39b $\displaystyle\int \sin^n u \cos^m u\, du = \frac{\sin^{n+1} u \cos^{m-1} u}{n+m} + \frac{m-1}{n+m} \int \sin^n u \cos^{m-2} u\, du$ if m

40 $\displaystyle\int u \sin u\, du = \sin u - u \cos u + C$

41 $\displaystyle\int u \cos u\, du = \cos u + u \sin u + C$

42 $\displaystyle\int u^n \sin u\, du = -u^n \cos u + n \int u^{n-1} \cos u\, du$

43 $\displaystyle\int u^n \cos u\, du = u^n \sin u - n \int u^{n-1} \sin u\, du$

FORMS INVOLVING $\sqrt{u^2 \pm a^2}$

44 $\displaystyle\int \sqrt{u^2 \pm a^2}\, du = \frac{u}{2} \sqrt{u^2 \pm a^2} \pm \frac{a^2}{2} \ln|u + \sqrt{u^2 \pm a^2}| + C$

45 $\displaystyle\int \frac{du}{\sqrt{u^2 \pm a^2}} = \ln|u + \sqrt{u^2 \pm a^2}| + C$

46 $\displaystyle\int \frac{\sqrt{u^2 + a^2}}{u}\, du = \sqrt{u^2 + a^2} - a \ln\left(\frac{a + \sqrt{u^2 + a^2}}{u}\right) + C$

47 $\displaystyle\int \frac{\sqrt{u^2 - a^2}}{u}\, du = \sqrt{u^2 - a^2} - a \sec^{-1} \frac{u}{a} + C$

48 $\displaystyle\int u^2 \sqrt{u^2 \pm a^2}\, du = \frac{u}{8} (2u^2 \pm a^2) \sqrt{u^2 \pm a^2} - \frac{a^4}{8} \ln|u + \sqrt{u^2 \pm a^2}| + C$

49 $\displaystyle\int \frac{u^2\, du}{\sqrt{u^2 \pm a^2}} = \frac{u}{2} \sqrt{u^2 \pm a^2} \mp \frac{a^2}{2} \ln|u + \sqrt{u^2 \pm a^2}| + C$

50 $\displaystyle\int \frac{du}{u^2 \sqrt{u^2 \pm a^2}} = \mp \frac{\sqrt{u^2 \pm a^2}}{a^2 u} + C$

51 $\displaystyle\int \frac{\sqrt{u^2 \pm a^2}}{u^2}\, du = -\frac{\sqrt{u^2 \pm a^2}}{u} + \ln|u + \sqrt{u^2 \pm a^2}| + C$

52 $\displaystyle\int \frac{du}{(u^2 \pm a^2)^{3/2}} = \frac{\pm u}{a^2 \sqrt{u^2 \pm a^2}} + C$

53 $\displaystyle\int (u^2 \pm a^2)^{3/2}\, du = \frac{u}{8} (2u^2 \pm 5a^2) \sqrt{u^2 \pm a^2} + \frac{3a^4}{8} \ln|u + \sqrt{u^2 \pm a^2}| + C$

FORMS INVOLVING $\sqrt{a^2 - u^2}$

54 $\displaystyle\int \sqrt{a^2 - u^2}\, du = \frac{u}{2} \sqrt{a^2 - u^2} + \frac{a^2}{2} \sin^{-1} \frac{u}{a} + C$

55 $\displaystyle\int \frac{\sqrt{a^2 - u^2}}{u}\, du = \sqrt{a^2 - u^2} - a \ln\left|\frac{a + \sqrt{a^2 - u^2}}{u}\right| + C$

56 $\displaystyle\int \frac{u^2\, du}{\sqrt{a^2 - u^2}} = -\frac{u}{2} \sqrt{a^2 - u^2} + \frac{a^2}{2} \sin^{-1} \frac{u}{a} + C$

57 $\displaystyle\int u^2 \sqrt{a^2 - u^2}\, du = \frac{u}{8} (2u^2 - a^2) \sqrt{a^2 - u^2} + \frac{a^4}{8} \sin^{-1} \frac{u}{a} + C$

58 $\displaystyle\int \frac{du}{u^2 \sqrt{a^2 - u^2}} = -\frac{\sqrt{a^2 - u^2}}{a^2 u} + C$

59 $\displaystyle\int \frac{\sqrt{a^2 - u^2}}{u^2}\, du = -\frac{\sqrt{a^2 - u^2}}{u} - \sin^{-1} \frac{u}{a} + C$

60 $\displaystyle\int \frac{du}{u \sqrt{a^2 - u^2}} = -\frac{1}{a} \ln\left|\frac{a + \sqrt{a^2 - u^2}}{u}\right| + C$

61 $\displaystyle\int \frac{du}{(a^2 - u^2)^{3/2}} = \frac{u}{a^2 \sqrt{a^2 - u^2}} + C$

62 $\displaystyle\int (a^2 - u^2)^{3/2}\, du = \frac{u}{8} (5a^2 - 2u^2) \sqrt{a^2 - u^2} + \frac{3a^4}{8} \sin^{-1} \frac{u}{a} + C$

EXPONENTIAL AND LOGARITHMIC FORMS

63 $\displaystyle\int u e^u\, du = (u - 1) e^u + C$

64 $\displaystyle\int u^n e^u\, du = u^n e^u - n \int u^{n-1} e^u\, du$

65 $\displaystyle\int \ln u\, du = u \ln u - u + C$

66 $\displaystyle\int u^n \ln u\, du = \frac{u^{n+1}}{n+1} \ln u - \frac{u^{n+1}}{(n+1)^2} + C$

67 $\displaystyle\int e^{au} \sin bu\, du = \frac{e^{au}}{a^2 + b^2} (a \sin bu - b \cos bu) + C$

68 $\displaystyle\int e^{au} \cos bu\, du = \frac{e^{au}}{a^2 + b^2} (a \cos bu + b \sin bu) + C$

(Continued inside back cover)

Calculus and Analytic Geometry

Calculus and

PRENTICE-HALL, INC., ENGLEWOOD CLIFFS, NEW JERSEY   07632

Analytic Geometry

C. H. Edwards, Jr.

David E. Penney

The University of Georgia, Athens

Library of Congress Cataloging in Publication Data

Edwards, C. H. (Charles Henry), (date)
    Calculus and analytic geometry.

    Bibliography: p. 835
    Includes index.
    1. Calculus.   2. Geometry, Analytic.   I. Penney,
David E.   II. Title.
QA303.E223   515'.15   81-8612
ISBN 0-13-111609-6   AACR2

Calculus and Analytic Geometry

C. H. Edwards, Jr. and David E. Penney

10 9 8 7 6 5 4 3

PRENTICE-HALL INTERNATIONAL, INC., *London*

PRENTICE-HALL OF AUSTRALIA PTY. LIMITED, *Sydney*

PRENTICE-HALL OF CANADA, LTD., *Toronto*

PRENTICE-HALL OF INDIA PRIVATE LIMITED, *New Delhi*

PRENTICE-HALL OF JAPAN, INC., *Tokyo*

PRENTICE-HALL OF SOUTHEAST ASIA PTE. LTD., *Singapore*

WHITEHALL BOOKS LIMITED, *Wellington, New Zealand*

Editorial/production supervision:
Eleanor Henshaw Hiatt

Interior design and cover design:
Walter A. Behnke

Illustrator: Eric G. Hieber

Editorial assistant: Susan Pintner

Manufacturing buyer: John Hall

Cover photograph: *Close to the Wind,*
a crystal sculpture designed by
Lloyd Atkins, Steuben Glass.

Contents

Preface

This book is a text for the standard calculus course for science, mathematics, and engineering students. It includes enough material for four quarters or three semesters. With regard to the selection, sequence, and treatment of mathematical topics, as well as notation and terminology, our approach is a traditional one. We do, however, place special emphasis on concrete examples, applications, and problems. These serve both to motivate and highlight the mathematical development and to demonstrate the remarkable power and versatility of calculus in the investigation of important scientific problems.

We wrote this book with five related objectives in constant view: *concreteness*, *readability*, *motivation*, *applicability*, and *accuracy*. The following paragraphs give more details about these aspects of the book. In addition, there are some comments about the organization of the book (designed to allow reasonable flexibility in its use) and about the problems, to which we devoted as much attention as to the text itself.

CONCRETENESS

The power of calculus is impressive in its precise answers to realistic questions and problems. In the necessary development of the theory, we keep in mind the central question, How does one actually *compute* it? We hope that a robust numerical flavor enhances the concreteness of our exposition. A propitious numerical computation often lends tangibility to the derivation of a general formula. For example:

- In Sections 3-8 and 5-5, we use the inverse-square law of gravitational attraction to calculate both the size of a "black hole" with the mass of the sun and the power a rocket engine must produce in order to put a satellite into earth orbit.

- In Sections 6-5 and 6-6 we solve simple differential equations to analyze the elimination of a drug from the bloodstream and of pollutants from a lake.
- In Section 10-5 the centripetal acceleration formula $a = r\omega^2$ is used to compute the angle at which a racetrack should be banked, and to deduce Kepler's third law of motion for the case of uniform circular motion.
- In Sections 12-3 and 12-9 we use geometric series to study safe drug dosage and to approximate accurately the period of oscillation of a pendulum.
- In Section 14-5 we apply multivariable maximum-minimum techniques to contrast the corporate profits that result from collusion and from competition in the marketplace.
- In Section 15-4 moments of inertia are used to compare the speeds with which different objects will roll down a hill.

In such examples we use realistic data, resulting in plausible numerical answers. Students who enjoy using hand-held calculators will find many exercises and examples to their taste, but because we show details of computations, a calculator is not required.

READABILITY

Difficulties in learning mathematics are often complicated by language difficulties. Our writing style stems from the observation that a crisp exposition, both intuitive and precise, makes mathematics more accessible—and hence more readily learned—with no loss of rigor. We hope our language will be clear and attractive to students and that they can and actually will read it. If so, this text will enable the instructor to spend more class time on the less routine aspects of teaching calculus.

MOTIVATION

The key to motivating effective study is stimulation of interest. Before asking the student to learn a new concept or technique, we try (whenever feasible) to make it clear how the knowledge gained will be worth the effort expended. For example, before we introduce the derivative in Section 1-5, we pave the way by showing in an earlier section how an ability to find slopes of tangent lines can be used to solve some interesting maximum-minimum problems. In theoretical discussions, especially, we try to provide an intuitive picture of the goal before we set off in pursuit of it. The student who reads Section 2-5, for example, will have a conceptual understanding of the statement and significance of the Mean Value Theorem before encountering its proof. Our exposition is centered around examples of the use of calculus to solve real problems of interest to real people. In selecting such problems for our examples and exercises, we took the view that stimulation of interest and effective teaching go hand in hand.

APPLICATIONS

We have included a somewhat wider range of scientific applications, from physical to economic, than one finds in many calculus books. But it is neither necessary nor desirable that the course cover all the applications in the book. Each section or subsection that may be omitted without loss of continuity is marked (both in the text and in the table of contents) with an asterisk. This provides flexibility for each instructor to steer his or her own path between theory and applications, or to direct the course toward one or another type of application.

Even if time does not permit classroom discussion of applications, they should be useful for supplementary reading. Its diverse applications are what attract many students to calculus, and realistic applications provide valuable motivation and reinforcement for all students. Section 1-1 contains a list of twenty *sample* applications that the student can anticipate for later study. Moreover, each applied example is self-contained, requiring little or no prior knowledge of the particular applied area.

PROBLEMS

For the students, the most important part of the book may be the problem sets. Our approximately 4400 problems have been winnowed carefully from a much larger initial number. We have included many of the old standbys that form an indispensable part of the calculus textbook tradition and a great many entirely new ones. The problem sets are graded carefully, beginning in most cases with routine computational exercises, then progressing to more challenging problems. To facilitate assigning homework after only a part of a section has been covered in class, both the routine computational problems (the first part of each problem set) and the less routine ones (the latter portion) are arranged in the approximate order of the presentation of topics within the section. Each chapter concludes with a generous selection of miscellaneous review problems. The answers to the odd-numbered problems are included in the back of the book. The complete solution of every third problem is available in a separate and optional solutions manual.

ORGANIZATION

The book's approximately 140 sections (exclusive of some introductory sections) are designed for coverage in one or two class days each, depending upon the length and pace of the course and the amount of time devoted to optional topics. The following comments on certain core topics may be useful to the instructor.

Limits and the Derivative   Although Chapter 1 is entitled "A Prelude to Differential Calculus," our objective is a quick start on calculus itself. Each of the necessary preliminaries is carefully motivated by showing how it will be needed. The derivative makes its first appearance in Section 1-5. In Section 1-8 we give an intuitive introduction to the limit concept, one sufficient for most purposes. Section 1-10 includes a rigorous treatment of limits and may be regarded as an optional section or as a reference.

Trigonometric Functions   We include an early introduction to the calculus of trigonometric functions (Section 2-9, which begins with a review of trigonometry). If the instructor prefers, this section may be deferred and treated as an introduction to Chapter 7. If so, the trigonometric problems and examples in the intervening chapters may be omitted with no loss of continuity.

Vectors   Students whose geometric background is minimal sometimes experience initial difficulty with vectors. For this reason, we first limit our discussion to vectors in the plane (Section 10-3). Vectors in space and the geometric interpretation of the scalar product are delayed until Section 13-1. Although it might be more efficient to treat plane vectors as a special case of space vectors, we have found the two-tiered approach to be more effective in practice.

Infinite Series   The chapter on infinite series (Chapter 12) may be delayed until after partial differentiation (Chapter 14) and multiple integrals (Chapter 15) if desired. Taylor's formula with remainder and some of its applications appear in Sections 11-2 and 11-5. The instructor who prefers to discuss Taylor's formula in the context of infinite series should insert these two sections between Sections 12-6 and 12-7.

Differential Equations   Many calculus instructors now feel that elementary differential equations and their applications should be introduced earlier than has been customary. In Section 6-6 we treat the linear first order differential equation with constant coefficients as an example of a separable equation; more general separable equations are taken up in Section 6-8. This permits early inclusion of some impressive applications. These sections may, however, be delayed until Chapter 17 ("Differential Equations") is begun. In this case Section 6-8 may be used as an introduction to Chapter 17, with Section 7-4 inserted just before Section 17-6. Alternatively, the whole of Chapter 17 (other than its final section) may be used at any point in the course after Chapter 8.

FLEXIBILITY

In writing this book we aimed at a presentation that accommodates the more common permutations in the order of coverage of the standard topics. With regard to the level of rigor, we favor an intuitive and conceptual treatment that is careful and precise in the formulation of definitions and the statements of theorems, some of whose proofs must be postponed until advanced calculus. The latter include, for example, the intermediate value and maximum value properties of continuous functions on closed intervals (Section 1-9). Certain other proofs that we include, but which may be omitted at the instructor's discretion, are placed at the ends of sections. In this way we leave ample room for variation in seeking the proper balance between rigor and intuition.

ACCURACY

We have taken considerable care to eliminate the glitches and misprints that sometimes appear in first editions. The first draft of our manuscript was scrutinized and dissected by an excellent group of reviewers. The revised draft was checked both by several of the original reviewers and by a new group; then we prepared the final draft for publication. Each sheet of galley proofs was corrected by each author and by five additional persons. Errors in the answers are especially disconcerting to students, so each problem was worked independently by at least two people. The many individuals who have helped us in the preparation of this text are listed on the acknowledgments page. In addition, we give special thanks to those students with whom the material in this book has been class-tested during the past ten years; their reactions have shaped it more than they know. We would appreciate our new readers informing us of any rough spots that remain despite all the assistance we have received.

C. H. E., JR.

D. E. P.

Acknowledgments

We have many people to thank for their assistance. The following professional colleagues reviewed various drafts of the manuscript, shared with us their insights and often their enthusiasm, and contributed more improvements than we could count.

Steve Batterson, Emory University
Jess Collins, McLennan Community College
Clement DeMayo, Providence College
Wayne Duncan, McLennan Community College
George Feissner, State University of New York, Cortland
Mark Hale, University of Florida, Gainesville
James E. Hall, University of Wisconsin, Madison
James Hurley, University of Connecticut, Storrs
John Jewett, Oklahoma State University
Joseph Krebs, Boston College, Chestnut Hill
Melvin Lax, California State University, Long Beach
Albert Liberi, Westchester Community College
Robert McFadden, Northern Illinois University
C. David Minda, University of Cincinnati
Richard Pellerin, Northern Virginia Community College, Annandale
Fred S. Roberts, Rutgers University, New Brunswick
William Wardlaw, United States Naval Academy, Annapolis
Michael Williams, Virginia Polytechnic Institute and State University

Carol Penney read critically every page of the final manuscript and worked every problem in the book. Those who individually corrected all galley proofs include Carol Brown and Alice Edwards.

We believe the finished book is ample evidence of the outstanding job by the staff of Prentice-Hall, Inc. Those whose personal contributions deserve special mention include Eleanor Henshaw Hiatt, production editor; Walter Behnke, designer; Eric G. Hieber, illustrator; and Robert Sickles and Harry Gaines, mathematics editors.

Prelude to Differential Calculus

1

Introduction

We live in a world of ceaseless change, filled with bodies in motion and with phenomena of ebb and flow. The principal object of the body of computational methods known as **calculus** is the analysis of problems of change and motion. This mathematical discipline stems from the seventeenth-century investigations of Isaac Newton (1642–1727) and Gottfried Wilhelm Leibniz (1646–1716), and it serves today as the quantitative language of science and technology. We list below some of the problems that you will learn to solve as you study this book. The first two problems will be discussed in this chapter, and the others will be covered in later chapters.

1 *The Fence Problem.* What is the maximum rectangular area that can be enclosed with a fence of perimeter 140 meters?

2 *The Refrigerator Problem.* The manager of an appliance store buys refrigerators at a wholesale price of $250 each. On the basis of past experience, the manager knows she can sell 20 refrigerators each month at $400 each and an additional one each month for each $3 reduction in selling price. What selling price will maximize the store's monthly profit?

3 A cork ball of specific gravity $\frac{1}{4}$ is thrown into water. How deep will it sink? (Section 3-4)

4 What is the maximum possible radius of a "black hole" with the same mass as the sun? (Section 3-8)

5 If you have enough "spring" in your legs to jump straight up 4 feet (on earth), could you blast off under your own power from an asteroid with a diameter of 3 miles? (Section 3-8)

6 How much power must a rocket engine produce in order to put a satellite into orbit around the earth? (Section 5-5)

7 If the Earth's population continues to grow at its present rate, when will it be "standing room only?" (Section 6-5)

8 Suppose that you deposit $100 each month in a savings account that pays 6% interest compounded continuously. How much will you have in the bank after 10 years? (Section 6-5)

9 The factories polluting a certain lake are ordered to cease immediately. How long will it take for natural processes to restore the lake to an acceptable level of purity? (Section 6-6)

10 According to newspaper accounts, it is possible to survive a free fall (without parachute) from a height of 20,000 feet. Can this be true? (Section 6-6)

11 How can a pendulum clock be used to determine the altitude of a mountain peak? (Section 7-4)

12 What is the best shape for the reflector in a solar heater? (Section 9-2)

13 At what angle should a racetrack curve be banked to best accommodate cars traveling at 150 mph? (Section 10-5)

14 How does a satellite of the earth use its thrusters to transfer from one circular orbit to another? (Section 10-7)

15 How do we know that $\pi = 3.14159265 \cdots$? (Section 11-5)

16 How often can a fixed dose of a drug be administered without producing a dangerous level of the drug in the patient's bloodstream? (Section 12-3)

17 What temperature can a mercury thermometer withstand before its bulb bursts? (Section 14-7)

18 How can two companies that make the same product conspire to maximize their total profits? (Section 14-5)

19 Which rolls down an inclined plane the fastest, a coin, a baseball, or a hoop? (Section 15-4)

20 How deep should a wine cellar be to keep the wine below 58°F throughout the year? (Section 17-7)

Our first chapter contains some preliminaries to your study of calculus: functions and their graphs, straight lines and slopes, and equations of circles and of parabolas. We introduce limits of functions and discuss how they can be used to find tangent lines to curves. This leads to the key concept of the derivative of a function and its interpretation as the rate of change of that function. Then we use these concepts to solve problems like the first two in the list above.

1-2
Real Numbers and Functions

The **real numbers** are already familiar to you; they are just those numbers ordinarily used in most measurements. The mass, speed, temperature, or charge of a body is measured by a real number. Real numbers can be represented by **terminating** or **nonterminating** decimal expansions. Any terminating decimal can be written in nonterminating form by adding zeros:

$$\tfrac{3}{8} = 0.375 = 0.375000000 \cdots.$$

Any **repeating** nonterminating decimal, such as

$$\tfrac{7}{22} = 0.31818181818 \cdots,$$

represents a **rational** number, one that is a quotient of two integers. The decimal expansion of an **irrational** number (one that is not rational), such as

$$\sqrt{2} = 1.414213562 \cdots$$

or

$$\pi = 3.141592653 \cdots,$$

is both nonterminating and nonrepeating.

The geometric representation of real numbers as points on the **real line** $\mathcal{R}$ is also familiar to you. Positive numbers lie to the right of zero and negative numbers lie to its left. The distance along the real line between zero and the number a is its **absolute value,** written $|a|$. Equivalently,

$$|a| = \begin{cases} a & \text{if } a \geqq 0; \\ -a & \text{if } a < 0. \end{cases} \tag{1}$$

1.1 The distance between a and b.

So it's always true that $|a| \geqq 0$, and $|a| = 0$ if and only if $a = 0$. The **distance** between the real numbers a and b is $|a - b| = |b - a|$, as shown in Figure 1.1.

It follows from (1) that, if b is a positive real number, then

$$|a| \leqq b \quad \text{if and only if} \quad -b \leqq a \leqq b. \tag{2}$$

This, in turn, implies the important

$$\boxed{\textit{Triangle Inequality:} \quad |a + b| \leqq |a| + |b|. \tag{3}}$$

Proof We add the inequalities $-|a| \leqq a \leqq |a|$ and $-|b| \leqq b \leqq |b|$. This gives us

$$-(|a| + |b|) \leqq a + b \leqq |a| + |b|.$$

Then, with the aid of (2), we may deduce (3).

INTERVALS

Suppose that S is a set of real numbers. Then we write $x \in S$ if and only if x is an element of S. The set S may be described by means of the notation

$$S = \{x : \cdots\},$$

where the ellipsis represents a condition that the real number x satisfies exactly when x belongs to S. The sets of real numbers that are most important in calculus are **intervals.** If $a < b$, then the **open interval** (a, b) is defined by

$$(a, b) = \{x : a < x < b\},$$

and the **closed interval** $[a, b]$ by

$$[a, b] = \{x : a \leqq x \leqq b\}.$$

Thus a closed interval contains its endpoints, while an open interval does not. We also use the **half-open intervals**

$$[a, b) = \{x : a \leqq x < b\}$$

and

$$(a, b] = \{x : a < x \leqq b\}.$$

We have shown examples of such intervals in Figure 1.2, as well as some **infinite intervals,** which have forms such as

$$[a, \infty) = \{x : x \geqq a\},$$

$$(-\infty, a] = \{x : x \leqq a\},$$

$$(a, \infty) = \{x : x > a\}, \qquad \text{and}$$

$$(-\infty, a) = \{x : x < a\}.$$

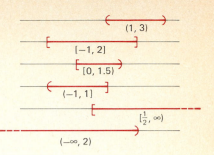

1.2 Some examples of intervals of real numbers.

An open interval containing a point is often called a **neighborhood** of that point. For example, if $\delta > 0$, then $(a - \delta, a + \delta)$ is a neighborhood of a with a as its midpoint.

The set of solutions of an inequality is often an interval or union of intervals, as in the following example.

EXAMPLE 1 Solve the inequality $|3 - 5x| < 2$.

Solution We use (2) with $<$ instead of $\leqq$; thus we transform the given inequality into

$$-2 < 3 - 5x < 2.$$

We subtract 3 from each member and find that

$$-5 < -5x < -1.$$

Multiplication by any negative number, such as $-\frac{1}{5}$, reverses inequalities. So $1 > x > \frac{1}{5}$, or

$$\tfrac{1}{5} < x < 1.$$

Thus the solution set is the open interval $(\frac{1}{5}, 1)$.

FUNCTIONS

The key to the mathematical analysis of a geometric or scientific situation is often the recognition of relationships between the variables that describe the situation. Such a relationship often takes the form of a formula that expresses one variable as a function of another. For example, the area A of a circle of radius r is given by $A = \pi r^2$. The volume V and surface area S of a sphere of radius r are given by $V = \frac{4}{3}\pi r^3$ and $S = 4\pi r^2$, respectively. After t seconds, a body that has been dropped from rest has fallen a distance $s = \frac{1}{2}gt^2$ feet and has speed $v = gt$ feet per second, where $g \approx 32$ ft/sec^2 is the gravitational acceleration. The volume V (in liters) of 3 grams of CO_2 at $27°C$ is given in terms of its pressure p (in atm) by $V = 1.68/p$. These are all examples of real-valued functions of a real variable.

> ***Definition of a Function***
> A real-valued **function** f defined on a set D of real numbers is a rule that assigns to each number x in D exactly one real number $f(x)$.

The set D of numbers x for which $f(x)$ is defined is called the **domain** (or **domain of definition**) of the function f. The number $f(x)$ is called the **value** of f at x. The set of all values $y = f(x)$ is called the **range** R of f; that is,

$$R = \{y : y = f(x) \quad \text{for some } x \in D\}.$$

We may indicate the domain and range of f by writing $f : D \rightarrow R$. The arrow is meant to suggest the "action" of f in assigning elements of the range R to elements of the domain D.

When we describe the function f by writing a formula $y = f(x)$, we call x the **independent variable** and y the **dependent variable**, because the value of y depends—through f—upon the choice of x. Not every function, however, is defined by means of a single formula. For instance, the assignment

$$f(x) = \begin{cases} x^2 & \text{if } x \geqq 0; \\ x & \text{if } x < 0, \end{cases}$$

determines a perfectly good function. Alternatively, the function in the following example is given by a verbal description rather than by a formula.

EXAMPLE 2 In 1980 the first-class postage rate was 15¢ for the first ounce plus 13¢ for each additional ounce or fraction thereof. If $g(w)$ denotes the postage in cents for w ounces, what is the range of the function g?

Solution First, $g(w) = 15$ if $w \in (0, 1]$. If $w > 1$ and n denotes the smallest integer that is no less than $w - 1$, then $g(w) = 15 + 13n$. This makes it clear that the range of g is the set

$$\{15, 28, 41, 54, 67, 80, 93, 106, \ldots\}.$$

What is the domain supposed to be when it is not specified? This is a common situation, and it occurs when we give a function f by writing a formula $y = f(x)$. If no domain is given, we assume that the domain D is the set of all real numbers x for which the expression $f(x)$ makes sense. For example, the domain of $f(x) = 1/x$ is the set of all nonzero real numbers.

EXAMPLE 3 Find the domain and range of $g(x) = 1/\sqrt{2x + 4}$.

Solution Recall that the symbol $\sqrt{x}$ denotes the *nonnegative* square root of x; it is defined only if x is itself nonnegative. So $\sqrt{2x + 4}$ is defined only if $2x + 4 \geqq 0$; that is, for $x \geqq -2$. In order that the reciprocal $1/\sqrt{2x + 4}$ be defined, we also require that $\sqrt{2x + 4} \neq 0$, and thus that $x \neq -2$. Hence the domain of g is the interval $D = (-2, +\infty)$. Its range is $R = (0, +\infty)$, the set of all positive real numbers.

A function f with domain D may be thought of as a machine that accepts a number x from D and then puts out (or displays, or prints) the number $f(x)$. Such a machine is the familiar pocket calculator with a $\sqrt{}$ button. When a nonnegative number x is entered and this button is pressed, the calculator displays the number $\sqrt{x}$.

EXAMPLE 4 Express the volume V of a cube as a function of its total surface area S.

Solution Suppose that e is the length of each edge of the cube. Then $V = e^3$ and $S = 6e^2$ (there are six square faces). Hence $e = \sqrt{S/6}$, and so

$$V = V(S) = \left(\frac{S}{6}\right)^{3/2}.$$

The domain of V is the open interval $(0, +\infty)$.

In the two examples that follow, we begin the solution of the fence problem and the refrigerator problem, the first two problems listed in Section 1-1.

EXAMPLE 5 In connection with the fence problem, write the area A of a rectangle of perimeter 140 as a function of its base x.

Solution As shown in Figure 1.3, let y denote the height of the rectangle. Then $A = xy$. But $2x + 2y = 140$, so $y = 70 - x$. Therefore $A = x(70 - x)$.

1.3 The rectangle of Example 5.

In addition to this formula, we should also specify the domain of the function A. Only values of $x > 0$ will produce actual rectangles, but we shall find it convenient to include the value $x = 0$ as well. This value of x corresponds to a "degenerate rectangle" with base 0 and height 70. For similar reasons, we have the restriction $y \geqq 0$. Since $y = 70 - x$, it follows that $x \leqq 70$. Thus the complete definition of our area function is

$$A(x) = x(70 - x), \qquad 0 \leqq x \leqq 70. \qquad (4)$$

Example 5 illustrates an important part of the solution of a typical problem involving applications of mathematics. The domain of a function is a necessary part of its definition, and for each function we must specify the domain of values of the independent variable. In applications, we use the values of the independent variable that are relevant to the problem at hand.

EXAMPLE 6 In connection with the refrigerator problem, express the monthly profit P as a function of the number x of refrigerators sold monthly.

Solution Let p be the selling price of each appliance. Then $P = x(p - 250)$ because the store manager buys them at a wholesale price of $250 each.

Now recall that 20 refrigerators per month can be sold at $400 each, but that an additional 1 per month can be sold for each $3 reduction in selling price. The number of $3 reductions (below $400) is $x - 20$, so

$$p = 400 - 3(x - 20) - 460 - 3x.$$

Hence

$$P = x(p - 250) = x(210 - 3x),$$

so that

$$P = 3x(70 - x).$$

Finally, we must find the relevant domain of values of x. Certainly $x \geq 0$. On the other hand, a negative profit would be unacceptable, so $x \leq 70$. Therefore the complete description of our function is

$$P(x) = 3x(70 - x), \qquad 0 \leq x \leq 70. \tag{5}$$

Though we are not yet prepared to complete the solution of either problem, you should note the similarity between them: The only difference between $A(x)$ in (4) and $P(x)$ in (5) is the constant 3. So the same value of x, whatever it may be, will maximize both $A(x)$ and $P(x)$. As different as the fence problem and the refrigerator problem seem at first glance, each—when reduced to essentials—amounts to the problem of finding the maximum value of $f(x) = x(70 - x)$ for $x \in [0, 70]$. We shall attack this problem in Section 1-4 after a review of rectangular coordinates in Section 1-3.

Much of the marvelous power and efficiency of calculus stems from the use of functional language to recognize such similarities between diverse types of applied problems, and from the exploitation of these similarities by the application of general mathematical methods.

1-2 PROBLEMS

In each of Problems 1–10, the given inequality describes a set of real numbers. Write this set in terms of intervals without using absolute value notation.

1 $2x - 7 < -3$.

2 $1 - 4x > 2$.

3 $-3 < 3x + 4 \leq 6$.

4 $-2 \leq \dfrac{5 - 2x}{3} \leq 2$.

5 $|5x + 3| \leq 4$.

6 $|1 - 3x| > 2$.

7 $1 < |7x - 1| < 3$.

8 $2 \leq |4 - 5x| \leq 4$.

9 $\dfrac{1}{2x + 1} > 3$.

10 $\dfrac{2}{7 - 3x} \leq -5$.

In each of Problems 11–25, the given formula, where meaningful, determines a function. Write the domain and range of this function.

11 $f(x) = 10 - x^2$.

12 $f(x) = x^3 + 5$.

13 $g(t) = \sqrt{t^2}$.

14 $g(t) = (\sqrt{t})^2$.

15 $f(x) = \sqrt{3x - 5}$.

16 $g(t) = \sqrt[3]{t + 4}$.

17 $f(t) = \sqrt{1 - 2t}$.

18 $g(x) = \dfrac{1}{(x + 2)^2}$.

19 $f(x) = \dfrac{2}{3 - x}$.

20 $g(t) = \left(\dfrac{2}{3 - t}\right)^{1/2}$.

21 $f(x) = \sqrt{x^2 + 9}$.

22 $h(z) = \dfrac{1}{\sqrt{4 - z^2}}$.

23 $f(x) = (4 - \sqrt{x})^{1/2}$.

24 $f(x) = \left(\dfrac{x + 1}{x - 1}\right)^{1/2}$.

25 $g(t) = \dfrac{t}{|t|}$.

26 Express the area A of a square as a function of its perimeter P.

27 Express the circumference C of a circle as a function of its area A.

28 Express the volume V of a sphere as a function of its surface area S.

29 If $0°C$ is the same as $32°F$, and a change of $1°C$ is the same as a change of $1.8°F$, express the Celsius temperature C as a function of the Fahrenheit temperature F.

30 A rectangular box has volume 125 and square base with edge x. Write its total surface area A as a function of x.

31 A rectangle with base x is inscribed in a circle of radius 2. Express the area A of the rectangle as a function of x.

32 An oil field containing 20 wells has been producing 4000 barrels of oil daily. Suppose that, for each new well that is drilled, the daily production of each well decreases by 5 barrels. Write the total daily production of the oil field as a function of the number x of new wells drilled.

The Coordinate Plane and Straight Lines

Imagine the flat, featureless, two-dimensional plane of Euclid's geometry. Install a copy of the real number line $\mathscr{R}$, with the line horizontal and the positive numbers to the right, as in Figure 1.4. Add another copy of $\mathscr{R}$ perpendicular to the first, with the two lines crossing where zero is located on each. The second line should have the positive numbers above and the negative numbers below the first line. The first line will be called the *x-axis* and the second is the *y-axis*.

With these added features, we call the plane the *coordinate plane*, since it's now possible to locate any point there by a pair of numbers, called the *coordinates of the point*. For if P is a point in the plane, draw perpendiculars, as shown in Figure 1.5, from P to the coordinate axes. One perpendicular meets the x-axis in the **x-coordinate** of P, called x_1. The other meets the y-axis in the **y-coordinate** y_1 of P. The pair of numbers (x_1, y_1) is called the **coordinate pair** for P, or just the **coordinates** of P. In compact notation, we speak of "the point $P(x_1, y_1)$." The numbers x_1 and y_1 are called the **abscissa** and **ordinate**, respectively, of the point P.

This coordinate system is called the **rectangular** coordinate system, or sometimes the **Cartesian** coordinate system (because its use in plane geometry was popularized, beginning in the 1630s, by the French mathematician and philosopher René Descartes (1596–1650)). The plane, thus coordinatized, is sometimes denoted by $\mathscr{R}^2$ because of the two copies of $\mathscr{R}$ that are used and is sometimes called the **Cartesian plane.**

Rectangular coordinates are easy to work with because $P(x_1, y_1)$ and $Q(x_2, y_2)$ denote the same point when and only when $x_1 = x_2$ and $y_1 = y_2$. Thus, when you know that P and Q are different points, you may conclude that P and Q have different abscissas or different ordinates (or both).

The point of symmetry $(0, 0)$ where the coordinate axes cross is called the **origin.** The points on the x-axis all have coordinates of the form $(x, 0)$, and, while the *real number* x is not the same as the *geometric point* $(x, 0)$, there are situations in which it is useful to think of the two as the same. Similar remarks can be made about the points $(0, y)$ that constitute the y-axis.

The following formula gives the distance $d(P_1, P_2)$ between the two points P_1 and P_2 in terms of their coordinates.

Distance Formula

The **distance** between $P_1(x_1, y_1)$ and $P_2(x_2, y_2)$ is

$$d(P_1, P_2) = \sqrt{(x_2 - x_1)^2 + (y_2 - y_1)^2}. \qquad (1)$$

Proof If $x_1 \neq x_2$ and $y_1 \neq y_2$, then Formula (1) follows from an application of the Pythagorean theorem. Use the right triangle with vertices P_1, P_2, and $P_3(x_2, y_1)$, shown in Figure 1.6.

If $x_1 = x_2$, then P_1 and P_2 lie in a vertical line, In this case,

$$d(P_1, P_2) = |y_2 - y_1| = \sqrt{(y_2 - y_1)^2}.$$

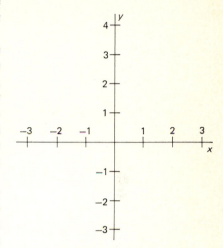

1.4 The coordinate plane.

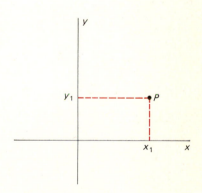

1.5 The point P has rectangular coordinates (x_1, y_1).

1.6 Use this triangle to deduce the distance formula.

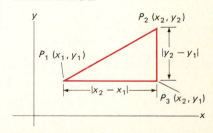

This agrees with Formula (1), because $x_1 = x_2$. The remaining case, in which $y_1 = y_2$, is similar.

EXAMPLE 1 Show that the triangle PQR with vertices $P(1, 0)$, $Q(5, 4)$, and $R(-2, 3)$ is a right triangle. (This triangle is illustrated in Figure 1.7.)

Solution The distance formula gives

$$a^2 = [d(P, Q)]^2 = (5 - 1)^2 + (4 - 0)^2 = 32,$$
$$b^2 = [d(P, R)]^2 = (-2 - 1)^2 + (3 - 0)^2 = 18,$$

and

$$c^2 = [d(Q, R)]^2 = (-2 - 5)^2 + (3 - 4)^2 = 50.$$

Since $a^2 + b^2 = c^2$, the *converse* of the Pythagorean theorem implies that PQR is a right triangle. The right angle is at P since P is opposite the longest side, QR.

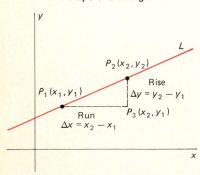

1.7 Is this a right triangle? See Example 1.

Another immediate application of the distance formula is an expression for the coordinates of the midpoint M of the line segment with end points P_1 and P_2. The coordinates of M are the *averages* of the corresponding coordinates of P_1 and P_2.

Midpoint Formula

The **midpoint** of the line segment with end points $P_1(x_1, y_1)$ and $P_2(x_2, y_2)$ is the point $M(\overline{x}, \overline{y})$ with coordinates

$$\overline{x} = \tfrac{1}{2}(x_1 + x_2), \qquad \overline{y} = \tfrac{1}{2}(y_1 + y_2). \tag{2}$$

Proof If you substitute the coordinates of P_1, P_2, and M into the distance formula, you will find that $d(P_1, M) = d(P_2, M)$. All that remains is to show that M is on the line segment $P_1 P_2$. We leave the details to you.

STRAIGHT LINES AND SLOPE

We need to define the slope of a straight line, a measure of its rate of rise or fall from left to right. Given a line L in the plane, with L not vertical, pick two points $P_1(x_1, y_1)$ and $P_2(x_2, y_2)$ on L. Consider the **increments** Δx and Δy (read "delta x" and "delta y") in the x- and y-coordinates from P_1 to P_2. These are defined to be

$$\Delta x = x_2 - x_1 \quad \text{and} \quad \Delta y = y_2 - y_1. \tag{3}$$

Then Δy is the **rise** from P_1 to P_2, and Δx is the **run** from P_1 to P_2. They are illustrated in Figure 1.8. The **slope** m of the nonvertical line L is then given by

$$m = \frac{\text{rise}}{\text{run}} = \frac{\Delta y}{\Delta x} = \frac{y_2 - y_1}{x_2 - x_1}. \tag{4}$$

1.8 The slope of a straight line.

You should note that if $P_3(x_3, y_3)$ and $P_4(x_4, y_4)$ are two other points on L, then the similarity of the triangles of Figure 1.9 implies that

$$\frac{y_4 - y_3}{x_4 - x_3} = \frac{y_2 - y_1}{x_2 - x_1}.$$

Hence the slope m as defined in Equation (4) does *not* depend upon the particular choice of P_1 and P_2.

If the line L is horizontal, then $\Delta y = 0$; thus Equation (4) gives $m = 0$. If L is vertical, then $\Delta x = 0$; the slope of L is *not defined* in this case. Thus we have the following statements.

Horizontal lines have slope zero;
vertical lines have no slope at all.

The concept of slope lets us write equations of straight lines. Let $P_0(x_0, y_0)$ be a fixed point on the nonvertical line L, and let $P(x, y)$ be any other point on L. We apply (4) with P and P_0 in place of P_2 and P_1 and find that

$$m = \frac{y - y_0}{x - x_0}$$

or

$$y - y_0 = m(x - x_0). \tag{5}$$

The last equation is the **point-slope equation** of the straight line L that passes through $P_0(x_0, y_0)$ and has slope m. A point $P(x, y)$ lies on L if and only if its coordinates satisfy Equation (5).

EXAMPLE 2   Write an equation for the straight line L through the points $P_1(1, -1)$ and $P_2(3, 5)$.

Solution   The slope of L is

$$m = \frac{5 - (-1)}{3 - 1} = 3.$$

We take P_1 as the fixed point. Then, with the aid of (5), the point-slope equation of L is

$$y + 1 = 3(x - 1),$$

or after simplification, $y - 3x + 4 = 0$.

Equation (5) can be written in the form

$$y = mx + b, \tag{6}$$

where the constant $b = y_0 - mx_0$. Since $y = b$ when $x = 0$, the **y-intercept** of L is the point $(0, b)$ of Figure 1.10. For this reason, Equation (6) is called the **slope-intercept equation** of the line with slope m and y-intercept b.

Suppose now that $B \neq 0$. The general linear equation

$$Ax + By = C \tag{7}$$

can be written in the form of Equation (6) by dividing by B. So the above equation represents a straight line with slope the coefficient of x *after solution*

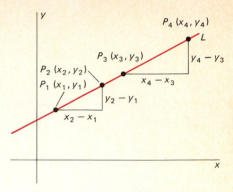

1.9   The result of the slope computation does not depend upon which two points on L are used.

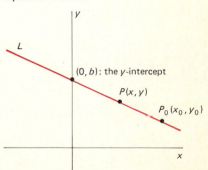

1.10   Finding the slope-intercept equation of a line.

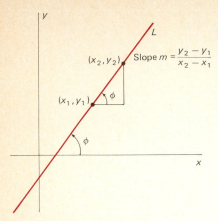

1.11 How is the angle of inclination ϕ related to the slope m?

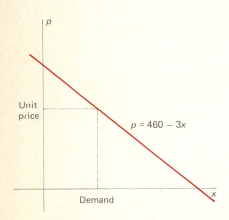

1.12 The demand curve for the refrigerator problem. See Example 4.

1.13 Illustration of the proof of Theorem 2.

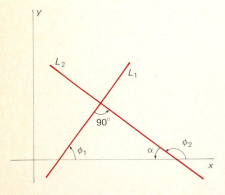

of the equation for y. If $B = 0$, then (7) reduces to the equation of a vertical line: $x = K$ (K a constant). If $A = 0$, then (7) reduces to the equation of a horizontal line: $y = H$ (H a constant).

If the line L is not horizontal, it must cross the x-axis. Then its **angle of inclination** is the angle $\phi \in (0°, 180°)$ from the positive x-axis to L, measured counterclockwise. The angle of inclination of a horizontal line is $\phi = 0°$. It is clear from Figure 1.11 that this angle ϕ and the slope m of a nonvertical line are related by the equation

$$m = \tan \phi. \qquad (8)$$

It is equally clear that two lines are parallel if and only if they have the same angle of inclination. So it follows from (8) that two parallel, nonvertical lines have the same slope, and that two lines with the same slope must be parallel.

Theorem 1 *Slopes of Parallel Lines*

Two nonvertical lines are parallel if and only if they have the same slope.

EXAMPLE 3 Write an equation of the line L that passes through the point $P(3, -2)$ and is parallel to the line L' with equation $x + 2y = 6$.

Solution When we solve the equation of L' for y, we get $y = -\frac{1}{2}x + 3$. So L' has slope $m = -\frac{1}{2}$. The point-slope equation of L is then

$$y + 2 = -\tfrac{1}{2}(x - 3),$$

or $x + 2y + 1 = 0$.

EXAMPLE 4 In our discussion of the refrigerator problem in Example 6 of Section 1-2, we found that the selling price p and the *demand x*, or the number of refrigerators x that could be sold monthly at unit price p, were related by the equation

$$p = 460 - 3x.$$

This is the equation of a straight line in the xp-plane, and it is shown in Figure 1.12. This line is the **demand curve** (or **line**) for the store's refrigerators. (Demand curves are usually *not* straight lines.)

Theorem 2 *Slopes of Perpendicular Lines*

Two lines L_1 and L_2 with slopes $m_1 = \tan \phi_1$ and $m_2 = \tan \phi_2$ are perpendicular if and only if

$$m_1 m_2 = -1. \qquad (9)$$

That is, the slope of each is the *negative reciprocal* of the slope of the other.

Proof If the two lines are perpendicular and each has slope, then neither is horizontal or vertical. Thus the situation is like the one illustrated in Figure 1.13. Then $\phi_1 + \alpha + 90° = 180° = \phi_2 + \alpha$, so it follows that $\phi_2 =$

CHAP. 1: **Prelude to Differential Calculus**

$\phi_1 + 90°$. Therefore

$$m_2 = \tan \phi_2 = \tan(\phi_1 + 90°)$$

$$= \frac{\sin(\phi_1 + 90°)}{\cos(\phi_1 + 90°)} = \frac{\cos \phi_1}{-\sin \phi_1} = -\frac{1}{\tan \phi_1} = -\frac{1}{m_1}.$$

Hence Equation (9) holds. In Problem 30 below, you will find an outline of a proof of the converse.

EXAMPLE 5 Write an equation of the line L through the point $P(3, -2)$ that is perpendicular to the line L' with equation $x + 2y = 6$.

Solution As we saw in Example 3, the slope of L' is $m' = -\frac{1}{2}$. By Theorem 2, the slope of L is $m = -1/m' = 2$, so L has the point-slope equation

$$y + 2 = 2(x - 3),$$

or $2x - y - 8 = 0$.

Finally, note that the *sign* of the slope m of the line L indicates whether L runs upward or downward from left to right. If $m > 0$, then, because $m = \tan \phi$, ϕ must be an acute angle. In this case, L "runs upward." If $m < 0$, then ϕ is obtuse, so that L "runs downward." Figure 1.14 shows the geometry behind these observations.

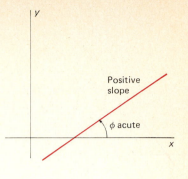

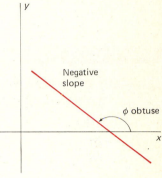

1.14 Positive and negative slope; effect on ϕ.

1-3   PROBLEMS

In Problems 1 and 2, plot the points A, B, and C, and then show that they are the vertices of a right triangle. Find the area of this triangle and also the midpoint of its hypotenuse.

1 $A(-2, -1)$, $B(2, 7)$, $C(4, -4)$.
2 $A(6, -1)$, $B(2, 3)$, $C(-3, -2)$.

In Problems 3 and 4, plot the points P, Q, and R to see whether they *appear* to lie on a single straight line. Then use slopes to determine whether or not they do.

3 $P(-2, 3)$, $Q(1, 2)$, $R(10, -1)$.
4 $P(-1, 2)$, $Q(6, 5)$, $R(11, 7)$.

In Problems 5–13, sketch the set of all points (x, y) that satisfy the given condition.

5 $x < y$.   **6** $y > 2x - 1$.
7 $x + y \geqq 1$.   **8** $x^2 + y^2 < 1$.
9 $(x - 3)^2 + (y - 4)^2 \leqq 25$.
10 $|x| = |y|$.   **11** $xy < 0$.
12 $|x| \leqq 1$ and $|y| \leqq 1$.   **13** $|x| + |y| = 1$.

In Problems 14–23, write an equation of the straight line L that is described.

14 L is vertical and has x-intercept 7.
15 L is horizontal and passes through $(3, -5)$.

16 L has x-intercept 2 and y-intercept -3.
17 L passes through $(2, -3)$ and $(5, 3)$.
18 L passes through $(-1, -4)$ and has slope $\frac{1}{2}$.
19 L passes through $(4, 2)$ and has angle of inclination $135°$.
20 L has slope 6 and y-intercept 7.
21 L passes through $(1, 5)$ and is parallel to the line with equation $2x + y = 10$.
22 L passes through $(-2, 4)$ and is perpendicular to the line with equation $x + 2y = 17$.
23 L is the perpendicular bisector of the line segment that has end points $(-1, 2)$ and $(3, 10)$.
24 Find the perpendicular distance from the point $(2, 1)$ to the line with equation $y = x + 1$.
25 Find the perpendicular distance between the parallel lines $y = 5x + 1$ and $y = 5x + 9$.
26 The points $A(-1, 6)$, $B(0, 0)$, and $C(3, 1)$ are three vertices of a parallelogram. Find the fourth vertex.
27 The Fahrenheit temperature F and the absolute temperature K satisfy a linear equation. Given that $K = 273$ when $F = 32$ and that $K = 373$ when $F = 212$, express K in terms of F. What is F when $K = 0$?
28 The length L (in centimeters) of a copper rod is a linear function of its Celsius temperature C. If $L = 124.942$ when $C = 20$ and $L = 125.134$ when $C = 110$, express L in terms of C.

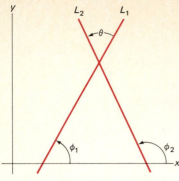

1.15 Figure for Problem 30.

29 A grocery owner finds that he can sell 980 gallons of milk each week at \$1.69 per gallon and 1220 gallons of milk each week at \$1.49. Assume a linear relationship between selling price and demand. How many gallons could he sell weekly at \$1.56 per gallon?

30 Refer to Figure 1.15, where θ is the angle between the lines L_1 and L_2, which have angles of inclination ϕ_1 and ϕ_2, respectively. (a) Show that $\theta = \phi_2 - \phi_1$. (b) Apply the trigonometric identity

$$\tan(\phi_2 - \phi_1) = \frac{\tan\phi_2 - \tan\phi_1}{1 + (\tan\phi_1)(\tan\phi_2)}$$

to show that if $m_1 m_2 = -1$, then it follows that L_1 and L_2 are perpendicular.

1-4

Graphs of Equations and Functions

We saw in the last section that the points (x, y) that satisfy a linear equation $Ax + By = C$ form a very simple set: a straight line. By contrast, the set of points (x, y) satisfying the equation

$$x^4 - 4x^3 + 3x^2 + 2x^2y^2 = y^2 + 4xy^2 - y^4$$

is the rather exotic curve of Figure 1.16. But both curves are examples of graphs.

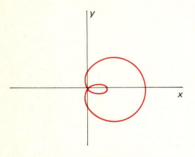

1.16 The graph of the equation
$x^4 - 4x^3 + 3x^2 + 2x^2y^2$
$\qquad = y^2 + 4xy^2 - y^4$.

> **Definition** *Graph of an Equation*
>
> The **graph** of an equation in the two variables x and y is the set of all points (x, y) that satisfy the equation.

For instance, the distance formula tells us that the graph of the equation

$$x^2 + y^2 = r^2 \qquad (1)$$

is the circle of radius r centered at the origin. More generally, the graph of the equation

$$(x - h)^2 + (y - k)^2 = r^2 \qquad (2)$$

is the circle of radius r with center (h, k). The relation between Equations (1) and (2) is a special case of the following principle.

> **Translation Principle**
>
> Suppose that the curve C is the graph of the equation $F(x, y) = 0$. (Here, $F(x, y)$ simply denotes an expression or formula involving the two variables x and y.) Let C' be the curve obtained by translating each point (x, y) of C to $(x + h, y + k)$. Then C' is the graph of the equation
>
> $$F(x - h, y - k) = 0. \qquad (3)$$

That is, the equation of C' is obtained from the equation of C by replacement of x by $x - h$ and of y by $y - k$. The reason is simple: (x, y) is a point on C' if and only if $(x - h, y - k)$ is a point on C.

In particular, Equation (2) can be obtained from Equation (1) by using the replacement of the translation principle. Note also that Equation (2) can be written in the form

$$x^2 + y^2 + ax + by = c. \tag{4}$$

Conversely, when we encounter an equation of this form, we can determine the center and radius of the circle it represents by using the elementary technique of "completing the square." To do so, note that

$$x^2 + ax = \left(x + \frac{a}{2}\right)^2 - \frac{a^2}{4},$$

which shows that $x^2 + ax$ can be made into a perfect square by adding to it the square of *half* the coefficient of x.

EXAMPLE 1 Find the center and radius of the circle represented by the equation

$$x^2 + y^2 - 4x + 6y = 12.$$

Solution We complete the square separately in each of the two variables. This gives us

$$(x^2 - 4x + 4) + (y^2 + 6y + 9) = 12 + 4 + 9,$$

or

$$(x - 2)^2 + (y + 3)^2 = 25.$$

So the circle has center $(2, -3)$ and radius 5.

The graph of a function is a special case of the graph of an equation.

> **Definition** *Graph of a Function*
> The **graph** of the function f is the graph of the equation $y = f(x)$.

It helps to remember that the graph of a function cannot intersect a *vertical* line in more than one point.

EXAMPLE 2 Construct the graph of the absolute value function $f(x) = |x|$.

Solution Recall that $|x| = -x$ if $x \leq 0$ and that $|x| = x$ for $x \geq 0$. So the graph of $y = |x|$ consists of the left half of the line $y = -x$, together with the right half of the line $y = x$, as shown in Figure 1.17.

EXAMPLE 3 Construct the graph of the parabola $y = x^2$.

Solution Here, we plot the points in a table of values.

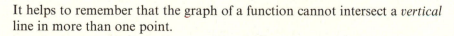

x	-3	-2	-1	0	1	2	3
$y = x^2$	9	4	1	0	1	4	9

When we draw a smooth curve through these points, we obtain the curve of Figure 1.18. The parabola $y = -x^2$ would look similar except that it

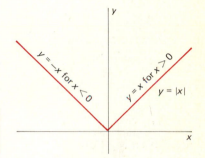

1.17 The graph of the absolute value function $y = |x|$.

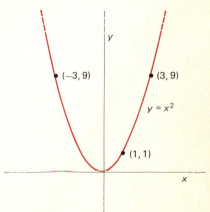

1.18 The graph of the parabola $y = x^2$.

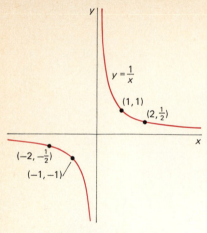

$y = \frac{1}{x}$

(1, 1)

$(2, \frac{1}{2})$

$(-2, -\frac{1}{2})$

$(-1, -1)$

1.19 The graph of the reciprocal function $y = 1/x$.

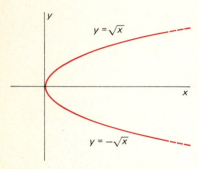

$y = \sqrt{x}$

$y = -\sqrt{x}$

1.20 The graph of the parabola $x = y^2$.

1.21 The graph of $f(x) = x(70 - x)$, $0 \leq x \leq 70$.

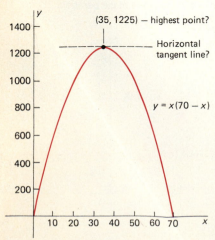

(35, 1225) — highest point?

Horizontal tangent line?

$y = x(70 - x)$

would open downward instead of upward. And more generally, the graph of the equation

$$y = ax^2 \tag{5}$$

is a parabola with its *vertex* at the origin. This parabola opens upward if $a > 0$ and downward if $a < 0$.

EXAMPLE 4 Sketch the graph of the reciprocal function $f(x) = 1/x$.

Solution We note that $|f(x)| = 1/|x|$ is very large when $|x|$ is very small, and conversely. Also, $f(x) = 1/x$ has the same sign as x. To get started with the graph, we can plot a few points, such as $(\pm 10, \pm 1/10)$, $(\pm 1, \pm 1)$, $(\pm 1/10, \pm 10)$. The rest of the information above suggests that the actual graph is much like the one of Figure 1.19.

EXAMPLE 5 Construct the graphs of the functions $y = \sqrt{x}$ and $y = -\sqrt{x}$.

Solution After plotting and connecting points as in Example 3, we obtain the parabola $y^2 = x$ (shown in Figure 1.20). Note that the parabola opens to the right. The upper half is the graph of $y = \sqrt{x}$, while the lower half is the graph of $y = -\sqrt{x}$.

In Section 1-2 we saw that the fence problem and the refrigerator problem can be reduced to the common problem of finding the maximum value of the function

$$f(x) = x(70 - x), \qquad 0 \leq x \leq 70.$$

We can begin our attack on the latter problem by graphing the function f. By plotting the points in the table below

x	0	10	20	30	40	50	60	70
$x(70 - x)$	0	600	1000	1200	1200	1000	600	0

and then sketching a smooth curve through them, we obtain the graph of f. It is shown in Figure 1.21.

In this figure we have used different scales along the two axes: a 10-unit interval on the x-axis has the same length as a 150-unit interval on the y-axis. In an applied problem in which there is no physical relation between the units used to measure the independent and dependent variables, there is no necessity to use the same scale on the two axes. As in this case, you may choose the scale for convenience in sketching the graph. Only in a geometric problem involving angles or distances will you find it important to use the same scale on the two axes.

The graph of Figure 1.21 seems to be symmetric about a vertical axis. This symmetry suggests strongly that its highest point is the point (35, 1225). For the moment, let us assume that this is so. What are the consequences of this assumption?

16

First, in Example 5 of Section 1-2, we saw that the area of a rectangle with perimeter 140 and base x is

$$A(x) = x(70 - x), \qquad 0 \leq x \leq 70.$$

A consequence of our assumption about the graph of f is then that the rectangle with perimeter 140 and *maximal* area is a *square* with base 35.

Next, in Example 6 of Section 1-2 (the refrigerator problem), we saw that if x refrigerators are sold monthly at a price of $p = 460 - 3x$ dollars each, then the monthly profit is

$$P(x) = 3x(70 - x) = 3f(x), \qquad 0 \leq x \leq 70.$$

The assumption that the maximum value of f is $f(35) = 1225$ then implies that the maximum profit of \$3675 monthly is made by selling 35 refrigerators each month at a price of \$355 each.

How can we be *certain* that the situation is indeed as it appears in Figure 1.21, where our visual inspection *suggests* that the highest point on the graph of $y = x(70 - x)$ is (35, 1225)? As a clue to the answer, notice that visual inspection also suggests that the tangent line to the curve at its highest point is horizontal. If we *knew* that the tangent line to a graph at its highest point must be horizontal, then our problem would reduce to that of showing that (35, 1225) is the only point of the graph in question where the tangent line is horizontal.

But what is meant by the **tangent line** to an arbitrary curve? We pursue this question in Section 1-5. The answer will open the door to the possibility of finding maximum and minimum values of arbitrary functions.

In the case of the particular function

$$f(x) = x(70 - x) = 70x - x^2$$

of current interest, we have an alternative method for nailing down its maximum value. We note that the curve $y = 70x - x^2$ is a translated parabola of the sort that opens downward. Indeed, the graph of any equation of the form

$$y = ax^2 + bx + c \qquad (a \neq 0) \tag{6}$$

can be recognized as a translated parabola by first completing the square in x and then by applying the translation principle. The size of $|a|$ determines the "width" of the parabola, which opens upward when $a > 0$ and downward when $a < 0$.

EXAMPLE 6   Show that the graph of $y = 70x - x^2$ is a translated parabola. Locate its vertex.

Solution   We complete the square in the variable x:

$$y = 70x - x^2 = -(x^2 - 70x)$$
$$= 1225 - (x^2 - 70x + 1225),$$

or

$$y - 1225 = -(x - 35)^2.$$

The translation principle therefore implies that our graph is shaped like the parabola $y = -x^2$, but with vertex $(35, 1225)$ rather than $(0, 0)$. Since the latter is the highest point of $y = -x^2$, the former must be the highest point of $y = 70x - x^2$.

Indeed, the fact that

$$f(x) = x(70 - x) = 1225 - (x - 35)^2$$

makes it clear that the maximum value of f is $f(35) = 1225$, because $(x - 35)^2 > 0$ except when $x = 35$.

The following example shows that the graph of a function need not be a "connected" curve.

EXAMPLE 7   Graph the function f whose formula is

$$f(x) = x - [x] - \tfrac{1}{2}.$$

Here, $[x]$ denotes the greatest integer that is less than or equal to x. For example, $[-10] = -10$, $[-4.2] = -5$, $[0] = 0$, and $[\pi] = 3$.

Solution   Think of x as temporarily fixed. If n is an integer such that $n \leqq x < n + 1$, then $[x] = n$. If so, then

$$f(x) = x - n - \tfrac{1}{2}.$$

Since $y = x - n - \tfrac{1}{2}$ has as its graph a straight line of slope 1, and $f(n) = n - n - \tfrac{1}{2} = -\tfrac{1}{2}$, it follows that the graph of f takes the form shown in Figure 1.22. This "sawtooth function" is our first example of a discontinuous function. The values of x where the value of $f(x)$ makes a "jump" are called **points of discontinuity** of the function f. Thus the points of discontinuity of the sawtooth function are the integers—as x approaches the integer n from the left, the value of $f(x)$ approaches $+\tfrac{1}{2}$, but it abruptly jumps to $-\tfrac{1}{2}$ when $x = n$. A precise definition of continuity and discontinuity for functions is given in Section 1-9.

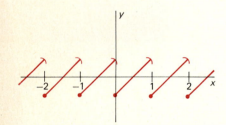

1.22   The sawtooth function $f(x) = x - [x] - \tfrac{1}{2}$.

1-4   PROBLEMS

Sketch the graphs of the functions in Problems 1–17.

1 $f(x) = 2 - 5x, \quad -1 \leqq x \leqq 1$.
2 $f(x) = 2 - 5x, \quad 0 \leqq x < 2$.
3 $f(x) = 10 - x^2$.
4 $f(x) = 1 + 2x^2$.
5 $f(x) = x^3$.
6 $f(x) = x^4$.

7 $f(x) = \sqrt{4 - x^2}$.
8 $f(x) = -\sqrt{9 - x^2}$.

9 $f(x) = \sqrt{x^2 - 9}$.
10 $f(x) = \dfrac{1}{1 - x}$.

11 $f(x) = \dfrac{1}{2x + 3}$.
12 $f(x) = \dfrac{1}{(2x + 3)^2}$.

13 $f(x) = \sqrt{1 - x}$.
14 $f(x) = \dfrac{1}{\sqrt{1 - x}}$.

15 $f(x) = \dfrac{1}{\sqrt{2x + 3}}$.
16 $f(x) = |2x - 2|$.

17 $f(x) = |x| + x$.

The graph of the equation $(x - h)^2 + (y - k)^2 = C$ is a circle if $C > 0$, is the point (h, k) if $C = 0$, and contains *no* points if $C < 0$ (why?). Find the graphs of the equations in Problems 18–22. In the case of a circle, give its center and radius.

18 $x^2 + y^2 - 2x + 4y + 1 = 0$.
19 $x^2 + y^2 - 6x + 8y = 0$.

20 $x^2 + y^2 - 2x + 2y + 2 = 0.$
21 $x^2 + y^2 + 2x + 6y + 20 = 0.$
22 $2x^2 + 2y^2 - 2x + 6y + 5 = 0.$

Sketch the translated parabolas in Problems 23–26 and indicate the vertex of each.

23 $y = x^2 + 2x + 4.$ **24** $2y = x^2 - 4x + 8.$
25 $y = 5x^2 + 20x + 23.$ **26** $y = x - x^2.$

Graph the functions given in Problems 27–31. Indicate each point of discontinuity.

27 $f(x) = 0$ if $x < 0$; $f(x) = 1$ if $x \geq 0.$
28 $f(x) = 1$ if x is an integer;   otherwise, $f(x) = 0.$

29 $f(x) = \dfrac{|x|}{x}$ if $x \neq 0$; $f(0) = 0.$

30 $f(x) = [x]$ (the greatest integer function).
31 $f(x) = [x] - x.$
32 In Problem 32 of Section 1-2, you were asked to express the daily production of a certain oil field as a function $P = f(x)$ of the number x of new oil wells that are drilled. Now construct the graph of f and use it to find the value of x that maximizes P.
33 If a ball is thrown straight upward with an initial velocity of 96 feet per second, then its height t seconds later is $y = 96t - 16t^2$ feet. Determine the maximum height it attains by constructing the graph of y as a function of t.

1-5

We suggested in the last section that we might locate the maximum value of a function f by finding the point (or points!) on its graph where the tangent line is horizontal. The typical picture of a maximum value (as in Figure 1.23) does suggest that the tangent should be horizontal there.

In our later studies, we shall find many ways to use the tangent line. For now, we need to decide how to *define* the tangent line L at an arbitrary point P of a curve $y = f(x)$. Our intuitive idea is this: The tangent line L should be the straight line through P that has the same direction as does the curve there. Since the direction of a line is determined by its slope, our plan for defining the tangent line amounts to finding a "slope-prediction" formula—one that will give the slope of the desired tangent line.

Our first example illustrates this approach in the case of one of the simplest of all nonstraight curves, the parabola $y = x^2$.

Tangent Lines and the Derivative

EXAMPLE 1   Determine the slope of the tangent line to the parabola $y = x^2$ at the point $P(a, a^2)$.

1.23 The maximum should occur where the tangent line is horizontal.

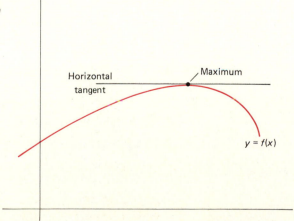

Horizontal tangent

Maximum

$y = f(x)$

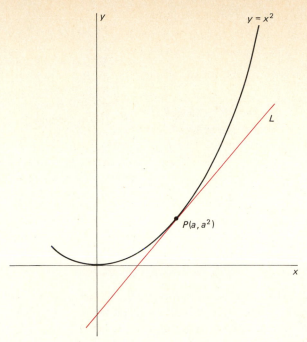

1.24 The tangent line at P should have the same direction as the curve does there.

Solution In Figure 1.24, we show the parabola $y = x^2$ with a typical point $P(a, a^2)$ marked on it. The figure also shows the result of a visual guess as to the position of the desired tangent line L. Our problem, then, is this: Find the slope of L.

To start with, we *can* compute the slope of a line that passes through two points. This line is the **secant line** K of Figure 1.25; it passes through $P(a, a^2)$

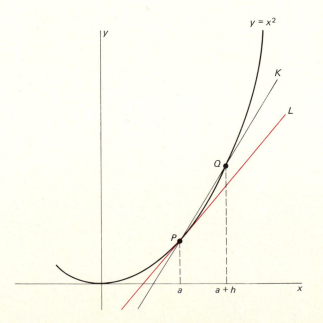

1.25 The secant line K passes through two points—P and Q—which we can use to compute its slope.

and the nearby point Q of the parabola. We take care to choose $Q \neq P$, though we *do* take Q quite close to P. Thus the abscissa of Q is not a; it is some nearby number, so it can be written in the form $a + h$, where h is some small nonzero number.

In effect, think of starting with the point $P(a, a^2)$, then changing the abscissa a of P by a small amount $h \neq 0$ to obtain a different number $a + h$, and finally using $x = a + h$ in the formula $y = x^2$ to get the ordinate of Q. Thus we can describe the point Q in terms of the single variable h, for we can write its coordinates in those terms: $Q = Q(a + h, (a + h)^2)$.

Since $Q \neq P$, we can use the definition of slope to compute the slope of the line K through P and Q. We denote this slope with the function notation $m(h)$ since the slope m is actually a function of h; if you change h, you change the line K, and thereby change the slope of K. Therefore

$$m(h) = \frac{(a + h)^2 - a^2}{(a + h) - a} = \frac{2ah + h^2}{h}, \tag{1}$$

or

$$m(h) = 2a + h. \tag{2}$$

Now imagine what happens as the point Q draws nearer and nearer to the point P. This corresponds to h approaching zero. The line K still passes through P and Q, but it pivots around the fixed point P. As h approaches zero, the secant line K comes more and more nearly into coincidence with the tangent line L. This phenomenon is suggested in Figure 1.26, which shows the secant line K almost coinciding with the tangent line L.

Thus the tangent line L is, by definition, the limiting position of the secant line K. To see precisely what this means, we look at what happens to

1.26 As $h \rightarrow 0$, Q approaches P, and K moves into coincidence with the tangent line L.

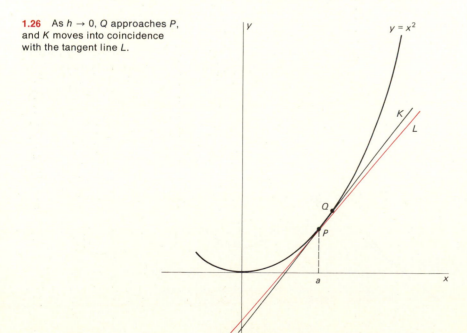

the slope of K as K pivots into coincidence with L:

> As h approaches zero,
> Q approaches P, and so
> K approaches L; meanwhile,
> *the slope of K approaches the slope of L.*

But we found above that the slope of the secant line K is

$$m(h) = 2a + h.$$

And it is quite clear that $m(h)$ approaches $2a$ as h approaches zero. We *must* define the slope m of the tangent line L by

$$m = 2a.$$

We summarize our discussion by saying that m is the **limit of $m(h)$** as h approaches zero, and we write

$$m = \lim_{h \to 0} m(h) = \lim_{h \to 0} (2a + h) = 2a. \qquad (3)$$

Now that we have the slope of the tangent line, we can write its equation. For instance, the point-slope formula for the tangent line to the parabola $y = x^2$ at the point $(1, 1)$ is

$$y - 1 = 2(x - 1);$$

that is,

$$y = 2x - 1.$$

The general case $y = f(x)$ is scarcely more complicated than the special case $y = x^2$. Suppose that $y = f(x)$ is given and that we want to find the slope of the tangent line L to its graph at the point $(a, f(a))$. As shown in Figure 1.27, let K be the secant line through P and a nearby point $Q =$

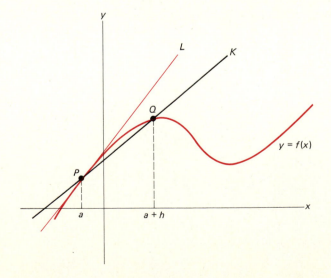

1.27 As $h \to 0$, $Q \to P$, and the slope of K approaches the slope of the tangent line L.

CHAP. 1: **Prelude to Differential Calculus**

$Q(a + h, f(a + h))$ on the graph. The slope of this secant line K is

$$m_{sec} = m(h) = \frac{f(a + h) - f(a)}{h} \qquad \text{(for } h \neq 0\text{).}} \qquad (4)$$

We now force Q to approach P along the curve by making h approach zero. If m_{sec} approaches a number m as h gets smaller and smaller approaching zero, then m is, *by definition*, the slope of the tangent line L at $P(a, f(a))$. In this case, we call m the **limit of m_{sec}** as h approaches zero, and we write

$$m = \lim_{h \to 0} m_{sec} = \lim_{h \to 0} \frac{f(a + h) - f(a)}{h}. \qquad (5)$$

The slope m in Equation (5) depends on both the function f and the number a. We indicate this by writing

$$m = f'(a) = \lim_{h \to 0} \frac{f(a + h) - f(a)}{h}. \qquad (6)$$

When we replace a by the more generic x, we see that we are really giving the definition of a new function:

$$f'(x) = \lim_{h \to 0} \frac{f(x + h) - f(x)}{h}. \qquad (7)$$

This new function f' is defined for all values of x such that the above limit exists. It is called the **derivative** of the original function f, and the process of computing the derivative f' (read "f prime") is called **differentiation** of f.

Our discussion leading up to Equation (7) shows that the derivative has the following important geometric interpretation:

$f'(x)$ is the slope of the line tangent to
the curve $y = f(x)$ at the point $(x, f(x))$.

This interpretation can be illustrated as shown in Figure 1.28.

There is a useful alternative notation for the derivative, and it originates from an early custom of writing Δx in place of h (because $\Delta x = h$ is an

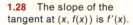

1.28 The slope of the tangent at $(x, f(x))$ is $f'(x)$.

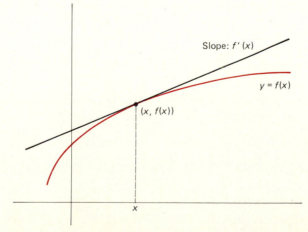

Slope: $f'(x)$

$y = f(x)$

$(x, f(x))$

x

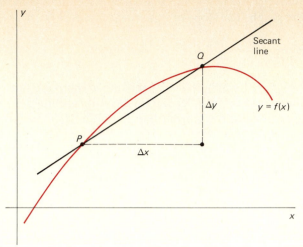

1.29 Genesis of the dy/dx notation.

increment in x) and $\Delta y = f(x + \Delta x) - f(x)$ for the resulting change (or increment) in y. The slope of the secant line K of Figure 1.29 is then

$$m_{sec} = \frac{\Delta y}{\Delta x}, \tag{8}$$

and the slope of the tangent line is

$$m = \frac{dy}{dx} = \lim_{\Delta x \to 0} \frac{\Delta y}{\Delta x}. \tag{9}$$

Hence, if $y = f(x)$, we often write

$$\frac{dy}{dx} = f'(x). \tag{10}$$

At present we can differentiate only very simple functions such as the one of Example 1, for which the meaning and existence of the limit in Equation (7) is obvious. In Chapter 2 we take up the systematic methods used to differentiate more complicated functions. These methods involve some deeper investigations of the concept of limit, which can be found in Section 1-8. Here, we mention only the following theorem, which tells how to differentiate a quadratic function with a graph that is a translated parabola.

Theorem *Differentiation of Quadratic Functions*
If $f(x) = ax^2 + bx + c$, then

$$f'(x) = 2ax + b. \tag{11}$$

Proof The definition of the derivative in (7) calls for us to substitute the given function f, and then to make algebraic simplifications until we can recognize the value of the limit as $h \to 0$. It will help to remember that x may

CHAP. 1: **Prelude to Differential Calculus**

be thought of as *constant* throughout such computations. We thus have

$$f'(x) = \lim_{h \to 0} \frac{f(x + h) - f(x)}{h}$$

$$= \lim_{h \to 0} \frac{[a(x + h)^2 + b(x + h) + c] - [ax^2 + bx + c]}{h}$$

$$= \lim_{h \to 0} \frac{2axh + bh + ah^2}{h}$$

$$= \lim_{h \to 0} (2ax + b + ah).$$

Thus we find that

$$f'(x) = 2ax + b,$$

since the value of ah approaches zero as $h \to 0$.

Note that the symbol $\lim_{h \to 0}$ signifies an operation to be performed, and so we continue to write it until the final step, when the operation is performed.

In geometric terms, the above theorem tells us this: The slope-prediction formula for curves of the form $y = ax^2 + bx + c$ is

$$m = 2ax + b. \tag{12}$$

EXAMPLE 2   Write equations for both the tangent line and the normal line to the parabola $y = 2x^2 - 3x + 5$ at the point $P(-1, 10)$.

Solution   We substitute $a = 2$, $b = -3$, and $x = -1$ into (12) and obtain $m = -7$ for the slope of the tangent line. The point-slope formula of the line is therefore

$$y - 10 = -7(x + 1)$$

or

$$y = 3 - 7x.$$

The **normal line** at P is, by definition, the straight line through P that is perpendicular to the tangent line. Its slope is $m' = -1/m = \frac{1}{7}$, so its point-slope formula is

$$y - 10 = \tfrac{1}{7}(x + 1).$$

Now we apply the derivative to wrap up our discussion of the refrigerator problem. Recall that the store manager has been purchasing refrigerators at a wholesale price of $250 each and has been selling 20 refrigerators per month at $400 each, for a monthly profit of $20(400 - 250) = \$3000$. Also, past experience indicates that an additional refrigerator can be sold each month for each $3 reduction in selling price. In Example 6 of Section 1-2, we found that, if x refrigerators per month are sold for $p = 460 - 3x$ dollars each, then the monthly profit will be

$$P(x) = 3x(70 - x)$$
$$= -3x^2 + 210x, \qquad 0 \leqq x \leqq 70.$$

This is the profit function that we seek to maximize.

Let us accept the intuitive fact—we shall give a rigorous proof in Chapter 2—that the maximum value of $P(x)$ occurs at a point where the tangent line to the graph of $y = P(x)$ is horizontal. Then we can apply the above theorem to find this maximum point. The slope at an arbitrary point $(x, P(x))$ is

$$m = P'(x) = -6x + 210.$$

We ask when $m = 0$, and we find that this happens when

$$-6x + 210 = 0,$$

or when $x = 35$. Thus, in agreement with the result we got by algebraic methods in Section 1-4, we find that the maximum profit is

$$P(35) = 3(35)(70 - 35) = 3675.$$

So just a bit of mathematics enables the store manager to add $675 to her monthly profit!

It is possible for a function to be defined everywhere but to fail to have a derivative—that is, not be differentiable—at some points in its domain. The example below shows this.

TERMINOLOGY The function f is said to be **differentiable** at the point x provided that the limit of Equation (7) (the definition of $f'(x)$) exists.

EXAMPLE 3 Show that the function $f(x) = |x|$ is *not* differentiable at $x = 0$.

Solution When $x = 0$, we have

$$\frac{f(x + h) - f(x)}{h} = \frac{|h|}{h} = \begin{cases} -1 & \text{if } h < 0; \\ +1 & \text{if } h > 0. \end{cases}$$

Thus there is no *single* value that the quotient

$$\frac{f(0 + h) - f(0)}{h}$$

approaches as $h \to 0$. This means that the limit in question does not exist, and so $f(x) = |x|$ is not differentiable at $x = 0$.

We have shown the graph of the function f of Example 3 in Figure 1.30. The graph has a sharp corner at the point $(0, 0)$, and this explains why there can be no tangent line there—no single straight line is a good approximation to the shape of the graph at $(0, 0)$. On the other hand, the figure makes it evident that $f'(x)$ exists if $x \neq 0$. Indeed,

$$f'(x) = \begin{cases} -1 & \text{if } x < 0; \\ +1 & \text{if } x > 0. \end{cases}$$

1.30 The graph of $f(x) = |x|$ has a corner point at $(0, 0)$.

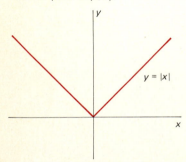

CHAP. 1: Prelude to Differential Calculus

Apply Equation (7), the definition of the derivative, to find $f'(x)$ for each of the functions in Problems 1–12. Then write an equation for the tangent line to the curve $y = f(x)$ at the point $(2, f(2))$.

1 $f(x) = 3x - 1.$

2 $f(x) = x^2 - x - 2.$

3 $f(x) = 2x^3 - 3x + 5.$

4 $f(x) = 70x - x^2.$

5 $f(x) = (x - 1)^2.$

6 $f(x) = x^3.$

7 $f(x) = \dfrac{1}{x}.$

8 $f(x) = x^4.$

9 $f(x) = \dfrac{1}{x^2}.$

10 $f(x) = x^2 + \dfrac{3}{x}.$

11 $f(x) = \dfrac{2}{x - 1}.$

12 $f(x) = \dfrac{x}{x - 1}.$

In the remaining problems, you should use Equation (11) to find the derivative of a quadratic function where necessary. In Problems 13–15, write equations for the tangent line and the normal line to the curve $y = f(x)$ at the point P given in the problem.

13 $y = x^2$; $P(-2, 4).$

14 $y = 5 - x - 2x^2$; $P(-1, 4).$

15 $y = 2x^2 + 3x - 5$; $P(2, 9).$

In Problems 16–18 find the point(s) of the curve $y = f(x)$ where the tangent line is horizontal.

16 $y = 10 - x^2.$

17 $y = 10x - x^2.$

18 $y = 3x^2 + 12x + 16.$

19 One of the two lines that pass through the point $(3, 0)$ and are tangent to the parabola $y = x^2$ is the x-axis itself. Find an equation for the *other* line.

20 Write an equation of the line through $(3, 0)$ that is normal to the parabola $y = x^2$.

21 Write equations for the two straight lines through the point $(2, 5)$ that are tangent to the parabola $y = 4x - x^2$.

22 Prove that the tangent line to the parabola $y = x^2$ at the point (x_0, y_0) intersects the x-axis at the point $(\frac{1}{2}x_0, 0)$.

23 In Problem 32 of Section 1-2, you were asked to express the daily production of a certain oil field as a function $P = f(x)$ of the number x of new oil wells drilled. Determine the maximum possible production by finding the point where $f'(x) = 0$.

24 If a projectile is fired at an angle of 45° from the horizontal, with initial position the origin in the xy-plane and with an initial velocity of $100\sqrt{2}$ feet per second, then its trajectory is the parabola $y = x - (x/25)^2$.
(a) How far does the projectile travel (horizontally) before it hits the ground?
(b) What is the maximum height above the ground that the projectile attains?

1-6

The Derivative and Rates of Change

In Section 1-5 we introduced the derivative of a function as the slope of the tangent line to its graph. Here we introduce the equally important interpretation of the derivative of a function as the function's rate of change with respect to the independent variable.

We begin with the instantaneous rate of change of a function whose independent variable is time t. Suppose that Q is a quantity that varies with time, and write $Q = f(t)$ for its value at time t. For instance, Q might be

1 The size of a population (such as rabbits, people, or bacteria);

2 The number of dollars in a bank account;

3 The volume of a balloon that is being inflated;

4 The amount of water in a reservoir with variable inflow and outflow; or

5 The amount of a certain chemical product produced in a reaction up to time t.

The change in Q from time t to time $t + \Delta t$ is the **increment**

$$\Delta Q = f(t + \Delta t) - f(t).$$

The **average rate of change** of Q (per unit of time) is, by definition, the quotient

$$\frac{\Delta Q}{\Delta t} = \frac{f(t + \Delta t) - f(t)}{\Delta t}. \tag{1}$$

We define the **instantaneous rate of change** of Q (per unit of time) to be the limit of this average rate as $\Delta t \to 0$. That is, the instantaneous rate of change is

$$\lim_{\Delta t \to 0} \frac{\Delta Q}{\Delta t} = \lim_{\Delta t \to 0} \frac{f(t + \Delta t) - f(t)}{\Delta t}. \tag{2}$$

But this latter limit is simply the derivative $f'(t)$. So we see that the instantaneous rate of change of $Q = f(t)$ is

$$\frac{dQ}{dt} = f'(t). \tag{3}$$

This concept of the instantaneous rate of change agrees with our intuitive idea of what it should be. If we think of Q as changing with time, but then suddenly, at time t, continuing in the direction of its graph at time t without subsequent curvature, the graph of Q would appear as shown in Figure 1.31. The broken line in that figure is meant to indicate how $Q = f(t)$ might normally behave. But if Q were to follow the straight line, that would be "change at a constant rate." Since the straight line is tangent to the graph of Q, we can interpret dQ/dt as the instantaneous rate of change of Q at time t. In brief,

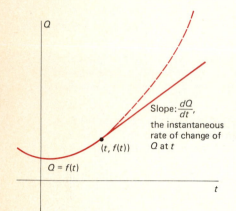

Slope: $\frac{dQ}{dt}$, the instantaneous rate of change of Q at t

$(t, f(t))$

$Q = f(t)$

1.31 The relation between the tangent line at $(t, f(t))$ and the instantaneous rate of change at t.

> The instantaneous rate of change of $Q = f(t)$ at time t is equal to the slope of the tangent line to the curve $Q = f(t)$ at the point $(t, f(t))$.

There are important additional conclusions that we can draw. Because a positive slope corresponds to a rising tangent line, and a negative slope to a falling tangent line, we see that

$$Q \text{ is increasing at time } t \text{ if } \frac{dQ}{dt} > 0;$$

$$Q \text{ is decreasing at time } t \text{ if } \frac{dQ}{dt} < 0.$$

Note this: It's clear how to define the phrase "$Q = f(t)$ is increasing *over the time interval* $a \leqq t \leqq b$." What we have here is a way to make precise what is meant by "$Q = f(t)$ is increasing *at time t*."

EXAMPLE 1 A cylindrical tank with vertical axis is initially filled with 200,000 gallons of water. This tank takes 50 minutes to empty after a drain in its bottom is opened. Suppose that the drain is opened at time $t = 0$. A consequence of Torricelli's law is that the volume V of water remaining in the tank after t minutes is

$$V = (200{,}000)\left(1 - \frac{t}{50}\right)^2,$$

or

$$V = 200{,}000 - 8000t + 80t^2 \qquad \text{(gallons)}.$$

Find the instantaneous rate at which the water is flowing out of the tank when $t = 30$ (minutes).

Solution We simply need to find the value of dV/dt when $t = 30$. But

$$\frac{dV}{dt} = -8000 + 160t$$

as long as the water is flowing, and the value of dV/dt at $t = 30$ is just -3200. The units for V are gallons, and t is measured in minutes in this problem. Therefore the units for "rate of change of V with respect to t" are gallons per minute. Thirty minutes after the drainhole is opened, the water is flowing out at 3200 gallons per minute.

The minus sign we obtained above means that V is a *decreasing* function of t; had dV/dt been a positive number, we would have concluded that the tank was being filled rather than emptied.

Our examples of functions have so far been restricted to those having either formulas or verbal descriptions. In scientific work we often work with a table of values obtained from observation or from empirical measurement. Our next example shows how the instantaneous rate of change of such a tabulated function can be estimated.

EXAMPLE 2 The table in Figure 1.32 gives the United States population P (in millions) at 10-year intervals in the nineteenth century. Estimate the instantaneous rate of population growth in 1850.

Solution We take $t = 0$ in 1800, so that $t = 50$ in 1850. In Figure 1.33 we have plotted the data of our example and added a freehand sketch of a smooth curve that fits these data.

However it may be obtained, a curve that fits the data ought to be a good approximation to the true graph of the unknown function $P = f(t)$. The instantaneous rate of change dP/dt is the slope of the tangent line at the point (50, 23.2). We draw the tangent as accurately as we can by visual inspection

t	Year	U.S. Pop. (millions)
0	1800	5.3
10	1810	7.2
20	1820	9.6
30	1830	12.9
40	1840	17.1
50	1850	23.2
60	1860	31.4
70	1870	38.6
80	1880	50.2
90	1890	62.9
100	1900	76.0

1.32 Data for Example 2.

1.33 A smooth curve that fits the data well.

SEC. 1-6: The Derivative and Rates of Change

and then measure the base and height of the triangle of Figure 1.33. Thus we approximate the slope of the tangent as

$$\frac{dP}{dt} \approx \frac{36}{51} \approx 0.71$$

millions of people per year (in 1850). Although there was no national census in 1851, we would expect the population of the United States then to have been approximately $23.2 + 0.7 = 23.9$ million people.

VELOCITY AND ACCELERATION

A particle moves along a straight line with **position function** $s = f(t)$. Thus we make the line of motion a coordinate axis: To so specify f is to have chosen an origin and a positive direction on the line of motion. Also, $f(t)$ is just the coordinate of the particle at time t.

Think of the time interval from t to $t + \Delta t$. The particle moves from $f(t)$ to $f(t + \Delta t)$. Its displacement is then the increment

$$\Delta s = f(t + \Delta t) - f(t).$$

The **average velocity** of the particle during this time interval is

$$\bar{v} = \frac{\text{Displacement}}{\text{Elapsed time}} = \frac{\Delta s}{\Delta t} = \frac{f(t + \Delta t) - f(t)}{\Delta t}. \tag{4}$$

We define the **instantaneous velocity** v of the particle at the time t to be the limit of this average velocity $\bar{v}$ as $\Delta t \to 0$. That is,

$$v = \lim_{\Delta t \to 0} \frac{\Delta s}{\Delta t} = \lim_{\Delta t \to 0} \frac{f(t + \Delta t) - f(t)}{\Delta t}. \tag{5}$$

We recognize the right-hand limit—it's the definition of the derivative of f at time t. So the velocity of the moving particle at time t is

$$v = \frac{ds}{dt} = f'(t). \tag{6}$$

Thus *velocity is instantaneous rate of change of position.* The velocity of a moving particle is positive or negative, depending on whether it is moving in the positive or negative direction along the line of motion. The *speed* is the absolute value $|v|$ of the particle's velocity.

For example, if a particle is projected straight upward from a height s_0 above the ground, at time $t = 0$, and with initial (upward) velocity v_0, then its height s above the ground at time t is

$$s = -\tfrac{1}{2}gt^2 + v_0 t + s_0. \tag{7}$$

Here g denotes the gravitational acceleration. Near the surface of the earth, $g \approx 32 \text{ ft/sec}^2$, or $g \approx 980 \text{ cm/sec}^2$. If we differentiate s, we obtain the velocity of the particle at time t:

$$v = \frac{ds}{dt} = -gt + v_0. \tag{8}$$

The **acceleration** of the particle is the instantaneous rate of change, or derivative, of its velocity,

$$a = \frac{dv}{dt} = -g. \tag{9}$$

Your intuition, and probably your experience, tell you that a body so projected upward will reach its maximum height at the instant that its velocity vanishes—when $v = 0$. (We shall see why this is true in Chapter 2.)

EXAMPLE 3 Find the maximum height attained by a ball that is thrown straight upward from the ground with an initial velocity of $v_0 = +96$ feet per second.

Solution We use Equation (8), with $v_0 = 96$ and $g = 32$. We find the velocity of the ball to be

$$v = -32t + 96.$$

Since $v = 0$ when $t = 3$, the ball attains its maximum height s_{max} at time $t = 3$ seconds. Then we use Equation (7) to find s_{max}:

$$s_{max} = -16(3)^2 + 96(3) = 144 \quad \text{(feet)}.$$

The derivative of any function—not merely a function of time—may be interpreted as its instantaneous rate of change with respect to the independent variable. If $y = f(x)$, then the **average rate of change** of y (per unit change in x) in the interval $[x, x + \Delta x]$ is the quotient

$$\frac{\Delta y}{\Delta x} = \frac{f(x + \Delta x) - f(x)}{\Delta x}.$$

The **instantaneous rate of change of y with respect to x** is the limit as $\Delta x \to 0$ of this average rate of change. That is,

$$\lim_{\Delta x \to 0} \frac{\Delta y}{\Delta x} = \frac{dy}{dx} = f'(x).$$

For instance, the rate of change of the area $A = \pi r^2$ of a circle with respect to its radius r is

$$\frac{dA}{dr} = 2\pi r.$$

1-6 PROBLEMS

In each of Problems 1–5, the position function $s = f(t)$ of a body moving in a line is given. Find its location s when its velocity v is zero.

1 $s = 100 - 16t^2$.
2 $s = -16t^2 + 160t + 25$.
3 $s = -16t^2 + 80t - 1$.
4 $s = 10t^2 + 50$.
5 $s = 100 - 20t - 5t^2$.

6 The Celsius temperature C is given in terms of Fahrenheit temperature F by $C = \frac{5}{9}(F - 32)$. Find the rate of change of C with respect to F and also the rate of change of F with respect to C.

7 Find the rate of change of the area A of a circle with respect to its circumference C.

8 A stone dropped into a pond (at $t = 0$) causes a circular ripple that travels out from the point of impact at 5 feet per second. At what rate is the area within this circle increasing when $t = 10$ (seconds)?

9 A car is traveling at 100 feet per second when the driver suddenly applies the brakes ($s = 0$, $t = 0$). The car's position function while skidding is

$$s = 100t - 5t^2.$$

How long and how far does it skid before coming to a stop?

10 A water bucket containing 10 gallons of water develops a leak when $t = 0$, and the volume of water in the bucket t seconds later is

$$V = 10\left(1 - \frac{t}{100}\right)^2.$$

(a) At what rate is water leaking out of the bucket after 1 minute has passed?
(b) When is the instantaneous rate of change of V equal to the average rate of change of V from $t = 0$ to $t = 100$ (seconds)?

11 A certain population of rodents numbers

$$P = 100(1 + 0.3t + 0.04t^2)$$

after t months.

(a) How long does it take this population to double?
(b) What is the rate of growth of the population when $P = 200$?

12 The following data describe the growth of the population P (in thousands) of a certain city during the 1970s. Use the graphical method of Example 2 to estimate its rate of growth in 1975.

Year	1970	1972	1974	1976	1978	1980
P	265	293	324	358	395	437

13 The following data give the distance s (in feet) traveled by an accelerating car (starting from rest) in the first t seconds. Use the graphical method of Example 2 to estimate its speed (in miles per hour) when $t = 20$ seconds and again when $t = 40$ seconds.

t	0	10	20	30	40	50	60
s	0	224	810	1655	2686	3850	5109

In Problems 14–16, use the fact that the derivative of $y = x^3$ is $dy/dx = 3x^2$.

14 Show that the rate of change of the volume of a cube with respect to its edge length is equal to half the surface area of the cube.

15 Show that the rate of change of the volume of a sphere with respect to its radius is equal to its surface area.

16 The height of a cylinder is equal to twice its radius. Show that the rate of change of its volume with respect to its radius is equal to its total surface area.

17 A spherical balloon with an initial radius of 5 inches begins to leak at $t = 0$, and its radius t seconds later is $r = (60 - t)/12$ inches. At what rate (in cubic inches per second) is air leaking out of the balloon when $t = 30$ (seconds)?

18 The volume V (in liters) of 3 grams of CO_2 at 27°C is given in terms of its pressure p (atm) by the formula $V = 1.68/p$. What is the rate of change of V with respect to p when $p = 2$ (atm)?

1-7

Combinations of Functions and Inverse Functions

Many important functions are built of simpler ones, and the results can be quite complex. Here we discuss some of the ways of combining functions to obtain new ones.

Suppose that f and g are functions, and that c is a fixed number. The **(scalar) multiple** cf, the **sum** $f + g$, the **difference** $f - g$, the **product** $f \cdot g$, and the **quotient** f/g are the new functions determined by the formulas

$$(cf)(x) = cf(x), \tag{1}$$

$$(f + g)(x) = f(x) + g(x), \tag{2}$$

$$(f - g)(x) = f(x) - g(x), \tag{3}$$

$$(f \cdot g)(x) = f(x) \cdot g(x), \quad \text{and} \tag{4}$$

$$\left(\frac{f}{g}\right)(x) = \frac{f(x)}{g(x)}. \tag{5}$$

The combinations in (2) through (4) are defined for every number x in both the domain of f and the domain of g. In (5), we must require that $g(x) \neq 0$.

For example, let $f(x) = x^2 + 1$ and $g(x) = x - 1$. Then

$$(3f)(x) = 3(x^2 + 1),$$

$$(f + g)(x) = (x^2 + 1) + (x - 1) = x^2 + x,$$

$$(f - g)(x) = (x^2 + 1) - (x - 1) = x^2 - x + 2,$$

$$(f \cdot g)(x) = (x^2 + 1)(x - 1) = x^3 - x^2 + x - 1, \qquad \text{and}$$

$$(f/g)(x) = \frac{x^2 + 1}{x - 1} \qquad (x \neq 1).$$

If $f(x) = \sqrt{1 - x}$ for $x \leq 1$ and $g(x) = \sqrt{1 + x}$ for $x \geq -1$, then the sum or product of f and g is defined where *both* f and g are defined. In this case, that domain would be the closed interval $[-1, 1]$.

You can build very complicated functions from simple functions using only the operations of Equations (1)–(5). For example, from the *identity* function $I(x) = x$ and the constant function $1(x) = 1$ (a useful abuse of notation), you can construct the function $f(x) = x^2 - 2x - 3$ by forming

$$f = I \cdot I - 2 \cdot I - 3.$$

In fact, you can "build" *any* polynomial in similar fashion, for

$$p(x) = a_n x^n + a_{n-1} x^{n-1} + \cdots + a_1 x + a_0$$

is the sum and product of constant functions and the identity function. By dividing the polynomial $p(x)$ by $q(x)$, you obtain also the **general rational function**

$$r(x) = \frac{p(x)}{q(x)}.$$

COMPOSITION OF FUNCTIONS

Some important functions cannot be constructed in any of the ways suggested above; for example, the function

$$k(x) = \sqrt{1 - x^2}, \qquad -1 \leq x \leq 1.$$

The graph of k is the upper half of the circle $x^2 + y^2 = 1$, and it is shown in Figure 1.34.

Certainly $f(x) = \sqrt{x}$ and $g(x) = 1 - x^2$ are simple functions. But they cannot be added, subtracted, multiplied, or divided to produce $k(x)$. On the other hand, if you start with x, let g act upon it,

$$x \xrightarrow{\;g\;} 1 - x^2$$

and next use the *output* of g as the *input* for f,

$$x \xrightarrow{\;g\;} 1 - x^2 \xrightarrow{\;f\;} \sqrt{1 - x^2},$$

then the *net* effect is that x is transformed into $\sqrt{1 - x^2}$. That is exactly how k is supposed to act on the input x.

When you think of a function as a sort of calculator, one that accepts values of x and produces values of the function, then the way we have

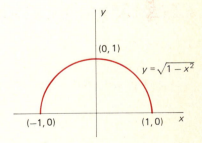

1.34 The graph of k.

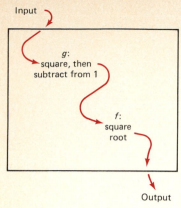

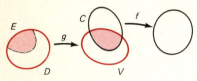

1.35 The action of $k(x) = \sqrt{1 - x^2}$.

1.36 The domain E of $f(g)$.

combined $f(x) = x^{1/2}$ and $g(x) = 1 - x^2$ to produce $k(x)$ could be represented in the way illustrated in Figure 1.35. Moreover, a natural way to write k in terms of f and g would be this:

$$k(x) = f(g(x)).$$

The expression $g(x)$ appears as the input or argument of f to signify that the *value* $g(x)$ is to be acted on by the function f. This is exactly what we need to produce the effect of the function k.

WARNING The order in which the two functions g and f are to be applied is this: *first g, then f*. Though the notation $k(x) = f(g(x))$ shows f appearing first, you should note upon close examination that *f is to be applied last*.

The way in which f and g are combined to form k is called the *composition* of f and g and is perhaps the most important way in which two functions can be combined. The reason is that many human activities are performed in the same way. Certain operations are to be performed in a certain order and each step acts on the result of the previous step: It's much like assembling an automobile, making a pizza, dancing, or learning calculus.

What about the domain of $f(g(x))$? Suppose that g has domain D and range V, and that f has domain C that at least partly overlaps V, as we have shown in Figure 1.36. Then the **composition** of f and g is denoted by $f(g)$, and its value at x is given by

$$(f(g))(x) = f(g(x)). \tag{6}$$

The domain E of $f(g)$ is the set of all numbers in D that give values in C.

Another common notation for the composition is $f \circ g$, with a small circle between f and g.

The composition notation is very different from the product notation. The product of the two functions f and g is written $f \cdot g$, or simply fg. But the composition of f and g may involve no multiplication whatsoever and is written $f(g)$. This is read "f of g," and its formula $f(g(x))$ is read "f of g of x."

A sure way to see that products differ from compositions is this: The product of f and g is the same whether f or g is written first. But when we reverse the order of composition, a different function is usually the result. With $f(x) = x^{1/2}$ and $g(x) = 1 - x^2$ as above, we find that

$$g(f(x)) = 1 - (\sqrt{x})^2 = 1 - x, \qquad x \geqq 0.$$

This is not at all the same as

$$f(g(x)) = \sqrt{1 - x^2}, \qquad -1 \leqq x \leqq 1.$$

EXAMPLE 1 Find $f(g)$ and $g(f)$, given $f(x) = x^3$ and $g(x) = x - 2$.

Solution First, $f(g)$ has the rule

$$f(g(x)) = (x - 2)^3.$$

The domain of g is all of $\mathcal{R}$. Since every possible value of g lies in the domain (also $\mathcal{R}$) of f, the domain of $f(g)$ is just $\mathcal{R}$.

Next, $g(f)$ has the rule

$$g(f(x)) = g(x^3) = x^3 - 2,$$

and its domain is also all of $\mathcal{R}$.

Note that the two possible results of composition in Example 1 are *not* the same function. In addition, neither is equal to the product of f and g:

$$(f \cdot g)(x) = f(x)g(x) = x^3(x - 2).$$

INVERSE FUNCTIONS

We often encounter pairs of functions that are natural opposites of one another, in the sense that each undoes the result of applying the other. A schematic of this phenomenon is shown in Figure 1.37, but actual examples may be more informative: Below, we give some pairs of functions that are inverses of each other. The domain of every function is, unless otherwise specified, the set of all real numbers for which its formula is meaningful.

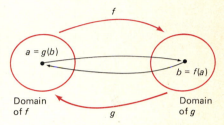

1.37 The action of inverse functions f and g.

EXAMPLE 2 The following are some pairs of inverse functions.

$$f(x) = x + 1 \quad \text{and} \quad g(x) = x - 1$$

("Adding 1" and "subtracting 1" are inverse operations; doing one undoes the other.)

$$f(x) = 2x \quad \text{and} \quad g(x) = \frac{x}{2}$$

("Doubling" and "halving" are inverse operations.)

$$f(x) = \frac{1}{x} \quad \text{and} \quad g(x) = \frac{1}{x}$$

(A function can be its own inverse.)

$$\begin{array}{ccc} f : (0, \infty) \to \mathcal{R} & & g : (0, \infty) \to \mathcal{R} \\ \text{by } f(x) = x^2 & \text{and} & \text{by } g(x) = \sqrt{x} \end{array}$$

("Squaring" and "taking the square root" are inverse operations when only positive numbers are involved.)

Each pair f and g of functions given above has the property that

$$f(g(x)) = x \quad \text{and} \quad g(f(x)) = x. \tag{7}$$

When this pair of equations holds for all values of x in the appropriate domains, then we say that f and g are *inverses* of each other.

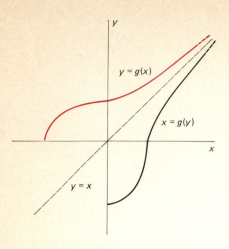

1.38 The graphs of two inverse functions are reflections of each other through the line $y = x$.

Definition *Inverse Functions*

Two functions $f : A \to B$ and $g : D \to E$ are said to be **inverse functions,** or **inverses** of each other, provided that

(i) The domain of each is the range of the other: $A = E$ and $D = B$;

and

(ii) $f(g(x)) = x$ for all x in D, and $g(f(x)) = x$ for all x in A.

Care is frequently required in specifying the domains of f and g to insure that Condition (i) of the definition is satisfied. For example, the rule $f(x) = x^2$ makes sense for all real x, but it is not the same as the function

$$f(x) = x^2, \qquad x \geq 0$$

that is the inverse of $g(x) = \sqrt{x},\ x \geq 0$. On the other hand, the functions

$$f(x) = x^3 \quad \text{and} \quad g(x) = x^{1/3}$$

are inverses of each other, with no restriction on their domains.

Condition (ii) of the definition above is the same as the statement that

$$f(x) = y \quad \text{if and only if} \quad g(y) = x$$

for all x in the domain of f and all y in the domain of g. That is, the graphs of the two equations

$$y = f(x) \quad \text{and} \quad x = g(y)$$

are *identical.* On the other hand, the graphs of the two equations

$$x = g(y) \quad \text{and} \quad y = g(x)$$

are *reflections* of one another across the line $y = x$. This is shown in Figure 1.38. The reason for this reflection phenomenon is that either of the last two equations may be obtained from the other simply by interchanging x and y; geometrically, this corresponds to interchanging the x-axis with the y-axis. This is the reason for the "one-dimensional mirror" along the line $y = x$. And hence:

Theorem 1 *Reflection Property of Inverse Functions*

Two functions f and g are inverse functions if and only if their graphs are reflections of each other across the line $y = x$.

We illustrate this property with the graphs of the pairs of inverse functions of Example 2, shown in Figures 1.39 through 1.42.

The reflection property above gives us a criterion for determining whether a given function has an inverse function. Suppose that f and g are inverses of each other. Then each vertical line intersects the graph of g in at most one point. It then follows from the reflection property that each horizontal line

1.39 The inverse function pair $f(x) = x + 1$ and $g(x) = x - 1$.

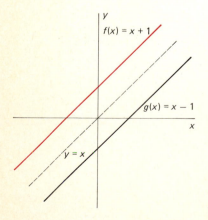

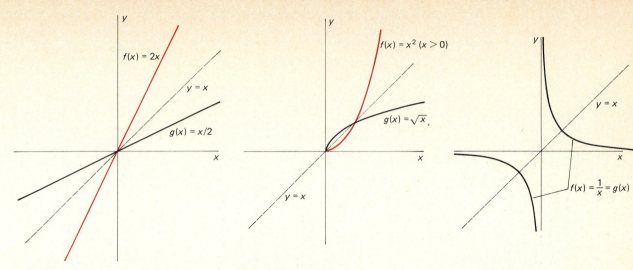

1.40 The inverse function pair $f(x) = 2x$ and $g(x) = x/2$.

1.41 The inverse function pair $f(x) = x^2$ $(x > 0)$ and $g(x) = \sqrt{x}$ $(x > 0)$.

1.42 The inverse of $f(x) = 1/x$ is $g(x) = 1/x = f(x)$.

intersects the graph of f in at most one point. Consequently f is a **one-to-one** function; that is, given y_1 in the range of f, there is only one number x_1 in the domain of f such that $f(x_1) = y_1$. By reasoning along such lines, one can construct a proof of the following theorem.

Theorem 2   *One-to-One Property of Inverse Functions*

The function f has an inverse function if and only if f is a one-to-one function.

If the function f defined by the formula $y = f(x)$ has an inverse function g, then we can attempt to find a formula for g by solving the equation $y = f(x)$ for $x = g(y)$.

EXAMPLE 3   The function

$$f(x) = \frac{1}{x^2 + 1}, \qquad x \geqq 0,$$

is one-to-one (because $0 \leqq a < b$ implies $0 \leqq a^2 < b^2$). Find its inverse function g.

Solution   We solve the equation $y = 1/(x^2 + 1)$ and obtain

$$x^2 + 1 = \frac{1}{y}$$

so that

$$x = \sqrt{\frac{1}{y} - 1} = \sqrt{\frac{1 - y}{y}} = g(y).$$

We take the positive root because $x \geq 0$. The domain of g is the range $(0, 1]$ of f. Hence, taking x as the independent variable, we find that g is the function

$$g(x) = \sqrt{\frac{1-x}{x}}, \qquad 0 < x \leq 1.$$

1-7 PROBLEMS

In Problems 1–5, find $f + g$, $f \cdot g$, and f/g, and give the domain of each of these new functions.

1 $f(x) = x + 1$, $g(x) = x^2 + 2x - 3$.

2 $f(x) = \dfrac{1}{x-1}$, $g(x) = \dfrac{1}{2x+1}$.

3 $f(x) = \sqrt{x}$, $g(x) = \sqrt{x-2}$.

4 $f(x) = \sqrt{x+1}$, $g(x) = \sqrt{5-x}$.

5 $f(x) = \sqrt{x^2-1}$, $g(x) = 1/\sqrt{4-x^2}$.

In Problems 6–10, find $f(g(x))$ and $g(f(x))$.

6 $f(x) = 1 - x^2$, $g(x) = 2x + 3$.

7 $f(x) = -17$, $g(x) = |x|$.

8 $f(x) = \sqrt{x^2 - 3}$, $g(x) = x^2 + 3$.

9 $f(x) = x^2 + 1$, $g(x) = \dfrac{1}{x^2+1}$.

10 $f(x) = x^3 - 4$, $g(x) = \sqrt[3]{x+4}$.

In Problems 11–15, find g such that $f(g(x)) = h(x)$.

11 $f(x) = 2x + 3$, $h(x) = 2x + 5$.

12 $f(x) = x + 1$, $h(x) = x^3$.

13 $f(x) = x^2$, $h(x) = x^4 + 1$.

14 $f(x) = \dfrac{1}{x}$, $h(x) = x^5$.

15 $f(x) = \sqrt{x-2}$, $h(x) = \sqrt{x+2}$.

In Problems 16–20, determine whether f and g are inverse functions.

16 $f(x) = 4x - 5$, $g(x) = \dfrac{x+5}{4}$.

17 $f(x) = 2x + 3$, $g(x) = \dfrac{x+3}{2}$.

18 $f(x) = x^2 + 3$, $g(x) = \sqrt{x+3}$.

19 $f(x) = \dfrac{1}{\sqrt{x+1}}$, $g(x) = \dfrac{1-x^2}{x^2}$.

20 $f(x) = x^3 + 1$, $g(x) = \sqrt[3]{x-1}$.

In Problems 21–26, find the inverse function of f.

21 $f(x) = 13x - 5$.

22 $f(x) = \sqrt{3x + 7}$.

23 $f(x) = 3x^3 + 1$.

24 $f(x) = x^2 - 4$, $0 \leq x \leq 2$.

25 $f(x) = (x^3 - 1)^{17}$, $0 \leq x \leq 1$.

26 $f(x) = \dfrac{x+1}{x-1}$.

27 Suppose that $f(x) = ax + b$ $(a \neq 0)$. Prove that f has an inverse function defined for all x.

28 Suppose that $f(x) = x^2 + bx + c$ with domain the set $\mathcal{R}$ of all real numbers. Prove that f does not have an inverse function.

29 Suppose that g_1 and g_2 are both inverse functions of f. Prove that $g_1 = g_2$.

30 The function f is called an **increasing function** provided that $x_1 < x_2$ implies $f(x_1) < f(x_2)$. Prove that every increasing function has an inverse function.

31 Apply the result of Problem 30 to show that $f(x) = x^3 + 2x$ has an inverse function.

1-8

Limits and the Limit Laws

In Section 1-5, we defined the derivative $f'(a)$ of the function f at the number a by

$$f'(a) = \lim_{h \to 0} \frac{f(a+h) - f(a)}{h}. \tag{1}$$

CHAP. 1: **Prelude to Differential Calculus**

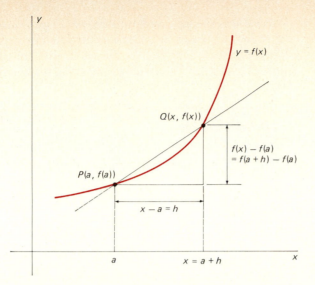

1.43 The derivative can be defined this way:

$$f'(a) = \lim_{x \to a} \frac{f(x) - f(a)}{x - a}.$$

The picture that motivated this definition appears in Figure 1.43, with $a + h$ relabeled as x (so that $h = x - a$). We see that x approaches a as h approaches zero, so (1) can be rewritten in the form

$$f'(a) = \lim_{x \to a} \frac{f(x) - f(a)}{x - a}. \qquad (2)$$

Thus the computation of $f'(a)$ amounts to the determination of the limit, as x approaches a, of the function

$$F(x) = \frac{f(x) - f(a)}{x - a}. \qquad (3)$$

To prepare for the differentiation of more complicated functions than we could handle in Section 1-5, we now turn our attention to the meaning of the statement

$$\lim_{x \to a} F(x) = L. \qquad (4)$$

This is read as "the limit of $F(x)$, as x approaches a, is L." We shall sometimes write (4) in the compact form $F(x) \to L$ as $x \to a$.

To discuss the limit of F at a is not to require that F be defined at a. Neither its value there, nor the possibility that it has no value at a, is of importance. All we require is that F be defined in some **deleted neighborhood** of a, meaning a set obtained by deleting the single point a from some neighborhood of a. Note that this is exactly the situation in the case of a function like (3) above, which is defined *except at a*.

The following statement gives the meaning of Equation (4) in intuitive language.

Idea of the Limit

We say that the number L is the *limit* of $F(x)$ as x approaches a provided that the number $F(x)$ can be made as close as one pleases to L by choosing x sufficiently near, though not equal to, the number a.

What this means, roughly, is that $F(x)$ gets closer and closer to L as x gets closer and closer to a. Once we decide how close to L we want $F(x)$ to be, it is necessary that $F(x)$ be that close to L for *all* x sufficiently close to (but not equal to) a. The actual definition of the limit makes this requirement precise.

Definition of the Limit

We say that the number L is the **limit** of $F(x)$ as x approaches a provided that, given any number $\varepsilon > 0$, there exists a number $\delta > 0$ such that

$$|F(x) - L| < \varepsilon$$

for all x such that

$$0 < |x - a| < \delta.$$

In this section we explore the "idea of the limit," mainly through the investigation of specific examples. We defer further discussion of the precise definition of the limit to Section 1-10.

EXAMPLE 1   Evaluate $\lim\limits_{x \to 2} x^2$.

Investigation   Note that we call this an "investigation" rather than a solution. Our method is useful for gleaning preliminary information about what the value of a limit might be. This method reinforces our intuitive guess as to what a limit ought to be, but it does not constitute a proof of the value of the limit.

We form a table of values of x^2, using values of x approaching $a = 2$. We use steadily approaching values of x since that makes the behavior exhibited in the table easier to understand. We use values of x that are "nice" in the decimal system since that, too, makes the behavior in the table of Figure 1.44 easier to follow.

In any case, examine the table—read down the column for x, since "down" is the table's direction for "approaches"—to see what happens to the values of x^2. In spite of the fact that we're using only rather special values of x, the table provides convincing evidence that

$$\lim\limits_{x \to 2} x^2 = 4.$$

Surely it is almost certain, and perhaps quite clear, that the limit of x^2 is 4 as x approaches 2.

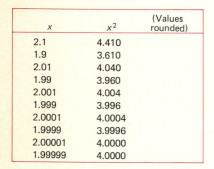

x	x^2	(Values rounded)
2.1	4.410	
1.9	3.610	
2.01	4.040	
1.99	3.960	
2.001	4.004	
1.999	3.996	
2.0001	4.0004	
1.9999	3.9996	
2.00001	4.0000	
1.99999	4.0000	

1.44   What happens to x^2 as x approaches 2?

IMPORTANT   *Note that we did not substitute the value $x = 2$ into the function $F(x) = x^2$ to obtain the value 4 of the limit.* While such substitution produces the correct answer in this particular case, in many important limits it produces either no answer at all or an incorrect answer.

EXAMPLE 2   Evaluate $\lim\limits_{x \to 3} \dfrac{x - 1}{x + 2}$.

CHAP. 1:   Prelude to Differential Calculus

Investigation Your natural guess is, of course, that the limit is $\frac{2}{5} = 0.4$. The data shown in Figure 1.45 reinforce this guess. If you experiment with other sequences of values of x approaching 3, the results will bear out the guess that the limit is, indeed, 0.4.

x	$\dfrac{x-1}{x+2}$	(Values rounded)
3.1	0.41176	
3.01	0.40120	
3.001	0.40012	
3.0001	0.40001	
3.00001	0.40000	
3.000001	0.40000	

1.45 Investigating the limit of Example 2.

EXAMPLE 3 Evaluate $\lim\limits_{x \to -4} \sqrt{x^2 + 9}$.

Investigation The evidence of the table in Figure 1.46 is certainly suggestive; it seems obvious that

$$\lim_{x \to -4} \sqrt{x^2 + 9} = 5.$$

EXAMPLE 4 Evaluate $\lim\limits_{x \to 0} \dfrac{\sqrt{x+25} - 5}{x}$.

Investigation Here the possibility of making a preliminary guess by substituting $x = 0$ is not open to us, because the formula is meaningless when $x = 0$. But the table of Figure 1.47 certainly indicates that

$$\lim_{x \to 0} \frac{\sqrt{x+25} - 5}{x} = \frac{1}{10}.$$

x	$\sqrt{x^2+9}$	(Values rounded)
−4.1	5.080354	
−4.01	5.008004	
−4.001	5.000800	
−4.0001	5.000080	
−4.00001	5.000008	
−4.000001	5.000001	
−4.0000001	5.000000	

1.46 The behavior of $\sqrt{x^2 + 9}$ as $x \to -4$.

In the next example the value obtained by substituting $x = a$ into $F(x)$ to find $\lim\limits_{x \to a} F(x)$ is incorrect.

EXAMPLE 5 Evaluate $\lim\limits_{x \to 0} F(x)$ where

$$F(x) = \begin{cases} 1 & \text{if } x \neq 0; \\ 0 & \text{if } x = 0. \end{cases}$$

x	$\dfrac{\sqrt{x+25} - 5}{x}$	(Values rounded)
10	0.09161	
1	0.09902	
0.1	0.09990	
0.01	0.09999	
0.001	0.10000	
0.0001	0.10000	

1.47 Numerical data for Example 4.

Solution The fact that $F(x) = 1$ for *every* value of x in any deleted neighborhood of 0 implies that

$$\lim_{x \to 0} F(x) = 1.$$

But the value of the limit is *not* equal to the value $F(0) = 0$.

EXAMPLE 6 Investigate $\lim\limits_{x \to 0} F(x)$ if

$$F(x) = \begin{cases} 1 & \text{if } x > 0; \\ -1 & \text{if } x < 0. \end{cases}$$

Solution There are positive values of x as close as you please to zero such that $F(x) = 1$, and negative values of x also as close as you please to zero such that $F(x) = -1$. Hence $F(x)$ *cannot* be made as close as one pleases to any single value of L merely by choosing x sufficiently close to zero. Hence the above limit *does not exist*.

Numerical investigations like those in Examples 1 through 4 provide us with an intuitive feeling for limits and often suggest the correct value of a limit. But most limit computations are based neither on merely suggestive numerical estimates nor on direct application of the definition of limit. Instead, such computations are most easily and naturally performed with the aid of the *limit laws* we give below. The proofs of these laws, which *are* based on the earlier definition of limit, are in Section 1-10.

THE LIMIT LAWS

Constant Law

$$\lim_{x \to a} C = C \qquad (C \text{ is a constant}). \tag{5}$$

In the following three laws, assume that

$$\lim_{x \to a} F(x) = L \quad \text{and that} \quad \lim_{x \to a} G(x) = M.$$

Addition Law

$$\lim_{x \to a} [F(x) \pm G(x)] = L \pm M. \tag{6}$$

(The limit of the sum is the sum of the limits; the limit of the difference is the difference of the limits.)

Product Law

$$\lim_{x \to a} [F(x)G(x)] = LM. \tag{7}$$

(The limit of the product is the product of the limits.)

Quotient Law

$$\lim_{x \to a} \frac{F(x)}{G(x)} = \frac{L}{M} \qquad \text{if } M \neq 0. \tag{8}$$

(The limit of the quotient is equal to the quotient of the limits, provided that the limit of the denominator isn't zero.)

In the next law, n is a positive integer and $a > 0$ if n is even.

Root Law

$$\lim_{x \to a} \sqrt[n]{x} = \sqrt[n]{a}. \tag{9}$$

The case $n = 1$ of the root law is the virtual tautology

$$\lim_{x \to a} x = a. \tag{10}$$

The next two examples show how the limit laws can be used to evaluate limits of polynomials and rational functions.

EXAMPLE 7

$$\lim_{x \to 3} (x^2 + 2x + 4) = \lim_{x \to 3} x^2 + \lim_{x \to 3} 2x + \lim_{x \to 3} 4$$

$$= \left(\lim_{x \to 3} x\right)^2 + 2\left(\lim_{x \to 3} x\right) + \lim_{x \to 3} 4$$

$$= (3)^2 + 2(3) + 4 = 19.$$

EXAMPLE 8

$$\lim_{x \to 3} \frac{2x + 5}{x^2 + 2x + 4} = \frac{2\left(\lim_{x \to 3} x\right) + \lim_{x \to 3} 5}{\lim_{x \to 3} (x^2 + 2x + 4)} = \frac{11}{19}.$$

The method of Examples 7 and 8 implies that limits of polynomials and rational functions *can* be evaluated by substitution. That is, if $p(x)$ and $q(x)$ are polynomials, then

$$\lim_{x \to a} p(x) = p(a), \tag{11}$$

and

$$\lim_{x \to a} \frac{p(x)}{q(x)} = \frac{p(a)}{q(a)} \qquad \text{if } q(a) \neq 0. \tag{12}$$

In general, the limit of a rational function will fail to exist at a point where its numerator is not zero but its denominator *is* zero.

EXAMPLE 9   Investigate $\displaystyle\lim_{x \to 1} \frac{1}{(x - 1)^2}$.

Solution   Since

$$\lim_{x \to 1} (x - 1)^2 = 0,$$

it follows that $1/(x - 1)^2$ can be made arbitrarily large by choosing x sufficiently close to 1. Hence $1/(x - 1)^2$ cannot approach any [finite] number as x approaches 1. Therefore the limit of Example 9 does not exist.

To evaluate limits of more complicated functions, such as roots of polynomials, we need to take limits of compositions of functions. The next limit law says that, under appropriate conditions, the limit of $f(g(x))$ as $x \to a$ may be found by substituting the limit of $g(x)$ as $x \to a$ into the function f.

Substitution Law

Suppose that

$$\lim_{x \to a} g(x) = L \quad \text{and that} \quad \lim_{x \to L} f(x) = f(L).$$

Then

$$\lim_{x \to a} f(g(x)) = f\left(\lim_{x \to a} g(x)\right) = f(L). \tag{13}$$

Thus the condition under which (13) holds is that the limit of the *outer* function f not only exists at $x = L$, but is in fact equal to the "expected" value, namely, $f(L)$.

EXAMPLE 10   Verify that

$$\lim_{x \to -4} \sqrt{x^2 + 9} = 5,$$

as our investigations in Example 3 suggest.

Solution   Let $g(x) = x^2 + 9$ and $f(x) = \sqrt{x}$. Then

$$f(g(x)) = \sqrt{x^2 + 9}.$$

We want to evaluate the limit of this composition as $x \to -4$. First note that

$$\lim_{x \to -4} g(x) = \lim_{x \to -4} (x^2 + 9) = 25$$

by Equation (11), or by direct application of the sum and product laws. In order to apply the substitution law, we therefore need to know that

$$\lim_{x \to 25} \sqrt{x} = f(25) = \sqrt{25}.$$

This follows from the root law, and so we may apply the substitution law; it gives

$$\lim_{x \to -4} \sqrt{x^2 + 9} = \lim_{x \to -4} f(g(x))$$

$$= \sqrt{\lim_{x \to -4} (x^2 + 9)} = \sqrt{25} = 5.$$

In general, you can use the method of Example 10 to show this: If n is a positive integer and the limit of $g(x)$ as $x \to a$ exists (and is positive if n is even), then

$$\lim_{x \to a} \sqrt[n]{g(x)} = \sqrt[n]{\lim_{x \to a} g(x)}. \tag{14}$$

Here is a useful special case. When m is a positive integer, take $g(x) = x^m$. Then Equation (14) yields

$$\lim_{x \to a} x^{m/n} = a^{m/n} \tag{15}$$

if $a > 0$.

Our discussion of limits began with the derivative. When we evaluate the limit

$$f'(x) = \lim_{h \to 0} \frac{f(x + h) - f(x)}{h},$$

it is important to remember that x plays the role of a *constant*, and h is the *variable* that is approaching zero.

EXAMPLE 11   Differentiate $f(x) = x/(x + 3)$.

Solution

$$f'(x) = \lim_{h \to 0} \frac{1}{h}[f(x+h) - f(x)]$$

$$= \lim_{h \to 0} \frac{1}{h}\left(\frac{x+h}{x+h+3} - \frac{x}{x+3}\right)$$

$$= \lim_{h \to 0} \frac{(x+h)(x+3) - x(x+h+3)}{h(x+h+3)(x+3)}$$

$$= \lim_{h \to 0} \frac{3h}{h(x+h+3)(x+3)}$$

$$= \frac{3}{\left(\lim\limits_{h \to 0}(x+h+3)\right)\left(\lim\limits_{h \to 0}(x+3)\right)}.$$

Therefore

$$f'(x) = \frac{3}{(x+3)^2}.$$

Another example illustrates an algebraic device often used in "preparing" functions before taking limits. This device can be applied when roots are present and resembles the simple computation

$$\frac{1}{\sqrt{3} - \sqrt{2}} = \frac{1}{\sqrt{3} - \sqrt{2}} \cdot \frac{\sqrt{3} + \sqrt{2}}{\sqrt{3} + \sqrt{2}}$$

$$= \frac{\sqrt{3} + \sqrt{2}}{3 - 2} = \sqrt{3} + \sqrt{2}.$$

EXAMPLE 12   Differentiate $f(x) = \sqrt{x}$.

Solution

$$f'(x) = \lim_{h \to 0} \frac{\sqrt{x+h} - \sqrt{x}}{h}.$$

To prepare the fraction for evaluation of the limit, we multiply the numerator and the denominator by $\sqrt{x+h} + \sqrt{x}$:

$$f'(x) = \lim_{h \to 0} \frac{\sqrt{x+h} - \sqrt{x}}{h} \cdot \frac{\sqrt{x+h} + \sqrt{x}}{\sqrt{x+h} + \sqrt{x}}$$

$$= \lim_{h \to 0} \frac{(x+h) - x}{h(\sqrt{x+h} + \sqrt{x})}$$

$$= \lim_{h \to 0} \frac{1}{\sqrt{x+h} + \sqrt{x}}.$$

Thus

$$f'(x) = \frac{1}{2\sqrt{x}}.$$

(In the final step we use the sum, quotient, and root laws.) In particular, if you take $x = 25$, you obtain the limit of Example 4.

A final property of limits that we will need is the fact that taking limits preserves inequalities between functions.

Squeeze Law

Suppose that $f(x) \leqq g(x) \leqq h(x)$ in some deleted neighborhood of a and also that

$$\lim_{x \to a} f(x) = L = \lim_{x \to a} h(x).$$

Then

$$\lim_{x \to a} g(x) = L$$

as well.

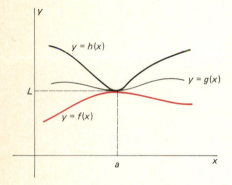

1.48 How the squeeze law works.

Figure 1.48 illustrates not only how and why the squeeze law works, but also suggests how it got its name. The idea is that $g(x)$ is trapped between $f(x)$ and $h(x)$; the latter both approach L, and so $g(x)$ must approach L as well.

The limit in the following example will play an important role when we differentiate trigonometric functions in Chapter 2.

EXAMPLE 13 Show that

$$\lim_{\theta \to 0} \frac{\sin \theta}{\theta} = 1.$$

Here $\sin \theta$ denotes the sine of an angle of θ in radians, not in degrees.

Solution A calculator set in radian mode provides us the numerical evidence of Figure 1.49. This table strongly suggests that the limit of $(\sin \theta)/\theta$ is 1 as $\theta \to 0$. To prove it, we examine Figure 1.50, which shows that

$$\text{area}(\triangle OPQ) < \text{area}(\text{sector } OPR) < \text{area}(\triangle ORS).$$

In terms of θ, this means that

$$\frac{1}{2} \sin \theta \cos \theta < \frac{1}{2} \theta < \frac{1}{2} \tan \theta = \frac{\sin \theta}{2 \cos \theta},$$

using the fact that the area of a circular sector in a circle of radius r is $A = \frac{1}{2} r^2 \theta$ if the sector is subtended by an angle of θ radians (here, we have $r = 1$). We divide each member of the above inequality by $\frac{1}{2} \sin \theta$, and obtain

$$\cos \theta < \frac{\theta}{\sin \theta} < \frac{1}{\cos \theta}.$$

We take reciprocals, which reverses the inequalities, and thus

$$\cos \theta < \frac{\sin \theta}{\theta} < \frac{1}{\cos \theta}.$$

θ	$\dfrac{\sin \theta}{\theta}$	(Values rounded)
1.0	0.84147	
0.1	0.99833	
0.01	0.99998	
0.001	1.00000	
0.0001	1.00000	

1.49 The numerical data suggest that $\lim_{\theta \to 0} \dfrac{\sin \theta}{\theta} = 1$.

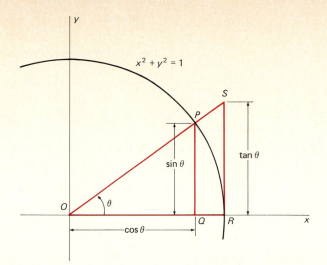

1.50 Finding the limit of Example 13.

Now we apply the squeeze law with

$$f(\theta) = \cos \theta, \qquad g(\theta) = \frac{\sin \theta}{\theta}, \quad \text{and} \quad h(\theta) = \frac{1}{\cos \theta}.$$

Since it is clear (why?) that $f(\theta)$ and $h(\theta)$ approach 1 as $\theta \to 0$, so does $g(\theta) = (\sin \theta)/\theta$. This geometrical argument shows that $(\sin \theta)/\theta \to 1$ for *positive* values of θ approaching zero. But the same result follows for negative values of θ because $\sin(-\theta) = -\sin \theta$.

1-8   PROBLEMS

In Problems 1–20, use the limit laws of this section to evaluate those limits that exist.

1 $\lim\limits_{x \to 0} (3x^2 + 7x - 12)$.

2 $\lim\limits_{x \to 2} (4x^2 - x + 5)$.

3 $\lim\limits_{x \to -1} (2x - x^5)$.

4 $\lim\limits_{x \to -2} (x^2 - 2)^5$.

5 $\lim\limits_{x \to 3} \dfrac{(x - 1)^7}{(2x - 5)^4}$.

6 $\lim\limits_{x \to 1} \dfrac{x + 1}{x^2 - x - 2}$.

7 $\lim\limits_{x \to -1} \dfrac{x + 1}{x^2 - x - 2}$.

8 $\lim\limits_{t \to 2} \dfrac{t^2 + 2t - 5}{t^3 - 2t}$.

9 $\lim\limits_{t \to 3} \dfrac{t^2 - 9}{t - 3}$.

10 $\lim\limits_{y \to 3} \dfrac{(1/y) - 1/3}{y - 3}$.

11 $\lim\limits_{x \to 3} (x^2 - 1)^{3/2}$.

12 $\lim\limits_{t \to -4} \sqrt{\dfrac{t + 8}{25 - t^2}}$.

13 $\lim\limits_{z \to 8} \dfrac{z^{2/3}}{z - \sqrt{2z}}$.

14 $\lim\limits_{t \to 2} \sqrt[3]{3t^3 + 4t - 5}$.

15 $\lim\limits_{x \to 0} \dfrac{\sqrt{x + 4} - 2}{x}$.

16 $\lim\limits_{h \to 0} \dfrac{1/(2 + h) - 1/2}{h}$.

17 $\lim\limits_{h \to 0} \dfrac{1/\sqrt{9 + h} - 1/3}{h}$.

18 $\lim\limits_{x \to 2} \dfrac{(x - 2)^4}{x^4 - 16}$.

19 $\lim\limits_{x \to 0} x \cos x$.

20 $\lim\limits_{x \to 0} x \sin \dfrac{1}{x}$.

Use a calculator to investigate numerically (as in Examples 1–4) the limits in Problems 21–25.

21 $\lim\limits_{x \to 0} \dfrac{\sqrt{x + 9} - 3}{x}$.

22 $\lim\limits_{x \to 4} \dfrac{x^{3/2} - 8}{x - 4}$.

23 $\lim\limits_{x \to 0} \dfrac{1 - \cos x}{x^2}$ (x in radians).

24 $\lim\limits_{h \to 0} (1 + h)^{1/h}$.

25 $\lim\limits_{x \to 0} \dfrac{3^x - 1}{x}$.

In each of Problems 26–30, find the derivative of $f(x)$.

26 $f(x) = \dfrac{1}{2x + 3}$.

27 $f(x) = \dfrac{1}{\sqrt{x}}$.

28 $f(x) = \sqrt{3x + 1}$.

29 $f(x) = \dfrac{x}{2x + 1}$.

30 $f(x) = \dfrac{1}{\sqrt{x+4}}$.

In each of Problems 31–35, use the result of Example 13 to find the given limit.

31 $\displaystyle\lim_{\theta \to 0} \dfrac{\theta^2}{\sin \theta}$.

32 $\displaystyle\lim_{\theta \to 0} \dfrac{\sin^2 \theta}{\theta^2}$.

33 $\displaystyle\lim_{\theta \to 0} \dfrac{1 - \cos \theta}{\theta^2}$.

34 $\displaystyle\lim_{\theta \to 0} \dfrac{\tan \theta}{\theta}$.

35 $\displaystyle\lim_{x \to 0} \dfrac{2x}{(\sin x) - x}$.

36 Apply the product law to prove that $\displaystyle\lim_{x \to a} x^n = a^n$ if n is a positive integer.

37 Suppose that $p(x) = b_n x^n + \cdots + b_1 x + b_0$ is a polynomial. Prove that $\displaystyle\lim_{x \to a} p(x) = p(a)$.

38 Use the root law and the substitution law to establish Equation (14).

39 Let $f(x) = [x]$ be the greatest integer function. For what values of a does $\displaystyle\lim_{x \to a} f(x)$ exist?

40 (a) Suppose that there exists a number M such that $|f(x)| \leq M$ for all x. Apply the squeeze law to show that $\displaystyle\lim_{x \to 0} x f(x) = 0$.

(b) Conclude that $\displaystyle\lim_{x \to 0} x^2 \cos(x^{-1/3}) = 0$. Supply the details.

41 Let $r(x) = p(x)/q(x)$ where $p(x)$ and $q(x)$ are polynomials. Suppose that $p(a) \neq 0$ and that $q(a) = 0$. Prove that $\displaystyle\lim_{x \to a} r(x)$ does not exist. (*Suggestion*: Consider $\displaystyle\lim_{x \to a} q(x) r(x)$.)

1-9

Continuous Functions

When we stated the substitution law for limits in Section 1-8, we needed for its validity the condition

$$\lim_{x \to a} f(x) = f(a) \tag{1}$$

on the outer function in the composition $f(g(x))$. (The law as originally stated has L in place of a, but of course this is unimportant.) A function that satisfies Equation (1) is said to be *continuous* at the number a.

> ### Definition of Continuity
> Suppose that the function f is defined in a neighborhood of a. We say that f is **continuous at** a provided that $\displaystyle\lim_{x \to a} f(x)$ exists and, moreover, that the value of this limit is $f(a)$; that is, Equation (1) holds.

Briefly, continuity of f at a means that the limit of f at a is equal to its value there. Another way to put it is this: The limit of f at a is its "expected" value—the value that you would assign if you knew the values of f in a deleted neighborhood of a and you knew f to be "predictable." Alternatively, continuity of f at a means this: When x is close to a, then $f(x)$ is close to $f(a)$.

We say simply that f is **continuous** if it is continuous at each point of its domain of definition. The definition of continuity in terms of limits makes precise the intuitive idea of a function with a graph that is a connected curve—connected in the sense that it can be traced with a "continuous" motion of the pencil without lifting the pencil from the paper.

If the function f is *not* continuous at a, we say that it is **discontinuous** there, or that a is a **discontinuity** of f. Intuitively, a discontinuity of f is a point where its graph has a "gap" or "jump" of some sort. For instance, the function

$$f(x) = \begin{cases} 1 & \text{if } x \neq 0; \\ 0 & \text{if } x = 0, \end{cases}$$

(its graph appears in Figure 1.51) jumps in value from 1 to 0 and back as x increases through 0. And f is not continuous at $x = 0$ because, as we noted in Example 5 of Section 1-8,

$$\lim_{x \to 0} f(x) = 1 \neq 0 = f(0);$$

the limit of f at 0 is not equal to its value there.

Also, the function

$$g(x) = \begin{cases} 1 & \text{if } x \geq 0; \\ -1 & \text{if } x < 0, \end{cases}$$

is discontinuous at $x = 0$. As its graph (in Figure 1.52) shows, there is a jump from -1 to 1 at 0. We saw in Example 6 of Section 1-8 that the limit of $g(x)$ as $x \to 0$ does not exist.

The graph of

$$h(x) = \begin{cases} 1/x & \text{if } x \neq 0; \\ 0 & \text{if } x = 0, \end{cases}$$

appears in Figure 1.53. This function has what might be called an "infinite discontinuity" at zero. Finally, the "sawtooth function" $f(x) = x - [x] - \frac{1}{2}$, which has the graph shown in Figure 1.22, is continuous at every *noninteger* point but discontinuous at every integer point.

We are more interested in functions that *are* continuous; most (though not all) of the functions that we study in calculus are continuous on their natural domains of definition. For example, a consequence of Equation (11) of Section 1-8 is this: *Every polynomial function*

$$p(x) = b_n x^n + b_{n-1} x^{n-1} + \cdots + b_1 x + b_0$$

is continuous at every number. And Equation (12) from Section 1-8 tells us that every rational function $f(x) = p(x)/q(x)$ is continuous wherever it is defined; that is, where $q(x) \neq 0$. If $q(a) = 0$ while $p(a) \neq 0$, so that $f(x) = p(x)/q(x)$ is not defined by its formula at $x = a$, it follows from Problem 41 in Section 1-8 that $\lim_{x \to a} f(x)$ does not exist. In such a case it is impossible to assign any value whatsoever to $f(a)$ to make f continuous at $x = a$.

The next theorem has a valuable consequence: Functions that are built by forming compositions of continuous functions are themselves continuous.

Theorem 1 *Continuity of Compositions*

The composition of two continuous functions is itself continuous. More precisely, if g is continuous at a and f is continuous at $g(a)$, then $f(g(x))$ is continuous at a.

Proof The substitution law for limits gives

$$\lim_{x \to a} f(g(x)) = f\left(\lim_{x \to a} g(x)\right) = f(g(a)),$$

as desired.

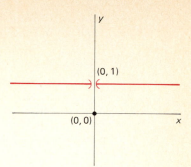

1.51 A discontinuity in f at $x = 0$.

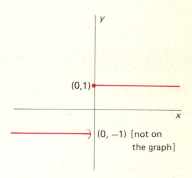

1.52 A discontinuity in g at $x = 0$.

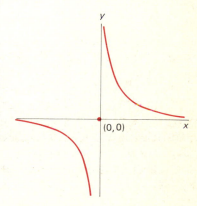

1.53 The function h has an infinite discontinuity at $x = 0$.

The following theorem is normally proved in a course in advanced calculus; it tells us that the inverse of a continuous function is continuous.

Theorem 2 *Continuity of Inverse Functions*

Let f be a continuous function with domain the open interval I. Suppose that f has an inverse function g with domain the open interval J. Then g is continuous (at each point of J).

For example, consider the nth-power function $f(x) = x^n$ (here, n is a positive integer). If n is odd, we take I as the whole real line $(-\infty, +\infty)$. If n is even, we take I to be the positive half-line $(0, +\infty)$. Then f has as its inverse function the nth-root function $g(x) = \sqrt[n]{x}$. The domain of g is the interval J, where $J = (-\infty, +\infty)$ if n is odd and $J = (0, +\infty)$ if n is even. Since $f(x) = x^n$ is continuous on I, Theorem 2 implies that $g(x) = \sqrt[n]{x}$ is continuous on J. Thus

$$\lim_{x \to a} \sqrt[n]{x} = \sqrt[n]{a},$$

with the stipulation that $a > 0$ if n is even. This proves the root law of Section 1-8.

We may combine this result with Theorem 1. Then we see that a root of a continuous function is continuous where defined. That is, the composition

$$g(x) = [f(x)]^{1/n}$$

is continuous at a if f is, assuming that $f(a) > 0$ if n is even (so that $\sqrt[n]{f(a)}$ will be defined.)

EXAMPLE 1 Show that the function

$$f(x) = \left(\frac{x - 7}{x^2 + 2x + 2}\right)^{2/3}$$

is continuous on the whole real line.

Solution Note first that the denominator

$$x^2 + 2x + 2 = (x + 1)^2 + 1$$

is never zero. Hence the rational function $r(x) = (x - 7)/(x^2 + 2x + 2)$ is defined and continuous everywhere. It then follows from Theorem 1 and the continuity of the cube root function that

$$f(x) = ([r(x)]^2)^{1/3}$$

is continuous everywhere. Hence (for example)

$$\lim_{x \to -1} \left(\frac{x - 7}{x^2 + 2x + 2}\right)^{2/3} = f(-1) = (-8)^{2/3} = 4.$$

CONTINUOUS FUNCTIONS
ON CLOSED INTERVALS

An applied problem often involves a function with a domain that is a closed interval. For example, in the fence problem we found that the area of a rectangle with perimeter 140 and base x is $f(x) = x(70 - x)$. Though the formula for f is meaningful for all x, only values of x in the closed interval $[0, 70]$ correspond to rectangles and are thereby pertinent to the stated problem.

In order to say what is meant by continuity of such a function at the endpoints of its interval of definition, we need the concept of **one-sided limits.** We say that the **right-hand limit** of $f(x)$ at a is L and write

$$\lim_{x \to a^+} f(x) = L,$$

provided that $f(x)$ can be made as close as we please to L by choosing x sufficiently close to *and to the right of a* (thus $x > a$). In short, $f(x)$ approaches L as x approaches *a from the right*. We define the **left-hand limit** $\lim\limits_{x \to a^-} f(x)$ similarly, except of course with x approaching a from the left.

EXAMPLE 2

$$\lim_{x \to 0} \sqrt{x}$$

does not exist because $\sqrt{x}$ is not defined in a deleted neighborhood of 0. What *is* true is that

$$\lim_{x \to 0^+} \sqrt{x} = 0.$$

EXAMPLE 3

$$\lim_{x \to 0^-} \frac{|x|}{x} = -1 \quad \text{while} \quad \lim_{x \to 0^+} \frac{|x|}{x} = +1.$$

EXAMPLE 4 The function $f(x) = \sqrt{x(2 - x)}$ is defined only on the closed interval $[0, 2]$. Evidently

$$\lim_{x \to 0^+} \sqrt{x(2 - x)} = 0 = \lim_{x \to 2^-} \sqrt{x(2 - x)}.$$

The function f defined on the closed interval $[a, b]$ is said to be **continuous on $[a, b]$** provided that it is continuous at each point of the open interval (a, b) *and* that

$$\lim_{x \to a^+} f(x) = f(a), \qquad \lim_{x \to b^-} f(x) = f(b).$$

These last two conditions mean that, at each endpoint, the value of the function is equal to its limit from within the interval.

The special importance of continuity of a function on a closed interval is this: Such a function has both the maximum value property and the intermediate value property. The two theorems that insure these properties are stated below and proved in courses in advanced calculus.

PRELIMINARY COMMENT   The value $f(c)$ is called the **maximum value** of $f(x)$ on $[a, b]$ if $f(c) \geqq f(x)$ for all x in $[a, b]$.

Theorem 3   *Maximum Value Property*

If f is continuous on the closed interval $[a, b]$, then f attains a maximum value $f(c)$ at some point c in $[a, b]$.

By applying this result to the function $g(x) = -f(x)$, we can see that $f(x)$ must also attain a minimum value at some point of the interval.

Earlier in this chapter we saw that both the fence problem and the refrigerator problem amount to finding the maximum value of $f(x) = x(70 - x)$ for x in $[0, 70]$. Now we see that it is the continuity of f on the closed interval $[0, 70]$ that implies that its maximum value exists and is attained at some point of that interval. The following example shows that a function that is not continuous may have no maximum value on a closed interval.

EXAMPLE 5   The function f defined on the closed interval $[0, 1]$ by

$$f(x) = \begin{cases} 1/x & \text{if} \quad x \in (0, 1]; \\ 1 & \text{if} \quad x = 0, \end{cases}$$

is not continuous on $[0, 1]$ because

$$\lim_{x \to 0^+} \frac{1}{x}$$

does not exist. This function does attain its minimum value of 1 at 0 and at 1, but attains no maximum value on $[0, 1]$ because $1/x$ can be made arbitrarily large by choosing x very close to zero (while keeping x positive).

For a variation on this example, the function $g(x) = 1/x$ with domain the *open* interval $(0, 1)$ attains neither a maximum nor a minimum value there.

We mentioned above that continuity of a function is related to the possibility of tracing its graph without lifting the pencil from the paper. The following theorem expresses this fact more precisely.

Theorem 4   *Intermediate Value Property*

Suppose that the function f is continuous on the closed interval $[a, b]$. Then $f(x)$ assumes every intermediate value between $f(a)$ and $f(b)$. Thus, if K is any number between $f(a)$ and $f(b)$, then there exists at least one point c in (a, b) such that $f(c) = K$.

In Figure 1.54 we show the graph of a fairly typical continuous function with domain the closed interval $[a, b]$. The number K is located on the y-axis, somewhere between $f(a)$ and $f(b)$ (in the figure, we have $f(a) < f(b)$, but that isn't essential). The horizontal line through K must cut the graph of f somewhere, and the x-coordinate of the point where graph and line meet is the value of c, the existence of which is assured by the Intermediate Value Property.

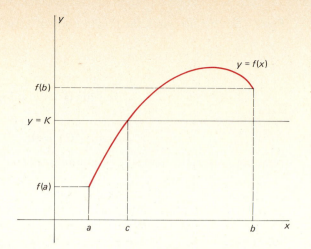

1.54 The continuous function f attains the intermediate value K at $x = c$.

Thus the Intermediate Value Property implies that each horizontal line meeting the y-axis between $f(a)$ and $f(b)$ must cut the graph of this continuous function f somewhere. This is a way of saying that the graph has no gaps and suggests that the idea of being able to trace such a graph without lifting the pencil from the paper is accurate.

EXAMPLE 6 The discontinuous function defined on $[-1, 1]$ by

$$f(x) = \begin{cases} 0 & \text{if } x < 0; \\ 1 & \text{if } x \geqq 0, \end{cases}$$

does *not* attain the intermediate value $\frac{1}{2}$. (See Figure 1.55.)

EXAMPLE 7 Show that the equation

$$x^3 + x - 1 = 0$$

has a solution between 0 and 1.

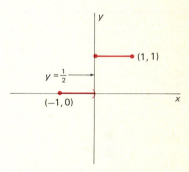

1.55 This discontinuous function does not have the Intermediate Value Property.

Solution The function $f(x) = x^3 + x - 1$ is continuous on $[0, 1]$, and $f(0) = -1$ while $f(1) = 1$. By the Intermediate Value Property, f must attain the value 0 at some point between $x = 0$ and $x = 1$.

CONTINUITY AND DIFFERENTIABILITY

We have seen that a wide variety of functions are continuous. Though there exist continuous functions so exotic that they are nowhere differentiable, the following theorem tells us that every differentiable function is continuous.

Theorem 5 *Differentiability Implies Continuity*

Suppose that f is defined in a neighborhood of a. If f is differentiable at a, then f is continuous at a.

Proof Since $f'(a)$ exists, the product law of limits yields this:

$$\lim_{x \to a} [f(x) - f(a)] = \lim_{x \to a} (x - a) \cdot \frac{f(x) - f(a)}{x - a}$$

$$= \left(\lim_{x \to a} (x - a) \right) \left(\lim_{x \to a} \frac{f(x) - f(a)}{x - a} \right)$$

$$= 0 \cdot f'(a) = 0.$$

Thus $\lim_{x \to a} f(x) = f(a)$, and so f is continuous at a.

Although it is difficult to describe a nowhere differentiable though everywhere continuous function, the absolute value function, $f(x) = |x|$, provides us a simple example of an everywhere continuous function that is not differentiable at the one point $x = 0$. (We saw this earlier, too, in Example 3 of Section 1-5.)

1-9 PROBLEMS

One-sided limits satisfy the limit laws stated in Section 1-8. Use those laws in Problems 1–8 to find those limits that exist.

1 $\lim\limits_{x \to 5^-} \sqrt{x(5 - x)}$.

2 $\lim\limits_{x \to 2^-} x\sqrt{4 - x^2}$.

3 $\lim\limits_{x \to 4^+} \sqrt{\dfrac{x - 4}{4x}}$.

4 $\lim\limits_{x \to -3^+} \sqrt{6 - x - x^2}$.

5 $\lim\limits_{x \to -4^+} \dfrac{1}{x^2 - 16}$.

6 $\lim\limits_{x \to 5^-} \dfrac{x - 5}{|x - 5|}$.

7 $\lim\limits_{x \to 3^-} \dfrac{\sqrt{x^2 - 6x + 9}}{x - 3}$.

8 $\lim\limits_{x \to 2^+} \dfrac{x - 2}{x^2 - 5x + 6}$.

In Problems 9–24, tell where the given function f is continuous. For each isolated point where f is not defined, tell whether f can be defined so as to be continuous there.

9 $f(x) = \dfrac{x}{(x + 3)^2}$.

10 $f(x) = \dfrac{x}{x^2 - 1}$.

11 $f(x) = \dfrac{x - 2}{x^2 - 4}$.

12 $f(x) = \dfrac{x + 1}{x^2 - x - 6}$.

13 $f(x) = \dfrac{1}{1 - |x|}$.

14 $f(x) = \dfrac{x + 3}{\sqrt{x - 3}}$.

15 $f(x) = \dfrac{x + 3}{\sqrt{x^2 - 9}}$.

16 $f(x) = \dfrac{x^2 + 5x + 6}{x + 2}$.

17 $f(x) = \sqrt{\dfrac{1 - x}{1 + x}}$.

18 $f(x) = \dfrac{|x - 1|}{(x - 1)^3}$.

19 $f(x) = \dfrac{x - 17}{|x - 17|}$.

20 $f(x) = \dfrac{\sqrt{x^2 - 4x + 4}}{x - 2}$.

21 $f(x) = [x]$ (the greatest integer function).

22 $f(x) = [x] - x + 1$.

23 $f(x) = \begin{cases} 0 & \text{if } x \text{ is rational;} \\ 1 & \text{if } x \text{ is irrational.} \end{cases}$

24 $f(x) = \begin{cases} 0 & \text{if } x \text{ is rational;} \\ x^2 & \text{if } x \text{ is irrational.} \end{cases}$

In each of Problems 25–30, tell whether the given function attains maximum or minimum values at points of the given interval.

25 $f(x) = |x|$; $x \in (-1, 1)$.

26 $f(x) = \dfrac{1}{\sqrt{x}}$; $x \in (0, 1]$.

27 $f(x) = \dfrac{1}{1 + |x|}$; $x \in [-1, 0)$.

28 $f(x) = 5 - x^2$; $x \in [-1, 2)$.

29 $f(x) = x^3 + x$; $x \in [-1, 1]$.

30 $f(x) = \dfrac{1}{1 + x^2}$; $x \in (-\infty, +\infty)$.

In each of the next three problems, apply the Intermediate Value Property to show that the given equation has a solution in the given interval.

31 $x^2 - 5 = 0$,   on $[2, 3]$.

32 $x^3 + x + 1 = 0$,   on $[-1, 0]$.

33 $x^3 - 3x^2 + 1 = 0$,   on $[0, 1]$.

34 Prove that the equation $f(x) = x^3 - 4x + 1 = 0$ has three distinct roots by calculating the values of $f(x)$ at $x = 0$, $\pm 1, \pm 2, \pm 3$ and then applying the Intermediate Value Property.

35 Apply the Intermediate Value Property to show that every positive number a has a square root. That is, given $a > 0$, there is a number x such that $x^2 = a$.

36 Apply the Intermediate Value Property to show that every real number has a cube root.

37 Show that there is a number x between 0 and $\pi/2$ (radians) such that $\cos x = x$.

38 Show that there is a number between $\pi/2$ and π (radians) such that $\tan x = -x$. (*Suggestion:* First sketch the graphs of $y = \tan x$ and $y = -x$.)

An Appendix: Proofs of the Limit Laws

Recall the definition of the limit:

$$\lim_{x \to a} F(x) = L$$

provided that, given $\varepsilon > 0$, there exists a $\delta > 0$ such that

$$0 < |x - a| < \delta \quad \text{implies} \quad |F(x) - L| < \varepsilon. \tag{1}$$

Note that the number $\varepsilon > 0$ comes *first. Then* a value of $\delta > 0$ must be found so that (1) holds. In order to prove that $F(x) \to L$ as $x \to a$, you must, in effect, be able to stop the first person you see on the street and ask him or her to pick a positive number ε at random. Then *you* must be able to respond with a positive number δ, which is chosen so that you can prove that (1) holds with your δ and his or her ε.

To do this you will ordinarily have to give an explicit method—a recipe—for producing the value of δ that works for a given value of ε. As the next few examples show, this method or recipe will depend on the particular function F involved and on the numbers a and L.

EXAMPLE 1   Prove that $\lim_{x \to 3} (2x - 1) = 5$.

Solution   Given $\varepsilon > 0$, we must find a $\delta > 0$ such that

$$|(2x - 1) - 5| < \varepsilon \quad \text{if} \quad 0 < |x - 3| < \delta.$$

Now

$$|(2x - 1) - 5| = |2x - 6| = 2|x - 3|.$$

So

$$0 < |x - 3| < \frac{\varepsilon}{2} \quad \text{implies} \quad |(2x - 1) - 5| < 2 \cdot \frac{\varepsilon}{2} = \varepsilon.$$

Hence, given $\varepsilon > 0$, it is sufficient to choose $\delta = \varepsilon/2$. This illustrates the observation that the required δ is generally a function of the given ε.

EXAMPLE 2   Prove that $\lim_{x \to 2} (3x^2 + 5) = 17$.

Solution   Given $\varepsilon > 0$, we must find $\delta > 0$ such that

$$0 < |x - 2| < \delta \quad \text{implies} \quad |(3x^2 + 5) - 17| < \varepsilon.$$

Now

$$|(3x^2 + 5) - 17| = |3x^2 - 12| = 3 \cdot |x + 2| \cdot |x - 2|.$$

Our problem, therefore, involves showing that $|x + 2| \cdot |x - 2|$ can be made as small as we please by choosing $|x - 2|$ sufficiently small. The idea is that $|x + 2|$ cannot be too large if $|x - 2|$ is fairly small. For instance, if $|x - 2| < 1$, then

$$|x + 2| = |(x - 2) + 4| \leq |x - 2| + 4 < 5.$$

Therefore

$$0 < |x - 2| < 1 \quad \text{implies} \quad |(3x^2 + 5) - 17| < 15 \cdot |x - 2|.$$

Consequently, let us choose δ to be the minimum of the two numbers 1 and $\varepsilon/15$. Then

$$0 < |x - 2| < \delta \quad \text{implies} \quad |(3x^2 + 5) - 17| < 15 \cdot \frac{\varepsilon}{15} = \varepsilon,$$

as desired.

EXAMPLE 3   Prove that

$$\lim_{x \to a} \frac{1}{x} = \frac{1}{a} \quad \text{if} \quad a \neq 0.$$

Solution   We take the liberty of considering only the case $a > 0$, since that will simplify the notation; the method is quite similar in the case $a < 0$. Now suppose that $\varepsilon > 0$ is given. We must find a number $\delta > 0$ such that

$$0 < |x - a| < \delta \quad \text{implies} \quad \left| \frac{1}{x} - \frac{1}{a} \right| < \varepsilon.$$

Now

$$\left| \frac{1}{x} - \frac{1}{a} \right| = \left| \frac{a - x}{ax} \right| = \frac{|x - a|}{a|x|}.$$

The idea is that $1/|x|$ cannot be too large if $|x - a|$ is fairly small. For example, if $|x - a| < a/2$, then $a/2 < x < 3a/2$. Therefore

$$|x| > \frac{a}{2}, \quad \text{so that} \quad \frac{1}{|x|} < \frac{2}{a}.$$

It follows that

$$\left| \frac{1}{x} - \frac{1}{a} \right| < \frac{2}{a^2} \cdot |x - a|$$

if $|x - a| < a/2$. Thus, if we choose δ to be the minimum of the two numbers $a/2$ and $a^2 \cdot \varepsilon/2$, then

$$0 < |x - a| < \delta \quad \text{implies} \quad \left| \frac{1}{x} - \frac{1}{a} \right| < \frac{2}{a^2} \cdot \frac{a^2 \varepsilon}{2} = \varepsilon,$$

as desired.

We are now ready to give proofs of the limit laws that we stated in Section 1-8.

Constant Law

If C is a constant, then

$$\lim_{x \to a} C = C.$$

Proof Since $|C - C| = 0$, we merely choose $\delta = 1$, regardless of the previously given value of $\varepsilon > 0$. Then, if $0 < |x - a| < \delta$, it is automatic that $|C - C| < \varepsilon$.

Addition Law

If

$$\lim_{x \to a} F(x) = L \quad \text{and} \quad \lim_{x \to a} G(x) = M,$$

then

$$\lim_{x \to a} [F(x) + G(x)] = L + M.$$

Proof Let $\varepsilon > 0$ be given. We must find $\delta > 0$ such that

$$0 < |x - a| < \delta \quad \text{implies} \quad |(F(x) + G(x)) - (L + M)| < \varepsilon.$$

Since L is the limit of $F(x)$ as $x \to a$, there is a number $\delta_1 > 0$ such that

$$0 < |x - a| < \delta_1 \quad \text{implies} \quad |F(x) - L| < \frac{\varepsilon}{2}.$$

Because M is the limit of $G(x)$ as $x \to a$, there is a number $\delta_2 > 0$ such that

$$0 < |x - a| < \delta_2 \quad \text{implies} \quad |G(x) - M| < \frac{\varepsilon}{2}.$$

So we let δ be the minimum of the two numbers δ_1 and δ_2. Then $0 < |x - a| < \delta$ implies that

$$|(F(x) + G(x)) - (L + M)| \leqq |F(x) - L| + |G(x) - M|$$

$$< \frac{\varepsilon}{2} + \frac{\varepsilon}{2} = \varepsilon,$$

as desired.

Product Law

If

$$\lim_{x \to a} F(x) = L \quad \text{and} \quad \lim_{x \to a} G(x) = M,$$

then

$$\lim_{x \to a} F(x) \cdot G(x) = L \cdot M.$$

Proof Given $\varepsilon > 0$, we must find a number $\delta > 0$ such that

$$0 < |x - a| < \delta \quad \text{implies} \quad |F(x) \cdot G(x) - L \cdot M| < \varepsilon.$$

But first, note that the triangle inequality gives the result

$$|F(x) \cdot G(x) - L \cdot M| = |F(x)G(x) - L \cdot G(x) + L \cdot G(x) - LM|$$
$$\leqq |G(x)| \cdot |F(x) - L| + |L| \cdot |G(x) - M|. \tag{2}$$

Since $\lim\limits_{x \to a} F(x) = L$, there exists $\delta_1 > 0$ such that

$$0 < |x - a| < \delta_1 \quad \text{implies} \quad |F(x) - L| < \frac{\varepsilon}{2(|M| + 1)}. \tag{3}$$

And since $\lim\limits_{x \to a} G(x) = M$, there is a number $\delta_2 > 0$ such that

$$0 < |x - a| < \delta_2 \quad \text{implies} \quad |G(x) - M| < \frac{\varepsilon}{2(|L| + 1)}. \tag{4}$$

Moreover, there is a *third* number $\delta_3 > 0$ such that

$$0 < |x - a| < \delta_3 \quad \text{implies} \quad |G(x) - M| < 1,$$

which in turn implies that

$$|G(x)| < |M| + 1. \tag{5}$$

We now choose δ to be the least of the three positive numbers δ_1, δ_2, and δ_3. Then we substitute (3), (4), and (5) into (2), and finally see that $0 < |x - a| < \delta$ implies

$$|F(x)G(x) - LM| < (|M| + 1) \cdot \frac{\varepsilon}{2(|M| + 1)} + |L| \cdot \frac{\varepsilon}{2(|L| + 1)}$$

$$< \frac{\varepsilon}{2} + \frac{\varepsilon}{2} = \varepsilon.$$

This establishes the product law. The use of $|M| + 1$ and $|L| + 1$ in the above denominators takes care of the possibility that L or M might be zero.

Substitution Law

If

$$\lim_{x \to a} g(x) = L \quad \text{and} \quad \lim_{x \to L} f(x) = f(L),$$

then

$$\lim_{x \to a} f(g(x)) = f(L).$$

Proof Let $\varepsilon > 0$ be given. We must find a number $\delta > 0$ such that

$$0 < |x - a| < \delta \quad \text{implies} \quad |f(g(x)) - f(L)| < \varepsilon.$$

Since $\lim\limits_{y \to L} f(y) = f(L)$, there exists $\delta_1 > 0$ such that

$$0 < |y - L| < \delta_1 \quad \text{implies} \quad |f(y) - f(L)| < \varepsilon. \tag{6}$$

And since $\lim\limits_{x \to a} g(x) = L$, we can choose $\delta > 0$ such that

$$0 < |x - a| < \delta \quad \text{implies} \quad |g(x) - L| < \delta_1,$$

or

$$|y - L| < \delta_1,$$

where $y = g(x)$. From (6) we see that

$$0 < |x - a| < \delta \quad \text{implies} \quad |f(g(x)) - f(L)| = |f(y) - f(L)| < \varepsilon,$$

exactly as desired.

Reciprocal Law

If

$$\lim_{x \to a} g(x) = L \quad \text{and} \quad L \neq 0,$$

then

$$\lim_{x \to a} \frac{1}{g(x)} = \frac{1}{L}.$$

Proof Let $f(x) = 1/x$. Then, as we saw in Example 3 above,

$$\lim_{x \to L} f(x) = \lim_{x \to L} \frac{1}{x} = \frac{1}{L} = f(L).$$

Hence the substitution law gives us the result

$$\lim_{x \to a} \frac{1}{g(x)} = \lim_{x \to a} f(g(x)) = f(L) = \frac{1}{L},$$

and this completes the proof.

Quotient Law

Suppose that

$$\lim_{x \to a} F(x) = L \quad \text{and} \quad \lim_{x \to a} G(x) = M \neq 0.$$

Then

$$\lim_{x \to a} \frac{F(x)}{G(x)} = \frac{L}{M}.$$

Proof It follows immediately from the product and reciprocal laws that

$$\lim_{x \to a} \frac{F(x)}{G(x)} = \lim_{x \to a} F(x) \cdot \frac{1}{G(x)}$$

$$= \left(\lim_{x \to a} F(x) \right) \cdot \left(\lim_{x \to a} \frac{1}{G(x)} \right) = L \cdot \frac{1}{M} = \frac{L}{M}.$$

Squeeze Law

Suppose that $f(x) \leqq g(x) \leqq h(x)$ in some deleted neighborhood of a and also that

$$\lim_{x \to a} f(x) = L = \lim_{x \to a} h(x).$$

Then

$$\lim_{x \to a} g(x) = L$$

as well.

Proof   Given $\varepsilon > 0$, we choose $\delta_1 > 0$ and $\delta_2 > 0$ such that

$$0 < |x - a| < \delta_1 \quad \text{implies} \quad |f(x) - L| < \varepsilon$$

and

$$0 < |x - a| < \delta_2 \quad \text{implies} \quad |h(x) - L| < \varepsilon.$$

If we now choose $\delta > 0$ as the smaller of the two numbers δ_1 and δ_2, then $0 < |x - a| < \delta$ implies that $f(x)$ and $h(x)$ are both points of the open interval $(L - \varepsilon, L + \varepsilon)$. So

$$L - \varepsilon < f(x) \leqq g(x) \leqq h(x) < L + \varepsilon.$$

Thus

$$0 < |x - a| < \delta \quad \text{implies} \quad |g(x) - L| < \varepsilon,$$

as desired. This completes the proof of the squeeze law.

1-10   PROBLEMS

In Problems 1–6, apply the definition of the limit to establish the given equality.

1 $\lim_{x \to a} x = a.$

2 $\lim_{x \to -3} (2x + 1) = -5.$

3 $\lim_{x \to -1} (2x^2 - 1) = 1.$

4 $\lim_{x \to a} \dfrac{1}{x^2} = \dfrac{1}{a^2}$ if $a \neq 0.$

5 $\lim_{x \to a} \dfrac{1}{x^2 + 1} = \dfrac{1}{a^2 + 1}.$

6 $\lim_{x \to a} \dfrac{1}{\sqrt{x}} = \dfrac{1}{\sqrt{a}}$ if $a > 0.$

7 Suppose that C is a constant and that $\lim_{x \to a} f(x) = L.$ Apply the definition of the limit to prove that

$$\lim_{x \to a} C \cdot f(x) = C \cdot L.$$

8 Suppose that $L \neq 0$ and that $\lim_{x \to a} f(x) = L.$ Use the method of Example 3 and the definition of the limit to show directly that

$$\lim_{x \to a} \frac{1}{f(x)} = \frac{1}{L}.$$

9 Use in this problem the algebraic identity

$$x^n - a^n = (x - a)(x^{n-1} + x^{n-2}a + \cdots + xa^{n-2} + a^{n-1}).$$

Show directly from the definition of the limit that $\lim_{x \to a} x^n = a^n$ if n is a positive integer.

10 Apply the identity

$$|\sqrt{x} - \sqrt{a}| = \frac{|x - a|}{\sqrt{x} + \sqrt{a}}$$

to show directly from the definition of the limit that $\lim_{x \to a} \sqrt{x} = \sqrt{a}$ if $a > 0.$

Use the list below as a guide to ideas that you may need to review.

1 Rational and irrational numbers
2 The real line
3 Absolute value of a real number
4 The triangle inequality
5 Intervals: open and closed
6 Function; domain and range of a function
7 Independent and dependent variable
8 The coordinate plane
9 Distance formula
10 Midpoint formula
11 The slope of a straight line
12 The point-slope equation of a line
13 The slope-intercept equation of a line
14 Angle of inclination of a line
15 Slope relationship between two perpendicular lines
16 Graph of an equation; graph of a function
17 Translation principle
18 The graph of $y = ax^2 + bx + c$

19 The tangent line to the graph of a function
20 The derivative of a function
21 The derivative of $f(x) = ax^2 + bx + c$
22 Average rate of change of a function
23 Instantaneous rate of change of a function
24 Position function; velocity and acceleration
25 Composition of functions
26 Inverse functions; one-to-one functions
27 The limit of $F(x)$ as x approaches a
28 The limit laws: constant, addition, product, quotient, root, substitution, squeeze
29 Polynomials and rational functions
30 Continuous functions
31 Continuity of compositions of functions and of the inverse of a continuous function
32 One-sided limits
33 Maximum Value Property
34 Intermediate Value Property
35 Differentiability implies continuity

MISCELLANEOUS PROBLEMS

Express in terms of intervals the solutions of the inequalities in Problems 1–3.

1 $-7 \leqq 1 - 4x < 3.$

2 $3 < |2x - 5| \leqq 6.$

3 $-2 \leqq \dfrac{3}{4x - 1} \leqq 4.$

In each of Problems 4–7, find the domain of the function whose formula is given.

4 $f(x) = \dfrac{x + 1}{x^2 - 2x}.$

5 $f(x) = \sqrt{2 - 3x}.$

6 $f(x) = \dfrac{1}{\sqrt{9 - x^2}}.$

7 $f(x) = \sqrt{(x - 2)(4 - x)}.$

8 The height of a circular cylinder is equal to its radius. Express its total surface area A (including both ends) as a function of its volume.

In each of Problems 9–14, write an equation of the straight line L described.

9 L passes through $(-3, 5)$ and $(1, 13)$.
10 L passes through $(4, -1)$ and has slope -3.
11 L has slope $\frac{1}{2}$ and y-intercept -5.
12 L passes through $(2, -3)$ and is parallel to $3x - 2y = 4$.

13 L passes through $(-3, 7)$ and also is perpendicular to $y - 2x = 10$.
14 L is the perpendicular bisector of the segment joining $(1, -5)$ and $(3, -1)$.

Sketch the graphs of the equations and functions given in Problems 15–23.

15 $2x - 5y = 7.$
16 $|x - y| = 1.$
17 $x^2 + y^2 = 2x.$
18 $x^2 + y^2 = 4y - 6x + 3.$
19 $y = 2x^2 - 4x - 1.$
20 $y = 4x - x^2.$

21 $f(x) = \dfrac{1}{x + 5}.$

22 $f(x) = \dfrac{1}{4 - x^2}.$

23 $f(x) = |x - 3|.$

24 Write an equation of the circle with center $(2, 3)$ that is tangent to the line $x + y + 3 = 0$.

In each of Problems 25–29, apply the definition of the derivative to find $f'(x)$.

25 $f(x) = 2x^2 + 3x - 4.$

26 $f(x) = x - x^3.$

27 $f(x) = \dfrac{1}{3 - x}.$

28 $f(x) = \sqrt{4 - 3x}.$

29 $f(x) = \dfrac{1}{\sqrt{2x - 1}}.$

30 Find the derivative of $f(x) = 3x - x^2 + |2x - 3|$ at the points where it is differentiable. Where is it *not* differentiable? Sketch the graph of f.

31 Write equations of the two straight lines through $(3, 4)$ that are tangent to the parabola $y = x^2$.

32 A ball thrown upward at 96 feet per second from a height of 112 feet (when $t = 0$) has height function $y = -16t^2 + 96t + 112$. (a) What is the ball's maximum height? (b) When and with what impact speed does it hit the ground?

33 A spaceship approaching touchdown on a distant planet has height y (meters) at time t (seconds) given by

$$y = 100 - 100t + 25t^2.$$

When, and with what speed, does it hit the ground?

34 A city had population $P = 100(1 + 0.04t + 0.003t^2)$ thousands, with t in years and $t = 0$ in 1970. (a) What was the rate of change of P in 1975? (b) What was the average rate of change of P from 1973 to 1978?

35 As a snowball with an initial radius of 12 centimeters melts, its radius decreases at a constant rate. It starts to melt when $t = 0$ and takes 12 hours to disappear. (a) What is its rate of change of volume when $t = 6$? (b) What is its average rate of change of volume from $t = 3$ to $t = 9$?

In each of Problems 36–39, find the inverse function of f.

36 $f(x) = 1 - 3x.$

37 $f(x) = \sqrt{5 - 2x}.$

38 $f(x) = \dfrac{1}{\sqrt{2x + 1}}.$

39 $f(x) = \dfrac{1}{(x^3 - 1)^5}.$

40 Suppose that $g(x) = 1 + \sqrt{x}$. Find f so that

$$f(g(x)) = 3 + 2\sqrt{x} + x.$$

41 Suppose that $f(x) = 1 + x^2$. Find g so that

$$f(g(x)) = 1 + x^2 - 2x^3 + x^4.$$

Use the limit laws of Section 1-8 to find the limits in Problems 42–48.

42 $\displaystyle\lim_{x \to 1} \frac{x^4 - 1}{x^2 + 2x - 3}.$

43 $\displaystyle\lim_{x \to 7} \frac{\sqrt{x + 2} - 3}{x - 7}.$

44 $\displaystyle\lim_{x \to 1^+} (x - \sqrt{x^2 - 1}).$

45 $\displaystyle\lim_{x \to -4} \frac{(1/\sqrt{13 + x}) - \frac{1}{3}}{x + 4}.$

46 $\displaystyle\lim_{x \to 1^+} \frac{1 - x}{|1 - x|}.$

47 $\displaystyle\lim_{x \to 2^+} \frac{2 - x}{\sqrt{4 - 4x + x^2}}.$

48 $\displaystyle\lim_{x \to 0} \frac{\sin 3x}{x}.$

Discuss the discontinuities of the functions given in Problems 49–52.

49 $f(x) = \dfrac{1 - x}{(2 - x)^2}.$

50 $f(x) = \dfrac{x^2 + x - 2}{x^2 + 2x - 3}.$

51 $f(x) = \dfrac{|x^2 - 1|}{x^2 - 1}.$

52 $f(x) = x \sin \dfrac{1}{x}.$

53 Apply the Intermediate Value Property to prove that the equation $x^5 + x = 1$ has a solution.

54 Apply the Intermediate Value Property to prove that the equation $x^3 - 3x^2 + 1 = 0$ has three different solutions.

55 Suppose that f is continuous at a and that $f(a) > 0$. Apply the definition of the limit to prove that there exists $h > 0$ such that

$$|x - a| < h \quad \text{implies} \quad f(x) > 0.$$

Differentiation

───────────────────────────────

2

Introduction

In Section 1-5 we introduced the derivative $f'(x)$ as the slope of the tangent line to the graph of the function f at the point $(x, f(x))$. More precisely, we were motivated by the geometry to *define* the tangent line to the graph at the point $P(a, f(a))$ to be the straight line through P with slope

$$m = \lim_{h \to 0} \frac{f(a + h) - f(a)}{h}. \tag{1}$$

This definition led to the introduction of a new function f', the derivative of f, defined by

$$f'(x) = \lim_{h \to 0} \frac{f(x + h) - f(x)}{h} \tag{2}$$

where this limit exists. In Section 1-6 we saw that the derivative can also be interpreted as an instantaneous rate of change. For example, if a particle moves along a line with position function $s = f(t)$, then the instantaneous rate of change of its position is its velocity $v = f'(t)$.

In Sections 1-5 and 1-8 we illustrated in several examples the process of differentiating a given function f by direct evaluation of the limit in (2) above. This involves carrying out these four steps.

1 Write the definition (2) of the derivative.
2 Substitute the expressions $f(x + h)$ and $f(x)$ as determined by the particular given function f.
3 Simplify the result by algebraic methods to make Step 4 possible.
4 Evaluate the limit—typically, by application of the limit laws.

For instance, we used such steps to deduce that:

$$\text{If} \quad f(x) = ax^2 + bx + c, \quad \text{then} \quad f'(x) = 2ax + b.$$

Even when the function f is rather simple, this process for computing f' directly from the definition of the derivative can be tedious. Also, Step 3 may require considerable ingenuity. Moreover, it would be very repetitious to continue to rely upon the above four-step process. To avoid tedium, we want a faster, easier, and shorter method.

That is the subject of this chapter: The development of systematic methods or "rules" for differentiating those functions that occur most frequently. These functions include polynomials, rational functions, compositions of such functions, and the trigonometric functions $\sin x$ and $\cos x$. Once these general differentiation rules have been established, they can be applied formally, almost mechanically, to compute derivatives. Only rarely will recourse to the definition of the derivative be required.

Thus our goal here is the routine, quick, and efficient computation of the derivatives of a wide variety of functions. With this computational facility, we shall be able to attack with success the optimization (maximum-minimum) and instantaneous rate problems that prompted our discussion of tangent lines, derivatives, and limits in Chapter 1.

We first recall from Section 1-5 the precise definition of the derivative.

> **Definition** *The Derivative*
> The **derivative** of the function f is the function f' defined by
>
> $$f'(x) = \lim_{h \to 0} \frac{f(x + h) - f(x)}{h} \tag{1}$$
>
> for all x such that this limit exists.

The last phrase implies, in particular, that f must be defined in a neighborhood of x.

We emphasized in Section 1-5 that we hold x fixed in (1) while h approaches zero. When we are specifically interested in the value of the derivative f' at $x = a$, we sometimes rewrite (1) in the form

$$f'(a) = \lim_{h \to 0} \frac{f(a + h) - f(a)}{h} = \lim_{x \to a} \frac{f(x) - f(a)}{x - a}. \tag{2}$$

The second limit here is obtained from the first by writing $x = a + h$ and $h = x - a$ and noting that $x \to a$ as $h \to 0$. The statement that these equivalent limits exist is sometimes abbreviated as "$f'(a)$ exists." In this case we say that the function f is **differentiable** at a. Finally, the process of finding the derivative f' is called **differentiation** of f.

When we interpreted the derivative in Section 1-6 as a rate of change, we found it useful to employ the dependent-independent variable notation

$$y = f(x), \qquad \Delta x = h, \qquad \Delta y = f(x + \Delta x) - f(x).$$

This led to the "differential notation"

$$\frac{dy}{dx} = \lim_{\Delta x \to 0} \frac{\Delta y}{\Delta x} = \lim_{\Delta x \to 0} \frac{f(x + \Delta x) - f(x)}{\Delta x} \tag{3}$$

for the derivative. When you use this notation, it is important to remember that the symbol dy/dx is simply an alternative notation for the derivative $f'(x)$. It is *not* defined as the quotient of two separate entities dy and dx (although we shall see in Chapter 3 that dx and dy can be defined separately and in such a way that dy divided by dx is actually equal to $f'(x)$).

A third notation is commonly used for the derivative $f'(x)$; it is $Df(x)$. Here, think of D as a "machine" that "operates" on the function f to produce the derivative function Df. Thus we can write the derivative of $y = f(x) = x^2$ in any of three ways:

$$f'(x) = \frac{dy}{dx} = Dx^2 = 2x.$$

These three notations for the derivative—the functional notation $f'(x)$, the differential notation dy/dx, and the operator notation $Df(x)$—are used interchangeably in mathematical and scientific writing, so familiarity with each is necessary.

DERIVATIVES OF POLYNOMIALS

Our first differentiation rule says that *the derivative of a constant function is identically zero*. This is geometrically obvious, since the graph of a constant function is a horizontal straight line and therefore has slope zero at each point.

Theorem 1 *Derivative of a Constant*

If $f(x) = C$ (a constant) for all x, then $f'(x) = 0$ for all x. That is,

$$\frac{dC}{dx} = DC = 0. \tag{4}$$

Proof   Since $f(x + h) = f(x) = C$, we see that

$$f'(x) = \lim_{h \to 0} \frac{f(x + h) - f(x)}{h}$$

$$= \lim_{h \to 0} \frac{C - C}{h} = \lim_{h \to 0} \frac{0}{h} = 0.$$

As motivation for the next rule, consider the following list of derivatives, all computed in Chapter 1.

$Dx = 1$	Special cases of the
$Dx^2 = 2x$	Theorem in Section 1-5
$Dx^3 = 3x^2$	(Section 1-5, Problem 6)
$Dx^{-1} = -x^{-2}$	(Section 1-5, Problem 7)
$Dx^{-2} = -2x^{-3}$	(Section 1-5, Problem 9)
$Dx^{1/2} = \frac{1}{2}x^{-1/2}$	(Section 1-8, Example 12)
$Dx^{-1/2} = -\frac{1}{2}x^{-3/2}$	(Section 1-8, Problem 27)

Each of these formulas fits the simple pattern

$$Dx^n = nx^{n-1}. \tag{5}$$

As an inference from the instances listed above, (5) is only a conjecture. But many discoveries in mathematics are made by detecting such apparent patterns, and later proving that they hold universally.

Eventually we shall see that Formula (5), called the **power rule,** is valid for all real numbers n. At this time we give a proof only for the case when the exponent n is a *positive integer*. We need the *binomial formula* from high-school algebra, which says that

$$(a + b)^n = a^n + na^{n-1}b + \frac{n(n - 1)}{1 \cdot 2} a^{n-2}b^2$$

$$+ \cdots + \frac{n(n - 1) \cdots (n - k + 1)}{1 \cdot 2 \cdot 3 \cdots k} a^{n-k}b^k$$

$$+ \cdots + nab^{n-1} + b^n \tag{6}$$

if n is a positive integer. The cases $n = 2$ and $n = 3$ are the familiar formulas

$$(a + b)^2 = a^2 + 2ab + b^2$$

and

$$(a + b)^3 = a^3 + 3a^2b + 3ab^2 + b^3.$$

There are $n + 1$ terms on the right-hand side of (6), but all we need to know here is the exact form of the first two terms, and the fact that all the others include b^2 as a factor.

Theorem 2 *Power Rule for n a Positive Integer*
If n is a positive integer and $f(x) = x^n$, then $f'(x) = nx^{n-1}$.

Proof The binomial formula gives

$$(x + h)^n = x^n + nx^{n-1}h + (n - 1 \qquad \text{terms each involving } h^2).$$

Therefore

$$f'(x) = \lim_{h \to 0} \frac{(x + h)^n - x^n}{h}$$

$$= \lim_{h \to 0} \left[nx^{n-1} + (n - 1 \qquad \text{terms each involving } h) \right]$$

$$= nx^{n-1},$$

because each of the $n - 1$ terms having h as a factor has limit zero.

For example, $Dx^{17} = 17x^{16}$ and $Dx^{1000} = 1000x^{999}$. Of course, we need not always use x as the independent variable. We may write the power rule with other symbols for the independent variable as

$$Dt^n = nt^{n-1}, \qquad Du^n = nu^{n-1}, \quad \text{and} \quad Dy^n = ny^{n-1}.$$

To use the power rule to differentiate polynomials, we need to know how to differentiate linear combinations. A **linear combination** of the functions f and g is a function of the form $af + bg$, where a and b are constants. It follows from the sum and product laws for limits that

$$\lim_{x \to c} \left[af(x) + bg(x) \right] = a \lim_{x \to c} f(x) + b \lim_{x \to c} g(x) \qquad (7)$$

provided that the two right-hand limits exist. Formula (7) is called the **linearity property** of the limit operation. It implies an analogous linearity property of differentiation.

Theorem 3 *Derivative of a Linear Combination*
If f and g are differentiable functions, then

$$D[af(x) + bg(x)] = aDf(x) + bDg(x). \qquad (8)$$

With $u = f(x)$ and $v = g(x)$, this takes the form

$$\frac{d(au + bv)}{dx} = a\frac{du}{dx} + b\frac{dv}{dx}. \qquad (8')$$

Proof   The linearity property of limits immediately gives

$$D[af(x) + bg(x)] = \lim_{h \to 0} \frac{[af(x + h) + bg(x + h)] - [af(x) + bg(x)]}{h}$$

$$= a\left(\lim_{h \to 0} \frac{f(x + h) - f(x)}{h}\right) + b\left(\lim_{h \to 0} \frac{g(x + h) - g(x)}{h}\right)$$

$$= aDf(x) + bDg(x).$$

Now take $a = c$ and $b = 0$ in Equation (8). The result is

$$D[cf(x)] = cDf(x) \quad \text{or} \quad \frac{d(cu)}{dx} = c\frac{du}{dx}. \tag{9}$$

That is, *the derivative of* a *constant multiple of a function is the same constant multiple of its derivative.*

Next, take $a = b = 1$ in Equation (8). We find that

$$D[f(x) + g(x)] = Dfx) + Dg(x) \quad \text{or} \quad \frac{d(u + v)}{dx} = \frac{du}{dx} + \frac{dv}{dx}. \tag{10}$$

Thus *the derivative of the sum of two functions is the sum of their derivatives.* Similarly, for differences we have

$$\frac{d(u - v)}{dx} = \frac{du}{dx} - \frac{dv}{dx}. \tag{11}$$

Repeated application of (10) to a sum of a finite number of differentiable functions gives

$$\frac{d(u_1 + u_2 + \cdots + u_n)}{dx} = \frac{du_1}{dx} + \frac{du_2}{dx} + \cdots + \frac{du_n}{dx}. \tag{12}$$

When we apply (9) and (12) and the power rule to the polynomial

$$p(x) = a_n x^n + a_{n-1}x^{n-1} + \cdots + a_2 x^2 + a_1 x + a_0,$$

we find the derivative *as fast as we can write it*; it is

$$p'(x) = na_n x^{n-1} + (n - 1)a_{n-1}x^{n-2} + \cdots + 2a_2 x + a_1. \tag{13}$$

With this result, it becomes a routine matter to write an equation for a tangent line to the graph of a polynomial.

EXAMPLE 1   Find the tangent line to the curve $y = 2x^3 - 7x^2 + 3x + 4$ at the point $(1, 2)$.

Solution   We compute the derivative as in (13) and find it to be

$$\frac{dy}{dx} = 6x^2 - 14x + 3.$$

We substitute $x = 1$ into the derivative and find that the slope of the tangent is $m = -5$. So the point-slope equation of the tangent line is

$$y - 2 = -5(x - 1).$$

EXAMPLE 2   The volume in cubic centimeters of a quantity of water varies with changing temperature T. For T between $0°C$ and $30°C$, the relationship is almost exactly

$$V = V_0(1 + \alpha T + \beta T^2 + \gamma T^3),$$

where V_0 is the volume at $0°C$, and the three constants have the values

$$\alpha = -0.06427 \times 10^{-3}, \qquad \beta = 8.5053 \times 10^{-6}, \qquad \gamma = -6.7900 \times 10^{-8}.$$

The rate of change of volume with respect to temperature is

$$\frac{dV}{dT} = V_0(\alpha + 2\beta T + 3\gamma T^2).$$

Suppose that $V_0 = 10^5$ (cc) and that $T = 20°C$. Then $V \approx 100,157$ (cc) and $dV/dT \approx 19.4$ cc/$°C$. We conclude that, at $T = 20°C$, this amount of water should increase in volume by approximately 19.4 cc for each rise of $1°C$ in temperature. By comparison, direct substitution into the above formula for V gives

$$V(21) - V(20) \approx 19.9 \quad \text{(cc)}.$$

THE PRODUCT AND QUOTIENT RULES

It might be natural to conjecture that the derivative of a product $f(x)g(x)$ is the product of the derivatives. This is *false*. For example, if $f(x) = g(x) = x$, then

$$D[f(x)g(x)] = Dx^2 = 2x$$

while

$$(Df(x))(Dg(x)) = (Dx)(Dx) = 1 \cdot 1 = 1.$$

In general, the derivative of a product is *not* merely the product of the derivatives. The following theorem tells what it *is*.

Theorem 4   *The Product Rule*

If f and g are differentiable at x, then fg is differentiable at x, and

$$D[f(x)g(x)] = f'(x)g(x) + f(x)g'(x). \tag{14}$$

With $u = f(x)$ and $v = g(x)$, this takes the form

$$\frac{d(uv)}{dx} = v\frac{du}{dx} + u\frac{dv}{dx}. \tag{14'}$$

When it is unmistakably clear what the independent variable is, we can write the product rule even more briefly:

$$(uv)' = u'v + uv'. \tag{14''}$$

Proof   We use an "add and subtract" device.

$$D[f(x)g(x)] = \lim_{h \to 0} \frac{f(x+h)g(x+h) - f(x)g(x)}{h}$$

$$= \lim_{h \to 0} \frac{f(x+h)g(x+h) - f(x)g(x+h) + f(x)g(x+h) - f(x)g(x)}{h}$$

$$= \left(\lim_{h \to 0} \frac{f(x+h) - f(x)}{h} \right) \left(\lim_{h \to 0} g(x+h) \right) + f(x) \left(\lim_{h \to 0} \frac{g(x+h) - g(x)}{h} \right)$$

$$= f'(x)g(x) + f(x)g'(x).$$

In this proof we used the sum and product laws for limits, the definitions of $f'(x)$ and $g'(x)$, and the fact that

$$\lim_{h \to 0} g(x + h) = g(x).$$

The last equation holds because g is differentiable and therefore continuous at x (by Theorem 5 in Section 1-9).

In words, the product rule says that *the derivative of a product is equal to the second factor times the derivative of the first factor, plus the first factor times the derivative of the second.*

EXAMPLE 3   Find the derivative of

$$f(x) = (1 - 4x^3)(3x^2 - 5x + 2)$$

without first multiplying the two factors.

Solution

$$D[(1 - 4x^3)(3x^2 - 5x + 2)]$$
$$= [D(1 - 4x^3)](3x^2 - 5x + 2) + (1 - 4x^3)[D(3x^2 - 5x + 2)]$$
$$= (-12x^2)(3x^2 - 5x + 2) + (1 - 4x^3)(6x - 5)$$
$$= -60x^4 + 80x^3 - 24x^2 + 6x - 5.$$

We can apply the product rule repeatedly to find the derivative of a product of three or more functions $u_1, u_2, \ldots, u_n$ of x. For example,

$$D[u_1 u_2 u_3] = D[(u_1 u_2) \cdot u_3]$$
$$= [D(u_1 u_2)] \cdot u_3 + (u_1 u_2) \cdot Du_3$$
$$= [(Du_1)u_2 + u_1(Du_2)]u_3 + (u_1 u_2)Du_3$$
$$= (Du_1)u_2 u_3 + u_1(Du_2)u_3 + u_1 u_2(Du_3).$$

The general result is

$$D(u_1 u_2 \cdots u_n) = (Du_1)u_2 \cdots u_n + u_1(Du_2)u_3 \cdots u_n$$
$$+ \cdots + u_1 u_2 \cdots u_{n-1}(Du_n), \tag{15}$$

where the last sum has one term for each of the n factors. It is easy to establish this result (Problem 46) if we use the following strategy for a *proof by induction on n*.

Principle of Mathematical Induction

Suppose it is required to prove true for all positive integers n a statement or formula involving the variable n. It is sufficient to show the following.

(i) The statement holds when $n = 1$.

(ii) The assumption that the statement holds for $n = k$ (a fixed but arbitrary positive integer) implies that it also holds for $n = k + 1$.

EXAMPLE 4   Use the principle of mathematical induction to establish the power rule for n a positive integer.

Solution   We are to show that $Dx^n = nx^{n-1}$ for all positive integers n. This statement is clearly true when $n = 1$; that is $Dx^1 = 1x^0$. Assume that it holds when $n = k$; that is, assume that $Dx^k = kx^{k-1}$ for some integer $k \geq 1$. The product rule then gives

$$Dx^{k+1} = D(x \cdot x^k) = (Dx) \cdot x^k + x \cdot (Dx^k).$$

Our assumption about the truth of the formula for $n = k$ now lets us continue these computations, and we find that

$$Dx^{k+1} = 1 \cdot x^k + x \cdot kx^{k-1} = x^k + kx^k = (k + 1)x^k.$$

This is the correct form of the power rule in the case $n = k + 1$. Thus we satisfy both conditions of the induction principle, and we conclude that $Dx^n = nx^{n-1}$ for all integral $n \geq 1$.

Our next theorem tells us how to find the derivative of the reciprocal of a function if we know the function's derivative.

Theorem 5   *The Reciprocal Rule*

If f is differentiable at x and $f(x) \neq 0$, then

$$D\frac{1}{f(x)} = -\frac{f'(x)}{[f(x)]^2}. \qquad (16)$$

With $u = f(x)$ the reciprocal rule takes the form

$$D\frac{1}{u} = -\frac{1}{u^2}\frac{du}{dx} \quad \text{or} \quad \left(\frac{1}{u}\right)' = -\frac{u'}{u^2}. \qquad (16')$$

Proof   As in the proof of Theorem 4, we use the limit laws, the definition of the derivative, and the fact that a function is continuous wherever it is differentiable. Here $f(x + h) \neq 0$ for h small, because $f(x) \neq 0$ and f is continuous at x. Therefore

$$D \frac{1}{f(x)} = \lim_{h \to 0} \frac{1}{h} \left(\frac{1}{f(x + h)} - \frac{1}{f(x)} \right)$$

$$= \lim_{h \to 0} \frac{f(x) - f(x + h)}{hf(x + h)f(x)}$$

$$= -\left(\lim_{h \to 0} \frac{1}{f(x + h)f(x)} \right) \left(\lim_{h \to 0} \frac{f(x + h) - f(x)}{h} \right)$$

$$= -\frac{f'(x)}{[f(x)]^2}.$$

For instance,

$$D \frac{1}{x^2 + 1} = -\frac{D(x^2 + 1)}{(x^2 + 1)^2} = -\frac{2x}{(x^2 + 1)^2}.$$

We now combine the reciprocal rule with the power rule for positive integral exponents to establish the power rule for negative integral exponents.

Theorem 6   *Power Rule for n a Negative Integer*

If n is a negative integer, then $Dx^n = nx^{n-1}$.

Proof   Let $m = -n$ so that m is a positive integer. Then

$$Dx^n = D \frac{1}{x^m} = -\frac{mx^{m-1}}{x^{2m}} = (-m)x^{(-m)-1} = nx^{n-1}.$$

This proof also shows that the rule of Theorem 6 holds exactly when the function being differentiated is defined: when $x \neq 0$. As an illustration, $Dx^{-7} = -7x^{-8}$ when $x \neq 0$.

Now we apply the product and reciprocal rules to get a rule for differentiation of the quotient of two functions.

Theorem 7   *The Quotient Rule*

If f and g are differentiable at x and $g(x) \neq 0$, then f/g is differentiable at x and

$$D \left(\frac{f(x)}{g(x)} \right) = \frac{f'(x)g(x) - f(x)g'(x)}{[g(x)]^2}. \tag{17}$$

With $u = f(x)$ and $v = g(x)$, this rule takes the forms

$$D \left(\frac{u}{v} \right) = \frac{v \dfrac{du}{dx} - u \dfrac{dv}{dx}}{v^2} \quad \text{and} \quad \left(\frac{u}{v} \right)' = \frac{u'v - uv'}{v^2}. \tag{17'}$$

Proof   We apply the product rule to the factorization

$$\frac{f(x)}{g(x)} = f(x) \cdot \frac{1}{g(x)}.$$

This gives

$$D\left(\frac{f(x)}{g(x)}\right) = (Df(x)) \cdot \frac{1}{g(x)} + f(x) \cdot D\frac{1}{g(x)}$$

$$= \frac{f'(x)}{g(x)} + f(x)\left(-\frac{g'(x)}{[g(x)]^2}\right)$$

$$= \frac{f'(x)g(x) - f(x)g'(x)}{[g(x)]^2}.$$

Note that the numerator in (17) is *not* the derivative of the product of f and g. And the minus sign means that the order of the terms in the numerator is important.

EXAMPLE 5   Find dz/dt if

$$z = \frac{1 - t^3}{1 + t^4}.$$

Solution   With t rather than x as the independent variable, the quotient rule gives

$$\frac{dz}{dt} = \frac{[D(1 - t^3)](1 + t^4) - (1 - t^3)[D(1 + t^4)]}{(1 + t^4)^2}$$

$$= \frac{(-3t^2)(1 + t^4) - (1 - t^3)(4t^3)}{(1 + t^4)^2} = \frac{t^6 - 4t^3 - 3t^2}{(1 + t^4)^2}.$$

2-2   PROBLEMS

Apply the differentiation rules of this section to find the derivatives of the functions given in Problems 1–26.

1 $f(x) = 3x^2 - x + 5.$ **2** $g(t) = 1 - 3t^2 - 2t^4.$

3 $f(x) = (2x + 3)(3x - 2).$

4 $g(x) = (2x^2 - 1)(x^3 + 2).$

5 $h(x) = (x + 1)^3.$ **6** $g(t) = (4t - 7)^2.$

7 $f(y) = y(2y - 1)(2y + 1).$

8 $f(x) = 4x^4 - \dfrac{1}{x^2}.$

9 $g(x) = \dfrac{1}{x + 1} - \dfrac{1}{x - 1}.$

10 $f(t) = \dfrac{1}{4 - t^2}.$ **11** $h(x) = \dfrac{3}{x^2 + x + 1}.$

12 $f(x) = \dfrac{1}{1 - (2/x)}.$

13 $g(t) = (t^2 + 1)(t^3 + t^2 + 1).$

14 $f(x) = (2x^3 - 3)(17x^4 - 6x + 2).$

15 $g(z) = \dfrac{1}{2z} - \dfrac{1}{3z^2}.$

16 $f(x) = \dfrac{2x^3 - 3x^2 + 4x - 5}{x^2}.$

17 $g(y) = 2y(3y^2 - 1)(y^2 + 2y + 3).$

18 $f(x) = \dfrac{x^2 - 4}{x^2 + 4}.$ **19** $g(t) = \dfrac{t - 1}{t^2 + 2t + 1}.$

20 $u(x) = \dfrac{1}{(x + 2)^2}.$ **21** $v(t) = \dfrac{1}{(t - 1)^3}.$

22 $h(x) = \dfrac{2x^3 + x^2 - 3x + 17}{2x - 5}.$

23 $g(x) = \dfrac{3x}{x^3 + 7x - 5}.$ **24** $f(t) = \dfrac{1}{[t + (1/t)]^2}.$

25 $g(x) = \dfrac{(1/x) - (2/x^2)}{(2/x^3) - (3/x^4)}.$

26 $f(x) = \dfrac{x^3 - [1/(x^2 + 1)]}{x^4 + [1/(x^2 + 1)]}.$

In each of Problems 27–31, find all points on the curve $y = f(x)$ where the tangent line is horizontal.

27 $y = x^3 + 7.$ **28** $y = x^4 - 8x^2 + 1.$

29 $y = 2x^3 - 3x^2 - 12x + 20.$

30 $y = x^6 - 3x^2 - 11.$ **31** $y = \dfrac{1}{1 + x^2}.$

In each of Problems 32–34, $s = f(t)$ is the position function of a point moving along a line. Determine the time intervals in which the velocity is positive and those in which it is negative.

32 $s = 2t^2 - 8t + 5$.

33 $s = 2t^3 + 3t^2 - 36t$.

34 $s = \dfrac{t}{t^2 + 1}$.

35 Apply the formula of Example 2 to answer the following two questions. (a) If 1000 cc of water at 0°C is heated, does it initially expand or contract? (b) What is the rate (in cc/°C) at which it contracts or expands?

36 Susan's weight in pounds is given by the formula $W = 2 \times 10^9/R^2$, where R is her distance in miles from the center of the earth. What is the rate of change of W with respect to R when $R = 3960$ miles? If Susan climbs a mountain, at what rate in ounces per (vertical) mile does her weight decrease?

2.1 The leaky tank (see Problem 37).

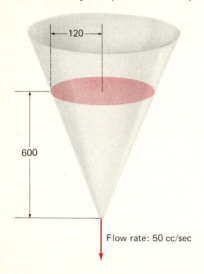

120

600

Flow rate: 50 cc/sec

37 The conical tank shown in Figure 2.1 has radius 160 cm and height 800 cm. Water is running out a small hole in the bottom of the tank. When the height h of water in the tank is 600 cm, what is the rate of change of its volume V with respect to h?

38 Find the intercepts of the straight line that is tangent to the curve $y = x^3 + x^2 + x$ at the point (1, 3).

39 Find the line through the point (1, 5) that is tangent to the curve $y = x^3$.

40 Find *two* lines through the point (2, 8) that are tangent to the curve $y = x^3$.

41 Show that no straight line can be tangent to the curve $y = x^2$ at two different points.

42 Find the two straight lines of slope -2 that are tangent to the curve $y = 1/x$.

43 Find the x-intercept of the line that is tangent to the curve $y = x^n$ at the point $P(x_0, y_0)$.

44 Show that the curve $y = x^5 + 2x$ has no horizontal tangent line. What is the smallest slope that a tangent to this curve can have?

45 Apply Equation (15) with $n = 3$ and $u_1 = u_2 = u_3 = f(x)$ to show that

$$D([f(x)]^3) = 3[f(x)]^2 f'(x).$$

46 Use the principle of mathematical induction to prove Equation (15).

47 Apply (15) to show that

$$D([f(x)]^n) = n[f(x)]^{n-1} f'(x)$$

if n is a positive integer and $f'(x)$ exists.

48 Use the result of Problem 47 to find $D(x^2 + x + 1)^{100}$.

49 Find $g'(x)$ if $g(x) = (x^3 - 17x + 35)^{17}$.

50 Find the constants a, b, c, and d if the curve $y = ax^3 + bx^2 + cx + d$ has horizontal tangent lines at the points (0, 1) and (1, 0).

2-3

Simple Maximum and Minimum Problems

In applications we often need to find the maximum or minimum value that a specified quantity can attain. The fence problem and the refrigerator problem stated in Section 1-1 are simple but typical examples of applied maximum-minimum problems. In Section 1-2 we saw that each of these two problems reduces to the purely mathematical problem of finding the maximum value attained by the function $f(x) = x(70 - x)$ on the closed interval $[0, 70]$.

In this section we discuss the general problem of finding the maximum and minimum values attained by a *continuous* function f on a *closed* interval $[a, b]$. By the Maximum Value Property of continuous functions (stated in Section 1-9), we know that f *does* have maximum and minimum values on $[a, b]$. The question, then, is this: Exactly where *are* these values located?

In Section 1-5 we solved the fence and refrigerator problems on the basis of the geometrically motivated assumption that the function $f(x) = x(70 - x)$ attains its maximum value on $[0, 70]$ at an interior point of that interval, one at which the tangent line is horizontal. Theorems 1 and 2 below provide a rigorous basis for the method we used there.

We say that the value $f(c)$ is a **local maximum value** of the function f if $f(x) \leq f(c)$ for all x sufficiently near c. More precisely, if this inequality holds for all x that are simultaneously in the domain of f and in some open interval containing c, then $f(c)$ is a local maximum of f. Similarly, we say that the value $f(c)$ is a **local minimum value** of f if $f(x) \geq f(c)$ for all x sufficiently near c. Thus a local maximum occurs where the value of f is at least as large as it is anywhere nearby; a local minimum occurs where the function is at least as small as at nearby points. As the graph of Figure 2.2 shows, a local maximum is a point such that no nearby points on the graph are higher, and a local minimum is one such that no nearby points on the curve are lower. A **local extremum** is a value of f that is either a local maximum or a local minimum.

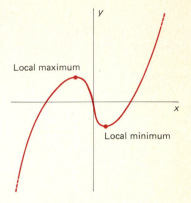

2.2 Local extrema.

Theorem 1   *Local Maxima and Minima*

If f is differentiable at c and is defined on an open interval containing c and $f(c)$ is either a local maximum value or a local minimum value of f, then $f'(c) = 0$.

Thus a local extremum of a *differentiable* function on an *open* interval can occur only at a point where the derivative is zero, and therefore where the tangent line to the graph is horizontal.

Proof   Suppose, for instance, that $f(c)$ is a local maximum value. The fact that $f'(c)$ exists means that the right-hand and left-hand limits

$$\lim_{h \to 0^+} \frac{f(c + h) - f(c)}{h} \quad \text{and} \quad \lim_{h \to 0^-} \frac{f(c + h) - f(c)}{h}$$

both exist and are equal to $f'(c)$.

If $h > 0$, then

$$\frac{f(c + h) - f(c)}{h} \leq 0$$

because $f(c) \geq f(c + h)$ for all small positive values of h. Hence, by a one-sided version of the squeeze property for limits (Section 1-8), the above inequality will be preserved when we take the limit as $h \to 0$. We thus find that

$$f'(c) = \lim_{h \to 0^+} \frac{f(c + h) - f(c)}{h} \leq \lim_{h \to 0^+} 0 = 0.$$

Similarly, in the case $h < 0$, we find that

$$\frac{f(c + h) - f(c)}{h} \geq 0.$$

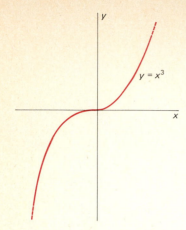

2.3 No extremum at $x = 0$, though the derivative is zero there.

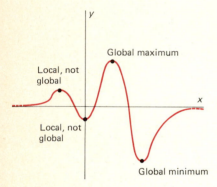

Global maximum

Local, not global

Local, not global

Global minimum

2.4 Some extrema are global, others are merely local.

So

$$f'(c) = \lim_{h \to 0^-} \frac{f(c + h) - f(c)}{h} \geq \lim_{h \to 0^-} 0 = 0.$$

Since both $f'(c) \leq 0$ and $f'(c) \geq 0$, we may conclude that $f'(c) = 0$. This establishes Theorem 1.

BEWARE   The converse of Theorem 1 is false. That is, the fact that $f'(c) = 0$ is *not enough* to imply that $f(c)$ is a local extremum. For example, consider the function $f(x) = x^3$. Its derivative $f'(x) = 3x^2$ vanishes at $x = 0$. But a glance at its graph, shown in Figure 2.3, shows us that $f(0) = 0$ is *not* a local extremum of x^3.

Thus the equation $f'(c) = 0$ is a *necessary* condition for $f(c)$ to be a local maximum or minimum value of f (for a function f differentiable on an open interval). It is *not a sufficient* condition. For $f'(x)$ can well be zero at points other than local maxima and minima. We shall give sufficient conditions for local maxima and minima in Chapter 3.

Now in most sorts of optimization problems, our interest is not in local maxima and minima as such, but rather in the global or *absolute* maximum and minimum values attained by a given continuous function. If f is a function with domain D, we call $f(c)$ the **absolute maximum** value of f on D provided that $f(c) \geq f(x)$ for *all* x in the domain D. Briefly, $f(c)$ is the largest value of f on D. It should be clear how to define the global or absolute minimum. The graph of Figure 2.4 illustrates some local and global extrema. Note that every global extremum is, of course, local; on the other hand, the graph shows some local extrema that are not global.

Theorem 2 below says that the absolute maximum and minimum values of the continuous function f on the closed interval $[a, b]$ both occur either at one of the end points a and b or at a critical point of f. The number c in the domain of the function f is called a **critical point** of f if either

(i) $f'(c) = 0$, or (ii) $f'(c)$ does not exist.

Theorem 2   *Absolute Maxima and Minima*
Suppose that $f(c)$ is the absolute maximum (or minimum) value of the continuous function f on the closed interval $[a, b]$. Then c is either a critical point of f or one of the end points a and b.

Proof   This follows almost immediately from Theorem 1. If c is not an end point of $[a, b]$, then $f(c)$ is a local extremum of f on the open interval (a, b). In this case Theorem 1 implies that $f'(c) = 0$, provided that f is differentiable at c.

As a consequence of Theorem 2, we can find the (absolute) maximum and minimum values of f on $[a, b]$ by first locating the critical points of f in (a, b) and then finding the value of f at each of these critical points *and* at the two end points. The largest of these values must then be the absolute maximum value of f, and the smallest will be its absolute minimum value. This procedure is called the **closed interval maximum-minimum method.**

EXAMPLE 1   Find the maximum and minimum values of

$$f(x) = 2x^3 - 3x^2 - 12x + 15$$

on the closed interval $[0, 3]$.

Solution   The derivative of f is

$$f'(x) = 6x^2 - 6x - 12 = 6(x - 2)(x + 1).$$

Hence the critical points of f are -1 and $+2$, but only the latter lies in $[0, 3]$. We evaluate f at this critical point and at the end points and obtain

$$f(0) = 15, \qquad f(2) = -5, \quad \text{and} \quad f(3) = 6.$$

Therefore the maximum value of $f(x)$ on $[0, 3]$ is $f(0) = 15$; the minimum is $f(2) = -5$.

In Example 1 the function f was differentiable everywhere. Example 2 illustrates the case of an extremum at a critical point at which the function is not differentiable.

EXAMPLE 2   Find the maximum and minimum values of the function $f(x) = 3 - |x - 2|$ on the interval $[1, 4]$.

Solution   If $x \leq 2$, then $x - 2 \leq 0$; in this case

$$f(x) = 3 + (x - 2) = x + 1.$$

If $x \geq 2$ then $x - 2 \geq 0$, and so

$$f(x) = 3 - (x - 2) = 5 - x.$$

Consequently the graph of f looks like the one shown in Figure 2.5. The only critical point of f in $[1, 4]$ is the point $x = 2$, because f is not differentiable there. (Why?) Evaluation of f at this critical point and at the two

2.5   Graph of the function of Example 2.

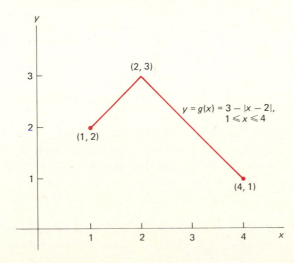

$y = g(x) = 3 - |x - 2|,$
$1 \leqslant x \leqslant 4$

endpoints yields

$$f(1) = 2, \qquad f(2) = 3, \quad \text{and} \quad f(4) = 1.$$

Thus the maximum value of $f(x)$ on $[1, 4]$ is $f(2) = 3$, and its minimum value is $f(4) = 1$.

APPLIED MAXIMUM-MINIMUM PROBLEMS

When we confront an applied maximum-minimum problem and choose to apply the closed interval maximum-minimum method, there is an important initial step: We must determine the quantity to be maximized or minimized. This quantity will be the dependent variable later in our solution.

This dependent variable must then be expressed as a function of an independent variable, one that "controls" the values of the dependent variable. If the domain of values of the independent variable—those that are pertinent to the applied problem—is a closed interval, then we may proceed with the closed interval maximum-minimum method. This plan of attack can be summarized in the following steps.

1 *Find the quantity to be maximized or minimized.* This quantity, which you should describe with a word or short phrase and label with a letter, will be your dependent variable. Since it is a *dependent* variable, it depends upon something else; that will be your independent variable. We call the independent variable x in what follows.

2 *Express the dependent variable as a function of the independent variable.* Use the conditions of the problem to write the dependent variable as a function of x. By all means, draw a figure or diagram and *label the variables*; this is frequently the best way to find the necessary relationships. Use auxiliary variables if they help; but do not use too many of them, for you must ultimately eliminate them. You *must* express the dependent variable in terms of a *single* independent variable x and various constants before you can compute a derivative. Find the domain of the function that is relevant to the stated problem as well as its formula. Force the domain to be a closed interval, if possible; if it is a bounded open interval, adjoin the endpoints.

3 *Apply calculus to find the critical points.* Compute the derivative $f'(x)$ of the function $f(x)$ found in Step 2 above. Use the derivative to find the critical points of type (i) (where $f'(x) = 0$) and of type (ii) (where $f'(x)$ does not exist).

4 *Identify the extrema.* Evaluate f at each critical point in your closed interval *and* at the two endpoints. The values you obtain will tell you which is the absolute maximum and which is the absolute minimum. Of course, either or both of these may occur at more than one point.

5 *Answer the question posed in the problem.* That is, interpret your results. The answer to the stated problem may be something other than merely the largest (or smallest) possible value of f. Give a precise answer to the specific question originally asked.

EXAMPLE 3   In one simple model of the spread of a contagious disease among members of a population of M people, the incidence of the disease, measured as the number of new cases per day, is given in terms of the number x of individuals already infected by

$$R(x) = kx(M - x) = kMx - kx^2$$

where k is some positive constant. How many individuals in the population are infected when the incidence R is the highest?

Solution   This problem is stated in such a way that Steps 1 and 2 above are already accomplished: The dependent variable is R, the independent variable is x, and $R(x)$ is the function given in the problem. The relevant domain of this function is $[0, M]$, since any number of people in the population, from none to all, may be infected.

Since $R'(x) = kM - 2kx$, we see that $x = M/2$ is the only critical point of R. Since

$$R(0) = 0, \qquad R\left(\frac{M}{2}\right) = \frac{1}{4} kM^2, \quad \text{and} \quad R(M) = 0,$$

we see that the incidence R is maximal when $x = M/2$. Hence the answer to the question posed is that the incidence is greatest when *half* the population is infected.

EXAMPLE 4   A piece of sheet metal is rectangular, 5 feet wide and 8 feet long. Equal squares are to be cut from its corners and the resulting piece of metal folded and welded to form a box with an open top, as shown in Figure 2.6. How should this be done to get a box of the largest possible volume?

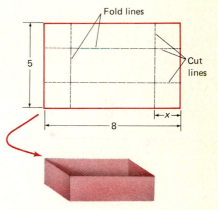

2.6   Making the box of Example 4.

Solution   The quantity to be maximized—the dependent variable—is the volume V of the box to be constructed. The shape and volume of the box are determined by the length x of the edge of each corner square removed. Hence x is a natural choice for the independent variable.

To write the volume as a function of x, note that the box will have height x, and its base will measure $8 - 2x$ feet by $5 - 2x$ feet. So

$$V = V(x) = x(5 - 2x)(8 - 2x) = 4x^3 - 26x^2 + 40x.$$

The procedure described in the problem will produce a box only if $0 < x < \frac{5}{2}$. But we make the domain the *closed* interval $[0, 2.5]$ in order to have the assurance that a maximum of $V(x)$ exists and to apply the closed interval maximum-minimum method. The values $x = 0$ and $x = \frac{5}{2}$ correspond to "degenerate" boxes.

Now we compute the derivative of $V(x)$.

$$\frac{dV}{dx} = 12x^2 - 52x + 40 = 4(3x - 10)(x - 1).$$

The critical points of V are now apparent: They occur when $x = 1$ and $x = \frac{10}{3}$. We discard the latter, for only $x = 1$ lies in the relevant interval $[0, 2.5]$. Here are the values of V we must examine:

$$V(0) = 0, \qquad V(1) = 18, \qquad V(2.5) = 0.$$

Thus the maximum value of $V(x)$ on $[0, 2.5]$ is $V(1) = 18$. The answer to the question posed is that the squares cut from the corners should be of edge length 1 foot each. The resulting box will measure 6 feet by 3 feet by 1 foot, and its volume will be 18 cubic feet.

Note that adjoining the endpoints 0 and 2.5 to the natural domain of V does not affect the answer, and makes the method of solution valid.

For our next application of the closed interval maximum-minimum method, let us consider a typical problem in business management. Suppose that x units of a certain product are to be manufactured at a total cost of $C(x)$ dollars. We make the following simple (and not always valid) assumption: The **cost function** $C(x)$ is the sum of two terms:

- a constant term a representing the fixed cost of acquiring and maintaining production facilities ("overhead"), and
- a variable term representing the additional cost of making x units at, for example, b dollars each.

Then $C(x)$ will be given by

$$C(x) = a + bx. \tag{1}$$

We also assume that the number of units that can be sold is a linear function of the selling price y, so that $x = m - ny$, where m and n are positive constants. The minus sign indicates, naturally enough, that an increase in selling price will cause a decrease in sales. If we solve this last equation for y, we get the **price function**

$$y = y(x) = A - Bx. \tag{2}$$

Here A and B are two constants that involve m and n; they result from solving the equation $x = m - ny$ for y.

What do we want to maximize? The profit, given here by the **profit function** $P(x)$, which is of course equal to the sales revenue minus the cost of production, so that

$$P(x) = x \cdot y(x) - C(x). \tag{3}$$

EXAMPLE 5 Suppose that the cost of publishing a certain book is $10,000 to set up the (annual) press run, plus $8 for each book actually printed. The publisher sold 7000 copies last year at $13 each, but this year sales dropped to 5000 copies when the price was raised to $15 per copy. Assume that as many as 10,000 copies can be printed in a single press run. How many copies should be printed, and what should be the selling price of each copy, in order to maximize the year's profit on this book?

Solution The dependent variable to be maximized is the profit P. As independent variable we choose the number x of copies to be printed, with $0 \leq x \leq 10,000$. Also, information in the problem tells us to take $a = 10,000$ and $b = 8$ in Equation (1), so that the cost function is

$$C(x) = 10,000 + 8x.$$

Now we substitute the data $x = 7000$ and $y = 13$ into Equation (2), as well as the data $x = 5000$ and $y = 15$. We obtain the equations

$$A - 7000B = 13 \quad \text{and} \quad A - 5000B = 15,$$

and their simultaneous solution is $A = 20$ and $B = 1/1000$. Hence the price function is

$$y = 20 - \frac{x}{1000},$$

and so the profit function is

$$P(x) = x\left(20 - \frac{x}{1000}\right) - (10{,}000 + 8x),$$

or

$$P(x) = 12x - \frac{x^2}{1000} - 10{,}000, \qquad x \text{ in } [0, 10{,}000].$$

Now

$$\frac{dP}{dx} = 12 - \frac{x}{500},$$

so the only critical point of the function $P(x)$ is at $x = 6000$. Since

$$P(0) = -10{,}000, \qquad P(6000) = 26{,}000, \quad \text{and} \quad P(10{,}000) = 10{,}000,$$

we see that the maximum possible profit of \$26,000 results from printing 6000 copies of the book. Each copy should be sold for \$14, since

$$y = 20 - \frac{6000}{1000} = 14.$$

EXAMPLE 6 (An inventory problem) A large department store sells 600 refrigerators each year, and currently orders them from the manufacturer on a quarterly basis; that is, 150 refrigerators are ordered each 3 months. Each order placed entails a fixed reorder cost of \$16, plus \$20 for each refrigerator ordered. In addition, the store incurs an annual carrying cost (warehouse storage charges) equal to \$30 multiplied by the average number of refrigerators on hand in inventory.

Assume that this average number on hand is equal to half the reorder lot size (the number ordered each time). What should be the reorder lot size in order to minimize the total reorder and carrying costs for the year? If this optimal reorder lot size is used, what annual savings will result, when compared with continued use of the present reorder lot size of 150?

Solution Let x denote the reorder lot size, so that $x = 150$ describes the store's current policy. Then the store must order $600/x$ lots per year. The cost of each reorder will be $16 + 20x$ dollars, and the average number of refrigerators on hand will be $x/2$. Therefore the annual reorder and carrying costs will be computed like this:

(reorder cost) = (cost per reorder)(number of reorders)

$$= (16 + 20x)\left(\frac{600}{x}\right).$$

(carrying cost) = (cost per refrigerator)(average number on hand)

$$= (30)\left(\frac{x}{2}\right).$$

Thus the total annual cost is

$$C(x) = 15x + (16 + 20x)\left(\frac{600}{x}\right) = 15x + \frac{9600}{x} + 12{,}000.$$

We assume at least one order each year and at least one refrigerator per order; if so, then C is defined on the closed interval $[1, 600]$. Hence we can use the closed interval maximum-minimum method. The derivative of C is

$$C'(x) = 15 - \frac{9600}{x^2}.$$

So $C'(x) = 0$ when $x \approx \pm 25.3$; we reject -25.3, and the value $x = 25.3$ cannot actually be attained in practice. We expect the greatest practical savings to occur either at $x = 25$ or at $x = 26$, unless it occurs at one of the two end points $x = 1$ and $x = 600$. Here are the values of the cost function at the 4 points of interest.

x	1	25	26	600
C	21,615	12,759.00	12,759.23	21,016

Thus the total reorder plus carrying cost is minimized by ordering 25 refrigerators twice each month. In comparison with the current quarterly reordering procedure, this will result in an annual saving of

$$C(150) - C(25) = 14,314 - 12,759 = 1555;$$

that is, $1555 every year.

These examples indicate that the closed interval maximum-minimum method is applicable to a wide range of particular problems. Indeed, applied optimization problems that initially look as different as refrigerators and diseases may turn out to have essentially identical mathematical models—the refrigerator problem of Chapter 1 and the disease problem of Example 3 above both reduced to maximizing a function of the form $f(x) = kx(a - x)$. This illustrates the power of generality that calculus exploits so impressively.

2-3 PROBLEMS

In each of Problems 1–15, find the maximum and minimum values attained by the given function on the indicated interval.

1 $f(x) = 3 - 2x$ on $[-1, 1]$.
2 $f(x) = x^2 - 4x + 3$ on $[0, 3]$.
3 $f(x) = 5 - 12x - 9x^2$ on $[-1, 1]$.
4 $f(x) = 2x^2 - 4x + 7$ on $[0, 2]$.
5 $f(x) = x^3 - 3x^2 - 9x + 5$ on $[-2, 4]$.
6 $f(x) = x^3 + x$ on $[-1, 2]$.
7 $f(x) = 3x^5 - 5x^3$ on $[-2, 2]$.
8 $f(x) = |2x - 3|$ on $[1, 2]$.
9 $f(x) = 5 + |7 - 3x|$ on $[1, 5]$.
10 $f(x) = |x + 1| + |x - 1|$ on $[-2, 2]$.
11 $f(x) = 50x^3 - 105x^2 + 72x$ on $[0, 1]$.

12 $f(x) = 2x + \dfrac{1}{2x}$ on $[1, 4]$.

13 $f(x) = \dfrac{x}{x + 1}$ on $[0, 3]$.

14 $f(x) = \dfrac{x}{1 + x^2}$ on $[0, 3]$.

15 $f(x) = \dfrac{1 - x}{x^2 + 3}$ on $[-2, 5]$.

16 A ball thrown straight upward with an initial velocity of 80 feet per second, starting at time $t = 0$ seconds, has a height of $-16t^2 + 80t$ feet after t seconds. What is the maximum height attained by the ball?

17 The sum of two positive numbers is 48. What is the smallest possible value of the sum of their squares?

18 What number between 0 and 1 exceeds its cube by the greatest amount?

19 The volume of 1 kg of water at a temperature T between $0°C$ and $30°C$ is approximately

$$V = 999.87 - (0.06426)T + (0.0085043)T^2$$
$$- (0.0000679)T^3$$

cubic centimeters. At what temperature does water have its maximum density?

20 What is the maximum possible area of a rectangle with base that lies on the x-axis and with two upper vertices that lie on the graph of the equation $y = 4 - x^2$?

21 A rectangular box has a square base whose edge has length at least 1 centimeter, and its total surface area is 600 square centimeters. What is the largest possible volume that such a box can have?

22 From 300 square inches of sheet metal, a rectangular box with square base and no top is to be made. No sheet metal will be wasted; you are allowed to order 5 pieces of any size and shape you wish to build the box, so long as it fits the conditions above. What is the greatest possible volume of such a box?

23 Three large squares of tin, each of edge length 1 meter, have 4 equal small squares cut from their corners. All 12 resulting small squares are to be the same size. The 3 large cross-shaped pieces are then folded to form boxes without tops, and the 12 small squares are used to make 2 small cubes. How should this be done to maximize the total volume of all 5 boxes?

24 A wire of length 100 centimeters is to be cut into 2 pieces. One piece is bent into a circle, the other into a square. How should the cut be made to maximize the sum of the areas of the square and the circle? How should it be done to minimize that sum?

25 A farmer has 600 meters of fencing with which he plans to enclose a rectangular divided pasture adjacent to a long existing wall. The plan is to build 1 fence parallel to the wall, 2 to form the ends of the enclosure, and a fourth to divide it. What is the maximum area that he can enclose in this way with the available amount of fencing?

26 A rectangular outdoor pen is to be added to an animal house with a "corner notch," as shown in Figure 2.7. If 85 meters of new fence is available, what should be the dimensions of the pen in order to maximize its area? No fence need be used along the wall of the animal house.

27 Suppose that a post office can accept a package for mailing by parcel post only if the sum of its length and girth (the circumference of its cross section) is at most 100 inches. What is the maximum volume of a rectangular box with square cross section that can be mailed?

28 Repeat Problem 27, except with a package that is cylindrical; its cross section is circular.

29 A printing company has 8 presses, each of which can print 3600 copies per hour. It costs $5.00 to set up each press for a run, and $10 + 6n$ dollars to run n presses for 1 hour. How many presses should be used in order to print 50,000 copies of a poster most profitably?

30 A farmer wants to hire workers to pick 900 bushels of beans. Each worker can pick 5 bushels per hour and is paid $1.00 per bushel. The farmer must also pay a supervisor $10 per hour while the picking is in progress, and he has miscellaneous additional expenses of $8 for each worker hired. How many workers should he hire in order to minimize the total cost? What will then be the cost per bushel picked?

31 The heating and cooling costs for a certain uninsulated house are $500 per year, but with $x \leq 10$ inches of insulation the costs would be $1000/(2 + x)$ dollars per year. It will cost $150 for each inch (thickness) of insulation installed. How many inches of insulation should be installed in order to minimize the *total* (initial plus annual) costs over a 10-year period? What will then be the annual savings resulting from this optimal insulation?

32 A concessionaire had been selling 5000 wieners each game night at 50¢ each. When she raised the price to 70¢ each, sales dropped to 4000 per night. If she has fixed costs of $1000 per night and each wiener costs her 25¢, what price will maximize her nightly profit?

33 A commuter train carries 600 passengers each day from a suburb to a city. It costs $1.50 per person to ride the train. It is found that 40 fewer people will ride the train for each 5¢ increase in the fare. What fare should be charged to make the largest possible revenue?

34 Find the shape of the cylinder of maximum volume that can be inscribed in a sphere of radius R. Show that the ratio of the cylinder's height to its radius is $\sqrt{2}$, and that the ratio of the sphere's volume to that of the cylinder is $\sqrt{3}$.

35 Find the dimensions of the right circular cylinder of greatest volume that can be inscribed in a right circular cone of radius R and height H.

36 In a circle of radius 1 is inscribed a trapezoid with the longer of its parallel sides coincident with a diameter of the circle. What is the maximum possible area of such a trapezoid? (*Suggestion:* A positive quantity is maximized when its square is maximized.)

2.7 The rectangular pen of Problem 26.

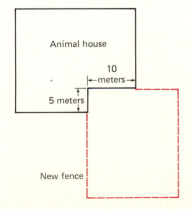

Animal house

10
←meters→

5 meters

New fence

Derivatives of Algebraic Functions

If f is a differentiable function, then the product rule gives

$$D[f(x)]^2 = D[f(x) \cdot f(x)]$$
$$= f'(x) \cdot f(x) + f(x) \cdot f'(x) = 2f(x)f'(x)$$

for the derivative of its square. Similarly, the derivative of its cube is

$$D[f(x)]^3 = D[(f(x))^2 \cdot f(x)]$$
$$= (D[f(x)]^2) \cdot f(x) + [f(x)]^2 \cdot f'(x)$$
$$= 2f(x)f'(x)f(x) + [f(x)]^2 f'(x)$$

or

$$D[f(x)]^3 = 3[f(x)]^2 f'(x).$$

These formulas for the derivative of a square or a cube of a differentiable function are special cases of the **generalized power rule.**

Theorem 1   *Generalized Power Rule*

If f is differentiable at x and r is a rational number, then

$$D[f(x)]^r = r[f(x)]^{r-1} f'(x) \qquad (1)$$

at all points where the right-hand side is meaningful: the points where $f(x) \neq 0$ if $r - 1 < 0$ and $f(x) > 0$ if an even root is involved.

In terms of the variable $u = f(x)$, we can write the generalized power rule in the form

$$D_x u^r = \frac{d(u^r)}{dx} = r u^{r-1} \frac{du}{dx}. \qquad (1')$$

The subscript x on D_x specifies that u^r is being differentiated with respect to x, rather than with respect to u (as would be understood if we wrote Du^n). When $u = f(x) = x$, we see that the above rule reduces to

$$Dx^r = rx^{r-1}, \qquad (2)$$

the power rule for rational exponents.

EXAMPLE 1   If $y = 5\sqrt{x^3} - \dfrac{2}{\sqrt[3]{x}}$

then

$$y = 5x^{3/2} - 2x^{-1/3},$$

so

$$\frac{dy}{dx} = 5\left(\frac{3}{2}\right)x^{1/2} - 2\left(-\frac{1}{3}\right)x^{-4/3}$$

$$= \frac{15}{2}x^{1/2} + \frac{2}{3}x^{-4/3} = \frac{15}{2}\sqrt{x} + \frac{2}{3\sqrt[3]{x^4}}$$

provided that $x > 0$.

EXAMPLE 2 With $f(x) = 2x^2 - 3x + 5$ and $r = \frac{1}{2}$, the generalized power rule gives

$$D\sqrt{2x^2 - 3x + 5} = \frac{1}{2}(2x^2 - 3x + 5)^{-1/2}D(2x^2 - 3x + 5)$$

$$= \frac{4x - 3}{2\sqrt{2x^2 - 3x + 5}}.$$

EXAMPLE 3 If $y = (5x + \sqrt[3]{(3x - 1)^4})^{10}$

then (1') with $u = 5x + (3x - 1)^{4/3}$ gives

$$\frac{dy}{dx} = 10u^9 \frac{du}{dx} = 10(5x + \sqrt[3]{(3x - 1)^4})^9(5 + 4\sqrt[3]{3x - 1}).$$

Proof of Theorem 1 for r an integer Suppose first that $r = n$, a *positive* integer. Then we can use the principle of mathematical induction much as in Example 4 of Section 2-2. The case $r = 1$ is just $Df(x) = f'(x)$. Assume the result is true in the case $r = n$; that is, that

$$D[f(x)]^n = n[f(x)]^{n-1}f'(x).$$

Then the product rule gives

$$D[f(x)]^{n+1} = D[(f(x))^n \cdot f(x)]$$
$$= (D[f(x)]^n) \cdot f(x) + [f(x)]^n \cdot f'(x)$$
$$= (n[f(x)]^{n-1}f'(x)) \cdot f(x) + [f(x)]^n \cdot f'(x)$$

and so

$$D[f(x)]^{n+1} = (n + 1)[f(x)]^n f'(x).$$

Since this is the generalized power rule for $r = n + 1$, the theorem now follows for positive integral values of r by induction.

Now suppose that $r = -n$, a *negative* integer. We combine the result of the case above with the reciprocal rule (Equation (16) in Section 2-1), and we find that

$$D[f(x)]^r = D\frac{1}{[f(x)]^n}$$

$$= -\frac{D[f(x)]^n}{[f(x)]^{2n}} = -\frac{n[f(x)]^{n-1}f'(x)}{[f(x)]^{2n}}$$

$$= -n\frac{f'(x)}{[f(x)]^{n+1}} = r[f(x)]^{r-1}f'(x).$$

The most difficult step in the proof of the generalized power rule for r a rational number p/q is the derivation of the formula

$$D[f(x)]^{1/q} = \frac{1}{q}[f(x)]^{(1/q)-1}f'(x), \tag{3}$$

which is the case $r = 1/q$ for q a nonzero integer. Since $[f(x)]^{1/q} = \sqrt[q]{f(x)}$, it is natural to call Formula (3) the **generalized root rule.** Let us assume it for now; then the proof of the generalized power rule with $r = p/q$ can be completed as follows.

$$D[f(x)]^r = D[f(x)]^{p/q} = D([f(x)]^{1/q})^p$$
$$= p([f(x)]^{1/q})^{p-1} \cdot D[f(x)]^{1/q}$$
$$= p[f(x)]^{(p-1)/q} \cdot \frac{1}{q}[f(x)]^{(1/q)-1}f'(x)$$
$$= \frac{p}{q}[f(x)]^{(p/q)-1}f'(x).$$

Thus

$$D[f(x)]^r = r[f(x)]^{r-1}f'(x)$$

for rational values of r (subject to the restrictions mentioned in the statement of the theorem).

In Theorem 3 below we shall establish Equation (3) for the special case $f(x) = x$; that is, the **root rule**

$$Dx^{1/q} = \frac{1}{q}x^{(1/q)-1}. \tag{4}$$

We give the proof of the general case of (3) with f an arbitrary differentiable function in Section 2-7. In Chapter 6 we shall see that the generalized power rule continues to hold even when r is an *irrational* number (such as $\sqrt{2}$ or π).

DERIVATIVES OF INVERSE FUNCTIONS

Recall from Section 1-7 that f and g are called inverse functions provided that $f(g(x)) = x$ and $g(f(x)) = x$ wherever these compositions are defined. Recall also that a given function f has an inverse function g if and only if f is one-to-one. The following theorem tells us how to differentiate g, provided that we already know how to differentiate f.

Theorem 2 *Differentiation of an Inverse Function*

Suppose that the one-to-one function f is differentiable with $f'(x) > 0$ for all x in its domain. Then its inverse function g is also differentiable, and

$$g'(x) = \frac{1}{f'(g(x))}. \tag{5}$$

COMMENT   This theorem is also true when the condition $f'(x) > 0$ is replaced by the condition $f'(x) < 0$. In Section 2-5 we shall see that the truth of either of these inequalities on an interval I implies that f must be one-to-one on I, and thus also that f^{-1} exists.

Formula (5) is easily remembered in differential notation. Let us write $x = f(y)$ and $y = g(x)$. Then $dy/dx = g'(x)$ and $dx/dy = f'(y)$. So (5) becomes the seemingly inevitable formula

$$\frac{dy}{dx} = \frac{1}{\dfrac{dx}{dy}}.$$

It is important to remember that dy/dx above is to be evaluated at x, while dx/dy is to be evaluated at the corresponding value of y; namely, $y = g(x)$.

Proof of Theorem 2   Figure 2.8 helps make it clear that

$$\text{if}\quad k = g(x + h) - g(x), \quad\text{then}\quad h = f(y + k) - f(y).$$

Hence

$$g'(x) = \lim_{h \to 0} \frac{g(x + h) - g(x)}{h}$$

$$= \lim_{k \to 0} \frac{k}{f(y + k) - f(y)} \qquad \begin{array}{l}\text{(Since } g \text{ is continuous by Theorem 2,}\\ \text{Section 1-9, } k \to 0 \text{ as } h \to 0.)\end{array}$$

$$= \frac{1}{\displaystyle\lim_{k \to 0} \frac{f(y + k) - f(y)}{k}} = \frac{1}{f'(y)} = \frac{1}{f'(g(x))}.$$

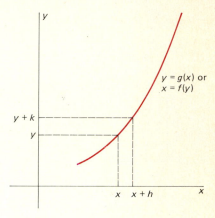

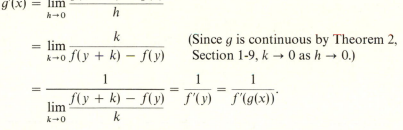

2.8 The relation between h and k in the proof of Theorem 2.

We can now use the fact that $f(x) = x^q$ and $g(x) = x^{1/q}$ are mutually inverse to help us compute the derivative of $x^{1/q}$.

Theorem 3   *The Root Rule*

$$Dx^{1/q} = \frac{1}{q} x^{(1/q) - 1}$$

if $x \neq 0$, and with the proviso that $x > 0$ if the positive integer q is even.

Proof   If $y = x^{1/q}$ and $x = y^q$, then (5′) gives

$$\frac{dy}{dx} = \frac{1}{\dfrac{dx}{dy}} = \frac{1}{qy^{q-1}} = \frac{1}{q(x^{1/q})^{q-1}},$$

so that

$$\frac{dy}{dx} = \frac{1}{q} x^{(1/q) - 1},$$

as desired.

2.9 The graph of the cube root function.

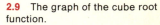
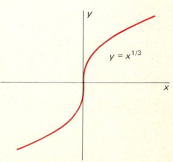

EXAMPLE 4   The root rule gives $Dx^{1/3} = \frac{1}{3} x^{-2/3}$, which increases without bound as x approaches zero. How is this information related to the graph of $y = x^{1/3}$? The graph appears in Figure 2.9. The derivative dy/dx does not exist at $x = 0$. So the curve does not have a tangent line (as defined in Section 1-5) at the origin. Nevertheless, in this case we may regard the

vertical line $x = 0$ as the tangent line to $y = x^{1/3}$ at $(0, 0)$. More generally, we say that

> The curve $y = f(x)$ has a **vertical tangent line** at the point $(a, f(a))$ provided that $|f'(x)| \to +\infty$ as $x \to a$.

EXAMPLE 5 (A sawmill problem) Suppose that you need to cut a beam with maximal rectangular cross section from a circular log of radius 1 foot. This is the geometric problem of finding the rectangle of greatest area that can be inscribed in a circle of radius 1.

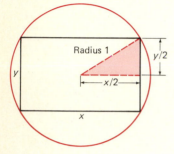

Solution Let x and y, as indicated in Figure 2.10, denote the base and height of the inscribed rectangle, respectively. Apply the Pythagorean theorem to the small shaded right triangle in the figure. This yields the equation

$$y = 2\left[1 - \left(\frac{x}{2}\right)^2\right]^{1/2} = [4 - x^2]^{1/2}.$$

Therefore the area of the inscribed rectangle is given by

$$A = xy = x[4 - x^2]^{1/2}.$$

2.10 A sawmill problem—Example 5.

The practical domain of definition of $A(x)$ is $(0, 2)$ and there is no harm in adjoining the end points, so we take $[0, 2]$ to be the domain.

Here is the derivative:

$$\frac{dA}{dx} = 1 \cdot [4 - x^2]^{1/2} + x \cdot \frac{1}{2}[4 - x^2]^{-1/2}(-2x) = \frac{4 - 2x^2}{[4 - x^2]^{1/2}}.$$

So the only critical point in $[0, 2]$ is $x = \sqrt{2}$. We evaluate A here and at the two end points and find that

$$A(0) = 0, \qquad A(\sqrt{2}) = 2, \quad \text{and} \quad A(2) = 0.$$

Hence the beam with rectangular cross section and maximal area is square, with edge length $\sqrt{2}$ feet.

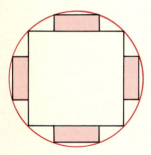

2.11 Cut four more beams after cutting one large one.

In Problem 23 at the conclusion of this section we ask you to maximize the cross-sectional area of the four planks that can be cut from the four pieces of log that remain after cutting out the square beam (see Figure 2.11).

EXAMPLE 6 We consider the reflection of a ray of light by a mirror M, as shown in Figure 2.12, which depicts a ray traveling from point A to point B via reflection in M at point P. We assume that the location of the point of reflection is such as to minimize the total distance $d_1 + d_2$ traveled by the light ray; this is an application of Fermat's *principle of least time* for the propagation of light. The problem: Find P.

2.12 Reflection at P of a light ray by a mirror M.

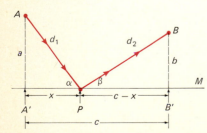

Solution Drop perpendiculars from A and B to the plane of the mirror M. Denote the feet of these perpendiculars by A' and B', as in Figure 2.12. Let a, b, c, and x denote (respectively) the lengths of the segments AA',

BB', $A'B'$, and $A'P$. Then $c - x$ is the length of the segment PB', and so—by the Pythagorean theorem—the distance to be minimized is

$$d_1 + d_2 = f(x) = (x^2 + a^2)^{1/2} + ([c - x]^2 + b^2)^{1/2}.$$

We may choose as the domain of f the interval $[0, c]$ because the minimum of f must occur somewhere within that interval. (To see this, examine the picture you get if x is *not* in that interval.)

Then

$$f'(x) = \frac{x}{(x^2 + a^2)^{1/2}} + \frac{(c - x)(-1)}{([c - x]^2 + b^2)^{1/2}}.$$

Since

$$f'(x) = \frac{x}{d_1} - \frac{c - x}{d_2},$$

we find that any horizontal tangent to the graph of f must occur over the point x determined by

$$\frac{x}{d_1} = \frac{c - x}{d_2}.$$

At such a point, $\cos \alpha = \cos \beta$, where α is the angle of the incident light ray and β is the angle of the reflected ray (these angles are also shown in Figure 2.12). But both α and β lie between 0 and $\pi/2$, and thus we find that $\alpha = \beta$. That is, the angle of incidence is equal to the angle of reflection, a familiar principle from physics.

2-4 PROBLEMS

Differentiate the functions given in Problems 1–20.

1 $f(x) = \sqrt{x^3 + 1}$.

2 $f(x) = \dfrac{1}{(x^4 + 3)^2}$.

3 $f(x) = \sqrt{2x^2 + 1}$.

4 $f(x) = \dfrac{x}{\sqrt{1 + x^4}}$.

5 $f(t) = \sqrt{2t^3}$.

6 $g(t) = \sqrt{1/(3t^5)}$.

7 $f(x) = (2x^2 - x + 7)^{3/2}$.

8 $g(z) = (3z^2 - 4)^{97}$.

9 $g(x) = \dfrac{1}{(x - 2x^3)^{4/3}}$.

10 $f(t) = (t^2 + [1 + t]^4)^5$.

11 $f(x) = x\sqrt{1 - x^2}$.

12 $g(x) = \sqrt{\dfrac{2x + 1}{x - 1}}$.

13 $f(t) = \sqrt{\dfrac{t^2 + 1}{t^2 - 1}}$.

14 $h(y) = \left(\dfrac{y + 1}{y - 1}\right)^{17}$.

15 $f(x) = \left(x - \dfrac{1}{x}\right)^3$.

16 $g(z) = \dfrac{z^2}{\sqrt{1 + z^2}}$.

17 $f(v) = \dfrac{\sqrt{v + 1}}{v}$.

18 $h(x) = \left(\dfrac{x}{1 + x^2}\right)^{5/3}$.

19 $f(x) = \sqrt[3]{1 - x^2}$.

20 $g(x) = \sqrt{x + \sqrt{x}}$.

21 Show that the rectangle with maximum perimeter that can be inscribed in a circle is a square.

22 Find the dimensions of the rectangle (with vertical and horizontal sides) of maximum area that can be inscribed in the ellipse $x^2/25 + y^2/9 = 1$ shown in Figure 2.13.

2.13 The ellipse of Problem 22.

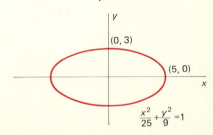

23 Find the dimensions that maximize the cross-sectional areas of the four planks that can be cut from the four pieces of the circular log of Example 5, the pieces that remain after cutting out a square beam (see Figure 2.11).

24 Assume that the strength of a rectangular beam is proportional to the product of its width and the *cube* of its depth. Find the dimensions of the strongest rectangular beam that can be cut from a cylindrical log of radius R.

25 A small island is 2 kilometers offshore in a large lake. A woman on the island can row her boat 10 kph (kilometers per hour) and can run 20 kph. Where should she land her boat in order to most quickly reach a village on the straight shoreline of the lake, and 6 kilometers down the shoreline from the point nearest the island?

26 A factory is located on one bank of a straight river that is 2000 meters wide. On the opposite bank but 4500 meters downstream is a power station from which the factory must draw its electricity. Assume that it costs three times as much per meter to lay an underwater cable as to lay a cable over the ground. What path should a cable connecting the power station to the factory take so as to minimize the cost of laying the cable?

27 A company has plants that are located (in an appropriate coordinate system) at the points $A(0, 1)$, $B(0, -1)$, and $C(3, 0)$. The company plans to construct a distribution center at the point $P(x, 0)$. What should be the value of x to minimize the sum of the distances of P from A, B, and C? (The object is to minimize transportation costs from the plants to the center.)

28 The sum of the surface areas of a cube and a sphere is 1000 square inches. What should be their dimensions in order to minimize the sum of their volumes? In order to maximize it?

29 Suppose that you are marooned in the desert: Your car has simultaneously run out of oil and water. You find yourself 2 kilometers north of the X River, which runs in an east-west direction. You are also 5 kilometers east of the Y Pipeline, which runs in a north-south direction. The

situation is shown in Figure 2.14, in which the units are kilometers. If you walk from your car to the X River to procure water, then to the Y Pipeline for some oil, and then back to your car, what should be your path to minimize the total length of your trip?

30 Light travels with speed c in air and with a somewhat slower speed v in water. (The constant c is approximately 3×10^{10} cm/sec; the ratio $n = c/v$, known as the **index of refraction** of water, depends upon the color of the light, but is approximately 1.33.) In Figure 2.15 we show the path of a light ray traveling from point A in air to point B in water, with what appears to be a sudden change in direction as the ray moves through the air-water interface.

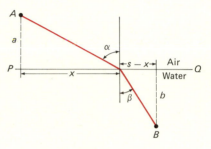

2.15 Snell's law gives the path of refracted light.

(a) Write the time T required for the ray to travel from A to B in terms of the variable x and the constants a, b, c, s, and v, all of which have been defined or are shown in the figure.

(b) Show that the equation $T'(x) = 0$ for minimizing T is equivalent to the condition

$$\frac{\sin \alpha}{\sin \beta} = \frac{c}{v} = n.$$

This is **Snell's law:** The ratio of the sines of the angles of incidence and refraction is equal to the index of refraction.

31 The mathematics of Snell's law is applicable to situations other than the refraction of light. Figure 2.16 shows a geological fault running from west to east, separating two towns at A and B. Assume that A is a miles north of the fault,

2.14 The X River and the Y Pipeline of Problem 29.

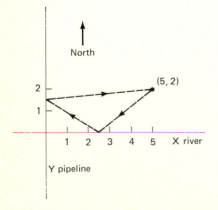

2.16 Illustration for Problem 31.

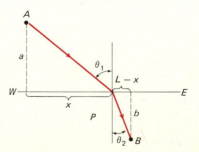

that B is b miles south of the fault, and that B is L miles east of A. We want to build a road from A to B. Because of differences in terrain, the cost of construction is C_1 (in millions of dollars per mile) north of the fault and C_2 south of it. Where should the point P be placed to minimize the total cost of the road?

(a) Using the notation of the figure, show that the cost is minimal when $C_1 \sin \theta_1 = C_2 \sin \theta_2$.

(b) Take the case $a = b = C_1 = 1$, $C_2 = 2$, and $L = 4$. Show that the equation in Part (a) is equivalent to

$$f(x) = 3x^4 - 24x^3 + 51x^2 - 32x + 64 = 0.$$

To approximate the desired solution of this equation, calcalute $f(0)$, $f(1)$, $f(2)$, $f(3)$, and $f(4)$. You should find that $f(3)$ is positive, while $f(4)$ is negative. Interpolate between $x = 3$ and $x = 4$ to approximate the root of the above equation.

32 Two noisy discothèques, one of them 4 times as noisy as the other, are located on opposite ends of a block 1000 feet long. What is the quietest point on the block between the two discos? Assume that the intensity of noise at a point removed from its source is proportional to the noisiness and inversely proportional to the square of the distance from the source.

The significance of the sign of the first derivative is simple but crucial:

f is increasing where $f'(x) > 0$;

f is decreasing where $f'(x) < 0$.

The Mean Value Theorem and Applications

Geometrically, this means that where $f'(x) > 0$, the graph of f is rising, as you scan it from left to right. Where $f'(x) < 0$, the graph is falling. We can make the terms *increasing* and *decreasing* more precise as follows.

Definition *Increasing and Decreasing Functions*

The function f is **increasing** on the interval I if, for each two numbers x_1 and x_2 in I with $x_1 < x_2$,

$$f(x_1) < f(x_2).$$

That is,

$$x_1 < x_2 \quad \text{implies} \quad f(x_1) < f(x_2).$$

The function f is **decreasing** on I provided that

$$x_1 < x_2 \quad \text{implies} \quad f(x_1) > f(x_2)$$

for any two points x_1 and x_2 in I.

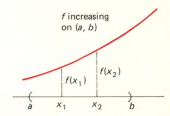

2.17 An increasing function; a decreasing function.

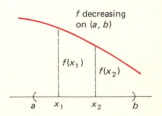

Figure 2.17 illustrates this definition.

Note that we speak of a function as increasing or decreasing *on an interval*, not at a single point. Nevertheless, if we consider the sign of f' at a single point, we get a useful intuitive picture of the significance of the sign of the derivative. This is because the derivative $f'(x)$ is the slope of the tangent line at the point $(x, f(x))$ on the graph of f. If $f'(x) > 0$, the tangent line has positive slope, and therefore rises as you scan from left to right. The graph looks much like its tangent line *at and near* the point of tangency, so the graph should also be rising where the derivative is positive.

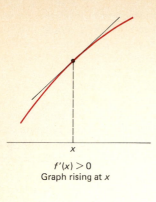

$f'(x) > 0$
Graph rising at x

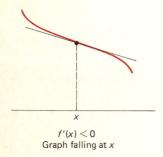

$f'(x) < 0$
Graph falling at x

2.18 A graph rising at x; a graph falling at x.

Similarly, we expect to see a falling graph where $f'(x)$ is negative. Figure 2.18 shows a pair of graphs illustrating this intuitive idea.

Though pictures of rising and falling graphs are quite suggestive, they provide no actual proof of the significance of the sign of the derivative. To establish the connection between the graph's rising or falling and the sign of the derivative, we need the **Mean Value Theorem,** stated below. This theorem is the principal theoretical tool of differential calculus, and we shall see that it has many important applications.

As an introduction to the Mean Value Theorem, we pose the following question. Suppose that P and Q are two points in the plane, with Q lying generally to the east of P, as shown in Figure 2.19. Is it possible to sail a boat from P to Q, sailing always roughly east, without *ever* (even for an instant) sailing in the exact direction from P to Q? That is, can we sail from P to Q without our instantaneous line of motion ever being parallel to the line PQ?

The Mean Value Theorem says that the answer to this question is No; there will always be at least one instant when we are sailing parallel to the line PQ, no matter which path we choose.

Here is a mathematical paraphrase: Let the path of the sailboat be the graph of a differentiable function, $y = f(x)$, with end points $P(a, f(a))$ and $Q(b, f(b))$. Then we are saying that there must be some point on this graph where the tangent line to the curve is parallel to the line PQ joining its end points. This is a *geometric interpretation* of the Mean Value Theorem.

But the slope of the tangent line at the point $(c, f(c))$, shown in Figure 2.20, is $f'(c)$, while the slope of the line PQ is

$$\frac{f(b) - f(a)}{b - a}.$$

We may think of this last quotient as the average (or *mean*) value of the slope of f. The Mean Value Theorem guarantees that there is a point c in (a, b) where the tangent line at $(c, f(c))$ is indeed parallel to the line PQ. In the language of algebra, there's a number c in (a, b) such that

$$f'(c) = \frac{f(b) - f(a)}{b - a}. \tag{1}$$

2.19 Can you sail from P to Q without ever sailing—at least for an instant—in the direction PQ (the direction of the arrow)?

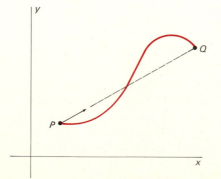

2.20 The sailboat problem in mathematical terminology.

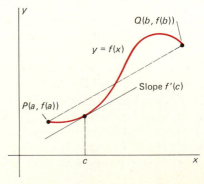

CHAP. 2: Differentiation

We first give a preliminary result, a lemma to expedite the proof of the Mean Value Theorem. This lemma is called **Rolle's theorem,** after the Frenchman Michel Rolle (1652–1719), who discovered it in 1690. Rolle studied the emerging subject of calculus in his youth but later renounced it; he argued that it was based on logical fallacies, and today he is remembered only for a single theorem that bears his name. It is ironic that his theorem plays an important role in the rigorous proofs of several calculus theorems.

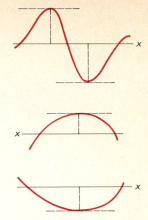

Rolle's Theorem

Suppose that the function f is continuous on the closed interval $[a, b]$ and is differentiable on its interior (a, b). If $f(a) = 0 = f(b)$, then $f'(c) = 0$ for some number c in (a, b).

2.21 The horizontal tangent is a consequence of Rolle's theorem.

Thus between each pair of zeros of a differentiable function there is *at least one* point where the tangent line is horizontal. Some possible pictures of the situation are indicated in Figure 2.21.

Proof Since f is continuous on $[a, b]$, it must attain both a maximum and a minimum value on $[a, b]$ (by the Maximum Value Property of Section 1-9). If f has any positive values at all, consider its maximum value $f(c)$. Since $f(a) = f(b) = 0$, c must be a point of (a, b). Since f is then differentiable at c, it follows from Theorem 1 of Section 2-3 that $f'(c) = 0$.

Similarly, if f has any negative values, we may consider its minimum value $f(c)$ and conclude that $f'(c) = 0$. If f has neither positive nor negative values, then f is identically zero on $[a, b]$, and it follows that $f'(c) = 0$ for *every* c in (a, b). Thus we see that the conclusion of Rolle's theorem is justified in every case.

EXAMPLE 1 Suppose that $f(x) = x^{1/2} - x^{3/2}$ on $[0, 1]$. Find a number c satisfying the conclusion of Rolle's theorem.

Solution Note that f is continuous on $[0, 1]$ and differentiable on $(0, 1)$; because of the presence of the term $x^{1/2}$, f is *not* differentiable at $x = 0$. Also, $f(0) = f(1) = 0$, so all hypotheses of Rolle's theorem are satisfied. Since

$$f'(x) = \tfrac{1}{2}x^{-1/2} - \tfrac{3}{2}x^{1/2} = \tfrac{1}{2}x^{-1/2}(1 - 3x),$$

we see that $f'(c) = 0$ for $c = \tfrac{1}{3}$.

EXAMPLE 2 Suppose that $f(x) = 1 - x^{2/3}$ on $[-1, 1]$. Then f satisfies the hypotheses of Rolle's theorem *except* for the fact that $f'(0)$ does not exist. It is clear from the graph of f (Figure 2.22) that there is *no* point where the tangent line is horizontal.

2.22 The function of Example 2; Rolle's hypotheses are not satisfied.

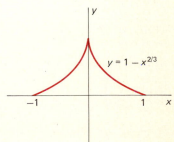

The Mean Value Theorem

Suppose that the function f is continuous on $[a, b]$ and differentiable on (a, b). Then

$$f(b) - f(a) = f'(c)(b - a) \qquad (2)$$

for some number c in (a, b).

Since Equation (2) is equivalent to Equation (1) above, the conclusion of the Mean Value Theorem is that there must be at least one point on the curve $y = f(x)$ where the tangent line is parallel to the line joining its end points $P(a, f(a))$ and $Q(b, f(b))$.

Motivation for the Proof We consider the "auxiliary function" ϕ suggested by Figure 2.23. The value $\phi(x)$ is by definition the vertical height difference over x of the point $(x, f(x))$ on the curve and the corresponding point on the line PQ. It appears that a point on the curve $y = f(x)$, where the tangent line is parallel to PQ, corresponds to a maximum or minimum of ϕ. It's clear also that $\phi(a) = \phi(b) = 0$, so Rolle's theorem can be applied to the function ϕ. So our plan for proving the Mean Value Theorem is this: First, we obtain a formula for the function ϕ. Second, we locate the point c such that $\phi'(c) = 0$. Finally, we show that this number c is exactly the number needed to satisfy Equation (2).

Proof of the Mean Value Theorem Since the line PQ passes through $(a, f(a))$ with slope

$$m = \frac{f(b) - f(a)}{b - a},$$

the point-slope formula for the equation of a straight line gives us the following equation for PQ:

$$y = y_{line} = f(a) + m(x - a).$$

Thus

$$\phi(x) = y_{curve} - y_{line} = f(x) - f(a) - m(x - a).$$

You may verify by direct substitution that $\phi(a) = \phi(b) = 0$. And, since ϕ is certainly continuous on $[a, b]$ and differentiable on (a, b), we may apply Rolle's theorem to it. Thus there is a point c somewhere in the open interval (a, b) at which $\phi'(c) = 0$. But

$$\phi'(x) = f'(x) - m = f'(x) - \frac{f(b) - f(a)}{b - a}.$$

So, since $\phi'(c) = 0$, we find that

$$0 = f'(c) - \frac{f(b) - f(a)}{b - a},$$

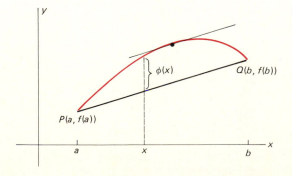

2.23 The construction of the auxiliary function ϕ.

CHAP. 2: Differentiation

and so

$$f(b) - f(a) = f'(c)(b - a).$$

CONSEQUENCES OF THE MEAN VALUE THEOREM

The first of these consequences is the *non*trivial converse of the trivial fact that the derivative of a constant function is identically zero. That is, we prove that there is *no* unknown exotic function that is nonconstant but which has derivative identically zero.

> **Corollary 1**   *Functions with Zero Derivatives*
>
> If $f'(x) = 0$ for all x in $[a, b]$, then f is a constant on $[a, b]$. That is, there is a constant C such that $f(x) = C$ for all x in $[a, b]$.

Proof   Apply the Mean Value Theorem to the function f on the interval $[a, x]$. The right-hand end point of the interval is indeed x, rather than b. We think of x as a "temporary constant," chosen in the range $a < x \leqq b$. We find that

$$f(x) - f(a) = f'(c) \cdot (x - a)$$

for some number c between a and x.

But f' is always zero on the interval $[a, b]$, and so $f'(c) = 0$. Thus $f(x) - f(a) = 0$, and so $f(x) = f(a)$.

Now we release x from the condition of being "temporarily constant" and observe that the result $f(x) = f(a)$ holds for *all* x in the range $a < x \leqq b$. That is, $f(x)$ has the fixed value $C = f(a)$ regardless of the choice of x in $[a, b]$. This establishes Corollary 1.

The ease and simplicity of the above proof suggests that the Mean Value Theorem is a powerful tool.

Corollary 1 is usually applied in a different but equivalent form, which we prove next.

> **Corollary 2**   *Functions with Equal Derivatives*
>
> Suppose that f and g are two differentiable functions such that $f'(x) = g'(x)$ for all x in the closed interval $[a, b]$. Then f and g differ by a constant on $[a, b]$. That is, there is a constant K such that
>
> $$f(x) = g(x) + K$$
>
> for all x in $[a, b]$.

Proof   Given the hypotheses, let $h(x) = f(x) - g(x)$. Then

$$h'(x) = f'(x) - g'(x) = 0$$

because $f' = g'$ on $[a, b]$. So, by Corollary 1, h is a constant K, on $[a, b]$.

That is, $f(x) - g(x) = K$ for all x in $[a, b]$, and so

$$f(x) = g(x) + K$$

for all x in $[a, b]$. This establishes Corollary 2.

The following consequence of the Mean Value Theorem verifies the remarks about increasing and decreasing functions with which we opened this section.

Corollary 3 *Increasing and Decreasing Functions*

Let f be a function that is continuous on $[a, b]$ and differentiable on (a, b). If $f'(x) > 0$ for all x in (a, b), then f is an increasing function on $[a, b]$. If $f'(x) < 0$ for all x in (a, b), then f is a decreasing function on $[a, b]$.

Proof Suppose, for example, that $f'(x) > 0$ for all x in (a, b). We need to show this: If x_1 and x_2 are points of $[a, b]$ with $x_1 < x_2$, then $f(x_1) < f(x_2)$. We apply the Mean Value Theorem to f, but on the interval $[x_1, x_2]$. This gives

$$f(x_2) - f(x_1) = f'(c)(x_2 - x_1)$$

for some c in (x_1, x_2). Since $x_2 > x_1$ and since, by hypothesis, $f'(c) > 0$, it follows that

$$f(x_2) - f(x_1) > 0, \quad \text{or} \quad f(x_2) > f(x_1),$$

as we wanted to show. The proof is similar in the case that f' is negative on (a, b).

EXAMPLE 3 Where is the function $f(x) = x^2 - 4x + 5$ increasing and where is it decreasing?

Solution The derivative of f is $f'(x) = 2x - 4$. It is clear that $f'(x) > 0$ if $x > 2$, while $f'(x) < 0$ if $x < 2$. Hence f is decreasing on $(-\infty, 2)$ and increasing on $(2, +\infty)$. Together with the fact that $f'(2) = 0$, this information is enough to obtain the sketch of the graph of f. It appears in Figure 2.24.

EXAMPLE 4 Show that the equation $x^3 + x - 1 = 0$ has exactly one (real) root.

Solution Let $f(x) = x^3 + x - 1$. Since $f(0) = -1 < 0$ and $f(1) = +1 > 0$, the Intermediate Value Property guarantees that $f(x)$ has *at least* one zero in $[0, 1]$. But

$$f'(x) = 3x^2 + 1 > 0$$

for all x, so Corollary 3 shows that f is an increasing function on the whole real line. Thus f cannot have more than one zero. (Why?)

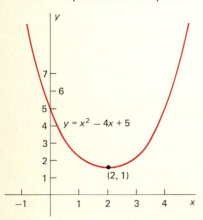

2.24 The parabola of Example 3.

In Problems 1–4, show that the given function f satisfies the hypotheses of Rolle's theorem on the indicated interval $[a, b]$, and find all numbers c in $[a, b]$ that satisfy the conclusion of the theorem.

1 $f(x) = x^2 - 2x$ on $[0, 2]$.
2 $f(x) = 9x^2 - x^4$ on $[-3, 3]$.

3 $f(x) = \dfrac{1 - x^2}{1 + x^2}$ on $[-1, 1]$.

4 $f(x) = 5x^{2/3} - x^{5/3}$ on $[0, 5]$.

In Problems 5–7, show that the given function f does not satisfy the conclusion of Rolle's theorem on the indicated interval. Which of the hypotheses does it fail to satisfy?

5 $f(x) = 1 - |x|$ on $[-1, 1]$.
6 $f(x) = 1 - (2 - x)^{2/3}$ on $[1, 3]$.
7 $f(x) = x^4 + x^2$ on $[0, 1]$.

In Problems 8–12, show that the given function f satisfies the hypotheses of the Mean Value Theorem on the indicated interval, and find all numbers c in $[a, b]$ that satisfy the conclusion of that theorem.

8 $f(x) = x^3$ on $[-1, 1]$.
9 $f(x) = 3x^2 + 6x - 5$ on $[-2, 1]$.
10 $f(x) = \sqrt{x - 1}$ on $[2, 5]$.
11 $f(x) = (x - 1)^{2/3}$ on $[1, 2]$.

12 $f(x) = x + \dfrac{1}{x}$ on $[1, 2]$.

In Problems 13–15, show that the given function satisfies neither the hypotheses nor the conclusion of the Mean Value Theorem on the indicated interval.

13 $f(x) = |x - 2|$ on $[1, 4]$.
14 $f(x) = 1 + |x - 1|$ on $[0, 3]$.
15 $f(x) = [x]$ (the greatest integer function) on $[-1, 1]$.

For each of the functions in Problems 16–20, determine the open intervals on which it is increasing and those on which it is decreasing.

16 $f(x) = x^2 + x + 1$.

17 $f(x) = 3x^5 - 5x^3$.

18 $f(x) = x^{3/2} - 3x^{1/3}$.

19 $f(x) = 4x + \dfrac{1}{x}$.

20 $f(x) = x\sqrt{8 - x^2}$.

In each of Problems 21–23, show that the given equation has exactly one root in the indicated interval.

21 $x^5 + 2x - 3 = 0$ in $[0, 1]$.
22 $x^{10} = 1000$ in $[1, 2]$.
23 $x^4 - 3x = 20$ in $[2, 3]$.
24 Show that the function $f(x) = x^{2/3}$ does not satisfy the hypotheses of the Mean Value Theorem on $[-1, 27]$, but nevertheless there is a number c in $(-1, 27)$ such that

$$f'(c) = \frac{f(27) - f(-1)}{27 - (-1)}.$$

25 Show that the function $f(x) = (1 + x)^{3/2} - \frac{3}{2}x - 1$ is increasing on $(0, +\infty)$. Conclude that $(1 + x)^{3/2} > 1 + \frac{3}{2}x$ for all $x > 0$.
26 Suppose that f' is constant on the interval $[a, b]$. Prove that f must be a linear function (a function which has a graph that is a straight line).
27 Suppose that $f'(x)$ is a polynomial of degree $n - 1$ on the interval $[a, b]$. Prove that $f(x)$ must be a polynomial of degree n on $[a, b]$.
28 Suppose that there are k different points of $[a, b]$ at which the differentiable function f vanishes (is zero). Show that f' must vanish at at least $k - 1$ points of $[a, b]$.
29 (a) Apply the Mean Value Theorem to $f(x) = \sqrt{x}$ on $[100, 101]$ to show that $\sqrt{101} = 10 + 1/2\sqrt{c}$ for some number c in $(100, 101)$.
(b) Show that if $100 < c < 101$, then $10 < \sqrt{c} < 10.5$, and use this fact to conclude from (a) that $10.0475 < \sqrt{101} < 10.0500$.
30 Show that the equation $x^7 + x^5 + x^3 + 1 = 0$ has exactly one real root.

2-6

Curve Sketching and the First Derivative Test

Suppose that f is a function that is differentiable on the open interval I and that f has a local extremum in I. Theorem 1 of Section 2-3 tells us this: That extremum must occur at the sort of critical point where $f'(x) = 0$. But if $f'(c) = 0$, it does not follow that there is a local extremum at c. What we need is a way of finding out, if $f'(c) = 0$, whether $f(c)$ is a local maximum value, a local minimum value, or neither.

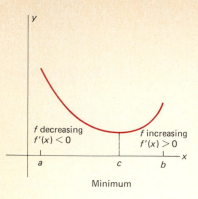

f decreasing
$f'(x) < 0$

f increasing
$f'(x) > 0$

a   c   b

Minimum

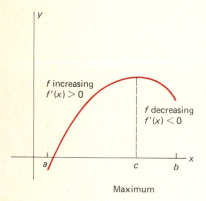

f increasing
$f'(x) > 0$

f decreasing
$f'(x) < 0$

a   c   b

Maximum

2.25 The first derivative test.

Figure 2.25 suggests how such criteria might be developed. If f is decreasing on the left of c and increasing on the right of c, then $f(c)$ should be a local minimum value. On the other hand, if f is increasing on the left of c and decreasing on its right, then $f(c)$ should be a local maximum. And we can tell, with the aid of Corollary 3 to the Mean Value Theorem, where the function f is increasing and where it is decreasing; such behavior is determined by the sign of $f'(x)$. Thus we obtain the following test for local maxima and minima.

Theorem 1   *The First Derivative Test*

Suppose that the function f is continuous on the open interval (a, b) and is differentiable there except possibly at c.

(i)   If $f'(x) < 0$ on (a, c) and $f'(x) > 0$ on (c, b), then $f(c)$ is the minimum value of f on (a, b).

(ii)   If $f'(x) > 0$ on (a, c) and $f'(x) < 0$ on (c, b), then $f(c)$ is the maximum value of f on (a, b).

(iii)   If $f'(x) > 0$ or if $f'(x) < 0$ for all x in (a, b) except for $x = c$, then $f(c)$ is neither a maximum nor a minimum value for f.

Thus $f(c)$ is a local extremum if the first derivative $f'(x)$ changes sign as x increases through c, and the direction of this sign change determines whether $f(c)$ is a local maximum or a local minimum. The best way to remember Parts (i) and (ii) of the first derivative test is simply to visualize Figure 2.25.

Proof   We shall prove Part (i); the proofs of the other two parts are similar. Let z be a point of (a, b) other than c. If $z < c$ then the fact that $f'(x) < 0$ on (z, c) implies that f is decreasing on $[z, c]$. So it follows that $f(z) > f(c)$. If $z > c$, the fact that $f'(x) > 0$ on (c, z) implies that f is increasing on $[c, z]$, so it follows that $f(c) < f(z)$. Thus $f(c) < f(z)$ for all z in (a, b) other than c, and this means that $f(c)$ is the minimum value of f on (a, b).

In Section 2-3 we discussed applied maximum-minimum problems in which the values of the dependent variable are given by a function defined on a *closed* interval. Sometimes, though, the function f describing the variable to be maximized or minimized is defined on an open interval (a, b), and we cannot "close" the interval by adjoining its end points; for instance, $f(x)$ may approach infinity as x approaches a or b. But if f has only a single critical point c in (a, b), then the first derivative test may tell us that $f(c)$ is the desired maximum or minimum value.

EXAMPLE 1   A cylindrical can with a volume of 125 cubic inches (about two quarts) is to be made by cutting its top and bottom out of metal squares and forming its curved surface by bending a rectangular sheet of metal to match its ends. What radius r and height h of the can will minimize the amount of material required?

Solution   We assume that the "corners" cut from the two squares, shown in Figure 2.26, are wasted, but that there is no other waste. As the figure shows, the total amount of sheet metal required is

$$A = 8r^2 + 2\pi rh.$$

The volume of the resulting can is then

$$V = \pi r^2 h = 125,$$

so that $h = 125/\pi r^2$. Hence A is given as a function of r by

$$A = 8r^2 + 2\pi r \cdot \frac{125}{\pi r^2} = 8r^2 + \frac{250}{r}.$$

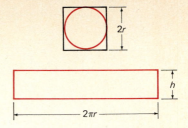

2.26   The parts to make the cylindrical can (Example 1).

Now r can have any positive value, so $A(r)$ is defined on the open interval $(0, +\infty)$. But $A \to \infty$ as $r \to 0$ and as $r \to \infty$. So we cannot use the closed interval maximum-minimum method. But we *can* use Theorem 1.

The first derivative of A is

$$\frac{dA}{dr} = 16r - \frac{250}{r^2} = \frac{16}{r^2}\left(r^3 - \frac{125}{8}\right),$$

so the only critical point in $(0, \infty)$ is at

$$r = \sqrt[3]{\tfrac{125}{8}} = \tfrac{5}{2}.$$

It is clear that $dA/dr < 0$ if $0 < r < \tfrac{5}{2}$, and that $dA/dr > 0$ if $r > \tfrac{5}{2}$. Therefore the first derivative test implies that the minimum value of $A(r)$ on $(0, \infty)$ is

$$A(\tfrac{5}{2}) = 8 \cdot (\tfrac{5}{2})^2 + \frac{250}{\tfrac{5}{2}} = 150.$$

Thus we minimize the amount of material required by making a can with radius $r = 2.5$ inches and height

$$h = \frac{125}{\pi(2.5)^2} \approx 6.37$$

inches. The total amount of material used is 150 square inches.

GRAPHS OF POLYNOMIALS

We can construct a reasonably accurate graph of the polynomial function

$$f(x) = a_n x^n + a_{n-1} x^{n-1} + \cdots + a_2 x^2 + a_1 x + a_0 \qquad (1)$$

by assembling the following information.

(a) The critical points of f. These include points on the graph where the tangent line is horizontal.
(b) The intervals on which f is increasing and those on which it is decreasing.
(c) The behavior of $f(x)$ as $x \to +\infty$ and as $x \to -\infty$.

To examine the behavior of f "at infinity," we write $f(x)$ in the form

$$f(x) = x^n \left(a_n + \frac{a_{n-1}}{x} + \cdots + \frac{a_1}{x^{n-1}} + \frac{a_0}{x^n}\right).$$

Thus we conclude that the behavior of $f(x)$ as $x \to \pm\infty$ is much the same as that of its *leading term* $a_n x^n$, because the terms with powers of x in the denominator all approach zero as $x \to \pm\infty$. In particular, if $a_n > 0$, then

$$\lim_{x \to \infty} f(x) = +\infty, \qquad (2)$$

meaning that $f(x)$ increases without bound as $x \to +\infty$. Also,

$$\lim_{x \to -\infty} f(x) = \begin{cases} +\infty & \text{if } n \text{ is even;} \\ -\infty & \text{if } n \text{ is odd.} \end{cases} \qquad (3)$$

If $a_n < 0$, reverse the signs on the right-hand sides of (2) and (3).

Every polynomial, including $f(x)$ above, is differentiable everywhere. So the critical points of $f(x)$ are the roots of the polynomial equation $f'(x) = 0$, or

$$n a_n x^{n-1} + (n-1)a_{n-1}x^{n-2} + \cdots + 2a_2 x + a_1 = 0. \qquad (4)$$

Sometimes one can find all (real) solutions of such an equation by factoring, and sometimes one must resort to numerical methods aided by calculator or computer. But suppose that we have somehow found *all* the solutions $c_1, c_2, \ldots, c_k$ of Equation (4). Then these solutions are the critical points of $f(x)$. If they are arranged in increasing order, as in Figure 2.27, they separate the x-axis into the finitely many open intervals

$$(-\infty, c_1), \quad (c_1, c_2), \quad (c_2, c_3), \ldots, \quad (c_{k-1}, c_k), \quad (c_k, +\infty)$$

that also appear in the figure. The Intermediate Value Property applied to $f'(x)$ tells us that f' can change sign only at the critical points of f, so that f' has only one sign on each of these open intervals. It is typical for $f'(x)$ to be positive on some and negative on the others. And it is easy to find the sign of f'; we need only substitute *any* convenient point of the interval into $f'(x)$.

Once we know the sign of f' on each of these intervals, we know where f is increasing and where it is decreasing. We then apply the first derivative test to find which of the critical values are local maxima, which are local minima, and which are neither—merely places where the tangent line is horizontal. With this information, the knowledge of the behavior of f as $x \to \pm\infty$, and the fact that f is continuous, we can sketch its graph. We plot the critical points $(c_i, f(c_i))$, and connect them with a smooth curve that is consistent with our other data.

It may also be helpful to plot the y-intercept $(0, f(0))$, and also any x-intercepts that are easy to find. But we recommend that (until inflection points are introduced in Chapter 3) you plot *only* these points—critical points and intercepts—and rely otherwise on the increasing-decreasing behavior of f.

EXAMPLE 2 Sketch the graph of $f(x) = x^3 - 27x$.

Solution Since the leading term is x^3, we see that

$$\lim_{x \to \infty} f(x) = +\infty \quad \text{and} \quad \lim_{x \to -\infty} f(x) = -\infty.$$

2.27 The zeros of f' divide the x-axis into intervals on each of which f' does not change sign.

CHAP. 2: Differentiation

Since

$$f'(x) = 3x^2 - 27 = 3(x + 3)(x - 3),$$

the critical points are $x = -3$ and $x = 3$. The corresponding points on the graph of f are $(-3, 54)$ and $(3, -54)$. The critical points separate the x-axis into the three open intervals

$$(-\infty, -3), \qquad (-3, 3), \quad \text{and} \quad (3, \infty).$$

The table below shows the behavior of f on each of those intervals.

Interval	$x + 3$	$x - 3$	$f'(x)$	f
$(-\infty, -3)$	Neg.	Neg.	Pos.	Increasing
$(-3, 3)$	Pos.	Neg.	Neg.	Decreasing
$(3, +\infty)$	Pos.	Pos.	Pos.	Increasing

We use this information to plot the critical points and the intercepts $(0, 0)$ and $(\pm 3\sqrt{3}, 0)$, and thus obtain the graph sketched in Figure 2.28.

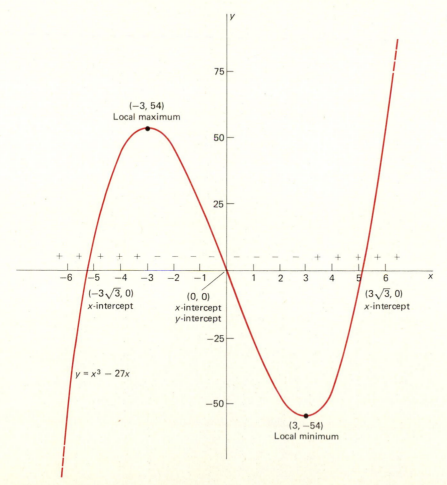

2.28 The function of Example 2.

In the figure we used plus and minus signs to mark the sign of $f'(x)$ in each interval. This makes it clear that $(-3, 54)$ is a local maximum and that $(3, -54)$ is a local minimum.

EXAMPLE 3 Sketch the graph of $f(x) = 8x^5 - 5x^4 - 20x^3$.

Solution Since

$$f'(x) = 40x^4 - 20x^3 - 60x^2 = 20x^2(x + 1)(2x - 3),$$

the critical points are -1, 0, and $\frac{3}{2}$. The behavior of f on the four open intervals determined by these critical points is indicated in the table that follows.

Interval	$x + 1$	x^2	$2x - 3$	$f'(x)$	f
$(-\infty, -1)$	Neg.	Pos.	Neg.	Pos.	Increasing
$(-1, 0)$	Pos.	Pos.	Neg.	Neg.	Decreasing
$(0, \frac{3}{2})$	Pos.	Pos.	Neg.	Neg.	Decreasing
$(\frac{3}{2}, \infty)$	Pos.	Pos.	Pos.	Pos.	Increasing

2.29 The graph for Example 3.

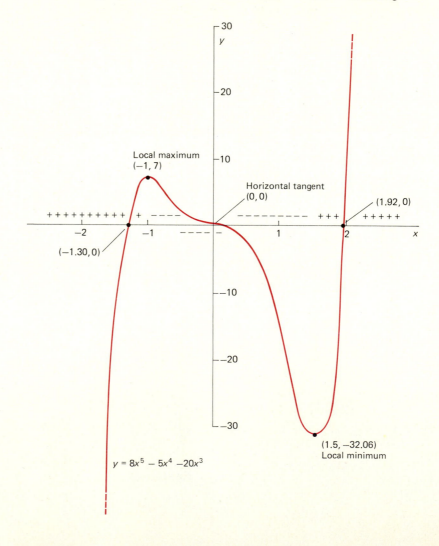

Local maximum
$(-1, 7)$

Horizontal tangent
$(0, 0)$

$(1.92, 0)$

$(-1.30, 0)$

$(1.5, -32.06)$
Local minimum

$y = 8x^5 - 5x^4 - 20x^3$

CHAP. 2: **Differentiation**

The points on the graph that correspond to the critical points are $(-1, 7)$, $(0, 0)$, and $(1.5, -32.06)$ (the last ordinate is an approximation).

We write $f(x)$ in the form

$$f(x) = x^3(8x^2 - 5x - 20);$$

we can use the quadratic formula to find the x-intercepts: $(-1.30, 0)$ and $(1.92, 0)$ (approximations again), and the origin $(0, 0)$. The latter is also the y-intercept. We apply the first derivative test to the increasing-decreasing behavior shown in the table, and thus see that $(-1, 7)$ is a local maximum, $(1.5, -32.06)$ is a local minimum, and $(0, 0)$ is neither. The graph looks like the one shown in Figure 2.29.

In our next example, the function is not a polynomial. Nevertheless, the methods of this section suffice for sketching its graph.

EXAMPLE 4   Sketch the graph of

$$f(x) = x^{2/3}(x^2 - 2x - 6) = x^{8/3} - 2x^{5/3} - 6x^{2/3}$$

Solution   The derivative of f is

$$f'(x) = \tfrac{8}{3}x^{5/3} - \tfrac{10}{3}x^{2/3} - \tfrac{12}{3}x^{-1/3} = \tfrac{2}{3}x^{-1/3}(4x^2 - 5x - 6)$$
$$= \tfrac{2}{3}x^{-1/3}(4x + 3)(x - 2).$$

The tangent line is horizontal at the two critical points $x = -\tfrac{3}{4}$ and $x = 2$. In addition, $|f'(x)| \to \infty$ as $x \to 0$, so $x = 0$ (a critical point because f is not differentiable there) is a point where the tangent line is vertical. These three critical points give the points $(-0.75, -3.25)$, $(0, 0)$, and $(2, -9.52)$ on the graph. (We use approximations where appropriate.) The following table summarizes our analysis of the sign of $f'(x)$ on each of the open intervals into which these critical points divide the x-axis.

Interval	$x^{-1/3}$	$4x + 3$	$x - 2$	$f'(x)$	f
$(-\infty, -\tfrac{3}{4})$	Neg.	Neg.	Neg.	Neg.	Decreasing
$(-\tfrac{3}{4}, 0)$	Neg.	Pos.	Neg.	Pos.	Increasing
$(0, 2)$	Pos.	Pos.	Neg.	Neg.	Decreasing
$(2, \infty)$	Pos.	Pos.	Pos.	Pos.	Increasing

The first derivative test now shows local minima at $(-0.75, -3.25)$ and $(2, -9.52)$, and a local maximum at $(0, 0)$. We note that the function is continuous everywhere: the only possible discontinuity is at $x = 0$, but $f(x)$ is close to zero when x is, so $\lim_{x \to 0} f(x) = 0 = f(0)$.

We use the quadratic formula to find the x-intercepts in addition to the origin; they are $(1 \pm \sqrt{7}, 0)$. We then plot the approximations $(-1.65, 0)$ and $(3.65, 0)$. Finally, we note that $f(x) \to \infty$ as $x \to \pm\infty$. So the graph has the shape shown in Figure 2.30.

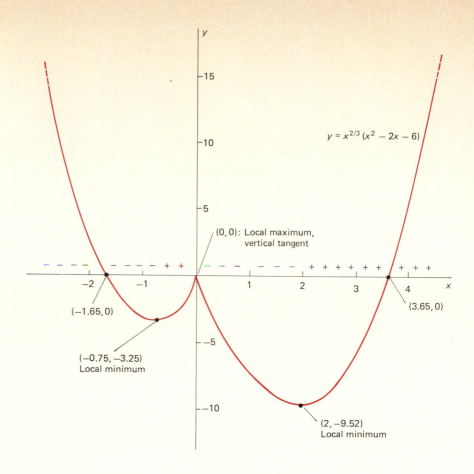

$y = x^{2/3} (x^2 - 2x - 6)$

(0,0): Local maximum, vertical tangent

(−1.65, 0)

(−0.75, −3.25) Local minimum

(3.65, 0)

(2, −9.52) Local minimum

2.30 The technique is valid for nonpolynomial functions, as in Example 4.

2-6 PROBLEMS

Sketch the graphs of the functions of Problems 1–18. Indicate the local maxima and minima of each function and the intervals on which it is increasing and decreasing.

1 $f(x) = 2x^2 - 3x - 9$.   **2** $f(x) = 6 - 5x - 6x^2$.

3 $f(x) = 2x^3 + 3x^2 - 12x$.

4 $f(x) = x^3 + 4x$.

5 $f(x) = 50x^3 - 105x^2 + 72x$.

6 $f(x) = x^3 - 3x^2 + 3x - 1$.

7 $f(x) = 3x^4 - 4x^3 - 12x^2 + 8$.

8 $f(x) = x^4 - 2x^2 + 1$.   **9** $f(x) = 3x^5 - 20x^3$.

10 $f(x) = 3x^5 - 25x^3 + 60x$.

11 $f(x) = 2x^3 + 3x^2 + 6x$.

12 $f(x) = x^4 - 4x^3$.   **13** $f(x) = 8x^4 - x^8$.

14 $f(x) = 1 - x^{1/3}$.   **15** $f(x) = x^{1/3}(4 - x)$.

16 $f(x) = x^{2/3}(x^2 - 16)$.   **17** $f(x) = x(x - 1)^{2/3}$.

18 $f(x) = x^{1/3}(2 - x)^{2/3}$.

In the following applied maximum-minimum problems, use the first derivative test to verify your solution.

19 Show that the rectangle with area 100 and minimum perimeter is a square.

20 Show that the rectangular solid with square base and volume 1000 that has the least total surface area is a cube.

21 A box with square base and open top is to have a volume of 62.5 cubic inches. Neglect the thickness of the material used to make the box, and find the dimensions that will minimize the amount of material used.

22 A tin can in the shape of a right circular cylinder is to have a volume of 16π cubic centimeters. What should be its radius r and height h in order to minimize its total surface area (including top and bottom)?

23 The metal used to make the top and bottom of a cylindrical can costs 4¢ per square inch, while the metal used for the sides costs 2¢ per square inch. The volume of the can is to be exactly 100 cubic inches. What should be the dimensions of the can to minimize its cost?

24 The pages of a book are each to contain 30 square inches of print, and each page must have 2-inch margins at top and bottom and 1-inch margins at each side. What is the minimum possible area of such a page?

25 What point or points on the curve $y = x^2$ are nearest the point $(0, 2)$? (*Suggestion:* The square of a distance is minimized exactly when the distance itself is minimized.)

26 What is the length of the shortest line segment lying wholly in the first quadrant with its end points on the coordinate axes and tangent to the graph of $y = 1/x$?

27 A rectangle has an area of 64 square inches. A straight line is to be drawn from one corner of the rectangle to the midpoint of one of the two more distant sides. What is the minimum possible length of such a line?

28 An oil can is to have volume 1000 cubic inches and is to be shaped like a cylinder with a flat bottom but with a hemispherical top. Neglect the thickness of the material of the can, and find the dimensions that will minimize the amount of material needed to make it.

29 Find the length L of the longest rod that can be carried

2.31 Carrying a rod around a corner (Problem 29).

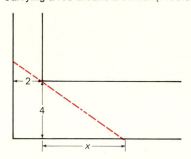

horizontally around a corner from a corridor 2 meters wide into one 4 meters wide. (*Suggestion: Minimize* the square of the length of the rod, indicated in Figure 2.31 by a dashed line.)

30 Find the length of the shortest ladder that will reach from the ground, over a wall 8 feet high, to the side of a building 1 foot behind the wall. That is, minimize the length $L = L_1 + L_2$ in Figure 2.32.

2.32 The ladder of Problem 30.

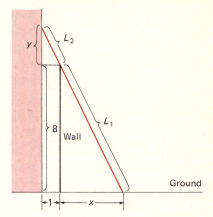

The Chain Rule

In Section 2-4 we showed that the generalized power rule

$$D[f(x)]^r = r[f(x)]^{r-1}f'(x) \tag{1}$$

holds if f is a differentiable function and r is an integer. Let us write $g(x) = x^r$, so that $g'(x) = rx^{r-1}$. Then (1) takes the form

$$D[g(f(x))] = g'(f(x)) \cdot f'(x). \tag{2}$$

The **chain rule** tells us that this last formula holds for *any* two differentiable functions f and g; g does *not* have to be a power function. Formula (2) is the last of the *general* differentiation rules in this book. All that remains is for you to learn the derivatives of some specific functions, such as the trigonometric functions of Section 2-9. Then you will be able to compute the derivative of any function in common use. Learning specific derivatives is important; note that (2) gives the derivative of the *composition* of two functions in terms of *their* derivatives.

Theorem *The Chain Rule*

Suppose that f is differentiable at x and that g is differentiable at $f(x)$. Then the composition $h = g(f)$ is differentiable at x, and its derivative there is

$$h'(x) = g'(f(x)) \cdot f'(x). \tag{3}$$

This is important: While the derivative of $h = g(f)$ is a product of the derivatives of f and g, these latter derivatives are evaluated at *different* points. For g' is evaluated at $f(x)$, while f' is evaluated at x. For a particular number $x = x_0$, (3) tells us that

$$h'(x_0) = g'(u_0)f'(x_0) \quad \text{where} \quad u_0 = f(x_0). \tag{4}$$

For example, if $h(x) = g(f(x))$, where f and g are differentiable functions, with

$$f(2) = 17, \quad f'(2) = -3, \quad \text{and} \quad g'(17) = 5,$$

then the chain rule gives

$$h'(2) = g'(f(2)) \cdot f'(2) = g'(17) \cdot f'(2) = (5)(-3) = -15.$$

If we write $u = f(x)$, so that $du/dx = f'(x)$, then the chain rule takes the form

$$D_x g(u) = g'(u)D_x u = g'(u)\frac{du}{dx}, \tag{5}$$

reminiscent of the generalized power rule in the form

$$D_x u^n = nu^{n-1}\frac{du}{dx}.$$

If also we write $y = g(u) = g(f(x)) = h(x)$, so that

$$\frac{dy}{dx} = h'(x) = g'(f(x))f'(x) = g'(u)f'(x),$$

and

$$\frac{dy}{du} = g'(u),$$

then the chain rule takes the form

$$\frac{dy}{dx} = \frac{dy}{du}\frac{du}{dx} \tag{6}$$

in differential notation.

Formula (6) is one that, once learned, is truly unlikely to be forgotten. Although dy/du and du/dx are *not* fractions—they are merely symbols representing the derivatives $g'(u)$ and $f'(x)$—it's just as though they *were* fractions, with the du in the first "cancelling" the du in the second. Of course, such "cancellation" no more proves the chain rule than cancelling d's proves that

$$\frac{dy}{dx} = \frac{y}{x} \quad \text{[an absurdity].}$$

It is nevertheless an excellent way to *remember* the chain rule. Such manipulations with differentials are so suggestive (even when invalid) that they played a substantial role in the early development of calculus in the seventeenth and eighteenth centuries. For thus were produced formulas that were later proved valid (as well as some formulas that were incorrect).

The differential notation of Formula (6) is especially fruitful; at the end of this section, we shall see that it not only suggests the chain rule, but even provides the idea for a correct proof of the chain rule.

EXAMPLE 1   A spherical balloon is being inflated, and its radius r is increasing at the rate of 0.2 inches per second at the instant when $r = 5$ inches. At what rate is the volume V of the balloon increasing at that instant?

Solution   Given $dr/dt = 0.2$ inches per second when $r = 5$ inches, we want to find dV/dt. Since the volume of the balloon is

$$V = \tfrac{4}{3}\pi r^3,$$

we see that $dV/dr = 4\pi r^2$. So the chain rule gives

$$\frac{dV}{dt} = \frac{dV}{dr}\frac{dr}{dt} = 4\pi r^2 \frac{dr}{dt}$$

$$= 4\pi(5)^2(0.2) \approx 62.83$$

cubic inches per second at the instant mentioned.

EXAMPLE 2   Imagine a spherical waterdrop falling through water vapor in the air. Suppose that the vapor adheres to its surface in such a way that the time rate of increase of the droplet's mass M is proportional to the droplet's surface area S. If the drop starts its fall with a radius that is effectively zero, and $r = 1$ mm after 20 seconds, when is the radius 3 mm?

Solution   We are given that

$$\frac{dM}{dt} = kS$$

where k is some constant that depends upon atmospheric conditions. Now

$$M = \tfrac{4}{3}\rho\pi r^3 \quad \text{and} \quad S = 4\pi r^2$$

where ρ is the density of water. Hence the chain rule gives

$$4\pi k r^2 = kS = \frac{dM}{dt} = \frac{dM}{dr}\frac{dr}{dt},$$

or

$$4\pi k r^2 = 4\pi\rho r^2 \frac{dr}{dt}.$$

This implies that

$$\frac{dr}{dt} = \frac{k}{\rho}, \qquad \text{a constant.}$$

So the radius of the droplet grows at a *constant* rate. Thus, if it takes 20 seconds for r to grow to 1mm, it will take 1 minute for it to grow to 3 mm.

PROOF OF THE GENERALIZED POWER RULE

We can now use the chain rule to complete the proof of the generalized power rule (1) for r any rational number. Recall from the discussion following Equation (3) in Section 2-4 that the only remaining case is the one with $r = 1/q$, the reciprocal of a positive integer q. Thus we want to prove that if f is a differentiable function, then

$$D[f(x)]^{1/q} = \frac{1}{q}[f(x)]^{(1/q)-1} \cdot f'(x) \tag{7}$$

at all points where the right-hand side is meaningful. Write

$$[f(x)]^{1/q} = u^{1/q} \quad \text{where} \quad u = f(x).$$

Then we know from the root rule (Theorem 3 in Section 2-4) that

$$D_u(u^{1/q}) = \frac{1}{q}u^{(1/q)-1}.$$

Hence the chain rule gives

$$D[f(x)]^{1/q} = D_x u^{1/q} = \frac{d(u^{1/q})}{du} \cdot \frac{du}{dx}$$

$$= \frac{1}{q}u^{(1/q)-1}\frac{du}{dx} = \frac{1}{q}[f(x)]^{(1/q)-1} \cdot f'(x),$$

thereby proving (7).

The key to successful application of the chain rule to find a derivative is the correct identification of the simpler functions f and g from which the composition $h = g(f)$ is built. To determine f and g, it may help to think of a hand-held calculator. Then the "outside" function g is represented by the *last* button you press in computing $h(x)$ for a particular x. The quantity $f(x)$ is what has been calculated just before pushing the last button. (We assume that the last button is something other than one of the arithmetic operations $+$, $-$, $\times$, or $\div$.) For if the last operation were multiplication (as an example), then you should be using the product rule. But if the last button is one such as *log*, *sin*, or x^2, then you are feeding the output from a previous calculation, $f(x)$, into a *last* function, and that function must be $g(x)$. In such cases the chain rule is needed to compute the derivative.

Of course, our "last button" analogy is intended only in a figurative sense. For example, if

$$h(x) = (1 + x^2)^{7/8},$$

the last functional operation performed in computing $h(x)$ is "raising to the power $\frac{7}{8}$," even though no calculator is likely to have a "$\frac{7}{8}$ power" button as such. So think of $h(x)$ as $g(f(x))$, where

$$g(u) = u^{7/8} \quad \text{and} \quad f(x) = 1 + x^2.$$

Then the chain rule gives

$$h'(x) = g'(u)f'(x) \qquad (u = 1 + x^2)$$

$$= \frac{7}{8}u^{-1/8}(2x) = \frac{7x}{4\sqrt[8]{1 + x^2}}.$$

As in this last example, it is often helpful to use explicitly the auxiliary variable u for the "inside" quantity $f(x)$ to which the "outside" function g is applied. This converts a complicated expression $h(x)$ into the simpler $g(u)$. The chain rule in the form of Equation (5),

$$D_x g(u) = g'(u)D_x u = g'(u)\frac{du}{dx}, \qquad (5)$$

can then be applied to produce the derivative $h'(x)$.

The use of the chain rule becomes clearer when it is applied to some functions other than the familiar algebraic functions of Section 2-4. In Section 2-9 we show that the derivatives of the sine and cosine functions (with x in radians) are

$$D \sin x = \cos x \quad \text{and} \quad D \cos x = -\sin x. \qquad (8)$$

If u is a differentiable function of x, then (5), first with $g(u) = \sin u$ and then with $g(u) = \cos u$, gives

$$D_x \sin u = (\cos u)\frac{du}{dx} \quad \text{and} \quad D_x \cos u = (-\sin u)\frac{du}{dx}. \qquad (9)$$

EXAMPLE 3 Differentiate $\cos \sqrt{x}$.

Solution $D \cos \sqrt{x} = -(\sin \sqrt{x})D_x(\sqrt{x}) = -(\sin \sqrt{x})/2\sqrt{x}.$

EXAMPLE 4 Differentiate

$$h(x) = \sin^2(2x - 1)^{3/2} = [\sin(2x - 1)^{3/2}]^2.$$

Solution Here the last operation, or outside function, is $g(u) = u^2$, where

$$u = f(x) = \sin(2x - 1)^{3/2}.$$

Hence

$$D \sin^2(2x - 1)^{3/2} = 2uD_x u$$
$$= 2[\sin(2x - 1)^{3/2}]D_x[\sin(2x - 1)^{3/2}]$$
$$= 2[\sin(2x - 1)^{3/2}][\cos(2x - 1)^{3/2}]D(2x - 1)^{3/2}$$
$$= 2[\sin(2x - 1)^{3/2}][\cos(2x - 1)^{3/2}]\tfrac{3}{2}(2x - 1)^{1/2}(2)$$
$$= 6(2x - 1)^{1/2}[\sin(2x - 1)^{3/2}][\cos(2x - 1)^{3/2}].$$

PROOF OF THE CHAIN RULE

Given $y = g(u)$ and $u = f(x)$, we want to compute the derivative

$$\frac{dy}{dx} = \lim_{\Delta x \to 0}\frac{\Delta y}{\Delta x} = \lim_{\Delta x \to 0}\frac{g(f(x + \Delta x)) - g(f(x))}{\Delta x}.$$

The differential form of the chain rule suggests the factorization

$$\frac{\Delta y}{\Delta x} = \frac{\Delta y}{\Delta u}\frac{\Delta u}{\Delta x}.$$

The product law of limits then gives

$$\frac{dy}{dx} = \lim_{\Delta x \to 0} \frac{\Delta y}{\Delta u}\frac{\Delta u}{\Delta x}$$

$$= \left(\lim_{\Delta u \to 0} \frac{\Delta y}{\Delta u}\right)\left(\lim_{\Delta x \to 0} \frac{\Delta u}{\Delta x}\right) = \frac{dy}{du}\frac{du}{dx}.$$

This will suffice to prove the chain rule *provided that*

$$\Delta u = f(x + \Delta x) - f(x)$$

is a *nonzero* quantity that approaches zero as $\Delta x \to 0$. Certainly $\Delta u \to 0$ as $\Delta x \to 0$ because f is differentiable and thus continuous. But it is quite possible that Δu *is* zero for some (even all) nonzero values of Δx. In such a case, the proposed factorization

$$\frac{\Delta y}{\Delta x} = \frac{\Delta y}{\Delta u}\frac{\Delta u}{\Delta x}$$

would involve the *invalid* step of division by zero.

On the other hand, this step *is* valid in case f is an increasing function, or even if it's known that f assumes no value twice in a small interval about x. Many functions have this latter property, and so the above attempt *is* a valid proof for a wide and important class of functions.

In addition, we can use the Mean Value Theorem to give a complete proof of the chain rule under the assumption that *the derivative g' of g is continuous at the point $f(x)$*. Although this additional hypothesis is not necessary for the truth of the chain rule, it will be satisfied in all our significant applications of it and eases the way along an otherwise stony path.

For this proof, we fix the number x, as if, again, x were a "temporary constant." Let $y_1 = f(x)$ and let $y_2 = f(x + h)$, where h is some small nonzero number. Then the definition of the derivative yields

$$Dg(f(x)) = \lim_{h \to 0} \frac{g(f(x + h)) - g(f(x))}{h} = \lim_{h \to 0} \frac{g(y_2) - g(y_1)}{h}. \tag{10}$$

2.33 Points used in proving the chain rule.

As indicated in Figure 2.33, we can apply the Mean Value Theorem to the function g on the closed interval $[y_1, y_2]$. We find that

$$g(y_2) - g(y_1) = (y_2 - y_1)g'(\bar{y}) \tag{11}$$

for some number $\bar{y}$ between y_1 and y_2. We are assuming that g is differentiable not only *at*, but also *near*, $f(x)$. This lets us apply the Mean Value Theorem to g on the interval $[y_1, y_2]$.

We also assume that f is differentiable at and near x, so that f is continuous on the closed interval with end points x and $x + h$. This means that we can apply the Intermediate Value Property to f there. We note that $\bar{y}$ is intermediate to the two values y_1 and y_2 of f and that the latter are indeed values of f, because $y_1 = f(x)$ and $y_2 = f(x + h)$. So there is a number $\bar{x}$ between x and $x + h$ such that $\bar{y} = f(\bar{x})$.

We have now prepared the way for our computation. We apply the definition of the derivative:

$$D_x g(f(x)) = \lim_{h \to 0} \frac{g(y_2) - g(y_1)}{h} \qquad \text{(By (10) above.)}$$

$$= \lim_{h \to 0} \frac{(y_2 - y_1)g'(\bar{y})}{h} \qquad \text{(By (11) above.)}$$

$$= \lim_{h \to 0} \frac{(y_2 - y_1)g'(f(\bar{x}))}{h}$$

$$= \lim_{h \to 0} \frac{f(x + h) - f(x)}{h} g'(f(\bar{x}))$$

$$= g(f(x))f'(x).$$

The last equality holds because as $h \to 0$, so that $\bar{x} \to x$, $f(\bar{x}) \to f(x)$ as well, because f is continuous at x. Thus $g'(f(\bar{x})) \to g'(f(x))$ by our assumption that g' is continuous at $f(x)$. This concludes the proof of the chain rule.

2-7 PROBLEMS

In each of Problems 1–8, identify two functions f and g such that $h(x) = g(f(x))$, and then apply the chain rule to find $h'(x)$.

1 $h(x) = \dfrac{1}{x^2 + 1}.$

2 $h(x) = \sqrt[3]{x + x^4}.$

3 $h(x) = \dfrac{1}{\sqrt{x^2 + 25}}.$

4 $h(x) = \sqrt{\dfrac{x}{(x^2 + 1)}}.$

5 $h(x) = \sqrt[3]{(x - 1)^5}.$

6 $h(x) = \dfrac{(x + 1)^{10}}{(x - 1)^{10}}.$

7 $h(x) = \cos(x^2 + 1).$

8 $h(x) = \sin(\cos x).$

Find dy/dx in Problems 9–12.

9 $y = \dfrac{1}{1 + u^2}$ and $u = x^2 + 1.$

10 $y = (1 - u^2)^{1/2}$ and $u = \dfrac{1}{x + 1}.$

11 $y = u^{1/2}$ and $u = \dfrac{x + 1}{x - 1}.$

12 $y = (1 + \sqrt{u})^{1/2}$ and $u = 1 + \sqrt{x}.$

Use the derivatives $D \sin x = \cos x$ and $D \cos x = -\sin x$ to differentiate the functions in Problems 13–18.

13 $f(x) = x \sin x.$

14 $f(x) = x^{10} \cos 10x.$

15 $g(t) = \sin(\sqrt{t^4 + 1}).$

16 $h(z) = \dfrac{\sin z^2}{z^2}.$

17 $f(x) = [\cos(\cos x)]^2.$

18 $g(t) = (\sin^2 t)(\cos^2 t).$

19 Given: $h(z) = f(g(z))$, $g(0) = 1$, $g'(0) = 5$, and $f'(1) = 2$. Find $h'(0)$.

20 Find $f'(-1)$ given $f(y) = h(g(y))$, $g(-1) = 2$, $h'(2) = -1$, and $g'(-1) = 7$.

21 Given: $G(t) = f(h(t))$, $h(1) = 4$, $f'(4) = 3$, and $h'(1) = -6$. Find $G'(1)$.

22 Suppose that $f(0) = 0$ and that $f'(0) = 1$. Calculate the value of the derivative of $f(f(f(x)))$ at $x = 0$.

23 Given: $\phi(t) = h(g(f(t)))$. Apply the chain rule successively to show that

$$\phi'(t) = h'(g(f(t))) \cdot g'(f(t)) \cdot f'(t).$$

Note that with $z = h(y)$, $y = g(x)$, and $x = f(t)$, this is

$$\frac{dz}{dt} = \frac{dz}{dy} \frac{dy}{dx} \frac{dx}{dt}$$

in differential notation.

24 Let $y = f(x)$ and $x = g(y)$ be differentiable functions that are inverses of each other. Then $g(f(x)) \equiv x$. Differentiate both sides of this identity, with the aid of the chain rule, to show that

$$\frac{dy}{dx} \frac{dx}{dy} = 1.$$

25 Air is being pumped into a spherical balloon in such a way that its radius is increasing at the rate of $dr/dt = 1$ centimeter per second. What is the time rate of increase, in cubic centimeters per second, of the balloon's volume when its radius r is 10 cm?

26 Suppose that the air is being pumped into the balloon of Problem 25 at the constant rate of 200π cc per second. What is the time rate of increase of the radius when $r = 5$ cm?

27 Air is escaping from a spherical balloon at the constant rate of 300π cc/sec. What is the radius of the balloon when its radius is decreasing at 3 cm/sec?

28 A spherical hailstone is losing mass by melting uniformly over its surface as it falls. At a certain time, its radius is 2 cm and its volume is decreasing at the rate of 0.1 cm³/sec. How fast is its radius decreasing at that time?

29 A spherical snowball is melting in such a way that the rate of decrease of its volume is proportional to its surface area. At 10 A.M. its volume was 500 cubic inches, and at

11 A.M. its volume was 250 cubic inches. When did the snowball finish melting? (See Example 2.)

In Chapter 6 we shall define a function $L = L(x)$, called the *natural logarithm function*, such that $L'(x) = 1/x$ for all $x > 0$. Find $f'(x)$ in Problems 30–38.

30 $f(x) = L(17x)$.   **31** $f(x) = L(x^{17})$.

32 $f(x) = [L(x)]^{17}$.   **33** $f(x) = L(x^2 + 1)$.

34 $f(x) = L\left(\dfrac{1}{x}\right)$.   **35** $f(x) = \dfrac{1}{L(x)}$.

36 $f(x) = L(L(x))$.   **37** $f(x) = L(\sin x)$.

38 $f(x) = \sin(L(x))$.

2-8

Implicit Differentiation

An equation in two variables x and y may have one or more solutions for y in terms of x or for x in terms of y. These solutions are functions that we say are **implicitly defined** by the equation. In this section we discuss the differentiation of such functions and the use of their derivatives in solving certain optimization problems.

For example, the equation $y^2 - x = 0$ of a parabola implicitly defines two continuous functions of x:

$$y = \sqrt{x} \quad \text{and} \quad y = -\sqrt{x}.$$

Each has domain the half line $x \geqq 0$. The graphs of these two functions are the upper and lower branches of the parabola of Figure 2.34. The whole parabola cannot be the graph of a function of x because no vertical line can meet the graph of a function in more than one point.

The equation of the unit circle, $x^2 + y^2 = 1$, implicitly defines four functions (among others):

$$y = +\sqrt{1 - x^2} \quad \text{for} \quad x \text{ in } [-1, 1],$$
$$y = -\sqrt{1 - x^2} \quad \text{for} \quad x \text{ in } [-1, 1],$$
$$x = +\sqrt{1 - y^2} \quad \text{for} \quad y \text{ in } [-1, 1], \quad \text{and}$$
$$x = -\sqrt{1 - y^2} \quad \text{for} \quad y \text{ in } [-1, 1].$$

The four graphs are highlighted against the four unit circles in Figure 2.35.

The equation $2x^2 + 2y^2 + 3 = 0$ implicitly defines *no* function, because this equation has no real solution (x, y).

In advanced calculus one studies conditions that will guarantee when an implicitly defined function is actually differentiable. Here we shall proceed on the assumption that our implicitly defined functions are differentiable at most points in their domains. (The functions that have the graphs shown in Figure 2.35 are *not* differentiable at the endpoints of their domains.) This is a reasonable assumption whenever the equation we are working with is itself "nice," such as a polynomial in two variables.

When we assume differentiability, we can use the chain rule to differentiate the given equation, thinking of x as the independent variable. We can

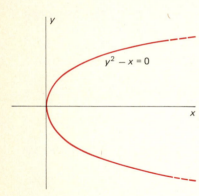

2.34 The parabola $y^2 - x = 0$.

2.35 Continuous functions defined implicitly by $x^2 + y^2 = 1$.

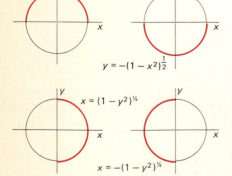

then solve the resulting equation for the derivative $dy/dx = f'(x)$ of the implicitly defined function. This process is called **implicit differentiation.**

EXAMPLE 1   Use implicit differentiation to find the derivative of a differentiable function $y = f(x)$ implicitly defined by the equation $x^2 + y^2 = 1$.

Solution   The equation $x^2 + y^2 = 1$ is to be thought of as an *identity* that implicitly defines y as a function of x. Since $x^2 + y^2$ is then a function of x, it has the same derivative as the constant function 1 on the other side of the identity. Thus we may differentiate both sides of $x^2 + y^2 = 1$ with respect to x and equate the results. We obtain

$$2x + 2y\frac{dy}{dx} = 0.$$

In this step, we must remember that y is a function of x, so that $D_x(y^2) = 2yD_xy$.

Then we solve for dy/dx:

$$\frac{dy}{dx} = -\frac{x}{y}. \tag{1}$$

It may be surprising to see a formula for dy/dx containing both x *and* y, but such a formula is frequently just as useful as one containing only x. For example, the above formula tells us that the slope of the tangent line to the circle $x^2 + y^2 = 1$ at the point $(1/\sqrt{5}, 2/\sqrt{5})$ is

$$\left.\frac{dy}{dx}\right|_{(1/\sqrt{5},\, 2/\sqrt{5})} = -\frac{1/\sqrt{5}}{2/\sqrt{5}} = -\frac{1}{2}.$$

We introduce here a self-explanatory notational device for specifying where dy/dx is evaluated.

Note that if $y = \pm\sqrt{1 - x^2}$, then

$$\frac{dy}{dx} = \frac{-x}{\pm\sqrt{1 - x^2}} = -\frac{x}{y},$$

in agreement with Equation (1) above. Thus (1) simultaneously gives us the derivatives of both the functions $y = +\sqrt{1 - x^2}$ and $y = -\sqrt{1 - x^2}$ defined implicitly by the equation $x^2 + y^2 = 1$.

EXAMPLE 2   The folium of Descartes is the graph of the equation

$$x^3 + y^3 = 3xy. \tag{2}$$

This curve was originally proposed by René Descartes (1596–1650) as a challenge to Pierre Fermat (1601–1665) to find its tangent line at an arbitrary point. The graph of the folium of Descartes appears in Figure 2.36; it has the line $x + y + 1 = 0$ as an asymptote. We indicate in a problem how this graph is constructed. Here we want to find the slope of its tangent line.

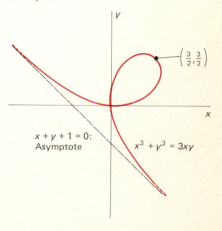

2.36   The folium of Descartes; see Example 2.

Solution   We use implicit differentiation (rather than the ad hoc methods with which Fermat met Descartes' challenge). We differentiate both sides of (1) and find that

$$3x^2 + 3y^2 \frac{dy}{dx} = 3y + 3x \frac{dy}{dx}.$$

We solve for the derivative:

$$\frac{dy}{dx} = \frac{y - x^2}{y^2 - x}. \tag{3}$$

For instance, at the point $(\frac{3}{2}, \frac{3}{2})$ on the folium, the slope of the tangent line is

$$\frac{dy}{dx}\bigg|_{(3/2,3/2)} = \frac{(\frac{3}{2}) - (\frac{3}{2})^2}{(\frac{3}{2})^2 - (\frac{3}{2})} = -1,$$

and this result agrees with our intuition about the figure. At the point $(\frac{2}{3}, \frac{4}{3})$, Equation (3) gives

$$\frac{dy}{dx}\bigg|_{(2/3,4/3)} = \frac{(\frac{4}{3}) - (\frac{2}{3})^2}{(\frac{4}{3})^2 - (\frac{2}{3})} = \frac{4}{5}.$$

Thus the equation of the tangent line at this point is

$$y - \tfrac{4}{3} = \tfrac{4}{5}(x - \tfrac{2}{3}), \quad \text{or} \quad 4x - 5y + 4 = 0.$$

MAXIMUM-MINIMUM PROBLEMS WITH CONSTRAINTS

In applied maximum-minimum problems, the quantity Q to be maximized or minimized often appears initially as a function of two variables, $Q = Q(x, y)$, with x and y related by an equation of the form

$$C(x, y) = K, \tag{4}$$

where K is some constant. That is, we are to find the maximum or minimum value of $Q(x, y)$ subject to the side condition or *constraint* that x and y satisfy Equation (4). For example, the problem of minimizing the total surface area of a box with a square base (as shown in Figure 2.37) required to have volume $V = 1000$ is that of minimizing $A = 2x^2 + 4xy$ subject to the constraint $x^2y = 1000$.

In Section 2-3 our method of solving such a problem was to solve the constraint equation (4) for y as a function of x and then substitute $y = y(x)$ into the expression $Q(x, y)$, the quantity to be maximized or minimized. This produced a function $Q = Q(x, y(x)) = f(x)$ of the single variable x, to which we then applied methods of calculus. The constraint equation was thus used to eliminate the "auxiliary" variable y.

Sometimes, however, it is difficult or even impossible to solve the constraint equation (4) for y as a function of x. If so, an alternative method is available. It involves implicit differentiation and is called the **method of auxiliary variables.** Here, one works with both x as the independent variable *and* the auxiliary variable y, now thought of as a function of x. Begin by differentiating $Q(x, y)$ and $C(x, y)$ implicitly. Then

$$\frac{dQ}{dx} = 0 \tag{5}$$

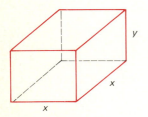

2.37   A box with square base of edge x and with height y.

at the maximum-minimum point being sought, while

$$\frac{dC}{dx} = 0 \qquad (6)$$

also holds, the result of implicit differentiation of the constraint equation $C(x, y) = K$ (a constant).

It is always easy to eliminate dy/dx from Equations (5) and (6); we use this technique in the examples below. The result is a single equation in x and y. The final step is to solve this equation and the constraint equation (4) simultaneously. The resulting solutions (x, y) are the points corresponding to horizontal tangents in our earlier method.

EXAMPLE 3   Use the method of auxiliary variables to solve the sawmill problem (Example 5 in Section 2-4), the problem of finding the rectangle of largest area inscribed in the unit circle $x^2 + y^2 = 1$.

Solution   Let (x, y) denote the first quadrant vertex of the inscribed rectangle, as in Figure 2.38. We want to maximize the area

$$A = 4xy$$

subject to the constraint

$$C = x^2 + y^2 = 1.$$

Equations (5) and (6) above now take the particular forms

$$\frac{dA}{dx} = 4y + 4x\frac{dy}{dx} = 0$$

and

$$\frac{dC}{dx} = 2x + 2y\frac{dy}{dx} = 0.$$

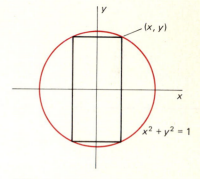

2.38   The sawmill problem.

The cases $x = 0$ and $y = 0$ do not produce maxima, so we may assume throughout that $x > 0$ and $y > 0$. With this assumption, the two equations above become

$$\frac{dy}{dx} = -\frac{y}{x} \quad \text{and} \quad \frac{dy}{dx} = -\frac{x}{y},$$

respectively. Thus

$$-\frac{y}{x} = -\frac{x}{y}, \quad \text{or} \quad x^2 = y^2.$$

The only first-quadrant solution of the equations

$$x^2 + y^2 = 1 \quad \text{and} \quad x^2 = y^2$$

is the point $(1/\sqrt{2}, 1/\sqrt{2})$, which determines a square with edge length $\sqrt{2}$. This is, of course, the same solution as the one we found in Example 5 of Section 2-4.

EXAMPLE 4   One common nuclear reactor design involves a cylindrical core with radius r and height h. For a certain mixture of moderator and fissionable material, the reactor will reach critical mass and thus produce a self-sustaining nuclear reaction if r and h (in centimeters) satisfy the equation

$$\frac{\lambda^2}{r^2} + \frac{\pi^2}{h^2} = B^2 \tag{7}$$

where $\lambda = 2.405$ and $B = 0.04$. Find the dimensions for which the critical mass volume $V = \pi r^2 h$ is minimal; that is, the least expensive reactor.

Solution Let us regard h as a function of r. Then

$$\frac{dV}{dr} = 2\pi rh + \pi r^2 \frac{dh}{dr}.$$

We also differentiate the constraint condition (7) and obtain

$$-\frac{2\lambda^2}{r^3} - \frac{2\pi^2}{h^3}\frac{dh}{dr} = 0.$$

We write the equation $dV/dr = 0$ and find that

$$2\pi rh + \pi r^2 \frac{dh}{dr} = 0.$$

We solve the last two equations for dh/dr:

$$\frac{dh}{dr} = -\frac{2h}{r} = -\frac{\lambda^2 h^3}{\pi^2 r^3},$$

and it follows that

$$h^2 = \frac{2\pi^2 r^2}{\lambda^2}.$$

We substitute this value of h^2 in Equation (7), and it's now easy to solve for

$$r = \frac{\lambda\sqrt{3}}{B\sqrt{2}} = \frac{(2.405)\sqrt{3}}{(0.04)\sqrt{2}} \approx 73.638$$

centimeters, or about 2.42 feet. This is the radius of the minimal volume critical mass core. The corresponding height is

$$h = \frac{\pi\sqrt{2}r}{\lambda} \approx \frac{\pi\sqrt{2}(73.638)}{2.405} \approx 136.04$$

centimeters, or about 4.46 feet.

2-8 PROBLEMS

In Problems 1–10, find dy/dx by implicit differentiation.

1 $x^2 - y^2 = 1.$
2 $xy = 1.$
3 $16x^2 + 25y^2 = 400.$
4 $x^3 + y^3 = 1.$
5 $x^{1/2} + y^{1/2} = 1.$
6 $x^2 + xy + y^2 = 9.$
7 $x^{2/3} + y^{2/3} = 1.$
8 $(x - 1)y^2 = x + 1.$
9 $x^2(x - y) = y^2(x + y).$
10 $(x^2 + y^2)^2 = 4xy.$

In Problems 11–14, find equations of the tangent and normal lines to the graph of the given equation at the indicated point.

11 $x^3 + y^3 = 2$; (1, 1).
12 $x^3 + 2xy + y^2 = 9$; (1, 2).
13 $x^4 + y^4 = 2$; (1, −1).
14 $y^3 + x + y + x^3 = 20$; (−2, 3).

CHAP. 2: Differentiation

Once dy/dx has been found by implicit differentiation, the second derivative d^2y/dx^2 can be found by differentiating dy/dx with respect to x, still regarding y as a function of x. Do this in Problems 15–18.

15 $x^2 - y^2 = 1$. **16** $x^3 + y^3 = 1$.

17 $x^{1/3} + y^{1/3} = 1$. **18** $x^2 - 3xy + 2y^2 = 5$.

19 The graph of the equation $x^2 - xy + y^2 = 9$ is the "rotated ellipse" shown in Figure 2.39. Find the tangent lines to this curve at the two points where it intersects the y-axis, and show that these two lines are parallel.

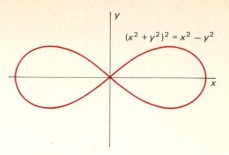

2.40 The lemniscate of Problem 22.

23 The lemniscate $(x^2 + y^2)^2 = 4xy$ looks like the figure-8 curve shown in Figure 2.41. Use the method of auxiliary variables to find the points on it that are farthest from the origin (maximizing, as usual, the *square* of the distance).

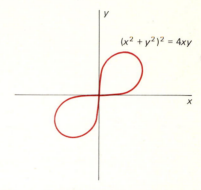

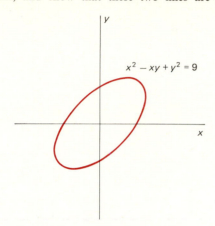

2.39 The rotated ellipse of Problem 19.

20 Find the points on the curve $x^2 - xy + y^2 = 9$ where the tangent line is horizontal ($dy/dx = 0$); then find the points where it is vertical ($dx/dy = 0$).

21 Find the points of the curve $x^2 - xy + y^2 = 9$ that are closest to and farthest from the origin. (*Suggestion:* Maximize the quantity $d^2 = x^2 + y^2$ subject to the constraint $x^2 - xy + y^2 = 9$.)

22 The lemniscate with equation $(x^2 + y^2)^2 = x^2 - y^2$ is shown in Figure 2.40. Find by implicit differentiation the four points on the lemniscate where the tangent line is horizontal. Also find the two points where the tangent line is vertical; that is, where $dx/dy = 1/(dy/dx) = 0$.

2.41 Another lemniscate (see Problem 23).

Use the method of auxiliary variables to solve the following applied maximum-minimum problems.

24 Problem 21, Section 2-3. **25** Problem 22, Section 2-3.
26 Problem 24, Section 2-3. **27** Problem 34, Section 2-3.
28 Problem 22, Section 2-4 (see Figure 2.13).
29 Problem 28, Section 2-4. **30** Problem 20, Section 2-6.
31 Problem 21, Section 2-6. **32** Problem 22, Section 2-6.
33 Problem 23, Section 2-6. **34** Problem 28, Section 2-6.

2-9

In elementary trigonometry, the six trigonometric functions of an acute angle θ in a right triangle are defined as ratios between pairs of sides of the triangle. As in Figure 2.42,

Derivatives of Sines and Cosines

$$\cos \theta = \frac{\text{adj}}{\text{hyp}}, \quad \sin \theta = \frac{\text{opp}}{\text{hyp}}, \quad \tan \theta = \frac{\text{opp}}{\text{adj}},$$

$$\sec \theta = \frac{\text{hyp}}{\text{adj}}, \quad \csc \theta = \frac{\text{hyp}}{\text{opp}}, \quad \cot \theta = \frac{\text{adj}}{\text{opp}}.$$

(1)

2.42

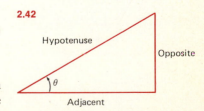

We generalize these definitions to *directed* angles of arbitrary size in the following way. Suppose that the initial side of the angle θ is the positive

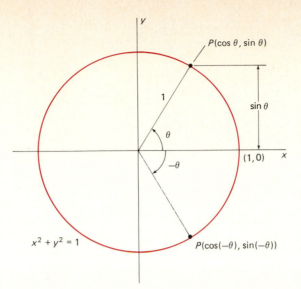

2.43 Defining the trigonometric functions by means of the unit circle.

x-axis, so its vertex is at the origin. The angle is **directed** if a direction of rotation from its initial side to its terminal side is specified. We call θ a **positive angle** if this rotation is counterclockwise and a **negative angle** if this rotation is clockwise.

Let $P(x, y)$ be the point at which the terminal side of θ intersects the *unit* circle $x^2 + y^2 = 1$. Then we define

$$\cos \theta = x, \qquad \sin \theta = y, \qquad \tan \theta = \frac{y}{x},$$

$$\sec \theta = \frac{1}{x}, \qquad \csc \theta = \frac{1}{y}, \qquad \cot \theta = \frac{x}{y}. \tag{2}$$

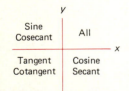

Positive in quadrants shown

2.44 The signs of the trigonometric functions.

2.45 Effect of replacing θ by $-\theta$ in sine and cosine.

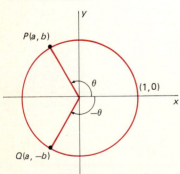

Of course we assume that $x \neq 0$ in the case of $\tan \theta$ and $\sec \theta$ and that $y \neq 0$ in the case of $\cot \theta$ and $\csc \theta$. If the angle θ is positive and acute, then it is clear from Figure 2.43 that the definitions in (2) agree with the right triangle definitions in (1) in terms of the coordinates of P. A glance at the figure also shows which of the functions are positive for angles in each of the four quadrants. The diagram in Figure 2.44 summarizes this information.

In this section we discuss primarily the two most basic trigonometric functions, the sine and the cosine. From (2) we see immediately that

$$\tan \theta = \frac{\sin \theta}{\cos \theta}, \qquad \sec \theta = \frac{1}{\cos \theta},$$

$$\cot \theta = \frac{\cos \theta}{\sin \theta}, \qquad \csc \theta = \frac{1}{\sin \theta}. \tag{3}$$

Next, we compare the angles θ and $-\theta$ in Figure 2.45. We see that

$$\cos(-\theta) = \cos \theta \quad \text{and} \quad \sin(-\theta) = -\sin \theta. \tag{4}$$

CHAP. 2: **Differentiation**

The equation $x^2 + y^2 = 1$ of the unit circle translates immediately into the **fundamental identity of trigonometry,**

$$\cos^2\theta + \sin^2\theta = 1. \qquad (5)$$

In Problems 34 and 35 following this section we outline derivations of the **addition formulas**

$$\sin(\alpha + \beta) = \sin \alpha \cos \beta + \cos \alpha \sin \beta, \qquad (6)$$

$$\cos(\alpha + \beta) = \cos \alpha \cos \beta - \sin \alpha \sin \beta. \qquad (7)$$

When we take $\alpha = \theta = \beta$ in (6) and (7), we obtain the **double-angle formulas**

$$\sin 2\theta = 2 \sin \theta \cos \theta, \qquad (8)$$

$$\cos 2\theta = \cos^2\theta - \sin^2\theta \qquad (9)$$

$$= 2 \cos^2\theta - 1 \qquad (9a)$$

$$= 1 - 2 \sin^2\theta, \qquad (9b)$$

where (9a) and (9b) are obtained from (9) by use of the fundamental identity in (5).

If we solve (9a) for $\cos^2\theta$ and (9b) for $\sin^2\theta$, we get the **half-angle formulas**

$$\cos^2\theta = \tfrac{1}{2}(1 + \cos 2\theta), \qquad (10)$$

$$\sin^2\theta = \tfrac{1}{2}(1 - \cos 2\theta). \qquad (11)$$

Radians	Degrees
0	0
$\pi/6$	30
$\pi/4$	45
$\pi/3$	60
$\pi/2$	90
$2\pi/3$	120
$3\pi/4$	135
$5\pi/6$	150
π	180
$3\pi/2$	270
2π	360
4π	720

2.46 Some radian-degree conversions.

2.47 Area of a sector and arc length for a circle.

RADIAN MEASURE

In elementary mathematics, angles are frequently measured in **degrees,** with 360 degrees in one complete revolution. In calculus it is more convenient, and often imperative, to measure angles in **radians.** The radian measure of an angle is the length of the arc it subtends in (that is, the arc it cuts out of) the unit circle, where the angle's vertex is at the circle's center. Since the circumference of the complete unit circle is 2π and its central angle is $360°$, it follows that

$$2\pi \text{ radians} = 360°,$$

$$180° = \pi \text{ radians} \approx 3.14159 \text{ radians.} \qquad (12)$$

Using (12) we can easily convert back and forth between radians and degrees:

$$1 \text{ radian} = \frac{180°}{\pi} \approx 57°17'44.8'', \qquad (12a)$$

$$1° = \frac{\pi}{180} \text{ radians} \approx 0.01745 \text{ radians.} \qquad (12b)$$

The table in Figure 2.46 shows radian-degree conversions for some frequently occurring angles.

Now consider an angle of θ radians placed at the center of a circle of radius r, as in Figure 2.47. Denote by s the length of the arc subtended by θ, and by A the area of the sector of the circle bounded by this angle. Then

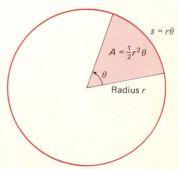

the evident proportions

$$\frac{s}{2\pi r} = \frac{A}{\pi r^2} = \frac{\theta}{2\pi}$$

give the formulas

$$s = r\theta \qquad (\theta \text{ in radians}) \tag{13}$$

and

$$A = \tfrac{1}{2} r^2 \theta \qquad (\theta \text{ in radians}). \tag{14}$$

Definitions (2) above refer to trigonometric functions of *angles* rather than trigonometric functions of *numbers*. Suppose that t is a real number. Then the number $\sin t$ is, by definition, the sine of an angle of t radians—remembering that a positive angle is directed counterclockwise from the positive x-axis, while a negative angle is directed clockwise. Briefly, $\sin t$ is the sine of an angle of t *radians*. The other trigonometric functions of the number t have similar definitions. Hence, when we write $\sin t$, $\cos t$, and so on, with t a real number, we *always* intend reference to an angle of t *radians*.

When we need to refer to the sine of an angle of t *degrees*, we will henceforth write $\sin t°$. The point is that $\sin t$ and $\sin t°$ are quite different functions of the variable t. For example, you will find that

$$\sin 1° \approx 0.0175 \quad \text{and} \quad \sin 30° = 0.5000$$

with your calculator set in degree mode. But when it is set radian mode, it will report that

$$\sin 1 \approx 0.8415 \quad \text{and} \quad \sin 30 \approx -0.9880.$$

The relationship between the functions $\sin t$ and $\sin t°$ is this:

$$\sin t° = \sin\left(\frac{\pi t}{180}\right). \tag{15}$$

The distinction extends even to most programming languages. In FORTRAN, the function SIN is the radian sine function above, and you write $\sin t°$ in the form SIND (T).

An angle of 2π radians corresponds to 1 revolution. This implies that the sine and cosine functions have **period** 2π, meaning that

$$\begin{aligned} \sin(t + 2\pi) &= \sin t, \\ \cos(t + 2\pi) &= \cos t. \end{aligned} \tag{16}$$

It follows from (16) that

$$\sin(t + 2n\pi) = \sin t, \qquad \cos(t + 2n\pi) = \cos t \tag{17}$$

for any integer n. This periodicity of the sine and cosine functions is evident in their graphs, which are shown in Figure 2.48.

From the definitions of $\sin t$ and $\cos t$, we see that

$$\sin 0 = 0, \qquad \sin \frac{\pi}{2} = 1, \qquad \sin \pi = 0,$$

$$\cos 0 = 1, \qquad \cos \frac{\pi}{2} = 0, \qquad \cos \pi = -1.$$

CHAP. 2: Differentiation

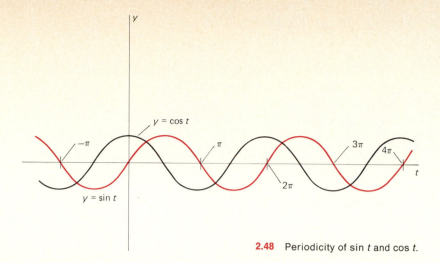

2.48 Periodicity of sin t and cos t.

The trigonometric functions of $\pi/6$, $\pi/4$, and $\pi/3$ are easy to read from the well-known triangles of Figure 2.49.

DIFFERENTIATION OF sin *u* AND cos *u*

In Example 13 of Section 1-8, we showed that

$$\lim_{\theta \to 0} \frac{\sin \theta}{\theta} = 1. \tag{18}$$

2.49

From this we obtain

$$\lim_{\theta \to 0} \frac{1 - \cos \theta}{\theta} = \lim_{\theta \to 0} \frac{1 - \cos \theta}{\theta} \cdot \frac{1 + \cos \theta}{1 + \cos \theta}$$

$$= \lim_{\theta \to 0} \frac{\sin^2 \theta}{\theta(1 + \cos \theta)}$$

$$= \left(\lim_{\theta \to 0} \frac{\sin \theta}{\theta} \right) \left(\lim_{\theta \to 0} \frac{\sin \theta}{1 + \cos \theta} \right),$$

so that

$$\lim_{\theta \to 0} \frac{1 - \cos \theta}{\theta} = 0. \tag{19}$$

In the last step we used the fact that $\sin \theta$ and $\cos \theta$ are continuous at $\theta = 0$ (as is evident from their definitions).

These trigonometric limits are all that we need to differentiate $\sin t$:

$$D \sin t = \lim_{h \to 0} \frac{\sin(t + h) - \sin t}{h}$$

$$= \lim_{h \to 0} \frac{(\sin t \cos h + \sin h \cos t) - \sin t}{h}$$

$$= (\cos t) \left(\lim_{h \to 0} \frac{\sin h}{h} \right) - (\sin t) \left(\lim_{h \to 0} \frac{1 - \cos h}{h} \right).$$

We apply (18) and (19), and it follows that

$$D \sin t = \cos t. \tag{20}$$

To combine this last formula with the chain rule, suppose that $y = \sin u$ where u is a differentiable function of x. Then

$$\frac{dy}{dx} = \frac{dy}{du} \cdot \frac{du}{dx} = (\cos u) \frac{du}{dx},$$

and so

$$D_x \sin u = (\cos u) \frac{du}{dx}. \tag{21}$$

As an application of the addition formula for the sine and cosine, we see that

$$\cos x = \sin\left(\frac{\pi}{2} - x\right) \quad \text{and} \quad \sin x = \cos\left(\frac{\pi}{2} - x\right).$$

Hence (21) gives

$$D \cos x = D_x \sin\left(\frac{\pi}{2} - x\right)$$

$$= \left[\cos\left(\frac{\pi}{2} - x\right)\right]\left[D_x\left(\frac{\pi}{2} - x\right)\right]$$

$$= -\cos\left(\frac{\pi}{2} - x\right).$$

That is,

$$D \cos x = -\sin x. \tag{22}$$

In combination with the chain rule (as above), we obtain

$$D_x \cos u = (-\sin u) \frac{du}{dx}. \tag{23}$$

From (21) we obtain

$$D \sin t^\circ = D_t \sin\left(\frac{\pi t}{180}\right) = \frac{\pi}{180} \cos\left(\frac{\pi t}{180}\right),$$

or

$$D \sin t^\circ \approx (0.01745)\cos t^\circ.$$

The necessity of using the approximate value 0.01745 here, and indeed its very presence, is one reason that radians instead of degrees are used in the calculus of trigonometric functions.

Examples 3 and 4 of Section 2-7 illustrate the use of Formulas (21) and (23) in the formal differentiation of functions involving sines and cosines. The examples below illustrate the use of trigonometric functions in applied problems.

EXAMPLE 1   A rocket is launched vertically and is tracked by an observing station located on the ground 5 miles from the launch pad. Suppose that the elevation angle θ of the line of sight to the rocket is increasing at 3° per second when $\theta = 60^\circ$. What is the velocity of the rocket at this instant?

Solution First we convert the given data from degrees to radians. We are given that

$$\frac{d\theta}{dt} = 3\frac{\text{deg}}{\text{sec}} \times \frac{\pi \text{ rad}}{180 \text{ deg}} = \frac{\pi \text{ rad}}{60 \text{ sec}} \qquad \text{when } \theta = 60° = \frac{\pi}{3} \text{ radians.}$$

From Figure 2.50 we see that the height y (in miles) of the rocket is

$$y = 5 \tan \theta.$$

Hence its velocity is

$$\frac{dy}{dt} = D_t\left(5\frac{\sin \theta}{\cos \theta}\right)$$

$$= 5\frac{(\cos \theta)(\cos \theta) - (\sin \theta)(-\sin \theta)}{\cos^2\theta} \cdot \frac{d\theta}{dt} = 5(\sec^2\theta)\frac{d\theta}{dt}.$$

Since $\sec(\pi/3) = 2$, the velocity of the rocket is

$$\frac{dy}{dt} = 5(2)^2\left(\frac{\pi}{60}\right) = \frac{\pi}{3}$$

miles per second, or about 3770 mph, at the time when $\theta = 60°$.

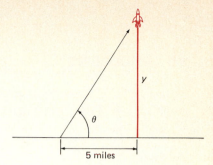

2.50 Tracking an ascending rocket.

EXAMPLE 2 Find the length of the longest rod that can be carried horizontally around the corner from a hall 2 meters wide into one that is 4 meters wide.

Solution The desired length will actually be the *minimum* length L of the dashed line in Figure 2.51, which represents the rod being carried around the corner. From the two similar triangles in the figure we see that

$$L = L_1 + L_2 = 4 \csc \theta + 2 \sec \theta,$$

or

$$L = L(\theta) = \frac{4}{\sin \theta} + \frac{2}{\cos \theta}.$$

The domain of L is the open interval $0 < \theta < \pi/2$. Clearly $L \to +\infty$ as either $\theta \to 0^+$ or $\theta \to (\pi/2)^-$.

We differentiate:

$$\frac{dL}{d\theta} = -\frac{4 \cos \theta}{\sin^2\theta} + \frac{2 \sin \theta}{\cos^2\theta} = \frac{2 \sin^3\theta - 4 \cos^3\theta}{\sin^2\theta \cos^2\theta},$$

so $dL/d\theta = 0$ when

$$2 \sin^3\theta = 4 \cos^3\theta,$$

$$\tan \theta = \sqrt[3]{2},$$

or $\theta \approx 0.90$ radians.

It's clear that $dL/d\theta < 0$ when $\theta < 0.90$, and that $dL/d\theta > 0$ when $\theta > 0.90$. So the graph of L looks as indicated in Figure 2.52. This means that the minimum value of L, and therefore the maximum length of the rod in question, is about

$$L(0.90) = \frac{4}{\sin(0.90)} + \frac{2}{\cos(0.90)},$$

or approximately 8.32 meters.

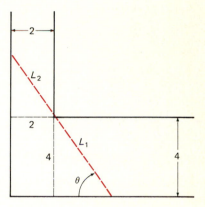

2.51 Carrying a rod around a corner.

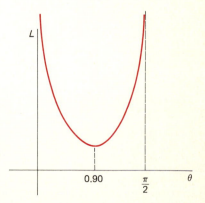

2.52 Graph of $L(\theta)$ (Example 2).

2-9   PROBLEMS

1 Express the following angles in radian measure.

(a) $40°$; (b) $-270°$; (c) $315°$; (d) $210°$;
(e) $-150°$.

2 Express the following angles, given in radian measure, in degrees.

(a) $\dfrac{\pi}{10}$; (b) $\dfrac{2\pi}{5}$; (c) 3π; (d) $\dfrac{15\pi}{4}$; (e) $\dfrac{23\pi}{60}$.

3 Evaluate the six trigonometric functions of x at the following values.

(a) $x = -\dfrac{\pi}{3}$; (b) $x = \dfrac{3\pi}{4}$;

(c) $x = \dfrac{7\pi}{6}$; (d) $x = \dfrac{5\pi}{3}$.

4 Find all numbers x such that

(a) $\sin x = 0$; (b) $\sin x = 1$; (c) $\sin x = -1$.

5 Find all numbers x such that

(a) $\cos x = 0$; (b) $\cos x = 1$; (c) $\cos x = -1$.

6 Find all numbers x such that

(a) $\tan x = 0$; (b) $\tan x = 1$; (c) $\tan x = -1$.

7 Suppose that $\tan x = \tfrac{3}{4}$ and that $\sin x$ is negative. Find the values of the other five trigonometric functions at x.

8 Suppose that $\csc x = -\tfrac{5}{3}$ and that $\cos x$ is positive. Find the values of the other five trigonometric functions at x.

9 Deduce from the fundamental identity $\cos^2\theta + \sin^2\theta = 1$ and the definitions of the other four trigonometric functions the identities

(a) $1 + \tan^2\theta = \sec^2\theta$; (b) $1 + \cot^2\theta = \csc^2\theta$.

10 Deduce from the addition formulas for the sine and cosine the addition formula for the tangent,

$$\tan(x + y) = \frac{\tan x + \tan y}{1 - \tan x \tan y}.$$

11 Derive from $D \sin x = \cos x$ and $D \cos x = -\sin x$ the following formulas for the derivatives of the other trigonometric functions.

(a) $D \tan x = \sec^2 x$.   (c) $D \sec x = \sec x \tan x$.
(b) $D \cot x = -\csc^2 x$.   (d) $D \csc x = -\csc x \cot x$.

Use $\lim\limits_{\theta \to 0} (\sin\theta)/\theta = 1$ to find the limits in Problems 12–20.

12 $\lim\limits_{\theta \to 0} \dfrac{\sin 2\theta}{\theta}$.

13 $\lim\limits_{\theta \to 0} \dfrac{1 - \cos\theta}{\theta \sin\theta}$.

14 $\lim\limits_{\theta \to 0} \dfrac{\sin^2\theta}{\theta}$.

15 $\lim\limits_{x \to 0} \dfrac{\tan x}{x}$.

16 $\lim\limits_{x \to 0} \dfrac{\tan 2x}{3x}$.

17 $\lim\limits_{x \to 0} x \cot 3x$.

18 $\lim\limits_{x \to 0} \dfrac{x - \tan x}{\sin x}$.

19 $\lim\limits_{x \to 0} \dfrac{1}{x^2}\sin^2\left(\dfrac{x}{2}\right)$.

20 $\lim\limits_{x \to 0} \dfrac{\sin 2x}{\sin 5x}$.

Find dy/dx in Problems 21–33.

21 $y = \sin 2x$.

22 $y = \cos 5x$.

23 $y = \sin(x^3)$.

24 $y = \sin^3 2x$.

25 $y = \sin^2(\sqrt{x})$.

26 $y = \dfrac{\cos 2x}{x}$.

27 $y = x^2 \cos(3x^2 - 1)$.

28 $y = \sin^3(x^4)$.

29 $y = (\sin 2x)(\cos 3x)$.

30 $y = \dfrac{x}{\sin 3x}$.

31 $y = \dfrac{\cos 3x}{\sin 5x}$.

32 $y = \sqrt{\cos \sqrt{x}}$.

33 $y = \sin(\cos \sqrt{1 - x^2})$.

34 The points $A(\cos\theta, -\sin\theta)$, $B(1, 0)$, $C(\cos\phi, \sin\phi)$, and $D(\cos(\theta + \phi), \sin(\theta + \phi))$ are shown in Figure 2.53; all are points on the unit circle. Deduce from the fact that line segments AC and BD have equal lengths (because they subtend the same angle $\theta + \phi$) that

$$\cos(\theta + \phi) = \cos\theta \cos\phi - \sin\theta \sin\phi.$$

2.53   Deriving the cosine addition formula (Problem 34).

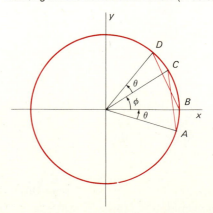

35 (a) Use the triangles shown in Figure 2.54 to deduce that

$$\sin\left(\theta + \frac{\pi}{2}\right) = \cos\theta \quad \text{and} \quad \cos\left(\theta + \frac{\pi}{2}\right) = -\sin\theta.$$

(b) Derive the addition formula for the sine from (a) and the addition formula for the cosine of Problem 34.

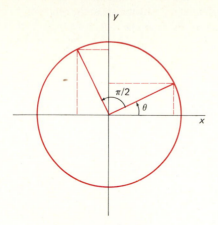

2.54 The identities of Problem 35.

36 Differentiate the identity $\sin 2x = 2 \sin x \cos x$ to discover a formula for $\cos 2x$ in terms of $\sin x$ and $\cos x$.

37 If a projectile is fired from ground level with initial velocity v_0 and inclination angle α, then its horizontal range (ignoring air resistance) is $R = \frac{1}{16}v_0^2 \sin\alpha \cos\alpha$. What value of α maximizes R?

38 Find the largest possible area A of a rectangle inscribed in the unit circle by maximizing A as a function of the angle θ indicated in Figure 2.55.

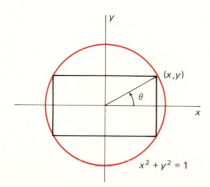

2.55 Rectangle inscribed in the unit circle.

39 An observer on the ground sights an approaching plane that is flying at a constant speed and at an altitude of 20,000 feet. From his point of view, the plane's angle of elevation is increasing at 0.5° per second when the angle is 60°. What is the plane's speed?

40 A plane flying at an altitude of 25,000 feet has an inoperative airspeed indicator. In order to determine his speed, the pilot sights a fixed point on the ground. At the moment when the angle of depression (from the horizontal) is 65°, he observes that this angle is increasing at 1.5° per second. What is the plane's speed?

41 A water trough is to be made from a long strip of tin 6 feet wide by bending up at an angle θ a 2-foot strip on each side, as shown in Figure 2.56. What should be the value of the angle θ to maximize the cross-sectional area, and thus the volume, of the trough?

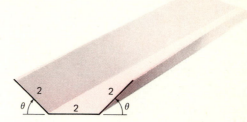

2.56 The water trough of Problem 41.

42 A circular area of radius 20 meters is surrounded by a walkway, and a light is placed atop a lamp post at the center. At what height should the light be placed to most strongly illuminate the walkway? The intensity I of illumination of a surface is given by $I = (k \sin\theta)/D^2$ where D is the distance from the light source to the surface, θ is the angle at which light strikes the surface, and k is a positive constant.

43 Find the minimum possible volume V of a cone in which a sphere of given radius R is inscribed. Minimize V as a function of the angle θ indicated in Figure 2.57.

2.57 Finding the smallest cone containing a fixed sphere.

44 A very long rectangular piece of paper is 20 centimeters wide. The bottom right-hand corner is folded along the crease shown as a heavy line in Figure 2.58, so that the corner

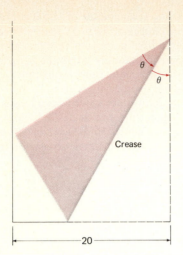

2.58 Fold a piece of paper; make the crease of minimal length.

just touches the left-hand side of the page. How should this be done so that the crease is as short as possible?

45 Find the maximum possible area A of the trapezoid inscribed in a semicircle of radius 1, as shown in Figure 2.59. Begin by expressing A as a function of θ.

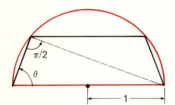

Trapezoid inscribed in a circle.

46 A 6-sided beam is to be cut from a circular log with diameter 30 cm, so that its cross section is as shown in Figure 2.60; the beam is highly symmetrical, with only two different angles, as the figure shows. Show that the area of the cross

2.60 Hexagonal beam cut from a circular log.

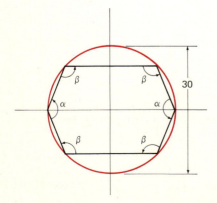

section is maximal when it is a regular hexagon, with equal sides and angles (corresponding to $\alpha = \beta = 2\pi/3$). Note that $\alpha + 2\beta = 2\pi$. (Why?)

47 Consider a circular arc of given length s and with its end points on the x-axis. Show that the area A bounded by this arc and the x-axis is maximal when the circular arc is in the shape of a semicircle. (*Suggestion:* Express A in terms of the angle θ subtended by the arc at the center of the circle, as shown in Figure 2.61. Show that A is maximal when $\theta = \pi$.)

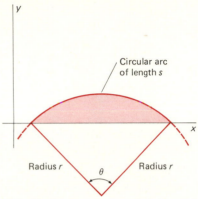

2.61 Maximum area bounded by a circular arc and its chord.

48 A hiker starting at a point P on a straight road wants to get to a forest cabin that is 2 kilometers from the point Q, which is 3 kilometers down the road from P, as shown in Figure 2.62. He can walk 8 kph along the road, but only 3 kph through the forest. How far down the road should he walk first, before setting off through the forest (straight for the cabin)? He wants to minimize the time required to reach the cabin. Use as independent variable the angle between the road and the path he takes through the forest.

2.62 Quickest path to the cabin in the forest.

49 (a) Set your calculator in *degree* mode (not radian mode) and then fill in the values in the following table.

θ	10	5	1	0.5	0.1	0.05	0.01
$(\sin \theta°)/\theta$							

(b) Now use the fact that $\lim\limits_{\theta \to 0} (\sin \theta)/\theta = 1$ to *prove* that

$$\lim_{\theta \to 0} \frac{\sin \theta°}{\theta} \approx 0.01745.$$

Differentiation Formulas

$$D_x(cu) = c\frac{du}{dx}$$

$$D_x(u + v) = \frac{du}{dx} + \frac{dv}{dx}$$

$$D_x(uv) = u\frac{dv}{dx} + v\frac{du}{dx}$$

$$D_x\frac{u}{v} = \frac{vD_x u - uD_x v}{v^2}$$

$$D_x(u^r) = ru^{r-1}\frac{du}{dx}$$

$$D_x \sin u = (\cos u)\frac{du}{dx}$$

$$D_x \cos u = (-\sin u)\frac{du}{dx}$$

$$D_x g(u) = g'(u)\frac{du}{dx}$$

Use the list below as a guide to ideas that you may need to review.

1 The derivative
2 Differential, function, and operator notation for derivatives
3 The binomial formula
4 The power rule
5 Linearity of differentiation
6 The product rule
7 The quotient rule
8 Principle of mathematical induction
9 Local minimum and local maximum
10 $f'(c) = 0$ as a necessary condition for local extremum
11 Critical point
12 Closed interval maximum-minimum method
13 Generalized power rule
14 Differentiation of inverse functions
15 Vertical tangent lines
16 Increasing and decreasing functions
17 The significance of the sign of the derivative
18 Rolle's theorem
19 The Mean Value Theorem
20 Consequences of the Mean Value Theorem
21 The first derivative test
22 Graphing of polynomials
23 The chain rule
24 Implicitly defined functions
25 Implicit differentiation
26 Maximum-minimum problems with constraints
27 Method of auxiliary variables
28 Definitions of the trigonometric functions
29 The fundamental identity of trigonometry
30 Addition, double-angle, and half-angle formulas
31 Radian measure of angles
32 Differentiation of sine and cosine

MISCELLANEOUS PROBLEMS

Find dy/dx in Problems 1–35.

1 $y = x^2 + \dfrac{3}{x^2}$.

2 $y^2 = x^2$.

3 $y = \sqrt{x} + \dfrac{1}{\sqrt[3]{x}}$.

4 $y = (x^2 + 4x)^{5/2}$.

5 $y = (x - 1)^7(3x + 2)^9$.

6 $y = \dfrac{x^4 + x^2}{x^2 + x + 1}$.

7 $y = \left(3x - \dfrac{1}{2x^2}\right)^4$.

8 $y = x^{10} \sin 10x$.

9 $xy = 9$.

10 $y = \sqrt{1/5x^6}$.

11 $y = \dfrac{1}{\sqrt{(x^3 - x)^3}}$.

12 $y = \sqrt[3]{2x + 1}\sqrt[5]{3x - 2}$.

13 $y = \dfrac{1}{1 + u^2}$, where $u = \dfrac{1}{1 + x^2}$.

14 $x^3 = \sin^2 y$.

15 $y = (\sqrt{x} + \sqrt[3]{2x})^{7/3}$.

16 $y = \sqrt{3x^5 - 4x^2}$.

17 $y = \dfrac{u + 1}{u - 1}$, $u = \sqrt{x + 1}$.

18 $y = \sin(2 \cos 3x)$.

19 $x^2 y^2 = x + y$.

20 $y = \sqrt{1 + \sin \sqrt{x}}$.

21 $y = \sqrt{x + \sqrt{2x + \sqrt{3x}}}$.

22 $y = \dfrac{x + \sin x}{x^2 + \cos x}$.

23 $x^{1/3} + y^{1/3} = 4$.

24 $x^3 + y^3 = xy$.

25 $y = (1 + 2u)^3, \quad u = \dfrac{1}{(1 + x)^3}$.

26 $y = \cos^2(\sin^2 x)$.

27 $y = \sqrt{\dfrac{\sin^2 x}{1 + \cos x}}$.

28 $y = (1 + \sqrt{x})^3(1 - 2\sqrt[3]{x})^4$.

29 $y = \dfrac{\cos 2x}{\sqrt{\sin 3x}}$.

30 $x^3 - x^2 y + xy^2 - y^3 = 4$.

31 $y = \sin^3 2x \cos^2 3x$.

32 $y = [1 + (2 + 3x)^{-3/2}]^{2/3}$.

33 $y = \sin^5\left(x + \dfrac{1}{x}\right)$.

34 $\sqrt{x + y} = \sqrt[3]{x - y}$.
35 $y = \cos^3 \sqrt[3]{x^4 + 1}$.

In each of Problems 36–39, find the tangent line to the given curve at the indicated point.

36 $y = \dfrac{x + 1}{x - 1}; \quad (0, -1)$.

37 $x = \sin 2y; \quad (1, \pi/4)$.
38 $x^2 - 3xy + 2y^2 = 0; \quad (2, 1)$.
39 $y^3 = x^2 + x; \quad (0, 0)$.
40 If a hemispherical bowl with radius 1 foot is filled with water to a depth of x inches, the volume of liquid in the bowl turns out to be $V = (\pi/3)(36x^2 - x^3)$ cubic inches. If the water flows out a hole at the bottom of the bowl at the rate of 36π cubic inches per second, how fast is x decreasing when $x = 6$ inches?
41 Falling sand forms a conical sandpile with height which always remains twice its radius r while both are increasing. If sand is falling onto the pile at the rate of 25π cubic feet per minute, how fast is r increasing when $r = 5$ feet?
42 Show that the equation $x^5 + x = 5$ has exactly one real root.

Sketch the graphs of the functions in Problems 43–47. Indicate the local maxima and minima of each function and the intervals on which it is increasing or decreasing.

43 $f(x) = x^2 - 6x + 4$.
44 $f(x) = 2x^3 - 3x^2 - 36x$.
45 $y = 3x^5 - 5x^3 + 60x$.
46 $y = x^{1/2}(3 - x)$.
47 $y = x^{1/3}(1 - x)$.
48 The period T of oscillation (in seconds) of a pendulum of length L (in feet) is given by $T = 2\pi(L/32)^{1/2}$. What is the rate of change of T with respect to L when $L = 4$ (feet)?

49 What is the rate of change of the volume $V = \frac{4}{3}\pi r^3$ of a sphere with respect to its surface area $S = 4\pi r^2$?
50 What is an equation for the straight line through $(1, 0)$ that is tangent to the graph of $h(x) = x + 1/x$ at a point in the first quadrant?
51 A rocket is launched vertically upward from a point two miles west of an observer on the ground. What is the rocket's speed when the angle of elevation (from the horizontal) of the observer's line of sight to the rocket is $50°$ and is increasing at $5°$ per second?
52 An oil field containing 20 wells has been producing 4000 barrels of oil daily. Suppose that, for each new well that is drilled, the daily production of each well decreases by 5 barrels. How many new wells should be drilled in order to maximize the total daily production of the oil field?
53 A triangle is inscribed in a circle of radius R with one side of the triangle coincident with a diameter of the circle. What, in terms of R, is the maximum possible area of such a triangle?
54 Five rectangular pieces of sheet metal measure 210 by 336 centimeters each. Equal squares are to be cut from all their corners, and the resulting 5 cross-shaped pieces of metal are to be folded and welded to form 5 boxes without tops. The 20 little squares left over are to be assembled in groups of 4 into 5 larger squares, and these 5 larger squares are to be assembled into a cubical box with no top. What is the maximum possible total volume of the 6 boxes that can be so obtained?
55 A mass of clay of volume V is formed into two spheres. For what distribution of clay is the total surface area of the two spheres a maximum? A minimum?
56 A right triangle has legs of lengths 3 and 4 meters. What is the maximum possible area of a rectangle inscribed in the triangle in the "obvious way"—that is, with one corner at the triangle's right angle, the adjacent sides of the rectangle lying on the triangle's legs, and the opposite corner on the triangle's hypotenuse?
57 What is the maximum possible volume of a right circular cone inscribed in a sphere of radius R?
58 A farmer has 400 feet of fencing with which to build a rectangular corral. He will use part or even all of an existing straight wall 100 feet long as part of the perimeter of the corral. What is the maximum area that he can enclose?
59 Suppose that we need to manufacture a cylindrical pot, without top, with a volume of 1 cubic foot. The cylindrical part of the pot is to be made of aluminum, the bottom of copper. Copper is 5 times as expensive as aluminum. What dimensions will minimize the cost of the pot?
60 Three sides of a trapezoid have side length L, a constant. What should be the length of the fourth side if the trapezoid is to have maximum area?
61 A rectangle of fixed perimeter P is rotated about one of its sides to generate a cylinder. Of all such possible rectangles, which generates the cylinder of greatest volume?

62 A small right circular cone is inscribed in a larger one, as shown in Figure 2.63. The larger cone has fixed radius R and fixed altitude H. What is the largest fraction of the volume of the larger cone that the smaller one can occupy?

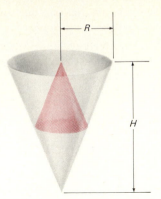

2.63 A small cone inscribed in a larger one.

63 A wall 8 feet high stands 4 feet away from a tall building. What is the length of the shortest ladder that will lean over the wall and touch the building? Choose as independent variable the angle that the ladder makes with the ground.

64 Show by the method of auxiliary variables that, if $Q(x, y)$ is the point of the curve $y = f(x)$ that is closest to the point $P(x_0, y_0)$ not on the curve, then

$$f'(x) = -\frac{x - x_0}{y - y_0}$$

at Q. Conclude that the segment PQ is perpendicular to the tangent line to the curve at Q. (*Suggestion:* Minimize the square of the distance PQ.)

65 Use the result of Problem 64 to show that the minimum distance from the point (x_0, y_0) to a point of the straight line $Ax + By + C = 0$ is

$$\frac{|Ax_0 + By_0 + C|}{\sqrt{A^2 + B^2}}.$$

66 A racetrack is to be built in the shape of two parallel and equal straightaways connected by semicircles on each end. The length of the track, one lap, is to be exactly 5 kilometers. What should be its design to maximize the rectangular area within it, indicated by shading in Figure 2.64?

2.64 Design the racetrack to maximize the shaded area.

67 Two towns are located near the straight shore of a lake. Their nearest distances to points on the shore are 1 mile and 2 miles, respectively, and these points on the shore are 6 miles apart. Where should a fishing pier be located to minimize the total amount of paving necessary to build a straight road from each town to the pier?

68 A hiker finds herself in a forest 2 kilometers from a long straight road. She wants to walk to her cabin 10 km away in the forest, and it's also 2 kilometers from the road. She can walk 8 kph along the road, but only 3 kph through the forest. So she decides to walk first to the road, then along the road, and finally through the forest to the cabin (see Figure 2.65). What should be the angle θ shown in the figure in order to minimize the time required for the hiker to reach her cabin? How much time is saved in comparison with the straight route through the forest?

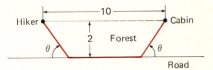

2.65 The hiker's quickest path to the cabin.

69 When an arrow is shot from the origin with initial velocity v_0 and initial angle of inclination α (from the horizontal x-axis, which represents the ground), then its trajectory is the curve

$$y = mx - \frac{16}{v_0^2}(1 + m^2)x^2$$

where $m = \tan \alpha$. (a) Find the maximum height reached by the arrow (in terms of m and v_0). (b) For what m (and hence, for what α) does the arrow travel the greatest horizontal distance?

70 A projectile is fired with initial velocity v_0 and angle of elevation θ from the base of a plane inclined at 45° from the horizontal, as shown in Figure 2.66. The range of the

2.66 A projectile fired uphill.

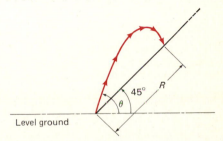

projectile, measured up this slope, is given by the formula

$$R = \frac{v_0^2\sqrt{2}}{16}(\cos\theta\sin\theta - \cos^2\theta).$$

What value of θ maximizes R?

71 The amount of water that flows over the spillway of a certain dam is given by $V = bx(2g[H - x])^{1/2}$, where V is the number of cubic feet of water that flow across per second, b is the width of the spillway, x is the depth of the water as it flows over the spillway, g (the acceleration of gravity) is 32 feet per second per second, and H is the water head in feet. Under normal circumstances, b is fixed, and H remains relatively constant over time intervals that are not too long. Assuming that they remain fixed, what value of x maximizes V?

Applications of Derivatives and Antiderivatives

3

Introduction

In this chapter we exploit the differentiation techniques of Chapter 2 to illustrate the diverse applications of the derivative concept. In Section 1-5 we saw that the derivative $f'(x)$ could be interpreted as the slope of the tangent to $y = f(x)$ at $(x, f(x))$. The maximum-minimum and curve-sketching problems of Chapter 2 are applications of this interpretation.

The applications of Chapter 3 range from solving equations that are not susceptible to elementary algebra (Section 3-4) to finding the size of a "black hole" (Section 3-8). Most scientific applications of derivatives stem from the rate of change interpretation that we introduced in Section 1-6. From this point of view, $f'(x)$ is the (instantaneous) rate of change of the dependent variable y or $f(x)$ with respect to the independent variable x. For example, the velocity v of a particle moving along a straight line is the rate of change of its position $s = f(t)$; that is, $v = ds/dt$. Its acceleration a is the rate of change of its velocity: $a = dv/dt$.

In Section 3-7 we introduce the operation of **antidifferentiation**—the inverse of differentiation. If $g'(x) = f(x)$, then the function g is called an **antiderivative** of f. The importance of antiderivatives results partly from the fact that scientific laws often specify the *rates of change* of quantities. The quantities themselves are then found by antidifferentiation. For example, in Section 3-8 we begin with the given acceleration of a moving particle. From this we find its velocity and then its position function, using successive antidifferentiation.

The methods of calculus are important not only in engineering and the physical sciences but in the biological, behavioral, and economic sciences as well. In Section 3-9 we discuss the use of elementary calculus in the analysis of business management decisions.

Increments and Differentials

Sometimes we need a quick and simple estimate of the change in $f(x)$ that results from a given change in x. Write y for $f(x)$, and suppose that the change in the independent variable x is the *increment* Δx, so that x changes from its original value x to the new value $x + \Delta x$. The actual change in the value of y is the *increment* Δy, computed by subtracting the "old" value of y from its "new" value:

$$\Delta y = f(x + \Delta x) - f(x). \tag{1}$$

The increments Δx and Δy are represented geometrically in Figure 3.1.

Now we compare the actual increment Δy with the change that *would* occur in y if it continued to change at the *fixed* rate $f'(x)$ as x changes to $x + \Delta x$. This hypothetical change in y is the **differential**

$$dy = f'(x)\,\Delta x. \tag{2}$$

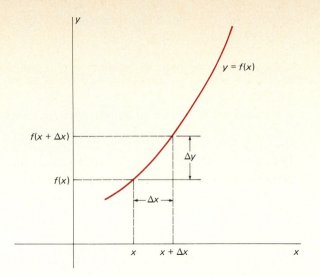

3.1 The increments Δx and Δy.

As Figure 3.2 shows, dy is the change in height of a point that moves along the tangent line at the point $(x, f(x))$, rather than along the curve $y = f(x)$.

Think of x as fixed. Then (2) shows that the differential dy is a *linear* function of the increment Δx. For this reason, dy is called the **linear approximation** to the actual increment Δy. We can approximate $f(x + \Delta x)$ by writing dy in place of Δy:

$$f(x + \Delta x) = y + \Delta y \approx y + dy.$$

Since $y = f(x)$ and $dy = f'(x)\,\Delta x$, this gives the **linear approximation formula**

$$f(x + \Delta x) \approx f(x) + f'(x)\,\Delta x. \tag{3}$$

The idea is that this last approximation is a "good" one, at least when Δx is relatively small.

3.2 The estimate dy of the actual increment Δy.

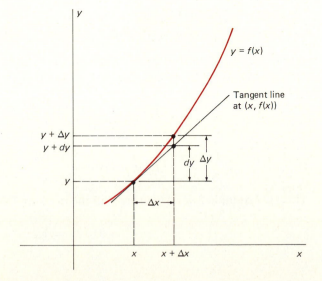

EXAMPLE 1   Use the linear approximation formula to estimate $(122)^{2/3}$. Note that $(125)^{2/3} = 25$.

Solution   Since we plan to estimate a particular value of $x^{2/3}$, we take $f(x) = x^{2/3}$. Thus, given the fact that $f(125) = 25$, we seek an estimate of $f(122)$.

Now $f'(x) = \frac{2}{3}x^{-1/3}$, so we also have the exact value

$$f'(125) = \tfrac{2}{3}(125)^{-1/3} = \tfrac{2}{15}.$$

We take $x = 125$ and $\Delta x = -3$ in the linear approximation formula (3) and find that

$$(122)^{2/3} = f(122) \approx f(125) + (-3)f'(125)$$
$$= 25 + (-3)(\tfrac{2}{15});$$

that is, $(122)^{2/3} \approx 24.6$. The actual value of $(122)^{2/3}$ is about 24.59838 (the digits given are correct), so Formula (3) gives us a rather good estimate.

EXAMPLE 2   A hemispherical bowl of radius 10 inches is filled with water to a depth of x inches. The volume of water in the bowl is given by the formula

$$V = \frac{\pi}{3}(30x^2 - x^3)$$

(V is in cubic inches). Suppose that you *measure* the depth of water in the bowl, and you find it to be 5 inches with a maximum possible measurement error of $\frac{1}{16}$ inch. Estimate the maximum error in the calculated volume of water in the bowl.

Solution   The error ΔV in the calculated volume V caused by an error Δx in the measured depth x is approximately equal to the differential

$$dV = \frac{dV}{dx}\,\Delta x = \frac{\pi}{3}(60x - 3x^2)\,\Delta x = \pi(20x - x^2)\,\Delta x.$$

We take $x = 5$ and $\Delta x = \pm\frac{1}{16}$, and obtain

$$dV = \pi[(20)(5) - 5^2](\pm\tfrac{1}{16}) \approx \pm 14.73$$

cubic inches. With $x = 5$, the formula for V gives $V(5) \approx 654.50$ cubic inches, but we see now that this may be in error by as much as 15 cubic inches either way.

The **relative error** in an estimated or measured value is defined by

$$\text{relative error} = \frac{\text{error}}{\text{value}}. \tag{4}$$

Thus, in Example 2, a relative error in the measured depth x of

$$\frac{\Delta x}{x} = \frac{\frac{1}{16}}{5} = 0.0125 = 1.25\%$$

leads to a relative error in the estimated volume of

$$\frac{dV}{V} = \frac{14.73}{654.50} = 0.0225 = 2.25\%.$$

The relationship between these two relative errors is of some interest. The formulas for dV and V above give

$$\frac{dV}{V} = \frac{\pi(20x - x^2)\,\Delta x}{\frac{1}{3}\pi(30x^2 - x^3)} = \frac{3(20 - x)}{30 - x} \cdot \frac{\Delta x}{x}.$$

In particular, when $x = 5$, this gives

$$\frac{dV}{V} = (1.80)\frac{\Delta x}{x}.$$

Hence, in order to estimate the volume of water in the bowl with a relative error of at most 0.5%, we would need to measure the depth with a relative error of at most $(0.5\%)/(1.8)$, or about 0.3%.

Now we take up the question of how closely the differential dy approximates the actual increment Δy. It is apparent in Figure 3.2 that the smaller is Δx, the closer are the corresponding points on the curve $y = f(x)$ and its tangent line. Since the difference in the heights of two such points is the value of $\Delta y - dy$ determined by a particular choice of Δx, we conclude that $\Delta y - dy \to 0$ as $\Delta x \to 0$.

But even more is true: As $\Delta x \to 0$, the difference $\Delta y - dy$ is small *even in comparison with* Δx. For

$$\frac{\Delta y - dy}{\Delta x} = \frac{f(x + \Delta x) - f(x) - f'(x)\,\Delta x}{\Delta x}$$

$$= \frac{f(x + \Delta x) - f(x)}{\Delta x} - f'(x);$$

that is,

$$\frac{\Delta y - dy}{\Delta x} = \varepsilon(\Delta x) \qquad (5)$$

where, by the definition of the derivative $f'(x)$, we see that $\varepsilon(\Delta x)$ is a function of Δx that approaches zero as $\Delta x \to 0$. We multiply both sides of (5) by Δx, write ε for $\varepsilon(\Delta x)$, and obtain

$$\Delta y = dy + \varepsilon\,\Delta x \qquad (6)$$

where $\varepsilon \to 0$ as $\Delta x \to 0$. If Δx is "very small," so that ε is also "very small," we might well describe the product $\varepsilon\,\Delta x = \Delta y - dy$ as "very *very* small."

EXAMPLE 3 Suppose that $y = x^3$. Verify that $\varepsilon = (\Delta y - dy)/\Delta x$ approaches zero as $\Delta x \to 0$.

Solution Simple computations give

$$\Delta y = (x + \Delta x)^3 - x^3 = 3x^2\,\Delta x + 3x(\Delta x)^2 + (\Delta x)^3$$

and $dy = 3x^2 \, \Delta x$. Hence

$$\varepsilon = \frac{\Delta y - dy}{\Delta x} = \frac{3x(\Delta x)^2 + (\Delta x)^3}{\Delta x} = 3x \, \Delta x + (\Delta x)^2,$$

which obviously approaches zero as $\Delta x \to 0$.

DIFFERENTIALS

The linear approximation formula (3) is often written with dx in place of Δx, in the form

$$f(x + dx) \approx f(x) + f'(x) \, dx. \tag{7}$$

In this case dx is an independent variable, called the **differential** of x, while x is fixed. Thus the differentials of x and y are defined by

$$dx = \Delta x \quad \text{and} \quad dy = f'(x) \, \Delta x = f'(x) \, dx. \tag{8}$$

With this definition it follows immediately that

$$\frac{dy}{dx} = \frac{f'(x) \, dx}{dx} = f'(x),$$

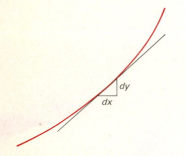

3.3 The slope of the tangent line as the ratio of the infinitesimals *dy* and *dx*.

in agreement with the notation we have been using. Indeed, Leibniz originated differential notation by visualizing infinitesimal increments dx and dy as in Figure 3.3, with their ratio dy/dx the slope of the tangent line.

Differential notation also provides us with a convenient way to write derivative formulas. For suppose that $z = g(u)$, so that $dz = g'(u) \, du$. For particular choices of the function g, we get the formulas

$$d(u^n) = nu^{n-1} \, du, \tag{9}$$

$$d(\sin u) = (\cos u) \, du, \tag{10}$$

and so on. This allows us to write differentiation rules in "differential form," without the need to identify the independent variable. The sum, product, and quotient rules take the forms

$$d(u + v) = du + dv, \tag{11}$$

$$d(uv) = u \, dv + v \, du, \quad \text{and} \tag{12}$$

$$d\left(\frac{u}{v}\right) = \frac{v \, du - u \, dv}{v^2}. \tag{13}$$

If $u = f(x)$ and $z = g(u)$, we may substitute $du = f'(x) \, dx$ into the formula $dz = g'(u) \, du$. This gives

$$dz = g'(f(x)) \cdot f'(x) \, dx. \tag{14}$$

This is the differential form of the chain rule

$$Dg(f(x)) = g'(f(x))f'(x).$$

Thus the chain rule appears here as though it were the result of mechanical manipulations of the differential notation. This compatibility with the chain rule is one reason for the extraordinary usefulness of differential notation in calculus.

In Problems 1–5, calculate Δy and dy with the indicated values of x and Δx.

1 $f(x) = 3x + 4$, $x = 2$, $\Delta x = 1$.
2 $f(x) = x^2$, $x = 10$, $\Delta x = 1$.
3 $f(x) = \sqrt{x}$, $x = 10$, $\Delta x = 1$.
4 $f(x) = x^{3/2}$, $x = 4$, $\Delta x = 0.1$.

5 $f(x) = \dfrac{1}{x^2 + 1}$, $x = 0$, $\Delta x = -0.2$.

In Problems 6–12, write dy in terms of x and dx.

6 $y = (x^2 + 1)^{100}$.   **7** $y = (4x - x^2)^{3/2}$.

8 $y = 8x^3(x^2 + 9)^{1/2}$.   **9** $y = \dfrac{x + 1}{x - 1}$.

10 $y = \sin(x^2)$.   **11** $y = x^2 \cos\sqrt{x}$.

12 $y = \dfrac{x}{\sin 2x}$.

In Problems 13–17, compute the differential of each side of the given equation, regarding x and y as *dependent* variables (as if both were functions of some third unspecified variable). Then solve for dy/dx.

13 $x^2 + y^2 = 1$.   **14** $x^{2/3} + y^{2/3} = 4$.
15 $x^3 + y^3 = 3xy$.   **16** $(x^2 + y^2)^2 = x^2 - y^2$.
17 $x \sin y = 1$.

In Problems 18–26, estimate the indicated number by linear approximation, as in Example 1.

18 $\sqrt{98}$.   **19** $\sqrt[3]{1005}$.
20 $\sqrt[3]{62}$.   **21** $(26)^{3/2}$.
22 $\sqrt[5]{30}$.   **23** $\sqrt[4]{17}$.
24 $\sqrt[10]{1000}$.   **25** $\sin(0.1)$.
26 $\cos(0.1)$.

In Problems 27–30, find Δy and dy, and then verify that $\varepsilon = (\Delta y - dy)/\Delta x \to 0$ as $\Delta x \to 0$.

27 $y = x^2 - x$.   **28** $y = x^3 + 2x$.

29 $y = \dfrac{1}{x}$.   **30** $y = \dfrac{x}{x + 1}$.

In Problems 31–35, estimate by linear approximation the change in the indicated quantity.

31 The volume $V = s^3$ of a cube, if its side length s is increased from 5 inches to 5.1 inches.
32 The area of a circle, if its radius is decreased from 10 to 9.8 cm.
33 The volume of a sphere, if its radius is increased from 5 to 5.1 inches.
34 The volume $V = 1000/P$ cubic inches of a gas, if the pressure P is decreased from 100 to 99 pounds per square inch.
35 The period of oscillation $T = 2\pi\sqrt{L/32}$ of a pendulum, if its length L is increased from 2 feet to 2 feet plus 1 inch.
36 The equatorial radius of the earth is approximately 3960 miles. Suppose that a wire is wrapped tightly around the earth at the equator. How much must this wire be lengthened if it is to be strung on poles 10 feet above the ground, all the way around the earth?
37 The radius of a spherical ball is measured as 10 inches, with a maximum error of $\frac{1}{16}$ inch. What is the maximum resulting error in its calculated volume?
38 With what accuracy must the radius of the ball of Problem 37 be measured to insure an error of at most 1 cubic inch in its calculated volume?
39 The radius of a hemispherical dome is measured as 100 meters with a maximum error of 1 cm. What is the maximum resulting error in its calculated surface area?
40 With what accuracy must the radius of the dome of Problem 39 be measured in order to have a percentage error of at most 0.01% in its calculated surface area?

3-3 Related Rates

A **related rates problem** involves two or more quantities that vary with time and an equation that expresses some relationship between them. Typically, the values of these quantities at some instant are given, together with all their time rates of change but one. The problem is usually to find the time rate of change that is *not* given, at some instant specified in the problem. One common method for solving such a problem is to begin by implicit differentiation of the equation that relates the given quantities.

For example, suppose that x and y are each functions of time such that

$$x^2 + y^2 = a^2 \qquad (a \text{ is a constant}).$$

Differentiate both sides of this equation *with respect to time t*. This produces the equation

$$2x\frac{dx}{dt} + 2y\frac{dy}{dt} = 0.$$

If the values of x, y, and dx/dt at a certain instant t are known, then this last equation can be solved for the value of dy/dt at time t. Note that it is *not* necessary to know x and y as functions of t. Indeed, it is typical for a related rates problem to contain insufficient information to express x and y as functions of t.

EXAMPLE 1   A rocket is launched vertically and is tracked by a radar station, which is located on the ground 3 miles from the launch site. What is the vertical speed of the rocket at the instant when its distance from the radar station is 5 miles and this distance is increasing at the rate of 5000 miles per hour?

Solution   Figure 3.4 illustrates this situation. We denote the altitude of the rocket by y and its distance from the radar station by z. We are given that

$$\frac{dz}{dt} = 5000 \quad (\text{mph}) \quad \text{when} \quad z = 5 \quad (\text{miles}).$$

We want to find the value of dy/dt at this instant.

We apply the Pythagorean theorem to the right triangle in the figure and obtain

$$y^2 + 9 = z^2$$

as a relation between y and z. From this we see that $y = 4$ when $z = 5$. Implicit differentiation now gives

$$2y\frac{dy}{dt} = 2z\frac{dz}{dt}.$$

We substitute $y = 4$, $z = 5$, and $dz/dt = 5000$. We thus find that

$$\frac{dy}{dt} = 6250 \qquad (\text{mph})$$

at the instant in question.

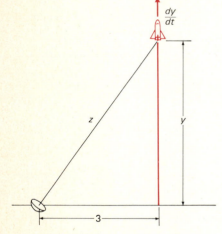

3.4   The rocket of Example 1.

Example 1 illustrates the following steps in the solution of a typical related rates problem of the sort that involves a geometrical situation.

1 Draw a diagram and label as variables the various quantities involved in the problem.
2 Record the values of the variables and their rates of change, as given in the problem.
3 Read an equation that relates the important variables of the problem from the diagram.
4 Differentiate this equation implicitly with respect to time t.

5 Substitute the given numerical data into the resulting equation, and then solve for the unknown.

WARNING The most common error to be avoided is the premature substitution of the given data, before rather than after implicit differentiation. If, in Example 1, we had substituted $z = 5$ to begin with, our equation would have been $y^2 + 9 = 25$, and implicit differentiation would have given the absurd result that $dy/dt = 0$.

In the following example, we use similar triangles (rather than the Pythagorean theorem) to discover the needed relation between the variables.

EXAMPLE 2 A man 6 feet tall walks with a speed of 8 feet per second away from a street light atop an 18-foot pole. How fast is the tip of his shadow moving along the ground when he is 100 feet from the light pole?

Solution Let x be the man's distance from the pole and z the distance of the tip of his shadow from the base of the pole. Note that, though x and z are functions of t, we do *not* attempt to obtain explicit formulas for either.

We are given that $dx/dt = +8$ (ft/sec), and we want to find dz/dt when $x = 100$ (feet). We equate ratios of corresponding sides of the two similar triangles of Figure 3.5 and find that

$$\frac{z}{18} = \frac{z - x}{6}.$$

Thus

$$2z = 3x.$$

Implicit differentiation now gives

$$2\frac{dz}{dt} = 3\frac{dx}{dt}.$$

We substitute $dx/dt = 8$, and find that

$$\frac{dz}{dt} = \frac{3}{2} \cdot \frac{dx}{dt} = \frac{3}{2} \cdot (8) = 12.$$

So the tip of the man's shadow is moving at 12 feet per second.

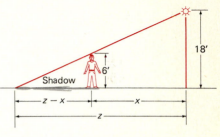

3.5 The moving shadow.

Example 2 is somewhat unusual in that the answer is independent of the man's distance from the light pole—the given value $x = 100$ is superfluous. The next example is a related rates problem with two relationships between the variables, not quite so unusual.

EXAMPLE 3 Two radar stations at A and B, with B 6 miles east of A, are tracking a ship. At a certain instant, the ship is 5 miles from A, and this distance is increasing at the rate of 28 miles per hour. At the same instant the ship is also 5 miles from B, while this distance is increasing at only 4 miles per hour. Where is the ship, how fast is it moving, and in what direction is it moving?

3.6 Radar stations tracking a ship.

Solution With the distances indicated in Figure 3.6, we find—again with the aid of the Pythagorean theorem—that

$$x^2 + y^2 = u^2 \quad \text{and} \quad (6 - x)^2 + y^2 = v^2.$$

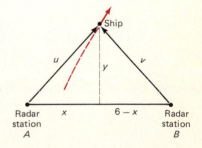

We are given that $u = v = 5$, that $du/dt = 28$, and that $dv/dt = 4$ at the instant in question. Since the ship is equally distant from A and B, it is clear that $x = 3$. (Why?) So $y = 4$. Thus the ship is 3 miles east and 4 miles north of A.

We differentiate implicitly the two equations above, and we obtain

$$2x\frac{dx}{dt} + 2y\frac{dy}{dt} = 2u\frac{du}{dt}$$

and

$$-2(6-x)\frac{dx}{dt} + 2y\frac{dy}{dt} = 2v\frac{dv}{dt}.$$

When we substitute the numerical data given or deduced, we find that

$$3\frac{dx}{dt} + 4\frac{dy}{dt} = 140$$

and

$$-3\frac{dx}{dt} + 4\frac{dy}{dt} = 20.$$

These equations are easy to solve; $dx/dt = dy/dt = 20$. Therefore the ship is sailing northeast at a speed of $\sqrt{(20)^2 + (20)^2} = 20\sqrt{2}$ miles per hour ... *if* the figure is correct! For a mirror along the line AB will reflect *another* ship, 3 miles east and 4 miles *south* of A, sailing *southeast* at a speed of $20\sqrt{2}$ miles per hour.

The lesson? Figures are important, helpful, often essential—and potentially misleading. Try to avoid taking anything for granted when you draw a figure. In this example there would be no real problem, for both radar stations would be able to tell whether the ship was generally to the north or to the south.

3-3 PROBLEMS

In each of Problems 1–5, a relation between x and y is given, together with the values at a particular instant of three of the four quantities x, y, dx/dt, and dy/dt. Find the value of the one not given.

1 $x^2 + y^2 = 13$; $x = 3$, $\dfrac{dx}{dt} = -4$, $\dfrac{dy}{dt} = 6$.

2 $y^2 = x^3 + 17$; $x = 2$, $y = -5$, $\dfrac{dx}{dt} = 10$.

3 $x^3 + 3xy + y^3 = 15$; $x = 1$, $y = 2$, $\dfrac{dy}{dt} = -3$.

4 $\sin^2 x + \cos^2 y = 1$; $x = y = \dfrac{\pi}{4}$, $\dfrac{dx}{dt} = 0.1$.

5 $2x \sin y = y \cos x$; $x = \dfrac{\pi}{6}$, $y = \dfrac{\pi}{3}$, $\dfrac{dy}{dt} = -1$.

6 Water is being collected from a block of ice with a square base. The water is produced because the ice is melting in such a way that the edge of the base of the block is decreasing at 2 inches per hour, while the block's height is decreasing at 3 inches per hour. What is the rate of flow of water into the collecting pan when the base has edge length 20 inches and the block's height is 15 inches?

7 Sand being emptied from a hopper at the rate of 10 cubic feet per second forms a conical pile whose height is always twice its radius. At what rate is the radius of the pile increasing when its height is 5 feet?

8 Suppose that water is being emptied from a spherical tank of radius 10 feet. If the depth of water in the tank is 5 feet and is decreasing at the rate of 3 feet per second, at what rate is the radius of the top surface of the water decreasing?

9 A circular oil slick of uniform thickness is caused by a spill of one cubic meter of oil. The thickness of the oil slick is decreasing at the rate of 0.1 centimeter per hour. At what rate is the radius of the slick increasing when it is 8 meters?

10 Suppose that an ostrich 5 feet tall is walking at a speed of 4 feet per second directly toward a street light 10 feet high. How fast is the tip of the ostrich's shadow moving along the ground? At what rate is the ostrich's shadow decreasing in length?

11 A 10-foot ladder is leaning against a wall. The bottom of the ladder begins to slide away from the wall at a speed of 1 mile per hour. How fast is the top of the ladder moving when it is 4 feet above the ground?

12 Two ships are sailing toward the same small island. One ship, the *Pinta*, is east of the island and is sailing due west at 15 miles per hour. The other ship, the *Niña*, is north of the island and is sailing due south at 20 miles per hour. At a certain time the *Pinta* is 30 miles from the island and the *Niña* is 40 miles from it. Are the two ships drawing closer together or farther apart at that time? At what rate?

13 At time $t = 0$, a single-engine military jet is flying due east at 12 miles per minute. At the same altitude and 208 miles directly ahead of it, still at time $t = 0$, is a commercial jet, flying due north at 8 miles per minute. When are the two planes closest to each other? What is the minimum distance between them?

14 A ship with a long anchor chain is anchored in 11 fathoms of water. The anchor chain is being wound in at the rate of 10 fathoms per minute, causing the ship to move toward the spot directly over where the anchor is resting on the ocean's bottom. The hawsehole—the point of contact between ship and chain—is located 1 fathom above the waterline. At what speed is the ship moving when there are exactly 13 fathoms of chain still out?

15 A water tank is in the shape of a cone with vertical axis and vertex downward. The tank's radius is 3 feet and the tank is 5 feet high. At first the tank is full of water, but at time $t = 0$ (in seconds), a small hole at the vertex is opened, and the water begins to drain. When the height of water in the tank has dropped to 3 feet, the water is flowing out at 0.02 cubic feet per second. At what rate, in feet per second, is the water level dropping then?

16 When a spherical tank with a radius of 10 feet contains water with a maximum depth of y feet, the volume of water in the tank is $V = (\pi/3)(30y^2 - y^3)$. If the tank is being filled at the rate of 200 gallons per minute, how fast is the water level rising when $y = 5$ (feet)? (*Note:* One gallon is approximately 0.1337 cubic feet.)

17 A water bucket is shaped like the frustum of a cone with height 2 feet and lower and upper base radii 6 inches and 12 inches, respectively. Water is leaking from the bucket at 10 cubic inches per minute. At what rate is the water level falling when the depth of water in the bucket is 1 foot? (*Note:* The volume of a conical frustum with height h and base radii a and b is

$$V = \frac{\pi}{3}h(a^2 + ab + b^2).)$$

18 Suppose that the radar stations at A and B of Example 3 are now 12.6 miles apart. At a certain instant, the ship is 10.4 miles from A, and its distance from A is increasing at 19.2 miles per hour. At the same instant, its distance from B is 5 miles, and this distance is decreasing at 0.6 miles per hour. Find the location, speed, and direction of motion of the ship.

19 An airplane climbing at an angle of $45°$ passes directly over a ground radar station at an altitude of 1 mile. A later reading shows that the distance from the radar station to the plane is 5 miles and is increasing at 7 miles per minute. What is the speed (in mph) of the plane?

20 The water tank of Problem 16 is completely full when a plug at its bottom is pulled. According to Torricelli's law, the water drains in such a way that

$$\frac{dV}{dt} = -k\sqrt{y},$$

where k is a positive empirical constant.
(a) Find dy/dt as a function of y.
(b) Find the depth of water when the water level is falling the *least* rapidly. (You will need to compute the derivative of dy/dt with respect to y.)

The quadratic formula is one tool for obtaining an exact solution of any second degree equation $ax^2 + bx + c = 0$. Formulas are known for the exact solutions of third and fourth degree equations, but they are so complicated that they are rarely used. And it can be proved that there is *no* general formula possible for giving the roots of an arbitrary fifth degree (or higher) polynomial equation in terms of algebraic operations. Thus the exact solution (for all its roots) of an equation such as

$$x^5 - 3x^3 + x^2 - 23x + 19 = 0$$

may be quite difficult, or even—as a practical matter—impossible.

Successive Approximations and Newton's Method

But calculus gives us a way to find accurate *approximations* to the solutions of quite general equations. These "approximation methods" are applicable to a wide variety of algebraic equations, and to equations involving trigonometric functions and the logarithmic and exponential functions of Chapter 6.

A typical approximation method involves starting with an initial "guess" or estimate x_0 of the desired solution x_*. Next, the equation and x_0 are used to construct a better approximation x_1. Repetition of this second step, which is known as **iteration,** gives an infinite sequence

$$x_0, x_1, x_2, x_3, \ldots, x_n, x_{n+1}, \ldots$$

of generally improving approximations to the actual root x_*. In this way we get arbitrarily accurate approximations of x_* provided that this sequence of approximations (denoted for brevity by $\{x_n\}_{n=0}^{\infty}$) *converges* to x_* in the sense of the following definition.

Definition *Limit of a Sequence*

We say that the sequence $\{x_n\}_{n=0}^{\infty}$ **converges** to x_* or has **limit** x_*, and we write

$$\lim_{n \to \infty} x_n = x_*, \tag{1}$$

provided that x_n can be made as close as we please to x_* by choosing n sufficiently large. That is, given any $\varepsilon > 0$, there exists an integer N such that

$$n \geqq N \quad \text{implies} \quad |x_n - x_*| < \varepsilon. \tag{2}$$

For example, with $\varepsilon = 1/10^{k+1}$, Condition (2) may be thought of as saying this: x_n and x_* agree to k decimal places if $n \geqq N$. If this accuracy suffices for our purposes, then x_n is a satisfactory approximation to the actual root x_*.

LINEAR INTERPOLATION

Any approximation method requires an initial approximation x_0 to the solution of the equation $f(x) = 0$ that we wish to solve. If a and b are two points such that $f(a)$ and $f(b)$ differ in sign, then we know from the Intermediate Value Property (Section 1-9) that the equation $f(x) = 0$ has at least one solution in (a, b) (assuming that f is continuous on $[a, b]$). If, in addition, either $f'(x) > 0$ on (a, b) or $f'(x) < 0$ on (a, b), then either f is increasing on $[a, b]$ or f is decreasing there, and it follows that the equation $f(x) = 0$ has *exactly one* solution x_* in (a, b).

A common way to obtain an initial approximation x_0 to x_* is **linear interpolation.** In this method we choose x_0 as the point where the line segment joining $(a, f(a))$ and $(b, f(b))$ meets the x-axis. From the similar triangles in Figure 3.7, we obtain the proportion

$$\frac{x_0 - a}{|f(a)|} = \frac{b - x_0}{|f(b)|}. \tag{3}$$

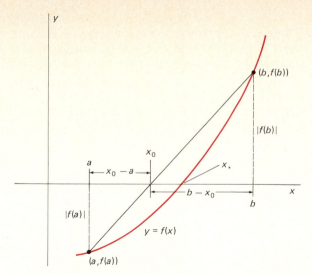

3.7 Obtaining x_0 by linear interpolation.

It is then easy to solve this equation for

$$x_0 = \frac{a|f(b)| + b|f(a)|}{|f(a)| + |f(b)|}.$$ (4)

In practice it will probably be more convenient to set up the proportion in (3) than to remember Formula (4).

EXAMPLE 1   A large cork ball has radius 1 foot, and its density is $\frac{1}{4}$ that of water. Archimedes' law of buoyancy implies that, when the ball floats in water, $\frac{1}{4}$ of its volume ($\frac{1}{4}$ of $4\pi/3$, which is $\pi/3$) is submerged. Find an equation determining the depth x to which the cork ball sinks, and use linear interpolation to make an initial approximation to its solution.

Solution   The volume of a spherical segment of height x and radius r is given by the known formula

$$V = \frac{\pi x}{6} (3r^2 + x^2).$$

From the right triangle of Figure 3.8, we see that

$$r^2 + (1 - x)^2 = 1^2,$$

so that

$$r^2 = 2x - x^2.$$

We combine Archimedes' law and the above volume formula and thus obtain

$$\frac{\pi x^2}{3} (3 - x) = \frac{\pi}{3}.$$

So we must solve the equation

$$f(x) = x^3 - 3x^2 + 1 = 0.$$ (5)

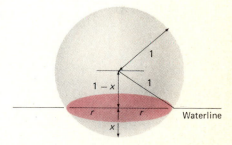

3.8 The floating cork ball.

This last equation has no rational solutions, so we cannot expect to solve it by entirely elementary methods. We note, however, that $f(0) = 1$ and $f(1) = -1$, so there is a root between 0 and 1. This is the solution we seek, since it is physically evident that the desired depth is between 0 and 1. With $a = 0$, $b = 1$, $f(a) = 1$, and $f(b) = -1$, the proportion in (3) is

$$\frac{x_0 - 0}{1} = \frac{1 - x_0}{1},$$

with solution $x_0 = \frac{1}{2}$. This will be our initial approximation to the root of Equation (5) that lies in $(0, 1)$.

THE METHOD OF REPEATED SUBSTITUTION

To generate a sequence of improving approximations to the actual root x_* of (5), we proceed as follows. First, we rewrite (5) in the form $3x^2 = x^3 + 1$, or

$$x = \sqrt{\frac{x^3 + 1}{3}}. \qquad (6)$$

We obtain x_1 by substituting x_0 in the right-hand side of (6):

$$x_1 = \sqrt{\frac{x_0^3 + 1}{3}} \approx 0.6124.$$

Then we find x_2 by substituting x_1 in like fashion:

$$x_2 = \sqrt{\frac{x_1^3 + 1}{3}} \approx 0.6402.$$

We continue in this way, obtaining each new approximation x_{n+1} by substituting x_n into the right-hand side of (6), and we obtain the values

$x_0 = 0.5$	$x_5 = 0.6523$
$x_1 = 0.6124$	$x_6 = 0.6526$
$x_2 = 0.6402$	$x_7 = 0.6527$
$x_3 = 0.6487$	$x_8 = 0.6527$
$x_4 = 0.6514$	

(the values are not exact, but are rounded to 4-place accuracy). Since x_8 is obtained by substitution of x_7 in the right-hand side of (6), the fact that $x_7 = x_8 = 0.6527$ to 4 decimal places means that

$$0.6527 = \sqrt{\frac{(0.6527)^3 + 1}{3}}$$

to this accuracy. Thus our root is $x_* = 0.6527$ to 4 decimal places, and so the cork ball with radius 1 foot sinks to a depth of 0.6527 feet, or about 7.8 inches.

The above computation illustrates the **method of repeated substitution.** It works well with many equations of interest. All we must do is write the given equation $f(x) = 0$ in the form

$$x = G(x).$$

Then we make a reasonably good initial guess x_0 and iterate:

$$x_1 = G(x_0), \qquad x_2 = G(x_1), \qquad x_3 = G(x_2),$$

and so on. We describe this iteration with the convenient notation

$$x_{n+1} = G(x_n) \qquad \text{for} \quad n = 1, 2, 3, \ldots. \tag{7}$$

That is, the approximation x_{n+1} of stage $n + 1$ is obtained by substitution of x_n into the right-hand side of the equation $x = G(x)$. We may stop when two successive approximations agree to the desired number of decimal places. In Figures 3.9 and 3.10, we show what's happening in the method of repeated substitution.

But sometimes the method of repeated substitution fails. Had we written Equation (5) in the equivalent form

$$x = \sqrt[3]{3x^2 - 1}$$

and begun with the same initial values $x_0 = 0.5$, we would have obtained the successive values

$x_0 = \ \ \ 0.5$	$x_3 = -0.1875$	$x_6 = 1.5059$
$x_1 = -0.6300$	$x_4 = -0.9635$	$x_7 = 1.7970$
$x_2 = \ \ \ 0.5754$	$x_5 = \ \ \ 1.2131$	$x_8 = 2.0557$

It seems we're getting nowhere! (Actually the sequence $\{x_n\}$ *does* converge, but *not* to the desired root in $(0, 1)$, and it converges exceedingly slowly to its limit.) We show at the end of this section a simple criterion for success of the method of repeated substitution; it is that $|G'(x)| < 1$ near the desired root. As a practical matter, one seldom checks this condition. For if the method succeeds, there's no problem. If it fails, we try an alternative, such as Newton's method.

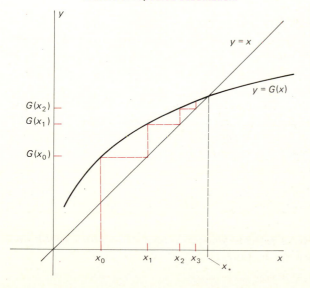

3.9 Convergence ("staircase mode") in the method of repeated substitution.

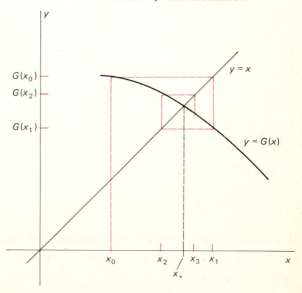

3.10 Convergence ("spiral mode") in the method of repeated substitution.

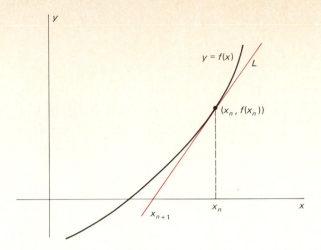

3.11 Geometry of the iteration of Newton's method.

NEWTON'S METHOD

We illustrate Newton's method of constructing a rapidly convergent sequence of successive approximations to a root x_* of the equation $f(x) = 0$ in Figure 3.11. The tangent line at $(x_n, f(x_n))$ is used to construct a better approximation x_{n+1} to x_* as follows. Start at the point x_n on the x-axis. Go vertically up (or down) to the point $(x_n, f(x_n))$ on the curve $y = f(x)$. Then follow the tangent line L there to the point where it intersects the x-axis. That point will be x_{n+1}.

Here is a formula for x_{n+1}. We obtain it by computing the slope of the line L in two ways: from the derivative and from the two-point definition of slope. Thus

$$f'(x_n) = \frac{f(x_n) - 0}{x_n - x_{n+1}},$$

and we solve easily for x_{n+1}:

$$x_{n+1} = x_n - \frac{f(x_n)}{f'(x_n)}. \tag{8}$$

This equation is the **iterative formula** for Newton's method, so called because in about 1669, Newton introduced an algebraic procedure (rather than the geometric construction above) that is equivalent to the iterative use of Equation (8). Newton's first example was the cubic equation $x^3 - 2x - 5 = 0$, for which he found the root $x_* \approx 2.0946$ (as we ask you to do in Problem 8).

Suppose now that we are using Newton's method to solve an equation. If we reach the point at which $x_{n+1} = x_n$ to the desired accuracy, then by Equation (8) it follows that x_n is very nearly a solution of the equation $f(x) = 0$. So, as in the method of repeated substitution, we simply continue iteration until we obtain the desired accuracy.

EXAMPLE 2 Use Newton's method to find $\sqrt{2}$ accurate to eight decimal places.

Solution More generally, consider the square root of the positive number a as the positive root of the equation

$$f(x) = x^2 - a = 0.$$

Since $f'(x) = 2x$, Equation (8) gives the iterative formula

$$x_{n+1} = x_n - \frac{x_n^2 - a}{2x_n} = \frac{1}{2}\left(x_n + \frac{a}{x_n}\right). \qquad (9)$$

The ancient Babylonians used this method, averaging the approximations x_n and a/x_n to $\sqrt{a}$ to get a better approximation. With $a = 2$ and initial guess $x_0 = 1.5$, Formula (9) gives the values

$$x_0 = 1.5$$
$$x_1 = 1.41666667$$
$$x_2 = 1.41421569$$
$$x_3 = 1.41421356$$
$$x_4 = 1.41421356$$

Thus $\sqrt{2} = 1.41421356$ to 8 decimal places. The very rapid convergence here is an important characteristic of Newton's method. As a general rule (with some exceptions), each iteration doubles the number of decimal places of accuracy.

EXAMPLE 3 Use Newton's method to solve the cork ball equation

$$f(x) = x^3 - 3x^2 + 1 = 0.$$

Solution Here $f'(x) = 3x^2 - 6x$, so the iterative Formula (8) becomes

$$x_{n+1} = x_n - \frac{x_n^3 - 3x_n^2 + 1}{3x_n^2 - 6x_n}. \qquad (10)$$

With $x_0 = 0.5$, this gives the values

$$x_0 = 0.5$$
$$x_1 = 0.6667$$
$$x_2 = 0.6528$$
$$x_3 = 0.6527$$
$$x_4 = 0.6527$$

Thus we obtain the root $x_* = 0.6527$ as before, with only half the number of iterations required in the method of repeated substitution.

The equation $x^3 - 3x^2 + 1 = 0$ has two additional solutions, a consequence of applying the Intermediate Value Property to $f(x) = x^3 - 3x^2 + 1$ on the intervals $[-1, 0]$, $[0, 1]$, and $[2, 3]$. The starting values of x_0 shown in the table below are estimated by linear interpolation.

x	$f(x)$	
-1	-3	
0	1	$x_0 = -0.25$
1	-1	$x_0 = +0.50$
2	-3	
3	1	$x_0 = +2.75$

Beginning with $x_0 = -0.25$ and, subsequently, with $x_0 = 2.75$, the iteration in (10) produces the two sequences

$$
\begin{array}{ll}
x_0 = -0.25 & x_0 = 2.75 \\
x_1 = -0.7222 & x_1 = 2.8939 \\
x_2 = -0.5626 & x_2 = 2.8795 \\
x_3 = -0.5331 & x_3 = 2.8794 \\
x_4 = -0.5321 & x_4 = 2.8794 \\
x_5 = -0.5321 &
\end{array}
$$

Thus the other two roots are -0.5321 and 2.8794 (to 4 decimal places).

Iterative methods are particularly well adapted to use with programmable calculators. With a calculator that is not programmable, the computations can be simplified a bit by replacing $f'(x_n)$ by the *constant* $f'(x_0)$ in the denominator of (8). This gives the "modified Newton's method"

$$
x_{n+1} = x_n - \frac{f(x_n)}{f'(x_0)}. \tag{11}
$$

For example, with $f(x) = x^3 - 3x^2 + 1$ and $x_0 = 0.5$, as in Example 3, (11) simplifies to

$$
x_{n+1} = x_n + \tfrac{4}{9}(x_n^3 - 3x_n^2 + 1).
$$

This iteration gives the values

$$
\begin{array}{ll}
x_0 = 0.5 & x_4 = 0.6526 \\
x_1 = 0.6667 & x_5 = 0.6527 \\
x_2 = 0.6502 & x_6 = 0.6527 \\
x_3 = 0.6531 &
\end{array}
$$

We need a couple more iterations, but each step is a little simpler.

We conclude this section with the theorem that validates the method of repeated substitution.

Theorem   *Method of Repeated Substitution*

Let G be a differentiable function on $[a, b]$ such that

(i) $a \leq G(x) \leq b$ for all x in $[a, b]$, and
(ii) $0 < |G'(x)| \leq K < 1$ for all x in $[a, b]$.

Let x_0 be any number in $[a, b]$. Then the sequence $\{x_n\}_{n=0}^{\infty}$ obtained by the iteration

$$
x_{n+1} = G(x_n) \quad \text{for} \quad n \geq 0
$$

converges to a root x_* of the equation

$$
x = G(x).
$$

Proof   Condition (i) assures us that there *is* a root x_*. In fact, it is unique because $G'(x) \neq 1$.

Next, note that

$$\left| x_1 - x_* \right| \leqq K \left| x_0 - x_* \right|.$$

This follows from (ii), since the Mean Value Theorem gives

$$\left| x_1 - x_* \right| = \left| G(x_0) - G(x_*) \right| = \left| G'(c_1) \right| \cdot \left| x_0 - x_* \right|.$$

Now assume, as part of a proof by induction, that

$$\left| x_n - x_* \right| \leqq K^n \left| x_0 - x_* \right|.$$

By the same sort of argument, we find that

$$\left| x_{n+1} - x_* \right| \leqq K^{n+1} \left| x_0 - x_* \right|.$$

But then

$$\left| x_n - x_* \right| \leqq K^n \left| x_0 - x_* \right| \qquad \text{for all } n,$$

and the last expression on the right approaches zero as $n \to \infty$ because $|K| < 1$. So the sequence $\{x_n\}$ approaches x_*, and this completes the proof.

3-4   PROBLEMS

In Problems 1–5, use the method of repeated substitution to find the root of the given equation in the indicated interval, accurate to three decimal places. Choose x_0 by linear interpolation.

1 $x^2 + 3x - 1 = 0$, $[0, 1]$.
2 $x^3 + 4x - 1 = 0$, $[0, 1]$.
3 $x - \frac{1}{4} \sin x - 1 = 0$, $[1, 2]$.
4 $x + \frac{1}{5} \cos x - 1 = 0$, $[0, 1]$.
5 $x^6 + 7x^2 - 4 = 0$, $[-1, 0]$.

In Problems 6–10, use Newton's method to find the root of the given equation in the indicated interval, accurate to four decimal places. Choose x_0 by linear interpolation.

6 $x^2 - 3 = 0$, $[1, 2]$ (to find $\sqrt{3}$).
7 $x^3 - 10 = 0$, $[2, 3]$ (to find $\sqrt[3]{10}$).
8 $x^3 - 2x - 5 = 0$, $[2, 3]$ (Newton's own example).
9 $x^5 - 5x - 10 = 0$, $[1, 2]$.
10 $x^{10} - 1000 = 0$, $[1, 2]$ (to find $\sqrt[10]{1000}$).
11 (a) Show that Newton's method applied to the equation $x^3 - a = 0$ yields the iteration

$$x_{n+1} = \frac{1}{3} \left(2x_n + \frac{a}{x_n^2} \right)$$

for approximating the cube root of a.
(b) Use this iteration to find $\sqrt[3]{2}$ accurate to five decimal places.

12 (a) Show that Newton's method yields the iteration

$$x_{n+1} = \frac{1}{k} \left[(k-1)x_n + \frac{a}{(x_n)^{k-1}} \right]$$

for approximating the kth root of the positive number a.
(b) Use this iteration to find $\sqrt[10]{100}$ accurate to five decimal places.

13 With many inexpensive hand-held calculators, you can calculate square roots directly, but not higher roots because there are no keys for exponentiation. Such a calculator can nevertheless be used to compute roots. For example, if $x = \sqrt[17]{a}$ for some positive number a, then $x^{17} = a$. So $x^{16} = a/x$, and thus

$$x = \sqrt{\sqrt{\sqrt{\sqrt{\sqrt{a/x}}}}}.$$

Use this formula and repeated substitution to find $\sqrt[17]{10}$ accurate to four decimal places.

14 Modify the method of Problem 13 in order to approximate $\sqrt[3]{5}$ accurate to four decimal places using only the square-root key (and the four arithmetic keys).
15 Repeat Problem 14 for $\sqrt[7]{10}$.
16 Show that Newton's method applied to the equation $1/x - a = 0$ yields the iteration $x_{n+1} = 2x_n - a(x_n)^2$, and

thus provides a method for approximating the reciprocal $1/a$ without performing any divisions. The reason for using such a method is that in most high-speed computers, the operation of division is *much* more time-consuming than even several additions and multiplications.

17 In Problem 31 of Section 2-4, we dealt with the problem of joining with least expense two points lying on opposite sides of a geological fault. This problem led to the equation

$$f(x) = 3x^4 - 24x^3 + 51x^2 - 32x + 64 = 0.$$

Use Newton's method to find the root of this equation that lies in (3, 4), accurate to four decimal places.

18 A moon of a certain planet has an elliptical orbit with eccentricity $\frac{1}{2}$, and its period of revolution about the planet is 100 days. If the moon is at the position $(a, 0)$ when $t = 0$, then, as illustrated in Figure 3.12, its central angle θ after

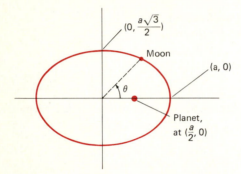

3.12 Elliptical orbit of Problem 18.

t days is given by *Kepler's equation*

$$\frac{2\pi t}{100} = \theta - \frac{1}{2}\sin\theta.$$

Use Newton's method to solve for θ when $t = 17$ (days). Take $\theta_0 = 1.5$ (radians), and calculate the first two approximations θ_1 and θ_2. Express θ_2 in degrees as well.

19 A great problem of Archimedes was that of using a plane to cut a sphere into two segments with volumes having a given (preassigned) ratio. Archimedes showed that the volume of a segment of height h of a sphere of radius a is

$$V = \frac{1}{3}\pi h^2(3a - h).$$

If a plane at distance x from the center of the unit sphere cuts it into two segments, one with twice the volume of the other, show that

$$3x^3 - 9x + 2 = 0.$$

Then use Newton's method to find x accurate to four decimal places.

20 The equation

$$f(x) = x^3 - 4x + 1 = 0$$

has three distinct real roots. Locate them by calculating the value of f for $x = -3, -2, -1, 0, 1, 2,$ and 3. Then use Newton's method to approximate each of the three roots to four-place accuracy.

21 The equation $x + \tan x = 0$ is important in a variety of applications—for example, in the study of the diffusion of heat. It has a sequence $\alpha_1, \alpha_2, \alpha_3, \ldots$ of positive roots, with the nth one slightly larger than $(n - \frac{1}{2})\pi$. Use Newton's method to compute α_1 and α_2 to three-place accuracy.

3-5

Higher Derivatives and Concavity

We saw in Section 2-5 that the sign of the first derivative f' tells whether the graph of the function is rising or falling. In this section we shall see that the sign of the *second* derivative of f, the derivative of f', tells which way the curve $y = f(x)$ is *bending*, upward or downward.

HIGHER DERIVATIVES

The **second derivative** of f is denoted by f'', and its value at x is

$$f''(x) = D(f'(x)) = D(Df(x)) = D^2f(x).$$

The derivative of f'' is the **third derivative** f''' of f, with

$$f'''(x) = D(f''(x)) = D(D^2f(x)) = D^3f(x).$$

The third derivative is also denoted by $f^{(3)}$. More generally, the result of starting with the function f and differentiating n times in succession is the **nth derivative** $f^{(n)}$ of f, with $f^{(n)}(x) = D^nf(x)$.

If $y = f(x)$, then the first n derivatives may be written as

$$D_x y, \qquad D_x^2 y, \qquad D_x^3 y, \qquad \ldots, \qquad D_x^n y$$

or

$$y', \qquad y'', \qquad y''', \qquad \ldots, \qquad y^{(n)},$$

or finally as

$$\frac{dy}{dx}, \qquad \frac{d^2 y}{dx^2}, \qquad \frac{d^3 y}{dx^3}, \qquad \ldots, \qquad \frac{d^n y}{dx^n}$$

in differential notation. The history of the curious placement of superscripts in the differential notation for higher derivative involves the metamorphosis

$$\frac{d}{dx}\left(\frac{dy}{dx}\right) \longrightarrow \frac{d}{dx}\frac{dy}{dx} \longrightarrow \frac{(d)^2 y}{(dx)^2} \longrightarrow \frac{d^2 y}{dx^2}.$$

EXAMPLE 1   Find the first four derivatives of

$$f(x) = 2x^3 + \frac{1}{x^2} + 16x^{7/2}.$$

Solution

$$f'(x) = 6x^2 - \frac{2}{x^3} + 56x^{5/2},$$

$$f''(x) = 12x + \frac{6}{x^4} + 140x^{3/2},$$

$$f'''(x) = 12 - \frac{24}{x^5} + 210x^{1/2}, \qquad \text{and}$$

$$f^{(4)}(x) = \frac{120}{x^6} + \frac{105}{x^{1/2}}.$$

The following example shows how higher derivatives of implicitly defined functions may be found.

EXAMPLE 2   Find the second derivative y'' of the function $y = y(x)$ if

$$x^2 - xy + y^2 = 9.$$

Solution   Implicit differentiation of the given equation with respect to x gives us

$$2x - y - xy' + 2yy' = 0,$$

so

$$y' = \frac{y - 2x}{2y - x}.$$

We obtain y'' by differentiating implicitly again, using the quotient rule. After that, we make substitutions for y' by using the result just found.

$$y'' = D_x\left(\frac{y - 2x}{2y - x}\right)$$

$$= \frac{(y' - 2)(2y - x) - (y - 2x)(2y' - 1)}{(2y - x)^2}$$

$$= \frac{3xy' - 3y}{(2x - y)^2}$$

$$= \frac{3x[(y - 2x)/(2y - x)] - 3y}{(2y - x)^2}.$$

Thus

$$y'' = -\frac{6(x^2 - xy + y^2)}{(2y - x)^3}.$$

CONCAVITY AND THE SECOND DERIVATIVE TEST

Now we investigate the significance of the *sign* of the second derivative. Suppose first that $f''(x) > 0$ on the interval I. Then $f'(x)$ is an increasing function on I because *its* derivative $f''(x)$ is positive. Thus, as we scan the graph $y = f(x)$ from left to right, we see the tangent line turning in a counter-clockwise direction, as illustrated in Figure 3.13. This situation may be described by saying that the curve $y = f(x)$ is **bending upward.** Note that a curve can be bending upward without rising, as illustrated in Figure 3.14.

On the other hand, if $f''(x) < 0$ on the interval I, then $f'(x)$ is decreasing on I, so the tangent line turns clockwise as x increases. In this case we say that the curve $y = f(x)$ is **bending downward.** Figures 3.15 and 3.16 illustrate two ways this can happen.

A comparison of Figures 3.13 and 3.14 with Figures 3.15 and 3.16 suggests that the question of whether the curve $y = f(x)$ is bending upward or downward is closely related to the question of whether it lies above or below

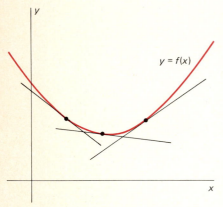

3.13 The graph is bending upward ("concave upward").

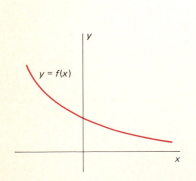

3.14 Another graph bending upward (concave upward).

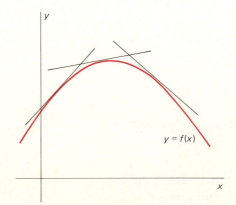

3.15 A graph bending downward ("concave downward").

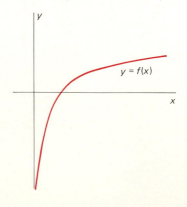

3.16 Another graph bending downward (concave downward).

its tangent lines. The latter question refers to the important property of *concavity*.

Definition of Concavity

Suppose that the function f is differentiable at the point a and that L is the tangent line to $y = f(x)$ at $(a, f(a))$. Then the function f (or its graph) is

(i) **Concave upward** at a if on some open interval containing a, the graph of f lies *above* L;
(ii) **Concave downward** at a if on some open interval containing a, the graph of f lies *below* L.

The following theorem establishes the connection between concavity and the sign of the second derivative. That connection is the one suggested by our discussion of bending.

Theorem 1   *Test for Concavity*

Suppose that f' is differentiable on an open interval containing a. Then f is

(i) Concave upward at a if $f''(a) > 0$;
(ii) Concave downward at a if $f''(a) < 0$.

Proof We give the proof for Case (i), illustrated by Figure 3.17. Write $y_{curve} = f(x)$ for the curve and

$$y_{line} = f(a) + f'(a)(x - a)$$

for the tangent line at $(a, f(a))$. Then it is enough to show that

$$y_{curve} - y_{line} = f(x) - f(a) - f'(a)(x - a)$$

is positive for all $x \neq a$ sufficiently near a. We apply the Mean Value Theorem to f on the interval $[a, x]$. Thus we find that

$$f(x) - f(a) = f'(c)(x - a)$$

where c is some number between a and x. If we substitute this information into the previous equation, we see that

$$y_{curve} - y_{line} = f'(c)(x - a) - f'(a)(x - a)$$
$$= [f'(c) - f'(a)](x - a)$$
$$= \frac{f'(c) - f'(a)}{c - a}(c - a)(x - a).$$

Now $c - a$ and $x - a$ have the same sign because c is between a and x. So all we must do is show that

$$\frac{f'(c) - f'(a)}{c - a} > 0$$

for x—and, therefore, for c—sufficiently near a.

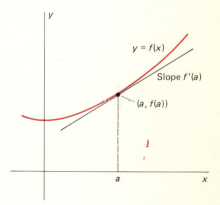

3.17 Case (i) of Theorem 1.

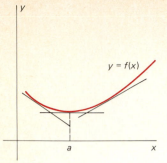

$f''(a) > 0$;
tangent turning counterclockwise;
graph concave up;
local minimum at a.

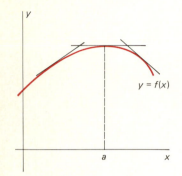

$f''(a) < 0$;
tangent turning clockwise;
graph concave down;
local maximum at a.

3.18 The second derivative test (Theorem 2).

3.19 No conclusion possible if $f'(a) = 0 = f''(a)$.

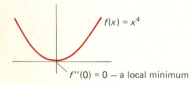

$f''(0) = 0$ — a local minimum

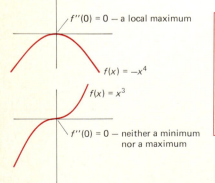

$f''(0) = 0$ — a local maximum

$f''(0) = 0$ — neither a minimum nor a maximum

But this last inequality follows from the hypothesis that $f''(a) > 0$. For, because $f''(a)$ exists, it is the limit of the above fraction; moreover, since that limit is positive, the fraction itself must be positive for all values of c sufficiently near a.

If we need to deal with values of $x < a$, the same proof works, except that we apply the Mean Value Theorem to f on the interval $[x, a]$ instead. Hence we have established Theorem 1.

Now suppose that, in addition to the differentiability hypothesis of Theorem 1, $f'(a) = 0$, so that a is a critical point of the sort where the tangent line is *horizontal*. Then f has a local minimum at a if, near a, its graph lies above this horizontal tangent line; by contrast, f has a local maximum at a if, near a, its graph lies below this horizontal tangent line. These observations together with Theorem 1 yield the following *sufficient* condition for a local extremum.

Theorem 2 *Second Derivative Test*

Suppose that the function f is differentiable on an open interval containing the critical point a, where $f'(a) = 0$. Then

(i) $f(a)$ is a local minimum value if $f''(a) > 0$;

(ii) $f(a)$ is a local maximum value if $f''(a) < 0$.

Rather than memorizing conditions (i) and (ii) verbatim, it is easier and more reliable to remember the second derivative test by visualizing the graphs shown in Figure 3.18.

Note that the second derivative test says nothing about what happens if $f''(a) = 0$. Consider the three functions $f(x) = x^4$, $f(x) = -x^4$, and $f(x) = x^3$. For each of these, $f'(0) = 0$ and $f''(0) = 0$. Their graphs, shown in Figure 3.19, demonstrate that *anything* can happen at such a point.

Similarly, the test for concavity (Theorem 1) says nothing about the case $f''(a) = 0$. A point where the second derivative vanishes *may or may not* be a point where the function changes concavity, from concave upward on one side to concave downward on the other. But if $(a, f(a))$ is a point where concavity reverses, whether or not $f''(a) = 0$, then it is called an **inflection point** of the function, or of its graph.

Theorem 3 *Inflection Point Test*

The point a is an inflection point of the continuous function f provided there is an open interval I containing a such that, for points x in I, either

(i) $f''(x) > 0$ if $x < a$ and $f''(x) < 0$ if $x > a$, or

(ii) $f''(x) < 0$ if $x < a$ and $f''(x) > 0$ if $x > a$.

The fact that a point where the second derivative changes sign is an inflection point follows immediately from Theorem 1 and the definition of

CHAP. 3: **Applications of Derivatives and Antiderivatives**

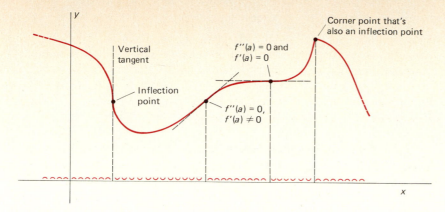

Vertical tangent

Inflection point

$f''(a) = 0$ and $f'(a) = 0$

$f''(a) = 0$, $f'(a) \neq 0$

Corner point that's also an inflection point

3.20 Some inflection points.

inflection point. At the inflection point a itself, either $f''(a) = 0$ or $f''(a)$ does not exist. Some of the various possibilities are indicated in Figure 3.20. Note how we have marked the intervals of upward concavity and of downward concavity with small cups opening upward and downward, respectively.

The tests for concavity, local extrema, and inflection points in Theorems 1 through 3 enable us to flesh out the curve-sketching techniques of Section 2-6.

EXAMPLE 3 Sketch the graph of $f(x) = 8x^5 - 5x^4 - 20x^3$, indicating local extrema, inflection points, and concave structure.

Solution We sketched this curve in Example 3 of Section 2-6; see Figure 2.29 for the graph. In that example we found the first derivative to be

$$f'(x) = 40x^4 - 20x^3 - 60x^2 = 20x^2(x + 1)(2x - 3),$$

so the critical points are $x = -1, 0,$ and $\frac{3}{2}$. The second derivative is

$$f''(x) = 160x^3 - 60x^2 - 120x = 160x(x^2 - \tfrac{3}{8}x - \tfrac{3}{4}).$$

Since $f''(-1) = -100 < 0$, $f''(0) = 0$, and $f''(\frac{3}{2}) = 225 > 0$, the second derivative test tells us that f has a local maximum at $x = -1$ and a local minimum at $x = \frac{3}{2}$. The second derivative test is not enough to determine the behavior of f at $x = 0$.

We use the quadratic formula to solve the equation

$$x^2 - \tfrac{3}{8}x - \tfrac{3}{4} = 0,$$

and thus we find that the roots of $f''(x) = 0$ are $x = 0$ and

$$x = \frac{(\tfrac{3}{8}) \pm \sqrt{(\tfrac{3}{8})^2 + 3}}{2} \approx -0.70, 1.07.$$

We take these two approximations as sufficiently exact, and write

$$f''(x) = 160x(x + 0.70)(x - 1.07).$$

This lets us analyze the concave structure of f in a manner like our analysis of increasing-decreasing behavior in Section 2-6. To do this we construct the following table.

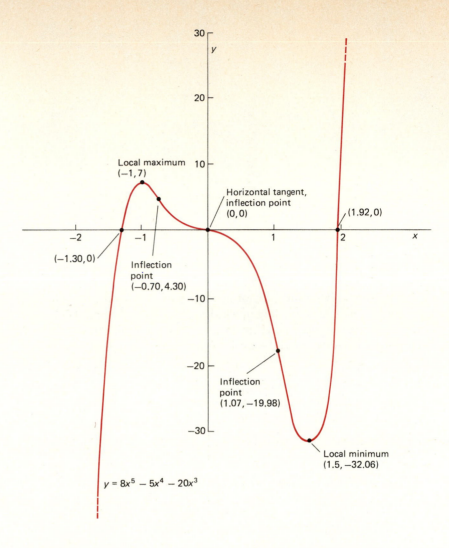

Local maximum
(−1,7)

Horizontal tangent,
inflection point
(0,0)

(1.92,0)

(−1.30,0)

Inflection
point
(−0.70,4.30)

Inflection
point
(1.07,−19.98)

Local minimum
(1.5,−32.06)

$y = 8x^5 - 5x^4 - 20x^3$

3.21 The graph of the function of Example 3.

Interval	x + 0.70	160x	x − 1.07	f''(x)	f
$(-\infty, -0.70)$	Neg.	Neg.	Neg.	Neg.	Concave Down
$(-0.70, 0)$	Pos.	Neg.	Neg.	Pos.	Concave Up
$(0, 1.07)$	Pos.	Pos.	Neg.	Neg.	Concave Down
$(1.07, +\infty)$	Pos.	Pos.	Pos.	Pos.	Concave Up

From the table we see that the direction of concavity of f changes at each of the points $x = -0.70$, $x = 0$, and $x = 1.07$. So these three points are indeed inflection points. This information is shown in the graph sketched in Figure 3.21.

EXAMPLE 4 Sketch the graph of $f(x) = 4x^{1/3} + x^{4/3}$, indicating local extrema, inflection points, and concave structure.

Solution Since

$$f'(x) = \tfrac{4}{3}x^{-2/3} + \tfrac{4}{3}x^{1/3} = \tfrac{4}{3}x^{-2/3}(x + 1),$$

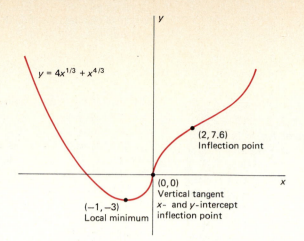

$y = 4x^{1/3} + x^{4/3}$

(2, 7.6)
Inflection point

(0, 0)
Vertical tangent
x- and y-intercept
inflection point

(−1, −3)
Local minimum

3.22 The graph of the function of Example 4.

the critical points are $x = -1$ (where the tangent line is horizontal) and $x = 0$ (where the tangent line is vertical). Since

$$f''(x) = -\tfrac{8}{9}x^{-5/3} + \tfrac{4}{9}x^{-2/3} = \tfrac{4}{9}x^{-5/3}(x - 2),$$

the possible inflection points are $x = 0$ and $x = 2$. The signs of $f'(x)$ and $f''(x)$ are easily analyzed as in the following tables.

Interval	$f'(x)$	f
$(-\infty, -1)$	Neg.	Decreasing
$(-1, 0)$	Pos.	Increasing
$(0, +\infty)$	Pos.	Increasing

Interval	$f''(x)$	f
$(-\infty, 0)$	Pos.	Concave Up
$(0, 2)$	Neg.	Concave Down
$(2, +\infty)$	Pos.	Concave Up

We note in addition that $f(x) \to \infty$ as $x \to \pm\infty$, and we use all this information to connect the points $(-1, -3)$, $(0, 0)$, and $(2, 6\sqrt[3]{2})$ with the smooth curve of Figure 3.22.

3-5 PROBLEMS

Calculate the first three derivatives of the functions in Problems 1–10.

1 $f(x) = x^3 - 2x.$

2 $f(x) = (x + 1)^{100}.$

3 $g(t) = \dfrac{1}{t} - \dfrac{1}{2t + 1}.$

4 $h(y) = \sqrt{3y - 1}.$

5 $f(t) = 2t^{3/2} - 3t^{4/3}.$

6 $g(x) = \dfrac{1}{x^2 + 9}.$

7 $h(t) = \dfrac{t + 2}{t - 2}.$

8 $f(z) = \sqrt[3]{z} + \dfrac{3}{\sqrt[5]{z}}.$

9 $g(x) = \sqrt[3]{5 - 4x}.$

10 $g(t) = \dfrac{8}{(3 - t)^{3/2}}.$

In each of Problems 11–15, calculate y' and y'', assuming that y is defined implicitly as a function of x by the given equation.

11 $x^2 - y^2 = 1.$

12 $y^3 + 2y - 1 = 3x.$

13 $x^2 + 4xy + 3y^2 = 5.$

14 $\sqrt{x} + \sqrt{y} = 1.$

15 $x^3 + y^3 = 3xy.$

Sketch the graphs of the functions in Problems 16–30, indicating all critical points and inflection points. Apply the second derivative test at each critical point. Show the correct concave structure in your sketches.

16 $f(x) = 12x - x^3.$

17 $f(x) = 2x^3 - 3x^2 - 12x + 3.$

18 $f(x) = 3x^4 - 4x^3 - 5.$

19 $f(x) = 6 + 8x^2 - x^4.$

20 $f(x) = 3x^5 - 5x^3$.

21 $f(x) = 3x^4 - 4x^3 - 12x^2 - 1$.

22 $f(x) = 3x^5 - 25x^3 + 60x$.

23 $f(x) = x^3(x - 1)^4$.

24 $f(x) = (x - 1)^2(x + 2)^3$.

25 $f(x) = 1 + x^{1/3}$. **26** $f(x) = 2 - (x - 3)^{1/3}$.

27 $f(x) = x^{1/2}(3 + x)$. **28** $f(x) = x^{2/3}(5 - 2x)$.

29 $f(x) = x^{1/3}(4 - x)$. **30** $f(x) = x^{1/3}(6 - x)^{2/3}$.

31 Find all the nonzero derivatives of $f(x) = (x + 1)^5$.

32 Suppose that $f(x) = x^n$, where n is a positive integer. Show by induction that $f^{(n)}(x) = n! = n(n - 1) \cdots 3 \cdot 2 \cdot 1$.

33 Conclude from the result of Problem 32 that, if $f(x)$ is a polynomial of degree n, then $f^{(k)}(x) \equiv 0$ ("is identically zero") if $k > n$.

34 (a) Calculate the first four derivatives of $f(x) = \sin x$. (b) Conclude that $D^{n+4} \sin x = D^n \sin x$ if n is a positive integer.

35 Suppose that $z = g(y)$ and that $y = f(x)$. Show that

$$\frac{d^2z}{dx^2} = \frac{d^2z}{dy^2}\left(\frac{dy}{dx}\right)^2 + \frac{dz}{dy}\frac{d^2y}{dx^2}.$$

36 Prove that the graph of a quadratic polynomial has no inflection points.

37 Prove that the graph of a cubic polynomial (one of degree three) has exactly one inflection point.

38 Prove that the graph of a polynomial function of degree four has either no inflection points or exactly two of them.

39 Suppose that the pressure P (in atmospheres), volume V (in cubic centimeters), and temperature T (in degrees Kelvin) of n moles of carbon dioxide satisfies van der Waals' equation

$$\left(P + \frac{n^2a}{V^2}\right)\left(V - nb\right) = nRT$$

where a, b, and R are empirical constants. The following experiment was carried out in order to find the values of these constants.

One mole of CO_2 was compressed at the constant temperature $T = 304°K$. The measured pressure-volume (PV) data were then plotted as in Figure 3.23, with the PV curve showing a *horizontal* inflection point at $V = 128.1$, $P = 72.8$. Use this information to calculate a, b, and R. (*Suggestion:* Solve van der Waals' equation for P and then calculate dP/dV and d^2P/dV^2.)

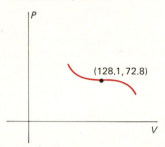

3.23 A problem involving van der Waals' equation.

(128.1, 72.8)

3-6

Curve Sketching and Asymptotes

Before we turn to scientific applications of differentiation, we want to extend the limit concept to include infinite limits and limits at infinity. This extension of the limit concept adds a final weapon to our arsenal of curve-sketching techniques, the notion of an **asymptote** to a curve—a straight line that the curve approaches arbitrarily closely (in a sense soon to be specified).

We say that $f(x)$ **increases without bound,** or **becomes infinite,** as x approaches a, and write

$$\lim_{x \to a} f(x) = \infty, \tag{1}$$

provided that $f(x)$ can be made arbitrarily large by choosing x sufficiently close (though not equal) to a. More precisely, (1) means that, given $M > 0$, there exists $\delta > 0$ such that

$$0 < |x - a| < \delta \quad \text{implies} \quad f(x) > M.$$

For example, it is apparent that

$$\lim_{x \to 2} \frac{1}{(x - 2)^2} = +\infty$$

because $(x - 2)^2$ is positive and is approaching zero as $x \to 2$.

CHAP. 3: Applications of Derivatives and Antiderivatives

The statement that $f(x)$ **decreases without bound,** or **becomes negatively infinite,** as $x \to a$, written

$$\lim_{x \to a} f(x) = -\infty, \qquad (2)$$

has an analogous definition.

One-sided versions of (1) and (2) also make sense. For instance, if n is an *odd* integer, then it is apparent that

$$\lim_{x \to 2^-} \frac{1}{(x-2)^n} = -\infty \quad \text{while} \quad \lim_{x \to 2^+} \frac{1}{(x-2)^n} = +\infty$$

because $(x-2)^n$ is negative when x is to the left of 2 and positive when x is to the right of 2.

We say that the line $x = a$ is a **vertical asymptote** for the curve $y = f(x)$ provided that

$$\lim_{x \to a} |f(x)| = \infty. \qquad (3)$$

The geometric significance of a vertical asymptote is illustrated by the graphs of $y = 1/(x-1)$ and $y = 1/(x-1)^2$ in Figures 3.24 and 3.25. In each case, as $x \to 1$ and $f(x) \to \pm\infty$, the point $(x, f(x))$ on the curve approaches the vertical asymptote $x = 1$.

The most typical occurrence of a vertical asymptote is for a rational function $f(x) = p(x)/q(x)$ at a point $x = a$ where $q(a) = 0$ but $p(a) \neq 0$. Several examples appear below.

In Section 2-6 we mentioned infinite limits at infinity in connection with the behavior of a polynomial as $x \to \pm\infty$. There is also such a thing as a finite limit at infinity. We say that $f(x)$ **approaches the number L as x increases without bound,** and write

$$\lim_{x \to \infty} f(x) = L, \qquad (4)$$

provided that $|f(x) - L|$ can be made arbitrarily small by choosing x sufficiently large. That is, given $\varepsilon > 0$, there exists $M > 0$ such that

$$x > M \quad \text{implies} \quad |f(x) - L| < \varepsilon. \qquad (5)$$

The statement that

$$\lim_{x \to -\infty} f(x) = L$$

has a definition of similar form: merely replace the condition $x > M$ by the condition $x < -M$.

The analogues for limits at infinity of the limit laws of Section 1-8 all hold, including in particular the sum, product, and quotient laws. In addition, it is not difficult to show that, if

$$\lim_{x \to \infty} f(x) = L \quad \text{and} \quad \lim_{x \to \infty} g(x) = \pm\infty,$$

then

$$\lim_{x \to \infty} \frac{f(x)}{g(x)} = 0.$$

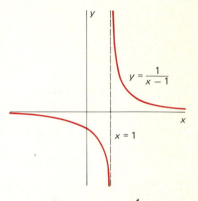

3.24 The graph of $y = \dfrac{1}{x-1}$.

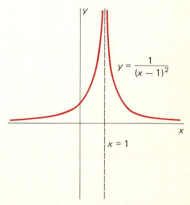

3.25 The graph of $y = 1/(x-1)^2$.

It follows from this result that

$$\lim_{x \to \infty} \frac{1}{x^k} = 0 \tag{6}$$

for any choice of the positive rational number k.

Using (6) and the limit laws, limits at infinity of rational functions are easy to evaluate. The general method is this: First divide each term in both numerator and denominator by the highest power of x that appears in any of the terms. Then apply the limit laws.

EXAMPLE 1 Find

$$\lim_{x \to \infty} f(x) \quad \text{if} \quad f(x) = \frac{3x^3 - x}{2x^3 + 7x^2 - 4}.$$

Solution

$$\lim_{x \to \infty} \frac{3x^3 - x}{2x^3 + 7x^2 - 4} = \lim_{x \to \infty} \frac{3 - 1/x^2}{2 + 7/x - 4/x^3}$$

$$= \frac{\lim_{x \to \infty} (3 - 1/x^2)}{\lim_{x \to \infty} (2 + 7/x - 4/x^3)}$$

$$= \frac{3 - 0}{2 + 0 - 0} = \frac{3}{2}.$$

Note that the same computation, but with $x \to -\infty$, also gives the result $\lim_{x \to -\infty} f(x) = \frac{3}{2}$.

EXAMPLE 2 Find $\lim_{x \to \infty} (\sqrt{x + a} - \sqrt{x})$.

Solution We use the familiar "divide and multiply" trick.

$$\lim_{x \to \infty} (\sqrt{x + a} - \sqrt{x}) = \lim_{x \to \infty} (\sqrt{x + a} - \sqrt{x}) \cdot \frac{\sqrt{x + a} + \sqrt{x}}{\sqrt{x + a} + \sqrt{x}}$$

$$= \lim_{x \to \infty} \frac{a}{\sqrt{x + a} + \sqrt{x}} = 0.$$

The geometric meaning of the statement $\lim_{x \to \infty} f(x) = L$ is that the point $(x, f(x))$ on the curve $y = f(x)$ approaches the horizontal line $y = L$ as $x \to \infty$. In particular, with the numbers M and ε of Condition (5), the part of the curve with $x > M$ lies between the horizontal lines $y = L - \varepsilon$ and $y = L + \varepsilon$ (see Figure 3.26). We therefore say that the line $y = L$ is a **horizontal asymptote** for the curve $y = f(x)$ if either

$$\lim_{x \to \infty} f(x) = L \quad \text{or} \quad \lim_{x \to -\infty} f(x) = L.$$

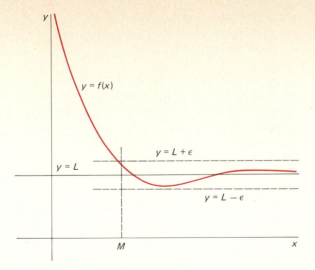

3.26 Geometry of the definition of horizontal asymptote.

EXAMPLE 3 Sketch the graph of $f(x) = x/(x - 2)$. Indicate any asymptotes.

Solution First we note that $x = 2$ is a vertical asymptote because $|f(x)| \to \infty$ as $x \to 2$. Also,

$$\lim_{x \to \pm \infty} \frac{x}{x - 2} = \lim_{x \to \pm \infty} \frac{1}{1 - (2/x)} = 1.$$

So $y = 1$ is a horizontal asymptote. The first two derivatives of f are readily found to be

$$f'(x) = -\frac{2}{(x - 2)^2} \quad \text{and} \quad f''(x) = \frac{4}{(x - 2)^3}.$$

Hence the function f has neither critical points nor inflection points. It is decreasing and concave downward on the interval $(-\infty, 2)$ and decreasing and concave upward on $(2, \infty)$. We show its graph in Figure 3.27.

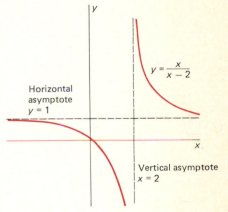

3.27 The graph for Example 3.

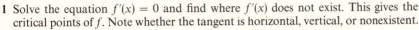

The curve-sketching techniques of Sections 2-6 and 3-5 and those of the present section can be summarized in the form of a list of steps. If you follow these, loosely rather than rigidly, you should obtain a rather accurate sketch of the graph of the curve $y = f(x)$.

1 Solve the equation $f'(x) = 0$ and find where $f'(x)$ does not exist. This gives the critical points of f. Note whether the tangent is horizontal, vertical, or nonexistent.

2 Determine the intervals on which f is increasing and those on which it is decreasing.

3 Solve the equation $f''(x) = 0$ and find where $f''(x)$ does not exist. These will be the *possible* inflection points of the graph.

4 Determine the intervals on which f is concave upward and those on which it is concave downward.

5 Find the y-intercept (if any) and the x-intercepts of the graph.

6 Plot and label the critical points, possible inflection points, and intercepts.

7 Determine the asymptotes (if any), discontinuities and behavior of f near them, and the behavior of f as $x \to \pm \infty$.

8 Finally, join the plotted points with a curve that is consistent with the information you have accumulated. Remember here that corner points are rare and that straight pieces of graph are even rarer.

Of course, you may follow these steps in any convenient order and omit any that present formidable computational difficulties. Many examples will require fewer than the eight steps; see Example 3. But our next example requires them all.

EXAMPLE 4 Sketch the graph of

$$f(x) = \frac{2 + x - x^2}{(x - 1)^2}.$$

Solution We notice immediately that

$$\lim_{x \to 1} f(x) = \infty,$$

so the line $x = 1$ is a vertical asymptote. Also

$$\lim_{x \to \pm\infty} \frac{2 + x - x^2}{(x - 1)^2} = \lim_{x \to \pm\infty} \frac{(2/x^2) + (1/x) - 1}{(1 - (1/x))^2} = -1,$$

so the line $y = -1$ is a horizontal asymptote.

Next, we apply the quotient rule and simplify to find that

$$f'(x) = \frac{x - 5}{(x - 1)^3}.$$

So the only critical point is $x = 5$, and we plot the point $(5, f(5)) = (5, -9/8)$ on a convenient coordinate plane. We note that there is a horizontal tangent there. And while we are inspecting the first derivative, we find the increasing-decreasing behavior of f:

Interval	$f'(x)$	f
$(-\infty, 1)$	Pos.	Increasing
$(1, 5)$	Neg.	Decreasing
$(5, \infty)$	Pos.	Increasing

After some simplification, we find the second derivative to be

$$f''(x) = \frac{14 - 2x}{(x - 1)^4}.$$

The only possible inflection point is at $x = 7$, corresponding to the point $(7, -\frac{10}{9})$ on the graph. The concave structure of the curve can be deduced with the aid of the next table.

Interval	$f''(x)$	f
$(-\infty, 1)$	Pos.	Concave Upward
$(1, 7)$	Pos.	Concave Upward
$(7, \infty)$	Neg.	Concave Downward

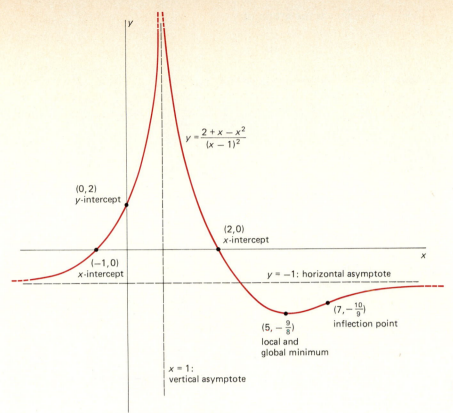

$$y = \frac{2 + x - x^2}{(x-1)^2}$$

(0, 2)
y-intercept

(2, 0)
x-intercept

(−1, 0)
x-intercept

y = −1: horizontal asymptote

$(7, -\frac{10}{9})$
inflection point

$(5, -\frac{9}{8})$
local and
global minimum

x = 1:
vertical asymptote

3.28 Graph of the function of Example 4.

Finding the intercepts is no problem at all; we plot them, sketch the asymptotes, and finally draw the graph with the aid of the two tables above. The result appears in Figure 3.28.

Asymptotes may be inclined as well as horizontal or vertical. The non-vertical line $y = mx + b$ is an **asymptote** for the curve $y = f(x)$ provided that

$$\lim_{x \to \pm \infty} [f(x) - (mx + b)] = 0. \qquad (7)$$

This condition means that, as $x \to \pm \infty$, the vertical distance between the point $(x, f(x))$ on the curve and the point $(x, mx + b)$ on the line approaches zero.

If $f(x) = p(x)/q(x)$ is a rational function with the degree of p greater by 1 than that of q, then by long division of $q(x)$ into $p(x)$, we find that $f(x)$ has the form

$$f(x) = mx + b + g(x)$$

where

$$\lim_{x \to \pm \infty} g(x) = 0.$$

Thus $y = mx + b$ is an asymptote for $y = f(x)$.

EXAMPLE 5   Sketch the graph of

$$f(x) = \frac{x^2 + x - 1}{x - 1}.$$

Solution   The long division suggested above takes this form:

$$\begin{array}{r} x + 2 \\ x - 1 \overline{)\, x^2 + x - 1} \\ \underline{x^2 - x} \\ 2x - 1 \\ \underline{2x - 2} \\ 1 \end{array}$$

Thus

$$f(x) = x + 2 + \frac{1}{x - 1}.$$

So $y = x + 2$ is an asymptote for the curve. Also,

$$\lim_{x \to 1} |f(x)| = \infty,$$

so $x = 1$ is a vertical asymptote. The first two derivatives of f are

$$f'(x) = 1 - \frac{1}{(x - 1)^2} = \frac{x(x - 2)}{(x - 1)^2}$$

and

$$f''(x) = \frac{2}{(x - 1)^3}.$$

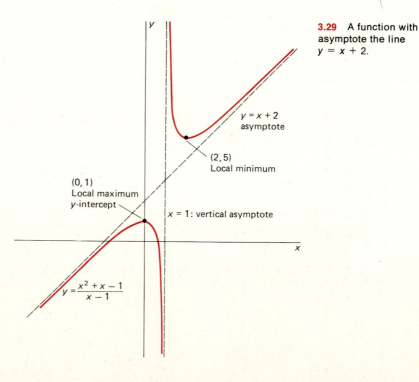

3.29 A function with asymptote the line $y = x + 2$.

$y = x + 2$
asymptote

(2, 5)
Local minimum

(0, 1)
Local maximum
y-intercept

$x = 1$: vertical asymptote

$y = \dfrac{x^2 + x - 1}{x - 1}$

It follows that f has critical points at $x = 0$ and at $x = 2$, but no inflection points. The sign of $f'(x)$ tells us that f is increasing on $(-\infty, 0)$ and on $(2, \infty)$ and decreasing on $(0, 1)$ and on $(1, 2)$. And examination of $f''(x)$ reveals that f is concave downward on $(-\infty, 1)$ and concave upward on $(1, \infty)$. In particular, $f(0) = 1$ is a local maximum value and $f(2) = 5$ is a local minimum value. So the graph of f looks much like the one in Figure 3.29.

3-6 PROBLEMS

Investigate the limits in Problems 1–16.

1 $\lim\limits_{x \to \infty} \dfrac{x}{x + 1}$.

2 $\lim\limits_{x \to -\infty} \dfrac{x^2 + 1}{x^2 - 1}$.

3 $\lim\limits_{x \to 1} \dfrac{x^2 + x - 2}{x - 1}$.

4 $\lim\limits_{x \to 1} \dfrac{x^2 - x - 2}{x - 1}$.

5 $\lim\limits_{x \to \infty} \dfrac{2x^2 - 1}{x^2 - 3x}$.

6 $\lim\limits_{x \to -\infty} \dfrac{x^2 + 3x}{x^3 - 5}$.

7 $\lim\limits_{x \to -1} \dfrac{x^2 + 2x + 1}{(x + 1)^2}$.

8 $\lim\limits_{x \to \infty} \dfrac{5x^3 - 2x + 1}{7x^3 + 4x^2 - 2}$.

9 $\lim\limits_{x \to 4} \dfrac{x - 4}{\sqrt{x - 2}}$.

10 $\lim\limits_{x \to \infty} \dfrac{2x + 1}{x - x^{3/2}}$.

11 $\lim\limits_{x \to -\infty} \dfrac{8 - \sqrt[3]{x}}{2 + x}$.

12 $\lim\limits_{x \to \infty} \dfrac{2x^2 - 17}{x^3 - 2x + 27}$.

13 $\lim\limits_{x \to \infty} \sqrt{\dfrac{4x^2 - x}{x^2 + 9}}$.

14 $\lim\limits_{x \to -\infty} \dfrac{\sqrt[3]{x^3 - 8x + 1}}{3x - 4}$.

15 $\lim\limits_{x \to -\infty} (\sqrt{x^2 + 2x} - x)$.

16 $\lim\limits_{x \to -\infty} (2x - \sqrt{4x^2 - 5x})$.

Sketch the graphs of the functions in Problems 17–32, indicating all critical points, inflection points, asymptotes, and concave structure.

17 $f(x) = \dfrac{2}{x - 3}$.

18 $f(x) = \dfrac{4}{5 - x}$.

19 $f(x) = \dfrac{3}{(x + 2)^2}$.

20 $f(x) = \dfrac{-4}{(3 - x)^2}$.

21 $f(x) = \dfrac{1}{(2x - 3)^3}$.

22 $f(x) = \dfrac{x + 1}{x - 1}$.

23 $f(x) = \dfrac{x^2}{x^2 + 1}$.

24 $f(x) = \dfrac{2x}{x^2 + 1}$.

25 $f(x) = \dfrac{1}{x^2 - 9}$.

26 $f(x) = \dfrac{x}{4 - x^2}$.

27 $f(x) = \dfrac{1}{x^2 + x - 6}$.

28 $f(x) = \dfrac{2x^2 + 1}{x^2 - 2x}$.

29 $f(x) = x + \dfrac{1}{x}$.

30 $f(x) = 2x + \dfrac{1}{x^2}$.

31 $f(x) = \dfrac{x^2}{x - 1}$.

32 $f(x) = \dfrac{2x^3 - 5x^2 + 4x}{x^2 - 2x + 1}$.

33 Suppose that $f(x) = x^2 + (2/x)$. Note that

$$\lim\limits_{x \to \pm\infty} (f(x) - x^2) = 0,$$

so the curve $y = f(x)$ approaches the parabola $y = x^2$ as $x \to \pm\infty$. Use this observation to make an accurate sketch of the graph of f.

34 Use the method of Problem 33 to make an accurate sketch of the graph of $f(x) = x^3 - 12/(x - 1)$.

3-7
Antiderivatives

The language of change is the natural language for the statement of most scientific laws and principles. For instance, Newton's law of cooling says that the *rate of change* of the temperature T of a body is proportional to the difference between T and the temperature of the surrounding medium. That is,

$$\frac{dT}{dt} = k(A - T) \tag{1}$$

where k is a positive constant and A—frequently assumed constant too—is the surrounding temperature. Similarly, the *rate of change* of a population P with constant birth and death rates is proportional to the size of the population:

$$\frac{dP}{dt} = kP \qquad (k \text{ constant}). \tag{2}$$

Torricelli's law says that the *rate of change* of the volume V of water in a draining tank is proportional to the square root of the water's depth y; that is,

$$\frac{dV}{dt} = -k\sqrt{y} \qquad (k \text{ a constant}). \tag{3}$$

Thus mathematical models of real-world situations frequently involve equations containing *derivatives* of unknown functions. Equations such as (1) through (3) above are called **differential equations.**

The simplest kind of differential equation has the form

$$\frac{dy}{dx} = f(x), \tag{4}$$

where f is a given (known) function and the function y of x is unknown. The process of finding a function from its derivative is the opposite of differentiation and is thus called **antidifferentiation.** If G is a function with derivative f, or $G'(x) = f(x)$, then G is called an **antiderivative** of f, and we write

$$G = D^{-1}f \quad \text{or} \quad G(x) = D^{-1}f(x). \tag{5}$$

The reason for this inverse notation is that the operations of differentiation (via D) and antidifferentiation (via D^{-1}) are inverses of each other—with one important proviso.

The proviso is that if a function has one antiderivative, then it has many antiderivatives. By contrast, a function can have only one derivative. While x^3 is an antiderivative of $3x^2$, so are the functions $x^3 + 17$, $x^3 + \pi$, and $x^3 - \sqrt[5]{2}$. Indeed, $x^3 + C$ is an antiderivative of $3x^2$ for *any* choice of the constant C.

More generally, if $G(x)$ is an antiderivative of $f(x)$, then so is $G(x) + C$ for any constant C. The converse of this statement is less obvious but equally true. That is, if $G(x)$ is one antiderivative of $f(x)$, then *every* antiderivative of $f(x)$ is of the form $G(x) + C$. This follows immediately from Corollary 2 in Section 2-5, according to which two functions with the same derivative differ only by a constant. Therefore, if $G(x)$ is one antiderivative of $f(x)$, then the *most general antiderivative* of $f(x)$ has the form

$$D^{-1}f(x) = G(x) + C.$$

Every differentiation formula yields an immediate antidifferentiation formula. We ordinarily write antidifferentiation formulas in terms of these most general antiderivatives. Thus

$$D^{-1}x^r = \frac{x^{r+1}}{r+1} + C \quad \text{if} \quad r \neq -1; \tag{6}$$

$$D^{-1}\cos x = \sin x + C; \tag{7}$$

$$D^{-1}\sin x = -\cos x + C. \tag{8}$$

Each such formula may be checked by differentiation of the right-hand side. This is the sure-fire way to check any antidifferentiation: To verify that G is an antiderivative of f, compute G' and see whether it is equal to f.

The linearity of the operation of differentiation implies immediately that, if the functions f and g have antiderivatives and a is a constant, then

$$D^{-1}(af) = aD^{-1}f + C \tag{9}$$

and

$$D^{-1}(f + g) = D^{-1}f + D^{-1}g + C. \tag{10}$$

That is, if $D^{-1}f$ and $D^{-1}(af)$ are particular antiderivatives of f and af, respectively, then (9) holds for some constant C; the meaning of the constant C in (10) is similar. For example,

$$D^{-1}\left(x^3 + 3\sqrt{x} - \frac{4}{x^2}\right) = D^{-1}(x^3 + 3x^{1/2} - 4x^{-2})$$

$$= D^{-1}(x^3) + 3D^{-1}(x^{1/2}) - 4D^{-1}(x^{-2}) + C$$

$$= \frac{1}{4}x^4 + 2x^{3/2} + \frac{4}{x} + C,$$

using (6) and (10).

A common technique of antidifferentiation is the application of the chain rule in reverse. The chain rule in the form

$$D_x g(u) = g'(u)\frac{du}{dx}$$

yields the antidifferentiation formula

$$D_x^{-1}\left(g'(u)\frac{du}{dx}\right) = g(u) + C, \tag{11}$$

where u denotes a differentiable function of x. Combination of this formula with our basic differentiation formulas gives these results:

$$D_x^{-1}\left(u^r\frac{du}{dx}\right) = \frac{u^{r+1}}{r+1} + C \quad \text{if} \quad r \neq -1, \tag{12}$$

$$D_x^{-1}\left((\cos u)\frac{du}{dx}\right) = \sin u + C, \tag{13}$$

and

$$D_x^{-1}\left((\sin u)\frac{du}{dx}\right) = -\cos u + C. \tag{14}$$

The key to applying such formulas as (12) through (14) is to determine the "inside function" $u = u(x)$. The following clues will help in spotting u.

1 The function u is inside some other function, as the above formulas suggest.
2 The derivative du/dx is a factor of the expression you are trying to antidifferentiate.

EXAMPLE 1   Find $D^{-1}((x^2 + 1)^{10} \cdot 2x)$.

Solution   It's clear that the inside function u is $u(x) = x^2 + 1$, since that puts the problem into the form

$$D^{-1}\left(u^{10}\frac{du}{dx}\right) = \frac{u^{11}}{11} + C.$$

Thus

$$D^{-1}((x^2 + 1)^{10} \cdot 2x) = \tfrac{1}{11}(x^2 + 1)^{11} + C.$$

The resubstitution of $x^2 + 1$ for u is essential, since the symbol u appears nowhere in the original problem.

EXAMPLE 2   Find $D^{-1}(x(1 + x^2)^{1/2})$.

Solution   Be willing to experiment. The required factor du/dx will be $2x$ if we choose $u = 1 + x^2$. The term x is present, but not $2x$. As a general rule, if the error is only a constant *multiplicative* factor (here, we have x rather than $2x$), the process will still work. So we cheerfully proceed by letting $u = 1 + x^2$.

The expression to be antidifferentiated involves $u^{1/2}$ as its most important factor. So it's plausible that the antiderivative resembles $u^{3/2}$. Try $u^{3/2}$ to see how well it works!

$$\begin{aligned}
D_x u^{3/2} &= D(1 + x^2)^{3/2} \\
&= \tfrac{3}{2}(1 + x^2)^{1/2}(2x) \\
&= 3x(1 + x^2)^{1/2}.
\end{aligned}$$

The resulting derivative is three times as large as we want. So by linearity we see that

$$D_x \tfrac{1}{3}u^{3/2} = x(1 + x^2)^{1/2}.$$

Thus

$$D^{-1}(x(1 + x^2)^{1/2}) = \tfrac{1}{3}(1 + x^2)^{3/2} + C.$$

The method above is rather a "cut and fit" technique. A cleaner but longer approach is to adjust the constants in advance. In Example 2, this alternative would look like this:

$$D^{-1}(x(1 + x^2)^{1/2}) = D^{-1}(x \cdot u^{1/2}) = D^{-1}\left(\frac{1}{2}(2x)u^{1/2}\right)$$

$$= \frac{1}{2}D^{-1}\left(u^{1/2}\frac{du}{dx}\right) = \frac{1}{2} \cdot \frac{u^{3/2}}{\frac{3}{2}} + C$$

$$= \frac{1}{3}u^{3/2} + C = \frac{1}{3}(1 + x^2)^{3/2} + C.$$

EXAMPLE 3   Find $D^{-1}(\sin x \cos x)$.

Solution   We take $u = \sin x$, so that $du/dx = \cos x$. Then we obtain

$$D^{-1}(\sin x \cos x) = D_x^{-1}\left(u\frac{du}{dx}\right)$$

$$= \frac{1}{2}u^2 + C = \frac{1}{2}\sin^2 x + C.$$

Alternatively, with $u = \cos x$ and $du/dx = -\sin x$, we obtain

$$D^{-1}(\sin x \cos x) = D_x^{-1}\left(-u\frac{du}{dx}\right)$$

$$= -\frac{1}{2}u^2 + C = -\frac{1}{2}\cos^2 x + C.$$

If both answers are correct, then $\frac{1}{2}\sin^2 x$ and $-\frac{1}{2}\cos^2 x$ must differ by a constant K. What is the value of K?

EXAMPLE 4   Find $D^{-1}\sin^2 x$.

Solution   Here it is important to remember the trigonometric identity

$$\sin^2 x = \tfrac{1}{2}(1 - \cos 2x).$$

With its aid, we find that

$$D^{-1}\sin^2 x = \tfrac{1}{2}D^{-1}(1 - \cos 2x)$$
$$= \tfrac{1}{2}(x - \tfrac{1}{2}\sin 2x) + C.$$

Since the original problem involved the sine of x rather than of $2x$, it's usually considered more tasteful to rewrite the answer in like terms. Thus we find that

$$D^{-1}\sin^2 x = \tfrac{1}{2}(x - \sin x \cos x) + C.$$

EXAMPLE 5   Find $D^{-1}(\sin^4 x \cos x)$.

Solution   We take $u = \sin x$ as in Example 3 and so

$$D^{-1}(\sin^4 x \cos x) = D_x^{-1}\left(u^4\frac{du}{dx}\right)$$

$$= \frac{1}{5}u^5 + C = \frac{1}{5}\sin^5 x + C.$$

EXAMPLE 6   Find $D^{-1}(x^3 \cos(x^4))$.

Solution   Here the inside function is $u = x^4$, with $du/dx = 4x^3$, so that $x^3 = \frac{1}{4}(du/dx)$. Hence

$$D^{-1}(x^3 \cos(x^4)) = D_x^{-1}\left[\frac{1}{4}(\cos u)\frac{du}{dx}\right]$$

$$= \frac{1}{4}\sin u + C = \frac{1}{4}\sin(x^4) + C.$$

*THE SUSPENSION BRIDGE

This is a simple example of the use of antiderivatives to explain a non-trivial physical phenomenon—the hanging suspension bridge. We assume that the bridge supports a horizontally uniform load. This is close to the situation in an ordinary suspension bridge, such as New York's Verrazano Narrows bridge, shown in Figure 3.30. The weight of the cables will be neglected in comparison with the weight of the roadway itself, and the latter may be assumed horizontally uniform.

Now suppose that the shape of the cable from which the bridge is suspended is the graph of the function $y = f(x)$. Our task is to find the function f: We really want an explicit formula for it, as if we planned on designing and building our own suspension bridge. If we can find f, we have determined the shape of the supporting cable.

3.30 The Verrazano Narrows Bridge. (Photograph courtesy of H. Armstrong Roberts, Philadelphia; and DePascale & Associates, Athens, Georgia)

As suggested in Figure 3.31, we assume that the solution is symmetric about the y-axis if that axis is arranged to pass through the lowest point of the graph. We concentrate on finding the formula for $y = f(x)$ for $x \geqq 0$, and in fact consider the part of the cable (or graph) that lies above the interval $[0, x]$ on the x-axis.

A consequence of elementary physics is this: The vertical and horizontal forces exerted on the segment of cable above $[0, x]$ must balance, or else the cable would fly off into space. Let T_0 (an unknown constant) denote the tension force exerted on the left end of this portion of the cable, and let T_x be the tension force exerted on the right end. We are correct in using the last subscript x, since the value of T_x *does* depend on x. What's pulling on the segment to produce T_0 is the part of the cable to the left of the y-axis and other structures, such as the towers from which the cable is hung. The force T_x is produced by the cable to the right of x.

Now T_0 is a horizontal force, since we are assuming that it is measured at the lowest point of the cable, at which $x = 0$ and the tangent to the cable is horizontal. The force T_x is inclined at the angle α, where $\tan \alpha = dy/dx = f'(x)$. The reason for this is that the direction of T_x must be the same as the cable's direction where it's measured.

Finally, the assumption that the load on the cable is horizontally uniform means that the weight supported by our portion of cable over $[0, x]$ is given by $W = \rho x$, where ρ is a constant. This weight is a *vertical* force; the horizontal component is zero.

We know that the forces on the segment of cable over $[0, x]$ balance, so we equate the vertical forces to get the first equation below and the horizontal forces to get the second.

$$T_x \sin \alpha = \rho x, \tag{15}$$

$$T_x \cos \alpha = T_0. \tag{16}$$

Now $dy/dx = \tan \alpha$, so we divide the first equation by the second and find that

$$\frac{dy}{dx} = \tan \alpha = \frac{\rho}{T_0} x.$$

Since ρ/T_0 is a constant, a single antidifferentiation gives us the equation

$$y = \frac{\rho}{2T_0} x^2 + C$$

3.31 Suspending cable for a suspension bridge.

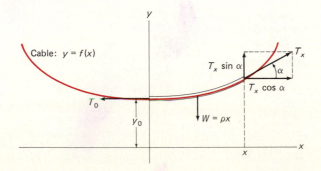

for some constant C. We denote by y_0 the minimum height of the cable (above the x-axis in Figure 3.31), and it follows that $C = y_0$. Thus

$$y = f(x) = \frac{\rho}{2T_0} x^2 + y_0 \tag{17}$$

is the equation of the cable. We have found that the hanging cable with uniform horizontal load takes the shape of a *parabola*.

3-7 PROBLEMS

Find the most general antiderivatives of the functions in Problems 1–25.

1 $f(x) = 3x^2 + 2x + 1$.

2 $g(t) = 3t^4 + 5t - 6$.

3 $h(x) = 1 - 2x^2 + 3x^3$.

4 $j(t) = -\frac{1}{t^2}$.

5 $f(x) = \frac{3}{x^3} + 2x^{3/2} - 1$.

6 $h(x) = x^{5/2} - \frac{5}{x^4} - \sqrt{x}$.

7 $f(t) = \frac{3}{2}t^{1/2} + 7$.

8 $h(x) = \frac{2}{x^{3/4}} - \frac{3}{x^{2/3}}$.

9 $f(x) = \sqrt[3]{x^2} + \frac{4}{\sqrt[4]{x^5}}$.

10 $g(x) = (x + 1)^4$.

11 $h(t) = (2t - 1)^5$.

12 $f(x) = 2x\sqrt{x^2 + 1}$.

13 $g(x) = 3x^2(x^3 + 7)^{5/3}$.

14 $h(t) = \sqrt{t + 1}$.

15 $f(x) = \frac{x}{\sqrt{x^2 + 1}}$.

16 $g(t) = \frac{t^3 + 3t^2 - 1}{\sqrt{t}}$.

17 $f(x) = \left(\sqrt[3]{x} + \frac{1}{\sqrt[3]{x}}\right)^2$.

18 $h(t) = t(1 - t^2)^{10}$.

19 $g(x) = \cos 2x$.

20 $f(x) = \sin^5 x \cos x$.

21 $h(x) = x \cos x^2$.

22 $j(x) = \frac{\cos \sqrt{x}}{\sqrt{x}}$.

23 $f(t) = \cos(2t + 1)$.

24 $g(x) = \cos^3 x \sin x$.

25 $f(x) = \sin 2x \cos 2x$.

Problems 26–28 deal with a suspension cable that supports a bridge with a horizontal load of ρ tons per linear foot. We choose the origin $(0, 0)$ as the lowest point of the cable, so its shape is that given by Equation (17) with $y_0 = 0$ and with T_0 in tons.

26 (a) Deduce from Equations (15) and (16) that the tension T_x at the point $(x, \rho x^2/2T_0)$ is given by

$$(T_x)^2 = (T_0)^2 + \rho^2 x^2.$$

(b) Eliminate T_0 to show that if $x \neq 0$, then

$$T_x = \frac{\rho x}{2y}(x^2 + 4y^2)^{1/2}.$$

27 Suppose that the bridge of Problem 26 is 200 feet long and weighs 200 tons, and that the two suspension towers are each 20 feet high. Use the results of Problem 26 to calculate:
(a) The tension T_0 (in tons) at the lowest point;
(b) The tension T_{100} at the points where the cable is attached to the towers; and
(c) The angle between the cable and the tower at the point of attachment.

28 Repeat the computations of Problem 27 in the case that the two suspension towers are each 40 feet high. Compare your answers with those of the previous problem. You should observe that increasing the height of the towers decreases the the tension in the supporting cable.

29 Consider a population of $P(t)$ small animals, with initial population $P(0) = 100$, with a rate of growth given by $dP/dt = \sqrt{P}$.
(a) Show that $P(t) = (t + 20)^2/4$. (*Suggestion:* Antidifferentiate the equation

$$\frac{1}{\sqrt{P}} \frac{dP}{dt} = 1,$$

and find the constant C that arises by using the information that $P = 100$ when $t = 0$.)
(b) What is the population when $t = 10$?
(c) When is $P(t) = 200$?

30 Suppose that a motorboat is traveling at $v = 40$ feet per second when its motor is cut off at time $t = 0$. Thereafter its deceleration due to water resistance is given by $dv/dt = -kv^2$ (k is a positive constant).
(a) Antidifferentiate the equation

$$-\frac{1}{v^2} \frac{dv}{dt} = k$$

to show that the boat's speed after t seconds is $v = 40/(1 + 40kt)$ feet per second.
(b) If the boat's speed after 10 seconds is 20 feet per second, how long does it take to slow to 5 feet per second?

31 Consider a cylindrical water tank with vertical sides and cross-sectional area A, so the volume of water in the tank is $V = Ay$ when the depth is y. Suppose that the initial volume of water is V_0 and that it takes T minutes for the tank to drain completely after the plug is pulled at time $t = 0$. Apply Torricelli's Law (Equation (3)) and the antidifferentiation method of Problem 29 to show that the volume of water in the tank after t minutes is $V = V_0[1 - (t/T)]^2$.

Velocity and Acceleration

Antidifferentiation is the tool that enables us, in many important cases, to analyze the motion of a particle (or "mass point") in terms of the forces acting upon it. We consider first the motion of a particle moving along a straight line under the influence of a *constant* accelerating force. If we regard the line of motion as the *x*-axis, then—as in Section 1-6—the motion of the particle is described by its **position function**

$$x = f(t) \qquad (1)$$

giving its *x*-coordinate at time t. Generally the function $f(t)$ will be unknown to start with, and our problem will be to find a formula for it, using such data as the initial position, the initial velocity, and the constant acceleration of the particle.

Recall from Section 1-6 that the *velocity* $v(t)$ of the moving particle is the derivative of its position function,

$$v = f'(t) \quad \text{or} \quad v = \frac{dx}{dt}, \qquad (2)$$

and that its *acceleration* $a(t)$ is the derivative of its velocity,

$$a = v'(t) = f''(t) \quad \text{or} \quad a = \frac{dv}{dt} = \frac{d^2x}{dt^2}. \qquad (3)$$

So if the acceleration is *constant*, we begin with the equation

$$\frac{dv}{dt} = a \qquad (a \text{ constant}), \qquad (4)$$

and antidifferentiate both sides to obtain

$$v = at + C. \qquad (5)$$

The constant C is ordinarily evaluated by substituting $t = 0$ into *both* sides of (5); this gives

$$v_0 = v(0) = a \cdot 0 + C = C,$$

so C turns out to be the *initial velocity* v_0. Hence

$$\frac{dx}{dt} = v = at + v_0. \qquad (6)$$

Another antidifferentiation gives

$$x = \tfrac{1}{2}at^2 + v_0 t + C. \qquad (7)$$

Here we should mention that the constants of antidifferentiation in Equations (5) and (7) have nothing to do with one another, despite the usual practice of denoting both by the same letter C. So we must, as before, evaluate the constant C in Equation (7), and—also as before—we do so by substituting

$t = 0$ into both sides of (7). This gives us

$$x_0 = x(0) = \tfrac{1}{2}a(0)^2 + v_0 \cdot (0) + C,$$

so that $C = x_0$, the *initial position* of the particle. And thus the position function of our particle is

$$x = \tfrac{1}{2}at^2 + v_0 t + x_0. \qquad (8)$$

WARNING Formulas (6) and (8) are valid only in the case of *constant* acceleration a. They do *not* apply to problems in which the acceleration varies.

EXAMPLE 1   The skid marks made by an automobile indicate that its brakes were fully applied for a distance of 225 feet before it came to a stop. Suppose it is known that the car in question has a constant deceleration of 50 ft/sec^2 under the conditions we describe. How fast was the car going when its brakes were applied?

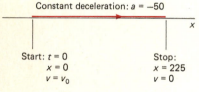

Constant deceleration: $a = -50$

Start: $t = 0$
  $x = 0$
  $v = v_0$

Stop:
  $x = 225$
  $v = 0$

3.32 Skid marks 225 feet long.

Solution   The introduction of a convenient coordinate system is often crucial to the successful solution of a physical problem. Here we take the x-axis as positively oriented in the direction of motion of the car. We choose the origin so that $x_0 = 0$ when $t = 0$, as indicated in Figure 3.32. In this coordinate system, the car's velocity v will be a decreasing function of time t, so that $a = -50$ (ft/sec^2) rather than $a = +50$. Hence Equations (6) and (8) take the forms

$$v = -50t + v_0,$$
$$x = -25t^2 + v_0 t.$$

The fact that the skid marks are 225 feet long tells us that $x = 225$ when the car came to a stop; that is, when $v = 0$. From the first of the equations above, we find that $t = v_0/50$ when $v = 0$, and substitution of this value of t together with the corresponding value $x = 225$ into the second equation gives

$$225 = -25\left(\frac{v_0}{50}\right)^2 + v_0\left(\frac{v_0}{50}\right).$$

When we solve this equation for v_0, we find that

$$v_0 = (100 \cdot 225)^{1/2} = 150$$

feet per second, or about 102 mph, was the velocity of the car when the brakes were first applied.

VERTICAL MOTION WITH CONSTANT GRAVITATIONAL ACCELERATION

One common application of Equations (4), (6), and (8) involves vertical motion near the earth's surface. A particle in such motion is subject to a downward acceleration denoted by g, which is about equal to 32 ft/sec^2. If we neglect air resistance, we may assume that this acceleration of gravity is the only outside influence on the moving particle; moreover, if the motion involved is fairly close to the earth's surface, we may also assume that g remains constant. (If you need more accurate values for g, you may use

CHAP. 3:   Applications of Derivatives and Antiderivatives

$g = 32.16$ ft/sec² in the fps system, $g = 980$ cm/sec² in the cgs system, or $g = 9.80$ m/sec² in the mks system.)

Since we deal with vertical motion here, it is natural to choose the y-axis as the coordinate system for position. If we choose the upward direction as the positive direction, then the effect of gravity on the particle is to *decrease* its height and also to *decrease* its velocity $v = dy/dt$, so we see that the particle's acceleration is

$$a = \frac{dv}{dt} = -g = -32 \text{ (ft/sec²)}.$$

Equations (6) and (8) then become

$$v = -32t + v_0 \qquad\qquad (6')$$

and

$$y = -16t^2 + v_0 t + y_0. \qquad\qquad (8')$$

Here y_0 is the initial height of the particle in feet and v_0 its initial velocity in feet per second.

EXAMPLE 2   A ball is thrown straight downward from the top of a tall building, with an initial speed of 30 feet per second. Suppose that the ball strikes the ground with a speed of 190 feet/second. How tall is the building?

Solution   We set up the coordinate system illustrated in Figure 3.33, with ground level corresponding to $y = 0$, with the ball thrown at time $t = 0$, and with the positive direction being the upward direction.

Since the height y of the ball is a decreasing function of time, the velocity $v = dy/dt$ is negative. So the data of the problem amount to this: $v_0 = -30$ and $v = -190$ when $y = 0$. We want to find y_0. Equations (6') and (8') give

$$y = -16t^2 - 30t + y_0,$$
$$v = -32t - 30.$$

If we use the fact that $v = -190$ when $y = 0$, the second equation will tell us when the ball hits the ground:

$$-190 = -32t - 30,$$

so that the solution $t = 5$ means that the ball strikes the ground 5 seconds after it's thrown. The other of the two equations may then be used, for we know that $y = 0$ when $t = 5$, and we find that

$$0 = (-16)(25) - (30)(5) + y_0,$$

so that $y_0 = 550$. Thus the building is 550 feet high.

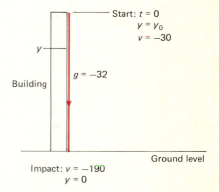

3.33   Ball thrown down from the top of a building.

*NEWTON'S INVERSE-SQUARE LAW OF GRAVITATION

Since we have already used whatever calculus is necessary in deriving the two key formulas in (6) and (8), such problems as the last—though interesting and even important—challenge only one's algebraic skills. Let us turn to some more exciting problems, involving motion of objects (meteors? spacecraft?) at considerable distances from the earth (and from other astronomical bodies). To do so, we need first to mention Newton's second law of motion and his inverse-square law of gravitation. According to the law of motion, the force F acting on a particle of mass m is proportional to the acceleration a that it produces:

$$F = ma. \tag{9}$$

According to the law of gravitation, the gravitational force of attraction between two point masses m and M located at a distance r apart is given by

$$F = \frac{GMm}{r^2} \tag{10}$$

where G is a certain empirical constant. The formula also applies if either of the two masses is not merely a point mass, but a homogeneous sphere; in this case, the distance r is measured between the centers of the spheres. (We shall derive this result in Chapter 16.)

Let M denote the mass of the earth and R its radius. We obtain the gravitational acceleration $a = g$ of a particle of mass m at the earth's surface by using Equations (9) and (10) above simultaneously:

$$ma = mg = \frac{GMm}{r^2} = \frac{GMm}{R^2},$$

so that

$$g = \frac{GM}{R^2}. \tag{11}$$

We shall use Equation (11) from time to time to remove the necessity for the actual determination of G in our problems and examples.

We mentioned that Equation (10) holds for point masses, and that we shall later show that it also holds for spheres. It appears that Newton was aware of the inverse-square law of gravitation for point masses before 1670, but was unable to show that it applies to massive objects (such as spherical planets) until some years later. This may have been the reason for the delay until 1687 of the publication of his *Philosophiae Naturalis Principia Mathematica* (*Mathematical Principles of Natural Philosophy*), the founding document of modern exact science.

EXAMPLE 3 If a woman has enough "spring" in her legs to jump vertically to a height of 4.00 feet on the earth, how high could she jump on the moon? Use the fact that the mass $\bar{M}$ and the radius $\bar{R}$ of the moon are given in terms of the mass M and radius R of the earth by $\bar{M} \approx (0.0123)M$ and $\bar{R} \approx (0.2725)R$. Because her mass is presumably the same anywhere in the universe, it's also reasonable to assume that she attains the same initial velocity before liftoff on the moon as on the earth.

Solution First we need to find the initial velocity v_0 required to jump 4.00 feet high on the earth. Then we can calculate how high she can jump on the moon with the same initial velocity.

CHAP. 3: **Applications of Derivatives and Antiderivatives**

On the earth, we use the standard equations

$$y = -16t^2 + v_0 t + y_0 \qquad (8')$$

and

$$v = -32t + v_0. \qquad (6')$$

In the woman's jump on the earth, we have $v = v_0$ when $t = 0$, $y_0 = 0$ (with $y = 0$ at ground level), and $v = 0$ when $y = 4.00$. So we substitute $v = 0$ into (6'), and find that the time required to reach her maximum height of 4.00 feet is $t = v_0/32$ seconds.

Hence Equation (8') gives

$$4 = -16\left(\frac{v_0}{32}\right)^2 + v_0\left(\frac{v_0}{32}\right) = \frac{v_0^2}{64}.$$

So $v_0 = +16$ feet per second.

Now we shift our attention to the moon. In order to find the correct versions of Equations (6') and (8') for the moon, we need to find the gravitational acceleration on its surface. But Equation (11) tells us that, if $\bar{g}$ denotes the gravitational acceleration for the moon,

$$\bar{g} = \frac{G\bar{M}}{(\bar{R})^2} = G\frac{(0.0123)M}{(0.2725R)^2} = (0.165643)\frac{GM}{R^2}$$

$$= (0.165643)g = (0.165643)(32) = 5.3$$

feet per second per second (approximately).

On the moon, then,

$$\frac{dv}{dt} = -5.3,$$

$$v = \frac{dy}{dt} = (-5.3)t + v_0, \qquad (6'')$$

and

$$y = (-2.65)t^2 + v_0 t + y_0. \qquad (8'')$$

In the moon jump, we have $y_0 = 0$ and $v_0 = 16$, from our previous work with the earth jump. In addition, $v = 0$ at maximum height y. So (6'') gives

$$0 = (-5.3)t + 16$$

then, and thus $t \approx 3.02$ seconds at her maximum height y_{max}. We substitute this value of t into (8''), and obtain

$$y_{max} \approx (-2.65)(3.02)^2 + (16)(3.02) \approx 24.15$$

feet as the height she can jump on the moon.

*ESCAPE VELOCITY

As an example of linear motion with *variable* acceleration, we now calculate the "escape velocity" from the earth—the minimum initial velocity with which an object must be launched straight upward from the earth's surface so that it will continue forever to move away from the earth.

The condition that the upward motion will continue forever becomes the relation

$$v = \frac{dy}{dt} > 0 \qquad \text{for all} \quad t \geq 0.$$

We will use $y = y(t)$ to denote the distance from the object to the earth's *center* because of the way that Newton's law of gravitation is phrased. Our launch site, though, will be the earth's *surface*, where we take $y_0 = y(0) = 6370$ km, or 6,370,000 meters. The solution to the escape velocity problem will be the initial velocity $v_0 = v(0)$ that is just sufficient to insure that $v = v(t)$ is never zero or negative.

Now imagine the launched object as it's shown in Figure 3.34, with mass m, at distance $y = y(t)$ from the earth's center at some time $t > 0$, with velocity $v = v(t)$ then. The only force acting on the mass m is the pull of the earth's gravity, given by Equation (10), Newton's law of gravitation, as

$$F = -\frac{GMm}{y^2}.$$

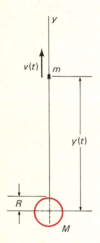

3.34 Escape velocity.

The resulting acceleration of m is then given, with the *same* values of F and m, by Newton's second law of motion:

$$F = ma. \tag{9}$$

We eliminate F by equating the last two right-hand sides. The minus sign in the gravitation equation belongs there because the force is acting opposite to the direction of increasing values of y. No minus sign is needed in the law of motion, because the sign of the acceleration a is built in automatically. And so

$$ma = m\frac{dv}{dt} = -\frac{GMm}{y^2},$$

which we simplify to

$$\frac{dv}{dt} = -\frac{GM}{y^2}.$$

To solve this differential equation, we use the chain rule:

$$\frac{dv}{dt} = \frac{dv}{dy}\frac{dy}{dt} = v\frac{dv}{dy}.$$

Thus

$$v\frac{dv}{dy} = -\frac{GM}{y^2}.$$

That is,

$$D_y\left(\frac{1}{2}v^2\right) = D_y\left(\frac{GM}{y}\right).$$

Note that we think of y as the independent variable, rather than the more natural t. It follows immediately that

$$\frac{1}{2}v^2 = \frac{GM}{y} + C.$$

CHAP. 3: **Applications of Derivatives and Antiderivatives**

To evaluate the constant C we use the fact that when $t = 0$, $v = v_0$ and $y = R$ (the earth's radius). Thus

$$\frac{1}{2} v_0^2 = \frac{GM}{R} + C,$$

and substitution of

$$C = \frac{1}{2} v_0^2 - \frac{GM}{R}$$

into the equation

$$\frac{1}{2} v^2 = \frac{GM}{y} + C$$

tells us that

$$v^2 = v_0^2 + 2GM \left(\frac{1}{y} - \frac{1}{R} \right).$$

In particular,

$$v^2 > v_0^2 - \frac{2GM}{R}.$$

Therefore v will remain positive provided that

$$v_0^2 \geqq \frac{2GM}{R}.$$

So the escape velocity for the earth is given by

$$v_0 = \sqrt{\frac{2GM}{R}} = \left(2 \frac{GM}{R^2} R \right)^{1/2} = \sqrt{2gR}. \qquad (12)$$

With $g = 9.8$ m/sec^2 and $R = 6.37 \times 10^6$ meters, this gives $v_0 \approx 11174$ meters per second, about 36,660 ft/sec, about 24,995 mph, or about 6.94 miles per second.

The point is that escape velocity is *finite* and relatively easy to attain, as you might deduce from the existence of the photograph of Figure 3.35.

EXAMPLE 4   Suppose that you are stranded—your rocket engine has failed—on an asteroid of diameter 3 miles, with density equal to that of the earth. If you have enough spring in your legs to jump 4 feet straight up on earth while wearing your present life-support equipment, can you blast off from this asteroid using leg power alone?

Solution   By the computation shown in the first steps of the solution to Example 3, we find that an initial velocity of $v_0 = 16$ feet per second is required to jump 4 feet high on earth. We must determine if this initial velocity is adequate for escape from the asteroid.

Let r denote the ratio of the radius of the asteroid to that of the earth, so that

$$r = \frac{1.5}{3960} = \frac{1}{2640}.$$

Also, the mass M_a and radius R_a of the asteroid are given by

$$M_a = r^3 M \quad \text{and} \quad R_a = rR$$

in terms of the mass M and radius R of the earth.

We found in our discussion of escape velocity that it is given by

$$v_0 = \sqrt{2gR} = \left(\frac{2GM}{R}\right)^{1/2} \tag{12}$$

where g is surface gravity and R the radius of the planet. The derivation was independent of the fact that we were discussing the Earth; the above formula applies to the moon, to Mars, to the sun, or to any asteroid that's roughly spherical, so long as the mass is M and the radius is R. So we find that escape velocity from the asteroid is

$$v_a = \left(\frac{2GM_a}{R_a}\right)^{1/2} = \left(\frac{2Gr^3 M}{rR}\right)^{1/2}$$

$$= r\sqrt{\frac{2GM}{R}} = rv_0,$$

CHAP. 3:   **Applications of Derivatives and Antiderivatives**

where v_0 is escape velocity from the earth. We found that v_0 is about 36,660 feet per second for the earth, so

$$v_a = rv_0 = \frac{36660}{2640} = 13.9 \quad \text{(ft/sec)}.$$

Since the escape velocity from the asteroid is less than the available initial velocity—16 feet per second—you can indeed jump right off this asteroid using your leg power alone.

*BLACK HOLES

Black holes have been in the news frequently in recent years. A black hole is a body that is so dense that escape velocity from its surface exceeds the speed of light, $c = 3.00 \times 10^8$ meters per second. According to the theory of relativity, no material body can travel even as fast as light, so nothing—no radiation, no light, no spaceship—can ever escape from a black hole.

From Equation (12) we see that the relationship between the mass M and radius R of a black hole is

$$c \leqq v_0 = \left(\frac{2GM}{R}\right)^{1/2}$$

so that

$$R \leqq \frac{2GM}{c^2}. \tag{13}$$

For the earth, with mass M_e and radius R_e, we know that the gravitational acceleration at the surface is

$$g = \frac{GM_e}{R_e^2} \approx 9.80$$

meters/sec^2, so

$$GM_e \approx 3.98 \times 10^{14} \; m^3/\text{sec}^2.$$

Hence if the earth were sufficiently compressed to form a black hole, its radius would be

$$R = \frac{2GM_e}{c^2} = \frac{2(3.98 \times 10^{14})}{(3.00 \times 10^8)^2} \approx 8.84 \times 10^{-3}$$

meters, or about 0.88 centimeters.

The mass of the sun is $M_s \approx 329{,}340 M_e$, so if the sun were compressed enough to be a black hole, its radius would be

$$R = \frac{2GM_s}{c^2} = \frac{2GM_e}{c^2} \frac{M_s}{M_e} \approx (8.84 \times 10^{-3})(329{,}340),$$

which turns out to be about 2.91×10^3 meters, or about 2.91 kilometers. The actual radius of the sun is close to 695,300 kilometers.

In each of Problems 1–6, find a function $y = f(x)$ satisfying the given differential equation and the prescribed initial conditions.

1 $\dfrac{dy}{dx} = 2x + 1; \quad y(0) = 3.$

2 $\dfrac{dy}{dx} = (x - 2)^3; \quad y(2) = 1.$

3 $\dfrac{dy}{dx} = \sqrt{x}; \quad y(4) = 0.$

4 $\dfrac{dy}{dx} = \dfrac{1}{x^2}; \quad y(1) = 5.$

5 $\dfrac{dy}{dx} = \dfrac{1}{\sqrt{x + 2}}; \quad y(2) = -1.$

6 $\dfrac{dy}{dx} = x\sqrt{x^2 + 9}; \quad y(-4) = 0.$

In Problems 7–13, find the position function $x(t)$ of a moving particle with the given acceleration $a(t)$, initial position $x_0 = x(0)$, and initial velocity $v_0 = v(0)$.

7 $a(t) = 50, \quad v_0 = 10, \quad x_0 = 20.$

8 $a(t) = -20, \quad v_0 = -15, \quad x_0 = 5.$

9 $a(t) = 3t, \quad v_0 = 5, \quad x_0 = 0.$

10 $a(t) = 2t + 1, \quad v_0 = -7, \quad x_0 = 4.$

11 $a(t) = 4(t + 3)^2, \quad v_0 = -1, \quad x_0 = 1.$

12 $a(t) = \dfrac{3}{\sqrt{t + 4}}, \quad v_0 = -1, \quad x_0 = 1.$

13 $a(t) = \dfrac{1}{(t + 1)^3}, \quad v_0 = 0, \quad x_0 = 0.$

14 Find the solution of the differential equation

$$\frac{dy}{dx} = 2xy^2$$

such that $y(0) = 1$. (*Suggestion:* Rewrite this differential equation in the form

$$\frac{1}{y^2}\frac{dy}{dx} = 2x; \quad \text{that is,} \quad D_x\!\left(-\frac{1}{y}\right) = D_x(x^2).$$

Then antidifferentiate with respect to x. Finally, substitute $x = 0$ and $y = 1$ to evaluate the constant that appears when you antidifferentiate.)

Use the method of Problem 14 to solve the differential equations in Problems 15–20.

15 $\dfrac{dy}{dx} = y^2; \quad y(0) = 1.$

16 $\dfrac{dy}{dx} = \sqrt{y}; \quad y(0) = 4.$

17 $\dfrac{dy}{dx} = \dfrac{1}{4y^3}; \quad y(0) = 1.$

18 $\dfrac{dy}{dx} = \dfrac{1}{x^2 y}; \quad y(1) = 2.$

19 $\dfrac{dy}{dx} = \sqrt{xy}; \quad y(0) = 4.$

20 $\dfrac{dy}{dx} = \dfrac{x}{y}; \quad y(3) = 5$

21 A ball is dropped from the top of a building that is 400 feet high. How long does it take to reach the ground? With what velocity does the ball strike the ground?

22 A car's brakes are applied when it is going 60 mph and provide a constant deceleration of 40 ft/sec². How far does the car travel before coming to a stop?

23 A ball is thrown straight upward from ground level with an initial speed of 160 feet per second. What is the maximum height that the ball attains? How long does it remain aloft?

24 A ball is thrown straight down from the top of a tall building. The initial speed of the ball is 25 feet per second. It hits the ground with a speed of 153 feet per second. How tall is the building?

25 A baseball is thrown straight downward with an initial speed of 40 feet per second from the top of the Washington Monument, 555 feet high. How long does it take to reach the ground, and with what speed does it strike the ground?

26 A spacecraft is free-falling toward the moon's surface at a speed of 1000 miles per hour. Its retrorockets, when fired, provide a deceleration of 20,000 miles per hour per hour. At what height above the surface should the retrorockets be turned on to insure a "soft touchdown" ($v = 0$ at impact)? Ignore the moon's gravitational field.

27 A bomb is dropped from a balloon hovering at an altitude of 800 feet. Directly below at point P, a projectile is fired upward toward the bomb, exactly 2 seconds after the bomb is released. With what initial speed should the projectile be fired in order to hit the bomb at an altitude of exactly 400 feet?

28 A car traveling at 60 mph skids 176 feet after its brakes are applied. What is the (constant) deceleration provided by the car's brakes?

29 The following table gives the masses and radii of several planetary bodies as multiples of the mass M_e and radius R_e of the earth. Calculate the escape velocity in miles per hour, kilometers per hour, and meters per second, from the surface of each of these bodies.

Body	M/M_e	R/R_e
Earth	1.0	1.0
Moon	0.0123	0.272
Mars	0.107	0.538
Jupiter	315.	11.3
Sun	329,400.	109.
Ganymede	0.0206	0.392
(Jovian satellite)		

30 Use the data of Problem 29 to determine how high the woman of Example 3 could jump on Jupiter.

31 What would the radius of Jupiter be if it were compressed just enough to become a black hole?

32 What would the radius of Ganymede be if it were compressed enough to become a black hole?

33 An object dropped 16 feet above the earth's surface strikes the ground in 1 second. How long would it take an object to fall to the sun's surface from an initial height of 16 feet?

---

*3-9

Applications to Economics

The methods of calculus have developed historically in close connection with physical applications such as those of Section 3-8. In recent years these methods have become important in the biological, behavioral, and economic sciences as well. In this section we discuss some applications of elementary calculus to decision problems that arise in the management of typical manufacturing and marketing operations.

We start with the *cost function* $C(x)$ introduced in Section 2-3; $C(x)$ is the total cost of manufacturing x units of a certain item. Once x items have been produced, the average cost per unit of producing Δx additional units is

$$\frac{\Delta C}{\Delta x} = \frac{C(x + \Delta x) - C(x)}{\Delta x},$$

which is approximately $C'(x)$ if Δx is fairly small. The derivative $C'(x)$ of the cost function is called the **marginal cost function.** Since

$$C(x + 1) - C(x) \approx C'(x) \tag{1}$$

by the linear approximation formula (Equation (3) of Section 3-2), we see that *the marginal cost $C'(x)$ is approximately equal to the cost of producing a single additional unit* (after x units have already been produced). This value generally differs from the **average cost** $u(x)$ of producing one unit, defined by $u(x) = C(x)/x$.

In order to minimize its costs or maximize its profits (not always the same thing), a company's managers need to know the specific form of its cost function. One realistic assumption is that the marginal cost is a linear function of x:

$$C'(x) = 2Ax + B. \tag{2}$$

(The reason for the factor 2 will appear in a moment.) The coefficient of x in (2) would be zero if the unit cost per additional item were constant. But it sometimes happens that the marginal cost increases as the total production x increases. For example, at a high level of production it may be necessary for the company to pay a premium for scarce raw materials or for additional energy sources. Similarly, a high level of production may necessitate the payment of overtime wages, thereby increasing unit labor costs.

If we antidifferentiate (2) we obtain the cost function

$$C(x) = Ax^2 + Bx + C_0. \tag{3}$$

The constant C_0 has a natural interpretation—it is the **setup cost** or **fixed cost** that the company incurs for such expenses as maintenance of facilities even if no items are produced.

Now we turn our attention to the marketing of the items being manufactured. As in Section 2-3, we assume that the company's market experience indicates that the number x of units that can be sold is a linear function of the selling price p per unit,

$$x = m - np. \tag{4}$$

When we solve (4) for p, we get the **price function**

$$p = p(x) = a - bx, \tag{5}$$

also known as the **demand function**—the price p corresponds to a **demand** for x units. The company's total income is given by the **revenue function**

$$R(x) = xp(x), \tag{6}$$

the number of units sold multiplied by the selling price of each. The company's **profit function** is

$$P(x) = R(x) - C(x). \tag{7}$$

The derivatives $R'(x)$ and $P'(x)$ are called the **marginal revenue** and **marginal profit,** respectively. By approximations similar to (1), they correspond to the revenue and profit that result from the sale of a single additional item, given that x items have already been sold. On the basis of maximum-minimum theory, we expect the maximum profit to occur at a critical point of the sort where

$$P'(x) = R'(x) - C'(x) = 0,$$

or $R'(x) = C'(x)$. Thus there may be a maximum profit when the marginal cost and marginal revenue are equal—when the approximate cost of producing an additional item is equal to the approximate additional revenue resulting from its sale.

EXAMPLE 1   A calculator company has been making 1500 programmable calculators monthly at a total cost of \$76,750 and selling them at \$57 each for a monthly profit of $(1500)(\$57) - \$76,750 = \$8750$. An analysis of its operations shows that its cost of producing x calculators in a month is

$$C(x) = (0.003)x^2 + 30x + 25{,}000,$$

and that the number of calculators that can be sold monthly for p dollars each is

$$x = 30{,}000 - 500p.$$

How many calculators should the company make, and at what price should it sell them, in order to maximize its monthly profit?

Solution   When we solve the equation $x = 30{,}000 - 500p$, we obtain the price function

$$p(x) = 60 - (0.002)x.$$

Hence the company's revenue function is

$$R(x) = xp(x) = 60x - (0.002)x^2,$$

and its profit function is

$$P(x) = R(x) - C(x)$$
$$= [60x - (0.002)x^2] - [(0.003)x^2 + 30x + 25,000];$$

that is,

$$P(x) = 30x - (0.005)x^2 - 25,000.$$

The marginal profit is

$$P'(x) = 30 - (0.010)x,$$

so $P'(x) = 0$ only when $x = 3000$. Since $P'(x) > 0$ when $x < 3000$ and $P'(x) < 0$ when $x > 3000$, the first derivative test shows that $x = 3000$ maximizes the monthly profit. These 3000 calculators should be sold at

$$p(3000) = 60 - (0.002)(3000) = \$54$$

each. The resulting monthly profit will be

$$P(3000) = 30(3000) - (0.005)(3000)^2 - 25,000 = \$20,000.$$

Thus the company can more than double its monthly profits by doubling production and simultaneously decreasing its selling price by $3. The following table gives a comparison of the two situations.

x	Price $p(x)$	Revenue $R(x)$	Cost $C(x)$	Profit $P(x)$	Average Cost $u(x) = C(x)/x$
1500	$57	$85,500	$76,750	$8,750	$51.17
3000	$54	$162,000	$142,000	$20,000	$47.33

IMPROVED TECHNOLOGY AND DECREASED AVERAGE COST

Our discussion preceding Example 1 contained the implicit assumption of a fixed situation regarding production facilities and available technology. When production is increased over a period of time, however, a significant reduction in marginal cost due to accumulated experience and improved technology is the usual rule. For example, calculators that cost several hundred dollars each when introduced in the early 1970s are now available at a very low price. Studies of the histories of a wide variety of products—from large airplanes to small calculators—indicate that a doubling of production typically results in a decrease of 20% to 30% in the cost per unit. These studies suggest a marginal cost function of the form

$$C'(x) = \frac{k}{x^\alpha}, \tag{8}$$

where $0 < \alpha < 1$.

We antidifferentiate the above equation to find the cost function:

$$C(x) = \frac{kx^{1-\alpha}}{1-\alpha}, \tag{9}$$

under the assumption that the production level x is so large that the constant of antidifferentiation (the setup cost) is negligible in comparison with the right-hand side of (9), and may therefore be ignored.

EXAMPLE 2 Calculate the percentage decrease in average cost $u(x) = C(x)/x$ that results from doubling x, taking the values $\alpha = \frac{1}{2}$ and $\alpha = \frac{1}{3}$ in (9).

Solution With $\alpha = \frac{1}{2}$ we have

$$\frac{u(2x)}{u(x)} = \frac{C(2x)/2x}{C(x)/x} = \frac{2k(2x)^{1/2}/2x}{2k(x)^{1/2}/x} = \frac{\sqrt{2}}{2} \approx 0.71;$$

thus the average cost is about 71% as much when $2x$ units have been produced as when x units have been produced, a decrease of 29% in the average cost. With $\alpha = \frac{1}{3}$ we have

$$\frac{u(2x)}{u(x)} = \frac{C(2x)/2x}{C(x)/x} = \frac{\frac{3}{2}k(2x)^{2/3}/2x}{\frac{3}{2}k(x)^{2/3}/x} = \frac{2^{2/3}}{2} \approx 0.79,$$

which amounts to a decrease of 21% in the average cost.

3-9 PROBLEMS

In Problems 1–5, find the following quantities for the given cost function $C(x)$:
(a) The cost of producing 400 items;
(b) The average cost of each of the first 400 items produced;
(c) The marginal cost when $x = 400$, and the exact cost of producing the 401st item.

1 $C(x) = 1000 + 3x.$
2 $C(x) = 500 + 50x + (0.002)x^2.$
3 $C(x) = (0.004)x^2 + 40x + 8000.$
4 $C(x) = 5000 + \sqrt{x}.$

5 $C(x) = 700 + 5x + \dfrac{100}{\sqrt{x}}.$

6 Minimize the average cost with the cost function of Problem 2.

7 Minimize the average cost with the cost function of Problem 3.

8 A small company makes x radios daily at a cost of $C(x) = 125 + 30x + 2x^{3/2}$ dollars. What daily production level will minimize the average cost?

9 Show that when the average cost is minimal, the average cost and the marginal cost are equal.

10 A company makes x watches weekly that can be sold for $35 each, at a cost of

$$C(x) = 2000 + 15x + (0.02)x^2$$

dollars. What production level will maximize its weekly profit?

11 Suppose now that the price function for the watches of Problem 10 is $p = 40 - (0.01)x$. How many watches should the company make weekly and at what price should it sell them in order to maximize its weekly profit?

12 Now assume that the company of Problem 10 has been selling 200 watches weekly for $40 each. Market research indicates that it can sell an additional 50 watches weekly for each reduction of $1 (below $40) in the selling price. How many watches should it make weekly and at what price should it sell them in order to maximize its weekly profit?

13 If the cost function is $C(x) = Bx + C_0$ and the price function is $p(x) = a - bx$ with $a > B$, show that there exists a value of x that maximizes the profit function.

14 At a theater seating 1000 persons, 500 tickets were sold daily at $4 each. When the price was reduced to $3 each, 700 tickets were sold daily. Assuming that the price function is a linear function, what price will maximize its daily revenue?

15 The marginal cost of producing a certain item is $50 + (0.002)x$ dollars, and the average cost for 100 items is $60. Find the cost function and the average cost for each of the first 200 items.

16 A car manufacturer has a weekly cost function of $C(x) = 1000 + 5x$ thousand dollars. The manufacturer has been breaking even by making 1000 cars weekly and selling them at $6000 each. Market research indicates that each rebate to the customer of $50 per car will stimulate the sales of 100 additional cars weekly. What rebate should it offer in order to maximize its weekly profits?

17 (a) With the cost function of Equation (9), show that $u(2x)/u(x) = 2^{-\alpha}$.
(b) Assume the cost function of Equation (9) with $\alpha = 0.6$. Suppose that the average cost of a new product is $100 in the first year of production and that the production level doubles every year. After how long will the average cost be $12.50? (*Suggestion:* Compute $100(2^{-\alpha})^n$ for $n = 1, 2, 3, \ldots$.)

Use the list below as a guide to ideas that you may need to review.

1 The increment Δy
2 The differential dy
3 The linear approximation formula
4 Differentiation rules in differential form
5 Solution of related rates problems
6 Linear interpolation
7 The method of repeated substitution
8 Newton's method
9 Calculation of higher derivatives
10 Concave upward and concave downward functions
11 Test for concavity
12 The second derivative test
13 Inflection points
14 Inflection point test

15 Infinite limits
16 Vertical asymptotes
17 Limits as $x \to \pm\infty$
18 Horizontal asymptotes
19 Curve-sketching strategy
20 Antidifferentiation and antiderivatives
21 The most general antiderivative of a function
22 Antidifferentiation formulas
23 Antidifferentiation using the chain rule
24 Velocity and acceleration
25 Solution of problems with constant acceleration
26 Newton's inverse-square law of gravitation
27 Marginal and average cost
28 Marginal revenue and marginal profit
29 Price or demand function
30 Maximizing the profit

MISCELLANEOUS PROBLEMS

In Problems 1 and 2, calculate Δy and dy with the indicated values of x and Δx.

1 $y = x^{2/3}$, $x = 125$, $\Delta x = -2$.
2 $y = \sqrt{x^2 + 9}$, $x = 4$, $\Delta x = 0.1$.

In Problems 3–5 estimate the indicated number by linear approximation.

3 $\sqrt[3]{123}$.   4 $\sqrt[5]{33}$.   5 $\sqrt[4]{80}$.

6 A rectangular block with square base is being squeezed in such a way that its height y is decreasing at the rate of 2 inches per minute while its volume remains constant. At what rate is the edge x of its base increasing when $x = 30$ inches and $y = 20$ inches?

7 Air is being pumped into a spherical balloon at the constant rate of 10 in^3/sec. At what rate is the surface area of the balloon increasing when its radius is 5 inches?

8 A ladder 10 feet long is leaning against a wall. If the bottom of the ladder slides away from the wall at the constant rate of 1 mile per hour, how fast (in mph) is the top of the ladder moving when it is 1/100 foot above the ground?

9 A water tank has the shape of an inverted cone (axis vertical and vertex downward) with a top radius of 5 feet and a height of 10 feet. The water is flowing out of the tank, through a hole at the vertex, at the rate of 50 ft^3/min. What is the time rate of change of the depth of the water in the tank at the instant when it is 6 feet deep?

10 Plane A is flying west toward an airport at an altitude of 2 miles; plane B is flying south toward the same airport at an altitude of 3 miles. When both planes are 2 miles (ground distance) from the airport, the speed of plane A is 500 mph and the distance between the two planes is decreasing at 600 mph. What is the speed of plane B?

11 A water tank is shaped so that the volume of water in the tank is $V = 2y^{3/2}$ in^3 when its depth is y inches. If water flows out a hole at the bottom at the rate of $3\sqrt{y}$ in^3/min, at what rate does the water level in the tank fall? Can you think of a practical application of such a water tank?

12 Use Newton's method to find the root of the equation $x^5 + x - 75 = 0$ accurate to three decimal places.

13 Find the depth to which a wooden ball with radius 2 feet sinks in water if its density is $\frac{1}{3}$ that of water. A useful formula appears in Problem 19, Section 3-4.

14 The equation $x^2 + 1 = 0$ has no real solutions. Try finding a solution by using Newton's method, and report what happens. Use the initial estimate $x_0 = 2$.

15 At the beginning of Section 3-4, we mentioned the fifth-degree equation

$$x^5 - 3x^3 + x^2 - 23x + 19 = 0,$$

which has at least one and no more than five distinct real solutions. Find all real solutions by Newton's method.

16 The equation $\tan x = 1/x$ has a sequence $\alpha_1, \alpha_2, \alpha_3, \ldots$ of positive roots, with α_n a bit more than $(n-1)\pi$. Use Newton's method to compute α_1 and α_2 to three-place accuracy.

In each of Problems 17–20, calculate y' and y'' under the assumption that y is defined implicitly as a function of x by the given equation.

17 $x^{1/3} + y^{1/3} = 1$.   **18** $2x^2 - 3xy + 5y^2 = 25$.
19 $y^5 - 4y + 1 = \sqrt{x}$.   **20** $\sin(xy) = xy$.

Sketch the graphs of the functions in Problems 21–30, indicating all critical points, inflection points, asymptotes, and concave structure.

21 $f(x) = x^4 - 32x$.   **22** $f(x) = 18x^2 - x^4$.
23 $f(x) = x^6 - 2x^4$.   **24** $f(x) = x\sqrt{x-3}$.
25 $f(x) = x\sqrt[3]{4-x}$.   **26** $f(x) = \dfrac{x-1}{x+2}$.
27 $f(x) = \dfrac{x^2+1}{x^2-4}$.   **28** $f(x) = \dfrac{x}{x^2-x-2}$.
29 $f(x) = \dfrac{2x^2}{x^2-x-2}$.   **30** $f(x) = \dfrac{x^3}{x^2-1}$.

Find the most general antiderivatives of the functions given in Problems 31–38.

31 $f(x) = \dfrac{x^5 - 2x + 5}{x^3}$.   **32** $f(x) = \dfrac{(1+\sqrt{x})^5}{\sqrt{x}}$.
33 $f(x) = \dfrac{x+1}{\sqrt[3]{x^2+2x}}$.   **34** $f(x) = x^{3/2}(1-\sqrt{x})^2$.
35 $f(x) = (\cos x)\sqrt{1 + \sin x}$.
36 $f(x) = x\sin(1+x^2)$.
37 $f(x) = \sqrt{x}(1+2x^{3/2})^{-2}$.
38 $f(x) = (1-3x)^9$.

In each of Problems 39–43, find a function $y = f(x)$ satisfying the given differential equation and the prescribed initial condition.

39 $\dfrac{dy}{dx} = (2x+1)^5$, $y(0) = 2$.

40 $\dfrac{dy}{dx} = \dfrac{2}{\sqrt{x+5}}$, $y(4) = 3$.

41 $\dfrac{dy}{dx} = \dfrac{1}{\sqrt[3]{x}}$, $y(1) = 1$.

42 $\dfrac{dy}{dx} = \dfrac{1}{\sqrt{y}}$, $y(0) = 1$.

43 $\dfrac{dy}{dx} = (x+1)\sqrt{y}$, $y(1) = 1$.

44 If a car starts from rest with an acceleration of 8 ft/sec², how far has it traveled by the time it reaches a speed of 60 mph?

45 Consider a planet where a ball dropped from a height of 20 feet hits the ground in 2 seconds. If a ball is dropped from the top of a 200-foot building on this planet, how long will it take to hit the ground? With what speed will it hit?

46 A man can throw a ball from the earth's surface straight upward to a maximum height of 144 feet. How high could he throw it on the planet of Problem 45?

47 Suppose that a car skids 44 feet if it is going 30 mph when the brakes are applied. Assuming the same constant deceleration, how far will it skid if it is going 60 mph when the brakes are applied?

48 (a) Deduce from the discussion of escape velocity in Section 3-8 that, if a projectile is launched upward from the earth's surface with an initial velocity v_0 that is less than the escape velocity, then the maximum distance from the center of the earth that the projectile will attain is

$$y_{\max} = \frac{2gR^2}{2gR - v_0^2},$$

where g and R are the surface gravity and radius of the earth (as in Section 3-8).
(b) With what initial velocity v_0 (in mph) must such a projectile be launched to yield a maximum height of 100 miles above the earth's surface?
(c) Find the maximum distance from the earth's center (in terms of earth radii) attained by a projectile launched from the earth's surface with 90% of escape velocity.

49 Suppose that it costs $1 + (0.0003)v^{3/2}$ dollars per mile to operate a truck at v mph. If there are additional costs (including, for example, the driver's pay) of \$10 per hour, what speed will minimize the total cost of a 1000-mile trip?

50 Water is being poured into the conical tank of Problem 9 at the rate of 50 ft³/min, and is draining out the hole at the bottom at the rate of $10\sqrt{y}$ ft³/min (here, y is the depth of the water in the tank).
(a) At what rate is the water level rising when $y = 5$ (feet)?
(b) Suppose that the tank is initially empty, water is poured in at 25 ft³/min, and water continues to drain out at $10\sqrt{y}$ ft³/min. What is the maximum depth attained by the water?

51 Sketch the curve $y^2 = x(x-1)(x-2)$, indicating that it consists of two pieces—one bounded and the other unbounded—and has two horizontal tangent lines, three vertical tangent lines, and two inflection points. (*Suggestion:* Note that the curve is symmetric about the x-axis, and begin by determining the intervals in which $x(x-1)(x-2)$ is positive. Compute y' and y'' by implicit differentiation.)

The Integral

4

Introduction

The earlier chapters have dealt with **differential calculus,** which is one of two closely related parts of *the* calculus. Differential calculus is centered on the concept of the *derivative*. Recall that the original motivation for the derivative was the problem of defining tangent lines to graphs of functions and calculating the slopes of such lines. By contrast, the importance of the derivative stems from its applications to diverse problems that may seem, upon initial inspection, to have little connection with tangent lines to graphs.

Integral calculus centers around the concept of the *integral*. The definition of the integral is motivated by the problem of defining and calculating the area of the region lying between the graph of a positive-valued function f and the x-axis over a closed interval $[a, b]$. The area of the region R of Figure 4.1 is given by the **integral** of f from a to b, denoted by the symbol $\int_a^b f(x)\, dx$. But the importance of the integral, like that of the derivative, is due to its applications in many problems that may appear unrelated to its original motivation.

The principal theorem of this chapter is the *Fundamental Theorem of Calculus* in Section 4-4. It provides a vital connection between the operations of differentiation and integration and also a method of computing values of integrals. In this chapter we concentrate on the definition and the basic computational properties of the integral, and reserve applications—other than those closely associated with area computations—to Chapter 5.

Elementary Area Computations

Perhaps everyone's first contact with the concept of area is the formula $A = bh$ for the area A of a rectangle as the product of its base b and its height h. We next learn that the area of a triangle is half the product of its base and height. This follows because any triangle can be split into two

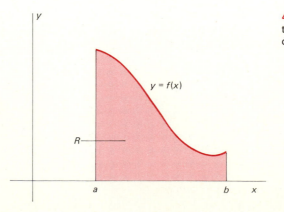

4.1 The problem of computing the area of R motivates integral calculus.

right triangles, and every right triangle is exactly half a rectangle (see Figure 4.2).

Given the formula $A = \frac{1}{2}bh$ for the area of a triangle, we can find the area of an arbitrary polygonal figure (a plane set bounded by a closed "curve" that consists of straight line segments). The reason is that any polygonal figure can be divided into nonoverlapping triangles, as shown in Figure 4.3, and the area of the polygonal figure is then the sum of the areas of these triangles. This approach to area dates back several thousand years to the ancient civilizations of Egypt and Babylonia.

The ancient Greeks began the investigation of areas of *curvilinear* figures in the fourth and fifth centuries B.C.; the approach they devised ultimately led—much later—to the integral as described in the introduction. We take it as obvious on intuitive grounds that every reasonably nice plane set S bounded by closed curves has an area $a(S)$, a nonnegative real number with the following properties:

I. If the set S is contained in the set T, then $a(S) \leqq a(T)$.

II. If S and T are nonoverlapping sets—they intersect only in points of their boundary curves, if at all—then the area of their union $S \cup T$ is $a(S \cup T) = a(S) + a(T)$.

III. If the sets S and T are congruent (have the same size and shape), then $a(S) = a(T)$.

To investigate the area of a curvilinear set S, we might inscribe in it a sequence of polygons $P_1, P_2, P_3, \ldots$ with increasing areas that gradually fill or "exhaust" the set S. We can then attempt to compute the limit $\lim_{n \to \infty} a(P_n)$ of the areas of these inscribed polygons. Recall from the definition of the limit of a sequence (Section 3-4) that

$$A = \lim_{n \to \infty} a(P_n)$$

means that $a(P_n)$ can be made as close to A as we please by choosing n sufficiently large.

Suppose that we also circumscribe about S a sequence of polygons $Q_1, Q_2, Q_3, \ldots$ with decreasing areas that gradually "close down" on the set S, and attempt to compute the limit $\lim_{n \to \infty} a(Q_n)$ of the areas of these circumscribed polygons. Our earlier inscribed polygons and these circumscribed polygons should enable us, as Figure 4.4 suggests, to obtain arbitrarily good estimates of the actual area $a(S)$. For by Property I of area above, we see that

$$a(P_n) \leqq a(S) \leqq a(Q_n).$$

By the squeeze law for limits of sequences (analogous to the squeeze law for limits of functions, stated in Section 1-8), it then follows that

$$\lim_{n \to \infty} a(P_n) \leqq a(S) \leqq \lim_{n \to \infty} a(Q_n),$$

provided that both these limits exist. If, moreover, the values of these two limits turn out to be equal, then the area of the set S we started with must be given by

$$a(S) = \lim_{n \to \infty} a(P_n) = \lim_{n \to \infty} a(Q_n). \tag{1}$$

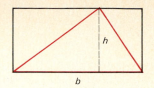

4.2 The formula for the area of a triangle follows with the aid of this figure.

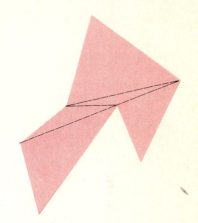

4.3 Every polygon can be represented as the union of nonoverlapping triangles.

4.4 Estimating the area $a(S)$ of the set S by using areas of inscribed and circumscribed polygons.

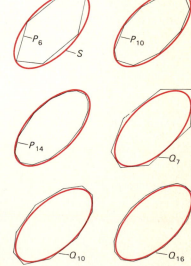

THE NUMBER π

In the third century B.C., Archimedes, the greatest mathematician of antiquity, used an approach similar to that outlined above to derive the famous estimate

$$\tfrac{223}{71} = 3\tfrac{10}{71} < \pi < 3\tfrac{1}{7} = \tfrac{22}{7}.$$

The number π is sometimes defined as the ratio of the circumference of a circle to its diameter and sometimes as the ratio of the area of a circle to the square of its radius. It may also be defined to be the area of the unit circle $x^2 + y^2 \leq 1$. With this last definition, the problem of computing π is that of finding the area of the unit circle.

Let P_n and Q_n be n-sided regular polygons, with P_n inscribed in the unit circle and Q_n circumscribed about it, as shown in Figure 4.5. Since the polygons are regular, all their sides and angles are equal, so all we need to find is the area of *one* of the triangles that we've shown making up P_n and one of those making up Q_n.

Let α_n be the central angle subtended by *half* of one of the sides. The angle α_n is the same whether we work with P_n or with Q_n. In degrees,

$$\alpha_n = \frac{360°}{2n} = \frac{180°}{n}.$$

We can read various dimensions and proportions from Figure 4.5; we see that the area of P_n is given by

$$a(P_n) = n \cdot 2 \cdot \frac{1}{2} \sin \alpha_n \cos \alpha_n = \frac{n}{2} \sin\left(\frac{360°}{n}\right) \tag{2}$$

and that

$$a(Q_n) = n \cdot 2 \cdot \frac{1}{2} \tan \alpha_n = n \tan\left(\frac{180°}{n}\right). \tag{3}$$

We substitute selected values of n into Formulas (2) and (3), and so obtain the entries of the table shown in Figure 4.6. Since $a(P_n) \leq \pi \leq a(Q_n)$ for all n, we see that $\pi \approx 3.14159$ to 5 decimal places.

SUMMATION NOTATION

For convenient application of the method of inscribed and circumscribed polygons to compute areas of rather general figures, we need a compact notation for sums of many numbers. We use the symbol $\sum_{i=1}^{n} a_i$ to abbreviate the sum of the n numbers $a_1, a_2, \ldots, a_n$, so that

$$\sum_{i=1}^{n} a_i = a_1 + a_2 + \cdots + a_n.$$

The Σ (Greek capital sigma) indicates the sum of the terms a_i as the **summation index** i runs through integer values from 1 to n. For example, the sum of the squares of the first ten positive integers is

$$\sum_{i=1}^{10} i^2 = 1^2 + 2^2 + \cdots + 10^2$$

$$= 1 + 4 + \cdots + 100 = 385.$$

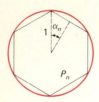
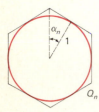

4.5 Estimating π by using inscribed and circumscribed regular polygons and the unit circle.

n	$a(P_n)$	$a(Q_n)$
6	2.598076	3.464102
12	3.000000	3.215390
24	3.105829	3.159660
48	3.132629	3.146086
96	3.139350	3.142715
180	3.140955	3.141912
360	3.141433	3.141672
720	3.141553	3.141613
1440	3.141583	3.141598
2880	3.141590	3.141594
5760	3.141592	3.141593

4.6 Data in estimating π (rounded to six-place accuracy).

CHAP. 4: The Integral

The particular letter used for the summation index is immaterial; we may write

$$\sum_{i=1}^{10} i^2 = \sum_{k=1}^{10} k^2 = \sum_{r=1}^{10} r^2 = 385.$$

The simple rules of summation

$$\sum_{i=1}^{n} ca_i = c\left(\sum_{i=1}^{n} a_i\right) \tag{4}$$

and

$$\sum_{i=1}^{n} (a_i + b_i) = \sum_{i=1}^{n} a_i + \sum_{i=1}^{n} b_i \tag{5}$$

are easy to verify by writing out each sum in full.

The sum of the kth powers of the first n positive integers,

$$\sum_{i=1}^{n} i^k = 1^k + 2^k + 3^k + \cdots + n^k,$$

often occurs in area computations. The values of this sum for $k = 1, 2, 3$, and 4 are given by the following formulas. They may be established by mathematical induction (see Problems 27–30).

$$\sum_{i=1}^{n} i = \frac{n}{2}(n + 1) = \frac{n^2}{2} + \frac{n}{2}. \tag{6}$$

$$\sum_{i=1}^{n} i^2 = \frac{n}{6}(n + 1)(2n + 1) = \frac{n^3}{3} + \frac{n^2}{2} + \frac{n}{6}. \tag{7}$$

$$\sum_{i=1}^{n} i^3 = \left[\frac{n}{2}(n + 1)\right]^2 = \frac{n^4}{4} + \frac{n^3}{2} + \frac{n^2}{4}. \tag{8}$$

$$\sum_{i=1}^{n} i^4 = \frac{n^5}{5} + \frac{n^4}{2} + \frac{n^3}{3} - \frac{n}{30}. \tag{9}$$

In fact, in the general case, it's known that

$$\sum_{i=1}^{n} i^k = \frac{n^{k+1}}{k + 1} + \frac{n^k}{2} + \text{(lower powers of } n\text{)}. \tag{10}$$

EXAMPLE 1 Use the summation formulas above to evaluate the sum

$$\sum_{i=1}^{10} (2i^3 - 3i^2 + 7) = 6 + 11 + 34 + \cdots + 1707.$$

Solution

$$\sum_{i=1}^{10} (2i^3 - 3i^2 + 7) = 2\left(\sum_{i=1}^{10} i^3\right) - 3\left(\sum_{i=1}^{10} i^2\right) + 7\left(\sum_{i=1}^{10} 1\right)$$

$$= 2[\tfrac{10}{2}(10 + 1)]^2 - 3 \cdot \tfrac{10}{6}(10 + 1)(20 + 1) + 7 \cdot 10$$

$$= 2(3025) - 3(385) + 70 = 4965.$$

EXAMPLE 2 Evaluate

$$\lim_{n \to \infty} \frac{1^2 + 2^2 + \cdots + n^2}{n^3}.$$

Solution Using Formula (7) we obtain

$$\lim_{n \to \infty} \frac{1^2 + 2^2 + \cdots + n^2}{n^3} = \lim_{n \to \infty} \frac{1}{n^3}\left(\frac{n^3}{3} + \frac{n^2}{2} + \frac{n}{6}\right)$$

$$= \lim_{n \to \infty} \left(\frac{1}{3} + \frac{1}{2n} + \frac{1}{6n^2}\right) = \frac{1}{3}.$$

We use the fact that the terms $1/2n$ and $1/6n^2$ approach zero as $n \to \infty$.

AREAS UNDER GRAPHS

Let f be a continuous, positive-valued function on $[a, b]$, and suppose that we want to calculate the **area A under the graph of f from a to b.** That is, A is the area of the region R that is bounded by the curve $y = f(x)$, the x-axis, and the vertical lines $x = a$ and $x = b$. This is the region shown earlier in Figure 4.1.

Our plan is to get high and low (but close) estimates of $A = a(R)$ by using certain special inscribed and circumscribed polygons. These polygons will be unions of rectangles with vertical sides, each having its base on the x-axis. We use such polygons because they are particularly easy to work with.

We start with a fixed positive integer n. Given n, we **partition,** or subdivide, the interval $[a, b]$ into n equal subintervals by choosing points x_0, $x_1, x_2, \ldots, x_n$ such that

$$a = x_0 < x_1 < x_2 < \cdots < x_{n-1} < x_n = b,$$

with the length of the ith subinterval $[x_{i-1}, x_i]$ given by

$$\Delta x = x_i - x_{i-1} = \frac{b - a}{n}$$

for each i ($1 \leq i \leq n$). The point x_i will be

$$x_i = x_0 + i\,\Delta x = a + \frac{i}{n}(b - a).$$

Symbol	Pronounced
$x_i^\flat$	x_i-flat
$x_i^\#$	x_i-sharp

Now let $x_i^\flat$ be a point of $[x_{i-1}, x_i]$ at which f attains its minimum value, so that $f(x_i^\flat)$ is the minimum value of $f(x)$ on the ith subinterval $[x_{i-1}, x_i]$. The rectangle with base $[x_{i-1}, x_i]$ and height $f(x_i^\flat)$ then lies *within* the region R. (See Figure 4.7.) The union of these rectangles for $i = 1, 2, \ldots, n$ is the **inscribed rectangular polygon P_n** associated with our partition of $[a, b]$ into n equal subintervals. Its area is the sum of the areas of the n rectangles with base length Δx and heights $f(x_1^\flat), f(x_2^\flat), \ldots, f(x_n^\flat)$. So

$$a(P_n) = \sum_{i=1}^{n} f(x_i^\flat)\,\Delta x. \tag{11}$$

Next, let $x_i^\#$ be a point of $[x_{i-1}, x_i]$ such that $f(x_i^\#)$ is the maximum value of $f(x)$ on the ith subinterval $[x_{i-1}, x_i]$. For each i ($1 \leq i \leq n$), consider the rectangle with base $[x_{i-1}, x_i]$, height $f(x_i^\#)$, and area $f(x_i^\#)\,\Delta x$. (See Figure 4.8.) The union of these n rectangles contains the region R under the graph; it is the **circumscribed rectangular polygon Q_n** associated with the partition of $[a, b]$ into n equal subintervals. Its area is

$$a(Q_n) = \sum_{i=1}^{n} f(x_i^\#)\,\Delta x. \tag{12}$$

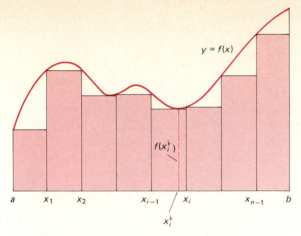

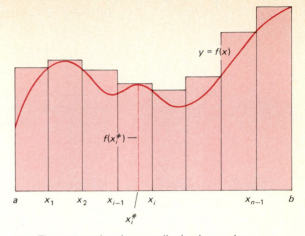

4.7 The area of the inscribed polygon P_n is an under-estimate of the area $A = a(R)$.

4.8 The rectangular circumscribed polygon gives an overestimate of the area $A = a(R)$.

Since the region R contains P_n but lies within Q_n, we conclude from (11) and (12) that its area A satisfies the inequalities

$$\sum_{i=1}^{n} f(x_i^\flat)\,\Delta x \leqq A \leqq \sum_{i=1}^{n} f(x_i^\#)\,\Delta x. \tag{13}$$

We could therefore attempt to estimate the value of A by computing (for a fixed value of n) the values of the two sums (11) and (12).

Figure 4.9 suggests that if n is very large, and hence Δx is very small, then the areas $a(P_n)$ and $a(Q_n)$ of the inscribed and circumscribed rectangular polygons will be very close to $A = a(R)$. For the difference in the heights of the inscribed and circumscribed rectangles over $[x_{i-1}, x_i]$ is $f(x_i^\#) - f(x_i^\flat)$, which should be small when Δx is small; so, intuitively, the difference between the two sums in (13) will be small when n is very large. Indeed, the assumption that f is continuous is enough to insure that this is so; in Section 4-3 we shall state a theorem which implies that

$$A = \lim_{n \to \infty} \sum_{i=1}^{n} f(x_i^\flat)\,\Delta x = \lim_{n \to \infty} \sum_{i=1}^{n} f(x_i^\#)\,\Delta x. \tag{14}$$

In this section we are proceeding on the basis of an intuitive concept of area. When the necessary groundwork has been laid, we shall see that the area under the graph of f from a to b must be *defined* by means of (14). The

4.9 The areas of the inscribed and circumscribed polygons should be nearly equal when Δx is small.

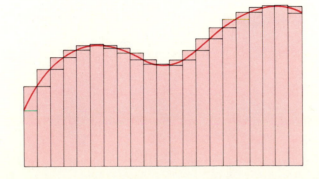

following two examples illustrate the use of (14) for the explicit computation of areas.

EXAMPLE 3   Find the area A under the graph of $f(x) = x^4$ from $x = 0$ to $x = b > 0$.

Solution   Since the two limits in (14) are equal, it is unnecessary to work with both inscribed and circumscribed rectangular polygons. We shall use only circumscribed figures, as shown in Figure 4.10. First we subdivide the interval $[0, b]$ into n equal subintervals, each with length $\Delta x = b/n$. To do so we must have $x_i = i \, \Delta x = ib/n$. Since $f(x) = x^4$ is an increasing function on $[0, b]$, f attains its maximum value on the ith subinterval $[x_{i-1}, x_i]$ at the right-hand end point $x_i^\# = x_i = ib/n$. Hence

$$\sum_{i=1}^{n} f(x_i^\#) \, \Delta x = \sum_{i=1}^{n} \left(\frac{ib}{n}\right)^4 \left(\frac{b}{n}\right) = \frac{b^5}{n^5} \sum_{i=1}^{n} i^4.$$

Now we use Formula (9) for $\sum_{i=1}^{n} i^4$. This gives a formula for the area of Q_n, the circumscribed polygon made up of n rectangles:

$$a(Q_n) = \frac{b^5}{n^5} \left(\frac{n^5}{5} + \frac{n^4}{2} + \frac{n^3}{3} - \frac{n}{30}\right)$$

$$= b^5 \left(\frac{1}{5} + \frac{1}{2n} + \frac{1}{3n^2} - \frac{1}{30n^4}\right).$$

Finally, we take the limit as $n \to \infty$. Since the terms $1/2n$, $1/3n^2$, and $1/30n^4$ approach zero as $n \to \infty$, we find that

$$A = \lim_{n \to \infty} \sum_{i=1}^{n} f(x_i^\#) \, \Delta x$$

$$= \lim_{n \to \infty} b^5 \left(\frac{1}{5} + \frac{1}{2n} + \frac{1}{3n^2} - \frac{1}{30n^4}\right) = \frac{b^5}{5}.$$

Note that the answer is *not* an approximation; it is *exact*.

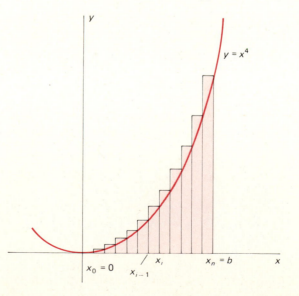

4.10   An area computation using circumscribed rectangles.

EXAMPLE 4   Find the area A under the graph of $f(x) = 100 - 3x^2$ from 1 to 5.

Solution   The function $f(x) = 100 - 3x^2$ is decreasing on $[1, 5]$ and therefore attains its minimum value on $[x_{i-1}, x_i]$ at the right-hand end point x_i. Hence $x_i^{\flat} = x_i$, and it is a bit more convenient to use inscribed rectangular polygons as indicated in Figure 4.11. We subdivide $[1, 5]$ into n subintervals each having length $\Delta x = 4/n$. Then

$$x_i^{\flat} = x_i = 1 + i\,\Delta x = 1 + \frac{4i}{n}.$$

And so

$$\sum_{i=1}^{n} f(x_i^{\flat})\,\Delta x = \sum_{i=1}^{n} \left[100 - 3\left(1 + \frac{4i}{n}\right)^2 \right]\left(\frac{4}{n}\right)$$

$$= \sum_{i=1}^{n} \left[97 - \frac{24i}{n} - \frac{48i^2}{n^2} \right]\left(\frac{4}{n}\right)$$

$$= \frac{388}{n} \sum_{i=1}^{n} 1 \;-\; \frac{96}{n^2} \sum_{i=1}^{n} i \;-\; \frac{192}{n^3} \sum_{i=1}^{n} i^2$$

$$= \frac{388}{n}(n) - \frac{96}{n^2}\left(\frac{n^2}{2} + \frac{n}{2}\right) - \frac{192}{n^3}\left(\frac{n^3}{3} + \frac{n^2}{2} + \frac{n}{6}\right)$$

$$= 276 - \frac{144}{n} - \frac{32}{n^2}.$$

We applied Formulas (6) and (7) in the next-to-last step. Consequently, the desired area is

$$A = \lim_{n \to \infty} \left(276 - \frac{144}{n} - \frac{32}{n^2} \right) = 276$$

since the terms $144/n$ and $32/n^2$ approach zero as $n \to \infty$.

4.11   An area computation using inscribed rectangles.

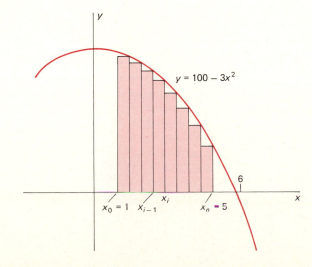

4-2 PROBLEMS

Use Formulas (6) through (9) to find the sums in Problems 1–10.

1 $\displaystyle\sum_{i=1}^{10} (4i - 3).$

2 $\displaystyle\sum_{j=1}^{8} (5 - 2j).$

3 $\displaystyle\sum_{i=1}^{10} (3i^2 + 1).$

4 $\displaystyle\sum_{k=1}^{6} (2k - 3k^2).$

5 $\displaystyle\sum_{r=1}^{8} (r - 1)(r + 2).$

6 $\displaystyle\sum_{i=1}^{5} (i^3 - 3i + 2).$

7 $\displaystyle\sum_{i=1}^{6} (i^4 - i^3).$

8 $\displaystyle\sum_{k=1}^{10} (2k - 1)^2.$

9 $\displaystyle\sum_{i=1}^{1000} i^2.$

10 $\displaystyle\sum_{i=1}^{100} i^3.$

Use the method of Example 2 to evaluate the limits in Problems 11–13.

11 $\displaystyle\lim_{n\to\infty} \frac{1 + 2 + \cdots + n}{n^2}.$

12 $\displaystyle\lim_{n\to\infty} \frac{1^3 + 2^3 + \cdots + n^3}{n^4}.$

13 $\displaystyle\lim_{n\to\infty} \frac{1^4 + 2^4 + \cdots + n^4}{n^5}.$

Use Formulas (6) through (9) to derive compact formulas in terms of n for the sums in Problems 14–16.

14 $\displaystyle\sum_{i=1}^{n} (2i - 1).$

15 $\displaystyle\sum_{i=1}^{n} (2i - 1)^2.$

16 $\displaystyle\sum_{i=1}^{n} (n^2 - i^2)^2.$

In each of Problems 17–26, use the method of Examples 3 and 4 to find the area under the graph of f from a to b.

17 $f(x) = 2x + 5; \quad a = 0, b = 3.$
18 $f(x) = 13 - 3x; \quad a = 0, b = 4.$
19 $f(x) = x^2; \quad a = 0, b = 1.$
20 $f(x) = x^3; \quad a = 0, b = 4.$
21 $f(x) = 3x^2 + 2; \quad a = 1, b = 5.$
22 $f(x) = 10 - x^2; \quad a = 1, b = 3.$
23 $f(x) = 3x^2 + 5x + 2; \quad a = 2, b = 4.$
24 $f(x) = 4x^3 + 2x; \quad a = 0, b = 3.$
25 $f(x) = x^2; \quad a = 0, b = b.$

26 $f(x) = x^3; \quad a = 0, b = b.$
27 Establish Formula (6) by mathematical induction.
28 Establish Formula (7) by mathematical induction.
29 Derive Formula (6) by adding the equations

$$\sum_{i=1}^{n} i = 1 + 2 + \cdots + n$$

and

$$\sum_{i=1}^{n} i = n + (n - 1) + \cdots + 1.$$

30 Write the n equations obtained by substituting the values $k = 1, 2, 3, \ldots, n$ into the identity

$$(k + 1)^3 - k^3 = 3k^2 + 3k + 1.$$

Add these n equations, and use their sum to deduce Formula (7) from Formula (6).

31 Derive the formula $A = \frac{1}{2}bh$ for the area of a right triangle by using circumscribed rectangular polygons to find the area under the graph of $f(x) = hx/b$ from 0 to b.

In Problems 32 and 33, let A denote the area and C the circumference of a circle of radius r and A_n and C_n denote the area and perimeter, respectively, of a regular n-sided polygon inscribed in this circle.

32 Show that

$$A_n = nr^2 \sin\left(\frac{\pi}{n}\right) \cos\left(\frac{\pi}{n}\right)$$

and that

$$C_n = 2nr \sin\left(\frac{\pi}{n}\right).$$

33 Deduce that $A = \frac{1}{2}rC$ by taking the limit of A_n/C_n as $n \to \infty$. Then given that $A = \pi r^2$, conclude from this that $C = 2\pi r$.

34 Use the method of Example 2 to deduce from Formula (10) that

$$\lim_{n\to\infty} \frac{1^k + 2^k + \cdots + n^k}{n^{k+1}} = \frac{1}{k + 1}.$$

35 Use circumscribed rectangular polygons and the result of Problem 34 to show that the area under the graph of $f(x) = x^k$ (k is a positive integer) from 0 to b is $A = b^{k+1}/(k+1)$.

In the previous section we saw that the area A under the graph from $x = a$ to $x = b$ of the *continuous, positive-valued* function f satisfies the inequalities

Riemann Sums and the Integral

$$\sum_{i=1}^{n} f(x_i^{\flat}) \, \Delta x \leqq A \leqq \sum_{i=1}^{n} f(x_i^{\#}) \, \Delta x, \tag{1}$$

where $f(x_i^{\flat})$ and $f(x_i^{\#})$ are the minimum and maximum values of f on the ith subinterval $[x_{i-1}, x_i]$ of a partition of $[a, b]$ into n equal subintervals of length Δx. The two approximating sums in (1) are both of the form

$$\sum_{i=1}^{n} f(x_i^{*}) \, \Delta x, \tag{2}$$

where x_i^{*} denotes an arbitrary point of the ith subinterval $[x_{i-1}, x_i]$. Sums of the form of (2) appear as approximations in a wide range of applications and also constitute the basis for the definition of the integral. Motivated by our discussion of area in Section 4-2, we want to define the integral of f from a to b as some sort of limit as $\Delta x \to 0$ of a sum such as (2).

We start with a function f on $[a, b]$ *not* necessarily either continuous or positive-valued. A **partition** P of $[a, b]$ is a collection of subintervals

$$[x_0, x_1], \quad [x_1, x_2], \quad [x_2, x_3], \quad \ldots, \quad [x_{n-1}, x_n]$$

of $[a, b]$ such that

$$a = x_0 < x_1 < x_2 < \cdots < x_{n-1} < x_n = b.$$

The **mesh** of the partition P is the largest of the lengths

$$\Delta x_i = x_i - x_{i-1}$$

of the subintervals in P and is denoted by $|P|$. To get a sum such as (2) we need a point x_i^{*} of the ith subinterval for each i, $1 \leqq i \leqq n$. A collection of points

$$S = \{x_1^{*}, x_2^{*}, x_3^{*}, \ldots, x_n^{*}\}$$

with x_i^{*} in $[x_{i-1}, x_i]$ (for each i) is called a **selection** for the partition P.

Definition *Riemann Sum*

Let f be a function defined on the interval $[a, b]$. If P is a partition of $[a, b]$ and S is a selection for P, then the **Riemann sum** for f determined by P and S is

$$R = \sum_{i=1}^{n} f(x_i^{*}) \, \Delta x_i. \tag{3}$$

We also say that this Riemann sum is **associated with** the partition P.

The German mathematician G. F. B. Riemann (1826–1866) provided a rigorous definition of the integral. Various special types of "Riemann sums"

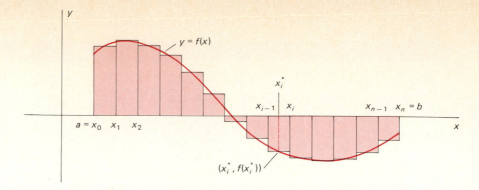

4.12 A geometric interpretation of the Riemann sum (3).

had appeared in area and volume computations since the time of Archimedes, but it was Riemann who framed the definition above in its full generality.

The Riemann sum (3) can be interpreted geometrically, as in Figure 4.12. On each subinterval $[x_{i-1}, x_i]$, we construct a rectangle with width Δx_i and "height" $f(x_i^*)$; if $f(x_i^*) > 0$, then this rectangle stands above the x-axis, while if $f(x_i^*) < 0$, it lies below the x-axis. The Riemann sum R is then the sum of the **signed** areas of these rectangles—that is, the sum of the areas of those rectangles that lie above the x-axis, *minus* the sum of the areas of those that lie below the x-axis.

If the widths Δx_i of these rectangles are all very small, that is, the mesh $|P|$ is very small, then it appears that the Riemann sum R ought to closely approximate the area from a to b under $y = f(x)$ above the x-axis, minus the area over $y = f(x)$ below the x-axis. This suggests that the integral of f from a to b be defined by taking the limit of the Riemann sum as the mesh $|P|$ approaches zero:

$$I = \lim_{|P| \to 0} \sum_{i=1}^{n} f(x_i^*) \, \Delta x_i. \tag{4}$$

The actual definition of the integral is obtained by saying precisely what this limit means. It means that the Riemann sum can be made as close to the number I as we like by taking the mesh of the partition P sufficiently small.

Definition   *The (Definite) Integral*

The **(definite) integral of the function f from a to b** is the number

$$I = \lim_{|P| \to 0} \sum_{i=1}^{n} f(x_i^*) \, \Delta x_i \tag{4}$$

provided that this limit exists, in which case we say that f is **integrable** on $[a, b]$. The meaning of (4) is that, for each number $\varepsilon > 0$, there exists a number $\delta > 0$ such that

$$\left| I - \sum_{i=1}^{n} f(x_i^*) \, \Delta x_i \right| < \varepsilon$$

for every Riemann sum associated with every partition P of $[a, b]$ with $|P| < \delta$.

The customary notation for the integral of f from a to b, due to the German mathematician and philosopher G. W. Leibniz (1646–1716), is

$$I = \int_a^b f(x)\,dx = \lim_{|P| \to 0} \sum_{i=1}^n f(x_i^*)\,\Delta x_i. \qquad (5)$$

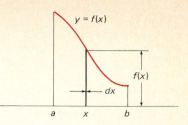

4.13 Origin of Liebniz's notation for the integral.

Considering I as the area under $y = f(x)$ from a to b, Leibniz first thought of a narrow strip with height $f(x)$ and "infinitesimally" small width dx (see Figure 4.13), so that its area would be the product $f(x)\,dx$. He regarded the integral as a sum of areas of such strips, denoting this sum by the elongated capital S (for *sum*) that appears as the integral symbol in (5).

We shall see that this integral notation is not only highly suggestive, but also exceedingly useful in manipulations with integrals. The numbers a and b are called the **lower limit** and **upper limit,** respectively, of the integral; they are merely the end points of the interval of integration. The function $f(x)$ that appears between the integral sign and dx is sometimes called the **integrand.** The symbol dx following $f(x)$ in (5) should be thought of as simply an indication of what the independent variable is. Like the index in a summation, x is merely a dummy variable, and we may write

$$\int_a^b f(x)\,dx = \int_a^b f(t)\,dt = \int_a^b f(u)\,du.$$

The definition above applies only if $a < b$, but it is convenient to include the cases $a = b$ and $a > b$ as well. The integral is *defined* in these cases by

$$\int_a^a f(x)\,dx = 0 \qquad (6)$$

and

$$\int_a^b f(x)\,dx = -\int_b^a f(x)\,dx \qquad (7)$$

provided the right-hand integral exists. Thus *interchanging the limits of integration reverses the sign of the integral.*

Not all functions are integrable. For example, if $\lim_{x \to c} f(x) = \infty$ for some point c in $[a, b]$, and $[x_{k-1}, x_k]$ is the subinterval of the partition P that contains c, then the Riemann sum (3) can be made arbitrarily large by choosing x_k^* sufficiently close to c. For our purposes, however, it is quite sufficient to know that every continuous function is integrable. The following theorem is proved in advanced calculus.

> **Theorem 1** *Existence of the Integral*
>
> If the function f is continuous on $[a, b]$, then f is integrable on $[a, b]$.

Although we omit the details, it is not difficult to show that the definition of the integral can be reformulated in terms of sequences of Riemann sums, as follows.

The function f is integrable on $[a, b]$ with integral I if and only if

$$\lim_{n \to \infty} R_n = I \qquad (8)$$

for every sequence $\{R_n\}_1^\infty$ of Riemann sums associated with a sequence of partitions $\{P_n\}_1^\infty$ of $[a, b]$ such that $|P_n| \to 0$ as $n \to \infty$.

This reformulation is advantageous, because it is easier to visualize a specific sequence of Riemann sums than the vast totality of all possible Riemann sums. In the case of a continuous function f that is known (by Theorem 1) to be integrable, the situation can be simplified even further by using only Riemann sums associated with partitions consisting of sub-intervals all having the same length,

$$\Delta x_1 = \Delta x_2 = \cdots = \Delta x_n = \frac{b - a}{n} = \Delta x.$$

Such a partition of $[a, b]$ into equal subintervals is called a **regular** partition of $[a, b]$.

Any Riemann sum associated with a regular partition can be written in the form

$$\sum_{i=1}^{n} f(x_i^*) \, \Delta x, \qquad (9)$$

where the absence of a subscript in Δx signifies that the sum is associated with a regular partition. In such a case, the conditions $|P| \to 0$, $\Delta x \to 0$, and $n \to \infty$ are equivalent, so the integral of a *continuous* function can be defined quite simply by

$$\int_a^b f(x) \, dx = \lim_{n \to \infty} \sum_{i=1}^{n} f(x_i^*) \, \Delta x = \lim_{\Delta x \to 0} \sum_{i=1}^{n} f(x_i^*) \, \Delta x. \qquad (10)$$

For continuous functions, two special sums are of particular importance. As in Section 4-2, let $f(x_i^?)$ and $f(x_i^\#)$ denote the minimum and maximum values, respectively, of f on the ith subinterval $[x_{i-1}, x_i]$ of a regular partition P of $[a, b]$. If $S = \{x_1^*, x_2^*, \ldots, x_n^*\}$ is any selection for P, then it follows that

$$\sum_{i=1}^{n} f(x_i^?) \, \Delta x \leqq \sum_{i=1}^{n} f(x_i^*) \, \Delta x \leqq \sum_{i=1}^{n} f(x_i^\#) \, \Delta x. \qquad (11)$$

The sums on the right and left in (11) are called the **upper Riemann sum** and **lower Riemann sum,** respectively, for f associated with the partition P. The following example illustrates how integrals can sometimes be computed by "trapping" arbitrary Riemann sums between upper and lower Riemann sums.

EXAMPLE 1 Use Riemann sums to compute $\int_0^b x^2 \, dx, b > 0$.

Solution We start with a regular partition P_n of $[0, b]$ into n equal sub-intervals by means of the points

$$0 = x_0 < x_1 < x_2 < \cdots < x_{n-1} < x_n = b$$

where $x_i = bi/n$. If $x_i^* \in [x_{i-1}, x_i]$, then

$$(x_{i-1})^2 \le (x_i^*)^2 \le (x_i)^2,$$

so

$$\frac{b^2(i-1)^2}{n^2} \le (x_i^*)^2 \le \frac{b^2 i^2}{n^2}$$

for $i = 1, 2, \ldots, n$. We multiply each term by $\Delta x = b/n$, then sum from $i = 1$ to $i = n$. This yields

$$\frac{b^3}{n^3} \sum_{i=1}^{n} (i-1)^2 \le \sum_{i=1}^{n} (x_i^*)^2 \, \Delta x \le \frac{b^3}{n^3} \sum_{i=1}^{n} i^2.$$

But Formula (7) in Section 4-2 gives

$$\sum_{i=1}^{n} i^2 = \frac{n^3}{3} + \frac{n^2}{2} + \frac{n}{6}$$

and

$$\sum_{i=1}^{n} (i-1)^2 = \left(\sum_{i=1}^{n} i^2 \right) - n^2 = \frac{n^3}{3} - \frac{n^2}{2} + \frac{n}{6}.$$

It follows that

$$b^3 \left(\frac{1}{3} - \frac{1}{2n} + \frac{1}{6n^2} \right) \le \sum_{i=1}^{n} (x_i^*)^2 \, \Delta x \le b^3 \left(\frac{1}{3} + \frac{1}{2n} + \frac{1}{6n^2} \right).$$

We take limits as $n \to \infty$, and the squeeze law for limits of sequences gives

$$\int_0^b x^2 \, dx = \lim_{n \to \infty} \sum_{i=1}^{n} (x_i^*)^2 \, \Delta x = \frac{b^3}{3}.$$

BASIC PROPERTIES OF INTEGRALS

Elementary proofs of the properties stated below are outlined in Problems 32–34 at the end of this section. We assume that f is integrable on $[a, b]$.

Integral of a Constant

$$\int_a^b c \, dx = c(b - a).$$

4.14 The integral of a constant is the area of a rectangle.

This is intuitively obvious, since the area represented by the integral is simply a rectangle with base $b - a$ and height c (see Figure 4.14).

Constant Multiple Property

$$\int_a^b cf(x) \, dx = c \int_a^b f(x) \, dx.$$

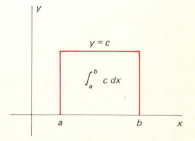

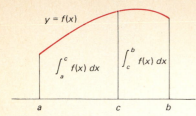

4.15 The way the interval union property works.

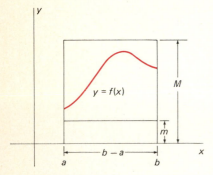

4.16 Plausibility of the comparison property.

Thus a constant c can be "moved out from under the integral sign."

Interval Union Property

If $a < c < b$ then

$$\int_a^b f(x)\,dx = \int_a^c f(x)\,dx + \int_c^b f(x)\,dx.$$

Figure 4.15 indicates the plausibility of this property.

Comparison Property

If $m \leq f(x) \leq M$ for all x in $[a, b]$, then

$$m(b - a) \leq \int_a^b f(x)\,dx \leq M(b - a).$$

The plausibility of this property is indicated in Figure 4.16. Note that m and M are not necessarily the minimum and maximum values of $f(x)$ on $[a, b]$.

EVALUATION OF INTEGRALS

Fortunately, we shall seldom find it necessary to evaluate an integral by the tedious use of Riemann sums as in Example 1. Theorem 2 below gives an easy way of finding $\int_a^b f(x)\,dx$, provided only that an antiderivative of f can be found.

To motivate this important result, consider a function $G(t)$ that gives the value at time t of some variable quantity. For example, $G(t)$ might be

- the size of a population (of bacteria, animals, or people);
- the volume of water in a tank;
- the amount of capital reserves a bank has on hand; or
- the position of a particle moving along a line.

Suppose that the time rate of change of this quantity, $G'(t) = f(t)$, is given. It would be common to want to find the change $G(b) - G(a)$ from time $t = a$ to time $t = b$.

We start with a regular partition P of $[a, b]$ into n equal subintervals $[t_{i-1}, t_i]$, each of length Δt. First we estimate the change in $G(t)$ corresponding to the ith subinterval. The linear approximation formula of Section 3-2 gives

$$G(t_i) - G(t_{i-1}) \approx G'(t_{i-1})\,\Delta t = f(t_{i-1})\,\Delta t.$$

Then summation gives

$$\begin{aligned}
G(b) - G(a) &= \left[G(t_n) - G(t_{n-1})\right] + \left[G(t_{n-1}) - G(t_{n-2})\right] \\
&\quad + \left[G(t_{n-2}) - G(t_{n-3})\right] + \cdots \\
&\quad + \left[G(t_2) - G(t_1)\right] + \left[G(t_1) - G(t_0)\right] \\
&= \sum_{i=1}^n \left[G(t_i) - G(t_{i-1})\right].
\end{aligned}$$

Therefore

$$G(b) - G(a) \approx \sum_{i=1}^{n} f(t_{i-1})\,\Delta t.$$

But the right-hand sum is a Riemann sum for the integral of f from a to b, so we now suspect that

$$G(b) - G(a) = \int_a^b f(t)\,dt.$$

The proof of the following theorem consists of making the above discussion precise by using the Mean Value Theorem in place of the linear approximation formula.

Theorem 2 *Evaluation of Integrals*

If G is an antiderivative of the continuous function f on the interval $[a, b]$, then

$$\int_a^b f(x)\,dx = G(b) - G(a). \tag{12}$$

Proof For each n, let P_n be the regular partition of $[a, b]$ into n equal subintervals, each of length $\Delta x = (b - a)/n$. Then, as we know,

$$\int_a^b f(x)\,dx = \lim_{n \to \infty} R_n$$

for any sequence $\{R_n\}$ of Riemann sums associated with the partitions $\{P_n\}$. To establish Equation (12), it therefore is enough to show that, for each n, there is a Riemann sum R_n associated with P_n such that

$$R_n = G(b) - G(a).$$

To do this, we apply the Mean Value Theorem to the differentiable function G on the ith subinterval $[x_{i-1}, x_i]$ of P_n. This yields a number x_i^* in that interval such that

$$G(x_i) - G(x_{i-1}) = G'(x_i^*)(x_i - x_{i-1}) = f(x_i^*)\,\Delta x.$$

We sum this equation from $i = 1$ to $i = n$, and find that

$$\sum_{i=1}^{n} [G(x_i) - G(x_{i-1})] = \sum_{i=1}^{n} f(x_i^*)\,\Delta x.$$

But we recognize the left-hand side as $G(b) - G(a)$, as in the computation preceding the statement of the theorem. Moreover, the right-hand side is a Riemann sum R_n for f associated with the partition P_n. Thus $G(b) - G(a) = R_n$, so it follows that

$$\int_a^b f(x)\,dx = \lim_{n \to \infty} R_n = G(b) - G(a),$$

and the proof of Theorem 2 is complete.

The difference $G(b) - G(a)$ is customarily abbreviated to $[G(x)]_a^b$, so Theorem 2 says that

$$\int_a^b f(x)\,dx = \Big[G(x)\Big]_a^b \tag{13}$$

if G is any antiderivative of the continuous function f. Thus, if we can find the antiderivative G, we can instantly evaluate the integral *without* recourse to the paraphernalia of limits of Riemann sums.

EXAMPLE 2 Some immediate applications of Theorem 2 are shown below.

$$\int_0^2 x^5 \, dx = \left[\frac{x^6}{6}\right]_0^2 = \frac{64}{6} - 0 = \frac{32}{3}.$$

$$\int_1^9 (2x - x^{-1/2} - 3) \, dx = \left[x^2 - 2\sqrt{x} - 3x\right]_1^9 = 52.$$

$$\int_0^1 x^2(x^3 + 1)^{1/3} \, dx = \left[\tfrac{1}{4}(x^3 + 1)^{4/3}\right]_0^1 = \tfrac{1}{4}(2^{4/3} - 1).$$

$$\int_0^{\pi/2} \sin 2x \, dx = \left[-\tfrac{1}{2}\cos 2x\right]_0^{\pi/2} = 1.$$

EXAMPLE 3 Evaluate $\int_a^b x^n \, dx$ (n is a positive integer).

Solution $D^{-1}x^n = x^{n+1}/(n + 1)$, so Theorem 2 gives

$$\int_a^b x^n \, dx = \left[\frac{x^{n+1}}{n+1}\right]_a^b = \frac{b^{n+1} - a^{n+1}}{n+1}.$$

Contrast this with the computation in Example 1.

EXAMPLE 4 Suppose that an animal population $P(t)$ initially numbers $P(0) = 100$, and it is observed that its rate of growth after t months is given by the formula

$$P'(t) = 10 + t + (0.06)t^2.$$

What is the population after 10 months?

Solution By Theorem 2 and the discussion preceding it, we know that

$$P(10) - P(0) = \int_0^{10} P'(t) \, dt$$

$$= \int_0^{10} (10 + t + 0.06t^2) \, dt$$

$$= \left[10t + \tfrac{1}{2}t^2 + 0.02t^3\right]_0^{10} = 170.$$

Thus $P(10) = 100 + 170 = 270$ individuals.

4-3 PROBLEMS

In each of Problems 1–8, compute the Riemann sum $\sum\limits_{i=1}^{n} f(x_i^*) \Delta x$ for the indicated function and a regular partition of the given interval into n equal subintervals. Do this in three ways:
(a) With $x_i^* = x_i$, the right-hand end point of the ith subinterval $[x_{i-1}, x_i]$;
(b) With $x_i^* = x_{i-1}$;
(c) With x_i^* the midpoint of $[x_{i-1}, x_i]$.

1 $f(x) = x^2$ on $[0, 1]$; $n = 5$.

2 $f(x) = x^3$ on $[0, 1]$; $n = 5$.

3 $f(x) = \dfrac{1}{x}$ on $[1, 6]$; $n = 5$.

4 $f(x) = \sqrt{x}$ on $[0, 5]$; $n = 5$.
5 $f(x) = 2x + 1$ on $[1, 4]$; $n = 6$.
6 $f(x) = x^2 + 2x$ on $[1, 4]$; $n = 6$.
7 $f(x) = x^3 - 3x$ on $[1, 4]$; $n = 5$.
8 $f(x) = 1 + 2\sqrt{x}$ on $[2, 3]$; $n = 5$.

Use the method of Example 1 to evaluate the integrals in Problems 9 and 10.

9 $\int_0^2 x\,dx.$ **10** $\int_0^4 x^3\,dx.$

In Problems 11–14, evaluate the given integral by computing

$$\lim_{n\to\infty} \sum_{i=1}^{n} f(x_i^*)\,\Delta x$$

for a regular partition of the interval of integration. Take $x_i^* = x_i$ in each case.

11 $\int_0^3 (2x + 1)\,dx.$ **12** $\int_1^5 (4 - 3x)\,dx.$

13 $\int_0^3 (3x^2 + 1)\,dx.$ **14** $\int_0^4 (x^3 - x)\,dx.$

Apply Theorem 2 to evaluate the integrals in Problems 15–24.

15 $\int_0^1 (x^2 - 3x + 4)\,dx.$

16 $\int_1^4 \sqrt{3t}\,dt.$

17 $\int_1^9 \left(\sqrt{x} - \dfrac{2}{\sqrt{x}}\right)dx.$

18 $\int_2^3 \dfrac{du}{u^2}.$ $\left(\text{Note the abbreviation for } \int_2^3 \dfrac{1}{u^2}\,du.\right)$

19 $\int_1^4 \dfrac{x^2 - 1}{\sqrt{x}}\,dx.$ **20** $\int_1^4 (t^2 - 2)\sqrt{t}\,dt.$

21 $\int_0^3 x\sqrt{x^2 + 4}\,dx.$ **22** $\int_0^{\pi/2} \cos 2x\,dx.$

23 $\int_0^\pi \sin^2 x\,dx.$ **24** $\int_0^\pi \sin^2 x \cos x\,dx.$

In each of Problems 25–30, evaluate the given limit by first recognizing the indicated sum as a Riemann sum associated with a regular partition of $[0, 1]$ and then evaluating the corresponding integral.

25 $\lim_{n\to\infty} \sum_{i=1}^{n} \left(\dfrac{2i}{n} - 1\right)\dfrac{1}{n}.$

26 $\lim_{n\to\infty} \sum_{i=1}^{n} \dfrac{i^2}{n^3}.$

27 $\lim_{n\to\infty} \dfrac{1 + 2 + 3 + \cdots + n}{n^2}.$

28 $\lim_{n\to\infty} \dfrac{1^3 + 2^3 + 3^3 + \cdots + n^3}{n^4}.$

29 $\lim_{n\to\infty} \dfrac{\sqrt{1} + \sqrt{2} + \sqrt{3} + \cdots + \sqrt{n}}{n\sqrt{n}}.$

30 $\lim_{n\to\infty} \sum_{i=1}^{n} \dfrac{1}{n} \sin \dfrac{\pi i}{n}.$

31 Evaluate the integral $\int_0^5 \sqrt{25 - x^2}\,dx$ by interpreting it as the area under the graph of an appropriate function.

32 Use sequences of Riemann sums to establish the constant multiple property.

33 Use sequences of Riemann sums to establish the interval union property of the integral. Note that if R_n' and R_n'' are Riemann sums for f on the intervals $[a, c]$ and $[c, b]$ respectively, then $R_n = R_n' + R_n''$ is a Riemann sum for f on $[a, b]$.

34 Use Riemann sums to establish the comparison property for integrals. Show first that, if $m \leq f(x) \leq M$ for all x in $[a, b]$, then

$$m(b - a) \leq R \leq M(b - a)$$

for every Riemann sum R for f on $[a, b]$.

35 Let $f(x) = x$, and let $\{x_0, x_1, x_2, \ldots, x_n\}$ be an arbitrary partition of the closed interval $[a, b]$. For each $i\,(1 \leq i \leq n)$, let $x_i^* = (x_{i-1} + x_i)/2$. Then show that

$$\sum_{i=1}^{n} x_i^* \,\Delta x_i = \tfrac{1}{2}b^2 - \tfrac{1}{2}a^2.$$

Explain why this computation proves that

$$\int_a^b x\,dx = \dfrac{b^2 - a^2}{2}.$$

36 Suppose that the population of a city in 1960 was 125 (thousand), and that its rate of growth t years later was $P'(t) = 8 + (0.5)t + (0.03)t^2$ (thousand) per year. What was its population in 1980?

37 Suppose that a tank initially contains 1000 gallons of water, and that the rate of change of its volume after draining t minutes is $V'(t) = (0.8)t - 40$ gallons per minute. How much water does the tank contain after it has been draining a half hour?

4-4

Newton and Leibniz are generally credited with the invention of calculus in the latter part of the seventeenth century. Actually, others had earlier calculated areas essentially equivalent to integrals, and tangent line slopes essentially equivalent to derivatives. The great accomplishment of Newton and Leibniz was the discovery and computational exploitation of the inverse relationship between differentiation and integration. This relationship is

The Fundamental Theorem of Calculus

embodied in the **Fundamental Theorem of Calculus.** One part of this theorem is Theorem 2 in Section 4-3—in order to evaluate $\int_a^b f(x)\,dx$, it suffices to find an antiderivative of f. The other part of the fundamental theorem tells us that every continuous function has an antiderivative.

THE AVERAGE VALUE OF A FUNCTION

As a preliminary to the proof of the Fundamental Theorem of Calculus, and also as a matter of independent interest, we investigate the notion of *average value.* For example, let the temperature T during a particular 24-hour day be described by $T = f(t)$, $t \in [0, 24]$. We might define the average temperature $\bar{T}$ for the day as the (ordinary arithmetical) average of the hourly temperatures:

$$\bar{T} = \frac{1}{24} \sum_{i=1}^{24} f(i) = \frac{1}{24} \sum_{i=1}^{24} f(t_i)$$

where $t_i = i$. If we subdivided the day into n equal subintervals $[t_{i-1}, t_i]$ rather than into twenty-four 1-hour intervals, we would obtain the more general average

$$\bar{T} = \frac{1}{n} \sum_{i=1}^{n} f(t_i).$$

The larger n is, the closer would we expect $\bar{T}$ to be to the "true" average temperature for the entire day. It is therefore plausible to define the true average temperature by letting n increase without bound. This gives

$$\bar{T} = \lim_{n \to \infty} \frac{1}{n} \sum_{i=1}^{n} f(t_i).$$

The right-hand side resembles a Riemann sum, and we can make it into a Riemann sum by introducing the factor $\Delta t = (b - a)/n$ where $a = 0$ and $b = 24$. Then

$$\bar{T} = \lim_{n \to \infty} \frac{1}{b - a} \sum_{i=1}^{n} f(t_i) \frac{b - a}{n}$$

$$= \frac{1}{b - a} \lim_{n \to \infty} \sum_{i=1}^{n} f(t_i) \Delta t$$

$$= \frac{1}{b - a} \int_a^b f(t)\,dt = \frac{1}{24} \int_0^{24} f(t)\,dt$$

under the assumption that f is continuous, so that the Riemann sums converge to the integral as $n \to \infty$. This example motivates the following definition.

Definition *Average Value of a Function*
Suppose that the function f is integrable on $[a, b]$. Then the **average value** $\bar{y}$ of $y = f(x)$ on $[a, b]$ is

$$\bar{y} = \frac{1}{b - a} \int_a^b f(x)\,dx. \tag{1}$$

For example, the average value of $f(x) = x^2$ on $[0, 2]$ is

$$\bar{y} = \frac{1}{2} \int_0^2 x^2 \, dx = \frac{1}{2}\left[\frac{x^3}{3}\right]_0^2 = \frac{4}{3}.$$

EXAMPLE 1   The mean daily temperature in degrees Fahrenheit in Athens, Georgia, t months after July 15, is closely approximated by

$$T = 61 + 18 \cos \frac{\pi t}{6}.$$

Find the average temperature between September 15 ($t = 2$) and December 15 ($t = 5$).

Solution   Definition (1) gives

$$\bar{T} = \frac{1}{5 - 2} \int_2^5 \left(61 + 18 \cos \frac{\pi t}{6}\right) dt$$

$$= \frac{1}{3}\left[61t + \frac{6(18)}{\pi} \sin \frac{\pi t}{6}\right]_2^5 \approx 57°\text{F}.$$

The following theorem tells us that every continuous function on a closed interval *attains* its average value at some point of the interval.

> **Average Value Theorem**
> If f is continuous on $[a, b]$, then
>
> $$f(\bar{x}) = \frac{1}{b - a} \int_a^b f(x) \, dx \qquad (2)$$
>
> for some point $\bar{x}$ of $[a, b]$.

Proof   Let $m = f(c)$ be the minimum value of f on $[a, b]$, and let $M = f(d)$ be its maximum value there. Then, by the comparison property of Section 4-3,

$$m = f(c) \leqq \bar{y} = \frac{1}{b - a} \int_a^b f(x) \, dx \leqq f(d) = M.$$

Since f is continuous, we can now apply the Intermediate Value Property. Because the number $\bar{y}$ is between the two values m and M of f, $\bar{y}$ itself must be a value of f; specifically, $\bar{y} = f(\bar{x})$ for some number $\bar{x}$ between a and b. This gives (2).

THE FUNDAMENTAL THEOREM

We state the Fundamental Theorem of Calculus in two parts. The first part is the fact that every continuous function f has an antiderivative. In particular, an antiderivative of f is obtained by integrating f in a certain way. Intuitively, in the case $f(x) > 0$, we let $A(x)$ denote the area under the graph of f from a to x. We shall prove that $A'(x) = f(x)$. We show the function $A(x)$

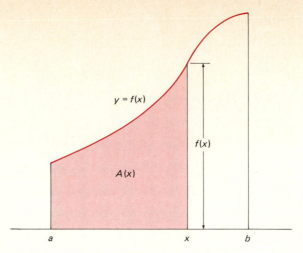

4.17 The area function A is an antiderivative of f.

geometrically in Figure 4.17. More precisely, we define the function A by

$$A(x) = \int_a^x f(t)\, dt,$$

where we use the dummy variable t in the integrand to avoid confusion with the upper limit x.

The Fundamental Theorem of Calculus

Let f be a continuous function defined on $[a, b]$.
Part 1: If the function F is defined on $[a, b]$ by

$$F(x) = \int_a^x f(t)\, dt \tag{3}$$

then F is an antiderivative of f. That is, $F'(x) = f(x)$ for x in (a, b).
Part 2: If G is any antiderivative of f on $[a, b]$, then

$$\int_a^b f(x)\, dx = \left[G(x) \right]_a^b = G(b) - G(a). \tag{4}$$

Proof of Part 1 By the definition of the derivative,

$$F'(x) = \lim_{h \to 0} \frac{F(x + h) - F(x)}{h}$$

$$= \lim_{h \to 0} \frac{1}{h} \left[\int_a^{x+h} f(t)\, dt - \int_a^x f(t)\, dt \right].$$

But

$$\int_a^{x+h} f(t)\, dt = \int_a^x f(t)\, dt + \int_x^{x+h} f(t)\, dt$$

by the interval union property (Section 4-3). Thus

$$F'(x) = \lim_{h \to 0} \frac{1}{h} \int_x^{x+h} f(t)\, dt.$$

The average value theorem above tells us that

$$\frac{1}{h} \int_x^{x+h} f(t)\, dt = f(\bar{t})$$

for some number $\bar{t}$ in $[x, x + h]$. Finally, we note that $\bar{t} \to x$ as $h \to 0$. Thus, since f is continuous, we see that

$$F'(x) = \lim_{h \to 0} \frac{1}{h} \int_x^{x+h} f(t)\, dt$$

$$= \lim_{h \to 0} f(\bar{t}) = \lim_{\bar{t} \to x} f(\bar{t}) = f(x).$$

Hence the function F defined by (3) is, indeed, an antiderivative of f.

Proof of Part 2 Here we apply Part 1 to give a second proof of Theorem 2 in Section 4-3. If G is *any* antiderivative of f, then—because it and the function F of Part 1 are both antiderivatives of f—we know that

$$G(x) = F(x) + C$$

on $[a, b]$ for some constant C. To evaluate C, we substitute $x = a$, and obtain

$$C = G(a) - F(a) = G(a)$$

because

$$F(a) = \int_a^a f(t)\, dt = 0.$$

Hence $G(x) = F(x) + G(a)$, or

$$F(x) = G(x) - G(a)$$

for all x in $[a, b]$. With $x = b$ this gives

$$G(b) - G(a) = F(b) = \int_a^b f(x)\, dx,$$

which establishes Equation (4).

Sometimes the Fundamental Theorem of Calculus is interpreted as a statement to the effect that differentiation and integration are *inverse processes.* Part 1 may be written in the form

$$\frac{d}{dx}\left(\int_a^x f(t)\, dt \right) = f(x)$$

if f is continuous. That is, if we first integrate the function f (with *variable* upper limit of integration x) and then differentiate with respect to x, the result is the function f again. So differentiation "cancels" the effect of integration.

Moreover, Part 2 of the fundamental theorem may be written in the form

$$\int_a^x G'(t)\, dt = G(x) - G(a),$$

if we assume that G' is continuous. If so, this equation means that if we first differentiate the function G and then integrate the result from a to x, the result can differ from the original function $G(x)$ by, at worst, the constant $G(a)$. If a is chosen so that $G(a) = 0$, this means that integration "cancels" the effect of differentiation.

For our first application of the Fundamental Theorem of Calculus, we use it to establish the **linearity property** of the integral:

$$\int_a^b [Af(x) + Bg(x)]\, dx = A \int_a^b f(x)\, dx + B \int_a^b g(x)\, dx \tag{5}$$

if A and B are constants and the functions f and g are continuous on $[a, b]$. For example,

$$\int_0^1 (3x^2 - 2x\sqrt{x^2 + 1})\, dx = 3 \int_0^1 x^2\, dx - 2 \int_0^1 x\sqrt{x^2 + 1}\, dx.$$

Thus linearity enables us to "divide and conquer," to split a relatively complicated integral into simpler ones.

The linearity of integration follows—via the Fundamental Theorem of Calculus—from the linearity of differentiation. The linearity formula (5) is a unified statement of the following two properties of integration.

$$\int_a^b Cf(x)\, dx = C \int_a^b f(x)\, dx, \tag{6}$$

$$\int_a^b [f(x) + g(x)]\, dx = \int_a^b f(x)\, dx + \int_a^b g(x)\, dx. \tag{7}$$

Formula (6) is merely the constant multiple property of Section 4-3. To prove (7), let $F(x)$ be an antiderivative of $f(x)$ and let $G(x)$ be an antiderivative of $g(x)$. Then

$$D[F(x) + G(x)] = f(x) + g(x),$$

because differentiation is linear. This means that $F + G$ is an antiderivative of $f + g$, which implies that

$$
\begin{aligned}
\int_a^b [f(x) + g(x)]\, dx &= [F(x) + G(x)]_a^b \\
&= [F(b) + G(b)] - [F(a) + G(a)] \\
&= [F(b) - F(a)] + [G(b) - G(a)] \\
&= [F(x)]_a^b + [G(x)]_a^b \\
&= \int_a^b f(x)\, dx + \int_a^b g(x)\, dx.
\end{aligned}
$$

Note that we have invoked the Fundamental Theorem of Calculus twice in this derivation.

One consequence of linearity is the fact that integration preserves inequalities between functions. That is, if f and g are continuous functions with $f(x) \leqq g(x)$ for all x in $[a, b]$, then

$$\int_a^b f(x)\, dx \leqq \int_a^b g(x)\, dx. \tag{8}$$

To prove this, let $H(x) = g(x) - f(x)$. Then $H(x) \geqq 0$ on $[a, b]$, so it follows from the comparison property of Section 4-3 that

$$\int_a^b H(x)\, dx \geqq 0.$$

So

$$\int_a^b g(x)\, dx - \int_a^b f(x)\, dx = \int_a^b H(x)\, dx \geqq 0,$$

which is equivalent to (8).

The verification of the following consequence of (8) is left for the problems:

$$\left| \int_a^b f(x)\, dx \right| \leq \int_a^b |f(x)|\, dx. \tag{9}$$

Examples 2 and 3 of Section 4-3 illustrate the use of Part 2 of the fundamental theorem in the evaluation of integrals. Additional examples appear in the problems and in the following section. The example below illustrates the necessity of splitting an integral into a sum of integrals when its integrand has different antiderivative formulas on different intervals.

EXAMPLE 2   Evaluate $\int_{-1}^{2} |x^3 - x|\, dx$.

Solution   We note that $x^3 - x \geq 0$ on $[-1, 0]$, that $x^3 - x \leq 0$ on $[0, 1]$, and that $x^3 - x \geq 0$ on $[1, 2]$. So we write

$$\int_{-1}^{2} |x^3 - x|\, dx = \int_{-1}^{0} (x^3 - x)\, dx + \int_0^1 (x - x^3)\, dx + \int_1^2 (x^3 - x)\, dx$$

$$= \left[\tfrac{1}{4}x^4 - \tfrac{1}{2}x^2 \right]_{-1}^{0} + \left[\tfrac{1}{2}x^2 - \tfrac{1}{4}x^4 \right]_0^1 + \left[\tfrac{1}{4}x^4 - \tfrac{1}{2}x^2 \right]_1^2$$

$$= \tfrac{1}{4} + \tfrac{1}{4} + [2 - (-\tfrac{1}{4})] = \tfrac{11}{4} = 2.75.$$

4–4 PROBLEMS

In each of Problems 1–5, find the average value of the given function on the indicated interval.

1 $f(x) = x^3$ on $[0, 5]$.

2 $f(x) = \dfrac{1}{\sqrt{x}}$ on $[1, 4]$.

3 $f(x) = \sqrt{x + 1}$ on $[0, 3]$.

4 $f(x) = \sin 2x$ on $[0, \pi/2]$.

5 $f(x) = \cos^2 x$ on $[0, \pi]$.

Evaluate the integrals in Problems 6–20.

6 $\int_{-1}^{2} (4 - 3x + 2x^2)\, dx$.

7 $\int_{-1}^{3} dx$ (Here, dx stands for $1\, dx$.)

8 $\int_1^2 (y^5 - 1)\, dy$.

9 $\int_1^4 \dfrac{dx}{\sqrt{9x^3}}$.

10 $\int_{-1}^{1} (x^3 + 2)^2\, dx$.

11 $\int_1^3 \dfrac{3t - 5}{t^4}\, dt$.

12 $\int_{-2}^{-1} \dfrac{x^2 - x + 3}{\sqrt[3]{x}}\, dx$.

13 $\int_0^{\pi} \sin x \cos x\, dx$.

14 $\int_{-1}^{2} |x|\, dx$.

15 $\int_1^2 \left(t - \dfrac{1}{2t} \right)^2 dt$.

16 $\int_{-1}^{1} \dfrac{x^2 - 4}{x + 2}\, dx$.

17 $\int_0^{\sqrt{\pi}} x \cos x^2\, dx$.

18 $\int_0^2 |x - \sqrt{x}|\, dx$.

19 $\int_{-2}^{2} |x^2 - 1|\, dx$.

20 $\int_0^{\pi/3} \sin 3x\, dx$.

In each of Exercises 21–25, apply the first part of the Fundamental Theorem of Calculus to find the derivative of the given function.

21 $f(x) = \int_{-1}^{x} (t^2 + 1)^{17}\, dt$.

22 $g(t) = \int_0^t (x^2 + 25)^{1/2}\, dx$.

23 $h(z) = \int_2^z (u - 1)^{1/3}\, du$.

24 $A(x) = \int_1^x \dfrac{dt}{t}$.

25 $f(x) = \int_x^{10} \left(t + \dfrac{1}{t} \right) dt$. $\left(Suggestion: \text{ Use the fact that} \right.$

$\left. \int_a^b g(t)\, dt = -\int_b^a g(t)\, dt. \right)$

In Problems 26–28, differentiate the function f by first writing $f(x)$ in the form $g(u)$, where u is the upper limit of the integral.

26 $f(x) = \int_0^{x^2} \sqrt{1 + t^3}\, dt.$

27 $f(x) = \int_1^{3x} \sin(t^2)\, dt.$

28 $f(x) = \int_0^{\sin x} \sqrt{1 - t^2}\, dt.$

29 The Fundamental Theorem of Calculus *seems* to say that

$$\int_{-1}^{1} \frac{dx}{x^2} = \left[-\frac{1}{x} \right]_{-1}^{1} = -2,$$

in apparent contradiction to the fact that $1/x^2$ is always positive. What's wrong here?

30 Note that $\pm f(x) \le |f(x)|$, and thereby deduce Inequality (9) from Inequality (8).

31 Apply Part 2 of the Fundamental Theorem of Calculus to establish the interval union property (Section 4-3) of the integral.

32 Show that the average rate of change

$$\frac{f(b) - f(a)}{b - a}$$

of the function f on $[a, b]$ is equal to the average value of its derivative on $[a, b]$.

33 If a ball is dropped from a height of 400 feet, find its average height and its average velocity between the time it is dropped and the time it strikes the ground.

34 Find the average value on $[0, 10]$ of the animal population $P(t) = 100 + 10t + \frac{1}{2}t^2 + (0.02)t^3$ of Example 4 in Section 4-3.

35 Suppose that a 5000-gallon water tank takes 10 minutes to drain, and after t minutes the amount of water left in the tank is $V(t) = 50(10 - t)^2$ gallons. What is the average amount of water in the tank while it is draining?

36 On a certain day the temperature t hours past midnight was

$$T(t) = 80 + 10 \sin\left(\frac{\pi}{12}(t - 10) \right).$$

What was the average temperature between noon and 6 P.M.?

4-5

Integration by Substitution

We can write the computational part of the Fundamental Theorem of Calculus in the form

$$\int_a^b f(x)\, dx = \left[D^{-1}f(x) \right]_a^b. \qquad (1)$$

Because of this relationship between integration and antidifferentiation, it is customary to write

$$\int f(x)\, dx = D^{-1}f(x) = F(x) + C \qquad (2)$$

if $F'(x) = f(x)$. The expression $\int f(x)\, dx$, with no limits on the integral symbol, is called the **indefinite integral** of the function f, in contrast with the **definite integral**, which has upper and lower limits. Thus the indefinite integral of f is simply the most general antiderivative of f, and indefinite integration is simply antidifferentiation. For example,

$$\int (3x^2 - 4)\, dx = x^3 - 4x + C, \qquad \text{and}$$

$$\int (t - \cos t)\, dt = \tfrac{1}{2}t^2 - \sin t + C.$$

Note that the indefinite integral is a *function* (actually, a collection of functions), while the definite integral is a *number*. The relationship between definite and indefinite integration is obtained by rewriting (1) as

$$\int_a^b f(x)\, dx = \left[\int f(x)\, dx \right]_a^b. \qquad (3)$$

In the notation of indefinite integrals, the antidifferentiation formulas (6) through (10) of Section 3-7 take the forms

$$\int af(x)\,dx = a\int f(x)\,dx \qquad (a \text{ is a constant}), \tag{4}$$

$$\int [f(x) + g(x)]\,dx = \int f(x)\,dx + \int g(x)\,dx, \tag{5}$$

$$\int x^r\,dx = \frac{x^{r+1}}{r+1} + C \qquad (\text{if } r \neq -1), \tag{6}$$

$$\int \cos x\,dx = \sin x + C, \tag{7}$$

$$\int \sin x\,dx = -\cos x + C. \tag{8}$$

A common sort of indefinite integral takes the form

$$\int f(g(x))g'(x)\,dx.$$

If we write $u = g(x)$, then the differential of u is $du = g'(x)\,dx$, so a purely mechanical substitution gives the *tentative* formula

$$\int f(g(x))g'(x)\,dx = \int f(u)\,du. \tag{9}$$

One of the beauties of differential notation is that Formula (9) is not only plausible, but in fact true—with the understanding that u is to be replaced by $g(x)$ after the indefinite integration on the right-hand side has been performed. Indeed, (9) is merely an indefinite integral version of the chain rule. For if $F'(x) = f(x)$, then

$$D_x F(g(x)) = F'(g(x))g'(x) = f(g(x))g'(x)$$

by the chain rule, so that

$$\int f(g(x))g'(x)\,dx = \int F'(g(x))g'(x)\,dx$$

$$= \int D[F(g(x))]\,dx = F(g(x)) + C$$

$$= F(u) + C \qquad (u = g(x))$$

$$= \int f(u)\,du.$$

Formula (9) is the basis for the powerful technique of indefinite **integration by substitution.** It may be used whenever the integrand function is recognizable in the form $f(g(x))g'(x)$.

EXAMPLE 1   Find $\int x^2\sqrt{x^3 + 9}\,dx$.

Solution   Note that x^2 is, to within a constant factor, the derivative of $x^3 + 9$. We can therefore substitute

$$u = x^3 + 9, \qquad du = 3x^2\,dx.$$

The constant factor 3 can be supplied if we compensate by multiplying the integral by $\frac{1}{3}$. This gives

$$\int x^2 \sqrt{x^3 + 9}\, dx = \frac{1}{3} \int (x^3 + 9)^{1/2} 3x^2\, dx$$

$$= \frac{1}{3} \int u^{1/2}\, du = \frac{1}{3} \frac{u^{3/2}}{\frac{3}{2}} + C$$

$$= \frac{2}{9} u^{3/2} + C = \frac{2}{9}(x^3 + 9)^{3/2} + C.$$

Another valid way to carry out the same substitution is to solve $du = 3x^2\, dx$ for $x^2\, dx = \frac{1}{3} du$, and then write

$$\int (x^3 + 9)^{1/2} x^2\, dx = \int u^{1/2} \cdot \tfrac{1}{3}\, du = \tfrac{1}{3} \int u^{1/2}\, du.$$

Three items worth noting appear upon examination of the solution to Example 1:

- The differential dx along with the rest of the integrand is "transformed" or replaced in terms of u and du.
- Once the actual integration has been performed, the constant C of integration is added.
- A final resubstitution is necessary to write the answer in terms of the original variable x.

The method of integration by substitution can also be used with definite integrals. Only one additional step is required—evaluation of the final antiderivative at the original limits of integration. For example, the substitution of Example 1 gives

$$\int_0^3 x^2 \sqrt{x^3 + 9}\, dx = \tfrac{1}{3} \int_*^{**} u^{1/2}\, du$$

$$= \tfrac{1}{3} \left[\tfrac{2}{3} u^{3/2} \right]_*^{**}$$

$$= \tfrac{2}{9} \left[(x^3 + 9)^{3/2} \right]_0^3$$

$$= \tfrac{2}{9} [216 - 27] = 42.$$

The limits * and ** on u are so indicated because they were not calculated—there was no need to know them—and because it would be only a coincidence if they were the same as the limits on x.

But sometimes it is more convenient to determine the limits of integration with respect to the new variable u. Under the above substitution $u = x^3 + 9$, the lower limit $x = 0$ corresponds to $u = 9$, and the upper limit $x = 3$ corresponds to $u = 36$. Hence we may alternatively write

$$\int_0^3 x^2 \sqrt{x^3 + 9}\, dx = \tfrac{1}{3} \int_9^{36} u^{1/2}\, du$$

$$= \tfrac{1}{3} \left[\tfrac{2}{3} u^{3/2} \right]_9^{36} = 42.$$

The following theorem tells how to transform the limits $x = a$ and $x = b$ under the substitution $u = g(x)$. The new lower limit is $u = g(a)$ and the new upper limit is $u = g(b)$, *whether or not $g(b)$ is greater than $g(a)$.*

> **Theorem** *Definite Integration by Substitution*
>
> Suppose that the function g has a continuous derivative on $[a, b]$, and that f is continuous on $[g(a), g(b)]$. Let $u = g(x)$. Then
>
> $$\int_a^b f(g(x))g'(x)\,dx = \int_{g(a)}^{g(b)} f(u)\,du. \qquad (10)$$

Proof Choose an antiderivative F of f, so that $F = D^{-1}f$ or $F' = f$. Then, by the chain rule,

$$D_x^{-1}[f(g(x))g'(x)] = F(g(x)).$$

So

$$\int_a^b f(g(x))g'(x)\,dx = \Big[\,F(g(x))\,\Big]_a^b = F(g(b)) - F(g(a))$$

$$= \Big[\,F(u)\,\Big]_{u=g(a)}^{g(b)} = \int_{g(a)}^{g(b)} f(u)\,du.$$

We used the fundamental theorem to obtain the first and last equalities in this argument.

EXAMPLE 2 Evaluate $\displaystyle\int_4^9 \frac{\sqrt{x}\,dx}{(30 - x^{3/2})^2}$.

Solution Note that $30 - x^{3/2}$ is nonzero on $[4, 9]$, so the integrand is continuous. We substitute

$$u = 30 - x^{3/2}, \quad \text{so that} \quad du = -\tfrac{3}{2}x^{1/2}\,dx$$

or $\sqrt{x}\,dx = -\tfrac{2}{3}du$. To change the limits, note that

$$\text{if } x = 4, \quad \text{then } u = 22, \qquad \text{and}$$
$$\text{if } x = 9, \quad \text{then } u = 3.$$

Hence our substitution gives

$$\int_4^9 \frac{\sqrt{x}\,dx}{(30 - x^{3/2})^2} = \int_{22}^3 \frac{-\tfrac{2}{3}du}{u^2}$$

$$= \frac{2}{3}\int_3^{22} \frac{du}{u^2} = \frac{2}{3}\left[\,-\frac{1}{u}\,\right]_3^{22}$$

$$= \frac{2}{3}\left(-\frac{1}{22} + \frac{1}{3}\right) = \frac{19}{99}.$$

EXAMPLE 3 Evaluate $\displaystyle\int_0^{\pi/4} \sin^3 2t \cos 2t\,dt$.

Solution We substitute

$$u = \sin 2t, \quad \text{so that} \quad du = 2\cos 2t\,dt.$$

Then

$$u = 0 \quad \text{when } t = 0, \quad \text{and} \quad u = 1 \text{ when } t = \frac{\pi}{4}.$$

Hence

$$\int_0^{\pi/4} \sin^3 2t \cos 2t\,dt = \tfrac{1}{2}\int_0^1 u^3\,du$$

$$= \tfrac{1}{2}\left[\tfrac{1}{4}u^4\right]_0^1 = \tfrac{1}{8}.$$

Evaluate the integrals in Problems 1–24.

1 $\int (4x - 3)^5 \, dx.$

2 $\int x\sqrt{x^2 - 1} \, dx.$

3 $\int x\sqrt{2 - 3x^2} \, dx.$

4 $\int 3t(1 - 2t^2)^{10} \, dt.$

5 $\int \dfrac{x^2 \, dx}{(x^3 + 5)^4}.$

6 $\int \dfrac{t \, dt}{\sqrt{2t^2 + 1}}.$

7 $\int x^2 \sqrt[3]{2 - 4x^3} \, dx.$

8 $\int \dfrac{(x + 1) \, dx}{(x^2 + 2x + 5)^2}.$

9 $\int \sin \dfrac{t}{3} \, dt.$

10 $\int \dfrac{\cos \sqrt{x}}{\sqrt{x}} \, dx.$

11 $\int_{-1}^{-2} \dfrac{dt}{(t + 3)^3}.$

12 $\int_0^4 x\sqrt{x^2 + 9} \, dx.$

13 $\int_0^4 \dfrac{dx}{\sqrt{2x + 1}}.$

14 $\int_{-1}^1 \dfrac{(x + 1) \, dx}{\sqrt{x^2 + 2x + 2}}.$

15 $\int_0^8 t(t + 1)^{1/2} \, dt.$ (*Suggestion:* Try $u = t + 1$.)

16 $\int_0^{\pi/2} \sin x \cos x \, dx.$

17 $\int_0^{\pi/6} \sin 2x \cos^3 2x \, dx.$

18 $\int_0^{\sqrt{\pi}} x \sin \dfrac{x^2}{2} \, dx.$

19 $\int_0^{\pi/2} (1 + \sin t)^{3/2} \cos t \, dt.$

20 $\int_1^4 \dfrac{(1 + \sqrt{x})^4}{\sqrt{x}} \, dx.$

21 $\int \dfrac{x^3 - 1}{(x^4 - 4x)^{2/3}} \, dx.$

22 $\int \dfrac{1}{x^3} \left(1 + \dfrac{1}{x^2} \right)^{5/3} dx.$

23 $\int (2 - t^2)\sqrt[4]{6t - t^3} \, dt.$

24 $\int \dfrac{2 - x^2}{(x^3 - 6x + 1)^5} \, dx.$

In each of Problems 25–28, verify the given formula by differentiating the right-hand side.

25 $\int \dfrac{dx}{x^2\sqrt{x^2 + a^2}} = -\dfrac{\sqrt{x^2 + a^2}}{a^2 x} + C.$

26 $\int (x^2 - a^2)^{3/2} \, dx = \dfrac{x}{4}(x^2 - a^2)^{3/2} -$

$$\dfrac{3}{4} a^2 \int \sqrt{x^2 - a^2} \, dx.$$

27 $\int \sin^3 x \, dx = \tfrac{1}{3} \cos^3 x - \cos x + C.$

28 $\int \dfrac{x \, dx}{(1 - x^2)^2} = \dfrac{x^2}{2(1 - x^2)} + C.$

29 Substitute $u = 1 - x^2$ to show that

$$\int \dfrac{x \, dx}{(1 - x^2)^2} = \dfrac{1}{2(1 - x^2)} + C.$$

Is this answer consistent with the formula of Problem 28? Explain.

30 Substitute $\sin^3 x = (\sin x)(1 - \cos^2 x)$ to derive the formula of Problem 27.

31 Suppose that f is an **odd** function, meaning that $f(-x) = -f(x)$ for all x. Then substitute $u = -x$ in the integral $\int_{-a}^0 f(x) \, dx$ to show that

$$\int_{-a}^a f(x) \, dx = 0.$$

32 If f is an **even** function, meaning that $f(-x) = f(x)$ for all x, use the method of Problem 31 to show that

$$\int_{-a}^a f(x) \, dx = 2 \int_0^a f(x) \, dx.$$

4-6

Computing Areas by Integration

In Section 4-2 we discussed the area A under the graph of a positive-valued continuous function f on the interval $[a, b]$. This discussion motivated our definition in Section 4-3 of the integral of f from a to b, with the result that

$$A = \int_a^b f(x) \, dx \tag{1}$$

by definition.

Here we consider the more general problem of finding the area of a region that is bounded by the graphs of *two* functions. Suppose that the functions f and g are continuous on $[a, b]$, and that $f(x) \geq g(x)$ for all x in $[a, b]$. We are interested in the area A of the region R shown in Figure

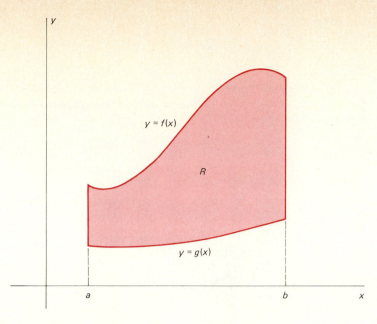

4.18 A region between two graphs.

4.18, in which R is bounded by the graphs of f and g and the vertical lines $x = a$ and $x = b$.

In order to approximate A, we consider a regular partition of $[a, b]$ into n equal subintervals, each with length Δx. If ΔA_i denotes the area of the region between the graphs of f and g and lying above the ith subinterval $[x_{i-1}, x_i]$, and x_i^* is an arbitrary number chosen in that subinterval (all this for $i = 1, 2, \ldots, n$), then ΔA_i is approximately equal to the area of a rectangle with height $f(x_i^*) - g(x_i^*)$ and width Δx (see Figure 4.19). Hence

$$\Delta A_i \approx [f(x_i^*) - g(x_i^*)] \, \Delta x,$$

4.19 A partition of $[a, b]$ divides R into vertical strips that we approximate with rectangular strips.

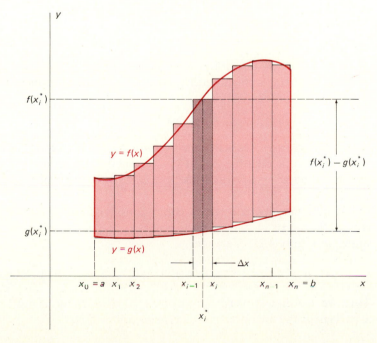

and so

$$A = \sum_{i=1}^{n} \Delta A_i \approx \sum_{i=1}^{n} [f(x_i^*) - g(x_i^*)] \, \Delta x.$$

We introduce the function $h(x) = f(x) - g(x)$, and observe that A is given approximately by a Riemann sum for h associated with our regular partition of $[a, b]$:

$$A \approx \sum_{i=1}^{n} h(x_i^*) \, \Delta x.$$

Intuition and reason both suggest that this approximation can be made arbitrarily accurate by choosing n sufficiently large (and hence $\Delta x = (b - a)/n$ sufficiently small). We would therefore conclude that

$$A = \lim_{\Delta x \to 0} \sum_{i=1}^{n} h(x_i^*) \, \Delta x$$

$$= \int_a^b h(x) \, dx = \int_a^b [f(x) - g(x)] \, dx.$$

Since our discussion is based on an intuitive concept rather than on a logical definition of area, it does *not* constitute a proof of the formula above. It does, however, provide justification for the following *definition* of the area in question.

Definition   *The Area Between Two Curves*

Let f and g be continuous with $f(x) \geqq g(x)$ for x in $[a, b]$. Then the **area** A of the region bounded by the curves $y = f(x)$ and $y = g(x)$ and the vertical lines $x = a$ and $x = b$ is

$$A = \int_a^b [f(x) - g(x)] \, dx. \qquad (2)$$

Note that our earlier Formula (1) is the case $g(x) \equiv 0$ of (2). On the other hand, if $f(x) \equiv 0$ and $g(x) \leqq 0$ on $[a, b]$, then (2) reduces to

$$A = -\int_a^b g(x) \, dx \quad \text{or} \quad \int_a^b g(x) \, dx = -A.$$

In this case the area lies below the x-axis, as in Figure 4.20. Thus the integral from a to b of a negative-valued function is the *negative* of the area of the region bounded by its graph, the x-axis, and the vertical lines $x = a$ and $x = b$.

More generally, consider a continuous function f with a graph that crosses the x-axis at finitely many points $c_1, c_2, \ldots, c_k$ between a and b, as shown in Figure 4.21. We write

$$\int_a^b f(x) \, dx = \int_a^{c_1} f(x) \, dx + \int_{c_1}^{c_2} f(x) \, dx + \cdots + \int_{c_k}^b f(x) \, dx.$$

Thus we see that $\int_a^b f(x) \, dx$ is equal to the area under $y = f(x)$ *above* the x-axis *minus* the area over $y = f(x)$ *below* the x-axis.

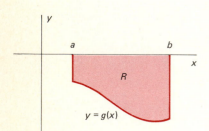

4.20 The integral gives the negative of the geometric area for a region lying below the x-axis.

CHAP. 4:   **The Integral**

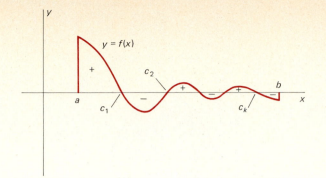

4.21 The integral computes area above *x*-axis *minus* area below *x*-axis.

The following *heuristic* (suggestive, though not rigorous) way of setting up integral formulas like (2) is sometimes useful. Consider the vertical strip of area lying above the interval $[x, x + dx]$ that is shaded in Figure 4.22. We think of the length dx of the interval $[x, x + dx]$ as being so small that this strip may be regarded as a rectangle with width dx and height $f(x) - g(x)$, so that its area is

$$dA = [f(x) - g(x)] \, dx.$$

Think now of the region over $[a, b]$ between $y = f(x)$ and $y = g(x)$ as made up of many such strips. Its area may be thought of as a sum of areas of such rectangles. If we write $\int$ for *sum*, this gives the formula

$$A = \int dA = \int_a^b [f(x) - g(x)] \, dx.$$

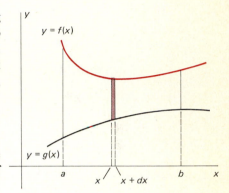

4.22 Heuristic approach to setting up area integrals.

This heuristic approach bypasses the subscript notation associated with Riemann sums. Nevertheless, it *is not and should not be regarded as a complete* derivation of the formula. It is best used only as a convenient memory device. For instance, in the figures for illustrative examples, we shall often show a strip of width dx as a visual aid in properly setting up the correct integral.

EXAMPLE 1 Find the area A of the region R that is bounded by the line $y = x$ and the parabola $y = 6 - x^2$.

Solution The region R appears in Figure 4.23. It is clear that we should use Formula (2), taking $f(x) = 6 - x^2$ and $g(x) = x$. The limits a and b will be the x-coordinates of the two points of intersection of the line and the parabola. To find a and b, we therefore equate $f(x)$ and $g(x)$ and solve the resulting equation for x:

$$x = 6 - x^2;$$
$$x^2 + x - 6 = 0;$$
$$(x - 2)(x + 3) = 0;$$
$$x = -3, 2.$$

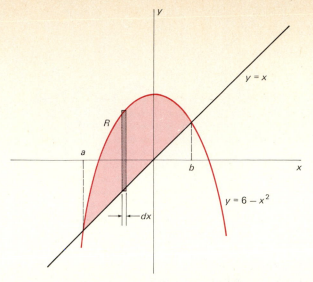

4.23 The region R of Example 1.

Thus $a = -3$ and $b = 2$, so Formula (2) gives

$$A = \int_{-3}^{2} (6 - x^2 - x)\, dx$$

$$= \left[6x - \frac{x^3}{3} - \frac{x^2}{2} \right]_{-3}^{2} = \frac{125}{6}.$$

As in the following example, it is sometimes necessary to subdivide a region before applying Formula (2).

EXAMPLE 2   Find the area A of the region R bounded by the line $y = x$ and the parabola $y^2 = 6 - x$.

Solution   The region R is shown in Figure 4.24. The points of intersection $(-3, -3)$ and $(2, 2)$ are found by equating $y = x$ and $y = \pm\sqrt{6 - x}$ and

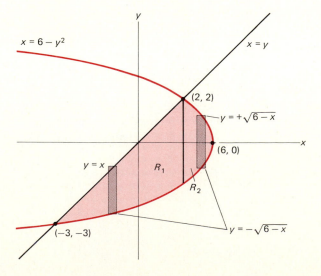

4.24 In Example 2, we split the region R into R_1 and R_2.

CHAP. 4:   **The Integral**

then solving for x. The lower boundary of R is given by $y = -\sqrt{6-x}$ on $[-3, 6]$. But the upper boundary of R is given by

$$y = x \quad \text{on} \quad [-3, 2],$$
$$y = +\sqrt{6-x} \quad \text{on} \quad [2, 6].$$

We must therefore subdivide R into the two regions R_1 and R_2 as indicated in Figure 4.24. Then Formula (2) gives

$$
\begin{aligned}
A &= \int_{-3}^{2} [x - (-\sqrt{6-x})]\, dx + \int_{2}^{6} [\sqrt{6-x} - (-\sqrt{6-x})]\, dx \\
&= \int_{-3}^{2} (x + \sqrt{6-x})\, dx + 2 \int_{2}^{6} \sqrt{6-x}\, dx \\
&= \left[\tfrac{1}{2}x^2 - \tfrac{2}{3}(6-x)^{3/2} \right]_{-3}^{2} + 2 \left[-\tfrac{2}{3}(6-x)^{3/2} \right]_{2}^{6} \\
&= (2 - \tfrac{16}{3}) - (\tfrac{9}{2} - 18) + 2(0) - 2(-\tfrac{16}{3}) = \tfrac{125}{6}.
\end{aligned}
$$

The region of Example 2 appears simpler if it is considered to be bounded by graphs of functions of y rather than functions of x. Figure 4.25 shows a region R that is bounded by the curves $x = f(y)$ and $x = g(y)$, with $f(y) \geq g(y)$ for y in $[c, d]$, and by the horizontal lines $y = c$ and $y = d$. To approximate the area A of R, we start with a regular partition of $[c, d]$ into n subintervals, each with length Δy. We choose a point y_i^* in the ith subinterval $[y_{i-1}, y_i]$ for each $i = 1, 2, \ldots, n$. The strip of R lying opposite $[y_{i-1}, y_i]$ is approximated by a rectangle with width Δy and length $f(y_i^*) - g(y_i^*)$. Hence

$$A \approx \sum_{i=1}^{n} [f(y_i^*) - g(y_i^*)]\, \Delta y.$$

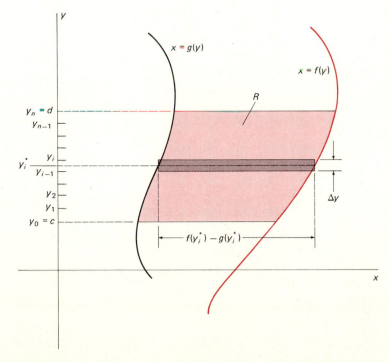

4.25 Finding area using an integral with respect to y.

Recognition of this sum as a Riemann sum for the integral

$$\int_c^d [f(y) - g(y)]\, dy$$

motivates the following definition.

> **Definition** *The Area Between Two Curves*
>
> Let f and g be continuous with $f(y) \geqq g(y)$ for y in $[c, d]$. Then the **area** A of the region bounded by the curves $x = f(y)$ and $x = g(y)$ and the horizontal lines $y = c$ and $y = d$ is
>
> $$A = \int_c^d [f(y) - g(y)]\, dy. \qquad (3)$$

Comparison of Example 2 with the one following illustrates the advantage of choosing the "right" variable of integration—the one that makes the resulting computation simpler.

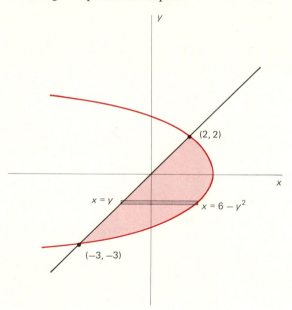

4.26 Recomputation of the area of Example 2.

EXAMPLE 3 Integrate with respect to y to find the area of the region R of Example 2.

Solution We see from Figure 4.26 that Formula (3) applies with $f(y) = 6 - y^2$ and $g(y) = y$ for y in $[-3, 2]$. This gives

$$A = \int_{-3}^2 [6 - y^2 - y]\, dy = \left[6y - \frac{y^3}{3} - \frac{y^2}{2} \right]_{-3}^2 = \frac{125}{6}.$$

EXAMPLE 4 Use calculus to derive the formula $A = \pi r^2$ for the area of a circle of radius r.

Solution We begin with the *definition* (in Section 4-2) of the number π as the area of the *unit* circle $x^2 + y^2 \leqq 1$. Then, with the aid of Figure 4.27,

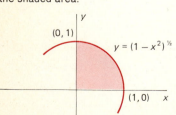

4.27 The number π is four times the shaded area.

CHAP. 4: **The Integral**

we may write

$$\pi = 4 \int_0^1 \sqrt{1 - x^2}\, dx, \tag{4}$$

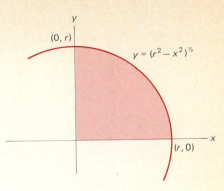

because the integral in (4) is, by Formula (1), the area of the first quadrant of the unit circle. We apply Formula (1) to the first quadrant of the circle of radius r of Figure 4.28, and we find the total area A of that circle to be

$$A = 4 \int_0^r \sqrt{r^2 - x^2}\, dx = 4r \int_0^r \sqrt{1 - (x/r)^2}\, dx$$

$$= 4r \int_0^1 \sqrt{1 - u^2}\, r\, du \quad \left(\text{Substitution:}\ u = \frac{x}{r},\, dx = r\, du. \right)$$

$$= 4r^2 \int_0^1 \sqrt{1 - u^2}\, du = 4r^2 \int_0^1 \sqrt{1 - x^2}\, dx.$$

So, by Equation (4) above, $A = \pi r^2$.

4.28 The shaded area can be written as an integral.

4-6 PROBLEMS

In each of Problems 1–22, sketch the region bounded by the given curves and then find its area.

1 $y = 0$ and $y = 25 - x^2$.
2 $y = x^2$ and $y = 4$.
3 $y = x^2$ and $y = 8 - x^2$.
4 $x = 0$ and $x = 16 - y^2$.
5 $x = y^2$ and $x = 25$.
6 $x = y^2$ and $x = 32 - y^2$.
7 $y = x^2$ and $y = 2x$.
8 $y = x^2$ and $x = y^2$.
9 $y = x^2$ and $y = x^3$.
10 $y = 2x^2$ and $y = 5x - 3$.
11 $x = 4y^2$ and $x + 12y + 5 = 0$.
12 $y = x^2$ and $y = 3 + 5x - x^2$.
13 $x = 3y^2$ and $x = -y^2 + 12y - 5$.
14 $y = x^2$ and $y = 4(x - 1)^2$.
15 $x = y^2 - 2y - 2$ and $x = -2y^2 + y + 4$.
16 $y = x^4$ and $y = 32 - x^4$.
17 $y = x^3$ and $y = 32\sqrt{x}$.
18 $y = x^3$ and $y = 2x - x^2$.
19 $y = x^2$ and $y = x^{2/3}$.
20 $y^2 = x$ and $y^2 = 2(x - 3)$.
21 $y = x^3$ and $y = 2x^3 + x^2 - 2x$.
22 $y = x^3$, $x + y = 0$, and $y = x + 6$.
23 The *ellipse* $x^2/a^2 + y^2/b^2 = 1$ is shown in Figure 4.29. Use the method of Example 4 to show that the area of the region it bounds is $A = \pi ab$, a pleasant generalization of the area formula for the circle.
24 Let A and B be the points of intersection of the parabola $y = x^2$ and the line $y = x + 2$, and let C be the point on the parabola where the tangent line is parallel to the line

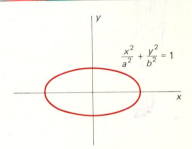

4.29 The ellipse of Problem 23.

$y = x + 2$. Show that the area of the parabolic segment (shown in Figure 4.30) that is cut off from the parabola by the line is four-thirds the area of the triangle ABC.

4.30 The parabolic segment of Problem 24.

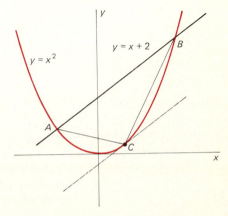

25 Find the area of the unbounded region that is shaded in Figure 4.31, regarding it as the limit as $b \to \infty$ of the region bounded by $y = 1/x^2$, $y = 0$, $x = 1$, and $x = b > 1$.

4.31 The unbounded region of Problem 25.

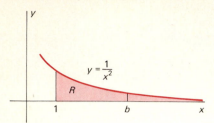

26 Find the total area of the bounded regions that are bounded by the x-axis and the curve $y = 2x^3 - 2x^2 - 12x$.

27 Suppose that the quadratic function $f(x) = px^2 + qx + r$ is never negative on $[a, b]$. Show that the area under the graph of f from a to b is

$$A = \frac{h}{3}\left[f(a) + 4f(m) + f(b)\right]$$

where $h = \frac{1}{2}(b - a)$ and $m = \frac{1}{2}(a + b)$. (*Suggestion:* By horizontal translation of this region, it may be assumed that $a = -h$, $m = 0$, and $b = h$.)

4-7

Numerical Integration

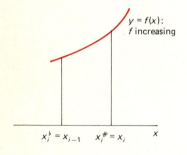

4.32 If f is increasing or decreasing, its extrema occur at the endpoints of the interval $[x_{i-1}, x_i]$.

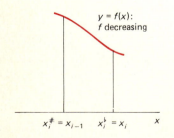

The Fundamental Theorem of Calculus, $\int_a^b f(x)\,dx = [G(x)]_a^b$, can be used to evaluate the integral only if a convenient formula for the antiderivative $G(x)$ of $f(x)$ can be found. But there are rather simple functions with antiderivatives that are not elementary functions. An **elementary function** is one that can be expressed in terms of polynomial, trigonometric, exponential, and logarithmic functions by means of finite combinations of sums, differences, products, quotients, roots, and function composition.

The problem is that elementary functions can have nonelementary antiderivatives. For example, it is known that the elementary function 2^{-x^2} has no elementary antiderivative. Consequently, we cannot use the Fundamental Theorem of Calculus to evaluate an integral such as

$$\int_0^1 2^{-x^2}\,dx.$$

In this section we discuss the use of Riemann sums to numerically approximate integrals that cannot conveniently be evaluated exactly, whether or not nonelementary functions are involved. Given a continuous function f on $[a, b]$ with an integral that is to be approximated, consider a regular partition P of $[a, b]$ into n subintervals, each with length $\Delta x = (b - a)/n$. Recall from Section 4-3 that the value of the integral lies between the upper and lower Riemann sums for f associated with the partition P:

$$\sum_{i=1}^{n} f(x_i^\flat)\,\Delta x \leq \int_a^b f(x)\,dx \leq \sum_{i=1}^{n} f(x_i^\#)\,\Delta x, \tag{1}$$

where $f(x_i^\flat)$ and $f(x_i^\#)$ are the minimum and maximum values, respectively, of f on the ith subinterval $[x_{i-1}, x_i]$ of the partition.

Relation (1) is particularly easy to apply when f is either an increasing function or a decreasing function on $[a, b]$. For if f is increasing on $[a, b]$, then the minimum value of f on the ith subinterval $[x_{i-1}, x_i]$ occurs at x_{i-1}, so that $x_i^\flat = x_{i-1}$. It is equally clear that $x_i^\# = x_i$. If f is instead a decreasing function on $[a, b]$, then $x_i^\flat = x_i$ and $x_i^\# = x_{i-1}$. The two cases are illustrated in Figure 4.32.

The *left-endpoint approximation* L_n and the *right-endpoint approximation* R_n to the definite integral $\int_a^b f(x)\,dx$, associated with the regular partition of $[a, b]$ into n equal subintervals, are the Riemann sums

$$L_n = \sum_{i=1}^{n} f(x_{i-1})\,\Delta x \qquad (2)$$

and

$$R_n = \sum_{i=1}^{n} f(x_i)\,\Delta x. \qquad (3)$$

The notation for L_n and R_n may be simplified a bit by writing y_i for $f(x_i)$.

Definition *Endpoint Approximations*

The **left-endpoint approximation** L_n and the **right-endpoint approximation** R_n to $\int_a^b f(x)\,dx$ with $\Delta x = (b - a)/n$ are

$$L_n = (\Delta x)(y_0 + y_1 + y_2 + \cdots + y_{n-1}) \qquad (2)$$

and

$$R_n = (\Delta x)(y_1 + y_2 + y_3 + \cdots + y_n). \qquad (3)$$

So Inequality (1) above amounts to this:

$$L_n \leqq \int_a^b f(x)\,dx \leqq R_n \qquad (4)$$

for an increasing function f, and

$$R_n \leqq \int_a^b f(x)\,dx \leqq L_n \qquad (5)$$

for a decreasing function f.

We also have an error estimate available. Note that

$$|R_n - L_n| = \left| \sum_{i=1}^{n} [f(x_i) - f(x_{i-1})]\,\Delta x \right|$$

$$= (\Delta x)|f(x_1) - f(x_0) + f(x_2) - f(x_1) + \cdots + f(x_n) - f(x_{n-1})|$$

$$= (\Delta x)|f(x_n) - f(x_0)|.$$

Thus

$$|R_n - L_n| = \frac{b - a}{n}\,|f(b) - f(a)|. \qquad (6)$$

Equation (6) enables us to estimate the error when we approximate the integral of an increasing (or a decreasing) function by either R_n or L_n. From a geometric point of view, Equation (6) is related to the observation that $|R_n - L_n|$ is the sum of the areas of the shaded small rectangles of Figure 4.33, and that these small rectangles can be moved horizontally to form a vertical stack. The stack, also shown in Figure 4.33, is a rectangle with base $\Delta x = (b - a)/n$ and height $|f(b) - f(a)|$.

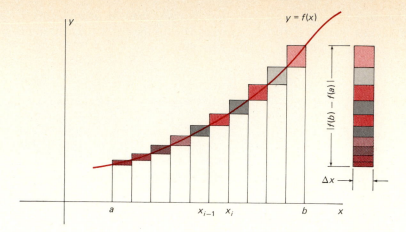

4.33 Error in the endpoint approximation.

Note also that Equation (6) can be used to determine how large n must be in order for L_n and R_n to approximate $\int_a^b f(x)\,dx$ with a given degree of accuracy. For example, to obtain three-place accuracy, it suffices to choose n sufficiently large that

$$\frac{b-a}{n}\left|f(b)-f(a)\right| < 0.0005.$$

We illustrate these numerical techniques by approximating the integral

$$\int_0^1 \frac{4}{1+x^2}\,dx. \tag{7}$$

In Chapter 7 we shall study the inverse tangent function $y = \arctan x$ (y is the angle between $-\pi/2$ and $\pi/2$ with tangent x), and we shall show there that $D \arctan x = 1/(1 + x^2)$. This implies that

$$\int_0^1 \frac{1}{1+x^2}\,dx = \left[\arctan x\right]_0^1 = \arctan 1 - \arctan 0 = \frac{\pi}{4}.$$

Hence the true value of the integral in (7) is

$$\pi = 3.14159\ 265358\ 97932\ \ldots .$$

This last fact will provide us with a check on the accuracy of our approximations.

EXAMPLE 1 Calculate the left-endpoint and right-endpoint approximations to the integral (7) with $n = 10$ and $\Delta x = 0.1$.

Solution Note first that $f(x) = 4/(1 + x^2)$ is a decreasing function on $[0, 1]$, so L_{10} will be a high estimate of the integral and R_{10} a low estimate. We use a hand-held calculator and note in writing only the final value of each approximation, because writing the individual terms and (subsequently) their sum would vitiate the eight- or nine-place internal accuracy of such a calculator. Since

$$x_0 = 0, \ x_1 = 0.1, \ x_2 = 0.2, \ldots, \ x_9 = 0.9, \ x_{10} = 1,$$

we obtain

$$L_{10} = (0.1)\left[\frac{4}{1.00} + \frac{4}{1.01} + \frac{4}{1.04} + \cdots + \frac{4}{1.81}\right]$$

$$\approx 3.239925990 \approx 3.24$$

and

$$R_{10} = (0.1)\left[\frac{4}{1.01} + \frac{4}{1.04} + \frac{4}{1.09} + \cdots + \frac{4}{2.00}\right]$$

$$\approx 3.039925990 \approx 3.04.$$

Now π lies between R_{10} and L_{10}, so these computations show that $\pi = 3.14 \pm 0.10$ if rounded to two decimal places (3.14 is the two-place average of R_{10} and L_{10}). Since $f(0) = 4$ and $f(1) = 2$, Equation (6) gives

$$|R_n - L_n| = \frac{2}{n}.$$

In order to achieve the accuracy $|R_n - L_n| \leqq 0.0005$, we would therefore need a partition with $n \geqq 4000$ subintervals.

The average $T_n = \frac{1}{2}(L_n + R_n)$ of the left-endpoint and right-endpoint approximations is called the *trapezoidal approximation* to $\int_a^b f(x)\,dx$ associated with the partition of $[a, b]$ into n equal subintervals. Written in full, we see that

$$T_n = \frac{1}{2}(L_n + R_n) = \frac{\Delta x}{2}\sum_{i=1}^{n}\left[f(x_{i-1}) + f(x_i)\right]$$

or

$$T_n = \frac{\Delta x}{2}\left[f(x_0) + 2f(x_1) + 2f(x_2) + \cdots + 2f(x_{n-2}) + 2f(x_{n-1}) + f(x_n)\right].$$

Note the 1–2–2–$\cdots$–2–2–1 pattern of the coefficients.

Definition *Trapezoidal Approximation*

The **trapezoidal approximation** to $\int_a^b f(x)\,dx$ with $\Delta x = (b - a)/n$ is

$$T_n = \frac{\Delta x}{2}(y_0 + 2y_1 + 2y_2 + \cdots + 2y_{n-1} + y_n). \qquad (8)$$

Figure 4.34 shows where T_n gets its name. The partition points $x_0, x_1, \ldots, x_n$ are used to build trapezoids from the x-axis to the function's graph. The trapezoid over the ith subinterval $[x_{i-1}, x_i]$ has **altitude** Δx and its parallel

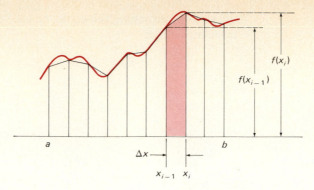

4.34 Geometry of the trapezoidal approximation.

bases have lengths $f(x_{i-1})$ and $f(x_i)$. So its area is

$$\frac{\Delta x}{2}[f(x_{i-1}) + f(x_i)].$$

A comparison of this with (8) shows that T_n is merely the sum of the areas of the n trapezoids shown in Figure 4.34.

EXAMPLE 2   Calculate the trapezoidal approximation to the integral in (7) with $n = 10$ and $\Delta x = 0.1$.

Solution   Since we have already calculated L_{10} and R_{10} in Example 1, we can simply write

$$T_{10} = \tfrac{1}{2}(L_{10} + R_{10})$$
$$\approx \tfrac{1}{2}(3.239925990 + 3.039925990)$$

so

$$T_{10} \approx 3.139925990.$$

Had L_{10} and R_{10} not been available, we would have written

$$T_{10} = \frac{0.1}{2}\left[\frac{(1)4}{1.00} + \frac{(2)4}{1.01} + \cdots + \frac{(2)4}{1.81} + \frac{(1)4}{2.00}\right]$$

$$\approx 3.139925990.$$

Another useful approximation to $\int_a^b f(x)\,dx$ is the *midpoint approximation* M_n. It is the Riemann sum obtained by choosing the point x_i^* in $[x_{i-1}, x_i]$ to be its midpoint $m_i = \tfrac{1}{2}(x_{i-1} + x_i)$. Thus

$$M_n = \sum_{i=1}^{n} f(m_i)\,\Delta x = (\Delta x)[f(m_1) + f(m_2) + \cdots + f(m_n)]. \qquad (9)$$

It is sometimes convenient to write $m_i = x_{i-1/2}$ and $f(m_i) = y_{i-1/2}$. With this convention we frame our next definition.

Definition   *Midpoint Approximation*

The **midpoint approximation** to $\int_a^b f(x)\,dx$ with $\Delta x = (b-a)/n$ is

$$M_n = (\Delta x)(y_{1/2} + y_{3/2} + y_{5/2} + \cdots + y_{n-1/2}). \qquad (9)$$

EXAMPLE 3 Calculate the midpoint approximation to the integral in (7) with $n = 10$ and $\Delta x = 0.1$.

Solution We still have the integral

$$\int_0^1 \frac{4}{1 + x^2}\, dx \qquad (7)$$

to be approximated. But now $m_1 = 0.05$, $m_2 = 0.15$, $m_3 = 0.25, \ldots$, and $m_{10} = 0.95$. So

$$M_{10} = (0.1)\left[\frac{4}{1.0025} + \frac{4}{1.0225} + \frac{4}{1.0625} + \cdots + \frac{4}{1.9025}\right]$$

$$\approx 3.142425986 \approx 3.14.$$

Thus M_{10} gives π rounded to two decimal places.

The midpoint approximation (9) is sometimes called the **tangent line** approximation, because the area of the rectangle with base $[x_{i-1}, x_i]$ and height $f(m_i)$ is also the area of another approximating figure. As shown in Figure 4.35, we draw a tangent segment to the graph of f, tangent at the point $(m_i, f(m_i))$ of its graph, and use that segment for one side of a trapezoid, somewhat like the method of the trapezoidal rule. This trapezoid and the rectangle mentioned above have the same area, and so the value of M_n is the sum of the areas of trapezoids like the one of Figure 4.35.

When we compare the results of Examples 2 and 3, we see that the midpoint approximation $M_{10} \approx 3.1424$ is somewhat closer to $\pi \approx 3.1416$ than is the trapezoidal approximation $T_{10} \approx 3.1399$. The trapezoids shown in Figure 4.36 suggest why this should happen. Moreover, the area of the trapezoid associated with the midpoint approximation is generally closer to the true value of

$$\int_{x_{i-1}}^{x_i} f(x)\, dx$$

than is the area of the trapezoid associated with the trapezoidal approximation. Figure 4.36 also shows this, in that the midpoint error E_m, which is crosshatched in the figure, is generally smaller than the trapezoidal error E_t, which is dotted in the figure. Figure 4.36 also indicates that if $y = f(x)$ is concave downward, then M_n will be an overestimate and T_n will be an underestimate of $\int_a^b f(x)\, dx$. If the graph is concave upward, then the situation will be reversed.

Such observations motivate the consideration of a *weighted* average of M_n and T_n, with M_n weighted more heavily than T_n, for further improving our numerical estimates of the definite integral. The particular weighted average

$$S_{2n} = \tfrac{1}{3}(2M_n + T_n) = \tfrac{2}{3}M_n + \tfrac{1}{3}T_n, \qquad (10)$$

is called *Simpson's approximation* to $\int_a^b f(x)\, dx$. The reason for the subscript $2n$ is that it's most natural to regard S_{2n} as being associated with a partition of $[a, b]$ into an *even* number $2n$ of equal subintervals with the endpoints

$$a = x_0 < x_1 < x_2 < \cdots < x_{2n-2} < x_{2n-1} < x_{2n} = b.$$

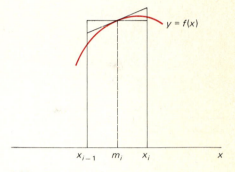

4.35 The midpoint or tangent rule.

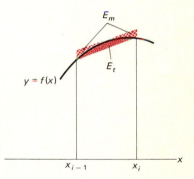

4.36 Comparing the midpoint rule error E_m with the trapezoidal rule error E_t.

The midpoint and trapezoidal approximations associated with n subintervals

$$[x_0, x_2], [x_2, x_4], [x_4, x_6], \ldots, [x_{2n-4}, x_{2n-2}], [x_{2n-2}, x_{2n}]$$

that all have the same length $2 \, \Delta x$ can then be written in the form

$$M_n = (2 \, \Delta x)[y_1 + y_3 + y_5 + \cdots + y_{2n-1}]$$

and

$$T_n = \frac{2 \, \Delta x}{2}[y_0 + 2y_2 + 2y_4 + \cdots + 2y_{2n-2} + y_{2n}].$$

We substitute these formulas for M_n and T_n into (10), and find—after a bit of algebra—that

$$S_{2n} = \frac{\Delta x}{3}[y_0 + 4y_1 + 2y_2 + 4y_3 + 2y_4 + \cdots + 2y_{2n-2} + 4y_{2n-1} + y_{2n}].$$

$$(11)$$

Definition *Simpson's Approximation*

Simpson's approximation to $\int_a^b f(x) \, dx$ with $\Delta x = (b - a)/2n$, associated with a partition of $[a, b]$ into an even number $2n$ of equal subintervals, is the sum S_{2n} given by Equation (11).

Note the 1–4–2–4–2–$\cdots$–4–2–4–1 pattern of coefficients in Simpson's approximation.

EXAMPLE 4 Calculate Simpson's approximation S_{10} to the integral

$$\int_0^1 \frac{4}{1 + x^2} \, dx. \tag{7}$$

Solution Working with 10 subintervals of $[0, 1]$ and $\Delta x = 0.1$, we obtain

$$S_{10} = \frac{0.1}{3}\left[\frac{(1)4}{1.00} + \frac{(4)4}{1.01} + \frac{(2)4}{1.04} + \frac{(4)4}{1.09} + \cdots + \frac{(2)4}{1.64} + \frac{(4)4}{1.81} + \frac{(1)4}{2.00}\right],$$

or

$$S_{10} \approx 3.141592616.$$

Simpson's approximation correctly gives the first *seven* decimal places of π!

EXAMPLE 5 Calculate Simpson's approximation S_{20} to the above integral (7).

Solution Since we have already calculated T_{10} and M_{10} in Examples 2 and 3, we can use (10):

$$S_{20} = \tfrac{2}{3}M_{10} + \tfrac{1}{3}T_{10}$$
$$\approx \tfrac{2}{3}(3.142425986) + \tfrac{1}{3}(3.139925990)$$

or

$$S_{20} \approx 3.141592654,$$

which is the correct value of π rounded to *nine* decimal places!

Although we have defined Simpson's approximation S_{2n} as a weighted average of the midpoint and trapezoidal approximations it has an important interpretation in terms of **parabolic approximations** to the curve $y = f(x)$. Starting with the partition of $[a, b]$ into $2n$ equal subintervals, we define the parabolic function

$$p_i(x) = A_i + B_i x + C_i x^2$$

on $[x_{2i-2}, x_{2i}]$. By solving three equations in three unknowns, the coefficients A_i, B_i, and C_i can be chosen so that $p_i(x)$ agrees with $f(x)$ at the three points x_{2i-2}, x_{2i-1}, and x_{2i}. A routine, albeit tedious, algebraic computation (see Problem 27 in Section 4-6) then shows that

$$\int_{x_{2i-2}}^{x_{2i}} p_i(x)\, dx = \frac{\Delta x}{3} (y_{2i-2} + 4y_{2i-1} + y_{2i}).$$

We now approximate $\int_a^b f(x)\, dx$ by replacing f by p_i on the interval $[x_{2i-2}, x_{2i}]$ for $i = 1, 2, 3, \ldots, n$. This gives

$$\int_a^b f(x)\, dx = \sum_{i=1}^{n} \int_{x_{2i-2}}^{x_{2i}} f(x)\, dx$$

$$\approx \sum_{i=1}^{n} \int_{x_{2i-2}}^{x_{2i}} p_i(x)\, dx$$

$$= \sum_{i=1}^{n} \frac{\Delta x}{3} (y_{2i-2} + 4y_{2i-1} + y_{2i})$$

$$= \frac{\Delta x}{3} (y_0 + 4y_1 + 2y_2 + 4y_3 + \cdots + 4y_{2n-3}$$

$$+ 2y_{2n-2} + 4y_{2n-1} + y_{2n}).$$

Thus the parabolic approximation described above results in Simpson's approximation S_{2n} to $\int_a^b f(x)\, dx$.

The numerical methods of this section are especially useful for approximating integrals of functions that are available only in graphical or in tabular form. This is often the case with functions derived from empirical data or from experimental measurements.

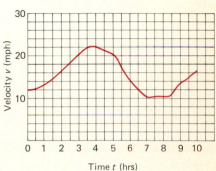

4.37 Velocity graph for the submarine of Example 6.

EXAMPLE 6 Suppose that the graph of Figure 4.37 shows the velocity recorded by instruments on board a nuclear submarine traveling under the polar ice cap directly toward the North Pole. Use the trapezoidal approximation and Simpson's approximation to estimate the distance $s = \int_a^b v(t)\, dt$ traveled by the submarine during the 10-hour period from $t = 0$ to $t = 10$.

Solution   We read the following data from the graph.

t	0	1	2	3	4	5	6	7	8	9	10	hrs
v	13	14	17	21	23	21	15	11	11	14	17	mph

Using the trapezoidal approximation with $n = 10$ and $\Delta x = 1$, we obtain

$$s = \int_0^{10} v(t)\,dt$$

$$\approx \tfrac{1}{2}[13 + 2(14 + 17 + 21 + 23 + 21 + 15 + 11 + 11 + 14) + 17]$$

$$= 162 \text{ miles.}$$

Using Simpson's approximation with $2n = 10$ and $\Delta x = 1$, we obtain

$$s = \int_0^{10} v(t)\,dt$$

$$\approx \tfrac{1}{3}[13 + 4(14) + 2(17) + 4(21) + 2(23) + 4(21)$$

$$+ 2(15) + 4(11) + 2(11) + 4(14) + 17]$$

$$= 162 \text{ miles.}$$

The submarine has traveled about 162 miles during this 10-hour time period.

*ERROR ESTIMATES

If f is either increasing ($f'(x) > 0$) on $[a, b]$ or decreasing ($f'(x) < 0$) on $[a, b]$, then $\int_a^b f(x)\,dx$ lies between the end-point approximations L_n and R_n. If f is either concave upward ($f''(x) > 0$) on $[a, b]$ or concave downward ($f''(x) < 0$) on $[a, b]$, then this integral lies between the midpoint and trapezoidal approximations M_n and T_n. So in these four cases the value of the integral can—in principle—be approximated with any desired degree of accuracy by choosing n sufficiently large.

In addition, there are *error estimates* that can be used to predict in advance the maximum possible error in a particular approximation. Here we shall simply state and briefly illustrate two such error estimates, and defer further discussion until Section 11-3.

Trapezoidal Error Estimate

Suppose that f'' is continuous on $[a, b]$ and that $|f''(x)| \leq M$ for all x in $[a, b]$. Then the difference between T_n and $\int_a^b f(x)\,dx$ is at most $M(b - a)^3/12n^2$.

Simpson's Error Estimate

Suppose that $f^{(4)}$ is continuous on $[a, b]$ and that $|f^{(4)}(x)| \leq M$ for all x in $[a, b]$. Then the difference between S_n [n even] and $\int_a^b f(x)\,dx$ is at most $M(b - a)^5/180n^4$.

For example, beginning with the integrand $f(x) = 4/(1 + x^2)$ of the integral in (7), repeated differentiation gives the second and fourth derivatives as

$$f''(x) = \frac{4(6x^2 - 2)}{(1 + x^2)^3} \quad \text{and} \quad f^{(4)}(x) = \frac{96(5x^4 - 10x^2 + 1)}{(1 + x^2)^5}.$$

By application of the closed interval maximum-minimum method of Section 2-3, it can be verified (see Example 2 in Section 11-3) that

$$|f''(x)| \leq 8 \quad \text{and} \quad |f^{(4)}(x)| \leq 96$$

for x in $[0, 1]$. Hence with $n = 10$ as in Example 2, the maximum error in the trapezoidal approximation is

$$\frac{(8)(1)^3}{12(10)^2} \approx 0.006666667,$$

so the actual value of π, according to the integral (7), is

$$3.139925990 \pm 0.006666667;$$

that is, π lies between

$$3.133259323 \approx 3.133 \quad \text{and} \quad 3.146592657 \approx 3.147.$$

With $n = 10$ as in Example 4, the maximum error in Simpson's approximation is

$$\frac{(96)(1)^5}{(180)(10)^4} \approx 0.000053333$$

so the actual value of π, according to the integral (7), is

$$3.141592616 \pm 0.000053333;$$

thus π lies between

$$3.141539283 \approx 3.1415 \quad \text{and} \quad 3.141645949 \approx 3.1417.$$

It is fairly typical of numerical integration that the approximations obtained, as in Examples 2 and 4, are actually much more accurate than the above error estimates suggest.

4-7 PROBLEMS

In Problems 1–5, calculate the right-endpoint and left-endpoint approximations to the given integral; use the indicated number of subintervals and round answers to two decimal places. Also, use Equation (6) to determine how large n must be in order that $|R_n - L_n| < 0.01$.

1 $\int_1^2 x^2 \, dx, \quad n = 5.$

2 $\int_0^1 \sqrt{x} \, dx, \quad n = 5.$

3 $\int_0^1 \sqrt{1 + x^3} \, dx, \quad n = 4.$

4 $\int_0^1 \dfrac{dx}{1 + x}, \quad n = 10.$

5 $\int_0^1 \dfrac{\sin x}{x} \, dx, \quad n = 10 \quad \left(\text{Make the integrand continuous} \right.$

by assuming its value at $x = 0$ is $\lim\limits_{x \to 0} \dfrac{\sin x}{x} = 1. \Big)$

6 Calculate the midpoint approximation to the integral of Problem 4 with $n = 10$, rounding the answer to four decimal places.

7 Repeat Problem 6, but with the integral of Problem 5.

In Problems 8–12, calculate
(a) The trapezoidal approximation T_n, and
(b) Simpson's approximation S_n
to the given integral, rounding answers to four decimal places.

8 $\int_1^3 x^3 \, dx, \quad n = 8.$

9 $\int_0^2 \dfrac{dx}{1 + x^3}, \quad n = 8.$

10 $\int_1^4 \sqrt{1 + x^4} \, dx, \quad n = 6.$

11 $\int_0^1 \sqrt{2 - x^2} \, dx, \quad n = 10.$

12 $\int_0^1 \dfrac{\sin x}{x} \, dx, \quad n = 10 \quad \text{(See the note to Problem 5.)}$

In Problems 13 and 14, calculate
(a) The trapezoidal approximation, and
(b) Simpson's approximation,
to $\int_a^b f(x)\,dx$, where f is the given tabulated function.

13

x	$a = 1.00$	1.25	1.50	1.75	2.00	2.25	$2.50 = b$
$f(x)$	3.43	2.17	0.38	1.87	2.65	2.31	1.97

14

x	$a = 0$	1	2	3	4	5	6	7	8	9	$10 = b$
$f(x)$	23	8	-4	12	35	47	53	50	39	29	5

15 The graph in Figure 4.38 shows the measured rate of water flow (gallons/minute) into a tank during a 10-minute period. Using 10 subintervals in each case, estimate the total amount of water flowing into the tank during this period using
(a) the trapezoidal approximation, and
(b) Simpson's approximation.

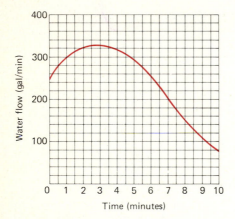

4.38 Water flow graph for Problem 15.

16 The graph in Figure 4.39 shows the daily mean temperatures recorded during a winter month at a certain location. Using 10 subintervals in each case, estimate the average temperature during the month using
(a) the trapezoidal approximation, and
(b) Simpson's approximation.

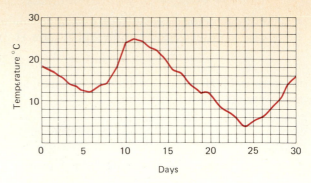

4.39 Temperature graph for Problem 16.

Problems 17–21 deal with the natural logarithm of 2, denoted by ln 2. In Chapter 6 we shall see that the value of ln 2 is given by the integral

$$\ln 2 = \int_1^2 \frac{1}{x}\,dx. \qquad (*)$$

The correct 7-place value is $\ln 2 \approx 0.6931472$.

17 Calculate the endpoint approximations L_n and R_n to the integral denoted by $(*)$; take $n = 10$.
18 Use Equation (6) to determine how large n must be in order to guarantee that L_n and R_n differ by at most 0.0005.
19 Tell why the integral $(*)$ lies between the midpoint approximation M_n and the trapezoidal approximation T_n. Then calculate these approximations with $n = 10$ subintervals, rounding the answers to 6 decimal places.
20 Use the trapezoidal error estimate to determine how large n must be in order to guarantee that T_n differs from ln 2 by at most 0.0005.
21 Use the Simpson's error estimate to determine how large n must be in order to guarantee that S_n differs from ln 2 by at most 0.000005.
22 The base for natural logarithms is a certain number e. In Chapter 6 we shall see that $\int_1^e \frac{1}{x}\,dx = 1$. Approximate the integrals

$$\int_1^{2.7} \frac{1}{x}\,dx \quad \text{and} \quad \int_1^{2.8} \frac{1}{x}\,dx$$

with sufficient accuracy to show that $2.7 < e < 2.8$.

CHAPTER 4 REVIEW: Definitions, Concepts, Results

Use the list below as a guide to ideas that you may need to review.

1 Properties of area
2 Summation notation
3 The area under the graph of f from a to b
4 Inscribed and circumscribed rectangular polygons
5 A partition of $[a, b]$
6 The mesh of a partition
7 A Riemann sum associated with a partition
8 The (definite) integral of f from a to b
9 Existence of the integral of a continuous function
10 The integral as the limit of a sequence of Riemann sums

11 Regular partition; $\displaystyle\lim_{\Delta x \to 0} \sum_{i=1}^{n} f(x_i^*)\, \Delta x$

12 Upper and lower Riemann sums

13 The constant multiple, interval union, and comparison properties of integrals

14 Evaluation of definite integrals using antiderivatives

15 The average value of $f(x)$ on the interval $[a, b]$

16 The average value theorem

17 The Fundamental Theorem of Calculus

18 Linearity of integration

19 Indefinite integral

20 The method of integration by substitution

21 Transforming the limits in definite integration by substitution

22 The area between $y = f(x)$ and $y = g(x)$ by integration with respect to x

23 The area between $x = f(y)$ and $x = g(y)$ by integration with respect to y

24 The right-endpoint and left-endpoint approximations

25 The trapezoidal approximation

26 The midpoint approximation

27 Simpson's approximation

28 Error estimates for the trapezoidal and Simpson's approximations

MISCELLANEOUS PROBLEMS

Find the sums in Problems 1–4.

1 $\displaystyle\sum_{i=1}^{100} 17$.

2 $\displaystyle\sum_{k=1}^{100} \left(\frac{1}{k} - \frac{1}{k+1}\right)$.

3 $\displaystyle\sum_{n=1}^{10} (3n - 2)^2$.

4 $\displaystyle\sum_{n=1}^{16} \sin \frac{n\pi}{2}$.

In each of Problems 5–7, find the limit of the given Riemann sum associated with a regular partition of the indicated interval $[a, b]$. First express it as an integral from a to b, and then evaluate that integral.

5 $\displaystyle\lim_{n \to \infty} \sum_{i=1}^{n} \frac{\Delta x}{\sqrt{x_i^*}}$; $[1, 2]$.

6 $\displaystyle\lim_{n \to \infty} \sum_{i=1}^{n} [(x_i^*)^2 - 3x_i^*]\Delta x$; $[0, 3]$.

7 $\displaystyle\lim_{n \to \infty} \sum_{i=1}^{n} 2\pi x_i^* \sqrt{1 + (x_i^*)^2}\, \Delta x$; $[0, 1]$.

8 Evaluate

$$\lim_{n \to \infty} \frac{1}{n^{11}} (1^{10} + 2^{10} + 3^{10} + \cdots + n^{10})$$

by expressing this limit as an integral from 0 to 1.

9 Use Riemann sums to prove that, if $f(x) \equiv c$ (a constant), then $\int_a^b f(x)\, dx = c(b - a)$.

10 Use Riemann sums to prove that, if f is continuous on $[a, b]$ and $f(x) \geq 0$ for all x in $[a, b]$, then $\int_a^b f(x)\, dx \geq 0$.

11 Use the comparison property of integrals (Section 4-3) to prove that $\int_a^b f(x)\, dx > 0$ if f is a continuous function with $f(x) > 0$ on $[a, b]$.

Evaluate the integrals in Problems 12–25.

12 $\displaystyle\int_0^1 (1 - x^2)^3\, dx$.

13 $\displaystyle\int \left(\sqrt{2x} - \frac{1}{\sqrt{3x^3}}\right) dx$.

14 $\displaystyle\int \frac{(1 - \sqrt[3]{x})^2}{\sqrt{x}}\, dx$.

15 $\displaystyle\int \frac{4 - x^3}{2x^2}\, dx$.

16 $\displaystyle\int_0^1 \frac{dt}{(3 - 2t)^2}$.

17 $\displaystyle\int \sqrt{x} \cos x^{3/2}\, dx$.

18 $\displaystyle\int_0^2 x^2 \sqrt{9 - x^3}\, dx$.

19 $\displaystyle\int \frac{1}{t^2} \sin \frac{1}{t}\, dt$.

20 $\displaystyle\int_0^2 \frac{2t + 1}{\sqrt{t^2 + t}}\, dt$.

21 $\displaystyle\int \frac{\sqrt[3]{u}}{(1 + u^{4/3})^3}\, du$.

22 $\displaystyle\int_0^{\pi/4} \frac{\sin t}{\sqrt{\cos t}}\, dt$.

23 $\displaystyle\int_1^4 \frac{(1 + \sqrt{t})^2}{\sqrt{t}}\, dt$.

24 $\displaystyle\int \frac{\sqrt[3]{1 - (1/u)}}{u^2}\, du$.

25 $\displaystyle\int \frac{\sqrt{4x^2 - 1}}{x^4}\, dx$.

Find the areas of the regions bounded by the curves given in Problems 26–32.

26 $y = x^3$ and the lines $x = -1$ and $y = 1$.

27 $y = x^4$ and $y = x^5$.

28 $y^2 = x$ and $3y^2 = x + 6$.

29 $y = x^4$ and $y = 2 - x^2$.

30 $y = x^4$ and $y = 2x^2 - 1$.

31 $y = (x - 2)^2$ and $y = 10 - 5x$.

32 $y = x^{2/3}$ and $y = 2 - x^2$.

33 Evaluate the integral $\int_0^2 \sqrt{2x - x^2}\, dx$ by interpreting it as the area of a region.

34 Repeat Problem 33 for the integral $\int_1^5 \sqrt{6x - 5 - x^2}\, dx$.

35 Find a function $f(x)$ such that

$$x^2 = 1 + \int_1^x \sqrt{1 + [f(t)]^2}\, dt$$

for all $x > 1$. (*Suggestion:* Differentiate both sides of the equation with the aid of the Fundamental Theorem of Calculus.)

36 Show that $G'(x) = \phi(h(x))h'(x)$ if

$$G(x) = \int_a^{h(x)} \phi(t)\, dt.$$

37 Use right- and left-endpoint approximations to estimate $\int_0^1 \sqrt{1 + x^2}\, dx$ with error not exceeding 0.05.

38 Calculate the trapezoidal and Simpson's approximations to $\int_0^\pi \sqrt{1 - \cos x}\, dx$ with six subintervals. For comparison, use the half-angle identity to calculate the exact value of this integral.

39 Calculate the midpoint and trapezoidal approximations to $\int_1^2 \dfrac{dx}{x + x^2}$ with $n = 5$ subintervals. Explain why the exact value of the integral lies between these two approximations.

In Problems 40–42, let $\{x_0, x_1, x_2, \cdots, x_n\}$ be a partition of $[a, b]$ where $0 < a < b$.

40 For $i = 1, 2, 3, \ldots, n$, let x_i^* be determined by

$$[x_i^*]^2 = \tfrac{1}{3}[(x_{i-1})^2 + x_{i-1}x_i + (x_i)^2].$$

Show first that $x_{i-1} < x_i^* < x_i$, and then use the algebraic identity

$$(c - d)(c^2 + cd + d^2) = c^3 - d^3$$

to show that

$$\sum_{i=1}^n (x_i^*)^2\, \Delta x_i = \tfrac{1}{3}(b^3 - a^3).$$

Explain why this computation proves that

$$\int_a^b x^2\, dx = \tfrac{1}{3}(b^3 - a^3).$$

41 Let $x_i^* = \sqrt{x_{i-1}x_i}$ for $i = 1, 2, 3, \ldots, n$. Show that

$$\sum_{i=1}^n \frac{\Delta x_i}{(x_i^*)^2} = \frac{1}{a} - \frac{1}{b}.$$

Then explain why this computation proves that

$$\int_a^b \frac{dx}{x^2} = \frac{1}{a} - \frac{1}{b}.$$

42 Define x_i^* by means of the equation

$$(x_i^*)^{1/2}(x_i - x_{i-1}) = \tfrac{2}{3}[(x_i)^{3/2} - (x_{i-1})^{3/2}].$$

Show that $x_{i-1} < x_i^* < x_i$, and then use this selection for the given partition to prove that

$$\int_a^b \sqrt{x}\, dx = \tfrac{2}{3}(b^{3/2} - a^{3/2}).$$

Applications of the Integral

5

Introduction: Setting Up Integral Formulas

In Section 4-3 we defined the integral of the function f from a to b as a limit of Riemann sums. Specifically, one begins with a partition P of the interval $[a, b]$. A selection of numbers x_i^* in each subinterval of P produces a Riemann sum associated with the partition P; this is a number of the form

$$\sum_{i=1}^{n} f(x_i^*)\, \Delta x_i.$$

Finally, the integral of f on $[a, b]$ is defined to be the limit of such sums as the mesh $|P|$ approaches zero. That is,

$$\int_a^b f(x)\, dx = \lim_{|P| \to 0} \sum_{i=1}^{n} f(x_i^*)\, \Delta x_i. \tag{1}$$

The wide applicability of the definite integral arises from the fact that many geometric and physical quantities can be approximated arbitrarily closely by Riemann sums of the sort that appear in the above definition. Such approximations lead to integral formulas for the computation of these quantities.

For example, our discussion in Sections 4-2 and 4-6 of the area from a to b under the graph of the positive-valued continuous function f can be summarized like this. Begin with a regular partition of $[a, b]$ into n subintervals, each with length $\Delta x = (b - a)/n$. For each i ($i = 1, 2, 3, \ldots, n$), select a point x_i^* in the ith subinterval $[x_{i-1}, x_i]$. Let ΔA_i denote the area below the graph of f over this subinterval. Then this area is given approximately by

$$\Delta A_i \approx f(x_i^*)\, \Delta x.$$

Figure 5.1 shows why this approximation should be good and why it should improve as the mesh of the partition approaches zero.

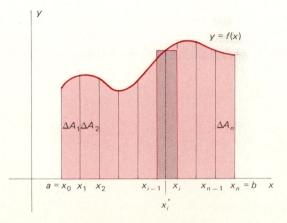

5.1 Approximating area by means of a Riemann sum.

It follows that the total area A under the graph of f is given approximately by

$$A = \sum_{i=1}^{n} \Delta A_i \approx \sum_{i=1}^{n} f(x_i^*) \, \Delta x. \tag{2}$$

The point is that the approximating sum on the right is a Riemann sum for f on $[a, b]$. Moreover,

(i) It is intuitively evident that the Riemann sum in (2) approaches the actual area A as $n \to \infty$;

(ii) By the definition of the integral, this Riemann sum approaches $\int_a^b f(x) \, dx$ as $n \to \infty$.

These observations justify the *definition* of the area A by means of the formula

$$A = \int_a^b f(x) \, dx. \tag{3}$$

This justification of the area formula (3) illustrates the following general method of setting up integral formulas. Suppose that we want to compute a certain quantity Q, where Q is associated with the interval $[a, b]$ in such a way that subintervals of $[a, b]$ correspond to specific portions of Q. Also assume that if $\Delta Q_1, \Delta Q_2, \Delta Q_3, \ldots, \Delta Q_n$ are the portions of Q corresponding to the subintervals of a partition of $[a, b]$ into n subintervals, then

$$Q = \sum_{i=1}^{n} \Delta Q_n. \tag{4}$$

Such assumptions hold, for instance, if Q is:

- an *area* lying over the interval $[a, b]$;
- the *distance* traveled by a moving particle during the time interval $a \le t \le b$;
- the *number of births* occurring in a certain population during the time interval $a \le t \le b$;
- the *volume* of water flowing into a tank during the time interval $a \le t \le b$;   or
- the *work* done by a force in moving a particle from the point $x = a$ to the point $x = b$.

For example, if Q is the volume of water flowing into a tank during the time interval $[a, b]$, then the portion ΔQ_i of Q corresponding to the subinterval $[t_{i-1}, t_i]$ is the volume of water that flows into the tank during this subinterval of time. In any such situation it will generally turn out that the quantity Q can be computed by evaluating a certain integral,

$$Q = \int_a^b f(x) \, dx \quad \text{or} \quad Q = \int_a^b f(t) \, dt, \tag{5}$$

depending on whether x or t is the independent variable. The key step in "setting up" such an integral formula is finding the particular function f to be integrated.

The usual way of finding f is to approximate the portion ΔQ_i of the quantity Q that corresponds to the subinterval $[x_{i-1}, x_i]$. We are still thinking of a regular partition of $[a, b]$ into n equal subintervals of length Δx, or into subintervals $[t_{i-1}, t_i]$ of length Δt in case t is the independent variable. Suppose we find that

$$\Delta Q_i \approx f(x_i^*) \, \Delta x \qquad (6)$$

for some point x_i^* in $[x_{i-1}, x_i]$. What follows now is the approximation

$$Q = \sum_{i=1}^{n} \Delta Q_i \approx \sum_{i=1}^{n} f(x_i^*) \, \Delta x. \qquad (7)$$

The right-hand sum in (7) is a Riemann sum that approaches the integral $\int_a^b f(x) \, dx$ as $n \to \infty$. If it is also evident—for geometric or physical reasons, for example—that this Riemann sum must approach the quantity Q as $n \to \infty$, then (7) justifies our setting up the integral formula

$$Q = \int_a^b f(x) \, dx. \qquad (5)$$

The crucial step above is the determination of the particular function f in the approximation (6): $\Delta Q_i \approx f(x_i^*) \, \Delta x$. If Q is the area under a curve, then (as we have seen) $f(x)$ should be the height of the curve. If Q is the distance traveled by a moving particle, then (as we shall see below) $f(t)$ should be its velocity function. If Q is the volume of water flowing into a tank during a specified time interval, then $f(t)$ should be the rate of water flow (for instance, in gallons per minute). This chapter is devoted largely to ways of setting up integral formulas like (5) in a variety of specific situations. The applications that we consider include computations of distances, volumes, work, fluid pressure, lengths of curves, areas of surfaces of revolution, and centers of mass.

EXAMPLE 1 Suppose that water is pumped into an initially empty tank. It is known that the rate of flow of water into the tank after t minutes is $50 - t$ gallons per minute. How much water flows into the tank during the first half hour?

Solution We want to compute the amount Q of water that flows into the tank during the time interval $[0, 30]$. Think of a regular partition of $[0, 30]$ into n equal subintervals of length $\Delta t = 30/n$ each.

Next choose a point t_i^* in the ith subinterval $[t_{i-1}, t_i]$. The rate of water flow between time t_{i-1} and time t_i is approximately $50 - t_i^*$ gallons per minute, so the amount ΔQ_i of water that flows into the tank during this subinterval of time is approximately given by

$$\Delta Q_i \approx (50 - t_i^*) \, \Delta t$$

(in gallons). Therefore the total amount Q we seek is given approximately by

$$Q = \sum_{i=1}^{n} \Delta Q_i \approx \sum_{i=1}^{n} (50 - t_i^*) \, \Delta t$$

(in gallons, still). We recognize the Riemann sum on the right, and—most important—we see that the function involved is $f(t) = 50 - t$. Hence we may conclude that

$$Q = \int_0^{30} (50 - t)\, dt = \left[50t - \tfrac{1}{2}t^2 \right]_0^{30} = 1050$$

gallons.

DISTANCE AND VELOCITY

Consider a particle traveling along a (directed) straight line with velocity $v = f(t)$ at time t. We want to compute the **net distance** s it travels between time $t = a$ and time $t = b$; that is, the distance between its initial position at time $t = a$ and its final position at time $t = b$.

We start with a regular partition of $[a, b]$ into n equal subintervals, each of length $\Delta t = (b - a)/n$. If t_i^* is an arbitrary point of the ith subinterval $[t_{i-1}, t_i]$, then during this subinterval the velocity of the particle is approximately $f(t_i^*)$. Hence the distance Δs_i traveled by the particle between time $t = t_{i-1}$ and $t = t_i$ is given approximately by

$$\Delta s_i \approx f(t_i^*)\, \Delta t.$$

Consequently the net distance s is given approximately by

$$s = \sum_{i=1}^{n} \Delta s_i \approx \sum_{i=1}^{n} f(t_i^*)\, \Delta t. \qquad (8)$$

We recognize the right-hand approximation as a Riemann sum for the integral $\int_a^b f(t)\, dt$, and we reason that this approximation should approach the actual net distance s as $n \to \infty$. Thus we conclude that

$$s = \int_a^b f(t)\, dt = \int_a^b v\, dt. \qquad (9)$$

Note that the *net distance is the definite integral of the velocity.*

If the velocity function $v = f(t)$ has both positive and negative values for t in $[a, b]$, then forward and backward distances cancel when we compute the net distance s given by the integral formula in (9). The **total distance** traveled, irrespective of direction, may be found by evaluating the integral of the *absolute value* of the velocity:

$$\text{total distance} = \int_a^b |f(t)|\, dt = \int_a^b |v|\, dt. \qquad (10)$$

This integral may be computed by integrating separately over the subintervals where v is positive and those where v is negative and then adding the absolute values of the results.

EXAMPLE 2 Suppose that the velocity of a moving particle is $v = 30 - 2t$ feet per second. Find both the net distance and the total distance it travels between the times $t = 0$ and $t = 20$ seconds.

Solution For net distance, we use Formula (9) and find that

$$s = \int_0^{20} (30 - 2t)\, dt = \left[30t - t^2 \right]_0^{20} = 200$$

feet. To find the total distance traveled, we note that v is positive if $t < 15$, while v is negative if $t > 15$. Since

$$\int_0^{15} (30 - 2t)\, dt = \left[30t - t^2 \right]_0^{15} = 225 \qquad \text{feet}$$

and

$$\int_{15}^{20} (2t - 30)\, dt = \left[t^2 - 30t \right]_{15}^{20} = 25 \qquad \text{feet,}$$

we see that the particle travels 225 feet forward, then 25 feet backward, for a total distance of 250 feet.

The derivation of Formula (9) above, based on the intuitive reasoning that the approximation in (8) *ought* to approach s as $n \to \infty$, is a "plausible justification" of the distance formula rather than a rigorous proof of it. Plausible justifications of this sort will suffice for most of our purposes, but on occasion we shall need a rigorous proof of an integral formula. A rigorous proof of distance formula (9) can be given as follows.

Begin as before with a regular partition of $[a, b]$ into n equal subintervals of length Δt. Let $f(t_i^\flat)$ and $f(t_i^\#)$ be the minimum and maximum values, respectively, of the velocity function for t in the ith subinterval $[t_{i-1}, t_i]$. Then the distance Δs_i traveled by the particle during this subinterval satisfies the inequalities

$$f(t_i^\flat)\, \Delta t \leq \Delta s_i \leq f(t_i^\#)\, \Delta t.$$

Since

$$s = \sum_{i=1}^{n} \Delta s_i,$$

addition of these inequalities for $i = 1, 2, 3, \ldots, n$ gives

$$\sum_{i=1}^{n} f(t_i^\flat)\, \Delta t \leq s \leq \sum_{i=1}^{n} f(t_i^\#)\, \Delta t.$$

Now we note that each of the last two sums is a Riemann sum for f on $[a, b]$. So, assuming that f is continuous, both sums approach $\int_a^b f(t)\, dt$ as $n \to \infty$. The squeeze law for limits therefore gives

$$\int_a^b f(t)\, dt \leq s \leq \int_a^b f(t)\, dt,$$

which establishes Formula (9).

5-1 PROBLEMS

In each of Problems 1–10, compute both the net distance and the total distance traveled between times $t = a$ and $t = b$ by a particle moving along a line with the given velocity function $v = f(t)$.

1 $v = -32$; $a = 0, b = 10$.

2 $v = 2t + 10$; $a = 1, b = 5$.

3 $v = 4t - 25$; $a = 0, b = 10$.

4 $v = |2t - 5|$; $a = 0, b = 5$.

5 $v = 4t^3$; $a = -2, b = 3$.

6 $v = t - \dfrac{1}{t^2}$; $a = \dfrac{1}{10}$, $b = 1$.

7 $v = \sin 2t$; $a = 0$, $b = \dfrac{\pi}{2}$.

8 $v = \cos 2t; \quad a = 0, \quad b = \dfrac{\pi}{2}.$

9 $v = \cos \pi t; \quad a = -1, \quad b = 1.$
10 $v = \sin t + \cos t; \quad a = 0, b = \pi.$

In each of Problems 11–15, x_i^* denotes an arbitrary point of the ith subinterval $[x_{i-1}, x_i]$ of a regular partition of the indicated interval $[a, b]$ into n equal subintervals of length Δx. Evaluate the given limit by computing the value of the appropriate related integral.

11 $\displaystyle\lim_{n \to \infty} \sum_{i=1}^{n} 2x_i^* \, \Delta x; \quad a = 0, b = 1.$

12 $\displaystyle\lim_{n \to \infty} \sum_{i=1}^{n} \dfrac{\Delta x}{(x_i^*)^2}; \quad a = 1, b = 2.$

13 $\displaystyle\lim_{n \to \infty} \sum_{i=1}^{n} (\sin \pi x_i^*) \, \Delta x; \quad a = 0, b = 1.$

14 $\displaystyle\lim_{n \to \infty} \sum_{i=1}^{n} [3(x_i^*)^2 - 1] \, \Delta x; \quad a = -1, b = 3.$

15 $\displaystyle\lim_{n \to \infty} \sum_{i=1}^{n} x_i^* \sqrt{(x_i^*)^2 + 9} \, \Delta x; \quad a = 0, b = 4.$

16 If a particle is thrown straight upward from the ground with an initial velocity of 160 feet per second, then its velocity after t seconds is $v = -32t + 160$ ft/sec, and it attains its maximum height when $t = 5$ sec (and $v = 0$). Use Formula (9) to compute this maximum height. Check your answer by the method of Section 3-8.

17 Suppose that the rate of flow of water into an initially empty tank is $100 - 3t$ gal/min after t minutes have elapsed.

How much water flows into the tank between time $t = 10$ minutes and time $t = 20$ minutes?

18 Suppose that the birth rate in a certain city t years after 1960 was $13 + t$ thousands of births per year. Set up and evaluate an appropriate integral to compute the number of births that occurred between 1960 and 1980.

19 Assume that the city of Problem 18 had a death rate of $5 + t/2$ thousands per year t years after 1960. If the population of the city was 125,000 in 1960, what was its population in 1980? Consider both births and deaths in working this problem.

20 The average daily rainfall in a certain locale is $r(t)$ inches per day at time $t, 0 \leq t \leq 365$. Begin with a regular partition of the interval $[0, 365]$ and derive the formula $R = \int_0^{365} r(t) \, dt$ for the average total annual rainfall.

21 Suppose, in connection with Problem 20, that

$$r(t) = a - b \cos\left(\dfrac{2\pi t}{365}\right)$$

where a and b are constants to be determined. If the average daily rainfall on January 1 ($t = 0$) is 0.1 inches per day and the average daily rainfall on July 1 ($t = 182.5$) is 0.5 inches per day, what is the average total annual rainfall in this locale?

22 Suppose that the rate of flow of water into a tank is $r(t)$ gal/min at time t. Use the method by which we established Formula (9) at the end of this section to derive rigorously the formula $Q = \int_a^b r(t) \, dt$ for the amount of water that flows into the tank between times $t = a$ and $t = b$.

Volumes by the Method of Cross Sections

In this section we show how to use integrals to calculate the volumes of certain solids or regions in space. We begin with the intuitive idea that volume has properties analogous to the properties of area listed in Section 4-2. That is, we assume that every reasonably nice solid region R has a volume $v(R)$, a nonnegative real number with the following properties:

I. If the solid region R is contained in the solid region S, then $v(R) \leq v(S)$.

II. If R and S are nonoverlapping solid regions, then the volume of their union is $v(R \cup S) = v(R) + v(S)$.

III. If the solid regions R and S are congruent (have the same size and shape), then $v(R) = v(S)$.

The **method of cross sections** is a way of computing the volume of a solid that is conveniently described in terms of its cross sections in planes perpendicular to a fixed *reference line L*. This reference line will ordinarily be either the x-axis or the y-axis. In a specific problem we attempt to choose as reference line that axis which makes the resulting computations simpler.

We first consider the case in which L is the x-axis. Suppose that the solid R with volume $V = v(R)$ that we want to calculate lies opposite the

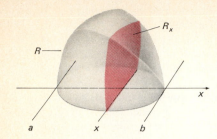

5.2 R_x is the cross section of R in the plane perpendicular to the x-axis at x.

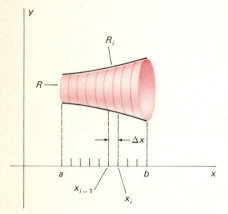

5.3 Planes through the partition points $x_0, x_1, \ldots, x_n$ partition the solid R into slabs $R_1, R_2, \ldots, R_n$.

interval $[a, b]$ on the x-axis. That is, a plane perpendicular to the x-axis intersects the solid if and only if this plane meets the x-axis in a point of $[a, b]$, as indicated in Figure 5.2. Let R_x denote the intersection of R with the perpendicular plane that meets the x-axis at the point x of $[a, b]$. We call R_x the (plane) **cross section** of the solid R at x.

This situation is especially nice if all the cross sections of R are congruent to one another and, in addition, are parallel translates of each other. In this case the solid R is called a **cylinder** with **bases** R_a and R_b and height $h = b - a$. If R_a and R_b are circles, then R is the familiar **circular cylinder.**

In order to calculate volumes by integration, we need a fourth property of volume adjoined to the three listed above:

IV. The volume of any cylinder, circular or not, is the product of its height and the area of its base.

This is a generalization of the familiar formula $V = \pi r^2 h$ for the volume of a circular cylinder with radius r and height h.

Now for each x in $[a, b]$, let $A(x)$ denote the area of the cross section R_x of our original solid R:

$$A(x) = a(R_x). \tag{1}$$

We shall assume that the shape of R is nice enough so that this **cross-sectional area function** A is continuous and, therefore, integrable.

To set up an integral formula for $V = v(R)$, we begin with a regular partition of $[a, b]$ into n equal subintervals each with length $\Delta x = (b - a)/n$. Let R_i denote the slab or slice of the solid R that lies opposite the ith subinterval $[x_{i-1}, x_i]$, as shown in Figure 5.3. We denote the volume of this ith slice of R by $\Delta V_i = v(R_i)$, so that

$$V = \sum_{i=1}^{n} \Delta V_i.$$

To approximate ΔV_i, we select an arbitrary point x_i^* in $[x_{i-1}, x_i]$, and consider the *cylinder* C_i with height Δx and whose base is the cross section $R_{x_i^*}$ of R at x_i^*. Figure 5.4 suggests that if Δx is small then $v(C_i)$ is a good

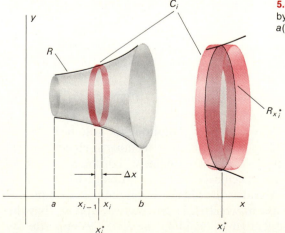

5.4 The slab R_i is approximated by the cylinder C_i of volume $a(R_{x_i^*}) \Delta x$.

CHAP. 5: **Applications of the Integral**

approximation to $\Delta V_i = v(R_i)$,

$$\Delta V_i \approx v(C_i) = a(R_{x_i^*}) \, \Delta x = A(x_i^*) \, \Delta x,$$

using Property (IV) of volume.

Then we add the volumes of these approximating cylinders for $i = 1$, $2, 3, \ldots, n$. We find that

$$V = \sum_{i=1}^{n} \Delta V_i \approx \sum_{i=1}^{n} A(x_i^*) \, \Delta x.$$

We recognize the approximating sum on the right as a Riemann sum that approaches $\int_a^b A(x) \, dx$ as $n \to \infty$. But this sum should also approach the actual volume V as $n \to \infty$. This justifies the following *definition* of the volume of a solid R in terms of its cross-sectional area function $A(x)$.

Definition *Volume by Cross Sections*

If the solid R lies opposite the interval $[a, b]$ on the x-axis and has continuous cross-sectional area function $A(x)$, then its volume $V = v(R)$ is

$$V = \int_a^b A(x) \, dx. \tag{2}$$

Formula (2) is sometimes called **Cavalieri's principle,** after the Italian mathematician Bonaventura Cavalieri (1598–1647), who systematically exploited the fact that the volume of a solid is determined by the areas of its cross sections perpendicular to a given reference line.

In the case of a solid R that lies opposite the interval $[c, d]$ on the y-axis, we denote by $A(y)$ the area of its cross section R_y in the plane that is perpendicular to the y-axis at the point y of $[c, d]$ (see Figure 5.5). A similar discussion, starting with a regular partition of $[c, d]$, leads to the volume formula

$$V = \int_c^d A(y) \, dy. \tag{3}$$

SOLIDS OF REVOLUTION

An important special case of Formula (2) gives the volume of a **solid of revolution.** For example, let the solid R be obtained by revolving around the x-axis the region under the graph of $y = f(x)$ over the interval $[a, b]$, where $f(x) \geqq 0$. The region and the resulting solid are shown in Figure 5.6.

Since the solid R is obtained by revolution, each cross section of R at x is a circular disk of radius $f(x)$. The cross-sectional area function is then $A(x) = \pi[f(x)]^2$, so Formula (2) gives us

$$V = \int_a^b \pi[f(x)]^2 \, dx \tag{4}$$

for the **volume of a solid of revolution around the x-axis.**

By a similar argument, if the region bounded by the curve $x = g(y)$, the y-axis, and the horizontal lines $y = c$ and $y = d$ is rotated about the

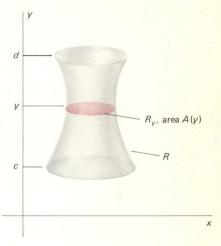

5.5 $A(y)$ is the area of the cross section R_y of R in the plane perpendicular to the y-axis at the point y.

5.6 Volume of revolution about the x-axis.

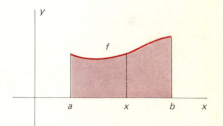

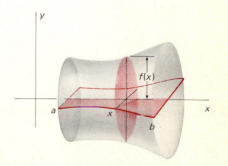

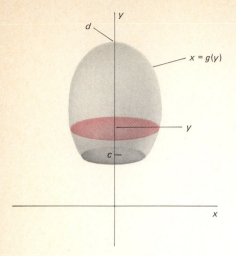

5.7 Volume of revolution about the y-axis.

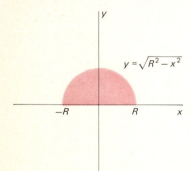

5.8 A sphere generated by rotation of a semicircular region.

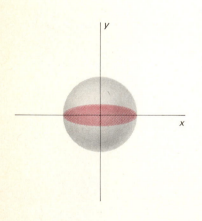

y-axis, then the volume of the resulting solid of revolution (Figure 5.7) is

$$V = \int_c^d \pi [g(y)]^2 \, dy. \tag{5}$$

EXAMPLE 1 Use the method of cross sections to verify the familiar formula $V = \frac{4}{3}\pi R^3$ for the volume of a sphere of radius R.

Solution We think of this sphere as the solid of revolution obtained by revolving the semicircular plane region of Figure 5.8 around the x-axis. This is the region bounded above by the semicircle $y = \sqrt{R^2 - x^2}$ ($-R \le x \le R$) and below by the interval $[-R, R]$ on the x-axis. To use Formula (4), we take $f(x) = \sqrt{R^2 - x^2}$, $a = -R$, and $b = R$. This gives

$$V = \int_{-R}^R \pi \left(\sqrt{R^2 - x^2}\right)^2 dx = \pi \int_{-R}^R (R^2 - x^2) \, dx$$

$$= \pi \left[R^2 x - \tfrac{1}{3} x^3 \right]_{-R}^R = \tfrac{4}{3}\pi R^3.$$

EXAMPLE 2 Use the method of cross sections to verify the familiar formula $V = \frac{1}{3}\pi r^2 h$ for the volume of a cone with base radius r and height h.

Solution We may think of the cone as the solid of revolution obtained by revolving the plane region bounded by the y-axis and the lines $y = h$ and $x = ry/h$ around the y-axis (as in Figure 5.9). Then Formula (5) with $g(y) = ry/h$ gives

$$V = \int_0^h \pi \left(\frac{ry}{h}\right)^2 dy = \frac{\pi r^2}{h^2} \int_0^h y^2 \, dy$$

$$= \frac{\pi r^2}{h^2} \left[\frac{1}{3} y^3 \right]_0^h = \frac{1}{3} \pi r^2 h.$$

EXAMPLE 3 Find the volume of the "wedge" that is cut from a circular cylinder with unit radius *and* height by a plane that passes through a diameter of the bottom base of the cylinder and through a point on the circumference of its top.

Solution The cylinder and wedge are shown in Figure 5.10. To form such a wedge, fill a cylindrical glass with cider and then drink slowly until half

5.9 Generating a cone by rotation (see Example 2).

5.10 The wedge of Example 3.

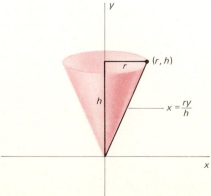

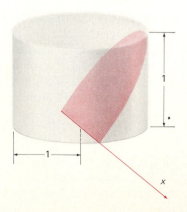

CHAP. 5: **Applications of the Integral**

the bottom of the glass is exposed; the remaining cider forms the wedge of Example 3.

We choose as reference line and x-axis the line through the "edge of the wedge"—the original diameter of the base of the cylinder. It's easy to verify with similar triangles that each cross section of the wedge perpendicular to the x-axis is an isosceles right triangle. One of these triangles is shown in Figure 5.11. We denote by y the (equal) base and height of this triangle.

In order to determine the cross-sectional area function $A(x)$, we must express y in terms of x. Figure 5.12 shows the unit circular base of the original cylinder. We apply the Pythagorean theorem to the right triangle in this figure and thus find that $y = \sqrt{1 - x^2}$. Hence

$$A(x) = \tfrac{1}{2}y^2 = \tfrac{1}{2}(1 - x^2),$$

so Formula (2) gives

$$V = \int_{-1}^{1} A(x)\,dx = 2 \int_{0}^{1} A(x)\,dx \qquad \text{(by symmetry)}$$

$$= 2 \int_{0}^{1} \tfrac{1}{2}(1 - x^2)\,dx = \left[x - \tfrac{1}{3}x^3 \right]_{0}^{1} = \tfrac{2}{3}$$

for the volume of the wedge.

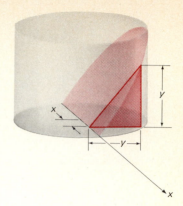

5.11   A cross section of the wedge—an isosceles triangle.

A useful habit is the checking of answers for plausibility whenever convenient. For example, we may compare a given solid with one whose volume is already known. Since the volume of the original cylinder in Example 3 is π, we have found that the wedge occupies the fraction

$$\frac{V_{wedge}}{V_{cyl}} = \frac{\tfrac{2}{3}}{\pi} \approx 21\%$$

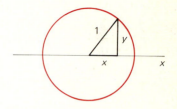

5.12   The base of the cylinder.

of the volume of the cylinder. A glance at Figure 5.10 indicates that this is plausible. An error in our computation would likely have given an implausible answer, thereby revealing the existence of the error.

The wedge of Example 3 has an ancient history. Its volume was first calculated in the third century B.C. by Archimedes, to whom also is originally due the formula $V = \tfrac{4}{3}\pi r^3$ for the volume of a sphere. His work on the wedge is found in a manuscript that was rediscovered in 1906 after having been lost for many centuries. Archimedes used a method of exhaustion for volumes similar to that discussed for areas in Section 4-2.

5.13   The region between two positive graphs is rotated about the x-axis; cross sections are annular rings.

Sometimes we need to calculate the volume of a solid generated by revolution of a plane region lying between two given curves. Suppose that $f(x) > g(x) > 0$ for x in the interval $[a, b]$, and that the solid R is generated by revolving the region between $y = f(x)$ and $y = g(x)$ about the x-axis. Then the cross section at x is an **annular ring** bounded by two circles, as shown in Figure 5.13. The ring has inner radius $g(x)$ and outer radius $f(x)$, so our formula for the cross-sectional area of R at x is

$$A(x) = \pi[f(x)]^2 - \pi[g(x)]^2.$$

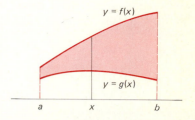

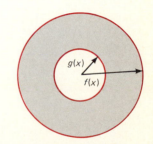

Thus, by Formula (2), the volume V of R is given by

$$V = \int_{a}^{b} \pi([f(x)]^2 - [g(x)]^2)\,dx. \tag{6}$$

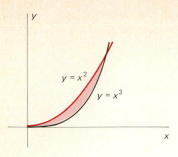

5.14 The plane region bounded by the graphs of $y = x^2$ and $y = x^3$.

Similarly, if $f(y) > g(y) > 0$ for y in $[c, d]$, then the volume of the solid obtained by revolving the region between $x = f(y)$ and $x = g(y)$ about the y-axis is

$$V = \int_c^d \pi([f(y)]^2 - [g(y)]^2)\, dy. \tag{7}$$

EXAMPLE 4 Consider the region shown in Figure 5.14, bounded by the curves $y = x^2$ and $y = x^3$ over the interval $0 \leq x \leq 1$. If this region is rotated about the x-axis, then Formula (6) gives the volume swept out as

$$V = \int_0^1 \pi(x^4 - x^6)\, dx = \frac{2\pi}{35}.$$

If the same region is rotated about the y-axis, then Formula (7) gives the volume of the resulting solid as

$$V = \int_0^1 \pi(y^{2/3} - y)\, dy = \frac{\pi}{10}.$$

5-2 PROBLEMS

In each of Problems 1–10, find the volume of the solid that is generated by revolving about the indicated axis the plane region bounded by the given curves.

1 $y = x^2$, $y = 0$, $x = 1$; about the x-axis.
2 $y = \sqrt{x}$, $y = 0$, $x = 4$; about the x-axis.
3 $y = x^2$, $y = 4$, $x = 0$ (first quadrant only); about the y-axis.
4 $y = 1/x$, $y = 0$, $x = \frac{1}{10}$, $x = 1$; about the x-axis.
5 $y = \sin x$ on $[0, \pi]$, $y = 0$; about the x-axis.
6 $y = 9 - x^2$, $y = 0$; about the x-axis.
7 $y = x^2$, $x = y^2$; about the x-axis.
8 $y = x^2$, $y = 4x$; about the line $x = 5$.
9 $y = x^2$, $y = 8 - x^2$; about the x-axis.
10 $x = y^2$, $x = y + 6$; about the y-axis.
11 Find the volume of the solid generated by revolving the region bounded by the parabolas $y^2 = x$ and $y^2 = 2(x - 3)$ about the x-axis.

5.15 The ellipse of Problems 12 and 13.

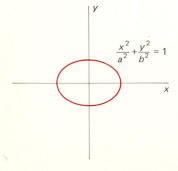

12 Find the volume of the ellipsoid that is generated by revolving the ellipse $x^2/a^2 + y^2/b^2 = 1$ around the x-axis (see Figure 5.15).
13 Repeat Problem 12, except revolve the ellipse around the y-axis.
14 Find the volume of the unbounded solid that is generated by revolving the unbounded region of Figure 5.16 around the x-axis. This is the region between the graph of $y = 1/x^2$ and the x-axis for $x \geq 1$. (*Method:* Compute the volume from $x = 1$ to $x = b$ (where $b > 1$) and then find the limit of this volume as $b \to \infty$.)

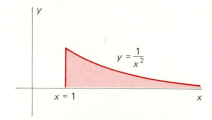

5.16 The unbounded plane region of Problem 14.

15 The base of a certain solid is a circle with diameter AB of length $2a$. Find the volume of the solid if each cross section perpendicular to AB is a square.
16 Repeat Problem 15 if each cross section is a semicircle rather than a square.
17 Repeat Problem 15 if each cross section of the solid is an equilateral triangle.

18 The base of a certain solid is the region in the xy-plane bounded by the parabolas $y = x^2$ and $x = y^2$. Find the volume of this solid if every cross section perpendicular to the x-axis is a square with base in the xy-plane.

19 The paraboloid generated by revolving around the x-axis the region under the parabola $y^2 = 2px$, $0 \le x \le h$, is shown in Figure 5.17. Show that the volume of the paraboloid is one-half that of the indicated circumscribed cylinder.

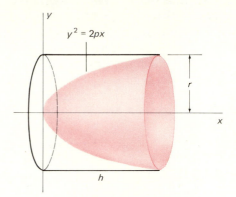

5.17 The paraboloid and cylinder of Problem 19.

20 A pyramid has height h and square base with area A. Show that its volume is $V = \frac{1}{3}Ah$. Note that each cross section parallel to the base is a square.

21 Repeat Problem 20, except that the base is a triangle with area A.

22 Find the volume of the solid that remains after a hole of radius 3 is bored through the center of a solid sphere of radius 5.

23 Two horizontal circular cylinders each have radius a, and their axes intersect at right angles. Find the volume of their solid of intersection. (*Suggestion*: First show that each horizontal cross section of the solid is a square. This is another problem solved by Archimedes in the long-lost manuscript.)

24 Figure 5.18 shows a "spherical segment" of height h cut

5.18 A spherical segment.

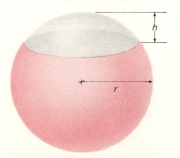

off from a sphere of radius r by a horizontal plane. Show that its volume is $V = \frac{1}{3}\pi h^2 (3r - h)$.

25 A doughnut-shaped solid called a *torus* is generated by revolving around the y-axis the circular disk

$$(x - b)^2 + y^2 \le a^2$$

centered at the point $(b, 0)$, where $0 < a < b$. Show that the volume of this torus is $V = 2\pi^2 a^2 b$. (*Suggestion*: Note that each cross section perpendicular to the y-axis is an annular ring, and recall that

$$\int_0^a \sqrt{a^2 - y^2}\, dy = \frac{1}{4}\pi a^2$$

because the integral represents the area of a quarter-circle of radius a.)

26 The summit of a hill is 100 feet higher than the surrounding level terrain, and each horizontal cross section of the hill is circular. The following table gives the radius r for selected values of the height h above the surrounding terrain. Use Simpson's approximation to estimate the volume of the hill.

h (feet)	0	25	50	75	100
r (feet)	60	55	50	35	0

27 (Newton's Wine Barrel) Consider a barrel with the shape of the solid generated by revolving around the x-axis the region under the parabola $y = R - kx^2$, $-h/2 \le x \le h/2$.
(a) Show that the radius of each end of the barrel is $r = R - \delta$, where $4\delta = kh^2$.
(b) Then show that the volume of the barrel is

$$V = \frac{1}{3}\pi h(2R^2 + r^2 - \frac{6}{15}\delta^2).$$

28 (The Clepsydra, or Water Clock) Consider a water tank whose surface is generated by revolving the curve $y = kx^4$ (k is a constant) around the y-axis.
(a) Compute $V(y)$, the volume of water in the tank as a function of its depth y.
(b) Suppose that water drains from the tank through a small hole at its bottom. Use the chain rule and Torricelli's law (Equation (3) in Section 3-7) to show that the water level in this tank falls at a *constant* rate.

29 Let V be the volume generated by revolving around the x-axis the area under the graph of the continuous and non-negative function f from $x = a$ to $x = b$. Let $f(x_i^b)$ and $f(x_i^\#)$ be the minimum and maximum values of f on the ith subinterval of a regular partition of $[a, b]$. Deduce from Properties I through IV of volume that

$$\sum_{i=1}^{n} \pi[f(x_i^b)]^2\, \Delta x \le V \le \sum_{i=1}^{n} \pi[f(x_i^\#)]^2\, \Delta x.$$

Explain why this implies the validity of Formula (4).

Volumes by the Method of Cylindrical Shells

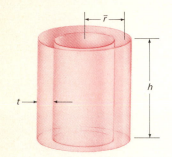

5.19 A cylindrical shell.

The method of cross sections of Section 5-2 is a technique of approximating a solid with a stack of thin slabs or slices. In the case of a solid of revolution, these slices are circular disks or annular rings. The **method of cylindrical shells** is a second way of computing volumes of solids of revolution. It is a technique of approximating a solid of revolution with a collection of thin cylindrical shells, and it sometimes leads to simpler computations.

A **cylindrical shell** is a region bounded by two concentric circular cylinders of the same height h. If, as in Figure 5.19, the inner cylinder has radius r_1 and the outer one has radius r_2, then we can write $\bar{r} = (r_1 + r_2)/2$ for the **average radius** of the cylindrical shell and $t = r_2 - r_1$ for its **thickness.** We then get the volume of the cylindrical shell by subtracting the volume of the inner cylinder from that of the outer one. So the shell has volume

$$V = \pi r_2^2 h - \pi r_1^2 h = 2\pi \frac{r_1 + r_2}{2}(r_2 - r_1)h = 2\pi \bar{r} t h. \tag{1}$$

In words, the volume of the shell is the product of 2π, its average radius, its thickness, and its height.

Now suppose that we want to find the volume V of revolution generated by revolving around the y-axis the region under $y = f(x)$ from $x = a$ to $x = b$. We assume, as in Figure 5.20, that $0 \leq a < b$ and that $f(x)$ is nonnegative on $[a, b]$. The solid will resemble the one also shown in Figure 5.20.

To find V, we begin with a regular partition of $[a, b]$ into n equal subintervals each of length $\Delta x = (b - a)/n$. Let x_i^* denote the midpoint of the ith subinterval $[x_{i-1}, x_i]$, and consider the rectangle with base $[x_{i-1}, x_i]$ and height $f(x_i^*)$. When this rectangle is revolved about the y-axis, it sweeps out a cylindrical shell like the one in Figure 5.20, with average radius x_i^*,

5.20 A solid of revolution—note the hole through its center—and a way to approximate it with nested cylindrical shells.

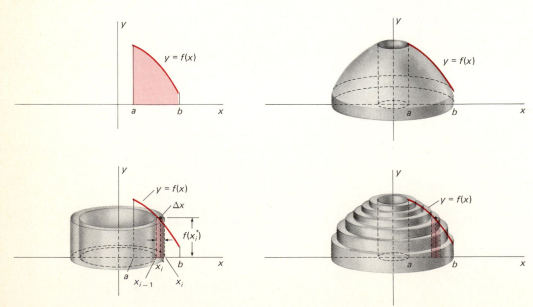

thickness Δx, and height $f(x_i^*)$. This cylindrical shell approximates the solid with volume ΔV_i that is obtained by revolving the region under $y = f(x)$ and over $[x_{i-1}, x_i]$, and thus Formula (1) gives

$$\Delta V_i \approx 2\pi x_i^* f(x_i^*) \, \Delta x.$$

We add the volumes of the n cylindrical shells determined by the partition. This sum should approximate V, since—as Figure 5.20 suggests—the union of the shells should approximate the solid of revolution. Thus we obtain the approximation.

$$V = \sum_{i=1}^{n} \Delta V_i \approx \sum_{i=1}^{n} 2\pi x_i^* f(x_i^*) \, \Delta x.$$

This approximation to V is a Riemann sum that approaches $\int_a^b 2\pi x f(x) \, dx$ as $\Delta x \to 0$, so it appears that the volume of our solid of revolution is given by

$$V = \int_a^b 2\pi x f(x) \, dx. \qquad (2)$$

A complete discussion would require a proof that this formula gives the same volume as that *defined* by the method of cross sections in Section 5-2 (see the appendix to this chapter).

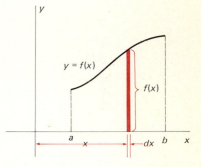

It is more reliable to learn how to set up integral formulas than merely to memorize such formulas. A useful heuristic device for setting up Formula (2) is to picture the very narrow rectangular strip of area shown in Figure 5.21. When this strip is revolved around the y-axis, it produces a thin cylindrical shell of radius x, height $f(x)$, and thickness dx, so its volume, if denoted by dV, can be written $dV = 2\pi x f(x) \, dx$. We think of V as a sum of very many such volumes, nested to form the solid of revolution. This makes it natural to write

5.21

$$V = \int dV = \int_a^b 2\pi x f(x) \, dx.$$

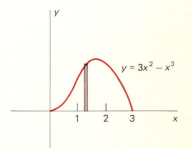

EXAMPLE 1 Find the volume of the solid generated by revolving around the y-axis the region under $y = 3x^2 - x^3$ from $x = 0$ to $x = 3$ (this region is shown in Figure 5.22).

Solution Here it would not be practical to use the method of cross sections, because a cross section perpendicular to the y-axis is an annular ring and finding its inner and outer radii would require solution of the equation $y = 3x^2 - x^3$ for x in terms of y. We would prefer to avoid this troublesome task, and Formula (2) provides us with an alternative: We take $f(x) = 3x^2 - x^3$, $a = 0$, and $b = 3$. It immediately follows that

5.22 The region of Example 1— rotate it about the y-axis.

$$V = \int_0^3 2\pi x(3x^2 - x^3) \, dx = 2\pi \int_0^3 (3x^3 - x^4) \, dx$$

$$= 2\pi \left[\frac{3}{4} x^4 - \frac{1}{5} x^5 \right]_0^3 = \frac{243\pi}{10}.$$

EXAMPLE 2 Find the volume of the solid that remains after boring a hole of radius a through the center of a solid sphere of radius $r > a$.

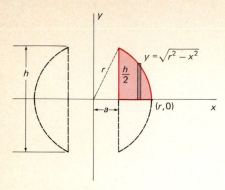

5.23 Middle cross section of the sphere-with-hole.

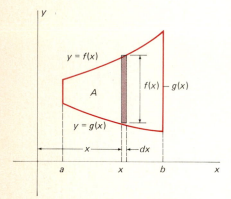

5.24 The region A between the graphs of f and g over $[a, b]$ is to be rotated about the y-axis.

5.25 The region A is to be rotated about the x-axis.

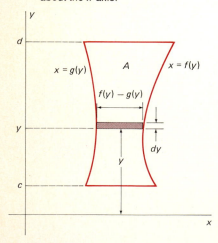

Solution We think of the sphere of radius r as being generated by revolving the right half of the circular disk $x^2 + y^2 \leq r^2$ around the y-axis, and we think of the hole as vertical and with its centerline coincident with part of the y-axis. Then the upper *half* of our solid is generated by revolving the region shaded in Figure 5.23 around the y-axis—this is the region under the graph of $y = \sqrt{r^2 - x^2}$ (and over the x-axis) from $x = a$ to $x = r$. The volume of the entire sphere-with-hole is then double that of the upper half, and Formula (2) gives

$$V = 2 \int_a^r 2\pi x \sqrt{r^2 - x^2}\, dx = 4\pi \left[-\tfrac{1}{3}(r^2 - x^2)^{3/2} \right]_a^r,$$

so that

$$V = \frac{4\pi}{3}(r^2 - a^2)^{3/2}.$$

A way to check this answer is to test it in some extreme cases. If $a = 0$, which corresponds to boring no hole at all, our result reduces to the volume $V = \tfrac{4}{3}\pi r^3$ of the whole sphere. If $a = r$, which corresponds to using a drill bit as large as the sphere, then $V = 0$; this, too, is correct.

Now let A denote the region between the curves $y = f(x)$ and $y = g(x)$ over the interval $[a, b]$, where $0 \leq a < b$ and $g(x) \leq f(x)$ for x in $[a, b]$. Such a region is shown in Figure 5.24. When A is rotated about the y-axis, it generates a solid of revolution. Suppose that we want to find the volume V of this solid. A development similar to that of Formula (2) leads to the approximation

$$V \approx \sum_{i=1}^{n} 2\pi x_i^* [f(x_i^*) - g(x_i^*)]\, \Delta x,$$

from which we may conclude that

$$V = \int_a^b 2\pi x [f(x) - g(x)]\, dx. \qquad (3)$$

The method of cylindrical shells is also an effective way to compute volumes of solids of revolution about the x-axis. Figure 5.25 shows the region A, bounded by the curves $x = f(y)$ and $x = g(y)$ for $c \leq y \leq d$ and by the horizontal lines $y = c$ and $y = d$. Let V be the volume obtained by revolving the region A about the x-axis. To compute V, we begin with a regular partition of $[c, d]$ into n equal subintervals of length Δy each. Let y_i^* denote the midpoint of the ith subinterval $[y_{i-1}, y_i]$. Then the volume of the cylindrical shell with average radius y_i^*, height $f(y_i^*) - g(y_i^*)$, and thickness Δy is

$$V_i = 2\pi y_i^* [f(y_i^*) - g(y_i^*)]\, \Delta y.$$

We add the volumes of these cylindrical shells and thus obtain the approximation

$$V \approx \sum_{i=1}^{n} 2\pi y_i^* [f(y_i^*) - g(y_i^*)]\, \Delta y.$$

We recognize the right-hand side as a Riemann sum for an integral with respect to y from c to d, and so conclude that the volume of the solid of

CHAP. 5: **Applications of the Integral**

revolution is given by

$$V = \int_c^d 2\pi y [f(y) - g(y)] \, dy. \tag{4}$$

EXAMPLE 3 Consider the region in the first quadrant that is bounded by the curves $y = x^2$ and $y = x^3$, shown in Figure 5.26. Use the method of cylindrical shells to compute the volumes of the solids obtained by revolving this region first about the y-axis, then instead about the x-axis.

Solution For the volume of revolution around the y-axis, Formula (3) can be applied with $f(x) = x^2$ and $g(x) = x^3$. This gives

$$V = \int_0^1 2\pi x (x^2 - x^3) \, dx = 2\pi \left[\frac{x^4}{4} - \frac{x^5}{5} \right]_0^1 = \frac{\pi}{10}.$$

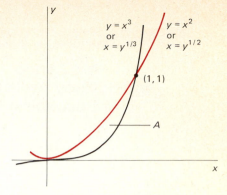

5.26 The region of Example 3.

For the volume of revolution around the x-axis, we use Formula (4) with $f(y) = y^{1/3}$ and $g(y) = y^{1/2}$:

$$V = \int_0^1 2\pi y [y^{1/3} - y^{1/2}] \, dy$$

$$= 2\pi \left[\frac{3}{7} y^{7/3} - \frac{2}{5} y^{5/2} \right]_0^1 = \frac{2\pi}{35}.$$

These answers are the same, of course, as those we obtained using the method of cross sections in Example 4 of Section 5-2.

5-3 PROBLEMS

In each of Problems 1–12, use the method of cylindrical shells to find the volume of the solid generated by revolving about the indicated axis the region bounded by the given curves.

1 $y = x^2$, $y = 0$, $x = 2$; about the y-axis.
2 $x = y^2$, $x = 4$; about the y-axis.
3 $y = 25 - x^2$, $y = 0$; about the y-axis.
4 $y = 2x^2$, $y = 8$; about the y-axis.
5 $y = x^2$, $y = 8 - x^2$; about the y-axis.
6 $x = 9 - y^2$, $x = 0$; about the x-axis.
7 $x = y$, $x + 2y = 3$, $y = 0$; about the x-axis.
8 $y = x^2$, $y = 2x$; about the line $y = 5$.
9 $y = 2x^2$, $y^2 = 4x$; about the x-axis.
10 $y = 3x - x^2$, $y = 0$; about the y-axis.
11 $y = 4x - x^3$, $y = 0$; about the y-axis.
12 $x = y^3 - y^4$, $x = 0$; about the line $y = -2$.
13 Verify the formula for the volume of a cone by using the method of cylindrical shells. Apply the method to the figure generated by revolving the triangle with vertices $(0, 0)$, $(r, 0)$, and $(0, h)$ around the y-axis.
14 Use the method of cylindrical shells to compute the volume of the paraboloid of Problem 19 in Section 5-2.
15 Use the method of cylindrical shells to find the volume of the ellipsoid obtained by revolving the ellipse $(x/a)^2 + (y/b)^2 = 1$ around the y-axis.
16 Use the method of cylindrical shells to derive the formula given in Problem 24 of Section 5-2 for the volume of a spherical segment.

17 Use the method of cylindrical shells to compute the volume of the torus of Problem 25 in Section 5-2. (*Suggestion:* Substitute u for $x - b$ in the integral given by Formula (2).)
18 Find the volume of the solid generated by revolving the region bounded by the curves $y = x^2$ and $y = x + 2$ (a) about the line $x = -2$; (b) about the line $x = 3$.
19 Find the volume of the solid generated by revolving the circular disk $x^2 + y^2 \leq a^2$ about the line $x = -a$.
20 (a) Verify by differentiation that

$$\int x \sin x \, dx = \sin x - x \cos x + C.$$

(b) Find the volume of the solid obtained by revolving about the y-axis the area under $y = \sin x$ from $x = 0$ to $x = \pi$.
21 In Example 2 of this section, we found that the volume remaining when a hole of radius a is bored through the center of a sphere of radius $r > a$ is

$$V = \frac{4\pi}{3} (r^2 - a^2)^{3/2}.$$

(a) Express the volume V above *without* use of the hole radius a, but in its place the hole height h. (*Suggestion:* Use the right triangle shown in Figure 5.23.)
(b) What is remarkable about the answer to Part (a)?

Arc Length and Surface Area of Revolution

A **smooth arc** is the graph of a smooth function defined on a closed interval; a **smooth function** f on $[a, b]$ is one which has derivative f' that is continuous on $[a, b]$. The continuity of f' rules out the possibility of corner points on the graph of f, points where the direction of the tangent line changes abruptly. The graphs of the functions $f(x) = |x|$ and $g(x) = x^{2/3}$ are shown in Figure 5.27; neither is smooth because each has a corner point at the origin.

To investigate the length of a smooth arc, we begin with the length of a straight line segment, which is simply the distance between its endpoints. Then, given a smooth arc C, we pose the following question: If C were a thin wire and we straightened it without stretching it, how long would the resulting straight wire be? The answer is what we would call the *length* of C.

To approximate the length s of the smooth arc C, we can inscribe in C a polygonal arc—one made up of straight line segments—and then calculate the length of this polygonal arc. We proceed in the following way, under the assumption that C is the graph of a smooth function f on the closed interval $[a, b]$. Consider a regular partition of $[a, b]$ into n subintervals each having length Δx. Let P_i denote the point $(x_i, f(x_i))$ on the arc C corresponding to the *i*th subdivision point x_i. Our polygonal arc "inscribed in" C is then the union of the line segments $P_0P_1, P_1P_2, P_2P_3, \ldots, P_{n-1}P_n$. So our approximation to the length s of C is

$$s \approx \sum_{i=1}^{n} |P_{i-1}P_i|, \tag{1}$$

the sum of the lengths of these line segments (see Figure 5.28).

The length of $P_{i-1}P_i$ is

$$|P_{i-1}P_i| = [(x_i - x_{i-1})^2 + (f(x_i) - f(x_{i-1}))^2]^{1/2}.$$

We apply the Mean Value Theorem to the function f on the interval $[x_{i-1}, x_i]$,

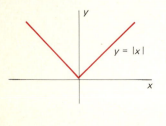

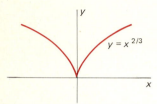

5.27 Graphs having corner points.

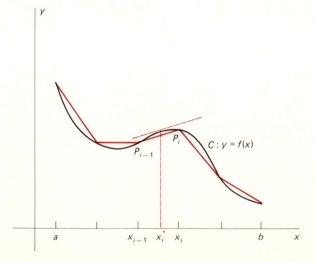

5.28 A polygonal arc inscribed in the smooth curve C.

and thereby conclude the existence of a point x_i^* in this interval such that

$$f(x_i) - f(x_{i-1}) = f'(x_i^*)(x_i - x_{i-1}).$$

Hence

$$|P_{i-1}P_i| = \left[1 + \left(\frac{f(x_i) - f(x_{i-1})}{x_i - x_{i-1}}\right)^2\right]^{1/2}(x_i - x_{i-1})$$

$$= [1 + [f'(x_i^*)]^2]^{1/2}\,\Delta x$$

where $\Delta x = x_i - x_{i-1}$.

We next substitute this expression for $|P_{i-1}P_i|$ into (1), which gives the approximation

$$s \approx \sum_{i=1}^{n} \sqrt{1 + [f'(x_i^*)]^2}\,\Delta x.$$

This sum is a Riemann sum for the function $\sqrt{1 + [f'(x)]^2}$ on $[a, b]$, and therefore (because f' is continuous) approaches the integral

$$\int_a^b \sqrt{1 + [f'(x)]^2}\,dx$$

as $\Delta x \to 0$. But our approximation ought also to approach the actual length s as $\Delta x \to 0$. On this basis we *define* the **length** s of the smooth arc C to be

$$s = \int_a^b \sqrt{1 + [f'(x)]^2}\,dx = \int_a^b \sqrt{1 + \left(\frac{dy}{dx}\right)^2}\,dx. \qquad (2)$$

In the case of a smooth arc given as a graph $x = g(y)$ for y in $[c, d]$, a similar discussion starting with a regular partition of $[c, d]$ leads to the formula

$$s = \int_c^d \sqrt{1 + [g'(y)]^2}\,dy = \int_c^d \sqrt{1 + \left(\frac{dx}{dy}\right)^2}\,dy \qquad (3)$$

for its length. The length of a more general curve, such as a circle, can be computed by subdividing it into finitely many smooth arcs, and then applying to each of these arcs whichever of Formulas (2) or (3) is required.

There is a convenient symbolic device that we can employ to remember Formulas (2) and (3) simultaneously. We think of two nearby points $P(x, y)$ and $Q(x + dx, y + dy)$ on the smooth arc C, and denote by ds the length of the arc joining P and Q. Imagine that P and Q are so close together that ds is, for all practical purposes, equal to the length of the straight line segment PQ. Then the Pythagorean theorem, applied to the small right triangle in Figure 5.29, gives

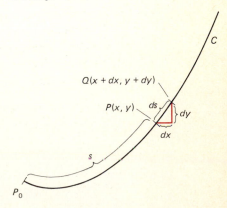

5.29 Heuristic development of the arc length formula.

$$ds = \sqrt{(dx)^2 + (dy)^2} \qquad (4)$$

$$= \sqrt{1 + \left(\frac{dy}{dx}\right)^2}\,dx \qquad (4')$$

$$= \sqrt{1 + \left(\frac{dx}{dy}\right)^2}\,dy. \qquad (4'')$$

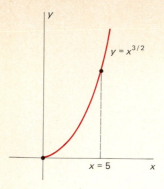

5.30 The semicubical parabola of Example 1.

Thinking of the whole length s of C as the sum of small pieces like ds, we write

$$s = \int ds. \tag{5}$$

Then formal (symbolic) substitution of Expressions (4') and (4'') for ds in Equation (5) yields Formulas (2) and (3); only the limits of integration remain to be determined.

EXAMPLE 1 Find the length of the "semicubical parabola" $y = x^{3/2}$ on $[0, 5]$, shown in Figure 5.30.

Solution We first compute the integrand of Formula (2):

$$\left[1 + \left(\frac{dy}{dx}\right)^2\right]^{1/2} = \left[1 + \left(\frac{3}{2}x^{1/2}\right)^2\right]^{1/2}$$

$$= \left(1 + \frac{9}{4}x\right)^{1/2} = \frac{1}{2}(4 + 9x)^{1/2}.$$

Hence the length of the arc $y = x^{3/2}$ over the interval $[0, 5]$ is

$$s = \int_0^5 \tfrac{1}{2}(4 + 9x)^{1/2}\, dx = \left[\tfrac{1}{27}(4 + 9x)^{3/2}\right]_0^5 = \tfrac{335}{27}.$$

EXAMPLE 2 A manufacturer needs to make corrugated metal sheets 36 inches wide with cross sections that are to have the form of the curve

$$y = \tfrac{1}{2}\sin \pi x, \qquad 0 \le x \le 36,$$

shown in Figure 5.31. What is the width of the flat sheets the manufacturer should use to produce these corrugated sheets?

Solution Since $D\tfrac{1}{2}\sin(\pi x) = (\pi/2)[\cos(\pi x)]$, Formula (2) tells us that the length of $y = \tfrac{1}{2}\sin(\pi x)$ over $[0, 36]$ is

$$s = \int_0^{36} \sqrt{1 + (\pi/2)^2 \cos^2(\pi x)}\, dx$$

$$= 36 \int_0^1 \sqrt{1 + (\pi/2)^2 \cos^2(\pi x)}\, dx.$$

It turns out that the right-hand integral cannot be evaluated in terms of elementary functions. Because of this, we cannot apply the Fundamental Theorem of Calculus. So we estimate its value with the aid of Simpson's approximation. With $n = 6$ subintervals, we find that

$$\int_0^1 \sqrt{1 + (\pi/2)^2 \cos^2(\pi x)}\, dx \approx 1.46$$

inches. Therefore the manufacturer should use flat sheets each having width approximately $(36)(1.46) \approx 52.6$ inches.

0   2   4   34   36

5.31 The corrugated sheet in the shape of $y = \tfrac{1}{2}\sin(\pi x)$.

CHAP. 5: **Applications of the Integral**

AREA OF SURFACES OF REVOLUTION

A **surface of revolution** is one obtained by revolving an arc or curve about an axis lying in its plane. The surface of a cylinder or of a sphere and the curved surface of a cone are important examples of surfaces of revolution.

Our basic approach to the area of such a surface is this: First we inscribe a polygonal arc in the curve to be revolved. We then regard the area of the surface generated by revolving the polygonal arc as an approximation to the surface generated by revolving the original curve. Since a surface generated by revolving a polygonal arc about an axis consists of frusta (sections) of cones, we can calculate its area in a reasonably simple way.

This approach to surface area originated with Archimedes. For example, it's the method he used to establish the formula $A = 4\pi r^2$ for the surface area of a sphere of radius r.

We shall need the formula

$$A = 2\pi \bar{r} L \qquad (6)$$

for the area of a frustum of a cone with average radius $\bar{r} = \frac{1}{2}(r_1 + r_2)$ and *slant height L*, as shown in Figure 5.32. Formula (6) follows from the formula

$$A = \pi r L \qquad (7)$$

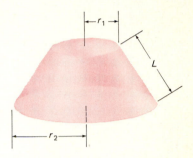

5.32 A frustum of a cone; the "slant height" is L.

for the area of a conical surface with base radius r and slant height L, as shown in Figure 5.33. It is easy to derive Formula (7) by "unrolling" the conical surface onto a sector of a circle of radius L (also in Figure 5.33), because the area of this sector is

$$A = \frac{2\pi r}{2\pi L} \cdot \pi L^2 = \pi r L.$$

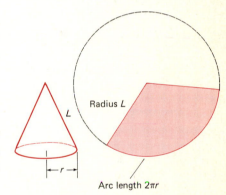

5.33 Surface area of cone: Cut along L, then unroll the cone onto the circular sector.

5.34 Derivation of Formula (6).

To derive Formula (6) from Formula (7), we think of the frustum as the lower section of a cone with slant height $L_2 = L + L_1$, as indicated in Figure 5.34. Then subtraction of the area of the upper conical section from that of the entire cone gives

$$A = \pi r_2 L_2 - \pi r_1 L_1 = \pi r_2 (L + L_1) - \pi r_1 L_1$$

$$= \pi (r_2 - r_1) L_1 + \pi r_2 L$$

for the area of the frustum. But the similar right triangles of Figure 5.34 yield the proportion

$$\frac{r_1}{L_1} = \frac{r_2}{L_2} = \frac{r_2}{L + L_1},$$

from which we find that $(r_2 - r_1)L_1 = r_1 L$. Hence the area of our frustum is

$$A = \pi r_1 L + \pi r_2 L = 2\pi \bar{r} L$$

where $\bar{r} = \frac{1}{2}(r_1 + r_2)$. So we have verified Formula (6).

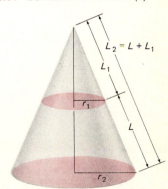

Now suppose that the surface S has area A and is generated by revolving around the x-axis the smooth arc $y = f(x)$, $a \leqq x \leqq b$; suppose also that $f(x)$ is never negative on $[a, b]$. To approximate A we begin with a regular partition of $[a, b]$ into n equal subintervals of length Δx. As in our discussion of arc length leading to Formula (2), let P_i denote the point $(x_i, f(x_i))$ on the arc. Then, as before, the line segment $P_{i-1}P_i$ has length

$$L_i = |P_{i-1}P_i| = [1 + (f'(x_i^*))^2]^{1/2}\, \Delta x$$

for some point x_i^* in the ith subinterval $[x_{i-1}, x_i]$.

The conical frustum obtained by revolving the segment $P_{i-1}P_i$ around the x-axis has slant height L_i and, as shown in Figure 5.35, average radius

$$\bar{r}_i = \tfrac{1}{2}[f(x_{i-1}) + f(x_i)].$$

Since $\bar{r}_i$ lies between the values $f(x_{i-1})$ and $f(x_i)$, the Intermediate Value Property (Section 1-9) yields a point x_i^{**} in $[x_{i-1}, x_i]$ such that $\bar{r}_i = f(x_i^{**})$. By Formula (6), the area of this conical frustum is, therefore,

$$2\pi \bar{r}_i L_i = 2\pi f(x_i^{**})[1 + (f'(x_i^*))^2]^{1/2}\, \Delta x.$$

We add the areas of these conical frusta for $i = 1, 2, 3, \ldots, n$; this gives the approximation

$$A \approx \sum_{i=1}^{n} 2\pi f(x_i^{**})[1 + (f'(x_i^*))^2]^{1/2}\, \Delta x.$$

If x_i^* and x_i^{**} were the *same* point of the ith subinterval $[x_{i-1}, x_i]$, then this approximation would be a Riemann sum for the integral

$$\int_a^b 2\pi f(x)\sqrt{1 + [f'(x)]^2}\; dx.$$

Even though the numbers x_i^* and x_i^{**} are generally unequal, it still follows (from a result stated in the appendix to this chapter) that our approximation approaches the above integral as $\Delta x \to 0$.

We therefore *define* the **area** A of the surface generated by revolving the smooth arc $y = f(x)$, $a \le x \le b$, around the x-axis by the integral formula

$$A = \int_a^b 2\pi f(x)\sqrt{1 + [f'(x)]^2}\; dx. \tag{8}$$

If we write $y = f(x)$ and $ds = \sqrt{1 + (dy/dx)^2}\; dx$, as in Formula (4'), then Formula (8) can be abbreviated to

$$A = \int 2\pi y\, ds \qquad (x\text{-axis}). \tag{9}$$

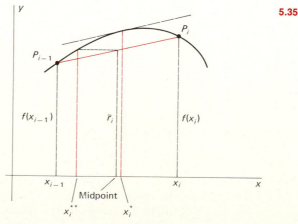

5.35

This abbreviated formula is conveniently remembered by thinking of $dA = 2\pi y\,ds$ as the area of the narrow frustum obtained by revolving the tiny arc ds around the x-axis in a circle of radius y, as in Figure 5.36.

If our smooth arc being revolved around the x-axis is given by $x = g(y)$, $c \le y \le d$, then an approximation based on a regular partition of $[c, d]$ leads to the area formula

$$A = \int_c^d 2\pi y \sqrt{1 + [g'(y)]^2}\, dy. \tag{10}$$

Note that Formula (10) can be obtained by making the formal substitution $ds = \sqrt{1 + (dx/dy)^2}\, dy$ of Formula (4″) in the abbreviated Formula (9) for surface area of revolution around the x-axis and then inserting the limits c and d.

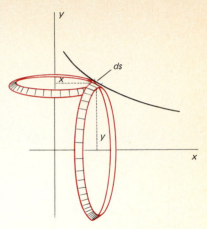

5.36 The arc ds may be revolved about either the x-axis or the y-axis.

Now let us consider the surface generated by revolving our smooth arc around the y-axis rather than the x-axis. In Figure 5.36 we see that the average radius of the narrow frustum obtained by revolving the tiny arc ds is now x instead of y. This suggests the abbreviated formula

$$A = \int 2\pi x\, ds \quad (y\text{-axis}) \tag{11}$$

for a surface area of revolution about the y-axis. If the smooth arc is given by $y = f(x), a \le x \le b$, then the symbolic substitution $ds = \sqrt{1 + (dy/dx)^2}\, dx$ gives

$$A = \int_a^b 2\pi x \sqrt{1 + [f'(x)]^2}\, dx. \tag{12}$$

But if our smooth arc is given by $x = g(y), c \le y \le d$, then the symbolic substitution $ds = \sqrt{1 + (dx/dy)^2}\, dy$ in (11) gives

$$A = \int_c^d 2\pi g(y) \sqrt{1 + [g'(y)]^2}\, dy. \tag{13}$$

It is possible to verify Formulas (12) and (13) by using approximations similar to the one leading to Formula (8).

Thus we have *four* formulas for areas of surfaces of revolution, summarized in Figure 5.37. Which one of these formulas is appropriate for computing the area of a given surface depends upon whether the smooth arc that generates it is described as the graph of a function of x or of a function of y, and upon whether this arc is to be revolved around the x-axis or around the y-axis. But memorizing the formulas in Figure 5.37 is unnecessary. We suggest that you instead remember the abbreviated Formulas (9) and (11) in conjunction with Figure 5.36, and make either the substitution

$$y = f(x), \quad ds = \sqrt{1 + [f'(x)]^2}\, dx$$

5.37 Area formulas for surfaces of revolution.

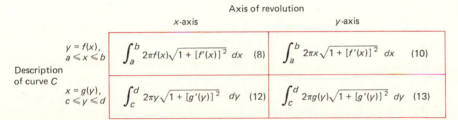

| | Axis of revolution | |
	x-axis	y-axis
$y = f(x),\ a \le x \le b$	$\int_a^b 2\pi f(x)\sqrt{1 + [f'(x)]^2}\,dx$ (8)	$\int_a^b 2\pi x\sqrt{1 + [f'(x)]^2}\,dx$ (10)
Description of curve C		
$x = g(y),\ c \le y \le d$	$\int_c^d 2\pi y\sqrt{1 + [g'(y)]^2}\,dy$ (12)	$\int_c^d 2\pi g(y)\sqrt{1 + [g'(y)]^2}\,dy$ (13)

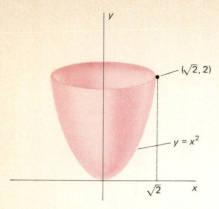

5.38 The paraboloid of Example 3.

or the substitution

$$x = g(y), \qquad ds = \sqrt{1 + [g'(y)]^2}\, dy,$$

depending upon the way in which the smooth arc is described.

EXAMPLE 3 Find the area of the paraboloid of Figure 5.38, generated by revolving the parabolic arc $y = x^2, 0 \leq x \leq \sqrt{2}$, around the y-axis.

Solution Following the suggestion that precedes the example, we get

$$A = \int 2\pi x\, ds = \int_a^b 2\pi x \sqrt{1 + \left(\frac{dy}{dx}\right)^2}\, dx$$

$$= \int_0^{\sqrt{2}} 2\pi x \sqrt{1 + (2x)^2}\, dx$$

$$= \left[\frac{\pi}{6}(1 + 4x^2)^{3/2}\right]_0^{\sqrt{2}} = \frac{13\pi}{3}.$$

Alternatively, we could describe our parabolic arc as $x = y^{1/2}, 0 \leq y \leq 2$, and get

$$A = \int 2\pi x\, ds = \int_c^d 2\pi x \sqrt{1 + \left(\frac{dx}{dy}\right)^2}\, dy$$

$$= \int_0^2 2\pi y^{1/2} \sqrt{1 + (\tfrac{1}{2}y^{-1/2})^2}\, dy$$

$$= \int_0^2 \pi \sqrt{1 + 4y}\, dy = \left[\frac{\pi}{6}(1 + 4y)^{3/2}\right]_0^2 = \frac{13\pi}{3}.$$

Note that the decision of which abbreviated formula—(9) or (11)—should be used is determined by the axis of revolution. By contrast, the decision whether the variable of integration should be x or y is determined by the way in which the smooth arc is given: as a function of x or as a function of y.

5-4 PROBLEMS

Find the lengths of the smooth arcs in Problems 1–8.

1 $y = \frac{2}{3}(x^2 + 1)^{3/2}$ from $x = 0$ to $x = 2$.

2 $x = \frac{2}{3}(y - 1)^{3/2}$ from $y = 1$ to $y = 5$.

3 $y = \dfrac{x^3}{6} + \dfrac{1}{2x}$ from $x = 1$ to $x = 3$.

4 $x = \dfrac{y^4}{8} + \dfrac{1}{4y^2}$ from $y = 1$ to $y = 2$.

5 $8x^2 y - 2x^6 = 1$ from $(1, \frac{3}{8})$ to $(2, \frac{129}{32})$.

6 $12xy - 4y^4 = 3$ from $(\frac{7}{12}, 1)$ to $(\frac{67}{24}, 2)$.

7 $y^3 = 8x^2$ from $(1, 2)$ to $(8, 8)$.

8 $(y - 3)^2 = 4(x + 2)^3$ from $(-1, 5)$ to $(2, 19)$.

In each of Problems 9–15, find the area of the surface of revolution that is generated by revolving the given curve about the indicated axis.

9 $y = \sqrt{x}, \quad 0 \leq x \leq 1$; the x-axis.

10 $y = x^3, 1 \leq x \leq 2$; the x-axis.

11 $y = \dfrac{x^5}{5} + \dfrac{1}{12x^3}$, $1 \leq x \leq 2$; the y-axis.

12 $x = \dfrac{y^4}{8} + \dfrac{1}{4y^2}$, $1 \leq y \leq 2$; the x-axis.

13 $y^3 = 3x, 0 \leq x \leq 9$; the y-axis.

14 $y = \frac{2}{3}x^{3/2}, 1 \leq x \leq 2$; the y-axis. (*Suggestion:* Make the substitution $u = 1 + x$.)

15 $y = (2x - x^2)^{1/2}, 0 \leq x \leq 2;$ the x-axis.

16 Show that the length of one arch of the sine curve $y = \sin x$ is equal to half the circumference of the ellipse $2x^2 + y^2 = 2$. (*Suggestion:* Substitute $x = \cos \theta$ into the arc length integral for the ellipse.)

17 Use Simpson's approximation with $n = 6$ subintervals to estimate the length of the sine arch of Problem 16.

18 Use Simpson's approximation with $n = 10$ subintervals to estimate the length of the parabola $y = x^2$ from $x = 0$ to $x = 1$.

19 Verify Formula (6) for the area of a conical frustum. Think of the frustum as being generated by revolving the line segment from $(r_1, 0)$ to (r_2, h) around the y-axis.

20 By considering a sphere of radius r as a surface of revolution, derive the formula $A = 4\pi r^2$ for its surface area.

21 Find the total length of the *astroid* shown in Figure 5.39. The equation of the graph there is $x^{2/3} + y^{2/3} = 1$.

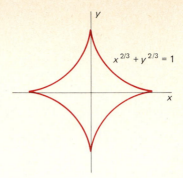

5.39 The astroid—see Problems 21 and 22.

22 Find the area of the surface that is generated by revolving the astroid of Problem 21 around the y-axis.

5-5 Force and Work

The concept of *work* is introduced to measure the cumulative effect of a force in moving a body from one position to another. In the simplest case, a particle is moved along a straight line by the action of a *constant* force. In this case, the work done by the force is defined to be the product of the force and the distance through which it acts:

$$\text{work} = (\text{constant force}) \times (\text{distance}). \qquad (1)$$

In this section we use the integral to generalize the above definition to the case in which a particle is moved along a straight line by a *variable* force. Given a **force function** $F(x)$ defined at each point x of the straight line segment $[a, b]$, we must first say what properties the work W done by the force F ought to have.

Work should be additive. That is, if we do work W_1 in moving a particle from a to c and do work W_2 in moving it from c to b, then the total work done in moving the particle from a to b should be

$$W = W_1 + W_2. \qquad (2)$$

Work should also have the following monotonicity property: If F_1 and F_2 are constants such that

$$F_1 \leq F(x) \leq F_2$$

for all x in $[a, b]$, then the work W done by the force F in moving a particle the distance Δx should satisfy the inequalities

$$F_1 \, \Delta x \leq W \leq F_2 \, \Delta x. \qquad (3)$$

We shall see that these two properties are enough to *determine* the way in which the work done by a variable force must be defined.

We begin with the usual regular partition of $[a, b]$ into n equal subintervals, each with length Δx. Let $F(x_i^?)$ and $F(x_i^\#)$ be the minimum and maximum values, respectively, of $F(x)$ on the ith subinterval $[x_{i-1}, x_i]$. By

Inequalities (3), the work W_i done by the force in moving the particle from x_{i-1} to x_i satisfies

$$F(x_i^\flat)\,\Delta x \leqq W_i \leqq F(x_i^\#)\,\Delta x.$$

If we add these inequalities for $i = 1, 2, 3, \ldots, n$, we find that

$$\sum_{i=1}^{n} F(x_i^\flat)\,\Delta x \leqq W \leqq \sum_{i=1}^{n} F(x_i^\#)\,\Delta x.$$

These two sums are the lower and upper Riemann sums for the function F on the interval $[a, b]$. If we know that F is continuous on $[a, b]$, it follows that (as $\Delta x \to 0$) both these sums approach the value of the integral $\int_a^b F(x)\,dx$. Consequently the squeeze law of limits leaves us no alternative but to *define* the **work** W done by the force $F(x)$ in moving a particle from $x = a$ to $x = b$ by

$$W = \int_a^b F(x)\,dx. \qquad (4)$$

The following heuristic way of setting up Formula (4) is useful in setting up integrals for work problems. Imagine that dx is so small a number that the value of F does not change appreciably on the tiny interval from x to $x + dx$. Then the work done by the force in moving a particle from x to $x + dx$ should be very close to

$$dW = F(x)\,dx.$$

Property (2) then suggests that we could obtain the total work W by adding these tiny elements of work, so that

$$W = \int dW = \int_a^b F(x)\,dx.$$

ELASTIC SPRINGS

Consider a spring with left end held fixed and right end free to move along the x-axis. We assume that the right end is at the origin $x = 0$ when the spring has its **natural length**; that is, when the spring is in its rest position, neither compressed nor stretched by outside forces.

According to **Hooke's law** for elastic springs, the force $F(x)$ that must be exerted on the spring to hold its right end at the point x is proportional to the displacement x of the right end from its rest position. That is,

$$F(x) = kx \qquad (5)$$

for some positive constant k. The constant k is a characteristic of the particular spring under study and is called the **spring constant.**

Figure 5.40 shows the arrangement of such a spring along the x-axis. The spring is being held with its right end at position x on the x-axis by a force $F(x)$. The figure shows the situation for $x > 0$, so that the spring is stretched. The force that the spring exerts on its right-hand end is directed to the left, so—as the figure shows—the external force $F(x)$ must act to the right. The right is the positive direction here, so $F(x)$ must be a positive number. Since x and $F(x)$ have the same sign, k must be positive too. You can mentally check that k is also positive in the case $x < 0$.

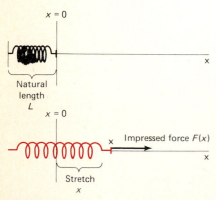

5.40 The stretch x is proportional to the impressed force F.

CHAP. 5:   **Applications of the Integral**

EXAMPLE 1   Suppose that a spring has a natural length of 1 foot and that a force of 10 pounds is required to hold it compressed to a length of 6 inches. How much work is done in stretching the spring from its natural length to a total length of 2 feet?

Solution   To move the free end from $x = 0$ (the natural length position) to $x = 1$ (stretched by 1 foot), we must exert a variable force $F(x)$ determined by Hooke's law. We are given that $F = -10$ pounds when $x = -\frac{1}{2}$ ft, so Equation (5), $F = kx$, implies that the spring constant for this spring is $k = 20$ lb/ft. Thus $F(x) = 20x$, and so—using Formula (4)—we find that the work done in stretching this spring in the manner given is

$$W = \int_0^1 20x \, dx = \left[10x^2\right]_0^1 = 10 \text{ ft-lb.}$$

*WORK AGAINST GRAVITY

According to Newton's law of gravitation, the force that must be exerted on a body to hold it at a distance r from the earth's center is inversely proportional to r^2. That is, if $F(r)$ denotes the holding force, then

$$F(r) = \frac{k}{r^2} \tag{6}$$

for some positive constant k. The value of this force at the earth's surface, where $r = R \approx 4000$ miles, is called the **weight** of the body.

Given the weight $F(R)$ of a particular body, we can find the corresponding value of k by using Formula (6):

$$k = R^2 F(R).$$

The work that must be done to lift the body vertically from the surface (where $r = R$) to a distance R_1 from the center of the earth is then

$$W = \int_R^{R_1} \frac{k}{r^2} \, dr. \tag{7}$$

If distance is measured in miles and force in pounds, then this integral gives the work in mile-pounds. This is a rather unconventional unit of work; we shall multiply by 5280 (feet per mile) to convert any such result into foot-pounds.

EXAMPLE 2   (Satellite Launch.)   How much work must be done in order to lift a 1000-pound satellite vertically from the surface of the earth to an orbit 1000 miles above the earth's surface? (Take $R = 4000$ miles as the radius of the earth.)

Solution   Since $F = 1000$ (pounds) when $r = R = 4000$ (miles), we find from Equation (6) that

$$k = (4000)^2(1000) = 16 \times 10^9 \text{ (mi}^2\text{-lb).}$$

Then by Formula (7) the work done is

$$W = \int_{4000}^{5000} \frac{k}{r^2}\, dr = \left[-\frac{k}{r} \right]_{4000}^{5000}$$

$$= (16 \times 10^9)\left(\frac{1}{4000} - \frac{1}{5000} \right) = 8 \times 10^5 \text{ mi-lb.}$$

We multiply by 5280 and write the answer as 4.224×10^9, or 4,224,000,000, foot-pounds.

We can also express the answer in terms of the power that the launch rocket must provide. **Power** is the rate at which work is performed. For instance, 1 **horsepower** is defined to be 33,000 foot-pounds per minute. If it is known that the ascent to orbit takes 15 minutes and that only 2% of the power generated by the rocket is effective in lifting the satellite (the rest is used in lifting the rocket and its fuel), we can convert the above answer to horsepower. The *average* power that the rocket engine must produce during the 15-minute ascent is

$$P = \frac{(50)(4.224 \times 10^9)}{(15)(33,000)} \approx 426,667 \text{ horsepower.}$$

The term 50 in the numerator comes from the fact that, because of the 2% "efficiency" of the rocket, the total power must be multiplied by $1/(0.02) = 50$.

WORK TO FILL A TANK

Examples 1 and 2 are applications of Formula (4) for calculating the work done by a variable force in moving a particle a certain distance. Another common type of force-work problem involves the summation of work done by constant forces that act through different distances. For example, consider the problem of pumping a fluid from ground level up into an above-ground tank, like the one shown in Figure 5.41.

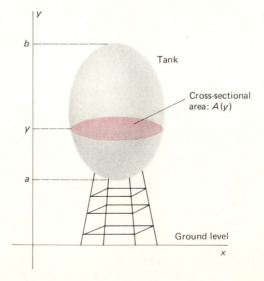

5.41 An above-ground tank.

CHAP. 5: **Applications of the Integral**

A convenient way to think of how the tank is filled is this: Think of thin horizontal layers of fluid, each lifted from the ground to its final position in the tank. No matter how the fluid actually behaves as the tank is filled, this simple way of thinking about the process gives a way to compute the work done in the filling process. When we think of filling the tank in this way, different layers of fluid must be lifted different distances to reach their eventual positions in the tank.

Suppose that the bottom of the tank is at height $y = a$ and its top at height $y = b$. Let $A(y)$ be the cross-sectional area of the tank at height y. Consider a regular partition of $[a, b]$ into n equal subintervals of length Δy. Then the volume of the horizontal slice of the tank that corresponds to the ith subinterval $[y_{i-1}, y_i]$ is

$$\Delta V_i = \int_{y_{i-1}}^{y_i} A(y)\, dy = A(y_i^*)\, \Delta y$$

for some number y_i^* in $[y_{i-1}, y_i]$; this is a consequence of the average value theorem for integrals (Section 4-4). If ρ is the density of the fluid (for instance, in lb/ft^3) then the force required to lift this slice is

$$F_i = \rho\, \Delta V_i = \rho A(y_i^*)\, \Delta y.$$

What about the distance through which this force must act? The fluid in question must be lifted from ground level to the level of the subinterval $[y_{i-1}, y_i]$, so every particle of the fluid is lifted at least the distance y_{i-1} and at most the distance y_i. Hence the work ΔW_i needed to lift this ith slice of fluid satisfies the inequalities $F_i y_{i-1} \le \Delta W_i \le F_i y_i$, or

$$\rho y_{i-1} A(y_i^*)\, \Delta y \le \Delta W_i \le \rho y_i A(y_i^*)\, \Delta y.$$

Now we add these inequalities for $i = 1, 2, 3, \ldots, n$, and find that the total work $W = \Sigma\, \Delta W_i$ satisfies the inequalities

$$\sum_{i=1}^{n} \rho y_{i-1} A(y_i^*)\, \Delta y \le W \le \sum_{i=1}^{n} \rho y_i A(y_i^*)\, \Delta y.$$

If the three points y_{i-1}, y_i^*, and y_i of $[y_{i-1}, y_i]$ were the same, then both the above sums would be Riemann sums for the function $f(y) = \rho y A(y)$ on $[a, b]$. Although the three points are not equal, it still follows—from a result stated in the appendix to this chapter—that both sums approach $\int_a^b \rho y A(y)\, dy$ as $\Delta y \to 0$. The squeeze law of limits therefore gives the formula

$$W = \int_a^b \rho y A(y)\, dy. \tag{8}$$

This is the work W done in pumping fluid of density ρ from the ground into a tank with horizontal cross-sectional area $A(y)$, located between heights $y = a$ and $y = b$ above the ground.

A quick heuristic way to set up Formula (8) is to think of a thin horizontal slice of fluid with volume $dV = A(y)\, dy$ and weight $\rho\, dV = \rho A(y)\, dy$. The work required to lift this slice a distance y is

$$dW = \rho y A(y)\, dy,$$

so the total work required to fill the tank is

$$W = \int dW = \int_a^b \rho y A(y)\, dy,$$

because the horizontal slices lie between $y = a$ and $y = b$.

EXAMPLE 3 Suppose that it took 20 years to construct the great pyramid of Khufu at Gizeh. This pyramid is 500 feet high and has a square base with edge length 750 feet. Suppose also that the pyramid is made of rock with density $\rho = 120$ pounds per cubic foot. Finally, suppose that each laborer did 160 foot-pounds per hour of effective work in lifting rocks from ground level to position in the pyramid and worked 12 hours daily for 330 days per year. How many laborers would have been required to construct this pyramid?

Solution We assume a constant labor force throughout the 20-year period of construction. We think of the pyramid as being made up of thin horizontal slabs of rock, each slab lifted from ground level to its ultimate height. Hence we can use Formula (8) to compute the work W required.

Figure 5.42 shows a vertical cross section of the pyramid. The horizontal cross section at height y is a square with edge s. From the similar triangles in Figure 5.42 we see that

$$\frac{s}{750} = \frac{500 - y}{500}, \quad \text{so that} \quad s = \frac{3}{2}(500 - y).$$

Hence the cross-sectional area at height y is

$$A(y) = \tfrac{9}{4}(500 - y)^2.$$

Formula (8) therefore gives

$$W = \int_0^{500} 120 \cdot y \cdot \tfrac{9}{4}(500 - y)^2 \, dy$$

$$= 270 \int_0^{500} (250{,}000y - 1000y^2 + y^3) \, dy$$

$$= 270 \left[125{,}000y^2 - \frac{1000}{3} y^3 + \frac{y^4}{4} \right]_0^{500},$$

so that $W \approx 1.406 \times 10^{12}$ ft-lb.

Since each laborer does

$$(160)(12)(330)(20) \approx 1.267 \times 10^7$$

foot-pounds of work, the construction of the pyramid would—under our assumptions—have required

$$\frac{1.406 \times 10^{12}}{1.267 \times 10^7},$$

or about 111,000, laborers.

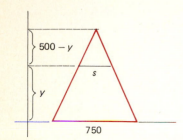

5.42 Vertical cross section of Khufu's pyramid.

5-5 PROBLEMS

In each of Problems 1–5, find the work done by the given force $F(x)$ in moving a particle along the x-axis from $x = a$ to $x = b$.

1 $F(x) = 10;\quad a = -2, b = 1.$

2 $F(x) = 3x - 1;\quad a = 1, b = 5.$

3 $F(x) = \dfrac{10}{x^2};\quad a = 1, b = 10.$

4 $F(x) = -3\sqrt{x};\quad a = 0, b = 4.$

5 $F(x) = \sin \pi x;\quad a = -1, b = 1.$

6 A spring has a natural length of 1 foot, and a force of 10 pounds is required to hold it stretched to a total length of 2 feet. How much work is done in compressing this spring from its natural length to a length of 6 inches?

7 A spring has a natural length of 2 feet, and a force of 15 pounds is required to hold it compressed to a length of 18 inches. How much work is done in stretching this spring from its natural length to a length of 3 feet?

8 Apply Formula (4) to compute the amount of work done in lifting a 100-pound weight a height of 10 feet, assuming this work is done against the constant force of gravity.

9 Compute the amount of work (in foot-pounds) done in lifting a 1000-pound weight from an orbit 1000 miles above the earth's surface to one 2000 miles above the earth's surface. Use the value of k given in Example 2.

10 A cylindrical tank is resting on the ground with its axis vertical; it has radius 5 feet and height 10 feet. Use Formula (8) to compute the amount of work done in filling this tank with water pumped in from ground level. (For the density of water, use $\rho = 62.4$ pounds per cubic foot.)

11 A conical tank is resting on its base—which is at ground level—and with its axis vertical. The tank has radius 5 feet and height 10 feet. Compute the work done in filling this tank with water ($\rho = 62.4 \text{ lb/ft}^3$) pumped in from ground level.

12 Repeat Problem 11, except that now the tank is upended: its vertex is at ground level, its base is 10 feet above the ground (but its axis is still vertical).

13 A tank with its lowest point 10 feet above ground has the shape of a cup obtained by rotating the parabola $y = \frac{1}{5}x^2$, $-5 \le x \le 5$, around the y-axis. The units on the coordinate axes are also in feet. How much work is done in filling this tank with oil weighing 50 pounds per cubic foot if the oil is pumped in from ground level?

14 Suppose that the tank of Figure 5.41 is filled with fluid of density ρ, and that all this fluid is to be pumped from the tank to the level $y = h$. Use Riemann sums as in the derivation of Formula (8) to obtain the formula

$$W = \int_a^b \rho(h - y)A(y)\,dy$$

for the work required to do this.

15 Use the formula of Problem 14 to find the amount of work done in pumping the water in the tank of Problem 10 to a height of 5 feet above the top of the tank.

16 Gasoline at a service station is stored in a cylindrical tank buried on its side, with the highest part of the tank 5 feet below the surface. The tank is 6 feet in diameter and 10 feet long. The density of gasoline is 45 pounds per cubic foot. Assume that the filler cap of each automobile gas tank is two feet above the ground. How much work is done in emptying all the gasoline from this tank, which is initially full, into automobiles?

17 Consider a spherical water tank of radius 10 feet, with center 50 feet above the ground. How much work is required to fill this tank by pumping water up from ground level? (*Suggestion:* It may simplify your computations to take $y = 0$ at the center of the tank and think of the distance each horizontal slice of water must be lifted.)

18 A hemispherical tank of radius 10 feet is located with its flat side down atop a tower 60 feet high. How much work is required to fill it with oil weighing 50 lb/ft³ if the oil is to be pumped into the tank from ground level?

19 Water is being drawn from a well 100 feet deep, using a bucket that scoops up 100 pounds of water. The bucket is being pulled up at the rate of 2 feet per second, but it has a hole in its bottom through which water leaks out at the rate of $\frac{1}{2}$ pound per second. How much work is done in pulling the bucket to the top? Neglect the weight of the bucket, the weight of the rope, and the work done in overcoming friction. (*Suggestion:* Take $y = 0$ at the level of water in the well, so that $y = 100$ at ground level. Let $\{y_0, y_1, y_2, \ldots, y_n\}$ be a subdivision of $[0,100]$ into n equal subintervals. Estimate the amount of work W_i required to raise the bucket from y_{i-1} to y_i. Then set up the sum $W = \Sigma W_i$ and proceed to the appropriate integral by letting $n \to \infty$.)

20 A rope 100 feet long and weighing $\frac{1}{4}$ pound per linear foot hangs from the edge of a very tall building. How much work is required to pull this rope to the top of the building?

21 Suppose we plug the hole in the leaking bucket of Problem 19. How much work is done in lifting the mended bucket, full of water, to the surface, using the rope of Problem 20?

22 Consider a volume V of gas in a cylinder fitted with a piston at one end, where the pressure p of the gas is a function $p(V)$ of its volume. Such a cylinder is shown in Figure 5.43.

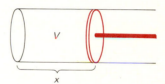

5.43 A cylinder fitted with a piston.

Let A be the area of the face of the piston. Then the force exerted on the piston by gas in the cylinder is $F = pA$. Assume that the gas expands from volume V_1 to volume V_2. Show that the work done by the force F is then given by

$$W = \int_{V_1}^{V_2} p(V)\,dV.$$

[*Suggestion:* If x is the length of the cylinder (from its fixed end to the face of the piston), then $F = Ap(Ax)$. Apply Formula (4) of this section, and make the substitution $V = Ax$ in the resulting integral.]

23 The pressure and volume of the steam in a small steam engine satisfy the condition $pV^{1.4} = c$ (constant). In one cycle, the steam expands from a volume $V_1 = 50 \text{ in}^3$ to $V_2 = 500 \text{ in}^3$ with an initial pressure of 200 lb/in². Use the formula of Problem 22 to compute the work (in foot-pounds) done by this engine in each such cycle.

Centroids of Plane Regions and Curves

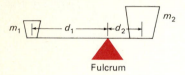

5.44 Law of the lever: The weights balance when $m_1 d_1 = m_2 d_2$.

According to the **law of the lever,** two masses m_1 and m_2 on opposite sides and at respective distances d_1 and d_2 from the fulcrum of a lever will balance provided that $m_1 d_1 = m_2 d_2$ (see Figure 5.44). Think of the x-axis as a lever with fulcrum at the origin. Then a more general form of the law of the lever states that masses $m_0, m_1, m_2, \ldots, m_n$ with coordinates $x_0, x_1, x_2, \ldots, x_n$ will balance provided that

$$\sum_{i=0}^{n} m_i x_i = m_0 x_0 + m_1 x_1 + \cdots + m_n x_n = 0. \tag{1}$$

Now consider arbitrary masses $m_1, m_2, \ldots, m_n$ at the points $x_1, x_2, \ldots, x_n$. Then a single particle with mass

$$m = m_1 + m_2 + \cdots + m_n = \sum_{i=1}^{n} m_i$$

and position $-\bar{x}$ will balance these n masses provided that

$$-m\bar{x} + \sum_{i=1}^{n} m_i x_i = 0;$$

that is, provided that

$$\bar{x} = \frac{1}{m} \sum_{i=1}^{n} m_i x_i. \tag{2}$$

Thus the original n masses act on the lever like a single particle of mass m located at the point $\bar{x}$. The point $\bar{x}$ is called the *center of mass*, and the $\sum_{i=1}^{n} m_i x_i$ that appears in (2) is called the *moment* about the origin.

Now consider a system of n particles with masses $m_1, m_2, \ldots, m_n$ located in the plane at the points $(x_1, y_1), (x_2, y_2), \ldots, (x_n, y_n)$. In analogy with the one-dimensional case above, we define the **moment** M_y of this system about the y-axis and its **moment** M_x about the x-axis by means of the equations

$$M_y = \sum_{i=1}^{n} m_i x_i \quad \text{and} \quad M_x = \sum_{i=1}^{n} m_i y_i. \tag{3}$$

The **center of mass** of this system of n particles is the point $(\bar{x}, \bar{y})$ with coordinates defined to be

$$\bar{x} = \frac{M_y}{m} \quad \text{and} \quad \bar{y} = \frac{M_x}{m} \tag{4}$$

where $m = m_1 + m_2 + \cdots + m_n$ is the sum of the masses. Thus $(\bar{x}, \bar{y})$ is the point where a single particle of mass m would have the same moments as the system about the two coordinate axes. In elementary physics it is shown that, if the xy-plane were a rigid and weightless plastic sheet with our n particles imbedded in it, then it would balance (horizontally) on the point of an icepick at the point $(\bar{x}, \bar{y})$.

As the number of particles we consider increases while their masses decrease in proportion, their aggregate more and more closely resembles a plane region of varying density. Let us first define the center of mass $(\bar{x}, \bar{y})$ and the moments about the coordinate axes of a thin plate or **lamina** of *constant* density ρ, one that occupies a bounded plane region R. Since ρ is constant, the numbers $\bar{x}$ and $\bar{y}$ should be independent of its actual value, so we will take $\rho = 1$ (for convenience) in our definition and computations. In this case $(\bar{x}, \bar{y})$ is called the **centroid of the plane region** R. We shall also define $M_y(R)$ and $M_x(R)$, the moments of the plane region R about the axes, with $\rho = 1$. The corresponding moments of a lamina of constant density $\rho \neq 1$ would then be $\rho M_y(R)$ and $\rho M_x(R)$.

Our definitions will be based upon the following two physical principles.

Symmetry Principle

If the region R is symmetric with respect to the line L—that is, R is carried onto itself when the plane is revolved through an angle of $180°$ about the axis L—then the centroid of R lies on L.

Additivity of Moments

If R is the union of the two nonoverlapping regions S and T, then

$$M_y(R) = M_y(S) + M_y(T)$$

and   (5)

$$M_x(R) = M_x(S) + M_x(T).$$

For example, the symmetry principle implies that the centroid of a rectangle is its geometric center: the intersection of its diagonals. We also assume that the moments of a rectangle R with area A and centroid $(\bar{x}, \bar{y})$ are $M_y(R) = A\bar{x}$ and $M_x(R) = A\bar{y}$. Knowing the centroid of a rectangle, our strategy will be to calculate the moments of a more general region by using additivity of moments and the integral, and then finally to define the centroid of this more general region by analogy with Equations (4).

Now suppose that the function f is continuous and nonnegative on $[a, b]$, and suppose also that the region R is the region between the graph of f and the x-axis for $a \leq x \leq b$. We begin with a regular partition of $[a, b]$ into n equal subintervals of length Δx and denote by x_i^* the midpoint of the ith subinterval $[x_{i-1}, x_i]$. As shown in Figure 5.45, the rectangle with base $[x_{i-1}, x_i]$ and height $f(x_i^*)$ has area $f(x_i^*)\,\Delta x$ and centroid $(x_i^*, \frac{1}{2}f(x_i^*))$. If P_n denotes the union of these rectangles for $i = 1, 2, 3, \ldots, n$, then—because moments are additive—the moments of the rectangular polygon P_n about the y-axis and the x-axis are

$$M_y(P_n) = \sum_{i=1}^{n} x_i^* f(x_i^*)\,\Delta x$$

and

$$M_x(P_n) = \sum_{i=1}^{n} \tfrac{1}{2}f(x_i^*) \cdot f(x_i^*)\,\Delta x,$$

respectively.

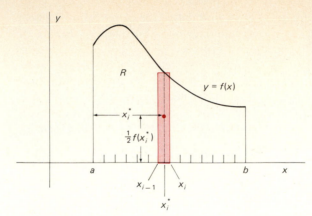

5.45 Locating the centroid of R by approximating R with rectangles.

We define the moments $M_y(R)$ and $M_x(R)$ of the region R itself by taking the limits of $M_y(P_n)$ and $M_x(P_n)$ as $\Delta x \to 0$. Since the two sums above are Riemann sums, their limits are the definite integrals

$$M_y(R) = \int_a^b xf(x)\,dx \qquad (6)$$

and

$$M_x(R) = \int_a^b \tfrac{1}{2}[f(x)]^2\,dx. \qquad (7)$$

Finally, we define the centroid $(\bar{x}, \bar{y})$ of R by

$$\bar{x} = \frac{M_y(R)}{A} \quad \text{and} \quad \bar{y} = \frac{M_x(R)}{A}, \qquad (8)$$

where $A = \int_a^b f(x)\,dx$ is the area of R. Thus

$$\bar{x} = \frac{1}{A} \int_a^b xf(x)\,dx \qquad (9)$$

and

$$\bar{y} = \frac{1}{A} \int_a^b \tfrac{1}{2}[f(x)]^2\,dx \qquad (10)$$

are the coordinates of the centroid of the region under $y = f(x)$ from $x = a$ to $x = b$.

By the symmetry principle, the centroid of a circular disk is its center. But the centroid of a semicircle is more interesting.

EXAMPLE 1   Find the centroid of the upper half D of the circular disk with center $(0, 0)$ and radius r.

5.46   The half-disk D of Example 1.

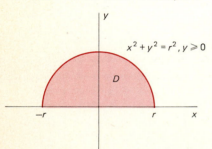

$x^2 + y^2 = r^2, y \geqslant 0$

Solution   By symmetry, the centroid of D lies on the y-axis, so $\bar{x} = 0$. Our semicircular disk lies under $y = \sqrt{r^2 - x^2}$ from $x = -r$ to $x = r$, as shown in Figure 5.46. So Formula (7) gives

$$M_x(D) = \int_{-r}^r \tfrac{1}{2}(\sqrt{r^2 - x^2})^2\,dx$$

$$= \tfrac{1}{2} \int_{-r}^r (r^2 - x^2)\,dx = \left[r^2 x - \frac{x^3}{3} \right]_0^r = \tfrac{2}{3}r^3.$$

Since $A = \frac{1}{2}\pi r^2$, Formula (8) gives

$$\bar{y} = \frac{M_x(D)}{A} = \frac{\frac{2}{3}r^3}{\pi r^2/2} = \frac{4r}{3\pi}.$$

Thus the centroid of D is the point $(0, 4r/3\pi)$. Note that our computed value for $\bar{y}$ has the dimensions of a length (because r is a length), as it should. Any answer having other dimensions would be suspect.

EXAMPLE 2 Find the centroid of the triangle T with vertices $(0, 0)$, $(1, 0)$, and $(0, 1)$.

Solution Figure 5.47 shows the triangle T. Note first that $\bar{x} = \bar{y}$ by symmetry. Formula (7) with $y = f(x) = 1 - x$ gives

$$M_y(T) = \int_0^1 x(1 - x)\, dx = \left[\tfrac{1}{2}x^2 - \tfrac{1}{3}x^3\right]_0^1 = \tfrac{1}{6}.$$

So

$$\bar{x} = \frac{M_y(T)}{A} = \frac{\frac{1}{6}}{\frac{1}{2}} = \frac{1}{3}.$$

Thus the centroid of T is $(\frac{1}{3}, \frac{1}{3})$.

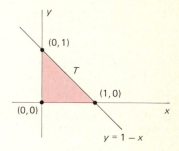

5.47 The triangle T of Example 2.

Additivity of moments can be used to define the moments and centroid of any plane region that is the union of a finite number of nonoverlapping regions shaped like the one shown in Figure 5.45. For example, suppose that $f(x) \geq g(x) \geq 0$ on $[a, b]$, and that R is the region between the graphs $y = f(x)$ and $y = g(x)$ from a to b. If R_f and R_g denote the regions under the graphs of f and g, respectively, then $M_y(R) + M_y(R_g) = M_y(R_f)$ by additivity of moments. Therefore

$$M_y(R) = M_y(R_f) - M_y(R_g)$$

$$= \int_a^b xf(x)\, dx - \int_a^b xg(x)\, dx \qquad \text{(by (6)),}$$

so that

$$M_y(R) = \int_a^b x[f(x) - g(x)]\, dx. \qquad (11)$$

Similarly,

$$M_x(R) = M_x(R_f) - M_x(R_g)$$

$$= \int_a^b \tfrac{1}{2}[f(x)]^2\, dx - \int_a^b \tfrac{1}{2}[g(x)]^2\, dx \qquad \text{(by (7)),}$$

and thus

$$M_x(R) = \int_a^b \tfrac{1}{2}([f(x)]^2 - [g(x)]^2)\, dx. \qquad (12)$$

We then define the centroid of R by means of Equations (8), with

$$A = \int_a^b [f(x) - g(x)]\, dx. \qquad (13)$$

An important theorem relating centroids and volumes of revolution is named for the Greek mathematician who stated it during the third century A.D.

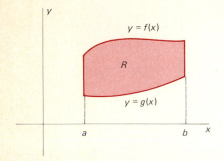

5.48 A region R between the graphs of two functions.

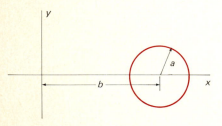

5.49 Rotate the circular disk about the y-axis to produce a torus (Example 4).

First Theorem of Pappus   *Volume of Revolution*

Suppose that a plane region R is revolved about an axis in its plane, generating a solid of revolution with volume V. Assume that the axis does not intersect the interior of R. Then V is the product of the area A of R and the distance traveled by the centroid of R.

Proof   (For the special case of a region like the one shown in Figure 5.48.) This is the region between the two graphs $y = f(x)$ and $y = g(x)$ from a to b, and we shall take the axis of revolution to be the y-axis. Then, in a revolution about the y-axis, the distance traveled by the centroid of R is $d = 2\pi\bar{x}$. By the method of cylindrical shells [see Formula (3) in Section 5-3] the volume of the solid generated is

$$V = \int_a^b 2\pi x[f(x) - g(x)]\, dx$$

$$= 2\pi M_y(R) \qquad\qquad [\text{by Formula (11)}]$$

$$= 2\pi\bar{x} \cdot A \qquad\qquad [\text{by Formula (8)}],$$

and thus $V = d \cdot A$, as desired.

EXAMPLE 3   Find the volume V of the sphere of radius r generated by revolving around the x-axis the semicircle D of Example 1.

Solution   The area of D is $A = \frac{1}{2}\pi r^2$, and we found in Example 1 that $\bar{y} = 4r/3\pi$. Hence Pappus's theorem gives

$$V = 2\pi\bar{y}A = 2\pi \cdot \frac{4r}{3\pi} \cdot \frac{\pi r^2}{2} = \frac{4}{3}\pi r^3.$$

EXAMPLE 4   Consider the circular disk of Figure 5.49, with radius a and center at the point $(b, 0)$; also, $0 < a < b$. Find the volume V of the solid torus generated by revolving this disk around the y-axis.

Solution   The centroid of the circle is its center $(b, 0)$, so $\bar{x} = b$. Hence this centroid is revolved through a distance $d = 2\pi b$. Consequently

$$V = d \cdot A = (2\pi b)(\pi a^2) = 2\pi^2 a^2 b.$$

Note that this result is dimensionally correct.

MOMENTS AND CENTROIDS OF CURVES

Moments and centroids of plane *curves* are defined in a way that is quite analogous to the method for plane regions, so we shall present this topic in less detail. The moments $M_y(C)$ and $M_x(C)$ of the curve C about the y- and x-axes, respectively, are defined to be

$$M_y(C) = \int x\, ds \quad \text{and} \quad M_x(C) = \int y\, ds.$$

The **centroid** $(\bar{x}, \bar{y})$ of C is then defined to be the point with coordinates

$$\bar{x} = \frac{1}{s} M_y(C) \quad \text{and} \quad \bar{y} = \frac{1}{s} M_x(C), \tag{15}$$

where s is the arc length of C.

The meaning of the integrals in (14) is that of the notation of Section 5-4. That is, ds is a symbol to be replaced (before evaluation of the integral) by either

$$ds = \sqrt{1 + \left(\frac{dy}{dx}\right)^2}\, dx \quad \text{or} \quad ds = \sqrt{1 + \left(\frac{dx}{dy}\right)^2}\, dy,$$

depending upon whether C is a smooth arc of the form $y = f(x)$ or one of the form $x = g(y)$. For example, if C is described by $y = f(x)$, $a \leqq x \leqq b$, then

$$M_y(C) = \int_a^b x\sqrt{1 + [f'(x)]^2}\, dx \tag{16}$$

and

$$M_x(C) = \int_a^b f(x)\sqrt{1 + [f'(x)]^2}\, dx. \tag{17}$$

EXAMPLE 5   Let J denote the upper half of the circle (not the *disk*) of radius r, defined by

$$y = \sqrt{r^2 - x^2}, \quad -r \leqq x \leqq r.$$

Find the centroid of J.

Solution   Note first that $M_y(J) = 0$ by symmetry. Now

$$\frac{dy}{dx} = -\frac{x}{\sqrt{r^2 - x^2}},$$

so

$$ds = \sqrt{1 + \frac{x^2}{r^2 - x^2}}\, dx = \frac{r\, dx}{\sqrt{r^2 - x^2}}.$$

Hence the second of Formulas (14) gives

$$M_x(J) = \int_{-r}^{r} \sqrt{r^2 - x^2}\, \frac{r\, dx}{\sqrt{r^2 - x^2}}$$

$$= \int_{-r}^{r} r\, dx = 2r^2.$$

Because $s = \pi r$, the coordinates of the centroid of J are

$$\bar{x} = \frac{M_y(J)}{s} = 0 \quad \text{and} \quad \bar{y} = \frac{M_x(J)}{s} = \frac{2r^2}{\pi r} = \frac{2r}{\pi}.$$

The first theorem of Pappus has an analogue for surface area of revolution.

> ***Second Theorem of Pappus***   *Area of Revolution*
> Let the plane curve C be revolved about an axis in its plane that does not intersect the curve. Then the area A of the surface of revolution generated is equal to the product of the length of C and the distance traveled by the centroid of C.

Proof (For the special case in which C is a smooth arc described by $y = f(x)$, $a \le x \le b$, and the axis of revolution is the y-axis.) The distance traveled by the centroid of C is $d = 2\pi\bar{x}$. By Formula (12) of Section 5-4, the area of the surface of revolution is

$$A = \int_a^b 2\pi x \sqrt{1 + [f'(x)]^2} \, dx$$

$$= 2\pi M_y(C) \qquad \text{(Formula (16))}$$

$$= 2\pi\bar{x}s = d \cdot s \qquad \text{(Formula (15))},$$

as desired.

EXAMPLE 6 Find the surface area A of the sphere of radius r generated by revolving the semicircular arc of Example 5 around the x-axis.

Solution Since we found that $\bar{y} = 2r/\pi$ and we know that $s = \pi r$, the second theorem of Pappus gives

$$A = 2\pi\bar{y}s = 2\pi\left(\frac{2r}{\pi}\right)(\pi r) = 4\pi r^2.$$

EXAMPLE 7 Find the surface area A of the torus of Example 4.

Solution Now we think of revolving the circle (*not* the disk) of radius a centered at the point $(0, b)$. Of course the centroid of the circle is its center $(0, b)$, as can be verified by computations like those in Example 5. Hence the distance traveled by the centroid is $d = 2\pi b$. Since the circumference of the circle is $s = 2\pi a$, the second theorem of Pappus gives

$$A = (2\pi b)(2\pi a) = 4\pi^2 ab.$$

5-6 PROBLEMS

In Problems 1–8, find the centroid of the plane region bounded by the given curves.

1. $y = 4 - x^2$ and $y = 0$.
2. $y = x^2$ and $y = 18 - x^2$.
3. $y = 3x^2$, $y = 0$, and $x = 1$.
4. $x = y^2$, $y = 0$, and $x = 4$ (*above* the x-axis).
5. $y = x$ and $y = 6 - x^2$.
6. $y = x^2$ and $y^2 = x$.
7. $y = x^2$ and $y = x^3$.
8. $y = \sin x$, $0 \le x \le \pi$, and $y = 0$.
9. Find the centroid of the first quadrant of the circular disk $x^2 + y^2 \le r^2$ by direct computation, as in Example 1.
10. Apply the first theorem of Pappus to find the centroid of the first quadrant of the circular disk $x^2 + y^2 \le r^2$, using the facts that $\bar{x} = \bar{y}$ by symmetry and that revolution of this quarter-disk about either coordinate axis gives a solid hemisphere with volume $V = \frac{2}{3}\pi r^3$.

11. Find the centroid of the arc consisting of the first quadrant portion of the circle $x^2 + y^2 = r^2$ by direct computation, as in Example 5.
12. Apply the second theorem of Pappus to find the centroid of the quarter-circular arc of Problem 11. Note that $\bar{x} = \bar{y}$ by symmetry, and that revolution of this arc about either coordinate axis gives a hemisphere with surface area $A = 2\pi r^2$.
13. Show by direct computation that the centroid of the triangle with vertices $(0, 0)$, $(r, 0)$, and $(0, h)$ is the point $(r/3, h/3)$. Verify that this point lies on the line from the vertex $(0, 0)$ to the midpoint of the opposite side of the triangle, and two-thirds of the way from the vertex to the midpoint.
14. Apply the first theorem of Pappus and the result of Problem 13 to verify the formula $V = \frac{1}{3}\pi r^2 h$ for the volume of the cone obtained by revolving the triangle around the y-axis.

15 Apply the second theorem of Pappus to show that the lateral surface area of the cone of Problem 14 is $A = \pi r L$ where $L = (r^2 + h^2)^{1/2}$ is the slant height of the cone.

16 (a) Use additivity of moments to find the centroid of the trapezoid shown in Figure 5.50.

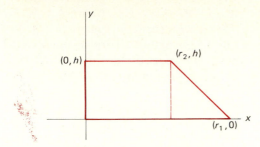

5.50 The trapezoid of Problem 16.

(b) Apply the first theorem of Pappus and the result of Part (a) to show that the volume of the conical frustum generated by revolving the trapezoid around the y-axis is

$$V = \frac{\pi}{3}(r_1^2 + r_1 r_2 + r_2^2)h.$$

17 Apply the second theorem of Pappus to show that the lateral surface area of the conical frustum in Problem 16 is $A = \pi(r_1 + r_2)L$, where

$$L = [(r_1 - r_2)^2 + h^2]^{1/2}$$

is its slant height.

18 Apply the second theorem of Pappus to verify that the curved surface area of a cylinder with height h and base radius r is $A = 2\pi r h$. Explain how this also follows from the result of Problem 17.

19 (a) Use additivity of moments to find the centroid of the plane region shown in Figure 5.51, which consists of a semicircle of radius a sitting atop a rectangle of width $2a$ and height b.

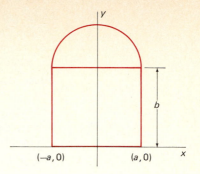

5.51 The plane region of Problem 19.

(b) Then apply the first theorem of Pappus to find the volume generated by rotating this region about the x-axis.

20 (a) Consider the plane region of Figure 5.52, bounded by $x^2 = 2py$, $x = 0$, and $y = h = r^2/2p$. Show that its area is $A = \frac{2}{3}rh$ and that the x-coordinate of its centroid is $\bar{x} = 3r/8$.

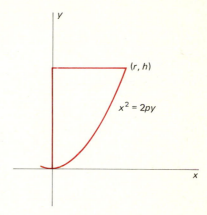

5.52 The region of Problem 20.

(b) Use Pappus's theorem and the result of Part (a) to show that the volume of a paraboloid of revolution with radius r and height h is $V = \frac{1}{2}\pi r^2 h$.

Additional Applications

We take up moments and centroids of solids in Chapter 15. But here we can quite simply compute moments for a special sort of solid, a solid of revolution. Since the centroid of such a solid will lie on its axis of symmetry, we shall need to compute only one coordinate of the centroid.

The moment of a mass particle in space about a *plane* is the product of its mass and its perpendicular distance from the plane. For solids (as for plane regions) we assume additivity of moments and symmetry; if a solid is symmetric with respect to a plane, then its centroid lies in that plane. For example, the centroid C of a solid cylinder lies on the axis of the cylinder and halfway between its two bases; thus the moment of the cylinder about a plane parallel to its bases is the product of its volume (we still assume, for simplicity, that density $\rho = 1$) and the distance from C to the plane.

Now suppose that f is a continuous and nonnegative function defined on the interval $[a, b]$ of the x-axis. Let S be the solid of revolution generated by revolving about the x-axis the region R that lies under the graph $y = f(x)$ for $a \leq x \leq b$. We think of S as a solid with constant density $\rho = 1$. We wish to define the moment of S about the plane that is perpendicular to the x-axis at the origin. Since it is natural to introduce a z-axis in space perpendicular to both the x-axis and the y-axis at the origin, we shall refer to this plane as the yz-plane.

To define the moment M_{yz} of S about the yz-plane, we begin with a regular partition of $[a, b]$ into n equal subintervals each of length Δx. Suppose that x_i^* is the midpoint of the ith subinterval $[x_{i-1}, x_i]$. In forming S by rotation, one "slice" of S is well approximated by the cylinder with axis $[x_{i-1}, x_i]$ and radius $f(x_i^*)$ shown in Figure 5.53. This cylinder has volume $\pi[f(x_i^*)]^2 \, \Delta x$ and centroid $(x_i^*, 0)$. Hence its moment about the yz-plane is $\pi x_i^*[f(x_i^*)]^2 \, \Delta x$.

By additivity of moments, the moment about the yz-plane of the union of these n cylinders (for $i = 1, 2, 3, \ldots, n$) is

$$\sum_{i=1}^{n} \pi x_i^*[f(x_i^*)]^2 \, \Delta x.$$

This sum is a Riemann sum for the function $\pi x[f(x)]^2$ on $[a, b]$. Therefore we define the **moment** M_{yz} of the solid of revolution S about the yz-plane by means of the equation

$$M_{yz} = \int_a^b \pi x[f(x)]^2 \, dx. \tag{1}$$

We then define the x-coordinate of the **centroid** of S to be

$$\bar{x} = \frac{M_{yz}}{V} = \frac{1}{V} \int_a^b \pi x[f(x)]^2 \, dx, \tag{2}$$

where V is the volume of S.

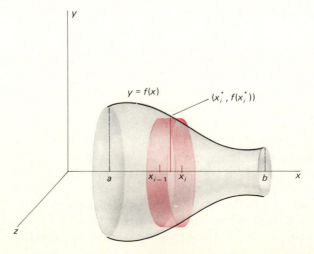

5.53 A solid of revolution is the union of almost-cylindrical slabs whose centroids are easy to locate.

y

$y = f(x)$

$(x_i^*, f(x_i^*))$

a x_{i-1} x_i b x

z

In the case of a solid generated by revolving a region $0 \leq x \leq g(y)$, $c \leq y \leq d$, around the y-axis, the roles of x and y are reversed, and the centroid lies on the y-axis. The **moment** M_{xz} of this solid about the xz-plane is defined to be

$$M_{xz} = \int_c^d \pi y [g(y)]^2 \, dy, \tag{3}$$

and the y-coordinate of its centroid is

$$\bar{y} = \frac{M_{xz}}{V} = \frac{1}{V} \int_c^d \pi y [g(y)]^2 \, dy. \tag{4}$$

EXAMPLE 1 Determine the centroid of the solid paraboloid shown in Figure 5.54, with "height" h, "base" a circle of radius r, and "vertex" at the origin, obtained by revolving the parabola $x^2 = py$ (with $p = r^2/h$) around the y-axis.

Solution By the method of cross sections, the volume of this solid paraboloid is

$$V = \int \pi x^2 \, dy = \int_0^h \pi p y \, dy$$

$$= \left[\tfrac{1}{2} \pi p y^2 \right]_0^h = \tfrac{1}{2} \pi p h^2 = \tfrac{1}{2} \pi r^2 h.$$

Formula (3) with $g(y) = \sqrt{py}$ gives

$$M_{xz} = \int_0^h \pi p y^2 \, dy = \left[\tfrac{1}{3} \pi p y^3 \right]_0^h = \tfrac{1}{3} \pi p h^3 = \tfrac{1}{3} \pi r^2 h^2.$$

Then

$$\bar{y} = \frac{M_{xz}}{V} = \frac{\pi r^2 h^2 / 3}{\pi r^2 h / 2} = \frac{2}{3} h.$$

Thus the centroid of a solid paraboloid lies on its axis two-thirds of the way from its vertex to its base.

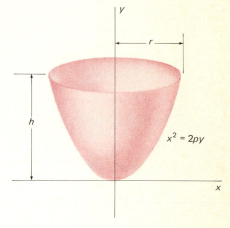

5.54 The parabolic solid of Example 1.

FORCE EXERTED BY A LIQUID

The **pressure** p at depth h in a liquid is the force per unit area exerted by the liquid at that depth, and is given by

$$p = \rho h \tag{5}$$

where ρ is the (weight) density of the liquid. For example, at a depth of 10 feet in water, for which $\rho = 62.4$ lb/ft^3, the pressure is $(62.4)(10) = 624$ pounds per square foot. Hence if a thin flat plate with area 5 square feet is suspended in a horizontal position at a depth of 10 feet in water, then the water exerts a downward force of $(624)(5) = 3120$ pounds on the top face of the plate and an equal upward force on its bottom face.

It is an important fact that at a given depth in a liquid the pressure is the same in all directions. But if a flat plate is submerged in a vertical position in a liquid, then the pressure on the face of the plate is not constant, because by (5) it increases with increasing depth. Consequently the total force exerted on the vertical plate must be computed by integration.

Consider a vertical flat plate submerged in a liquid of density ρ as indicated in Figure 5.55. The surface of the liquid is the line $y = c$, and the

5.55 A thin plate suspended vertically in a liquid.

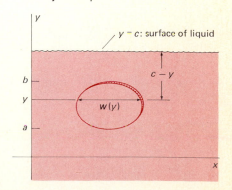

plate lies opposite the interval $a \leqq y \leqq b$. The width of the plate at depth $c - y$ is some function of y, which we denote by $w(y)$.

To compute the total force F exerted by the liquid on a face of this plate, we begin with a regular partition of $[a, b]$ into n equal subintervals of length Δy, and denote by y_i^* the midpoint of the subinterval $[y_{i-1}, y_i]$. The horizontal strip of the plate opposite this ith subinterval is approximated by a rectangle with width $w(y_i^*)$ and height Δy, and its average depth in the liquid is $c - y_i^*$. Hence the force ΔF_i exerted by the liquid on this strip of the plate is given approximately by

$$\Delta F_i \approx \rho(c - y_i^*)w(y_i^*)\,\Delta y, \tag{6}$$

and the total force on the entire plate is given approximately by

$$F = \sum_{i=1}^{n} \Delta F_i \approx \sum_{i=1}^{n} \rho(c - y_i^*)w(y_i^*)\,\Delta y.$$

We obtain the exact value of F by taking the limit of this Riemann sum as $\Delta y \to 0$:

$$F = \int_a^b \rho(c - y)w(y)\,dy. \tag{7}$$

EXAMPLE 2 A cylindrical tank 8 feet in diameter is lying on its side and is half full of oil with density $\rho = 75$ lb/ft^3. Find the force exerted by the oil on one end of the tank.

Solution We locate the y-axis as indicated in Figure 5.56, so that the surface of the oil is at the level $y = 0$. The oil corresponds to the interval $-4 \leqq y \leqq 0$, and from the right triangle in the figure we see that the width of the oil at depth $-y$ is $w(y) = 2\sqrt{16 - y^2}$. Hence Formula (7) gives

$$F = \int_{-4}^{0} 75(-y)(2\sqrt{16 - y^2})\,dy$$

$$= 75\left[\tfrac{2}{3}(16 - y^2)^{3/2}\right]_{-4}^{0} = 3200 \quad \text{lb.}$$

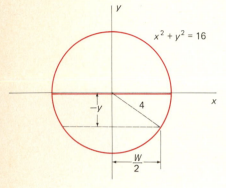

$x^2 + y^2 = 16$

5.56 View of one end of the oil tank.

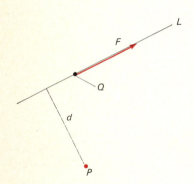

5.57 The moment of the force F about the point P.

5.58 View of a dam.

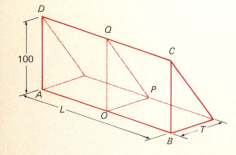

MOMENTS OF FORCES ON A DAM

This subsection is intended mainly for those who have already studied moments of forces in a physics course. Consider a force of magnitude F that acts at the point Q in the direction of the line L through Q shown in Figure 5.57. Recall that in physics the **moment** of the force F about the point P is defined to be

$$M = Fd$$

where d is the perpendicular distance from the point P to the line L.

If a force is distributed over an area, rather than being applied at a single point, then in general its moment about a given point must be computed by integration. We illustrate such a computation by analyzing a dam with the shape indicated in Figure 5.58. The vertical side $ABCD$ faces the water and has height 100 feet and length L feet. The dam is made of concrete of density 3.6ρ, where ρ denotes the density of water. We want to determine what the thickness T of the dam at its base should be in order that the weight W of the dam will balance the force F exerted by water 100 feet deep on the vertical face of the dam.

Figure 5.59 shows the vertical midsection OPQ of the dam. The point Z is the centroid of the dam, which by Problem 13 in Section 5-6 is the point $(T/3, 100/3)$ in the indicated xy-coordinate system. We assume that the weight W of the dam acts as a force directed downwards through Z, and that the force F exerted by the water acts as though it were concentrated along the segment OQ. We want to determine T so that the moments of F and W about the point P will be equal in magnitude; note that F acts clockwise around P while W acts in a counterclockwise direction.

The moment of W about P is

$$(3.6\rho)\left(\frac{100T}{2}\right)(L)\left(\frac{2T}{3}\right) = 120\rho L T^2.$$

To compute the moment of F about P, we begin with a regular partition of $[0, 100]$ into n equal subintervals of length Δy, and denote by y_i^* the midpoint of $[y_{i-1}, y_i]$. The force ΔF_i on the horizontal strip of the dam's face corresponding to $[y_{i-1}, y_i]$ is given by (6) with $c = 100$ and $w(y) = L$, so

$$\Delta F_i \approx \rho(100 - y_i^*)L\,\Delta y.$$

The moment ΔM_i of ΔF_i about P is therefore given approximately by

$$\Delta M_i \approx \rho y_i^*(100 - y_i^*)L\,\Delta y,$$

and the moment of F about P is obtained—again approximately—by adding these moments for $i = 1, 2, 3, \ldots, n$:

$$M \approx \sum_{i=1}^{n} \rho L(100y_i^* - [y_i^*]^2)\,\Delta y.$$

We take the limit as $\Delta y \to 0$, and this gives

$$M = \int_0^{100} \rho L(100y - y^2)\,dy$$

$$= \rho L\left[50y^2 - \tfrac{1}{3}y^3\right]_0^{100} = \tfrac{5}{3}\rho L \times 10^5.$$

5.59 Vertical cross section at the middle of the dam, as seen from the side.

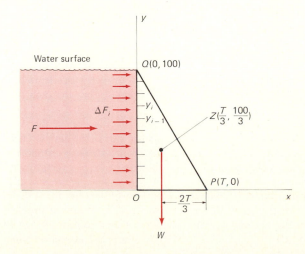

The condition that the moments of F and W balance about P can therefore be expressed in the form of the equation

$$120\rho LT^2 = \tfrac{5}{3}\rho L \times 10^5,$$

and so

$$T = \sqrt{\frac{500{,}000}{360}} \approx 37.27 \text{ feet.}$$

5-7 PROBLEMS

In each of Problems 1–8, find the centroid of the solid generated by revolving the region bounded by the given curves around the indicated axis.

1 $y^2 = 2x$, $x = 2$; the x-axis.
2 $y^2 = 3x$, $x = 0$, $y = 3$; the y-axis.
3 $y = 2x - x^2$, $y = 0$; the x-axis.
4 $x = 3y - y^2$, $x = 0$; the y-axis.
5 $y = x^2$, $y = 3x$; the x-axis.
6 $y = x^2$, $x = y^2$; the y-axis.
7 $x^2 - y^2 = 9$, $x = 5$; the x-axis.

8 $x = \dfrac{1}{y^2}$, $x = 0$, $y = 1$, $y = 2$; the y-axis.

9 Show that the centroid of a right circular cone lies on the axis of the cone and three-fourths of the way from the vertex to the base.
10 Show that the centroid of a solid hemisphere lies on its axis of symmetry, with its distance from the base equal to three-eighths the radius of the hemisphere.
11 Generalize the result of Problem 10 by finding the centroid of the upper half of the solid ellipsoid obtained by revolving the ellipse $(x/a)^2 + (y/b)^2 = 1$ around the y-axis.
12 Consider the first-quadrant region bounded by the coordinate axes and the parabola $y = h^2 - x^2$. If this region is revolved around the x-axis, show that the volume of the resulting solid is $V = \tfrac{8}{15}\pi h^5$ and that the x-coordinate of its centroid is $\bar{x} = \tfrac{5}{16}h$.
13 Suppose that n is a positive even integer; consider the solid that is generated by revolving the region bounded by the curve $y = x^n$ and the line $y = h$ around the y-axis. Show that the y-coordinate of its centroid is $\bar{y} = (n + 2)h/(2n + 2)$.
14 A water tank is in the shape of a solid of revolution, so that its horizontal cross sections are circular. The work W required to fill this tank with water pumped up from the ground is given by Formula (8) in Section 5-5. Show that $W = w\bar{y}$ where w is the weight of the water in the full tank and $\bar{y}$ is the y-coordinate of the centroid of that mass of water.

15 A water trough 10 feet long has a square cross section that is 2 feet wide. If the trough is full of water ($\rho = 62.4$ lb/ft³), find the force exerted by the water on one end of the trough.
16 Repeat Problem 15, where the cross section of the trough is an equilateral triangle with edge 3 feet.
17 Repeat Problem 15, where the cross section of the trough is a trapezoid 3 feet high, 2 feet wide at the bottom, and 4 feet wide at the top.
18 Find the force on one end of the cylindrical tank of Example 2 if it is filled with oil weighing 50 lb/ft³. Remember that

$$\int_0^a \sqrt{a^2 - y^2}\, dy = \frac{\pi a^2}{4}$$

because the integral represents the area of a quarter-circle of radius a.

In each of Problems 19–22, a gate in the vertical face of a dam is described. Find the total force of water on this gate if its top is 10 feet beneath the surface of the water.

19 A square of edge 5 feet with its top parallel to the water surface.
20 A circle with radius 3 feet.
21 An isosceles triangle 5 feet high and 8 feet wide at the top.
22 A semicircle with radius 4 feet with its diameter also its top edge and parallel to the water surface.
23 Suppose that the dam of Figure 5.58 is $L = 200$ feet long and $T = 30$ feet thick at its base. Find the force of water on the dam if the water is 100 feet deep and the *slanted* side of the dam faces the water.
24 Deduce from Formula (7) that the force exerted by fluid on a face of a vertically submerged flat plate with area A is $F = \rho h A$, where h is the depth (beneath the liquid's surface) of the centroid of the plate.
25 Apply the result of Problem 24 to find the force on the gate of (a) Problem 19; (b) Problem 20; (c) Problem 22 (see Example 1 in Section 5-6).

Appendix on Approximations and Riemann Sums

Several times in this chapter, our attempt to compute some quantity Q has led to the following situation. Beginning with a regular partition of an appropriate interval $[a, b]$ into n equal subintervals of length Δx, we have found an approximation A_n to Q of the form

$$A_n = \sum_{i=1}^{n} g(u_i)h(v_i)\,\Delta x \tag{1}$$

where u_i and v_i are two (generally different) points of the ith subinterval $[x_{i-1}, x_i]$. For example, in our discussion of surface area of revolution that precedes Formula (8) in Section 5-4, we found the approximation

$$\sum_{i=1}^{n} 2\pi f(u_i)[1 + (f'(v_i))^2]^{1/2}\,\Delta x \tag{2}$$

to the area of the surface generated by revolving the curve $y = f(x), a \leq x \leq b$, around the x-axis. (In Section 5-4, we wrote x_i^{**} for u_i and x_i^{*} for v_i.) Note that the expression in (2) is the same as the right-hand side of Equation (1); take $g(x) = 2\pi f(x)$ and $h(x) = [1 + (f'(x))^2]^{1/2}$.

In such a situation we observe that, if u_i and v_i were the *same* point x_i^{*} of $[x_{i-1}, x_i]$ for each i $(i = 1, 2, 3, \ldots, n)$, then Approximation (1) would be a Riemann sum for the function $g(x)h(x)$ on $[a, b]$. This leads us to suspect that

$$\lim_{\Delta x \to 0} \sum_{i=1}^{n} g(u_i)h(v_i)\,\Delta x = \int_a^b g(x)h(x)\,dx. \tag{3}$$

In Section 5-4, we assumed the truth of Equation (3) and concluded from Approximation (2) that surface area of revolution ought to be defined to be

$$A = \lim_{\Delta x \to 0} \sum_{i=1}^{n} 2\pi f(u_i)[1 + (f'(v_i))^2]^{1/2}\,\Delta x$$

$$= \int_a^b 2\pi f(x)[1 + (f'(x))^2]^{1/2}\,dx.$$

The following theorem guarantees that Equation (3) holds under only mild restrictions on the functions g and h.

Theorem 1 *A Generalization of Riemann Sums*

Suppose that h and the derivative g' are continuous on $[a, b]$. Then

$$\lim_{\Delta x \to 0} \sum_{i=1}^{n} g(u_i)h(v_i)\,\Delta x = \int_a^b g(x)h(x)\,dx \tag{3}$$

where u_i and v_i are arbitrary points of the ith subinterval of a regular partition of $[a, b]$ into n equal subintervals each of length Δx.

Proof Let M_1 and M_2 denote the maximum values on $[a, b]$ of $|g'(x)|$ and $|h(x)|$, respectively. Note that

$$\sum_{i=1}^{n} g(u_i)h(v_i)\,\Delta x = R_n + S_n$$

where

$$R_n = \sum_{i=1}^{n} g(v_i)h(v_i)\,\Delta x$$

is a Riemann sum approaching $\int_a^b g(x)h(x)\,dx$ as $\Delta x \to 0$, and

$$S_n = \sum_{i=1}^{n} [g(u_i) - g(v_i)]h(v_i)\,\Delta x.$$

To prove (3) it is sufficient to show that $S_n \to 0$ as $\Delta x \to 0$. The Mean Value Theorem gives

$$\begin{aligned}
|g(u_i) - g(v_i)| &= |g'(\bar{x}_i)| \cdot |u_i - v_i| \qquad (\bar{x}_i \text{ in } (u_i, v_i)) \\
&\leq M_1\,\Delta x
\end{aligned}$$

because u_i and v_i are both points of the interval $[x_{i-1}, x_i]$ of length Δx. Then

$$\begin{aligned}
|S_n| &\leq \sum_{i=1}^{n} |g(u_i) - g(v_i)| \cdot |h(v_i)|\,\Delta x \\
&\leq \sum_{i=1}^{n} M_1\,\Delta x \cdot M_2\,\Delta x \\
&= (M_1 M_2\,\Delta x) \sum_{i=1}^{n} \Delta x = M_1 M_2 (b - a)\,\Delta x,
\end{aligned}$$

from which it follows that $S_n \to 0$ as $\Delta x \to 0$, as desired.

As an application of Theorem 1, let us give a rigorous derivation of Formula (2) in Section 5-3,

$$V = \int_a^b 2\pi x f(x)\,dx, \qquad (4)$$

for the volume of the solid generated by revolving the region lying under $y = f(x)$, $a \leq x \leq b$, around the y-axis. Starting with the usual regular partition of $[a, b]$, let $f(x_i^{\flat})$ and $f(x_i^{\#})$ denote the minimum and maximum values of f on the ith subinterval $[x_{i-1}, x_i]$. Denote by x_i^* the midpoint of this subinterval. From Figure 5.60, we see that the part of our solid generated by revolving the region under $y = f(x)$, $x_{i-1} \leq x \leq x_i$, contains a cylindrical shell with average radius x_i^*, thickness Δx, and height $f(x_i^{\flat})$, and is contained in another cylindrical shell with the same average radius and thickness, but

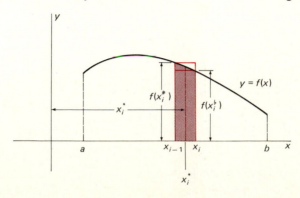

5.60 A careful estimate of the volume of a solid of revolution around the y-axis.

with height $f(x_i^\#)$. Hence the volume ΔV_i of this part of the solid satisfies the inequalities

$$2\pi x_i^* f(x_i^\flat)\, \Delta x \leqq \Delta V_i \leqq 2\pi x_i^* f(x_i^\#)\, \Delta x.$$

We add these inequalities for $i = 1, 2, 3, \ldots, n$, and find that

$$\sum_{i=1}^{n} 2\pi x_i^* f(x_i^\flat)\, \Delta x \leqq V \leqq \sum_{i=1}^{n} 2\pi x_i^* f(x_i^\#)\, \Delta x.$$

Since Theorem 1 implies that each of the last two sums approaches $\int_a^b 2\pi x f(x)\, dx$, the squeeze law of limits now implies Formula (4).

We shall occasionally need a generalization of Theorem 1 that involves the notion of a continuous function $F(x, y)$ of the two variables x and y. We say that F is *continuous* at the point (x_0, y_0) provided that the value $F(x, y)$ can be made arbitrarily close to $F(x_0, y_0)$ merely by choosing the point (x, y) sufficiently close to (x_0, y_0). We shall discuss continuity of functions of two variables in Chapter 14. Here it will suffice to accept the following facts: If $g(x)$ and $h(y)$ are continuous functions of the single variables x and y, respectively, then simple combinations such as

$$g(x) \pm h(y), \qquad g(x)h(y), \quad \text{and} \quad [(g(x))^2 + (h(y))^2]^{1/2}$$

are continuous functions of the two variables x and y.

Now consider a regular partition of $[a, b]$ into n equal subintervals of length Δx, and let u_i and v_i denote arbitrary points of the ith subinterval $[x_{i-1}, x_i]$. The following theorem—we omit the proof—tells us how to find the limit as $\Delta x \to 0$ of a sum of the form $\sum_{i=1}^{n} F(u_i, v_i)\, \Delta x$.

Theorem 2 *A Further Generalization*

Let $F(x, y)$ be continuous for x and y both in the interval $[a, b]$. Then, with the notation above,

$$\lim_{\Delta x \to 0} \sum_{i=1}^{n} F(u_i, v_i)\, \Delta x = \int_a^b F(x, x)\, dx. \tag{5}$$

Note that Theorem 1 is the special case $F(x, y) = g(x)h(y)$ of Theorem 2. Observe also that the integrand $F(x, x)$ on the right in Equation (5) is merely an ordinary function of the (single) variable x. As a formal matter, the integral corresponding to the sum in (5) is obtained by replacing the summation symbol with an integral symbol, changing both u_i and v_i to x, and replacing Δx by dx.

For instance, if the interval $[a, b]$ is $[0, 4]$, then

$$\lim_{\Delta x \to 0} \sum_{i=1}^{n} \sqrt{9u_i^2 + v_i^4}\, \Delta x = \int_0^4 \sqrt{9x^2 + x^4}\, dx$$

$$= \int_0^4 x\sqrt{9 + x^2}\, dx = \left[\tfrac{1}{3}(9 + x^2)^{3/2} \right]_0^4$$

$$= \tfrac{1}{3}\left[(25)^{3/2} - (9)^{3/2} \right] = \tfrac{98}{3}.$$

In each of Problems 1–7, u_i and v_i are arbitrary points of the ith subinterval of a regular partition of $[a, b]$ into n equal subintervals of length Δx each. Express the given limit as an integral from a to b, and then compute the value of this integral.

1 $\lim\limits_{\Delta x \to 0} \sum\limits_{i=1}^{n} u_i v_i \, \Delta x; \quad a = 0, b = 1.$

2 $\lim\limits_{\Delta x \to 0} \sum\limits_{j=1}^{n} (3u_j + 5v_j) \, \Delta x; \quad a = -1, b = 3.$

3 $\lim\limits_{\Delta x \to 0} \sum\limits_{i=1}^{n} u_i \sqrt{4 - v_i^2} \, \Delta x; \quad a = 0, b = 2.$

4 $\lim\limits_{\Delta x \to 0} \sum\limits_{i=1}^{n} \dfrac{u_i \, \Delta x}{\sqrt{16 + v_i^2}}; \quad a = 0, b = 3.$

5 $\lim\limits_{\Delta x \to 0} \sum\limits_{i=1}^{n} \sin u_i \cos v_i \, \Delta x; \quad a = 0, b = \dfrac{\pi}{2}.$

6 $\lim\limits_{\Delta x \to 0} \sum\limits_{i=1}^{n} \sqrt{\sin^2 u_i + \cos^2 v_i} \, \Delta x; \quad a = 0, b = \pi.$

7 $\lim\limits_{\Delta x \to 0} \sum\limits_{k=1}^{n} \sqrt{u_k^4 + v_k^7} \, \Delta x; \quad a = 0, b = 2.$

8 Explain how Theorem 1 applies to show that Formula (8) in Section 5-5 follows from the discussion preceding it in that section.

9 Use Theorem 1 to derive Formula (10) in Section 5-4.

CHAPTER 5 REVIEW: Definitions, Concepts, Results

Use the list below as a guide to ideas that you may need to review.

1 The general method of setting up an integral formula for a quantity by approximating it and then recognizing the approximation as a Riemann sum corresponding to the desired integral

2 Net distance traveled as the definite integral of velocity

3 Total distance as the integral of the absolute value of the velocity

4 The method of cross sections for computing volumes

5 The volume of a solid of revolution (around either the x-axis or the y-axis)

6 The method of cylindrical shells for computing volumes

7 The arc length of a smooth arc described either in the form $y = f(x)$, $a \leqq x \leqq b$, or in the form $x = g(y)$, $c \leqq y \leqq d$

8 The area of a frustum of a cone

9 The area of the surface of revolution generated by revolving a smooth arc, given in either of the forms $y = f(x)$ or $x = g(y)$, about either the x-axis or the y-axis

10 The work done by a force function in moving a particle along a straight line segment

11 Hooke's law and the work done in stretching or compressing an elastic spring

12 Work done against the force of gravity

13 Work done in filling a tank, or in pumping the liquid in a tank to another level

14 The center of mass of a finite system of particles

15 The symmetry principle (for centroids) and additivity of moments

16 The centroid of a plane region: definition and formulas

17 The moments of a plane region about the two coordinate axes

18 The centroid and moments of a smooth arc or curve

19 The first theorem of Pappus: volume of revolution

20 The second theorem of Pappus: surface area of revolution

21 The centroid of a solid of revolution

22 The force exerted by liquid on a face of a submerged plate

23 Recognition of integrals corresponding to limits of Riemann sums (Section 5-8)

MISCELLANEOUS PROBLEMS

In Problems 1–3, find both the net distance and the total distance traveled between times $t = a$ and $t = b$ by a particle moving along a line with the given velocity function $v = f(t)$.

1 $v = t^2 - t - 2; \quad a = 0, b = 3.$

2 $v = |t^2 - 4|; \quad a = 1, b = 4.$

3 $v = \pi \sin \dfrac{\pi}{2} (2t - 1); \quad a = 0, b = \dfrac{3}{2}.$

In Problems 4–8, u_i and v_i denote points of the ith subinterval of a regular partition of the indicated interval $[a, b]$ into n equal subintervals of length Δx. Evaluate the given limit by computing the value of an appropriate related integral.

4 $\lim\limits_{\Delta x \to 0} \sum\limits_{i=1}^{n} (3u_i^2 + 2) \, \Delta x; \quad a = -1, b = 2.$

5 $\lim\limits_{\Delta x \to 0} \sum\limits_{j=1}^{n} \sqrt{2v_j + 1} \, \Delta x; \quad a = 0, b = 12.$

6 $\displaystyle\lim_{\Delta x \to 0} \sum_{i=1}^{n} u_i \sqrt{v_i}\, \Delta x; \quad a = 0, b = 5.$

7 $\displaystyle\lim_{\Delta x \to 0} \sum_{i=1}^{n} \frac{u_i^3 - 1}{v_i^2}\, \Delta x; \quad a = 1, b = 2.$

8 $\displaystyle\lim_{\Delta x \to 0} \sum_{i=1}^{n} \sqrt{u_i + v_i^{5/2}}\, \Delta x; \quad a = 0, b = 1.$

9 Suppose that rainfall begins at time $t = 0$ and that the rate after t hours is $(t + 6)/12$ inches per hour. How many inches of rain fall during the first 12 hours?

10 The base of a certain solid is the region in the first quadrant that is bounded by the curves $y = x^3$ and $y = 2x - x^2$. Find its volume, if each cross section perpendicular to the x-axis is a square with one edge in the base of the solid.

11 Find the volume of the solid generated by revolving the first-quadrant region of Problem 10 about the x-axis.

12 Find the volume of the solid generated by revolving the region bounded by $y = 2x^4$ and $y = x^2 + 1$ around (a) the x-axis; (b) the y-axis.

13 A wire is made of copper (density 8.5 gm/cm³) and is shaped like a helix that spirals around the x-axis from $x = 0$ to $x = 20$ centimeters. Each cross section of this wire perpendicular to the x-axis is a circular disk of radius 0.25 cm. What is the total mass of the wire?

14 Derive the formula $V = \frac{1}{3}\pi h(r_1^2 + r_1 r_2 + r_2^2)$ for the volume of a frustum of a cone with height h and base radii r_1 and r_2.

15 Suppose that the point P lies on a line perpendicular to the xy-plane at the origin O, with $|OP| = h$. Consider the "elliptical cone" consisting of all points on line segments from P to points on and within the ellipse $(x/a)^2 + (y/b)^2 = 1$. Show that the volume of this elliptical cone is $V = \frac{1}{3}\pi abh$.

16 Figure 5.61 shows the region R bounded by the ellipse $(x/a)^2 + (y/b)^2 = 1$ and the line $x = a - h$, where $0 < h < a$. Revolution of R around the x-axis generates a "segment of an ellipsoid" with radius r, height h, and volume V. Show that

$$r^2 = \frac{b^2(2ah - h^2)}{a^2} \quad \text{and that} \quad V = \frac{1}{3}\pi r^2 h\, \frac{3a - h}{2a - h}.$$

5.61 A segment of an ellipsoid.

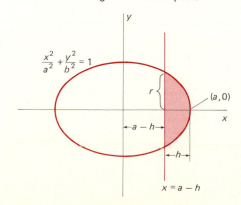

17 Figure 5.62 shows the region R bounded by the hyperbola $(x/a)^2 - (y/b)^2 = 1$ and the line $x = a + h$, where $h > 0$. Revolution of R around the x-axis generates a "segment of a hyperboloid" with radius r, height h, and volume V. Show that

$$r^2 = \frac{b^2(2ah + h^2)}{a^2} \quad \text{and that} \quad V = \frac{1}{3}\pi r^2 h\, \frac{3a + h}{2a + h}.$$

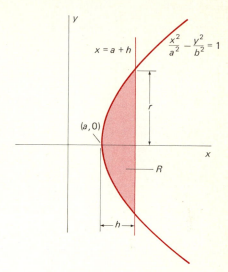

5.62 The region R of Problem 17.

In Problems 18–20, the function $f(x)$ is nonnegative and continuous for $x \geq 1$. When the region lying under $y = f(x)$ from $x = 1$ to $x = t$ is revolved around the indicated axis, the volume of the resulting solid is $V(t)$. Find the function $f(x)$.

18 $V(t) = \pi\left(1 - \dfrac{1}{t}\right);$ the x-axis.

19 $V(t) = \frac{1}{6}\pi[(1 + 3t)^2 - 16];$ the x-axis.

20 $V(t) = \frac{2}{9}\pi[(1 + 3t^2)^{3/2} - 8];$ the y-axis.

21 Use the integral formula of Problem 20 in Section 5-3 to find the volume of the solid generated by revolving the first-quadrant region bounded by $y = x$ and $y = \sin(\pi x/2)$ around the y-axis.

22 Use the method of cylindrical shells to find the volume of the solid generated by revolving the region bounded by $y = x^2$ and $y = x + 2$ around the line $x = -2$.

23 Find the length of the curve $y = \frac{1}{3}x^{3/2} - x^{1/2}$ from $x = 1$ to $x = 4$.

24 Find the area of the surface generated by revolving the curve of Problem 23 around (a) the x-axis; (b) the y-axis.

25 Find the length of the curve $x = \frac{3}{8}(y^{4/3} - 2y^{2/3})$ from $y = 1$ to $y = 8$.

26 Find the area of the surface generated by revolving the curve of Problem 25 around (a) the x-axis; (b) the y-axis.

27 Find the area of the surface generated by revolving the curve of Problem 23 around the line $x = 1$.

28 If $-r < a < b < r$, then a "spherical zone" of "height" $h = b - a$ is generated by revolving the circular arc

$$y = \sqrt{r^2 - x^2}, \qquad a \leqq x \leqq b$$

around the x-axis. Show that the area of this spherical zone is $A = 2\pi rh$, the same as that of a cylinder with radius r and height h.

29 Apply the result of Problem 28 to show that the surface area of a sphere of radius r is $A = 4\pi r^2$.

Find the centroids of the curves in Problems 30–33.

30 $y = \dfrac{x^5}{5} + \dfrac{1}{12x^3}$, $1 \leqq x \leqq 2$.

31 $x = \dfrac{y^4}{8} + \dfrac{1}{4y^2}$, $1 \leqq y \leqq 2$.

32 The curve of Problem 23.

33 The curve of Problem 25.

34 Find the centroid of the plane region in the first quadrant that is bounded by the curves $y = x^3$ and $y = 2x - x^2$.

35 Find the centroid of the plane region bounded by the curves $x = 2y^4$ and $x = y^2 + 1$.

36 Let T be the triangle with vertices $(0, 0)$, (a, b), and $(c, 0)$. Show that the centroid of the triangle T is the point of intersection of its medians.

37 Use the first theorem of Pappus to find the y-coordinate of the centroid of the upper half of the ellipse $(x/a)^2 + (y/b)^2 = 1$. Employ the facts that the area of this semiellipse is $A = \pi ab/2$, while the volume of the ellipsoid it generates when rotated about the x-axis is $V = \frac{4}{3}\pi ab^2$.

38 (a) Use the first theorem of Pappus to find the centroid of the first-quadrant part of the annular ring with boundary circles $x^2 + y^2 = a^2$ and $x^2 + y^2 = b^2$, with $0 < a < b$.
(b) Show that the limiting position of this centroid as $b \to a$ is the centroid of a quarter-circular arc, as found in Problem 12 of Section 5-6.

39 Let T be the triangle in the first quadrant with vertex $(0, 0)$ and opposite side L of length w joining the other two vertices (a, b) and (c, d), where $a > c > 0$ and $d > b > 0$. Let A denote the area of T, $\bar{y}$ the y-coordinate of the centroid of T, p the perpendicular distance from $(0, 0)$ to L, V the volume generated by revolving T around the x-axis, and S the surface area generated by revolving L around the x-axis. Then derive the formulas that follow in the order listed.

(a) $A = \frac{1}{2}(ad - bc)$;

(b) $\bar{y} = \frac{1}{3}(b + d)$;

(c) $V = \frac{1}{3}\pi(b + d)(ad - bc)$;

(d) $p = \dfrac{ad - bc}{w}$;

(e) $S = \pi(b + d)w$;

(f) $V = \frac{1}{3}pS$.

40 Show that the x-coordinate of the centroid of the segment of an ellipsoid described in Problem 16 is

$$\bar{x} = \frac{3(2a - h)^2}{4(3a - h)}.$$

41 Show that the x-coordinate of the centroid of the segment of a hyperboloid described in Problem 17 is

$$\bar{x} = \frac{3(2a + h)^2}{4(3a + h)}.$$

42 Find the y-coordinate of the centroid of the solid generated by revolving around the y-axis the region bounded by $y = x^2$ and $y = x$, in two ways:
(a) By integration.
(b) By additivity of moments, taking as known the centroids of a cone (Problem 9 in Section 5-7) and a solid paraboloid (Example 1 in Section 5-7).

43 Find the natural length L of a spring if five times as much work is required to stretch it from a length of 2 feet to a length of 5 feet, as is required to stretch it from a length of 2 feet to a length of 3 feet.

44 A steel beam weighing 1000 pounds hangs from a 50-foot cable; the cable weighs 5 pounds per linear foot. How much work is done in winding in 25 feet of the cable with a windlass?

45 A spherical tank of radius R (in feet) is initially full of oil weighing ρ pounds per cubic foot. Find the total work done in pumping all the oil from the sphere to a height of $2R$ above the top of the tank.

46 How much work is done by a colony of ants in building a conical anthill with height and diameter both 1 foot, using sand initially at ground level and with a density of 150 lb/ft³?

47 Below the surface of the earth the gravitational attraction is directly proportional to the distance from the earth's center. Suppose that a straight cylindrical hole with a radius of 1 foot is dug from the surface of the earth (radius 3960 miles) to its center. Assume that the earth has a uniform density of 350 lb/ft³. How much work, in ft-lb, is done in lifting a 1 pound weight from the bottom of this hole to its top?

48 How much work is done in digging the hole of Problem 47; that is, in lifting the material it at first contained to the earth's surface?

49 Suppose that a dam is shaped like a trapezoid with height 100 feet, 300 feet long at the top and 200 feet long at the bottom. When the water level behind the dam is level with its top, what is the total force that the water exerts on the dam?

50 In Problem 49, find the moment about the bottom of the dam of the force of water pressure.

51 Suppose that a dam has the same top and bottom lengths as in Problem 49 and the same vertical height of 100 feet, but that its face toward the water is slanted at an angle of $30°$ from the vertical. Now what is the total force of water pressure on the dam?

Exponential and Logarithmic Functions

6

Introduction

This introduction is a brief overview of the exponential and logarithmic functions, partly from the viewpoint of precalculus mathematics. It may simply be read (rather than studied in detail) as motivation for our systematic treatment of exponential and logarithmic functions, which begins in Section 6-2.

In elementary algebra a rational power of the positive real number a is defined in terms of integral roots and powers. There we learn that if p and q are integers (with $q > 0$), then $a^{p/q} = \sqrt[q]{a^p}$. The following **laws of exponents** are then established for all *rational* exponents r and s:

$$a^{r+s} = a^r a^s, \qquad (a^r)^s = a^{rs},$$

$$a^{-r} = \frac{1}{a^r}, \qquad (ab)^r = a^r b^r. \tag{1}$$

In applications we often need powers with irrational exponents as well as with rational ones. For example, consider a bacteria population $P(t)$ that increases by the same factor in any two time intervals of the same length and, in particular, doubles every hour. Suppose that the initial population is $P(0) = 1$ million. In 3 hours, P will increase by a factor of $2 \cdot 2 \cdot 2 = 2^3$, so that $P(3) = 2^3$ (million). If k is the factor by which P increases in $\frac{1}{3}$ hour, then in 1 hour P will increase by a factor of $k \cdot k \cdot k = k^3 = 2$, so $k = 2^{1/3}$ and $P(\frac{1}{3}) = 2^{1/3}$ (million). More generally, if p and q are positive integers, then $P(1/q) = 2^{1/q}$; in p/q hours, P will increase p times by a factor of $2^{1/q}$ each time, so $P(p/q) = (2^{1/q})^p = 2^{p/q}$. Thus $P(t) = 2^t$ if t is rational. But since time is not restricted to rational values alone, we surely ought to conclude that $P(t) = 2^t$ for *all* $t > 0$.

But what is meant by a number like $2^{\sqrt{2}}$ or 2^π? To find 2^π, we might proceed numerically, starting with the fact that π lies between the rational numbers 3.14159 and 3.14160. A calculator gives the (rounded) values

$$2^{3.14159} = 8.82496 \quad \text{and} \quad 2^{3.14160} = 8.82502.$$

From the correct assumption that 2^t is an increasing function of t, it follows that

$$8.82496 < 2^\pi < 8.82502,$$

so we conclude that $2^\pi \approx 8.8250$, rounded to four decimal places. (A calculator actually approximates 2^π by computing instead the number $2^{3.141592654}$; note that the latter exponent is rational.)

Let us investigate the function a^x in the case $a > 1$. We first note that $a^r > 1$ if r is a positive rational number. If $r < s$, then the laws of exponents give $a^r < a^r a^{s-r} = a^s$, so a^r is an *increasing* function of the rational exponent r. If we plot points on the graph of $y = a^x$ for rational values of x using a typical fixed value of a such as $a = 2$, we obtain a graph like the one in Figure 6.1. This graph is shown with a dotted curve to suggest that it is densely filled with "holes" corresponding to irrational values of x, for which we have not yet given a precise definition of a^x.

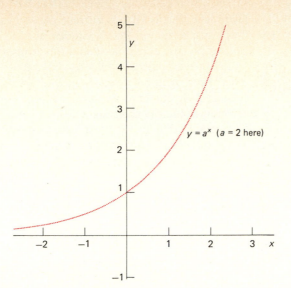

6.1 Graph of $y = a^x$ has "holes" if only rational values of x are used.

In Section 6-4 we shall show that the "holes" in this graph can be filled to obtain the graph of an increasing and continuous function f such that $f(r) = a^r$ for every rational number r. We therefore write $f(x) = a^x$ for all x, and call a^x the **exponential function with base a.** We shall verify that this function satisfies the laws of exponents (1), such as $a^{x+y} = a^x a^y$, for *all* values of the exponents x and y.

In elementary algebra the common logarithm function is introduced as the inverse function of the exponential function 10^x with base 10. That is, $y = \log_{10} x$ is the power to which 10 must be raised to obtain x, so that

$$y = \log_{10} x \quad \text{if and only if} \quad 10^y = x.$$

Similarly, the **base a logarithm function $\log_a x$** is the inverse function of the exponential function a^x with $a > 1$. That is,

$$y = \log_a x \quad \text{if and only if} \quad a^y = x. \tag{2}$$

It follows from the reflection property of inverse functions (Section 1-7) that the graph of $y = \log_a x$ is the reflection in the line $y = x$ of the graph of $y = a^x$, and therefore looks as shown in Figure 6.2. Since $a^0 = 1$, it follows that $\log_a 1 = 0$, so the intercepts in the figure are independent of the choice of a, at least if $a > 1$. Note also that $\log_a x$ is defined *only* for $x > 0$.

The inverse function relationship between $\log_a x$ and a^x can be used to deduce, from the laws of exponents (1), the following **laws of logarithms:**

$$\log_a xy = \log_a x + \log_a y,$$

$$\log_a \frac{1}{x} = -\log_a x,$$

$$\log_a \frac{x}{y} = \log_a x - \log_a y, \tag{3}$$

$$\log_a x^y = y \log_a x.$$

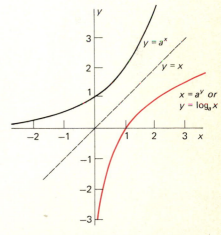

6.2 The graph of $x = a^y$ is the graph of the inverse function $\log_a x$ of the exponential function a^x—the case $a > 1$ is shown here.

We shall verify these laws of logarithms in Section 6-2.

There is a reason for our current pause in the development of integral calculus, more profound than a sudden need to have a precise definition of 2^{π}. The logarithmic and exponential functions are so important to calculus that further development would be difficult without some examination of their properties. We shall develop these properties systematically in Sections 6-2 through 6-4, but here we motivate that development by an informal investigation of their derivatives.

To compute the derivative of $\log_a x$, we begin with the definition of the derivative and then use the laws of logarithms to simplify. This gives us

$$D_x \log_a x = \lim_{h \to 0} \frac{\log_a(x + h) - \log_a x}{h}$$

$$= \lim_{h \to 0} \frac{1}{h} \log_a \left(1 + \frac{h}{x} \right)$$

$$= \frac{1}{x} \lim_{h \to 0} \frac{x}{h} \log_a \left(1 + \frac{h}{x} \right)$$

$$= \frac{1}{x} \lim_{k \to 0} \frac{1}{k} \log_a(1 + k) \qquad \left(k = \frac{h}{x} \right)$$

$$= \frac{1}{x} \lim_{k \to 0} \left[\log_a(1 + k)^{1/k} \right]$$

$$= \frac{1}{x} \log_a \left[\lim_{k \to 0} (1 + k)^{1/k} \right].$$

The last equality is valid provided that the logarithm function is continuous. If we also assume that the limit

$$e = \lim_{k \to 0} (1 + k)^{1/k} \tag{4}$$

exists, then our computations above show that

$$D_x \log_a x = \frac{\log_a e}{x}. \tag{5}$$

k	$(1 + k)^{1/k}$
0.1	2.594
0.01	2.705
0.001	2.717
0.0001	2.718
0.00001	2.718

6.3 Numerical estimate of the number e.

Let us investigate numerically the limit in (4). With a calculator we obtain the values in the table of Figure 6.3. The evidence suggests (although it doesn't prove) that $e \approx 2.718$ to 3 decimal places. This number e is one of the most important special numbers in mathematics, and its value to 15 places is

$$e \approx 2.718281828459045.$$

Since $\log_{10} e \approx 0.43429$, Formula (5) gives

$$D_x \log_{10} x \approx \frac{0.43429}{x}.$$

But $\log_e e = 1$ ("the power to which e must be raised to obtain e is 1"), so Formula (5) is simplest when we take $a = e$:

$$D_x \log_e x = \frac{1}{x}. \tag{6}$$

CHAP. 6: **Exponential and Logarithmic Functions**

For this reason, logarithms with base e are the most natural ones to use in calculus. Indeed, $\log_e x$ is called the *natural logarithm* of x. The natural logarithm function is denoted by ln, so that

$$\ln x = \log_e x. \tag{7}$$

Thus Formula (6) may also be written in the form

$$D \ln x = \frac{1}{x}. \tag{8}$$

The *natural exponential function* is e^x. Since e^x and $\ln x$ are inverse functions, we have

$$\ln(e^x) = x \tag{9}$$

for all x. If we assume that e^x is differentiable, we may differentiate both sides of Equation (9) using the chain rule on the left-hand side. This gives

$$\ln u = x \qquad \text{(with } u = e^x\text{)},$$

$$\frac{1}{u}\frac{du}{dx} = 1 \qquad \text{(by (8))},$$

$$\frac{du}{dx} = u,$$

and so

$$De^x = e^x. \tag{10}$$

Thus the natural exponential function e^x is its own derivative! In Sections 6-2 and 6-3, we shall substantiate the assumptions made here in deriving the basic differentiation formulas in (8) and (10).

In this chapter we shall see that the functions $\ln x$ and e^x play a vital role in the quantitative analysis of a wide range of natural phenomena— including population growth, radioactive decay, spread of epidemics, growth of investments, diffusion of pollutants, and motion with air resistance.

EXAMPLE   Consider the bacteria population discussed above, with $P(t) = 2^t$ (million). What is the rate of growth of the population after 5 hours?

Solution   First we write $2 = e^{\ln 2}$, a consequence of the fact that e^x and $\ln x$ are inverse functions. Thus we have

$$P(t) = e^{t \ln 2}.$$

Therefore

$$P'(t) = e^{t \ln 2}D(t \ln 2) = 2^t \ln 2.$$

Since $\ln 2 \approx 0.69315$, the rate of growth after 5 hours is

$$P'(5) = 2^5 \ln 2 \approx (0.69315)(32) \approx 22.18$$

million bacteria per hour.

The Natural Logarithm

We now begin our systematic development of the properties of exponential and logarithmic functions. Though we gave an informal overview of this material in Section 6-1, we begin here with a fresh start; in this section, we shall include the important details.

It is simplest to begin the development with the natural logarithm function. Guided by our earlier discussion, we want to define $\ln x$ for $x > 0$ so that

$$\ln 1 = 0 \quad \text{and} \quad D \ln x = \frac{1}{x}. \tag{1}$$

To do so, we recall part 1 of the Fundamental Theorem of Calculus (Section 4-4), according to which

$$D_x \left(\int_a^x f(t)\, dt \right) = f(x)$$

if f is a continuous function. In order that $\ln x$ satisfy Equations (1), we take $a = 1$ and $f(t) = 1/t$.

Definition *The Natural Logarithm*

The **natural logarithm** $\ln x$ of the positive number x is defined to be

$$\ln x = \int_1^x \frac{1}{t}\, dt. \tag{2}$$

Note that $\ln x$ is *not* defined for $x \leq 0$. Geometrically, $\ln x$ is the area under the graph of $y = 1/t$ from $t = 1$ to $t = x$ if $x > 1$ (this is the situation shown in Figure 6.4), the negative of this area if $0 < x < 1$, and 0 if $x = 1$. The fact that $D \ln x = 1/x$ follows immediately from the Fundamental Theorem of Calculus, as indicated above. By Theorem 5 in Section 1-9, the fact that the function $\ln x$ is differentiable implies that it is continuous for $x > 0$.

Since $D(\ln x) = 1/x > 0$ for $x > 0$, we see that $\ln x$ must be an increasing function. Because its second derivative

$$D^2(\ln x) = D\left(\frac{1}{x}\right) = -\frac{1}{x^2}$$

is negative for $x > 0$, it follows from Theorem 1 in Section 3-5 that the graph of $y = \ln x$ is everywhere concave downward. Below we shall use the laws of logarithms to show that

$$\lim_{x \to 0^+} \ln x = -\infty \tag{3}$$

and

$$\lim_{x \to \infty} \ln x = +\infty. \tag{4}$$

When we assemble all these facts, we see that the graph $y = \ln x$ has the shape shown in Figure 6.5.

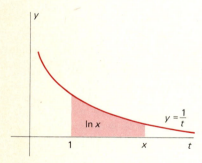

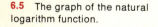

6.4 The natural logarithm function defined by means of an integral.

6.5 The graph of the natural logarithm function.

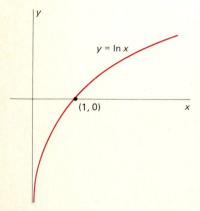

When we combine the formula $D \ln x = 1/x$ with the chain rule, we obtain the differentiation formula

$$D_x \ln u = \frac{D_x u}{u} = \frac{1}{u}\frac{du}{dx}. \tag{5}$$

Here, $u = u(x)$ denotes a positive-valued differentiable function of x.

EXAMPLE 1   If $y = \ln(x^2 + 1)$, then Formula (5) with $u = x^2 + 1$ gives

$$\frac{dy}{dx} = \frac{D_x(x^2 + 1)}{x^2 + 1} = \frac{2x}{x^2 + 1}.$$

If $y = \ln\sqrt{x^2 + 1}$, then with $u = \sqrt{x^2 + 1}$ we obtain

$$\frac{dy}{dx} = \frac{D_x\sqrt{x^2 + 1}}{\sqrt{x^2 + 1}} = \frac{1}{\sqrt{x^2 + 1}} \cdot \frac{x}{\sqrt{x^2 + 1}}$$

$$= \frac{x}{x^2 + 1}.$$

Note that these two derivatives are consistent with the fact that

$$\ln\sqrt{x^2 + 1} = \tfrac{1}{2}\ln(x^2 + 1)$$

(by a law of logarithms that we haven't yet rigorously proved).

The function $f(x) = \ln|x|$ is defined for all $x \neq 0$. If $x > 0$, then $|x| = x$ and, in this case,

$$f'(x) = D \ln x = \frac{1}{x}.$$

If $x < 0$, then $|x| = -x$ and so

$$f'(x) = D \ln(-x) = \frac{(-1)}{(-x)} = \frac{1}{x}.$$

In this computation, we used (5) with $u = -x$. Thus we have shown that

$$D \ln|x| = \frac{1}{x} \qquad (x \neq 0) \tag{6}$$

whether x is positive or negative.

When we combine (6) with the chain rule, we obtain the formula

$$D_x \ln|u| = \frac{D_x u}{u} = \frac{1}{u}\frac{du}{dx}, \tag{7}$$

valid wherever the differentiable function $u = g(x)$ is nonzero. This is equivalent to the integral formula

$$\int \frac{g'(x)}{g(x)}\,dx = \ln|g(x)| + C,$$

or simply

$$\int \frac{du}{u} = \ln|u| + C. \tag{8}$$

For example, with $u = x^2 - 1$, (8) gives

$$\int \frac{2x}{x^2 - 1}\, dx = \ln|x^2 - 1| + C = \begin{cases} \ln(x^2 - 1) + C & \text{if } |x| > 1; \\ \ln(1 - x^2) + C & \text{if } |x| < 1. \end{cases}$$

We now use our ability to differentiate logarithms to establish the laws of logarithms.

Theorem *Laws of Logarithms*

If x and y are positive numbers and r is a rational number, then:

$$\ln xy = \ln x + \ln y; \tag{9}$$

$$\ln\left(\frac{1}{x}\right) = -\ln x; \tag{10}$$

$$\ln\left(\frac{x}{y}\right) = \ln x - \ln y; \tag{11}$$

$$\ln(x^r) = r \ln x. \tag{12}$$

Proof of (9) We temporarily fix y; thus we may regard x as the independent variable and y as a constant. Then

$$D_x \ln xy = \frac{D_x(xy)}{xy} = \frac{y}{xy} = \frac{1}{x} = D_x \ln x.$$

Thus $\ln xy$ and $\ln x$ have the same derivative with respect to x. We antidifferentiate each and conclude that

$$\ln xy = \ln x + C$$

for some constant C. To evaluate C, we substitute $x = 1$ into both sides of the last equation. The fact that $\ln 1 = 0$ then implies that $C = \ln y$, and this establishes Equation (9).

Proof of (10) We differentiate $\ln(1/x)$:

$$D \ln\left(\frac{1}{x}\right) = \frac{(-1/x^2)}{(1/x)} = -\frac{1}{x} = D(-\ln x),$$

so $\ln(1/x)$ and $-\ln x$ have the same derivative. Hence antidifferentiation gives

$$\ln\left(\frac{1}{x}\right) = -\ln x + C$$

where C is a constant. We substitute $x = 1$ into this last equation. Since $\ln 1 = 0$, it follows that $C = 0$, and this proves (10).

Proof of (11) Since $x/y = x \cdot (1/y)$, Equation (11) follows immediately from Equations (9) and (10).

Proof of (12) We know that $Dx^r = rx^{r-1}$ if r is rational. So

$$D \ln x^r = \frac{rx^{r-1}}{x^r} = \frac{r}{x} = D(r \ln x).$$

Antidifferentiation then gives

$$\ln(x^r) = r \ln x + C$$

for some constant C. As before, substitution of $x = 1$ gives $C = 0$, which proves (12). In Section 6-4 we shall show that (12) holds whether or not r is rational.

Note that the proofs of (9), (10), and (12) are all quite similar—we differentiate the left-hand side, apply the fact that two functions with the same derivative (on an interval) differ by a constant C (on that interval), and evaluate C using the fact that $\ln 1 = 0$.

The laws of logarithms can often be used to simplify an expression prior to differentiating it, as in the following example.

EXAMPLE 2 Find dy/dx if

$$y = \ln \frac{\sqrt{x^2 + 1}}{\sqrt[3]{x^3 + 1}}.$$

Solution Immediate differentiation would require using the quotient rule and the chain rule (*several* times), all the while working with a complicated fraction. Our work is made much easier if we first use the laws of logarithms to simplify the formula for y:

$$y = \ln[(x^2 + 1)^{1/2}] - \ln[(x^3 + 1)^{1/3}]$$
$$= \tfrac{1}{2} \ln(x^2 + 1) - \tfrac{1}{3} \ln(x^3 + 1).$$

Finding dy/dx is now no trouble:

$$\frac{dy}{dx} = \frac{1}{2}\left(\frac{2x}{x^2 + 1}\right) - \frac{1}{3}\left(\frac{3x^2}{x^3 + 1}\right)$$

$$= \frac{x}{x^2 + 1} - \frac{x^2}{x^3 + 1}.$$

Now we establish the limits of $\ln x$ as $x \to 0^+$ and $x \to \infty$, as stated in (3) and (4). Since $\ln 2$ is the area under $y = 1/x$ from $x = 1$ to $x = 2$, we can inscribe and circumscribe a pair of rectangles, as shown in Figure 6.6, and conclude that

$$\tfrac{1}{2} < \ln 2 < 1.$$

6.6 Using rectangles to estimate ln 2.

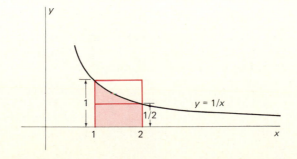

Law (12) of logarithms then gives

$$\ln(2^n) = n \ln 2 > \frac{n}{2}.$$

Next suppose that $M > 0$. Let n be an integer greater than $2M$. If $x > 2^n$, then—because $\ln x$ is an increasing function—it follows that

$$\ln x > \ln(2^n) > \frac{n}{2} > \frac{2M}{2} = M.$$

Thus we can make $\ln x$ as large as we please by choosing x sufficiently large. This proves (4): $\lim\limits_{x \to \infty} \ln x = \infty$. To prove (3), we use Law (10) of logarithms to write

$$\lim_{x \to 0^+} \ln x = -\lim_{x \to 0^+} \ln\left(\frac{1}{x}\right)$$

$$= -\lim_{y \to \infty} \ln y = -\infty,$$

by taking $y = 1/x$ and applying (4).

The graph of the natural logarithm function (Figure 6.5) suggests that, though $\ln x$ approaches infinity as x does, it does so rather slowly. Indeed, the function $\ln x$ increases more slowly than any positive integral power of x. By this statement, we mean that

$$\lim_{x \to \infty} \frac{\ln x}{x^n} = 0 \tag{13}$$

if n is a fixed positive integer. To prove this, note first that

$$\ln x = \int_1^x \frac{dt}{t} \leqq \int_1^x \frac{1}{\sqrt{t}} dt = 2(\sqrt{x} - 1),$$

because $1/t \leqq 1/\sqrt{t}$ if $t \geqq 1$. Hence if $x > 1$ then

$$0 < \frac{\ln x}{x^n} \leqq \frac{\ln x}{x} \leqq \frac{2}{\sqrt{x}} - \frac{2}{x}.$$

The last expression on the right approaches zero as $x \to \infty$. Equation (13) follows by the squeeze law of limits. Moreover, this argument shows that the positive integer n in (13) may be replaced by any rational number $k \geq 1$.

Because $\ln x$ is an increasing continuous function, the Intermediate Value Property implies that the curve $y = \ln x$ crosses the horizontal line $y = 1$ precisely once. This point of intersection has abscissa the important number $e \approx 2.71828$ mentioned in Section 6-1 (see Figure 6.7).

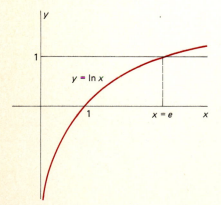

6.7 The fact that $\ln e = 1$ is expressed graphically here.

Definition of e

The number e is the unique number such that

$$\ln e = 1. \tag{14}$$

The letter e has been used to denote the number with natural logarithm 1 ever since this number was introduced by the Swiss mathematician Leonhard Euler (1707–1783); he used e for "exponential."

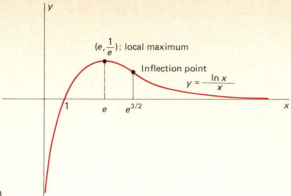

6.8 The graph for Example 3.

EXAMPLE 3 Sketch the graph of $f(x) = (\ln x)/x$, $x > 0$.

Solution First we compute the derivative

$$f'(x) = \frac{(1/x)(x) - (\ln x)(1)}{x^2} = \frac{1 - \ln x}{x^2}.$$

Thus the only critical point is where $\ln x = 1$; that is, where $x = e$. Since $f'(x) > 0$ if $x < e$ (the same as $\ln x < 1$) and $f'(x) < 0$ if $x > e$ (the same as $\ln x > 1$), we see that f is increasing if $x < e$ and decreasing if $x > e$. Hence f has a local maximum at $x = e$.

The second derivative of f is

$$f''(x) = \frac{(-1/x)(x^2) - (1 - \ln x)(2x)}{x^4} = \frac{2 \ln x - 3}{x^3},$$

so the only inflection point of the graph is where $\ln x = \frac{3}{2}$; that is, at $x = e^{3/2} \approx 4.48$.

Because

$$\lim_{x \to 0^+} \frac{\ln x}{x} = -\infty \quad \text{and} \quad \lim_{x \to \infty} \frac{\ln x}{x} = 0$$

—consequences of (3) and (13), respectively—we conclude that the graph of f looks like the one shown in Figure 6.8.

LOGARITHMS AND EXPERIMENTAL DATA

Certain empirical data can be explained by assuming that the observed dependent variable y is a **power** function of the independent variable x; that is, y is described by a mathematical model of the form

$$y = kx^m$$

where k and m are constants. If so, the laws of logarithms tell us that

$$\ln y = \ln k + m \ln x.$$

The experimenter then plots values of $\ln y$ versus values of $\ln x$. If the model assumed above is valid, the resulting data points will lie on a straight line with slope m and y-intercept $\ln k$, as shown in Figure 6.9. The usefulness of

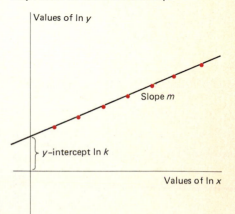

6.9 Plotting the logarithms of data may reveal a hidden relationship.

Planet	T (in days)	a (in 10^6 km)	ln T	ln a
Mercury	87.97	58	4.48	4.06
Venus	224.70	108	5.41	4.68
Earth	365.26	149	5.90	5.00
Mars	686.98	228	6.53	5.43
Jupiter	4332.59	778	8.37	6.66
Saturn	10759.20	1426	9.28	7.26

6.10 Data for Example 4.

this technique is that it is easy to see whether or not the data lie on a straight line, and—if they do—it is easy to measure the slope and y-intercept of the line and thus find the values k and m.

EXAMPLE 4 (Planetary Motion) The table of Figure 6.10 gives the period of revolution T and the major semiaxis a of the elliptical orbit of each of the first six planets about the sun, together with the logarithms of these numbers. If we plot ln T against ln a, it is immediately apparent that the resulting points lie on a straight line with slope $m = \frac{3}{2}$. Hence T and a satisfy an equation of the form $T = ka^{3/2}$, and so

$$T^2 = Ca^3.$$

This means that the square of the period T is proportional to the cube of the major semiaxis a. This is Kepler's third law of planetary motion, which he discovered empirically in 1619.

6-2 PROBLEMS

In each of Problems 1–15, find the derivative of the given function $f(x)$.

1 $\ln(3x - 1)$.

2 $\ln(4 - x^2)$.

3 $\ln\sqrt{1 + 2x}$.

4 $\ln[(1 + x)^3]$.

5 $\ln\sqrt[3]{x^3 - x}$.

6 $\ln(\sin^2 x)$.

7 $\cos(\ln x)$.

8 $(\ln x)^3$.

9 $\dfrac{1}{\ln x}$.

10 $\ln(\ln x)$.

11 $\ln[(2x + 1)^3(x^2 - 4)^4]$.

12 $\ln\sqrt{\dfrac{1 - x}{1 + x}}$.

13 $\ln\sqrt{\dfrac{4 - x^2}{9 + x^2}}$.

14 $\ln\dfrac{\sqrt{4x - 7}}{(3x - 2)^3}$.

15 $\ln\dfrac{\sqrt[4]{(1 - 5x)^3}}{\sqrt[3]{(1 + x^4)^5}}$.

Evaluate the indefinite integrals in Problems 16–22.

16 $\displaystyle\int \dfrac{dx}{3x + 5}$.

17 $\displaystyle\int \dfrac{x\,dx}{1 + 3x^2}$.

18 $\displaystyle\int \dfrac{x^2\,dx}{4 - x^3}$.

19 $\displaystyle\int \dfrac{(x + 1)\,dx}{2x^2 + 4x + 1}$.

20 $\displaystyle\int \dfrac{\cos x\,dx}{1 + \sin x}$.

21 $\displaystyle\int \dfrac{(\ln x)^2}{x}\,dx$.

22 $\displaystyle\int \dfrac{dx}{x \ln x}$.

23 Show that if $|x| \geq 1$, then

$$\ln(x + \sqrt{x^2 - 1}) = -\ln(x - \sqrt{x^2 - 1}).$$

24 Find a formula for $f^{(n)}(x)$ given $f(x) = \ln x$.

25 The heart rate R (in beats per minute) and weight W (in pounds) of various mammals were measured, with the results shown in Figure 6.11. Use the method of Example 4 to find a relation between the two of the form $R = kW^m$.

6.11 Data for Problem 25.

W	R
25	131
67	103
127	88
175	81
240	75
975	53

26 During the adiabatic expansion of a certain diatomic gas, its volume V (in liters) and its pressure P (in atmospheres) were measured, with the results shown in Figure 6.12. Use the method of Example 4 to find a relation between V and P of the form $P = kV^m$.

V	P
1.46	28.3
2.50	13.3
3.51	8.3
5.73	4.2
7.26	3.0

6.12 Data for Problem 26.

27 Substitute $y = x^p$ and then apply (13) to show that

$$\lim_{x \to \infty} \frac{\ln x}{x^p} = 0$$

if $0 < p < 1$.

28 Deduce from Problem 27 that

$$\lim_{x \to \infty} \frac{(\ln x)^k}{x} = 0 \quad \text{if} \quad k > 0.$$

29 Substitute $y = 1/x$ and then apply (13) to show that

$$\lim_{x \to 0^+} x^k \ln x = 0 \quad \text{if} \quad k > 0.$$

Use the limits in the preceding problems to sketch the graphs, for $x > 0$, of the equations given in Problems 31–33.

30 $y = x \ln x$.

31 $y = x^2 \ln x$.

32 $y = \sqrt{x} \ln x$.

33 $y = \dfrac{\ln x}{\sqrt{x}}$.

34 Problem 22 of Section 4-7 calls for showing, by numerical integration, that

$$\int_1^{2.7} \frac{dx}{x} < 1 < \int_1^{2.8} \frac{dx}{x}.$$

Explain carefully why this result shows that $2.7 < e < 2.8$.

35 If n moles of an ideal gas expand at *constant* temperature T, then its pressure and volume satisfy the equation $pV = nRT$ (R is a constant). With the aid of Problem 22 in Section 5-5, show that the work W done by the gas in expanding from volume V_1 to volume V_2 is

$$W = nRT \ln \frac{V_2}{V_1}.$$

36 "Gabriel's trumpet" is obtained by revolving the curve $y = 1/x$, $x \geq 1$, around the x-axis. Let A_b denote its surface area from $x = 1$ to $x = b$. Show that $A_b \geq 2\pi \ln b$, so—as a consequence—$A_b \to \infty$ as $b \to \infty$. Thus the surface area of Gabriel's trumpet is infinite. Is its volume finite or infinite?

37 According to the Prime Number Theorem, which was conjectured by the great German mathematician C. F. Gauss in 1792 (when he was 15 years old) but not proved until over a century later, the number of primes between the large positive numbers a and b ($a < b$) is given to a close approximation by the integral

$$\int_a^b \frac{dx}{\ln x}.$$

The midpoint and trapezoidal approximations with $n = 1$ subinterval provide an underestimate and an overestimate of the value of this integral. (Why?) Calculate them with $a = 90{,}000$ and $b = 100{,}000$. The actual number of primes in this range is 879.

6-3

The Exponential Function

We saw in Section 6-2 that the natural logarithm function $\ln x$ is continuous and increasing for $x > 0$, and that it attains arbitrarily large positive and negative values (because of Limits (3) and (4) in Section 6-2). It follows that $\ln x$ has an inverse function that is defined for all x. To see this, let y be any (fixed) real number whatsoever. If a and b are positive numbers such that $\ln a < y < \ln b$, then the Intermediate Value Property gives a number $x > 0$ (between a and b) such that $\ln x = y$. Because $\ln$ is an increasing function, there is only *one* such number x with $\ln x = y$. This inverse function to $\ln$ is called the *natural exponential function*, denoted by exp.

Definition *The Natural Exponential Function*
The (natural) **exponential function** exp is defined for all x by

$$\exp x = y \quad \text{if and only if} \quad \ln y = x. \tag{1}$$

Thus exp x is simply that (positive) number y whose natural logarithm is x. It is an immediate consequence of (1) that

$$\ln(\exp x) = x \quad \text{for all} \quad x, \tag{2}$$

and that

$$\exp(\ln y) = y \quad \text{for all} \quad y > 0. \tag{3}$$

It follows from the reflection property of inverse functions (Section 1-7) that the graphs $y = \exp x$ and $y = \ln x$ are reflections of each other in the line $y = x$, like the graphs in Figure 6.2. Therefore the graph of the exponential function looks like the one shown in Figure 6.13. In particular, exp x is positive-valued for all x, and

$$\exp 0 = 1, \tag{4}$$

$$\lim_{x \to \infty} \exp x = \infty, \quad \text{and} \tag{5}$$

$$\lim_{x \to -\infty} \exp x = 0. \tag{6}$$

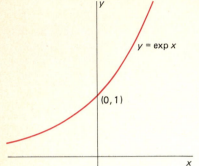

6.13 The graph of the exponential function exp.

These facts follow from the equation $\ln 1 = 0$ and from Limits (3) and (4) in Section 6-2.

Recall that the number $e \approx 2.71828$ was defined in Section 6-2 as that number which has natural logarithm 1. If r is any rational number, it follows that

$$\ln(e^r) = r \ln e = r.$$

But Equation (1) tells us that $\ln(e^r) = r$ if and only if

$$\exp r = e^r.$$

Thus exp x is equal to e^x (e to the power x) if x is a rational number. We therefore *define* e^x for x irrational as well as rational by

$$e^x = \exp x. \tag{7}$$

This is our first instance of irrational powers.

Equation (7) is the reason for calling exp the natural exponential function. With this notation, Equations (1) through (3) become

$$e^x = y \quad \text{if and only if} \quad \ln y = x, \tag{8}$$

$$\ln(e^x) = x \quad \text{for all } x, \quad \text{and} \tag{9}$$

$$e^{\ln x} = x \quad \text{for all } x > 0. \tag{10}$$

To justify Equation (7), we should show that powers of e satisfy the laws of exponents. We can do this immediately.

Theorem   Laws of Exponents

If x, x_1, and x_2 are real numbers and r is rational, then

$$e^{x_1} e^{x_2} = e^{x_1 + x_2}, \tag{11}$$

$$e^{-x} = \frac{1}{e^x}, \tag{12}$$

$$(e^x)^r = e^{rx}. \tag{13}$$

CHAP. 6: **Exponential and Logarithmic Functions**

Proof The laws of logarithms and (9) give

$$\ln(e^{x_1}e^{x_2}) = \ln e^{x_1} + \ln e^{x_2}$$
$$= x_1 + x_2 = \ln(e^{x_1 + x_2}),$$

so (11) follows from the fact that ln is an increasing, and hence a one-to-one, function. Similarly,

$$\ln([e^x]^r) = r \ln[e^x] = rx = \ln(e^{rx}).$$

So (13) follows in the same way. The proof of (12) is almost identical. In Section 6-4, we shall see that the restriction that r be rational in (13) is actually unnecessary; that is,

$$(e^x)^y = e^{xy}$$

for *all* real numbers x and y.

Because e^x is the inverse of the differentiable and increasing function $\ln x$, it follows, from Theorem 2 in Section 2-4 that e^x is differentiable and therefore also continuous. We may thus proceed to differentiate both sides of the equation (actually, the *identity*)

$$\ln(e^x) = x$$

with respect to x. Let $u = e^x$; the above equation becomes

$$\ln u = x,$$

and the derivatives must also be equal:

$$\frac{1}{u}\frac{du}{dx} = 1 \qquad \text{(since } u > 0\text{).}$$

So

$$\frac{du}{dx} = e^x;$$

that is,

$$De^x = e^x, \tag{14}$$

as we indicated in Section 6-1.

If u denotes a differentiable function of x, then (14) combined with the chain rule gives

$$D_x e^u = e^u \frac{du}{dx}. \tag{15}$$

The corresponding integration formula is

$$\int e^u \, du = e^u + C. \tag{16}$$

EXAMPLE 1 Find dy/dx if $y = e^{2 \sin x}$.

Solution With $u = 2 \sin x$, Formula (15) gives

$$\frac{dy}{dx} = e^{2 \sin x} D_x(2 \sin x) = 2e^{2 \sin x} \cos x.$$

EXAMPLE 2 Find $\int xe^{-3x^2}\,dx$.

Solution We substitute $u = -3x^2$. Since $du = -6x\,dx$, we have $x\,dx = -\frac{1}{6}du$, and we obtain

$$\int xe^{-3x^2}\,dx = -\frac{1}{6}\int e^u\,du$$

$$= -\frac{1}{6}e^u + C = -\frac{1}{6}e^{-3x^2} + C.$$

ORDER OF MAGNITUDE

The exponential function is remarkable for its high rate of increase with increasing x. In fact, e^x increases more rapidly as $x \to \infty$ than *any* fixed power of x. In the language of limits,

$$\lim_{x\to\infty}\frac{x^k}{e^x} = 0, \quad \text{or} \quad \lim_{x\to\infty}\frac{e^x}{x^k} = \infty \tag{17}$$

for any fixed $k > 0$. Since we have not yet defined x^k for k irrational, we prove (17) for the case k rational; once we know that (for $x > 1$) the power function x^k is an increasing function of k for all k, the general case will follow.

We begin by taking logarithms; we find that

$$\ln\left(\frac{e^x}{x^k}\right) = x - k\ln x = \left(\frac{x}{\ln x} - k\right)\ln x.$$

Since we know (from Formula (13) in Section 6-2) that $x/(\ln x) \to \infty$ as $x \to \infty$, this makes it clear that

$$\lim_{x\to\infty}\ln\left(\frac{e^x}{x^k}\right) = \infty.$$

Hence $e^x/x^k \to \infty$ as $x \to \infty$, so we have proved (17).

Formula (17) may be used to evaluate certain limits using the laws of limits. For example,

$$\lim_{x\to\infty}\frac{2e^x - 3x}{3e^x + x^3} = \lim_{x\to\infty}\frac{2 - 3xe^{-x}}{3 + x^3e^{-x}}$$

$$= \frac{2 - 0}{3 + 0} = \frac{2}{3}.$$

6.14 The graph of the function of Example 3.

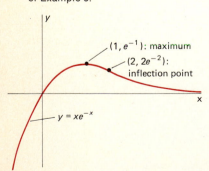

(1, e^{-1}): maximum

(2, $2e^{-2}$): inflection point

$y = xe^{-x}$

EXAMPLE 3 Sketch the graph of $f(x) = xe^{-x}$.

Solution From (17) we see that $f(x) \to 0$ as $x \to \infty$, while $f(x) \to -\infty$ as $x \to -\infty$. Since

$$f'(x) = e^{-x} - xe^{-x} = e^{-x}(1 - x),$$

the only critical point of f is $x = 1$, at which $y = e^{-1} \approx 0.37$. And since

$$f''(x) = -e^{-x}(1 - x) + e^{-x}(-1) = e^{-x}(x - 2),$$

the only inflection point is $x = 2$, where $y = 2e^{-2} \approx 0.27$. Hence the graph of f looks like Figure 6.14.

CHAP. 6: **Exponential and Logarithmic Functions**

THE NUMBER e AS A LIMIT

We now establish the following limit expression for the exponential function:

$$e^x = \lim_{n \to \infty} \left(1 + \frac{x}{n}\right)^n. \tag{18}$$

We begin by differentiating $\ln t$ using the definition of the derivative in combination with the fact that we already know that the derivative is $1/t$. Thus

$$\frac{1}{t} = D \ln t = \lim_{h \to 0} \frac{\ln(t + h) - \ln(t)}{h}$$

$$= \lim_{h \to 0} \ln \left[\left(1 + \frac{h}{t}\right)^{1/h}\right] \qquad \text{(by laws of logarithms)}$$

$$= \ln \left[\lim_{h \to 0} \left(1 + \frac{h}{t}\right)^{1/h}\right] \qquad \begin{array}{l}\text{(by continuity of the} \\ \text{logarithm function)}.\end{array}$$

The substitution $n = 1/h$ allows us to write

$$\frac{1}{t} = \ln \left[\lim_{n \to \infty} \left(1 + \frac{1}{nt}\right)^n\right];$$

then the substitution $x = 1/t$ gives

$$x = \ln \left[\lim_{n \to \infty} \left(1 + \frac{x}{n}\right)^n\right],$$

from which (18) follows, because $x = \ln y$ implies $e^x = y$. With $x = 1$, we obtain the following important expression of e as a limit:

$$e = \lim_{n \to \infty} \left(1 + \frac{1}{n}\right)^n. \tag{19}$$

*EXPONENTIALS AND THE HYDROGEN ATOM

In the modern quantum theory of atomic structure, the electron in a hydrogen atom does not travel in a well-defined orbit about the nucleus (the proton) of the atom. Instead it occupies a state known as an **orbital,** which may be visualized as a cloud of negative electricity surrounding the nucleus. The density of this "electron cloud" at a particular point is a measure of the probability that, at a given instant, the electron is near that point.

More precisely, a **spherically symmetric** orbital is described in terms of its **radial probability density function** $f(r)$, a function of the radius r, the distance from the nucleus. This function has the property that the probability $P(r, \Delta r)$ of finding the electron at a distance between r and $r + \Delta r$ from the nucleus is approximately

$$P(r, \Delta r) \approx f(r) \, \Delta r,$$

"approximately" in the sense that the error is small when compared with Δr. Here we are using an intuitive and natural concept of probability; you may think of $P(r, \Delta r)$ as the fraction of the time the electron spends between the spherical shells with radii r and $r + \Delta r$. Though the electron may be zipping about the nucleus, it's reasonable to regard the value of r that maximizes $f(r)$ as the most probable distance, or the "expected distance," of the electron from the nucleus.

For the normal ("unexcited") state of the hydrogen atom, called the $1s$-orbital, the density function is

$$f(r) = \frac{4r^2}{a_0^3} e^{-2r/a_0}.$$

The number a_0 is the **Bohr radius,** about 0.529 angstroms (1 angstrom is 10^{-10} meters). Because $f(r) > 0$ for all $r > 0$, it is theoretically possible for the electron to be anywhere in the universe! But the probability of finding the electron more than a few angstroms from the nucleus is *very* small.

To find the most probable distance of the electron from the nucleus, we differentiate $f(r)$ with the goal of eventually maximizing it:

$$f'(r) = \frac{4}{a_0^3}\left(2re^{-2r/a_0} - \frac{2r^2}{a_0} e^{-2r/a_0} \right)$$

$$= \frac{8}{a_0^4} r(a_0 - r)e^{-2r/a_0}.$$

Thus there are two critical points, $r = 0$ and $r = a_0$. Since $f(0) = 0$ and $f(r) \to 0$ as $r \to \infty$ (why?), we see that the graph of f looks like the one in Figure 6.15. In particular, $f(a_0)$ is the maximum value of $f(r)$, so the significance of the Bohr radius a_0 is this: It's the most probable distance of the electron from the nucleus in the $1s$-orbital of the hydrogen atom.

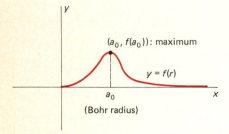

6.15 Graph of the radial probability density function.

6-3 PROBLEMS

In each of Problems 1–15, find the derivative of the given function $f(x)$.

1 e^{-3x}.

2 e^{x+1}.

3 $e^{-x^2/2}$.

4 $e^{-1/x}$.

5 $xe^{10 \ln x}$.

6 $x^3e^{4 - x^2}$.

7 $\dfrac{1 - x}{e^x}$.

8 $e^{\sqrt{x}} + e^{-\sqrt{x}}$.

9 $e^{(e^x)}$.

10 $\sqrt{e^{2x} + e^{-2x}}$.

11 $\sin(2e^x)$.

12 $\ln(x^2 + e^{-x})$.

13 $\dfrac{e^{-x}}{\ln(x + 1)}$.

14 $\cos(e^x + e^{-x})$.

15 $\dfrac{e^{\sqrt{x}}}{\sqrt{x}}$.

Evaluate the indefinite integrals in Problems 16–22.

16 $\displaystyle\int e^{2x+3}\, dx$.

17 $\displaystyle\int te^{-t^2/2}\, dt$.

18 $\displaystyle\int x^2 e^{1 - x^3}\, dx$.

19 $\displaystyle\int \frac{e^{\sqrt{x}}}{\sqrt{x}}\, dx$.

20 $\displaystyle\int \frac{e^{1/t}}{t^2}\, dt$.

21 $\displaystyle\int \frac{e^x}{1 + e^x}\, dx$.

22 $\displaystyle\int \exp(x + e^x)\, dx$.

Sketch the graphs of the equations in the next three problems. Show and label all critical points, inflection points, and asymptotes.

23 $y = x^2e^{-x}$.   **24** $y = x^3e^{-x}$.   **25** $y = e^{-x^2}$.

26 Find the area under $y = e^x$ from $x = 0$ to $x = 1$.

27 Find the volume generated by revolving the region of Problem 26 around the x-axis.

28 Find the volume generated by revolving the region R around the y-axis, where R is the plane figure bounded below by the x-axis, above by the graph of $y = e^{-x^2}$, and on the sides by the vertical lines at $x = 0$ and $x = 1$.

29 Find the length of the curve $y = \frac{1}{2}(e^x + e^{-x})$ from $x = 0$ to $x = 1$.

30 Find the area of the surface generated by revolving the curve of Problem 29 around the x-axis.

31 Show that the equation $e^{-x} = x - 1$ has a single solution, and use Newton's method to find it (to three-place accuracy).

32 If a plant releases an amount A of pollutant into a canal at time $t = 0$, then the resulting concentration of pollutant at time t in the water at a town on the canal at a distance x_0 from the plant is

$$C(t) = \frac{A}{\sqrt{k\pi t}} \exp\left(-\frac{x_0^2}{4kt}\right)$$

where k is a certain constant. Show that the maximum concentration at the town is

$$C_{max} = \frac{A}{x_0} \sqrt{\frac{2}{\pi e}}.$$

33 Sketch the graph of $f(x) = x^n e^{-x}$, $x \geq 0$ (n is a fixed but arbitrary positive integer). In particular, show that the maximum value of f is $f(n) = n^n e^{-n}$.

34 Approximate the number e as follows. First apply Simpson's approximation with $n = 2$ subintervals to the integral

$$\int_0^1 e^x \, dx = e - 1$$

to obtain the approximation $5e - 4\sqrt{e} - 7 \approx 0$. Then solve for e.

35 Suppose that $f(x) = x^n e^{-x}$, where n is a fixed but arbitrary positive integer. Conclude from Problem 33 that the numbers $f(n - 1)$ and $f(n + 1)$ are each less than $f(n) = n^n e^{-n}$. Deduce from this that

$$\left(1 + \frac{1}{n}\right)^n < e < \left(1 - \frac{1}{n}\right)^{-n}.$$

Substitute $n = 1024$ to show that $2.716 < e < 2.720$. Note that $1024 = 2^{10}$, so that a^{1024} can be computed easily with almost any calculator by entering a and then squaring a 10 times in succession.

36 Suppose that the quadratic equation $am^2 + bm + c = 0$ has the two real roots m_1 and m_2. Suppose that C_1 and C_2 are two arbitrary constants. Show that the function

$$y = y(x) = C_1 e^{m_1 x} + C_2 e^{m_2 x}$$

satisfies the differential equation $ay'' + by' + cy = 0$.

37 Use the result of Problem 36 to find a solution $y(x)$ of the differential equation $y'' + y' - 2y = 0$ such that $y(0) = 5$ and $y'(0) = 2$.

6-4

General Exponential and Logarithmic Functions

The natural exponential e^x and the natural logarithm $\ln x$ are often called the exponential and logarithm with *base e*. We now define general exponential and logarithm functions, having the forms a^x and $\log_a x$, with base a positive number $a \neq 1$. But it is now convenient to reverse the order of treatment in Sections 6-2 and 6-3, so we first consider the general exponential function.

If r is a rational number, then one of the laws of exponents (Equation (13) in Section 6-3) gives

$$a^r = (e^{\ln a})^r = e^{r \ln a}.$$

We therefore *define* arbitrary powers (rational *and* irrational) of the positive number a this way:

$$a^x = e^{x \ln a} \tag{1}$$

for all x. Then $f(x) = a^x$ is called the **exponential function with base a.** Note that $a^x > 0$ for all x and that $a^0 = e^0 = 1$ for all $a > 0$.

The *laws of exponents* for general exponentials follow almost immediately from Definition (1) and the laws of exponents for the natural exponential.

$$a^x a^y = a^{x+y}, \tag{2}$$

$$a^{-x} = \frac{1}{a^x}, \quad \text{and} \tag{3}$$

$$(a^x)^y = a^{xy} \tag{4}$$

for all x and y. To show the first formula,

$$a^x a^y = e^{x \ln a} e^{y \ln a} = e^{(x \ln a) + (y \ln a)}$$
$$= e^{(x+y) \ln a} = a^{x+y}.$$

To derive (4), note first from (1) that $\ln a^x = x \ln a$. Then

$$(a^x)^y = e^{y \ln(a^x)} = e^{xy \ln a} = a^{xy}.$$

Observe that this follows for all real numbers x and y, so the restriction that r be rational in the formula $(e^x)^r = e^{rx}$ (see Equation (13) in Section 6-3) has now been removed.

If $a > 1$ so that $\ln a > 0$, then Formulas (5) and (6) in Section 6-3 immediately give us the results

$$\lim_{a \to \infty} a^x = \infty \quad \text{and} \quad \lim_{x \to -\infty} a^x = 0. \tag{5}$$

The values of these two limits are interchanged if $0 < a < 1$, for then $\ln a < 0$.

Because

$$D_x a^x = D_x e^{x \ln a} = a^x \ln a \tag{6}$$

is positive for all x if $a > 1$, we see that—in this case—a^x is an increasing function of x. If $a > 1$, the graph of $y = a^x$ resembles that of $f(x) = e^x$, and in fact the graph of $y = 2^x$ appears in Figure 6.16. But if $0 < a < 1$, then $\ln a < 0$, and it then follows from (6) that a^x is a decreasing function. In this case, the graph of $y = a^x$ will look like the one shown in Figure 6.17.

If $u = u(x)$ is a differentiable function of x, then (6) combined with the chain rule gives

$$D_x a^u = a^u (\ln a) \frac{du}{dx}. \tag{7}$$

The corresponding integral formula is

$$\int a^u \, du = \frac{a^u}{\ln a} + C. \tag{8}$$

But rather than using these general formulas, it is generally simpler to rely upon Definition (1) alone, as in the following examples.

$$D3^{x^2} = De^{x^2 \ln 3} = e^{x^2 \ln 3}(2x \ln 3) = 3^{x^2}(2 \ln 3)x.$$

$$\int \frac{10^{\sqrt{x}} \, dx}{\sqrt{x}} = \int \frac{e^{\sqrt{x} \ln 10}}{\sqrt{x}} \, dx$$

$$= \int \frac{2e^u}{\ln 10} \, du \qquad \left(u = \sqrt{x} \ln 10, \quad du = \frac{\ln 10}{2\sqrt{x}} \, dx \right)$$

$$= \frac{2e^u}{\ln 10} + C = \frac{2(10^{\sqrt{x}})}{\ln 10} + C.$$

Whether or not the exponent r is rational, the **general power function** $f(x) = x^r$ is now defined for $x > 0$ by

$$x^r = e^{r \ln x}.$$

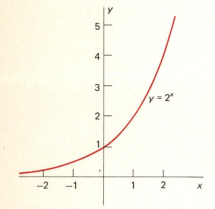

6.16 The graph of $y = 2^x$.

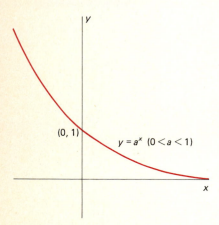

6.17 The graph of $y = a^x$ is decreasing and concave upward if $0 < a < 1$.

We may now prove the power rule for differentiation for an *arbitrary* (constant) exponent, as follows:

$$D_x x^r = D_x(e^{r \ln x}) = e^{r \ln x} D(r \ln x)$$

$$= x^r \cdot \frac{r}{x} = r x^{r-1}.$$

For example, we now know that

$$Dx^\pi = \pi x^{\pi - 1} \approx (3.14159) x^{2.14159}.$$

If $a > 1$, then the general exponential function a^x is continuous and increasing for all x, and attains all positive values (by an argument similar to that in the first paragraph of Section 6-3, using Limits (5) and (6) above). It therefore has an inverse function that is defined for all $x > 0$. This inverse function to a^x is called the **logarithm function with base a** and is denoted by $\log_a x$. Thus

$$y = \log_a x \quad \text{if and only if} \quad x = a^y. \tag{9}$$

Note that the logarithm function with base e is the natural logarithm function

$$\log_e x = \ln x.$$

On many calculators, the "log" button denotes common, or base 10, logarithms: $\log_{10} x = \log x$. Many computer programming languages, by contrast, use $\text{LOG}(x)$ to denote the natural logarithm function and use $\text{LOG10}(x)$ for the common logarithm. Computers use $\text{EXP}(x)$ for the natural exponential function, but the calculator button for this function is usually marked "e^x," and a button marked "exp" is sometimes used for entering the exponent in scientific notation.

The following *laws of logarithms* are easy to derive from the laws of exponents above.

$$\log_a xy = \log_a x + \log_a y, \tag{10}$$

$$\log_a \left(\frac{1}{x} \right) = -\log_a x, \tag{11}$$

$$\log_a x^y = y \log_a x. \tag{12}$$

These formulas hold for any positive base $a \neq 1$ and for all positive values of x and y; in the last formula, y may be negative or zero as well.

Logarithms with one base are related to logarithms with another base, and the relationship is most easily expressed by the formula

$$(\log_a b)(\log_b c) = \log_a c. \tag{13}$$

This formula holds for all values of a, b, and c that make sense; that is, the bases a and b are positive numbers other than 1 and c is positive. The proof of this formula is outlined in Problem 33. Note also how easy it is to remember Formula (13)—it is as if some arcane cancellation law held.

If we take $c = a$ in the above formula, this gives

$$(\log_a b)(\log_b a) = 1, \tag{14}$$

which in turn, with $b = e$, gives

$$\ln a = \frac{1}{\log_a e}. \tag{15}$$

Now let us differentiate the general logarithm function, using the fact that $\log_a x$ must be differentiable (by Theorem 2 in Section 2-4; $\log_a x$ is the inverse of the differentiable and increasing function a^x). We begin with the inverse function relation

$$a^{\log_a x} = x.$$

If we write $u = \log_a x$, then $a^u = x$, so that

$$D_x(a^u) = D_x x = 1.$$

Thus Formula (7) gives

$$a^u(\ln a)\frac{du}{dx} = 1,$$

so that

$$\frac{du}{dx} = \frac{1}{a^u \ln a} = \frac{1}{x \ln a}.$$

Then (15) gives

$$D \log_a x = \frac{\log_a e}{x}. \tag{16}$$

For example,

$$D \log_{10} x = \frac{\log_{10} e}{x} \approx \frac{0.4343}{x}.$$

If we now reason as in Formula (6) in Section 6-2, the chain rule yields the general formula

$$D_x \log_a|u| = \frac{\log_a e}{u}\frac{du}{dx} \qquad (u \neq 0) \tag{17}$$

if u is a differentiable function of x. For instance,

$$D_x \log_2 \sqrt{x^2 + 1} = \frac{1}{2} D_x \log_2(x^2 + 1)$$

$$= \frac{1}{2} \cdot \frac{\log_2 e}{x^2 + 1}(2x) \approx \frac{(1.4427)x}{x^2 + 1}.$$

We used here the fact that $\log_2 e = 1/(\ln 2)$ by (15).

LOGARITHMIC DIFFERENTIATION

The derivatives of certain functions are most conveniently found by first differentiating their logarithms. This process—called **logarithmic differentiation**—involves the following steps for finding $f'(x)$.

1 Given: $\qquad\qquad\qquad\qquad\qquad\qquad\qquad\qquad$ $y = f(x).$

2 Take *natural* logarithms, then simplify using laws of logarithms: $\qquad\qquad\qquad\qquad\qquad$ $\ln y = \ln f(x).$

3 Now differentiate with respect to x: $\qquad\qquad$ $\dfrac{1}{y}\dfrac{dy}{dx} = D_x[\ln f(x)].$

4 Finally, multiply both sides by $y = f(x)$: $\qquad$ $\dfrac{dy}{dx} = f(x)D_x[\ln f(x)].$

EXAMPLE 1 Find dy/dx if

$$y = \frac{\sqrt{(x^2 + 1)^3}}{\sqrt[3]{(x^3 + 1)^4}}.$$

Solution   The laws of logarithms give

$$\ln y = \ln \frac{(x^2 + 1)^{3/2}}{(x^3 + 1)^{4/3}} = \frac{3}{2} \ln(x^2 + 1) - \frac{4}{3} \ln(x^3 + 1).$$

So differentiation with respect to x gives

$$\frac{1}{y} \frac{dy}{dx} = \left(\frac{3}{2}\right) \frac{2x}{x^2 + 1} - \left(\frac{4}{3}\right) \frac{3x^2}{x^3 + 1}$$

$$= \frac{3x}{x^2 + 1} - \frac{4x^2}{x^3 + 1}.$$

Finally, to solve for dy/dx, we multiply both sides by

$$y = \sqrt{(x^2 + 1)^3}/\sqrt[3]{(x^3 + 1)^4},$$

and we obtain

$$\frac{dy}{dx} = \left(\frac{3x}{x^2 + 1} - \frac{4x^2}{x^3 + 1}\right) \frac{\sqrt{(x^2 + 1)^3}}{\sqrt[3]{(x^3 + 1)^4}}.$$

EXAMPLE 2   Find dy/dx if $y = x^{x+1}$.

Solution   If $y = x^{x+1}$, then

$$\ln y = \ln(x^{x+1}) = (x + 1) \ln x,$$

so

$$\frac{1}{y} \frac{dy}{dx} = (1) \ln x + (x + 1)\left(\frac{1}{x}\right)$$

$$= 1 + \frac{1}{x} + \ln x.$$

And now multiplication by $y = x^{x+1}$ gives

$$\frac{dy}{dx} = \left(1 + \frac{1}{x} + \ln x\right) x^{x+1}.$$

6-4   PROBLEMS

In each of Problems 1–14, find the derivative of the given function $f(x)$.

1 17^x.

2 $2^{\sqrt{x}}$.

3 $10^{1/x}$.

4 $3^{\sqrt{1-x^2}}$.

5 $2^{(2^x)}$.

6 $\log_2 x$.

7 $\log_3 \sqrt{x^2 + 4}$.

8 $\log_{10}(e^x)$.

9 $\log_3(2^x)$.

10 $\log_{10}(\log_{10} x)$.

11 $\log_2(\log_3 x)$.

12 $\pi^x + x^\pi + \pi^\pi$.

13 $\exp(\log_{10} x)$.

14 $\pi^{(x^3)}$.

Evaluate the integrals given in Problems 15–22.

15 $\int 3^{2x}\, dx$.

16 $\int x(10^{-x^2})\, dx$.

17 $\int \frac{2^{\sqrt{x}}}{\sqrt{x}}\, dx$.

18 $\int \frac{10^{1/x}}{x^2}\, dx$.

19 $\int x^2 7^{x^3+1} \, dx.$

20 $\int \dfrac{dx}{x \log_{10} x}.$

21 $\int \dfrac{\log_2 x}{x} \, dx.$

22 $\int (2^x) 3^{(2^x)} \, dx.$

In Problems 23–32, find dy/dx by logarithmic differentiation.

23 $y = \sqrt{(x^2 - 4)\sqrt{2x + 1}}.$

24 $y = \dfrac{\sqrt[3]{3 - x^2}}{\sqrt[4]{x^4 + 1}}.$

25 $y = 2^x.$

26 $y = x^x.$

27 $y = x^{\ln x}.$

28 $y = (1 + x)^{1/x}.$

29 $y = \sqrt[3]{\dfrac{(x + 1)(x + 2)}{(x^2 + 1)(x^2 + 2)}}.$

30 $y = \sqrt{x + 1}\sqrt[3]{x + 2}\sqrt[4]{x + 3}.$

31 $y = (\ln x)^{\sqrt{x}}.$

32 $y = (3 + 2^x)^x.$

33 Prove Formula (13). (*Suggestion:* Let $x = \log_a b$, $y = \log_b c$, and $z = \log_a c$. Then show that $a^z = a^{xy}$, and conclude that $z = xy$.)

34 Suppose that u and v are differentiable functions of x. Show by logarithmic differentiation that

$$D_x(u^v) = v(u^{v-1})\frac{du}{dx} + u^v(\ln u)\frac{dv}{dx}.$$

Interpret the two terms on the right in relation to each of the two special cases: (i) u is a constant; (ii) v is a constant.

35 Suppose that $a > 0$. Show that

$$\lim_{x \to \infty} a^{1/x} = 1$$

by examining $\ln(a^{1/x})$. It follows that

$$\lim_{n \to \infty} \sqrt[n]{a} = 1.$$

Test this conclusion by entering some positive number into your calculator and then pressing the square root key repeatedly. Tabulate the results of two such experiments.

36 Show that

$$\lim_{n \to \infty} \sqrt[n]{n} = 1$$

by showing that

$$\lim_{x \to \infty} x^{1/x} = 1.$$

Use the method of Problem 35.

37 Show that

$$\lim_{x \to \infty} \frac{x^x}{e^x} = \infty$$

by examining $\ln(x^x/e^x)$. Thus x^x increases faster than the exponential function e^x as $x \to \infty$.

38 (a) Show that the equation $2^x = x^{10}$ has the same solutions as the equation $(\ln x)/x = \frac{1}{10}\ln 2$.
(b) Conclude from the graph of $y = (\ln x)/x$ (Example 3 in Section 6-2) that the equation $2^x = x^{10}$ has exactly two solutions.
(c) Show that one of these two solutions is between 1 and 2, while the other is between 50 and 60. Then use Newton's method to approximate each solution with at least 2-place accuracy.

6-5

Natural Growth and Decay

Consider a population that numbers $P(t)$ persons—or animals, bacteria, or any sort of entity—at time t. We assume that this population has a constant birth rate β and a constant death rate δ. Roughly speaking, this means that, during any 1-year period, βP births and δP deaths occur.

But since P changes during the course of that year, some allowance must be made for changes in the number of births and the number of deaths. To be more precise, we think of a very brief time interval from t to $t + \Delta t$. For very small values of Δt, the value of $P = P(t)$ will change by such a small amount during the time interval $[t, t + \Delta t]$ that we can regard P as "almost" constant. We require that the numbers of births and deaths during this time interval be given with sufficient accuracy by the approximations

$$\text{number of births} \approx \beta P(t) \cdot \Delta t \quad \text{and} \quad \text{number of deaths} \approx \delta P(t) \cdot \Delta t. \tag{1}$$

What we mean when we say that the birth rate is β and the death rate is δ is this: The ratios to Δt of the errors in the above approximations both approach zero as $\Delta t \to 0$.

We would like to use the information in (1) to deduce, if possible, the form of the function $P(t)$ that describes our population. Our strategy begins with finding the **time rate of change** of P. Hence we consider the increment $\Delta P = P(t + \Delta t) - P(t)$ of P during the time interval $[t, t + \Delta t]$. Since ΔP is simply the number of births minus the number of deaths, we find from (1) that

$$\Delta P = P(t + \Delta t) - P(t) \approx \beta P(t)\, \Delta t - \delta P(t)\, \Delta t.$$

Therefore

$$\frac{\Delta P}{\Delta t} = \frac{P(t + \Delta t) - P(t)}{\Delta t} \approx (\beta - \delta)P(t).$$

The quotient on the left-hand side approaches the derivative $P'(t)$ as $\Delta t \to 0$ and, by the assumption following (1) above, it also approaches the right-hand side $(\beta - \delta)P(t)$. Hence, when we take the limit as $\Delta t \to 0$, we get the differential equation

$$P'(t) = (\beta - \delta)P(t)$$

or

$$\frac{dP}{dt} = kP \qquad \text{where } k = \beta - \delta. \tag{2}$$

This differential equation may be regarded as a *mathematical model* of the changing population.

The differential equation

$$\frac{dx}{dt} = kx \qquad (k \text{ a constant}) \tag{3}$$

serves as a mathematical model for an extraordinarily wide range of natural phenomena. It is easy to solve; we first write it in the form

$$\frac{1}{x}\frac{dx}{dt} = k;$$

that is,

$$D_t(\ln x) = D_t(kt).$$

We then antidifferentiate and obtain

$$\ln x = kt + C \qquad (\text{assuming } x > 0).$$

We apply the exponential function to both sides of this equation to solve for x. This gives

$$x = e^{\ln x} = e^{kt + C} = Ae^{kt}.$$

Here, $A = e^C$ is a constant that remains to be determined. But we see that A will simply be the value of x when $t = 0$; that is, $A = x(0) = x_0$. Thus the solution of the differential equation in (3) with the *initial value* $x(0) = x_0$ is

$$x(t) = x_0 e^{kt}. \tag{4}$$

As a result, Equation (3) is often called the **exponential growth equation,** or the **natural growth equation.** We see from (4) that with $x_0 > 0$, the solution

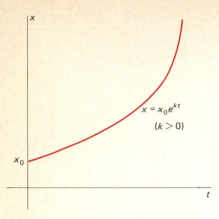

6.18 Solution of the exponential growth equation for $k > 0$.

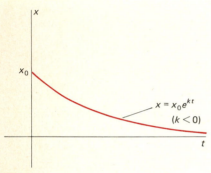

6.19 Solution of the exponential growth equation—now actually a *decay* equation—for the case $k < 0$.

$x(t)$ is an increasing function if $k > 0$ and a decreasing function if $k < 0$. These cases are shown in Figures 6.18 and 6.19, respectively. The remainder of this section is devoted to examples of natural phenomena for which this differential equation serves as a mathematical model.

POPULATION GROWTH

When we compare Equations (2), (3), and (4), we see that a population $P(t)$ with constant birth rate β and constant death rate δ is given by

$$P(t) = P_0 e^{kt} \tag{5}$$

where $P_0 = P(0)$ and $k = \beta - \delta$. If t is measured in years, then k is called the **annual growth rate**, whether it be positive, negative, or zero. Its value is often given as a percentage (which is its actual decimal value multiplied by 100). If k is not too large, this value is fairly close to the actual percentage increase (or decrease) of the population each year.

EXAMPLE 1 In mid-1977, the world population was 4.3 billion persons, and it was increasing then at the rate of a quarter of a million persons each day. Let us assume constant birth and death rates. We would like to find the answers to these questions:

(a) What is the annual growth rate k?
(b) How long will it take the world's population to double?
(c) What will the world's population be in the year 2000?
(d) When will the world's population be 50 billion (some demographers believe that this is the maximum for which the planet can supply food)?

Solution We measure the world's population $P(t)$ in billions and measure time in years. We shall take $t = 0$ to correspond to 1977, so that $P_0 = 4.3$. The fact that P is increasing by a quarter million, or 0.00025 billion, persons per day at time $t = 0$ means that

$$P'(0) = (0.00025)(365.25) \approx 0.0913$$

billion per year. From Equation (2) we now obtain

$$k = \frac{1}{P}\frac{dP}{dt}\bigg|_{t=0} = \frac{P'(0)}{P(0)} = \frac{0.0913}{4.3} \approx 0.0212.$$

Thus the population is growing at the rate of about 2.12% per year.

To find when the population will be 8.6 billion, we solve the equation $8.6 = (4.3)e^{0.0212t}$ for t, and find that $t \approx 32.7$ (years). Thus the world's population will double in approximately 33 years; that is, by the year 2010.

In the year 2000 the population will be

$$P(23) = (4.3)e^{(0.0212)(23)} \approx 7.0 \quad \text{(billion)}.$$

To find when the population will be 50 billion, we solve the equation $50 = (4.3)e^{0.0212t}$ for t, which turns out to be about 115.7. Thus the world population will reach 50 billion after approximately 116 years, in the year 2093.

RADIOACTIVE DECAY

Consider a sample of material that contains $N(t)$ atoms of a certain radioactive isotope at time t. It has been observed that a constant fraction of these radioactive atoms will spontaneously decay (into atoms of another element or another isotope of the same element) during each unit of time. Consequently the sample behaves exactly like a population with a constant death rate but with no births occurring. To write a model for $N(t)$, we use Equation (2) with N in place of P, with $k > 0$ in place of δ, and with $\beta = 0$. We thus get the differential equation

$$\frac{dN}{dt} = -kN. \tag{6}$$

From the solution (4) of Equation (3), with k replaced by $-k$, we conclude that

$$N(t) = N_0 e^{-kt} \tag{7}$$

where $N_0 = N(0)$, the number of radioactive atoms present in the sample at time $t = 0$.

The value of the *decay constant* k depends upon the particular isotope with which we are dealing. If k is large, then the isotope decays rapidly, while if k is near zero, the isotope decays quite slowly and thus may be a relatively persistent factor in its environment. The decay constant is often specified in terms of another empirical parameter, the **half-life** of the isotope, because this parameter is more convenient. The half-life τ of a radioactive isotope is the time required for *half* of it to decay. To find the relationship between k and τ, we set

$$t = \tau \quad \text{and} \quad N = \tfrac{1}{2}N_0$$

in Equation (7), so that

$$\tfrac{1}{2}N_0 = N_0 e^{-k\tau}.$$

When we solve for τ, we find that

$$\tau = \frac{\ln 2}{k}. \tag{8}$$

Note that the concept of half-life is meaningful; the value of τ depends *only* upon k and thus depends only on the radioactive isotope involved. It does *not* depend upon the amount of that isotope present.

EXAMPLE 2 The half-life of radioactive carbon C^{14} is about 5700 years. A specimen of charcoal found at Stonehenge turns out to contain 63% as much C^{14} as a sample of present-day charcoal. What is the age of the sample?

Solution The key to the method of *radiocarbon dating* is that living organic matter maintains a constant level of C^{14} by "breathing" air (or by consuming organic matter that does). Since air contains C^{14} along with the much more common stable isotope C^{12} of carbon, mostly in the gas CO_2, the same percentage permeates all life, because organic processes seem to make no distinction between the two isotopes. Of course, when a living organism dies, it ceases its metabolism of carbon, and the process of radioactive decay begins to deplete its C^{14} content. The fraction of C^{14} in the air remains roughly constant because new C^{14} is being generated constantly by the action of cosmic rays on nitrogen atoms in the upper atmosphere, and this generation has long been in a steady-state equilibrium with the loss of C^{14} through radioactive decay.

For our Stonehenge example, we take $t = 0$ as the time of "death" of the tree from which the charcoal was made. From Equation (8) we know that

$$k = \frac{\ln 2}{\tau} = \frac{\ln 2}{5700} \approx 0.0001216.$$

We are given that $N = (0.63)N_0$ now, so we solve the equation

$$(0.63)N_0 = N_0 e^{-kt}$$

with this value of k and thus find that

$$t = -\frac{\ln(0.63)}{0.0001216} \approx 3800$$

years. Thus the sample is about 3800 years old, and if it has any connection with the builders of Stonehenge, our computation suggests that this observatory, monument, or temple—whichever it may be—dates from almost 1800 B.C.

EXAMPLE 3 According to one cosmological theory, there were equal amounts of the two uranium isotopes U^{235} and U^{238} at the creation of the universe in the "big bang." At present there are 137.7 U^{238} atoms for each atom of U^{235}. Using the half-lives

4.51 billion years for U^{238},

0.71 billion years for U^{235},

calculate the age of the universe.

Solution Let $N_8(t)$ and $N_5(t)$ be the numbers of U^{238} and U^{235} atoms, respectively, at time t (in billions of years after the creation of the universe). Then

$$N_8(t) = N_0 e^{-kt} \quad \text{and} \quad N_5(t) = N_0 e^{-ct}$$

where N_0 is the original number of atoms of each isotope. Also

$$k = \frac{\ln 2}{4.51} \quad \text{and} \quad c = \frac{\ln 2}{0.71},$$

a consequence of Equation (8). We divide the equation for N_8 by the equation for N_5, and find that when t has the value corresponding to "now," then

$$137.7 = \frac{N_8}{N_5} = e^{(c-k)t},$$

and we solve this equation for

$$t = \frac{\ln 137.7}{[(1/0.71) - (1/4.51)] \ln 2} \approx 5.99.$$

Thus our estimate of the age of the universe is about 6 billion years, which is at least of the same order of magnitude as recent estimates of about 15 billion years.

CONTINUOUSLY COMPOUNDED INTEREST

Consider a savings account that is opened with an initial deposit of A_0 dollars and which thenceforth earns interest at the annual rate r. If there are $A(t)$ dollars in the account at time t and the interest is "compounded" at time $t + \Delta t$, this means that $rA(t) \, \Delta t$ dollars in interest are added to the account then. So

$$A(t + \Delta t) = A(t) + rA(t) \, \Delta t,$$

and thus

$$\frac{\Delta A}{\Delta t} = \frac{A(t + \Delta t) - A(t)}{\Delta t} = rA(t).$$

Continuous compounding of interest is the situation that results by taking the limit as $\Delta t \to 0$, so that

$$\frac{dA}{dt} = rA.$$

This is an exponential growth equation with solution

$$A(t) = A_0 e^{rt}. \tag{9}$$

For example, if $A_0 = \$1000$ is invested at 6% annual interest so that $r = 0.06$, then Equation (9) gives

$$A(1) = 1000e^{(0.06)(1)} = \$1061.84$$

for the amount present after one year if the interest is compounded continuously. Thus the "effective" interest rate is 6.184%. Most people are aware that the more often interest is compounded, the more rapidly their savings grow, but bank advertisements sometimes tend to overemphasize this advantage. For example, 6% compounded *monthly* multiplies your investment by

$$1 + \frac{0.06}{12} = 1.005$$

at the end of each month, so an initial investment would grow in one year to

$$(1000)(1.005)^{12} = \$1061.68,$$

only 16 cents less than would be yielded by continuous compounding.

*DRUG ELIMINATION

In many cases the amount $A(t)$ of a certain drug in the bloodstream, measured by the excess over the drug's natural level, will decline at a rate proportional to the current excess amount. That is,

$$\frac{dA}{dt} = -\lambda A \quad \text{so that} \quad A(t) = A_0 e^{-\lambda t}. \tag{10}$$

The parameter λ is called the *elimination constant* of the drug, and $T = 1/\lambda$ is called the *elimination time*.

EXAMPLE 4 The elimination time for alcohol varies from one person to another. If a person's "sobering time" $T = 1/\lambda$ is 2.5 hours, how long will it take for the excess bloodstream alcohol concentration to be reduced from 0.10% to 0.02%?

Solution We assume that the normal concentration of alcohol in the blood is zero, so that any amount is the excess amount. In this problem, we have $\lambda = 1/2.5 = 0.4$, so Equation (10) yields

$$0.02 = (0.10)e^{-(0.4)t}.$$

Thus

$$t = -\frac{\ln(0.2)}{0.4} \approx 4.02 \qquad \text{(hours)}.$$

*SALES DECLINE

Marketing studies for certain products show that if advertising for a particular product is halted and other market conditions remain unchanged —we mean such things as number and promotion of competing products, their prices, and so on—then the sales of the unadvertised product will decline at a rate that is proportional at any time to the current sales S. That is,

$$\frac{dS}{dt} = -\lambda S \quad \text{so that} \quad S(t) = S_0 e^{-\lambda t}. \tag{11}$$

As usual, S_0 denotes the "initial" value of the sales, which we take to be the sales in the last advertising month. If we take months as the natural units for time t, then $S(t)$ denotes sales t months after advertising is halted, and λ might be called the *sales decay constant*.

*LINGUISTICS

Consider a basic list of N_0 words in use in a given language at time $t = 0$. Let $N(t)$ denote the number of these words that are still in use at time t— those that have neither disappeared from the language nor been replaced by noncognates. According to one theory in linguistics, the rate of decrease of N is proportional to N. That is,

$$\frac{dN}{dt} = -\lambda N \quad \text{so that} \quad N(t) = N_0 e^{-\lambda t}. \tag{12}$$

If t is measured in millennia (standard in linguistics), $k = e^{-\lambda}$ is the fraction of the words in the original list that survive in use for 1000 years.

1 Suppose that $1000 is deposited in a savings account that pays 8% annual interest compounded continuously. At what rate (in dollars per year) is this account earning interest after 5 years? After 20 years?

2 (Population Growth) A certain city had a population of 25,000 in 1960 and a population of 30,000 in 1970. Assume that its population will continue to grow exponentially at a constant rate. What population can its city planners expect in the year 2000?

3 (Population Growth) In a certain culture of bacteria, the number of bacteria increased sixfold in 10 hours. How long did it take to double their number?

4 (Radiocarbon Dating) Carbon extracted from an ancient skull contained only $\frac{1}{6}$ as much radioactive C^{14} as carbon extracted from present-day bone. How old is the skull?

5 (Radiocarbon Dating) Carbon taken from a purported relic of the time of Christ contained 4.6×10^{10} atoms of C^{14} per gram. Carbon extracted from a present-day specimen of the same substance contained 5.0×10^{10} atoms of C^{14} per gram. Compute the approximate age of the relic. What is your opinion as to its authenticity?

6 An amount A is invested for t years at an annual interest rate r that is compounded n times over these years at equal intervals.

(a) Explain why the amount that has accrued after t years is

$$A_{r,n} = A\left(1 + \frac{rt}{n}\right)^n.$$

(b) Conclude from Limit (18) in Section 6-3 that

$$\lim_{n \to \infty} A_{r,n} = Ae^{rt}.$$

7 If an investment of A_0 dollars returns A_1 dollars after one year, the **effective annual interest rate** r is defined by means of the equation $A_1 = (1 + r)A_0$. Banks often advertise that they increase the effective interest rates on their customers' savings accounts by increasing the frequency of compounding. Calculate the effective annual interest rate if a 9% annual interest rate is compounded (a) quarterly; (b) monthly; (c) weekly; (d) daily; (e) continuously.

8 (Continuously Compounded Interest) Upon the birth of their first child, a couple deposited $5000 in a savings account that draws 6% interest compounded continuously. The interest payments are allowed to accumulate. How much will the account contain when the child is ready to go to college at age 18?

9 (Continuously Compounded Interest) Suppose that you discover in your attic an overdue library book on which your great-great-grandfather owed a fine of 30¢ 100 years ago. If an overdue fine grows exponentially at a 5% annual rate compounded continuously, how much would you have to pay if you returned the book today?

10 (Drug Elimination) Suppose that sodium pentobarbitol is used to anesthetize a dog; the dog is anesthetized when its bloodstream contains at least 45 milligrams of sodium pentobarbitol per kilogram of the dog's body weight. Suppose also that sodium pentobarbitol is eliminated exponentially from the dog's bloodstream, with a half-life of 5 hours. What single dose should be administered in order to anesthetize a 50-kilogram dog for 1 hour?

11 (Sales Decline) Advertising of a certain product has been discontinued; the company plans to resume advertising when sales have declined to 75% of their initial rate. (This phenomenon actually occurred when the Sony Corporation first introduced home videotape recorders in the United States in 1976.) If after 1 week without advertising, sales have declined to 95% of their initial rate, when should the company expect to resume advertising?

12 (Linguistics) The English language evolves in such a way that 77% of all words disappear (or are replaced by noncognates) every 1000 years. Of a basic list of words used by Chaucer in A.D. 1400, what percentage should we expect to find still in use today?

13 (Half-Life) The half-life of radioactive cobalt is 5.27 years. Suppose that a nuclear accident has left the level of cobalt radiation in a certain region at 100 times the level acceptable for human habitation. How long will it be before the region is again habitable? (Ignore the likely presence of other radioactive substances.)

14 Suppose that a mineral body, formed in an ancient cataclysm—perhaps the formation of the earth itself—originally contained the uranium isotope U^{238} (which has a half-life of 4.51×10^9 years) but no lead, the end product of the radioactive decay of U^{238}. If today the ratio of U^{238} atoms to lead atoms in the mineral body is 0.9, when did the cataclysm occur?

15 A certain moon rock was found to contain equal numbers of potassium and argon atoms. Assume that all the argon is the result of radioactive decay of potassium (its half-life is about 1.28×10^9 years) and that 1 of every 9 potassium atom disintegrations yields an argon atom. What is the age of the rock, measured from the time it contained only potassium?

16 If a body is cooling in a medium with constant temperature A, then—according to Newton's law of cooling—the rate of change of the temperature T of the body is proportional to $T - A$. A pitcher of buttermilk initially at $25°C$ is to be cooled by setting it out on the front porch, where the temperature is $0°C$. Suppose that the temperature of the buttermilk has dropped to $15°C$ after 20 minutes. When will it be at $5°C$?

17 When sugar is dissolved in water, the amount A that remains undissolved after t minutes satisfies the differential

equation $dA/dt = -kA \ (k > 0)$. If 25% of the sugar dissolves in 1 minute, how long does it take for half the sugar to dissolve?

18 The intensity of light I at a depth x meters below the surface of a lake satisfies the differential equation $dI/dx = -(1.4)I$.
(a) At what depth is the intensity half the intensity I_0 at the surface (where $x = 0$)?
(b) What is the intensity at a depth of 10 meters (as a fraction of I_0)?
(c) At what depth will the intensity be 1/100 what it is at the surface?

19 The barometric pressure p (in inches of mercury) at an altitude x miles above sea level satisfies the differential equation $dp/dx = -(0.2)p; \ p(0) = 29.92$.
(a) Calculate the barometric pressure at 10,000 feet and again at 30,000 feet.
(b) Without prior conditioning, few people can survive when the pressure drops to less than 15 inches of mercury. How high is that?

---

*6-6

Linear First Order Differential Equations and Applications

A **first order differential equation** is one in which only the first derivative (and not higher derivatives) of the dependent variable appears. It is called **linear** if it can be written in the form

$$\frac{dx}{dt} = ax + b, \tag{1}$$

where a and b denote functions of the independent variable t. In this section we discuss applications of the special case in which the coefficients a and b are *constants*.

To solve Equation (1) when the coefficients are constant, we write it as

$$\frac{1}{ax + b} \frac{dx}{dt} = 1.$$

If $ax + b > 0$, this says that

$$\frac{1}{a} D_t(\ln(ax + b)) = D_t t,$$

and antidifferentiation gives

$$\ln(ax + b) = at + C.$$

Then exponentiation gives

$$ax + b = Ke^{at}$$

where $K = e^C$. When we substitute $t = 0$ and denote the resulting value of x by x_0, we find that $K = ax_0 + b$. So

$$ax + b = (ax_0 + b)e^{at}.$$

Finally, we solve this equation for the solution $x = x(t)$ of Equation (1):

$$x(t) = \left(x_0 + \frac{b}{a}\right)e^{at} - \frac{b}{a}. \tag{2}$$

In this development, we have assumed that $ax + b > 0$, but this same formula gives the solution in the case $ax + b < 0$ as well (see Problem 11). In Problem 15 we outline a method by which Equation (1) may be solved when the coefficients a and b are functions of t rather than constants, but Solution (2) for the constant-coefficient case will be sufficient for the following applications.

POPULATION GROWTH WITH IMMIGRATION

Consider a population $P(t)$ with constant birth and death rates (β and δ, respectively), as in Section 6-5, but also with a constant immigration rate of I persons per year entering the country. To account for the immigration, our derivation of Equation (2) in Section 6-5 must be amended as follows:

$$P(t + \Delta t) - P(t) = \text{(births)} - \text{(deaths)} + \text{(immigrants)}$$
$$\approx \beta P(t)\,\Delta t - \delta P(t)\,\Delta t + I\,\Delta t,$$

so

$$\frac{P(t + \Delta t) - P(t)}{\Delta t} \approx (\beta - \delta)P(t) + I.$$

We take limits as $\Delta t \to 0$ and thus obtain the linear first-order differential equation

$$\frac{dP}{dt} = kP + I \tag{3}$$

with constant coefficients $k = \beta - \delta$ and I. According to Formula (2), the solution of (3) is

$$P(t) = P_0 e^{kt} + \frac{I}{k}(e^{kt} - 1). \tag{4}$$

The first term on the right-hand side is the effect of natural population growth; the second term is the effect of immigration.

EXAMPLE 1   Consider the U.S. population with $P_0 = 222$ million in 1980 ($t = 0$). Suppose that we ask about the effect of allowing immigration at the rate of half a million people per year for the next 20 years, assuming a natural growth rate of 1% annually, so that $k = 0.01$. Then

$$P_0 e^{kt} = 222e^{(0.01)(20)} \approx 271.2 \quad \text{(million)},$$

and

$$\frac{I}{k}(e^{kt} - 1) = \frac{0.5}{0.01}(e^{(0.01)(20)} - 1) \approx 11.1 \quad \text{(million)}.$$

Thus the effect of the immigration would be to increase the U.S. population in the year 2000 from 271.2 million to 282.3 million.

SAVINGS ACCOUNT WITH CONTINUOUS DEPOSITS

Consider the savings account of Section 6-5, containing A_0 dollars initially and earning interest at the annual rate r compounded continuously. In addition, we now suppose that deposits are added to this account at the rate of Q dollars per year. To simplify the mathematical model, we assume that these deposits are made continuously rather than (for instance) monthly. We may then regard the amount $A(t)$ in the account at time t as a "population" of dollars, having a natural annual growth rate r and with "immigration" (deposits) at the rate of Q dollars annually. Then by merely changing the notation in Equations (3) and (4), we get the differential equation

$$\frac{dA}{dt} = rA + Q \tag{5}$$

with solution

$$A(t) = A_0 e^{rt} + \frac{Q}{r}(e^{rt} - 1). \tag{6}$$

EXAMPLE 2 Suppose that you wish to arrange, at the time of her birth, for your daughter to have 20 thousand dollars available for college expenses at age 18. You plan to do so by making frequent small—essentially continuous—deposits in a savings account, at the rate of Q thousand dollars each year. This account accumulates 6% annual interest compounded continuously. What should Q be so that you may achieve your goal?

Solution With $A_0 = 0$ and $r = 0.06$, we are asking for the value of Q so that Equation (6) yields the result $A(18) = 20$. That is, we are to find Q so that

$$20 = \frac{Q}{0.06}(e^{(0.06)(18)} - 1).$$

When we solve this equation, we find that $Q = 0.61707$. Thus you should deposit $617.07 per year, or about $51.42 per month, in order to have $20,000 in the account after 18 years. You may wish to verify that your total deposits will be $11,107.23 and that the total interest accumulated will be $8892.77.

COOLING AND HEATING

According to Newton's law of cooling (or heating!), the time rate of change of the temperature T of a body is proportional to the difference between T and the temperature A of its surroundings, under the assumption that A is constant. We may translate this law into the language of differential equations by writing

$$\frac{dT}{dt} = -k(T - A). \tag{7}$$

Here k is a positive constant; the minus sign is needed to make $T'(t)$ negative when T exceeds A.

From Equation (2) we obtain the solution of Equation (7):

$$T(t) = A + (T_0 - A)e^{-kt}. \tag{8}$$

EXAMPLE 3   A 5-pound roast, initially at 50°F, is put into a 375°F-oven when $t = 0$; it's found that the temperature of the roast $T(t)$ (since T is a function of time t) is 125°F when $t = 75$ (minutes). When will the roast be medium rare, a temperature of 150°F?

Solution   Here we take $A = 375$ and $T_0 = 50$. Then Equation (8) tells us that

$$T = 375 - 325e^{-kt}.$$

We know that $T = 125$ when $t = 75$, and using this information, we find that

$$k = \frac{1}{75}\ln\frac{325}{250} \approx 0.0035.$$

So all we need do is solve the equation

$$150 = 375 - 325e^{-kt}.$$

We find that t is about 105, so the roast should remain in the oven for about another 30 minutes.

DIFFUSION OF INFORMATION

Let $N(t)$ denote the number of people (in a fixed population P) who by time t have heard a certain piece of information that is being spread by the mass media. Under certain common conditions, the time rate of increase of N will be proportional to the number of people who have not yet heard the piece of information. That is,

$$\frac{dN}{dt} = k(P - N). \tag{9}$$

If $N(0) = 0$, the solution to Equation (9) is

$$N(t) = P(1 - e^{-kt}). \tag{10}$$

If P and some value $N(t_1)$ are known, we can then solve for k and thereby determine $N(t)$ for all t.

EXAMPLE 4   (Elimination of Pollutants)   Consider a lake with a volume of 8 billion cubic feet and an initial pollutant concentration of 0.25%. Suppose that an inflowing river brings in 500 million cubic feet of water daily with a pollutant concentration of 0.05%, and that an outflowing river also removes 500 million cubic feet of the lake water daily. We make the simplifying assumption that the water in the lake, including that removed by the second river, is perfectly mixed at all times. If so, how long will it take to reduce the pollutant concentration in the lake to 0.10%?

Solution Let $x(t)$ denote the amount of pollutants in the lake after t days, in millions of cubic feet. Then $x_0 = (0.0025)(8000) = 20$. We want to know when $x = (0.001)(8000) = 8$.

We construct a mathematical model for this situation by estimating the increment in x during a short time interval Δt. Thus

$$x(t + \Delta t) - x(t) = \text{(pollutant in)} - \text{(pollutant out)}$$

$$\approx (0.0005)(500)\,\Delta t - \frac{x(t)}{8000}\, 500\,\Delta t$$

$$= \frac{1}{4}\,\Delta t - \frac{x}{16}\,\Delta t.$$

It follows that

$$\frac{dx}{dt} = \frac{1}{4} - \frac{x}{16}.$$

With $x_0 = 20$, Equation (2) gives the solution

$$x(t) = 4 + 16e^{-t/16}.$$

We can find when $x(t) = 8$ by solving the equation

$$8 = 4 + 16e^{-t/16};$$

this gives

$$t = 16 \ln 4 \approx 22.2 \qquad \text{(days)}.$$

MOTION WITH RESISTANCE

Here's a common situation: A body moves vertically near the surface of the earth. There are usually only two significant forces influencing the motion of the body:

> The force of gravity;
> The force of air resistance, which we assume
> is proportional to the speed of the body.

If the height y of the body is measured upward from ground level, then y is never negative and the value $y = 0$ corresponds to ground level. With this choice of coordinates, the gravitational force is given by $F_G = -mg$, where $g = 32$ ft/sec^2 if m is in slugs and F_G is in pounds. (Take $g = 9.8$ m/sec^2 if m is in kilograms and F_G is in newtons.)

The force of air resistance is

$$F_R = -kv \tag{11}$$

where k is a positive constant and $v = dy/dt$ is the velocity of the body. Note that the minus sign in Equation (11) makes F_R positive (an upward force) if the body is falling (v is negative), while it makes F_R negative (a downward force) if the body is rising (v is positive).

The net force acting on the body is then

$$F = F_R + F_G = -kv - mg,$$

and Newton's law of motion $F = m(dv/dt)$ yields the equation

$$m \frac{dv}{dt} = -kv - mg.$$

Thus

$$\frac{dv}{dt} = -\rho v - g \tag{12}$$

where $\rho = k/m > 0$.

Equation (12) is a linear first order differential equation, whose solution—given by Equation (2)—is

$$v(t) = \left(v_0 + \frac{g}{\rho}\right) e^{-\rho t} - \frac{g}{\rho}. \tag{13}$$

Here, $v_0 = v(0)$ is the initial velocity of the body. Note that

$$\lim_{t \to \infty} v(t) = -\frac{g}{\rho}. \tag{14}$$

Thus the speed of a body falling with air resistance does *not* increase indefinitely. Instead it approaches a *finite* limiting speed, or **terminal speed,**

$$v_\tau = \frac{g}{\rho} = \frac{mg}{k}. \tag{15}$$

This fact is what makes the parachute a practical invention; it even explains the occasional survival of people who fall without parachutes from high-flying airplanes.

We now rewrite Equation (13) in the form

$$\frac{dy}{dt} = (v_0 + v_\tau) e^{-\rho t} - v_\tau. \tag{16}$$

Antidifferentiation gives

$$y(t) = -\frac{1}{\rho}(v_0 + v_\tau) e^{-\rho t} - v_\tau t + C.$$

We substitute 0 for t, and let $y(0) = y_0$ denote the initial height of the body. We thus find that

$$C = y_0 + \frac{1}{\rho}(v_0 + v_\tau),$$

and so

$$y(t) = y_0 - v_\tau t + \frac{1}{\rho}(v_0 + v_\tau)(1 - e^{-\rho t}). \tag{17}$$

Equations (16) and (17) give the velocity v and height y of a body falling under the influence of gravity and air resistance. The formulas depend on the initial height y_0 of the body, its initial velocity v_0, and what might be called the "drag coefficient" ρ, the constant such that the acceleration due to air resistance is $a_R = -\rho v$. The formulas are also given in terms of the terminal speed v_τ.

For a person descending with the aid of a parachute, a typical value of ρ is 1.5, which corresponds to a terminal speed of $v_\tau \approx 21.3$ feet per second, or about 14.5 mph. Without a parachute, a typical value is $\rho = 0.15$, which yields a terminal speed of $v_\tau \approx 213$ feet per second, or about 145 mph. With an unbuttoned overcoat flapping in the wind in place of a parachute, an unlucky skydiver can increase ρ to perhaps as much as 0.5, which gives a terminal speed of $v_\tau \approx 64$ feet per second, or about 43.6 mph.

EXAMPLE 5   A man bails out of an airplane at an altitude of 10,000 feet, free falls for 20 seconds, and then opens his parachute. How long does it take him to reach the ground, assuming that $\rho = 0.15$ without parachute and that $\rho = 1.5$ with parachute?

Solution   First, we want to find his velocity after the 20 seconds of free fall. To do so, we substitute the initial data $y_0 = 10{,}000$ and $v_0 = 0$, as well as the given values $\rho = 0.15$ and $v_\tau = 213$, into Equation (16). We find that

$$v = 213(e^{-(0.15)(20)} - 1) \approx -202.4$$

feet per second. His height after 20 seconds of free fall is computed by substitution of the same values into Equation (17):

$$y = 10{,}000 - (213)(20) + \frac{1}{0.15}(213)(1 - e^{-(0.15)(20)}) \approx 7089 \qquad \text{(feet)}.$$

Now we move to the second stage of his descent. To find his height after t seconds of fall *with* parachute, we are actually working a new problem, with the initial data being the man's velocity and height after his 20 seconds of free fall. We substitute $y_0 = 7089$, $v_0 = -202.4$, $\rho = 1.5$, and $v_\tau = 21.3$ into Equation (17), and find that

$$y = 7089 - (21.3)t + \frac{1}{1.5}(-202.4 + 21.3)(1 - e^{-(1.5)t})$$

$$= 6968 - (21.3)t + (120.7)e^{-(1.5)t}.$$

To find the value of t at which $y = 0$, we must solve the equation

$$0 = 6968 - (21.3)t + (120.7)e^{-(1.5)t};$$

that is,

$$t = 327.1 + (5.668)e^{-(1.5)t}.$$

The method of repeated substitution (Section 3-4) is a natural way to solve this equation. For example, an initial guess of $t_0 = 300$ yields the successive approximations

$$t_1 = 327.1 + (5.668)e^{-(1.5)(300)} \approx 327.10 \qquad \text{(seconds)}$$

$$t_2 = 327.1 + (5.668)e^{-(1.5)(327.1)} \approx 327.10 \qquad \text{(seconds)}.$$

Thus the total time of descent is about $327 + 20 = 347$ seconds, or 5 minutes 47 seconds.

In each of Problems 1–4, use the method of derivation of Equation (2), rather than the equation itself, to find the solution of the given differential equation satisfying the indicated initial condition.

1 $\dfrac{dy}{dx} = y + 1$; $y(0) = 1$.

2 $\dfrac{dy}{dx} = 2 - y$; $y(0) = 3$.

3 $\dfrac{dy}{dx} = 2y - 3$; $y(0) = 2$.

4 $\dfrac{dy}{dx} = \dfrac{1}{4} - \dfrac{y}{16}$; $y(0) = 20$.

5 A certain city had a population of 1.5 million in 1980. Assume that it grows continuously at a 4% annual rate and also absorbs 50,000 newcomers per year. What will its population be in the year 2000?

6 A cake is removed from an oven at 210°F and left to cool at room temperature, which is 70°F. After 30 minutes the temperature of the cake is 140°F. When will it be 100°F? (*Suggestion:* Apply Newton's law of cooling.)

7 Payments are made on a mortgage (original loan) of P_0 dollars continuously at the constant rate of c dollars per month. Let $P(t)$ denote the principal (amount still owed) after t months, and let r denote the monthly interest rate paid by the borrower. (For example, $r = 0.06/12 = 0.005$ if the annual interest rate is 6%.) Derive the differential equation

$$\dfrac{dP}{dt} = rP - c, \quad P(0) = P_0.$$

8 An auto loan of $3600 is to be paid off continuously over a period of 36 months. Apply the result of Problem 7 to determine the monthly payment required if the annual interest rate is (a) 12%; (b) 18%.

9 A certain piece of dubious information about phenylethylamine in the drinking water began to spread one day in a city with a population of 100,000. Within a week, 10,000 people had heard this rumor. Assuming that the rate of increase of the number who have heard it is proportional to the number who have not yet heard it, how long will it be before half the population of the city has heard this piece of information?

10 A tank contains 1000 liters of a solution consisting of 50 pounds of salt dissolved in water. Pure water is pumped into the tank at the rate of 5 liters per second, and the mixture—kept uniform by stirring—is pumped out at the

same rate. After how many seconds will there remain only 10 pounds of salt in the tank?

11 Derive the solution (2) of Equation (1) under the assumption that $ax + b < 0$.

12 Suppose that a body moves through a resisting medium with resistance proportional to its velocity v, so that $dv/dt = -kv$.
(a) Show that its velocity and position at time t are given by

$$v = v_0 e^{-kt} \quad \text{and} \quad x = x_0 + \dfrac{v_0}{k}(1 - e^{-kt}).$$

(b) Conclude that the body travels only a *finite* distance v_0/k.

13 Suppose that a motorboat is moving at 40 feet per second when its motor suddenly quits, and that 10 seconds later the boat has slowed to 20 feet per second. Assume, as in Problem 12, that the resistance it encounters while coasting is proportional to its velocity. How far will the motor boat coast in all?

14 The acceleration of a certain sports car is proportional to the difference between 250 kph and the velocity of the sports car. If this machine can accelerate from rest to 100 kph in 10 seconds, how long will it take for it to accelerate from rest to 200 kph?

15 Consider the linear first order differential equation

$$x'(t) + p(t)x(t) = q(t)$$

with variable coefficients. Let $P(t)$ be an antiderivative of $p(t)$. Multiply both sides of the above equation by $e^{P(t)}$, and note that the left-hand side of the resulting equation is $D_t[e^{P(t)}x(t)]$. Conclude by antidifferentiation that

$$x(t) = e^{-P(t)}\left[\int e^{P(t)}q(t)\,dt + C\right].$$

16 Use the method of Problem 15 to derive the solution

$$x(t) = x_0 e^{-at} + b\,\dfrac{e^{ct} - e^{-at}}{a + c}$$

of the differential equation $dx/dt + ax = be^{ct}$ (under the assumption that $a + c \neq 0$).

17 A 30-year-old woman accepts an engineering position with a starting salary of $30,000 per year. Her salary S increases exponentially, with

$$S(t) = 30e^{(0.05)t}$$

thousand dollars after t years. Meanwhile, 12% of her salary is deposited continuously in a retirement account, which accumulates interest at a continuous annual rate of 6%.

(a) Estimate ΔA in terms of Δt to derive the following equation for the amount $A(t)$ in her retirement account at time t:

$$\frac{dA}{dt} - (0.06)A = (3.6)e^{(0.05)t}.$$

(b) Apply the result of Problem 16 to compute $A(40)$, the amount available for her retirement at age 70.

18 According to a newspaper account, a paratrooper survived a training jump from 1200 feet when his parachute failed to open but provided some resistance by flapping unopened in the wind. Allegedly he hit the ground at 100 mph after falling for 8 seconds. Test the accuracy of this account. (*Suggestion:* Find ρ (in Equation (12)) by assuming a terminal velocity of 100 mph. Then calculate the time required to fall 1200 feet.)

19 A ball for which $\rho = 0.1$ (in Equation (12)) is thrown straight upward from the ground at 160 feet per second.

(a) What is the maximum height reached by the ball, and when is that height attained?

(b) What is the total time that the ball spends in the air, and with what speed does it strike the ground?

20 Suppose that a body moving horizontally through a medium encounters a resistive force that is proportional to the *square* of the velocity, so that $dv/dt = -\rho v^2$. Show that

$$v(t) = \frac{v_0}{1 + v_0 \rho t} \quad \text{and} \quad x(t) = x_0 + \frac{1}{\rho} \ln(1 + v_0 \rho t).$$

Note that $\lim_{t \to \infty} x(t) = \infty$, in contrast with the result of Problem 12.

21 Suppose that the motorboat of Problem 13 encounters resistance proportional to v^2 (as in Problem 20). Starting with the same initial data as in Problem 13, find how far the boat has coasted after 5 minutes, and how fast it is then moving.

---

*6-7

Applications to Economics

Many economic variables typically exhibit—at least in the short run—the pattern of natural growth or decline that we discussed in Section 6-5. Consequently integrals involving exponential functions are frequently encountered in calculating such things as the amount of a commodity consumed over a period of time, the present value of income spread over a period of time, or the total cost of a project that is carried out during a period of monetary inflation.

CONSUMPTION OF A COMMODITY

Suppose that a certain commodity is consumed continuously from time $t = a$ to time $t = b$, with consumption rate $C(t)$ at time t. We take this to imply that, if $C_1 \le C(t) \le C_2$ during a time interval of length Δt, then the amount consumed during this interval satisfies

$$C_1 \, \Delta t \le (\text{amount consumed}) \le C_2 \, \Delta t. \tag{1}$$

To compute the total amount A consumed, we consider a regular partition of $[a, b]$ into n equal subintervals of length Δt. If $C(t_i^{\flat})$ and $C(t_i^{\#})$ are the minimum and maximum consumption rates, respectively, for t in the ith subinterval $[t_{i-1}, t_i]$, then by (1) the amount ΔA_i consumed during this subinterval satisfies the inequality

$$C(t_i^{\flat}) \, \Delta t \le \Delta A_i \le C(t_i^{\#}) \, \Delta t.$$

We add these inequalities for $i = 1, 2, 3, \ldots, n$ and find that

$$\sum_{i=1}^{n} C(t_i^{\flat}) \, \Delta t \le A \le \sum_{i=1}^{n} C(t_i^{\#}) \, \Delta t.$$

Both these sums are Riemann sums for the integral $\int_a^b C(t)\,dt$. We take the limits as $\Delta t \to 0$, and the squeeze law of limits therefore gives

$$A = \int_a^b C(t)\,dt. \tag{2}$$

Thus the total amount consumed is the integral of the consumption rate.

A typical case is that in which the consumption rate $C(t)$ *increases continuously at an annual rate of r*. This means that $dC/dt = rC$, and it follows from Equation (4) in Section 6-5 that

$$C(t) = C_0 e^{rt} \tag{3}$$

where $C_0 = C(0)$. Alternatively, Equation (3) may be taken as the *definition* of the statement that the consumption rate increases continuously at an annual rate r.

EXAMPLE 1   Assume that the known global reserves of petroleum amount to 6×10^{11} barrels, and that the current consumption rate is 2×10^{10} barrels per year. At this rate, should it remain constant, the known global reserves would be exhausted in 30 years.
(a) If the consumption rate continues to increase continuously at the rate of 3.5% annually, how long will the known global reserves last?
(b) How much would the consumption rate have to be decreased annually in order for the known global reserves of oil to last 50 more years?

Solution   With the world consumption rate of $C(t) = C_0 e^{rt}$ barrels per year t years from now, Formula (2) gives

$$A = \int_0^T C_0 e^{rt}\,dt = \frac{C_0}{r}(e^{rT} - 1) \tag{4}$$

for the amount A of oil consumed in T years.

For (a), we substitute the given values $A = 6 \times 10^{11}$, $C_0 = 2 \times 10^{10}$, and $r = 0.035$ into Equation (4) and solve for

$$T = \frac{1}{0.035}\ln\left(\frac{(0.035)(6 \times 10^{11})}{2 \times 10^{10}} + 1\right) \approx 20.51$$

years. Thus the 3.5% annual increase in consumption will cut by about 9.5 years the time the known reserves of oil will last.

Part (b) asks for the value of r such that Equation (4) is satisfied with $A = 6 \times 10^{11}$, $C_0 = 2 \times 10^{10}$, and $T = 50$; that is,

$$6 \times 10^{11} = \frac{2 \times 10^{10}}{r}(e^{50r} - 1).$$

We can solve this equation for r by rewriting it in the form

$$f(r) = e^{50r} - 30r - 1 = 0$$

and then using Newton's method (Section 3-4). We seek a *negative* value of r (a rate of *decrease*), so we begin with $r_0 = -0.02$. The iterative formula

$$r_{n+1} = r_n - \frac{f(r_n)}{f'(r_n)} = r_n - \frac{e^{50r_n} - 30r_n - 1}{50e^{50r_n} - 30}$$

then gives $r_1 = -0.0228$ and $r_2 = r_3 = -0.0225$. Thus the consumption rate would have to be *reduced* by 2.25% annually in order to extend the life of the known oil reserves to 50 years.

PRESENT VALUE OF FUTURE INCOME

A dollar in your hand today is worth more than a dollar to be received at some future date. For example, a dollar that you have today may be invested. If the interest rate is r, compounded continuously, then after t years the original dollar will have grown to e^{rt} dollars, as we saw in Section 6-5. On this basis, we define the **present value at continuous interest rate** r of a dollar to be received at time t (a "future dollar") to be e^{-rt} dollars. As an alternative motivation, we can think of inflation at a continuous rate of r annually, so that a dollar t years hence has the purchasing power of only e^{-rt} present dollars.

For example, if we use $r = 0.1$, corresponding to an annual interest rate of 10% or a like inflation rate, and take t to be 10 years, we find that the present value of a dollar to be paid to us 10 years from now is e^{-1} dollars, or about 37¢.

Now suppose that income is to be received from time $t = a$ to time $t = b$ at the rate of $f(t)$ dollars per year at time t. For instance, imagine yourself the lucky beneficiary of a trust fund, one that pays you $5000 tax-free dollars per year, starting in 5 years and lasting for 20 years. Then, in what follows, you would take $a = 5$, $b = 25$, and $f(t) \equiv 5000$.

We are interested in the total present value P of this income between times $t = a$ and $t = b$. Let t_i^* be a point of the ith subinterval $[t_{i-1}, t_i]$ of a regular partition of $[a, b]$ into n equal subintervals of length Δt each. Between times t_{i-1} and t_i, you will receive approximately $f(t_i^*) \, \Delta t$ dollars, and the present value ΔP_i of this part of your income will be given approximately by

$$\Delta P_i \approx e^{-rt_i^*} f(t_i^*) \, \Delta t.$$

It follows that P itself is given approximately by

$$P \approx \sum_{i=1}^{n} e^{-rt_i^*} f(t_i^*) \, \Delta t.$$

Since this approximation is a Riemann sum, we *define* the **present value** P of income to be received at the rate of $f(t)$ dollars per year from time $t = a$ to time $t = b$ to be

$$P = \int_a^b e^{-rt} f(t) \, dt, \tag{5}$$

where r is the interest or "discount rate."

For instance, using $r = 8\%$ and the above-mentioned values $a = 5$, $b = 25$, and $f(t) = 5000$ (constant), the present value of the 25-year trust fund we mentioned is

$$P = \int_5^{25} e^{-(0.08)t} (5000) \, dt$$

$$= \left[\frac{5000}{-0.08} e^{-(0.08)t} \right]_5^{25} = \$33,436.55.$$

By contrast, this trust fund will pay you a total of $20 \times \$5000 = \$100,000$ over its duration.

EXAMPLE 2 Suppose that you retire at age 65 with a life expectancy of 15 years. Your company retirement fund offers a choice of the following three plans:

Plan A: $10,000 per year paid continuously for life (which we assume to be for 15 years for the purpose of computing the present value of the income from plan A).

Plan B: $15,000 per year paid continuously for 10 years.

Plan C: $100,000 paid immediately.

Which of these plans is the best deal in terms of present value at 6% annually?

Solution The present value of plan A is given by Formula (5) as

$$P_A = \int_0^{15} (10,000)e^{-(0.06)t}\, dt$$

$$= (10,000)\left[\frac{e^{-(0.06)t}}{-0.06}\right]_0^{15} \approx \$98,905.$$

Similarly, the present value of plan B is

$$P_B = \int_0^{10} (15,000)e^{-(0.06)t}\, dt$$

$$= (15,000)\left[\frac{e^{-(0.06)t}}{-0.06}\right]_0^{10} \approx \$112,797.$$

Since $P_C = \$100,000$, it follows that plan B is the best deal and plan A the worst.

INFLATION OF COSTS

Consider a project that will require the expenditure or investment of funds during a period from time $t = a$ to time $t = b$, during which period continuous monetary inflation at the annual rate r occurs. If the function $g(t)$ gives the rate of expenditure in terms of *present value* dollars, then the actual rate of expenditure (in terms of dollars counted as they are spent) will be $e^{rt}g(t)$.

We can think of the project as "consuming" dollars at the rate of $C(t) = e^{rt}g(t)$ dollars per year at time t. Then Formula (2) gives the actual cost of the project as

$$C = \int_a^b e^{rt}g(t)\, dt. \tag{6}$$

The amount by which C exceeds the present value

$$P = \int_a^b g(t)\, dt \tag{7}$$

of the project may be thought of as inflation-generated cost overrun. A typical example would be the escalation in the cost of a construction project because of construction delays, during which inflation occurs.

EXAMPLE 3 Consider a solar power plant that would cost $1 billion if it could be built immediately and instantaneously. Suppose that the start of construction is delayed by 5 years because of government red tape and anti-trust suits filed against the consortium of sponsoring companies. Once construction has begun, it takes another 5 years to complete the power plant. Assume a 10% continuous annual inflation of construction costs during this 10-year period. What will be the actual cost of construction of the power plant?

Solution We assume that, during the 5-year period of construction, there is a constant rate of expenditures in terms of present-value dollars. This means that we use $g(t) = \frac{1}{5}$ (in billions of dollars per year). Then Formula (6) gives the total actual cost as

$$ C = \int_5^{10} \tfrac{1}{5} e^{(0.1)t} \, dt = \left[2e^{(0.1)t} \right]_5^{10} = 2(e - e^{1/2}), $$

or about $2.14 billion. Thus the 10-year delay results in more than doubling the apparent cost of the power plant.

6-7 PROBLEMS

In Problems 1–4, assume that the known global reserves of natural gas are 1.2×10^{15} cubic feet and that the current rate of consumption is 3×10^{13} cubic feet per year.

1 If the rate of consumption increases continously at 5% per year, how long will the known global reserves of natural gas last?

2 If the global reserves of natural gas are actually five times as large as now known, how long will they last with consumption increasing continuously at 5% per year?

3 If through conservation measures the rate of increase of consumption is reduced to 2% per year, how long will the *known* global reserves of natural gas last?

4 How much would the consumption rate have to be decreased in order for the known global reserves of natural gas to last for 60 years?

5 Suppose that you have signed a contract under which you will be paid $10,000 per year (on a continuous basis) during the next 5 years.

(a) What is the present value at 6% continuous interest rate of the $50,000 you will be paid under this contract?

(b) What is the present value of the $10,000 that you will be paid during the fifth year of this contract?

6 Suppose that you have signed a contract under which you will be paid a salary continuously for 10 years, starting at $10,000 per year and increasing steadily to $20,000 per year after 10 years. Thus your salary (in thousands of dollars) after t years will be $f(t) = 10 + t$.

(a) What is the present value (at 6% continuous annual interest) of this contract?

(b) What will be the total amount of dollars you actually receive during the 10-year period? Use the integral formula

$$ \int t e^{at} \, dt = \frac{e^{at}}{a^2}(at - 1) + C. $$

7 Suppose that you have $100,000 in book royalties payable now, but for income-tax reasons you desire to be paid at a constant rate for 10 years starting 5 years hence. At what annual rate should you be paid in order that the present value of the amount you receive will be the fair sum of $100,000? Assume 8% continuous annual interest rate.

8 Consider a nuclear carrier for which construction will require materials and services with current value of $1.5 billion. Suppose also that the annual rate of inflation is and remains at 7.5%.

(a) How much will the carrier cost if it is begun now and built over a 2-year period?

(b) Repeat Part (a), but assume a 3-year construction period.

(c) State carefully what assumptions you are making about the rate of construction in the other parts of this problem.

(d) How much will the carrier cost if its construction is delayed for 4 years for political reasons and then completed during the subsequent 2-year period?

9 Suppose that a small but rapidly growing company had a total income last year of $1 million and that production costs were $0.9 million, so that net profit was $0.1 million. Suppose also that during the foreseeable future its total income increases linearly by $0.2 million annually, but that its production costs increase exponentially at a continuous annual rate of 12%.

(a) What will be the total profits of the company during the next 5 years?

(b) What will be its profits for the subsequent 5 years?

A **separable** first order differential equation is one that can be written in the form

$$f(x)\frac{dx}{dt} = g(t), \qquad (1)$$

so that the variables may be "separated" by writing $f(x)\,dx = g(t)\,dt$. It is easy to solve this sort of differential equation by writing

$$\int f(x)\,dx = \int g(t)\,dt + C, \qquad (2)$$

provided that the antiderivatives $F(x) = D_x^{-1}f(x)$ and $G(t) = D_t^{-1}g(t)$ can be found. To see that (2) follows from (1), it is enough to note that (1) says that $D_t F(x(t)) = D_t G(t)$, which in turn implies that

$$F(x(t)) = G(t) + C. \qquad (3)$$

When a particular value of x is available, it may be used to determine the value of the constant of integration C. For instance, if $x_0 = x(0)$, then substitution of $t = 0$ into Equation (3) gives $C = F(x_0) - G(0)$. The problem of solving (1) subject to the initial condition $x(0) = x_0$ is called an **initial value problem,** and the resulting solution is called a **particular solution** of (1); in contrast, (3) is called the **general solution** of Equation (1).

EXAMPLE 1 Solve the initial value problem

$$\frac{dx}{dt} = 4xt^3, \qquad x(0) = 2.$$

Solution From

$$\int \frac{dx}{x} = \int 4t^3\,dt$$

we immediately obtain the general solution

$$\ln x = t^4 + C.$$

Substitution of $t = 0$ into *both* sides of the general solution gives $C = \ln 2$. Then we apply the natural exponential function to each side of the equation in order to solve for x; thus we find that

$$x = e^{t^4 + \ln 2} = 2e^{t^4}.$$

LIMITED POPULATIONS

In Section 6-5 we saw that a population $P(t)$ with constant birth rate β and constant death rate δ satisfies the differential equation

$$\frac{dP}{dt} = (\beta - \delta)P. \qquad (4)$$

This equation also applies when β and δ are variable. One important example is the case when β is a linear decreasing function of the population P, so that $\beta = \beta_0 - \beta_1 P$ where β_0 and β_1 are positive constants. If the death rate

$\delta = \delta_0$ remains constant, then (4) takes the form

$$\frac{dP}{dt} = (\beta_0 - \beta_1 P - \delta_0)P,$$

or

$$\frac{dP}{dt} = kP(M - P), \tag{5}$$

where $k = \beta_1$ and $M = (\beta_0 - \delta_0)/\beta_1$. We assume that $\beta_0 > \delta_0$ so that $M > 0$.

Equation (5) is called the **logistic equation.** If we assume that $P < M$, it may be solved as follows.

$$\int \frac{dP}{P(M - P)} = \int k \, dt;$$

$$\frac{1}{M} \int \left(\frac{1}{P} + \frac{1}{M - P} \right) dP = \int k \, dt;$$

$$\ln \left(\frac{P}{M - P} \right) = kMt + C.$$

Exponentiation gives

$$\frac{P}{M - P} = Ae^{kMt}$$

where $A = e^C$. We substitute $t = 0$ into both sides of this last equation, and this gives $A = P_0/(M - P_0)$. So

$$\frac{P}{M - P} = \frac{P_0 e^{kMt}}{M - P_0}.$$

This equation is easily solved for

$$P(t) = \frac{MP_0}{P_0 + (M - P_0)e^{-kMt}}. \tag{6}$$

Whereas we assumed in deriving (6) that $P < M$, this restriction is unnecessary, because we can verify by direct substitution into Equation (5) that $P(t)$ as given by (6) satisfies the logistic equation whether $P < M$ or $P \geqq M$.

If the initial population satisfies $P_0 < M$, then (6) shows that $P(t) < M$ for all $t > 0$ and also that

$$\lim_{t \to \infty} P(t) = M. \tag{7}$$

Thus a population that satisfies the logistic equation is *not* like a naturally growing population; it does *not* grow without bound. Instead, the population approaches the finite *limiting population M* as $t \to \infty$. But since

$$\frac{dP}{dt} = kP(M - P) > 0$$

in this case, we see that the population is steadily increasing. Moreover, differentiation with respect to t gives

$$\frac{d^2P}{dt^2} = \left[\frac{d}{dP} \left(\frac{dP}{dt} \right) \right] \left(\frac{dP}{dt} \right) = (kM - 2kP) \cdot kP(M - P).$$

So the graph of $P(t)$ has an inflection point where $P = M/2$. Therefore the graph of P has the shape of the lower curve in Figure 6.20. The population increases at an increasing rate until $P = M/2$, and thereafter increases at a decreasing rate.

If $P_0 > M$, then a similar analysis shows that $P(t)$ is a steadily decreasing function with a graph like the upper curve in Figure 6.20.

In 1845 the Belgian demographer Verhulst used the 1790 through 1840 population data for the United States to predict the U.S. population through the year 1930, under his assumption that it would continue to satisfy the logistic equation. With $P_0 = 3.9$ (in millions), $M = 197.3$ (in millions), and $k = 0.0001589$, Equation (6) gives the remarkable results shown in the table of Figure 6.21. Of course, the assumed limiting population of $M = 197.3$ millions has now been exceeded, so the population of the United States has *not* continued to satisfy the logistic equation in the decades following 1930.

EXAMPLE 2 Suppose that in 1910 the population of a certain country was 50 million and was growing at the rate of 750,000 people per year at that time. Suppose also that in 1940 its population was 100 million and was then growing at the rate of 1 million per year. Assume that this population satisfies the logistic equation. Determine both the limiting population M and the predicted population for the year 2000.

Solution We substitute the two given pairs of data into Equation (5) and find that

$$0.75 = 50k(M - 50),$$
$$1.00 = 100k(M - 100).$$

We solve simultaneously and find that $M = 200$ and $k = 0.0001$. Thus the limiting population of the country in question is 200 million. With these values of M and k and with $t = 0$ corresponding to the year 1940 (in which $P_0 = 100$), we find that, as a consequence of Equation (6), the population in the year 2000 will be

$$P = \frac{(100)(200)}{100 + (200 - 100)e^{-(0.0001)(200)(60)}},$$

or about 153.7 million people.

We describe next some situations that illustrate the varied circumstances in which the logistic equation is a satisfactory mathematical model.

1 *Limited Environment Situation.* A certain environment can support a population of at most M individuals. It's then reasonable to expect the growth rate $\beta - \delta$ (the combined birth and death rates) to be proportional to $M - P$, since we may think of $M - P$ as the "potential" for further expansion. Then $\beta - \delta = k(M - P)$, so that

$$\frac{dP}{dt} = (\beta - \delta)P = kP(M - P).$$

The classic example of a limited environment situation is a fruit fly population in a closed container.

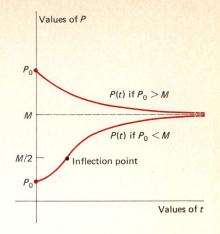

6.20 Graphs of the two solutions of the logistic equation.

Year	Actual U.S. population (millions)	Calculated population
1790	3.9	3.9
1800	5.3	5.3
1810	7.2	7.2
1820	9.6	9.7
1830	12.9	13.0
1840	17.1	17.4
1850	23.2	23.1
1860	31.4	30.2
1870	38.6	39.2
1880	50.2	49.9
1890	62.9	62.5
1900	76.0	76.6
1910	92.0	91.7
1920	106.5	107.1
1930	123.2	122.1

6.21 Actual United States population data compared with Verhulst's predictions.

2 *Competition Situation.* If the birth rate β is constant but the death rate δ is proportional to P, so that $\delta = \alpha P$, then

$$\frac{dP}{dt} = (\beta - \alpha P)P = kP(M - P).$$

This might be a reasonable working hypothesis in a study of a cannibalistic population, in which all deaths result from chance encounters between individuals. Of course, competition between individuals is not usually so deadly, nor its effects so immediate and so decisive.

3 *Joint Proportion Situation.* Let $P(t)$ denote the number of individuals in a constant susceptible population M that is infected with a certain contagious and incurable disease. The disease in question is spread by chance encounters. Then $P'(t)$ should be proportional to the product of the number P of individuals having the disease and the number $M - P$ of those not having it, so that

$$\frac{dP}{dt} = kP(M - P).$$

Again we discover that the mathematical model is the logistic equation. The mathematical description of the spread of a rumor in a population of M individuals is identical.

EXAMPLE 3 Suppose that at time $t = 0$, half of a population of 100,000 persons have heard a certain rumor, and that the number of those who have heard it is then increasing at the rate of 1000 persons per day. How long will it take for this rumor to spread to 80% of the population?

Solution Let us work in units of thousands of persons. Then we take $M = 100$ for the total fixed population. We substitute $M = 100$, $P'(0) = 1$, and $P_0 = 50$ into the logistic equation, and thereby obtain

$$1 = k(50)(100 - 50), \quad \text{so that} \quad k = 0.0004.$$

If t denotes the number of days until 80 thousand people have heard the rumor, then Equation (6) gives

$$80 = \frac{(50)(100)}{50 + (100 - 50)e^{-(0.04)t}},$$

so that t is approximately 34.66. Thus the rumor will have spread to 80% of the population in a little less than 35 days.

CHEMICAL REACTIONS

Consider a chemical reaction of the sort symbolized by

$$mA + nB \longrightarrow pC, \tag{8}$$

in which m molecules of reactant A combine with n molecules of reactant B to form p molecules of the product C (plus, perhaps, other products in which we have no interest). The simplest assumption concerning the rate at which this reaction takes place is that the rate of increase of the concentration $[C]$ of the product C is proportional to the product $[A] \cdot [B]$ of the concentrations of the reactants A and B. That is,

$$\frac{d[C]}{dt} = k[A][B] \qquad (k > 0). \tag{9}$$

In such cases we say that Reaction (8) satisfies the **law of mass action.**

Concentrations of the components in a chemical reaction are generally specified in moles per unit volume. One **mole** of a substance consists of 6.023×10^{23} molecules—the number of molecules present in an amount of the substance equal in mass (in grams) to its molecular weight. We consider only reactions taking place in a constant volume, which may then be taken as the unit of volume. The concentration of a substance is, in such a simple case, merely the number of moles of that substance present in the given volume.

Let $x(t)$ denote the concentration of the product C at time t. Suppose that the initial concentrations of the reactants A and B are a and b, respectively. Suppose also that $x(0) = 0$. Then the concentrations of A and B at time t are

$$[A] = a - \frac{m}{p} x \quad \text{and} \quad [B] = b - \frac{n}{p} x,$$

because it takes m molecules of A and n molecules of B to form p molecules of C. In this case, the law of mass action (9) for Reaction (8) becomes

$$\frac{dx}{dt} = k\left(a - \frac{m}{p} x\right)\left(b - \frac{n}{p} x\right). \tag{10}$$

EXAMPLE 4   Consider the chemical reaction

$$C_2H_4Br_2 + 3KI \longrightarrow KI_3 + \text{(other products)}$$

between ethylene bromide and potassium iodide. With $A = C_2H_4Br_2$, $B = KI$, and $C = KI_3$, we have $m = p = 1$ and $n = 3$. So Equation (10) becomes

$$\frac{dx}{dt} = k(a - x)(b - 3x). \tag{11}$$

Problems 16 and 17 deal with the solution of this separable differential equation.

6-8   PROBLEMS

In each of Problems 1–8, find the solution of the differential equation that satisfies the given initial condition.

1 $\dfrac{dx}{dt} = 2t - 1; \quad x(0) = 3.$

2 $\dfrac{dx}{dt} = 2x - 1; \quad x(0) = 1.$

3 $2x\dfrac{dx}{dt} = x^2 + 1; \quad x(0) = 2.$

4 $\dfrac{dx}{dt} = e^{t-x}; \quad x(0) = \ln 2.$

5 $\dfrac{dx}{dt} = (x - 3)^2; \quad x(0) = -3.$

6 $t\dfrac{dx}{dt} = x \ln x; \quad x(1) = e.$

7 $\dfrac{dx}{dt} = x^2 \sin t; \quad x(0) = 1.$

8 $\sqrt{xt}\dfrac{dx}{dt} = 1; \quad x(1) = 4.$

9 Suppose that when a certain lake is stocked with fish, the birth and death rates of the fish population $P(t)$ are each inversely proportional to $\sqrt{P}$, so that $dP/dt = k\sqrt{P}$ (k is a constant).
(a) Show that $P(t) = (\frac{1}{2}kt + \sqrt{P_0})^2$.
(b) If $P_0 = 100$ and after 6 months there are 169 fish in the lake, how many will there be after one year?

10 Consider a prolific breed of rabbits for which the birth and death rates are each proportional to the square of the population P, so that $dP/dt = kP^2$ $(k > 0)$.
(a) Show that

$$P(t) = \frac{P_0}{1 - kP_0 t}.$$

Note that $P(t) \to \infty$ as $t \to 1/kP_0$. This is "doomsday."
(b) Suppose that $P_0 = 2$ and that there are 4 rabbits after 3 months. When does doomsday occur?

Use the algebraic identity

$$\frac{1}{x + a} - \frac{1}{x + b} = \frac{b - a}{(x + a)(x + b)}$$

to find the solution $x(t)$ in Problems 11–13.

11 $\dfrac{dx}{dt} = x^2 - 1;$ $x(0) = 2.$

12 $\dfrac{dx}{dt} = 4 - x^2;$ $x(0) = 6.$

13 $\dfrac{dx}{dt} = x^2 + 3x + 2;$ $x(0) = 0.$

14 The data in the table of Figure 6.22 are given for a certain population $P(t)$ that satisfies the logistic equation.

Year	P (in millions)
1909	49.26
1910	50.00
1911	50.76
⋮	⋮
1939	99.02
1940	100.00
1941	101.02

6.22 Population data for Problem 14.

(a) What is the limiting population M? (*Suggestion:* Use the approximation

$$P'(t) \approx \frac{P(t + h) - P(t - h)}{2h}$$

with $h = 1$ to estimate the values of dP/dt when $P = 50$

and when $P = 100$. Then substitute these values into the logistic equation, and solve for k and M.
(b) Use the values of k and M found in part (a) to determine when $P = 150$. (*Suggestion:* Take $t = 0$ to correspond to the year 1940.)
15 A country contains 15,000 people who are susceptible to a spreading and contagious disease. At time $t = 0$, the number $N(t)$ of people who have the disease is 5000 and is then increasing at 500 cases per day. Assume that $N(t)$ satisfies the logistic equation. Determine how long it will take for another 5000 people to contract the disease.
16 Suppose that $b = 3a$ in Example 4 (the reaction between ethylene bromide and potassium iodide), so that Equation (11) takes the form $dx/dt = 3k(a - x)^2$. Derive the solution

$$x(t) = a\left(1 - \frac{1}{3akt + 1}\right).$$

Conclude that $\lim_{t \to \infty} x(t) = a$, and explain why this implies that *both* reactants are used up in the limit.
17 If $b \neq 3a$ in Equation (11) of Example 4, use the suggestion for Problems 11–13 to derive the solution

$$x(t) = ab \frac{1 - e^{k(b - 3a)t}}{3a - be^{k(b - 3a)t}}.$$

Conclude that $\lim_{t \to \infty} x(t) = a$ if $b > 3a$, while $\lim_{t \to \infty} x(t) = b/3$ if $b < 3a$. Explain these results in terms of the reaction continuing until *one* of the reactants is used up.
18 Consider the constant-volume gas reaction $A + B \longrightarrow 2C$. A typical example is the combination of hydrogen and chlorine to form the gas hydrogen chloride:

$$H_2 + Cl_2 \longrightarrow 2HCl.$$

Let $x(t)$ denote the concentration of the product C at time t, and suppose that $x(0) = 0$. Let a and b denote the initial concentrations of the reactants A and B.
(a) Show that the law of mass action for this reaction is $dx/dt = k(a - x/2)(b - x/2)$.
(b) Consider the case in which $a = b = 5$, and suppose that 5 moles of C are produced in the first 2 minutes of the reaction. What is the value of k? How long will it take to produce 8 moles of C?
(c) If $a = 10$ and $b = 5$, how long will it take to produce 8 moles of C?

CHAPTER 6 REVIEW: Definitions, Concepts, Results

Use the list below as a guide to concepts you may need to review.

1 The laws of exponents
2 The laws of logarithms
3 Definition of the natural logarithm
4 The graph of $y = \ln x$

5 Definition of the number e
6 Definition of the natural exponential function
7 The inverse function relationship between $\ln x$ and e^x
8 The graphs of $y = e^x$ and $y = e^{-x}$
9 Differentiation of $\ln u$ and e^u where u is a differentiable function of x

10 The order of magnitude of $(\ln x)/x^k$ and x^k/e^x as $x \to \infty$
11 The number e as a limit
12 Definition of general exponential and logarithm functions
13 Differentiation of a^u and $\log_a u$
14 Logarithmic differentiation
15 Solution of the differential equation $dx/dt = kx$

16 Natural population growth
17 Radioactive decay and radiocarbon dating
18 Solution of a linear first order differential equation with constant coefficients
19 Solution of separable first order differential equations
20 Evaluation of the constant of integration in an initial value problem

MISCELLANEOUS PROBLEMS

Differentiate the given function $f(x)$ in Problems 1–24.

1 $\ln 2\sqrt{x}$.

2 $e^{-2\sqrt{x}}$.

3 $\ln(x - e^x)$.

4 $10^{\sqrt{x}}$.

5 $\ln(2^x)$.

6 $\log_{10}(\sin x)$.

7 $x^3 e^{-1/x^2}$.

8 $x(\ln x)^2$.

9 $(\ln x)[\ln(\ln x)]$.

10 $\exp(10^x)$.

11 $2^{\ln x}$.

12 $\ln\left(\dfrac{e^x + e^{-x}}{e^x - e^{-x}}\right)$.

13 $e^{(x+1)/(x-1)}$.

14 $\ln(\sqrt{1 + x}\ \sqrt[3]{2 + x^2})$.

15 $\ln[(x - 1)/(3 - 4x^2)]^{3/2}$.

16 $\sin(\ln x)$.

17 $\exp([1 + \sin^2 x]^{1/2})$.

18 $\dfrac{x}{(\ln x)^2}$.

19 $\ln(3^x \sin x)$.

20 $(\ln x)^x$.

21 $x^{1/x}$.

22 $x^{\sin x}$.

23 $(\ln x)^{\ln x}$.

24 $(\sin x)^{\cos x}$.

Evaluate the indefinite integrals in Problems 25–36.

25 $\displaystyle\int \frac{dx}{1 - 2x}$.

26 $\displaystyle\int \frac{\sqrt{x}\ dx}{1 + x^{3/2}}$.

27 $\displaystyle\int \frac{(3 - x)\ dx}{1 + 6x - x^2}$.

28 $\displaystyle\int \frac{e^x - e^{-x}}{e^x + e^{-x}}\ dx$.

29 $\displaystyle\int \frac{\sin x\ dx}{2 + \cos x}$.

30 $\displaystyle\int \frac{e^{-1/x^2}}{x^3}\ dx$.

31 $\displaystyle\int \frac{10^{\sqrt{x}}}{\sqrt{x}}\ dx$.

32 $\displaystyle\int \frac{dx}{x(\ln x)^2}$.

33 $\displaystyle\int e^x \sqrt{1 + e^x}\ dx$.

34 $\displaystyle\int \frac{1}{x} \sqrt{1 + \ln x}\ dx$.

35 $\displaystyle\int 2^x 3^x\ dx$.

36 $\displaystyle\int \frac{dx}{x^{1/3}(1 + x^{2/3})}$.

Solve the initial value problems in Problems 37–44.

37 $\dfrac{dx}{dt} - 2t$; $x(0) = 17$.

38 $\dfrac{dx}{dt} = 2x$; $x(0) = 17$.

39 $\dfrac{dx}{dt} = e^t$; $x(0) = 2$.

40 $\dfrac{dx}{dt} = e^x$; $x(0) = 2$.

41 $\dfrac{dx}{dt} = 3x - 2$; $x(0) = 3$.

42 $\dfrac{dx}{dt} = x^2 t^2$; $x(0) = -1$.

43 $\dfrac{dx}{dt} = x \cos t$; $x(0) = \sqrt{2}$.

44 $\dfrac{dx}{dt} = \sqrt{x}$; $x(1) = 0$.

Sketch the graphs of the curves given in Problems 45–49.

45 $y = e^{-x}\sqrt{x}$.

46 $y = x - \ln x$.

47 $y = \sqrt{x} - \ln x$.

48 $y = x(\ln x)^2$.

49 $y = e^{-1/x}$.

50 Find the length of the curve $y = \frac{1}{2}x^2 - \frac{1}{4}\ln x$ from $x = 1$ to $x = e$.

51 A grain warehouse holds B bushels of grain, which is deteriorating in such a way that only $B \cdot 2^{-t/12}$ bushels will be salable after t months. Meanwhile the market price is increasing linearly; after t months it will be $2 + t/12$ dollars per bushel. After how many months should the grain be sold in order to maximize the revenue obtained?

52 Suppose that you have borrowed \$1000 at 10% annual interest compounded continuously to plant timber on a tract of land and agree to repay this loan when the timber is cut and sold. If the cut timber can be sold after t years for $800 \exp(\frac{1}{2}\sqrt{t})$ dollars, when should you cut and sell in order to maximize your profit?

53 Suppose that blood samples from 1000 students are to be tested for a certain disease that is known to occur in 1% of the population. Each test costs \$5, so it would cost \$5000 to test the samples individually. Suppose, however, that "lots" made up of x samples each are formed by pooling individual samples, and that these lots are tested first (for \$5 each). Only in case the test on a lot is positive—the probability of this is $1 - (0.99)^x$—will the x samples used to make up this lot be tested individually.
(a) Show that the total expected number of tests is

$$f(x) = \frac{1000}{x}\left[(1)(0.99)^x + (x + 1)(1 - (0.99)^x)\right]$$

$$= 1000 + \frac{1000}{x} - 1000(0.99)^x \qquad \text{if } x \geq 2.$$

(b) Show that the value of x that minimizes $f(x)$ is a root of the equation

$$x = \frac{(0.99)^{-x/2}}{[\ln(100/99)]^{1/2}}.$$

Since the denominator above is approximately $1/10$, it may be convenient to solve instead the simpler equation $x = 10(0.99)^{-x/2}$. Whichever you choose, the method of repeated substitution (beginning with $x_0 = 10$) is very effective.
(c) From the results above, deduce the cost of using this batch method to test the original 1000 samples.

54 Deduce from Problem 29 in Section 6-2 that

$$\lim_{x \to 0^+} x^x = 1.$$

55 Show that

$$\lim_{x \to 0} \frac{\ln(1 + x)}{x} = 1$$

by considering the value of $D \ln x$ for $x = 1$. Thus show that $\ln(1 + x) \approx x$ if x is very close to zero.

56 Show that

$$\lim_{h \to 0} \frac{1}{h}(a^h - 1) = \ln a$$

by considering the definition of the derivative of a^x at $x = 0$. Substitute $h = 1/n$ to obtain

$$\ln a = \lim_{n \to \infty} n(\sqrt[n]{a} - 1).$$

Approximate $\ln 2$ by taking $n = 1024 = 2^{10}$ and using only the square root key (10 times) on your calculator.

57 Suppose that the fish population in a lake is attacked by disease at time $t = 0$, with the result that $dP/dt = -3\sqrt{P}$ thereafter (t in weeks). How long will it take for all the fish in the lake to die, assuming that $P_0 = 900$?

58 A race car sliding along a level surface is decelerated by frictional forces proportional to its speed. Initially it is decelerated at 2 meters per second per second and travels a total of 1800 meters. What was its initial velocity? See Problem 12 in Section 6-6.

59 A home mortgage of $30,000 is to be paid off continuously over a period of 25 years. Apply the result of Problem 7 in Section 6-6 to determine the monthly payment required if the annual interest rate is (a) 8%; (b) 12%.

60 A powerboat weighs 32,000 pounds, and its motor provides a thrust of 5000 pounds. Assume that the water resistance is 100 pounds for each foot per second of the boat's speed. Then

$$1000 \frac{dv}{dt} = 5000 - 100v.$$

If the boat starts from rest, what is the maximum velocity that it can attain?

61 The temperature within a certain freezer is $-16°C$, and the room temperature is a constant $20°C$. At 11 P.M. the power goes off, and at 6 A.M. the next morning the temperature in the freezer has risen to $-10°C$. At what time will the temperature in the freezer reach the critical value of $0°C$ if the power remains off?

62 Suppose that the action of fluorocarbons depletes the ozone in the upper atmosphere by 0.25% annually, so that the amount A of ozone in the atmosphere satisfies the differential equation

$$\frac{dA}{dt} = -\frac{1}{400}A \qquad (t \text{ in years}).$$

(a) What percentage of the original amount A_0 of atmospheric ozone will remain 25 years from now?
(b) How long will it take for the amount of atmospheric ozone to be reduced to half its initial amount?

63 Suppose it is discovered experimentally that the amount x in grams of a solute that dissolves in a certain solvent in t seconds satisfies the logistic equation

$$\frac{dx}{dt} = (0.8)x - (0.0004)x^2.$$

(a) What is the maximum amount of the solute that will dissolve in this solvent?
(b) If $x = 50$ when $t = 0$, how long will it take for another 50 grams to dissolve?

64 A car starts from rest and travels along a straight road. Its engine provides a constant acceleration of a ft/sec^2. Air resistance and road friction cause a deceleration of ρ ft/sec^2 for every foot per second of the car's velocity.
(a) Show that after t seconds the car's velocity is

$$v = \frac{a}{\rho}(1 - e^{-\rho t}).$$

(b) If $a = 17.6$ ft/sec^2 and $\rho = 0.1$, find v when $t = 10$ (seconds), and also find the limiting velocity as $t \to \infty$. Give each answer in *miles per hour*.

65 (a) In the notation of Problem 18 in Section 6-8, show that the law of mass action for the reaction

$$2A + B \longrightarrow C$$

takes the form $dx/dt = k(a - 2x)(b - x)$.
(b) Let $a = 10$ and $b = 5$, and suppose that two moles of C are produced in the first 40 seconds. What is the value of k? How long will it take to produce 4 moles of C?
(c) Suppose that $a = b = 5$. How long will it take to produce two moles of C?

Trigonometric and Hyperbolic Functions

7

Introduction

The function f is called an **algebraic function** provided that $y = f(x)$ satisfies an equation of the form

$$a_n(x)y^n + a_{n-1}(x)y^{n-1} + \cdots + a_1(x)y + a_0(x) = 0$$

where the coefficients $a_0(x), a_1(x), \ldots, a_n(x)$ are polynomials in x. For example, since the equation $y^2 - p(x) = 0$ has the above form, we see that the square root of the polynomial $p(x)$ is an algebraic function. The equation $q(x)y - p(x) = 0$ also has the necessary form, and so we see that a rational function (a quotient of polynomials) is also an algebraic function.

A function that is *not* algebraic is said to be **transcendental.** The natural logarithm function $\ln x$ and the exponential function e^x are examples of transcendental functions. In this chapter we study the remaining transcendental functions of elementary character—the trigonometric and the hyperbolic functions. These functions have extensive scientific applications and also provide the basis for certain important methods of integration (these methods are discussed in Chapter 8).

Derivatives and Integrals of Trigonometric Functions

In Section 2-9 we saw that the derivatives of the sine and cosine functions are given by

$$D_x \sin u = (\cos u)\frac{du}{dx}, \qquad D_x \cos u = (-\sin u)\frac{du}{dx}, \tag{1}$$

where u denotes a differentiable function of x. Recall that when differentiating and integrating trigonometric functions, we use radian measure of angles exclusively.

The other four trigonometric functions—the tangent, cotangent, secant, and cosecant—may be defined in terms of the sine and cosine:

$$\tan x = \frac{\sin x}{\cos x}, \qquad \cot x = \frac{\cos x}{\sin x},$$

$$\sec x = \frac{1}{\cos x}, \qquad \csc x = \frac{1}{\sin x}. \tag{2}$$

These formulas are valid except where the denominators are zero. Thus $\tan x$ and $\sec x$ are undefined when x is an odd multiple of $\pi/2$, while $\cot x$ and $\csc x$ are undefined when x is an integral multiple of π. The graphs of the six trigonometric functions appear in Figure 7.1. There we show the sine and its reciprocal the cosecant in the same coordinate plane; we also pair the cosine with the secant and the tangent with the cotangent.

The graphs of Figure 7.1 suggest that all six trigonometric functions are periodic, and this is of course true. The sine and cosine functions have period 2π, and it follows from (2) that the other four must repeat their values in cycles of 2π as well. For example, $\tan(x + 2\pi) = \tan x$. But the **period** of the tangent and cotangent functions is the length of the shortest interval of repetition, so for those two functions the period is π, because $\tan(x + \pi) = \tan x$ for all (meaningful) values of x. This relation, too, is suggested in Figure 7.1. The secant and cosecant functions have period 2π.

If we begin with the derivatives in (1) and the definitions in (2), we can compute the derivatives of the latter four trigonometric functions. All we need is the quotient rule. The results are:

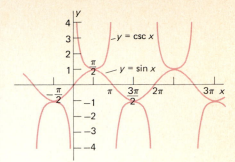

$$D_x \tan u = (\sec^2 u)\frac{du}{dx}, \qquad (3)$$

$$D_x \cot u = (-\csc^2 u)\frac{du}{dx}, \qquad (4)$$

$$D_x \sec u = (\sec u \tan u)\frac{du}{dx}, \qquad (5)$$

$$D_x \csc u = (-\csc u \cot u)\frac{du}{dx}. \qquad (6)$$

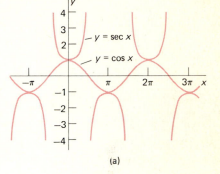

(a)

Here are the details of the first such derivation.

$$D \tan x = D \frac{\sin x}{\cos x} = \frac{(\cos x)(\cos x) - (\sin x)(-\sin x)}{\cos^2 x}$$

$$= \frac{\cos^2 x + \sin^2 x}{\cos^2 x} = \frac{1}{\cos^2 x} = \sec^2 x.$$

Combination of this result with the chain rule then yields Formula (3).

The patterns in Formulas (1) and (3) through (6) make them easy to memorize. Formulas (4) and (6) are the "cofunction analogues" of Formulas (3) and (5), respectively. And the derivative formulas for the cofunctions are the ones that involve minus signs.

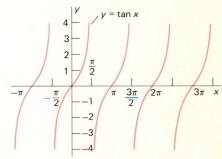

EXAMPLE 1 The following are some derivatives involving trigonometric functions.

(a) $D \tan(2x^3) = [\sec^2 2x^3]D_x(2x^3) = 6x^2 \sec^2 2x^3$.

(b) $D \cot^3(2x) = [3 \cot^2(2x)]D_x(\cot 2x)$
$\qquad = [3 \cot^2(2x)][-\csc^2(2x)]D_x(2x)$
$\qquad = -6 \cot^2 2x \csc^2 2x$.

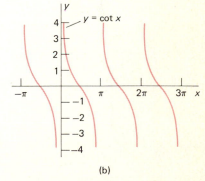

(b)

7.1 Graphs of the six trigonometric functions.

(c) $D \sec \sqrt{x} = (\sec \sqrt{x} \tan \sqrt{x})D_x(\sqrt{x}) = \dfrac{\sec \sqrt{x} \tan \sqrt{x}}{2\sqrt{x}}$.

(d) $D \csc^{1/2} x = \dfrac{1}{2}[\csc x]^{-1/2}D_x(\csc x)$

$\qquad = \dfrac{-\csc x \cot x}{2(\csc x)^{1/2}} = -\dfrac{1}{2}\csc^{1/2}x \cot x.$

By now we know that to every derivative formula there corresponds an integral formula. The integral versions of (1) and (3) through (6) are these:

$$\int \cos u \; du = \sin u + C, \tag{7}$$

$$\int \sin u \; du = -\cos u + C, \tag{8}$$

$$\int \sec^2 u \; du = \tan u + C, \tag{9}$$

$$\int \csc^2 u \; du = -\cot u + C, \tag{10}$$

$$\int \sec u \tan u \; du = \sec u + C, \tag{11}$$

$$\int \csc u \cot u \; du = -\csc u + C. \tag{12}$$

EXAMPLE 2 Compute the antiderivative: $\int \sec^2 3x \; dx$.

Solution With $u = 3x$, $du = 3 \; dx$, Formula (9) gives

$$\int \sec^2 3x \; dx = \tfrac{1}{3} \int \sec^2 u \; du$$

$$= \tfrac{1}{3} \tan u + C = \tfrac{1}{3} \tan 3x + C.$$

EXAMPLE 3 Find the antiderivative: $\displaystyle\int \frac{1}{\sqrt{x}} \csc \sqrt{x} \cot \sqrt{x} \; dx$.

Solution With $u = \sqrt{x}$, $du = \dfrac{dx}{2\sqrt{x}}$, Formula (12) gives

$$\int \frac{\csc \sqrt{x} \cot \sqrt{x}}{\sqrt{x}} \; dx = 2 \int \csc u \cot u \; du$$

$$= -2 \csc u + C = -2 \csc \sqrt{x} + C.$$

To integrate $\tan x$, the substitution $u = \cos x$, $du = -\sin x \; dx$ gives

$$\int \tan x \; dx = \int \frac{\sin x \; dx}{\cos x} = -\int \frac{du}{u}$$

$$= -\ln|u| + C,$$

and thus

$$\int \tan x \; dx = -\ln|\cos x| + C = \ln|\sec x| + C. \tag{13}$$

We use here the fact that $|\sec x| = 1/|\cos x|$.

Similarly,

$$\int \cot x \, dx = \ln|\sin x| + C = -\ln|\csc x| + C. \tag{14}$$

The integration of $\sec x$ may be accomplished by what amounts to an unmotivated trick:

$$\int \sec x \, dx = \int (\sec x) \frac{\tan x + \sec x}{\sec x + \tan x} \, dx$$

$$= \int \frac{\sec x \tan x + \sec^2 x}{\sec x + \tan x} \, dx$$

$$= \int \frac{du}{u} \qquad \text{(where } u = \sec x + \tan x)$$

$$= \ln|u| + C.$$

Therefore

$$\int \sec x \, dx = \ln|\sec x + \tan x| + C. \tag{15}$$

Similarly,

$$\int \csc x \, dx = -\ln|\csc x + \cot x| + C. \tag{16}$$

EXAMPLE 4   The substitution $u = \frac{1}{2}x$, $du = \frac{1}{2} \, dx$ gives

$$\int_0^{\pi/2} \sec \frac{x}{2} \, dx = 2 \int_0^{\pi/4} \sec u \, du$$

$$= 2 \Big[\ln|\sec u + \tan u| \Big]_0^{\pi/4} = 2 \ln(1 + \sqrt{2}).$$

To integrate $\sin^2 x$ and $\cos^2 x$ we use the half-angle identities (Formulas (11) and (12) in Section 2-9)

$$\sin^2 x = \tfrac{1}{2}(1 - \cos 2x), \qquad \cos^2 x = \tfrac{1}{2}(1 + \cos 2x). \tag{17}$$

For example,

$$\int \sin^2 x \, dx = \int \frac{1}{2}(1 - \cos 2x) \, dx$$

$$= \frac{x}{2} - \frac{1}{4} \sin 2x + C.$$

To integrate $\tan^2 x$ and $\cot^2 x$ we use the identities

$$1 + \tan^2 x = \sec^2 x, \qquad 1 + \cot^2 x = \csc^2 x. \tag{18}$$

The first of these follows from the fundamental identity $\cos^2 x + \sin^2 x = 1$ upon division of both sides by $\cos^2 x$. To obtain the second formula in (18), divide both sides of the fundamental identity by $\sin^2 x$.

EXAMPLE 5   Compute the antiderivative: $\int \cot^2 3x \, dx$.

Solution   By using the second identity in (18) we obtain

$$\cot^2 3x \, dx = \int (\csc^2 3x - 1) \, dx$$

$$= \int (\csc^2 u - 1)(\tfrac{1}{3} du) \qquad \text{(with } u = 3x)$$

$$= \tfrac{1}{3}(-\cot u - u) + C$$

$$= -\tfrac{1}{3} \cot 3x - x + C.$$

7-2   PROBLEMS

Find the derivatives of the functions given in Problems 1–20.

1 $\sin(2x + 3)$.

2 $\cos(x^2 - 1)$.

3 $\tan\left(\dfrac{2x}{3}\right)$.

4 $\cot \sqrt{x}$.

5 $\sec\left(\dfrac{1}{x}\right)$.

6 $\csc e^x$.

7 $\tan^3 4x$.

8 $\sec^2 \sqrt[3]{x}$.

9 $\ln(\sin x)$.

10 $e^{\cot 2x}$.

11 $\tan^2(\ln x)$.

12 $\sec(\csc x)$.

13 $\ln|\csc x + \cot x|$.

14 $\tan^3 x \sec^3 x$.

15 $\sin^3(\csc x)$.

16 $\dfrac{\cos^2 x}{1 - \sin x}$.

17 $\dfrac{\tan^2 x}{\sec x + 1}$.

18 $x^3 \tan^{3/2}\left(\dfrac{1}{x^2}\right)$.

19 $e^{\sin x} \sec x$.

20 $(\cot x) \ln|\cos x|$.

Find the integrals in Problems 21–40.

21 $\int \sec^2 \dfrac{x}{2} \, dx$.

22 $\int \csc^2 x \cot x \, dx$.

23 $\int \cos^2 3x \, dx$.

24 $\int \dfrac{\tan^2 \sqrt{x}}{\sqrt{x}} \, dx$.

25 $\int \dfrac{dx}{\csc x}$.

26 $\int \dfrac{dx}{\sin 2x \tan 2x}$.

27 $\int \dfrac{\sec x \tan x}{1 + \sec x} \, dx$.

28 $\int \tan 3x \sec^2 3x \, dx$.

29 $\int x \sec(x^2) \tan(x^2) \, dx$.

30 $\int \dfrac{\tan^2 x}{\sec x} \, dx$.

31 $\int \dfrac{e^{\sin 2x}}{\sec 2x} \, dx$.

32 $\int \dfrac{e^{\tan x}}{\cos^2 x} \, dx$.

33 $\int \dfrac{\sin x}{\cos^2 x} \, dx$.

34 $\int \dfrac{1}{x} \sec(\ln x) \, dx$.

35 $\int \dfrac{e^x \sec^2 e^x}{\tan e^x} \, dx$.

36 $\int \cos^2 2x \, dx$.

37 $\int \sin^5 x \cos x \, dx$.

38 $\int \tan^7 x \sec^2 x \, dx$.

39 $\int \dfrac{\sin 2x}{\cos^5 2x} \, dx$.

40 $\int \sin x \cos x \, dx$.

41 Find the critical points and inflection points of the function $f(x) = \tan x$, and verify that its graph looks as indicated in Figure 7.1.

42 The same as Problem 41, except for the function $g(x) = \sec x$.

43 Sketch the graph of $y = \sin x + \cos x$.

44 Sketch the graph of the function $f(x) = e^{-x} \cos x$ for $x \geqq 0$. This curve oscillates between the two curves $y = e^{-x}$ and $y = -e^{-x}$. Where are its local maxima and minima?

45 Find the area of the region between the curves $y = \tan^2 x$ and $y = \sec^2 x$ from $x = 0$ to $x = \pi/4$.

46 Find the volume generated by revolving the region under $y = \sec x$ from $x = 0$ to $x = \pi/4$ around the x-axis.

47 Find the volume generated by revolving the region under $y = \tan(\pi x^2/4)$ from $x = 0$ to $x = 1$ around the y-axis.

48 Find the length of the curve $y = \ln(\cos x)$ from $x = 0$ to $x = \pi/4$.

49 Show by evaluating the integral in two different ways that

$$\int \cot x \csc^2 x \, dx = -\tfrac{1}{2} \cot^2 x + C_1 = -\tfrac{1}{2} \csc^2 x + C_2.$$

Are these two answers consistent with one another?

50 Show by evaluating the integral in two different ways that

$$\int \sin x \cos x \, dx = \tfrac{1}{2}\sin^2 x + C_1 = -\tfrac{1}{2}\cos^2 x + C_2.$$

Are these two answers consistent with one another?

51 Sketch the graphs of $y = 1/x$ and $y = \tan x$ on the same coordinate plane. Then use Newton's method to find the least two positive solutions of the equation

$$\frac{1}{x} = \tan x.$$

7-3

Inverse Trigonometric Functions

Recall from Section 1-7 that if the function f is one-to-one on its domain of definition, then it has an inverse function f^{-1}. The domain of f^{-1} is the range of values of f, and

$$f^{-1}(x) = y \quad \text{if and only if} \quad f(y) = x. \tag{1}$$

For example, the exponential function e^x is increasing, and therefore one-to-one, on the whole real line, and it attains all positive values. Hence it has an inverse function, which we already know to be $\ln x$, and this inverse function is defined for all $x > 0$.

Here we want to define the inverses of the trigonometric functions. We must, however, confront the fact that the trigonometric functions fail to be one-to-one because each has period π or 2π. For example, $\sin y = \tfrac{1}{2}$ if y is either $\pi/6$ plus any multiple of 2π or $5\pi/6$ plus any multiple of 2π. Consequently we *cannot* define $y = \sin^{-1}x$, the inverse of the sine function, by saying simply that y is that value such that $\sin y = x$. There are *many* such values of y, and we must specify just which particular one of these is to be used.

We do this by suitably restricting the domain of the sine function. Since the function $\sin x$ is increasing on $[-\pi/2, \pi/2]$ and its range of values is $[-1, 1]$, for each x in $[-1, 1]$ there is a *single* y in $[-\pi/2, \pi/2]$ such that $\sin y = x$. This observation leads to the following definition of the **inverse sine (or arcsine) function,** denoted by $\sin^{-1}x$ or by arcsin x.

> **Definition**
>
> The **inverse sine function** is defined as follows:
>
> $$y = \sin^{-1}x \quad \text{if and only if} \quad \sin y = x \tag{2}$$
>
> where $-1 \leqq x \leqq 1$ and $-\pi/2 \leqq y \leqq \pi/2$.

Thus, if x is between -1 and $+1$ (inclusive), then $\sin^{-1}x$ is that number y between $-\pi/2$ and $\pi/2$ such that $\sin y = x$. Even more briefly, arcsin x is the angle nearest zero whose sine is x.

Note that the symbol -1 in the notation $\sin^{-1}x$ is *not an exponent*—it does *not* mean $(\sin x)^{-1}$.

It follows from the reflection property of inverse functions (Section 1-7) that the graph of $y = \sin^{-1}x$ is the reflection in the line $y = x$ of the graph of $y = \sin x$, $-\pi/2 \leqq x \leqq \pi/2$, and therefore looks like the graph in Figure 7.2.

7.2 The graph of $y = $ arcsin x.

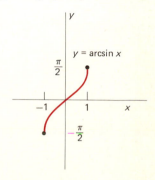

It follows from Equation (2) that

$$\sin(\sin^{-1}x) = x \quad \text{if} \quad -1 \le x \le 1, \tag{3a}$$

$$\sin^{-1}(\sin x) = x \quad \text{if} \quad -\frac{\pi}{2} \le x \le \frac{\pi}{2}. \tag{3b}$$

Because the derivative of $\sin x$ is positive for $-\pi/2 < x < \pi/2$, it follows from Theorem 2 in Section 2-4 that $\sin^{-1}x$ is differentiable on $(-1, 1)$. We may therefore proceed to differentiate both sides of Identity (3a), though we first write it in the form

$$\sin y = x$$

where $y = \sin^{-1}x$. This gives

$$(\cos y)\frac{dy}{dx} = 1.$$

So

$$\frac{dy}{dx} = \frac{1}{\cos y} = \frac{1}{\sqrt{1 - \sin^2 y}} = \frac{1}{\sqrt{1 - x^2}}.$$

We are correct in taking the positive square root in this computation because $\cos y > 0$ for $-\pi/2 < y < \pi/2$. Thus

$$D \sin^{-1}x = \frac{1}{\sqrt{1 - x^2}} \tag{4}$$

provided that $-1 < x < 1$. When we combine this result with the chain rule we get

$$D_x \sin^{-1}u = \frac{1}{\sqrt{1 - u^2}}\frac{du}{dx} \tag{5}$$

if u is a differentiable function with values in the interval $(-1, 1)$.

The definition of the inverse cosine function is similar, except that we begin by restricting the cosine function to the interval $[0, \pi]$, where it is a decreasing function. Thus the **inverse cosine** (or **arccosine**) **function** is defined by means of the rule

$$y = \cos^{-1}x \quad \text{if and only if} \quad \cos y = x \tag{6}$$

where $-1 \le x \le 1$ and $0 \le y \le \pi$.

We may compute the derivative of $\cos^{-1}x$, also written $\arccos x$, by differentiation of both sides of the identity

$$\cos(\cos^{-1}x) = x \qquad (-1 < x < 1).$$

The computations are similar to those for $D \sin^{-1}x$, and lead to the result

$$D \cos^{-1}x = -\frac{1}{\sqrt{1 - x^2}}.$$

CHAP. 7: **Trigonometric and Hyperbolic Functions**

And if u denotes a differentiable function with values in $(-1, 1)$, the chain rule then gives

$$D_x \cos^{-1} u = -\frac{1}{\sqrt{1 - u^2}}\frac{du}{dx}. \qquad (7)$$

The graph of $y = \cos^{-1}x$ is the reflection in the line $y = x$ of the graph of $y = \cos x$, $0 \leq x \leq \pi$, and is shown in Figure 7.3.

EXAMPLE 1 Suppose that $y = \sin^{-1}x^2$. Then (5) with $u = x^2$ gives

$$\frac{dy}{dx} = \frac{2x}{\sqrt{1 - x^4}}.$$

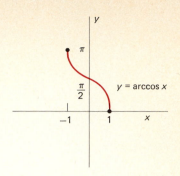

7.3 The graph of $y = \arccos x$.

The tangent function is increasing on the *open* interval $(-\pi/2, \pi/2)$—it is not defined at the end points $-\pi/2$ and $\pi/2$—and its range of values is the whole real line. We may in consequence define the **inverse tangent** (or **arctangent**) **function**, denoted by $\tan^{-1}x$ or by arctan x, as the inverse of $y = \tan x$, $-\pi/2 < x < \pi/2$.

Definition

The **inverse tangent function** is defined as follows:

$$y = \tan^{-1}x \quad \text{if and only if} \quad \tan y = x \qquad (8)$$

where $-\pi/2 < y < \pi/2$.

Since the tangent function attains all real values, $\tan^{-1}x$ is defined for all real numbers x; $\tan^{-1}x$ is that number y between $-\pi/2$ and $\pi/2$ such that $\tan y = x$. Alternatively, arctan x is the angle nearest zero whose tangent is x. The graph of $y = \tan^{-1}x$ is the reflection in the line $y = x$ of the graph of $y = \tan x$, $-\pi/2 < x < \pi/2$, and is shown in Figure 7.4.

It follows from (8) that

$$\tan(\tan^{-1}x) = x \qquad \text{for all } x, \qquad (9a)$$

$$\tan^{-1}(\tan x) = x \qquad \text{if } -\pi/2 < x < \pi/2. \qquad (9b)$$

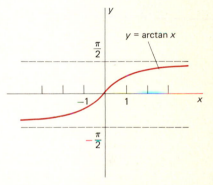

7.4 The inverse tangent function has domain all real x.

Because the derivative of $\tan x$ is positive for all x in the interval $(-\pi/2, \pi/2)$, it follows from Theorem 2 in Section 2-4 that $\tan^{-1}x$ is differentiable for all x. We may therefore differentiate both sides of the identity (9a). First we write that identity in the form

$$\tan y = x$$

where $y = \tan^{-1}x$. Then

$$(\sec^2 y)\frac{dy}{dx} = 1.$$

So

$$\frac{dy}{dx} = \frac{1}{\sec^2 y} = \frac{1}{1 + \tan^2 y} = \frac{1}{1 + x^2}.$$

Thus

$$D \tan^{-1}x = \frac{1}{1 + x^2},$$ (10)

and if u is any differentiable function of x, then

$$D_x \tan^{-1}u = \frac{1}{1 + u^2}\frac{du}{dx}.$$ (11)

The definition of the inverse cotangent function is similar, except that we begin by restricting the cotangent function to the interval $(0, \pi)$, where it is a decreasing function attaining all real values. Thus the **inverse cotangent** (or **arccotangent**) **function** is defined as

$$y = \cot^{-1}x \quad \text{if and only if} \quad \cot y = x$$ (12)

where x is any real number and $0 < y < \pi$. Then differentiation of the identity $\cot(\cot^{-1}x) = x$ leads, as in the derivation of (10), to

$$D \cot^{-1}x = -\frac{1}{1 + x^2}.$$

If u is a differentiable function of x, then the chain rule gives

$$D_x \cot^{-1}u = -\frac{1}{1 + u^2}\frac{du}{dx}.$$ (13)

EXAMPLE 2 A mountain climber on one edge of a deep canyon 800 feet wide sees a large rock fall from the opposite edge at time $t = 0$. As he watches the rock plummeting downward, he notices that his eyes first move slowly, then faster, then more slowly again. Let θ denote the angle of depression of his line of sight below the horizontal. At what angle θ would the rock seem to be moving the most rapidly; that is, when would $d\theta/dt$ be maximal?

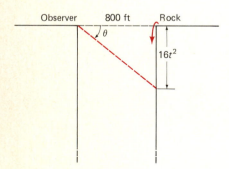

7.5 The falling rock.

Solution From our study of constant acceleration in Section 3-8, we know that the rock will fall $16t^2$ feet in the first t seconds. We refer to Figure 7.5 and see that the value of θ after t seconds will be

$$\theta = \theta(t) = \tan^{-1}\left(\frac{16t^2}{800}\right) = \tan^{-1}\left(\frac{t^2}{50}\right).$$

Hence

$$\frac{d\theta}{dt} = \frac{1}{1 + (t^2/50)^2}\cdot\frac{2t}{50} = \frac{100t}{t^4 + 2500}.$$

To find when $d\theta/dt$ is maximal, we find when *its* derivative is zero.

$$\frac{d}{dt}\left(\frac{d\theta}{dt}\right) = \frac{100(t^4 + 2500) - 100t(4t^3)}{(t^4 + 2500)^2}$$

$$= \frac{100(2500 - 3t^4)}{(t^4 + 2500)^2}.$$

CHAP. 7: **Trigonometric and Hyperbolic Functions**

So $d^2\theta/dt^2$ is zero when $3t^4 = 2500$; that is, when

$$t = \sqrt[4]{2500/3} \approx 5.37 \qquad \text{(seconds)}.$$

This is the value of t when $d\theta/dt$ is maximal, and at this time we have $t^2 = 50/\sqrt{3}$. So the angle at this time is

$$\theta = \arctan\left(\frac{1}{50} \cdot \frac{50}{\sqrt{3}}\right) = \arctan\left(\frac{1}{\sqrt{3}}\right) = \frac{\pi}{6}.$$

So the *apparent* speed of the falling rock is greatest when the climber's line of sight is $30°$ below the horizontal. The speed of the rock then is $32t$ with $t \approx 5.37$ and thus is about 172 feet per second.

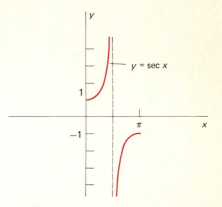

Figure 7.6 shows that the secant function is increasing on each of the intervals $[0, \pi/2)$ and $(\pi/2, \pi]$. On the union of these two intervals the secant function attains all real values y such that $|y| \geq 1$. We may therefore define the **inverse secant** (or **arcsecant**) **function**, denoted by $\sec^{-1}x$ or by arcsecant x, by restricting the secant function to the union of the intervals $[0, \pi/2)$ and $(\pi/2, \pi]$.

7.6 Restriction of the secant function to the union of the intervals $[0, \pi/2)$ and $(\pi/2, \pi]$.

Definition

The **inverse secant function** is defined as follows:

$$y = \sec^{-1}x \quad \text{if and only if} \quad \sec y = x \qquad (14)$$

where $|x| \geq 1$ and $0 \leq y \leq \pi$.

The graph of $y = \sec^{-1}x$ is the reflection in the line $y = x$ of the graph of $y = \sec x$, suitably restricted to the intervals $0 \leq x < \pi/2$ and $\pi/2 < x \leq \pi$. The graph appears in Figure 7.7. It follows from the definition above that

$$\sec(\sec^{-1}x) = x \qquad \text{if } |x| \geq 1, \qquad (15a)$$

$$\sec^{-1}(\sec x) = x \qquad \text{for } x \text{ in } [0, \pi/2) \cup (\pi/2, \pi]. \qquad (15b)$$

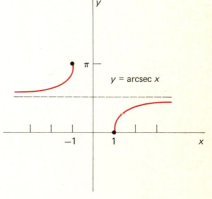

Following the now familiar pattern, we find $D \sec^{-1}x$ by differentiating both sides of (15a) in the form

$$\sec y = x$$

where $y = \sec^{-1}x$. This yields

7.7 The graph of $y = \sec^{-1}x = $ arcsec x.

$$(\sec y \tan y)\frac{dy}{dx} = 1,$$

and so

$$\frac{dy}{dx} = \frac{1}{\sec y \tan y} = \frac{1}{\pm x\sqrt{x^2 - 1}},$$

because $\tan y = \pm\sqrt{\sec^2 y - 1} = \pm\sqrt{x^2 - 1}$.

To obtain the correct choice of sign above, note what happens in the two cases $x > 1$ and $x < -1$. In the first case, $0 < y < \pi/2$ and $\tan y > 0$, so we choose the $+$ sign. If $x < -1$ then $\pi/2 < y < \pi$ and $\tan y < 0$, so we take the $-$ sign. Thus

$$D \sec^{-1} x = \frac{1}{|x|\sqrt{x^2 - 1}} \qquad (|x| > 1). \tag{16}$$

If u is a differentiable function of x with values that exceed 1 in magnitude, then by the chain rule we have

$$D_x \sec^{-1} u = \frac{1}{|u|\sqrt{u^2 - 1}} \frac{du}{dx}. \tag{17}$$

EXAMPLE 3 The function $\sec^{-1} e^x$ is defined if $x > 0$, for then $e^x > 1$. Then by (17)

$$D_x \sec^{-1} e^x = \frac{e^x}{|e^x|\sqrt{e^{2x} - 1}} = \frac{1}{\sqrt{e^{2x} - 1}}$$

because $|e^x| = e^x$ for all x.

The **inverse cosecant** (or **arccosecant**) function is the inverse of the function $y = \csc x$ where x is restricted to the union of the intervals $[-\pi/2, 0)$ and $(0, \pi/2]$. Thus

$$y = \csc^{-1} x \quad \text{if and only if} \quad \csc y = x \tag{18}$$

where $|x| \geqq 1$ and $-\pi/2 < y < \pi/2$. Its derivative formula, which has a derivation similar to that of (17) above, is

$$D_x \csc^{-1} u = -\frac{1}{|u|\sqrt{u^2 - 1}} \frac{du}{dx}. \tag{19}$$

It is noteworthy that the derivatives of the six inverse trigonometric functions are all simple *algebraic* functions. As a consequence, inverse trigonometric functions frequently appear when we integrate algebraic functions. Observe also that the derivatives of $\cos^{-1} x$, $\cot^{-1} x$, and $\csc^{-1} x$ differ only in sign from the derivatives of their respective cofunctions. For this reason only the arcsine, arctangent, and arcsecant functions are necessary for integration, and only these three are in common use. That is, one need commit to memory the integral formulas only for the latter three functions. They follow immediately from (5), (11), and (17), and may be written in the form below.

$$\int \frac{du}{\sqrt{1 - u^2}} = \sin^{-1} u + C, \tag{20}$$

$$\int \frac{du}{1 + u^2} = \tan^{-1} u + C, \tag{21}$$

$$\int \frac{du}{u\sqrt{u^2 - 1}} = \sec^{-1} |u| + C. \tag{22}$$

It is easy to verify that the absolute value on the right side in (22) follows from the one in (17). Remember also that, because $\sec^{-1}|u|$ is undefined unless $|u| \geq 1$, the definite integral

$$\int_a^b \frac{du}{u\sqrt{u^2 - 1}}$$

is meaningful only if the limits a and b are both at least 1 or both at most -1.

EXAMPLE 4   The substitution $u = x/2$, $du = \frac{1}{2}dx$ gives

$$\int \frac{dx}{\sqrt{4 - x^2}} = \int \frac{dx}{2\sqrt{1 - (x/2)^2}}$$

$$= \int \frac{du}{\sqrt{1 - u^2}} = \arcsin u + C$$

$$= \arcsin\left(\frac{x}{2}\right) + C.$$

EXAMPLE 5   The substitution $u = 3x$, $du = 3\, dx$ gives

$$\int \frac{dx}{1 + 9x^2} = \frac{1}{3}\int \frac{3\, dx}{1 + (3x)^2} = \frac{1}{3}\int \frac{du}{1 + u^2}$$

$$= \frac{1}{3}\tan^{-1}u + C = \frac{1}{3}\tan^{-1}3x + C.$$

EXAMPLE 6   The substitution $u = x\sqrt{2}$, $du = \sqrt{2}\, dx$ gives

$$\int_1^{\sqrt{2}} \frac{dx}{x\sqrt{2x^2 - 1}} = \int_{\sqrt{2}}^2 \frac{du}{u\sqrt{u^2 - 1}}$$

$$= \left[\sec^{-1}|u|\right]_{\sqrt{2}}^2 = \sec^{-1}2 - \sec^{-1}\sqrt{2}$$

$$= \frac{\pi}{3} - \frac{\pi}{4} = \frac{\pi}{12}.$$

7-3   PROBLEMS

Differentiate the functions in Problems 1–20.

1 $\sin^{-1}e^x$.

2 $\arctan\sqrt{x}$.

3 $\cos^{-1}x + \sec^{-1}\left(\frac{1}{x}\right)$.

4 $\cot^{-1}\left(\frac{1}{x^2}\right)$.

5 $\csc^{-1}x^2$.

6 $\arccos\left(\frac{1}{\sqrt{x}}\right)$.

7 $\dfrac{1}{\arctan x}$.

8 $(\arcsin x)^2$.

9 $\tan^{-1}(\ln x)$.

10 $\operatorname{arcsec}\sqrt{x^2 + 1}$.

11 $\tan^{-1}e^x + \cot^{-1}e^{-x}$.

12 $\exp(\arcsin x)$.

13 $\sin(\arctan x)$.

14 $\sec(\sec^{-1}e^x)$.

15 $\dfrac{e^x}{\arctan 3x}$.

16 $(\sin^{-1}2x^2)^{-2}$.

17 $\dfrac{\arctan x}{(1 + x^2)^2}$.

18 $\arcsin\left(\dfrac{x}{a}\right)$.

19 $\dfrac{1}{a}\tan^{-1}\left(\dfrac{x}{a}\right)$.

20 $\dfrac{1}{a}\sec^{-1}\left(\dfrac{x}{a}\right)$.

Evaluate or antidifferentiate, as appropriate, in Problems 21–35.

21 $\displaystyle\int_0^1 \dfrac{dx}{1+x^2}$.

22 $\displaystyle\int_0^{1/2} \dfrac{dx}{\sqrt{1-x^2}}$.

23 $\displaystyle\int_1^2 \dfrac{dx}{x\sqrt{x^2-1}}$.

24 $\displaystyle\int_{\sqrt{2}}^2 \dfrac{dx}{x\sqrt{x^2-1}}$.

25 $\displaystyle\int_0^3 \dfrac{dx}{9+x^2}$.

26 $\displaystyle\int_0^4 \dfrac{dx}{\sqrt{16-x^2}}$.

27 $\displaystyle\int \dfrac{dx}{\sqrt{1-4x^2}}$.

28 $\displaystyle\int \dfrac{dx}{9x^2+4}$.

29 $\displaystyle\int \dfrac{dx}{x\sqrt{x^2-25}}$.

30 $\displaystyle\int \dfrac{dx}{x\sqrt{4x^2-9}}$.

31 $\displaystyle\int \dfrac{e^x\,dx}{1+e^{2x}}$.

32 $\displaystyle\int \dfrac{x^2\,dx}{x^6+25}$.

33 $\displaystyle\int \dfrac{dx}{x\sqrt{x^6-25}}$.

34 $\displaystyle\int \dfrac{\sqrt{x}\,dx}{1+x^3}$.

35 $\displaystyle\int \dfrac{dx}{\sqrt{x(1-x)}}$.

36 Conclude from the formula $D\cos^{-1}x = -D\sin^{-1}x$ that $\sin^{-1}x + \cos^{-1}x = \pi/2$ if $0 \le x \le 1$.

37 Show that $D\sec^{-1}x = D\cos^{-1}(1/x)$ if $x \ge 1$, and conclude that $\sec^{-1}x = \cos^{-1}(1/x)$ if $x \ge 1$. This fact can be used to find arcsecants on a calculator that has the key for the arccosine function, usually written "inv cos" or "cos^{-1}."

38 (a) Deduce from the addition formula for tangents (Problem 10 in Section 2-9) that

$$\arctan x + \arctan y = \arctan\left(\frac{x+y}{1-xy}\right)$$

provided that $xy < 1$.

(b) Apply Part (a) to show that each of the following numbers is equal to $\pi/4$:

(i) $\arctan(\tfrac{1}{2}) + \arctan(\tfrac{1}{3})$;

(ii) $2\arctan(\tfrac{1}{3}) + \arctan(\tfrac{1}{7})$;

(iii) $\arctan(\tfrac{120}{119}) - \arctan(\tfrac{1}{239})$;

(iv) $4\arctan(\tfrac{1}{5}) - \arctan(\tfrac{1}{239})$.

39 A billboard *parallel* to a highway is to be 12 feet high, and its bottom will be 4 feet above the eye level of a passing motorist. How far from the highway should the billboard be placed in order to maximize the angle it subtends at the motorist's eyes?

40 Use inverse trigonometric functions to prove that the vertical angle subtended by a rectangular painting on a wall is greatest when the painting is hung with its center at the level of the observer's eyes.

41 Show that the circumference of a circle of radius a is $2\pi a$ by finding the length of the circular arc $y = \sqrt{a^2-x^2}$ from $x = 0$ to $x = a/\sqrt{2}$ and then multiplying by 8.

42 Find the volume generated by revolving the area under $y = 1/(1+x^4)$ from $x = 0$ to $x = 1$ around the y-axis.

43 Show that the area of the unbounded region lying between the x-axis and the curve $y = 1/(1+x^2)$ is finite by evaluating

$$\lim_{a\to\infty}\int_{-a}^a \frac{dx}{1+x^2}.$$

44 A building 250 feet high is equipped with an external elevator. The elevator starts at the top at time $t = 0$ and descends at the constant rate of 25 feet per second. You are watching the elevator from a window that is 100 feet above the ground, in a building 50 feet from the elevator. At what height does the elevator appear to you to be moving the fastest?

45 Suppose that the function f is defined for all x with $|x| > 1$, and has the property that $f'(x) = 1/(x\sqrt{x^2-1})$ for all such x.

(a) Explain why there exist two constants A and B such that

$$f(x) = \operatorname{arcsec} x + A \qquad \text{if } x > 1;$$
$$f(x) = -\operatorname{arcsec} x + B \qquad \text{if } x < -1.$$

(b) Determine the values of A and B so that $f(\pm 2) = 1$. Then sketch the graph of $y = f(x)$.

---

*7-4

Periodic Phenomena and Simple Harmonic Motion

As we saw in Chapter 6, exponential functions are important in the study of natural growth and decay processes—population growth, radioactive decay, and the like. The trigonometric functions are equally important in the analysis of periodic phenomena in nature, such as the ebb and flow of seasons and tides, and the oscillatory motions that occur in vibrating mechanical systems. These applications are so remarkably diverse partly because of the truth of the following theorem.

Theorem *The Equation* $x'' = -k^2 x$

The function $x = f(t)$ satisfies the second order differential equation

$$\frac{d^2 x}{dt^2} = -k^2 x \qquad (k \text{ constant}) \tag{1}$$

if and only if there exist constants A and B such that

$$f(t) = A \cos kt + B \sin kt. \tag{2}$$

Proof Suppose first that $x = f(t)$ has the form given in Equation (2). Then

$$\frac{dx}{dt} = -kA \sin kt + kB \cos kt \tag{3}$$

and

$$\begin{aligned}
\frac{d^2 x}{dt^2} &= -k^2 A \cos kt - k^2 B \sin kt \\
&= -k^2 (A \cos kt + B \sin kt) = -k^2 x.
\end{aligned}$$

Thus the function in (2) satisfies the given differential equation (1). This proves the "if" part of the theorem. Problem 21 is an outline of the "only if" part.

If a particle moves along the x-axis with acceleration given by Equation (1), then the theorem implies that its position function $x = f(t)$ is of the form of (2). Motion of this type is called **simple harmonic motion.** The constants A and B may be determined in terms of the particle's initial position $x_0 = f(0)$ and initial velocity $v_0 = f'(0)$, as follows. Substitution of $t = 0$ into (2) gives $A = x_0$, and substitution of $t = 0$ into (3) gives $B = v_0/k$. Hence the position function of the particle is

$$x = x_0 \cos kt + \frac{v_0}{k} \sin kt. \tag{4}$$

To analyze and to help us visualize the simple harmonic motion of a particle with the position function in (2), we choose constants C and α such that

$$C \sin k\alpha = -A \quad \text{and} \quad C \cos k\alpha = B. \tag{5}$$

Here is one way to find C and α. First, the sum of the squares of the left-hand sides above is C^2, and so $C^2 = A^2 + B^2$. When we form quotients using the two equations above, we find that $\tan k\alpha = -A/B$. And hence we take

$$C = \sqrt{A^2 + B^2} \quad \text{and} \quad \alpha = \frac{1}{k} \tan^{-1}(-A/B). \tag{6}$$

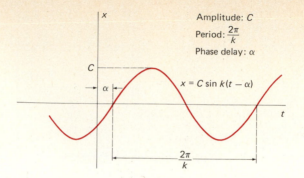

Amplitude: C

Period: $\dfrac{2\pi}{k}$

Phase delay: α

$x = C \sin k(t - \alpha)$

$\dfrac{2\pi}{k}$

7.8 Amplitude, period, and phase delay.

Then substitution of Equations (5) into (2) yields

$$x = (-C \sin k\alpha) \cos kt + (C \cos k\alpha) \sin kt$$
$$= C(\sin kt \cos k\alpha - \cos kt \sin k\alpha).$$

Thus

$$x = C \sin k(t - \alpha). \qquad (7)$$

With $k\beta = k\alpha + \pi/2$, the addition formula for cosines gives the alternative form

$$x = C \cos k(t - \beta). \qquad (8)$$

The graph of Equation (7) exhibits the same oscillatory behavior as the curve $x = \sin t$. But the effect of the coefficient C is to stretch the graph in the x-direction, so that x has maximum value $+C$ and minimum value $-C$. Because of the constant k, the function $f(t) = C \sin k(t - \alpha)$ has period $2\pi/k$ instead of 2π, since

$$f\left(t + \frac{2\pi}{k}\right) = C \sin k\left(t + \frac{2\pi}{k} - \alpha\right)$$
$$= C \sin(kt - k\alpha + 2\pi)$$
$$= C \sin k(t - \alpha) = f(t).$$

Finally, the effect of the constant α is to translate the graph α units to the right of the curve $x = C \sin kt$. When we assemble all this information, we see that the graph of (7) looks like the one in Figure 7.8.

Simple harmonic motion with the position function in (7) may be described as having

amplitude	C,
period	$\dfrac{2\pi}{k}$,
phase delay	α.

(9)

EXAMPLE 1 Sketch the graph of $x = 3 \sin(2t - \pi/2)$.

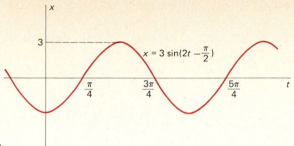

7.9 The graph for Example 1.

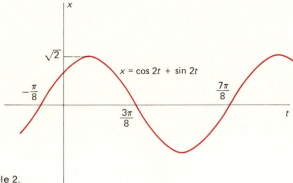

7.10 The function of Example 2.

Solution We write $x = 3 \sin 2(t - \pi/4)$, which shows that the function x has amplitude $C = 3$, period $2\pi/k = \pi$, and phase delay $\alpha = \pi/4$, so the graph looks like Figure 7.9.

EXAMPLE 2 Sketch the graph of $x = \cos 2t + \sin 2t$.

Solution Here we have the form (2) with $A = B = 1$ and $k = 2$. Equations (6) give

$$C = \sqrt{2} \quad \text{and} \quad \alpha = \frac{1}{2} \tan^{-1}(-1) = -\frac{\pi}{8}.$$

Thus our curve is $x = \sqrt{2} \sin 2(t + \pi/8)$, with amplitude $C = \sqrt{2}$, period $2\pi/k = \pi$, and phase delay $\alpha = -\pi/8$. Hence the graph looks like Figure 7.10.

Some typical examples of simple harmonic motion are the motion of a mass attached to a vibrating string, the motion of an air molecule in a sound wave, the motion of a point on a vibrating musical string (hence the term *harmonic*), and the motion of a piston in an engine.

A RECIPROCATING PISTON

Consider the piston of Figure 7.11, moving up and down on one end of a 6-foot shaft. The other end is attached at Q to a horizontal slotted arm fitted to a peg P on the rim of a wheel of radius 2 feet. Suppose that the

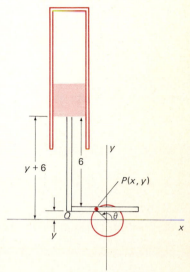

7.11 The reciprocating piston.

wheel begins with the point P at $\theta = \pi/4$ when $t = 0$ and rotates counter-clockwise at 5 radians per second. Then $\theta = 5t + \pi/4$ after t seconds, so the y-coordinate of the point Q after t seconds is

$$y = 2 \sin \theta = 2 \sin\left(5t + \frac{\pi}{4}\right)$$

$$= 2 \sin 5\left(t + \frac{\pi}{20}\right).$$

Thus the motion of the piston—essentially that of the point Q—is simple harmonic motion with amplitude $C = 2$ feet, period $2\pi/k = 2\pi/5$, and phase delay $\alpha = -\pi/20$.

MASS AND SPRING

Suppose that a mass m rests on a frictionless horizontal plane, and is attached to a spring of natural (unstretched) length L. We assume that the spring obeys Hooke's law: If the spring is stretched an amount x, then it exerts a restorative force $F = -cx$, where c is a positive constant (the *spring constant*). As shown in Figure 7.12, the spring has one end attached to a fixed wall and the other to the mass m.

Now suppose that the mass is set in motion with position function $x(t)$, measured from the equilibrium position $x = 0$ (corresponding to an unstretched spring). Combining Newton's law $F = ma = mx''(t)$ and Hooke's law $F = -cx$, we find that

$$mx''(t) = -cx(t),$$

or

$$x''(t) = -k^2 x(t)$$

where $k = \sqrt{c/m}$. Thus the position function of the mass satisfies differential equation (1), and so by the theorem above, its motion is simple harmonic:

$$x(t) = A \cos kt + B \sin kt,$$

where the constants A and B are determined by the initial position x_0 and the initial velocity v_0 of the mass (as in Equation (4)). From (9) we see that the period of motion of the mass is

$$\frac{2\pi}{k} = 2\pi \sqrt{\frac{m}{c}}, \tag{10}$$

which depends only upon the mass m and the spring constant c and is independent of x_0 and v_0. The **frequency** v of this simple harmonic motion —the number of cycles completed per second—is the reciprocal of the period:

$$v = \frac{k}{2\pi} = \frac{1}{2\pi} \sqrt{\frac{c}{m}}. \tag{11}$$

EXAMPLE 3   A mass-and-spring system is constructed as in the above discussion; the frequency turns out, by experiment, to be $v = 1/\pi$. The mass is then set into motion from the initial position $x_0 = 1$ with initial velocity $v_0 = 2$. Describe its subsequent motion.

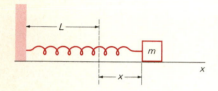

7.12   A simple mass-and-spring system.

CHAP. 7:   **Trigonometric and Hyperbolic Functions**

Solution By Equation (4) we know that

$$x(t) = \cos kt + \frac{2}{k} \sin kt.$$

From Equation (11) we see that $v = 1/\pi$ implies that $k = 2$, so

$$x(t) = \cos 2t + \sin 2t = \sqrt{2} \sin 2\left(t + \frac{\pi}{8}\right).$$

In the last step we used the result of Example 2; we see that the amplitude of the motion is $\sqrt{2}$ and the mass first reaches the position $x = \sqrt{2}$ when $t = \pi/8$ and first reaches the position $x = -\sqrt{2}$ when $t = 5\pi/8$.

THE SIMPLE PENDULUM

A simple pendulum consists of a mass m swinging to and fro on the end of a string (or, better, a "massless rod") of length L, as in Figure 7.13. We may specify the position of the mass at time t by giving the counter-clockwise angle $\theta = \theta(t)$ that the string or rod makes with the vertical. To analyze the motion of the mass m, we will apply the law of the *conservation of energy*, according to which the sum of the kinetic energy and the potential energy of m remains constant.

The distance along the circular arc from 0 to m is $s = L\theta$, so the speed of the mass is

$$v = \frac{ds}{dt} = L\frac{d\theta}{dt},$$

and its kinetic energy is

$$\text{K.E.} = \frac{1}{2}mv^2 = \frac{1}{2}m\left(\frac{ds}{dt}\right)^2 = \frac{1}{2}mL^2\left(\frac{d\theta}{dt}\right)^2$$

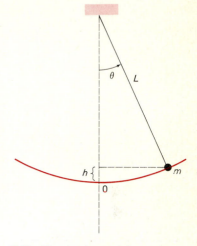

7.13 The simple pendulum.

We next choose as reference point the lowest point O reached by the mass (see Figure 7.13). Then its potential energy is the product of its weight mg (where g is gravitational acceleration) and its vertical height $h = L(1 - \cos \theta)$ above O, so that

$$\text{P.E.} = mgL(1 - \cos \theta).$$

The fact that the sum of K.E. and P.E. is a constant C therefore gives

$$\frac{1}{2}mL^2\left(\frac{d\theta}{dt}\right)^2 + mgL(1 - \cos \theta) = C.$$

We differentiate both sides of this identity with respect to t and find that

$$mL^2\left(\frac{d\theta}{dt}\right)\frac{d^2\theta}{dt^2} + mgL(\sin \theta)\frac{d\theta}{dt} = 0,$$

and so

$$\frac{d^2\theta}{dt^2} = -\frac{g}{L}\sin \theta \qquad (12)$$

after cancellation of the common factor $mL^2(d\theta/dt)$.

Equation (12) is a bit too difficult to solve here. But recall from Section 2-9 that $(\sin \theta)/\theta \to 1$ as $\theta \to 0$, so θ and $\sin \theta$ are approximately equal if θ is sufficiently small. In particular, examination of a table of sines or of some calculator values indicates that θ (in radians) and $\sin \theta$ agree to two decimal places when θ is at most $\pi/12$ (that is, $15°$). In a typical pendulum clock, for example, θ would certainly not exceed $15°$.

It therefore seems reasonable to simplify our mathematical model of the pendulum by replacing $\sin \theta$ by θ in Equation (12). This gives

$$\frac{d^2\theta}{dt^2} = -\frac{g}{L}\,\theta, \tag{13}$$

a differential equation of the form of (1) with θ instead of x and with $k = \sqrt{g/L}$. Its solution is of the form

$$\theta = A \cos kt + B \sin kt = C \sin k(t - \alpha).$$

Thus the angular motion of the pendulum is simple harmonic, and by (9) its period p is

$$p = \frac{2\pi}{k} = 2\pi\sqrt{\frac{L}{g}}. \tag{14}$$

Note that p depends only upon the length L of the pendulum and the gravitational acceleration g, but not upon either the mass m of the pendulum bob or the initial conditions. For example, using $g = 32.2$ ft/sec^2, we find that the period of oscillation of a 3-foot pendulum is

$$p = 2\pi\sqrt{\frac{3}{32.2}} \approx 1.92 \qquad \text{(seconds)}.$$

Recall that the gravitational acceleration g at a point on the earth's surface at distance R from its center is $g = GM/R^2$, where G is the gravitational constant and M is the earth's mass. If this value of g is substituted into Equation (14), the result is

$$p = 2\pi R\sqrt{\frac{L}{GM}}. \tag{15}$$

If the period of a pendulum of given length L is first measured at sea level —where R is known—and then measured atop a mountain, then Equation (15) may be used to compute the value of R at the mountain top and thereby find its height. Indeed, this method is one of the most accurate ways of determining altitude.

In 1672 the French astronomer Richer traveled to South America on a scientific mission. While there, he found that his pendulum clock, which had kept very accurate time in Paris, lost more than 2 minutes per day in French Guiana. Newton deduced from the behavior of Richer's clock that French Guiana must be farther than Paris from the center of the earth, and thereby Newton discovered the equatorial bulge of the earth. There are similar computations in the problems below.

In each of Problems 1–7 sketch the graph of the given function. Show at least one cycle.

1 $x = 2 \sin \pi t$.

2 $x = 2 \cos 3\pi t$.

3 $x = \sin \pi(t - 1)$.

4 $x = 2 \cos(2t - \pi)$.

5 $x = 3 \sin \dfrac{\pi}{2}(4t - 1)$.

6 $x = \sin t + \sqrt{3} \cos t$.

7 $x = 3 \sin \pi t + 4 \cos \pi t$.

8 In a certain city the average temperature $f(t)$ on the tth day of the year (with $t = 1$ on January 1) is given by a formula of the form $f(t) = A + B \sin k(t - \alpha)$.
(a) What must k be for the period of $f(t)$ to be 365 days?
(b) The minimum daily average temperature of 50°F occurs on January 20 and the maximum of 80°F occurs half a year later. What are the values of A, B, and α?
(c) On what days of the year is the average temperature 60°F? Repeat for 70°F.

Problems 9–15 deal with a mass on a spring, undergoing simple harmonic motion with position function $x(t)$.

9 What is the amplitude of the motion if its frequency is 2 hz (cycles per second), $x(0) = 1$, and $x'(0) = 0$?

10 What is the amplitude of the motion if its frequency is $2/\pi$ hz, $x(0) = 0$, and $x'(0) = -4$?

11 What is its amplitude if its period is π seconds and $x(0) = x'(0) = 1$?

12 What is the period of the motion if its amplitude is 4, $x(0) = 0$, and $x'(0) = 2$?

13 What is the maximum velocity of the mass if the amplitude of the motion is 5 and its period is 2 seconds?

14 What is the amplitude of the motion if its frequency is $1/\pi$ hz and the maximum velocity of the mass is 10?

15 Find the period and amplitude of the motion if $x(0) = 3$, $x'(0) = 4$, and $x''(0) = -3$.

16 Consider a piston as in Figure 7.11, except that the piston shaft is 5 feet long and the diameter of the wheel is 3 feet. Suppose that the wheel begins at time $t = 0$ with $\theta = \pi$ and rotates counterclockwise at 5 revolutions per second. Express the displacement of the piston as a function of t. What is the maximum velocity of the piston?

17 Two pendulums are of lengths L_1 and L_2, and—when located at the respective distances R_1 and R_2 from the center of the earth—have periods p_1 and p_2. Show that

$$\frac{p_1}{p_2} = \frac{R_1\sqrt{L_1}}{R_2\sqrt{L_2}}.$$

18 A certain pendulum clock keeps perfect time in Paris, where the radius of the earth is $R = 3956$ miles. But this clock loses 2 min 40 sec per day at a location on the equator. Use the result of Problem 17 to find the amount of the equatorial bulge of the earth.

19 A pendulum of length 100.10 inches, located at a point at sea level where the radius of the earth is $R = 3960$ miles, has the same period as does a pendulum of length 100.00 inches atop a nearby mountain. Use the result of Problem 17 to find the height of the mountain.

20 Most grandfather clocks have pendulums with lengths that are adjustable. One such clock loses 10 minutes per day when the length of the pendulum is 30 inches. With what length pendulum will this clock keep perfect time?

21 This is an outline of the proof of the "only if" part of the theorem in this section.
(a) Assuming that $f''(t) = -k^2 f(t)$, define

$$g(t) = f(t) - f(0) \cos kt - \frac{f'(0)}{k} \sin kt.$$

Then show that $g''(t) = -k^2 g(t)$ and that $g(0) = g'(0) = 0$.
(b) Let $H(t) = [g'(t)]^2 + k^2[g(t)]^2$. Show that $H'(t) \equiv 0$. Conclude that $H(t) \equiv 0$. Why does this complete the proof?

22 Suppose that the position function $x = x(t)$ satisfies the differential equation $x''(t) + k^2 x(t) = a$, where a and k are constants. Let $y(t) = x(t) - (a/k^2)$. Show that $y''(t) = -k^2 y(t)$. Conclude that

$$x(t) = \frac{a}{k^2} + A \cos kt + B \sin kt.$$

Thus $x(t)$ describes simple harmonic motion about the *equilibrium position* $x_e = a/k^2$.

23 Consider a mass m hanging on the end of a vertical spring, as in Figure 7.14. If $x(t)$ denotes the distance of the mass *below* its position $x = 0$ when the spring is unstretched,

7.14 Mass hanging from a spring—as in Problem 23.

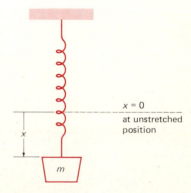

$x = 0$
at unstretched
position

m

show that Newton's law $F = ma$ yields

$$mx''(t) + cx(t) = mg$$

where c is the spring constant. Conclude from the result of Problem 22 that the mass undergoes simple harmonic motion with period $p = 2\pi\sqrt{m/c}$ about the equilibrium position $x_e = mg/c$.

24 Consider a floating cylindrical buoy with radius r, height h, and density $\rho \leqq \frac{1}{2}$ (recall that the density of water is 1 gm/cm³). The buoy is initially suspended at rest with its bottom at the top surface of the water and is released at time $t = 0$. Thereafter it is acted upon by two forces: a downward gravitational force equal to its weight $mg = \rho\pi r^2 hg$ and an upward force of buoyancy equal to the weight $\pi r^2 x g$ of water displaced, where $x = x(t)$ is the depth of the buoy's bottom beneath the surface at time t. Conclude from Problem 22 that the buoy undergoes simple harmonic motion about the equilibrium position $x_e = \rho h$ with period $p = 2\pi\sqrt{\rho h/g}$. Compute p and the amplitude of the motion if $\rho = 0.5$ gm/cm³, $h = 200$ cm, and $g = 980$ cm/sec².

7-5

Hyperbolic Functions

The **hyperbolic cosine** and the **hyperbolic sine** of the real number x are denoted by $\cosh x$ and $\sinh x$ and are defined to be

$$\cosh x = \frac{e^x + e^{-x}}{2}, \qquad \sinh x = \frac{e^x - e^{-x}}{2}. \tag{1}$$

These particular combinations of familiar exponentials are useful in certain applications of calculus and are also helpful in evaluating certain integrals. The other four hyperbolic functions—the hyperbolic tangent, cotangent, secant, and cosecant—are defined in terms of $\cosh x$ and $\sinh x$ by analogy with trigonometry:

$$\tanh x = \frac{\sinh x}{\cosh x} = \frac{e^x - e^{-x}}{e^x + e^{-x}},$$

$$\coth x = \frac{\cosh x}{\sinh x} = \frac{e^x + e^{-x}}{e^x - e^{-x}} \qquad (x \neq 0); \tag{2}$$

$$\text{sech } x = \frac{1}{\cosh x} = \frac{2}{e^x + e^{-x}},$$

$$\text{csch } x = \frac{1}{\sinh x} = \frac{2}{e^x - e^{-x}} \qquad (x \neq 0). \tag{3}$$

The trigonometric terminology and notation for these hyperbolic functions stem from the fact that they satisfy a list of identities that, apart from an occasional difference of sign, much resemble the familiar trigonometric identities:

$$\cosh^2 x - \sinh^2 x = 1; \tag{4}$$

$$1 - \tanh^2 x = \text{sech}^2 x; \tag{5}$$

$$\coth^2 x - 1 = \text{csch}^2 x; \tag{6}$$

$$\sinh(x + y) = \sinh x \cosh y + \cosh x \sinh y; \tag{7}$$

$$\cosh(x + y) = \cosh x \cosh y + \sinh x \sinh y; \tag{8}$$

$$\sinh 2x = 2 \sinh x \cosh x; \tag{9}$$

$$\cosh 2x = \cosh^2 x + \sinh^2 x; \tag{10}$$

$$\cosh^2 x = \tfrac{1}{2}(\cosh 2x + 1); \tag{11}$$

$$\sinh^2 x = \tfrac{1}{2}(\cosh 2x - 1). \tag{12}$$

Identities (4), (7), and (8) follow directly from the definition of cosh x and sinh x. For example,

$$\cosh^2 x - \sinh^2 x = \tfrac{1}{4}(e^x + e^{-x})^2 - \tfrac{1}{4}(e^x - e^{-x})^2$$
$$= \tfrac{1}{4}(e^{2x} + 2 + e^{-2x}) - \tfrac{1}{4}(e^{2x} - 2 + e^{-2x})$$
$$= 1.$$

The other identities above may be derived from (4), (7), and (8) in ways that parallel the derivations of the standard trigonometric identities.

The trigonometric functions are sometimes called *circular* functions because the point $(\cos \theta, \sin \theta)$ lies on the circle $x^2 + y^2 = 1$ for all θ. Similarly, Identity (4) above tells us that the point $(\cosh \theta, \sinh \theta)$ lies on the hyperbola $x^2 - y^2 = 1$, and this is the reason for the name *hyperbolic* functions (see Figure 7.15).

The graphs of $y = \cosh x$ and $y = \sinh x$ are easily constructed: Add (for cosh) or subtract (for sinh) the ordinates of the graphs of $y = \tfrac{1}{2}e^x$ and $y = \tfrac{1}{2}e^{-x}$. The graphs of the other four hyperbolic functions can then be constructed by dividing ordinates. The graphs of all six are shown in Figure 7.16.

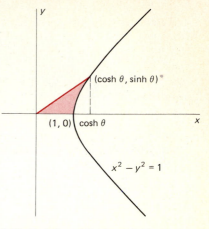

7.15 Relation of the hyperbolic cosine and hyperbolic sine to the hyperbola $x^2 - y^2 = 1$.

7.16 Graphs of the six hyperbolic functions.

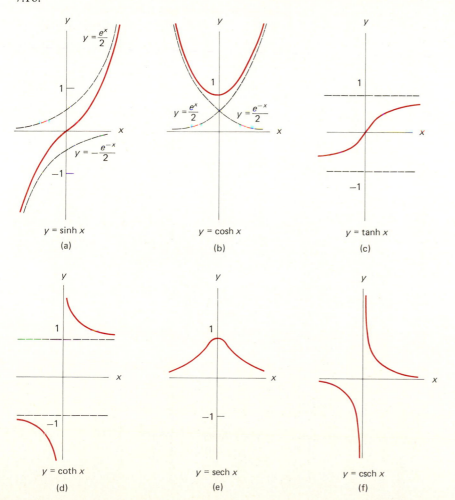

A striking difference is apparent in these graphs, a difference between the hyperbolic functions and the ordinary trigonometric functions: None of the hyperbolic functions is periodic. They do, however, have even-odd properties like the circular functions. The two functions cosh and sech are even, because

$$\cosh(-x) = \cosh x \quad \text{and} \quad \text{sech}(-x) = \text{sech } x$$

for all x. The other four hyperbolic functions are odd:

$$\sinh(-x) = -\sinh x, \qquad \tanh(-x) = -\tanh x,$$

and so on.

The formulas for the derivatives of the hyperbolic functions parallel those for the trigonometric functions, with occasional sign differences. For example,

$$D \cosh x = D(\tfrac{1}{2}e^x + \tfrac{1}{2}e^{-x})$$
$$= \tfrac{1}{2}e^x - \tfrac{1}{2}e^{-x} = \sinh x.$$

The chain rule then gives

$$D_x \cosh u = (\sinh u)\frac{du}{dx} \tag{13}$$

if u is a differentiable function of x. The other five differentiation formulas are

$$D_x \sinh u = (\cosh u)\frac{du}{dx}, \tag{14}$$

$$D_x \tanh u = (\text{sech}^2 u)\frac{du}{dx}, \tag{15}$$

$$D_x \coth u = -(\text{csch}^2 u)\frac{du}{dx}, \tag{16}$$

$$D_x \text{sech } u = (-\text{sech } u \tanh u)\frac{du}{dx}, \tag{17}$$

$$D_x \text{csch } u = (-\text{csch } u \coth u)\frac{du}{dx}. \tag{18}$$

Formula (14) is derived just as (13) is. Then Formulas (15) through (18) follow from (13) and (14) with the aid of the quotient rule and identities (5) and (6).

The antiderivative versions of the above differentiation formulas are the following integral formulas.

$$\int \sinh u \, du = \cosh u + C, \tag{19}$$

$$\int \cosh u \, du = \sinh u + C, \tag{20}$$

$$\int \text{sech}^2 u \, du = \tanh u + C, \tag{21}$$

$$\int \text{csch}^2 u \, du = -\coth u + C, \tag{22}$$

$$\int \text{sech } u \text{ tanh } u \, du = -\text{sech } u + C, \tag{23}$$

$$\int \text{csch } u \text{ coth } u \, du = -\text{csch } u + C. \tag{24}$$

The integrals in the following example illustrate the fact that simple hyperbolic integrals may be treated in much the same way as simple trigonometric integrals (Section 7-2).

EXAMPLE 1

(a) $\int \sinh^2 x \, dx = \int \frac{1}{2}(\cosh 2x - 1) \, dx = \frac{1}{4} \sinh 2x - \frac{1}{2}x + C.$

(b) $\int \cosh^2 x \sinh x \, dx = \frac{1}{3} \cosh^3 x + C.$

(c) $\int \frac{\text{sech}^2 \sqrt{x}}{\sqrt{x}} \, dx = 2 \int (\text{sech}^2 \sqrt{x}) \frac{dx}{2\sqrt{x}} = 2 \tanh \sqrt{x} + C.$

(d) $\int_0^1 \tanh^2 x \, dx = \int_0^1 (1 - \text{sech}^2 x) \, dx = \left[x - \tanh x \right]_0^1$

$$= 1 - \tanh 1 = 1 - \frac{e^1 - e^{-1}}{e^1 + e^{-1}}$$

$$= \frac{2}{e^2 + 1} \approx 0.238.$$

(e) $\int \coth x \, dx = \int \frac{\cosh x \, dx}{\sinh x} = \ln|\sinh x| + C.$

DOWNWARD MOTION WITH RESISTANCE

In Section 6-6 we discussed the vertical motion of a body of mass m under the influence of a gravitational force $F_G = -mg$ and a resistive force— such as air resistance—which we assumed proportional to its velocity $v = dy/dt$. In the case of a heavy and rapidly moving body, it is often more accurate to assume that the force of air resistance is proportional to the *square* of the velocity:

$$F_R = \pm kv^2 \qquad (k \text{ constant}, \quad k > 0). \tag{25}$$

Then Newton's law $F = ma$ gives

$$m \frac{dv}{dt} = F_G + F_R = -mg \pm kv^2,$$

so that

$$\frac{dv}{dt} = -g \pm \rho v^2 \tag{26}$$

where $\rho = k/m > 0$.

The choice of plus or minus sign in (26) depends upon the direction of motion. For downward motion F_R must be positive (directed upward, which we take as the positive direction), so we choose the plus sign:

$$\frac{dv}{dt} = -g + \rho v^2. \tag{26a}$$

In the case of upward motion we would take the minus sign.

EXAMPLE 2 Suppose that a body is dropped at time $t = 0$ from a height y_0 above the ground, so that its height function $y(t)$ must satisfy Equation (26a) and the initial conditions $y(0) = y_0$ and $y'(0) = 0$. Verify that the function

$$y(t) = y_0 - \frac{1}{\rho}\ln(\cosh t\sqrt{\rho g}) \tag{27}$$

meets these requirements.

Solution We differentiate the function $y(t)$ of Equation (27), and thus we find that

$$v = \frac{dy}{dt} = -\frac{1}{\rho}\frac{\sinh t\sqrt{\rho g}}{\cosh t\sqrt{\rho g}}\sqrt{\rho g} = -\sqrt{g/\rho}\,\tanh t\sqrt{\rho g}.$$

So

$$v^2 = \frac{g}{\rho}\tanh^2 t\sqrt{\rho g}.$$

Another differentiation gives

$$\frac{dv}{dt} = -g\,\text{sech}^2\, t\sqrt{\rho g} = -g(1 - \tanh^2 t\sqrt{\rho g})$$

$$= -g + \rho \cdot \frac{g}{\rho}\tanh^2 t\sqrt{\rho g} = -g + \rho v^2.$$

This shows that the function of (27) satisfies Equation (26a). The facts that $y(0) = y_0$ and $v(0) = 0$ follow from the facts that $\cosh 0 = 1$ and $\tanh 0 = 0$. Problem 30 in Section 7-6 will outline a systematic derivation of the solution in (27) from the differential equation (26a).

7-5 PROBLEMS

Find the derivatives of the functions in Problems 1–14.

1 $\cosh(3x - 2)$.

2 $\sinh\sqrt{x}$.

3 $x^2 \tanh\left(\dfrac{1}{x}\right)$.

4 $\text{sech}\, e^{2x}$.

5 $\coth^3 4x$.

6 $\ln \sinh 3x$.

7 $e^{\text{csch}\, x}$.

8 $\cosh \ln x$.

9 $\sin(\sinh x)$.

10 $\tan^{-1}(\tanh x)$.

11 $\sinh x^4$.

12 $\sinh^4 x$.

13 $\dfrac{1}{x + \tanh x}$.

14 $\cosh^2 x - \sinh^2 x$.

Find the integrals in Problems 15–28.

15 $\displaystyle\int x \sinh x^2\, dx$.

16 $\displaystyle\int \cosh^2 3u\, du$.

17 $\displaystyle\int \tanh^2 3x\, dx$.

18 $\displaystyle\int \frac{\text{sech}\sqrt{x}\,\tanh\sqrt{x}}{\sqrt{x}}\, dx$.

19 $\displaystyle\int \sinh^2 2x \cosh 2x\, dx$.

20 $\displaystyle\int \tanh 3x\, dx$.

21 $\displaystyle\int \frac{\sinh x}{\cosh^3 x}\, dx.$

22 $\displaystyle\int \sinh^4 x\, dx.$

23 $\displaystyle\int \coth x\, \operatorname{csch}^2 x\, dx.$

24 $\displaystyle\int \operatorname{sech} x\, dx.$

25 $\displaystyle\int \frac{\sinh x\, dx}{1 + \cosh x}.$

26 $\displaystyle\int \frac{\sinh \ln x}{x}\, dx.$

27 $\displaystyle\int \frac{dx}{(e^x + e^{-x})^2}.$

28 $\displaystyle\int \frac{e^x + e^{-x}}{e^x - e^{-x}}\, dx.$

29 Apply Definition (1) to prove Identity (7).

30 Derive Identities (5) and (6) from Identity (4).

31 Deduce Identities (10) and (11) from Identity (8).

32 Suppose that A and B are constants. Show that the function $x = A \cosh kt + B \sinh kt$ is a solution of the differential equation $x''(t) = k^2 x(t)$.

33 Find the length of the curve $y = \cosh x$ over the interval $[0, a]$.

34 Find the volume of the solid obtained by revolving around the x-axis the area under $y = \sinh x$ from $x = 0$ to $x = \pi$.

35 Show that the area $A(\theta)$ of the shaded sector in Figure 7.15 is $\theta/2$. This corresponds to the fact that the area of the sector of the unit circle between the positive x-axis and the radius to the point $(\cos \theta, \sin \theta)$ is $\theta/2$. (*Suggestion:* Note first that

$$A(\theta) = \tfrac{1}{2} \cosh \theta \sinh \theta - \int_1^{\cosh \theta} \sqrt{x^2 - 1}\, dx.$$

Then use the Fundamental Theorem of Calculus to show that $A'(\theta) = \tfrac{1}{2}$ for all θ.)

36 (a) Verify that $\lim\limits_{x \to \infty} \tanh x = 1$, as indicated in Figure 7.16.

(b) Deduce from the solution to Example 2 that a body falling from rest subject to a resistive force proportional to the square of its velocity (Equation (26a)) approaches the terminal speed $v_\tau = \sqrt{g/\rho}$.

---

*7-6 Inverse Hyperbolic Functions

A glance at Figure 7.16 in Section 7-5 shows that the hyperbolic sine and tangent functions are increasing on the whole real line. Thus each is one-to-one on the set of all real numbers. The hyperbolic contangent and cosecant functions are also one-to-one where they are defined; each of the latter has domain the set of all real numbers other than zero. Thus each of these four hyperbolic functions has an inverse function, whose domain is in each case the original function's range of values:

$$y = \sinh^{-1} x \quad \text{if } x = \sinh y, \qquad \text{all } x; \tag{1}$$

$$y = \tanh^{-1} x \quad \text{if } x = \tanh y, \qquad -1 < x < 1; \tag{2}$$

$$y = \coth^{-1} x \quad \text{if } x = \coth y, \qquad |x| > 1; \tag{3}$$

$$y = \operatorname{csch}^{-1} x \quad \text{if } x = \operatorname{csch} y, \qquad x \neq 0. \tag{4}$$

The hyperbolic cosine and secant functions are one-to-one on the half-line $x \geq 0$, where cosh is increasing and sech is decreasing. Hence we define

$$y = \cosh^{-1} x \quad \text{if } \quad x = \cosh y \tag{5}$$

where $x \geq 1$ and $y \geq 0$, and

$$y = \operatorname{sech}^{-1} x \quad \text{if } \quad x = \operatorname{sech} y \tag{6}$$

where $0 < x \leq 1$ and $y \geq 0$. The graphs of the six inverse hyperbolic functions may be obtained by reflecting through the line $y = x$ the graphs of the hyperbolic functions (restricted to the half-line $x \geq 0$ in the cases of cosh and sech). We show these graphs in Figure 7.17.

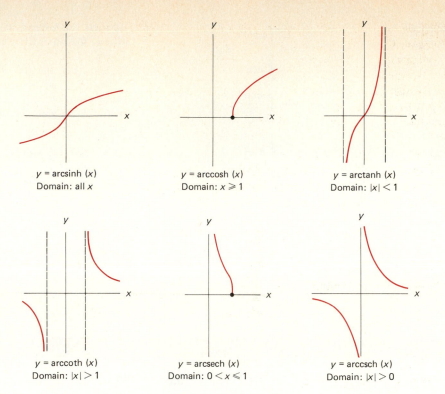

$y = \text{arcsinh}\,(x)$
Domain: all x

$y = \text{arccosh}\,(x)$
Domain: $x \geqslant 1$

$y = \text{arctanh}\,(x)$
Domain: $|x| < 1$

$y = \text{arccoth}\,(x)$
Domain: $|x| > 1$

$y = \text{arcsech}\,(x)$
Domain: $0 < x \leqslant 1$

$y = \text{arccsch}\,(x)$
Domain: $|x| > 0$

7.17 The inverse hyperbolic functions.

The derivatives of the six inverse hyperbolic functions are these:

$$D \sinh^{-1}x = \frac{1}{\sqrt{x^2 + 1}}, \tag{7}$$

$$D \cosh^{-1}x = \frac{1}{\sqrt{x^2 - 1}}, \tag{8}$$

$$D \tanh^{-1}x = \frac{1}{1 - x^2}, \tag{9}$$

$$D \coth^{-1}x = \frac{1}{1 - x^2}, \tag{10}$$

$$D \text{sech}^{-1}x = \frac{-1}{x\sqrt{1 - x^2}}, \tag{11}$$

$$D \text{csch}^{-1}x = \frac{-1}{|x|\sqrt{1 + x^2}}. \tag{12}$$

One can derive these formulas by the standard method of finding the derivative of an inverse function given the derivative of the function itself. The only requirement is that the inverse function is known in advance to be differentiable. For example, to differentiate $\tanh^{-1}x$, we begin with the relation

$$\tanh(\tanh^{-1}x) = x$$

which follows from Definition (2), and substitute $u = \tanh^{-1}x$. Then, since the equation above is actually an identity,

$$D_x \tanh u = Dx = 1,$$

so that

$$(\operatorname{sech}^2 u)\frac{du}{dx} = 1.$$

Thus

$$D \tanh^{-1}x = \frac{du}{dx} = \frac{1}{\operatorname{sech}^2 u} = \frac{1}{1 - \tanh^2 u}$$

$$= \frac{1}{1 - \tanh^2(\tanh^{-1}x)} = \frac{1}{1 - x^2}.$$

This establishes Formula (9). One can use similar methods to verify the other five derivative formulas above.

The hyperbolic functions are defined in terms of the natural exponential function e^x, so it is no surprise to find that their inverses may be expressed in terms of $\ln x$. In fact,

$$\sinh^{-1}x = \ln(x + \sqrt{x^2 + 1}) \qquad \text{for all } x; \qquad (13)$$

$$\cosh^{-1}x = \ln(x + \sqrt{x^2 - 1}) \qquad \text{for all } x \geqq 1; \qquad (14)$$

$$\tanh^{-1}x = \frac{1}{2}\ln\left(\frac{1 + x}{1 - x}\right) \qquad \text{for } |x| < 1; \qquad (15)$$

$$\coth^{-1}x = \frac{1}{2}\ln\left(\frac{x + 1}{x - 1}\right) \qquad \text{for } |x| > 1; \qquad (16)$$

$$\operatorname{sech}^{-1}x = \ln\left(\frac{1 + \sqrt{1 - x^2}}{x}\right) \qquad \text{if } 0 < x \leqq 1; \qquad (17)$$

$$\operatorname{csch}^{-1}x = \ln\left(\frac{1}{x} + \frac{\sqrt{1 + x^2}}{|x|}\right) \qquad \text{if } x \neq 0. \qquad (18)$$

Each of these identities may be established by showing that each side has the same derivative, and that the two sides also agree for at least one value of x in every interval in their respective domains. For example,

$$D \ln(x + \sqrt{x^2 + 1}) = \frac{1 + x(x^2 + 1)^{-1/2}}{x + (x^2 + 1)^{1/2}}$$

$$= \frac{1}{\sqrt{x^2 + 1}} = D \sinh^{-1}x.$$

So

$$\sinh^{-1}x = \ln(x + \sqrt{x^2 + 1}) + C.$$

But $\sinh^{-1}(0) = 0 = \ln(0 + \sqrt{0 + 1})$. This implies that $C = 0$ and thus establishes Formula (13). It is not quite so easy to show that $C = 0$ in the proofs of Formulas (16) and (18); see Problems 26 and 27.

Formulas (13) through (18) may be used to calculate inverse hyperbolic function values. This is convenient if you own a calculator with a repertoire that does not include the inverse hyperbolic functions.

EXAMPLE 1 Suppose that a body is dropped at time $t = 0$ from a height y_0 above the ground. Assume that it is acted on both by gravity and by air resistance proportional to v^2 (where v is its velocity). By Example 2 in Section 7-5, its height t seconds after release will be

$$y = y_0 - \frac{1}{\rho} \ln(\cosh t \sqrt{\rho g}).$$

Assume that $y_0 = 2000$ feet, that $g = 32$ ft/sec², and that $\rho = 0.001$. How long will it take for this body to fall to the ground?

Solution   If we set $y = 0$ in the above equation and then solve for t, we get

$$t = \frac{1}{\sqrt{\rho g}} \cosh^{-1}(e^{\rho y_0}).$$

Then Formula (14) gives

$$t = \frac{1}{\sqrt{\rho g}} \ln(e^{\rho y_0} + \sqrt{e^{2\rho y_0} - 1}).$$

We substitute the data given in the problem and find that

$$t = \frac{1}{\sqrt{0.032}} \ln(e^2 + \sqrt{e^4 - 1}) \approx 15.03$$

seconds—this is the time required for the body to fall to the ground.

The differentiation formulas in (7) through (12) may, in the usual way, be written as the following integral formulas.

$$\int \frac{du}{\sqrt{u^2 + 1}} = \sinh^{-1} u + C. \tag{19}$$

$$\int \frac{du}{\sqrt{u^2 - 1}} = \cosh^{-1} u + C. \tag{20}$$

$$\int \frac{du}{1 - u^2} = \begin{cases} \tanh^{-1} u + C & \text{if } |u| < 1; \tag{21a} \\ \coth^{-1} u + C & \text{if } |u| > 1; \tag{21b} \end{cases}$$

$$= \frac{1}{2} \ln \left| \frac{1 + u}{1 - u} \right| + C. \tag{21c}$$

$$\int \frac{du}{u\sqrt{1 - u^2}} = -\operatorname{sech}^{-1} |u| + C. \tag{22}$$

$$\int \frac{du}{u\sqrt{1 + u^2}} = -\operatorname{csch}^{-1} |u| + C. \tag{23}$$

The distinction between the two cases $|u| < 1$ and $|u| > 1$ in Formula (21) results from the fact that the inverse hyperbolic tangent is defined for $|x| < 1$, while the inverse hyperbolic cotangent is defined for $|x| > 1$.

EXAMPLE 2

$$\int \frac{dx}{\sqrt{4x^2 + 1}} = \frac{1}{2} \int \frac{du}{\sqrt{u^2 + 1}} \qquad [u = 2x, \quad dx = \tfrac{1}{2}\, du]$$

$$= \frac{1}{2} \sinh^{-1} 2x + C.$$

EXAMPLE 3

$$\int_0^{1/2} \frac{dx}{1 - x^2} = \left[\tanh^{-1} x \right]_0^{1/2}$$

$$= \frac{1}{2} \left[\ln \left| \frac{1 + x}{1 - x} \right| \right]_0^{1/2} = \frac{1}{2} \ln 3.$$

EXAMPLE 4

$$\int_2^5 \frac{dx}{1 - x^2} = \left[\coth^{-1} x \right]_2^5 = \frac{1}{2} \left[\ln \left| \frac{1 + x}{1 - x} \right| \right]_2^5$$

$$= \frac{1}{2} \left[\ln \left(\frac{6}{4} \right) - \ln 3 \right] = -\frac{1}{2} \ln 2.$$

THE HANGING CABLE

In Section 3-7 we investigated the shape of a flexible and inelastic cable supporting a load that is uniformly distributed horizontally, such as a supporting cable of a suspension bridge. We found that such a cable hangs in the shape of a parabola. Here, however, we investigate the shape of a cable that hangs of its own weight alone, which we assume is uniformly distributed along the cable, and therefore is not uniformly distributed horizontally.

Let w denote the density of the cable, measured in pounds per foot of length. Figure 7.18 shows the cable with lowest point P, and with a coordinate system set up just as in Section 3-7.

7.18 The hanging cable.

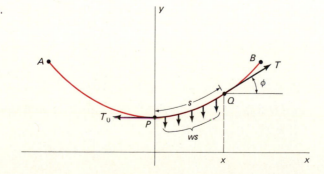

Now consider a section PQ of length s. We plan to obtain a differential equation for the shape $y = f(x)$ of the hanging cable by balancing horizontal and vertical components of the forces acting on PQ. These forces are:

T_0 the horizontal tension pulling on the cable at P;

T the tangential tension pulling on the cable at Q;   and

ws the force of gravity pulling downward on the section PQ.

When we equate the horizontal and vertical components we find that

$$T \cos \phi = T_0 \quad \text{and} \quad T \sin \phi = ws. \tag{24}$$

As Figure 7.18 shows, ϕ is the angle the tangent to the cable at Q makes with the horizontal.

We divide the second equation above by the first and find that

$$\frac{dy}{dx} = \tan \phi = \frac{T \sin \phi}{T \cos \phi} = \frac{ws}{T_0}. \tag{25}$$

Since s is a function of x, this differential equation is not as simple as we might hope. But after we differentiate both sides, we find that

$$\frac{d^2 y}{dx^2} = \frac{w}{T_0} \frac{ds}{dx}.$$

We know from our study of arc length (in Section 5-4) that

$$\frac{ds}{dx} = \sqrt{1 + \left(\frac{dy}{dx}\right)^2},$$

and so

$$\frac{d^2 y}{dx^2} = \frac{w}{T_0} \sqrt{1 + \left(\frac{dy}{dx}\right)^2}. \tag{26}$$

This is the differential equation that will give us the shape $y = y(x)$ of the hanging cable if we can manage to find the solution of the equation.

There is a standard method for solving a second order differential equation like the one above; it is used when the dependent variable y does not appear explicitly. We substitute

$$p = \frac{dy}{dx} \quad \text{and} \quad \frac{dp}{dx} = \frac{d^2 y}{dx^2}.$$

With Equation (26), this yields the simpler equation

$$\frac{dp}{dx} = \frac{w}{T_0} \sqrt{1 + p^2},$$

which we rewrite in the form

$$\frac{1}{\sqrt{1 + p^2}} \frac{dp}{dx} = \frac{w}{T_0}.$$

Since $D \sinh^{-1} x = 1/\sqrt{1 + x^2}$, integration of both sides of the above equation gives

$$\sinh^{-1} p = \frac{w}{T_0} x + C_1.$$

Now $p = 0$ when $x = 0$, because the cable has a horizontal tangent at its lowest point P. This tells us that $C_1 = \sinh^{-1}(0) = 0$, and thus that

$$\sinh^{-1}p = \frac{w}{T_0}x.$$

And so

$$\frac{dy}{dx} = p = \sinh\left(\frac{w}{T_0}x\right).$$

Finally, we integrate both sides of this last equation, and thus we find that

$$y = \frac{T_0}{w}\cosh\left(\frac{w}{T_0}x\right) + C_2.$$

If y_0 is the height of the point P (above the x-axis), then

$$C_2 = y_0 - \frac{T_0}{w},$$

and so the shape of the hanging cable is given by

$$y = \frac{T_0}{w}\cosh\left(\frac{w}{T_0}x\right) + y_0 - \frac{T_0}{w}. \tag{27}$$

We may choose the location of the x-axis so that $y_0 = T_0/w$, and then Equation (27) takes the simple form

$$y = \frac{T_0}{w}\cosh\left(\frac{wx}{T_0}\right), \tag{28}$$

or, if we let $a = T_0/w$, the form

$$y = a\cosh\left(\frac{x}{a}\right).$$

This curve is frequently called a *catenary*, from the Latin word *catena* (chain).

Problems 31–35 deal with the relationship between the length S of the cable, the distance $2L$ between the two points from which it is suspended (at equal heights), and the dip or sag H of the cable at its middle. All these lengths are shown in Figure 7.19.

7.19 Hanging cable of length S, sag H, suspended from two points at distance $2L$ apart.

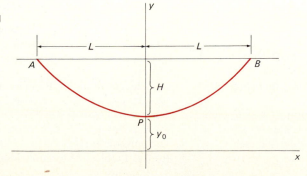

Find the derivatives of the functions in Problems 1–10.

1 $\sinh^{-1} 2x$.

2 $\cosh^{-1}(x^2 + 1)$.

3 $\tanh^{-1}\sqrt{x}$.

4 $\coth^{-1}\sqrt{x^2 + 1}$.

5 $\operatorname{sech}^{-1}\left(\dfrac{1}{x}\right)$.

6 $\operatorname{csch}^{-1} e^x$.

7 $(\sinh^{-1} x)^{3/2}$.

8 $\sinh^{-1}(\ln x)$.

9 $\ln(\tanh^{-1} x)$.

10 $\dfrac{1}{\tanh^{-1} 3x}$.

Evaluate the integrals in Problems 11–20.

11 $\displaystyle\int \frac{dx}{\sqrt{x^2 + 9}}$.

12 $\displaystyle\int \frac{dy}{\sqrt{4y^2 - 9}}$.

13 $\displaystyle\int_{1/2}^{1} \frac{dx}{4 - x^2}$.

14 $\displaystyle\int_{5}^{10} \frac{dx}{4 - x^2}$.

15 $\displaystyle\int \frac{dx}{x\sqrt{4 - 9x^2}}$.

16 $\displaystyle\int \frac{dx}{x\sqrt{x^2 + 25}}$.

17 $\displaystyle\int \frac{e^x\, dx}{\sqrt{e^{2x} + 1}}$.

18 $\displaystyle\int \frac{x\, dx}{\sqrt{x^4 - 1}}$.

19 $\displaystyle\int \frac{dx}{\sqrt{1 - e^{2x}}}$.

20 $\displaystyle\int \frac{\cos x\, dx}{\sqrt{1 + \sin^2 x}}$.

21 Establish Formula (7) for $D \sinh^{-1} x$.

22 Establish Formula (11) for $D \operatorname{sech}^{-1} x$.

23 Prove Formula (15) by differentiating both sides.

24 Establish Formula (13) by solving the equation

$$x = \sinh y = \tfrac{1}{2}(e^y - e^{-y})$$

for y in terms of x.

25 Establish Formula (16) by solving the equation

$$x = \coth y = \frac{e^y + e^{-y}}{e^y - e^{-y}}$$

for y in terms of x.

26 (a) Differentiate both sides of Formula (16) to show that they differ by a constant C. (b) Then prove that $C = 0$ by using the definition of $\coth x$ to show that $\coth^{-1} 2 = \tfrac{1}{2} \ln 3$.

27 (a) Differentiate both sides of Formula (18) to show that they differ by a constant C. (b) Then prove that $C = 0$ by using the definition of $\operatorname{csch} x$ to show that $\operatorname{csch}^{-1} 1 = \ln(1 + \sqrt{2})$.

28 Suppose that the body of Example 1 is dropped from the top of a 500-foot building. After how long and with what speed does it strike the ground?

29 Assume that $\rho = 0.075$ (in the formula of Example 1) for a paratrooper falling with open parachute. If he jumps from an altitude of 10,000 feet and opens his parachute immediately, after how long and with what speed will he land?

30 (a) Starting with the initial value problem

$$\frac{dv}{dt} = -g + \rho v^2, \qquad v(0) = 0,$$

first separate variables, then use Formula (9) to derive the solution

$$v = -\sqrt{\frac{g}{\rho}}\, \tanh t \sqrt{\rho g}.$$

(b) Integrate again to obtain Equation (27) in Section 7-5.

31 Consider a hanging cable, of the sort to which Equation (28) applies. Show that the length of the section lying above the interval $[0, x]$ is

$$s = \frac{T_0}{w} \sinh\left(\frac{wx}{T_0}\right).$$

32 Deduce from (28) that the sag of the cable at its middle is

$$H = \frac{T_0}{w}\left[\cosh\left(\frac{wL}{T_0}\right) - 1\right].$$

33 Deduce from Equation (24) that the tension in the cable satisfies the equation $T^2 = T_0^2 + w^2 s^2$.

34 Use the results of Problems 31 and 33 to show that the tension in the cable at the point (x, y) is

$$T = T_0 \cosh\left(\frac{wx}{T_0}\right) = wy,$$

the weight of a cable of the same density and of length y.

35 A high-voltage transmission line 205 feet long and weighing 1 pound per foot is strung between 2 towers 200 feet apart.

(a) Use the result of Problem 31 to find the minimum tension T_0. (*Suggestion:* Use Newton's method to solve for $u = 100/T_0$.)

(b) Use Problem 33 to find the maximum tension in the transmission line.

(c) Use Problem 32 to find the sag H at the middle.

Use the list below as a guide to concepts you may need to review.

1 The derivatives of the six trigonometric functions, and the corresponding integral formulas
2 The definitions of the six inverse trigonometric functions

3 The derivatives of the inverse trigonometric functions
4 The integral formulas corresponding to the derivatives of the inverse sine, tangent, and secant functions
5 The definitions and derivatives of the hyperbolic functions
6 The definitions of the inverse hyperbolic functions

MISCELLANEOUS PROBLEMS

Differentiate the functions in Problems 1–20.

1 $\sin \sqrt{x}$.

2 $x \cos \left(\dfrac{1}{x}\right)$.

3 $\tan(\ln x)$.

4 $\sec^2(3x + 1)$.

5 $(\csc 2x + \cot 2x)^5$.

6 $\sqrt{\tan x^2}$.

7 $\ln(\sec^2 3x)$.

8 $\cot e^{3x}$.

9 $e^{\arctan x}$.

10 $\ln(\sin^{-1} x)$.

11 $\arcsin \sqrt{x}$.

12 $x \sec^{-1} x^2$.

13 $\tan^{-1}(1 + x^2)$.

14 $\sin^{-1}\sqrt{1 - x^2}$.

15 $e^x \sinh e^x$.

16 $\ln \cosh x$.

17 $\tanh^2 3x + \operatorname{sech}^2 3x$.

18 $\sinh^{-1}\sqrt{x^2 - 1}$.

19 $\cosh^{-1}\sqrt{x^2 + 1}$.

20 $\tanh^{-1}(1 - x^2)$.

Evaluate the integrals in Problems 21–44.

21 $\displaystyle\int x \sec^2 x^2 \, dx$.

22 $\displaystyle\int e^x \sec e^x \tan e^x \, dx$.

23 $\displaystyle\int \csc^2 2x \cot 2x \, dx$.

24 $\displaystyle\int \frac{\cot^2\sqrt{x}}{\sqrt{x}} \, dx$.

25 $\displaystyle\int \frac{\tan^2(1/x)}{x^2} \, dx$.

26 $\displaystyle\int \tan^3 2x \sec^2 2x \, dx$.

27 $\displaystyle\int x \tan(x^2 + 1) \, dx$.

28 $\displaystyle\int \frac{e^x \, dx}{\sqrt{1 - e^x}}$.

29 $\displaystyle\int \frac{e^x \, dx}{\sqrt{1 - e^{2x}}}$.

30 $\displaystyle\int \frac{x \, dx}{1 + x^4}$.

31 $\displaystyle\int \frac{dx}{\sqrt{9 - 4x^2}}$.

32 $\displaystyle\int \frac{dx}{9 + 4x^2}$.

33 $\displaystyle\int \frac{x^2 \, dx}{1 + x^6}$.

34 $\displaystyle\int \frac{\cos x \, dx}{1 + \sin^2 x}$.

35 $\displaystyle\int \frac{dx}{x\sqrt{4x^2 - 1}}$.

36 $\displaystyle\int \frac{dx}{x\sqrt{x^4 - 1}}$.

37 $\displaystyle\int \frac{dx}{\sqrt{e^{2x} - 1}}$.

38 $\displaystyle\int x^2 \cosh x^3 \, dx$.

39 $\displaystyle\int \frac{\sinh \sqrt{x}}{\sqrt{x}} \, dx$.

40 $\displaystyle\int \operatorname{sech}^2(3x - 2) \, dx$.

41 $\displaystyle\int \frac{\tan^{-1} x}{1 + x^2} \, dx$.

42 $\displaystyle\int \frac{dx}{\sqrt{4x^2 - 1}}$.

43 $\displaystyle\int \frac{dx}{\sqrt{4x^2 + 9}}$.

44 $\displaystyle\int \frac{x \, dx}{\sqrt{x^4 + 1}}$.

45 Find the volume generated by revolving the region under $y = (1 - x^4)^{-1/2}$ from $x = 0$ to $x = 1/\sqrt{2}$ around the y-axis.

46 Find the volume generated by revolving the region under $y = (x^4 + 1)^{-1/2}$ from $x = 0$ to $x = 1$ around the y-axis.

47 Sketch the graph of the curve $y = 5 \cos 2x + 12 \sin 2x$. Show at least one complete cycle.

48 A mass on the end of a spring oscillates vertically with a period of 2π seconds. At $t = 0$ the mass is 4 feet from the equilibrium point ($y = 0$) and is approaching it at 3 feet per second. What is the amplitude of the motion of this mass?

49 Use Formulas (14) through (17) in Section 7-6 to show that

(a) $\coth^{-1} x = \tanh^{-1}\left(\dfrac{1}{x}\right)$;

(b) $\operatorname{sech}^{-1} x = \cosh^{-1}\left(\dfrac{1}{x}\right)$.

50 Show that $x''(t) = k^2 x(t)$ if

$$x(t) = A \cosh kt + B \sinh kt,$$

where A and B are constants. Determine A and B if

(a) $x(0) = 1, \quad x'(0) = 0$;
(b) $x(0) = 0, \quad x'(0) = 1$.

51 Use Newton's method to find the least positive solution of the equation $\cos x \cosh x = 1$. Begin by sketching the graphs $y = \cos x$ and $y = \operatorname{sech} x$.

52 (a) Verify by differentiation that

$$\int \sec x \, dx = \sinh^{-1}(\tan x) + C.$$

(b) Show similarly that

$$\int \operatorname{sech} x \, dx = \tan^{-1}(\sinh x) + C.$$

53 Deduce from Problems 31 and 32 in Section 7-6 that the length S, sag H, density w, and minimum tension T_0 of a hanging cable satisfy the equation

$$4wH^2 - wS^2 + 8HT_0 = 0.$$

54 A 300-foot length of cable weighing 2 pounds per foot is hung between 2 points at equal height. A 30-foot sag is then observed at the middle of the cable. (a) Use the result of Problem 53 to find the minimum tension T_0. (b) Find the distance $2L$ between the ends of the cable. (c) Find the maximum tension in the cable.

55 A 400-foot wire weighing 2 pounds per foot can withstand a tension of 5000 pounds. Use the results of Problem 53 and Problem 33 in Section 7-6 to determine the minimum sag that can be allowed in hanging this wire between 2 points of equal height.

56 A cable is 105 feet long and weighs 2 pounds per foot. It is hung with its ends at the same level, and the ends are 100 feet apart. Find the sag at its middle and the maximum and minimum tension in the cable.

57 Assume that the earth is a solid sphere of uniform density, with mass M and radius $R = 3960$ miles. For a particle of mass m *within* the earth at distance r from the center of the earth, the gravitational force attracting m toward the center is $F_r = -GM_r m/r^2$, where M_r is the mass of the part of the earth within a sphere of radius r.

(a) Show that $F_r = -(GMm/R^3)r$.

(b) Now suppose that a hole is drilled straight through the center of the earth, connecting two antipodal points on its surface. Let a particle of mass m be dropped at time $t = 0$ into this hole with initial speed zero, and let $r(t)$ be its distance from the center of the earth at time t. Conclude from Newton's second law and part (a) that $r''(t) = -k^2 r(t)$, where $k^2 = GM/R^3 = g/R$ and $g = 32.2$ feet per second per second.

(c) Deduce from Part (b) that the particle undergoes simple harmonic motion back and forth between the ends of the hole, with a period of about 84 minutes.

(d) With what speed (in miles per hour) does the particle pass through the earth's center?

Techniques of Integration

8

Introduction

In the past three chapters we have seen that many geometric and physical quantities can be expressed as definite integrals. The Fundamental Theorem of Calculus reduces the problem of calculating the definite integral $\int_a^b f(x)\,dx$ to that of finding an antiderivative $G(x)$ of $f(x)$; once this is accomplished, then

$$\int_a^b f(x)\,dx = \Big[G(x)\Big]_a^b = G(b) - G(a).$$

But as yet we have relied largely on trial-and-error methods for finding the required antiderivative $G(x)$. Sometimes a knowledge of elementary derivative formulas, perhaps in combination with a simple substitution, will allow us to integrate a given function. This approach can, however, be inefficient and time-consuming, especially in view of the following surprising fact: There exist simple-looking integrals, such as

$$\int e^{-x^2}\,dx, \qquad \int \frac{\sin x}{x}\,dx, \quad \text{and} \quad \int \sqrt{1 + x^4}\,dx,$$

that cannot be evaluated in terms of finite combinations of the familiar algebraic and elementary transcendental functions. For example, the antiderivative

$$H(x) = \int_0^x e^{-t^2}\,dt$$

of e^{-x^2} has no finite expression in terms of elementary functions. Any attempt to find such an expression will, therefore, inevitably be unsuccessful.

The presence of such examples indicates that we cannot hope to reduce integration to a routine process like differentiation. In fact, finding antiderivatives is an art, depending upon experience and practice. Nevertheless, there are a number of techniques whose systematic use can substantially reduce our dependence upon chance and intuition alone. This chapter deals with some of these systematic techniques of integration.

Integral Tables and Simple Substitutions

Integration would be a fairly simple matter if we had a list of integral formulas, an *integral table*, in which we could locate any integral that we ever needed to evaluate. But the diversity of integrals that we encounter in practice is too great for such an all-inclusive integral table to be practical. It is more sensible to print, or memorize, a short table of integrals of the sort seen frequently, and to learn techniques by which the range of applicability of this short table can be extended. We may begin with the following list of integrals (Figure 8.1), which are familiar from earlier chapters; each is equivalent to one of the basic derivative formulas.

A table of approximately 110 integral formulas appears inside the covers of this book. Still more extensive integral tables are readily available. For example, the volume of *Standard Mathematical Tables* edited by Samuel M. Selby and published annually by the Chemical Rubber Company in Cleveland, contains a list of over 700 integral formulas. But even such a lengthy table can be expected to include only a small fraction of the integrals we may need to evaluate. Thus it is necessary to learn techniques for deriving new formulas and for transforming a given integral either into one that's already familiar or into one appearing in an accessible table.

The principal such technique is the *method of substitution*, which we first considered in Section 4-5. Recall that, if

$$\int f(u)\,du = F(u) + C,$$

then

$$\int f(g(x))g'(x)\,dx = F(g(x)) + C.$$

Thus the substitution $u = g(x)$, $du = g'(x)\,dx$ transforms the integral $\int f(g(x))g'(x)\,dx$ into the simpler integral $\int f(u)\,du$. The key to making this simplification lies in spotting the composition $f(g(x))$ in the given integrand. In order to convert this integrand into a function of u alone, it is necessary that the remaining factor be a constant multiple of the derivative $g'(x)$ of the inside function $g(x)$. In this case we replace $f(g(x))$ by the simpler $f(u)$ and also $g'(x)\,dx$ by du. The preceding three chapters contain numerous illustrations of this method of substitution, and the problems for the present section provide an opportunity to review it.

$$\int u^n\,du = \frac{u^{n+1}}{n+1} + C \quad [n \neq -1] \quad (1)$$

$$\int \frac{du}{u} = \ln|u| + C \quad (2)$$

$$\int e^u\,du = e^u + C \quad (3)$$

$$\int \cos u\,du = \sin u + C \quad (4)$$

$$\int \sin u\,du = -\cos u + C \quad (5)$$

$$\int \sec^2 u\,du = \tan u + C \quad (6)$$

$$\int \csc^2 u\,du = -\cot u + C \quad (7)$$

$$\int \sec u \tan u\,du = \sec u + C \quad (8)$$

$$\int \csc u \cot u\,du = -\csc u + C \quad (9)$$

$$\int \frac{du}{\sqrt{1 - u^2}} = \sin^{-1} u + C \quad (10)$$

$$\int \frac{du}{1 + u^2} = \tan^{-1} u + C \quad (11)$$

$$\int \frac{du}{u\sqrt{u^2 - 1}} = \sec^{-1}|u| + C \quad (12)$$

8.1 A short table of integrals.

EXAMPLE 1 Find $\displaystyle\int \frac{x\,dx}{1 + x^4}$.

Solution Here it is not so clear what the inside function is. But comparison with the integral formula in (11) (in Figure 8.1) suggests that we try the substitution $u = x^2$ and its consequent $du = 2x\,dx$. We take advantage of the factor $x\,dx = \frac{1}{2}du$ that is available in the integrand and make these computations:

$$\int \frac{x\,dx}{1 + x^4} = \frac{1}{2}\int \frac{du}{1 + u^2}$$

$$= \frac{1}{2}\tan^{-1} u + C = \frac{1}{2}\tan^{-1} x^2 + C.$$

Note that the substitution $u = x^2$ would have been of little use had the integrand been either $1/(1 + x^4)$ or $x^2/(1 + x^4)$.

Example 1 illustrates the device of using a substitution to convert a given integral into a familiar one. Often an integral that does not appear in any integral table may be transformed into one that does by using the techniques of this chapter. In the following example we employ an appropriate substitution to "reconcile" the given integral with the standard integral formula

$$\int \frac{u^2 \, du}{\sqrt{a^2 - u^2}} = \frac{a^2}{2} \sin^{-1}\left(\frac{u}{a}\right) - \frac{u}{2}\sqrt{a^2 - u^2} + C,$$

which is Formula 56 (inside the front cover).

EXAMPLE 2 Find $\displaystyle\int \frac{\sin^2 x \cos x \, dx}{\sqrt{25 - 16 \sin^2 x}}$.

Solution   In order that $25 - 16\sin^2 x$ be equal to $a^2 - u^2$, we take $a = 5$ and $u = 4\sin x$, so that $du = 4\cos x \, dx$. This gives

$$\int \frac{\sin^2 x \cos x \, dx}{\sqrt{25 - 16\sin^2 x}} = \int \frac{(u/4)^2(1/4) \, du}{\sqrt{25 - u^2}}$$

$$= \frac{1}{64} \int \frac{u^2 \, du}{\sqrt{25 - u^2}}$$

$$= \frac{1}{64}\left(\frac{25}{2}\sin^{-1}\left(\frac{u}{5}\right) - \frac{u}{2}\sqrt{25 - u^2}\right) + C$$

$$= \frac{25}{128}\sin^{-1}\left(\frac{4}{5}\sin x\right) - \frac{\sin x}{32}\sqrt{25 - 16\sin^2 x} + C.$$

In Section 8-5 we shall see how integral formulas like the one applied in Example 2 can be derived.

8-2   PROBLEMS

Evaluate the integrals in Problems 1–30.

1 $\displaystyle\int (2 - 3x)^4 \, dx.$

2 $\displaystyle\int \frac{dx}{(1 + 2x)^2}.$

3 $\displaystyle\int x^2 \sqrt{2x^3 - 4} \, dx.$

4 $\displaystyle\int \frac{5t \, dt}{5 + 2t^2}.$

5 $\displaystyle\int \frac{3x \, dx}{\sqrt[3]{2x^2 + 3}}.$

6 $\displaystyle\int x \sec^2 x^2 \, dx.$

7 $\displaystyle\int \frac{\cot \sqrt{y} \csc \sqrt{y}}{\sqrt{y}} \, dy.$

8 $\displaystyle\int \sin \pi(2x + 1) \, dx.$

9 $\displaystyle\int (1 + \sin \theta)^5 \cos \theta \, d\theta.$

10 $\displaystyle\int \frac{\sin 2x}{4 + \cos 2x} \, dx.$

11 $\displaystyle\int e^{-\cot x} \csc^2 x \, dx.$

12 $\displaystyle\int \frac{e^{\sqrt{x+4}}}{\sqrt{x+4}} \, dx.$

13 $\displaystyle\int \frac{[\ln t]^{10}}{t} \, dt.$

14 $\displaystyle\int \frac{t \, dt}{\sqrt{1 - 9t^2}}.$

15 $\displaystyle\int \frac{dt}{\sqrt{1 - 9t^2}}.$

16 $\displaystyle\int \frac{e^{2x}}{1 + e^{2x}} \, dx.$

17 $\displaystyle\int \frac{e^{2x}}{1 + e^{4x}} \, dx.$

18 $\displaystyle\int \frac{e^{\tan^{-1}x}}{1 + x^2} \, dx.$

19 $\displaystyle\int \frac{3x \, dx}{\sqrt{1 - x^4}}.$

20 $\displaystyle\int \sin^3 2x \cos 2x \, dx.$

21 $\displaystyle\int \tan^4 3x \sec^2 3x \, dx.$

22 $\displaystyle\int \frac{dt}{1 + 4t^2}.$

23 $\displaystyle\int \frac{\cos \theta}{1 + \sin^2 \theta} \, d\theta.$

24 $\displaystyle\int \frac{\sec^2 \theta \, d\theta}{1 + \tan \theta}.$

25 $\displaystyle\int \frac{(1 + \sqrt{x})^4}{\sqrt{x}} \, dx.$

26 $\displaystyle\int t^{-1/3}\sqrt{t^{2/3} - 1} \, dt.$

27 $\displaystyle\int \frac{dt}{(1 + t^2)\arctan t}$.

28 $\displaystyle\int \frac{\sec 2x \tan 2x\, dx}{(1 + \sec 2x)^{3/2}}$.

29 $\displaystyle\int \frac{dx}{\sqrt{e^{2x} - 1}}$.

30 $\displaystyle\int \frac{x\, dx}{\sqrt{\exp(2x^2) - 1}}$.

In Problems 31–35, make the indicated substitution to evaluate the given integral.

31 $\displaystyle\int x^2\sqrt{x - 2}\, dx$, $\quad u = x - 2$.

32 $\displaystyle\int \frac{x^2\, dx}{\sqrt{x + 3}}$, $\quad u = x + 3$.

33 $\displaystyle\int \frac{x\, dx}{\sqrt{2x + 3}}$, $\quad u = 2x + 3$.

34 $\displaystyle\int x\sqrt[3]{x - 1}\, dx$, $\quad u = x - 1$.

35 $\displaystyle\int \frac{x\, dx}{\sqrt[3]{x + 1}}$, $\quad u = x + 1$.

In each of Problems 36–42, evaluate the given integral by first transforming it into one of those listed in the integral table inside the covers of this book.

36 $\displaystyle\int x\sqrt{4 - x^4}\, dx$.

37 $\displaystyle\int e^x\sqrt{9 + e^{2x}}\, dx$.

38 $\displaystyle\int \frac{\cos x\, dx}{(\sin^2 x)\sqrt{1 + \sin^2 x}}$.

39 $\displaystyle\int \frac{\sqrt{x^4 - 1}}{x}\, dx$.

40 $\displaystyle\int \frac{e^{3x}\, dx}{\sqrt{25 + 16e^{2x}}}$.

41 $\displaystyle\int \frac{(\ln x)^2}{x}\sqrt{1 + (\ln x)^2}\, dx$.

42 $\displaystyle\int x^8\sqrt{4x^6 - 1}\, dx$.

43 The substitution $u = x^2$, $x = \sqrt{u}$, and $dx = du/2\sqrt{u}$ appears to lead to this result:

$$\int_{-1}^{1} x^2\, dx = \tfrac{1}{2}\int_{1}^{1} \sqrt{u}\, du = 0.$$

Do you believe this result? If not, why not?

44 Use the fact that $x^2 + 4x + 5 = (x + 2)^2 + 1$ to evaluate

$$\int \frac{dx}{x^2 + 4x + 5}.$$

45 Use the fact that $1 - (x - 1)^2 = 2x - x^2$ to evaluate

$$\int \frac{dx}{\sqrt{2x - x^2}}.$$

8-3

Trigonometric Integrals

The substitution $u = \sin x$, $du = \cos x\, dx$ gives

$$\int \sin^3 x \cos x\, dx = \int u^3\, du$$

$$= \tfrac{1}{4}u^4 + C = \tfrac{1}{4}\sin^4 x + C.$$

This type of substitution can be used to evaluate an integral of the form

$$\int \sin^m x \cos^n x\, dx \tag{1}$$

in the first of the following two cases.

Case 1: At least one of the two numbers m and n is an *odd positive integer*. If so, the other may be any real number.

Case 2: Both m and n are *nonnegative even integers*.

Suppose, for example, that m is an odd positive integer. Then we split off one $\sin x$ factor and use the identity $\sin^2 x = 1 - \cos^2 x$ to express the remaining factor $\sin^{m-1} x$ in terms of $\cos x$, as follows:

$$\int \sin^m x \cos^n x\, dx = \int (1 - \cos^2 x)^{(m-1)/2} \cos^n x \sin x\, dx.$$

Since $(m - 1)/2$ is a nonnegative integer (because m is odd), the substitution $u = \cos x$, $du = -\sin x\, dx$ yields an integral of the form $\int f(u)\, du$ where $f(u)$ is a polynomial in u. If it is n that is odd, then we can split off one $\cos x$ factor and convert the remaining cosines into sines.

EXAMPLE 1

(a)
$$\int \sin^3 x \cos^2 x \, dx = \int (1 - \cos^2 x) \cos^2 x \sin x \, dx$$

$$= \int (u^4 - u^2) \, du \qquad\qquad (u = \cos x)$$

$$= \tfrac{1}{5} u^5 - \tfrac{1}{3} u^3 + C$$

$$= \tfrac{1}{5} \cos^5 x - \tfrac{1}{3} \cos^3 x + C.$$

(b)
$$\int \cos^5 x \, dx = \int (1 - \sin^2 x)^2 \cos x \, dx$$

$$= \int (1 - u^2)^2 \, du \qquad\qquad (u = \sin x)$$

$$= \int (1 - 2u^2 + u^4) \, du$$

$$= u - \tfrac{2}{3} u^3 + \tfrac{1}{5} u^5 + C$$

$$= \sin x - \tfrac{2}{3} \sin^3 x + \tfrac{1}{5} \sin^5 x + C.$$

In Case 2 of the sine-cosine integral in (1), with both m and n nonnegative even integers, we use the half-angle formulas

$$\sin^2 \theta = \tfrac{1}{2}(1 - \cos 2\theta) \qquad\qquad (2a)$$

and

$$\cos^2 \theta = \tfrac{1}{2}(1 + \cos 2\theta) \qquad\qquad (2b)$$

to halve the even powers of $\sin x$ and $\cos x$. Repetition of this process with the resulting powers of $\cos 2x$—if necessary—leads to integrals involving odd powers, and we have seen how to handle these in Case 1.

EXAMPLE 2 Use of Formulas (2a) and (2b) gives

$$\int \sin^2 x \cos^2 x \, dx = \int \tfrac{1}{2}(1 - \cos 2x)\tfrac{1}{2}(1 + \cos 2x) \, dx$$

$$= \tfrac{1}{4} \int (1 - \cos^2 2x) \, dx$$

$$= \tfrac{1}{4} \int (1 - \tfrac{1}{2}(1 + \cos 4x)) \, dx$$

$$= \tfrac{1}{8} \int (1 - \cos 4x) \, dx$$

$$= \tfrac{1}{8} x - \tfrac{1}{32} \sin 4x + C.$$

In the third step we have used Formula (2b) with $\theta = 2x$.

EXAMPLE 3 Here we apply Formula (2b), first with $\theta = 3x$, then with $\theta = 6x$.

$$\int \cos^4 3x \, dx = \int \tfrac{1}{4}(1 + \cos 6x)^2 \, dx$$

$$= \tfrac{1}{4} \int (1 + 2 \cos 6x + \cos^2 6x) \, dx$$

$$= \tfrac{1}{4} \int (\tfrac{3}{2} + 2 \cos 6x + \tfrac{1}{2} \cos 12x) \, dx$$

$$= \tfrac{3}{8} x + \tfrac{1}{12} \sin 6x + \tfrac{1}{96} \sin 12x + C.$$

An integral of the form

$$\int \tan^m x \sec^n x \, dx \qquad\qquad (3)$$

can be routinely evaluated in either of the following two cases.

Case 1: m is an *odd positive integer*.

Case 2: n is an *even positive integer*.

In Case 1, we split off the factor $\sec x \tan x$ to form, along with dx, the differential $\sec x \tan x \, dx$ of $\sec x$. We then use the identity $\tan^2 x = \sec^2 x - 1$ to convert the remaining even power of $\tan x$ into powers of $\sec x$. This prepares us for the substitution $u = \sec x$.

EXAMPLE 4

$$
\begin{aligned}
\int \tan^3 x \sec^3 x \, dx &= \int (\sec^2 x - 1)\sec^2 x \sec x \tan x \, dx \\
&= \int (u^4 - u^2) \, du \qquad\qquad (u = \sec x) \\
&= \tfrac{1}{5}u^5 - \tfrac{1}{3}u^3 + C \\
&= \tfrac{1}{5}\sec^5 x - \tfrac{1}{3}\sec^3 x + C.
\end{aligned}
$$

To evaluate Integral (3) in Case 2, we split off $\sec^2 x$ to form, along with dx, the differential of $\tan x$. Use of the identity $\sec^2 x = 1 + \tan^2 x$ to convert the remaining even power of $\sec x$ to powers of $\tan x$ then prepares us for the substitution $u = \tan x$.

EXAMPLE 5

$$
\begin{aligned}
\int \sec^6 x \, dx &= \int (1 + \tan^2 x)^2 \sec^2 x \, dx \\
&= \int (1 + u^2)^2 \, du \qquad\qquad (u = \tan x) \\
&= \int (1 + 2u^2 + u^4) \, du \\
&= u + \tfrac{2}{3}u^3 + \tfrac{1}{5}u^5 + C \\
&= \tan x + \tfrac{2}{3}\tan^3 x + \tfrac{1}{5}\tan^5 x + C.
\end{aligned}
$$

Similar methods are effective with integrals of the form

$$\int \csc^m x \cot^n x \, dx.$$

The method of Case 1 succeeds with the integral $\int \tan^n x \, dx$ only when n is an odd positive integer, but there is another approach that works equally well whether n be even *or* odd. We split off the factor $\tan^2 x$ and replace it by $\sec^2 x - 1$. This gives

$$
\begin{aligned}
\int \tan^n x \, dx &= \int \tan^{n-2} x(\sec^2 x - 1) \, dx \\
&= \int \tan^{n-2} x \sec^2 x \, dx - \int \tan^{n-2} x \, dx.
\end{aligned}
$$

We integrate what we can, and find that

$$\int \tan^n x \, dx = \frac{\tan^{n-1} x}{n-1} - \int \tan^{n-2} x \, dx. \qquad (4)$$

Equation (4) is our first example of a **reduction formula.** Its use reduces the original exponent from n to $n-2$. Repeated application of (4) leads eventually either to

$$\int \tan^2 x \, dx = \int (\sec^2 x - 1) \, dx$$
$$= \tan x - x + C$$

or to

$$\int \tan x \, dx = \int \frac{\sin x}{\cos x} \, dx$$
$$= -\ln|\cos x| + C.$$

EXAMPLE 6   Two applications of Formula (4) give

$$\int \tan^6 x \, dx = \tfrac{1}{5} \tan^5 x - \int \tan^4 x \, dx$$

$$= \tfrac{1}{5} \tan^5 x - \left(\tfrac{1}{3} \tan^3 x - \int \tan^2 x \, dx \right)$$

$$= \tfrac{1}{5} \tan^5 x - \tfrac{1}{3} \tan^3 x + \tan x - x + C.$$

Finally, in the case of a trigonometric integral involving tangents, cosecants, and so on, a last resort is to express the integrand in terms of sines and cosines. Simplification may then yield an integrable expression.

8-3   PROBLEMS

Evaluate the integrals in Problems 1–25.

1 $\int \sin^3 x \, dx.$

2 $\int \sin^4 x \, dx.$

3 $\int \sin^2 \theta \cos^3 \theta \, d\theta.$

4 $\int \sin^3 t \cos^3 t \, dt.$

5 $\int \cos^5 x \, dx.$

6 $\int \frac{\sin t}{\cos^3 t} \, dt.$

7 $\int \frac{\sin^3 x}{\sqrt{\cos x}} \, dx.$

8 $\int \sin^3 3\phi \cos^4 3\phi \, d\phi.$

9 $\int \sin^5 2z \cos^2 2z \, dz.$

10 $\int \sin^{3/2} x \cos^3 x \, dx.$

11 $\int \frac{\sin^3 4x}{\cos^2 4x} \, dx.$

12 $\int \cos^6 4\theta \, d\theta.$

13 $\int \sec^4 t \, dt.$

14 $\int \tan^3 x \, dx.$

15 $\int \cot^3 2x \, dx.$

16 $\int \tan \theta \sec^4 \theta \, d\theta.$

17 $\int \tan^5 2x \sec^2 2x \, dx.$

18 $\int \cot^3 x \csc^2 x \, dx.$

19 $\int \csc^6 2t \, dt.$

20 $\int \frac{\sec^4 t}{\tan^2 t} \, dt.$

21 $\int \frac{\tan^3 \theta}{\sec^4 \theta} \, d\theta.$

22 $\int \frac{\cot^3 x}{\csc^2 x} \, dx.$

23 $\int \frac{\tan^3 t}{\sqrt{\sec t}} \, dt.$

24 $\int \frac{1}{\cos^4 2x} \, dx.$

25 $\int \frac{\cot \theta}{\csc^3 \theta} \, d\theta.$

26 Derive a reduction formula, one analogous to Formula (4), for $\int \cot^n x \, dx$.

27 Find $\int \tan x \sec^4 x \, dx$ in two different ways. Then show that your two results are equivalent.

28 Find $\int \cot^3 x \, dx$ in two different ways. Then show that the results are equivalent.

Problems 29–32 are applications of the trigonometric identities

$$\sin\alpha\sin\beta = \tfrac{1}{2}[\cos(\alpha-\beta) - \cos(\alpha+\beta)],$$

$$\sin\alpha\cos\beta = \tfrac{1}{2}[\sin(\alpha-\beta) + \sin(\alpha+\beta)],$$

$$\cos\alpha\cos\beta = \tfrac{1}{2}[\cos(\alpha-\beta) + \cos(\alpha+\beta)].$$

29 Find $\int \sin 3x \cos 5x\, dx$.

30 Find $\int \sin 2x \sin 4x\, dx$.

31 Find $\int \cos x \cos 4x\, dx$.

32 Suppose that m and n are positive integers with $m \neq n$. Show that

(a) $\int_0^{2\pi} \sin mx \sin nx\, dx = 0$;

(b) $\int_0^{2\pi} \cos mx \sin nx\, dx = 0$;

(c) $\int_0^{2\pi} \cos mx \cos nx\, dx = 0$.

8-4

Integration by Parts

One reason for transforming one given integral into another is to produce an integral that is easier to evaluate. There are two general ways to accomplish such a transformation—the first is integration by substitution and the second is *integration by parts*.

The formula for integration by parts is a simple consequence of the product rule for derivatives:

$$D_x(uv) = v\frac{du}{dx} + u\frac{dv}{dx}.$$

If we write this formula in the form

$$u(x)v'(x) = D_x[u(x)v(x)] - v(x)u'(x), \tag{1}$$

then antidifferentiation gives

$$\int u(x)v'(x)\, dx = u(x)v(x) - \int v(x)u'(x)\, dx. \tag{2}$$

This is the formula for **integration by parts.** With $du = u'(x)\, dx$ and $dv = v'(x)\, dx$, Formula (2) becomes

$$\int u\, dv = uv - \int v\, du. \tag{3}$$

To apply the integration by parts formula to a given integral, we must first factor its integrand into two "parts," u and dv, the latter including the differential dx. We try to choose these parts in such a way that

(i) The antiderivative $v = \int dv$ is easy to find, and

(ii) The new integral $\int v\, du$ is easier to compute than the original integral $\int u\, dv$.

An effective strategy is to choose for dv the most complicated factor that can readily be integrated. The other part u then gets differentiated (rather than integrated) to find du.

We begin with two examples in which we have little flexibility in choosing the parts u and dv.

EXAMPLE 1 Find $\int \ln x\, dx$.

Solution Here there is little alternative to the natural choice $u = \ln x$ and $dv = dx$. Then $du = dx/x$ and $v = x$. It is helpful to systematize the procedure of integration by parts by writing u, dv, du, and v in an array like the one below:

$$u = \ln x \qquad\qquad dv = dx$$

$$du = \frac{dx}{x} \qquad\qquad v = x$$

The first line specifies one choice of u and dv; the second line is computed from the first. Then Formula (3) gives

$$\int \ln x \, dx = x \ln x - \int dx = x \ln x - x + C.$$

Comment 1: The constant of integration appears only at the last step. We know that, once we have found one antiderivative, any other may be obtained by adding a constant C to the one we have found.

Comment 2: In computing $v = \int dv$, we ordinarily take the constant of integration to be zero. Had we written $v = x + C_1$ in Example 1, the result would have been

$$\int \ln x \, dx = (x + C_1) \ln x - \int \left(1 + \frac{C_1}{x}\right) dx$$

$$= x \ln x + C_1 \ln x - (x + C_1 \ln x) + C$$

$$= x \ln x - x + C$$

as before, so introducing the extra constant C_1 makes no difference.

EXAMPLE 2 Find $\int \sin^{-1} x \, dx$.

Solution Let

$$u = \sin^{-1} x \qquad\qquad dv = dx$$

$$du = \frac{dx}{\sqrt{1 - x^2}} \qquad\qquad v = x.$$

Then Formula (3) gives

$$\int \sin^{-1} x \, dx = x \sin^{-1} x - \int \frac{x \, dx}{\sqrt{1 - x^2}}$$

$$= x \sin^{-1} x + \sqrt{1 - x^2} + C.$$

EXAMPLE 3 Find $\int x e^{-x} \, dx$.

Solution Here we have some flexibility. Suppose that we try

$$u = e^{-x} \qquad\qquad dv = x \, dx$$

$$du = -e^{-x} \, dx \qquad\qquad v = \tfrac{1}{2} x^2.$$

Then integration by parts gives

$$\int x e^{-x} \, dx = \tfrac{1}{2} x^2 e^{-x} + \tfrac{1}{2} \int x^2 e^{-x} \, dx.$$

The new integral on the right looks more troublesome than the one we started with! Let us begin anew with

$$u = x \qquad dv = e^{-x}\,dx$$
$$du = dx \qquad v = -e^{-x}.$$

Now integration by parts gives

$$\int xe^{-x}\,dx = -xe^{-x} + \int e^{-x}\,dx$$
$$= -xe^{-x} - e^{-x} + C.$$

Integration by parts can also be applied to definite integrals. We integrate Equation (1) from $x = a$ to $x = b$ and apply the Fundamental Theorem of Calculus. This gives

$$\int_a^b u(x)v'(x)\,dx = \int_a^b D_x[u(x)v(x)]\,dx - \int_a^b v(x)u'(x)\,dx$$
$$= \Big[u(x)v(x)\Big]_a^b - \int_a^b v(x)u'(x)\,dx.$$

In the notation of (3) this is

$$\int_{x=a}^{x=b} u\,dv = \Big[uv\Big]_{x=a}^{x=b} - \int_{x=a}^{x=b} v\,du, \qquad (4)$$

though we must not forget that u and v are functions of x. For example, with $u = x$ and $dv = e^{-x}\,dx$, as in Example 3, we obtain

$$\int_0^1 xe^{-x}\,dx = \Big[-xe^{-x}\Big]_0^1 + \int_0^1 e^{-x}\,dx$$
$$= -e^{-1} + \Big[-e^{-x}\Big]_0^1 = 1 - 2e^{-1}.$$

EXAMPLE 4 Find $\int x^2 e^{-x}\,dx$.

Solution If we choose $u = x^2$, then $du = 2x\,dx$, so we will reduce the exponent of x by this choice. With

$$u = x^2 \qquad dv = e^{-x}\,dx$$
$$du = 2x\,dx \qquad v = -e^{-x},$$

integration by parts gives

$$\int x^2 e^{-x}\,dx = -x^2 e^{-x} + 2\int xe^{-x}\,dx.$$

We substitute the result $\int xe^{-x}\,dx = -xe^{-x} - e^{-x}$ of Example 3, and this yields

$$\int x^2 e^{-x}\,dx = -x^2 e^{-x} - 2xe^{-x} - 2e^{-x} + C$$
$$= -(x^2 + 2x + 2)e^{-x} + C.$$

In effect, we have annihilated the original x^2 factor by integrating by parts twice in succession (since Example 3 was itself an integration by parts).

EXAMPLE 5 Find $\int e^{2x}\sin 3x\,dx$.

Solution This is another example in which repeated integration by parts succeeds, but with a different twist. Let

$$u = \sin 3x \qquad\qquad dv = e^{2x}\, dx$$
$$du = 3 \cos 3x\, dx \qquad\qquad v = \tfrac{1}{2}e^{2x}.$$

Then

$$\int e^{2x} \sin 3x\, dx = \tfrac{1}{2}e^{2x} \sin 3x - \tfrac{3}{2} \int e^{2x} \cos 3x\, dx.$$

At first it appears that little progress has been made, since the integral on the right seems just as difficult as the one on the left. We ignore this objection and try again. We apply integration by parts to the new integral. With

$$u = \cos 3x \qquad\qquad dv = e^{2x}\, dx$$
$$du = -3 \sin 3x\, dx \qquad\qquad v = \tfrac{1}{2}e^{2x},$$

we find that

$$\int e^{2x} \cos 3x\, dx = \tfrac{1}{2}e^{2x} \cos 3x + \tfrac{3}{2} \int e^{2x} \sin 3x\, dx.$$

When we substitute this information into the previous equation, we discover that

$$\int e^{2x} \sin 3x\, dx = \tfrac{1}{2}e^{2x} \sin 3x - \tfrac{3}{4}e^{2x} \cos 3x - \tfrac{9}{4} \int e^{2x} \sin 3x\, dx.$$

So we're back where we started. Or are we? In fact we are *not*, because we actually can *solve* the last equation for the desired integral! We move the right-hand integral above to the left-hand side of the equation; this gives

$$\tfrac{13}{4} \int e^{2x} \sin 3x\, dx = \tfrac{1}{4}e^{2x}(2 \sin 3x - 3 \cos 3x) + C_1,$$

and so

$$\int e^{2x} \sin 3x\, dx = \tfrac{1}{13}e^{2x}(2 \sin 3x - 3 \cos 3x) + C.$$

EXAMPLE 6 Find a reduction formula for $\int \sec^n x\, dx$.

Solution The idea is that n is a (large) positive integer, and that we want to express the given integral in terms of a lower power of $\sec x$. The easiest power of $\sec x$ to integrate is $\sec^2 x$, so we let

$$u = \sec^{n-2} x \qquad\qquad dv = \sec^2 x\, dx$$
$$du = (n-2) \sec^{n-2} x \tan x\, dx \qquad\qquad v = \tan x.$$

This gives

$$\int \sec^n x\, dx = \sec^{n-2} x \tan x - (n-2) \int \sec^{n-2} x \tan^2 x\, dx$$

$$= \sec^{n-2} x \tan x - (n-2) \int (\sec^{n-2} x)(\sec^2 x - 1)\, dx.$$

Hence

$$\int \sec^n x\, dx = \sec^{n-2} x \tan x - (n-2) \int \sec^n x\, dx + (n-2) \int \sec^{n-2} x\, dx.$$

We solve this last equation for the desired integral and find that

$$\int \sec^n x\, dx = \frac{\sec^{n-2} x \tan x}{n-1} + \frac{n-2}{n-1} \int \sec^{n-2} x\, dx. \tag{5}$$

This is the desired reduction formula. For example, if we take $n = 3$ in this formula, we find that

$$\int \sec^3 x \, dx = \frac{1}{2} \sec x \tan x + \frac{1}{2} \int \sec x \, dx$$

$$= \frac{1}{2} \sec x \tan x + \frac{1}{2} \ln|\sec x + \tan x| + C. \qquad (6)$$

In the last step we used Equation (15) of Section 7-2:

$$\int \sec x \, dx = \ln|\sec x + \tan x| + C.$$

The reason for using the reduction formula in (5) is that repeated applications must yield one of the two elementary integrals $\int \sec x \, dx$ and $\int \sec^2 x \, dx$.

8-4 PROBLEMS

Use integration by parts to compute the integrals in Problems 1–25.

1 $\int x e^{2x} \, dx.$

2 $\int x^2 e^{2x} \, dx.$

3 $\int t \sin t \, dt.$

4 $\int t^2 \sin t \, dt.$

5 $\int x \cos 3x \, dx.$

6 $\int x \ln x \, dx.$

7 $\int x^3 \ln x \, dx.$

8 $\int e^{3z} \cos 3z \, dz.$

9 $\int \tan^{-1} x \, dx.$

10 $\int x \tan^{-1} x \, dx.$

11 $\int y^{1/2} \ln y \, dy.$

12 $\int x \sec^2 x \, dx.$

13 $\int (\ln t)^2 \, dt.$

14 $\int t(\ln t)^2 \, dt.$

15 $\int x \sqrt{x + 3} \, dx.$

16 $\int x^3 \sqrt{1 - x^2} \, dx.$

17 $\int x^5 \sqrt{x^3 + 1} \, dx.$

18 $\int \sin^2 \theta \, d\theta.$

19 $\int \csc^3 \theta \, d\theta.$

20 $\int \sin(\ln t) \, dt.$

21 $\int x^2 \arctan x \, dx.$

22 $\int \ln(1 + x^2) \, dx.$

23 $\int \sec^{-1} \sqrt{x} \, dx.$

24 $\int x \tan^{-1} \sqrt{x} \, dx.$

25 $\int \tan^{-1} \sqrt{x} \, dx.$

26 Use the method of cylindrical shells to find the volume generated by revolving the area under $y = \cos x$, where $0 \leqq x \leqq \pi/2$, around the y-axis.

27 Find the volume of the solid obtained by revolving the area under $y = \ln x$ from $x = 1$ to $x = e$ around the x-axis.

28 Find the centroid of the area mentioned in Problem 27.

29 Find the centroid of the solid of Problem 27.

30 Find the centroid of the region under $y = e^{\sqrt{x}}$ from $x = 0$ to $x = 1$. (*Suggestion:* Substitute $u = \sqrt{x}$.)

Derive the reduction formulas given in Problems 31–36.

31 $\int x^n e^x \, dx = x^n e^x - n \int x^{n-1} e^x \, dx.$

32 $\int x^n e^{-x^2} \, dx = -\frac{1}{2} x^{n-1} e^{-x^2} + \frac{n-1}{2} \int x^{n-2} e^{-x^2} \, dx.$

33 $\int (\ln x)^n \, dx = x(\ln x)^n - n \int (\ln x)^{n-1} \, dx.$

34 $\int x^n \cos x \, dx = x^n \sin x - n \int x^{n-1} \sin x \, dx.$

35 $\int \sin^n x \, dx = -\frac{\sin^{n-1} x \cos x}{n} + \frac{n-1}{n} \int \sin^{n-2} x \, dx.$

36 $\int \cos^n x \, dx = \frac{\cos^{n-1} x \sin x}{n} + \frac{n-1}{n} \int \cos^{n-2} x \, dx.$

Use the reduction formulas in Problems 31–33 to evaluate the integrals in Problems 37–39.

37 $\int_0^1 x^3 e^x \, dx.$

38 $\int_0^1 x^5 e^{-x^2} \, dx.$

39 $\int_1^e (\ln x)^3 \, dx.$

40 Apply the reduction formula of Problem 35 to show that, for each positive integer n,

$$\int_0^{\pi/2} \sin^{2n} x \, dx = \frac{\pi}{2} \frac{1}{2} \frac{3}{4} \frac{5}{6} \cdots \frac{2n - 1}{2n}$$

and

$$\int_0^{\pi/2} \sin^{2n+1} x \, dx = \frac{2}{3} \frac{4}{5} \frac{6}{7} \cdots \frac{2n}{2n + 1}.$$

Trigonometric Substitution

The method of *trigonometric substitution* is often effective in dealing with integrals when the integrands involve expressions such as $(a^2 - u^2)^{1/2}$, $(u^2 - a^2)^{3/2}$, and $1/(a^2 + u^2)^2$. There are three basic trigonometric substitutions:

If the integral involves	then substitute	and use the identity
$a^2 - u^2$	$u = a \sin \theta$	$1 - \sin^2\theta = \cos^2\theta$
$a^2 + u^2$	$u = a \tan \theta$	$1 + \tan^2\theta = \sec^2\theta$
$u^2 - a^2$	$u = a \sec \theta$	$\sec^2\theta - 1 = \tan^2\theta$

By the substitution $u = a \sin \theta$ we mean, more precisely, the substitution

$$\theta = \sin^{-1}\frac{u}{a}, \qquad -\frac{\pi}{2} \le \theta \le \frac{\pi}{2},$$

where $|u| \le a$. Suppose, for example, that an integral involves the expression $(a^2 - u^2)^{1/2}$. Then this substitution yields

$$(a^2 - u^2)^{1/2} = \sqrt{a^2 - a^2 \sin^2\theta}$$
$$= \sqrt{a^2 \cos^2\theta} = a \cos \theta.$$

We take the positive square root because $\cos \theta \ge 0$ for $-\pi/2 \le \theta \le \pi/2$. Thus the troublesome factor $(a^2 - u^2)^{1/2}$ becomes $a \cos \theta$, and meanwhile, $du = a \cos \theta \, d\theta$. If the trigonometric integral that results from this substitution can be evaluated using the methods of Section 8-3, the result will normally involve $\theta = \sin^{-1}(u/a)$ and trigonometric functions of θ. The final step will be to express the answer in terms of the original variable. For this purpose the values of the various trigonometric functions can be read from the right triangle of Figure 8.2, which contains an angle θ such that $\sin \theta = u/a$ (if u is negative then θ is negative).

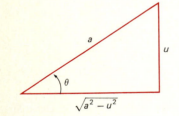

8.2 Reference triangle for the substitution $u = a \sin \theta$.

EXAMPLE 1 Evaluate $\displaystyle\int \frac{x^3 \, dx}{\sqrt{1 - x^2}}$ where $|x| < 1$.

Solution Here $a = 1$ and $u = x$, so we substitute $x = \sin \theta$, $dx = \cos \theta \, d\theta$. This gives

$$\int \frac{x^3 \, dx}{\sqrt{1 - x^2}} = \int \frac{\sin^3\theta \cos \theta \, d\theta}{\sqrt{1 - \sin^2\theta}}$$

$$= \int \sin^3\theta \, d\theta = \int (\sin \theta)(1 - \cos^2\theta) \, d\theta$$

$$= \tfrac{1}{3} \cos^3\theta - \cos \theta + C.$$

Since $\cos \theta = \sqrt{1 - \sin^2\theta} = \sqrt{1 - x^2}$, our final answer is

$$\int \frac{x^3 \, dx}{\sqrt{1 - x^2}} = \tfrac{1}{3}(1 - x^2)^{3/2} - \sqrt{1 - x^2} + C.$$

EXAMPLE 2   Verify by using trigonometric substitution that the area $A = \int_0^a \sqrt{a^2 - x^2} \, dx$ of the quarter-circle lying under $y = \sqrt{a^2 - x^2}$, $0 \leq x \leq a$, is $\tfrac{1}{4}\pi a^2$.

Solution   The substitution $x = a \sin \theta$, $dx = a \cos \theta \, d\theta$ gives

$$\int \sqrt{a^2 - x^2} \, dx = \int \sqrt{a^2 - a^2 \sin^2\theta} \, (a \cos \theta) \, d\theta$$

$$= \int a^2 \cos^2\theta \, d\theta = \frac{a^2}{2} \int (1 + \cos 2\theta) \, d\theta$$

$$= \frac{a^2}{2} \left(\theta + \frac{1}{2} \sin 2\theta \right) + C$$

$$= \frac{a^2}{2} (\theta + \sin \theta \cos \theta) + C.$$

(We used the identity $\sin 2\theta = 2 \sin \theta \cos \theta$ in the last step.) Now $\sin \theta = x/a$ and (from Figure 8.2, with $u = x$) $\cos \theta = (a^2 - x^2)^{1/2}/a$. Hence

$$A = \int_0^a \sqrt{a^2 - x^2} \, dx$$

$$= \frac{a^2}{2} \left[\sin^{-1} \frac{x}{a} + \frac{x(a^2 - x^2)^{1/2}}{a^2} \right]_0^a$$

$$= \frac{a^2}{2} \frac{\pi}{2} = \frac{1}{4} \pi a^2.$$

By the substitution $u = a \tan \theta$ in an integral involving $a^2 + u^2$ is meant the substitution

$$\theta = \tan^{-1} \frac{u}{a}, \qquad -\frac{\pi}{2} < \theta < \frac{\pi}{2}.$$

Note that in this case

$$\sqrt{a^2 + u^2} = \sqrt{a^2 + a^2 \tan^2\theta}$$

$$= \sqrt{a^2 \sec^2\theta} = a \sec \theta$$

under the assumption that $a > 0$. We take the positive square root here because $\sec \theta > 0$ for $-\pi/2 < \theta < \pi/2$. The values of the various trigonometric functions of θ under this substitution can be read from the right triangle of Figure 8.3, which shows a (positive or negative) acute angle θ such that $\tan \theta = u/a$.

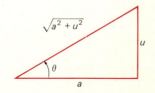

8.3   The substitution $u = a \tan \theta$.

EXAMPLE 3   Find $\displaystyle\int \frac{dx}{(4x^2 + 9)^2}$.

Solution The factor $4x^2 + 9$ corresponds to $u^2 + a^2$ with $u = 2x$ and $a = 3$. Hence the substitution $u = a \tan \theta$ amounts to

$$2x = 3 \tan \theta, \qquad x = \tfrac{3}{2} \tan \theta, \quad \text{and} \quad dx = \tfrac{3}{2} \sec^2\theta \, d\theta.$$

This gives

$$\int \frac{dx}{(4x^2 + 9)^2} = \int \frac{\tfrac{3}{2} \sec^2\theta \, d\theta}{(9 \tan^2\theta + 9)^2}$$

$$= \frac{3}{2} \int \frac{\sec^2\theta \, d\theta}{(9 \sec^2\theta)^2} = \frac{1}{54} \int \frac{d\theta}{\sec^2\theta}$$

$$= \frac{1}{54} \int \cos^2\theta \, d\theta = \frac{1}{108} (\theta + \sin \theta \cos \theta) + C.$$

The actual integration in the last step is the same as in Example 2. Now $\theta = \tan^{-1}(2x/3)$, and the triangle of Figure 8.4 gives

$$\sin \theta = \frac{2x}{\sqrt{4x^2 + 9}}, \qquad \cos \theta = \frac{3}{\sqrt{4x^2 + 9}}.$$

Hence

$$\int \frac{dx}{(4x^2 + 9)^2} = \frac{1}{108} \left[\tan^{-1}\left(\frac{2x}{3}\right) + \frac{2x}{\sqrt{4x^2 + 9}} \cdot \frac{3}{\sqrt{4x^2 + 9}} \right] + C$$

$$= \frac{1}{108} \tan^{-1}\left(\frac{2x}{3}\right) + \frac{x}{18(4x^2 + 9)} + C.$$

By the substitution $u = a \sec \theta$ in an integral involving $u^2 - a^2$ is meant the substitution

$$\theta = \sec^{-1}\frac{u}{a}, \qquad 0 \leq \theta \leq \pi,$$

where $|u| \geq a$ (because of the domain and range of the inverse secant function). Then

$$\sqrt{u^2 - a^2} = \sqrt{a^2 \sec^2\theta - a^2}$$

$$= \sqrt{a^2 \tan^2\theta} = \pm a \tan \theta.$$

Here we must take the plus sign if $u > a$, so that $0 < \theta < \pi/2$ and $\tan \theta > 0$. If $u < -a$, so that $\pi/2 < \theta < \pi$ and $\tan \theta < 0$, we take the minus sign above. In either case the values of the various trigonometric functions of θ can be read from the right triangle of Figure 8.5.

EXAMPLE 4   Find $\displaystyle\int \frac{\sqrt{x^2 - 25}}{x} \, dx, \quad x > 5.$

Solution   We substitute $x = 5 \sec \theta$, $dx = 5 \sec \theta \tan \theta \, d\theta$. Then

$$\sqrt{x^2 - 25} = \sqrt{25(\sec^2\theta - 1)} = +5 \tan \theta,$$

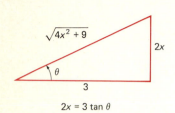

8.4 Reference triangle for Example 3.

$\sqrt{4x^2 + 9}$; $2x$; 3 ; θ

$2x = 3 \tan \theta$

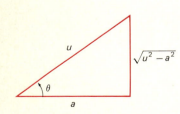

8.5 Reference triangle for the substitution $u = a \sec \theta$.

u ; $\sqrt{u^2 - a^2}$; a ; θ

because $x > 5$ implies $0 < \theta < \pi/2$, so that $\tan \theta > 0$. Hence this substitution gives

$$\int \frac{\sqrt{x^2 - 25}}{x}\, dx = \int \frac{5 \tan \theta}{5 \sec \theta}\, (5 \sec \theta \tan \theta\, d\theta)$$

$$= 5 \int \tan^2\theta\, d\theta = 5 \int (\sec^2\theta - 1)\, d\theta$$

$$= 5 \tan \theta - 5\theta + C$$

$$= \sqrt{x^2 - 25} - 5 \sec^{-1}\left(\frac{x}{5}\right) + C.$$

Hyperbolic substitutions may be used in a similar way, and to the same effect, as trigonometric substitutions. The three basic hyperbolic substitutions—which are not ordinarily memorized—are listed here for reference.

If the integral involves	use the substitution	and use the identity
$a^2 - u^2$	$u = a \tanh \theta$	$1 - \tanh^2\theta = \operatorname{sech}^2\theta$
$a^2 + u^2$	$u = a \sinh \theta$	$1 + \sinh^2\theta = \cosh^2\theta$
$u^2 - a^2$	$u = a \cosh \theta$	$\cosh^2\theta - 1 = \sinh^2\theta$

EXAMPLE 5   Find $\displaystyle\int \frac{dx}{\sqrt{x^2 - 1}}$, $x > 1$.

Solution   For purpose of comparison, we evaluate this integral both by trigonometric substitution and by hyperbolic substitution. The trigonometric substitution

$$x = \sec \theta, \qquad dx = \sec \theta \tan \theta\, d\theta, \qquad \tan \theta = \sqrt{x^2 - 1}$$

gives

$$\int \frac{dx}{\sqrt{x^2 - 1}} = \int \frac{\sec \theta \tan \theta\, d\theta}{\tan \theta}$$

$$= \int \sec \theta\, d\theta$$

$$= \ln|\sec \theta + \tan \theta| + C \qquad \text{(Equation (15), Section 7-2)}$$

$$= \ln|x + \sqrt{x^2 - 1}| + C.$$

With the hyperbolic substitution $x = \cosh \theta$, $dx = \sinh \theta\, d\theta$, we have

$$\sqrt{x^2 - 1} = \sqrt{\cosh^2\theta - 1} = \sinh \theta.$$

We take the positive square root here because $x > 1$ implies that $\theta = \cosh^{-1}x > 0$ and thus that $\sinh \theta > 0$. Hence

$$\int \frac{dx}{\sqrt{x^2 - 1}} = \int \frac{\sinh \theta}{\sinh \theta}\, d\theta = \int d\theta$$

$$= \theta + C = \cosh^{-1}x + C.$$

Our two results appear to differ, but Equation (14) in Section 7-6 shows that they are equivalent.

Use trigonometric substitutions to evaluate the integrals in Problems 1–20.

1 $\int \dfrac{\sqrt{1-x^2}}{x^2}\,dx.$

2 $\int \dfrac{\sqrt{1+x^2}}{x^2}\,dx.$

3 $\int \dfrac{\sqrt{x^2-1}}{x^2}\,dx.$

4 $\int x^3\sqrt{4-x^2}\,dx.$

5 $\int x^3\sqrt{9+4x^2}\,dx.$

6 $\int \dfrac{x^3\,dx}{\sqrt{x^2+25}}.$

7 $\int \dfrac{(1-4x^2)^{1/2}}{x}\,dx.$

8 $\int \dfrac{dx}{\sqrt{1+x^2}}.$

9 $\int \dfrac{dx}{\sqrt{9+4x^2}}.$

10 $\int \sqrt{1+4x^2}\,dx.$

11 $\int \dfrac{x^2\,dx}{\sqrt{25-x^2}}.$

12 $\int \dfrac{x^3\,dx}{\sqrt{25-x^2}}.$

13 $\int \dfrac{x^2\,dx}{\sqrt{1+x^2}}.$

14 $\int \dfrac{x^3\,dx}{\sqrt{1+x^2}}.$

15 $\int \dfrac{x^2\,dx}{\sqrt{4+9x^2}}.$

16 $\int (1-x^2)^{3/2}\,dx.$

17 $\int \dfrac{dx}{(1+x^2)^{3/2}}.$

18 $\int \dfrac{dx}{(4-x^2)^2}.$

19 $\int \dfrac{dx}{(4-x^2)^3}.$

20 $\int \dfrac{dx}{(4x^2+9)^3}.$

Use hyperbolic substitutions to evaluate the following integrals.

21 $\int \dfrac{dx}{\sqrt{25+x^2}}.$

22 $\int \sqrt{1+x^2}\,dx.$

23 $\int \dfrac{\sqrt{x^2-4}}{x^2}\,dx.$

24 $\int \dfrac{dx}{\sqrt{1+9x^2}}.$

25 $\int x^2\sqrt{1+x^2}\,dx.$

26 Compute the arc length of the parabola $y = x^2$ over the interval $0 \le x \le 1$.

27 Compute the area of the surface obtained by revolving the parabolic arc of Problem 26 around the x-axis.

28 Show that the length of one arch of the sine curve $y = \sin x$ is equal to half the circumference of the ellipse $x^2 + \frac{1}{2}y^2 = 1$. (*Suggestion:* Substitute $x = \cos\theta$ in the arc length integral for the ellipse.)

29 Compute the arc length of the curve $y = \ln x$ for $1 \le x \le 2$.

30 Compute the area of the surface obtained by revolving the curve of Problem 29 around the y-axis.

31 A torus is obtained by revolving the circle

$$(x-a)^2 + y^2 = b^2,$$

where $0 < a \le b$, around the y-axis. Show that its surface area is $4\pi^2 ab$.

32 Find the area under the curve $y = \sqrt{9+x^2}$ from $x = 0$ to $x = 4$.

33 Find the area of the surface obtained by revolving the curve $y = \sin x$, $0 \le x \le \pi$, around the x-axis.

34 An ellipsoid of revolution is obtained by revolving the ellipse $x^2/a^2 + y^2/b^2 = 1$ around the x-axis. Suppose that $a > b$, and show then that the ellipsoid has surface area

$$A = 2\pi ab\left[\frac{b}{a} + \frac{a}{c}\sin^{-1}\left(\frac{c}{a}\right)\right]$$

where $c^2 = a^2 - b^2$. Assume that $a \approx b$, so that $c \approx 0$ and $\sin^{-1}(c/a) \approx c/a$. Conclude that $A \approx 4\pi a^2$.

35 Suppose that $b > a$ for the ellipsoid of revolution of Problem 34. Show that its surface area is then

$$A = 2\pi ab\left[\frac{b}{a} + \frac{a}{c}\ln\left(\frac{b+c}{a}\right)\right]$$

where $c^2 = b^2 - a^2$. Assume the fact that $\ln(1+x) \approx x$ if $x \approx 0$, and thereby conclude that $A \approx 4\pi a^2$ if $a \approx b$.

8-6

Integrals Involving Quadratic Polynomials

An integral involving a square root or negative power of a quadratic expression $ax^2 + bx + c$ can often be simplified by the process of *completing the square*. The object is to convert $ax^2 + bx + c$ into a sum or difference of squares, either $u^2 \pm a^2$ or $a^2 - u^2$, so that the method of trigonometric substitution may be used. To see how this is done, suppose first that $a = 1$, so that the quadratic in question is of the form $x^2 + bx + c$. The sum $x^2 + bx$ of the first two terms can be "completed" to a perfect square by adding $b^2/4$, the square of half the coefficient of x, and in turn subtracting $b^2/4$ from the constant term c. This gives

$$x^2 + bx + c = \left(x^2 + bx + \frac{b^2}{4}\right) + \left(c - \frac{b^2}{4}\right)$$

$$= \left(x + \frac{1}{2}b\right)^2 + \left(c - \frac{b^2}{4}\right).$$

With $u = x + \frac{1}{2}b$, our result above is of the form $u^2 + A^2$ or $u^2 - A^2$ (depending on the sign of $c - b^2/4$). If we factor out the coefficient a to begin with, the general case is handled similarly:

$$ax^2 + bx + c = a\left(x^2 + \frac{b}{a}x + \frac{c}{a}\right).$$

EXAMPLE 1   Find $\displaystyle\int \frac{dx}{9x^2 + 6x + 5}$.

Solution   Our first step is completion of the square.

$$9x^2 + 6x + 5 = 9(x^2 + \tfrac{2}{3}x) + 5$$
$$= 9(x^2 + \tfrac{2}{3}x + \tfrac{1}{9}) - 1 + 5$$
$$= 9(x + \tfrac{1}{3})^2 + 4$$
$$= (3x + 1)^2 + 2^2.$$

Hence

$$\int \frac{dx}{9x^2 + 6x + 5} = \int \frac{dx}{(3x + 1)^2 + 4}$$

$$= \frac{1}{3} \int \frac{du}{u^2 + 4} \qquad (u = 3x + 1)$$

$$= \frac{1}{6} \int \frac{\frac{1}{2}\,du}{(u/2)^2 + 1}$$

$$= \frac{1}{6} \tan^{-1}\left(\frac{u}{2}\right) + C$$

$$= \frac{1}{6} \tan^{-1}\left(\frac{3x + 1}{2}\right) + C.$$

EXAMPLE 2   Find $\displaystyle\int \frac{dx}{\sqrt{9 + 16x - 4x^2}}$.

Solution   First we complete the square:

$$9 + 16x - 4x^2 = 9 - 4(x^2 - 4x)$$
$$= 9 - 4(x^2 - 4x + 4) + 16$$
$$= 25 - 4(x - 2)^2.$$

Hence

$$\int \frac{dx}{\sqrt{9 + 16x - 4x^2}} = \int \frac{dx}{\sqrt{25 - 4(x-2)^2}}$$

$$= \frac{1}{5} \int \frac{dx}{\sqrt{1 - \frac{4}{25}(x-2)^2}}$$

$$= \frac{1}{2} \int \frac{du}{\sqrt{1 - u^2}} \qquad \left(u = \frac{2(x-2)}{5} \right)$$

$$= \frac{1}{2} \sin^{-1} u + C$$

$$= \frac{1}{2} \sin^{-1} \left(\frac{2(x-2)}{5} \right) + C.$$

An integral involving a quadratic expression can sometimes be split into two simpler integrals. The next two examples illustrate this technique.

EXAMPLE 3 Find $\int \frac{(2x + 3)\,dx}{9x^2 + 6x + 5}$.

Solution Since $D(9x^2 + 6x + 5) = 18x + 6$, we write the above integral as the sum of two integrals, one of which has numerator $18x + 6$ in its integrand.

$$\int \frac{(2x + 3)\,dx}{9x^2 + 6x + 5} = \frac{1}{9} \int \frac{(18x + 6)\,dx}{9x^2 + 6x + 5} + \frac{7}{3} \int \frac{dx}{9x^2 + 6x + 5}.$$

The first integral on the right is a logarithm, and the second is given by Example 1. Thus

$$\int \frac{(2x + 3)\,dx}{9x^2 + 6x + 5} = \frac{1}{9} \ln(9x^2 + 6x + 5) + \frac{7}{18} \tan^{-1} \left(\frac{3x + 1}{2} \right) + C.$$

Alternatively, we first could complete the square in the denominator. The substitution $u = 3x + 1$, $x = \frac{1}{3}(u - 1)$, $dx = \frac{1}{3}\,du$ then gives

$$\int \frac{(2x + 3)\,dx}{(3x + 1)^2 + 4} = \int \frac{[\frac{2}{3}(u - 1) + 3]\frac{1}{3}\,du}{u^2 + 4}$$

$$= \frac{1}{9} \int \frac{2u\,du}{u^2 + 4} + \frac{7}{9} \int \frac{du}{u^2 + 4}$$

$$= \frac{1}{9} \ln(u^2 + 4) + \frac{7}{18} \tan^{-1} \left(\frac{u}{2} \right) + C$$

$$= \frac{1}{9} \ln(9x^2 + 6x + 5) + \frac{7}{18} \tan^{-1} \left(\frac{3x + 1}{2} \right) + C.$$

EXAMPLE 4 Find $\displaystyle\int \frac{2-x}{(2x-x^2)^2}\,dx.$

Solution Since $D_x(2x-x^2) = 2 - 2x$, we first write

$$\int \frac{(2-x)\,dx}{(2x-x^2)^2} = \frac{1}{2}\int \frac{(2-2x)\,dx}{(2x-x^2)^2} + \int \frac{dx}{(2x-x^2)^2}$$

$$= -\frac{1}{2(2x-x^2)} + \int \frac{dx}{(2x-x^2)^2}.$$

To find the remaining integral, we complete the square:

$$2x - x^2 = 1 - (x-1)^2.$$

This suggests the substitution $x - 1 = \sin\theta$, so that

$$dx = \cos\theta\,d\theta \quad\text{and}\quad 1 - (x-1)^2 = \cos^2\theta.$$

This substitution gives

$$\int \frac{dx}{(2x-x^2)^2} = \int \frac{dx}{(1-(x-1)^2)^2}$$

$$= \int \frac{\cos\theta\,d\theta}{(\cos^2\theta)^2} = \int \sec^3\theta\,d\theta$$

$$= \frac{1}{2}\sec\theta\tan\theta + \frac{1}{2}\int \sec\theta\,d\theta$$

(by Formula (6) in Section 8-4)

$$= \frac{1}{2}\sec\theta\tan\theta + \frac{1}{2}\ln|\sec\theta + \tan\theta| + C$$

$$= \frac{1}{2}\left(\frac{x-1}{2x-x^2}\right) + \frac{1}{2}\ln\left|\frac{x}{\sqrt{2x-x^2}}\right| + C.$$

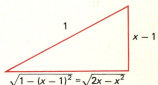

Here we have read the values of $\sec\theta$ and $\tan\theta$ in terms of x from the right triangle in Figure 8.6.

When we combine all our results above, we finally obtain the answer:

$$\int \frac{(2-x)\,dx}{(2x-x^2)^2} = -\frac{1}{2(2x-x^2)} + \frac{x-1}{2(2x-x^2)} + \frac{1}{2}\ln\left|\frac{x}{\sqrt{2x-x^2}}\right| + C.$$

8.6 The reference triangle for Example 4.

The method of Example 4 can be used to evaluate a general integral having the form

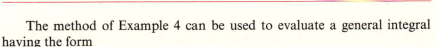

$$\int \frac{Ax+B}{(ax^2+bx+c)^n}\,dx \qquad\qquad (1)$$

where n is a positive integer. By splitting such an integral into two simpler ones and by completing the square in the quadratic expression in the denominator, the problem of evaluating the integral in (1) can be reduced to that of computing

$$\int \frac{du}{(a^2 \pm u^2)^n}. \tag{2}$$

If the sign in (2) is the plus sign, then the substitution $u = a \tan \theta$ transforms the integral into the form

$$\int \cos^m \theta \, d\theta$$

(see Problem 19). This integral can be handled by the methods of Section 8-3 or by use of the reduction formula

$$\int \cos^k \theta \, d\theta = \frac{\cos^{k-1}\theta \sin \theta}{k} + \frac{k-1}{k} \int \cos^{k-2}\theta \, d\theta$$

of Problem 36 in Section 8-4.

If the sign in (2) is the minus sign, then the substitution $u = a \sin \theta$ transforms the integral into the form

$$\int \sec^m \theta \, d\theta$$

(see Problem 20). This integral may be evaluated with the aid of the reduction formula

$$\int \sec^k \theta \, d\theta = \frac{\sec^{k-2}\theta \tan \theta}{k-1} + \frac{k-2}{k-1} \int \sec^{k-2}\theta \, d\theta$$

(Equation (5) in Section 8-4).

8–6 PROBLEMS

Find the integrals in Problems 1–18.

1 $\int \dfrac{dx}{x^2 + 4x + 13}.$

2 $\int \dfrac{dx}{\sqrt{2x - x^2}}.$

3 $\int \dfrac{dx}{3 + 2x - x^2}.$

4 $\int x\sqrt{8 + 2x - x^2}\, dx.$

5 $\int \dfrac{2x - 5}{x^2 + 2x + 2}\, dx.$

6 $\int \dfrac{2x - 1}{4x^2 + 4x - 15}\, dx.$

7 $\int \dfrac{x \, dx}{\sqrt{5 + 12x - 9x^2}}.$

8 $\int (3x - 2)\sqrt{9x^2 + 12x + 8}\, dx.$

9 $\int (7 - 2x)\sqrt{9 + 16x - 4x^2}\, dx.$

10 $\int \dfrac{2x + 3}{\sqrt{x^2 + 2x + 5}}\, dx.$

11 $\int \dfrac{x + 4}{(6x - x^2)^{3/2}}\, dx.$

12 $\int \dfrac{x - 1}{(x^2 + 1)^2}\, dx.$

13 $\int \dfrac{2x + 3}{(4x^2 + 12x + 13)^2}\, dx.$

14 $\int \dfrac{x^3}{(1 - x^2)^4}\, dx.$

15 $\int \dfrac{3x - 1}{x^2 + x + 1}\, dx.$

16 $\int \dfrac{3x - 1}{(x^2 + x + 1)^2}\, dx.$

17 $\int \dfrac{dx}{(x^2 - 4)^2}.$

18 $\int (x - x^2)^{3/2}\, dx.$

19 Show that the substitution $u = a \tan \theta$ gives

$$\int \frac{du}{(a^2 + u^2)^n} = \frac{1}{a^{2n-1}} \int \cos^{2n-2}\theta \, d\theta.$$

20 Show that the substitution $u = a \sin \theta$ gives

$$\int \frac{du}{(a^2 - u^2)^n} = \frac{1}{a^{2n-1}} \int \sec^{2n-1}\theta \, d\theta.$$

CHAP. 8: Techniques of Integration

In this section we show that every rational function can be integrated in terms of elementary functions. Recall that a rational function $R(x)$ is one expressible as a quotient of two polynomials. That is,

$$R(x) = \frac{P(x)}{Q(x)} \tag{1}$$

where $P(x)$ and $Q(x)$ are polynomials. The **method of partial fractions** involves decomposing $R(x)$ into a sum of terms:

$$R(x) = \frac{P(x)}{Q(x)} = p(x) + F_1(x) + F_2(x) + \cdots + F_k(x), \tag{2}$$

where $p(x)$ is a polynomial and each expression $F_i(x)$ is a fraction that can be integrated by the methods of earlier sections.

For example, one can verify (by finding a common denominator on the right) that

$$\frac{x^3 - 1}{x^3 + x} = 1 - \frac{1}{x} + \frac{x - 1}{x^2 + 1}. \tag{3}$$

It follows that

$$\int \frac{x^3 - 1}{x^3 + x}\, dx = \int \left(1 - \frac{1}{x} + \frac{x}{x^2 + 1} - \frac{1}{x^2 + 1} \right) dx$$

$$= x - \ln|x| + \frac{1}{2}\ln|x^2 + 1| - \tan^{-1}x + C.$$

Of course, the key to this simple integration lies in finding the decomposition given in Equation (3). That such a decomposition exists and the technique of finding it is what the method of partial fractions is about.

According to a theorem that is proved in advanced algebra, every rational function can be written in the form of (2) with each of the $F_i(x)$ being either a fraction of the form

$$\frac{A}{(ax + b)^n} \tag{4}$$

or one of the form

$$\frac{Bx + C}{(ax^2 + bx + c)^n}, \tag{5}$$

where the quadratic polynomial $ax^2 + bx + c$ is **irreducible,** meaning that it is not a product of linear factors with real coefficients. This is the same as saying that the equation $ax^2 + bx + c = 0$ has no real roots, and the quadratic formula tells us that this is the case exactly when $b^2 - 4ac < 0$.

Fractions of the forms in (4) and (5) are called **partial fractions,** and the sum in (2) is called the **partial-fraction decomposition** of $R(x)$. Thus (3) gives the partial-fraction decomposition of $(x^3 - 1)/(x^3 + x)$. A partial fraction of the form in (4) may be integrated immediately, and we saw in Section 8-6 how to integrate one of the form in (5).

The first step in finding the partial-fraction decomposition of $R(x)$ is finding the polynomial $p(x)$ in (2). It turns out that $p(x) \equiv 0$ provided that the degree of the numerator $P(x)$ is *less than* that of the denominator $Q(x)$; in this case the rational fraction $R(x) = P(x)/(Q(x)$ is called **proper.** If $R(x)$ is not proper, then $p(x)$ may be found by "long division" of $Q(x)$ into $P(x)$, as in the following example.

EXAMPLE 1   Find $\displaystyle\int \frac{x^3 + x^2 + x - 1}{x^2 + 2x + 2}\,dx.$

Solution   Long division of denominator into numerator is carried out as follows.

$$
\begin{array}{r}
x - 1 \qquad\qquad\; p(x) \quad\text{(quotient)} \\
x^2 + 2x + 2 \,\overline{)\, x^3 + x^2 + x - 1} \\
\underline{x^3 + 2x^2 + 2x} \\
-x^2 - x - 1 \\
\underline{-x^2 - 2x - 2} \\
x + 1 \quad r(x) \quad\text{(remainder)}
\end{array}
$$

As in arithmetic,

$$\text{fraction} = \text{quotient} + \frac{\text{remainder}}{\text{divisor}}.$$

Thus

$$\frac{x^3 + x^2 + x - 1}{x^2 + 2x + 2} = (x - 1) + \frac{x + 1}{x^2 + 2x + 2}.$$

And hence

$$\int \frac{x^3 + x^2 + x - 1}{x^2 + 2x + 2}\,dx = \int \left(x - 1 + \frac{x + 1}{x^2 + 2x + 2} \right) dx$$

$$= \tfrac{1}{2} x^2 - x + \tfrac{1}{2} \ln(x^2 + 2x + 2) + C.$$

By using long division as in Example 1, any rational function $R(x)$ can be written as a sum of a polynomial $p(x)$ and a *proper* rational function,

$$R(x) = p(x) + \frac{r(x)}{Q(x)}.$$

To see how to integrate an arbitrary rational function, we therefore need only see how to find the partial-fraction decomposition of a proper rational function.

To obtain such a decomposition, we first must factor the denominator $Q(x)$ into a product of linear factors (those of the form $ax + b$) and irreducible quadratic factors (those of the form $ax^2 + bx + c$ with $b^2 - 4ac < 0$). This is always possible in principle but may be quite difficult in practice. Once this factorization of $Q(x)$ has been found, the partial-fraction decomposition may be obtained by routine algebra (described below). Each linear or irreducible quadratic factor of $Q(x)$ leads to one or more partial fractions of the forms in (4) and (5).

LINEAR FACTORS

Let $R(x) = P(x)/Q(x)$ be a *proper* rational fraction, and suppose that the linear factor $ax + b$ occurs n times in the factorization of $Q(x)$. That is, $(ax + b)^n$ is the highest power of $ax + b$ that divides "evenly" into $Q(x)$. In this case we call n the **multiplicity** of the factor $ax + b$.

Rule 1 *Linear Factor Partial Fractions*

The part of the partial-fraction decomposition of $R(x)$ corresponding to the linear factor $ax + b$ of multiplicity n is a sum of n partial fractions, having the form

$$\frac{A_1}{ax + b} + \frac{A_2}{(ax + b)^2} + \cdots + \frac{A_n}{(ax + b)^n}, \tag{6}$$

where $A_1, A_2, \ldots, A_n$ are constants.

If *all* the factors of $Q(x)$ are linear, then the partial-fraction decomposition of $R(x)$ is a sum of expressions like (6). The situation is especially simple if each of these linear factors is *nonrepeated*—that is, if each has multiplicity $n = 1$. In this case Expression (6) reduces to its first term, and the partial-fraction decomposition of $R(x)$ is a sum of such terms. The solution to the following example illustrates how the constant numerators can be determined.

EXAMPLE 2   Find $\displaystyle\int \frac{4x^2 - 3x - 4}{x^3 + x^2 - 2x}\, dx.$

Solution   The rational function to be integrated is proper, so we immediately proceed to factor its denominator.

$$x^3 + x^2 - 2x = x(x^2 + x - 2)$$
$$= x(x - 1)(x + 2).$$

We are dealing with three nonrepeated linear factors, so the partial-fraction decomposition is of the form

$$\frac{4x^2 - 3x - 4}{x^3 + x^2 - 2x} = \frac{A}{x} + \frac{B}{x - 1} + \frac{C}{x + 2}.$$

To find the constants A, B, and C, we multiply both sides of this equation by the left-hand denominator $x(x - 1)(x + 2)$ and find thereby that

$$4x^2 - 3x - 4 = A(x - 1)(x + 2) + Bx(x + 2) + Cx(x - 1). \quad (7)$$

Then we collect coefficients of like powers on the right:

$$4x^2 - 3x - 4 = (A + B + C)x^2 + (A + 2B - C)x + (-2A).$$

Now two polynomials are equal only if the coefficients of corresponding powers of x are the same, and so we may conclude that

$$A + B + C = 4,$$
$$A + 2B - C = -3,$$
$$-2A = -4.$$

We solve these simultaneous equations and thus find that $A = 2$, $B = -1$, and $C = 3$.

There is an alternative way of finding A, B, and C, one that is especially convenient in the case of nonrepeated linear factors. Substitute the values $x = 0$, 1, and -2 (the zeros of the linear factors of the denominator) in turn into Equation (7). Substitution of $x = 0$ into (7) immediately gives $-4 = -2A$, so that $A = 2$. Substitution of $x = 1$ into (7) gives $-3 = 3B$, and so $B = -1$. Substitution of $x = -2$ gives $18 = 6C$, so $C = 3$.

With these values $A = 2$, $B = -1$, and $C = 3$, however obtained, we find that

$$\int \frac{4x^2 - 3x - 4}{x^3 + x^2 - 2x} \, dx = \int \left(\frac{2}{x} - \frac{1}{x - 1} + \frac{3}{x + 2} \right) dx$$

$$= 2 \ln|x| - \ln|x - 1| + 3 \ln|x + 2| + C.$$

Laws of logarithms allow us to write this antiderivative in the more compact form

$$\ln \left| \frac{x^2(x + 2)^3}{x - 1} \right| + C.$$

EXAMPLE 3 Find $\int \dfrac{x^3 - 4x - 1}{x(x - 1)^3} \, dx.$

Solution Here we have a linear factor of multiplicity $n = 3$. According to Rule 1 above, the partial-fraction decomposition of the integrand has the form

$$\frac{x^3 - 4x - 1}{x(x - 1)^3} = \frac{A}{x} + \frac{B}{x - 1} + \frac{C}{(x - 1)^2} + \frac{D}{(x - 1)^3}.$$

To find the constants A, B, C, and D, we multiply both sides of this equation by $x(x - 1)^3$ and obtain

$$x^3 - 4x - 1 = A(x - 1)^3 + Bx(x - 1)^2 + Cx(x - 1) + Dx.$$

We multiply, then collect coefficients on the right-hand side; this yields

$$x^3 - 4x - 1 = (A + B)x^3 + (-3A - 2B + C)x^2$$
$$+ (3A + B - C + D)x - A.$$

Then we equate coefficients of like powers of x on each side of this equation. We get the four equations

$$A + B \qquad\qquad = 1,$$
$$-3A - 2B + C \qquad = 0,$$
$$3A + B - C + D = -4,$$
$$-A \qquad\qquad = -1.$$

The last equation gives $A = 1$; then the first equation gives $B = 0$. Next, the second equation gives $C = 3$. Substitution of $A = 1$, $B = 0$, and $C = 3$ into the third equation finally gives $D = -4$. Hence

$$\int \frac{x^3 - 4x - 1}{x(x - 1)^3}\, dx = \int \left(\frac{1}{x} + \frac{3}{(x - 1)^2} - \frac{4}{(x - 1)^3} \right) dx$$

$$= \ln|x| - \frac{3}{x - 1} + \frac{2}{(x - 1)^2} + C.$$

QUADRATIC FACTORS

Suppose that $R(x) = P(x)/Q(x)$ is a proper rational fraction, and suppose that the irreducible linear factor $ax^2 + bx + c$ occurs n times in the factorization of $Q(x)$. That is, $(ax^2 + bx + c)^n$ is the highest power of $ax^2 + bx + c$ that divides evenly into $Q(x)$.

Rule 2 *Quadratic Factor Partial Fractions*

The part of the partial-fraction decomposition of $R(x)$ corresponding to the irreducible quadratic factor $ax^2 + bx + c$ of multiplicity n is a sum of n partial fractions, having the form

$$\frac{B_1 x + C_1}{ax^2 + bx + c} + \frac{B_2 x + C_2}{(ax^2 + bx + c)^2} + \cdots + \frac{B_n x + C_n}{(ax^2 + bx + c)^n}, \quad (8)$$

where $B_1, B_2, \ldots, B_n, C_1, C_2, \ldots, C_n$ are constants.

If $Q(x)$ has both linear and irreducible quadratic factors, then the partial-fraction decomposition of $R(x)$ is the sum of the expressions of the form in (6) corresponding to the linear factors, plus the sum of the expressions of the form in (8) corresponding to the quadratic factors. In the case of an irreducible quadratic factor of multiplicity $n = 1$, the expression (8) reduces to its first term, just as in the linear case.

EXAMPLE 4 Find $\displaystyle\int \frac{5x^3 - 3x^2 + 2x - 1}{x^4 + x^2}\, dx.$

Solution The denominator $x^4 + x^2 = x^2(x^2 + 1)$ has both a quadratic factor and a repeated linear factor. The partial-fraction decomposition takes the form

$$\frac{5x^3 - 3x^2 + 2x - 1}{x^4 + x^2} = \frac{A}{x} + \frac{B}{x^2} + \frac{Cx + D}{x^2 + 1}.$$

We multiply both sides by $x^4 + x^2$ and obtain

$$5x^3 - 3x^2 + 2x - 1 = Ax(x^2 + 1) + B(x^2 + 1) + (Cx + D)x^2$$
$$= (A + C)x^3 + (B + D)x^2 + Ax + B.$$

As before, we equate coefficients of like powers of x and obtain the four equations

$$
\begin{aligned}
A \quad\quad + C \quad\quad &= \quad 5, \\
B \quad\quad + D &= -3, \\
A \quad\quad\quad\quad\quad &= \quad 2, \\
B \quad\quad\quad\quad\quad &= -1.
\end{aligned}
$$

So $A = 2$, $B = -1$, $C = 3$, and $D = -2$. Thus

$$\int \frac{5x^3 - 3x^2 + 2x - 1}{x^4 + x^2}\,dx = \int \left(\frac{2}{x} - \frac{1}{x^2} + \frac{3x - 2}{x^2 + 1} \right) dx$$

$$= 2\ln|x| + \frac{1}{x} + \frac{3}{2}\int \frac{2x\,dx}{x^2 + 1} - 2\int \frac{dx}{x^2 + 1}$$

$$= 2\ln|x| + \frac{1}{x} + \frac{3}{2}\ln(x^2 + 1) - 2\tan^{-1} x + C.$$

EXAMPLE 5 Find $\displaystyle\int \frac{x^3 - 2x}{(x^2 + 2x + 2)^2}\,dx.$

Solution The repeated quadratic factor $x^2 + 2x + 2$ is irreducible because

$$b^2 - 4ac = (2)^2 - 4(1)(2) = -4 < 0.$$

Hence the partial-fraction decomposition of the integrand, as given by Rule 2, is

$$\frac{x^3 - 2x}{(x^2 + 2x + 2)^2} = \frac{Ax + B}{x^2 + 2x + 2} + \frac{Cx + D}{(x^2 + 2x + 2)^2}.$$

We multiply each side by $(x^2 + 2x + 2)^2$ and find that

$$x^3 - 2x = (Ax + B)(x^2 + 2x + 2) + (Cx + D)$$
$$= Ax^3 + (2A + B)x^2 + (2A + 2B + C)x + (2B + D).$$

And so

$$
\begin{aligned}
A \quad\quad\quad\quad\quad\quad &= \quad 1, \\
2A + B \quad\quad\quad\quad &= \quad 0, \\
2A + 2B + C \quad\quad &= -2, \\
2B \quad\quad + D &= \quad 0.
\end{aligned}
$$

We find that $A = 1$, $B = -2$, $C = 0$, and $D = 4$. So

$$\int \frac{x^3 - 2x}{(x^2 + 2x + 2)^2}\, dx = \int \frac{x - 2}{x^2 + 2x + 2}\, dx + 4 \int \frac{1}{(x^2 + 2x + 2)^2}\, dx$$

$$= \int \frac{x - 2}{(x + 1)^2 + 1}\, dx + 4 \int \frac{1}{[(x + 1)^2 + 1]^2}\, dx$$

$$= \int \frac{u - 3}{u^2 + 1}\, du + 4 \int \frac{1}{(u^2 + 1)^2}\, du \qquad (9)$$

where $u = x + 1$. The first integral here is simple to evaluate:

$$\int \frac{u - 3}{u^2 + 1}\, du = \frac{1}{2} \int \frac{2u}{u^2 + 1}\, du - 3 \int \frac{du}{u^2 + 1}$$

$$= \frac{1}{2} \ln(u^2 + 1) - 3 \tan^{-1} u + C$$

$$= \frac{1}{2} \ln(x^2 + 2x + 2) - 3 \tan^{-1}(x + 1) + C. \qquad (10)$$

For the second integral in (9), we use the substitution $u = \tan\theta$, so that $du = \sec^2\theta\, d\theta$. Then

$$4 \int \frac{du}{(u^2 + 1)^2} = 4 \int \frac{\sec^2\theta\, d\theta}{\sec^4\theta}$$

$$= 4 \int \cos^2\theta\, d\theta = 2 \int (1 + \cos 2\theta)\, d\theta$$

$$= 2(\theta + \sin\theta\cos\theta) + C$$

$$= 2 \tan^{-1} u + \frac{2u}{u^2 + 1} + C$$

$$= 2 \tan^{-1}(x + 1) + \frac{2(x + 1)}{x^2 + 2x + 2} + C. \qquad (11)$$

We have read the values of $\sin\theta$ and $\cos\theta$ in terms of u from the right triangle of Figure 8.7 and then resubstituted $x + 1$ for u. Finally, addition of our results in (10) and (11) gives the answer:

$$\int \frac{x^3 - 2x}{(x^2 + 2x + 2)^2}\, dx = \frac{1}{2} \ln(x^2 + 2x + 2) - \tan^{-1}(x + 1)$$

$$+ \frac{2(x + 1)}{x^2 + 2x + 2} + C.$$

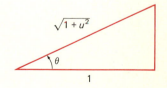

8.7 The reference triangle for Example 5.

Find the integrals in Problems 1–25.

1 $\displaystyle\int \frac{x - 1}{x + 1}\, dx.$

2 $\displaystyle\int \frac{2x^3 - 1}{x^2 + 1}\, dx.$

5 $\displaystyle\int \frac{dx}{x^2 - 4}.$

6 $\displaystyle\int \frac{x^4\, dx}{x^2 + 4x + 4}.$

3 $\displaystyle\int \frac{x^2 + 2x}{(x + 1)^2}\, dx.$

4 $\displaystyle\int \frac{2x - 4}{x^2 - x}\, dx.$

7 $\displaystyle\int \frac{x + 10}{2x^2 + 5x - 3}\, dx.$

8 $\displaystyle\int \frac{x + 1}{x^3 - x^2}\, dx.$

9 $\displaystyle\int \frac{x^2 + 1}{x^3 + 2x^2 + x}\, dx.$

10 $\displaystyle\int \frac{(x^2 + x)\, dx}{x^3 - x^2 - 2x}.$

11 $\displaystyle\int \frac{4x^3 - 7x}{x^4 - 5x^2 + 4}\, dx.$

12 $\displaystyle\int \frac{(2x^2 + 3)\, dx}{x^4 - 2x^2 + 1}.$

13 $\displaystyle\int \frac{x^2\, dx}{(x + 2)^3}.$

14 $\displaystyle\int \frac{(x^2 + x)\, dx}{(x^2 - 4)(x + 4)}.$

15 $\displaystyle\int \frac{dx}{x^3 + x}.$

16 $\displaystyle\int \frac{x^2\, dx}{x^2 + x + 1}.$

17 $\displaystyle\int \frac{x + 4}{x^3 + 4x}\, dx.$

18 $\displaystyle\int \frac{(x^2 + 1)\, dx}{x^3 + x^2 + x}.$

19 $\displaystyle\int \frac{x\, dx}{(x + 1)(x^2 + 1)}.$

20 $\displaystyle\int \frac{x^2 + 2}{(x^2 + 1)^2}\, dx.$

21 $\displaystyle\int \frac{(x^2 - 10)\, dx}{2x^4 + 9x^2 + 4}.$

22 $\displaystyle\int \frac{x^2\, dx}{x^4 - 1}.$

23 $\displaystyle\int \frac{(3x + 1)\, dx}{(x^2 + 2x + 5)^2}.$

24 $\displaystyle\int \frac{(x^2 + 4)\, dx}{(x^2 + 1)^2(x^2 + 2)}.$

25 $\displaystyle\int \frac{x^4 + 3x^2 - 4x + 5}{(x - 1)^2(x^2 + 1)}\, dx.$

In Problems 26–29, make a preliminary substitution before using the method of partial fractions.

26 $\displaystyle\int \frac{\cos\theta\, d\theta}{\sin^2\theta - \sin\theta - 6}.$

27 $\displaystyle\int \frac{e^{4t}\, dt}{(e^{2t} - 1)^3}.$

28 $\displaystyle\int \frac{\sec^2 t\, dt}{\tan^3 t + \tan^2 t}.$

29 $\displaystyle\int \frac{(1 + \ln t)\, dt}{t(3 + 2\ln t)^2}.$

In Problems 30 and 31, factor the denominator by first noting by inspection a root r of the denominator, and then employing long division by $x - r$. Finally, use the method of partial fractions to aid in finding the indicated antiderivative.

30 $\displaystyle\int \frac{dx}{x^3 + 8}.$

31 $\displaystyle\int \frac{x^4 + 2x^2}{x^3 - 1}\, dx.$

32 (a) Find constants a and b such that

$$x^4 + 1 = (x^2 + ax + 1)(x^2 + bx + 1).$$

(b) Show that

$$\int_0^1 \frac{x^2 + 1}{x^4 + 1}\, dx = \frac{\pi}{2\sqrt{2}}.$$

33 Factor $x^4 + x^2 + 1$ as in Problem 32, and then find

$$\int \frac{x^3 + 2x}{x^4 + x^2 + 1}\, dx.$$

34 Find the volume generated by revolving the area bounded by the following curve around the x-axis:

$$y^2 = x^2 \frac{1 - x}{1 + x}, \qquad 0 \leq x \leq 1.$$

---

*8-8

Rationalizing Substitutions

In the following example we make a substitution that eliminates the radical. This is an example of a *rationalizing* substitution.

EXAMPLE 1 Find $\displaystyle\int x^2\sqrt{x + 1}\, dx.$

Solution Let $u = x + 1$. Then $x = u - 1$ and $dx = du$. This substitution yields

$$\int x^2\sqrt{x + 1}\, dx = \int (u - 1)^2\sqrt{u}\, du$$

$$= \int (u^2 - 2u + 1)\sqrt{u}\, du$$

$$= \int (u^{5/2} - 2u^{3/2} + u^{1/2})\, du$$

$$= \tfrac{2}{7}u^{7/2} - \tfrac{4}{5}u^{5/2} + \tfrac{2}{3}u^{3/2} + C$$

$$= \tfrac{2}{7}(x + 1)^{7/2} - \tfrac{4}{5}(x + 1)^{5/2} + \tfrac{2}{3}(x + 1)^{3/2} + C.$$

The method of Example 1 succeeds with any integral of the form

$$\int p(x)\sqrt{ax + b}\, dx \qquad (1)$$

where $p(x)$ is a polynomial. In fact, either the substitution $u = ax + b$ or the substitution $u = (ax + b)^{1/2}$ may be used. More generally, the substitution $u^n = f(x)$ may succeed when the integrand involves $\sqrt[n]{f(x)}$. It *always* succeeds in the case of an integral of the form

$$\int p(x)\ \sqrt[n]{\frac{ax + b}{cx + d}}\, dx, \qquad (2)$$

where $p(x)$ is a polynomial. In particular, the substitution

$$u^n = \frac{ax + b}{cx + d}$$

converts (2) into the integral of a rational function of u (see Problem 21), which can then be integrated by the method of partial fractions.

EXAMPLE 2   Find $\displaystyle\int \frac{\sqrt{y}\, dy}{\sqrt{c - y}}$.

COMMENT   This integral arises in the solution of the problem of determining the shape of a wire joining two fixed points, down which a bead slides in the least possible time from the upper point O to the lower point P. This is the famous "brachistochrone problem" (from the Greek *brachistos*, shortest, and *chronos*, time).

We take the y-axis pointing downward from O, as shown in Figure 8.8. An analysis of the physics of the situation, together with an advanced technique from the calculus of variations, shows that the optimal shape $y = f(x)$ of the wire—the shape that gives the bead the least time of transit—satisfies the differential equation

$$y\left(1 + \left(\frac{dy}{dx}\right)^2\right) = c \qquad (c \text{ constant}),$$

so that

$$\frac{dy}{dx} = \frac{\sqrt{c - y}}{\sqrt{y}}.$$

Thus

$$x = x(y) = \int \frac{\sqrt{y}\, dy}{\sqrt{c - y}}.$$

So we can solve the brachistochrone problem if we can perform the above antidifferentiation, and thus obtain x as a function of y; this will give us the formula for the inverse of the desired function f.

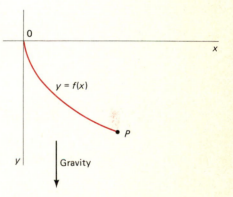

8.8   The brachistochrone problem: Find the shape of the graph of $y = f(x)$ to minimize the time it takes a bead to slide down the graph from O to P.

Solution We use the substitution suggested above for an integral of the form (2). If $u^2 = y/(c - y)$, then

$$y = \frac{cu^2}{1 + u^2} \quad \text{and} \quad dy = \frac{2cu}{(1 + u^2)^2} \, du.$$

Hence

$$x = \int \frac{2cu^2}{(1 + u^2)^2} \, du$$

$$= \int \frac{2c \tan^2\theta \, \sec^2\theta}{(1 + \tan^2\theta)^2} \, d\theta \qquad (u = \tan\theta)$$

$$= 2c \int \frac{\tan^2\theta}{\sec^2\theta} \, d\theta = 2c \int \sin^2\theta \, d\theta.$$

And so

$$x = c(\theta - \sin\theta \cos\theta). \qquad (3)$$

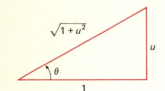

8.9 Reference triangle for the substitution $u = \tan\theta$.

The constant of integration is zero because $y = u = \theta = 0$ when $x = 0$. Now, with the aid of Figure 8.9, we retrace our steps.

$$x = \int \frac{\sqrt{y}}{\sqrt{c - y}} \, dy = c(\theta - \sin\theta \cos\theta)$$

$$= c \tan^{-1} u - \frac{cu}{u^2 + 1},$$

and therefore

$$x = c \tan^{-1}\left(\frac{\sqrt{y}}{\sqrt{c - y}}\right) - \sqrt{cy - y^2}.$$

This *is* a solution, but one whose form tells us little about the actual shape of the curve of quickest descent. In Section 10-1 we shall see what this curve looks like. For reference then, we record the fact that

$$y = \frac{cu^2}{1 + u^2} = \frac{c \tan^2\theta}{1 + \tan^2\theta} = c \sin^2\theta. \qquad (4)$$

Equations (3) and (4) express x and y in terms of the substitution variable θ. This is a "parametric" description of the curve of quickest descent.

In the case of an integral involving roots of the variable x, we may try the substitution $x = u^n$, with n chosen to be the smallest integer that will rationalize all the roots.

EXAMPLE 3 Find $\displaystyle\int \frac{dx}{x^{1/2} + x^{1/3}}$.

Solution The least common multiple of 2 and 3 (from the exponents $\frac{1}{2}$ and $\frac{1}{3}$) is 6, so we try the substitution $x = u^6$, $dx = 6u^5 \, du$. This gives

$$\int \frac{dx}{x^{1/2} + x^{1/3}} = \int \frac{6u^5 \, du}{u^3 + u^2} = 6 \int \frac{u^3 \, du}{u + 1}$$

$$= 6 \int \left(u^2 - u + 1 - \frac{1}{u + 1} \right) du \qquad \text{(by long division)}$$

$$= 2u^3 - 3u^2 + 6u - 6 \ln|u + 1| + C$$

$$= 2x^{1/2} - 3x^{1/3} + 6x^{1/6} - 6 \ln|1 + x^{1/6}| + C.$$

RATIONAL FUNCTIONS OF sin θ AND cos θ

In Section 8-3 we saw how to evaluate certain integrals of the form

$$\int R(\sin \theta, \cos \theta) \, d\theta, \qquad (5)$$

where $R(\sin \theta, \cos \theta)$ is a rational function of $\sin \theta$ and $\cos \theta$. That is, $R(\sin \theta, \cos \theta)$ is a quotient of polynomials in the two "variables" $\sin \theta$ and $\cos \theta$. The special substitution

$$u = \tan \frac{\theta}{2} \qquad (6)$$

can be used to find *any* integral of the Form (5).

In order to carry out the substitution in (6), we must express $\sin \theta$, $\cos \theta$, and $d\theta$ in terms of u and du. Note first that

$$\theta = 2 \tan^{-1} u, \quad \text{so that} \quad d\theta = \frac{2 \, du}{1 + u^2}. \qquad (7)$$

From the triangle of Figure 8.10, we see that

$$\sin \frac{\theta}{2} = \frac{u}{\sqrt{1 + u^2}}, \qquad \cos \frac{\theta}{2} = \frac{1}{\sqrt{1 + u^2}}.$$

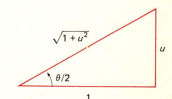

8.10 The special rationalizing substitution $u = \tan \dfrac{\theta}{2}$.

Hence

$$\sin \theta = 2 \sin \frac{\theta}{2} \cos \frac{\theta}{2} = \frac{2u}{1 + u^2}, \qquad (8)$$

$$\cos \theta = \cos^2 \frac{\theta}{2} - \sin^2 \frac{\theta}{2} = \frac{1 - u^2}{1 + u^2}. \qquad (9)$$

It should be clear that these substitutions will convert Integral (5) into an integral of a rational function of u. The latter can then be evaluated by the methods of Section 8-7.

In Section 7-2 we resorted to an unmotivated trick to show that

$$\int \sec \theta \, d\theta = \ln|\sec \theta + \tan \theta| + C.$$

We can now use the substitution $u = \tan(\theta/2)$ to integrate $\sec\theta$ more systematically.

EXAMPLE 4   Find $\int \sec\theta \, d\theta$.

Solution   Equations (7) through (9) give

$$\int \sec\theta \, d\theta = \int \frac{1}{\cos\theta} \, d\theta$$

$$= \int \frac{1 + u^2}{1 - u^2} \frac{2 \, du}{1 + u^2}$$

$$= \int \frac{2 \, du}{1 - u^2} = \int \left(\frac{1}{1 - u} + \frac{1}{1 + u} \right) du$$

$$= \ln \left| \frac{1 + u}{1 - u} \right| + C = \ln \left| \frac{1 + \tan(\theta/2)}{1 - \tan(\theta/2)} \right| + C.$$

Elementary trigonometric identities may now be used to show that this result is the same as our earlier one.

EXAMPLE 5   Suppose that $a > b > 0$. Find

$$\int \frac{d\theta}{a + b \cos\theta}.$$

Solution   We make the substitutions of Equations (7) through (9) and thus find that

$$\int \frac{d\theta}{a + b \cos\theta} = \int \frac{1}{a + b(1 - u^2)/(1 + u^2)} \frac{2 \, du}{1 + u^2}$$

$$= \int \frac{2 \, du}{(a + b) + (a - b)u^2}$$

$$= \frac{2}{a - b} \int \frac{du}{c^2 + u^2} \qquad \left(\text{where } c = \left(\frac{a + b}{a - b} \right)^{1/2} \right)$$

$$= \frac{2/c}{a - b} \tan^{-1}\left(\frac{u}{c} \right) + C.$$

Thus

$$\int \frac{d\theta}{a + b \cos\theta} = \frac{2}{\sqrt{a^2 - b^2}} \tan^{-1}\left(\sqrt{\frac{a - b}{a + b}} \tan\frac{\theta}{2} \right) + C. \qquad (10)$$

8-8   PROBLEMS

Find the integrals in Problems 1–20.

1 $\int x^3 \sqrt{3x - 2} \, dx.$

2 $\int x^3 \sqrt[3]{x^2 + 1} \, dx.$

3 $\int \frac{1}{1 + \sqrt{x}} \, dx.$

4 $\int \frac{dx}{x^{1/2} - x^{1/4}}.$

5 $\int \frac{x^3 \, dx}{(x^2 - 1)^{4/3}}.$

6 $\int x^2(x - 1)^{3/2} \, dx.$

7 $\int \frac{1 - \sqrt{x}}{1 + \sqrt[4]{x}} \, dx.$

8 $\int \frac{\sqrt[3]{x}}{1 + \sqrt{x}} \, dx.$

9 $\int \dfrac{x^5 \, dx}{\sqrt{x^3 + 1}}$.

10 $\int x^7 \sqrt[3]{x^4 + 1} \, dx$.

11 $\int \dfrac{dx}{1 + x^{2/3}}$.

12 $\int \sqrt{\dfrac{1 + x}{1 - x}} \, dx$.

13 $\int \dfrac{dx}{1 + \sqrt{x + 4}}$.

14 $\int \dfrac{\sqrt[3]{x + 1}}{x} \, dx$.

15 $\int \dfrac{d\theta}{1 + \sin \theta}$.

16 $\int \dfrac{d\theta}{(1 - \cos \theta)^2}$.

17 $\int \dfrac{d\theta}{\sin \theta + \cos \theta}$.

18 $\int \dfrac{d\phi}{\sin \phi + \cos \phi + 2}$.

19 $\int \dfrac{\sin \theta}{2 + \cos \theta} \, d\theta$.

20 $\int \dfrac{\sin \theta - \cos \theta}{\sin \theta + \cos \theta} \, d\theta$.

21 Show that if $p(x)$ is a polynomial, the substitution $u^n = (ax + b)/(cx + d)$ transforms the integral

$$\int p(x) \left(\dfrac{ax + b}{cx + d} \right)^{1/n} dx$$

into the integral of a rational function of u.

22 Rework Example 2 using the trigonometric substitution $y = c \sin^2\theta$ instead of the algebraic substitution used in the text.

23 Find the area bounded by the loop of the curve

$$y^2 = x^2(1 - x), \qquad 0 \leqq x \leqq 1.$$

24 Find the area bounded by the loop of the curve

$$y^2 = x^2 \dfrac{1 - x}{1 + x}, \qquad 0 \leqq x \leqq 1.$$

25 Find the length of the arc $y = 2\sqrt{x}$, $0 \leqq x \leqq 1$. (*Suggestion:* Substitute $x = \tan^2\theta$ in the arc length integral.)

26 Use the result of Example 5 to find

$$\int \dfrac{A + B \cos \theta}{a + b \cos \theta} \, d\theta \qquad (a > b > 0).$$

Do this by dividing the denominator into the numerator—think of the denominator as a polynomial in the variable $\cos \theta$.

27 Show that

$$\int \dfrac{d\theta}{a + b \cos \theta} = \dfrac{1}{\sqrt{b^2 - a^2}}$$

$$\times \ln \left| \dfrac{\sqrt{b - a} \tan(\theta/2) + \sqrt{b + a}}{\sqrt{b - a} \tan(\theta/2) - \sqrt{b + a}} \right| + C$$

if $0 < a < b$.

When you confront the problem of evaluating a particular integral, you first must decide which of the several methods of this chapter to try. There are only two *general* methods of integration—integration by substitution (Section 8-2) and integration by parts (Section 8-4). These are the analogues for integration of the chain rule and product rule, respectively, for differentiation.

Look first at the given integral to see if you spot a substitution that transforms it into an elementary or familiar integral or one likely to be found in an integral table. If the integrand is an unfamiliar product of two functions, one of which is easily differentiated and the other easily integrated, then an attempt to integrate by parts is indicated.

Beyond these two quite general methods, the chapter deals with a number of *special* methods. In the case of a conspicuously trigonometric integral, the simple "split-off" methods of Section 8-3 may succeed. Recall that reduction formulas (like Formula (5) and Problems 35 and 36 in Section 8-4) are available for integrating an integral power of a single trigonometric function. As a last resort, change to

sines and cosines, and remember that any rational function of $\sin \theta$ and $\cos \theta$ can be integrated using the substitution $u = \tan(\theta/2)$ of Section 8-8. The result of this substitution is a rational function of u.

Any integral of a rational function (quotient of two polynomials) can be evaluated by the method of partial fractions (Section 8-7). If the degree of the numerator is not less than that of the denominator—that is, if the rational function is not proper—first use long division to express it as the sum of a polynomial (easily integrated) and a proper rational fraction. Then decompose the latter into partial fractions. The partial fractions corresponding to linear factors are easily integrated, and those corresponding to irreducible quadratic factors can be integrated by completing the square and making (if necessary) a trigonometric substitution. As we explained in Section 8-6, the trigonometric integrals that result can always be evaluated.

In the case of an integral involving

$$\sqrt{ax^2 + bx + c},$$

you should first complete the square (Section 8-6) and then rationalize the integral by making an appropriate trigonometric substitution (Section 8-5); this will leave you with a trigonometric integral. Finally, an integral involving $\sqrt[n]{f(x)}$ is sometimes rationalized by the special substitution $u^n = f(x)$ of Section 8-8.

MISCELLANEOUS PROBLEMS

Evaluate the integrals in Problems 1–100.

1 $\displaystyle\int \frac{dx}{\sqrt{x}\,(1+x)}.$

2 $\displaystyle\int \frac{\sec^2 t\,dt}{1+\tan t}.$

3 $\displaystyle\int \sin x \sec x\,dx.$

4 $\displaystyle\int \frac{\csc x \cot x}{1+\csc^2 x}\,dx.$

5 $\displaystyle\int \frac{\tan\theta}{\cos^2\theta}\,d\theta.$

6 $\displaystyle\int \csc^4 x\,dx.$

7 $\displaystyle\int x \tan^2 x\,dx.$

8 $\displaystyle\int x^2 \cos^2 x\,dx.$

9 $\displaystyle\int x^5\sqrt{2-x^3}\,dx.$

10 $\displaystyle\int \frac{dx}{\sqrt{x^2+4}}.$

11 $\displaystyle\int \frac{x^2\,dx}{\sqrt{25+x^2}}.$

12 $\displaystyle\int \cos x\sqrt{4-\sin^2 x}\,dx$

13 $\displaystyle\int \frac{dx}{x^2-x+1}.$

14 $\displaystyle\int \sqrt{x^2+x+1}\,dx.$

15 $\displaystyle\int \frac{5x+31}{3x^2-4x+11}\,dx.$

16 $\displaystyle\int \frac{x^4+1}{x^2+2}\,dx.$

17 $\displaystyle\int \frac{d\theta}{5+4\cos\theta}.$

18 $\displaystyle\int \frac{\sqrt{x}}{1+x}\,dx.$

19 $\displaystyle\int \frac{\cos x\,dx}{\sqrt{4-\sin^2 x}}.$

20 $\displaystyle\int \frac{\cos 2x}{\cos x}\,dx.$

21 $\displaystyle\int \frac{\tan x}{\ln(\cos x)}\,dx.$

22 $\displaystyle\int \frac{x^7\,dx}{\sqrt{1-x^4}}.$

23 $\displaystyle\int \ln(1+x)\,dx.$

24 $\displaystyle\int x \sec^{-1} x\,dx.$

25 $\displaystyle\int \sqrt{x^2+9}\,dx.$

26 $\displaystyle\int \frac{x^2\,dx}{\sqrt{4-x^2}}.$

27 $\displaystyle\int \sqrt{2x-x^2}\,dx.$

28 $\displaystyle\int \frac{4x-2}{x^3-x}\,dx.$

29 $\displaystyle\int \frac{x^4\,dx}{x^2-2}.$

30 $\displaystyle\int \frac{\sec x \tan x\,dx}{\sec x+\sec^2 x}.$

31 $\displaystyle\int \frac{x\,dx}{(x^2+2x+2)^2}.$

32 $\displaystyle\int \frac{\sqrt[3]{x}\,dx}{\sqrt{x}+\sqrt[4]{x}}.$

33 $\displaystyle\int \frac{d\theta}{1+\cos 2\theta}.$

34 $\displaystyle\int \frac{\sec x}{\tan x}\,dx.$

35 $\displaystyle\int \sec^3 x \tan^3 x\,dx.$

36 $\displaystyle\int x^2 \tan^{-1} x\,dx.$

37 $\displaystyle\int x(\ln x)^3\,dx.$

38 $\displaystyle\int \frac{dx}{x\sqrt{1+x^2}}.$

39 $\displaystyle\int e^x\sqrt{1+e^{2x}}\,dx.$

40 $\displaystyle\int \frac{x\,dx}{\sqrt{4x-x^2}}.$

41 $\displaystyle\int \frac{dx}{x^3\sqrt{x^2-9}}.$

42 $\displaystyle\int \frac{x\,dx}{(7x+1)^{17}}.$

43 $\displaystyle\int \frac{4x^2+x+1}{4x^3+x}\,dx.$

44 $\displaystyle\int \frac{4x^3-x+1}{x^3+1}\,dx.$

45 $\displaystyle\int \tan^2 x \sec x\,dx.$

46 $\displaystyle\int \frac{x^2+2x+2}{(x+1)^3}\,dx.$

47 $\displaystyle\int \frac{x^4+2x+2}{x^5+x^4}\,dx.$

48 $\displaystyle\int \frac{8x^2-4x+7}{(x^2+1)(4x+1)}\,dx.$

49 $\displaystyle\int \frac{dx}{(x^3-1)^2}.$

50 $\displaystyle\int \frac{x\,dx}{x^4+4x^2+8}.$

51 $\displaystyle\int \frac{d\theta}{4+5\cos\theta}.$

52 $\displaystyle\int \frac{(1+x^{2/3})^{3/2}}{\sqrt[3]{x}}\,dx.$

53 $\displaystyle\int \frac{(\sin^{-1} x)^2}{\sqrt{1-x^2}}\,dx.$

54 $\displaystyle\int \frac{dx}{x^{3/2}(1+x^{1/3})}.$

55 $\displaystyle\int \tan^3 z\,dz.$

56 $\displaystyle\int \sin^2\omega \cos^4\omega\,d\omega.$

57 $\displaystyle\int \frac{xe^{x^2}\,dx}{1+e^{2x^2}}.$

58 $\displaystyle\int \frac{\cos^3 x}{\sqrt{\sin x}}\,dx.$

59 $\displaystyle\int x^3 e^{-x^2}\,dx.$

60 $\displaystyle\int \sin\sqrt{x}\,dx.$

61 $\displaystyle\int \frac{\arcsin x}{x^2}\,dx.$

62 $\displaystyle\int \sqrt{x^2-9}\,dx.$

63 $\displaystyle\int x^2\sqrt{1-x^2}\,dx.$

64 $\displaystyle\int x\sqrt{2x-x^2}\,dx.$

65 $\displaystyle\int \frac{(x-2)\,dx}{4x^2+4x+1}.$

66 $\displaystyle\int \frac{2x^2-5x-1}{x^3-2x^2-x+2}\,dx.$

67 $\displaystyle\int \frac{e^{2x}\,dx}{e^{2x}-1}.$

68 $\displaystyle\int \frac{\cos x\,dx}{\sin^2 x - 3\sin x + 2}.$

69 $\displaystyle\int \frac{2x^3 + 3x^2 + 4}{(x+1)^4}\,dx.$

70 $\displaystyle\int \frac{\sec^2 x\,dx}{\tan^2 x + 2\tan x + 2}.$

71 $\displaystyle\int \frac{x^3 + x^2 + 2x + 1}{x^4 + 2x^2 + 1}\,dx.$

72 $\displaystyle\int \frac{3 + \cos\theta}{2 - \cos\theta}\,d\theta.$

73 $\displaystyle\int x^5\sqrt{x^3 - 1}\,dx.$

74 $\displaystyle\int \frac{d\theta}{2 + 2\cos\theta + \sin\theta}.$

75 $\displaystyle\int \frac{\sqrt{1 + \sin x}}{\sec x}\,dx.$

76 $\displaystyle\int \frac{dx}{x^{2/3}(1 + x^{2/3})}.$

77 $\displaystyle\int \frac{\sin x}{\sin 2x}\,dx.$

78 $\displaystyle\int \sqrt{1 + \cos t}\,dt.$

79 $\displaystyle\int \sqrt{1 + \sin t}\,dt.$

80 $\displaystyle\int \frac{\sec^2 t\,dt}{1 - \tan^2 t}.$

81 $\displaystyle\int \ln(x^2 + x + 1)\,dx.$

82 $\displaystyle\int e^x \sin^{-1}(e^x)\,dx.$

83 $\displaystyle\int \frac{\arctan x\,dx}{x^2}.$

84 $\displaystyle\int \frac{x^2\,dx}{\sqrt{x^2 - 25}}.$

85 $\displaystyle\int \frac{x^3\,dx}{(x^2 + 1)^2}.$

86 $\displaystyle\int \frac{dx}{x\sqrt{6x - x^2}}.$

87 $\displaystyle\int \frac{(3x + 2)\,dx}{(x^2 + 4)^{3/2}}.$

88 $\displaystyle\int x^{3/2}\ln x\,dx.$

89 $\displaystyle\int \frac{(1 + \sin^2 x)^{1/2}}{\sec x\,\csc x}\,dx.$

90 $\displaystyle\int \frac{e^{\sqrt{\sin x}}\,dx}{\sec x\,\sqrt{\sin x}}.$

91 $\displaystyle\int xe^x \sin x\,dx.$

92 $\displaystyle\int x^2 \exp(x^{3/2})\,dx.$

93 $\displaystyle\int \frac{\tan^{-1} x}{(x-1)^3}\,dx.$

94 $\displaystyle\int \ln(1 + \sqrt{x})\,dx.$

95 $\displaystyle\int \frac{(2x + 3)\,dx}{\sqrt{3 + 6x - 9x^2}}.$

96 $\displaystyle\int \frac{d\theta}{2 + 2\sin\theta + \cos\theta}.$

97 $\displaystyle\int \frac{\sin^3\theta\,d\theta}{\cos\theta - 1}.$

98 $\displaystyle\int x^{3/2}\tan^{-1}(x^{1/2})\,dx.$

99 $\displaystyle\int \sec^{-1}\sqrt{x}\,dx.$

100 $\displaystyle\int x\left(\frac{1 - x^2}{1 + x^2}\right)^{1/2}\,dx.$

101 Find the area of the surface generated by revolving the curve $y = \cosh x$, $0 \le x \le 1$, around the x-axis.

102 Find the length of the curve $y = e^{-x}$, $0 \le x \le 1$.

103 (a) Find the area A_b of the surface generated by revolving the curve $y = e^{-x}$, $0 \le x \le b$, around the x-axis. (b) Find $\displaystyle\lim_{b\to\infty} A_b$.

104 (a) Find the area A_b of the surface generated by revolving the curve $y = 1/x$, $1 \le x \le b$, around the x-axis. (b) Find $\displaystyle\lim_{b\to\infty} A_b$.

105 Find the area of the surface generated by revolving the curve

$$y = \sqrt{x^2 - 1}, \qquad 1 \le x \le 2,$$

around the x-axis.

106 (a) Derive the reduction formula

$$\int x^m (\ln x)^n\,dx = \frac{x^{m+1}(\ln x)^n}{m + 1} - \frac{n}{m + 1}\int x^m (\ln x)^{n-1}\,dx.$$

(b) Find $\displaystyle\int_1^e x^3 (\ln x)^3\,dx.$

107 Derive the reduction formula

$$\int \sin^m x \cos^n x\,dx = -\frac{\sin^{m-1} x \cos^{n+1} x}{m + n}$$
$$+ \frac{m - 1}{m + n}\int \sin^{m-2} x \cos^n x\,dx.$$

108 Use the reduction formulas of Problem 107 above and Problem 36 in Section 8-4 to evaluate

$$\int_0^{\pi/2} \sin^6 x \cos^5 x\,dx.$$

109 Find the area bounded by the curve $y^2 = x^5(2 - x)$, $0 \le x \le 2$. (*Suggestion:* Substitute $x = 2\sin^2\theta$, then use Problem 40 in Section 8-4.)

110 Show that

$$\int_0^1 \frac{t^4(1 - t)^4}{1 + t^2}\,dt = \frac{22}{7} - \pi.$$

111 Evaluate $\displaystyle\int_0^1 t^4(1 - t)^4\,dt$, and then apply the result of Problem 110 to conclude that

$$\tfrac{22}{7} - \tfrac{1}{630} < \pi < \tfrac{22}{7} - \tfrac{1}{1260}.$$

Thus $3.1412 < \pi < 3.1421$.

112 Find the length of the curve $y = \frac{4}{5}x^{5/4}$, $0 \le x \le 1$.

113 Find the length of the curve $y = \frac{4}{3}x^{3/4}$, $1 \le x \le 4$.

114 An initially empty water tank is shaped like a cone with vertical axis, vertex at the bottom, depth of 9 feet, and top radius 4.5 feet. Beginning at time $t = 0$, water is poured into this tank at 50 cubic feet per minute. Meanwhile, water leaks out a hole at the bottom at the rate of $10\sqrt{y}$ cubic feet per minute—consistent with Torricelli's law. How long does it take to fill the tank?

115 (a) Evaluate $\displaystyle\int \frac{dx}{1 + e^x + e^{-x}}$.

(b) Explain why your substitution in (a) suffices to integrate any rational function of e^x.

116 (a) The equation $x^3 + x + 1 = 0$ has one real root r.

Use Newton's method to find it, accurate to at least two places.

(b) Use long division to find the irreducible quadratic factor of $x^3 + x + 1$.

(c) Use the factorization of part (b) to evaluate

$$\int_0^1 \frac{dx}{x^3 + x + 1}.$$

117 Find $\displaystyle\int \frac{dx}{1 + e^x}$.

Conic Sections and Polar Coordinates

9

Analytic Geometry and the Conic Sections

Plane analytic geometry, the main topic of this chapter, is the use of algebra and calculus to study the properties of curves in the plane. The ancient Greeks used deductive reasoning and the methods of axiomatic Euclidean geometry to study lines, circles, and the **conic sections** (parabolas, ellipses, and hyperbolas). The properties of conic sections have played an important role in diverse scientific applications since the 17th century, when Kepler discovered—and Newton explained—the fact that the orbits of planets and other bodies in the solar system are conic sections.

The French mathematicians Descartes and Fermat, working almost independently of one another, initiated analytic geometry in 1637. The central idea of analytic geometry is the correspondence between an equation $F(x, y) = 0$ and the set or **locus** (typically, a curve) of all those points (x, y) in the plane with coordinates that satisfy this equation.

The idea is this: Given a geometric locus or curve, its properties can be derived algebraically or analytically from its defining equation $F(x, y) = 0$. For example, suppose that the equation of a given curve turns out to be the linear equation

$$Ax + By = C, \tag{1}$$

where A, B, and C are constants with $B \neq 0$. This equation may be written in the form

$$y = mx + b \tag{2}$$

where $m = -A/B$ and $b = C/B$. But (2) is the slope-intercept equation (Section 1-3) of the straight line with slope m and y-intercept b. Hence the given curve is this straight line. In the following example we use this approach to show that a certain geometrically described locus is a particular straight line.

EXAMPLE 1 Show that the set of all points equally distant from the points $(1, 1)$ and $(5, 3)$ is the perpendicular bisector of the line segment joining these two points.

Solution The typical point $P(x, y)$ is equidistant from $(1, 1)$ and $(5, 3)$ if and only if

$$(x - 1)^2 + (y - 1)^2 = (x - 5)^2 + (y - 3)^2,$$

$$(x^2 - 2x + 1) + (y^2 - 2y + 1) = (x^2 - 10x + 25) + (y^2 - 6y + 9),$$

$$2x + y = 8. \tag{3}$$

Thus the given locus is the straight line in (3) with slope -2. The straight line through $(1, 1)$ and $(5, 3)$ has equation

$$y - 1 = \tfrac{1}{2}(x - 1) \tag{4}$$

and slope $\frac{1}{2}$. Since the product of the slopes of these two lines is -1, it follows (by Theorem 2 in Section 1-3) that they are perpendicular lines. If we solve Equations (3) and (4) simultaneously, we find that the intersection of these lines is, indeed, the midpoint (3, 2) of the given line segment. Thus the locus described is the perpendicular bisector of this line segment.

The circle with center (h, k) and radius r is described geometrically as the set of all points $P(x, y)$ whose distance from (h, k) is r. The distance formula then gives

$$(x - h)^2 + (y - k)^2 = r^2 \qquad (5)$$

as the equation of this circle. In particular, if $h = k = 0$, then (5) takes the simple form

$$x^2 + y^2 = r^2. \qquad (6)$$

We can see directly from this equation, without further reference to the definition, that a circle centered at the origin has the following symmetry properties:

- *Symmetry about the x-axis:* The equation of the curve is unaltered when y is replaced by $-y$.
- *Symmetry about the y-axis:* The equation of the curve is unaltered when x is replaced by $-x$.
- *Symmetry with respect to the origin:* The equation of the curve is unaltered when x is replaced by $-x$ and y is replaced by $-y$.
- *Symmetry about the 45°-line $y = x$:* The equation is unaltered when x and y are interchanged.

Equation (6) can be rewritten in each of the forms

$$y = \pm\sqrt{r^2 - x^2} \quad \text{and} \quad x = \pm\sqrt{r^2 - y^2}.$$

Since the expressions beneath the square roots must be nonnegative, these new equations show us that $|x| \leq r$ and $|y| \leq r$ for any point $P(x, y)$ of the circle of radius r centered at the origin. (This is evident from the geometry.)

The relationship between Equations (5) and (6) is an illustration of the translation principle of Section 1-4. Imagine a translation ("slide") of the plane that moves the point (x, y) to the new position $(x + h, y + k)$. Under such a translation, a curve C would be moved to a new curve. The equation of the new curve is easily obtained from the old equation—we simply replace x by $x - h$ and y by $y - k$. Conversely, we can recognize a translated circle from its equation: Any equation of the form

$$x^2 + y^2 + Ax + By + C = 0 \qquad (7)$$

can be rewritten in the form

$$(x - h)^2 + (y - k)^2 = p$$

by completing the square as in Example 1 of Section 1-4. Thus the graph of Equation (6) is either a circle (if $p > 0$), a single point (if $p = 0$), or empty (if $p < 0$). We use this approach in the following example to discover that the locus described is a particular circle.

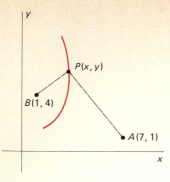

9.1 The locus of Example 2.

EXAMPLE 2 Determine the locus of a point $P(x, y)$ if its distance $|AP|$ from $A(7, 1)$ is twice its distance $|BP|$ from $B(1, 4)$.

Solution The points A, B, and P appear in Figure 9.1, along with a curve through P representing the given locus. Since

$$|AP|^2 = 4|BP|^2 \qquad \text{(from } |AP| = 2|BP|\text{),}$$

we get the equation

$$(x - 7)^2 + (y - 1)^2 = 4[(x - 1)^2 + (y - 4)^2].$$

Hence

$$3x^2 + 3y^2 + 6x - 30y + 18 = 0,$$

$$x^2 + y^2 + 2x - 10y = -6,$$

and so

$$(x + 1)^2 + (y - 5)^2 = 20.$$

Thus the locus is a circle with center $(-1, 5)$ and radius $r = \sqrt{20} = 2\sqrt{5}$.

CONIC SECTIONS

The phrase *conic sections* stems from the fact that these are the curves in which a plane intersects a cone. The cone used is a right circular cone with two "nappes" extending infinitely far in both directions, as in Figure 9.2. There are three types of conic sections, as illustrated in Figure 9.3. If the cutting plane is parallel to some generator of the cone, then the curve of intersection is a *parabola*. Otherwise it is either a single closed curve—an *ellipse*—or a *hyperbola* with two "branches."

In Section 13-5 we shall use the methods of three-dimensional analytic geometry to show that, if an appropriate xy-coordinate system is set up in the intersecting plane, then the equations of the three conic sections take the following forms.

Parabola: $\qquad\qquad y^2 = kx;$ $\qquad\qquad$ (8)

Ellipse: $\qquad \dfrac{x^2}{a^2} + \dfrac{y^2}{b^2} = 1;$ $\qquad\qquad$ (9)

Hyperbola: $\qquad \dfrac{x^2}{a^2} - \dfrac{y^2}{b^2} = 1.$ $\qquad\qquad$ (10)

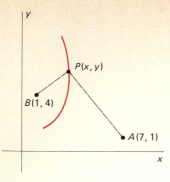

9.2 A cone of two nappes.

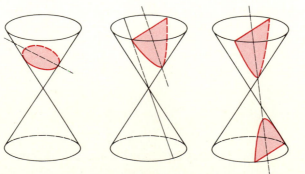

9.3 The conic sections.

Ellipse $\qquad\qquad$ Parabola $\qquad\qquad$ Hyperbola

In the following three sections, we shall discuss these conic sections on the basis of two-dimensional definitions that do not require the three-dimensional setting of a cone and an intersecting plane. The following example illustrates one such approach to the conic sections.

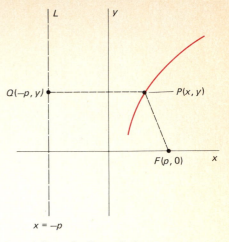

EXAMPLE 3   Let e be a given positive number (*not* to be confused with the natural logarithm base). Determine the locus of a point $P(x, y)$ if its distance from the fixed point $F(p, 0)$ is e times its distance from the vertical line L with equation $x = -p$. See Figure 9.4.

9.4   A figure for Example 3.

Solution   Let PQ be the perpendicular from P to the line L. Then the condition

$$\overline{PF} = e\overline{PQ}$$

takes the analytic form

$$\sqrt{(x - p)^2 + y^2} = e|x - (-p)|.$$

That is,

$$(x^2 - 2px + p^2) + y^2 = e^2(x^2 + 2px + p^2),$$

and so

$$x^2(1 - e^2) - 2p(1 + e^2)x + y^2 = -p^2(1 - e^2). \tag{11}$$

Case 1: e = 1.   Then (11) reduces to

$$y^2 = 4px. \tag{12}$$

We see upon comparison with Equation (8) that the locus of P is a *parabola* if $e = 1$.

Case 2: e < 1.   Dividing Equation (11) by $1 - e^2$ yields

$$x^2 - 2p\frac{1 + e^2}{1 - e^2}x + \frac{y^2}{1 - e^2} = -p^2.$$

We now complete the square in x, and get

$$\left(x - p\frac{1 + e^2}{1 - e^2}\right)^2 + \frac{y^2}{1 - e^2} = p^2\left[\left(\frac{1 + e^2}{1 - e^2}\right)^2 - 1\right] = a^2.$$

This equation takes the form

9.5   An ellipse: $e < 1$.

$$\frac{(x - h)^2}{a^2} + \frac{y^2}{b^2} = 1 \tag{13}$$

where

$$h = +p\frac{1 + e^2}{1 - e^2} \quad \text{and} \quad b^2 = a^2(1 - e^2). \tag{14}$$

When we compare Equations (9) and (13), we see that if $e < 1$, then the locus of P is an *ellipse* with $(0, 0)$ translated to $(h, 0)$, as illustrated in Figure 9.5.

Case 3: e > 1.   In this case, Equation (11) reduces to a translated version of Equation (10), so the locus of P is a *hyperbola*. The details, which are similar to those of Case 2, are left for Problem 25.

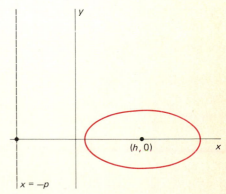

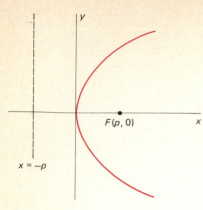

9.6 A parabola: $e = 1$.

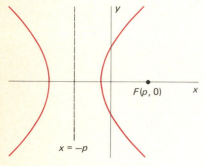

9.7 A hyperbola: $e > 1$.

Thus the locus in Example 3 is a *parabola* if $e = 1$, an *ellipse* if $e < 1$, and a *hyperbola* if $e > 1$. The number e is called the **eccentricity** of the conic section. The point $F(p, 0)$ is commonly called its **focus** in the parabolic case, Figure 9.6 shows the parabola of Case 1, and Figure 9.7 illustrates the hyperbola of Case 3.

If we begin with Equations (8) through (10), we can derive the general characteristics of the three conic sections shown in Figures 9.5 through 9.7. For example, in the case of the parabola with the equation in (8), we see that the curve passes through the origin, that $x \geq 0$ at each of its points, that $y \to \pm \infty$ as $x \to \infty$, and that the graph is symmetric about the x-axis (because the curve is unchanged when y is replaced by $-y$).

In the case of the ellipse with the equation in (9), we see that the graph must be symmetric about both axes. At each point (x, y) of the graph, we must have $|x| \leq a$ and $|y| \leq b$. The graph intersects the axes at the four points $(\pm a, 0)$ and $(0, \pm b)$.

Finally, the hyperbola with the equation in (10), or its alternative form

$$y = \pm \frac{b}{a} \sqrt{x^2 - a^2},$$

is symmetric about both axes. It meets the x-axis at the two points $(\pm a, 0)$, has one branch consisting of points with $x \geq a$, and has another branch where $x \leq -a$. And $y \to \pm \infty$ as $x \to \pm \infty$.

9-1 PROBLEMS

In each of Problems 1–6, write an equation of the specified straight line.

1 The line through the point $(1, -2)$ that is parallel to the line with equation $x + 2y = 5$.

2 The line through the point $(-3, 2)$ that is perpendicular to the line with equation $3x - 4y = 7$.

3 The line that is tangent to the circle $x^2 + y^2 = 25$ at the point $(3, -4)$.

4 The line that is tangent to the curve $y^2 = x + 3$ at the point $(6, -3)$.

5 The line that is perpendicular to the curve $x^2 + 2y^2 = 6$ at the point $(2, -1)$.

6 The perpendicular bisector of the line segment with end points $(-3, 2)$ and $(5, -4)$.

In each of Problems 7–10, find the center and radius of the circle described in the given equation.

7 $x^2 + 2x + y^2 = 4$.
8 $x^2 + y^2 - 4y = 5$.
9 $x^2 + y^2 - 4x + 6y = 3$.
10 $x^2 + y^2 + 8x - 6y = 0$.

In each of Problems 11–14, write the equation of the specified circle.

11 The circle with center $(-1, -2)$ that passes through the point $(2, 3)$.

12 The circle with center $(2, -2)$ that is tangent to the line $y = x + 4$.

13 The circle with center $(6, 6)$ that is tangent to the line $y = 2x - 4$.

14 The circle that passes through the points $(4, 6), (-2, -2)$, and $(5, -1)$. (*Suggestion:* Determine A, B, and C so that each of the three given points satisfies the equation $x^2 + y^2 + Ax + By + C = 0$.)

In each of Problems 15–20, derive the equation of the set of all points $P(x, y)$ satisfying the given condition. Then sketch the graph of the equation.

15 The point $P(x, y)$ is equally distant from the two points $(3, 2)$ and $(7, 4)$.

16 The distance from P to the point $(-2, 1)$ is one-half its distance from $(4, -2)$.

17 The point P is three times as far from the point $(-3, 2)$ as it is from the point $(5, 10)$.

18 The distance of P from the line $x = -3$ is equal to its distance from the point $(3, 0)$.

19 The sum of the distances of P from the points $(4, 0)$ and $(-4, 0)$ is 10.

20 The sum of the distances of P from the points $(3, 0)$ and $(-3, 0)$ is 10.

21 Find all lines through the point $(2, 1)$ that are tangent to the parabola $y = x^2$.

22 Find all lines through the point $(-1, 2)$ that are normal to the parabola $y = x^2$.

23 Find all lines that are normal to the curve $xy = 4$ and simultaneously parallel to the line $y = 4x$.

24 Find all lines that are tangent to the curve $y = x^3$ and are also parallel to the line $3x - y = 5$.

25 Suppose that $e > 1$. Show that Equation (11) of this section can be written in the form

$$\frac{(x-h)^2}{a^2} - \frac{y^2}{b^2} = 1,$$

thus showing that its graph is a hyperbola. Find a, b, and h in terms of p and e.

The Parabola

The case $e = 1$ of Example 3 in the previous section is motivation for this formal definition:

> **Definition** *The Parabola*
>
> A **parabola** is the set of all points P in the plane that are equally distant from a fixed point F (called the **focus** of the parabola) and a fixed line L (called its **directrix**) not containing P.

If the focus of the parabola is $F(p, 0)$ and its directrix is the line $x = -p$, $p > 0$, then we know from Equation (12) of Section 9-1 that the equation of the parabola is

$$y^2 = 4px. \tag{1}$$

When we replace x by $-x$, both in the equation and in the discussion that precedes it, we get the equation of a parabola with focus at $(-p, 0)$ and with directrix the vertical line $x = p$. The new parabola has equation

$$y^2 = -4px. \tag{2}$$

The old and new parabolas appear in Figure 9.8.

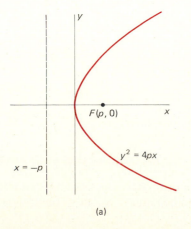

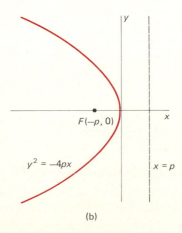

(a) (b)

9.8 Two parabolas with vertical directrices.

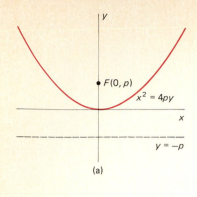

(a)

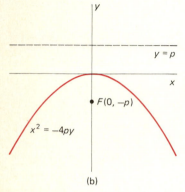

(b)

9.9 Two parabolas with horizontal directrices.

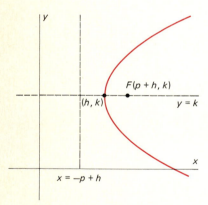

9.10 A translation of the parabola $x^2 = 4py$.

We could also interchange x and y in Equation (1). This would give the equation of a parabola that has focus at $(0, p)$ and directrix the *horizontal* line $y = -p$. This parabola opens upward. Its equation is

$$x^2 = 4py, \tag{3}$$

and it is shown in Figure 9.9(a).

Finally, we replace y with $-y$ in Equation (3). This gives the equation

$$x^2 = -4py \tag{4}$$

of a downward-opening parabola with focus $(0, -p)$ and directrix $y = p$. It is shown in Figure 9.9(b).

Each of the parabolas discussed above is symmetric about one of the coordinate axes. The line about which a parabola is symmetric is called its **axis.** The point on a parabola midway between its focus and its directrix is called its **vertex.** Note that the very definition of a parabola implies that the parabola contains its vertex. Each parabola we discussed in connection with Equations (1) through (4) has the origin as its vertex.

EXAMPLE 1   Determine the focus, directrix, axis, and vertex of the parabola $x^2 = 12y$.

Solution   We write the given equation as $x^2 = 4(3)y$. In this form it matches Equation (3) with $p = 3$. Hence the given parabola has its focus at $(0, 3)$, and its directrix is the horizontal line $y = -3$. The y-axis is the axis of symmetry, and the parabola opens upward from its vertex at the origin.

Suppose that we begin with the parabola of Equation (1), and translate its vertex to the new point (h, k). Then the translated parabola has equation

$$(y - k)^2 = 4p(x - h). \tag{1a}$$

The new parabola has focus $F(p + h, k)$ and its directrix is the vertical line $x = -p + h$, as indicated in Figure 9.10. Its axis is the horizontal line $y = k$.

We can obtain the translates of the other three parabolas (2) through (4) in the same way. If the vertex is moved from the origin to the point (h, k), then the three equations take these forms:

$$(y - k)^2 = -4p(x - h), \tag{2a}$$

$$(x - h)^2 = 4p(y - k), \quad \text{and} \tag{3a}$$

$$(x - h)^2 = -4p(y - k). \tag{4a}$$

Now note that Equations (1a) and (2a) both take the general form

$$y^2 + Ax + By + C = 0 \quad (A \neq 0), \tag{5}$$

while Equations (3a) and (4a) take the general form

$$x^2 + Ax + By + C = 0 \quad (B \neq 0). \tag{6}$$

What is significant about these last two equations is what they have in common: Each is linear in one of the coordinate variables and quadratic in the other. In fact, *any* such equation can be reduced to one of the standard forms in (1a) through (4a) by completing the square in the coordinate variable that appears quadratically. This means that the graph of any equation of the form of either (5) or (6) is a parabola. The features of the parabola can be read from the standard form of the equation.

EXAMPLE 2 Determine the graph of the equation

$$4y^2 - 8x - 12y + 1 = 0.$$

Solution This equation is linear in x and quadratic in y. We divide through by the coefficient of y^2 and then collect all terms involving y on one side of the equation:

$$y^2 - 3y = 2x - \tfrac{1}{4}.$$

Then we complete the square in the variable y and thus find that

$$y^2 - 3y + \tfrac{9}{4} = 2x - \tfrac{1}{4} + \tfrac{9}{4} = 2x + 2.$$

The final step is to write the terms on the right involving x in the form $4p(x - h)$:

$$(y - \tfrac{3}{2})^2 = 4(\tfrac{1}{2})(x + 1).$$

This equation has the form of Equation (1a), with $p = \tfrac{1}{2}$, $h = -1$, and $k = \tfrac{3}{2}$. Thus the graph is a parabola that opens to the right from the vertex at $(-1, \tfrac{3}{2})$. It focus is at $(-\tfrac{1}{2}, \tfrac{3}{2})$, its directrix is the vertical line $x = -\tfrac{3}{2}$, and its axis is the horizontal line $y = \tfrac{3}{2}$, as indicated in Figure 9.11.

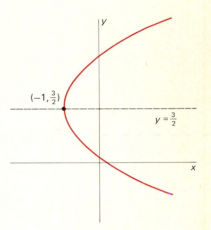

9.11 The parabola of Example 2.

An important property of the parabola, known as its **reflection property,** has many practical applications. This property involves the fact that the tangent line to the parabola $y^2 = 4px$ at any point $P(x_0, y_0)$ makes equal angles with the horizontal line through P and the line PF from P to the focus F of the parabola. That is, $\alpha = \beta$ in Figure 9.12. To verify this fact,

9.12 Reflection property of the parabola: $\alpha = \beta$.

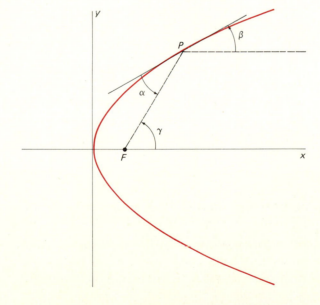

note first that the derivative of $y = 2\sqrt{px}$ is $dy/dx = \sqrt{p/x}$, so

$$\tan \beta = \sqrt{\frac{p}{x_0}}.$$

The slope of the line PF is

$$\tan \gamma = \frac{y_0 - 0}{x_0 - p} = \frac{2\sqrt{px_0}}{x_0 - p}.$$

Since $\alpha = \gamma - \beta$, it follows that

$$\tan \alpha = \tan(\gamma - \beta) = \frac{\tan \gamma - \tan \beta}{1 + \tan \gamma \tan \beta}$$

$$= \frac{[2\sqrt{px_0}/(x_0 - p)] - (\sqrt{p}/\sqrt{x_0})}{1 + [2\sqrt{px_0}/(x_0 - p)](\sqrt{p}/\sqrt{x_0})} \cdot \frac{(x_0 - p)\sqrt{x_0}}{(x_0 - p)\sqrt{x_0}}$$

$$= \frac{2\sqrt{p}\,x_0 - \sqrt{p}(x_0 - p)}{(x_0 - p)\sqrt{x_0} + 2\sqrt{px_0}\sqrt{p}} = \frac{\sqrt{p}(x_0 + p)}{\sqrt{x_0}(x_0 + p)}$$

$$= \sqrt{\frac{p}{x_0}} = \tan \beta.$$

Since α and β are acute angles, the fact that $\tan \alpha = \tan \beta$ implies that $\alpha = \beta$.

The reflection property is exploited in the design of parabolic mirrors. Such a mirror has the shape of the surface obtained by revolving a parabola around its axis of symmetry. Because of the law of reflection for light rays, a beam of incoming rays parallel to the axis will then be focused at the point T, as shown in Figure 9.13. The reflection property is also exploited "in reverse"—rays emanating from the focus will be reflected in a beam parallel to the axis, thus keeping the light beam intense. Parabolic mirrors are used in visual and radio telescopes, radar antennas, searchlights, automobile headlights, microphone systems, and solar heating devices.

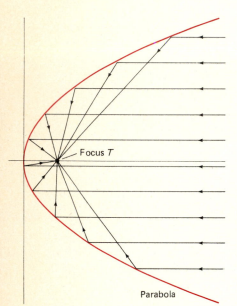

Focus T

Parabola

9.13 Incident rays parallel to the axis reflect through the focus.

Galileo discovered in the early 1600s that the trajectory of a projectile fired from a gun is a parabola (assuming that air resistance is ignored). Suppose that the projectile is fired at time $t = 0$ from the origin with initial velocity v_0, at an angle α of inclination from the horizontal x-axis. Then the initial velocity splits into the components

$$v_{0x} = v_0 \cos \alpha \quad \text{and} \quad v_{0y} = v_0 \sin \alpha$$

as indicated in Figure 9.14. The fact that the projectile continues to move horizontally with *constant* speed v_{0x}, together with Equation (8) in Section 3-8, implies that its x- and y-coordinates after t seconds are

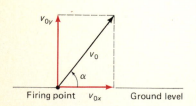

v_{0y}

v_0

α

Firing point $\quad v_{0x} \quad$ Ground level

9.14 Resolution of the initial velocity v_0 into horizontal and vertical components.

$$x = (v_0 \cos \alpha)t, \tag{7}$$

$$y = -\tfrac{1}{2}gt^2 + (v_0 \sin \alpha)t. \tag{8}$$

By substituting $t = x/(v_0 \cos \alpha)$ from (7) into (8) and then completing the square, we can derive (as in Problem 24) an equation of the form

$$y - M = -4p(x - \tfrac{1}{2}R)^2. \tag{9}$$

Here,

$$M = \frac{v_0^2 \sin^2\alpha}{2g} \tag{10}$$

is the maximum height attained by the projectile, and

$$R = \frac{v_0^2 \sin 2\alpha}{g} \tag{11}$$

is the **range** or horizontal distance it travels before returning to the ground. Thus the trajectory is the parabola shown in Figure 9.15.

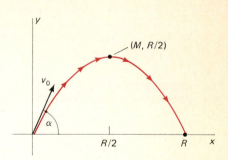

9.15 The trajectory of the projectile, showing its maximum altitude and its range.

9-2 PROBLEMS

In each of Problems 1–5, find the equation and sketch the graph of the parabola with vertex V and focus F.

1 $V(0, 0)$ and $F(3, 0)$.
2 $V(0, 0)$ and $F(0, -2)$.
3 $V(2, 3)$ and $F(2, 1)$.
4 $V(-1, -1)$ and $F(-3, -1)$.
5 $V(2, 3)$ and $F(0, 3)$.

In each of Problems 6–10, find the equation and sketch the graph of the parabola with the given focus and directrix.

6 $F(1, 2)$; $x = -1$.   **7** $F(0, -3)$; $y = 0$.
8 $F(1, -1)$; $x = 3$.   **9** $F(0, 0)$; $y = -2$.
10 $F(-2, 1)$; $x = -4$.

In each of Problems 11–18, sketch the parabola with the given equation. Show and label its vertex, focus, axis, and directrix.

11 $y^2 = 12x$.   **12** $x^2 = -8y$.
13 $y^2 = -6x$.   **14** $x^2 = 7y$.
15 $x^2 - 4x - 4y = 0$.
16 $y^2 - 2x + 6y + 15 = 0$.
17 $4x^2 + 4x + 4y + 13 = 0$.
18 $4y^2 - 12y + 9x = 0$.
19 Show that the point of the parabola $y^2 = 4px$ that is closest to its focus is its vertex.
20 Find the equation of a parabola with vertical axis that passes through the points $(2, 3)$, $(4, 3)$, and $(6, -5)$.
21 Show that the equation of the tangent line to the parabola $y^2 = 4px$ at the point (x_0, y_0) is

$$2px - y_0y + 2px_0 = 0.$$

Conclude that this tangent line intersects the x-axis at

the point $(-x_0, 0)$. This fact provides a quick method for constructing a tangent line to a parabola at a given point.
22 A comet has a parabolic orbit with the sun at its focus. When the comet is $100\sqrt{2}$ million miles from the sun, the line from the sun to the comet makes an angle of $45°$ with the axis of the parabola, as shown in Figure 9.16. What will be the minimum distance between the comet and the sun? (*Suggestion:* Write the equation of the parabola with the origin at the focus and then use the result of Problem 19.)

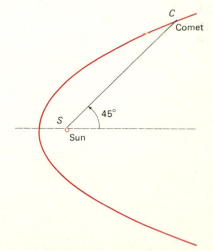

9.16 The comet of Problem 22 in parabolic orbit about the sun.

23 Suppose that the angle of Problem 22 increases from $45°$ to $90°$ in 3 days. How much longer will be required for the comet to reach its point of closest approach to the sun? Assume that the line segment from the sun to the comet sweeps out area at a constant rate (Kepler's second law).

24 Use Equations (7) and (8) to derive Equation (9) with the values of M and R given by Equations (10) and (11).

25 Deduce from Equation (11) that, given a fixed initial velocity v_0, the maximum range of the projectile is $R_{max} = v_0^2/g$ and is attained when $\alpha = 45°$.

In Problems 26–28, assume that a projectile is fired from the origin with initial velocity $v_0 = 160$ ft/sec and with initial angle of inclination α. Use $g = 32$ ft/sec^2.

26 If $\alpha = 45°$, find the range R of the projectile and the maximum height that it attains.

27 For what value(s) of α is the range $R = 400$ ft?

28 Find the range R of the projectile and the length of time it is in the air if (a) $\alpha = 30°$; (b) $\alpha = 60°$.

9-3

The Ellipse

An ellipse is a conic section with eccentricity e less than 1, as in Example 3 of Section 9-1.

> **Definition**   *The Ellipse*
>
> Suppose that $0 < e < 1$, and let F be a fixed point and L a fixed line not containing F. Then the **ellipse** with **eccentricity** e, **focus** F, and **directrix** L is the set of all points P such that the distance $|PF|$ is e times the (perpendicular) distance from P to the line L.

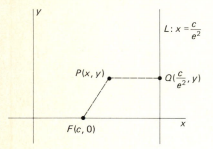

9.17 Ellipse: focus, F; directrix, L; eccentricity, e.

The equation of the ellipse is especially simple if F is the point $(c, 0)$ on the x-axis and L is the vertical line $x = c/e^2$. The case $c > 0$ is shown in Figure 9.17. If Q is the point $(c/e^2, y)$, then PQ is the perpendicular from $P(x, y)$ to L. The condition $|PF| = e|PQ|$ gives

$$(x - c)^2 + y^2 = e^2\left(x - \frac{c}{e^2}\right)^2,$$

$$x^2 - 2cx + c^2 + y^2 = e^2 x^2 - 2cx + \frac{c^2}{e^2},$$

$$x^2(1 - e^2) + y^2 = c^2\left(\frac{1}{e^2} - 1\right) = \frac{c^2}{e^2}(1 - e^2).$$

Thus

$$x^2(1 - e^2) + y^2 = a^2(1 - e^2)$$

where

$$a = \frac{c}{e}. \tag{1}$$

Division of both sides of the next-to-last equation above by $a^2(1 - e^2)$ gives

$$\frac{x^2}{a^2} + \frac{y^2}{a^2(1 - e^2)} = 1.$$

Finally, with the aid of the fact that $e < 1$, we may let

$$b^2 = a^2(1 - e^2) = a^2 - c^2. \tag{2}$$

Then the equation of the ellipse with focus $(c, 0)$ and directrix $x = c/e^2 = a/e$ takes the pleasant form

$$\frac{x^2}{a^2} + \frac{y^2}{b^2} = 1. \tag{3}$$

We see from Equation (3) that this ellipse is symmetric about both axes. Its x-intercepts are $(\pm a, 0)$, and its y-intercepts are $(0, \pm b)$. The points $(\pm a, 0)$ are called the **vertices** of the ellipse, and the line segment joining them is its **major axis.** The line segment joining its y-intercepts $(0, \pm b)$ is its **minor axis** (note from (2) that $b < a$). The alternative form

$$a^2 = b^2 + c^2 \tag{4}$$

of (2) is the Pythagorean relation for the right triangle of Figure 9.18. Indeed, visualization of this triangle is an excellent way to remember Equation (4). The numbers a and b are lengths of the major and minor **semiaxes,** respectively.

Since $a = c/e$, the directrix of the ellipse in (3) is $x = a/e$. If we had begun instead with focus $(-c, 0)$ and directrix $x = -a/e$, we would still have got Equation (3), because only the squares of a and c are involved in the derivation. Thus the ellipse in (3) has *two* foci, $(c, 0)$ and $(-c, 0)$, and *two* directrices, $x = a/e$ and $x = -a/e$. These are shown in Figure 9.19.

The larger the eccentricity $e < 1$ is, the more elongated is the ellipse. If $e = 0$ then Equation (2) gives $b = a$, so Equation (3) reduces to the equation of a circle of radius a. Thus a circle may be regarded as an ellipse with eccentricity zero. Compare the three cases shown in Figure 9.20.

EXAMPLE 1   Find the equation of the ellipse with foci $(\pm 3, 0)$ and vertices $(\pm 5, 0)$.

Solution   We are given that $c = 3$ and $a = 5$, so Equation (2) gives $b = 4$. Thus Equation (3) gives

$$\frac{x^2}{25} + \frac{y^2}{16} = 1.$$

This ellipse is shown in Figure 9.21.

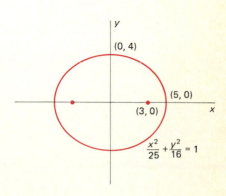

9.18   The parts of an ellipse.

9.19   The ellipse as a conic section: Two foci, two directrices.

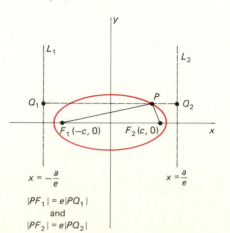

$x = -\dfrac{a}{e}$ $x = \dfrac{a}{e}$

$|PF_1| = e|PQ_1|$
and
$|PF_2| = e|PQ_2|$

9.20   Relation between eccentricity of an ellipse and its shape.

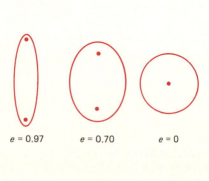

$e = 0.97$ $e = 0.70$ $e = 0$

9.21   The ellipse of Example 1.

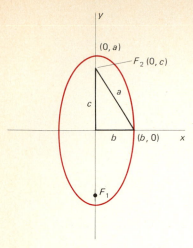

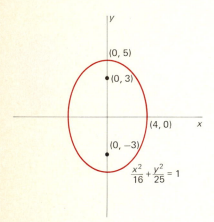

9.22 An ellipse with vertical major axis.

9.23 The ellipse of Example 2.

If our ellipse has its two foci on the y-axis, for example $F_1(0, -c)$ and $F_2(0, c)$, then its equation is

$$\frac{x^2}{b^2} + \frac{y^2}{a^2} = 1, \tag{5}$$

and it is still true that

$$a^2 = b^2 + c^2. \tag{4}$$

But now the major axis of length $2a$ is vertical, and the minor axis of length $2b$ is horizontal. The derivation of Equation (5) is similar to that of Equation (3); see Problem 23. Figure 9.22 shows the case of an ellipse with vertical major axis.

In practice, there is little chance of confusing Equations (3) and (5). The equation or given data will make it clear whether the major axis of the ellipse is horizontal or vertical. Just use the equation to read the ellipse's intercepts. The two that are farthest from the origin are the end points of the major axis, and the other two are the end points of the minor axis. The two foci lie on the major axis, each at distance c from the ellipse's center—which will be at the origin if the ellipse's equation has the form of either of Equations (3) or (5).

EXAMPLE 2   Sketch the graph of the equation

$$\frac{x^2}{16} + \frac{y^2}{25} = 1.$$

Solution   The x-intercepts are $(\pm 4, 0)$; the y-intercepts are $(0, \pm 5)$. So the major axis is vertical. We take $a = 5$ and $b = 4$ in Equation (4) and find that $c = 3$. The foci are thus at $(0, \pm 3)$. Hence this ellipse has the appearance of the one shown in Figure 9.23.

Any equation of the form

$$Ax^2 + Cy^2 + Dx + Ey + F = 0, \tag{6}$$

in which the coefficients A and C of the squared terms are *both nonzero* and have the *same sign*, may be reduced to the form

$$A(x - h)^2 + C(y - k)^2 = G$$

by completing the square in x and y. If $G < 0$, there are no points satisfying (6), and the graph is the empty set. If $G = 0$, there is exactly one point on the locus—the single point (h, k). And if $G > 0$, we can divide both sides of the last equation by G, and get an equation that resembles one of the two below:

$$\frac{(x - h)^2}{a^2} + \frac{(y - k)^2}{b^2} = 1, \tag{7a}$$

$$\frac{(x - h)^2}{b^2} + \frac{(y - k)^2}{a^2} = 1. \tag{7b}$$

Which equation to choose? The one consistent with the condition $a \geq b > 0$. Finally note that either of Equations (7) is the equation of a translated ellipse. Thus, apart from the exceptional cases noted, the graph of Equation (6) is an ellipse if $AC > 0$.

EXAMPLE 3 Determine the graph of the equation

$$3x^2 + 5y^2 - 12x + 30y + 42 = 0.$$

Solution We collect terms containing x, terms containing y, and complete the square in each variable. This gives

$$3(x^2 - 4x) + 5(y^2 + 6y) = -42,$$

$$3(x^2 - 4x + 4) + 5(y^2 + 6y + 9) = 15,$$

$$\frac{(x - 2)^2}{5} + \frac{(y + 3)^2}{3} = 1.$$

Thus the given equation is that of a translated ellipse. The center is at $(2, -3)$, the ellipse has horizontal major semiaxis of length $a = \sqrt{5}$, and the minor semiaxis has length $b = \sqrt{3}$. The distance from the center to each focus is $c = \sqrt{2}$, and the eccentricity is $e = c/a = \sqrt{\frac{2}{5}}$. This ellipse appears in Figure 9.24.

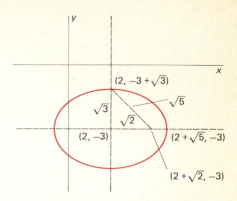

9.24 The ellipse of Example 3.

EXAMPLE 4 The orbit of the earth is an ellipse with the sun at one focus. The planet's maximum distance from the sun is 94.56 million miles and its minimum distance is 91.45 million miles. What are the major and minor semiaxes of the earth's orbit, and what is its eccentricity?

Solution As Figure 9.25 shows, we have

$$a + c = 94.56 \quad \text{and} \quad a - c = 91.45$$

with units in millions of miles. From these equations we conclude that $a = 93.00$, that $c = 1.56$, and then that

$$b = \sqrt{(93.00)^2 - (1.56)^2} \approx 92.99$$

million miles. Finally,

$$e = \frac{1.56}{93.00} \approx 0.017,$$

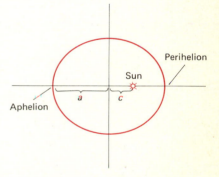

9.25 The orbit of the earth, with exaggerated eccentricity.

a number rather close to zero; this means that the earth's orbit is nearly circular. Indeed, the major and minor semiaxes are so nearly equal that, on any usual scale, the earth's orbit would appear to be a perfect circle. But the difference between uniform circular motion and the earth's actual motion has some important aspects, including the fact that the sun is 1.56 million miles off center, and—as we shall see in Chapter 10—the earth's orbital speed is not constant.

EXAMPLE 5 One of the most famous of all comets is Halley's comet, named for Edmund Halley (1656–1742, a disciple of Newton). By studying the records of the paths of earlier comets, Halley deduced that the comet of 1682 was the same one that had been sighted in 1607, in 1531, in 1456, and in 1066 (an omen at the Battle of Hastings). In 1682, he predicted that the comet would return in 1759, in 1835, and in 1910; he was correct each time. The period of Halley's comet is about 76 years; it can vary a couple of years in either direction because of perturbations of its orbit by Jupiter. The orbit of the comet is an ellipse with the sun at one focus. In terms of astronomical units (1 A.U. is the earth's mean distance from the sun), the major and minor semiaxes of this elliptical orbit are 18.09 A.U. and 4.56 A.U., respectively. What are the maximum and minimum distances from the sun of Halley's comet?

Solution   We are given that $a = 18.09$ (A.U.) and that $b = 4.56$ (A.U.), so

$$c = \sqrt{(18.09)^2 - (4.56)^2} \approx 17.51 \text{ A.U.}$$

Hence its maximum distance from the sun is $a + c \approx 35.60$ A.U., and its minimum distance is $a - c \approx 0.58$ A.U. The eccentricity of its orbit is

$$e = \frac{17.51}{18.09} \approx 0.97,$$

a very eccentric orbit.

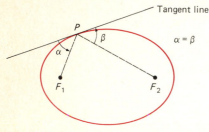

9.26   The reflection property: $\alpha = \beta$.

The *reflection property* of the ellipse is the fact that at a point P of an ellipse, the tangent line makes equal angles with the lines PF_1 and PF_2 from P to the two foci of the ellipse (see Figure 9.26). This property is the basis of the "whispering gallery" phenomenon, which can be observed, for instance, in the U.S. Senate whispering gallery. Suppose that the ceiling of a large room is shaped like half an ellipsoid of the sort obtained by revolving an ellipse about its major axis. Sound waves, like light waves, are reflected at equal angles of incidence and reflection. Thus, if two diplomats are standing in quiet conversation near one focus of the ellipsoid, a reporter standing near the other focus would be able to eavesdrop on their conversation even though standing fifty feet away, and even if the diplomats' conversation were inaudible to others in the same room.

An alternative definition of the ellipse with foci F_1 and F_2 and major axis of length $2a$ may be given as follows: It is the locus of a point P such that the sum of the distances $|PF_1|$ and $|PF_2|$ is the constant $2a$ (Problem 26). This fact gives a convenient way of drawing the ellipse, making use of two tacks placed at F_1 and F_2 and a string of length $2a$. This construction is illustrated in Figure 9.27.

9.27

9-3   PROBLEMS

In each of Problems 1–15, find an equation of the ellipse specified.

1 Vertices $(\pm 4, 0)$ and $(0, \pm 5)$.

2 Foci $(\pm 5, 0)$ and major semiaxis 13.

3 Foci $(0, \pm 8)$ and major semiaxis 17.

4 Center $(0, 0)$, vertical major axis 12, minor axis 8.

5 Foci $(\pm 3, 0)$, eccentricity 0.75.

6 Foci $(0, \pm 4)$, eccentricity $\frac{2}{3}$.

7 Center $(0, 0)$, horizontal major axis 20, eccentricity $\frac{1}{2}$.

8 Center $(0, 0)$, horizontal minor axis 10, eccentricity $\frac{1}{2}$.

9 Foci $(\pm 2, 0)$, directrices $x = \pm 8$.

10 Foci $(0, \pm 4)$, directrices $y = \pm 9$.

11 Center $(2, 3)$, horizontal axis 8, vertical axis 4.

12 Center $(1, -2)$, horizontal major axis 8, eccentricity $\frac{3}{4}$.

13 Foci $(-2, 1)$ and $(4, 1)$, major axis 10.

14 Foci $(-3, 0)$ and $(-3, 4)$, minor axis 6.

15 Foci $(-2, 2)$ and $(4, 2)$, eccentricity $\frac{1}{3}$.

Sketch the graphs of the equations in Problems 16–20. Indicate centers, foci, and lengths of axes.

16 $4x^2 + y^2 = 16$.

17 $4x^2 + 9y^2 = 144$.

18 $4x^2 + 9y^2 = 24x$.

19 $9x^2 + 4y^2 - 32y + 28 = 0$.

20 $2x^2 + 3y^2 + 12x - 24y + 60 = 0$.

21 The orbit of the comet Kahoutek is an ellipse of extreme eccentricity $e = 0.999925$; the sun is at one focus of this ellipse. The minimum distance between Kahoutek and the sun is 0.13 A.U. What is the maximum distance between Kahoutek and the sun?

22 The orbit of the planet Mercury is an ellipse with eccentricity $e = 0.206$. Its maximum and minimum distances from the sun are 0.467 and 0.307 A.U., respectively. What are the major and minor semiaxes of Mercury's orbit? Would you describe this orbit as "nearly circular"?

23 Derive Equation (5) for an ellipse if its foci lie on the y-axis.

24 Show that the tangent line to the ellipse $x^2/a^2 + y^2/b^2 = 1$ at the point $P(x_0, y_0)$ of that ellipse has equation

$$\frac{xx_0}{a^2} + \frac{yy_0}{b^2} = 1.$$

25 Use the result of Problem 24 to establish the reflection property of the ellipse. (*Suggestion:* Let m be the slope of the normal line to the ellipse at $P(x_0, y_0)$, and m_1 and m_2 be the slopes of the lines PF_1 and PF_2, respectively. Show that

$$\frac{m - m_1}{1 + m_1 m} = \frac{m_2 - m}{1 + m_2 m},$$

and use the identity for $\tan(A - B)$.)

26 Given $F_1(-c, 0)$ and $F_2(c, 0)$ and $a > c > 0$, show that the ellipse $x^2/a^2 + y^2/b^2 = 1$ (with $b^2 = a^2 - c^2$) is the locus of a point P such that $|PF_1| + |PF_2| = 2a$.

27 Find the equation of the ellipse with horizontal and vertical axes that passes through the points $(-1, 0)$, $(3, 0)$, $(0, 2)$ and $(0, -2)$.

28 Derive an equation for the ellipse with foci $(3, -3)$, $(-3, 3)$, and major axis of length 10. Note that the foci of this ellipse lie on neither a vertical line nor a horizontal line.

9-4

The Hyperbola

A hyperbola is defined in the same way as is an ellipse, except that the eccentricity e of a hyperbola is greater than 1.

> **Definition**   *The Hyperbola*
>
> Suppose that $e > 1$, and let F be a fixed point and L a fixed line not containing F. Then the **hyperbola** with **eccentricity** e, **focus** F, and **directrix** L is the set of all points P such that the distance $|PF|$ is e times the (perpendicular) distance from P to the line L.

As with the ellipse, the equation of a hyperbola is simplest if F is the point $(c, 0)$ on the x-axis and L is the vertical line $x = c/e^2$. The case $c > 0$ is shown in Figure 9.28. If Q is the point $(c/e^2, y)$, then PQ is the perpendicular from $P(x, y)$ to L. The condition $|PF| = e|PQ|$ gives

$$(x - c)^2 + y^2 = e^2 \left(x - \frac{c}{e^2} \right)^2,$$

$$x^2 - 2cx + c^2 + y^2 = e^2 x^2 - 2cx + \frac{c^2}{e^2},$$

$$(e^2 - 1)x^2 - y^2 = c^2 \left(1 - \frac{1}{e^2} \right) = \frac{c^2}{e^2}(e^2 - 1).$$

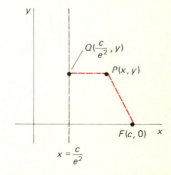

9.28 The definition of the hyperbola.

Thus

$$(e^2 - 1)x^2 - y^2 = a^2(e^2 - 1)$$

where

$$a = \frac{c}{e}. \tag{1}$$

Division of both sides of the next-to-last equation above by $a^2(e^2 - 1)$ gives

$$\frac{x^2}{a^2} - \frac{y^2}{a^2(e^2 - 1)} = 1.$$

To simplify this equation, we let

$$b^2 = a^2(e^2 - 1) = c^2 - a^2. \qquad (2)$$

This is permissible because $e > 1$. So the equation of the hyperbola with focus $(c, 0)$ and directrix $x = c/e^2 = a/e$ takes the form

$$\frac{x^2}{a^2} - \frac{y^2}{b^2} = 1. \qquad (3)$$

The minus sign on the left is the only difference between the equation of a hyperbola and that of an ellipse. Of course, Equation (2) differs from the relation $a^2 = b^2 + c^2$ for the case of the ellipse.

The hyperbola of Equation (3) is clearly symmetric about both coordinate axes, and has x-intercepts $(\pm a, 0)$. But it has no y-intercepts! If we rewrite Equation (3) in the form

$$y = \pm \frac{b}{a}\sqrt{x^2 - a^2}, \qquad (4)$$

then we see that there are points on the graph only if $|x| \geq a$. Hence the hyperbola has two **branches,** as shown in Figure 9.29. We also see from (4) that $|y| \to \infty$ as $|x| \to \infty$.

The x-intercepts $V_1(-a, 0)$ and $V_2(a, 0)$ are the **vertices** of the hyperbola, and the line segment joining them is its **transverse axis.** The line segment joining $W_1(0, -b)$ and $W_2(0, b)$ is its **conjugate axis.** The alternative form

$$c^2 = a^2 + b^2 \qquad (5)$$

of Equation (2) is the Pythagorean relation for the right triangle of Figure 9.29.

The lines $y = \pm bx/a$ that pass through the **center** $(0, 0)$ and the opposite vertices of the rectangle in Figure 9.30 are **asymptotes** of the two branches of the hyperbola in both directions. That is, if

$$y_1 = \frac{bx}{a} \quad \text{and} \quad y_2 = \frac{b}{a}\sqrt{x^2 - a^2},$$

then

$$\lim_{x \to \infty}(y_1 - y_2) = 0 = \lim_{x \to -\infty}(y_1 - (-y_2)). \qquad (6)$$

To verify the first limit, note that

$$\lim_{x \to +\infty} \frac{b}{a}(x - \sqrt{x^2 - a^2}) = \lim_{x \to +\infty} \frac{b}{a} \cdot \frac{(x - \sqrt{x^2 - a^2})(x + \sqrt{x^2 - a^2})}{x + \sqrt{x^2 - a^2}}$$

$$= \lim_{x \to +\infty} \frac{b}{a} \cdot \frac{a^2}{x + \sqrt{x^2 - a^2}} = 0.$$

Just as in the case of the ellipse, the hyperbola with focus $(c, 0)$ and directrix $x = a/e$ also has focus $(-c, 0)$ and directrix $x = -a/e$, as shown both in Figure 9.30 and in Figure 9.31. Since $c = ae$ by (1), the foci $(\pm ae, 0)$ and the directrices $x = \pm a/e$ take the same forms in terms of a and e for both the hyperbola $(e > 1)$ and the ellipse $(e < 1)$.

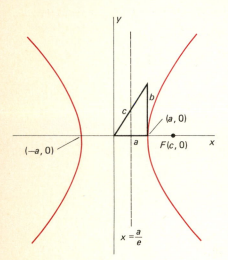

9.29 A hyperbola has two branches.

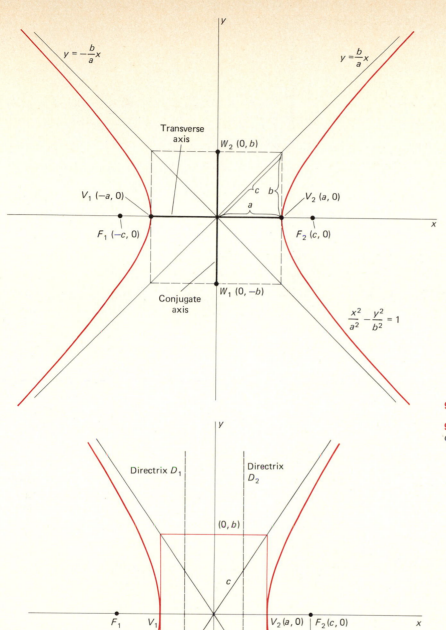

$y = -\dfrac{b}{a}x$

$y = \dfrac{b}{a}x$

Transverse axis

$W_2\ (0, b)$

$V_1\ (-a, 0)$

$V_2\ (a, 0)$

c b

a

$F_1\ (-c, 0)$

$F_2\ (c, 0)$

$W_1\ (0, -b)$

Conjugate axis

$\dfrac{x^2}{a^2} - \dfrac{y^2}{b^2} = 1$

9.30 The parts of a hyperbola.

9.31 Relations between the parts of a hyperbola.

Directrix D_1

Directrix D_2

$(0, b)$

c

F_1 V_1

$V_2\ (a, 0)$ $F_2\ (c, 0)$

$\dfrac{a}{e}$

a

ae

$y = \dfrac{b}{a}x$

$y = -\dfrac{b}{a}x$

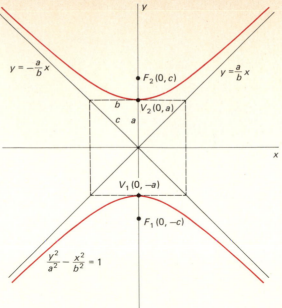

$$\frac{y^2}{a^2} - \frac{x^2}{b^2} = 1$$

9.32 The hyperbola of Equation (7) has horizontal directrices.

If we interchange x and y in Equation (3), we obtain the equation

$$\frac{y^2}{a^2} - \frac{x^2}{b^2} = 1. \tag{7}$$

This hyperbola has foci at $(0, \pm c)$. They, and its transverse axis, lie on the y-axis. Its asymptotes are $y = \pm ax/b$, and its graph is like the one in Figure 9.32.

When we studied the ellipse, we saw that its orientation—whether the major axis is horizontal or vertical—is determined by the relative sizes of a and b. In the case of the hyperbola, the situation is different, for the relative sizes of a and b make no such difference. The direction in which the hyperbola opens—horizontal as in Figure 9.31, or vertical as in Figure 9.32—is determined by the *signs* of the x^2 and y^2 terms.

9.33 The hyperbola of Example 1.

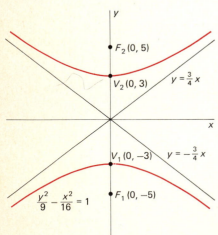

$$\frac{y^2}{9} - \frac{x^2}{16} = 1$$

EXAMPLE 1 Sketch the graph of the hyperbola with equation

$$\frac{y^2}{9} - \frac{x^2}{16} = 1.$$

Solution This is an equation of the form of (7), so the hyperbola opens vertically. Since $a = 3$ and $b = 4$, we find that $c = 5$ by using Equation (2): $b^2 = c^2 - a^2$. Thus the vertices are $(0, \pm 3)$, the foci are the two points $(0, \pm 5)$, and the asymptotes are the two lines $y = \pm 3x/4$. This hyperbola appears in Figure 9.33.

EXAMPLE 2 Find the equation of the hyperbola with foci $(\pm 10, 0)$ and asymptotes $y = \pm 4x/3$.

Solution Since $c = 10$, we have

$$a^2 + b^2 = 100 \quad \text{and} \quad \frac{b}{a} = \frac{4}{3}.$$

Thus $b = 8$ and $a = 6$, and the equation of the hyperbola is

$$\frac{x^2}{36} - \frac{y^2}{64} = 1.$$

As we noted in Section 9-3, any equation of the form

$$Ax^2 + Cy^2 + Dx + Ey + F = 0 \qquad (8)$$

with A and C nonzero can be reduced to the form

$$A(x - h)^2 + C(y - k)^2 = G$$

by completing the square in x and y. Now suppose that the coefficients A and C of the quadratic terms have *opposite signs*; for example, suppose that $A = p^2$ and $C = -q^2$. The equation above becomes

$$p^2(x - h)^2 - q^2(y - k)^2 = G. \qquad (9)$$

If $G = 0$, then factorization of the difference of squares on the left yields the equations

$$p(x - h) + q(y - k) = 0,$$
$$p(x - h) - q(y - k) = 0$$

of two straight lines through (h, k) with slopes $\pm p/q$. If $G \neq 0$, then division of Equation (9) by G gives an equation that looks either like

$$\frac{(x - h)^2}{a^2} - \frac{(y - k)^2}{b^2} = 1 \qquad \text{(if } G > 0\text{)}$$

or like

$$\frac{(y - k)^2}{a^2} - \frac{(x - h)^2}{b^2} = 1 \qquad \text{(if } G < 0\text{)}.$$

Thus if $AC < 0$ in Equation (8), the graph is either a pair of intersecting straight lines or a hyperbola.

EXAMPLE 3 Determine the graph of the equation

$$9x^2 - 4y^2 - 36x + 8y = 4.$$

Solution We collect the terms containing x, those containing y, and complete the square in each variable. We find that

$$9(x - 2)^2 - 4(y - 1)^2 = 36,$$

so

$$\frac{(x - 2)^2}{4} - \frac{(y - 1)^2}{9} = 1.$$

Hence the graph is a hyperbola with horizontal transverse axis and center $(2, 1)$. Since $a = 2$ and $b = 3$, $c = \sqrt{13}$; the vertices of the hyperbola are $(0, 1)$ and $(4, 1)$ and its foci are the two points $(2 \pm \sqrt{13}, 1)$. Its asymptotes are the two lines

$$y - 1 = \pm\tfrac{3}{2}(x - 2),$$

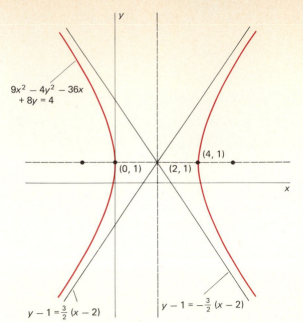

$$9x^2 - 4y^2 - 36x + 8y = 4$$

$(4, 1)$

$(0, 1)$ $(2, 1)$

$y - 1 = \frac{3}{2}(x - 2)$ $y - 1 = -\frac{3}{2}(x - 2)$

9.34 The hyperbola of Example 3, a translate of the hyperbola $x^2/4 - y^2/9 = 1$.

translates of the asymptotes $y = \pm \frac{3}{2}x$ of the hyperbola $x^2/4 - y^2/9 = 1$. Figure 9.34 shows the graph of the translated hyperbola.

The *reflection property* of the hyperbola takes the same form as for the ellipse. If P is a point on a hyperbola, then the two lines PF_1 and PF_2 from P to the two foci make equal angles with the tangent line at P. In Figure 9.35 this means that $\alpha = \beta$.

For an application of this reflection property, consider a mirror shaped like one branch of a hyperbola, reflectorized on the outside. Then an incoming light ray aimed toward one focus will be reflected toward the other

9.35 Reflection property of the hyperbola: $\alpha = \beta$.

9.36 How a hyperbolic mirror reflects a ray aimed at one focus; note that $\alpha = \beta$ again.

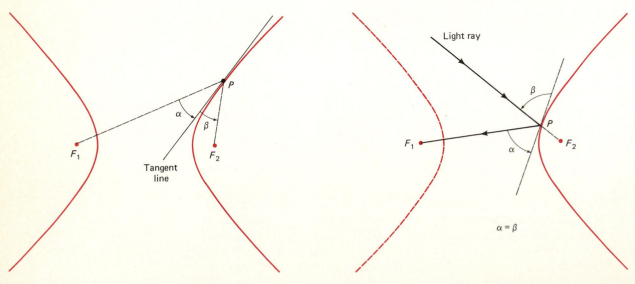

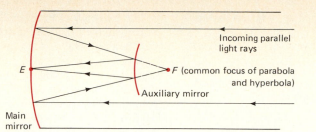

9.37 One sort of reflecting telescope: Main mirror parabolic, auxiliary mirror hyperbolic.

focus, as shown in Figure 9.36. Figure 9.37 indicates the design of a reflecting telescope that makes use of the reflection properties of the parabola and hyperbola. The parallel incoming light rays first are reflected by the parabola toward its focus at F. Then they are intercepted by an auxiliary hyperbolic mirror with foci at E and F and reflected into the eyepiece located at E.

The following example illustrates how hyperbolas are sometimes used to determine the position of ships at sea.

EXAMPLE 4   A ship lies at sea due east of point A on a long north-south coastline. Simultaneous signals are transmitted by a radio station at A and by one at B on the coast 200 miles south of A. The ship receives the signal from A 500 microseconds before it receives the one from B. Assume that the speed of radio signals is 980 feet per microsecond. How far out at sea is the ship?

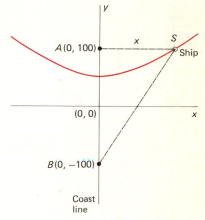

Solution   The situation is diagrammed in Figure 9.38.
The difference between the distances of the ship S from A and B is

$$|SB| - |SA| = (500)(980)$$

feet, or 92.8 miles. Thus (by Problem 24) the ship lies on a hyperbola with foci A and B, and with $c = 100$, we have

$$a = \tfrac{1}{2}(92.8) = 46.4 \quad \text{and} \quad b = \sqrt{(100)^2 - (46.4)^2} \approx 88.6.$$

In the coordinate system of Figure 9.38, the hyperbola has equation

$$\frac{y^2}{(46.4)^2} - \frac{x^2}{(88.6)^2} = 1.$$

9.38 Solving a navigation problem (Example 4).

We substitute $y = 100$, because the ship is due east of A. We find that the ship's distance from the coastline is $x \approx 169.1$ miles.

9-4   PROBLEMS

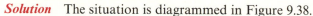

In each of Problems 1–14, find an equation of the hyperbola described.

1 Foci $(\pm 4, 0)$ and vertices $(\pm 1, 0)$.

2 Foci $(0, \pm 3)$ and vertices $(0, \pm 2)$.

3 Foci $(\pm 5, 0)$ and asymptotes $y = \dfrac{\pm 3x}{4}$.

4 Vertices $(\pm 3, 0)$ and asymptotes $y = \dfrac{\pm 3x}{4}$.

5 Vertices $(0, \pm 5)$ and asymptotes $y = \pm x$.
6 Vertices $(\pm 3, 0)$ and eccentricity $e = \tfrac{5}{3}$.
7 Foci $(0, \pm 6)$ and eccentricity $e = 2$.
8 Vertices $(\pm 4, 0)$ and passing through $(8, 3)$.
9 Foci $(\pm 4, 0)$ and directrices $x = \pm 1$.
10 Foci $(0, \pm 9)$ and directrices $y = \pm 4$.
11 Center $(2, 2)$, horizontal transverse axis of length 6, and eccentricity 2.
12 Center $(-1, 3)$, vertices at $(-4, 3)$ and $(2, 3)$, and foci at $(-6, 3)$ and $(4, 3)$.

13 Center $(1, -2)$, vertices $(1, 1)$ and $(1, -5)$, and asymptotes $3x - 2y - 7 = 0$ and $3x + 2y + 1 = 0$.

14 Focus $(8, -1)$ and asymptotes $3x - 4y - 13 = 0$ and $3x + 4y - 5 = 0$.

Sketch the graph of the equation given in each of Problems 15–20; indicate centers, foci, and asymptotes.

15 $x^2 - y^2 - 2x + 4y = 4$.

16 $x^2 - 2y^2 + 4x = 0$.

17 $y^2 - 3x^2 - 6y = 0$.

18 $x^2 - y^2 - 2x + 6y = 9$.

19 $9x^2 - 4y^2 + 18x + 8y = 31$.

20 $4y^2 - 9x^2 - 18x - 8y = 41$.

21 Show that the graph of the equation

$$\frac{x^2}{15 - c} - \frac{y^2}{c - 6} = 1$$

is:

(a) a hyperbola with foci $(\pm 3, 0)$ if $6 < c < 15$;

(b) an ellipse if $c < 6$.

Identify the graph in the case $c > 15$.

22 Establish that the tangent line to the hyperbola $x^2/a^2 - y^2/b^2 = 1$ at the point $P(x_0, y_0)$ has equation

$$\frac{xx_0}{a^2} - \frac{yy_0}{b^2} = 1.$$

23 Use the result of Problem 22 to establish the reflection property of the hyperbola. (See the suggestion for Problem 25 in Section 9-3.)

24 Suppose that $0 < a < c$, and let $b = (c^2 - a^2)^{1/2}$. Show that the hyperbola $x^2/a^2 - y^2/b^2 = 1$ is the locus of a point P such that the *difference* between the distances $|PF_1|$ and $|PF_2|$ is equal to $2a$, where F_1 and F_2 are the two foci of the hyperbola.

25 Derive an equation for the hyperbola with foci $(\pm 5, \pm 5)$ and vertices $(\pm 3/\sqrt{2}, \pm 3/\sqrt{2})$. Use the difference definition of a hyperbola that is implied by Problem 24.

26 Two radio signal stations at A and B lie on an east-west line with A 100 miles west of B. A plane is flying west on a line 50 miles north of the line AB. Radio signals are sent (traveling at 980 feet per microsecond) simultaneously from A and B, and the one sent from B arrives 400 microseconds before the one from A. Where is the plane?

27 Two radio signal stations are located as in Problem 26 and transmit radio signals that travel at the same speed as in that problem. In this problem, however, it is known only that the plane is generally somewhere north of the line AB, that the signal sent from B arrives 400 microseconds before the one sent from A, and that the signal sent from A and reflected by the plane takes a total of 600 microseconds to reach B. Where is the plane?

9-5

Rotation of Axes and Second-Degree Curves

In the previous three sections we have studied the second-degree equation

$$Ax^2 + Cy^2 + Dx + Ey + F = 0, \tag{1}$$

which contains no xy term. We found that its graph is always a conic section, apart from exceptional cases of the following types:

$$2x^2 + 3y^2 = -1 \qquad \text{(no locus)},$$

$$2x^2 + 3y^2 = 0 \qquad \text{(one point)},$$

$$(2x - 1)^2 = 0 \qquad \text{(a line)},$$

$$(2x - 1)^2 = 1 \qquad \text{(two parallel lines)},$$

$$x^2 - y^2 = 0 \qquad \text{(two intersecting lines)}.$$

We may therefore say that the graph of Equation (1) is a conic section, possibly **degenerate.** If either A or C is zero (but not both), then the graph is a parabola. It is an ellipse if $AC > 0$ (by the discussion of Equation (6) in Section 9-3), a hyperbola if $AC < 0$ (by the discussion of Equation (8) in Section 9-4).

Let us assume that $AC \neq 0$. Then we can determine the particular conic section represented by Equation (1) by completing squares; that is, we write (1) in the form

$$A(x - h)^2 + B(y - k)^2 = G. \tag{2}$$

This equation can be simplified further by a **translation of coordinates** to a new $\bar{x}\bar{y}$-coordinate system centered at the point (h, k) in the old xy-system. The geometry of this change of coordinates is shown in Figure 9.39. The relation between the old and new coordinates is

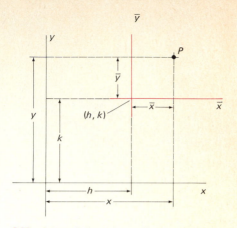

$$\left.\begin{array}{r} \bar{x} = x - h, \\ \bar{y} = y - k \end{array}\right\} \quad \text{or} \quad \begin{cases} x = \bar{x} + h, \\ y = \bar{y} + k. \end{cases} \tag{3}$$

In the new $\bar{x}\bar{y}$-coordinate system, Equation (2) takes the simpler form

$$A\bar{x}^2 + C\bar{y}^2 = G, \tag{2'}$$

from which it is clear whether we have an ellipse, a hyperbola, or a degenerate case.

9.39 A translation of coordinates.

We now turn to the general second-degree equation

$$Ax^2 + Bxy + Cy^2 + Dx + Ey + F = 0. \tag{4}$$

Note the presence of the "cross-product," or xy-, term. In order to recognize its graph, we need to change to a new $x'y'$-coordinate system obtained by a **rotation of axes.**

We get the $x'y'$-axes from the xy-axes by a rotation through an angle α in the counterclockwise direction. In the notation of Figure 9.40, we have

$$x = OQ = OP \cos(\phi + \alpha) \quad \text{and} \quad y = PQ = OP \sin(\phi + \alpha). \tag{5}$$

Similarly,

$$x' = OR = OP \cos \phi \quad \text{and} \quad y' = PR = OP \sin \phi. \tag{6}$$

Recall the addition formulas

$$\cos(\phi + \alpha) = \cos \phi \cos \alpha - \sin \phi \sin \alpha,$$

$$\sin(\phi + \alpha) = \sin \phi \cos \alpha + \cos \phi \sin \alpha.$$

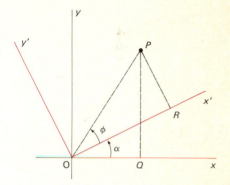

9.40 A rotation of coordinates through the angle α.

With the aid of these identities and the substitution of Equations (6) into Equations (5), we obtain this result:

Equations for Rotation of Axes

$$x = x' \cos \alpha - y' \sin \alpha,$$

$$y = x' \sin \alpha + y' \cos \alpha. \tag{7}$$

These equations express the old xy-coordinates of the point P in terms of its new $x'y'$-coordinates and the rotation angle α. The following example illustrates how Equations (7) may be used to transform the equation of a curve from xy-coordinates into the rotated $x'y'$-coordinates.

EXAMPLE 1 The xy-axes are rotated through an angle of $\alpha = 45°$. Find the equation of the curve $2xy = 1$ in the new coordinates x', y'.

Solution Since $\cos 45° = \sin 45° = 1/\sqrt{2}$, Equations (7) yield

$$x = \frac{x' - y'}{\sqrt{2}} \quad \text{and} \quad y = \frac{x' + y'}{\sqrt{2}}.$$

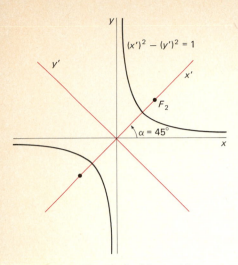

$(x')^2 - (y')^2 = 1$

y'

x'

F_2

$\alpha = 45°$

9.41 The graph of $2xy = 1$ is a hyperbola.

The original equation $2xy = 1$ thus becomes

$$(x')^2 - (y')^2 = 1.$$

So, in the $x'y'$-coordinate system, we have a hyperbola with $a = b = 1$, $c = \sqrt{2}$, and foci $(\pm\sqrt{2}, 0)$. In the original xy-coordinates, its foci are $(1, 1)$ and $(-1, -1)$, and its asymptotes are the x- and y-axes. This hyperbola is shown in Figure 9.41. A hyperbola of this form, one which has equation $xy = k$, is called a **rectangular** hyperbola (because its asymptotes are perpendicular).

Example 1 suggests that the cross-product term Bxy of Equation (4) may disappear upon rotation of the coordinate axes. One can, indeed, always choose an appropriate angle α of rotation so that, in the new coordinate system, there is no xy-term.

To determine the appropriate rotation angle, we substitute Equations (7) for x and y into the general second-degree equation in (4). We obtain the following new second-degree equation:

$$A'(x')^2 + B'x'y' + C'(y')^2 + D'x' + E'y' + F' = 0. \tag{8}$$

The new coefficients are given in terms of the old ones and the angle α by the following equations:

$$A' = A\cos^2\alpha + B\cos\alpha\sin\alpha + C\sin^2\alpha,$$

$$B' = B(\cos^2\alpha - \sin^2\alpha) + 2(C - A)\sin\alpha\cos\alpha,$$

$$C' = A\sin^2\alpha - B\sin\alpha\cos\alpha + C\cos^2\alpha,$$

$$D' = D\cos\alpha + E\sin\alpha, \tag{9}$$

$$E' = -D\sin\alpha + E\cos\alpha, \qquad\qquad \text{and}$$

$$F' = F.$$

Now suppose that an equation of the form of (4) is given, with $B \neq 0$. We simply choose α so that $B' = 0$ in the above list of new coefficients. Then Equation (8) will have no cross-product term, and we can identify and sketch the curve with little trouble in the $x'y'$-coordinate system. But is it really easy to choose such an angle α?

It is. We recall that

$$\cos 2\alpha = \cos^2\alpha - \sin^2\alpha \quad \text{and} \quad \sin 2\alpha = 2\sin\alpha\cos\alpha.$$

So the above equation for B' may be written

$$B' = B\cos 2\alpha + (C - A)\sin 2\alpha.$$

9.42 Finding $\sin\alpha$ and $\cos\alpha$, given $\cot 2\alpha = \dfrac{A - C}{B}$.

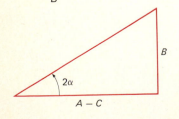

B

2α

$A - C$

We can cause B' to be zero by choosing α as that (unique) acute angle such that

$$\cot 2\alpha = \frac{A - C}{B}. \tag{10}$$

If we plan to use Equations (9) to calculate the coefficients in the transformed Equation (8), we shall need the values of $\sin\alpha$ and $\cos\alpha$ that follow from Equation (10). It is sometimes convenient to calculate these values directly from $\cot 2\alpha$, as follows. From the triangle in Figure 9.42, we can

440

read the numerical value of $\cos 2\alpha$. Since the cosine and cotangent are both positive in the first quadrant and both negative in the second quadrant, we give $\cos 2\alpha$ the same sign as $\cot 2\alpha$. Then we use the half-angle formulas to get $\sin \alpha$ and $\cos \alpha$:

$$\sin \alpha = \left(\frac{1 - \cos 2\alpha}{2}\right)^{1/2}, \qquad \cos \alpha = \left(\frac{1 + \cos 2\alpha}{2}\right)^{1/2}. \qquad (11)$$

Once we have the values of $\sin \alpha$ and $\cos \alpha$, we can compute the coefficients in the resulting Equation (8) by means of Equations (9). Alternatively, it's frequently simpler to get Equation (8) directly by substituting Equations (9), with the numerical values of $\sin \alpha$ and $\cos \alpha$ obtained as above, into Equation (4).

EXAMPLE 2 Determine the graph of the equation

$$73x^2 - 72xy + 52y^2 - 30x - 40y - 75 = 0.$$

Solution We begin with Equation (10), and find that $\cot 2\alpha = -\frac{7}{24}$, so that $\cos 2\alpha = -\frac{7}{25}$. Thus

$$\sin \alpha = \left(\frac{1 - (-\frac{7}{25})}{2}\right)^{1/2} = \frac{4}{5}, \qquad \cos \alpha = \left(\frac{1 + (-\frac{7}{25})}{2}\right)^{1/2} = \frac{3}{5}.$$

Then with $A = 73$, $B = -72$, $C = 52$, $D = -30$, $E = -40$, and $F = -75$, Equations (9) yield

$$A' = 25, \qquad\qquad\qquad D' = -50,$$
$$B' = 0 \text{ (this was the point)}, \qquad E' = 0,$$
$$C' = 100, \qquad\qquad\qquad F' = -75.$$

Consequently the equation in the new $x'y'$-coordinate system, obtained by rotation through an angle of $\alpha = \arcsin(\frac{4}{5}) \approx 53.13°$, is

$$25(x')^2 + 100(y')^2 - 50x' = 75.$$

Alternatively, we could have obtained this equation by substituting

$$x = \tfrac{3}{5}x' - \tfrac{4}{5}y', \qquad y = \tfrac{4}{5}x' + \tfrac{3}{5}y'$$

into the original equation.

By completing the square in x' we finally obtain

$$25(x' - 1)^2 + 100(y')^2 = 100,$$

which we put into the standard form

$$\frac{(x' - 1)^2}{4} + \frac{(y')^2}{1} = 1.$$

Thus the original curve is an ellipse with major semiaxis 2, minor semiaxis 1, and center $(1, 0)$ in the $x'y'$-coordinate system, as shown in Figure 9.43.

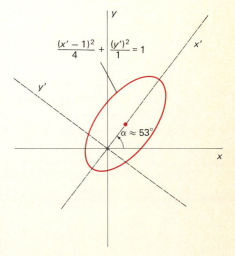

9.43 The ellipse of Example 2.

Example 2 illustrates the general procedure for finding the graph of a second-degree equation. First, if there is a cross-product (xy) term, rotate axes to eliminate it. Then translate axes (if necessary) to reduce the equation to the standard form of a parabola, ellipse, or hyperbola (or a degenerate case of one of these).

There *is* a test by which the nature of the curve may be discovered without actually carrying out the transformations described. This test stems from the fact (see Problem 14) that, whatever the angle α of rotation, Equations (9) imply that

$$(B')^2 - 4A'C' = B^2 - 4AC. \qquad (12)$$

Thus the **discriminant** $B^2 - 4AC$ is an *invariant* under any rotation of axes. If α is chosen so that $B' = 0$, then the left-hand side of Equation (12) is simply $-4A'C'$. Since A' and C' are the coefficients of the squared terms, our earlier discussion of Equation (1) now applies. It follows that the graph will be

(a) a *parabola* if $B^2 - 4AC = 0$,
(b) an *ellipse* if $B^2 - 4AC < 0$, and
(c) a *hyperbola* if $B^2 - 4AC > 0$.

Of course, degenerate cases may occur.

Here are some examples:

(a) $x^2 + 2xy + y^2 = 1$ is a (degenerate) parabola,
(b) $x^2 + xy + y^2 = 1$ is an ellipse, and
(c) $x^2 + 3xy + y^2 = 1$ is a hyperbola.

9-5 PROBLEMS

In each of Problems 1–12, first use the discriminant to classify the graph of the given equation. Then eliminate the cross-product term by an appropriate rotation of the coordinate axes. Finally, translate coordinates (if necessary) and sketch the curve in the original xy-coordinate system.

1 $x^2 + 2xy + y^2 = 2.$ **2** $x^2 + xy + y^2 = 3.$
3 $x^2 + 3xy + y^2 = 5.$ **4** $5x^2 - 6xy + 5y^2 = 8.$
5 $x^2 + 4xy - 2y^2 = 18.$
6 $x^2 - 2xy + y^2 = 2x + 2y.$
7 $x^2 - xy + y^2 - 2\sqrt{2}x + \sqrt{2}y = 0.$
8 $41x^2 - 24xy + 34y^2 + 20x - 140y + 125 = 0.$
9 $23x^2 - 72xy + 2y^2 + 140x + 20y = 75.$
10 $9x^2 + 24xy + 16y^2 - 170x - 60y + 245 = 0.$
11 $161x^2 + 480xy - 161y^2 - 510x - 272y = 0.$
12 $144x^2 - 120xy + 25y^2 - 65x - 156y = 169.$
13 Solve Equations (7) to show that the rotated coordinates (x', y') are given in terms of the original coordinates by

$$x' = x \cos \alpha + y \sin \alpha, \qquad y' = -x \sin \alpha + y \cos \alpha.$$

14 Use Equations (9) to verify Equation (12).
15 Show that the sum $A + C$ of the coefficients of x^2 and y^2 in Equation (4) is invariant under rotation. That is, show that $A' + C' = A + C$ for any rotation through an angle α.
16 Use Equations (9) to show that any rotation of axes

transforms the equation $x^2 + y^2 = r^2$ into the equation $(x')^2 + (y')^2 = r^2$.

17 Consider the equation

$$x^2 + Bxy - y^2 + Dx + Ey + F = 0.$$

Show that there is a rotation of axes such that $A' = 0 = C'$ in the resulting equation. (*Suggestion:* Find the angle α for which $A' = 0$. Then apply the result of Problem 15.) What can you conclude about the graph of the given equation?

18 Suppose that $B^2 - 4AC < 0$, so that the equation

$$Ax^2 + Bxy + Cy^2 = 1$$

represents an ellipse. Show that its area is

$$\pi ab = \frac{2\pi}{\sqrt{4AC - B^2}}$$

where a and b are the lengths of its semiaxes.
19 Show that the equation $27x^2 + 37xy + 17y^2 = 1$ represents an ellipse, and then find the points of the ellipse that are closest to and farthest from the origin.
20 Show that the equation $x^2 + 14xy + 49y^2 = 100$ represents a parabola, and then find the point of the parabola that is closest to the origin.

Polar Coordinates

A familiar way to locate a point in the plane is by specifying its rectangular coordinates (x, y)—that is, by giving its abscissa x and ordinate y relative to given perpendicular axes. In some problems it is more convenient to locate a point by means of its **polar coordinates.** The polar coordinates of a point give its position relative to a fixed reference point O (the **pole**) and a given ray beginning at O (the **polar axis**).

For convenience, we begin with a given xy-coordinate system, then take the origin as the pole and the nonnegative x-axis as the polar axis. Given the pole O and the polar axis, the point P with **polar coordinates** r and θ, written as the ordered pair (r, θ), is located as follows. First find the terminal side of the angle θ, given in radians, where θ is measured counterclockwise (if $\theta > 0$) from the positive x-axis (or polar axis) as its initial side. If $r \geqq 0$, then P is on this terminal side at the distance r from the origin. If $r < 0$, then P lies on the ray opposite the terminal side at the distance $|r| = -r$ from the pole. The **radial** coordinate r can be described as the *directed* distance of P from the pole along the terminal side of the angle θ. Thus, if r is positive, the point P lies in the same quadrant as θ, while if r is negative, then P lies in the opposite quadrant. If $r = 0$, the angle θ does not matter; the polar coordinates $(0, \theta)$ represent the origin whatever the **angular** coordinate θ might be. Of course, the origin or pole is the only point for which $r = 0$. The cases $r > 0$ and $r < 0$ are illustrated separately in Figure 9.44.

Polar coordinates differ from rectangular coordinates in that any point has more than one representation in polar coordinates. For example, the polar coordinates (r, θ) and $(-r, \theta + \pi)$ represent the same point P, as shown in Figure 9.45. More generally, this same point P has the polar coordinates $(r, \theta + n\pi)$ for any even integer n, as well as $(-r, \theta + n\pi)$ for any odd integer n. Thus the polar coordinate pairs

$$(2, \pi/3), \qquad (-2, 4\pi/3), \qquad (2, 7\pi/3), \quad \text{and} \quad (-2, -2\pi/3)$$

all represent the point P of Figure 9.46. The rectangular coordinates of P are $(1, \sqrt{3})$.

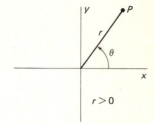

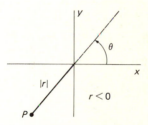

9.44 The difference between the two cases $r > 0$ and $r < 0$.

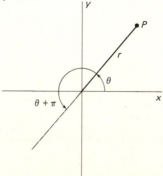

9.45 The polar coordinates (r, θ) and $(-r, \theta + \pi)$ represent the same point P.

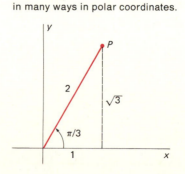

9.46 The point P can be described in many ways in polar coordinates.

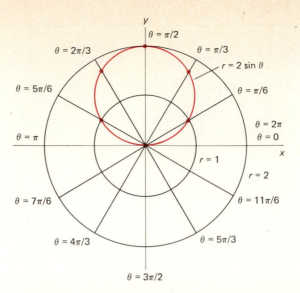

9.48 The graph of the polar equation $r = 2 \sin \theta$.

θ	r
0	0.00
$\pi/6$	1.00
$\pi/3$	1.73
$\pi/2$	2.00
$2\pi/3$	1.73
$5\pi/6$	1.00
π	0.00
$7\pi/6$	−1.00
$4\pi/3$	−1.73
$3\pi/2$	−2.00
$5\pi/3$	−1.73
$11\pi/6$	−1.00
2π	0.00
	(data rounded)

9.47 Values of $r = 2 \sin \theta$.

Some curves have simpler equations in polar coordinates than in rectangular coordinates; this is an important reason for the usefulness of polar coordinates. The **graph** of an equation in the polar coordinate variables r and θ is the set of all those points P such that P has a pair of polar coordinates (r, θ) satisfying the given equation. The graph of an equation $r = f(\theta)$ can be constructed by computing a table of values of r, then plotting the corresponding points (r, θ) on polar coordinate graph paper.

EXAMPLE 1 Construct the graph of the equation $r = 2 \sin \theta$.

Solution Figure 9.47 shows a table of values of r as a function of θ. The corresponding points are plotted in Figure 9.48, using the rays at multiples of $\pi/6$ and the circles (centered at the pole) of radii 1 and 2 to locate these points. A visual inspection of the smooth curve connecting these points suggests that it is a circle of radius 1. Let us assume for the moment that this is so. Note then that the point $P(r, \theta)$ moves *once around this circle counterclockwise* as θ increases from 0 to π, then moves around the circle *a second time* as θ increases from π to 2π. This is because the negative values of r for θ between π and 2π give (in this example) the same geometric points as the positive values of r found when θ is between 0 and π.

The verification that the graph of $r = 2 \sin \theta$ is the indicated circle illustrates the general procedure for transferring back and forth between polar and rectangular coordinates. For changing from polar to rectangular coordinates we use the basic relations

$$x = r \cos \theta, \qquad y = r \sin \theta \qquad (1)$$

that we read from the right triangle of Figure 9.49. In the opposite direction we have

$$r^2 = x^2 + y^2, \qquad \theta = \tan^{-1}\left(\frac{y}{x}\right) \text{ if } x \neq 0. \qquad (2)$$

For example, the first of Equations (2) transforms the rectangular equation $x^2 + y^2 = a^2$ of a circle into $r^2 = a^2$, or simply $r = a$. To trans-

9.49 Read Equations (1) and (2)—conversions between polar and rectangular coordinates—from this figure.

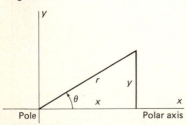

CHAP. 9: Conic Sections and Polar Coordinates

form the equation $r = 2 \sin \theta$ of Example 1 into rectangular coordinates, we first multiply both sides by r to get

$$r^2 = 2r \sin \theta.$$

Equations (1) and (2) now give

$$x^2 + y^2 = 2y.$$

Finally, after we complete the square, we have

$$x^2 + (y - 1)^2 = 1,$$

the rectangular coordinates equation of a circle with center $(0, 1)$ and radius 1. More generally, the graphs of the equations

$$r = 2a \sin \theta \quad \text{and} \quad r = 2a \cos \theta \qquad (3)$$

are circles of radius a centered, respectively, at the points $(0, a)$ and $(a, 0)$.

Substitution of Equations (1) transforms the rectangular equation $ax + by = c$ of a straight line into

$$ar \cos \theta + br \sin \theta = c.$$

Let us take $a = 1$ and $b = 0$. Then we see that the polar coordinates equation of the vertical line $x = c$ is $r = c \sec \theta$, as we can deduce directly from Figure 9.50.

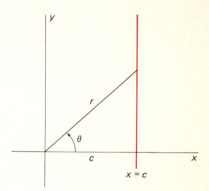

9.50 Finding the polar equation of the vertical line $x = c$.

EXAMPLE 2 Sketch the graph of the equation

$$r = 2 + 2 \sin \theta.$$

Solution If we scan the second column of the table in Figure 9.47, mentally adding 2 to each entry for r, we see that

- r increases from 2 to 4 as θ increases from 0 to $\pi/2$;
- r decreases from 4 to 2 as θ increases from $\pi/2$ to π;
- r decreases from 2 to 0 as θ increases from π to $3\pi/2$; and
- r increases from 0 to 2 as θ increases from $3\pi/2$ to 2π.

This information tells us that the graph resembles the curve shown in Figure 9.51. This heart-shaped graph is called a **cardioid.** The graphs of the equations

$$r = a(1 \pm \sin \theta) \quad \text{and} \quad r = a(1 \pm \cos \theta)$$

are all cardioids, differing only in size (determined by a), axis of symmetry (whether horizontal or vertical), and the direction in which the cusp at the pole points.

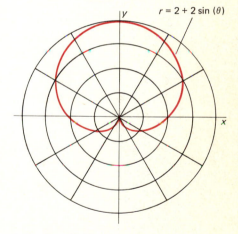

$r = 2 + 2 \sin (\theta)$

9.51 A cardioid.

EXAMPLE 3 Sketch the graph of the equation $r = 2 \cos 2\theta$.

Solution Rather than constructing a table of values of r as a function of θ and then plotting points, let us reason qualitatively. Note first that $r = 0$ if θ is an odd multiple of $\pi/4$. Also $|r| = 2$ if θ is an even multiple of $\pi/4$. We begin by plotting these points.

Then think of what happens as θ increases from one odd multiple of $\pi/4$ to the next. For instance, if θ goes from $\pi/4$ to $3\pi/4$, then 2θ increases from $\pi/2$ to $3\pi/2$. So $r = 2 \cos 2\theta$ decreases from 0 to -2, then increases to 0. The graph for such values of θ forms a loop, starting at the origin

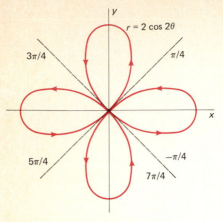

r = 2 cos 2θ

3π/4   π/4

x

5π/4   −π/4

7π/4

9.52 A four-leaved rose.

and lying between the rays $\theta = 5\pi/4$ and $\theta = 7\pi/4$. Something similar happens between any two consecutive odd multiples of $\pi/4$, so the entire graph consists of four loops (Figure 9.52). The arrows in the figure indicate the direction in which the point $P(r, \theta)$ moves along the curve as θ increases.

The curve of Example 3 is called a *four-leaved rose*. The equations $r = a \cos n\theta$ and $r = a \sin n\theta$ represent "roses" with $2n$ "leaves" or loops if n is even and $n \geq 2$, and with n loops if n is odd and $n \geq 3$.

The four-leaved rose exhibits several types of symmetry. The following are some *sufficient* conditions for symmetry in polar coordinates.

- *For symmetry about the x-axis:* The equation is unaltered when θ is replaced by $-\theta$.
- *For symmetry about the y-axis:* The equation is unaltered when θ is replaced by $\pi - \theta$.
- *For symmetry with respect to the origin:* The equation is unchanged when r is replaced by $-r$.

Since $\cos 2\theta = \cos(-2\theta) = \cos 2(\pi - \theta)$, the equation $r = 2 \cos 2\theta$ of the four-leaved rose satisfies the first two conditions above, and therefore its graph is symmetric about both the x-axis and the y-axis. Thus it is also symmetric about the origin. Nevertheless, this equation does *not* satisfy the third condition above, the condition for symmetry about the origin. This illustrates that while the conditions above are *sufficient* for the symmetries described, they are not *necessary* conditions.

Example 3 illustrates a peculiarity of graphs of polar equations, one that is caused by the fact that a single point has multiple representations in polar coordinates. For the point with polar coordinates $(2, \pi/2)$ clearly lies on the four-leaved rose, but these coordinates do *not* satisfy the equation $r = 2 \cos 2\theta$. This means that a point may have one pair of polar coordinates that satisfy a given equation and others that do not. Hence we must be careful to understand this: The graph of a polar equation consists of all those points having *at least one* polar coordinate representation satisfying the given equation.

9.53 Simultaneous solution of the two polar equations yields only *one* of the two obvious intersection points.

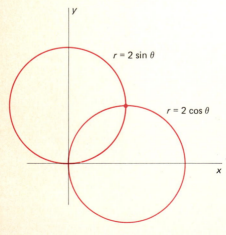

r = 2 sin θ

r = 2 cos θ

x

Another result of the multiplicity of polar coordinates is that the simultaneous solution of two polar equations does not always give all the points of intersection of their graphs. For instance, consider the circles $r = 2 \sin \theta$ and $r = 2 \cos \theta$ shown in Figure 9.53. The origin is clearly a point of intersection of these two circles. Its polar representation $(0, \pi)$ satisfies the equation $r = 2 \sin \theta$, and its representation $(0, \pi/2)$ satisfies the other equation $r = 2 \cos \theta$. But the origin has no *single* polar representation that satisfies both equations simultaneously! If we think of θ as increasing uniformly with time, then the moving points on the two circles pass through the origin at different times. Hence the origin cannot be discovered as a point of intersection of the two circles by solving their equations algebraically.

As a consequence of the phenomenon illustrated by the above example, the only way we can be certain of finding *all* points of intersection of two curves in polar coordinates is by graphing both curves.

EXAMPLE 4 Find all points of intersection of the graphs of the equations $r = 1 + \sin \theta$ and $r^2 = 4 \sin \theta$.

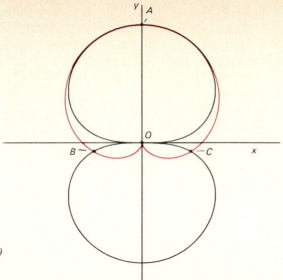

9.54 The cardioid $r = 1 + \sin\theta$ and the figure eight $r^2 = 4\sin\theta$ meet in four places.

Solution The graph of $r = 1 + \sin\theta$ is a scaled-down version of the cardioid of Example 2. In Problem 32 we ask you to show that the graph of $r^2 = 4\sin\theta$ is the figure-eight curve shown with the cardioid in Figure 9.54. The figure shows four points of intersection, called O, A, B, and C. Can we find all four points with algebra?

Given the two equations, we begin by eliminating r. Since

$$(1 + \sin\theta)^2 = r^2 = 4\sin\theta,$$

it follows that

$$\sin^2\theta - 2\sin\theta + 1 = 0,$$

$$(\sin\theta - 1)^2 = 0,$$

and thus that $\sin\theta = 1$. So θ must be an angle of the form $\pi/2 \pm 2n\pi$ where n is an integer. All points on the cardioid and all points on the figure-eight curve are produced by letting θ increase from 0 to 2π, so $\theta = \pi/2$ will produce all the solutions that we can possibly get by this method. The only such point is $A(2, \pi/2)$, and the other three points are not detected except by graphing the two equations.

9-6 PROBLEMS

1 Plot the points with the following polar coordinates and then find the rectangular coordinates of each.

(a) $(1, \pi/4)$ (b) $(-2, 2\pi/3)$ (c) $(1, -\pi/3)$

(d) $(3, 3\pi/2)$ (e) $(2, 9\pi/4)$ (f) $(-2, -7\pi/6)$

(g) $(2, 5\pi/6)$

2 Find two polar coordinate representations, one with $r > 0$ and the other with $r < 0$, for the points with these rectangular coordinates.

(a) $(-1, -1)$ (b) $(\sqrt{3}, -1)$ (c) $(2, 2)$

(d) $(-1, \sqrt{3})$ (e) $(\sqrt{2}, -\sqrt{2})$ (f) $(-3, \sqrt{3})$

In each of Problems 3–10, express the given rectangular equation in polar form.

3 $x = 4$.   **4** $y = 6$.

5 $x = 3y$.   **6** $x^2 + y^2 = 25$.

7 $xy = 1$.   **8** $x^2 - y^2 = 1$.

9 $y = x^2$.   **10** $x + y = 4$.

In each of Problems 11–18, express the given polar equation in rectangular coordinates.

11 $r = 3$.   **12** $\theta = 3\pi/4$.

13 $r = -5\cos\theta$.   **14** $r = \sin 2\theta$.

15 $r = 1 - \cos 2\theta$.

16 $r = 2 + \sin \theta$.

31 $r = 2 \sin 5\theta$ (five-leaved rose).

17 $r = 3 \sec \theta$.

18 $r^2 = \cos 2\theta$.

32 $r^2 = 4 \sin \theta$ (figure eight).

Sketch the graphs of the polar equations in Problems 19–32 and indicate any symmetry about either axis or the origin.

In Problems 33–38, find all points of intersection of the given curves.

19 $r = 2 \cos \theta$.

20 $r = 2 \sin \theta + 2 \cos \theta$ (circle).

21 $r = 1 + \cos \theta$ (cardioid).

22 $r = 1 - \sin \theta$ (cardioid).

23 $r = 2 + 4 \cos \theta$ (limaçon).

24 $r = 4 + 2 \cos \theta$ (limaçon).

25 $r^2 = 4 \sin 2\theta$ (lemniscate).

26 $r^2 = 4 \cos 2\theta$ (lemniscate).

27 $r = 2 \sin 2\theta$ (four-leaved rose).

28 $r = 3 \sin 3\theta$ (three-leaved rose).

29 $r = 3 \cos 3\theta$ (three-leaved rose).

30 $r = 3\theta$ (spiral of Archimedes).

33 $r = 2$ and $r = \cos \theta$.

34 $r = \sin \theta$ and $r^2 = 3 \cos^2 \theta$.

35 $r = \sin \theta$ and $r = \cos 2\theta$.

36 $r = 1 + \cos \theta$ and $r = 1 - \sin \theta$.

37 $r = 1 - \cos \theta$ and $r^2 = 4 \cos \theta$.

38 $r^2 = 4 \sin \theta$ and $r^2 = 4 \cos \theta$.

39 (a) The straight line L passes through the point with polar coordinates (p, α) and is perpendicular to the line segment joining the pole and the point (p, α). Write the polar coordinates equation of L.

(b) Show that the rectangular coordinates equation of L is $x \cos \alpha + y \sin \alpha = p$.

---

*9-7

Conic Sections in Polar Coordinates

In order to investigate orbits of satellites, such as planets or comets orbiting the sun or natural or artificial moons orbiting a planet, we need the equations of the conic sections in polar coordinates. Recall the focus-directrix definition of a conic section with eccentricity e. It is the locus of a point P whose distance from the focus is e times its distance from the directrix. This locus is an ellipse if $e < 1$, a parabola if $e = 1$, and a hyperbola if $e > 1$. We shall find that all three conic sections have the same general equation in polar coordinates.

To derive the polar equation of a conic section, suppose that its focus is the pole (origin) and that its directrix is the vertical line $x = -p$ (with $p > 0$). In the notation of Figure 9.55, the fact that $|OP| = e|PQ|$ tells us that

$$r = e(p + r \cos \theta),$$

so that

$$r = \frac{pe}{1 - e \cos \theta}.$$

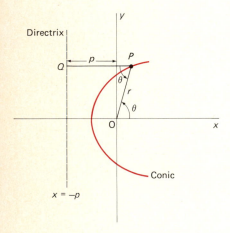

9.55 A conic section: $|OP| = e|PQ|$.

If the directrix is the line $x = +p$, then a similar computation (Problem 22) gives the same result, except that the denominator $1 - e \cos \theta$ is replaced by $1 + e \cos \theta$.

> **Polar Coordinates Equation of a Conic Section**
>
> The polar equation of the conic section with eccentricity e, focus O, and directrix $x = \pm p$ is
>
> $$r = \frac{pe}{1 \pm e \cos \theta}. \qquad (1)$$

Here $p > 0$, and we take either the plus sign in both $x = \pm p$ and Equation (1), or the minus sign in both.

If the directrix of the conic section is the horizontal line $y = \pm p$, then a similar computation (Problem 23) yields the polar coordinates equation

$$r = \frac{pe}{1 \pm e \sin \theta}. \qquad (2)$$

Now consider an ellipse given by Equation (1) for some choice of $e < 1$. Its vertices correspond to $\theta = 0$ and $\theta = \pi$. Hence it follows with the aid of Figure 9.56 that the length $2a$ of its major axis is

$$2a = \frac{pe}{1 - e} + \frac{pe}{1 + e} = \frac{2pe}{1 - e^2}.$$

From this equation we obtain the important relation

$$pe = a(1 - e^2). \qquad (3)$$

EXAMPLE 1   Sketch the graph of the equation

$$r = \frac{16}{5 - 3 \cos \theta}.$$

Solution   First we divide numerator and denominator by 5 and find that

$$r = \frac{\frac{16}{5}}{1 - \frac{3}{5} \cos \theta}.$$

Thus $e = \frac{3}{5}$ and $pe = \frac{16}{5}$. Equation (3) then implies that $a = 5$. Finally, $c = ae = 3$ and

$$b = \sqrt{a^2 - c^2} = 4.$$

So we have here an ellipse with major semiaxis $a = 5$, minor semiaxis $b = 4$, and center at $(3, 0)$ in rectangular coordinates. The ellipse is shown in Figure 9.57.

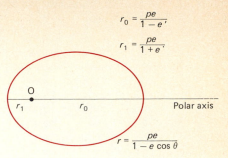

$$r_0 = \frac{pe}{1 - e},$$
$$r_1 = \frac{pe}{1 + e}.$$

$$r = \frac{pe}{1 - e \cos \theta}$$

9.56   Find the major axis of an ellipse given in polar form.

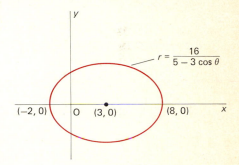

$$r = \frac{16}{5 - 3 \cos \theta}$$

9.57   The ellipse of Example 1.

In contrast with the rectangular coordinates situation discussed in Section 9-5, the polar coordinates equation of a rotated conic section is quite simple. Figure 9.58 shows an ellipse rotated through a counterclockwise angle α about its focus at the origin O. In the rotated system with polar coordinates (r', θ'), the ellipse has equation

$$r' = \frac{pe}{1 - e \cos \theta'}.$$

Since $r' = r$ and $\theta' = \theta - \alpha$, the equation of the ellipse in the original (unrotated) polar coordinate system is

$$r = \frac{pe}{1 - e \cos(\theta - \alpha)}. \qquad (4)$$

With $e \geqq 1$, Equation (4) represents a parabola or a hyperbola rotated through the angle α.

If we combine Equations (1) and (3), we find that the equation of an ellipse with eccentricity e, major semiaxis a, and focus at the origin is

$$r = \frac{a(1 - e^2)}{1 \pm e \cos \theta}. \qquad (5)$$

9.58   An ellipse rotated about one focus through the angle α.

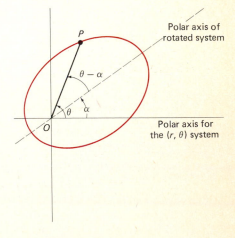

θ (degrees)	r_{par}	r_{ell}
180	0.500	0.500
135	0.586	0.585
90	1.000	0.999
60	2.000	1.998
45	3.414	3.403
30	7.464	7.409
0	∞	999.000
	[data rounded]	

9.59 Comparison between a parabola and an ellipse with $e = 0.999$.

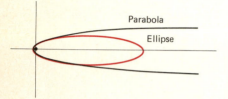

9.60 An eccentric ellipse resembles a parabola near their common focus.

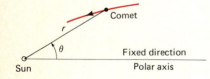

9.61 The comet of Example 2.

The limiting form of this equation as $e \to 0$ is the equation $r = a$ of a circle. Since $p \to \infty$ as $e \to 0$ in Equation (3), we may thus regard any circle as an ellipse of eccentricity zero and with directrix $x = \pm\infty$.

If we begin with Equation (1) and let $e \to 1$, the limiting form of the equation is the equation of a parabola:

$$r = \frac{p}{1 \pm \cos\theta}.$$

Let's take $p = 1$ and compare values of r for the parabola

$$r_{par} = \frac{1}{1 - \cos\theta}$$

and the ellipse

$$r_{ell} = \frac{0.999}{1 - (0.999)\cos\theta}.$$

The ellipse has eccentricity $e = 0.999$, and should be, in some sense, very like a parabola. The table of Figure 9.59 shows the result of our comparison. The parabola and the ellipse almost coincide for $\theta > 30°$. This sort of approximation of an ellipse by a parabola is useful in studying comets with highly eccentric elliptical orbits. The way in which the two orbits would seem to coincide is illustrated in Figure 9.60.

EXAMPLE 2   A certain comet is known to have a highly eccentric elliptical orbit with the sun at one focus. Its position is described in polar coordinates, as shown in Figure 9.61; the following measurements—made by radar—refer to the same figure:

$$r = 6 \text{ A.U.} \quad \text{(astronomical units)} \qquad \text{when} \quad \theta = 60°,$$
$$r = 2 \text{ A.U.} \qquad\qquad\qquad\qquad\qquad \text{when} \quad \theta = 90°.$$

Estimate the position of the comet at its point of closest approach to the sun.

Solution   First we write the equation of the orbit of the comet in the polar form suggested by Figure 9.61. Since the orbit is highly eccentric ($e \approx 1$) we assume that near the sun it is approximately the parabola

$$r = \frac{p}{1 - \cos(\theta - \alpha)}. \tag{6}$$

This parabola has its axis rotated through the angle α from the polar axis.

We know that the vertex of a parabola is its point closest to the focus. Hence the minimum distance of the comet from the sun will be $r = p/2$ when $\theta = \pi + \alpha$. Our problem, then, is to determine p and α.

Substitution of the given data — $r = 6$ when $\theta = \pi/3$ and $r = 2$ when $\theta = \pi/2$ — into Equation (6) gives us the two equations

$$6 = \frac{p}{1 - \cos(\pi/3 - \alpha)} \quad \text{and} \quad 2 = \frac{p}{1 - \cos(\pi/2 - \alpha)}.$$

CHAP. 9:   **Conic Sections and Polar Coordinates**

We divide the first of these equations by the second. This eliminates p, and we obtain

$$3 = \frac{1 - \cos(\pi/2 - \alpha)}{1 - \cos(\pi/3 - \alpha)} = \frac{1 - \sin\alpha}{1 - \frac{1}{2}\cos\alpha - \frac{1}{2}\sqrt{3}\sin\alpha}.$$

We simplify this last equation and find that α is a root of the equation

$$f(\alpha) = (1.598)\sin\alpha + (1.5)\cos\alpha - 2 = 0.$$

We can approximate the value of α by using the iterative process of Newton's method:

$$\alpha_{n+1} = \alpha_n - \frac{f(\alpha_n)}{f'(\alpha_n)}$$

$$= \alpha_n - \frac{(1.598)\sin\alpha_n + (1.5)\cos\alpha_n - 2}{(1.598)\cos\alpha_n - (1.5)\sin\alpha_n}.$$

The data of the problem suggest that α is a first-quadrant angle, so we use $\alpha_1 = 0.5$ (radians) as our first approximation. This leads to the sequence

$$\alpha_2 = 0.38, \qquad \alpha_3 = 0.39, \qquad \alpha_4 = \alpha_5 = 0.40,$$

and so α is about 0.40 radians, or about $23°$. Then

$$p = 6\left[1 - \cos\left(\frac{\pi}{3} - 0.40\right)\right] \approx 1.23$$

astronomical units. The closest approach of the comet to the sun is thus approximately

$$\tfrac{1}{2}(1.23)(93) \approx 57.2$$

million miles, since 1 A.U. is about 93 million miles.

9-7 PROBLEMS

In each of Problems 1–10, identify and sketch the conic section with the given equation.

1 $r = \dfrac{6}{1 + \cos\theta}.$

2 $r = \dfrac{6}{1 + 2\cos\theta}.$

3 $r = \dfrac{3}{1 - \sin\theta}.$

4 $r = \dfrac{8}{8 - 2\cos\theta}.$

5 $r = \dfrac{6}{2 - \sin\theta}.$

6 $r = \dfrac{12}{3 + 2\cos\theta}.$

7 $r = \dfrac{12}{2 - 3\cos\theta}.$

8 $r = \dfrac{12}{2 - 3\cos(\theta - \pi/3)}.$

9 $r = \dfrac{2}{1 - \sin\theta - \cos\theta}.$

10 $r = \dfrac{4}{2 - \sqrt{3}\cos\theta + \sin\theta}.$

In each of Problems 11–15, find a polar equation of the given conic section with focus at the origin.

11 Eccentricity 1, directrix $r = 3\sec\theta$.

12 Eccentricity 2, directrix $r = -4\sec\theta$.

13 Eccentricity $\frac{1}{3}$, directrix $r = 2\csc\theta$.

14 Eccentricity $\frac{2}{3}$, directrix $r = -3\csc\theta$.

15 Eccentricity 1, directrix $r(\cos\theta + \sin\theta) = 3$.

In each of Problems 16–20, find an equation in rectangular coordinates for the graph of the indicated problem above.

16 Problem 11.   **17** Problem 12.

18 Problem 13.   **19** Problem 14.

20 Problem 15.

21 Identify and sketch the graph of the equation

$$r = \frac{4}{2 - \sqrt{2}(\cos\theta + \sin\theta)}.$$

(*Suggestion:* Find A and α so that

$$\cos\theta + \sin\theta = A\cos(\theta - \alpha).$$

Then refer to Equation (4).)

22 Derive Equation (1) with the plus sign.

23 Derive Equation (2) with the minus sign.

24 A comet has a parabolic orbit with the sun at one focus. When the comet is 150 million miles from the sun, the sun-comet line makes an angle of 45° with the axis of the parabola. What will be the minimum distance between the comet and the sun?

25 A satellite has an elliptical orbit with the center of the earth (take its radius to be 4000 miles) as one focus.

The lowest point of its orbit is 500 miles above the North Pole, and the highest point is 5000 miles above the South Pole. What is the height of the satellite above the surface of the earth when the satellite crosses the equator?

26 Find the closest approach to the sun of a comet as in Example 2 of this section; assume that $r = 2.5$ A.U. when $\theta = 45°$ and that $r = 1$ A.U. when $\theta = 90°$.

9-8

Area in Polar Coordinates

The graph of the polar coordinates equation $r = f(\theta)$ may bound an area, as does the cardioid $r = 2(1 + \sin \theta)$ of Figure 9.51 (Section 9-6). To calculate this area, we may find it more convenient to work directly in polar coordinates, rather than change to rectangular coordinates.

To see how to set up an area integral in polar coordinates, we consider the region R of Figure 9.62. This region is bounded by the two radial lines $\theta = \alpha$ and $\theta = \beta$ and by the curve $r = f(\theta)$, $\alpha \leq \theta \leq \beta$. To approximate the area A of R, we begin with a regular partition

$$\alpha = \theta_0 < \theta_1 < \theta_2 < \cdots < \theta_n = \beta$$

of the interval $[\alpha, \beta]$ into n equal subintervals, each having length $\Delta\theta = (\beta - \alpha)/n$, and select a point θ_i^* in the ith subinterval $[\theta_{i-1}, \theta_i]$ for $i = 1, 2, \ldots, n$.

Let ΔA_i denote the area of the sector bounded by the lines $\theta = \theta_{i-1}$ and $\theta = \theta_i$, and the curve $r = f(\theta)$. We see from Figure 9.62 that, for small values of $\Delta\theta$, ΔA_i is approximately equal to the area of the *circular* sector with radius $r_i = f(\theta_i^*)$ and bounded by the same angle. That is,

$$\Delta A_i \approx \tfrac{1}{2}(r_i^*)^2 \, \Delta\theta = \tfrac{1}{2}[f(\theta_i^*)]^2 \, \Delta\theta.$$

We add the areas of these sectors for $i = 1, 2, \ldots, n$ and thereby find that

$$A = \sum_{i=1}^{n} \Delta A_i \approx \sum_{i=1}^{n} \tfrac{1}{2}[f(\theta_i^*)]^2 \, \Delta\theta.$$

9.62 We obtain the area formula from Riemann sums.

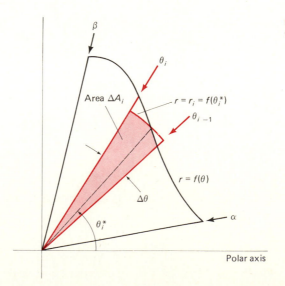

Polar axis

The right-hand sum is a Riemann sum for the integral $\frac{1}{2}\int_\alpha^\beta [f(\theta)]^2\,d\theta$. Hence the value of this integral is, assuming that f is continuous, the limit of the sum above as $\Delta\theta \to 0$. We therefore conclude that *the area A of the region R bounded by the lines $\theta = \alpha$ and $\theta = \beta$ and the curve $r = f(\theta)$ is*

$$A = \frac{1}{2}\int_\alpha^\beta [f(\theta)]^2\,d\theta. \tag{1}$$

The infinitesimal sector shown in Figure 9.63, with radius r, central angle $d\theta$, and area $dA = \frac{1}{2}r^2\,d\theta$, serves as a useful visual device for remembering Formula (1) in the abbreviated form

$$A = \frac{1}{2}\int_\alpha^\beta r^2\,d\theta. \tag{2}$$

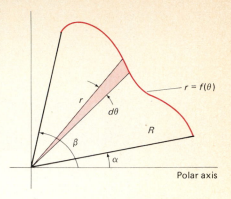

9.63 Heuristic derivation of the area formula in polar coordinates.

EXAMPLE 1 Find the area A bounded by the cardioid $r = 2(1 + \sin\theta)$, $0 \le \theta \le 2\pi$, shown in Figure 9.51.

Solution Formula (2) with $\alpha = 0$, $\beta = 2\pi$ gives

$$A = \frac{1}{2}\int_0^{2\pi} 4(1 + \sin\theta)^2\,d\theta$$

$$= 2\int_0^{2\pi} (1 + 2\sin\theta + \sin^2\theta)\,d\theta$$

$$= 2\int_0^{2\pi} (1 + 2\sin\theta + \frac{1}{2} - \frac{1}{2}\cos 2\theta)\,d\theta$$

$$= 2\left[\frac{3}{2}\theta - 2\cos\theta - \frac{1}{4}\sin 2\theta\right]_0^{2\pi} = 6\pi.$$

EXAMPLE 2 Find the area bounded by each loop of the limaçon $r = 1 + 2\cos\theta$ shown in Figure 9.64.

Solution The equation $1 + 2\cos\theta = 0$ has two solutions for θ between 0 and 2π: $\theta = 2\pi/3$ and $\theta = 4\pi/3$. The upper half of the outer loop of the limaçon corresponds to values of θ between 0 and $2\pi/3$, where r is positive. Since the curve is symmetric about the x-axis, we can find the total area A_1 bounded by the outer loop by integrating from 0 to $2\pi/3$ and then doubling:

$$A_1 = 2\int_0^{2\pi/3} \frac{1}{2}(1 + 2\cos\theta)^2\,d\theta$$

$$= \int_0^{2\pi/3} (1 + 4\cos\theta + 4\cos^2\theta)\,d\theta$$

$$= \int_0^{2\pi/3} (3 + 4\cos\theta + 2\cos 2\theta)\,d\theta$$

$$= \left[3\theta + 4\sin\theta + \sin 2\theta\right]_0^{2\pi/3} = 2\pi + \frac{3}{2}\sqrt{3}.$$

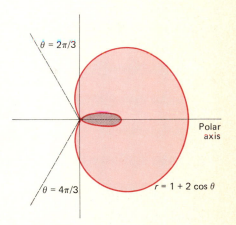

9.64 The limaçon of Example 2.

The inner loop of the limaçon corresponds to values of θ between $2\pi/3$ and $4\pi/3$, where r is negative. Hence the area A_2 bounded by the inner loop is

$$\frac{1}{2}\int_{2\pi/3}^{4\pi/3} (1 + 2\cos\theta)^2\,d\theta = \frac{1}{2}\left[3\theta + 4\sin\theta + \sin 2\theta\right]_{2\pi/3}^{4\pi/3}$$

$$= \pi - \frac{3}{2}\sqrt{3}.$$

The area of the region lying between the two loops of the limaçon is

$$A = A_1 - A_2 = (2\pi + \tfrac{3}{2}\sqrt{3}) - (\pi - \tfrac{3}{2}\sqrt{3}) = \pi + 3\sqrt{3}.$$

Now consider two curves $r = f(\theta)$ and $r = g(\theta)$, with $f(\theta) \geq g(\theta)$ for $\alpha \leq \theta \leq \beta$. Then the area of the region bounded by these two curves and the rays $\theta = \alpha$ and $\theta = \beta$ (shown in Figure 9.65) may be found by subtracting the area bounded by the inner curve from that bounded by the outer curve. That is, the area A between the curves is given by

$$A = \tfrac{1}{2}\int_\alpha^\beta [f(\theta)]^2 \, d\theta - \tfrac{1}{2}\int_\alpha^\beta [g(\theta)]^2 \, d\theta$$

$$= \tfrac{1}{2}\int_\alpha^\beta ([f(\theta)]^2 - [g(\theta)]^2) \, d\theta. \tag{3}$$

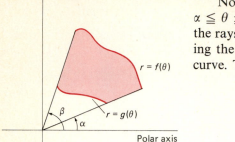

9.65 The area between the graphs of f and g.

With $r_2 = f(\theta)$ for the outer curve and $r_1 = g(\theta)$ for the inner curve, we get the abbreviated formula

$$A = \tfrac{1}{2}\int_\alpha^\beta (r_2^2 - r_1^2) \, d\theta \tag{4}$$

for the area of the region shown in Figure 9.65.

EXAMPLE 3 Find the area A of the region that lies within the limaçon $r = 1 + 2\cos\theta$ and outside the circle $r = 2$.

Solution The circle and the limaçon are shown in Figure 9.66, where the area A between them is shaded. The points of intersection of the circle and the limaçon are given by

$$1 + 2\cos\theta = 2, \quad \text{so} \quad \cos\theta = \tfrac{1}{2},$$

and the figure shows that we should choose the solutions $\theta = \pm\pi/3$. These two values of θ are the needed limits of integration, and when we use Formula (3) we find that

$$A = \tfrac{1}{2}\int_{-\pi/3}^{\pi/3} ((1 + 2\cos\theta)^2 - 2^2) \, d\theta$$

$$= \int_0^{\pi/3} (4\cos\theta + 4\cos^2\theta - 3) \, d\theta \qquad \text{(symmetry about the polar axis)}$$

$$= \int_0^{\pi/3} (4\cos\theta + 2\cos 2\theta - 1) \, d\theta$$

$$= \left[4\sin\theta + \sin 2\theta - \theta \right]_0^{\pi/3} = \frac{5}{2}\sqrt{3} - \frac{\pi}{3}.$$

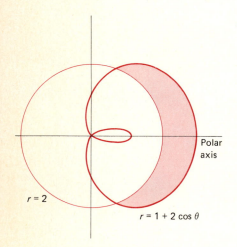

9.66 The region of Example 3.

9-8 PROBLEMS

In each of Problems 1–6, find the area bounded by the given curve.

1 $r = 2\cos\theta$.
2 $r = 4\sin\theta$.
3 $r = 1 + \cos\theta$.
4 $r = 2 - 2\sin\theta$.
5 $r = 2 - \cos\theta$.
6 $r = 3 + 2\sin\theta$.

In each of Problems 7–12, find the area bounded by one loop of the given curve.

7 $r = 2\cos 2\theta$.
8 $r = 3\sin 3\theta$.
9 $r = 2\cos 4\theta$.
10 $r = \sin 5\theta$.
11 $r^2 = 4\sin 2\theta$.
12 $r^2 = 4\cos 2\theta$.

CHAP. 9: Conic Sections and Polar Coordinates

In each of Problems 13–20, find the area of the region described.

13 Inside $r = 2 \sin \theta$ and outside $r = 1$.
14 Inside both $r = 4 \cos \theta$ and $r = 2$.
15 Inside both $r = \cos \theta$ and $r = \sqrt{3} \sin \theta$.
16 Inside $r = 2 + \cos \theta$ and outside $r = 2$.
17 Inside $r = 3 + 2 \sin \theta$ and outside $r = 4$.
18 Inside $r^2 = 2 \cos 2\theta$ and outside $r = 1$.
19 Inside both $r^2 = \cos 2\theta$ and $r^2 = \sin 2\theta$.
20 Inside the large loop and outside the small loop of $r = 1 - 2 \sin \theta$.
21 Find the area of the circle $r = \sin \theta + \cos \theta$ by integration in polar coordinates. Check the answer by identifying the given circle in rectangular coordinates, then using the familiar area formula.
22 Find the area of the region that lies interior to all three circles

$$r = 1, \qquad r = 2 \cos \theta, \quad \text{and} \quad r = 2 \sin \theta.$$

23 The *spiral of Archimedes*, shown in Figure 9.67, has the simple defining equation $r = a\theta$ (a is a constant). Let A_n denote the area bounded by the nth turn of the spiral, for $2\pi(n - 1) \le \theta \le 2\pi n$, and the portion of the polar axis joining its end points. For each $n \ge 2$, let $R_n = A_n - A_{n-1}$ denote the area between the $(n - 1)$st and the nth turns. Then derive the following results of Archimedes:

(a) $A_1 = \frac{1}{3}\pi(2\pi a)^2$,

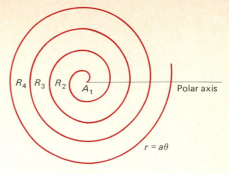

9.67 The spiral of Archimedes.

(b) $A_2 = \frac{7}{12}\pi(4\pi a)^2$,
(c) $R_2 = 6A_1$,
(d) $R_{n+1} = nR_2$ for $n \ge 2$.

24 The orbit of a certain comet approaching the sun is the parabola $r = 1/(1 - \cos \theta)$. The units for r are in astronomical units (recall that 1 A.U. is about 93 million miles). Suppose that it takes 15 days for the comet to travel from the position $\theta = 60°$ to the position $\theta = 90°$. How much longer will it require for the comet to reach its point of closest approach to the sun? Assume that the radius from the sun to the comet sweeps out area at a constant rate as the comet moves.

CHAPTER 9 REVIEW: Properties of Conic Sections

The parabola with focus $(p, 0)$ and directrix $x = -p$ has eccentricity $e = 1$ and equation $y^2 = 4px$. The table on the right compares the properties of an ellipse and a hyperbola, each with foci $(\pm c, 0)$ and major axis of length $2a$.

Use the list below as a guide to additional concepts that you may need to review.

1 Translation of coordinates
2 Equations and procedure for rotation of axes
3 Use of the discriminant to classify the graph of a second degree equation
4 The relationship between rectangular and polar coordinates
5 The graph of a polar coordinates equation
6 The equation of a conic section in polar coordinates
7 The area formula in polar coordinates

	Ellipse	*Hyperbola*
eccentricity	$e = \dfrac{c}{a} < 1$	$e = \dfrac{c}{a} > 1$
a, b, c relation	$a^2 = b^2 + c^2$	$c^2 = a^2 + b^2$
equation	$\dfrac{x^2}{a^2} + \dfrac{y^2}{b^2} = 1$	$\dfrac{x^2}{a^2} - \dfrac{y^2}{b^2} = 1$
vertices	$(\pm a, 0)$	$(\pm a, 0)$
y-intercepts	$(0, \pm b)$	none
directrices	$x = \pm \dfrac{a}{e}$	$x = \pm \dfrac{a}{e}$
asymptotes	none	$y = \pm \dfrac{b}{a} x$

MISCELLANEOUS PROBLEMS

Sketch the graphs of the equations in Problems 1–30, labeling centers, foci, and vertices in the case of conic sections.

1 $x^2 + y^2 - 2x + 2y = 2$.
2 $x^2 + y^2 = x + y$.
3 $x^2 + y^2 - 6x + 2y + 9 = 0$.
4 $y^2 = 4(x + y)$.
5 $x^2 = 8x - 2y - 20$.
6 $x^2 + 2y^2 - 2x + 8y + 8 = 0$.

7 $9x^2 + 4y^2 = 36x.$
8 $x^2 - y^2 = 2x - 2y - 1.$
9 $y^2 - 2x^2 = 4x + 2y + 3.$
10 $9y^2 - 4x^2 = 8x + 18y + 31.$
11 $x^2 + 2y^2 = 4x + 4y - 12.$
12 $x^2 + 2xy + y^2 + 1 = 0.$
13 $x^2 + 2xy - y^2 = 7.$ **14** $xy + 8 = 0.$
15 $3x^2 - 2xy + 3y^2 = 4.$ **16** $x^2 - 6xy + y^2 = 4.$
17 $9x^2 - 24xy + 16y^2 = 20x + 15y.$
18 $7x^2 + 48xy - 7y^2 + 25 = 0.$
19 $r = -2\cos\theta.$ **20** $\cos\theta + \sin\theta = 0.$

21 $r = \dfrac{1}{\sin\theta - \cos\theta}.$ **22** $r\sin^2\theta = \cos\theta.$

23 $r = 3\csc\theta.$ **24** $r = 2(\cos\theta - 1).$
25 $r^2 = 4\cos\theta.$ **26** $r\theta = 1.$

27 $r = 3 - 2\sin\theta.$ **28** $r = \dfrac{1}{1 + \cos\theta}.$

29 $r = \dfrac{4}{2 + \cos\theta}.$ **30** $r = \dfrac{4}{1 - 2\cos\theta}.$

In each of Problems 31–38, find the area of the region described.

31 Inside both $r = 2\sin\theta$ and $r = 2\cos\theta$.
32 Inside $r^2 = 4\cos\theta$.
33 Inside $r = 3 - 2\sin\theta$ and outside $r = 4$.
34 Inside $r^2 = 2\sin 2\theta$ and outside $r = 2\sin\theta$.
35 Inside $r = 2\sin 2\theta$ and outside $r = \sqrt{2}$.
36 Inside $r = 3\cos\theta$ and outside $r = 1 + \cos\theta$.
37 Inside $r = 1 + \cos\theta$ and outside $r = \cos\theta$.
38 Between the two loops of $r = 1 - 2\sin\theta$.
39 Find a polar coordinates equation of the circle that passes through the origin and is centered at the point with polar coordinates (p, α).
40 Find the equation of the parabola with focus the origin and directrix the line $y = x + 4$. Recall from Chapter 2, Miscellaneous Problem 65, that the distance from the point (x, y) to the line $Ax + By + C = 0$ is

$$\frac{|Ax + By + C|}{\sqrt{A^2 + B^2}}.$$

41 A "diameter" of an ellipse is a chord through its center. What are the maximum and minimum lengths of diameters of the ellipse with equation $x^2/a^2 + y^2/b^2 = 1$?
42 Use calculus to prove that the ellipse $x^2/a^2 + y^2/b^2 = 1$ is normal to the coordinate axis at each of its four vertices.
43 The parabolic arch of a bridge has base width b and height h at its center. Write its equation, choosing the origin on the ground at the left end of the arch.
44 Use methods of calculus to find the points on the ellipse $x^2/a^2 + y^2/b^2 = 1$ that are nearest to and farthest from:
(a) The center $(0, 0)$; (b) The focus $(c, 0)$.

45 Consider a line segment QR that contains a point P such that $|QP| = a$ and $|PR| = b$. Suppose that Q is constrained to move on the y-axis, while R must remain on the x-axis. Show that the locus of P is an ellipse.
46 Suppose that $a > 0$ and that F_1 and F_2 are two fixed points in the plane with $|F_1 F_2| > 2a$. Imagine a point P that moves in such a way that $|PF_2| = 2a + |PF_1|$. Show that the locus of P is one branch of a hyperbola with foci F_1 and F_2. Then—as a consequence—explain how to construct points on a hyperbola by drawing appropriate circles centered at its foci.
47 Let Q_1 and Q_2 be two points on the parabola $y^2 = 4px$. Let P be the point of the parabola at which the tangent line is parallel to $Q_1 Q_2$. Show that the horizontal line through P bisects the segment $Q_1 Q_2$.
48 Determine the locus of a point P such that the product of its distances from the two fixed points $F_1(-a, 0)$ and $F_2(a, 0)$ is a^2.
49 Find the eccentricity of the conic section

$$3x^2 - y^2 + 12x + 9 = 0$$

by writing its equation in polar coordinates.
50 Find the area bounded by the loop of the *strophoid* $r = \sec\theta - 2\cos\theta$ shown in Figure 9.68.

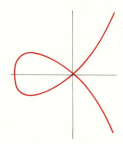

9.68 Strophoid.

51 Find the area bounded by the loop of the *folium of Descartes* $x^3 + y^3 = 3xy$ shown in Figure 2.36. (*Suggestion*: Change to polar coordinates, then substitute $u = \tan\theta$ to evaluate the area integral.)
52 Use the method of Problem 51 to find the area bounded by the first-quadrant loop (similar to the *folium*) of the curve $x^5 + y^5 = 5x^2 y^2$.
53 The asteroid Icarus has a period of revolution of 15 months in an elliptical orbit with the sun at one focus. Its distance from the sun ranges from 17 million to 183 million miles.
(a) Write the equation of its orbit in polar coordinates.
(b) Find how much time during one revolution Icarus spends in the portion of its orbit (nearest the sun) for which $-\pi/2 \le \theta \le \pi/2$. Assume Kepler's law of equal areas swept out in equal times by the sun-Icarus radius. (*Suggestion*: Use the substitution $u = \tan\theta/2$ of Section 8-8 to evaluate the integral $\int (1 + e\cos\theta)^{-2}\, d\theta$.)

Parametric Curves and
Vectors in the Plane

10

Parametric Curves

Until now we have encountered *curves* mainly as graphs of equations. An equation of the form $y = f(x)$ or one of the form $x = g(y)$ determines a curve by giving one of the coordinate variables explicitly as a function of the other. An equation of the form $F(x, y) = 0$ may also determine a curve, but here each variable is given implicitly as a function of the other.

Another important sort of curve is the trajectory of a point moving in the plane. The motion of the point can be described by giving its position $(x(t), y(t))$ at time t. Such a description involves expressing both the rectangular coordinate variables x and y as functions of a third variable, or **parameter,** t, rather than as functions of one another. In this context a *parameter* is an independent variable (not a constant, as is sometimes meant in popular usage). This approach motivates the following definition.

Definition Parametric Curve

A **parametric curve** C in the plane is a pair of functions

$$x = f(t), \qquad y = g(t), \tag{1}$$

that give x and y as continuous functions of the real number t (the parameter) in some interval I.

Each value of the parameter t gives a point $(f(t), g(t))$, and the set of all such points is the **graph** of the curve C. Often the distinction between the curve—the pair of **coordinate functions** f and g—and the graph is not made. Therefore we shall sometimes refer interchangeably to the curve and to its graph. The two equations in (1) are called the **parametric equations** of the curve.

In most cases the interval I will be a closed interval of the form $[a, b]$, and in such cases the two points $P_a(f(a), g(a))$ and $P_b(f(b), g(b))$ are called the **endpoints** of the curve C. If these two points coincide, then we say that the curve C is **closed.** If no distinct pair of values of t, except possibly for the values $t = a$ and $t = b$, give rise to the same point on the graph of the curve, then the curve C is non-self-intersecting, and we say that the curve is **simple.** These concepts are illustrated in Figure 10.1

10.1 Parametric curves may be simple or not, closed or not.

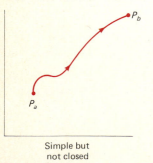

Simple but
not closed

Neither closed
nor simple

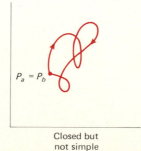

Closed but
not simple

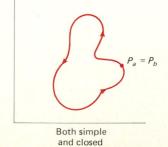

Both simple
and closed

The graph of a parametric curve may be sketched by plotting enough points to indicate its likely shape. Sometimes one can eliminate the parameter t and thus obtain an equation in x and y. This equation may give us more information about the shape of the curve.

EXAMPLE 1   Determine the graph of the curve

$$x = \cos t, \qquad y = \sin t, \qquad 0 \leq t \leq 2\pi.$$

Solution   Figure 10.2 shows a table of values of x and y that correspond to multiples of $\pi/4$ for the parameter t. These values give the eight points $(\pm 1, 0)$, $(0, \pm 1)$, and $(\pm 1/\sqrt{2}, \pm 1/\sqrt{2})$, all of which lie on the unit circle. This suggests that the graph is, in fact, the unit circle. To verify this, we note that the fundamental identity of trigonometry gives

$$x^2 + y^2 = \cos^2 t + \sin^2 t = 1,$$

so every point of the graph lies on the unit circle $x^2 + y^2 = 1$. Conversely, the point of the circle with angular (polar) coordinate t is the point $(\cos t, \sin t)$ of the graph. Thus the graph is precisely the unit circle.

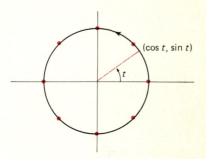

t	x	y
0	1	0
$\pi/4$	$1/\sqrt{2}$	$1/\sqrt{2}$
$\pi/2$	0	1
$3\pi/4$	$-1/\sqrt{2}$	$1/\sqrt{2}$
π	-1	0
$5\pi/4$	$-1/\sqrt{2}$	$-1/\sqrt{2}$
$3\pi/2$	0	-1
$7\pi/4$	$1/\sqrt{2}$	$-1/\sqrt{2}$
2π	1	0

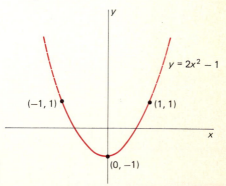

10.2   Table of values for Example 1 and the graph.

What is lost in the above process is the information about how the graph is produced as t goes from 0 to 2π. But this is easy to determine by inspection. As t increases from 0 to 2π, the point $(\cos t, \sin t)$ starts at $(1, 0)$ and travels counterclockwise around the circle, ending at $(1, 0)$ when $t = 2\pi$. The direction of motion is indicated by the arrow in Figure 10.2.

A given figure in the plane may be the graph of different curves. Speaking more loosely, a given curve may have different **parametrizations.** For example, the graph of the curve

$$x = \frac{1 - t^2}{1 + t^2}, \qquad y = \frac{2t}{1 + t^2}, \qquad -\infty < t < \infty$$

also lies on the unit circle, because we find that $x^2 + y^2 = 1$ here, too. If t begins at 0 and increases (decreases), then the point $P(x(t), y(t))$ begins at $(1, 0)$ and travels along the upper (lower) half of the circle. As $t \to \pm\infty$, the point P approaches the point $(-1, 0)$. Thus the graph consists of the unit circle with the single point $(-1, 0)$ deleted. A slight modification of Example 1,

$$x = \cos t, \qquad y = \sin t, \qquad -\pi < t < \pi,$$

is a different parametrization of this graph.

EXAMPLE 2   Eliminate the parameter to determine the graph of the parametric curve

$$x = t - 1, \qquad y = 2t^2 - 4t + 1, \qquad 0 \leq t \leq 2.$$

Solution   We substitute $t = x + 1$ (from the equation for x) into the equation for y. This yields

$$y = 2(x + 1)^2 - 4(x + 1) + 1 = 2x^2 - 1$$

for $-1 \leq x \leq 1$. Thus the graph of the given curve is a portion of the parabola $y = 2x^2 - 1$, as shown in Figure 10.3. As t increases from 0 to 2, the point $(t - 1, 2t^2 - 4t + 1)$ travels along the parabola from $(-1, 1)$ to $(1, 1)$.

10.3   The curve of Example 2 is part of a parabola.

The same parabolic arc can be reparametrized with

$$x = \sin t, \qquad y = 2 \sin^2 t - 1.$$

Now, as t increases, the point $(\sin t, 2 \sin^2 t - 1)$ travels back and forth along the parabola between the two points $(-1, 1)$ and $(1, 1)$, rather like the bob of a pendulum.

Examples 1 and 2 are parametric curves in which we can eliminate the parameter and thus obtain an explicit equation $y = f(x)$. Conversely, any explicit curve $y = f(x)$ can be viewed as a parametric curve by writing

$$x = t, \qquad y = f(t),$$

with t running through the values in the original domain of f. For example, the straight line

$$y - y_0 = m(x - x_0)$$

with slope m and passing through the point (x_0, y_0) is parametrized by

$$x = t, \qquad y = y_0 + m(t - t_0).$$

An even simpler parametrization is obtained with $x - x_0 = t$. This gives

$$x = x_0 + t, \qquad y = y_0 + mt.$$

The use of parametric equations $x = x(t)$ and $y = y(t)$ to describe a curve is most advantageous when elimination of the parameter is either impossible or would lead to an equation $y = f(x)$ considerably more complicated than the original parametric equations. This often happens when the curve is a geometric locus or the path of a point moving under specified conditions.

EXAMPLE 3   The curve traced by a point P on the edge of a rolling circle is called a **cycloid.** The circle is to roll along a straight line without slipping or stopping. You will see a cycloid if you watch a patch of bright paint on the tire of a bicycle crossing your path from left to right. Find parametric equations for the cycloid if the line along which the circle rolls is the x-axis, the circle is above the x-axis but always tangent to it, and the point P begins at the origin.

Solution   Evidently the cycloid consists of a series of arches. As parameter t we take the angle (in radians) through which the circle has turned since it began with P at the origin. This is the angle TCP in Figure 10.4.

The distance the circle has rolled is $|OT|$, so this is also the length of the circumference subtended by the angle TCP. Thus $|OT| = at$ if a is the radius of the circle, and so the center C of the rolling circle has coordinates (at, a) at time t. The right triangle CPQ of Figure 10.4 provides us with the relations

$$at - x = a \sin t \quad \text{and} \quad a - y = a \cos t.$$

Therefore the parametric equations of the cycloid, the path of the moving point P, are

$$x = a(t - \sin t), \qquad y = a(1 - \cos t). \tag{2}$$

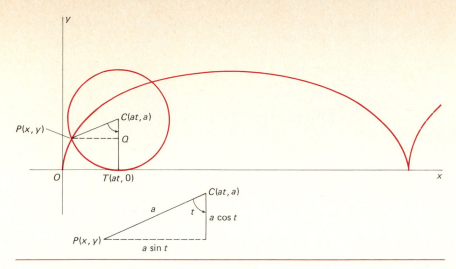

10.4 The cycloid.

In Example 2 of Section 8-8 (Equations (3) and (4)) we obtained the parametrization

$$x = c(\theta - \sin\theta\cos\theta), \qquad y = c\sin^2\theta \qquad (3)$$

of the brachistochrone—the curve of quickest descent. But the substitution $t = 2\theta$, $a = c/2$, together with the identities $\sin 2\theta = 2\sin\theta\cos\theta$ and $\sin^2\theta = \frac{1}{2}(1 - \cos 2\theta)$, transform Equations (3) into Equations (2). In June 1696 John Bernoulli proposed the brachistochrone problem as a public challenge, with a six-month deadline (later extended to Easter 1697 at Leibniz's request). Newton received Bernoulli's challenge on January 29, 1697 (new style), probably after a hard day at the office, and the next day communicated his solution—the curve of minimal descent time is an arc of an inverted cycloid—to the Royal Society of London.

A curve given in polar coordinates by the equation $r = f(\theta)$ can be viewed as a parametric curve with parameter θ. To see this, we recall that the equations $x = r\cos\theta$ and $y = r\sin\theta$ allow us to change from polar to rectangular coordinates. We replace r with $f(\theta)$, and this gives the parametric equations

$$x = f(\theta)\cos\theta, \qquad y = f(\theta)\sin\theta, \qquad (4)$$

expressing x and y in terms of the parameter θ.

For instance, the spiral of Archimedes shown in Figure 10.5 has the polar coordinates equation $r = a\theta$. Equations (4) give the spiral the parametrization

$$x = a\theta\cos\theta, \qquad y = a\theta\sin\theta.$$

10.5 The spiral of Archimedes.

TANGENT LINES TO PARAMETRIC CURVES

The parametric curve $x = f(t)$, $y = g(t)$ is called **smooth** if the derivatives $f'(t)$ and $g'(t)$ are continuous and never simultaneously zero. In some neighborhood of each point of its graph, a smooth parametric curve can be described either in the form $y = F(x)$ or in the form $x = G(y)$. To see why this is so, suppose (for example) that $f'(t) > 0$ on the interval I. Then

$f(t)$ is an increasing function on I, and therefore has an inverse function $t = \phi(x)$ there. Substitution of $t = \phi(x)$ into the equation $y = g(t)$ then gives

$$y = g(\phi(x)) = F(x).$$

We can use the chain rule to compute the slope dy/dx of the tangent line to a smooth parametric curve. Differentiation of $y = F(x)$ with respect to t yields

$$\frac{dy}{dt} = \frac{dy}{dx}\frac{dx}{dt},$$

so

$$\frac{dy}{dx} = \frac{dy/dt}{dx/dt} = \frac{g'(t)}{f'(t)} \tag{5}$$

at any point where $f'(t) \neq 0$. The tangent line is vertical at a point where $f'(t) = 0$ but $g'(t) \neq 0$.

Equation (5) gives $y' = dy/dx$ as a function of t. Another differentiation with respect to t, again using the chain rule, results in the formula

$$\frac{dy'}{dt} = \frac{dy'}{dx}\frac{dx}{dt}.$$

So

$$\frac{d^2y}{dx^2} = \frac{dy'}{dx} = \frac{dy'/dt}{dx/dt}. \tag{6}$$

EXAMPLE 4   Calculate dy/dx and d^2y/dx^2 for the cycloid with the parametric equations in (2).

Solution   We begin with

$$x = a(t - \sin t), \qquad y = a(1 - \cos t). \tag{2}$$

Then Equation (5) gives

$$\frac{dy}{dx} = \frac{dy/dt}{dx/dt} = \frac{a \sin t}{a(1 - \cos t)} = \frac{\sin t}{1 - \cos t}. \tag{7}$$

This derivative is zero when t is an odd multiple of π, so the tangent line is horizontal at the midpoint of each arch of the cycloid. The endpoints of the arches correspond to even multiples of π, where both the numerator and the denominator of (7) vanish. These are isolated points (called *cusps*) at which the cycloid fails to be a smooth curve.

Next, Equation (6) yields

$$\frac{d^2y}{dx^2} = \frac{(\cos t)(1 - \cos t) - (\sin t)(\sin t)}{(1 - \cos t)^2} \div a(1 - \cos t)$$

$$= -\frac{1}{a(1 - \cos t)^2}.$$

Since $d^2y/dx^2 < 0$ for all t (except for the isolated even multiples of π), this shows that each arch of the cycloid is concave downward, as shown in Figure 10.4.

The slope dy/dx can also be computed in terms of polar coordinates. Given a polar coordinates curve $r = f(\theta)$, we use the parametrization

$$x = f(\theta) \cos \theta, \qquad y = f(\theta) \sin \theta$$

of Equations (4). Then Equation (5), with θ in place of t, gives

$$\frac{dy}{dx} = \frac{f'(\theta) \sin \theta + f(\theta) \cos \theta}{f'(\theta) \cos \theta - f(\theta) \sin \theta} \tag{8}$$

or, alternatively,

$$\frac{dy}{dx} = \frac{r' \sin \theta + r \cos \theta}{r' \cos \theta - r \sin \theta} \qquad \left(\text{where } r' = \frac{dr}{d\theta} \right). \tag{9}$$

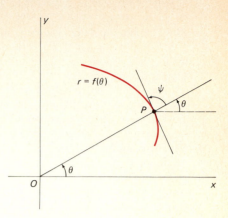

Formula (9) has the following useful consequence. Let ψ denote the angle between the tangent line at P and the radius OP (extended) from the origin, as in Figure 10.6. Then

$$\cot \psi = \frac{1}{r} \frac{dr}{d\theta}. \tag{10}$$

10.6 The interpretation of the angle ψ [see Formula (10)].

In Problem 26 we indicate how Formula (10) can be derived from Formula (9).

EXAMPLE 5 Consider the *logarithmic spiral* with polar equation $r = e^{\theta}$. Show that $\psi = 45°$ at every point of the spiral, and write the equation of its tangent line at the point $(e^{\pi/2}, \pi/2)$.

Solution Since $dr/d\theta = e^{\theta}$, Equation (10) tells us that $\cot \psi = e^{\theta}/e^{\theta} = 1$. Thus $\psi = 45°$. When $\theta = \pi/2$, Equation (9) gives

$$\frac{dy}{dx} = \frac{e^{\pi/2} \sin(\pi/2) + e^{\pi/2} \cos(\pi/2)}{e^{\pi/2} \cos(\pi/2) - e^{\pi/2} \sin(\pi/2)} = -1.$$

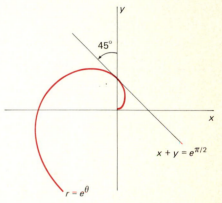

Since $x = 0$ and $y = e^{\pi/2}$ when $\theta = \pi/2$, it follows that the equation of the desired tangent line is

$$y - e^{\pi/2} = -x, \quad \text{or} \quad x + y = e^{\pi/2}.$$

The line and the spiral appear in Figure 10.7.

10.7 The angle ψ is always 45° for the logarithmic spiral.

10-1 PROBLEMS

In Problems 1–12, eliminate the parameter and then sketch the curve.

1 $x = t + 1$, $y = 2t - 1$.
2 $x = t^2 + 1$, $y = 2t^2 - 1$.
3 $x = t^2$, $y = t^3$.
4 $x = \sqrt{t}$, $y = 3t - 2$.
5 $x = t + 1$, $y = 2t^2 - t - 1$.
6 $x = t^2 + 3t$, $y = t - 2$.
7 $x = e^t$, $y = 4e^{2t}$.
8 $x = 2e^t$, $y = 2e^{-t}$.
9 $x = 5 \cos t$, $y = 3 \sin t$.
10 $x = 2 \cosh t$, $y = 3 \sinh t$.
11 $x = \sec t$, $y = \tan t$.
12 $x = \cos 2t$, $y = \sin t$.

In Problems 13–17, write the equation of the tangent line to the given parametric curve at the point corresponding to the given value of t.

13 $x = 2t^2 + 1$, $y = 3t^3 + 2$; $t = 1$.
14 $x = \cos^3 t$, $y = \sin^3 t$; $t = \pi/4$.
15 $x = t \sin t$, $y = t \cos t$; $t = \pi/2$.
16 $x = e^t$, $y = e^{-t}$; $t = 0$.

17 $x = \dfrac{3t}{1 + t^3}$, $y = \dfrac{3t^2}{1 + t^3}$; $t = 1$.

In each of Problems 18–21, find the angle ψ between the radius OP and the tangent line at the point P that corresponds to the given value of θ.

18 $r = 1/\theta$, $\theta = 1$.
19 $r = e^{\theta \sqrt{3}}$, $\theta = \pi/2$.

---

20 $r = 1 - \cos\theta$, $\theta = \pi/3$. **21** $r = \sin 3\theta$, $\theta = \pi/6$.

22 Eliminate t to determine the graph of the curve $x = t^2$, $y = t^3 - 3t$, $-2 \le t \le 2$. Note that $t = \sqrt{x}$ if $t > 0$, while $t = -\sqrt{x}$ if $t < 0$. Your sketch should contain a loop and should be symmetric about the x-axis.

23 The curve C is determined by the parametric equations $x = e^{-t}$, $y = e^{2t}$. Calculate dy/dx and d^2y/dx^2 directly from these parametric equations. Conclude that C is concave upward at every point. Then sketch C.

24 The graph of the folium of Descartes with rectangular equation $x^3 + y^3 = 3xy$ appears in Figure 10.8. Parametrize its loop as follows: Let P be the point of intersection of the line $y = tx$ with the loop, then solve for the coordinates x and y of P in terms of t.

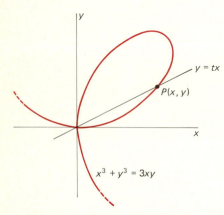

10.8 The loop of the folium of Descartes (Problem 24).

25 Parametrize the parabola $y^2 = 4px$ by expressing x and y as functions of the slope m of the tangent line at the point $P(x, y)$ of the parabola.

26 Let P be a point of the curve with polar equation $r = f(\theta)$, and let ψ be the angle between the extended radius OP and the tangent line at P. Let α be the angle of inclination of this tangent line, measured counterclockwise from the horizontal. Then $\psi = \alpha - \theta$. Verify Formula (10) by substituting $\tan\alpha = dy/dx$ from Formula (9) and $\tan\theta = y/x = (\sin\theta)/(\cos\theta)$ into the identity

$$\cot\psi = \frac{1}{\tan(\alpha - \theta)} = \frac{1 + \tan\alpha\,\tan\theta}{\tan\alpha - \tan\theta}.$$

27 Let P_0 be the highest point of the circle of Figure 10.4—the circle that generates the cycloid of Example 3. Show that the line through P_0 and the point P of the cyloid (P is shown in Figure 10.4) is tangent to the cycloid at P. This fact gives a geometric construction of the tangent line to the cycloid.

28 A circle of radius b rolls without slipping inside a circle of radius $a > b$. The path of a point P fixed on the circumference of the rolling circle is called a *hypocycloid*. Let P begin its journey at $A(a, 0)$, shown in Figure 10.9, and

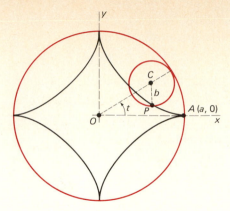

10.9 The hypocycloid of Problem 28.

let t be the angle AOC, where O is the center of the large circle and C the center of the rolling circle. Show that the coordinates of P are given by the parametric equations

$$x = (a - b)\cos t + b\cos\left(\frac{a - b}{b}t\right),$$

$$y = (a - b)\sin t - b\sin\left(\frac{a - b}{b}t\right).$$

29 If $b = a/4$ in Problem 28, show that the parametric equations of the hypocycloid reduce to

$$x = a\cos^3 t, \qquad y = a\sin^3 t.$$

30 (a) Show that the hypocycloid $x = \cos^3 t$, $y = \sin^3 t$ is the graph of the equation $x^{2/3} + y^{2/3} = 1$.
(b) Find the points of the hypocycloid where its tangent line is either horizontal or vertical and the intervals on which it is concave upward and those on which it is concave downward.
(c) Sketch this hypocycloid.

31 Consider a point P on the spiral of Archimedes, the curve shown in Figure 10.10 with polar equation $r = a\theta$.

10,10 The segment PQ is tangent to the spiral (a result of Archimedes).

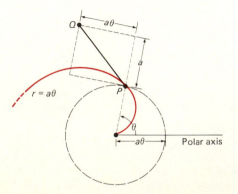

Archimedes viewed the path of P as compounded of two motions, one with speed a directly away from the origin O and the other a circular motion with unit angular speed around O. This suggests Archimedes' result that the line PQ in the figure is tangent to the spiral at P. Prove that this is indeed true.

Integral Computations with Parametric Curves

In Chapter 5 we discussed the computation of a variety of geometric quantities associated with the graph $y = f(x)$ of a nonnegative function on the interval $[a, b]$. These included:

Area under
the curve:
$$A = \int_a^b y \, dx. \tag{1}$$

Volume of
revolution around
the x-axis:
$$V_x = \int_a^b \pi y^2 \, dx. \tag{2a}$$

Volume of
revolution around
the y-axis:
$$V_y = \int_a^b 2\pi x y \, dx. \tag{2b}$$

Arc length of
the curve:
$$s = \int_0^s ds = \int_a^b [1 + (y')^2]^{1/2} \, dx. \tag{3}$$

Area of surface
of revolution
about x-axis:
$$A_x = \int_a^b 2\pi y \, ds. \tag{4a}$$

Area of surface
of revolution
about y-axis:
$$A_y = \int_a^b 2\pi x \, ds. \tag{4b}$$

Of course we substitute $y = f(x)$ in each of these integrals before integrating from $x = a$ to $x = b$.

We now want to compute these same quantities for a smooth parametric curve

$$x = f(t), \qquad y = g(t), \qquad \alpha \leq t \leq \beta.$$

To insure that this parametric curve looks like the graph of a function defined on some interval $[a, b]$ of the x-axis, we assume that $f(t)$ *is either an increasing or a decreasing function of t on $[\alpha, \beta]$, and that $g(t) \geq 0$.* If $a = f(\alpha)$ and $b = f(\beta)$, then the curve will be traced from left to right as t increases, while if $a = f(\beta)$ and $b = f(\alpha)$, it is traced from right to left. See Figure 10.11.

Any one of the quantities A, V_x, V_y, s, A_x, and A_y may then be computed by making the substitutions

$$x = f(t), \qquad y = g(t),$$
$$dx = f'(t)\,dt, \qquad dy = g'(t)\,dt, \qquad \text{and} \tag{5}$$
$$ds = \sqrt{[f'(t)]^2 + [g'(t)]^2}\,dt$$

in the appropriate one of the integral formulas in (1) through (4) above. In the case of Formulas (1) and (2), which involve dx, we then integrate

10.11 Tracing a parametrized curve.

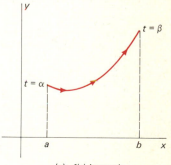

(a): $f(t)$ increasing

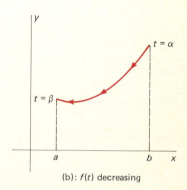

(b): $f(t)$ decreasing

from $t = \alpha$ to $t = \beta$ in the case shown in part (a) of Figure 10.11, and from $t = \beta$ to $t = \alpha$ in the case shown in part (b). Thus the proper choice of limits on t corresponds to traversing the curve *from left to right*. For example,

$$A = \int_{\alpha}^{\beta} g(t)f'(t)\, dt \quad \text{if} \quad f(\alpha) < f(\beta),$$

while

$$A = \int_{\beta}^{\alpha} g(t)f'(t)\, dt \quad \text{if} \quad f(\beta) < f(\alpha).$$

The validity of this method of evaluating integrals (1) and (2) follows from the theorem on integration by substitution in Section 4-5.

In the case of Formulas (3) and (4), which involve ds, we integrate from $t = \alpha$ to $t = \beta$ in either case. To see why this is so, recall from Formula (5) in Section 10-1 that $dy/dx = g'(t)/f'(t)$ if $f'(t) \neq 0$ on $[\alpha, \beta]$. Hence

$$s = \int_{a}^{b} \sqrt{1 + \left(\frac{dy}{dx}\right)^2}\, dx$$

$$= \int_{f^{-1}(a)}^{f^{-1}(b)} \sqrt{1 + \left[\frac{g'(t)}{f'(t)}\right]^2}\, f'(t)\, dt.$$

Since $f'(t) > 0$ if $f(\alpha) = a$ and $f(\beta) = b$, while $f'(t) < 0$ if $f(\alpha) = b$ and $f(\beta) = a$, it follows that

$$s = \int_{\alpha}^{\beta} \sqrt{1 + \left[\frac{g'(t)}{f'(t)}\right]^2}\, |f'(t)|\, dt$$

and so

$$s = \int_{\alpha}^{\beta} \sqrt{[f'(t)]^2 + [g'(t)]^2}\, dt$$

$$= \int_{\alpha}^{\beta} \sqrt{\left(\frac{dx}{dt}\right)^2 + \left(\frac{dy}{dt}\right)^2}\, dt. \tag{6}$$

This formula, derived under the assumption that $f'(t) \neq 0$ on $[\alpha, \beta]$, may be taken as the *definition* of arc length for an arbitrary smooth parametric curve. Similarly, the area of a surface of revolution is defined for smooth parametric curves as the result of first making substitutions (5) in Formula (4a) or (4b) and then integrating from $t = \alpha$ to $t = \beta$.

The infinitesimal triangle of Figure 10.12 serves as a convenient device for remembering the substitution

$$ds = \sqrt{[f'(t)]^2 + [g'(t)]^2}\, dt.$$

When the Pythagorean theorem is applied to this right triangle, it leads to the symbolic manipulation

$$ds = \sqrt{(dx)^2 + (dy)^2}$$

$$= \sqrt{\left(\frac{dx}{dt}\right)^2 + \left(\frac{dy}{dt}\right)^2}\, dt$$

$$= \sqrt{[f'(t)]^2 + [g'(t)]^2}\, dt. \tag{7}$$

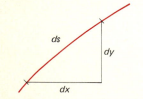

10.12 Nearly a right triangle for small dx and dy.

CHAP. 10: **Parametric Curves and Vectors in the Plane**

EXAMPLE 1 Use the parametrization $x = a \cos t$, $y = a \sin t$ of the circle with center O and radius a to find the volume V and surface area A of the sphere obtained by revolving this circle around the x-axis.

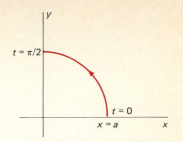

10.13 The quarter circle of Example 1.

Solution Half of the sphere is obtained by revolving the first quadrant of the circle, which is shown in Figure 10.13. The left-to-right direction along the curve is from $t = \pi/2$ to $t = 0$, so Formula (2a) gives

$$V = 2\pi \int_{\pi/2}^{0} (a \sin t)^2 (-a \sin t \, dt)$$

$$= 2\pi a^3 \int_{0}^{\pi/2} (1 - \cos^2 t) \sin t \, dt$$

$$= 2\pi a^3 \left[-\cos t + \tfrac{1}{3} \cos^3 t \right]_{0}^{\pi/2} = \tfrac{4}{3}\pi a^3.$$

The arc length differential for the parametrized curve is

$$ds = \sqrt{(-a \sin t)^2 + (a \cos t)^2} \, dt = a \, dt.$$

Hence Formula (4a) gives

$$A = 2 \int_{0}^{\pi/2} 2\pi (a \sin t)(a \, dt) = 4\pi a^2 \int_{0}^{\pi/2} \sin t \, dt$$

$$= 4\pi a^2 \left[-\cos t \right]_{0}^{\pi/2} = 4\pi a^2.$$

Of course the results of Example 1 are familiar. By contrast, the following example requires the new methods of this section.

EXAMPLE 2 Find the area under and the arc length of the cycloidal arch

$$x = a(t - \sin t), \qquad y = a(1 - \cos t), \qquad 0 \leq t \leq 2\pi.$$

Solution We shall use the identity $1 - \cos t = 2 \sin^2(t/2)$ and the result of Problem 40 in Section 8-4,

$$\int_{0}^{\pi} \sin^{2n} u \, du = \pi \frac{1}{2} \frac{3}{4} \frac{5}{6} \cdots \frac{2n-1}{2n}.$$

Since $dx = a(1 - \cos t) \, dt$ and the left-to-right direction along the curve is from $t = 0$ to $t = 2\pi$, Formula (1) gives

$$A = \int_{0}^{2\pi} a(1 - \cos t) \cdot a(1 - \cos t) \, dt$$

$$= a^2 \int_{0}^{2\pi} (1 - \cos t)^2 \, dt = 4a^2 \int_{0}^{2\pi} \sin^4 \frac{t}{2} \, dt$$

$$= 8a^2 \int_{0}^{\pi} \sin^4 u \, du \qquad\qquad \left(u = \frac{t}{2} \right)$$

$$= 8a^2 \cdot \pi \cdot \frac{1}{2} \cdot \frac{3}{4} = 3\pi a^2$$

for the area under one arch of the cycloid. The arc length differential is

$$ds = \sqrt{a^2(1 - \cos t)^2 + (a \sin t)^2} \, dt$$

$$= a\sqrt{2(1 - \cos t)} \, dt = 2a \sin\left(\frac{t}{2}\right) dt,$$

so Formula (3) gives

$$s = \int_0^{2\pi} 2a \sin \frac{t}{2}\, dt = \left[-4a \cos \frac{t}{2} \right]_0^{2\pi} = 8a$$

for the length of one arch of the cycloid.

PARAMETRIC POLAR COORDINATES

Suppose that a parametric curve is determined by giving its polar coordinates

$$r = r(t), \qquad \theta = \theta(t), \qquad \alpha \leq t \leq \beta$$

as functions of the parameter t. Then this curve is described in rectangular coordinates by means of the parametric equations

$$x(t) = r(t) \cos \theta(t), \qquad y(t) = r(t) \sin \theta(t), \qquad \alpha \leq t \leq \beta,$$

giving x and y as functions of t. These latter parametric equations may then be used in integral formulas (1) through (4).

To compute ds, we first calculate the derivatives

$$\frac{dx}{dt} = (\cos \theta) \frac{dr}{dt} - (r \sin \theta) \frac{d\theta}{dt},$$

$$\frac{dy}{dt} = (\sin \theta) \frac{dr}{dt} + (r \cos \theta) \frac{d\theta}{dt}.$$

Upon substituting these expressions for dx/dt and dy/dt into (7) and making algebraic simplifications, we find that the arc length differential in parametric polar coordinates is

$$ds = \sqrt{\left(\frac{dr}{dt}\right)^2 + \left(r \frac{d\theta}{dt}\right)^2}\, dt. \tag{8}$$

In the case of a curve with explicit polar coordinate equation $r = f(\theta)$, we may use θ rather than t as the parameter. Then (8) takes the simpler form

$$ds = \sqrt{\left(\frac{dr}{d\theta}\right)^2 + r^2}\, d\theta. \tag{9}$$

EXAMPLE 3 Find the perimeter s of the cardioid $r = 1 + \cos \theta$; also find the surface area A generated by revolving it around the x-axis.

Solution This cardioid is shown in Figure 10.14. Formula (9) gives

$$ds = \sqrt{(-\sin \theta)^2 + (1 + \cos \theta)^2}\, d\theta$$

$$= \sqrt{2(1 + \cos \theta)}\, d\theta = \sqrt{4 \cos^2 \left(\frac{\theta}{2}\right)}\, d\theta.$$

Hence $ds = 2 \cos(\theta/2)\, d\theta$ on the upper half of the cardioid, where $0 \leq \theta \leq \pi$ and $\cos(\theta/2) \geq 0$. Therefore

$$s = 2 \int_0^{\pi} 2 \cos \frac{\theta}{2}\, d\theta = 8 \left[\sin \frac{\theta}{2} \right]_0^{\pi} = 8.$$

10.14 The cardioid of Example 3.

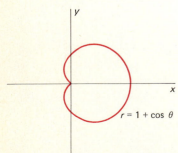

$r = 1 + \cos \theta$

CHAP. 10: **Parametric Curves and Vectors in the Plane**

The surface area of revolution is

$$A = \int 2\pi y \, ds = \int_{\theta=0}^{\pi} 2\pi(r\sin\theta)\,ds$$

$$= \int_0^{\pi} 2\pi(1+\cos\theta)(\sin\theta)\cdot 2\cos\frac{\theta}{2}\,d\theta$$

$$= 16\pi \int_0^{\pi} \cos^4\frac{\theta}{2}\sin\frac{\theta}{2}\,d\theta$$

$$= 16\pi\left[-\frac{2}{5}\cos^5\frac{\theta}{2}\right]_0^{\pi} = \frac{32\pi}{5}.$$

10-2 PROBLEMS

In each of Problems 1–6, find the area of the region that lies between the given parametric curve and the x-axis.

1 $x = t^3,\ y = 2t^2 + 1;\ \ -1 \le t \le 1.$
2 $x = e^{3t},\ y = e^{-t};\ \ 0 \le t \le \ln 2.$
3 $x = \cos t,\ y = \sin^2 t;\ \ 0 \le t \le \pi.$
4 $x = 2 - 3t,\ y = e^{2t};\ \ 0 \le t \le 1.$
5 $x = \cos t,\ y = e^t;\ \ 0 \le t \le \pi.$
6 $x = 1 - e^t,\ y = 2t + 1;\ \ 0 \le t \le 1.$

In Problems 7–10, find the volume obtained by revolving the region indicated in the given problem above around the x-axis.

7 Problem 1.   **8** Problem 2.
9 Problem 3.   **10** Problem 5.

In Problems 11–16, find the arc length of the given curve.

11 $x = 2t,\ y = \frac{2}{3}t^{3/2};\ \ 5 \le t \le 12.$
12 $x = \frac{1}{2}t^2,\ y = \frac{1}{3}t^3;\ \ 0 \le t \le 1.$

13 $x = \sin t - \cos t,\ y = \sin t + \cos t;\ \ \dfrac{\pi}{4} \le t \le \dfrac{\pi}{2}.$

14 $x = e^t \sin t,\ y = e^t \cos t;\ \ 0 \le t \le \pi.$
15 $r = e^{\theta/2};\ \ 0 \le \theta \le 4\pi.$
16 $r = \theta;\ \ 2\pi \le \theta \le 4\pi.$

In Problems 17–22, find the area of the surface of revolution generated by revolving the given curve around the indicated axis.

17 $x = 1 - t,\ y = 2\sqrt{t},\ 1 \le t \le 4;$ the x-axis.

18 $x = 2t^2 + \dfrac{1}{t},\ y = 8t^{1/2},\ 1 \le t \le 2;$ the x-axis.

19 $x = t^3,\ y = 2t + 3,\ -1 \le t \le 1;$ the y-axis.
20 $x = 2t + 1,\ y = t^2 + t,\ 0 \le t \le 3;$ the y-axis.
21 $r = 4\sin\theta;$ the x-axis.

22 $r = e^{\theta},\ 0 \le \theta \le \dfrac{\pi}{2};$ the y-axis.

23 Find the volume generated by revolving the region under the cycloidal arch of Example 2 around the x-axis.
24 Find the area of the surface generated by revolving the cycloidal arch of Example 2 around the x-axis.
25 Use the parametrization $x = a\cos t,\ y = b\sin t$ to find
(a) the area bounded by the ellipse $x^2/a^2 + y^2/b^2 = 1$;
(b) the volume of the ellipsoid generated by revolving this ellipse around the x-axis.
26 Find the area bounded by the loop of the parametric curve $x = t^2,\ y = t^3 - 3t$ of Problem 22 in Section 10-1.
27 Use the parametrization $x = t\cos t,\ y = t\sin t$ of the Archimedean spiral to find the arc length of the first full turn of this spiral (corresponding to $0 \le t \le 2\pi$).
28 The circle $(x - b)^2 + y^2 = a^2$ with radius $a < b$ and center $(b, 0)$ can be parametrized by

$$x = b + a\cos t, \qquad y = a\sin t, \qquad 0 \le t \le 2\pi.$$

Find the surface area of the torus obtained by revolving this circle around the y-axis.
29 The *astroid* (or four-cusped hypocycloid) $x^{2/3} + y^{2/3} = 1$ arose in connection with Problems 28–30 of Section 10-1 and is shown in Figure 10.9 at the end of that section. It has the parametrization

$$x = a\cos^3 t, \qquad y = a\sin^3 t, \qquad 0 \le t \le 2\pi.$$

Find the area of the region bounded by the astroid.
30 Find the total length of the astroid of Problem 29.
31 Find the area of the surface obtained by revolving the astroid of Problem 29 around the x-axis.
32 Find the area of the surface generated by revolving the lemniscate $r^2 = 2a^2\cos 2\theta$ around the y-axis. (*Suggestion:* Use Formula (9) and note that $r\,dr = -2a^2\sin 2\theta\,d\theta$.)

Vectors in the Plane

A physical quantity such as length, temperature, or mass can be specified in terms of a single real number, its magnitude. Such a quantity is called a **scalar.** Other physical entities such as force and velocity possess both magnitude and direction; these entities are called **vector quantities,** or simply *vectors.*

For example, to specify the velocity of a moving point in the plane, we must give both the rate at which it moves (its speed) and the direction of that motion. The combination is the **velocity vector** of the moving point. It is convenient to represent this velocity vector by an arrow, located at the current position of the moving point on its trajectory, as shown in Figure 10.15.

Although the arrow or directed line segment carries the desired information—both magnitude (its length) and direction—it is a pictorial object rather than a mathematical object. We shall see that the following formal definition of a vector captures the essence of magnitude combined with direction.

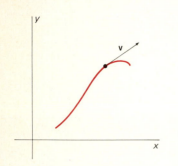

10.15 A velocity vector may be represented by an arrow.

Definition of Vector

A **vector** $\mathbf{v}$ in the Cartesian plane is an ordered pair of real numbers, of the form $\langle a, b \rangle$. We write $\mathbf{v} = \langle a, b \rangle$ and call a and b the **components** of the vector $\mathbf{v}$.

The directed line segment $\overrightarrow{OP}$ from the origin O to the point $P(a, b)$ is one geometric representation of the vector $\mathbf{v}$. For this reason, the vector $\mathbf{v} = \langle a, b \rangle$ is called the **position vector** of the point $P(a, b)$. In fact, the relationship between $\mathbf{v} = \langle a, b \rangle$ and $P(a, b)$ is so close that, in certain contexts, it is convenient to confuse the two deliberately—to regard $\mathbf{v}$ and P as the same mathematical object. In other contexts the distinction between the two is important; in particular, *any* directed line segment with the same direction and magnitude (length) as $\overrightarrow{OP}$ is a representation of $\mathbf{v}$, as Figure 10.16 suggests. What is important about the vector $\mathbf{v}$ is usually not *where* it is, but how long it is and which way it points.

The magnitude associated with the vector $\mathbf{v} = \langle a, b \rangle$, called its **length,** is denoted by $v = |\mathbf{v}|$ and is defined to be

$$v = |\mathbf{v}| = |\langle a, b \rangle| = \sqrt{a^2 + b^2}. \tag{1}$$

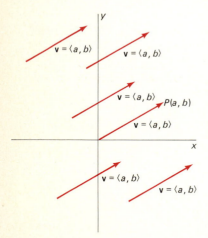

10.16 All these arrows represent the same vector $\mathbf{v} = \langle a, b \rangle$.

The notation $v = |\mathbf{v}|$ is used because the length of a vector is in some ways analogous to the absolute value of a real number.

The only vector with length zero is the **zero vector** with both components zero, denoted by $\mathbf{0} = \langle 0, 0 \rangle$. The zero vector is also unique in having no specific direction.

The operations of addition and multiplication of real numbers have analogues for vectors. We shall define each of these operations of "vector algebra" in terms of components of vectors and then give a geometric interpretation in terms of arrows.

Definition *Addition of Vectors*

The **sum u + v** of the two vectors $\mathbf{u} = \langle u_1, u_2 \rangle$ and $\mathbf{v} = \langle v_1, v_2 \rangle$ is the vector

$$\mathbf{u} + \mathbf{v} = \langle u_1 + v_1, u_2 + v_2 \rangle. \tag{2}$$

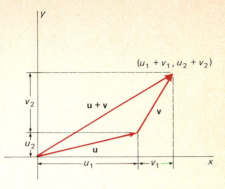

10.17 The triangle law, a geometric interpretation of vector addition.

Thus we add vectors by adding corresponding components; that is, by "componentwise addition." The geometric interpretation of vector addition is the **triangle law of addition,** illustrated in Figure 10.17, where the labeled lengths indicate why this interpretation is valid. An equivalent interpretation is the **parallelogram law of addition** illustrated in Figure 10.18.

It is natural to write $2\mathbf{u} = \mathbf{u} + \mathbf{u}$. If $\mathbf{u} = \langle u_1, u_2 \rangle$, then

$$2\mathbf{u} = \mathbf{u} + \mathbf{u} = \langle u_1, u_2 \rangle + \langle u_1, u_2 \rangle = \langle 2u_1, 2u_2 \rangle.$$

This suggests that multiplication of a vector by a scalar (real number) also be defined in a componentwise manner.

Definition *Multiplication of a Vector by a Scalar*

If $\mathbf{u} = \langle u_1, u_2 \rangle$ and c is a real number, then the **scalar multiple** $c\mathbf{u}$ is the vector

$$c\mathbf{u} = \langle cu_1, cu_2 \rangle. \tag{3}$$

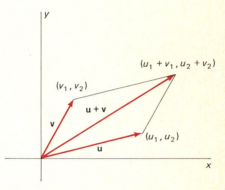

10.18 The parallelogram law for vector addition.

Note that

$$|c\mathbf{u}| = \sqrt{(cu_1)^2 + (cu_2)^2} = |c|\sqrt{(u_1)^2 + (u_2)^2} = |c| \cdot |\mathbf{u}|.$$

Thus the length of $c\mathbf{u}$ is $|c|$ times that of $\mathbf{u}$.

The negative of the vector $\mathbf{u}$ is the vector

$$-\mathbf{u} = (-1)\mathbf{u} = \langle -u_1, -u_2 \rangle$$

10.19 The vector $c\mathbf{u}$ may have the same direction as $\mathbf{u}$ or the opposite direction.

with the same length as $\mathbf{u}$ but the opposite direction. We say that the two nonzero vectors $\mathbf{u}$ and $\mathbf{v}$ have

 (i) The **same direction** if $\mathbf{u} = c\mathbf{v}$ for some $c > 0$;
 (ii) **Opposite directions** if $\mathbf{u} = c\mathbf{v}$ for some $c < 0$.

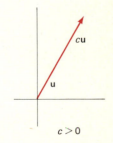

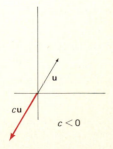

The geometric interpretation of scalar multiplication is that $c\mathbf{u}$ is the vector with length $|c| \cdot |\mathbf{u}|$, with the same direction as $\mathbf{u}$ if $c > 0$ but the opposite direction if $c < 0$. See Figure 10.19.

The **difference u − v** of the vectors $\mathbf{u} = \langle u_1, u_2 \rangle$ and $\mathbf{v} = \langle v_1, v_2 \rangle$ is defined to be

$$\mathbf{u} - \mathbf{v} = \mathbf{u} + (-\mathbf{v}) = \langle u_1 - v_1, u_2 - v_2 \rangle. \tag{4}$$

If we think of $\langle u_1, u_2 \rangle$ and $\langle v_1, v_2 \rangle$ as position vectors of the points P and Q, respectively, then $\mathbf{u} - \mathbf{v}$ may be represented by the arrow $\overrightarrow{QP}$ from Q

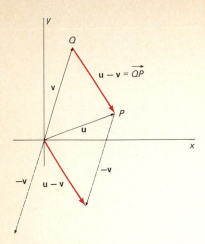

10.20 Geometric interpretation of the difference **u** − **v**.

to P. We may therefore write

$$\mathbf{u} - \mathbf{v} = \overrightarrow{OP} - \overrightarrow{OQ} = \overrightarrow{QP},$$

as illustrated in Figure 10.20.

EXAMPLE 1   Suppose that $\mathbf{u} = \langle 4, -3 \rangle$ and $\mathbf{v} = \langle -2, 3 \rangle$. Find $|\mathbf{u}|$ and the vectors $\mathbf{u} + \mathbf{v}$, $\mathbf{u} - \mathbf{v}$, $3\mathbf{u}$, $-2\mathbf{v}$, and $2\mathbf{u} + 4\mathbf{v}$.

Solution

$$|\mathbf{u}| = \sqrt{4^2 + (-3)^2} = \sqrt{25} = 5.$$

$$\mathbf{u} + \mathbf{v} = \langle 4 + (-2), (-3) + 3 \rangle = \langle 2, 0 \rangle.$$

$$\mathbf{u} - \mathbf{v} = \langle 4 - (-2), (-3) - 3 \rangle = \langle 6, -6 \rangle.$$

$$3\mathbf{u} = \langle 3 \cdot (4), 3 \cdot (-3) \rangle = \langle 12, -9 \rangle.$$

$$-2\mathbf{v} = \langle -2 \cdot (-2), -2 \cdot (3) \rangle = \langle 4, -6 \rangle.$$

$$2\mathbf{u} + 4\mathbf{v} = \langle 2 \cdot (4) + 4 \cdot (-2), 2 \cdot (-3) + 4 \cdot (3) \rangle = \langle 0, 6 \rangle.$$

Many familiar algebraic properties of real numbers carry over to the following analogous properties of vector addition and scalar multiplication. Let $\mathbf{a}$, $\mathbf{b}$, and $\mathbf{c}$ be vectors and r and s be numbers. Then

$$\begin{align} &\text{(i) } \mathbf{a} + \mathbf{b} = \mathbf{b} + \mathbf{a}; \\ &\text{(ii) } \mathbf{a} + (\mathbf{b} + \mathbf{c}) = (\mathbf{a} + \mathbf{b}) + \mathbf{c}; \\ &\text{(iii) } r(\mathbf{a} + \mathbf{b}) = r\mathbf{a} + r\mathbf{b}; \\ &\text{(iv) } (r + s)\mathbf{a} = r\mathbf{a} + s\mathbf{a}; \\ &\text{(v) } (rs)\mathbf{a} = r(s\mathbf{a}) = s(r\mathbf{a}). \end{align} \tag{5}$$

These identities are verified by working with components. For instance, if $\mathbf{a} = \langle a_1, a_2 \rangle$ and $\mathbf{b} = \langle b_1, b_2 \rangle$, then

$$\begin{align} r(\mathbf{a} + \mathbf{b}) &= r\langle a_1 + b_1, a_2 + b_2 \rangle \\ &= \langle r(a_1 + b_1), r(a_2 + b_2) \rangle \\ &= \langle ra_1 + rb_1, ra_2 + rb_2 \rangle \\ &= \langle ra_1, ra_2 \rangle + \langle rb_1, rb_2 \rangle \\ &= r\mathbf{a} + r\mathbf{b}. \end{align}$$

The proofs of the other four identities in (5) are left as exercises.

THE UNIT VECTORS i AND j

A **unit** vector is one with length 1. If $\mathbf{a} = \langle a_1, a_2 \rangle \neq \mathbf{0}$ then

$$\mathbf{u} = \frac{\mathbf{a}}{|\mathbf{a}|} \tag{6}$$

is the unit vector having the same direction as $\mathbf{a}$, because

$$|\mathbf{u}| = \sqrt{\left(\frac{a_1}{|\mathbf{a}|}\right)^2 + \left(\frac{a_2}{|\mathbf{a}|}\right)^2} = \frac{1}{|\mathbf{a}|}\sqrt{a_1^2 + a_2^2} = 1.$$

CHAP. 10:   **Parametric Curves and Vectors in the Plane**

For example, if $\mathbf{a} = \langle 3, -4 \rangle$, then $|\mathbf{a}| = 5$. Thus $\langle \frac{3}{5}, -\frac{4}{5} \rangle$ is a unit vector that has the same direction as $\mathbf{a}$.

Two particular unit vectors play a special role. They are the vectors $\mathbf{i} = \langle 1, 0 \rangle$ and $\mathbf{j} = \langle 0, 1 \rangle$. The first points in the positive x-direction, the second in the positive y-direction. Together they provide a useful alternative notation for vectors. For if $\mathbf{a} = \langle a_1, a_2 \rangle$, then

$$\mathbf{a} = \langle a_1, 0 \rangle + \langle 0, a_2 \rangle = a_1 \langle 1, 0 \rangle + a_2 \langle 0, 1 \rangle = a_1 \mathbf{i} + a_2 \mathbf{j}. \qquad (7)$$

Thus every vector is a **linear combination** of $\mathbf{i}$ and $\mathbf{j}$. The usefulness of this notation is based on the fact that such linear combinations of $\mathbf{i}$ and $\mathbf{j}$ may be manipulated as if they were ordinary sums. For example, if

$$\mathbf{a} = a_1 \mathbf{i} + a_2 \mathbf{j} \quad \text{and} \quad \mathbf{b} = b_1 \mathbf{i} + b_2 \mathbf{j},$$

then

$$\begin{aligned} \mathbf{a} + \mathbf{b} &= (a_1 \mathbf{i} + a_2 \mathbf{j}) + (b_1 \mathbf{i} + b_2 \mathbf{j}) \\ &= (a_1 + b_1) \mathbf{i} + (a_2 + b_2) \mathbf{j}. \end{aligned}$$

Also,

$$c\mathbf{a} = c(a_1 \mathbf{i} + a_2 \mathbf{j}) = (ca_1) \mathbf{i} + (ca_2) \mathbf{j}.$$

EXAMPLE 2 Suppose that $\mathbf{a} = 2\mathbf{i} - 3\mathbf{j}$ and $\mathbf{b} = 3\mathbf{i} + 4\mathbf{j}$. Express $5\mathbf{a} - 3\mathbf{b}$ in terms of $\mathbf{i}$ and $\mathbf{j}$.

Solution $5\mathbf{a} - 3\mathbf{b} = 5(2\mathbf{i} - 3\mathbf{j}) - 3(3\mathbf{i} + 4\mathbf{j})$

$$= (10 - 9)\mathbf{i} + (-15 - 12)\mathbf{j} = \mathbf{i} - 27\mathbf{j}.$$

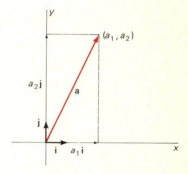

10.21 Resolution of $\mathbf{a} = \langle a_1, a_2 \rangle$ into its horizontal and vertical components.

Formula (7) expresses the vector $\mathbf{u} = \langle a_1, a_2 \rangle$ as the sum of a horizontal vector $a_1 \mathbf{i}$ and a vertical vector $a_2 \mathbf{j}$, as Figure 10.21 shows. The decomposition, or "resolution," of a vector into its horizontal and vertical components is an important technique in the study of vector quantities. For example, a force $\mathbf{F}$ may be decomposed into its horizontal and vertical components $F_1 \mathbf{i}$ and $F_2 \mathbf{j}$, respectively. The physical effect of the single force $\mathbf{F}$ is the same as the combined effect of the separate forces $F_1 \mathbf{i}$ and $F_2 \mathbf{j}$. (This is an instance of the empirically verifiable parallelogram law of addition of forces.) Because of this decomposition, many two-dimensional problems can be reduced to one-dimensional problems, the latter solved, and the two results combined (again by vector methods) to give the solution of the original problem.

10.22 Rotate $\mathbf{a}$ 90° counterclockwise to obtain $\mathbf{a}_\perp$.

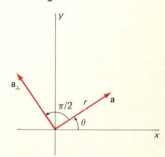

PERPENDICULAR VECTORS AND THE DOT PRODUCT

Suppose that the vector $\mathbf{a} = a_1 \mathbf{i} + a_2 \mathbf{j}$ is the position vector of the point P with polar coordinates (r, θ). Then $a_1 = r \cos \theta$ and $a_2 = r \sin \theta$ because P has rectangular coordinates (a_1, a_2). Consequently

$$\mathbf{a} = (r \cos \theta)\mathbf{i} + (r \sin \theta)\mathbf{j}. \qquad (8)$$

Let $\mathbf{a}_\perp$ denote the perpendicular vector with $|\mathbf{a}_\perp| = |\mathbf{a}|$ obtained by rotating $\mathbf{a}$ through a counterclockwise angle of $\pi/2$ (see Figure 10.22). We replace θ

by $\theta + \pi/2$ in Formula (8), and find that

$$\mathbf{a}_\perp = r \cos\left(\theta + \frac{\pi}{2}\right)\mathbf{i} + r \sin\left(\theta + \frac{\pi}{2}\right)\mathbf{j}$$

$$= -(r \sin\theta)\mathbf{i} + (r \cos\theta)\mathbf{j}.$$

Thus

$$\mathbf{a}_\perp = -a_2\mathbf{i} + a_1\mathbf{j}. \tag{9}$$

That is, $\mathbf{a}_\perp$ is obtained from $\mathbf{a}$ by first interchanging the components of $\mathbf{a}$ and then changing the sign of the new first component. We can use (9) to establish a criterion for an arbitrary vector to be perpendicular to $\mathbf{a}$. Two vectors $\mathbf{a}$ and $\mathbf{b}$ are called **perpendicular** if, upon representing them as position vectors $\mathbf{a} = \overrightarrow{OP}$ and $\mathbf{b} = \overrightarrow{OQ}$, the segments OP and OQ are perpendicular.

Theorem   *Test for Perpendicular Vectors*

The two nonzero vectors $\mathbf{a} = \langle a_1, a_2 \rangle$ and $\mathbf{b} = \langle b_1, b_2 \rangle$ are perpendicular if and only if

$$a_1 b_1 + a_2 b_2 = 0. \tag{10}$$

Proof   The vector $\mathbf{b}$ is perpendicular to $\mathbf{a}$ if and only if $\mathbf{b}$ is a scalar multiple of the particular perpendicular vector $\mathbf{a}_\perp = -a_2\mathbf{i} + a_1\mathbf{j}$. If

$$\mathbf{b} = c\mathbf{a}_\perp = c(-a_2\mathbf{i} + a_1\mathbf{j}) = -ca_2\mathbf{i} + ca_1\mathbf{j}$$

then

$$b_1 = -ca_2 \quad \text{and} \quad b_2 = ca_1,$$

so

$$a_1 b_1 + a_2 b_2 = a_1(-ca_2) + a_2(ca_1) = 0.$$

Conversely, suppose that $a_1 b_1 + a_2 b_2 = 0$. Since either $a_1 \neq 0$ or $a_2 \neq 0$, assume, for example, that $a_1 \neq 0$. Then $a_1 b_1 + a_2 b_2 = 0$ implies that $b_1 = -a_2 b_2/a_1$, so that

$$\mathbf{b} = \left(\frac{-a_2 b_2}{a_1}\right)\mathbf{i} + b_2\mathbf{j}$$

$$= \frac{b_2}{a_1}(-a_2\mathbf{i} + a_1\mathbf{j}) = c\mathbf{a}_\perp$$

where $c = b_2/a_1$. Thus $\mathbf{b}$ is perpendicular to $\mathbf{a}$.

The particular combination of the components of $\mathbf{a}$ and $\mathbf{b}$ in (10) has important applications and is called the *dot product* of $\mathbf{a}$ and $\mathbf{b}$.

Definition   *The Dot Product of Two Vectors*

The **dot product** $\mathbf{a} \cdot \mathbf{b}$ of the two vectors $\mathbf{a} = \langle a_1, a_2 \rangle$ and $\mathbf{b} = \langle b_1, b_2 \rangle$ is the real number

$$\mathbf{a} \cdot \mathbf{b} = a_1 b_1 + a_2 b_2. \tag{11}$$

Note that the dot product (sometimes called the *scalar product*) of two vectors is a *scalar*. For example,

$$\langle 2, -3 \rangle \cdot \langle -4, 2 \rangle = (2)(-4) + (-3)(2) = -14.$$

The following algebraic properties of the dot product are easy to verify by componentwise calculations.

$$
\begin{array}{lll}
\text{(i)} & \mathbf{a} \cdot \mathbf{a} = |\mathbf{a}|^2; & \\
\text{(ii)} & \mathbf{a} \cdot \mathbf{b} = \mathbf{b} \cdot \mathbf{a}; & \\
\text{(iii)} & \mathbf{a} \cdot (\mathbf{b} + \mathbf{c}) = \mathbf{a} \cdot \mathbf{b} + \mathbf{a} \cdot \mathbf{c}; & (12) \\
\text{(iv)} & (r\mathbf{a}) \cdot \mathbf{b} = r(\mathbf{a} \cdot \mathbf{b}) = \mathbf{a} \cdot (r\mathbf{b}). &
\end{array}
$$

For example,

$$\mathbf{a} \cdot \mathbf{a} = (a_1)^2 + (a_2)^2 = |\mathbf{a}|^2$$

and

$$
\begin{aligned}
\mathbf{a} \cdot (\mathbf{b} + \mathbf{c}) &= \langle a_1, a_2 \rangle \cdot \langle b_1 + c_1, b_2 + c_2 \rangle \\
&= a_1(b_1 + c_1) + a_2(b_2 + c_2) \\
&= (a_1 b_1 + a_2 b_2) + (a_1 c_1 + a_2 c_2) = \mathbf{a} \cdot \mathbf{b} + \mathbf{a} \cdot \mathbf{c}.
\end{aligned}
$$

Since $\mathbf{i} \cdot \mathbf{i} = \mathbf{j} \cdot \mathbf{j} = 1$ and $\mathbf{i} \cdot \mathbf{j} = \mathbf{j} \cdot \mathbf{i} = 0$, application of Properties (12) gives

$$
\begin{aligned}
\mathbf{a} \cdot \mathbf{b} &= (a_1 \mathbf{i} + a_2 \mathbf{j}) \cdot (b_1 \mathbf{i} + b_2 \mathbf{j}) \\
&= (a_1 b_1) \mathbf{i} \cdot \mathbf{i} + (a_1 b_2 + a_2 b_1) \mathbf{i} \cdot \mathbf{j} + (a_2 b_2) \mathbf{j} \cdot \mathbf{j} \\
&= a_1 b_1 + a_2 b_2,
\end{aligned}
$$

in accord with the definition of $\mathbf{a} \cdot \mathbf{b}$. Thus dot products of vectors expressed in terms of $\mathbf{i}$ and $\mathbf{j}$ can be computed by "ordinary" algebra.

The test for perpendicularity is best remembered in terms of the dot product.

Corollary *Test for Perpendicular Vectors*

The two nonzero vectors $\mathbf{a}$ and $\mathbf{b}$ are perpendicular if and only if $\mathbf{a} \cdot \mathbf{b} = 0$.

EXAMPLE 3 Determine whether or not the following pairs of vectors are perpendicular.

(a) $\langle 3, 4 \rangle$ and $\langle 8, -6 \rangle$;   (b) $\langle 2, 3 \rangle$ and $\langle 4, -3 \rangle$.

Solution (a) $\langle 3, 4 \rangle \cdot \langle 8, -6 \rangle = 24 - 24 = 0$, so these two vectors *are* perpendicular.

(b) $\langle 2, 3 \rangle \cdot \langle 4, -3 \rangle = 8 - 9 = -1 \neq 0$, so $\langle 2, 3 \rangle$ and $\langle 4, -3 \rangle$ are *not* perpendicular.

In Section 13-1, we shall see that $\mathbf{a} \cdot \mathbf{b} = |\mathbf{a}| |\mathbf{b}| \cos \theta$, where θ is the angle between the vectors $\mathbf{a}$ and $\mathbf{b}$. The case $\theta = \pi/2$ gives the corollary above.

10-3 PROBLEMS

In each of Problems 1–8 find $|\mathbf{a}|$, $|-2\mathbf{b}|$, $|\mathbf{a} - \mathbf{b}|$, $\mathbf{a} + \mathbf{b}$, and $3\mathbf{a} - 2\mathbf{b}$. Also determine whether or not $\mathbf{a}$ and $\mathbf{b}$ are perpendicular.

1 $\mathbf{a} = \langle 1, -2 \rangle$, $\mathbf{b} = \langle -3, 2 \rangle$.
2 $\mathbf{a} = \langle 3, 4 \rangle$, $\mathbf{b} = \langle -4, 3 \rangle$.
3 $\mathbf{a} = \langle -2, -2 \rangle$, $\mathbf{b} = \langle -3, -4 \rangle$.
4 $\mathbf{a} = -2\langle 4, 7 \rangle$, $\mathbf{b} = -3\langle -4, -2 \rangle$.
5 $\mathbf{a} = \mathbf{i} + 3\mathbf{j}$, $\mathbf{b} = 2\mathbf{i} - 5\mathbf{j}$.
6 $\mathbf{a} = 2\mathbf{i} - 5\mathbf{j}$, $\mathbf{b} = \mathbf{i} - 6\mathbf{j}$.
7 $\mathbf{a} = 4\mathbf{i}$, $\mathbf{b} = -7\mathbf{j}$.
8 $\mathbf{a} = -\mathbf{i} - \mathbf{j}$, $\mathbf{b} = 2\mathbf{i} + 2\mathbf{j}$.

In each of Problems 9–12, find a unit vector $\mathbf{u}$ with the same direction as the given vector $\mathbf{a}$. Express $\mathbf{u}$ in terms of $\mathbf{i}$ and $\mathbf{j}$. Also find a unit vector $\mathbf{v}$ with the direction opposite that of $\mathbf{a}$.

9 $\mathbf{a} = \langle -3, -4 \rangle$. **10** $\mathbf{a} = \langle 5, -12 \rangle$.
11 $\mathbf{a} = 8\mathbf{i} + 15\mathbf{j}$. **12** $\mathbf{a} = 7\mathbf{i} - 24\mathbf{j}$.

In each of Problems 13–16, find the vector $\mathbf{a}$, expressed in terms of $\mathbf{i}$ and $\mathbf{j}$, that is represented by the arrow $\overrightarrow{PQ}$ in the plane.

13 $P = (3, 2)$, $Q = (3, -2)$.
14 $P = (-3, 5)$, $Q = (-3, 6)$.
15 $P = (-4, 7)$, $Q = (4, -7)$.
16 $P = (1, -1)$, $Q = (-4, -1)$.

In each of Problems 17–20, determine whether or not the given vectors $\mathbf{a}$ and $\mathbf{b}$ are perpendicular.

17 $\mathbf{a} = \langle 6, 0 \rangle$, $\mathbf{b} = \langle 0, -7 \rangle$.
18 $\mathbf{a} = 3\mathbf{j}$, $\mathbf{b} = 3\mathbf{i} - \mathbf{j}$.
19 $\mathbf{a} = 2\mathbf{i} - \mathbf{j}$, $\mathbf{b} = 4\mathbf{j} + 8\mathbf{i}$.
20 $\mathbf{a} = 8\mathbf{i} + 10\mathbf{j}$, $\mathbf{b} = 15\mathbf{i} - 12\mathbf{j}$.
21 Find a vector that has the same direction as $5\mathbf{i} - 7\mathbf{j}$ and (a) three times its length; (b) one-third its length.
22 Find a vector that has the opposite direction from $-3\mathbf{i} + 5\mathbf{j}$ and (a) four times its length; (b) one-fourth its length.
23 Find a vector of length 5 with (a) the same direction as $7\mathbf{i} - 3\mathbf{j}$; (b) the opposite direction from $8\mathbf{i} + 5\mathbf{j}$.
24 For what numbers c are the vectors $\langle c, 2 \rangle$ and $\langle c, -8 \rangle$ perpendicular?
25 For what numbers c are the vectors $2c\mathbf{i} - 4\mathbf{j}$ and $3\mathbf{i} + c\mathbf{j}$ perpendicular?
26 Given the three points $A(2, 3)$, $B(-5, 7)$, and $C(1, -5)$,

verify by direct computation of the vectors and their sum that $\overrightarrow{AB} + \overrightarrow{BC} + \overrightarrow{CA} = \mathbf{0}$.

In each of Problems 27–32, give a componentwise proof of the indicated property of vector algebra. Take $\mathbf{a} = \langle a_1, a_2 \rangle$ and $\mathbf{b} = \langle b_1, b_2 \rangle$ throughout.

27 $\mathbf{a} + (\mathbf{b} + \mathbf{c}) = (\mathbf{a} + \mathbf{b}) + \mathbf{c}$.
28 $(r + s)\mathbf{a} = r\mathbf{a} + s\mathbf{a}$.
29 $(rs)\mathbf{a} = r(s\mathbf{a})$.
30 $\mathbf{a} \cdot \mathbf{b} = \mathbf{b} \cdot \mathbf{a}$.
31 $(r\mathbf{a}) \cdot \mathbf{b} = r(\mathbf{a} \cdot \mathbf{b})$.
32 If $\mathbf{a} + \mathbf{b} = \mathbf{a}$, then $\mathbf{b} = \mathbf{0}$.

In Problems 33–35, assume the following fact: If an airplane flies with velocity vector $\mathbf{v}_a$ relative to the air and the velocity vector of the wind is $\mathbf{w}$, then the velocity vector of the plane relative to the ground is

$$\mathbf{v}_g = \mathbf{v}_a + \mathbf{w}.$$

The **apparent velocity** vector is $\mathbf{v}_a$, while $\mathbf{v}_g$ is called the **true velocity** vector.

33 Suppose that the wind is blowing from the northeast at 50 mph, and the pilot wishes to fly due east at 500 mph. What should the plane's apparent velocity vector be?
34 Repeat Problem 33 with the phrase *due east* replaced by *due west*.
35 Repeat Problem 33 in the case that the pilot wishes to fly northwest at 500 mph.
36 In the triangle ABC, let M_1 and M_2 be the midpoints of AB and AC, respectively. Show that $\overrightarrow{M_1M_2} = \frac{1}{2}\overrightarrow{BC}$. Conclude that the line segment joining the midpoints of two sides of a triangle is parallel to the third side.
37 Prove that the diagonals of a parallelogram $ABCD$ bisect each other. (*Suggestion:* If M_1 and M_2 denote the midpoints of AC and BD, respectively, show that $\overrightarrow{OM_1} = \overrightarrow{OM_2}$.)
38 Use vectors to prove that the midpoints of the four sides of an arbitrary quadrilateral are the vertices of a parallelogram.
39 Show that the vector $\mathbf{n} = a\mathbf{i} + b\mathbf{j}$ is perpendicular to the line with equation $ax + by + c = 0$. (*Suggestion:* If $P_1(x, y)$ and $P_2(\bar{x}, \bar{y})$ are two points on the line, show that $\mathbf{n} \cdot \overrightarrow{P_1P_2} = 0$.)

10-4

Motion and Vector-Valued Functions

We now use vectors to study the motion of a point in the plane. If the coordinates of the moving point at time t are given by the parametric equations $x = f(t)$, $y = g(t)$, then the vector

$$\mathbf{r}(t) = f(t)\mathbf{i} + g(t)\mathbf{j} \quad \text{or} \quad \mathbf{r} = x\mathbf{i} + y\mathbf{j} \tag{1}$$

is called the **position vector** of the point. Equation (1) determines a **vector-valued function** $\mathbf{r} = \mathbf{r}(t)$ that associates with the number t the vector $\mathbf{r}(t)$. In the bracket notation for vectors, a vector-valued function is an ordered pair of real-valued functions: $\mathbf{r}(t) = \langle f(t), g(t) \rangle$.

Much of the calculus of (ordinary) real-valued functions applies to vector-valued functions. To begin with, the **limit** of a vector-valued function $\mathbf{r}$ is defined as follows:

$$\lim_{t \to a} \mathbf{r}(t) = \left\langle \lim_{t \to a} f(t), \lim_{t \to a} g(t) \right\rangle$$

$$= \mathbf{i} \left(\lim_{t \to a} f(t) \right) + \mathbf{j} \left(\lim_{t \to a} g(t) \right), \qquad (2)$$

provided that the limits in the latter two expressions exist. Thus we take limits of vector-valued functions by taking limits of their component functions.

We say that $\mathbf{r} = \mathbf{r}(t)$ is **continuous** at the number a provided that

$$\lim_{t \to a} \mathbf{r}(t) = \mathbf{r}(a).$$

This amounts to saying that $\mathbf{r}$ is continuous at a if and only if its component functions f and g are continuous at a.

The derivative $\mathbf{r}'(t)$ of the vector-valued function $\mathbf{r}(t)$ is defined in much the same way as the derivative of a real-valued function. Specifically,

$$\mathbf{r}'(t) = \lim_{h \to 0} \frac{\mathbf{r}(t + h) - \mathbf{r}(t)}{h}, \qquad (3)$$

provided that this limit exists. A geometric note: Figure 10.23 suggests that $\mathbf{r}'(t)$ will be tangent to the curve swept out by $\mathbf{r}$ if $\mathbf{r}'(t)$ is attached at the point of evaluation.

Our next theorem tells us the simple *but important* fact that $\mathbf{r}'(t)$ can be calculated by componentwise differentiation. We shall also denote derivatives by

$$\mathbf{r}'(t) = D_t \mathbf{r}(t) = \frac{d\mathbf{r}}{dt}.$$

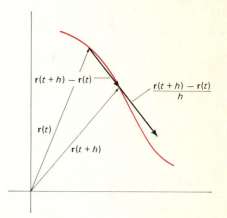

10.23 Geometry of the derivative of a vector-valued function.

Theorem 1 *Componentwise Differentiation*

Suppose that

$$\mathbf{r}(t) = \langle f(t), g(t) \rangle = f(t)\mathbf{i} + g(t)\mathbf{j},$$

where both f and g are differentiable functions. Then

$$\mathbf{r}'(t) = \langle f'(t), g'(t) \rangle = f'(t)\mathbf{i} + g'(t)\mathbf{j}. \qquad (4)$$

That is, if $\mathbf{r} = x\mathbf{i} + y\mathbf{j}$, then

$$\frac{d\mathbf{r}}{dt} = \frac{dx}{dt}\mathbf{i} + \frac{dy}{dt}\mathbf{j}.$$

Proof We simply take the limit in Equation (3) by taking limits of the components. We find that

$$\mathbf{r}'(t) = \lim_{h \to 0} \frac{1}{h} [\mathbf{r}(t + h) - \mathbf{r}(t)]$$

$$= \lim_{h \to 0} \frac{1}{h} [f(t + h)\mathbf{i} + g(t + h)\mathbf{j} - f(t)\mathbf{i} - g(t)\mathbf{j}]$$

$$= \left(\lim_{h \to 0} \frac{f(t + h) - f(t)}{h}\right)\mathbf{i} + \left(\lim_{h \to 0} \frac{g(t + h) - g(t)}{h}\right)\mathbf{j}$$

$$= f'(t)\mathbf{i} + g'(t)\mathbf{j}.$$

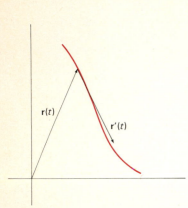

10.24 The derivative vector *is* tangent to the curve at the point of evaluation.

In Equation (5) of Section 10-1, we saw that the tangent line to the parametric curve $x = f(t)$, $y = g(t)$ has slope $g'(t)/f'(t)$ and therefore is parallel to the line through $(0, 0)$ and $(f'(t), g'(t))$. Hence, by Theorem 1, the derivative vector $\mathbf{r}'(t) = \langle f'(t), g'(t) \rangle$ may be visualized as a tangent vector to the curve at the point $(f(t), g(t))$. See Figure 10.24.

Our next theorem tells us that the formulas for computing derivatives of sums and products of vector-valued functions are formally similar to those for real-valued functions.

Theorem 2 *Differentiation Formulas*

Let $\mathbf{u}(t)$ and $\mathbf{v}(t)$ be differentiable vector-valued functions. Let $h(t)$ be a differentiable real-valued function, and let c be a (constant) scalar. Then

(i) $D_t[\mathbf{u}(t) + \mathbf{v}(t)] = \mathbf{u}'(t) + \mathbf{v}'(t)$;

(ii) $D_t[c\mathbf{u}(t)] = c\mathbf{u}'(t)$;

(iii) $D_t[h(t)\mathbf{u}(t)] = h'(t)\mathbf{u}(t) + h(t)\mathbf{u}'(t)$;   and

(iv) $D_t[\mathbf{u}(t) \cdot \mathbf{v}(t)] = \mathbf{u}'(t) \cdot \mathbf{v}(t) + \mathbf{u}(t) \cdot \mathbf{v}'(t)$.

Proof We shall prove Part (iv) and leave the other parts as exercises. If

$$\mathbf{u}(t) = f_1(t)\mathbf{i} + f_2(t)\mathbf{j} \quad \text{and} \quad \mathbf{v}(t) = g_1(t)\mathbf{i} + g_2(t)\mathbf{j},$$

then

$$\mathbf{u}(t) \cdot \mathbf{v}(t) = f_1(t)g_1(t) + f_2(t)g_2(t).$$

Hence the ordinary product rule gives

$$D_t[\mathbf{u}(t) \cdot \mathbf{v}(t)] = D_t[f_1(t)g_1(t) + f_2(t)g_2(t)]$$

$$= [f_1'(t)g_1(t) + f_2'(t)g_2(t)] + [f_1(t)g_1'(t) + f_2(t)g_2'(t)]$$

$$= \mathbf{u}'(t) \cdot \mathbf{v}(t) + \mathbf{u}(t) \cdot \mathbf{v}'(t).$$

Now we can discuss the motion of a point with position vector $\mathbf{r}(t) = f(t)\mathbf{i} + g(t)\mathbf{j}$. Its **velocity vector** $\mathbf{v}(t)$ and **acceleration vector** $\mathbf{a}(t)$ are defined by

$$\mathbf{v}(t) = \mathbf{r}'(t) = f'(t)\mathbf{i} + g'(t)\mathbf{j},$$

$$\mathbf{v} = \frac{d\mathbf{r}}{dt} = \frac{dx}{dt}\mathbf{i} + \frac{dy}{dt}\mathbf{j}; \qquad \left.\right\} \quad (5)$$

$$\mathbf{a}(t) = \mathbf{v}'(t) = f''(t)\mathbf{i} + g''(t)\mathbf{j},$$

$$\mathbf{a} = \frac{d^2\mathbf{r}}{dt^2} = \frac{d^2x}{dt^2}\mathbf{i} + \frac{d^2y}{dt^2}\mathbf{j}. \quad \Bigg\} \quad (6)$$

The moving point also has a **speed** $v(t)$ and **scalar acceleration** $a(t)$. These are just the lengths of the corresponding velocity and acceleration vectors; that is,

$$v(t) = |\mathbf{v}(t)| = \sqrt{\left(\frac{dx}{dt}\right)^2 + \left(\frac{dy}{dt}\right)^2} \quad (7)$$

and

$$a(t) = |\mathbf{a}(t)| = \sqrt{\left(\frac{d^2x}{dt^2}\right)^2 + \left(\frac{d^2y}{dt^2}\right)^2}. \quad (8)$$

Do *not* assume that the scalar acceleration a and the derivative dv/dt of the speed are equal. Though this holds part of the time for one-dimensional motion, it is almost *never* true for two-dimensional motion.

EXAMPLE 1 A moving particle has position vector

$$\mathbf{r}(t) = t\mathbf{i} + t^2\mathbf{j}.$$

Find the velocity and acceleration vectors and the speed and the scalar acceleration when $t = 2$.

Solution Since $\mathbf{r}(2) = 2\mathbf{i} + 4\mathbf{j}$, the particle is at the point $(2, 4)$ on the parabola $y = x^2$ when $t = 2$. Its velocity vector is

$$\mathbf{v}(t) = \mathbf{i} + 2t\mathbf{j}.$$

So $\mathbf{v}(2) = \mathbf{i} + 4\mathbf{j}$ and $v(2) = |\mathbf{v}(2)| = \sqrt{17}$. Note that the *velocity* $\mathbf{v}(2)$ is a *vector*, while the *speed* $v(2)$ is a *scalar*.

The acceleration vector is $\mathbf{a} = 2\mathbf{j}$, so the acceleration of the particle is the (constant) vector $2\mathbf{j}$. In particular, when $t = 2$, its acceleration is $2\mathbf{j}$ and its scalar acceleration is $a = 2$. Figure 10.25 shows the trajectory of the particle with the vectors $\mathbf{v}(2)$ and $\mathbf{a}(2)$ attached at the point corresponding to $t = 2$.

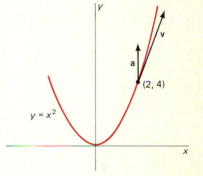

10.25 The velocity and acceleration vectors at $t = 2$ (Example 1).

If $\mathbf{r}(t) = f(t)\mathbf{i} + g(t)\mathbf{j}$ is continuous, then—by taking limits componentwise—we get

$$\int_a^b \mathbf{r}(t)\, dt = \lim_{\Delta t \to 0} \sum_{i=1}^{n} \mathbf{r}(t_i^*)\, \Delta t$$

$$= \mathbf{i}\left(\lim_{\Delta t \to 0} \sum_{i=1}^{n} f(t_i^*)\, \Delta t\right) + \mathbf{j}\left(\lim_{\Delta t \to 0} \sum_{i=1}^{n} g(t_i^*)\, \Delta t\right).$$

This gives the *definition*

$$\int_a^b \mathbf{r}(t)\, dt = \mathbf{i}\left(\int_a^b f(t)\, dt\right) + \mathbf{j}\left(\int_a^b g(t)\, dt\right). \quad (9)$$

Thus *a vector-valued function is integrated componentwise.*

If $\mathbf{R}(t)$ is an antiderivative of $\mathbf{r}(t)$, which means that $\mathbf{R}'(t) = \mathbf{r}(t)$, we find that the Fundamental Theorem of Calculus takes the form

$$\int_a^b \mathbf{r}(t)\, dt = \Big[\, \mathbf{R}(t)\, \Big]_a^b = \mathbf{R}(b) - \mathbf{R}(a). \tag{10}$$

Vector integration is the basis for one method of navigation. If a submarine is cruising beneath the icecap at the North Pole, and thus can use neither visual nor radio methods to determine its position, here is an alternative: Build a sensitive gyroscope-accelerometer combination and install it in the submarine. The device continuously measures the sub's acceleration vector, starting at the time $t = 0$ when its position $\mathbf{r}(0)$ and velocity $\mathbf{v}(0)$ are known. Since $\mathbf{v}'(t) = \mathbf{a}(t)$, Equation (10) gives

$$\int_0^t \mathbf{a}(t)\, dt = \Big[\, \mathbf{v}(t)\, \Big]_0^t = \mathbf{v}(t) - \mathbf{v}(0),$$

so that

$$\mathbf{v}(t) = \mathbf{v}(0) + \int_0^t \mathbf{a}(t)\, dt.$$

Thus the velocity at every time t is known. Similarly, because $\mathbf{r}'(t) = \mathbf{v}(t)$, a second integration gives

$$\mathbf{r}(t) = \mathbf{r}(0) + \int_0^t \mathbf{v}(t)\, dt$$

for the position at every time t. On-board computers can be programmed to carry out these integrations (perhaps using Simpson's approximation), and thus continuously provide the crew with the submarine's position and velocity.

Indefinite integrals of vector-valued functions may also be computed. If $\mathbf{R}'(t) = \mathbf{r}(t)$, then every antiderivative of $\mathbf{r}(t)$ is of the form $\mathbf{R}(t) + \mathbf{C}$ for some constant vector $\mathbf{C}$. We therefore write

$$\int \mathbf{r}(t)\, dt = \mathbf{R}(t) + \mathbf{C} \qquad \text{if} \quad \mathbf{R}'(t) = \mathbf{r}(t). \tag{11}$$

EXAMPLE 2 Suppose that a moving point has initial position $\mathbf{r}(0) = 2\mathbf{i}$, initial velocity $\mathbf{v}(0) = \mathbf{i} - \mathbf{j}$, and acceleration $\mathbf{a}(t) = 2\mathbf{i} + 6t\mathbf{j}$. Find its position and velocity at time t.

Solution Since $\mathbf{a}(t) = \mathbf{v}'(t)$, Equation (11) gives

$$\mathbf{v}(t) = \int \mathbf{a}(t)\, dt + \mathbf{C} = \int (2\mathbf{i} + 6t\mathbf{j})\, dt + \mathbf{C},$$

$$= 2t\mathbf{i} + 3t^2\mathbf{j} + \mathbf{C}.$$

To evaluate $\mathbf{C}$, we use the fact that $\mathbf{v}(0)$ is known; we substitute $t = 0$ into both sides of the last equation and find that $\mathbf{C} = \mathbf{v}(0) = \mathbf{i} - \mathbf{j}$. So

$$\mathbf{v}(t) = (2t\mathbf{i} + 3t^2\mathbf{j}) + (\mathbf{i} - \mathbf{j}) = (2t + 1)\mathbf{i} + (3t^2 - 1)\mathbf{j}.$$

Next,

$$\mathbf{r}(t) = \int \mathbf{v}(t)\,dt + \mathbf{C}^* = (t^2 + t)\mathbf{i} + (t^3 - t)\mathbf{j} + \mathbf{C}^*.$$

Again we substitute $t = 0$, and find that $\mathbf{C}^* = \mathbf{r}(0) = 2\mathbf{i}$. Hence

$$\mathbf{r}(t) = (t^2 + t + 2)\mathbf{i} + (t^3 - t)\mathbf{j}$$

is the position vector of the moving point.

In Problems 1–8, find the values of $\mathbf{r}'(t)$ and $\mathbf{r}''(t)$ for the indicated value of t.

1 $\mathbf{r}(t) = 3\mathbf{i} - 2\mathbf{j}; \quad t = 1.$
2 $\mathbf{r}(t) = t^2\mathbf{i} - t^3\mathbf{j}; \quad t = 2.$
3 $\mathbf{r}(t) = e^{2t}\mathbf{i} + e^{-t}\mathbf{j}; \quad t = 0.$
4 $\mathbf{r}(t) = \mathbf{i}\cos t + \mathbf{j}\sin t; \quad t = \pi/4.$
5 $\mathbf{r}(t) = 3\mathbf{i}\cos 2\pi t + 3\mathbf{j}\sin 2\pi t; \quad t = \frac{3}{4}.$
6 $\mathbf{r}(t) = 5\mathbf{i}\cos t + 4\mathbf{j}\sin t; \quad t = \pi.$
7 $\mathbf{r}(t) = \mathbf{i}\sec t + \mathbf{j}\tan t; \quad t = 0.$
8 $\mathbf{r}(t) = (2t + 3)\mathbf{i} + (6t - 5)\mathbf{j}; \quad t = 10.$

Calculate the integrals in Problems 9–12.

9 $\displaystyle\int_0^{\pi/4} (\mathbf{i}\sin t + 2\mathbf{j}\cos t)\,dt.$

10 $\displaystyle\int_1^e \left(\frac{1}{t}\mathbf{i} - \mathbf{j}\right) dt.$

11 $\displaystyle\int_0^2 t^2(1 + t^3)^{3/2}\mathbf{i}\,dt.$

12 $\displaystyle\int_0^1 (\mathbf{i}e^t - \mathbf{j}te^{-t^2})\,dt.$

In Problems 13–15, apply Theorem 2 to compute the derivative $D_t[\mathbf{u}(t) \cdot \mathbf{v}(t)]$.

13 $\mathbf{u}(t) = 3t\mathbf{i} - \mathbf{j}, \mathbf{v}(t) = 2\mathbf{i} - 5t\mathbf{j}.$
14 $\mathbf{u}(t) = t\mathbf{i} + t^2\mathbf{j}, \mathbf{v}(t) = t^2\mathbf{i} - t\mathbf{j}.$
15 $\mathbf{u}(t) = \mathbf{i}\cos t + \mathbf{j}\sin t, \mathbf{v}(t) = \mathbf{i}\sin t - \mathbf{j}\cos t.$

In Problems 16–20, find $\mathbf{v}(t)$ and $\mathbf{r}(t)$ corresponding to the given values of the acceleration $\mathbf{a}(t)$, initial velocity $\mathbf{v}_0$, and initial position $\mathbf{r}_0$.

16 $\mathbf{a} = \mathbf{0}, \mathbf{r}_0 = 2\mathbf{i} + 3\mathbf{j}, \mathbf{v}_0 = -2\mathbf{j}.$
17 $\mathbf{a} = 2\mathbf{j}, \mathbf{r}_0 = \mathbf{0}, \mathbf{v}_0 = \mathbf{i}.$
18 $\mathbf{a} = \mathbf{i} - \mathbf{j}, \mathbf{v}_0 = \mathbf{i} + \mathbf{j}, \mathbf{r}_0 = \mathbf{j}.$
19 $\mathbf{a} = t\mathbf{i} + t^2\mathbf{j}, \mathbf{r}_0 = \mathbf{i}, \mathbf{v}_0 = \mathbf{0}.$
20 $\mathbf{a} = -\mathbf{i}\sin t + \mathbf{j}\cos t, \mathbf{r}_0 = \mathbf{i}, \mathbf{v}_0 = \mathbf{j}.$

21 Suppose that the vector-valued functions $\mathbf{u}(t)$ and $\mathbf{v}(t)$ both have limits as $t \to a$. Prove that

(i) $\lim_{t\to a} (\mathbf{u}(t) + \mathbf{v}(t)) = \lim_{t\to a} \mathbf{u}(t) + \lim_{t\to a} \mathbf{v}(t);$

(ii) $\lim_{t\to a} (\mathbf{u}(t) \cdot \mathbf{v}(t)) = \left(\lim_{t\to a} \mathbf{u}(t)\right) \cdot \left(\lim_{t\to a} \mathbf{v}(t)\right)$

22 Prove Part (i) of Theorem 2.
23 Prove Part (ii) of Theorem 2.
24 Suppose that both the vector-valued function $\mathbf{r}(t)$ and the real-valued function $h(t)$ are differentiable. Deduce the chain rule for vector-valued functions,

$$D_t[\mathbf{r}(h(t))] = h'(t)\mathbf{r}'(h(t)),$$

in componentwise fashion from the ordinary chain rule.
25 A point moves with constant speed, so that its velocity vector $\mathbf{v}$ satisfies the condition $v^2 = \mathbf{v} \cdot \mathbf{v} = C$ (a constant). Prove that the velocity and acceleration vectors for the point are always perpendicular to each other.
26 A point moves on a circle with center at the origin. Use the dot product to show that the position and velocity vectors of the moving point are always perpendicular.
27 A point moves on the hyperbola $x^2 - y^2 = 1$, with position vector

$$\mathbf{r}(t) = \mathbf{i}\cosh \omega t + \mathbf{j}\sinh \omega t$$

(the number ω is a constant). Show that $\mathbf{a}(t) = c\mathbf{r}(t)$ where c is a positive constant. What sort of external force would produce this sort of motion?
28 Suppose that a point moves on the ellipse $x^2/a^2 + y^2/b^2 = 1$ with position vector $\mathbf{r}(t) = \mathbf{i}a\cos \omega t + \mathbf{j}b\sin \omega t$, ω a constant. Show that $\mathbf{a}(t) = c\mathbf{r}(t)$ where c is a negative constant. To what sort of external force $\mathbf{F}(t)$ does this motion correspond?
29 A point moves in the plane with constant acceleration vector $\mathbf{a} = a\mathbf{j}$. Show that its path is a parabola or a straight line.
30 Suppose that a particle is subject to no force, so its acceleration vector $\mathbf{a}(t)$ is identically zero. Prove that the particle travels with constant speed along a straight line (Newton's first law of motion).

Projectiles and Uniform Circular Motion

Suppose that a projectile is launched from the point (x_0, y_0), with y_0 denoting its initial height above the surface of the earth. Let α be the angle of inclination from the horizontal of its initial velocity vector $\mathbf{v}_0$, as in Figure 10.26. Then its initial position vector is

$$\mathbf{r}_0 = x_0\mathbf{i} + y_0\mathbf{j}, \tag{1a}$$

and Equation (8) in Section 10-3 gives

$$\mathbf{v}_0 = (v_0 \cos \alpha)\mathbf{i} + (v_0 \sin \alpha)\mathbf{j} \tag{1b}$$

where $v_0 = |\mathbf{v}_0|$ is the initial speed of the projectile.

We suppose that the motion takes place sufficiently close to the surface that we may assume the earth is flat and gravity perfectly uniform. Then, if we also ignore air resistance, the acceleration of the projectile is

$$\mathbf{a} = \frac{d\mathbf{v}}{dt} = -g\mathbf{j}$$

where $g \approx 32$ ft/sec^2. Antidifferentiation gives

$$\mathbf{v} = -gt\mathbf{j} + \mathbf{C}_1.$$

Put $t = 0$ into both sides of this last equation. This shows that $\mathbf{C}_1 = \mathbf{v}_0$, and thus that

$$\mathbf{v} = \frac{d\mathbf{r}}{dt} = -gt\mathbf{j} + \mathbf{v}_0.$$

Another antidifferentiation gives

$$\mathbf{r} = -\tfrac{1}{2}gt^2\mathbf{j} + \mathbf{v}_0 t + \mathbf{C}_2.$$

Now substitution of $t = 0$ gives $\mathbf{C}_2 = \mathbf{r}_0$, so the position vector of the projectile at time t is

$$\mathbf{r}(t) = -\tfrac{1}{2}gt^2\mathbf{j} + \mathbf{v}_0 t + \mathbf{r}_0. \tag{2}$$

Equations (1) now give

$$\mathbf{r}(t) = [(v_0 \cos \alpha)t + x_0]\mathbf{i} + [-\tfrac{1}{2}gt^2 + (v_0 \sin \alpha)t + y_0]\mathbf{j},$$

so the parametric equations of the trajectory of the projectile are

$$x = (v_0 \cos \alpha)t + x_0, \tag{3}$$

$$y = -\tfrac{1}{2}gt^2 + (v_0 \sin \alpha)t + y_0. \tag{4}$$

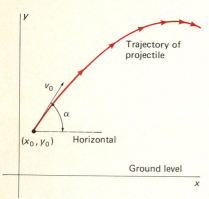

10.26 Trajectory of a projectile launched at the angle α.

EXAMPLE 1 An airplane is flying horizontally at an altitude of 1600 feet, in order to pass directly over snowbound cattle on the ground. Its speed is a constant 150 mph (220 ft/sec). At what angle of sight ϕ (between

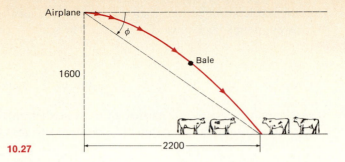

10.27

the horizontal and the direct line to the target) should a bale of hay be released in order to hit the target?

Solution   See Figure 10.27. We take $x_0 = 0$ where the bale of hay is released at time $t = 0$. Then $y_0 = 1600$ (feet), $v_0 = 220$ (ft/sec), and $\alpha = 0$. Then Equations (3) and (4) give

$$x = 220t, \qquad y = -16t^2 + 1600.$$

From the second of these equations we find that $t = 10$ (seconds) when the bale of hay hits the ground ($y = 0$). It has then traveled a horizontal distance of

$$x = (220)(10) = 2200$$

feet. Hence the required angle of sight is

$$\phi = \tan^{-1}\left(\frac{1600}{2200}\right) \approx 36^\circ.$$

UNIFORM CIRCULAR MOTION

Consider a point that moves around the circle with center $(0, 0)$ and radius r at a constant angular speed of ω radians per second. If its initial position is the point $(r, 0)$, then its position vector at time t is

$$\mathbf{r} = \mathbf{i}r\cos\omega t + \mathbf{j}r\sin\omega t. \tag{5}$$

We differentiate to find the velocity vector:

$$\mathbf{v} = -\mathbf{i}r\omega\sin\omega t + \mathbf{j}r\omega\cos\omega t.$$

Note that $\mathbf{v} \cdot \mathbf{r} = 0$, so that $\mathbf{v}$ is tangent to the circle. This is shown in Figure 10.28. The speed of the moving point is

$$v = |\mathbf{v}| = r\omega. \tag{6}$$

A second differentiation gives the acceleration vector

$$\frac{d\mathbf{v}}{dt} = \mathbf{a} = -\mathbf{i}r\omega^2\cos\omega t - \mathbf{j}r\omega^2\sin\omega t,$$

and we note that $\mathbf{a}$ is a scalar multiple of $\mathbf{r}$; indeed,

$$\mathbf{a} = -\omega^2\mathbf{r}.$$

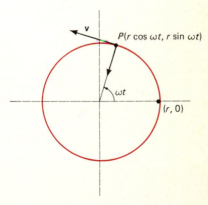

10.28   Uniform circular motion.

Thus the acceleration vector, when we think of it as located at the endpoint of $\mathbf{r}$, always points back toward the origin. The scalar acceleration is

$$a = |\mathbf{a}| = \omega^2|\mathbf{r}| = r\omega^2, \tag{7}$$

proportional to the radius of the circle and to the *square* of the angular velocity.

If the moving point is a particle of mass m, then the **central** force (one directed towards the origin) required to produce this motion is

$$F = ma = mr\omega^2 = \frac{mv^2}{r}, \tag{8}$$

since $v = r\omega$. In the examples below and in the problems, we give some of the many applications of Formula (8).

EXAMPLE 2 A circular racetrack has a radius of $\frac{1}{4}$ mile. At what angle α must the track be banked in order that a car can travel around it at 150 mph (220 ft/sec) with no danger of skidding? The idea is to assume that the track is coated with oil and water, so there's no friction between the tires and the surface of the track.

Solution The forces involved are shown in Figure 10.29. The two forces acting on the car are the gravitational force mg and a normal force $\mathbf{N}$ exerted on the car by the track. The vertical component of $\mathbf{N}$ must balance the gravitational force:

$$|\mathbf{N}| \cos\alpha = mg.$$

The horizontal component of $\mathbf{N}$ supplies the force of Equation (8):

$$|\mathbf{N}| \sin\alpha = \frac{mv^2}{r}.$$

If we divide the second of these two equations by the first, we eliminate $|\mathbf{N}|$. Thus we discover that

$$\tan\alpha = \frac{v^2}{gr}. \tag{9}$$

Now we use the data of the problem—that $v = 220$ (ft/sec) and $r = 1320$ (the number of feet in a quarter mile). We take $g = 32$, and Equation (9) tells us that α is about 49°. Indeed, the *rated speed* of a circular curve in a road banked at angle α is the speed v given by Equation (9).

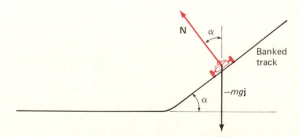

10.29 Racing on a circular banked track.

CHAP. 10: **Parametric Curves and Vectors in the Plane**

Suppose now that our particle of mass m is a natural or an artificial satellite in a circular orbit around a body of mass M. The latter might be either the sun or a planet. According to Newton's law of gravitation, the force that M exerts on m has the scalar value

$$F = G\frac{Mm}{r^2} \qquad (G \text{ is a constant}).$$

We equate this value of F with that in Equation (8), which leads to

$$F = \frac{GMm}{r^2} = \frac{mv^2}{r}. \qquad (10)$$

If the satellite has period of revolution T, then its (constant) speed is $v = 2\pi r/T$. A consequence of Equation (10) is

$$T^2 = kr^3 \qquad (11)$$

where k is the constant $4\pi^2/GM$. Thus, for satellites in circular orbits about the *same* body, *the square of the period of revolution is proportional to the cube of the radius of the orbit*. This is Kepler's third law for the case of uniform circular motion.

EXAMPLE 3 A communications relay satellite is to be placed in a circular orbit about the earth and to have a period of revolution of 24 hours. This is a *synchronous* orbit, in which the satellite appears to be stationary in the sky. Assume that the earth's natural moon has a period of 27.32 days in a circular orbit of radius 239,000 miles. What should be the radius of the satellite's orbit?

Solution Equation (11), when applied to the moon, yields $(27.32)^2 = k(239,000)^3$. For the stationary satellite that has period $T = 1$ (day), it yields $(1)^2 = kr^3$ where r is the radius of the synchronous orbit. To eliminate k, we divide the second of these equations by the first, and we find that

$$r^3 = \frac{(239,000)^3}{(27.32)^2}.$$

Thus r is approximately 26,350 miles. The radius of the earth is about 3960 miles, so the satellite will be 22,390 miles above the surface.

10-5 PROBLEMS

Problems 1–6 deal with a projectile fired from the origin (so that $x_0 = y_0 = 0$) with initial speed v_0 and initial angle of inclination α. The *range* R of the projectile is the horizontal distance it travels before returning to the ground.

1 If $\alpha = 45°$, what value of v_0 gives a range of 1 mile?

2 If $\alpha = 60°$ and $R = 1$ mile, what is the maximum height attained by the projectile?

3 Deduce from Equations (3) and (4) the fact that $R = \frac{1}{16}v_0^2 \sin\alpha \cos\alpha$.

4 Given the initial speed v_0, find the angle α that maximizes the range R. Use the result of Problem 3.

5 Suppose that $v_0 = 160$ feet per second. Find the max-

imum height y_m and the range R of the projectile if (a) $\alpha = 30°$; (b) $\alpha = 45°$; (c) $\alpha = 60°$.

6 The projectile of Problem 5 is to be fired at a target 600 feet away, and there is a hill 300 feet high midway between the gun site and this target. At what initial angle of inclination should the projectile be fired?

7 A projectile is to be fired horizontally from the top of a 400-foot cliff at a target 1 mile from the base of the cliff. What should be the initial velocity of the projectile?

8 A bomb is dropped (initial speed zero) from a helicopter hovering at a height of 800 feet. A projectile is fired from a gun located on the ground 800 feet to the west of the point directly beneath the helicopter. The intent is for the projectile to intercept the bomb at a height of exactly 400 feet. If the projectile is fired at the same instant that the bomb is dropped, what should be its initial velocity and angle of inclination?

9 Suppose, more realistically, that the projectile of Problem 8 is fired one second after the bomb is dropped. What should be its initial velocity and angle of inclination?

In Problems 10–14, use the value $G = 6.673 \times 10^{-11}$ for the gravitational constant. This is its approximate value in mks units.

10 The acceleration of gravity at the surface of the earth is 32.17 ft/sec², and the radius of the earth is 3960 miles. Calculate the mass of the earth in kilograms.

11 Assume that the earth's orbit around the sun is circular (this is very nearly true) with a radius of 1.496×10^{11} meters. Assume also that the period of the earth is 365.26 days. Calculate the mass of the sun and its ratio to the mass of the earth.

12 Given the fact that Jupiter's period of (almost) circular revolution around the sun is 11.86 years, calculate the distance of Jupiter from the sun.

13 Given the period 27.32 days of the moon's almost circular motion around the earth, compute the distance (in miles) to the moon.

14 Jupiter's moon Ganymede has a period of revolution of 7.166 (Earth) days in an orbit with a radius of 1,070,000 kilometers. What is the mass of Jupiter, both in kilograms and as a multiple of the mass of the earth?

15 Suppose that the circular racetrack of Example 2 has a radius of 1 mile. At what angle should it be banked in order to allow cars to travel at 200 mph?

16 A circular racetrack is like the one of Example 2 except that it is *not* banked. Instead, the necessary centripetal force

$mr\omega^2$ is supplied by a force of resistance to sliding that is equal to one-fourth of the weight $W = mg$ of the car. How fast can a car travel without skidding around this track?

17 At carnivals one sometimes sees a motorcyclist riding around the interior of a vertical cylinder. The motorcycle is acted upon by three forces, as shown in Figure 10.30:

> A normal force $N = mr\omega^2$ that supplies the necessary centripetal acceleration,
>
> The weight $W = mg$ of the motorcycle, and
>
> A force $R = kN = W$ of sliding resistance.

If the radius of the cylinder is $r = 25$ feet and the proportionality constant k is 0.4, what should the speed v of the motorcycle be?

10.30 A daring motorcyclist.

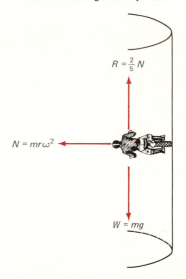

$$R = \tfrac{2}{5} N$$

$$N = mr\omega^2$$

$$W = mg$$

10-6

Curvature and Acceleration

The word *curvature* has an intuitive meaning that we need to make precise. Most people would agree that a straight line does not curve at all, while a circle of small radius seems to curve more than a circle of large radius (see Figure 10.31). This judgment may be based on a feeling that curvature is "rate of change of direction." The direction of a curve is determined by its velocity vector, so you would expect the idea of curvature to have something to do with the rate at which the velocity vector is turning.

10.31 The intuitive idea of curvature.

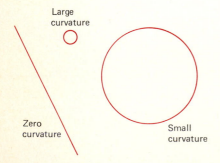

Large curvature

Zero curvature

Small curvature

Let $\mathbf{r}(t) = x(t)\mathbf{i} + y(t)\mathbf{j}$, $a \leq t \leq b$, be the position vector of a smooth curve with nonzero velocity vector $\mathbf{v}(t) = \mathbf{r}'(t)$. The curve's **unit tangent vector** at the point $\mathbf{r}(t)$ is the unit vector

$$\mathbf{T}(t) = \frac{\mathbf{v}(t)}{|\mathbf{v}(t)|} = \frac{\mathbf{v}(t)}{v(t)} \qquad (1)$$

where $v(t) = |\mathbf{v}(t)|$ is the speed. Now denote by ϕ the angle of inclination of $\mathbf{T}$, measured counterclockwise from the positive x-axis, as in Figure 10.32. Then

$$\mathbf{T} = \mathbf{i} \cos \phi + \mathbf{j} \sin \phi. \qquad (2)$$

The **principal unit normal vector** to the curve is the unit vector

$$\mathbf{N} = \pm \mathbf{T}_\perp = \pm(-\mathbf{i} \sin \phi + \mathbf{j} \cos \phi) \qquad (3)$$

where $\mathbf{T}_\perp$ is the vector obtained by rotating $\mathbf{T}$ through a counterclockwise angle of 90° (compare this with Equation (9) in Section 10-3). We choose the sign of $\mathbf{N}$ so that $\mathbf{N}$ points in the direction in which the curve is bending. Specifically, if $\mathbf{T}$ points to the right (its $\mathbf{i}$-component is positive), then we take the plus sign in (3) if the curve is concave upward and the minus sign if it is concave downward. If $\mathbf{T}$ points to the left, the choice of signs is reversed. The relationship between $\mathbf{T}$, $\mathbf{N}$, and the direction of curvature is shown in Figure 10.33.

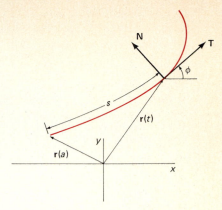

10.32   The unit tangent vector **T**.

EXAMPLE 1   The parabola $y = x^2$ can be parametrized by $x = t$, $y = t^2$. Find $\mathbf{T}$ and $\mathbf{N}$ at the point $(1, 1)$.

Solution   The position vector is $\mathbf{r}(t) = t\mathbf{i} + t^2\mathbf{j}$, so $\mathbf{v}(t) = \mathbf{i} + 2t\mathbf{j}$. The speed is $v(t) = (1 + 4t^2)^{1/2}$, so Equation (1) yields

$$\mathbf{T}(t) = \frac{\mathbf{i} + 2t\mathbf{j}}{\sqrt{1 + 4t^2}}.$$

By substituting $t = 1$, we find that the unit tangent vector at $(1, 1)$ is

$$\mathbf{T} = \frac{1}{\sqrt{5}}\mathbf{i} + \frac{2}{\sqrt{5}}\mathbf{j}.$$

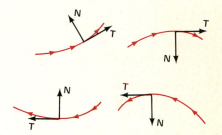

10.33   Direction of the principal unit normal vector **N**.

Since the parabola is concave upward at $(1, 1)$, the plus sign is the correct one to choose in Equation (3), and the principal unit normal vector is thus

$$\mathbf{N} = -\frac{2}{\sqrt{5}}\mathbf{i} + \frac{1}{\sqrt{5}}\mathbf{j}.$$

The **curvature** at a point of a curve, denoted by the lower-case Greek letter kappa, is defined to be

$$\kappa = \left| \frac{d\phi}{ds} \right|, \qquad (4)$$

the absolute value of the rate of change of the angle ϕ with respect to arc length s. We measure s along the curve from its initial point $\mathbf{r}(a)$ to the point $\mathbf{r}(t)$, so that

$$s = \int_a^t \sqrt{[x'(u)]^2 + [y'(u)]^2} \, du = \int_a^t v(u) \, du \qquad (5)$$

by Formula (6) in Section 10-2. The Fundamental Theorem of Calculus then implies that

$$v = \frac{ds}{dt}. \qquad (6)$$

We define the curvature κ in terms of $d\phi/ds$ rather than $d\phi/dt$ because the latter depends not only upon the shape of the curve but also upon the

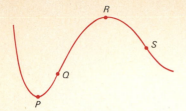

10.34 The curvature is large at P and R, small at Q and S.

speed of the moving point $\mathbf{r}(t)$. For a straight line the angle ϕ is a constant, so the curvature given by Equation (4) is zero. If you imagine a point moving with constant speed along a curve, the curvature is greatest at points where ϕ changes most rapidly, such as the points P and R on the curve of Figure 10.34. The curvature is least at points such as Q and S, where ϕ is changing the least rapidly.

We need to derive a formula that is effective in computing the curvature of a *smooth* parametric curve $x = x(t)$, $y = y(t)$. First we note that

$$\phi = \tan^{-1}\left(\frac{dy}{dx}\right) = \tan^{-1}\left(\frac{y'(t)}{x'(t)}\right)$$

provided $x'(t) \neq 0$. Hence

$$\frac{d\phi}{dt} = \frac{y''x' - y'x''}{(x')^2} \div \left(1 + \left(\frac{y'}{x'}\right)^2\right) = \frac{x'y'' - x''y'}{(x')^2 + (y')^2}.$$

Since $v = ds/dt > 0$, Equation (4) gives

$$\kappa = \left|\frac{d\phi}{ds}\right| = \left|\frac{d\phi}{dt}\frac{dt}{ds}\right| = \frac{1}{v}\left|\frac{d\phi}{dt}\right|;$$

thus

$$\kappa = \frac{|x'y'' - x''y'|}{[(x')^2 + (y')^2]^{3/2}} = \frac{|x'y'' - x''y'|}{v^3}. \tag{7}$$

At a point where $x'(t) = 0$, we know that $y'(t) \neq 0$ because the curve is smooth. Thus we will obtain the same result if we begin with the equation $\phi = \cot^{-1}(x'/y')$.

An explicitly described curve $y = f(x)$ may be regarded as a parametric curve $x = x$, $y = f(x)$. Then $x' = 1$ and $x'' = 0$, so Equation (7)—with x in place of t as the parameter—becomes

$$\kappa = \frac{|y''|}{[1 + (y')^2]^{3/2}} = \frac{|d^2y/dx^2|}{[1 + (dy/dx)^2]^{3/2}}. \tag{8}$$

EXAMPLE 2 Show that the curvature at each point of a circle of radius a is $\kappa = 1/a$.

Solution With the familiar parametrization $x = a \cos t$, $y = a \sin t$ of such a circle centered at the origin, we have

$$x' = -a \sin t, \qquad y' = a \cos t,$$

and

$$x'' = -a \cos t, \qquad y'' = -a \sin t.$$

Hence (7) gives

$$\kappa = \frac{|(-a \sin t)(-a \sin t) - (-a \cos t)(a \cos t)|}{[(-a \sin t)^2 + (a \cos t)^2]^{3/2}} = \frac{a^2}{a^3} = \frac{1}{a}.$$

Alternatively, we could have used Formula (8); our point of departure would be the equation $x^2 + y^2 = a^2$ of the same circle, and we would compute y' and y'' by implicit differentiation (see Problem 21).

Suppose that P is a point on a parametrized curve where $\kappa \neq 0$. Consider the circle that is tangent to the curve at the point P and which has the

CHAP. 10: **Parametric Curves and Vectors in the Plane**

same curvature there. The center of the circle is to lie on the concave side of the curve; that is, on the side toward which the normal vector **N** points. This circle is called the **osculating circle (or circle of curvature)** of the curve at the given point because it touches the curve so closely there (*osculum* is the Latin word for *kiss*). Let ρ be the radius of the osculating circle and let γ be the position vector of its center; $\gamma = \overrightarrow{OC}$ where C is the center of the osculating circle. Then ρ is called the **radius of curvature** of the curve at the point P, and γ is called the (vector) **center of curvature** of the curve at P. See Figure 10.35.

Example 2 implies that the radius of curvature is

$$\rho = \frac{1}{\kappa}, \tag{9}$$

and the fact that $|\mathbf{N}| = 1$ implies that the position vector of the center of curvature is

$$\gamma = \mathbf{r} + \rho\mathbf{N} \qquad (\mathbf{r} = \overrightarrow{OP}). \tag{10}$$

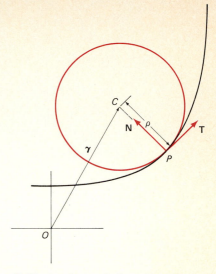

10.35 Osculating circle, radius of curvature, center of curvature.

EXAMPLE 3 Determine the radius of curvature and the center of curvature of the parabola $y = x^2$ at the point (1, 1).

Solution   By Equation (8),

$$\kappa = \frac{|2|}{(1 + 4x^2)^{3/2}},$$

so $\kappa = 2/(5\sqrt{5})$ and $\rho = (5\sqrt{5})/2$ at the point (1, 1).

Now for the vector center of curvature: In Example 1 we found the unit tangent and normal vectors to the same curve at the same point:

$$\mathbf{T} = \frac{1}{\sqrt{5}}\mathbf{i} + \frac{2}{\sqrt{5}}\mathbf{j} \quad \text{and} \quad \mathbf{N} = -\frac{2}{\sqrt{5}}\mathbf{i} + \frac{1}{\sqrt{5}}\mathbf{j}.$$

Hence Equation (10) gives the center of curvature as

$$\gamma = \langle 1, 1 \rangle + \frac{5\sqrt{5}}{2}\left\langle -\frac{2}{\sqrt{5}}, \frac{1}{\sqrt{5}} \right\rangle = \left\langle -4, \frac{7}{2} \right\rangle.$$

The equation of the osculating circle to the parabola at (1, 1) is therefore

$$(x + 4)^2 + (y - \tfrac{7}{2})^2 = \rho^2 = \tfrac{125}{4}.$$

The parabola and this osculating circle are shown in Figure 10.36.

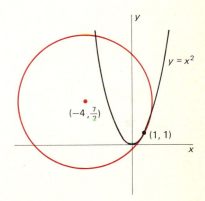

10.36 The osculating circle of Example 3.

There is an important relationship between the curvature κ, the unit normal vector **N**, and the rate of change of the unit tangent vector **T** with respect to arc length. We begin with Equation (2), and differentiate each side with respect to arc length s. This gives

$$\frac{d\mathbf{T}}{ds} = (-\mathbf{i}\sin\phi + \mathbf{j}\cos\phi)\frac{d\phi}{ds} = (\pm\mathbf{N})(\pm\kappa).$$

To determine the correct sign, we first consider the case in which **T** points to the right. From our analysis of the signs in Equation (3) it follows that

$$-\mathbf{i}\sin\phi + \mathbf{j}\cos\phi = +\mathbf{N}$$

if the curve is concave upward, while

$$-\mathbf{i} \sin \phi + \mathbf{j} \cos \phi = -\mathbf{N}$$

if it is concave downward. But the definition of κ shows that $d\phi/ds = +\kappa$ > 0 if the curve is concave upward, while $d\phi/ds = -\kappa < 0$ if it is concave downward. Thus we take both plus signs in the one case and both minus signs in the other. In either case we obtain

$$\frac{d\mathbf{T}}{ds} = \kappa \mathbf{N}. \tag{11}$$

The result of the analysis is the same if $\mathbf{T}$ points to the left.

NORMAL AND TANGENTIAL COMPONENTS OF ACCELERATION

We may apply Equation (11) to analyze the meaning of the acceleration vector of a moving particle with velocity vector $\mathbf{v}$ and speed v. Then Equation (1) gives $\mathbf{v} = v\mathbf{T}$, so the acceleration vector of the particle is

$$\mathbf{a} = \frac{d\mathbf{v}}{dt} = \frac{dv}{dt}\mathbf{T} + v\frac{d\mathbf{T}}{dt}$$

$$= \frac{dv}{dt}\mathbf{T} + v\frac{d\mathbf{T}}{ds}\frac{ds}{dt}.$$

Since $ds/dt = v$, Equation (11) gives us the formula

$$\mathbf{a} = \frac{dv}{dt}\mathbf{T} + \kappa v^2 \mathbf{N}. \tag{12}$$

Since $\mathbf{T}$ and $\mathbf{N}$ are unit vectors tangent and normal to the curve, respectively, Equation (12) provides a *decomposition of the acceleration vector* into its components tangent and normal to the trajectory. Such a decomposition is illustrated in Figure 10.37. The **tangential component**

$$a_T = \frac{dv}{dt} \tag{13}$$

is the particle's rate of change of speed, while the **normal component**

$$a_N = \kappa v^2 = \frac{v^2}{\rho} \tag{14}$$

measures the rate of change of its direction of motion.

The results of Section 10-5, on uniform circular motion (motion with constant speed v around a circle of radius r), follow directly from Equation (12). Because v is constant, $dv/dt = 0$, and so $\mathbf{a} = \kappa v^2 \mathbf{N}$. But $\mathbf{N}$ is a unit vector directed toward the center of the circle, and $\kappa = 1/r$ by Example 2 of this section. Hence the centripetal acceleration is

$$|\mathbf{a}| = \frac{v^2}{r} = r\omega^2,$$

where $\omega = v/r$ is the angular speed.

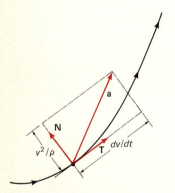

10.37 Resolution of acceleration **a** into tangential and normal components.

CHAP. 10: **Parametric Curves and Vectors in the Plane**

As an application of Equation (12), think of a train moving along a straight track with constant speed v, so that $a_T = 0 = a_N$ (the latter because $\kappa = 0$ for a straight line). Suppose that at time $t = 0$, the train enters a circular curve of radius ρ. At that instant, it will be *suddenly* subjected to a normal acceleration of magnitude v^2/ρ. A passenger in the train will experience a sudden jerk. If v is large, the stresses may be so great as to damage the track or to derail the train. It is for exactly this reason that railroads are not built with curves shaped like arcs of circles, but with *approach curves* in which the curvature, and hence the normal acceleration, build up smoothly.

EXAMPLE 4 A particle moves with parametric equations

$$x = \tfrac{3}{2}t^2, \qquad y = \tfrac{4}{3}t^3.$$

Find the tangential and normal components of its acceleration vector when $t = 1$.

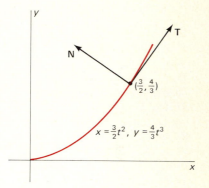

Solution The trajectory, **N**, and **T** appear in Figure 10.38; **N** and **T** are shown attached at the point of evaluation, at which $t = 1$. The particle has position vector $\mathbf{r} = \tfrac{3}{2}t^2\mathbf{i} + \tfrac{4}{3}t^3\mathbf{j}$, and velocity vector $\mathbf{v} = 3t\mathbf{i} + 4t^2\mathbf{j}$. Hence its speed is

$$v = \sqrt{9t^2 + 16t^4},$$

from which we calculate

10.38 The moving particle of Example 4.

$$a_T = \frac{dv}{dt} = \frac{9t + 32t^3}{\sqrt{9t^2 + 16t^4}}.$$

Thus $v = 5$ and $a_T = \tfrac{41}{5}$ when $t = 1$.

Since $x' = 3t$, $y' = 4t^2$, $x'' = 3$, and $y'' = 8t$, we can use Equation (7) to find the curvature at $t = 1$:

$$\kappa = \frac{|x'y'' - x''y'|}{v^3} = \frac{|(3)(8) - (3)(4)|}{(5)^3} = \frac{12}{125}.$$

Hence

$$a_N = \kappa v^2 = (\tfrac{12}{125})(5)^2 = \tfrac{12}{5}$$

when $t = 1$. As a check (Problem 22), you might compute **T** and **N** when $t = 1$ and verify that

$$\tfrac{41}{5}\mathbf{T} + \tfrac{12}{5}\mathbf{N} = \mathbf{a} = 3\mathbf{i} + 8\mathbf{j}.$$

10-6 PROBLEMS

In Problems 1–6, find the curvature of the given curve at the indicated point.

1 $y = x^3$; at $(0, 0)$.
2 $y = x^3$; at $(-1, -1)$.
3 $y = \cos x$; at $(0, 1)$.
4 $x = t - 1, y = t^2 + 3t + 2$; where $t = 2$.

5 $x = 5 \cos t, y = 4 \sin t$; where $t = \pi/4$.

6 $x = 5 \cosh t, y = 3 \sinh t$; where $t = 0$.

In Problems 7–10, find the point or points of the given curve at which the curvature is a maximum.

7 $y = e^x$.
8 $y = \ln x$.
9 $x = 5 \cos t, y = 3 \sin t$.
10 $xy = 1$.

For each of the curves in Problems 11–15, find the unit tangent and normal vectors at the indicated point.

11 $y = x^3$; at $(-1, -1)$.
12 $x = t^3, y = t^2$; at $(-1, 1)$.
13 $x = 3 \sin 2t, y = 4 \cos 2t$; where $t = \pi/6$.
14 $x = t - \sin t, y = 1 - \cos t$; where $t = \pi/2$.
15 $x = \cos^3 t, y = \sin^3 t$; where $t = 3\pi/4$.

The position vector of a moving particle is given in each of Problems 16–20. Find the tangential and normal components of the acceleration vector.

16 $\mathbf{r}(t) = 3\mathbf{i} \sin \pi t + 3\mathbf{j} \cos \pi t$.
17 $\mathbf{r}(t) = (2t + 1)\mathbf{i} + (3t^2 - 1)\mathbf{j}$.
18 $\mathbf{r}(t) = \mathbf{i} \cosh 3t + \mathbf{j} \sinh 3t$.
19 $\mathbf{r}(t) = \mathbf{i}t \cos t + \mathbf{j}t \sin t$.
20 $\mathbf{r}(t) = \mathbf{i}e^t \sin t + \mathbf{j}e^t \cos t$.
21 Use Formula (8) to compute the curvature of the circle with equation $x^2 + y^2 = a^2$.
22 Verify the equation $\frac{41}{5}\mathbf{T} + \frac{12}{5}\mathbf{N} = 3\mathbf{i} + 8\mathbf{j}$ mentioned at the end of Example 4.

In each of Problems 23–25, find the equation of the osculating circle for the given curve at the indicated point.

23 $y = 1 - x^2$; at $(0, 1)$.
24 $y = e^x$; at $(0, 1)$.
25 $xy = 1$; at $(1, 1)$.
26 (a) Deduce from Equation (11) that

$$\mathbf{N} = \frac{d\mathbf{T}/dt}{|d\mathbf{T}/dt|}.$$

(b) Use the above formula to compute $\mathbf{N}$ for the circle $x = a \sin \omega t, y = a \cos \omega t$.
27 A particle moves under the influence of a force that is always perpendicular to its direction of motion. Show that the speed of the particle must be constant.

28 Deduce from Equation (12) that

$$\kappa = \frac{\sqrt{a^2 - a_T^2}}{v^2} = \frac{[(x'')^2 + (y'')^2 - (v')^2]^{1/2}}{(x')^2 + (y')^2}$$

where each prime denotes differentiation with respect to t.
29 Apply the formula of Problem 28 to calculate the curvature of the curve

$$x = \cos t + t \sin t, \qquad y = \sin t - t \cos t.$$

30 Find the curvature and center of curvature of the folium of Descartes $x^3 + y^3 = 3xy$ at the point $(\frac{3}{2}, \frac{3}{2})$. Begin by calculating dy/dx and d^2y/dx^2 by implicit differentiation.
31 Determine the constants $A, B, C, D, E,$ and F so that the curve

$$y = Ax^5 + Bx^4 + Cx^3 + Dx^2 + Ex + F$$

does, simultaneously, *all* the following:

(i) Joins the two points $(0, 0)$ and $(1, 1)$;
(ii) Has slope 0 at $(0, 0)$ and slope 1 at $(1, 1)$;
(iii) Has curvature 0 at both $(0, 0)$ and $(1, 1)$.

The curve in question is shown in color in Figure 10.39. Why would this be a good curve to join the railway tracks, which are shown in black in the figure?

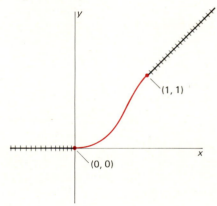

10.39 Connecting two railroad tracks (Problem 31).

---

*10-7

Orbits of Planets and Satellites

Ancient Greek astronomers and mathematicians developed an elaborate mathematical model to account for the complicated motions of the sun, moon, and six planets (then known) as viewed from Earth. A combination of uniform circular motions was used to describe the motion of each body about the Earth—if the Earth is placed at the origin, then each body *does* orbit the Earth.

In this system, it was typical for a planet P to travel uniformly around a small circle (the *epicycle*) with center C, which in turn traveled uniformly

CHAP. 10: **Parametric Curves and Vectors in the Plane**

around a circle centered at the Earth, labeled E in Figure 10.40. The radii of the circles and the angular speeds of P and C around them were chosen to match the observed motion of the planet as closely as possible. For greater accuracy, one could use secondary circles. In fact, several circles were required for each body in the solar system. The theory of epicycles reached its definitive form in Ptolemy's *Almagest* of the second century A.D.

In 1543 Copernicus altered Ptolemy's approach by placing the center of each primary circle at the sun rather than at the Earth. This change was of much greater philosophical than mathematical importance. For, contrary to popular belief, this *heliocentric system* was *not* simpler than Ptolemy's geocentric system; indeed, Copernicus's system actually required more circles.

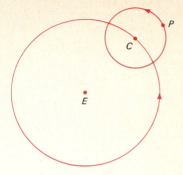

10.40 The small circle is the epicycle.

It was Johann Kepler (1571–1630) who finally got rid of all these circles. On the basis of a detailed analysis of a lifetime of planetary observations by the Danish astronomer Tycho Brahe, Kepler stated the following three propositions, now known as **Kepler's laws of planetary motion:**

I. The orbit of each planet is an ellipse with the sun at one focus.
II. The radius vector from the sun to a planet sweeps out area at a constant rate.
III. The *square* of the period of revolution of a planet is proportional to the *cube* of the major semiaxis of its elliptical orbit.

In his *Principia Mathematica* (1687) Newton showed that Kepler's laws follow from the basic principles of mechanics ($F = ma$, and so on) and the inverse-square law of gravitational attraction. His success in using mathematics to explain natural phenomena ("I now demonstrate the frame of the System of the World") inspired confidence that the universe could be understood, and perhaps even mastered. This new confidence permanently altered humanity's perception of itself and of its place in the scheme of things.

In this section we show how Kepler's laws can be derived as indicated above. To begin with, set up a coordinate system in which the sun is located at the origin in the plane of the planet's motion. Let $r = r(t)$ and $\theta = \theta(t)$ be the polar coordinates at time t of the planet as it moves in its orbit about the sun. We want first to split the planet's position, velocity, and acceleration vectors $\mathbf{r}$, $\mathbf{v}$, and $\mathbf{a}$ into *radial* and *transverse* components. To do so, we introduce at each point (r, θ) of the plane (the origin excepted) the *unit* vectors

$$\mathbf{u}_r = \mathbf{i} \cos \theta + \mathbf{j} \sin \theta, \qquad \mathbf{u}_\theta = -\mathbf{i} \sin \theta + \mathbf{j} \cos \theta. \tag{1}$$

If we substitute $\theta = \theta(t)$, then $\mathbf{u}_r$ and $\mathbf{u}_\theta$ become functions of t. The **radial** unit vector $\mathbf{u}_r$ always points directly away from the origin; the **transverse** unit vector $\mathbf{u}_\theta$ is obtained from $\mathbf{u}_r$ by a 90° counterclockwise rotation, as shown in Figure 10.41.

In Problem 6 we ask you to verify, by componentwise differentiation of Equations (1), that

$$\frac{d\mathbf{u}_r}{dt} = \mathbf{u}_\theta \frac{d\theta}{dt} \quad \text{and} \quad \frac{d\mathbf{u}_\theta}{dt} = -\mathbf{u}_r \frac{d\theta}{dt}. \tag{2}$$

The position vector $\mathbf{r}$ points directly away from the origin and has length $|\mathbf{r}| = r$, so

$$\mathbf{r} = r\mathbf{u}_r. \tag{3}$$

10.41 The radial and transverse unit vectors $\mathbf{u}_r$ and $\mathbf{u}_\theta$.

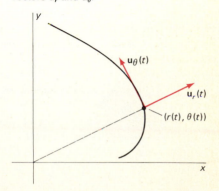

Differentiation of both sides of Equation (3) with respect to t gives the result

$$\mathbf{v} = \frac{d\mathbf{r}}{dt} = \mathbf{u}_r \frac{dr}{dt} + r \frac{d\mathbf{u}_r}{dt}.$$

We use the first equation in (2) and find that the planet's velocity vector is

$$\mathbf{v} = \mathbf{u}_r \frac{dr}{dt} + r \frac{d\theta}{dt} \mathbf{u}_\theta. \tag{4}$$

Thus we have expressed the velocity $\mathbf{v}$ in terms of the radial vector $\mathbf{u}_r$ and the transverse vector $\mathbf{u}_\theta$.

We differentiate this last equation, and find that

$$\mathbf{a} = \frac{d\mathbf{v}}{dt} = \left(\mathbf{u}_r \frac{d^2r}{dt^2} + \frac{dr}{dt} \frac{d\mathbf{u}_r}{dt} \right) + \left(\frac{dr}{dt} \frac{d\theta}{dt} \mathbf{u}_\theta + r \frac{d^2\theta}{dt^2} \mathbf{u}_\theta + r \frac{d\theta}{dt} \frac{d\mathbf{u}_\theta}{dt} \right).$$

Then, by using Equations (2) and collecting the coefficients of $\mathbf{u}_r$ and $\mathbf{u}_\theta$ (Problem 7), we obtain the decomposition

$$\mathbf{a} = \left[\frac{d^2r}{dt^2} - r \left(\frac{d\theta}{dt} \right)^2 \right] \mathbf{u}_r + \left[\frac{1}{r} \frac{d}{dt} \left(r^2 \frac{d\theta}{dt} \right) \right] \mathbf{u}_\theta \tag{5}$$

of the acceleration vector into its radial and transverse components.

Let M denote the mass of the sun and m the mass of the orbiting planet. The inverse-square law of gravitation in vector form is

$$\mathbf{F} = m\mathbf{a} = -\frac{GMm}{r^2} \mathbf{u}_r,$$

so the acceleration of the planet *also* is given by

$$\mathbf{a} = -\frac{\mu}{r^2} \mathbf{u}_r \tag{6}$$

where $\mu = GM$. We equate the transverse components of (5) and (6), and thus obtain

$$\frac{1}{r} \frac{d}{dt} \left(r^2 \frac{d\theta}{dt} \right) = 0.$$

We antidifferentiate both sides and find that

$$r^2 \frac{d\theta}{dt} = h \qquad (h \text{ is a constant}). \tag{7}$$

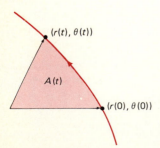

10.42 Area swept out by the radius vector.

We know from Section 9-8 that, if $A(t)$ denotes the area swept out by the planet's radius vector between time 0 and time t (see Figure 10.42), then

$$A(t) = \int \frac{1}{2} r^2 \, d\theta = \int_0^t \frac{1}{2} r^2 \frac{d\theta}{dt} \, dt.$$

Now we apply the Fundamental Theorem of Calculus, which yields

$$\frac{dA}{dt} = \frac{1}{2} r^2 \frac{d\theta}{dt}. \tag{8}$$

CHAP. 10: **Parametric Curves and Vectors in the Plane**

When we compare Equations (7) and (8), we see that

$$\frac{dA}{dt} = \frac{h}{2}. \qquad (9)$$

Since $h/2$ is a constant, we have derived Kepler's second law: The radius vector from sun to planet sweeps out area at a constant rate.

Now we derive Kepler's first law. Choose coordinate axes so that at time $t = 0$ the planet is on the polar axis and is at its closest point of approach to the sun, with initial position vector $\mathbf{r}_0$ and initial velocity vector $\mathbf{v}_0$, as in Figure 10.43. By Equation (4),

$$\mathbf{v}_0 = r_0\, \theta'(0)\mathbf{u}_\theta = v_0\mathbf{j} \qquad (10)$$

because, when $t = 0$, $\mathbf{u}_\theta = \mathbf{j}$ and $dr/dt = 0$ (because r is minimal then).

From Equation (6) we have

$$\frac{d\mathbf{v}}{dt} = -\frac{\mu}{r^2}\,\mathbf{u}_r = -\frac{\mu}{r^2}\left(-\frac{1}{d\theta/dt}\frac{d\mathbf{u}_\theta}{dt}\right)$$

$$= \frac{\mu}{r^2}\frac{d\mathbf{u}_\theta}{d\theta/dt\; dt} = \frac{\mu}{h}\frac{d\mathbf{u}_\theta}{dt},$$

by using the second equation in (2) in the first step and then Equation (7).

Next we antidifferentiate and find that

$$\mathbf{v} = \frac{\mu}{h}\,\mathbf{u}_\theta + \mathbf{C}.$$

To find $\mathbf{C}$, we substitute $t = 0$; this yields $\mathbf{u}_\theta = \mathbf{j}$ and $\mathbf{v} = v_0\mathbf{j}$. We get

$$\mathbf{C} = v_0\mathbf{j} - \frac{\mu}{h}\,\mathbf{j} = \left(v_0 - \frac{\mu}{h}\right)\mathbf{j}.$$

Consequently

$$\mathbf{v} = \frac{\mu}{h}\,\mathbf{u}_\theta + \left(v_0 - \frac{\mu}{h}\right)\mathbf{j}. \qquad (11)$$

Now we take the dot product of each side of Equation (11) with the unit vector $\mathbf{u}_\theta$. Remember that

$$\mathbf{u}_\theta \cdot \mathbf{v} = r\frac{d\theta}{dt} \quad \text{and} \quad \mathbf{u}_\theta \cdot \mathbf{j} = \cos\theta,$$

consequences of Equations (4) and (1), respectively. The result is

$$r\frac{d\theta}{dt} = \frac{\mu}{h} + \left(v_0 - \frac{\mu}{h}\right)\cos\theta.$$

Since $r(d\theta/dt) = h/r$ by Equation (7), we have

$$\frac{h}{r} = \frac{\mu}{h} + \left(v_0 - \frac{\mu}{h}\right)\cos\theta.$$

We solve for r, and find that

$$r = \frac{h^2/\mu}{1 + [(v_0 h/\mu) - 1]\cos\theta}.$$

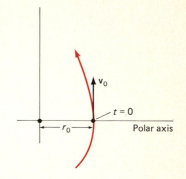

10.43 Setup of the coordinate system for the derivation of Kepler's first law.

Next, from Equations (7) and (10) we see that

$$h = r_0 v_0,$$ (12)

so finally we may write the equation of the planet's orbit:

$$r = \frac{(r_0 v_0)^2 / \mu}{1 + (r_0 v_0^2 / \mu - 1) \cos \theta} = \frac{pe}{1 + e \cos \theta}.$$ (13)

From Equation (1) in Section 9-7, we see that the orbit of the planet is a conic section with eccentricity

$$e = \frac{r_0 v_0^2}{GM} - 1.$$ (14)

Because the nature of a conic section is determined by its eccentricity, we see that the planet's orbit is:

A circle if $r_0 v_0^2 = GM$,

An ellipse if $GM < r_0 v_0^2 < 2GM$,

A parabola if $r_0 v_0^2 = 2GM$, (15)

A hyperbola if $r_0 v_0^2 > 2GM$.

This comprehensive description of the situation is a generalization of Kepler's first law. Of course, the elliptical case is the one that holds for planets in the solar system.

Let us consider further the case in which the orbit is an ellipse with major semiaxis a and minor semiaxis b. (The case in which the ellipse is actually a circle, with $a = b$, will be a special case of this discussion.) By Equation (3) of Section 9-7, the constant

$$pe = \frac{h^2}{GM}$$

used in (13) satisfies the equations

$$pe = a(1 - e^2) = a\left(1 - \frac{a^2 - b^2}{a^2}\right) = \frac{b^2}{a}.$$ (16)

We equate these two expressions for pe and find that

$$h^2 = GM \frac{b^2}{a}.$$

Now let T denote the period of revolution of the planet in its elliptical orbit. Then from Equation (9) we see that the area of the ellipse is $A = \frac{1}{2}hT = \pi ab$, and thus that

$$T^2 = \frac{4\pi^2 a^2 b^2}{h^2} = \frac{4\pi^2 a^2 b^2}{GMb^2 / a},$$

so that

$$T^2 = \gamma a^3$$ (17)

where $\gamma = 4\pi^2 / GM$. This is Kepler's third law.

CHAP. 10: **Parametric Curves and Vectors in the Plane**

EXAMPLE 1 The period of revolution of Mercury in its elliptical orbit about the sun is 87.97 days, while that of Earth is 365.26 days. Compute the major semiaxis (in astronomical units) of the orbit of Mercury.

Solution The major semiaxis of Earth's orbit is, by definition, 1 A.U. So Equation (17) gives us the value of the constant $\gamma = (365.26)^2$ (day^2/au^3). Hence the major semiaxis of the orbit of Mercury is

$$a = \left(\frac{T^2}{\gamma}\right)^{1/3} = \left(\frac{(87.97)^2}{(365.26)^2}\right)^{1/3} \approx 0.387 \text{ A.U.}$$

As yet we have considered only planets in orbit about the sun. But Kepler's laws and the equations of this section apply to bodies in orbit about any common central mass, so long as they move solely under the influence of its gravitational attraction. Examples include satellites (artificial or natural) orbiting the Earth or the moons of Jupiter. All we need to know is the value of the constant $\mu = GM$ for the central body. For Earth, we can compute μ by beginning with the values

$$g = \frac{GM}{R^2} = 32.15$$

ft/sec^2 for surface gravitational acceleration and the radius $R = 3960$ (miles) of Earth—which we assume to be spherical. Then

$$\mu = GM = R^2g = (3960)^2\left(\frac{32.15}{5280}\right) \approx 95{,}500 \text{ mi}^3/\text{sec}^2.$$

EXAMPLE 2 A rocket is in a circular orbit about the earth with orbital radius $r_0 = 4500$ miles. Its thrusters, when activated, provide a tangential acceleration of 0.1 miles per second per second. We want to insert the rocket into an elliptical orbit with perigee (closest distance) $r_0 = 4500$ miles and apogee (farthest distance) $r_1 = 26{,}300$ miles. How long must the thrusters burn to accomplish this?

Solution The problem is to determine by how much the speed v_0 of the rocket, when at the perigee of the desired elliptical orbit (see Figure 10.44), exceeds the speed v_c of the rocket in its original circular orbit.

From our study of uniform circular motion (Section 10-5), we know that the acceleration in a circular orbit with radius r_0 is

$$a = \frac{v_c^2}{r_0} = \frac{GM}{r_0^2},$$

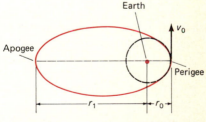

10.44 The transfer orbit of Example 2.

so the rocket's original speed—in the circular orbit—is

$$v_c = \sqrt{\frac{GM}{r_0}} = \sqrt{\frac{95{,}500}{4500}} \approx 4.61$$

miles per second.

We may use Equation (12), that $h = r_0v_0$, to find v_0 once we know h. And we can obtain h from Equation (16):

$$pe = \frac{h^2}{GM} = a(1 - e^2),$$

after we find a and e for the desired ellipse. Recall that the focus of an ellipse is at the distance ae from its center. Thus

$$r_0 = a(1 - e) = 4500 \quad \text{and} \quad r_1 = a(1 + e) = 26,300$$

(the units are miles). After we solve these simultaneous equations for $a = 15,400$ (miles) and $e = 0.71$, the previous equation gives

$$h = \sqrt{(95,500)(15,400)(1 - [0.71]^2)} \approx 27,091.$$

Thus insertion into the elliptical orbit requires an initial speed of

$$v_0 = \frac{h}{r_0} = \frac{27,091}{4500} \approx 6.02$$

miles per second. The thrusters must be activated for long enough to increase the rocket's speed by $v_0 - v_c$, or approximately 1.41 miles per second. With an acceleration of 0.1 mi/sec^2, this will require a burn time of 14.1 seconds.

The ellipse of Example 2 is a **transfer orbit** between the original circular orbit and the orbit of a "stationary satellite" at $r = 26,300$ miles. In Problem 13 we ask you to continue Example 2 to determine how to complete the transfer to the stationary satellite's orbit.

10-7 PROBLEMS

For each of the parametric curves in Problems 1–5, express its velocity and acceleration vectors in terms of $\mathbf{u}_r$ and $\mathbf{u}_\theta$; that is, in the form $A(t)\mathbf{u}_r + B(t)\mathbf{u}_\theta$.

1 $r = a, \theta = t.$
2 $r = 2\cos t, \theta = t.$
3 $r = t, \theta = t.$
4 $r = e^t, \theta = 2t.$
5 $r = 3\sin 4t, \theta = 2t.$
6 Derive both equations in (2) by differentiation of Equations (1).
7 Derive Equation (5) by differentiating Equation (4).
8 Consider a body in an elliptical orbit with major and minor semiaxes a and b and period of revolution T.
(a) Deduce from Equation (4) that $v = r(d\theta/dt)$ when the body is nearest to and farthest from its focus.
(b) Then apply Kepler's second law to conclude that $v = 2\pi ab/rT$ at the body's nearest and farthest points.

In each of Problems 9–12, apply the equation of Part (b) of Problem 8 to compute the speed (in miles per second) of the given body at the nearest and farthest points of its orbit. Use the value 1 A.U. = 92,956,000 miles for the major semiaxis of the Earth's orbit.

9 The planet Mercury: $a = 0.387$ A.U., $e = 0.206$, $T = 87.97$ days.
10 The planet Earth: $e = 0.0167$, $T = 365.26$ days.
11 The moon: $a = 238,900$ miles, $e = 0.055$, $T = 27.32$ days.
12 An earth satellite: $a = 10,000$ miles, $e = 0.5$.
13 Continue Example 2 as follows. (a) Apply Kepler's third law to find how long it will take the rocket to travel from perigee to apogee in the elliptical transfer orbit. (b) Then compute, as in Example 2, how long the thrusters should burn at apogee to insert the rocket into a circular orbit with a radius of 26,300 miles.

CHAPTER 10 REVIEW: Definitions and Concepts

Use the list below as a guide to concepts that you may need to review.

1 Definition of a parametric curve; of a smooth parametric curve
2 The slope of the tangent line to a smooth parametric curve (both in rectangular and in polar coordinates)
3 Integral computations with parametric curves (Formulas (1) through (4) in Section 10-2)
4 Arc length of a parametric curve
5 Vectors: their definition, length, addition, multiplication by scalars, and dot product
6 Test for perpendicular vectors

7 Vector-valued functions, velocity and acceleration vectors

8 Componentwise differentiation and integration of vector-valued functions

9 The equations of motion of a projectile

10 The speed and the scalar acceleration of a particle in uniform circular motion

11 Definitions of unit tangent vector and principal unit normal vector

12 Definition and computation of curvature of an explicit or parametric curve

13 Normal and tangential components of acceleration

14 Kepler's three laws of planetary motion

15 The radial and transverse unit polar vectors

16 Polar decomposition of velocity and acceleration vectors

17 Outline of the derivation of Kepler's laws from Newton's law of gravitation

MISCELLANEOUS PROBLEMS

In Problems 1–5, eliminate the parameter and sketch the curve.

1 $x = 2t^3 - 1, y = 2t^3 + 1$.
2 $x = \cosh t, y = \sinh t$.
3 $x = 2 + \cos t, y = 1 - \sin t$.
4 $x = \cos^4 t, y = \sin^4 t$.
5 $x = 1 + t^2, y = t^3$.

In Problems 6–10, write the equation of the tangent line to the given curve at the indicated point.

6 $x = t^2, y = t^3$; $t = 1$.
7 $x = 3 \sin t, y = 4 \cos t$; $t = \pi/4$.
8 $x = e^t, y = e^{-t}$; $t = 0$.
9 $r = \theta$; $\theta = \pi/2$.
10 $r = 1 + \sin \theta$; $\theta = \pi/3$.

In each of Problems 11–14, find the area of the region between the given curve and the x-axis.

11 $x = 2t + 1, y = t^2 + 3$; $-1 \leq t \leq 2$.
12 $x = e^t, y = e^{-t}$; $0 \leq t \leq 10$.
13 $x = 3 \sin t, y = 4 \cos t$; $0 \leq t \leq \pi/2$.
14 $x = \cosh t, y = \sinh t$; $0 \leq t \leq 1$.

In each of Problems 15–19, find the arc length of the given curve.

15 $x = t^2, y = t^3$; $0 \leq t \leq 1$.
16 $x = \ln(\cos t), y = t$; $0 \leq t \leq \pi/4$.
17 $x = 2t, y = t^3 + 1/3t$; $1 \leq t \leq 2$.
18 $r = \sin \theta$; $0 \leq \theta \leq \pi$.
19 $r = \sin^3(\theta/3)$; $0 \leq \theta \leq \pi$.

In each of Problems 20–24, find the area of the surface generated by revolving the given curve around the x-axis.

20 $x = t^2 + 1, y = 3t$; $0 \leq t \leq 2$.
21 $x = 4t^{1/2}, y = t^3/3 + 1/2t^2$; $1 \leq t \leq 4$.
22 $r = 4 \cos \theta$.
23 $r = e^{\theta/2}$; $0 \leq \theta \leq \pi$.
24 $x = e^t \cos t, y = e^t \sin t$; $0 \leq t \leq \pi/2$.

25 Consider the rolling circle of radius a that was used to generate the cycloid in Example 3 of Section 10-1. Suppose that this circle is the rim of a disk, and let Q be a point of this disk at distance $b < a$ from the center. Find parametric equations for the curve traced by Q as the circle rolls along the x-axis; assume that Q begins at the point $(0, a - b)$. Sketch this curve, which is called a *trochoid*.

26 If the smaller circle of Problem 28 in Section 10-1 rolls around the *outside* of the larger circle, the path of the point P is called an *epicycloid*. Show that it has parametric equations

$$x = (a + b) \cos t - b \cos \left(\frac{a + b}{b} t \right),$$

$$y = (a + b) \sin t - b \sin \left(\frac{a + b}{b} t \right).$$

27 Suppose that $b = a$ in Problem 26. Show that the epicycloid is then the cardioid $r = 2a(1 - \cos \theta)$ translated a units to the right.

28 Find the area of the surface generated by revolving lemniscate $r^2 = 2a^2 \cos 2\theta$ around the x-axis.

29 Find the volume generated by revolving around the y-axis the area under the cycloid

$$x = a(t - \sin t), \qquad y = a(1 - \cos t), \qquad 0 \leq t \leq 2\pi.$$

30 Show that the length of one arch of the hypocycloid of Problem 28 in Section 10-1 is $s = 8b(a - b)/a$.

31 Let ABC be an isosceles triangle with $AB = AC$. If M is the midpoint of BC, use the dot product to show that AM and BC are perpendicular. (*Suggestion:* Show first that $\vec{BC} = \vec{AC} - \vec{AB}$ and $\vec{AM} = \frac{1}{2}(\vec{AC} + \vec{AB})$.)

32 Use the dot product to show that the diagonals of a rhombus (a parallelogram with all four sides equal) are perpendicular to each other.

33 The acceleration of a certain particle is $\mathbf{a} = \mathbf{i} \sin t - \mathbf{j} \cos t$. Assume that the particle begins at time $t = 0$ at the

point $(0, 1)$, and has initial velocity $\mathbf{v}_0 = -\mathbf{i}$. Show that its path is a circle.

34 A particle moves in an attracting central force field, with force proportional to the distance from the origin. This implies that the particle's acceleration is given by $\mathbf{a} = -\omega^2\mathbf{r}$, where $\mathbf{r}$ is the position vector of the particle. Assume that the particle's initial position is $\mathbf{r}_0 = p\mathbf{i}$ and that its initial velocity is $\mathbf{v}_0 = q\omega\mathbf{j}$. Show that the trajectory of the particle is the ellipse $x^2/p^2 + y^2/q^2 = 1$. (*Suggestion:* Apply the theorem of Section 7-4.)

35 At time $t = 0$, a target is 160 feet from a gun and is moving directly away from it with a constant speed of 80 ft/sec. If the muzzle velocity of the gun is 320 ft/sec, at what angle of elevation α should it be fired in order to strike the moving target?

36 Suppose that a gun with muzzle velocity v_0 is located at the foot of a hill with a 30° slope. At what angle of elevation (from the horizontal) should the gun be fired in order to maximize its range, as measured up the hill?

37 Find the points on the curve $y = \sin x$ where the curvature is maximal and those where it is minimal.

38 The right branch of the hyperbola $x^2 - y^2 = 1$ is parametrized by $x = \cosh t$, $y = \sinh t$. Find the point where its curvature is maximal.

39 Find the vectors $\mathbf{N}$ and $\mathbf{T}$ at the point of $x = t \cos t$, $y = t \sin t$ where $t = \pi/2$.

40 Find the points on the ellipse $x^2/a^2 + y^2/b^2 = 1$ (with $a > b$) where the curvature is maximal and those where it is minimal.

41 Suppose that the curve $r = f(\theta)$ is given in polar coordinates. Put $r' = f'(\theta)$ and $r'' = f''(\theta)$. Show that its curvature is given by

$$\kappa = \frac{|r^2 + 2(r')^2 - rr''|}{(r^2 + (r')^2)^{3/2}}.$$

42 Use the formula of Problem 41 to calculate the curvature $\kappa(\theta)$ at the point (r, θ) of the Archimedean spiral $r = \theta$. Then show that $\kappa(\theta) \to 0$ as $\theta \to \infty$.

43 A railway curve is to join two straight tracks, one extending westward from $(-1, -1)$, the other extending

eastward from $(1, 1)$. Determine A, B, and C so that the curve $y = Ax + Bx^3 + Cx^5$ joins $(-1, -1)$ and $(1, 1)$ and so that the slope and curvature of this connecting curve are zero at both its end points.

44 Assume that a body has an elliptical orbit $r = pe/(1 + e \cos \theta)$ and satisfies Kepler's second law in the form $r^2(d\theta/dt) = h$ (constant). Then deduce from Equation (5) in Section 10-7 that its acceleration vector $\mathbf{a}$ is equal to $(k/r^2)\mathbf{u}_r$ for some constant k. This shows that Newton's inverse-square law of gravitation follows from Kepler's first and second laws.

45 Consider the water surface in a cylindrical bucket that is rotating with angular speed ω. Assume that the water has reached steady state with all the water particles rotating at the same angular speed as the bucket. A water particle of mass m at the water surface is subject to two forces, shown in Figure 10.45: A force $\mathbf{N}$ perpendicular to the surface, and the weight mg of the water particle. The vertical components balance because of the steady-state situation, and the horizontal component of $\mathbf{N}$ supplies the necessary centripetal force $mx\omega^2$. Deduce that a vertical cross section $y = f(x)$ of the water surface is a parabola.

10.45 Cross section of water in a rotating bucket.

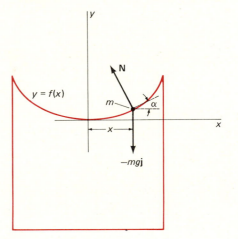

Indeterminate Forms, Taylor's Formula, and Improper Integrals

11

Indeterminate Forms and L'Hôpital's Rule

An *indeterminate form* is a certain type of expression with a limit that is not evident by inspection. If

$$\lim_{x \to a} f(x) = 0 = \lim_{x \to a} g(x),$$

then we say that the quotient $f(x)/g(x)$ has the **indeterminate form** $0/0$ at $x = a$. For example, to differentiate the trigonometric functions (Section 2-9), we needed to know that $\lim_{x \to 0} (\sin x)/x = 1$. Because $(\sin x)/x$ has the indeterminate form $0/0$ at $x = 0$, we had to use a geometric argument to find its limit (Example 13 in Section 1-8). Indeed, something of this sort happens whenever we compute a derivative, because the quotient

$$\frac{f(x) - f(a)}{x - a},$$

whose limit as $x \to a$ is the derivative $f'(a)$, has the indeterminate form $0/0$ at $x = a$.

Sometimes the limit of an indeterminate form can be found by a special algebraic manipulation or construction, as in our earlier computations of derivatives. Often, however, it is more convenient to apply a rule that appeared in the first calculus textbook ever published, in 1696, by the Marquis de l'Hôpital. L'Hôpital was a French nobleman who had hired the Swiss mathematician John Bernoulli as his calculus tutor, and "l'Hôpital's rule" is actually due to Bernoulli.

Theorem 1 *L'Hôpital's Rule*

Suppose that the functions f and g are differentiable in a deleted neighborhood of the point a, and that $g'(x)$ is nonzero in that neighborhood. Suppose also that

$$\lim_{x \to a} f(x) = 0 = \lim_{x \to a} g(x).$$

Then

$$\lim_{x \to a} \frac{f(x)}{g(x)} = \lim_{x \to a} \frac{f'(x)}{g'(x)} \tag{1}$$

provided that the limit on the right either exists (as a finite number) or is $\pm \infty$.

In essence, the rule says that if $f(x)/g(x)$ has the indeterminate form $0/0$ at $x = a$, then—subject to a few mild restrictions—the form has the same limit at $x = a$ as does the quotient $f'(x)/g'(x)$ of *derivatives*.

EXAMPLE 1   Find $\lim\limits_{x \to 0} \dfrac{e^x - 1}{\sin 2x}$.

Solution   The above fraction has the indeterminate form $0/0$ at $x = 0$. The numerator and denominator are clearly differentiable in a deleted neighborhood of $x = 0$, and the derivative of the denominator is certainly nonzero if the neighborhood is small enough (if $0 < |x| < \pi/4$). So l'Hôpital's rule applies, and

$$\lim_{x \to 0} \frac{e^x - 1}{\sin 2x} = \lim_{x \to 0} \frac{e^x}{2 \cos 2x} = \frac{1}{2}.$$

If it turns out that the quotient $f'(x)/g'(x)$ is again indeterminate, then l'Hôpital's rule can be applied a second (or third, or . . .) time, as in the following example. When the rule is applied repeatedly, however, the conditions for its applicability must be checked each time.

EXAMPLE 2   Find $\lim\limits_{x \to 1} \dfrac{1 - x + \ln x}{1 + \cos \pi x}$.

Solution

$$\lim_{x \to 1} \frac{1 - x + \ln x}{1 + \cos \pi x}$$

$$= \lim_{x \to 1} \frac{-1 + 1/x}{-\pi \sin \pi x} \qquad \text{(still of the form 0/0)}$$

$$= \lim_{x \to 1} \frac{x - 1}{\pi x \sin \pi x} \qquad \text{(algebraic simplification)}$$

$$= \lim_{x \to 1} \frac{1}{\pi \sin \pi x + \pi^2 x \cos \pi x} \qquad \text{(l'Hôpital's rule again)}$$

$$= -\frac{1}{\pi^2} \qquad \text{(by inspection)}.$$

Since the final limit exists, so do the previous ones, because the existence of the right-hand limit in (1) implies the existence of the left-hand limit.

When you need to apply l'Hôpital's rule repeatedly in this way, you need only keep differentiating the numerator and denominator separately until at least one of them has a nonzero finite limit. At that point you can recognize the limit of the quotient by inspection, as in the final step of Example 2.

EXAMPLE 3 $\lim\limits_{x \to 0} \dfrac{\sin x}{x + x^2} = \lim\limits_{x \to 0} \dfrac{\cos x}{1 + 2x} = 1.$

The point of Example 3 is to issue a warning. Had we *erroneously* attempted to apply l'Hôpital's rule to the second limit in Example 3, we would then have obtained the *wrong* answer 0. It is an oversimplification to say that l'Hôpital's rule "works when you need it and fails when you don't need it," but there is still much truth in that statement.

In order to prove l'Hôpital's rule, we need a generalization of the Mean Value Theorem due to the French mathematician Augustin-Louis Cauchy (1789–1857). He used it in the early nineteenth century to give rigorous proofs of several calculus results not previously established firmly.

Theorem 2 *Cauchy's Mean Value Theorem*

Suppose that the functions f and g are continuous on the closed interval $[a, b]$ and differentiable on (a, b). Suppose also that $g'(x)$ is never zero on (a, b). Then there exists a number c in (a, b) such that

$$\frac{f(b) - f(a)}{g(b) - g(a)} = \frac{f'(c)}{g'(c)}. \tag{2}$$

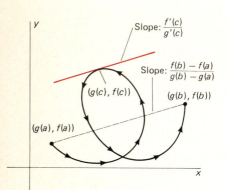

11.1 The idea of Cauchy's Mean Value Theorem. The proof given in the text suffices to establish the conclusion in (2) under the stronger hypotheses that, for each x in (a, b), $g(x)$ lies between $g(a)$ and $g(b)$, and $f'(x)$ and $g'(x)$ are never simultaneously zero.

To see that this theorem is a generalization of the (ordinary) Mean Value Theorem, take $g(x) = x$. Then the conclusion (2) reduces to the fact that

$$\frac{f(b) - f(a)}{b - a} = f'(c)$$

for some number c in (a, b).

Equation (2) also has a geometric interpretation like that of the ordinary Mean Value Theorem. Think of $x = g(t)$, $y = f(t)$, for t in $[a, b]$, as the parametric equations of a curve C in the xy-plane (as in Figure 11.1). The left-hand side of Equation (2) is the slope of the line joining the endpoints of this curve, and $f'(c)/g'(c)$ is the slope of the line tangent to C at the point $(g(c), f(c))$. Thus Cauchy's Mean Value Theorem says that there is a point of the parametric curve where the tangent line is parallel to the line joining the endpoints of C. Of course, this is exactly what the ordinary Mean Value Theorem says for an explicitly described curve $y = f(x)$. The proof of Cauchy's Mean Value Theorem will be deferred until the end of this section.

PROOF OF L'HÔPITAL'S RULE

Suppose that $f(x)/g(x)$ has the indeterminate form $0/0$ at $x = a$. We may use the continuity of f and g to allow the assumption that $f(a) = 0 = g(a)$. That is, we simply define $f(a)$ and $g(a)$ to be zero in case their values at $x = a$ are not originally given.

Now we restrict our attention to values of x in a fixed deleted neighborhood of a on which both f and g are differentiable. Choose one such value of x and hold it temporarily constant. Then apply Cauchy's Mean Value Theorem on the interval $[a, x]$. (If $x < a$, use the interval $[x, a]$ instead.) We find that there is a number z between a and x that behaves as c does in Equation (2). Hence, by virtue of Equation (2), we obtain the equation

$$\frac{f(x)}{g(x)} = \frac{f(x) - f(a)}{g(x) - g(a)} = \frac{f'(z)}{g'(z)}.$$

Now z depends upon x, but z is trapped between x and a, so that z must approach a as x does (see Figure 11.2). We conclude that

$$\lim_{x \to a} \frac{f(x)}{g(x)} = \lim_{z \to a} \frac{f'(z)}{g'(z)} = \lim_{x \to a} \frac{f'(x)}{g'(x)},$$

under the assumption that the right-hand limit exists.

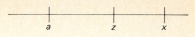

11.2 One case in the proof of l'Hôpital's rule.

INDETERMINATE FORMS INVOLVING ∞

L'Hôpital's rule has several variations. In addition to the fact that the limit in Equation (1) is allowed to be infinite, the real number a in l'Hôpital's rule may be replaced by either $+\infty$ or $-\infty$. For example,

$$\lim_{x \to \infty} \frac{f(x)}{g(x)} = \lim_{x \to \infty} \frac{f'(x)}{g'(x)} \qquad (3)$$

provided the other hypotheses are satisfied. In particular, to use Equation (3) we must first verify that

$$\lim_{x \to \infty} f(x) = 0 = \lim_{x \to \infty} g(x)$$

and that the right-hand limit in Equation (3) exists. The proof of this version of l'Hôpital's rule is outlined in Problem 43.

L'Hôpital's rule may also be used when $f(x)/g(x)$ has the **indeterminate form** ∞/∞. This means that

$$\lim_{x \to a} f(x) \quad \text{is either } +\infty \text{ or } -\infty,$$

and that

$$\lim_{x \to a} g(x) \quad \text{is either } +\infty \text{ or } -\infty.$$

The proof of this extension of the rule is more difficult, and will be omitted. [For a proof, see, for example, A. E. Taylor and W. R. Mann, *Advanced Calculus* (Lexington, Mass.: Xerox, 1972), p. 115.]

EXAMPLE 4

$$\lim_{x \to \infty} \frac{x^{1/3}}{\ln x} = \lim_{x \to \infty} \frac{\frac{1}{3} x^{-2/3}}{1/x} = \lim_{x \to \infty} \frac{1}{3} x^{1/3} = +\infty.$$

If $\lim_{x \to a} f(x) = 0$ and $\lim_{x \to a} g(x) = \infty$, we say that the product $f(x)g(x)$ has the **indeterminate form** $0 \cdot \infty$ at $x = a$. To find the limit of $f(x)g(x)$ at $x = a$, we can change the problem to one of the forms $0/0$ or ∞/∞ in this way:

$$f(x)g(x) = \frac{f(x)}{1/g(x)} \quad \text{or} \quad f(x)g(x) = \frac{g(x)}{1/f(x)}.$$

Then l'Hôpital's rule can be applied.

EXAMPLE 5 Find $\displaystyle \lim_{x \to \infty} x \ln \left(\frac{x - 1}{x + 1} \right)$.

Solution   This has the indeterminate form $\infty \cdot 0$, so we write

$$\lim_{x \to \infty} x \ln \left(\frac{x-1}{x+1} \right) = \lim_{x \to \infty} \frac{\ln[(x-1)/(x+1)]}{1/x}.$$

The latter limit has the form 0/0, so we can apply l'Hôpital's rule. First we note that

$$D \ln [(x-1)/(x+1)] = \frac{2}{x^2 - 1}.$$

Thus

$$\lim_{x \to \infty} x \ln \left(\frac{x-1}{x+1} \right) = \lim_{x \to \infty} \frac{2/(x^2 - 1)}{-1/x^2}$$

$$= \lim_{x \to \infty} \frac{-2}{1 - (1/x^2)} = -2.$$

If $\lim\limits_{x \to a} f(x) = \infty = \lim\limits_{x \to a} g(x)$, then we say that $f(x) - g(x)$ has the **indeterminate form** $\infty - \infty$. To then evaluate

$$\lim_{x \to a} (f(x) - g(x)),$$

we try by algebraic manipulation to convert $f(x) - g(x)$ into a form of type 0/0 or ∞/∞, to which l'Hôpital's rule can then be applied. If $f(x)$ or $g(x)$ is a fraction, we sometimes can do this by finding a common denominator. In most cases, however, less obvious methods are required. Example 6 illustrates the technique of finding a common denominator, while Example 7 illustrates a factoring technique that is sometimes effective.

EXAMPLE 6

$$\lim_{x \to 0} \left(\frac{1}{x} - \frac{1}{\sin x} \right) = \lim_{x \to 0} \frac{(\sin x) - x}{x \sin x} \qquad \text{(form 0/0)}$$

$$= \lim_{x \to 0} \frac{(\cos x) - 1}{\sin x + x \cos x} \qquad \text{(\emph{still} 0/0)}$$

$$= \lim_{x \to 0} \frac{-\sin x}{2 \cos x - x \sin x} = 0.$$

EXAMPLE 7

$$\lim_{x \to \infty} (\sqrt{x^2 + 3x} - x) = \lim_{x \to \infty} x \left(\sqrt{1 + \frac{3}{x}} - 1 \right) \qquad \text{(form } \infty \cdot 0)$$

$$= \lim_{x \to \infty} \frac{\sqrt{1 + 3/x} - 1}{1/x} \qquad \text{(form 0/0 now)}$$

$$= \lim_{x \to \infty} \frac{\frac{1}{2}(1 + 3/x)^{-1/2}(-3/x^2)}{-1/x^2}$$

$$= \lim_{x \to \infty} \frac{\frac{3}{2}}{\sqrt{1 + 3/x}} = \frac{3}{2}.$$

THE INDETERMINATE FORMS 0^0, ∞^0, AND 1^∞

Suppose that we need to find the limit of a quantity

$$y = [f(x)]^{g(x)}$$

where the limits of f and g as $x \to a$ are such that one of the indeterminate forms 0^0, ∞^0, or 1^∞ is produced. We first compute the natural logarithm:

$$\ln y = \ln([f(x)]^{g(x)}) = g(x) \ln f(x).$$

For each of the three indeterminate cases mentioned, $g(x) \ln f(x)$ has the form $0 \cdot \infty$, so we can use our previous methods to find $L = \lim_{x \to a} \ln y$. Then

$$\lim_{x \to a} [f(x)]^{g(x)} = \lim_{x \to a} e^{\ln y} = e^L,$$

because the exponential function is continuous.

EXAMPLE 8 Let k be a constant. Find $\lim\limits_{x \to 0} (1 + kx)^{1/x}$.

Solution Here we have the indeterminate form 1^∞; we therefore let $y = (1 + kx)^{1/x}$. Then

$$\ln y = \ln((1 + kx)^{1/x}) = \frac{1}{x} \ln(1 + kx).$$

Now we have a $0 \cdot \infty$ form that we can convert immediately to a $0/0$ form, so

$$\lim_{x \to 0} \ln y = \lim_{x \to 0} \frac{\ln(1 + kx)}{x}$$

$$= \lim_{x \to 0} \frac{k/(1 + kx)}{1} = k.$$

Therefore $\lim\limits_{x \to 0} (1 + kx)^{1/x} = e^k$.

Proof of Cauchy's Mean Value Theorem Suppose that f and g are functions satisfying the hypotheses of the theorem. We consider the parametric curve $x = g(t)$, $y = f(t)$, for t in $[a, b]$. Such a curve C is shown in Figure 11.3.

The motivation here is the same as for the proof of the ordinary Mean Value Theorem in Section 2-5. We introduce an auxiliary function $\phi(t)$, the difference of the heights above the point $x = g(t)$ of the curve and the line segment joining the endpoints $(g(a), f(a))$ and $(g(b), f(b))$ of C. The equation of this line is

$$y = f(a) + \frac{f(b) - f(a)}{g(b) - g(a)} [x - g(a)].$$

As Figure 11.3 suggests, the formula for the auxiliary function ϕ is

$$\phi(t) = f(t) - f(a) - \frac{f(b) - f(a)}{g(b) - g(a)} [g(t) - g(a)].$$

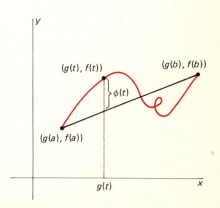

11.3 How the function $\phi(t)$ is defined. This figure shows a self-intersection permitted by the weaker hypotheses given in the caption of Figure 11.1.

Since it is evident that $\phi(a) = 0 = \phi(b)$, we may apply Rolle's theorem to ϕ on $[a, b]$. This gives us a point c in (a, b) such that $\phi'(c) = 0$. Since

$$\phi'(t) = f'(t) - \frac{f(b) - f(a)}{g(b) - g(a)}\, g'(t),$$

it follows that

$$f'(c) - \frac{f(b) - f(a)}{g(b) - g(a)}\, g'(c) = 0.$$

Since $g'(c) \neq 0$, this equation is equivalent to Equation (2), so we have completed the proof of Cauchy's Mean Value Theorem.

11-1 PROBLEMS

Find the limits in Problems 1–35.

1 $\displaystyle \lim_{x \to 1} \frac{x - 1}{x^2 - 1}.$

2 $\displaystyle \lim_{x \to \infty} \frac{3x - 4}{2x - 5}.$

3 $\displaystyle \lim_{x \to \infty} \frac{2x^2 - 1}{5x^2 + 3x}.$

4 $\displaystyle \lim_{x \to 0} \frac{e^{3x} - 1}{x}.$

5 $\displaystyle \lim_{x \to 0} \frac{\sin x^2}{x}.$

6 $\displaystyle \lim_{x \to 0^+} \frac{1 - \cos \sqrt{x}}{x}.$

7 $\displaystyle \lim_{x \to 1} \frac{x - 1}{\sin x}.$

8 $\displaystyle \lim_{x \to 0} \frac{1 - \cos x}{x^3}.$

9 $\displaystyle \lim_{x \to 0} \frac{e^x - x - 1}{x^2}.$

10 $\displaystyle \lim_{z \to \pi/2} \frac{1 + \cos 2z}{1 - \sin 2z}.$

11 $\displaystyle \lim_{u \to 0} \frac{u \tan^{-1} u}{1 - \cos u}.$

12 $\displaystyle \lim_{x \to 0} \frac{x - \arctan x}{x^3}.$

13 $\displaystyle \lim_{x \to \infty} \frac{\ln x}{x^{1/10}}.$

14 $\displaystyle \lim_{r \to \infty} \frac{e^r}{(r + 1)^4}.$

15 $\displaystyle \lim_{x \to 10} \frac{\ln(x - 9)}{x - 10}.$

16 $\displaystyle \lim_{t \to \infty} \frac{t^2 + 1}{t \ln t}.$

17 $\displaystyle \lim_{x \to 0} \frac{e^x + e^{-x} - 2}{x \sin x}.$

18 $\displaystyle \lim_{x \to (\pi/2)^-} \frac{\tan x}{\ln \cos x}.$

19 $\displaystyle \lim_{x \to 0^+} x \ln x.$

20 $\displaystyle \lim_{x \to \pi/2} (\tan x)(\cos 3x).$

21 $\displaystyle \lim_{x \to \pi} (x - \pi) \csc x.$

22 $\displaystyle \lim_{x \to \infty} e^{-x^2}(x - \sin x).$

23 $\displaystyle \lim_{x \to 0^+} \left(\frac{1}{\sqrt{x}} - \frac{1}{\sin x} \right).$

24 $\displaystyle \lim_{x \to 0} \left(\frac{1}{x} - \frac{1}{e^x - 1} \right).$

25 $\displaystyle \lim_{x \to 1} \left(\frac{x}{x^2 + x - 2} - \frac{1}{x - 1} \right).$

26 $\displaystyle \lim_{x \to \infty} (\sqrt{x + 1} - \sqrt{x}).$

27 $\displaystyle \lim_{x \to 0} \left(\frac{1}{x} - \frac{1}{\ln(1 + x)} \right).$

28 $\displaystyle \lim_{x \to \infty} (\sqrt{x^2 + x} - \sqrt{x^2 - x}).$

29 $\displaystyle \lim_{x \to \infty} (\sqrt[3]{x^3 + 2x + 5} - x).$

30 $\displaystyle \lim_{x \to 0^+} x^x.$

31 $\displaystyle \lim_{x \to 0^+} x^{\sin x}.$

32 $\displaystyle \lim_{x \to \infty} \left(1 + \frac{1}{x} \right)^x.$

33 $\displaystyle \lim_{x \to \infty} [\ln x]^{1/x}.$

34 $\displaystyle \lim_{x \to \infty} \left(1 - \frac{1}{x^2} \right)^x.$

35 $\displaystyle \lim_{x \to 0} \left(\frac{\sin x}{x} \right)^{1/x^2}.$

36 Sketch the graph of the function $y = x^{1/x}$, $x > 0$.

37 Suppose that f is a twice-differentiable function. Use l'Hôpital's rule to prove that:

(a) $\displaystyle \lim_{h \to 0} \frac{f(x + h) - f(x - h)}{2h} = f'(x);$

(b) $\displaystyle \lim_{h \to 0} \frac{f(x + h) - 2f(x) + f(x - h)}{h^2} = f''(x).$

38 Suppose that n is a fixed positive integer. Show that

$$\lim_{x \to \infty} \left(\sqrt[n]{x^n + a_1 x^{n-1} + a_2 x^{n-2} + \cdots + a_{n-1} x + a_n} - x \right) = \frac{a_1}{n}.$$

39 In Section 6-6, we saw that if a body falls under the influence both of gravity and of air resistance proportional to its velocity v (so that $dv/dt = -\rho v - g$), then its velocity at time t is

$$v(t) = \left(v_0 + \frac{g}{\rho} \right) e^{-\rho t} - \frac{g}{\rho}.$$

Show that $\lim\limits_{\rho \to 0} v(t) = v_0 - gt$, the velocity that the falling body would have if there were no air resistance.

40 In Section 6-6, we saw that the height at time t of the body of Problem 39 is

$$y(t) = y_0 - \frac{gt}{\rho} + \frac{1}{\rho}\left(v_0 + \frac{g}{\rho}\right)(1 - e^{-\rho t}).$$

Show that $\lim\limits_{\rho \to 0} y(t) = -\frac{1}{2}gt^2 + v_0 t + y_0$, the height it would have at time t if there were no air resistance.

41 According to Problem 34 in Section 8-5, the surface area of the ellipsoid obtained by revolving the ellipse $x^2/a^2 + y^2/b^2 = 1$ $(a > b)$ around the x-axis is

$$A = 2\pi a b\left(\frac{b}{a} + \frac{a}{c}\sin^{-1}\left(\frac{c}{a}\right)\right)$$

where $c^2 = a^2 - b^2$. Use l'Hôpital's rule to show that $\lim\limits_{b \to a} A = 4\pi a^2$, the area of a sphere of radius a.

42 Consider a long thin rod with heat conductivity k and coinciding with the x-axis. Suppose that at time $t = 0$ the temperature at x is $A/2\varepsilon$ if $-\varepsilon \leq x \leq \varepsilon$ and is zero if $|x| > \varepsilon$.

Then it turns out that the temperature $T(x, t)$ of the rod at the point x at time $t > 0$ is given by

$$T(x, t) = \frac{A}{\varepsilon\sqrt{4\pi kt}} \int_0^\varepsilon e^{-(x-u)^2/4kt}\, du.$$

Use l'Hôpital's rule to show that

$$\lim_{\varepsilon \to 0} T(x, t) = \frac{A}{\sqrt{4\pi kt}}\, e^{-x^2/4kt}.$$

This is the temperature resulting from an initial "hot spot" at the origin.

43 Establish the 0/0 version of l'Hôpital's rule in the case $a = \infty$. (*Suggestion:* Let $F(t) = f(1/t)$ and $G(t) = g(1/t)$. Then show that

$$\lim_{x \to \infty} \frac{f(x)}{g(x)} = \lim_{t \to 0} \frac{F(t)}{G(t)}$$

$$= \lim_{t \to 0} \frac{F'(t)}{G'(t)} = \lim_{x \to \infty} \frac{f'(x)}{g'(x)},$$

using l'Hôpital's rule in the case $a = 0$.)

Taylor's Formula and Polynomial Approximations

The definitions of the various elementary transcendental functions leave it unclear how to compute their values precisely, except at a few isolated points. For example, $\ln x = \int_1^x dt/t$ by definition, so obviously $\ln 1 = 0$; but no other value of $\ln x$ is obvious. The exponential function is the inverse of the function $\ln x$, so it is clear that $e^0 = 1$, but it is not at all clear how to compute e^x for $x \neq 0$. Indeed, even such an innocent-looking expression as $\sqrt{x}$ is not computable (precisely, and in a finite number of steps) unless x happens to be the square of a rational number.

On the other hand, *any* value of a polynomial

$$P(x) = c_0 + c_1 x + c_2 x^2 + \cdots + c_n x^n$$

with known coefficients $c_0, c_1, c_2, \ldots, c_n$ is easy to calculate; only addition and multiplication are required. The goal of this section is to use the fact that polynomial values are so readily computable to help us calculate values of functions like $\ln x$ and e^x.

Suppose that we want to calculate (or at least approximate closely) a specific value $f(x_0)$ of a given function f. It would suffice to find a polynomial $P(x)$ with a graph which is very close to that of f on some interval containing x_0. For then we could use the value $P(x_0)$ as an approximation to the actual value of $f(x_0)$. Once we know how to find such an approximating polynomial $P(x)$, we can then ask how accurately $P(x_0)$ approximates the desired value $f(x_0)$.

The simplest example of polynomial approximation is the linear approximation

$$f(x) \approx f(a) + f'(a)(x - a)$$

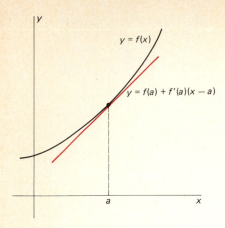

11.4 The tangent line at $(a, f(a))$ is the best linear approximation to f at a.

obtained by writing $\Delta x = x - a$ in Equation (3) of Section 3-2. The graph of the first-degree polynomial

$$P_1(x) = f(a) + f'(a)(x - a) \qquad (1)$$

is the tangent line to the curve $y = f(x)$ at the point $(a, f(a))$; see Figure 11.4. Note that this first-degree polynomial agrees with f and its first derivative at $x = a$; that is, $P_1(a) = f(a)$ and $P_1'(a) = f'(a)$.

For example, suppose that $f(x) = \ln x$ and that $a = 1$. Then $f(1) = 0$ and $f'(1) = 1$, so $P_1(x) = x - 1$. Hence we expect that $\ln x \approx x - 1$ for x near 1. With $x = 1.1$, we find that

$$P_1(1.1) = 0.100000 \quad \text{while} \quad \ln(1.1) \approx 0.095310 \qquad \text{(rounded)}.$$

The error in this approximation is about 5%.

To better approximate $\ln x$ near $x = 1$, let us look for a second-degree polynomial

$$P_2(x) = c_0 + c_1 x + c_2 x^2$$

that has not only the same value and the same first derivative as does f at $x = 1$ but also has the same second derivative there: $P_2''(1) = f''(1) = -1$. To satisfy these conditions, we must have

$$P_2(1) = c_2 + c_1 + c_0 = 0,$$

$$P_2'(1) = 2c_2 + c_1 = 1,$$

$$P_2''(1) = 2c_2 = -1.$$

When we solve these equations we find that $c_0 = -\frac{3}{2}, c_1 = 2$, and $c_2 = -\frac{1}{2}$, so

$$P_2(x) = -\tfrac{1}{2}x^2 + 2x - \tfrac{3}{2}.$$

With $x = 1.1$ we find that $P_2(1.1) = 0.095000$, which is accurate to three decimal places because $\ln(1.1) \approx 0.095310$. The graph of $y = -\frac{1}{2}x^2 + 2x - \frac{3}{2}$ is a parabola through $(1, 0)$ with the same slope *and curvature* there as $y = \ln x$; see Figure 11.5.

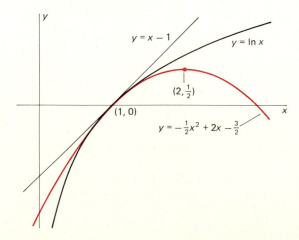

11.5 The linear and parabolic approximations to $y = \ln x$ at the point $(1, 0)$.

The tangent line and the parabola used in these computations illustrate the general approach to polynomial approximation. In order to approximate the function $f(x)$ near $x = a$, we look for an nth-degree polynomial

$$P_n(x) = c_0 + c_1 x + c_2 x^2 + \cdots + c_n x^n$$

such that its value at a and the values of its first n derivatives at a agree with the corresponding values of f. That is, we require that

$$P_n(a) = f(a),$$
$$P_n'(a) = f'(a),$$
$$P_n''(a) = f''(a), \qquad\qquad (2)$$
$$\vdots$$
$$P_n^{(n)}(a) = f^{(n)}(a).$$

These $n + 1$ conditions can be used to determine the values of the $n + 1$ coefficients $c_0, c_1, \ldots, c_n$.

The algebra involved is much simpler, however, if we begin with $P_n(x)$ expressed as a nth degree polynomial in powers of $x - a$ rather than in powers of x:

$$P_n(x) = b_0 + b_1(x - a) + b_2(x - a)^2 + \cdots + b_n(x - a)^n. \qquad (3)$$

Then substitution of $x = a$ in (3) yields $b_0 = P_n(a) = f(a)$ by the first condition in (2). Substitution of $x = a$ into

$$P_n'(x) = b_1 + 2b_2(x - a) + 3b_3(x - a)^2 + \cdots + nb_n(x - a)^{n-1}$$

yields $b_1 = P_n'(a) = f'(a)$ by the second condition in (2). Next, substitution of $x = a$ into

$$P_n''(x) = 2b_2 + 3 \cdot 2b_3(x - a) + \cdots + n(n - 1)b_n(x - a)^{n-2}$$

yields $2b_2 = P_n''(a) = f''(a)$, so that $b_2 = \frac{1}{2}f''(a)$.

We continue this process to find $b_3, b_4, \ldots, b_n$. In general, the constant term in the kth derivative $P_n^{(k)}(x)$ is $k!b_k$, since it is the kth derivative of the kth degree term $b_k(x - a)^k$ in $P_n(x)$:

$$P_n^{(k)}(x) = k!b_k + (\text{powers of } x - a).$$

So when we substitute $x = a$ into $P_n^{(k)}(x)$, we find that

$$k!b_k = P_n^{(k)}(a) = f^{(k)}(a),$$

and thus that

$$b_k = \frac{f^{(k)}(a)}{k!} \qquad\qquad (4)$$

for $k = 1, 2, 3, \ldots, n$.

Indeed, Formula (4) also holds if we use the common convention that $0! = 1$ and agree that the zeroth derivative $g^{(0)}$ of the function g is just g itself. With such conventions, our computations establish the following theorem.

> **Theorem 1** *The nth-Degree Taylor Polynomial*
> Suppose that the first n derivatives of the function $f(x)$ exist at $x = a$. Let $P_n(x)$ be the nth-degree polynomial
>
> $$P_n(x) = \sum_{k=0}^{n} \frac{f^{(k)}(a)}{k!}(x-a)^k \tag{5}$$
>
> $$= f(a) + f'(a)(x-a)$$
>
> $$+ \frac{f''(a)}{2!}(x-a)^2 + \cdots + \frac{f^{(n)}(a)}{n!}(x-a)^n.$$
>
> Then the values of $P_n(x)$ and its first n derivatives agree, at $x = a$, with the values of f and its first n derivatives there. That is, the equations in (2) hold.

The polynomial given in Equation (5) is called the **nth degree Taylor polynomial of the function f at the point $x = a$.** Note that $P_n(x)$ is a polynomial in powers of $x - a$ rather than in powers of x. In order to use $P_n(x)$ effectively for the approximation of $f(x)$, for x near a, we shall need to be able to compute the values of the derivatives $f(a)$, $f'(a)$, $f''(a)$, and so on, all the way to $f^{(n)}(a)$.

The line $y = P_1(x)$ is simply the tangent line to $y = f(x)$ at $(a, f(a))$. Since the curvature of $y = f(x)$ at $(a, f(a))$ is

$$\kappa = \frac{|f''(a)|}{(1 + [f'(a)]^2)^{3/2}},$$

we see that $y = P_2(x)$ has the same slope and the same curvature at $(a, f(a))$ as does $y = f(x)$. The third-degree Taylor polynomial $y = P_3(x)$ has not only the same slope and curvature, but also the same rate of change of curvature at $x = a$ as $y = f(x)$. These observations suggest the possibility that, the larger n is, the more closely $P_n(x)$ will approximate $f(x)$ for x near a.

EXAMPLE 1 Find the nth-degree Taylor polynomial of $f(x) = \ln x$ at $a = 1$.

Solution The first few derivatives of $f(x) = \ln x$ are

$$f'(x) = \frac{1}{x}, \qquad f''(x) = -\frac{1}{x^2}, \qquad f'''(x) = \frac{2}{x^3},$$

$$f^{(4)}(x) = -\frac{3!}{x^4}, \qquad f^{(5)}(x) = \frac{4!}{x^5}.$$

The pattern should be clear:

$$f^{(k)}(x) = (-1)^{k-1}\frac{(k-1)!}{x^k} \qquad \text{for } k \geqq 1.$$

Hence $f^{(k)}(1) = (-1)^{k-1}(k-1)!$, and so Equation (5) gives

$$P_n(x) = (x-1) - \frac{1}{2}(x-1)^2 + \frac{1}{3}(x-1)^3 - \frac{1}{4}(x-1)^4 + \cdots + \frac{(-1)^{n-1}}{n}(x-1)^n.$$

With $n = 2$ we obtain the quadratic polynomial
$$P_2(x) = (x - 1) - \tfrac{1}{2}(x - 1)^2 = -\tfrac{1}{2}x^2 + 2x - \tfrac{3}{2};$$
we saw its graph earlier in Figure 11.5. With the third degree Taylor polynomial
$$P_3(x) = (x - 1) - \tfrac{1}{2}(x - 1)^2 + \tfrac{1}{3}(x - 1)^3$$
we can go a step further in approximating $\ln(1.1)$—the value $P_3(1.1) = 0.095333 \ldots$ is correct to four decimal places.

EXAMPLE 2 Find the nth degree Taylor polynomial for $f(x) = e^x$ at $a = 0$.

Solution This is the easiest of all Taylor polynomials to compute, because $f^{(k)}(x) = e^x$ for all $k \geq 0$. Hence $f^{(k)}(0) = 1$ for all $k \geq 0$, and so
$$P_n(x) = 1 + x + \frac{x^2}{2!} + \frac{x^3}{3!} + \cdots + \frac{x^n}{n!}.$$

The first few Taylor polynomials of the exponential function at $a = 0$ are therefore
$$P_0(x) = 1,$$
$$P_1(x) = 1 + x,$$
$$P_2(x) = 1 + x + \tfrac{1}{2}x^2,$$
$$P_3(x) = 1 + x + \tfrac{1}{2}x^2 + \tfrac{1}{6}x^3,$$
$$P_4(x) = 1 + x + \tfrac{1}{2}x^2 + \tfrac{1}{6}x^3 + \tfrac{1}{24}x^4,$$
$$P_5(x) = 1 + x + \tfrac{1}{2}x^2 + \tfrac{1}{6}x^3 + \tfrac{1}{24}x^4 + \tfrac{1}{120}x^5.$$

The table of Figure 11.6 shows how these polynomials approximate $f(x) = e^x$ for $x = 0.1$ and for $x = 0.5$. Note that—at least for these two values of

11.6 Approximating $y = e^x$ with Taylor polynomials at $a = 0$.

$x = 0.1$

n	$P_n(x)$	e^x	$e^x - P_n(x)$
0	1.000000	1.105171	0.105171
1	1.100000	1.105171	0.005171
2	1.105000	1.105171	0.000171
3	1.105167	1.105171	0.000004
4	1.105171	1.105171	0.000000085
5	1.105171	1.105171	0.000000001

$x = 0.5$

n	$P_n(x)$	e^x	$e^x - P_n(x)$
0	1.000000	1.648721	0.648721
1	1.500000	1.648721	0.148721
2	1.625000	1.648721	0.023721
3	1.645833	1.648721	0.002888
4	1.648438	1.648721	0.000284
5	1.648698	1.648721	0.000023

[all data rounded]

x—the closer x is to $a = 0$, the more rapidly $P_n(x)$ appears to approach $f(x)$ as n increases.

The closeness with which $P_n(x)$ approximates $f(x)$ is measured by the difference

$$R_n(x) = f(x) - P_n(x),$$

for which

$$f(x) = P_n(x) + R_n(x). \qquad (6)$$

This difference $R_n(x)$ is called the **nth degree remainder for $f(x)$ at $x = a$.** It is the *error* made if the value $f(x)$ is replaced by the approximation $P_n(x)$.

The theorem that lets us estimate the error or remainder $R_n(x)$ is called **Taylor's formula,** after Brook Taylor, a follower of Newton, who introduced "Taylor polynomials" in an article published in 1715. The particular expression for $R_n(x)$ that we give next is called the *Lagrange form* for the remainder because it first appeared in 1797 in a book written by the French mathematician Joseph Louis Lagrange (1736–1813).

Theorem 2 *Taylor's Formula*

Suppose that the $(n + 1)$st derivative of the function f exists on an interval containing the points a and b. Then

$$f(b) = f(a) + f'(a)(b - a) + \frac{f''(a)}{2!}(b - a)^2 + \frac{f^{(3)}(a)}{3!}(b - a)^3$$

$$+ \cdots + \frac{f^{(n)}(a)}{n!}(b - a)^n + \frac{f^{(n+1)}(z)}{(n+1)!}(b - a)^{n+1} \qquad (7)$$

for some number z between a and b.

The proof of Taylor's formula is given at the end of this section. If we write $b = x$ in (7), we get the nth degree **Taylor's formula with remainder** at $x = a$,

$$f(x) = f(a) + f'(a)(x - a) + \frac{f''(a)}{2!}(x - a)^2 + \frac{f^{(3)}(a)}{3!}(x - a)^3$$

$$+ \cdots + \frac{f^{(n)}(a)}{n!}(x - a)^n + \frac{f^{(n+1)}(z)}{(n+1)!}(x - a)^{n+1} \qquad (8)$$

where z is some number between a and x. Thus the nth degree remainder term is

$$R_n(x) = \frac{f^{(n+1)}(z)}{(n+1)!}(x - a)^{n+1}, \qquad (9)$$

which is easy to remember—it's the same as the *last* term of $P_{n+1}(x)$, except that $f^{(n+1)}(a)$ is replaced by $f^{(n+1)}(z)$.

Ordinarily the exact value of z is unknown. One effective way to skirt this difficulty is to estimate $f^{(n+1)}(z)$; what we usually seek is an overestimate, a number M such that

$$\left| f^{(n+1)}(z) \right| \leq M$$

for *all* z between *a* and *x*. If we can find such a number *M*, then

$$|f(x) - P_n(x)| = |R_n(x)| \leqq \frac{M|x - a|^{n+1}}{(n + 1)!}. \tag{10}$$

The fact that $(n + 1)!$ grows very rapidly as *n* increases is often helpful in showing that $|R_n(x)|$ is small.

For a particular *x*, we may be able to show that $\lim_{n \to \infty} R_n(x) = 0$. It will then follow that

$$f(x) = \lim_{n \to \infty} P_n(x).$$

In such a case, we can then approximate $f(x)$ with *any* desired degree of accuracy simply by choosing *n* sufficiently large.

EXAMPLE 3 Estimate the accuracy of the approximation $\ln(1.1) \approx 0.095333 \cdots$ obtained in Example 1.

Solution   Recall that if $f(x) = \ln x$, then

$$f^{(k)}(x) = (-1)^{k-1} \frac{(k - 1)!}{x^k},$$

so that

$$f^{(k)}(1) = (-1)^{k-1}(k - 1)!.$$

Hence the third degree Taylor formula *with remainder* at $a = 1$ is

$$\ln x = (x - 1) - \frac{1}{2}(x - 1)^2 + \frac{1}{3}(x - 1)^3 + \frac{(-1)^3 3!}{4!z^4}(x - 1)^4$$

for some *z* between $a = 1$ and *x*. With $x = 1.1$, this gives

$$\ln(1.1) = 0.095333 - \frac{0.0001}{4z^4}$$

with *z* between $a = 1$ and $x = 1.1$. The largest possible numerical value of the remainder term is obtained with $z = 1$,

$$\frac{0.0001}{4(1)^4} = 0.000025.$$

It follows that

$$0.095308 < \ln(1.1) < 0.095333,$$

so we can say that $\ln(1.1) = 0.0953$ to four-place accuracy.

EXAMPLE 4 Use a Taylor polynomial for $f(x) = e^x$ at $a = 0$ (as in Example 2) to approximate the number *e* to five-place accuracy.

Solution   Since $f^{(k)}(x) = e^x$ for all *k*, the *n*th degree Taylor polynomial with remainder is

$$e^x = 1 + x + \frac{x^2}{2!} + \frac{x^3}{3!} + \cdots + \frac{x^n}{n!} + \frac{e^z}{(n + 1)!}x^{n+1}$$

for some *z* between $a = 0$ and *x*. With $x = 1$ we find that

$$e = 2 + \frac{1}{2!} + \frac{1}{3!} + \cdots + \frac{1}{n!} + \frac{e^z}{(n + 1)!}$$

with z between 0 and 1. Thus $e^0 < e^z < e^1$. Since we already know (Problem 34 in Section 6-2) that $e < 3$, it follows that $1 < e^z < 3$. We can therefore achieve five-place accuracy by choosing n so large that $3/(n+1)! < 0.000005$; that is, so that

$$(n + 1)! > \frac{3}{0.000005} = 600,000.$$

With either a table of factorials or a calculator, we find that the first factorial greater than 600,000 is $10! = 3,628,800$. We therefore take $n = 9$ and find that

$$e = 2 + \frac{1}{2!} + \frac{1}{3!} + \cdots + \frac{1}{9!} + R_9 = 2.7182815 + R_9$$

where

$$0 < R_9 = \frac{e^z}{10!} < \frac{3}{10!} = 0.000000827.$$

Thus $2.7182815 < e < 2.7182824$, and so the familiar $e = 2.71828$ is indeed the correct value rounded to five places.

Because of the factor $(x - a)^{n+1}$ in the remainder term $R_n(x)$, we see that (with n fixed) the closer x is to a, the better $P_n(x)$ approximates $f(x)$. To approximate $f(x)$ with the least labor, we naturally choose a as the nearest point to x at which we already know the value of f and the required derivatives of f. For example, to compute the sine of $5°$ (which is $\pi/36$ in radians), we choose $a = 0$. But to compute the sine of $50°$ ($5\pi/18$ radians), it's better to use $a = \pi/4$.

EXAMPLE 5 Show that the values of $\sin x$ for angles between $40°$ and $50°$ can be computed with three-place accuracy by using the approximation

$$\sin x \approx \frac{1}{\sqrt{2}}\left[1 + \left(x - \frac{\pi}{4}\right) - \frac{1}{2}\left(x - \frac{\pi}{4}\right)^2\right].$$

Solution Let $f(x) = \sin x$. Since $f'(x) = \cos x$, $f''(x) = -\sin x$, $f^{(3)}(x) = -\cos x$, the second degree Taylor polynomial with remainder for $\sin x$ at $a = \pi/4$ is

$$\sin x = \sin\frac{\pi}{4} + \left[\cos\frac{\pi}{4}\right]\left(x - \frac{\pi}{4}\right) - \frac{1}{2!}\left[\sin\frac{\pi}{4}\right]\left(x - \frac{\pi}{4}\right)^2 + R_2(x)$$

$$= \frac{1}{\sqrt{2}}\left[1 + \left(x - \frac{\pi}{4}\right) - \frac{1}{2}\left(x - \frac{\pi}{4}\right)^2\right] + R_2(x)$$

where

$$R_2(x) = -\frac{\cos z}{3!}\left(x - \frac{\pi}{4}\right)^3$$

for some number z between $\pi/4$ and x. Now $|\cos x| \leq 1$ for any z, and

$$\left|x - \frac{\pi}{4}\right| \leq \frac{\pi}{36} < 0.1$$

for the angles in question. Hence

$$|R_2(x)| < \frac{(0.1)^3}{6} \approx 0.000167 < 0.0002.$$

Thus the given polynomial will indeed give three-place accuracy, exactly as desired. For example,

$$\sin 50° = \sin\left(\frac{\pi}{4} + \frac{\pi}{36}\right)$$

$$\approx \frac{1}{\sqrt{2}}\left(1 + \frac{\pi}{36} - \frac{1}{2}\left(\frac{\pi}{36}\right)^2\right) = 0.766$$

(to three-place accuracy).

PROOF OF TAYLOR'S FORMULA

Although the motivation is not so obvious, the proof of Taylor's formula is a generalization of the proof of the Mean Value Theorem. We are going to apply Rolle's theorem to the auxiliary function ϕ, which we define like this:

$$\phi(x) = f(x) - P_n(x) - K(x - a)^{n+1}, \tag{11}$$

where the constant K is chosen so that $\phi(b) = 0$. (To accomplish this, we merely substitute $x = b$ into Equation (11), set $\phi(b) = 0$, and solve for K.) Then Rolle's theorem gives us a number z_1 between a and b such that $\phi'(z_1) = 0$.

Now note that

$$\phi'(a) = \phi''(a) = \cdots = \phi^{(n)}(a) = 0, \tag{12}$$

because $f^{(k)}(a) = P_n^{(k)}(a)$ for $k \leq n$, and for each k the kth derivative of $K(x - a)^{n+1}$ is a constant multiple of $(x - a)^{n-k+1}$.

We now apply Rolle's theorem to the function $\phi'(x)$ on $[a, z_1]$ and obtain a number z_2 between a and z_1 such that $\phi''(z_2) = 0$. This means that we can again apply Rolle's theorem to $\phi''(x)$ on the interval $[a, z_2]$.

In fact, because of the relations in (12) above, we may continue in this fashion, successively applying Rolle's theorem to the functions ϕ', ϕ'', ..., $\phi^{(n-1)}$. We obtain a sequence $z_1, z_2, \ldots, z_n$, shown in Figure 11.7. In the nth step we obtain z_n between a and z_{n-1}, and thus between a and b, such that $\phi^{(n)}(z_n) = 0$.

Now for one final application of Rolle's theorem, to the function $\phi^{(n)}$ on $[a, z_n]$. We obtain a point z between a and b such that $\phi^{(n+1)}(z) = 0$. But from Equation (11) we see that

$$\phi^{(n+1)}(x) = f^{(n+1)}(x) - (n + 1)!K.$$

The reason is that $P_n(x)$ is an nth degree polynomial whose $(n + 1)$st derivative is therefore identically zero. Because $\phi^{(n+1)}(z) = 0$, it follows that the

11.7 Result of repeated application of Rolle's theorem.

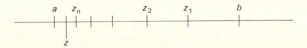

value of the constant K is

$$K = \frac{f^{(n+1)}(z)}{(n+1)!}.$$

We finally obtain Formula (7), our goal, by substituting this value of K and $x = b$ into Equation (11). We use the fact that $\phi(b) = 0$ and merely solve for $f(b)$. The result is Formula (7).

11-2 PROBLEMS

The case $a = 0$ of Taylor's formula with remainder is often called Maclaurin's formula, after the Scottish mathematician Colin Maclaurin, who used it as a basic tool in a calculus book he published in 1742. In Problems 1–10, find Maclaurin's formula with remainder for the given function and given value of n.

1 $f(x) = e^{-x};\quad n = 5.$ **2** $f(x) = \sin x;\quad n = 4.$

3 $f(x) = \cos x;\quad n = 4.$ **4** $f(x) = \dfrac{1}{1-x};\quad n = 4.$

5 $f(x) = \sqrt{1+x};\quad n = 3.$
6 $f(x) = \ln(1+x);\quad n = 4.$
7 $f(x) = \tan x;\quad n = 3.$
8 $f(x) = \arctan x;\quad n = 2.$
9 $f(x) = \sin^{-1} x;\quad n = 2.$
10 $f(x) = x^3 - 3x^2 + 5x - 7;\quad n = 4.$

In Problems 11–16, find the Taylor polynomial with remainder using the given values of a and n.

11 $f(x) = e^x;\quad a = 1, n = 4.$
12 $f(x) = \cos x;\quad a = \pi/4, n = 3.$
13 $f(x) = \sin x;\quad a = \pi/6, n = 3.$
14 $f(x) = \sqrt{x};\quad a = 100, n = 3.$
15 $f(x) = 1/(x-4)^2;\quad a = 5, n = 5.$
16 $f(x) = \tan x;\quad a = \pi/4, n = 4.$

In Problems 17–20, determine the number of decimal places of accuracy the given approximation formula yields for $|x| \le 0.1$.

17 $e^x \approx 1 + x + \frac{1}{2}x^2 + \frac{1}{6}x^3 + \frac{1}{24}x^4.$

18 $\sin x \approx x - \dfrac{x^3}{6} + \dfrac{x^5}{120}.$

19 $\ln(1+x) \approx x - \frac{1}{2}x^2 + \frac{1}{3}x^3 - \frac{1}{4}x^4.$
20 $\sqrt{1+x} \approx 1 + \frac{1}{2}x - \frac{1}{8}x^2.$

21 Show that the approximation in Problem 17 gives e^x to within 0.001 if $|x| \le 0.5$. Then compute $\sqrt[3]{e}$ accurate to two decimal places.

22 For what values of x is the approximation $\sin x \approx x - \frac{1}{6}x^3$ accurate to five decimal places?

23 (a) Show that values of the cosine function for angles between $40°$ and $50°$ can be computed with five-place accuracy by means of the approximation

$$\cos x \approx \frac{\sqrt{2}}{2}\left[1 - \left(x - \frac{\pi}{4}\right) - \frac{1}{2}\left(x - \frac{\pi}{4}\right)^2 + \frac{1}{6}\left(x - \frac{\pi}{4}\right)^3\right].$$

(b) Show that this approximation yields eight-place accuracy for angles between $44°$ and $46°$.

In Problems 24–28, calculate the indicated number with the required accuracy using Taylor's formula for an appropriate function at the given value of a.

24 $\sin 10°;\quad a = 0$, four decimal places.
25 $\cos 35°;\quad a = \pi/6$, four decimal places.
26 $e^{1/4};\quad a = 0$, four decimal places.
27 $\sin 62°;\quad a = \pi/3$, six decimal places.
28 $\sqrt{105};\quad a = 100$, three decimal places.
29 Show that the Taylor polynomial of degree $2n$ for $f(x) = \cos x$ at $a = 0$ is

$$\sum_{k=0}^{n} (-1)^k \frac{x^{2k}}{(2k)!} = 1 - \frac{x^2}{2!} + \frac{x^4}{4!} - \cdots + (-1)^n \frac{x^{2n}}{(2n)!}.$$

30 Show that the Taylor polynomial of degree $2n+1$ for $f(x) = \sin x$ at $a = 0$ is

$$\sum_{k=0}^{n} (-1)^k \frac{x^{2k+1}}{(2k+1)!} = x - \frac{x^3}{3!} + \frac{x^5}{5!} - \cdots + (-1)^n \frac{x^{2n+1}}{(2n+1)!}.$$

*11-3

Error Estimates for Numerical Integration Rules

In Section 4-7 we introduced several numerical integration rules for approximating the integral $\int_a^b f(x)\, dx$—the right and left endpoint approximations, the midpoint and trapezoidal approximations, and Simpson's approximation. The latter two of these are the most commonly used.

CHAP. 11: **Indeterminate Forms, Taylor's Formula, and Improper Integrals**

Here we give error estimates for the trapezoidal approximation and Simpson's approximation that are rather like the remainder term in Taylor's formula.

Let the points $a = x_0 < x_1 < \cdots < x_{n-1} < x_n = b$ partition the interval $[a, b]$ into n equal subintervals, each with length $\Delta x = (b - a)/n$. Let $y_i = f(x_i)$ for $i = 0, 1, 2, \ldots, n$. Recall that the trapezoidal approximation T_n and Simpson's approximation S_n to $\int_a^b f(x)\, dx$ are defined as follows:

$$T_n = \frac{\Delta x}{2} (y_0 + 2y_1 + 2y_2 + \cdots + 2y_{n-1} + y_n) \tag{1}$$

and

$$S_n = \frac{\Delta x}{3} (y_0 + 4y_1 + 2y_2 + \cdots + 2y_{n-2} + 4y_{n-1} + y_n). \tag{2}$$

In the case of Simpson's approximation, the number n of subintervals must be even.

For such approximations to be of real value, we must have some knowledge of the error involved. Let us denote these errors by ET_n and ES_n, respectively; the error terms are defined by means of the equations

$$\int_a^b f(x)\, dx = T_n + ET_n \tag{3}$$

and

$$\int_a^b f(x)\, dx = S_n + ES_n. \tag{4}$$

The following two theorems give expressions for ET_n and ES_n that often allow us to estimate them effectively.

Theorem 1 *Trapezoidal Error Estimate*

If f'' is continuous on $[a, b]$, then

$$ET_n = -\frac{(b - a)^3}{12n^2} f''(z) \tag{5}$$

for some number z in $[a, b]$. That is, for some such z,

$$\int_a^b f(x)\, dx = T_n - \frac{(b - a)^3}{12n^2} f''(z). \tag{6}$$

Theorem 2 *Simpson's Error Estimate*

If $f^{(4)}$ is continuous on $[a, b]$, then

$$ES_n = -\frac{(b - a)^5}{180n^4} f^{(4)}(z) \tag{7}$$

for some z in $[a, b]$. That is, for some such z,

$$\int_a^b f(x)\, dx = S_n - \frac{(b - a)^5}{180n^4} f^{(4)}(z). \tag{8}$$

The proofs of Theorems 1 and 2 are based on the same general idea as the proof of Taylor's formula, that of approximating a function with a polynomial. As we indicated in Section 4-7, the trapezoidal approximation results from approximating $f(x)$ on each subinterval $[x_{i-1}, x_i]$ with a linear polynomial, while Simpson's approximation results from approximating $f(x)$ on each subinterval $[x_{2i-2}, x_{2i}]$ with a second-degree polynomial. Detailed proofs of Theorems 1 and 2 may be found in most introductory numerical analysis books [for example, S. D. Conte and C. de Boor, *Elementary Numerical Analysis* (New York: McGraw-Hill, 1972) pp. 284–92].

If we let M_2 denote the maximum value of $|f''(z)|$ for z in $[a, b]$ and let M_4 denote the maximum value of $|f^{(4)}(z)|$ there, then Theorems 1 and 2 yield the error estimates stated in Section 4-7:

$$\left| \int_a^b f(x)\, dx - T_n \right| \le \frac{(b-a)^3}{12n^2} M_2 \qquad (9)$$

and

$$\left| \int_a^b f(x)\, dx - S_n \right| \le \frac{(b-a)^5}{180n^4} M_4. \qquad (10)$$

If the value of either M_2 or M_4 is known, then we may attain any desired degree of accuracy by choosing n sufficiently large.

EXAMPLE 1   Use the trapezoidal error estimate with $n = 10$ to find upper and lower bounds for the value of

$$\int_0^1 e^{-x^2}\, dx.$$

Solution   It is a sound idea to estimate the error first, so that we will know how many decimal places to retain in our computations. With $f(x) = e^{-x^2}$, we have

$$f'(x) = -2xe^{-x^2},$$
$$f''(x) = (4x^2 - 2)e^{-x^2}, \qquad \text{and}$$
$$f^{(3)}(x) = 4x(3 - 2x^2)e^{-x^2}.$$

We have calculated the third derivative in order to find the maximum and minimum values of $f''(z)$ for z in $[0, 1]$. Since $f^{(3)}(x) \ge 0$ for $0 \le x \le 1$, we see that $f''(z)$ lies between $f''(0) = -2$ and $f''(1) = 2/e \approx 0.736$. Thus

$$-0.736 < -f''(z) < 2.$$

We multiply each part of this inequality by

$$\frac{(b-a)^3}{12n^2} = \frac{1}{1200}$$

and find that

$$-0.00061 < ET_{10} < 0.00167.$$

This suggests that we compute T_{10} to five decimal places. We obtain

$$T_{10} = \frac{0.1}{2} \left(e^{-(0.0)^2} + 2e^{-(0.1)^2} + \cdots + 2e^{-(0.9)^2} + e^{-(1.0)^2} \right)$$

$$= 0.74621$$

to five places. We get our upper and lower bounds by subtracting 0.00061 and adding 0.00167, respectively, to T_{10}. Thus we find that

$$0.74560 < \int_0^1 e^{-x^2} \, dx < 0.74788.$$

Since $|ET_{10}| < 0.00167$, a more crude estimate would be

$$\int_0^1 e^{-x^2} \, dx = 0.74621 \pm 0.00167.$$

Incidentally, to five-place accuracy, the true value of the integral is 0.74682.

EXAMPLE 2   We saw in Section 4-7 that

$$\frac{\pi}{4} = \int_0^1 \frac{dx}{1 + x^2}.$$

Use this fact and Simpson's approximation to approximate the number π accurate to five decimal places.

Solution   If

$$\frac{\pi}{4} = \int_0^1 \frac{dx}{1 + x^2} = S_n + ES_n,$$

then

$$\pi = 4S_n + 4ES_n.$$

To bracket the value of π in an interval of length less than 0.000005, we therefore need to choose n so large that

$$|8ES_n| < 0.000005, \quad \text{and thus} \quad |ES_n| < 0.000000625.$$

To find the smallest suitable value of n, we first estimate $f^{(4)}(z)$ for z in $[0, 1]$. We begin with $f(x) = 1/(1 + x^2)$; successive differentiation gives

$$f'(x) = \frac{-2x}{(1 + x^2)^2},$$

$$f''(x) = \frac{6x^2 - 2}{(1 + x^2)^3},$$

$$f^{(3)}(x) = \frac{24x(1 - x^2)}{(1 + x^2)^4},$$

$$f^{(4)}(x) = \frac{24(5x^4 - 10x^2 + 1)}{(1 + x^2)^5}, \quad \text{and}$$

$$f^{(5)}(x) = \frac{-240x(3x^4 - 10x^2 + 3)}{(1 + x^2)^6}.$$

The roots of $f^{(5)}(x) = 0$ are $x = 0$, $\pm 1/\sqrt{3}$, and $\pm \sqrt{3}$. We need check only the one of these that lies in the interval $(0, 1)$:

$$f^{(4)}(1/\sqrt{3}) = -10.125.$$

The only other critical points of $f^{(4)}(x)$ are the end points $x = 0$ and $x = 1$, and there we find that $f^{(4)}(0) = 24$, $f^{(4)}(1) = -3$. Thus

$$|f^{(4)}(z)| \leq 24$$

for z in $[0, 1]$. We therefore must choose n so that

$$\frac{(1)^5(24)}{180 n^4} < 0.000000625,$$

which is equivalent to $n > 21.49$. Hence we take $n = 22$ and find that the maximum possible error is

$$|ES_{22}| \leq \frac{(1)(24)}{(180)(22)^4} \leq 0.00000057.$$

Simpson's approximation with $n = 22$ is

$$S_{22} = \frac{\frac{1}{22}}{3}\left(\frac{1}{1 + (0)^2} + \frac{4}{1 + (\frac{1}{22})^2} + \frac{2}{1 + (\frac{2}{22})^2}\right.$$
$$\left. + \cdots + \frac{2}{1 + (\frac{20}{22})^2} + \frac{4}{1 + (\frac{21}{22})^2} + \frac{1}{1 + (1)^2}\right),$$

so that

$$S_{22} \approx 0.78539816.$$

When we use our (high) estimate of $|ES_{22}|$ we write

$$S_{22} - 0.00000057 < \frac{\pi}{4} = \int_0^1 \frac{dx}{1 + x^2} < S_{22} + 0.00000057.$$

We replace S_{22} by 0.78539816 and then multiply by 4; this finally gives

$$3.14159036 < \pi < 3.14159492.$$

Thus π actually is 3.14159 to five decimal places.

EXACT APPROXIMATIONS

The fourth derivative of a cubic (third degree) polynomial $P_3(x)$ is identically zero, so Simpson's approximation will be exact rather than a mere estimate. Thus, with $n = 2$,

$$\int_a^b P_3(x)\, dx = \frac{b - a}{6}\left[P_3(a) + 4P_3(m) + P_3(b)\right] \qquad (11)$$

where $m = (a + b)/2$.

For a typical application of (11), suppose that the cross-sectional area function of a solid is a third degree polynomial (or one of lower degree).

Then Equation (11) gives us an easy-to-apply formula for its volume:

$$V = \frac{H}{6}(A_0 + 4A_m + A_1).\qquad(12)$$

Here, $H = b - a$ is the height of the solid, $A_0 = A(a)$ and $A_1 = A(b)$ are the areas of its ends, and $A_m = A(m)$ is the area of a cross section through its middle. Formula (12) is called the *prismoidal formula* for such solids (see Figure 11.8).

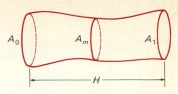

11.8 Using the prismoidal formula for volume.

For example, for a sphere with radius r, the prismoidal formula (12) gives

$$V = \frac{2r}{6}(0 + 4\pi r^2 + 0) = \frac{4}{3}\pi r^3.$$

11-3   PROBLEMS

In Problems 1–5, use the trapezoidal error estimate, with the indicated number n of subintervals, to find upper and lower bounds for the value of the integral. Then compare these with the exact value.

1 $\int_0^{10} x\, dx;\quad n = 2.$

2 $\int_1^2 \sqrt{x}\, dx;\quad n = 5.$

3 $\int_0^1 e^{-x}\, dx;\quad n = 10.$

4 $\int_1^2 \ln x\, dx;\quad n = 10.$

5 $\int_0^\pi \sin x\, dx;\quad n = 8.$

In Problems 6–10, use Simpson's error estimate, with the indicated number n of subintervals, to find upper and lower bounds for the value of the integral. Then compare these with the exact value.

6 $\int_0^{10} x^3\, dx;\quad n = 2.$

7 $\int_1^2 x^{3/2}\, dx;\quad n = 6.$

8 $\int_0^1 e^{-x}\, dx;\quad n = 10.$

9 $\int_1^2 \ln x\, dx;\quad n = 10.$

10 $\int_0^\pi \sin x\, dx;\quad n = 8.$

In Problems 11–14, determine a value of n such that the trapezoidal approximation T_n differs from the true value of the given integral by less than the indicated allowable error E.

11 $\int_0^4 x^2\, dx;\quad E = 0.001.$

12 $\int_0^1 \sqrt{1 + x}\, dx;\quad E = 0.005.$

13 $\int_0^1 e^{-x}\, dx;\quad E = 0.005.$

14 $\int_0^\pi \sin x\, dx;\quad E = 0.0001.$

In Problems 15–18, determine a value of n such that Simpson's approximation S_n differs from the true value of the given integral by no more than the maximum allowable error E given in the problem.

15 $\int_0^5 x^4\, dx;\quad E = 0.001.$

16 $\int_0^1 \frac{dx}{\sqrt{1 + x}};\quad E = 0.005.$

17 $\int_1^2 \ln x\, dx;\quad E = 0.0005.$

18 $\int_0^1 e^{-x}\, dx;\quad E = 0.0001.$

In Problems 19–22, use the prismoidal formula to calculate the exact volume of the solid described.

19 A right circular cone with radius r and height h.

20 A frustum of a cone with height h and base radii r and R.

21 The paraboloid obtained by revolving the parabola $y = \sqrt{x}, 0 \leq x \leq a$, around the x-axis.

22 The ellipsoid obtained by revolving the ellipse $x^2/a^2 + y^2/b^2 = 1$ around the x-axis.

23 Use the trapezoidal rule to calculate $\ln 2 = \int_1^2 dx/x$ with error less than 0.002, thereby finding $\ln 2$ accurate to two decimal places.

24 Repeat Problem 23, except use Simpson's approximation.

25 Use the trapezoidal approximation to calculate

$$\int_0^1 \sqrt{1 + x^3}\, dx$$

with error less than 0.005. Begin by showing that if $f(x) = \sqrt{1 + x^3}$, then $|f''(x)| \leq 4$ for all x in $[0, 1]$.

26 Suppose that $P_4(x)$ is a fourth-degree polynomial. Show that ES_2 is a computable *constant*. Hence explain how to use Simpson's error estimate with $n = 2$ to find the *exact value* of $\int_a^b P_4(x)\, dx$.

27 Use Simpson's approximation with error estimate and $n = 2$ to compute the exact value of $\int_0^1 x^4\, dx$. (*Suggestion:* See Problem 26.)

28 Use Simpson's approximation with error estimate and $n = 2$ to compute the exact volume of the solid obtained by revolving around the x-axis the area bounded by the x-axis and the parabola $y = 25 - x^2$.

Improper Integrals

To show the existence of the definite integral, we have relied until now upon Theorem 1 in Section 4-3. This theorem states that the integral $\int_a^b f(x)\,dx$ exists provided that the function f is *continuous* on the closed (*bounded*) interval $[a, b]$. Certain applications of calculus, however, lead naturally to the formulation of integrals in which either

(a) The interval of integration is not bounded; it has one of the forms

$$[a, +\infty), \qquad (-\infty, a], \quad \text{or} \quad (-\infty, +\infty); \qquad \text{or}$$

(b) The integrand has an infinite discontinuity at some point c:

$$\lim_{x \to c} f(x) = \pm\infty.$$

An example of Case (a) is the integral

$$\int_1^\infty \frac{dx}{x^2}.$$

A geometric interpretation of this integral is the area of the unbounded region (shaded in Figure 11.9) that lies between the curve $y = 1/x^2$ and the x-axis and to the right of the vertical line $x = 1$. An example of Case (b) is the integral

$$\int_0^1 \frac{dx}{\sqrt{x}}.$$

This integral may be interpreted as the area of the unbounded region (shaded in Figure 11.10) that lies under the curve $y = 1/\sqrt{x}$ from $x = 0$ to $x = 1$.

Such integrals are called **improper** integrals. The natural interpretation of an improper integral is the area of an unbounded region. The surprise is that such an area can nevertheless be finite, and this section is meant to show you how to find such areas—that is, how to evaluate improper integrals.

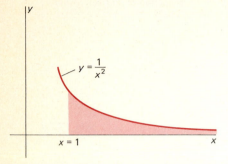

11.9 The shaded area cannot be measured using our earlier techniques.

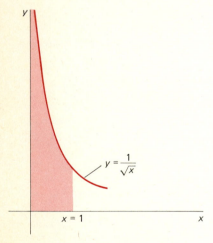

11.10 Another area that must be measured with an improper integral.

11.11 The shaded area $A(t)$ exists provided f is continuous.

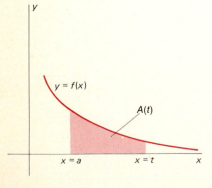

INFINITE LIMITS OF INTEGRATION

Suppose that the function f is continuous and nonnegative on the unbounded interval $[a, +\infty)$. Then, for any fixed $t > a$, the area $A(t)$ of the region under $y = f(x)$ from $x = a$ to $x = t$ (shaded in Figure 11.11) is given by the (ordinary) definite integral

$$A(t) = \int_a^t f(x)\,dx.$$

Suppose now that we let $t \to +\infty$, and find that the limit of $A(t)$ exists. Then we may regard this limit as the area of the unbounded region lying under $y = f(x)$ and over $[a, +\infty)$. For f continuous on $[a, +\infty)$, we therefore *define*

$$\int_a^\infty f(x)\,dx = \lim_{t \to \infty} \int_a^t f(x)\,dx \qquad (1)$$

provided that this limit exists (as a finite number). In this case we say that the improper integral on the left **converges**; otherwise, we say that it **diverges.** If $f(x)$ is *nonnegative* on $[a, +\infty)$, then the limit in Equation (1) either exists or is infinite, and in the latter case we write

$$\int_a^\infty f(x)\, dx = +\infty$$

and say that the above improper integral **diverges to infinity.**

If the function f has both positive and negative values on $[a, +\infty)$, then the improper integral can diverge *by oscillation;* that is, without diverging to infinity. This occurs with $\int_0^\infty \sin x\, dx$, because it is easy to verify that $\int_0^t \sin x\, dx$ is 0 if t is an even multiple of π but is 2 if t is an odd multiple of π. Thus $\int_0^t \sin x\, dx$ oscillates between 0 and 2 as $t \to \infty$, and so the limit in (1) does not exist.

We handle an infinite lower limit of integration similarly: We define

$$\int_{-\infty}^b f(x)\, dx = \lim_{t \to -\infty} \int_t^b f(x)\, dx \qquad (2)$$

provided the limit exists. If the function f is continuous on the whole real line, we define

$$\int_{-\infty}^\infty f(x)\, dx = \int_{-\infty}^c f(x)\, dx + \int_c^\infty f(x)\, dx \qquad (3)$$

for any convenient choice of c, provided that both improper integrals on the right converge. Note that $\int_{-\infty}^\infty f(x)\, dx$ is *not* necessarily equal to $\lim_{t\to\infty} \int_{-t}^t f(x)\, dx$ (see Problem 28).

It makes no difference what value of c is used in (3) because, if $c < d$, then

$$\int_{-\infty}^c f(x)\, dx + \int_c^\infty f(x)\, dx = \int_{-\infty}^c f(x)\, dx + \int_c^d f(x)\, dx + \int_d^\infty f(x)\, dx$$

$$= \int_{-\infty}^d f(x)\, dx + \int_d^\infty f(x)\, dx,$$

under the assumption that the limits involved all exist.

EXAMPLE 1   Investigate the improper integrals:

$$\text{(a)} \quad \int_1^\infty \frac{dx}{x^2}; \qquad \text{(b)} \quad \int_{-\infty}^0 \frac{dx}{\sqrt{1-x}}.$$

Solution

$$\text{(a)} \quad \int_1^\infty \frac{dx}{x^2} = \lim_{t\to\infty} \int_1^t \frac{dx}{x^2}$$

$$= \lim_{t\to\infty} \left[-\frac{1}{x} \right]_1^t = \lim_{t\to\infty} \left(-\frac{1}{t} + 1 \right) = 1.$$

Thus this improper integral converges to 1, and this is the area of the region shown in Figure 11.9.

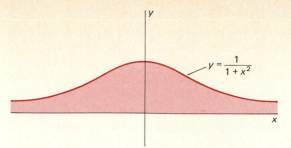

11.12 The area measured by the integral of Example 2.

(b) $\displaystyle\int_{-\infty}^{0} \frac{dx}{\sqrt{1-x}} = \lim_{t\to-\infty} \int_{t}^{0} \frac{dx}{\sqrt{1-x}}$

$\displaystyle\qquad = \lim_{t\to-\infty} \left[-2\sqrt{1-x}\right]_{t}^{0} = \lim_{t\to-\infty} (2\sqrt{1-t}-2) = +\infty.$

Thus the second improper integral of the example diverges to $+\infty$.

EXAMPLE 2 Investigate the improper integral

$$\int_{-\infty}^{\infty} \frac{1}{1+x^2}\, dx.$$

Solution The choice of $c = 0$ in Equation (3) gives

$$\int_{-\infty}^{\infty} \frac{dx}{1+x^2} = \int_{-\infty}^{0} \frac{dx}{1+x^2} + \int_{0}^{\infty} \frac{dx}{1+x^2}$$

$$= \lim_{s\to-\infty} \int_{s}^{0} \frac{dx}{1+x^2} + \lim_{t\to+\infty} \int_{0}^{t} \frac{dx}{1+x^2}$$

$$= \lim_{s\to-\infty} \left[\tan^{-1}x\right]_{s}^{0} + \lim_{t\to+\infty} \left[\tan^{-1}x\right]_{0}^{t}$$

$$= \lim_{s\to-\infty} (-\tan^{-1}s) + \lim_{t\to+\infty} (\tan^{-1}t)$$

$$= \frac{\pi}{2} + \frac{\pi}{2} = \pi.$$

The shaded area of Figure 11.12 is a geometric interpretation of the integral of Example 2.

11.13 An improper integral of the second type: $f(x) \to \infty$ as $x \to b^-$.

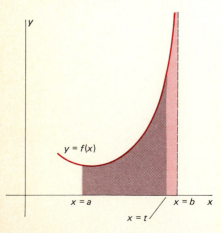

INFINITE INTEGRANDS

Suppose that the function f is continuous and nonnegative on $[a, b)$ but that $\lim_{x\to b^-} f(x) = \infty$. The graph of such a function appears in Figure 11.13. The area $A(t)$ of the region lying under $y = f(x)$ from $x = a$ to $x = t < b$ is the value of the (ordinary) definite integral

$$A(t) = \int_{a}^{t} f(x)\, dx.$$

If the limit of $A(t)$ as $t \to b^-$ exists, then this limit may be regarded as the area of the (unbounded) region under $y = f(x)$ from $x = a$ to $x = b$. For

CHAP. 11: Indeterminate Forms, Taylor's Formula, and Improper Integrals

f continuous on $[a, b)$, we therefore *define*

$$\int_a^b f(x)\, dx = \lim_{t \to b^-} \int_a^t f(x)\, dx, \tag{4}$$

provided that this limit exists (as a finite number), in which case we say that the improper integral on the left *converges*; otherwise, we say that it *diverges*. If

$$\int_a^b f(x)\, dx = \lim_{t \to b^-} \int_a^t f(x)\, dx = \infty,$$

then we say that the improper integral *diverges to infinity*.

If f is continuous on $(a, b]$ but the limit of $f(x)$ as $x \to a^+$ is infinite, then we *define*

$$\int_a^b f(x)\, dx = \lim_{t \to a^+} \int_t^b f(x)\, dx \tag{5}$$

provided that the limit exists. If f is continuous at every point of $[a, b]$ except for the point c in (a, b) and one or both one-sided limits at c are infinite, then we *define*

$$\int_a^b f(x)\, dx = \int_a^c f(x)\, dx + \int_c^b f(x)\, dx \tag{6}$$

provided that both improper integrals on the right converge.

EXAMPLE 3   Investigate the improper integrals:

(a) $\quad \displaystyle\int_0^1 \frac{dx}{\sqrt{x}};$ (b) $\quad \displaystyle\int_1^2 \frac{dx}{(x-2)^2}.$

Solution

(a) The integrand $1/\sqrt{x}$ becomes infinite as $x \to 0^+$, so

$$\int_0^1 \frac{dx}{\sqrt{x}} = \lim_{t \to 0^+} \int_t^1 \frac{dx}{\sqrt{x}}$$

$$= \lim_{t \to 0^+} \left[2\sqrt{x} \right]_t^1 = \lim_{t \to 0^+} 2(1 - \sqrt{t}) = 2.$$

Thus the area of the unbounded region shown in Figure 11.10 is 2.

(b) Here the integrand becomes infinite as x approaches the right-hand endpoint, so

$$\int_1^2 \frac{dx}{(x-2)^2} = \lim_{t \to 2^-} \int_1^t \frac{dx}{(x-2)^2}$$

$$= \lim_{t \to 2^-} \left[-\frac{1}{x-2} \right]_1^t = \lim_{t \to 2^-} \left(-1 - \frac{1}{t-2} \right) = +\infty.$$

Hence this improper integral diverges to infinity. It follows that the improper integral

$$\int_1^3 \frac{dx}{(x-2)^2} = \int_1^2 \frac{dx}{(x-2)^2} + \int_2^3 \frac{dx}{(x-2)^2}$$

also diverges, because not both of the right-hand improper integrals converge. (It turns out that the second one also diverges to $+\infty$.)

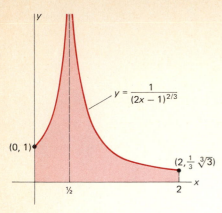

11.14 The region of Example 4.

EXAMPLE 4 Investigate the improper integral

$$\int_0^2 \frac{dx}{(2x-1)^{2/3}}.$$

Solution This improper integral corresponds to the region shaded in Figure 11.14. The integrand function has an infinite discontinuity at the point $c = \frac{1}{2}$ within the interval of integration, so we write

$$\int_0^2 \frac{dx}{(2x-1)^{2/3}} = \int_0^{1/2} \frac{dx}{(2x-1)^{2/3}} + \int_{1/2}^2 \frac{dx}{(2x-1)^{2/3}}$$

and investigate separately the two improper integrals on the right. We find that

$$\int_0^{1/2} \frac{dx}{(2x-1)^{2/3}} = \lim_{t \to (1/2)^-} \int_0^t \frac{dx}{(2x-1)^{2/3}}$$

$$= \lim_{t \to (1/2)^-} \left[\tfrac{3}{2}(2x-1)^{1/3} \right]_0^t$$

$$= \lim_{t \to (1/2)^-} \tfrac{3}{2}\left[(2t-1)^{1/3} - (-1)^{1/3} \right] = \tfrac{3}{2},$$

and

$$\int_{1/2}^2 \frac{dx}{(2x-1)^{2/3}} = \lim_{t \to (1/2)^+} \int_t^2 \frac{dx}{(2x-1)^{2/3}}$$

$$= \lim_{t \to (1/2)^+} \left[\tfrac{3}{2}(2x-1)^{1/3} \right]_t^2$$

$$= \lim_{t \to (1/2)^+} \tfrac{3}{2}\left[(3)^{1/3} - (2t-1)^{1/3} \right] = \tfrac{3}{2}\sqrt[3]{3}.$$

Therefore

$$\int_0^2 \frac{dx}{(2x-1)^{2/3}} = \frac{3}{2}(1 + \sqrt[3]{3}).$$

Special functions in advanced mathematics are frequently defined by means of improper integrals. An important example is the **gamma function** $\Gamma(t)$ that the prolific Swiss mathematician Leonhard Euler (1707–1783) introduced to "interpolate" the factorial function $n!$. The gamma function is defined for all real numbers $t > 0$ by

$$\Gamma(t) = \int_0^\infty x^{t-1} e^{-x}\, dx. \tag{7}$$

This definition gives a continuous function of t such that

$$\Gamma(n + 1) = n! \tag{8}$$

if n is a positive integer (see Problems 29 and 30).

EXAMPLE 5 Find the volume V of the unbounded solid obtained by revolving around the y-axis the region under the curve $y = e^{-x^2}$, $x \geqq 0$.

Solution The region in question is shown in Figure 11.15. Let V_t denote the volume generated by revolving the shaded part of this region between

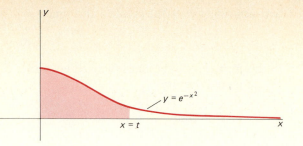

11.15 *All* the region under the graph is rotated about the y-axis; this produces an unbounded solid.

$x = 0$ and $x = t$. Then the method of cylindrical shells gives

$$V_t = \int_0^t 2\pi x e^{-x^2}\, dx.$$

We get the whole volume V by letting $t \to \infty$:

$$V = \lim_{t \to \infty} V_t = \int_0^\infty 2\pi x e^{-x^2}\, dx$$
$$= \lim_{t \to \infty} \left[-\pi e^{-x^2} \right]_0^t = \pi.$$

*ESCAPE VELOCITY

We saw in Section 5-5 how to compute the work W_r required to lift a body of mass m from the surface of a planet (of mass M and radius R) to a distance r from the center of the planet. According to Equation (7) of that section, the answer is

$$W_r = \int_R^r \frac{GMm}{x^2}\, dx.$$

So the work required to move the mass m "infinitely far" from the planet is

$$W = \lim_{r \to \infty} W_r = \int_R^\infty \frac{GMm}{x^2}\, dx = \lim_{r \to \infty} \left[-\frac{GMm}{x} \right]_R^r = \frac{GMm}{R}.$$

Suppose that the mass is projected straight upward from the planet's surface with initial velocity v_0, as in Jules Verne's novel *From the Earth to the Moon* (1865), in which a spaceship was fired from an immense cannon. Then the initial kinetic energy $\frac{1}{2}mv_0^2$ is available to supply this work—by conversion into potential energy. From the equation $\frac{1}{2}mv_0^2 = GMm/R$, we find that

$$v_0 = \sqrt{\frac{2GM}{R}}.$$

This provides an alternative derivation of the escape velocity formula (Equation (12) in Section 3-8).

*PRESENT VALUE OF A PERPETUITY

Consider a "perpetual annuity," under which you and your heirs (and theirs, ad infinitum) will be paid A dollars annually. By Equation (5) of Section 6-7, the **present value** of the amount you (and your heirs) receive between time $t = 0$ (the present) and time $t = T$ is

$$P_T = \int_0^T Ae^{-rt}\, dt.$$

In this formula, r is the continuous interest rate. So the total present value of the perpetual annuity is

$$P = \lim_{T \to \infty} P_T = \int_0^\infty Ae^{-rt}\, dt = \lim_{T \to \infty} \left[-\frac{A}{r} e^{-rt} \right]_0^T = \frac{A}{r}.$$

For example, at an interest rate of 10% ($r = 0.10$), you should be able to purchase a perpetuity paying $A = \$10,000$ annually for $P = \$10,000/(0.10)$ or $\$100,000$.

11-4 PROBLEMS

Determine whether or not the improper integrals in Problems 1–24 converge, and evaluate those that do converge.

1 $\displaystyle\int_4^\infty \frac{dx}{x^{3/2}}.$

2 $\displaystyle\int_1^\infty \frac{dx}{x^{2/3}}.$

3 $\displaystyle\int_0^4 \frac{dx}{x^{3/2}}.$

4 $\displaystyle\int_0^8 \frac{dx}{x^{2/3}}.$

5 $\displaystyle\int_1^\infty \frac{dx}{x + 1}.$

6 $\displaystyle\int_3^\infty \frac{dx}{\sqrt{x + 1}}.$

7 $\displaystyle\int_5^\infty \frac{dx}{(x - 1)^{3/2}}.$

8 $\displaystyle\int_0^4 \frac{dx}{\sqrt{4 - x}}.$

9 $\displaystyle\int_0^9 \frac{dx}{(9 - x)^{3/2}}.$

10 $\displaystyle\int_0^3 \frac{dx}{(x - 3)^2}.$

11 $\displaystyle\int_{-\infty}^{-2} \frac{dx}{(x + 1)^3}.$

12 $\displaystyle\int_{-\infty}^0 \frac{dx}{\sqrt{4 - x}}.$

13 $\displaystyle\int_{-1}^8 \frac{1}{x^{1/3}}\, dx.$

14 $\displaystyle\int_{-4}^5 \frac{1}{(x + 4)^{2/3}}\, dx.$

15 $\displaystyle\int_2^\infty \frac{dx}{(x - 1)^{1/3}}.$

16 $\displaystyle\int_{-\infty}^\infty \frac{x\, dx}{(x^2 + 4)^{3/2}}.$

17 $\displaystyle\int_{-\infty}^\infty \frac{x}{x^2 + 4}\, dx.$

18 $\displaystyle\int_0^\infty e^{-(x+1)}\, dx.$

19 $\displaystyle\int_0^1 \frac{e^{\sqrt{x}}}{\sqrt{x}}\, dx.$

20 $\displaystyle\int_0^2 \frac{x}{x^2 - 1}\, dx.$

21 $\displaystyle\int_1^\infty \frac{dx}{x \ln x}.$

22 $\displaystyle\int_0^\infty \sin^2 x\, dx.$

23 $\displaystyle\int_0^\infty xe^{-2x}\, dx.$

24 $\displaystyle\int_0^\infty e^{-x} \sin x\, dx.$

Problems 25–27 deal with *Gabriel's horn*, the surface obtained by revolving the curve $y = 1/x$, $x \geq 1$, around the x-axis (see Figure 11.16).

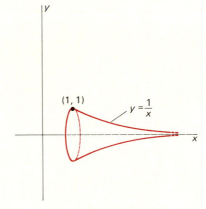

11.16 Gabriel's horn.

25 Show that the area under the curve $y = 1/x$, $x \geq 1$, is infinite.

26 Show that the volume of revolution enclosed by Gabriel's horn is finite, and compute it.

27 Show that the surface area of Gabriel's horn is infinite. (*Suggestion:* If A_t denotes the surface area from $x = 1$ to $x = t$, show that

$$A_t = \int_1^t \frac{2\pi}{x} \sqrt{1 + (1/x^4)}\, dx > 2\pi \ln t.)$$

28 Show that

$$\int_{-\infty}^\infty \frac{1 + x}{1 + x^2}\, dx$$

diverges, but that

$$\lim_{t \to \infty} \int_{-t}^t \frac{1 + x}{1 + x^2}\, dx = \pi.$$

29 Let n be a fixed positive integer. Begin with Integral (7) defining the gamma function and integrate by parts to show that $\Gamma(n + 1) = n\Gamma(n)$.

30 (a) Show that $\Gamma(1) = 1$. (b) Use the results of part (a) and Problem 29 to prove by mathematical induction that $\Gamma(n + 1) = n!$.

31 Use the substitution $x = e^{-u}$ and the fact that $\Gamma(n + 1) = n!$ to show that, if m and n are fixed but arbitrary positive integers, then

$$\int_0^1 x^m (\ln x)^n \, dx = \frac{n!(-1)^n}{(m + 1)^{n+1}}.$$

32 Consider a perpetual annuity under which you and your heirs will be paid at the rate of $(10 + t)$ thousand per year t years hence. Thus you will receive \$20 thousand 10 years hence, your heir will receive \$110 thousand 100 years hence, and so on. Show that the present value at an interest rate of 10% of this perpetuity is

$$P = \int_0^\infty (10 + t)e^{-t/10} \, dt,$$

and then evaluate this improper integral.

33 A "semi-infinite" uniform rod occupies the nonnegative x-axis ($x \geq 0$) and has linear density ρ; that is, a segment of length dx has mass $\rho \, dx$. Show that the force of gravitational attraction that the rod exerts on a point mass m at $(-a, 0)$ is

$$F = \int_0^\infty \frac{Gm\rho \, dx}{(a + x)^2} = \frac{Gm\rho}{a}.$$

34 A rod of linear density ρ occupies the entire y-axis. A point mass m is located at $(a, 0)$ on the x-axis, as indicated in Figure 11.17. Show that the total (horizontal) gravitational force that the rod exerts on m is

$$F = \int_{-\infty}^\infty \frac{Gm\rho \cos \theta}{r^2} \, dy = \frac{2Gm\rho}{a},$$

where $r^2 = a^2 + y^2$ and $\cos \theta = \dfrac{a}{r}$.

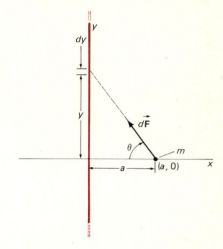

11.17 Gravitational attraction exerted on a point mass by an infinite rod.

11-5

Further Applications of Taylor's Formula

Suppose that f is a function with continuous derivatives of *all* orders in a neighborhood of the point a, and that n is an arbitrary positive integer. We know by Taylor's formula that

$$f(x) = P_n(x) + R_n(x) \tag{1}$$

where

$$P_n(x) = f(a) + f'(a)(x - a) + f''(a)\frac{(x - a)^2}{2!}$$

$$+ \cdots + f^{(n)}(a)\frac{(x - a)^n}{n!} \tag{2}$$

and

$$R_n(x) = \frac{f^{(n+1)}(z)}{(n + 1)!}(x - a)^{n+1} \tag{3}$$

where z is some number between a and x.

    Now suppose that, for some particular *fixed* value of x, we can show that

$$\lim_{n \to \infty} R_n(x) = 0. \tag{4}$$

Then it follows from Equation (1) that

$$f(x) = \lim_{n \to \infty} P_n(x) = \lim_{n \to \infty} \left(\sum_{k=0}^{n} \frac{f^{(k)}(a)}{k!} (x - a)^k \right). \tag{5}$$

The right-hand member of Equation (5) is the limit (as $n \to \infty$) of the sum of $n + 1$ terms. Because of this, we use the very natural notation

$$f(x) = \sum_{k=0}^{\infty} \frac{f^{(k)}(a)}{k!} (x - a)^k$$

$$= f(a) + f'(a)(x - a) + \frac{f''(a)}{2!} (x - a)^2 + \cdots$$

$$+ \frac{f^{(n)}(a)}{n!} (x - a)^n + \cdots. \tag{6}$$

The "infinite series" in Equation (6) is called the **Taylor series** of the function f at $x = a$. We shall take up infinite series in Chapter 12; for the moment, we simply may regard Equation (6) as a convenient and suggestive notation equivalent to Equation (5). Note that the nth degree Taylor polynomial $P_n(x)$ is the sum of the first $n + 1$ terms of the Taylor series.

Now suppose that Equation (4) holds—that, for a particular value of x, $R_n(x) \to 0$ as $n \to \infty$. Then, by Equation (5), we can compute the value of $f(x)$ with any desired accuracy by adding enough terms of the Taylor series of f at a. For example, from the problems and examples of Section 11-2, we know the following Taylor formulas for the exponential and trigonometric functions:

$$e^x = 1 + x + \frac{x^2}{2!} + \frac{x^3}{3!} + \cdots + \frac{x^n}{n!} + \frac{e^z}{(n + 1)!} x^{n+1}.$$

$$\cos x = 1 - \frac{x^2}{2!} + \frac{x^4}{4!} - \cdots + (-1)^n \frac{x^{2n}}{(2n)!} + (-1)^{n+1} \frac{\cos z}{(2n + 2)!} x^{2n+2}.$$

$$\sin x = x - \frac{x^3}{3!} + \frac{x^5}{5!} - \cdots + (-1)^n \frac{x^{2n+1}}{(2n + 1)!} + (-1)^{n+1} \frac{\cos z}{(2n + 3)!} x^{2n+3}.$$

In each case, z is a number between 0 and x.

Since z is between 0 and x, it follows that $0 < e^z \leqq e^{|x|}$ in Taylor's formula for e^x. In the formulas for the sine and cosine functions, $0 \leqq |\cos z| \leqq 1$. Therefore the fact that

$$\lim_{n \to \infty} \frac{x^n}{n!} = 0$$

for all x (see Problem 17) implies that $\lim_{n \to \infty} R_n(x) = 0$ in all three cases above. This gives the following Taylor series:

$$e^x = \sum_{n=0}^{\infty} \frac{x^n}{n!} = 1 + x + \frac{x^2}{2!} + \frac{x^3}{3!} + \frac{x^4}{4!} + \cdots, \tag{7}$$

$$\cos x = \sum_{n=0}^{\infty} (-1)^n \frac{x^{2n}}{(2n)!} = 1 - \frac{x^2}{2!} + \frac{x^4}{4!} - \frac{x^6}{6!} + \cdots, \tag{8}$$

$$\sin x = \sum_{n=0}^{\infty} (-1)^n \frac{x^{2n+1}}{(2n+1)!} = x - \frac{x^3}{3!} + \frac{x^5}{5!} - \frac{x^7}{7!} + \cdots. \qquad (9)$$

We shall discuss these series further in Section 12-7.

*THE NUMBER π

In Section 4-2 we described how Archimedes used polygons inscribed in and circumscribed about the unit circle to show that $3\frac{10}{71} < \pi < 3\frac{1}{7}$. With the aid of large electronic computers, π has been computed to as many as one million decimal places. We describe now some of the methods that have been used for such computations. [For a chronicle of mankind's perennial fascination with the number π, see Howard Eves, *An Introduction to the History of Mathematics*, 4th ed. (Boston: Allyn & Bacon, 1976), pp. 96–102.]

We begin with the elementary algebraic identity

$$\frac{1}{1+x} = 1 - x + x^2 - x^3 + \cdots + (-1)^{k-1}x^{k-1} + \frac{(-1)^k x^k}{1+x}, \qquad (10)$$

which can be verified by multiplying both sides by $1 + x$. We substitute t^2 for x and $n + 1$ for k, and thus find that

$$\frac{1}{1+t^2} = 1 - t^2 + t^4 - \cdots + (-1)^n t^{2n} + \frac{(-1)^{n+1} t^{2n+2}}{1+t^2}.$$

Since $D_t \tan^{-1} t = 1/(1 + t^2)$, integration of both sides of this last equation from $t = 0$ to $t = x$ gives

$$\tan^{-1}x = x - \frac{x^3}{3} + \frac{x^5}{5} - \cdots + (-1)^n \frac{x^{2n+1}}{2n+1} + R_{2n+1} \qquad (11)$$

where

$$\left| R_{2n+1} \right| = \left| \int_0^x \frac{t^{2n+2}}{1+t^2}\,dt \right| \leq \left| \int_0^x t^{2n+2}\,dt \right| = \frac{|x|^{2n+3}}{2n+3}. \qquad (12)$$

This estimate of the error makes it clear that

$$\lim_{n \to \infty} R_{2n+1} = 0$$

if $|x| \leq 1$. Hence we obtain the Taylor series for the inverse tangent function:

$$\tan^{-1}x = \sum_{n=0}^{\infty} (-1)^n \frac{x^{2n+1}}{2n+1} = x - \frac{x^3}{3} + \frac{x^5}{5} - \frac{x^7}{7} + \cdots, \qquad (13)$$

valid for $-1 \leq x \leq 1$.

If we substitute $x = 1$ into Equation (13), we obtain *Leibniz's series*

$$\frac{\pi}{4} = 1 - \frac{1}{3} + \frac{1}{5} - \frac{1}{7} + \cdots.$$

Though this is a beautiful series, it is not an effective way to compute π. But the error estimate in (12) shows that Formula (11) is effective for the calculation of $\tan^{-1}x$ if $|x|$ is small. For example, if $x = \frac{1}{5}$, the fact that

$$\frac{1}{(9)(5)^9} \approx 0.000000057$$

implies that the approximation

$$\alpha = \tan^{-1}\left(\tfrac{1}{5}\right) \approx \tfrac{1}{5} - \tfrac{1}{3}\left(\tfrac{1}{5}\right)^3 + \tfrac{1}{5}\left(\tfrac{1}{5}\right)^5 - \tfrac{1}{7}\left(\tfrac{1}{5}\right)^7$$

is accurate to six decimal places.

Let us begin with $\alpha = \tan^{-1}(\tfrac{1}{5})$. The addition formula for the tangent function can be used to show (Problem 8) that

$$\tan\left(\frac{\pi}{4} - 4\alpha\right) = -\frac{1}{239}.$$

Hence

$$\frac{\pi}{4} = 4\tan^{-1}\left(\frac{1}{5}\right) - \tan^{-1}\left(\frac{1}{239}\right). \tag{14}$$

In 1706 John Machin used Formula (14) to calculate the first hundred decimal places of π; in Problem 16 we ask you to use it to show that $\pi = 3.14159$ to five decimal places. In 1844 the lightning calculator Zacharias Dase, a German, computed the first 200 decimal places of π using the related formula

$$\frac{\pi}{4} = \tan^{-1}\left(\frac{1}{2}\right) + \tan^{-1}\left(\frac{1}{5}\right) + \tan^{-1}\left(\frac{1}{8}\right), \tag{15}$$

and you might enjoy verifying this formula (see Problem 9). A recent computation of 1 million decimal places of π used the formulas

$$\frac{\pi}{4} = 12\tan^{-1}\left(\frac{1}{18}\right) + 8\tan^{-1}\left(\frac{1}{57}\right) - 5\tan^{-1}\left(\frac{1}{239}\right)$$

$$= 6\tan^{-1}\left(\frac{1}{8}\right) + 2\tan^{-1}\left(\frac{1}{57}\right) + \tan^{-1}\left(\frac{1}{239}\right).$$

Although no practical application is likely to require more than 10 or 12 decimal places of π, these computations provide dramatic evidence of the power of Taylor's formula.

NUMERICAL INTEGRATION WITH TAYLOR POLYNOMIALS

As an alternative to the use of the trapezoidal rule or Simpson's approximation to compute $\int_a^b f(x)\,dx$, it is sometimes more convenient to substitute for the integrand $f(x)$ the Taylor formula for f at a:

$$f(x) = P_n(x) + \frac{f^{(n+1)}(z)}{(n+1)!}(x-a)^{n+1}$$

(for some number z between a and x). Then

$$\int_a^b f(x)\,dx = \int_a^b P_n(x)\,dx + E_n$$

with

$$|E_n| = \left|\int_a^b f^{(n+1)}(z)\frac{(x-a)^{n+1}}{(n+1)!}\,dx\right| \leq \frac{M(b-a)^{n+2}}{(n+2)!}, \tag{16}$$

where M is the maximum of $|f^{(n+1)}(x)|$ for x in $[a, b]$. Finding the integral of the polynomial $P_n(x)$ is simple, and (16) provides an upper bound on the error.

EXAMPLE 2 Substitute the eighth-degree Taylor polynomial of $\sin x$ to compute

$$\int_0^1 \frac{\sin x}{x}\, dx.$$

Solution The eighth-degree Taylor polynomial with remainder for $\sin x$ at $a = 0$ is

$$\sin x = x - \frac{x^3}{3!} + \frac{x^5}{5!} - \frac{x^7}{7!} + \frac{x^9}{9!} \cos z$$

with z between 0 and x. We divide both sides by x, then integrate:

$$\int_0^1 \frac{\sin x}{x}\, dx = \int_0^1 \left(1 - \frac{x^2}{3!} + \frac{x^4}{5!} - \frac{x^6}{7!}\right) dx + E$$

$$= 1 - \frac{1}{3!3} + \frac{1}{5!5} - \frac{1}{7!7} + E \approx 0.946083 + E.$$

Now

$$E = \int_0^1 \frac{x^8}{9!} \cos z\, dx \leqq \int_0^1 \frac{x^8}{9!}\, dx = \frac{1}{9!9} \approx 0.0000003.$$

Therefore we find that

$$\int_0^1 \frac{\sin x}{x}\, dx = 0.946083$$

to six decimal places, with considerably less work involved than would be necessary to achieve such accuracy with Simpson's approximation.

There's a bonus. The original integral is actually, in some sense, improper: The integrand is an indeterminate form at the endpoint $x = 0$. Notice how easily the Taylor series method avoids this potential difficulty.

TAYLOR'S FORMULA AND THE SECOND DERIVATIVE TEST

Here is a demonstration of the theoretical power of Taylor's formula: We use it to give a quick proof of the second derivative maximum-minimum test for a function that has a continuous second derivative. Suppose that $f'(a) = 0$ but that $f''(a) \neq 0$. Then the second degree Taylor formula for f at a is

$$f(x) = f(a) + \tfrac{1}{2} f''(z)(x - a)^2$$

with z somewhere between a and x. Therefore $f(x) - f(a)$ has the same sign as $f''(z)$. If x is sufficiently close to a to insure that $f''(z)$ has the same sign as $f''(a)$, it follows that $f(x) > f(a)$ if $f''(a) > 0$, while $f(x) < f(a)$ if $f''(a) < 0$. Therefore $f(a)$ is a local minimum value if $f''(a) > 0$; it is a local maximum value if $f''(a) < 0$.

11-5 PROBLEMS

In each of Problems 1–7, find the Taylor series (Equation (6)) of the given function $f(x)$ at the indicated point a.

1 $f(x) = \ln(1 + x)$; $a = 0$.

2 $f(x) = \dfrac{1}{1 - x}$; $a = 0$. **3** $f(x) = e^{-x}$; $a = 0$.

4 $f(x) = \cosh x$; $a = 0$. **5** $f(x) = \ln x$; $a = 1$.

6 $f(x) = \sin x$; $a = \dfrac{\pi}{2}$. **7** $f(x) = \cos x$; $a = \dfrac{\pi}{4}$.

8 Beginning with $\alpha = \tan^{-1}(\frac{1}{5})$, use the addition formula

$$\tan(A + B) = \frac{\tan A + \tan B}{1 - \tan A \tan B}$$

to show in turn that: (a) $\tan 2\alpha = \frac{5}{12}$; (b) $\tan 4\alpha = \frac{120}{119}$; and (c) $\tan(\pi/4 - 4\alpha) = -\frac{1}{239}$.

9 Apply the addition formula for the tangent function to verify Formula (15).

In Problems 10–15, substitute an appropriate Taylor formula with remainder for the integrand in order to compute the given integral accurate to the indicated number of decimal places.

10 $\displaystyle\int_0^1 e^{-x^2}\, dx$; two decimal places.

11 $\displaystyle\int_0^1 \frac{1 - e^{-x}}{x}\, dx$; two decimal places.

12 $\displaystyle\int_0^{1/2} \sin x^2\, dx$; five decimal places.

13 $\displaystyle\int_0^{1/2} \frac{\sin x}{\sqrt{x}}\, dx$; five decimal places.

14 $\displaystyle\int_0^1 e^{\sqrt{x}}\, dx$; two decimal places.

15 $\displaystyle\int_0^{1/2} \frac{\tan^{-1} x}{x}\, dx$; three decimal places.

16 Use Formulas (11), (12), and (14) to show that $\pi = 3.14159$ to five decimal places. (*Suggestion:* Compute $\arctan(\frac{1}{5})$ and $\arctan(\frac{1}{239})$ each with error less than 0.5×10^{-7}. You can do this by applying Formula (11) with $n = 4$ and with $n = 1$. Carry out your computations to seven decimal places and keep track of errors.)

17 Prove that $\lim\limits_{n \to \infty} x^n/n! = 0$ if x is a fixed real number. (*Suggestion:* Choose an integer k with $k > 2|x|$, and let $L = |x|^k/k!$. Then show that $|x|^n/n! < L/2^{n-k}$ if $n > k$.)

18 (a) Suppose that t is a positive real number but is otherwise arbitrary. Integrate both sides of (10) from $x = 0$ to $x = t$ to show that

$$\ln(1 + t) = t - \frac{t^2}{2} + \frac{t^3}{3} - \cdots + (-1)^{k-1}\frac{t^k}{k} + R_k$$

where $|R_k| \leq \dfrac{t^{k+1}}{k + 1}$.

(b) Conclude that

$$\ln(1 + t) = t - \frac{t^2}{2} + \frac{t^3}{3} - \cdots = \sum_{k=1}^{\infty} (-1)^{k-1}\frac{t^k}{k}$$

if $0 < t < 1$.

19 Use the result of Problem 18(a) to calculate $\ln(\frac{4}{3})$ accurate to three decimal places.

20 Suppose that f is a function with a continuous third derivative. If $f'(a) = f''(a) = 0$ but $f^{(3)}(a) \neq 0$, prove that $f(a)$ is *neither* a local maximum nor a local minimum value. Begin by writing the third degree Taylor expansion of f at $x = a$.

CHAPTER 11 REVIEW: Definitions, Concepts, Results

Use the list below as a guide to concepts that you may need to review.

1 L'Hôpital's rule and the indeterminate forms $0/0$, ∞/∞, $0 \cdot \infty$, $\infty - \infty$, 0^0, ∞^0, and 1^∞

2 Cauchy's Mean Value Theorem

3 The nth degree Taylor polynomial of the function f at the point $x = a$

4 Taylor's formula with remainder

5 Use of Taylor's formula to approximate values of functions and to evaluate integrals

6 Error estimate for the trapezoidal approximation and Simpson's approximation

7 Definition and evaluation of the improper integrals

$$\int_a^{\infty} f(x)\, dx \quad \text{and} \quad \int_{-\infty}^{b} f(x)\, dx$$

8 Definition and evaluation of the improper integral $\int_a^b f(x)\, dx$, where f has an infinite discontinuity at the point c of $[a, b]$

9 Taylor series

MISCELLANEOUS PROBLEMS

Find the limits in Problems 1–15.

1 $\lim_{x \to 2} \dfrac{x-2}{x^2-4}$.

2 $\lim_{x \to 0} \dfrac{\sin 2x}{x}$.

3 $\lim_{x \to \pi} \dfrac{1+\cos x}{(x-\pi)^2}$.

4 $\lim_{x \to 0} \dfrac{x-\sin x}{x^3}$.

5 $\lim_{t \to 0} \dfrac{t \tan^{-1} t - \sin^2 t}{t^6}$.

6 $\lim_{x \to \infty} \dfrac{\ln(\ln x)}{\ln x}$.

7 $\lim_{x \to 0} (\cot x) \ln(1+x)$.

8 $\lim_{x \to 0^+} (e^{1/x}-1) \tan x$.

9 $\lim_{x \to 0} \left(\dfrac{1}{x^2} - \dfrac{1}{1-\cos x} \right)$.

10 $\lim_{x \to \infty} \left(\dfrac{x^2}{x+2} - \dfrac{x^3}{x^2+3} \right)$.

11 $\lim_{x \to \infty} (\sqrt{x^2-x+1} - \sqrt{x})$.

12 $\lim_{x \to \infty} x^{1/x}$.

13 $\lim_{x \to \infty} (e^{2x}-2x)^{1/x}$.

14 $\lim_{x \to \infty} (1-e^{-x^2})^{1/x^2}$.

15 $\lim_{x \to \infty} x \left[\left(1+\dfrac{1}{x} \right)^x - e \right]$. (*Suggestion:* Let $u = 1/x$ and take the limit as $u \to 0^+$.)

Determine whether or not the improper integrals in Problems 16–22 converge, and evaluate those that do converge.

16 $\displaystyle\int_{-4}^{0} \dfrac{dx}{\sqrt{x+4}}$.

17 $\displaystyle\int_{2}^{6} \dfrac{dx}{(x-2)^{3/2}}$.

18 $\displaystyle\int_{-\infty}^{\infty} \dfrac{x\,dx}{\sqrt{x^2+9}}$.

19 $\displaystyle\int_{e}^{\infty} \dfrac{dx}{x(\ln x)^2}$.

20 $\displaystyle\int_{-\infty}^{\infty} \sin x\,dx$.

21 $\displaystyle\int_{0}^{\infty} x^2 e^{-x}\,dx$.

22 $\displaystyle\int_{0}^{2} \dfrac{dx}{(x-1)^{2/3}}$.

In Problems 23–31, find the Taylor formula with remainder for $f(x)$ with the given values of a and n.

23 $f(x) = e^{x+1}$; $a = 0, n = 5$.
24 $f(x) = xe^x$; $a = 0, n = 3$.
25 $f(x) = e^{-x^2}$; $a = 0, n = 3$.
26 $f(x) = \sqrt[3]{x-8}$; $a = 0, n = 3$.
27 $f(x) = \sqrt[3]{x}$; $a = 125, n = 3$.
28 $f(x) = \tan^{-1} x$; $a = 1, n = 3$.
29 $f(x) = x^{-1/2}$; $a = 1, n = 4$.
30 $f(x) = (x+2)^{3/2}$; $a = 2, n = 3$.
31 $f(x) = x^4$; $a = 1, n = 5$.

In each of Problems 32–37, find the Taylor series of $f(x)$ at the indicated point $x = a$.

32 $f(x) = e^{2x}$; $a = 0$.
33 $f(x) = \sinh x$; $a = 0$.

34 $f(x) = \dfrac{1}{(1-x)^2}$; $a = 0$.

35 $f(x) = \dfrac{1}{x}$; $a = 1$.

36 $f(x) = \cos x$; $a = \dfrac{\pi}{2}$.

37 $f(x) = \sin x$; $a = \dfrac{\pi}{4}$.

38 Use Taylor's formula to calculate $\sqrt[3]{120}$ accurate to three decimal places.

39 Use Taylor's formula to calculate $\sqrt[10]{1000}$ accurate to five decimal places.

In Problems 40–41, determine the number of decimal places of accuracy the given approximation formula yields for $|x| \le 1$.

40 $\cos x \approx 1 - \frac{1}{2}x^2 + \frac{1}{24}x^4$.
41 $\tan^{-1} x \approx x - \frac{1}{3}x^3$.
42 Show that the approximation

$$\sin x \approx x - \dfrac{x^3}{3!} + \dfrac{x^5}{5!} - \dfrac{x^7}{7!} + \dfrac{x^9}{9!}$$

can be used to calculate $\sin x$ accurate to four decimal places for all values of x that correspond to angles between $0°$ and $90°$.

43 Show that the approximation of Problem 42 is accurate to 15 decimal places for values of x that correspond to angles between $0°$ and $10°$.

44 According to Problem 35 in Section 8-5, the surface area of the ellipsoid obtained by revolving the ellipse $x^2/a^2 + y^2/b^2 = 1$ ($a < b$) around the x-axis is

$$A = 2\pi ab \left(\dfrac{b}{a} + \dfrac{a}{c} \ln \left(\dfrac{b+c}{a} \right) \right)$$

where $c^2 = b^2 - a^2$. Use l'Hôpital's rule to show that $\lim_{b \to a} A = 4\pi a^2$.

45 Given: $\displaystyle\int_{0}^{\infty} e^{-x^2}\,dx = \frac{1}{2}\sqrt{\pi}$. Deduce that $\Gamma(\frac{1}{2}) = \sqrt{\pi}$.

46 Recall that $\Gamma(x+1) = x\Gamma(x)$ if $x > 0$. Suppose that n is a positive integer. Use Problem 45 to establish that

$$\Gamma\left(n + \dfrac{1}{2} \right) = \dfrac{1 \cdot 3 \cdot 5 \cdots (2n-1)}{2^n} \sqrt{\pi}.$$

47 (a) Suppose that $k > 1$. Show by integration by parts that

$$\int_0^\infty x^k e^{-x^2}\, dx = \frac{k-1}{2} \int_0^\infty x^{k-2} e^{-x^2}\, dx.$$

(b) Suppose that n is a positive integer. Show that

$$\int_0^\infty x^{n-1} e^{-x^2}\, dx = \frac{1}{2} \Gamma\!\left(\frac{n}{2}\right).$$

48 Differentiate both sides of the identity

$$\frac{1}{1-x} = 1 + x + x^2 + \cdots + x^n + \frac{x^{n+1}}{1-x}$$

to obtain the result

$$\frac{1}{(1-x)^2} = 1 + 2x + 3x^2 + \cdots + nx^{n-1} + R_n$$

where, assuming that $|x| < 1$, $R_n \to 0$ as $n \to \infty$. Thus if $-1 < x < 1$ then

$$\frac{1}{(1-x)^2} = \sum_{n=0}^\infty (n+1)x^n.$$

49 Substitute t^2 for x in the first identity of Problem 48, and then integrate from $t = 0$ to $t = x$ to obtain the result that

$$\tanh^{-1} x = x + \frac{x^3}{3} + \cdots + \frac{x^{2n+1}}{2n+1} + R_n$$

where $R_n \to 0$ as $n \to \infty$, under the assumption that $|x| < 1$.

Thus if $-1 < x < 1$ then

$$\tanh^{-1} x = \sum_{n=0}^\infty \frac{x^{2n+1}}{2n+1}.$$

50 Use the formula $\ln 2 = \ln \frac{5}{4} + 2 \ln \frac{6}{5} + \ln \frac{10}{9}$ to calculate $\ln 2$ accurate to three decimal places. See Problem 18(a) in Section 11-5 for the Taylor's formula for $\ln(1 + x)$.

51 Prove as follows that the number e is irrational. First suppose to the contrary that $e = p/q$ where p and q are positive integers. Write

$$\frac{p}{q} = e = 1 + 1 + \frac{1}{2!} + \frac{1}{3!} + \cdots + \frac{1}{q!} + R_q$$

where $0 < R_q < 3/(q+1)!$. (Why?) Then show that multiplication of both sides of this equation by $q!$ would lead to the contradiction that one side of the result is an integer but the other side is not.

52 In a letter of 1672 that he wrote to a person who wanted a quick method for gauging the content of a wine barrel (volume, not quality), Newton mentioned that if a symmetric barrel has the shape obtained by revolving a parabolic arc, then its volume is exactly

$$V = \pi H(\tfrac{2}{3}R^2 + \tfrac{1}{3}r^2 - \tfrac{2}{15}\delta^2).$$

In this formula, H is the height of the barrel, R is the radius of its midsection, r is the radius of each end, and $\delta = R - r$. Use Simpson's approximation with error estimate to establish Newton's formula.

Infinite Series

12

Introduction

In the fifth century B.C., the Greek philosopher Zeno proposed the following paradox: In order for a runner to travel a given distance, the runner must first travel halfway, then half the remaining distance, then half the distance that still remains, and so on ad infinitum. But, Zeno argued, it is clearly impossible for a runner to accomplish these infinitely many steps in a finite period of time, so motion from one point to another must be impossible.

Zeno's paradox suggests the infinite subdivision of $[0, 1]$ indicated in Figure 12.1. There is one subinterval of length $1/2^n$ for each integer $n = 1, 2, 3, \ldots$. If the length of the interval is the sum of the lengths of the subintervals into which it is divided, then it would appear that

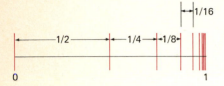

12.1 Subdivision of an interval to illustrate Zeno's paradox.

$$1 = \frac{1}{2} + \frac{1}{4} + \frac{1}{8} + \frac{1}{16} + \cdots + \frac{1}{2^n} + \cdots,$$

with infinitely many terms somehow adding up to 1. On the other hand, the formal infinite sum

$$1 + 2 + 3 + \cdots + n + \cdots$$

of all the positive integers seems meaningless—it does not appear to add up to *any* (finite) value.

The question is this: What, if anything, is meant by the sum of an *infinite* collection of numbers? This chapter explores conditions under which an *infinite* sum

$$a_1 + a_2 + a_3 + \cdots + a_n + \cdots,$$

known as an *infinite series, is* meaningful. We shall discuss methods for computing the sum of an infinite series, and applications of the algebra and calculus of infinite series. Infinite series are important in science and mathematics because many functions either arise most naturally in the form of infinite series, or have infinite series representations (such as the Taylor series of Section 11-5) that are useful for numerical computations.

Infinite Sequences

An (**infinite**) **sequence** of real numbers is a function whose domain of definition is the set of all positive integers. Thus if s is a sequence, then to each positive integer n there corresponds a real number $s(n)$. Ordinarily, a sequence is most conveniently described by listing its values in order, beginning with $s(1)$:

$$s(1), s(2), s(3), \ldots, s(n), \ldots.$$

With subscript notation rather than functional notation, we may write

$$s_1, s_2, s_3, \ldots, s_n, \ldots \tag{1}$$

for this list of values. The values in this list are the **terms** of the sequence; s_1 is the first term, s_2 the second term, s_n the **nth term.**

We use the notation $\{s_n\}_1^\infty$, or simply $\{s_n\}$, as abbreviation for the **ordered** list in (1), and we may refer to the sequence by saying simply "the sequence $\{s_n\}$." When a particular sequence is so described, the nth term s_n is generally (though not always) given by a formula in terms of its subscript n. In this case, listing the first few terms of the sequence often helps us to see it more concretely.

EXAMPLE 1 The following table lists explicitly the first four terms of each of several sequences.

$\{s_n\}_1^\infty$	$s_1, s_2, s_3, s_4, \dots$
$\left\{\dfrac{1}{n}\right\}_1^\infty$	$1, \dfrac{1}{2}, \dfrac{1}{3}, \dfrac{1}{4}, \dots$
$\left\{\dfrac{1}{10^n}\right\}_1^\infty$	$0.1, 0.01, 0.001, 0.0001, \dots$
$\left\{\dfrac{1}{n!}\right\}_1^\infty$	$1, \dfrac{1}{2}, \dfrac{1}{6}, \dfrac{1}{24}, \dots$
$\left\{\dfrac{(-1)^n}{n(n+1)}\right\}_1^\infty$	$-\dfrac{1}{2}, \dfrac{1}{6}, -\dfrac{1}{12}, \dfrac{1}{20}, \dots$
$\left\{\sin\dfrac{n\pi}{2}\right\}_1^\infty$	$1, 0, -1, 0, \dots$
$\left\{1 + (-1)^n\right\}_1^\infty$	$0, 2, 0, 2, \dots$

EXAMPLE 2 The Fibonacci sequence $\{F_n\}$ is defined as follows:

$$F_1 = 1, \qquad F_2 = 1, \quad \text{and} \quad F_{n+1} = F_{n-1} + F_n \qquad \text{for } n \geq 2.$$

The first 10 terms of the Fibonacci sequence are

$$1, 1, 2, 3, 5, 8, 13, 21, 34, 55.$$

This is a *recursively defined* sequence—after the initial values are given, each term is defined in terms of its predecessors.

The limit of a sequence was defined in Section 3-4 in much the same way as the limit of an ordinary function (Section 1-8).

Definition *Limit of a Sequence*

We say that the sequence $\{s_n\}$ **converges** to the real number L, or has **limit** L, and we write

$$\lim_{n \to \infty} s_n = L, \tag{2}$$

provided that s_n can be made as close to L as we please by choosing n sufficiently large. That is, given any number $\varepsilon > 0$, there exists an integer N such that

$$|s_n - L| < \varepsilon \qquad \text{for all} \quad n \geq N. \tag{3}$$

EXAMPLE 3 Show that $\lim_{n \to \infty} \dfrac{1}{n} = 0$.

Solution We need to show this: To each positive number ε, there corresponds an integer N such that, for all $n \geqq N$,

$$\left|\frac{1}{n} - 0\right| = \frac{1}{n} < \varepsilon.$$

It suffices to pick any fixed integer $N > 1/\varepsilon$. Then $n \geqq N$ implies $1/n \leqq 1/N < \varepsilon$, as desired.

The limit laws stated in Section 1-8 for limits of functions have natural analogues for limits of sequences. Their proofs are based on techniques similar to those used in Section 1-10.

Theorem 1 *Limit Laws for Sequences*

If the limits $\lim\limits_{n \to \infty} a_n = A$ and $\lim\limits_{n \to \infty} b_n = B$ exist (so that A and B are real numbers), then:

(a) $\lim\limits_{n \to \infty} ca_n = cA$ (c any real number);

(b) $\lim\limits_{n \to \infty} (a_n + b_n) = A + B$;

(c) $\lim\limits_{n \to \infty} a_n b_n = AB$;

(d) $\lim\limits_{n \to \infty} \dfrac{a_n}{b_n} = \dfrac{A}{B}$.

In this last case we must assume also that $B \neq 0$ and that $b_n \neq 0$ for all sufficiently large n.

Theorem 2 *Substitution Law for Sequences*

If $\lim\limits_{n \to \infty} a_n = A$ and the function $f(x)$ is continuous at $x = A$, then

$$\lim_{n \to \infty} f(a_n) = f(A).$$

Theorem 3 *Squeeze Law for Sequences*

If $a_n \leqq b_n \leqq c_n$ for all n and

$$\lim_{n \to \infty} a_n = L = \lim_{n \to \infty} c_n,$$

then $\lim\limits_{n \to \infty} b_n = L$ also.

These theorems can be used to compute limits of many sequences formally, without recourse to the definition. For example, if k is a positive integer and c is a constant, then Example 3 and the product law (Theorem 1c) give

$$\lim_{n \to \infty} \frac{c}{n^k} = c \cdot 0 \cdot 0 \cdots 0 = 0.$$

EXAMPLE 4 Show that $\displaystyle\lim_{n\to\infty}\frac{(-1)^n\cos n}{n^2}=0$.

Solution This result follows from the squeeze law and the fact that $1/n^2\to 0$ as $n\to\infty$, because

$$-\frac{1}{n^2}\leqq\frac{(-1)^n\cos n}{n^2}\leqq\frac{1}{n^2}.$$

EXAMPLE 5 Show that if $a>0$, then $\displaystyle\lim_{n\to\infty}\sqrt[n]{a}=1$.

Solution We apply the substitution law with $f(x)=a^x$ and $A=0$. Since $1/n\to 0$ as $n\to\infty$ and f is continuous at $x=0$, this gives

$$\lim_{n\to\infty}a^{1/n}=a^0=1.$$

EXAMPLE 6 The limit laws and the continuity of $\sqrt{x}$ at $x=4$ yield

$$\lim_{n\to\infty}\sqrt{\frac{4n-1}{n+1}}=\sqrt{\lim_{n\to\infty}\frac{4-(1/n)}{1+(1/n)}}=\sqrt{4}=2.$$

EXAMPLE 7 Show that if $|r|<1$, then $\displaystyle\lim_{n\to\infty}r^n=0$.

Solution Since $|r^n|=|(-r)^n|$, we may assume that $0<r<1$. Then $1/r=1+a$ with $a>0$, so the binomial formula yields

$$\frac{1}{r^n}=(1+a)^n=1+na+(\text{positive terms})>1+na,$$

so

$$0<r^n<\frac{1}{1+na}.$$

Since $1/(1+na)\to 0$ as $n\to\infty$, the squeeze law implies that $r^n\to 0$ as $n\to\infty$.

Let $f(x)$ be a function defined for every real number $x\geqq 1$, and $\{a_n\}$ a sequence such that $f(n)=a_n$ for every positive integer n. Then it follows from the definitions of limits of functions and sequences that

$$\text{if}\quad\lim_{x\to\infty}f(x)=L\quad\text{then}\quad\lim_{n\to\infty}a_n=L.\qquad(4)$$

Note that the converse of the above statement is generally false. For example,

$$\lim_{n\to\infty}\sin\pi n=0\quad\text{but}\quad\lim_{x\to\infty}\sin\pi x\quad\text{does not exist.}$$

Because of (4) we can use **l'Hôpital's rule for sequences:** If $a_n=f(n)$, $b_n=g(n)$ and $f(x)/g(x)$ has the indeterminate form ∞/∞ as $x\to\infty$, then

$$\lim_{n\to\infty}\frac{a_n}{b_n}=\lim_{x\to\infty}\frac{f(x)}{g(x)}=\lim_{x\to\infty}\frac{f'(x)}{g'(x)},\qquad(5)$$

provided that the right-hand limit exists.

EXAMPLE 8 Show that $\lim\limits_{n \to \infty} \dfrac{\ln n}{n} = 0$.

Solution The function $(\ln x)/x$ is defined for all $x \geq 1$ and agrees with the given sequence when $x = n$, an integer. Since $(\ln x)/x$ has the indeterminate form ∞/∞ as $x \to \infty$, l'Hôpital's rule gives

$$\lim_{n \to \infty} \frac{\ln n}{n} = \lim_{x \to \infty} \frac{\ln x}{x} = \lim_{x \to \infty} \frac{1/x}{1} = 0.$$

EXAMPLE 9 Show that $\lim\limits_{n \to \infty} \sqrt[n]{n} = 1$.

Solution First we note that

$$\ln \sqrt[n]{n} = \frac{\ln n}{n} \to 0 \text{ as } n \to \infty$$

by Example 8. By the substitution law with $f(x) = e^x$, this gives

$$\lim_{n \to \infty} \sqrt[n]{n} = \lim_{n \to \infty} \exp(\ln \sqrt[n]{n}) = e^0 = 1.$$

EXAMPLE 10 Find $\lim\limits_{n \to \infty} \dfrac{3n^3}{e^{2n}}$.

Solution We apply l'Hôpital's rule repeatedly, though we must be careful at each intermediate step to verify that we still have an indeterminate form. Thus we find that

$$\lim_{n \to \infty} \frac{3n^3}{e^{2n}} = \lim_{x \to \infty} \frac{3x^3}{e^{2x}} = \lim_{x \to \infty} \frac{9x^2}{2e^{2x}}$$

$$= \lim_{x \to \infty} \frac{18x}{4e^{2x}} = \lim_{x \to \infty} \frac{18}{8e^{2x}} = 0.$$

BOUNDED MONOTONE SEQUENCES

The set of all *rational* numbers has by itself all the most familiar elementary algebraic properties of the entire real number system. In order to guarantee the existence of irrational numbers, we must assume in addition a "completeness property" of the real numbers. Otherwise the real line might have "holes" where the irrational numbers ought to be. One way of stating this completeness property involves the convergence of an important type of sequence.

The sequence $\{a_n\}_1^\infty$ is called **monotone increasing** if

$$a_1 \leq a_2 \leq a_3 \leq \cdots \leq a_n \leq \cdots$$

and **monotone decreasing** if

$$a_1 \geq a_2 \geq a_3 \geq \cdots \geq a_n \geq \cdots.$$

It is **monotone** if it is *either* monotone increasing *or* monotone decreasing. The sequence $\{a_n\}$ is **bounded** if there is a number M such that $|a_n| \leq M$ for all n. The following assertion may be taken as an axiom for the real number system.

> **Bounded Monotone Sequence Property**
> Every bounded monotone infinite sequence converges (that is, has a finite limit).

Suppose, for example, that the monotone increasing sequence $\{a_n\}_1^\infty$ is bounded above by a number M, meaning that $a_n \leq M$ for all n. Since it is also bounded below (by a_1, for instance), the bounded monotone sequence property implies that $\lim_{n \to \infty} a_n = A$ for some real number $A \leq M$. If the monotone increasing sequence $\{a_n\}$ is *not* bounded above (by any number), then it follows that $\lim_{n \to \infty} a_n = \infty$ (Problem 38). These two alternatives are illustrated in Figure 12.2.

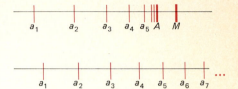

12.2 A bounded increasing sequence and an increasing sequence that is not bounded above.

EXAMPLE 11 Investigate the sequence $\{a_n\}$ defined recursively by

$$a_1 = \sqrt{2}, \qquad a_{n+1} = \sqrt{2 + a_n} \qquad \text{for} \quad n \geq 1.$$

Solution   The first four terms of $\{a_n\}$ are

$$\sqrt{2}, \ \sqrt{2 + \sqrt{2}}, \ \sqrt{2 + \sqrt{2 + \sqrt{2}}}, \ \sqrt{2 + \sqrt{2 + \sqrt{2 + \sqrt{2}}}}.$$

If $\lim_{n \to \infty} a_n$ exists, then this limit would seem to be the natural interpretation of the infinite expression

$$\sqrt{2 + \sqrt{2 + \sqrt{2 + \sqrt{2 + \cdots}}}}.$$

We shall apply the bounded monotone sequence property to show that this limit does indeed exist. Note first that $a_1 < 2$. If we assume inductively that $a_n < 2$, it then follows that

$$(a_{n+1})^2 = 2 + a_n < 4, \quad \text{so} \quad a_{n+1} < 2.$$

Thus the sequence $\{a_n\}$ is bounded above by the number 2. Since it is clear that $a_n > 1$ also, we can write $a_n = 2 - \varepsilon_n$ with $0 < \varepsilon_n < 1$. Then

$$\begin{aligned}
a_{n+1} = \sqrt{2 + a_n} &= \sqrt{4 - \varepsilon_n} \\
&> \sqrt{4 - \varepsilon_n - (3\varepsilon_n - \varepsilon_n^2)} \\
&= \sqrt{4 - 4\varepsilon_n + \varepsilon_n^2},
\end{aligned}$$

so that $a_{n+1} > 2 - \varepsilon_n = a_n$. And thus $\{a_n\}$ is a monotone increasing sequence.

Therefore the bounded monotone sequence property tells us that the sequence $\{a_n\}$ has a limit A. It does not tell us what the number A is. But now that we know that $A = \lim_{n \to \infty} a_n$ exists, we can write

$$\lim_{n \to \infty} a_{n+1} = \lim_{n \to \infty} \sqrt{2 + a_n},$$

$$A = \sqrt{2 + A},$$

and thus

$$A^2 = 2 + A.$$

The roots of this equation are -1 and 2. It is clear that $A > 0$, so we finally conclude that

$$\lim_{n \to \infty} a_n = 2.$$

To indicate what the bounded monotone sequence property has to do with the "completeness" of the real numbers, we outline in Problem 42 a proof (using this property) of the existence of the number $\sqrt{2}$. In Problems 43 and 44, we outline a proof of the equivalence of the bounded monotone sequence property and another common statement of the completeness of the real numbers —the *least upper bound property*.

12-2 PROBLEMS

In Problems 1–35 determine whether or not the sequence $\{a_n\}$ converges, and find its limit if it does converge.

1 $a_n = \dfrac{2n}{5n - 3}.$

2 $a_n = \dfrac{1 - n^2}{2 + 3n^2}.$

3 $a_n = \dfrac{n^3 - n + 7}{2n^3 + n^2}.$

4 $a_n = \dfrac{n^3}{10n^2 + 1}.$

5 $a_n = 1 + (\tfrac{9}{10})^n.$

6 $a_n = 2 - (-\tfrac{1}{2})^n.$

7 $a_n = 1 + (-1)^n.$

8 $a_n = \dfrac{1 + (-1)^n}{\sqrt{n}}.$

9 $a_n = \dfrac{1 + (-1)^n \sqrt{n}}{(\tfrac{3}{2})^n}.$

10 $a_n = \dfrac{\sin n}{3^n}.$

11 $a_n = \dfrac{\sin^2 n}{\sqrt{n}}.$

12 $a_n = \sqrt{\dfrac{2 + \cos n}{n}}.$

13 $a_n = n \sin \pi n.$

14 $a_n = n \cos \pi n.$

15 $a_n = \pi^{-(\sin n)/n}.$

16 $a_n = 2^{\cos n\pi}.$

17 $a_n = \dfrac{\ln n}{\sqrt{n}}.$

18 $a_n = \dfrac{\ln 2n}{\ln 3n}.$

19 $a_n = \dfrac{(\ln n)^2}{n}.$

20 $a_n = n \sin\left(\dfrac{1}{n}\right).$

21 $a_n = \dfrac{\tan^{-1} n}{n}.$

22 $a_n = \dfrac{n^3}{e^{n/10}}.$

23 $a_n = \dfrac{2^n + 1}{e^n}.$

24 $a_n = \dfrac{\sinh n}{\cosh n}.$

25 $a_n = \left(1 + \dfrac{1}{n}\right)^n.$

26 $a_n = \sqrt[n]{2n + 5}.$

27 $a_n = \left(\dfrac{n - 1}{n + 1}\right)^n.$

28 $a_n = (0.001)^{-1/n}.$

29 $a_n = \sqrt[n]{2^{n+1}}.$

30 $a_n = \left(1 - \dfrac{2}{n^2}\right)^n.$

31 $a_n = \left(\dfrac{2}{n}\right)^{3/n}.$

32 $a_n = (-1)^n \sqrt[n]{n^2 + 1}.$

33 $a_n = \left(\dfrac{2 - n^2}{3 + n^2}\right)^n.$

34 $a_n = \dfrac{(\tfrac{2}{3})^n}{1 - \sqrt[n]{n}}.$

35 $a_n = \dfrac{(\tfrac{2}{3})^n}{(\tfrac{1}{2})^n + (\tfrac{9}{10})^n}.$

36 Suppose that $\lim\limits_{n \to \infty} a_n = A$. Prove that $\lim\limits_{n \to \infty} |a_n| = |A|$.

37 Prove that $\{(-1)^n a_n\}$ diverges if $\lim\limits_{n \to \infty} a_n = A \neq 0$.

38 Suppose that $\{a_n\}$ is a monotone increasing sequence that is not bounded. Prove that $\lim\limits_{n \to \infty} a_n = \infty$.

39 Suppose that $A > 0$. Given x_1 arbitrary, define the sequence $\{x_n\}$ recursively by

$$x_{n+1} = \frac{1}{2}\left(x_n + \frac{A}{x_n}\right).$$

Prove that if $L = \lim\limits_{n \to \infty} x_n$ exists, then $L = \sqrt{A}$.

40 Let $\{F_n\}$ be the Fibonacci sequence of Example 2. Assume that $\tau = \lim\limits_{n \to \infty} (F_{n+1}/F_n)$ exists; show that $\tau = \frac{1}{2}(1 + \sqrt{5})$. (*Suggestion:* Substitute $F_{n+1} = F_n + F_{n-1}$ in the equation $\tau = \lim\limits_{n \to \infty} (F_{n+1}/F_n)$. Deduce that τ satisfies the equation $\tau^2 = \tau + 1$, and note that $\tau > 0$.)

41 Let the sequence $\{a_n\}$ be defined recursively by

$$a_1 = 2, \qquad a_{n+1} = \tfrac{1}{2}(a_n + 4) \qquad \text{for} \quad n \geq 1.$$

(a) Show by induction on n that $a_n < 4$ for each n and that $\{a_n\}$ is a monotone increasing sequence.
(b) Find the limit of this sequence.

42 For each $n = 1, 2, 3, \ldots$, let a_n be the largest multiple of $1/10^n$ such that $a_n^2 \leq 2$. (a) Prove that $\{a_n\}$ is a bounded and monotone increasing sequence, so that $A = \lim\limits_{n \to \infty} a_n$ exists. (b) Show that $A^2 > 2$ implies that $a_n^2 > 2$ for n sufficiently large. (c) Show that $A^2 < 2$ implies that $a_n^2 < B$ for some $B < 2$ and all sufficiently large n. (d) Conclude that $A^2 = 2$.

Problems 43 and 44 deal with the **least upper bound property** of the real numbers: If a nonempty set S of real numbers has an upper bound, then it has a least upper bound. The number M is an **upper bound** for S if $x \leq M$ for all x in S. The upper bound L of S is a **least upper bound** of S if no number smaller than L is an upper bound for S.

43 Show that the least upper bound property implies the bounded monotone sequence property. (*Suggestion:* If

$\{a_n\}$ is a bounded and monotone increasing sequence, and A is the least upper bound of $\{a_n\}$, show that $\lim_{n \to \infty} a_n = A$.)

44 Show that the bounded monotone sequence property implies the least upper bound property. (*Suggestion:* For each n, let a_n be the least multiple of $1/10^n$ that is an upper bound of the set S. Show that $\{a_n\}$ is a bounded and monotone decreasing sequence and then that $A = \lim_{n \to \infty} a_n$ is a least upper bound for S.)

12-3

Convergence of Infinite Series

An **infinite series** is an expression of the form

$$\sum_{n=1}^{\infty} a_n = a_1 + a_2 + a_3 + \cdots + a_n + \cdots \tag{1}$$

where $\{a_n\}$ is an infinite sequence of real numbers. The number a_n is called the **nth term** of the series. The symbol $\sum_{n=1}^{\infty} a_n$ is simply an abbreviation—compact notation—for the right-hand side of (1).

To say what is meant by such an infinite sum, we introduce the partial sums of the infinite series (1). The **nth partial sum** S_n of the series is the sum of its first n terms:

$$S_n = a_1 + a_2 + a_3 + \cdots + a_n. \tag{2}$$

Thus each infinite series has associated with it an infinite **sequence of partial sums**

$$S_1, S_2, S_3, \ldots, S_n, \ldots.$$

We define the sum of the infinite series to be the limit of its sequence of partial sums, provided that this limit exists.

Definition *Sum of an Infinite Series*

We say that the infinite series

$$\sum_{n=1}^{\infty} a_n \text{ converges (or is convergent)}$$

with **sum** S provided that the limit of its sequence of partial sums,

$$S = \lim_{n \to \infty} S_n, \tag{3}$$

exists (and is finite). Otherwise, we say that the series **diverges** (or is **divergent**). If a series diverges then it has no sum.

Thus an infinite sum is a limit of finite sums,

$$S = \sum_{n=1}^{\infty} a_n = \lim_{N \to \infty} \sum_{n=1}^{N} a_n,$$

provided that this limit exists.

EXAMPLE 1   Show that the series

$$\sum_{n=1}^{\infty} \left(\tfrac{1}{2}\right)^n = \tfrac{1}{2} + \tfrac{1}{4} + \tfrac{1}{8} + \tfrac{1}{16} + \cdots$$

converges, and find its sum.

Solution   The first four partial sums are

$$S_1 = \tfrac{1}{2}, \qquad S_2 = \tfrac{3}{4}, \qquad S_3 = \tfrac{7}{8}, \qquad S_4 = \tfrac{15}{16}.$$

It seems likely that $S_n = (2^n - 1)/2^n$, and indeed this follows easily by induction on n, since

$$\frac{2^n - 1}{2^n} + \frac{1}{2^{n+1}} = \frac{2^{n+1} - 2 + 1}{2^{n+1}} = \frac{2^{n+1} - 1}{2^{n+1}}.$$

Hence the sum of the given series is

$$S = \lim_{n \to \infty} S_n = \lim_{n \to \infty} \frac{2^n - 1}{2^n} = \lim_{n \to \infty} \left(1 - \frac{1}{2^n}\right) = 1.$$

EXAMPLE 2   Show that the series

$$\sum_{n=1}^{\infty} (-1)^{n+1} = 1 - 1 + 1 - 1 + 1 - 1 + \cdots$$

diverges.

Solution   The sequence of partial sums of the given series is

$$1, 0, 1, 0, 1, 0, 1, \ldots,$$

and this sequence has no limit. Therefore the series diverges.

EXAMPLE 3   Show that the infinite series

$$\sum_{n=1}^{\infty} \frac{1}{4n^2 - 1} = \frac{1}{3} + \frac{1}{15} + \frac{1}{35} + \frac{1}{63} + \cdots$$

converges, and find its sum.

Solution   We need a formula for the nth partial sum S_n so that we can evaluate its limit as $n \to \infty$. To find such a formula, we begin with the observation that

$$a_n = \frac{1}{4n^2 - 1} = \frac{1}{2}\left(\frac{1}{2n - 1} - \frac{1}{2n + 1}\right).$$

It follows that

$$S_n = \frac{1}{2}\left(1 - \frac{1}{3}\right) + \frac{1}{2}\left(\frac{1}{3} - \frac{1}{5}\right) + \frac{1}{2}\left(\frac{1}{5} - \frac{1}{7}\right)$$

$$+ \cdots + \frac{1}{2}\left(\frac{1}{2n - 1} - \frac{1}{2n + 1}\right)$$

$$= \frac{1}{2}\left(1 - \frac{1}{2n + 1}\right) = \frac{n}{2n + 1}.$$

And hence

$$\sum_{n=1}^{\infty} \frac{1}{4n^2 - 1} = \lim_{n \to \infty} \frac{n}{2n + 1} = \frac{1}{2}.$$

The sum for S_n in Example 3 is called a *telescoping* sum and provides us with a way to find the sums of certain series. The series in Examples 1 and 2 are examples of a much more common sort of series, the *geometric series*.

Definition *Geometric Series*

The series $\sum\limits_{n=0}^{\infty} a_n$ is said to be a **geometric series** if each term after the first is a fixed multiple of the term before it; that is, if there is a number r—called the **ratio** of the series—such that

$$a_{n+1} = ra_n$$

for all $n \geq 0$.

Thus every geometric series takes the form

$$a_0 + ra_0 + r^2 a_0 + r^3 a_0 + \cdots = \sum_{n=0}^{\infty} r^n a_0. \qquad (4)$$

Note that it is convenient to begin the summation at $n = 0$, and thus we regard the sum

$$S_n = a_0(1 + r + r^2 + \cdots + r^n)$$

of the first $n + 1$ terms as the nth partial sum of the series.

Theorem 1 *Sum of a Geometric Series*

If $|r| < 1$ then the geometric series in (4) converges, and its sum is

$$S = \sum_{n=0}^{\infty} r^n a_0 = \frac{a_0}{1 - r}. \qquad (5)$$

If $|r| \geq 1$ and $a_0 \neq 0$, then the geometric series diverges.

Proof If $r = 1$, then $S_n = (n + 1)a_0$, so the series certainly diverges. If $r = -1$, then it diverges by an argument like the one in Example 2. So we suppose that $r \neq 1$ and $r \neq -1$. Then the elementary identity

$$\frac{1}{1 - r} = 1 + r + r^2 + \cdots + r^n + \frac{r^{n+1}}{1 - r}$$

follows after multiplying each side by $1 - r$. Hence

$$S_n = a_0 \left(\frac{1}{1 - r} - \frac{r^{n+1}}{1 - r} \right).$$

If $|r| < 1$, then $r^{n+1} \to 0$ as $n \to \infty$, by Example 7 in Section 12-2. So in this case

$$S = \lim_{n \to \infty} a_0 \left(\frac{1}{1 - r} - \frac{r^{n+1}}{1 - r} \right) = \frac{a_0}{1 - r}.$$

If $|r| > 1$, then $\lim\limits_{n \to \infty} r^{n+1}$ does not exist, and so $\lim\limits_{n \to \infty} S_n$ does not exist. This establishes the theorem.

For example, with $a_0 = 1$ and $r = -\frac{1}{2}$, we find that

$$1 - \tfrac{1}{2} + \tfrac{1}{4} - \tfrac{1}{8} + \cdots = \sum_{n=0}^{\infty} \left(-\tfrac{1}{2}\right)^n$$

$$= \frac{1}{1 - \left(-\frac{1}{2}\right)} = \frac{2}{3}.$$

The following theorem says that the operations of addition and multiplication by a constant can be carried out termwise in the case of *convergent* series. Since the sum of an infinite series is the limit of its sequence of partial sums, this theorem follows immediately from the limit laws for sequences (Theorem 1 in Section 12-2).

Theorem 2 *Termwise Addition and Multiplication*

If the series $A = \Sigma a_n$ and $B = \Sigma b_n$ converge to the indicated sums and c is a constant, then the series $\Sigma(a_n + b_n)$ and $\Sigma c a_n$ also converge, with sums

(i) $\Sigma(a_n + b_n) = A + B$;
(ii) $\Sigma c a_n = cA$.

The geometric series formula in (5) may be used to find the rational number represented by a given repeating infinite decimal. For example,

$$0.55555\cdots = \frac{5}{10} + \frac{5}{100} + \frac{5}{1000} + \cdots$$

$$= \sum_{n=0}^{\infty} \frac{5}{10}\left(\frac{1}{10}\right)^n = \frac{\frac{5}{10}}{1 - \left(\frac{1}{10}\right)} = \frac{5}{9}.$$

In a more complicated example, we may wish to use the "termwise algebra" of Theorem 2:

$$0.728\,28\,28\,28\ldots = \frac{7}{10} + \frac{28}{10^3} + \frac{28}{10^5} + \frac{28}{10^7} + \cdots$$

$$= \frac{7}{10} + \frac{28}{10^3}\left(1 + \frac{1}{10^2} + \frac{1}{10^4} + \cdots\right)$$

$$= \frac{7}{10} + \frac{28}{1000}\sum_{n=0}^{\infty}\left(\frac{1}{10^2}\right)^n = \frac{7}{10} + \frac{28}{1000}\left(\frac{1}{1 - \frac{1}{100}}\right)$$

$$= \frac{7}{10} + \frac{28}{1000}\left(\frac{100}{99}\right) = \frac{7}{10} + \frac{28}{990} = \frac{721}{990}.$$

It should be apparent that this technique can be generalized to show that every repeating infinite decimal represents a rational number; consequently, the decimal expansions of irrational numbers such as π, e, and $\sqrt{2}$ must be nonrepeating as well as infinite. Conversely, if p and q are integers with

$q \neq 0$, then long division of q into p yields a repeating decimal expansion for the rational number p/q because such a division can yield at each stage only q possible different remainders.

The following theorem is often useful in showing that a given series does *not* converge.

Theorem 3 *The nth Term Test for Divergence*

If $\lim\limits_{n \to \infty} a_n \neq 0$, then the infinite series Σa_n diverges.

Here the hypothesis $\lim\limits_{n \to \infty} a_n \neq 0$ is to have its most liberal interpretation—that the limit does not exist *or*, should it exist, is not equal to zero.

Proof We want to show under the stated hypothesis that the series Σa_n diverges. It suffices to show that, *if* the series Σa_n does converge, then $\lim\limits_{n \to \infty} a_n = 0$. So suppose that Σa_n converges with sum $S = \lim\limits_{n \to \infty} S_n$, where

$$S_n = a_1 + a_2 + a_3 + \cdots + a_n.$$

We note that $a_n = S_n - S_{n-1}$, so that

$$\lim_{n \to \infty} a_n = \lim_{n \to \infty} (S_n - S_{n-1})$$

$$= \lim_{n \to \infty} S_n - \lim_{n \to \infty} S_{n-1} = S - S = 0.$$

Consequently, if $\lim\limits_{n \to \infty} a_n \neq 0$, then the series Σa_n diverges.

For example, the series

$$\sum_{n=1}^{\infty} (-1)^{n-1} n^2 = 1 - 4 + 9 - 16 + 25 - \cdots$$

diverges because $\lim\limits_{n \to \infty} a_n$ does not exist, while the series

$$\sum_{n=1}^{\infty} \frac{n}{3n+1} = \frac{1}{4} + \frac{2}{7} + \frac{3}{10} + \frac{4}{13} + \cdots$$

diverges because

$$\lim_{n \to \infty} \frac{n}{3n+1} = \frac{1}{3} \neq 0.$$

WARNING   The converse of Theorem 3 is false. The condition $\lim\limits_{n \to \infty} a_n = 0$ is necessary *but not sufficient* for convergence of the series Σa_n. That is, a series may satisfy the condition $\lim\limits_{n \to \infty} a_n = 0$ and yet diverge. An important example of a divergent series with terms that approach zero is the *harmonic series*

$$\sum_{n=1}^{\infty} \frac{1}{n} = 1 + \frac{1}{2} + \frac{1}{3} + \frac{1}{4} + \frac{1}{5} + \cdots. \tag{6}$$

> **Theorem 4**   *The harmonic series diverges.*

Proof   Since each term of the harmonic series is positive, its sequence of partial sums $\{S_n\}$ is monotone increasing. We shall prove that $\lim\limits_{n \to \infty} S_n = \infty$, and thus that the harmonic series diverges, by showing that there are arbitrarily large partial sums. Consider the following grouping of terms:

$$\sum_{n=1}^{\infty} \frac{1}{n} = 1 + \frac{1}{2} + \left(\frac{1}{3} + \frac{1}{4} \right) + \left(\frac{1}{5} + \cdots + \frac{1}{8} \right)$$

$$+ \left(\frac{1}{9} + \cdots + \frac{1}{16} \right) + \left(\frac{1}{17} + \cdots + \frac{1}{32} \right) + \cdots$$

$$= S_2 + (S_4 - S_2) + (S_8 - S_4) + (S_{16} - S_8) + (S_{32} - S_{16}) + \cdots.$$

The sum of each group of terms within parentheses is greater than $\frac{1}{2}$ because

$$S_{2n} - S_n = \frac{1}{n+1} + \frac{1}{n+2} + \cdots + \frac{1}{2n}$$

$$> \frac{1}{2n} + \frac{1}{2n} + \cdots + \frac{1}{2n} = \frac{1}{2}.$$

Hence

$$S_{2^k} = S_2 + (S_4 - S_2) + (S_8 - S_4) + \cdots + (S_{2^k} - S_{2^{k-1}})$$
$$> \tfrac{3}{2} + \tfrac{1}{2} + \tfrac{1}{2} + \cdots + \tfrac{1}{2} \qquad\qquad (k \text{ terms})$$
$$= \tfrac{3}{2} + \tfrac{1}{2}(k-1) = 1 + \tfrac{1}{2}k.$$

If $n > 2^k$, then $S_n > S_{2^k} > 1 + \frac{1}{2}k$. Since $1 + \frac{1}{2}k$ can be made arbitrarily large, it follows that

$$\lim_{n \to \infty} S_n = \infty.$$

Since its sequence of partial sums has no (finite) limit, the harmonic series therefore diverges.

If the sequence of partial sums of the series Σa_n diverges to infinity, then we say that the series itself **diverges to infinity,** and we write

$$\Sigma a_n = \infty.$$

The series $\Sigma(-1)^{n+1}$ of Example 2 is one that diverges but does not diverge to infinity. In the nineteenth century it was common to say that such a series was "divergent by oscillation;" today we say merely that it diverges.

Our proof of Theorem 4 shows that $\Sigma 1/n = \infty$. But the partial sums of the harmonic series diverge to infinity very slowly. If N_A denotes the smallest integer such that

$$\sum_{n=1}^{N_A} \frac{1}{n} \geq A,$$

then it is known that

$$N_5 = 83$$

$$N_{10} = 12,367;$$

(This can be verified with the aid of a programmable calculator.)

$$N_{20} = 272,400,600;$$

$$N_{100} \approx 1.5 \times 10^{43};$$

$$N_{1000} \approx 1.1 \times 10^{434}.$$

Thus you would need to add more than a quarter of a billion terms of the harmonic series to get a partial sum over 20. At this point the next few terms would each be approximately $0.000000004 = 4 \times 10^{-9}$. The number of terms you'd have to add to reach 1000 is far larger than the estimated number of elementary particles in the entire universe (10^{80}). If you enjoy such large numbers, see the article *Partial Sums of Infinite Series, and How They Grow* by R. P. Boas, Jr., in the American Mathematical Monthly, 84 (1977), 237–48.

The following theorem says that if two infinite series have the same terms from some point on, then either both converge or both diverge. The proof is left to Problem 33.

Theorem 5 *Series that are Eventually the Same*

If there exists a positive integer k such that $a_n = b_n$ for all $n > k$, then the series Σa_n and Σb_n either both converge or both diverge.

It follows that a *finite* number of terms can be changed, deleted from, or adjoined to an infinite series without altering its convergence or divergence (although the *sum* of a convergent series will be changed by such alterations). In particular, taking $b_n = 0$ for $n \le k$ and $b_n = a_n$ for $n > k$, we see that the series $\sum_{n=1}^{\infty} a_n$ and the series $\sum_{n=k+1}^{\infty} a_n$ obtained by deleting its first k terms either both converge or both diverge.

*SAFE DRUG DOSAGE

As a typical application of Formula (5) for the sum of a geometric series, we use it to determine the minimum permissible time between successive doses of a certain drug administered to a patient. Suppose that the patient receives a fixed dose every t_0 hours, and that a single dose of this particular drug results in an immediate bloodstream concentration of C_0. As we saw in Section 6-5, the concentration t hours after administration of a single dose is $C_0 e^{-kt}$, where k is the elimination constant of the drug. The concentration resulting from n successive doses, immediately before taking the $(n + 1)$ st dose, is then

$$C_0 e^{-kt_0} + C_0 e^{-2kt_0} + \cdots + C_0 e^{-nkt_0}.$$

It is therefore natural to expect the *residual concentration* of this drug—resulting from an extended regime of dosage at the prescribed level—to be

$$R = \sum_{n=1}^{\infty} C_0 e^{-nkt_0} = C_0 e^{-kt_0} \sum_{n=0}^{\infty} (e^{-kt_0})^n.$$

We take $r = e^{-kt_0}$ in Formula (5), and find the residual concentration to be

$$R = \frac{C_0 e^{-kt_0}}{1 - e^{-kt_0}}. \tag{7}$$

This will be the expected bloodstream concentration before each new dose.

Now let C_S denote the highest concentration that is safe (or otherwise desirable). In order that the concentration immediately *after* each dose not exceed C_S, we require that

$$C_0 + R = C_0 \left(1 + \frac{e^{-kt_0}}{1 - e^{-kt_0}} \right) \leq C_S.$$

After we solve for t_0, we find that this requirement amounts to

$$t_0 \geq \frac{1}{k} \ln \left(\frac{C_S}{C_S - C_0} \right). \tag{8}$$

The quantity on the right in (8) is, then, the minimum permissible time between successive doses.

EXAMPLE 4 Suppose that the drug under consideration is alcohol, and that the contemplated dose of one ounce produces a bloodstream concentration of $C_0 = 0.05\%$ (a volume concentration). How often can this dose be taken if the bloodstream concentration is not to exceed $C_S = 0.15\%$? Assume an elimination constant that corresponds to half of the blood alcohol being eliminated in 1 hour.

Solution We are given that $C_0 e^{-k \cdot 1} = \frac{1}{2} C_0$, so the elimination constant is $k = \ln 2$. From (8) we therefore obtain

$$t_0 = \frac{1}{\ln 2} \ln \left(\frac{0.15}{0.15 - 0.05} \right) \approx 0.585 \quad \text{(hours)},$$

or about 35 minutes.

12-3 PROBLEMS

In Problems 1–20, determine whether each given infinite series converges or diverges. If it converges, find its sum.

1 $1 + \frac{1}{3} + \frac{1}{9} + \cdots + (\frac{1}{3})^n + \cdots.$

2 $1 + e^{-1} + e^{-2} + \cdots + e^{-n} + \cdots.$

3 $1 + 3 + 5 + 7 + \cdots + (2n - 1) + \cdots.$

4 $\frac{1}{2} + \frac{1}{\sqrt{2}} + \frac{1}{\sqrt[3]{2}} + \cdots + \frac{1}{\sqrt[n]{2}} + \cdots.$

5 $1 - 2 + 4 - 8 + \cdots + (-2)^{n+1} + \cdots.$

6 $1 - \frac{1}{4} + \frac{1}{16} - \cdots + (-\frac{1}{4})^n + \cdots.$

7 $4 + \frac{4}{3} + \frac{4}{9} + \frac{4}{27} + \cdots + \frac{4}{3^n} + \cdots.$

8 $\frac{1}{3} + \frac{2}{9} + \frac{4}{27} + \cdots + \frac{2^{n-1}}{3^n} + \cdots.$

9 $1 + (1.01) + (1.01)^2 + \cdots + (1.01)^n + \cdots.$

10 $1 + \frac{1}{\sqrt{2}} + \frac{1}{\sqrt[3]{3}} + \cdots + \frac{1}{\sqrt[n]{n}} + \cdots.$

11 $\sum_{n=0}^{\infty} \frac{(-1)^n n}{n + 1}.$

12 $\sum_{n=1}^{\infty} \left(\frac{e}{10} \right)^n.$

13 $\sum_{n=0}^{\infty} (-1)^n \left(\frac{3}{e} \right)^n.$

14 $\sum_{n=0}^{\infty} \frac{3^n - 2^n}{4^n}.$

15 $\sum\limits_{n=1}^{\infty} (\sqrt{2})^{1-n}$.

16 $\sum\limits_{n=1}^{\infty} \left(\dfrac{2}{n} - \dfrac{1}{2^n} \right)$.

17 $\sum\limits_{n=1}^{\infty} \dfrac{n}{10n + 17}$.

18 $\sum\limits_{n=1}^{\infty} \dfrac{\sqrt{n}}{\ln(n + 1)}$.

19 $\sum\limits_{n=1}^{\infty} (5^{-n} - 7^{-n})$.

20 $\sum\limits_{n=0}^{\infty} \dfrac{1}{1 + (\frac{9}{10})^n}$.

In Problems 21–25, find the rational number represented by the given repeating decimal.

21 $0.474747\ldots.$

22 $0.252525\ldots.$

23 $0.123123123\ldots.$

24 $0.337733773377\ldots.$

25 $3.141591415914159\ldots.$

In each of Problems 26–30, use the method of Example 3 to find a formula for the nth partial sum S_n, and then compute the sum of the given infinite series.

26 $\sum\limits_{n=1}^{\infty} \dfrac{1}{n(n + 1)}$.

27 $\sum\limits_{n=1}^{\infty} \ln\left(\dfrac{n + 1}{n} \right)$.

28 $\sum\limits_{n=0}^{\infty} \dfrac{4}{16n^2 - 8n - 3}$.

29 $\sum\limits_{n=2}^{\infty} \dfrac{2}{n^2 - 1}$.

30 $\dfrac{1}{1 \cdot 3} - \dfrac{1}{2 \cdot 4} + \dfrac{1}{3 \cdot 5} - \dfrac{1}{4 \cdot 6} + \cdots.$

31 Prove: If Σa_n diverges and c is a nonzero constant, then $\Sigma c a_n$ diverges.

32 Suppose that Σa_n converges and that Σb_n diverges. Prove that $\Sigma (a_n + b_n)$ diverges.

33 Let S_n and T_n denote the nth partial sums of Σa_n and Σb_n, respectively. Suppose that $a_n = b_n$ for all $n > k$. Show that $S_n - T_n = S_k - T_k$ for $n > k$. Hence prove Theorem 5.

34 Suppose that $0 < x \leq 1$. Integrate both sides of the identity

$$\frac{1}{1 + t} = 1 - t + t^2 - \cdots + (-1)^n t^n + \frac{(-1)^{n+1} t^{n+1}}{1 + t}$$

from $t = 0$ to $t = x$ to show that

$$\ln(1 + x) = x - \frac{x^2}{2} + \frac{x^3}{3} - \cdots + (-1)^n \frac{x^{n+1}}{n + 1} + R_n$$

where $\lim\limits_{n \to \infty} R_n = 0$. Hence conclude that

$$\ln(1 + x) = \sum_{n=1}^{\infty} (-1)^{n+1} \frac{x^n}{n}$$

if $0 < x \leq 1$.

35 Criticize the following "proof" that $2 = 1$. Substitution of $x = 1$ into the result of Problem 34 gives the fact that

$$\ln 2 = 1 - \tfrac{1}{2} + \tfrac{1}{3} - \tfrac{1}{4} + \cdots.$$

If

$$S = 1 + \tfrac{1}{2} + \tfrac{1}{3} + \tfrac{1}{4} + \cdots,$$

then

$$\ln 2 = S - 2(\tfrac{1}{2} + \tfrac{1}{4} + \tfrac{1}{6} + \tfrac{1}{8} + \cdots) = S - S = 0.$$

Hence $2 = e^{\ln 2} = e^0 = 1$.

36 A ball has *bounce coefficient* $r < 1$ if, when it is dropped from height h, it bounces back to a height of rh. Suppose that such a ball is dropped from the initial height a and subsequently bounces infinitely many times. Use a geometric series to show that the total up-and-down distance it travels in all its bouncing is

$$D = a \frac{1 + r}{1 - r}.$$

37 A ball with bounce coefficient $r = 0.64$ (see Problem 36) is dropped from an initial height of $a = 4$ feet. Use a geometric series to compute the total time required for it to complete its infinitely many bounces. The time required for a ball to drop h feet (from rest) is $\sqrt{2h/g}$ seconds where $g = 32$ ft/sec^2.

38 Suppose that the government spends \$1 billion, and that each recipient spends 90% of the dollars he or she receives. In turn, the secondary recipients spend 90% of the dollars they receive, and so on. How much total spending results from the original injection of \$1 billion into the economy?

39 A tank initially contains a mass M_0 of air. Each stroke of a vacuum pump removes 5% of the air in the container. Compute (a) the mass M_n of air remaining in the tank after n strokes of the pump; (b) $\lim\limits_{n \to \infty} M_n$.

40 It requires 1 day for half of a dose of a certain drug to be eliminated from the bloodstream. The maximum safe concentration of the drug is three times that resulting from a single dose. How often can doses of this drug be taken?

41 A pane of glass of a certain material reflects half the incident light, absorbs a fourth, and transmits a fourth. A window is made of two panes of this glass separated by a small space, as shown in Figure 12.3. How much of the incident light I is transmitted by the double window?

12.3 Diagram of the double-pane window of Problem 41.

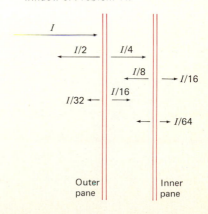

The Integral Test

Given an infinite series Σa_n, it is the exception rather than the rule when a nice formula for its nth partial sum S_n can be found and used directly to determine whether the series converges or diverges. There are, however, several *convergence tests* that involve the *terms* of an infinite series rather than its partial sums. Such a test will (when successful) tell us whether or not the series converges. Once we know that a series Σa_n does converge, it is then a separate matter actually to find its sum S. It may be necessary to approximate S by adding up sufficiently many terms; in this case we shall need to know how many terms are required for the desired accuracy.

In this section and the following one, we concentrate our attention on **positive term series**—that is, those with terms that are all positive. If $a_n > 0$ for all n, then

$$S_1 < S_2 < S_3 < \cdots < S_n < \cdots,$$

so the sequence of partial sums $\{S_n\}$ is monotone increasing. Hence there are just two possibilities. If the sequence $\{S_n\}$ is *bounded*—there exists a number M such that $S_n \leqq M$ for all n—then the bounded monotone sequence property (Section 12-2) implies that $S = \lim_{n \to \infty} S_n$ exists, so the series Σa_n *converges*. Otherwise it diverges to infinity (by Problem 38 in Section 12-2).

A similar alternative holds for improper integrals. Suppose that the function f is continuous and positive-valued for $x \geqq 1$. Then it follows (Problem 21) that the improper integral

$$\int_1^\infty f(x)\, dx = \lim_{b \to \infty} \int_1^b f(x)\, dx \tag{1}$$

either converges (the limit is a real number) or diverges to infinity (the limit is $+\infty$). This analogy between positive-term series and improper integrals of positive functions is the key to the **integral test.** We compare the behavior of the series Σa_n with that of the improper integral in (1), where f is an appropriately selected function. (Among other things, we require that $f(n) = a_n$ for all $n \geqq 1$. For example, in the case of the harmonic series $\Sigma 1/n$, we would take $f(x) = 1/x$.)

> **Theorem 1** *The Integral Test*
>
> Suppose that Σa_n is a positive-term series and that f is a positive-valued, decreasing, continuous function for $x \geqq 1$. If $f(n) = a_n$ for all integers $n \geqq 1$, then the series and the improper integral
>
> $$\sum_{n=1}^\infty a_n \quad \text{and} \quad \int_1^\infty f(x)\, dx$$
>
> either both converge or both diverge.

Proof   Since f is a decreasing function, the rectangular polygon with area

$$S_n = a_1 + a_2 + \cdots + a_n$$

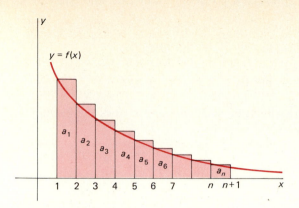

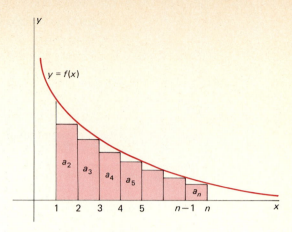

12.4 Underestimating the partial sums with an integral.

12.5 Overestimating the partial sums with an integral.

shown in Figure 12.4 contains the region under $y = f(x)$ from $x = 1$ to $x = n + 1$. Hence

$$\int_1^{n+1} f(x)\, dx \leqq S_n. \tag{2}$$

Similarly, the rectangular polygon with area

$$S_n - a_1 = a_2 + a_3 + \cdots + a_n$$

shown in Figure 12.5 is contained in the region under $y = f(x)$ from $x = 1$ to $x = n$. Hence

$$S_n - a_1 \leqq \int_1^n f(x)\, dx. \tag{3}$$

Suppose first that the improper integral $\int_1^\infty f(x)\, dx$ diverges (to infinity). Then

$$\lim_{n \to \infty} \int_1^{n+1} f(x)\, dx = \infty,$$

so it follows from (2) that $\lim_{n \to \infty} S_n = \infty$ also, and hence the infinite series Σa_n likewise diverges.

On the other hand, suppose that the improper integral $\int_1^\infty f(x)\, dx$ converges, with (finite) value I. Then (3) implies that

$$S_n \leqq a_1 + \int_1^n f(x)\, dx \leqq a_1 + I,$$

so the monotone increasing sequence $\{S_n\}_1^\infty$ is bounded. Thus the infinite series $\Sigma a_n = \lim_{n \to \infty} S_n$ converges also. Hence we have shown that the infinite series and the improper integral either both converge or both diverge.

EXAMPLE 1 The integral test gives another proof that the harmonic series

$$\sum_{n=1}^{\infty} \frac{1}{n} = 1 + \frac{1}{2} + \frac{1}{3} + \cdots + \frac{1}{n} + \cdots$$

diverges. The function $f(x) = 1/x$ is positive, continuous, and decreasing for $x \geq 1$, and

$$\int_1^\infty \frac{dx}{x} = \lim_{b \to \infty} \int_1^b \frac{dx}{x} = \lim_{b \to \infty} \left[\ln x \right]_1^b$$

$$= \lim_{b \to \infty} (\ln b - \ln 1) = +\infty.$$

Thus the improper integral diverges, and therefore so does the harmonic series.

The harmonic series is the case $p = 1$ of the **p-series**

$$\sum_{n=1}^\infty \frac{1}{n^p} = 1 + \frac{1}{2^p} + \frac{1}{3^p} + \cdots + \frac{1}{n^p} + \cdots . \tag{4}$$

Whether the p-series converges or diverges depends upon the value of p.

EXAMPLE 2 Show that the p-series converges if $p > 1$ but diverges if $0 < p < 1$.

Solution If $p > 0$ but $p \neq 1$, then the function $f(x) = 1/x^p$ satisfies the conditions of the integral test, and

$$\int_1^\infty \frac{dx}{x^p} = \lim_{b \to \infty} \int_1^b \frac{dx}{x^p} = \lim_{b \to \infty} \left[\frac{-1}{(p-1)x^{p-1}} \right]_1^b$$

$$= \lim_{b \to \infty} \frac{1}{p-1} \left(1 - \frac{1}{b^{p-1}} \right).$$

If $p > 1$ then

$$\int_1^\infty \frac{dx}{x^p} = \frac{1}{p-1} < \infty,$$

so the integral and the series both converge. But if $0 < p < 1$, then

$$\int_1^\infty \frac{dx}{x^p} = \lim_{b \to \infty} \frac{1}{1-p} (b^{1-p} - 1) = \infty,$$

and in this case the integral and the series both diverge.

For example, the series

$$\sum_{n=1}^\infty \frac{1}{n^2} = 1 + \frac{1}{2^2} + \frac{1}{3^2} + \frac{1}{4^2} + \cdots + \frac{1}{n^2} + \cdots$$

converges $(p = 2)$, while the series

$$\sum_{n=1}^\infty \frac{1}{\sqrt{n}} = 1 + \frac{1}{\sqrt{2}} + \frac{1}{\sqrt{3}} + \frac{1}{\sqrt{4}} + \cdots + \frac{1}{\sqrt{n}} + \cdots$$

diverges $(p = \frac{1}{2})$.

Now suppose that the positive-term series Σa_n converges by the integral test, and we wish to approximate its sum by adding up sufficiently many of its terms. The difference between the sum S and the nth partial sum S_n is the **remainder**

$$R_n = S - S_n = a_{n+1} + a_{n+2} + a_{n+3} + \cdots . \qquad (5)$$

This remainder is the error made when the actual sum S is estimated by using in its place the partial sum S_n.

Theorem 2 *Integral Test Remainder Estimate*

Suppose that the infinite series and improper integral

$$\sum_{n=1}^{\infty} a_n \quad \text{and} \quad \int_1^{\infty} f(x)\, dx$$

satisfy the hypotheses of the integral test, and that both converge. Then

$$\int_{n+1}^{\infty} f(x)\, dx \leqq R_n \leqq \int_n^{\infty} f(x)\, dx \qquad (6)$$

where R_n is the remainder given in (5).

Proof   From Figure 12.6 we see that

$$\int_k^{k+1} f(x)\, dx \leqq a_k \leqq \int_{k-1}^{k} f(x)\, dx$$

for $k = n + 1, n + 2, \ldots$. We add these inequalities for all such values of k, and the result is (6), since

$$\sum_{k=n+1}^{\infty} \int_k^{k+1} f(x)\, dx = \int_{n+1}^{\infty} f(x)\, dx$$

and

$$\sum_{k=n+1}^{\infty} \int_{k-1}^{k} f(x)\, dx = \int_n^{\infty} f(x)\, dx.$$

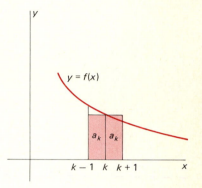

12.6 Establishing the integral test remainder estimate.

In Section 12-7 we shall see that the exact sum of the p-series with $p = 2$ is $\pi^2/6$, thus giving the beautiful formula

$$\frac{\pi^2}{6} = 1 + \frac{1}{2^2} + \frac{1}{3^2} + \frac{1}{4^2} + \cdots . \qquad (7)$$

EXAMPLE 3   Approximate the number π by applying the integral test remainder estimate with $n = 50$ to the series in Equation (7).

Solution   The sum of the first 50 terms in (7) is, to the accuracy shown,

$$\sum_{n=1}^{50} \frac{1}{n^2} = 1.6251327.$$

You can add these 50 terms one by one on a pocket calculator in a few minutes, but this is precisely the sort of thing a programmable calculator is really good for.
    Because

$$\int_a^{\infty} \frac{dx}{x^2} = \lim_{b \to \infty} \left[-\frac{1}{x} \right]_a^b = \frac{1}{a},$$

the integral test remainder estimate with $f(x) = 1/x^2$ gives $\frac{1}{51} \leq R_{50} \leq \frac{1}{50}$, so

$$\frac{1}{51} + 1.6251327 \leq \frac{\pi^2}{6} \leq \frac{1}{50} + 1.6251327.$$

We multiply through by 6, extract the square root, and round to four-place accuracy. The result is that

$$3.1414 < \pi < 3.1418.$$

With $n = 200$, a programmable calculator gives

$$\sum_{n=1}^{200} \frac{1}{n^2} = 1.6399465,$$

so

$$\frac{1}{201} + 1.6399465 \leq \frac{\pi^2}{6} \leq \frac{1}{200} + 1.6399465.$$

This leads to the inequality $3.14158 < \pi < 3.14161$, and it follows that $\pi = 3.1416$ rounded to four decimal places.

EXAMPLE 4 Show that the series

$$\sum_{n=2}^{\infty} \frac{1}{n(\ln n)^2}$$

converges, and find how many terms you would need to add to find its sum accurate to within 0.01.

Solution We take $f(x) = 1/x(\ln x)^2$ and find that

$$\int_2^{\infty} \frac{dx}{x(\ln x)^2} = \lim_{b \to \infty} \left[-\frac{1}{\ln x} \right]_2^b = \frac{1}{\ln 2} < \infty,$$

so the series does converge by the integral test. We want to choose n sufficiently large that $R_n < 0.01$. The right-hand inequality in (6) gives

$$R_n \leq \int_n^{\infty} \frac{dx}{x(\ln x)^2} = \frac{1}{\ln n}.$$

Hence we need

$$\frac{1}{\ln n} \leq 0.01,$$

$$\ln n \geq 100,$$

and thus

$$n \geq e^{100} \approx 2.7 \times 10^{43}.$$

This is a far larger number of terms than any conceivable computer could add within the expected lifetime of the universe. But accuracy to within 0.05 would require only that $n \geq e^{20} \approx 4.85 \times 10^8$, fewer than half a billion terms—well within the range of a modern computer.

In Problems 1–20, use the integral test to test the given series for convergence.

1 $\displaystyle\sum_{n=1}^{\infty} \frac{n}{n^2 + 1}$.

2 $\displaystyle\sum_{n=1}^{\infty} \frac{n}{e^{n^2}}$.

3 $\displaystyle\sum_{n=1}^{\infty} \frac{1}{\sqrt{n+1}}$.

4 $\displaystyle\sum_{n=1}^{\infty} \frac{1}{(n+1)^{4/3}}$.

5 $\displaystyle\sum_{n=1}^{\infty} \frac{1}{n^2 + 1}$.

6 $\displaystyle\sum_{n=1}^{\infty} \frac{1}{n(n+1)}$.

7 $\displaystyle\sum_{n=2}^{\infty} \frac{1}{n \ln n}$.

8 $\displaystyle\sum_{n=1}^{\infty} \frac{\ln n}{n}$.

9 $\displaystyle\sum_{n=1}^{\infty} \frac{1}{2^n}$.

10 $\displaystyle\sum_{n=1}^{\infty} \frac{n}{e^n}$.

11 $\displaystyle\sum_{n=1}^{\infty} \frac{n^2}{e^n}$.

12 $\displaystyle\sum_{n=1}^{\infty} \frac{1}{17n - 13}$.

13 $\displaystyle\sum_{n=1}^{\infty} \frac{\ln n}{n^2}$.

14 $\displaystyle\sum_{n=1}^{\infty} \frac{n+1}{n^2}$.

15 $\displaystyle\sum_{n=1}^{\infty} \frac{n}{n^4 + 1}$.

16 $\displaystyle\sum_{n=1}^{\infty} \frac{1}{n^3 + n}$.

17 $\displaystyle\sum_{n=1}^{\infty} \frac{1}{n(n+1)(n+2)}$.

18 $\displaystyle\sum_{n=1}^{\infty} \ln\left(\frac{n+1}{n}\right)$.

19 $\displaystyle\sum_{n=1}^{\infty} \ln\left(1 + \frac{1}{n^2}\right)$.

20 $\displaystyle\sum_{n=1}^{\infty} \frac{\sqrt[n]{2}}{n^2}$.

21 Suppose that the function f is continuous and positive-valued for $x \geqq 1$. Let $b_n = \int_1^n f(x)\, dx$ for $n = 1, 2, 3, \ldots$. (a) Suppose that the monotone increasing sequence $\{b_n\}$ is bounded, so that $B = \lim_{n \to \infty} b_n$ exists. Prove that $\int_1^\infty f(x)\, dx = B$. (b) Prove that if the sequence $\{b_n\}$ is not bounded, then $\int_1^\infty f(x)\, dx = \infty$.

22 Show that the series

$$\sum_{n=2}^{\infty} \frac{1}{n(\ln n)^p}$$

converges if $p > 1$ and diverges if $p \leqq 1$.

23 Use the integral test remainder estimate to find $\displaystyle\sum_{n=1}^{\infty} \frac{1}{n^5}$ accurate to three decimal places.

24 Using the integral test remainder estimate, how many terms are needed to approximate $\displaystyle\sum_{n=1}^{\infty} \frac{1}{n^{3/2}}$ with two-place accuracy?

25 Deduce from Inequalities (2) and (3) with $f(x) = 1/x$ that

$$\ln n \leqq 1 + \frac{1}{2} + \frac{1}{3} + \cdots + \frac{1}{n} \leqq 1 + \ln n.$$

If a computer adds 1 million terms of the harmonic series $\Sigma 1/n$ per second, how long will it take for the partial sum to reach 50?

26 (a) Let

$$c_n = \left(1 + \frac{1}{2} + \cdots + \frac{1}{n}\right) - \ln n.$$

Deduce from Problem 25 that $0 \leqq c_n \leqq 1$ for all n. (b) Note that

$$\int_n^{n+1} \frac{1}{x}\, dx \geqq \frac{1}{n+1};$$

conclude that the sequence $\{c_n\}$ is monotone decreasing. Therefore the sequence $\{c_n\}$ converges. The number

$$\gamma = \lim_{n \to \infty} \left(1 + \frac{1}{2} + \cdots + \frac{1}{n} - \ln n\right) \approx 0.57722$$

is known as **Euler's constant.**

27 It is known that

$$\sum_{n=1}^{\infty} \frac{1}{n^4} = \frac{\pi^4}{90}.$$

Use the integral test remainder estimate and the first ten terms of this series to show that $\pi = 3.1416$ rounded to four decimal places.

12-5

With the integral test we attempt to determine whether or not an infinite series converges by comparing it with an improper integral. The methods of this section involve comparing the terms of the *positive-term* series Σa_n with those of another series Σb_n whose convergence or divergence is already

Comparison Tests for Positive-Term Series

known. We have already developed two families of *reference series* for the role of the known series Σb_n; these are the geometric series of Section 12-3 and the *p*-series of Section 12-4. They are well adapted for our new purposes because their convergence or divergence is quite easy to determine: The geometric series Σr^n converges if $|r| < 1$ and diverges if $|r| \geq 1$, and the *p*-series $\Sigma 1/n^p$ converges for $p > 1$ and diverges if $0 < p \leq 1$.

Let Σa_n and Σb_n be positive-term series. Then we say that the series Σb_n **dominates** the series Σa_n provided that $a_n \leq b_n$ for all n. The following theorem says that the positive-term series Σa_n converges if it is dominated by a convergent series, and diverges if it dominates a divergent series.

Theorem 1 *Comparison Test*

Suppose that Σa_n and Σb_n are positive-term series. Then

(i) Σa_n converges if Σb_n converges and $a_n \leq b_n$ for all n;
(ii) Σa_n diverges if Σb_n diverges and $a_n \geq b_n$ for all n.

Proof Denote the *n*th partial sums of the series Σa_n and Σb_n by S_n and T_n, respectively. Then $\{S_n\}$ and $\{T_n\}$ are monotone increasing sequences. To prove Part (i), suppose that Σb_n converges, so that $T = \lim_{n \to \infty} T_n$ exists (T is a real number). Then the fact that $a_n \leq b_n$ for all n implies that $S_n \leq T_n \leq T$ for all n. Thus the sequence $\{S_n\}$ of partial sums of Σa_n is bounded and monotone increasing and therefore converges. Thus Σa_n converges.

Part (ii) is just a restatement of Part (i). If the series Σa_n converged, then the fact that Σa_n dominates Σb_n would imply—by Part (i) with a_n and b_n interchanged—that Σb_n converged. But Σb_n diverges, so it follows that Σa_n must also diverge.

We know by Theorem 5 in Section 12-3 that the convergence or divergence of an infinite series is not affected by the addition or deletion of finitely many terms. Consequently the conditions $a_n \leq b_n$ and $a_n \geq b_n$ in the two parts of the comparison test need only be assumed to hold for all $n \geq k$, where k is some fixed positive integer. Thus we can say that the series Σa_n converges if it is "eventually dominated" by a convergent series.

EXAMPLE 1 Since

$$\frac{1}{n(n + 1)(n + 2)} < \frac{1}{n^3}$$

for all $n \geq 1$, the series

$$\sum_{n=1}^{\infty} \frac{1}{n(n + 1)(n + 2)} = \frac{1}{1 \cdot 2 \cdot 3} + \frac{1}{2 \cdot 3 \cdot 4} + \frac{1}{3 \cdot 4 \cdot 5} + \cdots$$

is dominated by the series $\Sigma 1/n^3$, which is a convergent *p*-series with $p = 3$. Hence Part (i) of the comparison test shows that the series $\Sigma 1/n(n + 1)(n + 2)$ converges.

EXAMPLE 2 Since

$$\frac{1}{\sqrt{2n-1}} > \frac{1}{\sqrt{2n}}$$

for all $n \geq 1$, the series

$$\sum_{n=1}^{\infty} \frac{1}{\sqrt{2n-1}} = 1 + \frac{1}{\sqrt{3}} + \frac{1}{\sqrt{5}} + \frac{1}{\sqrt{7}} + \cdots$$

dominates the series

$$\sum_{n=1}^{\infty} \frac{1}{\sqrt{2n}} = \frac{1}{\sqrt{2}} \sum_{n=1}^{\infty} \frac{1}{n^{1/2}}.$$

But $\Sigma 1/n^{1/2}$ is a divergent p-series with $p = \frac{1}{2}$, and a constant multiple of a divergent series diverges, so Part (ii) of the comparison test shows that the series $\Sigma 1/\sqrt{2n-1}$ diverges.

EXAMPLE 3 Test the series

$$\sum_{n=0}^{\infty} \frac{1}{n!} = 1 + \frac{1}{1!} + \frac{1}{2!} + \frac{1}{3!} + \cdots$$

for convergence.

Solution We note first that if $n \geq 1$ then

$$n! = n(n-1)(n-2)\cdots 3 \cdot 2 \cdot 1$$
$$\geq 2 \cdot 2 \cdot 2 \cdots 2 \cdot 2 \cdot 1; \qquad \text{(same number of factors)}$$

that is, $n! \geq 2^{n-1}$ for $n \geq 1$. Thus

$$\frac{1}{n!} \leq \frac{1}{2^{n-1}} \qquad \text{for} \quad n \geq 1,$$

so the series $\displaystyle\sum_{n=0}^{\infty} \frac{1}{n!}$ is dominated by the series

$$1 + \sum_{n=1}^{\infty} \frac{1}{2^{n-1}} = 1 + \sum_{n=0}^{\infty} \frac{1}{2^n},$$

which is a convergent geometric series. Hence the series $\displaystyle\sum_{n=0}^{\infty} \frac{1}{n!}$ converges.

From the Taylor series of the exponential function (Section 11-5) we know that its sum is the number $e = 2.71828\ldots$. Thus

$$e = 1 + \frac{1}{1!} + \frac{1}{2!} + \frac{1}{3!} + \cdots + \frac{1}{n!} + \cdots.$$

Suppose that Σa_n is a series such that $a_n \to 0$ as $n \to \infty$, so that (in connection with the nth term divergence test of Section 12-3) Σa_n has at least a chance of converging. How do we choose an appropriate series Σb_n to compare it with? A good idea is to pick b_n as a *simple* function of n, simpler than a_n, such that a_n and b_n approach zero at the same rate as $n \to \infty$. If the formula for a_n is a fraction, we can try discarding all but the terms of largest magnitude in its numerator and denominator. For example, if

$$a_n = \frac{3n^2 + n}{n^4 + \sqrt{n}},$$

then we reason that n is small in comparison with $3n^2$ and $\sqrt{n}$ is small in comparison with n^4 when n is quite large. This suggests that we pick $b_n = 3n^2/n^4 = 3/n^2$. The series $\Sigma 3/n^2$ converges ($p = 2$), but when we attempt to compare Σa_n and Σb_n, we find that $a_n \geqq b_n$ (rather than $a_n \leqq b_n$). Consequently the comparison test does not apply immediately—the fact that Σa_n dominates a convergent series does *not* imply that Σa_n itself converges. The following theorem provides a convenient way of handling such a situation.

Theorem 2   Limit Comparison Test

Suppose that Σa_n and Σb_n are positive-term series. If the limit

$$L = \lim_{n \to \infty} \frac{a_n}{b_n}$$

exists and $0 < L < \infty$, then either both series converge or both series diverge.

Proof   Choose two fixed positive numbers P and Q such that $P < L < Q$. Then $P < a_n/b_n < Q$ for all n sufficiently large, and so

$$Pb_n < a_n < Qb_n$$

for all n sufficiently large. If Σb_n converges, then Σa_n is eventually dominated by the convergent series $Q\Sigma b_n$, so Part (i) of the comparison test implies that Σa_n also converges. If Σb_n diverges, then Σa_n eventually dominates the divergent series $P\Sigma b_n$, so Part (ii) of the comparison test implies that Σa_n also diverges. Thus the convergence of either series implies the convergence of the other.

EXAMPLE 4   With

$$a_n = \frac{3n^2 + n}{n^4 + \sqrt{n}} \quad \text{and} \quad b_n = \frac{1}{n^2}$$

(motivated by the discussion preceding Theorem 2), we find that

$$\lim_{n \to \infty} \frac{a_n}{b_n} = \lim_{n \to \infty} \frac{3n^4 + n^3}{n^4 + \sqrt{n}} = \lim_{n \to \infty} \frac{3 + 1/n}{1 + n^{-7/2}} = 3.$$

Since $\Sigma 1/n^2$ is a convergent p-series ($p = 2$), the limit comparison test tells us that the series

$$\sum_{n=1}^{\infty} \frac{3n^2 + n}{n^4 + \sqrt{n}}$$

also converges.

EXAMPLE 5 Test for convergence: $\displaystyle\sum_{n=1}^{\infty} \frac{1}{2n + \ln n}$.

Solution Since $\lim_{n\to\infty} (\ln n)/n = 0$ by Example 8 in Section 12-2, we note that $\ln n$ is small compared with $2n$ when n is quite large. We therefore take $a_n = 1/(2n + \ln n)$ and—ignoring the constant coefficient 2—we take $b_n = 1/n$. Then we find that

$$\lim_{n\to\infty} \frac{a_n}{b_n} = \lim_{n\to\infty} \frac{n}{2n + \ln n} = \lim_{n\to\infty} \frac{1}{2 + (\ln n)/n} = \frac{1}{2}.$$

Since the harmonic series $\Sigma 1/n$ diverges, it follows that the given series Σa_n also diverges.

It is important to realize that if $L = \lim(a_n/b_n)$ is either 0 or ∞, then the limit comparison test does not apply, and no conclusion is possible. For instance, if $a_n = 1/n^2$ and $b_n = 1/n$, then $\lim(a_n/b_n) = 0$. But in this case Σa_n converges while Σb_n diverges.

We close our discussion of positive-term series with the observation that the sum of a convergent *positive*-term series is not altered by grouping or rearranging its terms. For example, let Σa_n be a convergent positive-term series and consider

$$\sum_{n=1}^{\infty} b_n = (a_1 + a_2 + a_3) + (a_4) + (a_5 + a_6) + \cdots.$$

That is, the new series has $b_1 = a_1 + a_2 + a_3, b_2 = a_4, b_3 = a_5 + a_6$, and so on. Then every partial sum T_n of Σb_n is equal to some partial sum S_{n_i} of Σa_n. Since $\{S_n\}$ is a monotone increasing sequence with limit $S = \Sigma a_n$, it follows easily that $\{T_n\}$ is a monotone increasing sequence with the same limit. Thus $\Sigma b_n = S$ also. The argument is more subtle if terms of Σa_n are "moved out of place," as in

$$\sum_{n=1}^{\infty} c_n = (a_1 + a_3 + a_2) + (a_5 + a_7 + a_4) + (a_9 + a_{11} + a_6) + \cdots,$$

but the same conclusion holds: Any rearrangement of a convergent *positive*-term series also converges, and to the same sum.

Similarly, it is easy to prove that any grouping or rearrangement of a divergent positive-term series also diverges. But these observations all fail in the case of infinite series having both positive and negative terms. For example, the series $\Sigma(-1)^n$ diverges, but it has the convergent regrouping

$$(-1 + 1) + (-1 + 1) + (-1 + 1) + \cdots = 0 + 0 + 0 + \cdots = 0.$$

It follows from Problem 34 in Section 12-3 that

$$\ln 2 = 1 - \tfrac{1}{2} + \tfrac{1}{3} - \tfrac{1}{4} + \tfrac{1}{5} - \cdots,$$

but the rearrangement

$$1 + \tfrac{1}{3} - \tfrac{1}{2} + \tfrac{1}{5} + \tfrac{1}{7} - \tfrac{1}{4} + \tfrac{1}{9} + \tfrac{1}{11} - \tfrac{1}{6} + \cdots$$

converges to $\tfrac{3}{2} \ln 2$. The above series for $\ln 2$ even has rearrangements that diverge.

Use comparison tests to determine whether the infinite series in Problems 1–25 converge or diverge.

1 $\displaystyle\sum_{n=1}^{\infty} \frac{1}{n^2 + n + 1}$.

2 $\displaystyle\sum_{n=1}^{\infty} \frac{n^3 + 1}{n^4 + 2}$.

3 $\displaystyle\sum_{n=1}^{\infty} \frac{1}{n + \sqrt{n}}$.

4 $\displaystyle\sum_{n=1}^{\infty} \frac{1}{n + n^{3/2}}$.

5 $\displaystyle\sum_{n=1}^{\infty} \frac{1}{1 + 3^n}$.

6 $\displaystyle\sum_{n=1}^{\infty} \frac{10n^2}{n^4 + 1}$.

7 $\displaystyle\sum_{n=2}^{\infty} \frac{10n^2}{n^3 - 1}$.

8 $\displaystyle\sum_{n=1}^{\infty} \frac{n^2 - n}{n^4 + 2}$.

9 $\displaystyle\sum_{n=1}^{\infty} \frac{1}{\sqrt{37n^3 + 3}}$.

10 $\displaystyle\sum_{n=1}^{\infty} \frac{1}{\sqrt{n^2 + 1}}$.

11 $\displaystyle\sum_{n=1}^{\infty} \frac{\sqrt{n}}{n^2 + n}$.

12 $\displaystyle\sum_{n=1}^{\infty} \frac{1}{3 + 5^n}$.

13 $\displaystyle\sum_{n=2}^{\infty} \frac{1}{\ln n}$.

14 $\displaystyle\sum_{n=2}^{\infty} \frac{1}{n - \ln n}$.

15 $\displaystyle\sum_{n=1}^{\infty} \frac{\sin^2 n}{n^2 + 1}$.

16 $\displaystyle\sum_{n=1}^{\infty} \frac{\cos^2 n}{3^n}$.

17 $\displaystyle\sum_{n=1}^{\infty} \frac{n + 2^n}{n + 3^n}$.

18 $\displaystyle\sum_{n=0}^{\infty} \frac{1}{2^n + 3^n}$.

19 $\displaystyle\sum_{n=2}^{\infty} \frac{1}{n^2 \ln n}$.

20 $\displaystyle\sum_{n=1}^{\infty} \frac{1}{n^{1 + \sqrt{n}}}$.

21 $\displaystyle\sum_{n=1}^{\infty} \frac{\ln n}{n^2}$.

22 $\displaystyle\sum_{n=1}^{\infty} \frac{\tan^{-1} n}{n}$.

23 $\displaystyle\sum_{n=1}^{\infty} \frac{\sin^2(1/n)}{n^2}$.

24 $\displaystyle\sum_{n=1}^{\infty} \frac{e^{1/n}}{n}$.

25 $\displaystyle\sum_{n=2}^{\infty} \frac{\ln n}{e^n}$.

26 (a) Show that $\ln n < n^{1/8}$ for all n sufficiently large. (b) Conclude that $\Sigma 1/(\ln n)^8$ diverges.

27 Prove that $\Sigma(a_n/n)$ converges if Σa_n is a convergent positive-term series.

28 Suppose that Σa_n is a convergent positive-term series and $\{c_n\}$ is a sequence of positive numbers such that $\lim_{n \to \infty} c_n = 0$. Prove that $\Sigma a_n c_n$ converges.

29 Use the result of Problem 28 to prove that, if Σa_n and Σb_n are convergent positive-term series, then $\Sigma a_n b_n$ converges.

30 Show that the series

$$\sum_{n=1}^{\infty} \frac{1}{1 + 2 + 3 + \cdots + n}$$

converges.

31 Use the result of Problem 26 in Section 12-4 to show that the series

$$\sum_{n=1}^{\infty} \left(1 + \frac{1}{2} + \frac{1}{3} + \cdots + \frac{1}{n}\right)^{-1}$$

diverges.

12-6

Alternating Series and Absolute Convergence

In Sections 12-4 and 12-5, we concentrated on positive-term series. Now we discuss infinite series that have both positive and negative terms. An important example is a series with terms that are alternately positive and negative. An **alternating series** is an infinite series of the form

$$\sum_{n=1}^{\infty} (-1)^{n+1} a_n = a_1 - a_2 + a_3 - a_4 + \cdots \qquad (1)$$

or of the form $\displaystyle\sum_{n=1}^{\infty} (-1)^n a_n$, where $a_n > 0$ for all n. For example, the series

$$\sum_{n=1}^{\infty} \frac{(-1)^{n+1}}{n} = 1 - \frac{1}{2} + \frac{1}{3} - \frac{1}{4} + \cdots$$

is an alternating series. The following theorem shows that this series converges because the sequence of absolute values of its terms is monotone decreasing with limit zero.

> **Theorem 1** *Alternating Series Test*
>
> If $a_n \geqq a_{n+1} > 0$ for all n and $\lim\limits_{n \to \infty} a_n = 0$, then the alternating series in (1) converges.

Proof We consider first the "even" partial sums $S_2, S_4, S_6, \ldots, S_{2n}$. We may write

$$S_{2n} = (a_1 - a_2) + (a_3 - a_4) + \cdots + (a_{2n-1} - a_{2n}).$$

Since $a_k - a_{k+1} \geqq 0$ for all k, we see that the sequence $\{S_{2n}\}$ is monotone increasing. Also, since

$$S_{2n} = a_1 - (a_2 - a_3) - \cdots - (a_{2n-2} - a_{2n-1}) - a_{2n},$$

we see that $S_{2n} \leqq a_1$ for all n, so the monotone increasing sequence $\{S_{2n}\}$ is bounded above. Hence the limit

$$S = \lim_{n \to \infty} S_{2n}$$

exists by the bounded monotone sequence property of Section 12-2. And it follows that

$$0 < S < a_1$$

because the even partial sums S_{2n} are positive and bounded by a_1. It remains only to see that the "odd" partial sums $S_1, S_3, S_5, \ldots$ also converge to S. But $S_{2n+1} = S_{2n} + a_{2n+1}$ and $\lim\limits_{n \to \infty} a_{2n+1} = 0$, so

$$\lim_{n \to \infty} S_{2n+1} = \lim_{n \to \infty} S_{2n} + \lim_{n \to \infty} a_{2n+1} = S.$$

Thus $\lim\limits_{n \to \infty} S_n = S$, and therefore the series converges.

EXAMPLE 1 The series

$$\sum_{n=1}^{\infty} \frac{(-1)^{n+1}}{2n-1} = 1 - \frac{1}{3} + \frac{1}{5} - \frac{1}{7} + \cdots$$

satisfies the conditions of Theorem 1, and therefore converges. The alternating series test does not tell us what its sum is, but we saw in Section 11-5 that the sum of this series is $\pi/4$.

EXAMPLE 2 The series

$$\sum_{n=1}^{\infty} (-1)^{n+1} \frac{n}{2n-1} = 1 - \frac{2}{3} + \frac{3}{5} - \frac{4}{7} + \cdots$$

is an alternating series, and

$$a_n = \frac{n}{2n-1} > \frac{n+1}{2n+1} = a_{n+1}$$

for all n because $n(2n+1) > (2n-1)(n+1) = 2n^2 + n - 1$ if $n \geqq 1$. But $\lim\limits_{n \to \infty} a_n = \frac{1}{2} \neq 0$, so the alternating series test does not apply. In fact this series diverges by the nth term divergence test.

If a series converges by the alternating series test, the following theorem shows how to approximate its sum with any desired degree of accuracy—*if* you have some method (a computer?) for adding enough of its terms.

Theorem 2   *Alternating Series Error Estimate*

Suppose that the series $\Sigma(-1)^{n+1}a_n$ satisfies the conditions of the alternating series test and therefore converges. Let S denote its sum. Denote by $R_n = S - S_n$ the error made in replacing S with the nth partial sum S_n. Then R_n has the same sign as the next term $(-1)^{n+2}a_{n+1}$, and

$$0 < |R_n| < a_{n+1}. \tag{2}$$

Proof   We note that

$$R_n = S - S_n$$
$$= (-1)^{n+2}a_{n+1} + (-1)^{n+3}a_{n+2} + (-1)^{n+4}a_{n+3} + \cdots$$
$$= (-1)^n(a_{n+1} - a_{n+2} + a_{n+3} - \cdots).$$

So

$$|R_n| = a_{n+1} - a_{n+2} + a_{n+3} - \cdots$$

is an alternating series with first term a_{n+1}. Hence Inequality (2) follows from the fact that the sum of a convergent alternating series lies between zero and its first term, as we saw in the proof of Theorem 1.

EXAMPLE 3   In Section 11-5 we saw that

$$e^x = \sum_{n=0}^{\infty} \frac{x^n}{n!}$$

and thus that

$$\frac{1}{e} = e^{-1} = 1 - 1 + \frac{1}{2!} - \frac{1}{3!} + \frac{1}{4!} - \cdots.$$

Use this alternating series to compute e^{-1} accurate to four decimal places.

Solution   We want $|R_n| < 1/(n+1)! \leqq 0.00005$. The least value of n for which this inequality holds is $n = 7$. Then

$$e^{-1} = 1 - 1 + \frac{1}{2!} - \frac{1}{3!} + \frac{1}{4!} - \frac{1}{5!} + \frac{1}{6!} - \frac{1}{7!} + R_7$$

$$= 0.367857 + R_7.$$

(We are carrying six places because we want four-place accuracy in the final answer.) Now Inequality (2) gives

$$0 < R_7 < \frac{1}{8!} < 0.000025.$$

Thus

$$0.367857 < e^{-1} < 0.367882.$$

If we take reciprocals, we find that

$$2.718263 < e < 2.718448.$$

Thus $e^{-1} = 0.3679$ rounded to four places and also $e = 2.718$ rounded to three places.

ABSOLUTE CONVERGENCE

The series

$$\sum_{n=1}^{\infty} \frac{(-1)^{n+1}}{n} = 1 - \frac{1}{2} + \frac{1}{3} - \frac{1}{4} + \cdots$$

converges, but if we simply add the absolute values of its terms, we get the *divergent* series

$$1 + \tfrac{1}{2} + \tfrac{1}{3} + \tfrac{1}{4} + \cdots.$$

On the other hand, the series

$$\sum_{n=0}^{\infty} \frac{(-1)^n}{2^n} = 1 - \frac{1}{2} + \frac{1}{4} - \frac{1}{8} + \cdots$$

has the property that the associated positive-term series

$$1 + \tfrac{1}{2} + \tfrac{1}{4} + \tfrac{1}{8} + \tfrac{1}{16} + \cdots$$

converges. Our next theorem tells us this: If a series of *positive* terms converges, then we may insert minus signs in front of any of the terms—every other one, for instance—and the resulting series must also converge.

Theorem 3 *Absolute Convergence Implies Convergence*
If the series $\Sigma |a_n|$ converges, then so does the series Σa_n.

Proof Suppose that the series $\Sigma |a_n|$ converges. Note that

$$0 \leq a_n + |a_n| \leq 2|a_n|$$

for all n. Let $b_n = a_n + |a_n|$, and then it follows from the comparison test that the positive-term series Σb_n converges; it is dominated by the convergent series $\Sigma 2|a_n|$. Since it is easily verified that the termwise difference of two convergent series also converges, we now see that the series

$$\Sigma a_n = \Sigma (b_n - |a_n|) = \Sigma b_n - \Sigma |a_n|$$

converges.

Thus we have another convergence test: Given the series Σa_n, test $\Sigma |a_n|$ for convergence. If the latter series converges, then so does the former. This phenomenon motivates us to make the following definition.

Definition *Absolute Convergence*
The series Σa_n is said to **converge absolutely** (and is called **absolutely convergent**) provided that the series

$$\Sigma |a_n| = |a_1| + |a_2| + |a_3| + \cdots + |a_n| + \cdots$$

converges.

Thus Theorem 3 says simply that *every absolutely convergent series is convergent*. The two examples preceding Theorem 3 show that a convergent series may either converge absolutely or fail to do so:

$$1 - \tfrac{1}{2} + \tfrac{1}{4} - \tfrac{1}{8} + \tfrac{1}{16} - \cdots$$

is an absolutely convergent series because

$$1 + \tfrac{1}{2} + \tfrac{1}{4} + \tfrac{1}{8} + \tfrac{1}{16} + \cdots$$

converges, while

$$1 - \tfrac{1}{2} + \tfrac{1}{3} - \tfrac{1}{4} + \tfrac{1}{5} - \cdots$$

is an example of a series that, though it converges, does *not* converge absolutely. A series that converges but does not converge absolutely is said to be **conditionally convergent.**

Consequently the terms *absolutely convergent*, *conditionally convergent*, and *divergent* are simultaneously all-inclusive and mutually exclusive.

Note that there is some advantage in the application of Theorem 3, since to apply it we test the *positive*-term series $\Sigma |a_n|$ for convergence; we have a variety of tests, such as comparison tests or the integral test, designed for use on positive-term series.

Note also that absolute convergence of the series Σa_n means convergence of *another* series $\Sigma |a_n|$, and the two sums generally will differ. For example, with $a_n = (-\tfrac{1}{3})^n$ the formula for the sum of a geometric series gives

$$\sum_{n=0}^{\infty} a_n = \sum_{n=0}^{\infty} \left(-\tfrac{1}{3}\right)^n = \frac{1}{1 - (-\tfrac{1}{3})} = \frac{3}{4}$$

while

$$\sum_{n=0}^{\infty} |a_n| = \sum_{n=0}^{\infty} \left(\tfrac{1}{3}\right)^n = \frac{1}{1 - (\tfrac{1}{3})} = \frac{3}{2}.$$

EXAMPLE 4 Discuss the convergence of the series

$$\sum_{n=1}^{\infty} \frac{\cos n}{n^2} = \cos 1 + \frac{\cos 2}{4} + \frac{\cos 3}{9} + \cdots.$$

Solution Let $a_n = (\cos n)/n^2$. Then

$$|a_n| = \frac{|\cos n|}{n^2} \le \frac{1}{n^2}$$

for all $n \ge 1$. Hence the positive-term series $\Sigma |a_n|$ converges by the comparison test, since it is dominated by the convergent *p*-series $\Sigma (1/n^2)$. Thus the given series is absolutely convergent, and therefore (by Theorem 3) it converges.

One reason for the importance of absolute convergence is the fact (proved in advanced calculus) that the terms of an absolutely convergent series may be regrouped or rearranged without changing the sum of the series. As we remarked at the end of Section 12-5, this is *not* true of conditionally convergent series.

THE RATIO AND ROOT TESTS

Each of our next two convergence tests involves a way of measuring the rate of growth or decrease of the sequence $\{a_n\}$ of terms in order to determine whether the series Σa_n diverges or converges absolutely.

Theorem 4 *The Ratio Test*

Suppose that the limit

$$\rho = \lim_{n \to \infty} \left| \frac{a_{n+1}}{a_n} \right| \tag{3}$$

either exists or is infinite. Then the infinite series Σa_n with nonzero terms

  (i) Converges absolutely if $\rho < 1$;
  (ii) Diverges if $\rho > 1$.

If $\rho = 1$, the ratio test is inconclusive.

Proof If $\rho < 1$, pick a (fixed) number r with $\rho < r < 1$. Then (3) implies that there exists an integer N such that $|a_{n+1}| \leqq r|a_n|$ for all $n \geqq N$. It follows that

$$|a_{N+1}| \leqq r|a_N|,$$

$$|a_{N+2}| \leqq r|a_{N+1}| \leqq r^2|a_N|,$$

$$|a_{N+3}| \leqq r|a_{N+2}| \leqq r^3|a_N|,$$

and—in general—that

$$|a_{N+k}| \leqq r^k|a_N| \qquad \text{for} \quad k \geqq 0.$$

Hence the series

$$|a_N| + |a_{N+1}| + |a_{N+2}| + \cdots$$

is dominated by the geometric series

$$|a_N|(1 + r + r^2 + \cdots),$$

and the latter converges because $r < 1$. Thus the series $\Sigma |a_n|$ converges, and so the series Σa_n converges absolutely.

If $\rho > 1$, then (3) implies that there exists a positive integer N such that $|a_{n+1}| > |a_n|$ for all $n \geqq N$. It follows that $|a_n| > |a_N| > 0$ for all $n > N$. Thus the sequence $\{a_n\}$ cannot approach zero as $n \to \infty$, and consequently— by the nth term divergence test—the series Σa_n diverges.

To see that Σa_n may either converge or diverge if $\rho = 1$, consider the divergent series $\Sigma(1/n)$ and the convergent series $\Sigma(1/n^2)$, for both of which $\rho = 1$:

$$\lim_{n \to \infty} \left| \frac{1/(n + 1)}{1/n} \right| = \lim_{n \to \infty} \frac{n}{n + 1} = 1,$$

$$\lim_{n \to \infty} \left| \frac{1/(n + 1)^2}{1/n^2} \right| = \lim_{n \to \infty} \frac{n^2}{(n + 1)^2} = 1.$$

EXAMPLE 5   Consider the series

$$\sum_{n=1}^{\infty} \frac{(-1)^n 2^n}{n!} = -2 + \frac{4}{2!} - \frac{8}{3!} + \frac{16}{4!} - \cdots.$$

Then

$$\rho = \lim_{n \to \infty} \left| \frac{a_{n+1}}{a_n} \right| = \lim_{n \to \infty} \left| \frac{(-1)^{n+1} 2^{n+1} / (n+1)!}{(-1)^n 2^n / n!} \right|$$

$$= \lim_{n \to \infty} \frac{2}{n+1} = 0.$$

Since $\rho < 1$, the series converges absolutely.

EXAMPLE 6   Test the following series for convergence.

$$\text{(a)} \quad \sum_{n=1}^{\infty} \frac{n}{2^n}; \qquad \text{(b)} \quad \sum_{n=1}^{\infty} \frac{3^n}{n^2}.$$

Solution   (a) We have

$$\rho = \lim_{n \to \infty} \left| \frac{a_{n+1}}{a_n} \right| = \lim_{n \to \infty} \frac{(n+1)/2^{n+1}}{n/2^n} = \lim_{n \to \infty} \frac{n+1}{2n} = \frac{1}{2}.$$

Since $\rho < 1$, this series converges absolutely.

(b) Here we have

$$\rho = \lim_{n \to \infty} \left| \frac{a_{n+1}}{a_n} \right| = \lim_{n \to \infty} \frac{3^{n+1}/(n+1)^2}{3^n/n^2}$$

$$= \lim_{n \to \infty} \frac{3n^2}{(n+1)^2} = 3.$$

Since $\rho > 1$, this series diverges.

> **Theorem 5**   *The Root Test*
>
> Suppose that the limit
>
> $$\rho = \lim_{n \to \infty} \sqrt[n]{|a_n|} \qquad (4)$$
>
> either exists or is infinite. Then the infinite series Σa_n
>
>   (i) Converges absolutely if $\rho < 1$;
> (ii) Diverges if $\rho > 1$.
>
> If $\rho = 1$, the root test is inconclusive.

Proof   If $\rho < 1$, pick a (fixed) number r such that $\rho < r < 1$. Then $|a_n|^{1/n} < r$, and hence $|a_n| < r^n$, for n sufficiently large. Thus the series $\Sigma |a_n|$ is eventually dominated by the convergent geometric series Σr^n. Therefore $\Sigma |a_n|$ converges, and so the series Σa_n converges absolutely.

   If $\rho > 1$, then $|a_n|^{1/n} > 1$, and hence $|a_n| > 1$, for n sufficiently large. Therefore the nth term divergence test implies that the series Σa_n diverges.

The ratio test is generally simpler to apply than the root test and there-fore is ordinarily the one to try first. But there are certain series for which the root test succeeds while the ratio test fails.

EXAMPLE 7 Consider the series

$$\sum_{n=0}^{\infty} \frac{1}{2^{n+(-1)^n}} = \frac{1}{2} + \frac{1}{1} + \frac{1}{8} + \frac{1}{4} + \frac{1}{32} + \frac{1}{16} + \cdots.$$

Then $a_{n+1}/a_n = \frac{1}{2}$ if n is even while $a_{n+1}/a_n = \frac{1}{8}$ if n is odd. So the limit required for the ratio test does not exist. But

$$\lim_{n\to\infty} |a_n|^{1/n} = \lim_{n\to\infty} \left(\frac{1}{2^{n+(-1)^n}}\right)^{1/n}$$

$$= \lim_{n\to\infty} \frac{1}{2}\left(\frac{1}{2^{(-1)^n/n}}\right) = \frac{1}{2},$$

so the given series converges by the root test. (Its convergence also follows from the fact that it is a rearrangement of the positive-term geometric series $\Sigma 1/2^n$.)

12-6 PROBLEMS

Determine whether or not the alternating series in Problems 1–10 converge.

1 $\displaystyle\sum_{n=1}^{\infty} \frac{(-1)^{n+1}}{n^2}.$

2 $\displaystyle\sum_{n=1}^{\infty} \frac{(-1)^{n+1}n}{3n+2}.$

3 $\displaystyle\sum_{n=1}^{\infty} (-1)^n \frac{n}{n^2+1}.$

4 $\displaystyle\sum_{n=2}^{\infty} (-1)^n \frac{n}{\ln n}.$

5 $\displaystyle\sum_{n=2}^{\infty} (-1)^n \frac{1}{\ln n}.$

6 $\displaystyle\sum_{n=1}^{\infty} \frac{(-1)^n}{\sqrt{2n+1}}.$

7 $\displaystyle\sum_{n=0}^{\infty} \frac{(-1)^n}{n!}.$

8 $\displaystyle\sum_{n=1}^{\infty} (-1)^n \frac{(1.01)^n}{n^4+1}.$

9 $\displaystyle\sum_{n=1}^{\infty} \frac{(-1)^n}{n^{1/n}}.$

10 $\displaystyle\sum_{n=1}^{\infty} (-1)^n \frac{n!}{(2n)!}.$

Determine whether the series in Problems 11–32 converge absolutely, converge conditionally, or diverge.

11 $\displaystyle\sum_{n=1}^{\infty} \frac{(-1)^{n+1}}{2^n}.$

12 $\displaystyle\sum_{n=1}^{\infty} \frac{n}{n^2+1}.$

13 $\displaystyle\sum_{n=1}^{\infty} \frac{(-1)^n \ln n}{n}.$

14 $\displaystyle\sum_{n=1}^{\infty} \frac{1}{n^n}.$

15 $\displaystyle\sum_{n=1}^{\infty} \left(\frac{10}{n}\right)^n.$

16 $\displaystyle\sum_{n=1}^{\infty} \frac{3^n}{n!n}.$

17 $\displaystyle\sum_{n=0}^{\infty} \frac{(-10)^n}{n!}.$

18 $\displaystyle\sum_{n=1}^{\infty} (-1)^n \frac{n!}{n^n}.$

19 $\displaystyle\sum_{n=1}^{\infty} (-1)^n \left(\frac{n}{n+1}\right)^n.$

20 $\displaystyle\sum_{n=1}^{\infty} \frac{n!n^2}{(2n)!}.$

21 $\displaystyle\sum_{n=1}^{\infty} \left(\frac{\ln n}{n}\right)^n.$

22 $\displaystyle\sum_{n=0}^{\infty} (-1)^n \frac{2^{3n}}{7^n}.$

23 $\displaystyle\sum_{n=0}^{\infty} (-1)^n(\sqrt{n+1}-\sqrt{n}).$

24 $\displaystyle\sum_{n=1}^{\infty} n(\tfrac{3}{4})^n.$

25 $\displaystyle\sum_{n=2}^{\infty} \left[\ln\left(\frac{1}{n}\right)\right]^n.$

26 $\displaystyle\sum_{n=0}^{\infty} \frac{(n!)^2}{(2n)!}.$

27 $\displaystyle\sum_{n=1}^{\infty} (-1)^n \frac{3^n}{n(2^n+1)}.$

28 $\displaystyle\sum_{n=1}^{\infty} (-1)^n \frac{\tan^{-1}n}{n}.$

29 $\displaystyle\sum_{n=1}^{\infty} (-1)^n \frac{n!}{1\cdot3\cdot5\cdots(2n-1)}.$

30 $\displaystyle\sum_{n=1}^{\infty} \frac{1\cdot3\cdot5\cdots(2n-1)}{1\cdot4\cdot7\cdots(3n-2)}.$

31 $\displaystyle\sum_{n=0}^{\infty} \frac{(n+2)!}{3^n(n!)^2}.$

32 $\displaystyle\sum_{n=1}^{\infty} (-1)^n \frac{n^n}{3^{n^2}}.$

In each of Problems 33–36, find the least positive integer n such that S_n (the sum of the first n terms of the series) approximates the sum S of the given series accurate to three decimal places; that is, so that $|R_n| = |S - S_n| < 0.0005$.

33 $\displaystyle\sum_{n=1}^{\infty} \frac{(-1)^{n+1}}{n}.$

34 $\displaystyle\sum_{n=1}^{\infty} \frac{(-1)^{n+1}}{n^2}.$

35 $\displaystyle\sum_{n=0}^{\infty} \frac{(-1)^n}{3^n}.$

36 $\displaystyle\sum_{n=1}^{\infty} \frac{(-1)^n}{n^n}.$

In each of Problems 37–40, approximate the sum of the given series accurate to three decimal places.

37 $e^{-1/2} = \displaystyle\sum_{n=0}^{\infty} \frac{(-1)^n}{n!2^n}.$

38 $\cos 1 = \displaystyle\sum_{n=0}^{\infty} \frac{(-1)^n}{(2n)!}.$

39 $\ln(1.1) = \displaystyle\sum_{n=1}^{\infty} (-1)^{n+1} \frac{(0.1)^n}{n}.$

40 $\displaystyle\sum_{n=1}^{\infty} \frac{(-1)^{n+1}}{n^5}.$

41 Approximate the sum of the series

$$\frac{\pi^2}{12} = \sum_{n=1}^{\infty} \frac{(-1)^{n+1}}{n^2}$$

with error less than 0.01. Use the corresponding partial sum and error estimate to verify that $3.13 < \pi < 3.15$.

42 Show that $\Sigma|a_n|$ diverges if the series Σa_n diverges.

43 Give an example of two convergent series Σa_n and Σb_n such that $\Sigma a_n b_n$ diverges.

44 (a) Suppose that r is a (fixed) number with $|r| < 1$. Use the ratio test to show that the series

$$\sum_{n=0}^{\infty} nr^n$$

converges. Let S denote its sum.
(b) Show that

$$(1 - r)S = \sum_{n=1}^{\infty} r^n.$$

Show how to conclude that

$$\sum_{n=0}^{\infty} nr^n = \frac{r}{(1-r)^2}.$$

12-7

Power Series

Up to this point we have concentrated on infinite series with *constant* terms. Much of the practical importance of infinite series, however, stems from the fact that many functions have useful representations as infinite series with *variable* terms. For example, from Theorem 1 of Section 12-3 (with x in place of r), we know that the geometric series

$$\frac{1}{1-x} = 1 + x + x^2 + \cdots + x^n + \cdots \tag{1}$$

represents (that is, *converges to*) the function $f(x) = 1/(1-x)$ when $|x| < 1$. In Section 11-5 we derived the Taylor series

$$e^x = \sum_{n=0}^{\infty} \frac{x^n}{n!} = 1 + x + \frac{x^2}{2!} + \frac{x^3}{3!} + \cdots, \tag{2}$$

$$\cos x = \sum_{n=0}^{\infty} (-1)^n \frac{x^{2n}}{(2n)!} = 1 - \frac{x^2}{2!} + \frac{x^4}{4!} - \cdots, \tag{3}$$

and

$$\sin x = \sum_{n=0}^{\infty} (-1)^n \frac{x^{2n+1}}{(2n+1)!} = x - \frac{x^3}{3!} + \frac{x^5}{5!} - \cdots. \tag{4}$$

There we used Taylor's formula with remainder to show that these series converge (for all x) to the functions e^x, $\cos x$, and $\sin x$, respectively. Thus we did not then require the convergence tests of this chapter.

The infinite series above all have the form

$$\sum_{n=0}^{\infty} a_n x^n = a_0 + a_1 x + a_2 x^2 + \cdots + a_n x^n + \cdots \tag{5}$$

with constant *coefficients* $a_0, a_1, a_2, \ldots$. An infinite series of this form is called a **power series in** (powers of) x. In order that the first terms of the two sides of Equation (5) agree, we adopt here the convention that $x^0 = 1$ even if $x = 0$.

The power series (5) obviously converges when $x = 0$. In general, it will converge for some nonzero values of x and diverge for others. Because of the way in which powers of x are involved, the ratio test is particularly effective in deciding for which values of x a power series converges. Assume that the limit

$$\rho = \lim_{n \to \infty} \left| \frac{a_{n+1}}{a_n} \right| \tag{6}$$

exists. This is the limit that we need if we want to apply the ratio test to the series Σa_n of constants. To apply the ratio test to the power series of (5), we write $u_n = a_n x^n$ and compute the limit

$$\lim_{n \to \infty} \left| \frac{u_{n+1}}{u_n} \right| = \lim_{n \to \infty} \left| \frac{a_{n+1} x^{n+1}}{a_n x^n} \right| = \rho |x|. \tag{7}$$

If $\rho = 0$, we see that $\Sigma a_n x^n$ converges absolutely for all x. If $\rho = \infty$, we see that $\Sigma a_n x^n$ diverges for all $x \neq 0$. If ρ is a positive real number, we see from (7) that $\Sigma a_n x^n$ converges absolutely for all x such that $\rho |x| < 1$; that is, when

$$|x| < r = \frac{1}{\rho}. \tag{8}$$

In this case the ratio test also implies that $\Sigma a_n x^n$ diverges if $|x| > r$, but the test is inconclusive when $x = \pm r$. We therefore have proved the following theorem, under the additional hypothesis that the limit in (6) exists. In Problems 43 and 44, we outline a proof that does not require this additional hypothesis.

Theorem 1 *Convergence of Power Series*

If $\Sigma a_n x^n$ is a power series, then either:

  (i) The series converges absolutely for all x; or
  (ii) The series converges only when $x = 0$; or
  (iii) There exists a number $r > 0$ such that $\Sigma a_n x^n$ converges absolutely if $|x| < r$ and diverges if $|x| > r$.

The number r of Case (iii) is called the **radius of convergence** of the power series $\Sigma a_n x^n$. We shall write $r = \infty$ in Case (i) and $r = 0$ in Case (ii). The set of all real numbers x for which the series converges is called its **interval of convergence**. If $0 < r < \infty$, then the interval of convergence is one of the intervals

$$(-r, r), \qquad (-r, r], \qquad [-r, r), \quad \text{or} \quad [-r, r].$$

When we substitute either of the end points $x = \pm r$ into the series $\Sigma a_n x^n$, we obtain an infinite series with constant terms whose convergence must be determined separately. Since these will be numerical series, the earlier tests of this chapter are appropriate.

EXAMPLE 1 Find the interval of convergence of the series

$$\sum_{n=1}^{\infty} \frac{x^n}{n \cdot 3^n}.$$

Solution With $u_n = x^n/(n \cdot 3^n)$ we find that

$$\lim_{n \to \infty} \left| \frac{u_{n+1}}{u_n} \right| = \lim_{n \to \infty} \left| \frac{x^{n+1}/([n+1] \cdot 3^{n+1})}{x^n/(n \cdot 3^n)} \right|$$

$$= \lim_{n \to \infty} \frac{n|x|}{3(n+1)} = \frac{|x|}{3}.$$

Now $|x|/3 < 1$ provided that $|x| < 3$, so the ratio test implies that the given series converges absolutely if $|x| < 3$ and diverges if $|x| > 3$. When $x = 3$ we have the divergent harmonic series $\Sigma(1/n)$, and when $x = -3$ we have the convergent alternating series $\Sigma(-1)^n/n$. Thus the interval of convergence of the given power series is $[-3, 3)$.

EXAMPLE 2 Find the interval of convergence of

$$\sum_{n=0}^{\infty} \frac{2^n x^n}{n!}.$$

Solution With $u_n = 2^n x^n/n!$, we find that

$$\lim_{n \to \infty} \left| \frac{u_{n+1}}{u_n} \right| = \lim_{n \to \infty} \left| \frac{2^{n+1} x^{n+1}/(n+1)!}{2^n x^n/n!} \right|$$

$$= \lim_{n \to \infty} \frac{2|x|}{n+1} = 0$$

for all x. Hence the ratio test implies that the power series converges for all x, and its interval of convergence is $(-\infty, +\infty)$, the whole real line.

EXAMPLE 3 Find the interval of convergence of the series

$$\sum_{n=0}^{\infty} n^n x^n.$$

Solution With $u_n = n^n x^n$, we find that

$$\lim_{n \to \infty} \left| \frac{u_{n+1}}{u_n} \right| = \lim_{n \to \infty} \left| \frac{(n+1)^{n+1} x^{n+1}}{n^n x^n} \right|$$

$$= \lim_{n \to \infty} (n+1)\left(1 + \frac{1}{n}\right)^n |x| = +\infty$$

for all $x \neq 0$, because $\lim_{n \to \infty} (1 + 1/n)^n = e$. Thus the given series diverges for all $x \neq 0$, and its interval of convergence consists of the single point $x = 0$.

EXAMPLE 4 Use the ratio test to verify that the Taylor series for $\cos x$ in (3) above converges for all x.

Solution With $u_n = (-1)^n x^{2n}/(2n)!$ we find that

$$\lim_{n \to \infty} \left| \frac{u_{n+1}}{u_n} \right| = \lim_{n \to \infty} \left| \frac{(-1)^{n+1} x^{2n+2}/(2n+2)!}{(-1)^n x^{2n}/(2n)!} \right|$$

$$= \lim_{n \to \infty} \frac{x^2}{(2n+1)(2n+2)} = 0$$

for all x, so the series converges for all x. Note, however, that the ratio test tells us only that the series converges to *some* number, *not* necessarily the particular number $\cos x$. The argument of Section 11-5 (using the Taylor formula with remainder) is required to establish that the sum of the series is actually $\cos x$.

An infinite series of the form

$$\sum_{n=0}^{\infty} a_n (x - c)^n = a_0 + a_1 (x - c) + a_2 (x - c)^2 + \cdots \qquad (9)$$

(where c is a constant) is called a **power series in** (powers of) $x - c$. By the same reasoning that leads to Theorem 1, with x^n replaced by $(x - c)^n$ throughout, we conclude that either:

(i) The series (9) coverges absolutely for all x;
(ii) The series converges only when $x - c = 0$—that is, when $x = c$;   or
(iii) There exists a number $r > 0$ such that $\Sigma a_n (x - c)^n$ converges absolutely if $|x - c| < r$ and diverges if $|x - c| > r$.

As in the case of a power series with $c = 0$, the number r is called its **radius of convergence,** and the **interval of convergence** of the series $\Sigma a_n (x - c)^n$ is the set of all numbers x for which it converges. As before, when $0 < r < \infty$ the convergence of the series at the endpoints $x = c - r$ and $x = c + r$ of its interval of convergence must be checked separately.

EXAMPLE 5 Determine the interval of convergence of the series

$$\sum_{n=1}^{\infty} (-1)^n \frac{(x - 2)^n}{(4^n)\sqrt{n}}.$$

Solution We let $u_n = (-1)^n (x - 2)^n / [(4^n)\sqrt{n}]$. Then

$$\lim_{n \to \infty} \left| \frac{u_{n+1}}{u_n} \right| = \lim_{n \to \infty} \left| \frac{(-1)^{n+1}(x - 2)^{n+1}/[(4^{n+1})\sqrt{n+1}]}{(-1)^n (x - 2)^n / [(4^n)\sqrt{n}]} \right|$$

$$= \lim_{n \to \infty} \frac{|x - 2|}{4} \left(\frac{n}{n+1} \right)^{1/2} = \frac{|x - 2|}{4}.$$

Hence the given series converges when $|x - 2|/4 < 1$; that is, when $|x - 2| < 4$, so the radius of convergence is $r = 4$. Since $c = 2$, the series converges when $-2 < x < 6$ and diverges if either $x < -2$ or $x > 6$. When $x = -2$ the series reduces to the divergent p-series $\Sigma(1/\sqrt{n})$, and when $x = 6$ it reduces to the convergent alternating series $\Sigma(-1)^n/\sqrt{n}$. Thus the interval of convergence of the given power series is $(-2, 6]$.

POWER SERIES REPRESENTATION OF FUNCTIONS

Power series provide us with an important tool for computing (or approximating) values of functions. Suppose that the series $\Sigma a_n x^n$ converges to the value $f(x)$; that is,

$$f(x) = a_0 + a_1 x + a_2 x^2 + \cdots + a_n x^n + \cdots$$

for each x in the interval of convergence of the power series. Then we call $\Sigma a_n x^n$ a **power series representation** of $f(x)$. For example, the geometric series Σx^n in (1) is a power series representation of the function $f(x) = 1/(1 - x)$ on the interval $(-1, 1)$.

In Section 11-5 we saw how Taylor's formula with remainder can often be used to find a power series representation of a given function. Recall that the nth degree Taylor's formula for $f(x)$ at $x = a$ is

$$f(x) = f(a) + f'(a)(x - a) + \frac{f''(a)}{2!}(x - a)^2 + \cdots$$

$$+ \frac{f^{(n)}(a)}{n!}(x - a)^n + R_n(x). \tag{10}$$

The remainder $R_n(x)$ is given by

$$R_n(x) = \frac{f^{(n+1)}(c)}{(n+1)!}(x - a)^{n+1}$$

where c is some number between a and x. If we let $n \to +\infty$ in Formula (10), we obtain the following theorem.

Theorem 2 *Taylor Series Representation*

Suppose that the function f has derivatives of all orders on some interval containing a and also that

$$\lim_{n \to \infty} R_n(x) = 0 \tag{11}$$

for each x in that interval. Then

$$f(x) = \sum_{n=0}^{\infty} \frac{f^{(n)}(a)}{n!}(x - a)^n \tag{12}$$

for each x in the interval.

The power series in (12) is the **Taylor series** of the function f *at $x = a$* (or *in powers of $x - a$, or with center a*). If $a = 0$ we obtain the power series

$$f(x) = \sum_{n=0}^{\infty} \frac{f^{(n)}(0)}{n!} x^n$$

$$= f(0) + f'(0)x + \frac{f''(0)}{2!} x^2 + \cdots, \tag{13}$$

which is often called the **Maclaurin series** of f. Thus the power series in Equations (2) through (4) at the beginning of this section are the Maclaurin series of the functions e^x, $\cos x$, and $\sin x$, respectively.

Upon replacing x by $-x$ in the Maclaurin series for e^x, we obtain

$$e^{-x} = 1 - x + \frac{x^2}{2!} - \frac{x^3}{3!} + \cdots + (-1)^n \frac{x^n}{n!} + \cdots .$$

Let us add the series for e^x and e^{-x}. This gives

$$\cosh x = \frac{e^x + e^{-x}}{2}$$

$$= \frac{1}{2}\left(1 + x + \frac{x^2}{2!} + \frac{x^3}{3!} + \frac{x^4}{4!} + \cdots\right)$$

$$+ \frac{1}{2}\left(1 - x + \frac{x^2}{2!} - \frac{x^3}{3!} + \frac{x^4}{4!} - \cdots\right),$$

so that

$$\cosh x = 1 + \frac{x^2}{2!} + \frac{x^4}{4!} + \frac{x^6}{6!} + \cdots .$$

Similarly,

$$\sinh x = x + \frac{x^3}{3!} + \frac{x^5}{5!} + \frac{x^7}{7!} + \cdots .$$

Note the resemblance with the series for $\sin x$ and $\cos x$.

Upon replacement of x by $-x^2$ in the series for e^x, we obtain

$$e^{-x^2} = \sum_{n=0}^{\infty} (-1)^n \frac{x^{2n}}{n!} = 1 - x^2 + \frac{x^4}{2!} - \frac{x^6}{3!} + \cdots .$$

Because this power series converges to e^{-x^2} for all x, it must be the Maclaurin series for e^{-x^2} (see Problem 40). Think how tedious it would be to compute the derivatives of e^{-x^2} needed to write its Maclaurin series directly from (13).

The following example gives one of the most famous and useful of all series, the *binomial series*, which was discovered by Newton in the 1660s. It is the infinite series generalization of the (finite) binomial formula of elementary algebra.

EXAMPLE 6 Suppose that α is a nonzero real number. Show that the Maclaurin series of $f(x) = (1 + x)^\alpha$ is

$$(1 + x)^\alpha = 1 + \sum_{n=1}^{\infty} \frac{\alpha(\alpha - 1)(\alpha - 2)\cdots(\alpha - n + 1)}{n!} x^n$$

$$= 1 + \alpha x + \frac{\alpha(\alpha - 1)}{2!} x^2 + \frac{\alpha(\alpha - 1)(\alpha - 2)}{3!} x^3 + \cdots . \quad (14)$$

Also determine the interval of convergence of this **binomial series**.

Solution To derive the series itself, we simply list the derivatives of $f(x) = (1 + x)^\alpha$:

$$f(x) = (1 + x)^\alpha,$$

$$f'(x) = \alpha(1 + x)^{\alpha-1},$$

$$f''(x) = \alpha(\alpha - 1)(1 + x)^{\alpha-2},$$

$$f^{(3)}(x) = \alpha(\alpha - 1)(\alpha - 2)(1 + x)^{\alpha-3},$$

$$\vdots$$

$$f^{(n)}(x) = \alpha(\alpha - 1)(\alpha - 2) \cdots (\alpha - n + 1)(1 + x)^{\alpha-n}.$$

Thus

$$f^{(n)}(0) = \alpha(\alpha - 1)(\alpha - 2) \cdots (\alpha - n + 1).$$

Substitution of this value of $f^{(n)}(0)$ into the Maclaurin series formula of (13) gives the binomial series in (14).

To determine the interval of convergence of the binomial series, we let

$$u_n = \frac{\alpha(\alpha - 1)(\alpha - 2) \cdots (\alpha - n + 1)}{n!} x^n.$$

We find that

$$\lim_{n \to \infty} \left| \frac{u_{n+1}}{u_n} \right| = \lim_{n \to \infty} \left| \frac{\alpha(\alpha - 1)(\alpha - 2) \cdots (\alpha - n)x^{n+1}/(n + 1)!}{\alpha(\alpha - 1)(\alpha - 2) \cdots (\alpha - n + 1)x^n/n!} \right|$$

$$= \lim_{n \to \infty} \left| \frac{(\alpha - n)x}{n + 1} \right| = |x|.$$

Hence the ratio test shows that the binomial series converges absolutely if $|x| < 1$ and diverges if $|x| > 1$. Its convergence at the end points $x = \pm 1$ depends upon the value of α; we shall not pursue this. Problem 41 outlines a proof that the sum of the binomial series is actually $(1 + x)^\alpha$ if $|x| < 1$.

If $\alpha = k$, a positive integer, then the coefficient of x^n in (14) vanishes for $n > k$, and the binomial series reduces to the binomial formula

$$(1 + x)^k = \sum_{n=0}^{k} \frac{k!}{n!(k - n)!} x^n.$$

Otherwise (14) is an infinite series. For example, with $\alpha = \frac{1}{2}$, we obtain

$$\sqrt{1 + x} = 1 + \frac{1}{2}x + \frac{(\frac{1}{2})(-\frac{1}{2})}{2!}x^2 + \frac{(\frac{1}{2})(-\frac{1}{2})(-\frac{3}{2})}{3!}x^3$$

$$+ \frac{(\frac{1}{2})(-\frac{1}{2})(-\frac{3}{2})(-\frac{5}{2})}{4!}x^4 + \cdots$$

$$= 1 + \frac{1}{2}x - \frac{1}{8}x^2 + \frac{1}{16}x^3 - \frac{5}{128}x^4 + \cdots. \tag{15}$$

If we replace x by $-x$ and take $\alpha = -\frac{1}{2}$, we get the series

$$\frac{1}{\sqrt{1-x}} = 1 + \frac{(-\frac{1}{2})}{1!}(-x) + \frac{(-\frac{1}{2})(-\frac{3}{2})}{2!}(-x)^2$$

$$+ \cdots + \frac{1 \cdot 3 \cdot 5 \cdots (2n-1)}{n! \cdot 2^n} x^n + \cdots,$$

which in summation notation takes the form

$$\frac{1}{\sqrt{1-x}} = 1 + \sum_{n=1}^{\infty} \frac{1 \cdot 3 \cdot 5 \cdots (2n-1)}{2 \cdot 4 \cdot 6 \cdots (2n)} x^n. \tag{16}$$

Sometimes it is inconvenient to compute the repeated derivatives of a function in order to find its Taylor series. Another common method of deriving new power series is by the differentiation and integration of known power series. Suppose that a power series representation of the function $f(x)$ is known. Then the following theorem (we leave its proof to advanced calculus) says that the function $f(x)$ may be differentiated by separately differentiating the individual terms in its power series. That is, the power series obtained by "termwise" differentiation converges to the derivative $f'(x)$. Similarly, a function can be integrated by termwise integration of its power series.

Theorem 3 *Termwise Differentiation and Integration*

Suppose that the function f has a power series representation

$$f(x) = \sum_{n=0}^{\infty} a_n x^n = a_0 + a_1 x + a_2 x^2 + a_3 x^3 + \cdots$$

with nonzero radius of convergence r. Then f is differentiable on $(-r, r)$ and

$$f'(x) = \sum_{n=1}^{\infty} n a_n x^{n-1} = a_1 + 2a_2 x + 3a_3 x^2 + \cdots. \tag{17}$$

Also

$$\int_0^x f(t)\, dt = \sum_{n=0}^{\infty} \frac{a_n x^{n+1}}{n+1} = a_0 x + \frac{a_1 x^2}{2} + \frac{a_2 x^3}{3} + \cdots \tag{18}$$

for each x in $(-r, r)$. Moreover, the power series in (17) and (18) have the same radius of convergence r.

EXAMPLE 7 Termwise differentiation of the geometric series for $f(x) = 1/(1-x)$ yields

$$\frac{1}{(1-x)^2} = D_x\left(\frac{1}{1-x}\right)$$

$$= D_x(1 + x + x^2 + x^3 + \cdots)$$

$$= 1 + 2x + 3x^2 + 4x^3 + \cdots.$$

Thus

$$\frac{1}{(1-x)^2} = \sum_{n=1}^{\infty} nx^{n-1} = \sum_{n=0}^{\infty} (n+1)x^n.$$

The series converges to $1/(1-x)^2$ if $-1 < x < 1$.

EXAMPLE 8 Replacement of x by $-t$ in the geometric series above gives

$$\frac{1}{1+t} = 1 - t + t^2 - t^3 + \cdots + (-1)^n t^n + \cdots.$$

Since $D_t \ln(1+t) = 1/(1+t)$, termwise integration from $t = 0$ to $t = x$ now gives

$$\ln(1+x) = \int_0^x \frac{dt}{1+t}$$

$$= \int_0^x (1 - t + t^2 - \cdots + (-1)^n t^n + \cdots)\, dt;$$

$$\ln(1+x) = x - \frac{x^2}{2} + \frac{x^3}{3} - \frac{x^4}{4} + \cdots + (-1)^{n-1}\frac{x^n}{n} + \cdots \qquad (19)$$

if $|x| < 1$.

EXAMPLE 9 Find a power series representation for the arctangent function.

Solution Since $D_t \tan^{-1} t = 1/(1+t^2)$, termwise integration of the series

$$\frac{1}{1+t^2} = 1 - t^2 + t^4 - t^6 + t^8 - \cdots$$

gives

$$\tan^{-1} x = \int_0^x \frac{dt}{1+t^2} = \int_0^x (1 - t^2 + t^4 - t^6 + t^8 - \cdots)\, dt.$$

Therefore

$$\tan^{-1} x = x - \frac{x^3}{3} + \frac{x^5}{5} - \frac{x^7}{7} + \frac{x^9}{9} - \cdots \qquad (20)$$

if $-1 < x < 1$.

EXAMPLE 10 Find a power series representation for the arcsine function.

Solution First we substitute t^2 for x in Equation (16). This yields

$$\frac{1}{\sqrt{1-t^2}} = 1 + \sum_{n=1}^{\infty} \frac{1 \cdot 3 \cdot 5 \cdots (2n-1)}{2 \cdot 4 \cdot 6 \cdots (2n)} t^{2n}$$

if $|t| < 1$. Since $D_t \sin^{-1} t = 1/(1-t^2)^{1/2}$, termwise integration of this series from $t = 0$ to $t = x$ gives

$$\sin^{-1} x = \int_0^x \frac{dt}{\sqrt{1-t^2}}$$

$$= x + \sum_{n=1}^{\infty} \frac{1 \cdot 3 \cdot 5 \cdots (2n-1)}{2 \cdot 4 \cdot 6 \cdots (2n)} \cdot \frac{x^{2n+1}}{2n+1} \qquad (21)$$

if $|x| < 1$. Problem 42 shows how to use this series for $\sin^{-1}x$ to derive the series

$$\frac{\pi^2}{6} = 1 + \frac{1}{2^2} + \frac{1}{3^2} + \frac{1}{4^2} + \cdots + \frac{1}{n^2} + \cdots$$

that was used in Example 3 of Section 12-4 to approximate the number π.

Theorem 3 has the important consequence that, if the two power series $\Sigma a_n x^n$ and $\Sigma b_n x^n$ both converge and $\Sigma a_n x^n = \Sigma b_n x^n$ for all $|x| < r$ $(r > 0)$, then $a_n = b_n$ for all n. In particular, the Taylor series of a function is its unique power series representation (if any). See Problem 40.

12-7 PROBLEMS

Find the interval of convergence of each of the power series in Problems 1–20.

1 $\displaystyle\sum_{n=1}^{\infty} \frac{x^n}{n}$.

2 $\displaystyle\sum_{n=0}^{\infty} \frac{(-1)^n x^n}{n^2 + 1}$.

3 $\displaystyle\sum_{n=1}^{\infty} (-1)^n n^2 x^n$.

4 $\displaystyle\sum_{n=1}^{\infty} n! x^n$.

5 $\displaystyle\sum_{n=1}^{\infty} \frac{(-1)^n x^{2n}}{2n - 1}$.

6 $\displaystyle\sum_{n=1}^{\infty} \frac{n x^n}{5^n}$.

7 $\displaystyle\sum_{n=0}^{\infty} (5x - 3)^n$.

8 $\displaystyle\sum_{n=1}^{\infty} \frac{(2x - 1)^n}{n^4 + 16}$.

9 $\displaystyle\sum_{n=1}^{\infty} \frac{2^n (x - 3)^n}{n^2}$.

10 $\displaystyle\sum_{n=1}^{\infty} \frac{n!}{n^n} x^n$ (do not test the endpoints; it diverges at each).

11 $\displaystyle\sum_{n=1}^{\infty} \frac{(2n)!}{n!} x^n$.

12 $\displaystyle\sum_{n=1}^{\infty} \frac{1 \cdot 3 \cdot 5 \cdots (2n + 1)}{n!} x^n$ (do not test the endpoints; it diverges at each).

13 $\displaystyle\sum_{n=1}^{\infty} \frac{n^3 (x + 1)^n}{3^n}$.

14 $\displaystyle\sum_{n=1}^{\infty} (-1)^n \frac{(x - 2)^n}{n^2}$.

15 $\displaystyle\sum_{n=1}^{\infty} \frac{(3 - x)^n}{n^3}$.

16 $\displaystyle\sum_{n=1}^{\infty} (-1)^n \frac{10^n}{n!} (x - 10)^n$.

17 $\displaystyle\sum_{n=1}^{\infty} \frac{n!}{2^n} (x - 5)^n$.

18 $\displaystyle\sum_{n=1}^{\infty} \frac{(-1)^n}{n \cdot 10^n} (x - 2)^n$.

19 $\displaystyle\sum_{n=0}^{\infty} x^{(2n)}$.

20 $\displaystyle\sum_{n=0}^{\infty} \left(\frac{x^2 + 1}{5}\right)^n$.

In each of Problems 21–30, use power series established in this section to find a power series representation of the given function. Then determine the radius of convergence of the resulting series.

21 $f(x) = x^2 e^{-3x}$

22 $f(x) = \dfrac{1}{10 + x}$.

23 $f(x) = \sin x^2$.

24 $f(x) = \cos^2 x = \frac{1}{2}(1 + \cos 2x)$.

25 $f(x) = \sqrt[3]{1 - x}$.

26 $f(x) = (1 + x^2)^{3/2}$.

27 $f(x) = (1 + x)^{-3}$.

28 $f(x) = \dfrac{1}{\sqrt{9 + x^3}}$.

29 $f(x) = \dfrac{\ln(1 + x)}{x}$.

30 $f(x) = \dfrac{x - \tan^{-1}x}{x^3}$.

In each of Problems 31–36, find a power series representation for the given function $f(x)$ using termwise integration.

31 $f(x) = \displaystyle\int_0^x \sin t^3 \, dt$.

32 $f(x) = \displaystyle\int_0^x \frac{\sin t}{t} \, dt$.

33 $f(x) = \displaystyle\int_0^x e^{-t^3} \, dt$.

34 $f(x) = \displaystyle\int_0^x \frac{\tan^{-1}t}{t} \, dt$.

35 $f(x) = \displaystyle\int_0^x \frac{1 - e^{-t^2}}{t^2} \, dt$.

36 $\tanh^{-1}x = \displaystyle\int_0^x \frac{dt}{1 - t^2}$.

37 Deduce from the arctangent series (Example 9) that

$$\pi = \frac{6}{\sqrt{3}} \sum_{n=0}^{\infty} \frac{(-1)^n}{2n + 1} \left(\frac{1}{3}\right)^n.$$

Then use this alternating series to show that $\pi = 3.14$ accurate to two decimal places.

38 Substitute the Maclaurin series for $\sin x$, and then assume the validity of termwise integration of the resulting series, to derive the formula

$$\int_0^\infty e^{-t} \sin xt \, dt = \frac{x}{1 + x^2} \qquad (|x| < 1).$$

Use the fact that $\int_0^\infty t^n e^{-t} \, dt = \Gamma(n + 1) = n!$ (Section 11-4).

39 (a) Deduce from the Maclaurin series for e^t that

$$\frac{1}{x^x} = \sum_{n=0}^{\infty} \frac{(-1)^n}{n!} (x \ln x)^n.$$

(b) Assuming the validity of termwise integration of the series in (a), use the integral formula of Problem 31 in

Section 11-4 to conclude that

$$\int_0^1 \frac{dx}{x^x} = \sum_{n=1}^{\infty} \frac{1}{n^n}.$$

40 Suppose that $f(x)$ is represented by the power series $\sum_{n=0}^{\infty} a_n x^n$ for all x in some neighborhood of $x = 0$. Show by repeated termwise differentiation of the series, substituting $x = 0$ each time, that $a_n = f^{(n)}(0)/n!$. Thus the only power series in x that represents a given function at and near $x = 0$ is its Maclaurin series.

41 (a) Consider the binomial series

$$f(x) = \sum_{n=0}^{\infty} \frac{\alpha(\alpha - 1)(\alpha - 2) \cdots (\alpha - n + 1)}{n!} x^n,$$

which converges (to *something*) if $|x| < 1$. Compute the derivative $f'(x)$ by termwise differentiation, and show that it satisfies the differential equation $(1 + x)f'(x) = \alpha f(x)$, so that

$$\frac{f'(x)}{f(x)} = \frac{\alpha}{1 + x}.$$

(b) Antidifferentiate both sides of the last equation in Part (a) to obtain $f(x) = C(1 + x)^{\alpha}$ for some constant C. Finally, show that $C = 1$. Thus the binomial series converges to $(1 + x)^{\alpha}$ if $|x| < 1$.

42 (a) Show by direct integration that

$$\int_0^1 \frac{\sin^{-1} x}{\sqrt{1 - x^2}} \, dx = \frac{\pi^2}{8}.$$

(b) Use the result of Problem 40 in Section 8-4 to show that

$$\int_0^1 \frac{x^{2n+1}}{\sqrt{1 - x^2}} \, dx = \frac{2 \cdot 4 \cdot 6 \cdots (2n)}{1 \cdot 3 \cdot 5 \cdots (2n + 1)}.$$

(c) Substitute the series of Example 10 for $\sin^{-1} x$ into the integral of Part (a), and then use the integral of Part (b) to integrate termwise. Conclude that

$$\int_0^1 \frac{\sin^{-1} x}{\sqrt{1 - x^2}} \, dx = 1 + \frac{1}{3^2} + \frac{1}{5^2} + \cdots.$$

(d) Note that

$$\sum_{n=1}^{\infty} \frac{1}{n^2} = \sum_{n=1}^{\infty} \frac{1}{(2n - 1)^2} + \sum_{n=1}^{\infty} \frac{1}{(2n)^2}.$$

Use this information and Parts (a) and (c) to show that

$$\sum_{n=1}^{\infty} \frac{1}{n^2} = \frac{\pi^2}{6}.$$

43 Prove that, if the power series $\sum a_n x^n$ converges for $x = x_0 \neq 0$, then it converges absolutely for all x such that $|x| < |x_0|$. (*Suggestion:* Conclude from the fact that $\lim_{n \to \infty} a_n x_0^n = 0$ that $|a_n x^n| \le |x/x_0|^n$ for n sufficiently large. Thus the series $\sum |a_n x^n|$ is eventually dominated by the geometric series $\sum |x/x_0|^n$, which converges if $|x| < |x_0|$.)

44 Suppose that the power series $\sum a_n x^n$ converges for some but not all nonzero values of x. Let S be the set of all real numbers x for which the series converges absolutely.

(a) Conclude from Problem 43 that the set S is bounded above.

(b) Let r be the least upper bound of the set S (see Problem 44 in Section 12-2). Then show that $\sum a_n x^n$ converges absolutely if $|x| < r$ and diverges if $|x| > r$. Explain why this proves Theorem 1 without the additional hypothesis that $\lim_{n \to \infty} |a_{n+1}/a_n|$ exists.

12-8

Power Series Computations

In Section 11-5 we discussed the use of Taylor polynomials to approximate numerical values of functions and of integrals. Power series are often used for these same purposes. In the case of an *alternating* power series, it may be simpler to work with the alternating series error estimate (Theorem 2 in Section 12-6) than with a Taylor formula remainder term.

EXAMPLE 1   Use the binomial series

$$\sqrt{1 + x} = 1 + \tfrac{1}{2}x - \tfrac{1}{8}x^2 + \tfrac{1}{16}x^3 - \tfrac{5}{128}x^4 + \cdots$$

to approximate $\sqrt{105}$ and estimate the accuracy.

Solution   The series above is, after the first term, an alternating series. Hence

$$\sqrt{105} = \sqrt{100 + 5} = 10\sqrt{1 + 0.05}$$
$$= 10(1 + \tfrac{1}{2}(0.05) - \tfrac{1}{8}(0.05)^2 + \tfrac{1}{16}(0.05)^3 + E)$$
$$\approx 10(1.02469531 + E) = 10.2469531 + 10E$$

where E is negative and

$$|E| < \frac{5}{128}(0.05)^4 < 0.0000003.$$

Consequently

$$10.246950 < \sqrt{105} < 10.246954,$$

and so $\sqrt{105} = 10.24695$ to five decimal places.

EXAMPLE 2   Approximate

$$\int_0^1 \frac{1 - \cos x}{x^2}\, dx$$

accurate to five decimal places.

Solution   We replace $\cos x$ by its Maclaurin series and get

$$\int_0^1 \frac{1 - \cos x}{x^2}\, dx = \int_0^1 \frac{1}{x^2}\left(\frac{x^2}{2!} - \frac{x^4}{4!} + \frac{x^6}{6!} - \cdots\right) dx$$

$$= \int_0^1 \left(\frac{1}{2!} - \frac{x^2}{4!} + \frac{x^4}{6!} - \frac{x^6}{8!} + \cdots\right) dx$$

$$= \frac{1}{2!} - \frac{1}{4!3} + \frac{1}{6!5} - \frac{1}{8!7} + \cdots.$$

Since this last series is an alternating series, it follows that

$$\int_0^1 \frac{1 - \cos x}{x^2}\, dx = \frac{1}{2!} - \frac{1}{4!3} + \frac{1}{6!5} + E$$

$$\approx 0.486389 + E,$$

where E is negative and

$$|E| < \frac{1}{8!7} < 0.000004.$$

Therefore

$$\int_0^1 \frac{1 - \cos x}{x^2}\, dx \approx 0.48639$$

rounded to five decimal places.

EXAMPLE 3   The binomial series with $\alpha = \frac{1}{3}$ gives

$$(1 + x^2)^{1/3} = 1 + \tfrac{1}{3}x^2 - \tfrac{1}{9}x^4 + \tfrac{5}{81}x^6 - \tfrac{10}{243}x^8 + \cdots,$$

which alternates after its first term. Use the first five terms of this series to estimate the value of

$$\int_0^{1/2} \sqrt[3]{1 + x^2}\, dx.$$

Solution Termwise integration of the above series yields

$$\int_0^{1/2} \sqrt[3]{1 + x^2}\, dx = \int_0^{1/2} \left(1 + \frac{x^2}{3} - \frac{x^4}{9} + \frac{5x^6}{81} - \frac{10x^8}{243} + \cdots \right) dx$$

$$= \left[x + \frac{1}{9}x^3 - \frac{1}{45}x^5 + \frac{5}{567}x^7 - \frac{10}{2187}x^9 + \cdots \right]_0^{1/2}$$

$$= \frac{1}{2} + \frac{1}{9}\left(\frac{1}{2}\right)^3 - \frac{1}{45}\left(\frac{1}{2}\right)^5 + \frac{5}{567}\left(\frac{1}{2}\right)^7$$

$$- \frac{10}{2187}\left(\frac{1}{2}\right)^9 + \cdots .$$

Since this last series is an alternating series, it follows that

$$\int_0^{1/2} \sqrt[3]{1 + x^2}\, dx = \tfrac{1}{2} + \tfrac{1}{9}(\tfrac{1}{2})^3 - \tfrac{1}{45}(\tfrac{1}{2})^5 + \tfrac{5}{567}(\tfrac{1}{2})^7 + E$$

$$\approx 0.513263 + E$$

where E is negative and

$$|E| < \tfrac{10}{2187}(\tfrac{1}{2})^9 < 0.000009.$$

Therefore

$$0.513254 < \int_0^{1/2} \sqrt[3]{1 + x^2}\, dx < 0.513263,$$

so

$$\int_0^{1/2} \sqrt[3]{1 + x^2}\, dx = 0.51326 \pm 0.00001.$$

THE ALGEBRA OF POWER SERIES

The following theorem, which we state without proof, says that power series may be added and multiplied much like polynomials. The guiding principle is that of collecting coefficients of like powers of x.

Theorem 1 *Adding and Multiplying Power Series*

Let $\Sigma a_n x^n$ and $\Sigma b_n x^n$ be power series with nonzero radii of convergence. Then

$$\sum_{n=0}^{\infty} a_n x^n + \sum_{n=0}^{\infty} b_n x^n = \sum_{n=0}^{\infty} (a_n + b_n)x^n \qquad (1)$$

and

$$\left(\sum_{n=0}^{\infty} a_n x^n \right)\left(\sum_{n=0}^{\infty} b_n x^n \right) = \sum_{n=0}^{\infty} c_n x^n$$

$$= a_0 b_0 + (a_0 b_1 + a_1 b_0)x$$

$$+ (a_0 b_2 + a_1 b_1 + a_2 b_0)x^2 + \cdots \qquad (2)$$

where

$$c_n = a_0 b_n + a_1 b_{n-1} + \cdots + a_{n-1} b_1 + a_n b_0. \qquad (3)$$

Series (1) and (2) converge for any x that lies interior to the intervals of convergence of both $\Sigma a_n x^n$ and $\Sigma b_n x^n$.

Thus, if $\Sigma a_n x^n$ and $\Sigma b_n x^n$ are power series representations of the functions $f(x)$ and $g(x)$, respectively, then the product power series $\Sigma c_n x^n$ found by "ordinary multiplication" and collection of terms is a power series representation of the function $f(x)g(x)$. This fact can also be used to divide one power series by another, *provided* it is known in advance that the quotient has a power series representation.

EXAMPLE 4   Assume that the tangent function has a power series representation $\tan x = \Sigma a_n x^n$. Use the Maclaurin series for $\sin x$ and $\cos x$ to find a_0, a_1, a_2, and a_3.

Solution   We multiply series to obtain

$$\sin x = \tan x \cos x$$

$$= (a_0 + a_1 x + a_2 x^2 + a_3 x^3 + \cdots)\left(1 - \frac{x^2}{2} + \frac{x^4}{24} - \cdots\right)$$

$$= a_0 + a_1 x + \left(a_2 - \frac{1}{2} a_0\right)x^2 + \left(a_3 - \frac{1}{2} a_1\right)x^3 + \cdots.$$

Since

$$\sin x = x - \tfrac{1}{6}x^3 + \tfrac{1}{120}x^5 - \cdots,$$

comparison of coefficients gives the equations

$$
\begin{array}{rcl}
a_0 & = & 0, \\
a_1 & = & 1, \\
-\tfrac{1}{2}a_0 + a_2 & = & 0, \\
-\tfrac{1}{2}a_1 + a_3 & = & -\tfrac{1}{6}.
\end{array}
$$

Then we find that $a_0 = 0$, $a_1 = 1$, $a_2 = 0$, and $a_3 = \tfrac{1}{3}$. So

$$\tan x = x + \tfrac{1}{3}x^3 + \cdots.$$

Things are not always as they first appear; the continuation of this series is

$$\tan x = x + \tfrac{1}{3}x^3 + \tfrac{2}{15}x^5 + \tfrac{17}{315}x^7 + \cdots.$$

For the general form of the nth coefficient a_n, see page 204 of K. Knopp's *Theory and Application of Infinite Series* (Hafner, 1971). You may check that the first few terms given above agree with the result of the ordinary long division indicated below.

$$(1 - \tfrac{1}{2}x^2 + \tfrac{1}{24}x^4 - \cdots)\overline{)x - \tfrac{1}{6}x^3 + \tfrac{1}{120}x^5 - \cdots}$$
$$\quad\quad\quad\quad x + \tfrac{1}{3}x^3 + \tfrac{2}{15}x^5 + \cdots$$

POWER SERIES AND INDETERMINATE FORMS

According to Theorem 3 in Section 12-7, a power series is differentiable and therefore continuous within its interval of convergence. It follows that

$$\lim_{x \to c} \sum_{n=0}^{\infty} a_n(x - c)^n = a_0. \tag{4}$$

The following two examples illustrate the use of this fact to find the limit of an indeterminate form $f(x)/g(x)$ by first substituting power series representations for $f(x)$ and $g(x)$.

EXAMPLE 5 Find $\displaystyle\lim_{x\to 0} \frac{\sin x - \tan^{-1}x}{x^2 \ln(1 + x)}$.

Solution The power series of Equations (4), (19), and (20) in Section 12-7 give

$$\sin x - \tan^{-1}x = (x - \tfrac{1}{6}x^3 + \tfrac{1}{120}x^5 - \cdots) - (x - \tfrac{1}{3}x^3 + \tfrac{1}{5}x^5 - \cdots)$$
$$= \tfrac{1}{6}x^3 - \tfrac{23}{120}x^5 + \cdots$$

and

$$x^2 \ln(1 + x) = x^2(x - \tfrac{1}{2}x^2 + \tfrac{1}{3}x^3 - \cdots)$$
$$= x^3 - \tfrac{1}{2}x^4 + \tfrac{1}{3}x^5 - \cdots.$$

Hence

$$\lim_{x\to 0} \frac{\sin x - \tan^{-1}x}{x^2 \ln(1 + x)} = \lim_{x\to 0} \frac{\tfrac{1}{6}x^3 - \tfrac{23}{120}x^5 + \cdots}{x^3 - \tfrac{1}{2}x^4 + \cdots}$$

$$= \lim_{x\to 0} \frac{\tfrac{1}{6} - \tfrac{23}{120}x^2 + \cdots}{1 - \tfrac{1}{2}x + \cdots} = \frac{1}{6}.$$

EXAMPLE 6 Find $\displaystyle\lim_{x\to 1} \frac{\ln x}{x - 1}$.

Solution We first replace x by $x - 1$ in the power series for $\ln(1 + x)$ used in Example 5. This gives us

$$\ln x = (x - 1) - \tfrac{1}{2}(x - 1)^2 + \tfrac{1}{3}(x - 1)^3 - \cdots.$$

Hence

$$\lim_{x\to 1} \frac{\ln x}{x - 1} = \lim_{x\to 1} \frac{(x - 1) - \tfrac{1}{2}(x - 1)^2 + \tfrac{1}{3}(x - 1)^3 - \cdots}{x - 1}$$

$$= \lim_{x\to 1} (1 - \tfrac{1}{2}(x - 1) + \tfrac{1}{3}(x - 1)^2 - \cdots) = 1.$$

The method of Examples 5 and 6 provides a useful alternative to l'Hôpital's rule, especially when repeated differentiation of numerator and denominator is inconvenient or too time-consuming.

12-8 PROBLEMS

In each of Problems 1–5, use an infinite series to approximate the indicated number accurate to three decimal places.

1 $\sqrt[3]{65}$. **2** $\sqrt[4]{630}$.

3 $\sin(0.5)$. **4** $e^{-0.2}$.

5 $\tan^{-1}(0.5)$.

In each of Problems 6–10, use an infinite series to approximate the value of the given integral accurate to three decimal places.

6 $\displaystyle\int_0^1 \frac{\sin x}{\sqrt{x}}\, dx$.

7 $\displaystyle\int_0^1 \frac{\tan^{-1}x}{x}\, dx$.

8 $\int_0^1 \sin x^2 \, dx$.

9 $\int_0^{0.1} \dfrac{\ln(1+x)}{x} \, dx$.

10 $\int_0^{1/2} \dfrac{dx}{\sqrt{1+x^4}}$.

In each of Problems 11–16, use power series to evaluate the given limit.

11 $\lim\limits_{x \to 0} \dfrac{1+x-e^x}{x^2}$.

12 $\lim\limits_{x \to 0} \dfrac{x - \sin x}{x^3 \cos x}$.

13 $\lim\limits_{x \to 0} \dfrac{1 - \cos x}{x(e^x - 1)}$.

14 $\lim\limits_{x \to 0} \dfrac{e^x - e^{-x} - 2x}{x - \arctan x}$.

15 $\lim\limits_{x \to 0} \left(\dfrac{1}{x} - \dfrac{1}{\sin x} \right)$.

16 $\lim\limits_{x \to 1} \dfrac{\ln(x^2)}{x - 1}$.

17 Derive the geometric series representation of $1/(1-x)$ by finding $a_0, a_1, a_2, \ldots$ such that

$$(1-x)(a_0 + a_1 x + a_2 x^2 + a_3 x^3 + \cdots) = 1.$$

18 Derive the first five coefficients in the binomial series for $\sqrt{1+x}$ by finding $a_0, a_1, a_2, a_3,$ and a_4 such that

$$(a_0 + a_1 x + a_2 x^2 + a_3 x^3 + a_4 x^4 + \cdots)^2 = 1 + x.$$

19 Use the method of Example 4 to find the coefficients $a_0, a_1, a_2, a_3,$ and a_4 in the series

$$\sec x = \dfrac{1}{\cos x} = \sum_{n=0}^{\infty} a_n x^n.$$

20 Multiply the geometric series for $1/(1-x)$ and the series for $\ln(1-x)$ to show that, if $|x| < 1$, then

$$-\dfrac{1}{1-x}\ln(1-x) = x + \left(1 + \dfrac{1}{2}\right)x^2 + \left(1 + \dfrac{1}{2} + \dfrac{1}{3}\right)x^3$$

$$+ \left(1 + \dfrac{1}{2} + \dfrac{1}{3} + \dfrac{1}{4}\right)x^4 + \cdots.$$

21 Take the logarithmic series

$$\ln(1 + x) = x - \dfrac{x^2}{2} + \dfrac{x^3}{3} - \dfrac{x^4}{4} + \cdots$$

as known. Find the first four coefficients in the series for e^x by finding $a_0, a_1, a_2,$ and a_3 so that

$$1 + x = e^{\ln(1+x)} = \sum_{n=0}^{\infty} a_n \left(x - \dfrac{x^2}{2} + \dfrac{x^3}{3} - \dfrac{x^4}{4} + \cdots \right)^n.$$

This is exactly how the power series for e^x was first discovered (by Newton)!

22 Use the method of Example 4 to show that

$$\dfrac{x}{\sin x} = 1 + \dfrac{1}{6}x^2 + \dfrac{7}{360}x^4 + \cdots.$$

23 Show that long division of power series gives

$$\dfrac{2 + x}{1 + x + x^2} = 2 - x - x^2 + 2x^3 - x^4 - x^5$$

$$+ 2x^6 - x^7 - x^8 + \cdots.$$

Show that the radius of convergence of this series is $r = 1$.

---

*12-9
The Simple Pendulum Again

In Section 7-4 we saw that the period of oscillation T of a simple pendulum of length L, as in Figure 12.7, is *approximately* $2\pi(L/g)^{1/2}$, where g is the acceleration of gravity. If L is in feet, we take $g = 32.2$ ft/sec^2.

Now we can use infinite series to compute the actual value of T to any desired degree of accuracy. We begin as we did in Section 7-4, but refine our methods when we reach the point at which, earlier, we employed a crude approximation. The kinetic and potential energies of the mass m on the end of the pendulum are, respectively,

$$\text{K.E.} = \tfrac{1}{2}mL^2(d\theta/dt)^2 \quad \text{and} \quad \text{P.E.} = mgL(1 - \cos\theta).$$

As in Figure 12.7, the angle θ is the angle of deviation of the pendulum from the vertical.

We suppose that the pendulum is released from rest from the initial angle $\theta = \alpha$. Since total mechanical energy is conserved (we assume there is no friction), we can write the equation

$$\frac{1}{2} mL^2 \left(\frac{d\theta}{dt}\right)^2 + mgL(1 - \cos\theta) = mgL(1 - \cos\alpha).$$

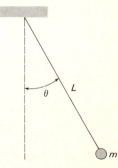

12.7 The simple pendulum.

Thus

$$dt = \pm \sqrt{\frac{L}{2g}} \, \frac{d\theta}{\sqrt{\cos\theta - \cos\alpha}}.$$

One-fourth of a period is required for the pendulum to swing from $\theta = \alpha$ to $\theta = 0$. Therefore

$$T = 4 \left(\frac{L}{2g}\right)^{1/2} \int_0^\alpha \frac{d\theta}{\sqrt{\cos\theta - \cos\alpha}}. \tag{1}$$

Thus our problem—finding the period T—reduces to the evaluation of this integral.

First we use the trigonometric identity $\cos\theta = 1 - 2\sin^2(\theta/2)$, which gives

$$T = 2 \left(\frac{L}{g}\right)^{1/2} \int_0^\alpha \frac{d\theta}{\sqrt{k^2 - \sin^2(\theta/2)}} \tag{2}$$

where $k = \sin(\alpha/2)$. Next, we use the substitution $u = (1/k)\sin(\theta/2)$, which yields

$$T = 4 \left(\frac{L}{g}\right)^{1/2} \int_0^1 \frac{du}{\sqrt{(1 - u^2)(1 - k^2 u^2)}}.$$

Finally, the substitution $u = \sin\phi$ gives

$$T = 4 \left(\frac{L}{g}\right)^{1/2} \int_0^{\pi/2} \frac{d\phi}{\sqrt{1 - k^2 \sin^2\phi}}. \tag{3}$$

The integral of Equation (3) is known as an *elliptic integral*, because a similar integral arises in the evaluation of the arc length of an ellipse. To evaluate it, we use the binomial series

$$\frac{1}{\sqrt{1 - x}} = \sum_{n=0}^{\infty} \frac{1 \cdot 3 \cdot 5 \cdots (2n - 1)}{2 \cdot 4 \cdot 6 \cdots (2n)} x^n$$

of Section 12-7 (as in Formula (16) there, the coefficient is 1 when $n = 0$). With $x = k^2 \sin^2\phi$,

$$\frac{1}{\sqrt{1 - k^2 \sin^2\phi}} = \sum_{n=0}^{\infty} \frac{1 \cdot 3 \cdot 5 \cdots (2n - 1)}{2 \cdot 4 \cdot 6 \cdots (2n)} k^{2n} \sin^{2n}\phi.$$

We substitute this series into the integral of Equation (3), then integrate term by term. The result is

$$T = 4 \left(\frac{L}{g}\right)^{1/2} \sum_{n=0}^{\infty} \frac{1 \cdot 3 \cdot 5 \cdots (2n - 1)}{2 \cdot 4 \cdot 6 \cdots (2n)} k^{2n} \int_0^{\pi/2} \sin^{2n}\phi \, d\phi.$$

But

$$\int_0^{\pi/2} \sin^{2n}\phi \, d\phi = \frac{\pi}{2} \frac{1 \cdot 3 \cdot 5 \cdots (2n - 1)}{2 \cdot 4 \cdot 6 \cdots (2n)}$$

by Problem 40 of Section 8-4. Consequently

$$\begin{aligned}
T &= 2\pi \left(\frac{L}{g}\right)^{1/2} \sum_{n=0}^{\infty} \left(\frac{1 \cdot 3 \cdot 5 \cdots (2n - 1)}{2 \cdot 4 \cdot 6 \cdots (2n)}\right)^2 k^{2n} \\
&= 2\pi \left(\frac{L}{g}\right)^{1/2} \left[1 + \left(\frac{1}{2}\right)^2 k^2 + \left(\frac{1 \cdot 3}{2 \cdot 4}\right)^2 k^4 + \left(\frac{1 \cdot 3 \cdot 5}{2 \cdot 4 \cdot 6}\right)^2 k^6 + \cdots\right]. \tag{4}
\end{aligned}$$

Let us denote by $S(k)$ the infinite series factor in Equation (4). We find that our earlier estimate $T \approx 2\pi(L/g)^{1/2}$ can now be replaced by

$$T = 2\pi \left(\frac{L}{g}\right)^{1/2} S(k). \qquad (5)$$

We can check the approximation $T \approx 2\pi(L/g)^{1/2}$ by verifying that $S(k)$ is near 1 for very small initial angles α. We take $\alpha = 10°$, so that $k = \sin 5°$. Then

$$S(k) \geq 1 + \tfrac{1}{4}k^2 \approx 1.00190.$$

To get an upper bound for $S(k)$, we note that the coefficients in the series for $S(k)$ are all at most $\tfrac{1}{4}$, so

$$S(k) \leq 1 + \frac{1}{4}k^2 + \frac{1}{4}k^4 + \frac{1}{4}k^6 + \cdots$$

$$= 1 + \frac{1}{4}k^2(1 + k^2 + k^4 + k^6 + \cdots)$$

$$= 1 + \frac{1}{4}k^2\left(\frac{1}{1-k^2}\right) = \frac{4 - 3k^2}{4 - 4k^2}.$$

Hence

$$S(k) \leq \frac{4 - 3k^2}{4 - 4k^2} \approx 1.00191$$

when $k = \sin 5°$.

Thus $S(k) = 1.0019$ to four decimal places, so the approximation $T \approx 2\pi(L/g)^{1/2}$ is about 0.19% low. In a well-designed pendulum clock, the lessening of the angular amplitude due to friction tends to compensate for this increased period and helps the clock to keep time more accurately.

12-9 PROBLEMS

1 Consider two pendulum clocks with identical working mechanisms. The period of oscillation of the pendulum of one clock is 0.19% greater than the period of the other. If the two clocks initially agree, by how much will they disagree when the slower clock indicates that exactly one week has elapsed?

2 Figure 12.8 shows a parabolic suspension cable for a bridge of total span $2S$ and height H at each end.
(a) Show that the total length L of the cable is

$$L = 2\int_0^S \left(1 + \frac{4H^2}{S^4}x^2\right)^{1/2} dx.$$

12.8 The parabolic supporting cable of a suspension bridge.

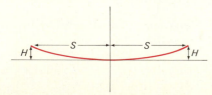

(b) Then use the binomial series to show that

$$L = 2S\left(1 + \frac{2H^2}{3S^2} - \frac{2H^4}{5S^4} + \frac{4H^6}{7S^6} - \cdots\right).$$

3 A single-span suspension bridge across the Strait of Messina (5 miles wide) between Italy and Sicily is currently in the planning stage. The plans include suspension towers 1250 feet high at each end. Use the series of Problem 2 to estimate the length of the suspension cable for this proposed bridge.

4 Consider the power series

$$S(x) = \sum_{n=0}^{\infty} a_n x^n,$$

with positive and nonincreasing coefficients; thus $a_n \geq a_{n+1} > 0$ for all $n \geq 0$. Let

$$S_{n-1}(x) = a_0 + a_1 x + a_2 x^2 + \cdots + a_{n-1}x^{n-1}.$$

Use the geometric series to show that, if $0 < x < 1$, then

$$S_{n-1}(x) + a_n x^n < S(x) < S_{n-1}(x) + \frac{a_n x^n}{1 - x}.$$

Explain how this result may be used to compute $S(x)$ with any desired degree of accuracy.

5 Apply the result of Problem 4 to show that, for a simple pendulum with initial angle $\alpha = 90°$, $S(k) = 1.18$ to two decimal places. Thus the actual period of such a pendulum would be 18% larger than $2\pi(L/g)^{1/2}$.

6 Use the result of Problem 4 and the first five terms of the Maclaurin series for e^x to compute $e^{1/3}$ accurate to three decimal places.

7 (a) Consider the ellipse with equation $x^2/a^2 + y^2/b^2 = 1$ and eccentricity $\varepsilon = (a^2 - b^2)^{1/2}/a$. Use the parametrization $x = a \sin \phi$, $y = b \cos \phi$ to show that the circum-

ference of this ellipse is

$$S = 4a \int_0^{\pi/2} \sqrt{1 - \varepsilon^2 \sin^2 \phi} \, d\phi.$$

(b) Expand the integrand as a binomial series and integrate termwise to obtain

$$S = 2\pi a \left(1 - \frac{1}{2^2} \varepsilon^2 - \frac{1^2 \cdot 3}{2^2 \cdot 4^2} \varepsilon^4 - \frac{1^2 \cdot 3^2 \cdot 5}{2^2 \cdot 4^2 \cdot 6^2} \varepsilon^6 - \cdots \right).$$

(c) Suppose that $a = 10$ and $\varepsilon = \frac{1}{2}$. Use the series in Part (b) to show that $S = 58.7$ to one decimal place.

CHAPTER 12 REVIEW: Definitions, Concepts, Results

Use the list below as a guide to concepts that you may need to review.

1 Limit of a sequence (definition)
2 The limit laws for sequences
3 The bounded monotone sequence property
4 Sum of an infinite series (definition)
5 Sum of a geometric series (formula)
6 The nth term test for divergence
7 Divergence of the harmonic series
8 The integral test.
9 Convergence of p-series
10 The comparison and limit comparison tests

11 The alternating series test
12 Absolute convergence (definition *and* the fact that it implies convergence)
13 The ratio test
14 The root test
15 Power series; radius and interval of convergence
16 Taylor series of the elementary transcendental functions
17 The binomial series
18 Termwise differentiation and integration of power series
19 Use of power series to approximate values of functions and integrals
20 Product of two power series
21 Use of power series to evaluate indeterminate forms

MISCELLANEOUS PROBLEMS

In Problems 1–15, determine whether or not the sequence $\{a_n\}$ converges, and find the limit if it does converge.

1 $a_n = \dfrac{n^2 + 1}{n^2 + 4}$.

2 $a_n = \dfrac{8n - 7}{7n - 8}$.

3 $a_n = 10 - (0.99)^n$.

4 $a_n = n \sin \pi n$.

5 $a_n = \dfrac{1 + (-1)^n \sqrt{n}}{n + 1}$.

6 $a_n = \sqrt{\dfrac{1 + (-\frac{1}{2})^n}{n + 1}}$.

7 $a_n = \dfrac{\sin 2n}{n}$.

8 $a_n = 2^{-(\ln n)/n}$.

9 $a_n = (-1)^{\sin(n\pi/2)}$.

10 $a_n = \dfrac{(\ln n)^3}{n^2}$.

11 $a_n = \dfrac{1}{n} \sin \dfrac{1}{n}$.

12 $a_n = \dfrac{n - e^n}{n + e^n}$.

13 $a_n = \dfrac{\sinh n}{n}$.

14 $a_n = \left(1 + \dfrac{2}{n} \right)^{2n}$.

15 $a_n = \sqrt[n]{2n^2 + 1}$.

Determine whether each infinite series in Problems 16–30 converges or diverges.

16 $\displaystyle\sum_{n=1}^{\infty} \dfrac{(n^2)!}{n^n}$.

17 $\displaystyle\sum_{n=1}^{\infty} (-1)^n \dfrac{\ln n}{n^2}$.

18 $\displaystyle\sum_{n=0}^{\infty} \dfrac{3^n}{2^n + 4^n}$.

19 $\displaystyle\sum_{n=0}^{\infty} \dfrac{n!}{e^{n^2}}$.

20 $\displaystyle\sum_{n=1}^{\infty} \dfrac{\sin(1/n)}{n^{3/2}}$.

21 $\displaystyle\sum_{n=0}^{\infty} \dfrac{(-2)^n}{3^n + 1}$.

22 $\displaystyle\sum_{n=1}^{\infty} 2^{-(2/n^2)}$.

23 $\displaystyle\sum_{n=2}^{\infty} \dfrac{(-1)^n n}{(\ln n)^3}$.

24 $\displaystyle\sum_{n=1}^{\infty} \dfrac{(-1)^n}{\sqrt[n]{10}}$.

25 $\displaystyle\sum_{n=1}^{\infty} \dfrac{\sqrt{n} + \sqrt[3]{n}}{n^2 + n^3}$.

26 $\displaystyle\sum_{n=1}^{\infty} (-1)^n \dfrac{1}{n^{[1 + (1/n)]}}$.

27 $\displaystyle\sum_{n=1}^{\infty} (-1)^n \dfrac{\tan^{-1} n}{\sqrt{n}}$.

28 $\displaystyle\sum_{n=1}^{\infty} n \sin \dfrac{1}{n}$.

29 $\displaystyle\sum_{n=3}^{\infty} \dfrac{1}{n(\ln n)(\ln \ln n)}$.

30 $\displaystyle\sum_{n=3}^{\infty} \frac{1}{n(\ln n)(\ln \ln n)^2}$.

Find the interval of convergence of each of the power series in Problems 31–40.

31 $\displaystyle\sum_{n=0}^{\infty} \frac{2^n x^n}{n!}$.

32 $\displaystyle\sum_{n=0}^{\infty} \frac{(3x)^n}{2^{n+1}}$.

33 $\displaystyle\sum_{n=1}^{\infty} \frac{(x-1)^n}{n(3^n)}$.

34 $\displaystyle\sum_{n=0}^{\infty} \frac{(2x-3)^n}{4^n}$.

35 $\displaystyle\sum_{n=1}^{\infty} \frac{(-1)^n x^n}{4n^2 - 1}$.

36 $\displaystyle\sum_{n=0}^{\infty} \frac{(2x-1)^n}{n^2 + 1}$.

37 $\displaystyle\sum_{n=0}^{\infty} \frac{n! x^{2n}}{10^n}$.

38 $\displaystyle\sum_{n=2}^{\infty} \frac{x^n}{\ln n}$.

39 $\displaystyle\sum_{n=0}^{\infty} \frac{1+(-1)^n}{2(n!)} x^n$.

40 $\displaystyle\sum_{n=1}^{\infty} \left(1 + \frac{1}{n}\right)^n (x-1)^n$.

Find the set of values of x for which each series in Problems 41–43 converges.

41 $\displaystyle\sum_{n=1}^{\infty} (x-n)^n$.

42 $\displaystyle\sum_{n=1}^{\infty} (\ln x)^n$.

43 $\displaystyle\sum_{n=0}^{\infty} \frac{e^{nx}}{n!}$.

44 Find the rational number that has repeated decimal expansion $2.7\ 1828\ 1828\ 1828 \cdots$.

45 Criticize the following argument:

If $x = 1 + 2 + 4 + 8 + \cdots$, then $1 + 2x = 1 + 2 + 4 + 8 + \cdots$, so $2x + 1 = x$, and hence $x = -1$.

46 Prove that Σa_n^2 converges if Σa_n is a convergent positive-term series.

47 Let the sequence $\{a_n\}$ be defined recursively by

$$a_1 = 1, \qquad a_{n+1} = 1 + \frac{1}{1 + a_n} \qquad \text{if } n \geq 1.$$

The limit of the sequence $\{a_n\}$ is the value of the *continued fraction*

$$1 + \cfrac{1}{2 + \cfrac{1}{2 + \cfrac{1}{2 + \cdots}}}.$$

Assuming that $A = \lim_{n \to \infty} a_n$ exists, show that $A = \sqrt{2}$.

48 Let $\{F_n\}_{n=1}^{\infty}$ be the Fibonacci sequence of Example 2 in Section 12-2.

(a) Show that $0 < F_n \leq 2^n$ for all $n \geq 1$, and hence conclude that the power series $F(x) = \displaystyle\sum_{n=1}^{\infty} F_n x^n$ converges if $|x| < \frac{1}{2}$.

(b) Show that $(1 - x - x^2)F(x) = x$, so that $F(x) = x/(1 - x - x^2)$.

49 We say that the "infinite product" indicated by

$$\prod_{n=1}^{\infty} (1 + a_n) = (1 + a_1)(1 + a_2)(1 + a_3) \cdots$$

converges provided that the infinite series

$$S = \sum_{n=1}^{\infty} \ln(1 + a_n)$$

converges, in which case the value of the infinite product is e^S. Use the integral test to show that

$$\prod_{n=1}^{\infty} \left(1 + \frac{1}{n}\right)$$

diverges.

50 (See Problem 49.) Show that the infinite product

$$\prod_{n=1}^{\infty} \left(1 + \frac{1}{n^2}\right)$$

converges, and use the integral test remainder estimate to approximate its value. The actual value of this infinite product is known to be $(\sinh \pi)/\pi \approx 3.67608$.

In each of Problems 51–55, use infinite series to approximate the indicated number accurate to three decimal places.

51 $\sqrt[5]{1.5}$.

52 $\ln(1.2)$.

53 $\displaystyle\int_0^1 e^{-x^2}\, dx$.

54 $\displaystyle\int_0^{1/2} \sqrt[3]{1 + x^4}\, dx$.

55 $\displaystyle\int_0^{1/2} \frac{1 - e^{-x}}{x}\, dx$.

56 Substitute the Maclaurin series for $\sin x$ into that for e^x to obtain

$$e^{\sin x} = 1 + x + \tfrac{1}{2}x^2 - \tfrac{1}{8}x^4 + \cdots.$$

57 Substitute the Maclaurin series for the cosine and then integrate termwise to derive the formula

$$\int_0^{\infty} e^{-t^2} \cos 2xt\, dt = \frac{\sqrt{\pi}}{2} e^{-x^2}.$$

Use the formula for $\displaystyle\int_0^{\infty} t^{2n} e^{-t^2}\, dt$ given in Miscellaneous Problem 47 in Chapter 11. It should be noted that the validity of this improper termwise integration is subject to verification.

58 Show that

$$\tanh^{-1} x = \int_0^x \frac{dt}{1 - t^2} = \sum_{n=0}^{\infty} \frac{x^{2n+1}}{2n + 1}$$

if $|x| < 1$.

59 Show that

$$\sinh^{-1}x = \int_0^x \frac{dt}{(1+t^2)^{1/2}}$$

$$= \sum_{n=0}^{\infty} (-1)^n \frac{1\cdot 3\cdot 5\cdots(2n-1)}{2\cdot 4\cdot 6\cdots(2n)}\frac{x^{2n+1}}{2n+1}$$

if $|x| < 1$.

60 Suppose that $\tan y = \Sigma a_n y^n$. Determine a_0, a_1, a_2, and a_3 by substituting the inverse tangent series (Equation (20) in Section 12-7) into the equation

$$x = \tan(\tan^{-1}x) = \Sigma a_n(\tan^{-1}x)^n.$$

61 According to *Stirling's series*, the value of $n!$ for n large is given to a close approximation by

$$n! \approx \sqrt{2\pi n}\left(\frac{n}{e}\right)^n e^{\mu(n)}$$

where

$$\mu(n) = \frac{1}{12n} - \frac{1}{360n^3} + \frac{1}{1260n^5}.$$

Substitute $\mu(n)$ into Maclaurin's series for e^x to show that

$$e^{\mu(n)} = 1 + \frac{1}{12n} + \frac{1}{288n^2} - \frac{139}{51840n^3} + \cdots.$$

Can you show that the next term is $-571/(2{,}488{,}320n^4)$?

Vectors, Curves, and Surfaces in Space

13

Rectangular Coordinates and Three-Dimensional Vectors

In the first twelve chapters we have discussed many aspects of the calculus of functions of a *single* variable. The geometry of such functions is two-dimensional because the graph of a function of a single variable is a curve in the plane. Most of the remainder of this book deals with the calculus of functions of *several* (two or more) independent variables. The geometry of functions of two variables is three-dimensional, because the graphs of such functions are surfaces in space.

Rectangular coordinates in the plane may be generalized in a natural way to rectangular coordinates in space. A point in space is determined by giving its location relative to three mutually perpendicular **coordinate axes** passing through the origin O. We shall always draw the x-, y-, and z-axes as shown in Figure 13.1, with arrows indicating the positive direction along each axis. With this configuration of axes our rectangular coordinate system is **right-handed:** If the curled fingers of the right hand point in the direction of a 90° rotation from the positive x-axis to the positive y-axis, then the thumb points in the direction of the positive z-axis. If the x- and y-axes were interchanged, then we would have a left-handed coordinate system. These two coordinate systems are different. So are L-alanine and D-alanine, shown in Figure 13.2; you can digest the former but not the latter. In this book we shall use right-handed coordinate systems exclusively.

The three coordinate axes taken in pairs determine three **coordinate planes:**

- the (horizontal) xy-plane, where $z = 0$;
- the (vertical) yz-plane, where $x = 0$; and
- the (vertical) xz-plane, where $y = 0$.

The point P in space is said to have **rectangular coordinates** (x, y, z) if (see Figure 13.3):

- x is its signed distance from the yz-plane;
- y is its signed distance from the xz-plane; and
- z is its signed distance from the xy-plane.

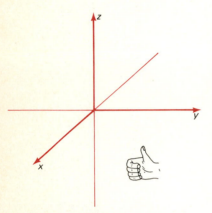

13.1 The right-handed rectangular coordinate system.

13.2 The stereoisomers of the amino acid alanine are physically and biologically different (L is for "levo" and D is for "dextro").

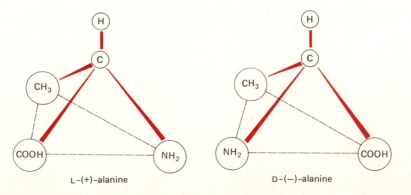

L–(+)–alanine   D–(−)–alanine

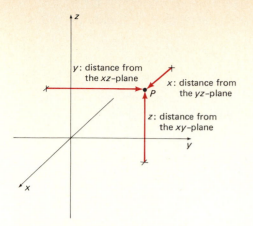

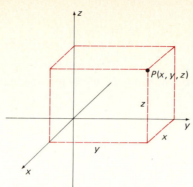

13.3 Locating the point P with rectangular coordinates.

13.4 "Completing the box" to show P with the illusion of the third dimension.

In this case we may describe the location of the point P by simply calling it "the point $P(x, y, z)$." There is a one-to-one correspondence between ordered triples (x, y, z) of real numbers and points P in space; this correspondence is called a **rectangular coordinate system** in space. In Figure 13.4 the point P is located in the **first octant**—the eighth of space in which all three rectangular coordinates are positive.

If we apply the Pythagorean theorem to the right triangles P_1QR and P_1RP_2 in Figure 13.5, we get

$$|P_1P_2|^2 = |RP_2|^2 + |P_1R|^2 = |RP_2|^2 + |QR|^2 + |P_1Q|^2$$
$$= (x_1 - x_2)^2 + (y_1 - y_2)^2 + (z_1 - z_2)^2.$$

Thus the **distance formula** for the **distance** $|P_1P_2|$ between the points P_1 and P_2 is

$$|P_1P_2| = \sqrt{(x_1 - x_2)^2 + (y_1 - y_2)^2 + (z_1 - z_2)^2}. \qquad (1)$$

13.5 The distance between P_1 and P_2 is the length of the long diagonal of the box.

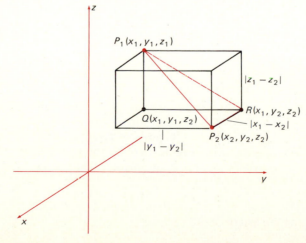

For example, the distance between the points $P_1(1, 3, -2)$ and $P_2(4, -3, 1)$ is

$$|P_1P_2| = \sqrt{(4 - 1)^2 + (-3 - 3)^2 + (1 + 2)^2}$$
$$= \sqrt{54} \approx 7.34847.$$

In Problem 43 we ask you to apply the distance formula in (1) to show that the **midpoint** M of the line segment joining $P_1(x_1, y_1, z_1)$ and $P_2(x_2, y_2, z_2)$ is

$$M(\tfrac{1}{2}[x_1 + x_2], \tfrac{1}{2}[y_1 + y_2], \tfrac{1}{2}[z_1 + z_2]). \qquad (2)$$

The **graph** of an equation in three variables x, y, and z is the set of all points in space with rectangular coordinates that satisfy the equation. In general, the graph of an equation in three variables will be a *two-dimensional surface* in $\mathscr{R}^3$ (three-dimensional space with rectangular coordinates). For example, let $C(h, k, l)$ be a fixed point. Then the graph of the equation

$$(x - h)^2 + (y - k)^2 + (z - l)^2 = r^2 \qquad (3)$$

is the set of all points $P(x, y, z)$ at distance r from the fixed point C. This means that Equation (3) is the equation of **sphere with radius r and center** $C(h, k, l)$. Moreover, given an equation of the form

$$x^2 + y^2 + z^2 + Ax + By + Cz + D = 0,$$

we can attempt—by completing the square in each variable—to write it in the form of Equation (3), and thereby show that its graph is a sphere.

EXAMPLE 1   Determine the graph of the equation

$$x^2 + y^2 + z^2 + 4x + 2y - 6z - 2 = 0.$$

Solution   We complete the squares, and the equation takes the form

$$(x^2 + 4x + 4) + (y^2 + 2y + 1) + (z^2 - 6z + 9) = 16.$$

That is,

$$(x + 2)^2 + (y + 1)^2 + (z - 3)^2 = 4^2.$$

Thus the given equation has as its graph a sphere with radius 4 and center $(-2, -1, 3)$.

13.6 The segment OP is a realization of the vector $\mathbf{v} = \overrightarrow{OP}$.

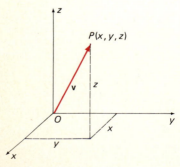

VECTORS IN SPACE

The discussion of vectors in the plane in Section 10-3 may be repeated almost verbatim for vectors in space. The major difference is that a vector in space has three components rather than two. The vector determined by the point $P(x, y, z)$ is its **position vector** $\mathbf{v} = \overrightarrow{OP} = \langle x, y, z \rangle$, which is represented pictorially (Figure 13.6) by the arrow from the origin O to P (or by any parallel translate of this arrow). The distance formula in (1) gives

$$|\mathbf{v}| = \sqrt{x^2 + y^2 + z^2} \qquad (4)$$

for the **length** of the vector $\mathbf{v} = \langle x, y, z \rangle$.

CHAP. 13:   Vectors, Curves, and Surfaces in Space

The vector $\overrightarrow{AB}$ represented by the arrow (Figure 13.7) from $A(a_1, a_2, a_3)$ to $B(b_1, b_2, b_3)$ is defined to be

$$\overrightarrow{AB} = \langle b_1 - a_1, b_2 - a_2, b_3 - a_3 \rangle,$$

and its length is simply the distance between the two points A and B.

We define addition and scalar multiplication of vectors exactly as in Section 10-3, taking into account that our vectors now have three components instead of two. The **sum** of the vectors $\mathbf{a} = \langle a_1, a_2, a_3 \rangle$ and $\mathbf{b} = \langle b_1, b_2, b_3 \rangle$ is the vector

$$\mathbf{a} + \mathbf{b} = \langle a_1 + b_1, a_2 + b_2, a_3 + b_3 \rangle. \tag{5}$$

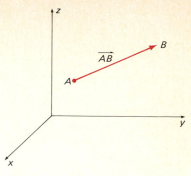

13.7 The segment AB is an instance of the vector $\overrightarrow{AB}$.

Since $\mathbf{a}$ and $\mathbf{b}$ lie in a plane (though perhaps not the xy-plane) if their initial points coincide, addition of vectors obeys the same **parallelogram law** as in the two-dimensional case (see Figure 13.8).

If c is a real number then the **scalar multiple** $c\mathbf{a}$ is the vector

$$c\mathbf{a} = \langle ca_1, ca_2, ca_3 \rangle. \tag{6}$$

The length of $c\mathbf{a}$ is $|c|$ times that of $\mathbf{a}$, and $c\mathbf{a}$ has the same direction as $\mathbf{a}$ if $c > 0$ but the opposite direction if $c < 0$. The following algebraic properties of vector addition and scalar multiplication are easy to establish; they follow from computations with components, just as in Section 10-3.

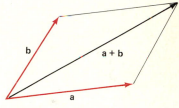

13.8 The parallelogram law for addition of vectors.

$$\begin{aligned}
&\text{(i)} &\mathbf{a} + \mathbf{b} &= \mathbf{b} + \mathbf{a}, \\
&\text{(ii)} &\mathbf{a} + (\mathbf{b} + \mathbf{c}) &= (\mathbf{a} + \mathbf{b}) + \mathbf{c}, \\
&\text{(iii)} &r(\mathbf{a} + \mathbf{b}) &= r\mathbf{a} + r\mathbf{b}, \\
&\text{(iv)} &(r + s)\mathbf{a} &= r\mathbf{a} + s\mathbf{a}, \\
&\text{(v)} &(rs)\mathbf{a} &= r(s\mathbf{a}) = s(r\mathbf{a}).
\end{aligned} \tag{7}$$

A **unit vector** is one with length 1. Any space vector can be expressed in terms of the three **basic unit vectors**

$$\mathbf{i} = \langle 1, 0, 0 \rangle, \qquad \mathbf{j} = \langle 0, 1, 0 \rangle, \qquad \mathbf{k} = \langle 0, 0, 1 \rangle.$$

When located with their initial points at the origin, these basic unit vectors form a right-handed triple of vectors pointing in the positive directions along the three coordinate axes (as shown in Figure 13.9).

Any space vector $\mathbf{a} = \langle a_1, a_2, a_3 \rangle$ can be written as

$$\mathbf{a} = a_1\mathbf{i} + a_2\mathbf{j} + a_3\mathbf{k}$$

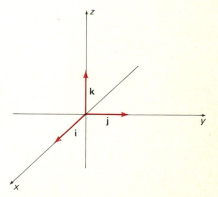

13.9 The basic unit vectors $\mathbf{i}$, $\mathbf{j}$, and $\mathbf{k}$.

in terms of the basic unit vectors. As in the two-dimensional case, the usefulness of this representation is that algebraic operations involving vectors may be carried out simply by collecting coefficients of $\mathbf{i}$, $\mathbf{j}$, and $\mathbf{k}$. For example,

$$\begin{aligned}
\mathbf{a} + \mathbf{b} &= (a_1\mathbf{i} + a_2\mathbf{j} + a_3\mathbf{k}) + (b_1\mathbf{i} + b_2\mathbf{j} + b_3\mathbf{k}) \\
&= (a_1 + b_1)\mathbf{i} + (a_2 + b_2)\mathbf{j} + (a_3 + b_3)\mathbf{k}.
\end{aligned}$$

The **dot product** of the two vectors

$$\mathbf{a} = a_1\mathbf{i} + a_2\mathbf{j} + a_3\mathbf{k} \quad \text{and} \quad \mathbf{b} = b_1\mathbf{i} + b_2\mathbf{j} + b_3\mathbf{k}$$

is also defined almost exactly as before: Multiply corresponding components, then add the results. Thus

$$\mathbf{a} \cdot \mathbf{b} = a_1 b_1 + a_2 b_2 + a_3 b_3. \tag{8}$$

If $a_3 = 0 = b_3$, then one may think of $\mathbf{a}$ and $\mathbf{b}$ as vectors in the xy-plane. Then the above definition reduces to the one given in Section 10-3 for the dot product of plane vectors. The three-dimensional dot product has the same list of properties as the two-dimensional dot product, and all those below can be established routinely by working with components.

$$\begin{align}
&\text{(i)} & \mathbf{a} \cdot \mathbf{a} &= |\mathbf{a}|^2, \\
&\text{(ii)} & \mathbf{a} \cdot \mathbf{b} &= \mathbf{b} \cdot \mathbf{a}, \\
&\text{(iii)} & \mathbf{a} \cdot (\mathbf{b} + \mathbf{c}) &= \mathbf{a} \cdot \mathbf{b} + \mathbf{a} \cdot \mathbf{c}, \\
&\text{(iv)} & (r\mathbf{a}) \cdot \mathbf{b} &= r(\mathbf{a} \cdot \mathbf{b}) = \mathbf{a} \cdot (r\mathbf{b}).
\end{align} \tag{9}$$

The significance of the dot product resides in its geometric interpretation. Let the vectors $\mathbf{a}$ and $\mathbf{b}$ be represented by the position vectors $\overrightarrow{OP}$ and $\overrightarrow{OQ}$, respectively. Then the angle θ between $\mathbf{a}$ and $\mathbf{b}$ is the angle at O in the triangle OPQ of Figure 13.10. We say that $\mathbf{a}$ and $\mathbf{b}$ are **parallel** if $\theta = 0$ or if $\theta = \pi$, and that $\mathbf{a}$ and $\mathbf{b}$ are **perpendicular** if $\theta = \pi/2$. For convenience, we regard the zero vector $\mathbf{0}$ as both parallel to *and* perpendicular to *every* vector.

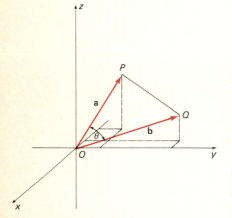

13.10 The angle θ between the vectors $\mathbf{a}$ and $\mathbf{b}$.

> **Theorem** *Interpretation of the Dot Product*
> If θ is the angle between the vectors $\mathbf{a}$ and $\mathbf{b}$, then
>
> $$\mathbf{a} \cdot \mathbf{b} = |\mathbf{a}|\,|\mathbf{b}| \cos \theta. \tag{10}$$

Proof If either $\mathbf{a} = \mathbf{0}$ or $\mathbf{b} = \mathbf{0}$, then Equation (10) follows immediately. If the vectors $\mathbf{a}$ and $\mathbf{b}$ are parallel, then $\mathbf{b} = t\mathbf{a}$ with either $t > 0$ and $\theta = 0$ or $t < 0$ and $\theta = \pi$. In either case, both sides of (10) reduce to $t|\mathbf{a}|^2$.

So we turn to the general case in which the vectors $\mathbf{a} = \overrightarrow{OP}$ and $\mathbf{b} = \overrightarrow{OQ}$ are nonzero and nonparallel. Then

$$\begin{align}
|\overrightarrow{PQ}|^2 = |\mathbf{a} - \mathbf{b}|^2 &= (\mathbf{a} - \mathbf{b}) \cdot (\mathbf{a} - \mathbf{b}) \\
&= \mathbf{a} \cdot \mathbf{a} - \mathbf{a} \cdot \mathbf{b} - \mathbf{b} \cdot \mathbf{a} + \mathbf{b} \cdot \mathbf{b};
\end{align}$$

application of the law of cosines to the triangle OPQ of Figure 13.10 yields

$$|\overrightarrow{PQ}|^2 = |\mathbf{a}|^2 + |\mathbf{b}|^2 - 2|\mathbf{a}|\,|\mathbf{b}| \cos \theta.$$

Comparison of these two expressions for $|\overrightarrow{PQ}|^2$ now gives (10).

This theorem tells us that the angle between the nonzero vectors $\mathbf{a}$ and $\mathbf{b}$ is given by

$$\cos \theta = \frac{\mathbf{a} \cdot \mathbf{b}}{|\mathbf{a}|\,|\mathbf{b}|}. \tag{11}$$

Note that this immediately implies the perpendicularity test of Section 10-3: *The two vectors $\mathbf{a}$ and $\mathbf{b}$ are perpendicular ($\theta = \pi/2$) if and only if $\mathbf{a} \cdot \mathbf{b} = 0$.*

EXAMPLE 2 Find the angles in the triangle whose vertices are $A(2, -1, 0)$, $B(5, -4, 3)$, and $C(1, -3, 2)$, shown in Figure 13.11.

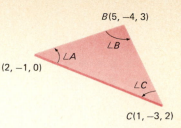

13.11 The triangle of Example 2.

Solution We apply Equation (10) with $\theta = \angle A$, $\mathbf{a} = \overrightarrow{AB} = \langle 3, -3, 3 \rangle$, and $\mathbf{b} = \overrightarrow{AC} = \langle -1, -2, 2 \rangle$. This yields

$$\angle A = \cos^{-1}\left(\frac{\langle 3, -3, 3 \rangle \cdot \langle -1, -2, 2 \rangle}{\sqrt{27}\sqrt{9}}\right)$$

$$= \cos^{-1}\left(\frac{9}{\sqrt{27}\sqrt{9}}\right) \approx 54.74°.$$

Similarly,

$$\angle B = \cos^{-1}\left(\frac{\overrightarrow{BA} \cdot \overrightarrow{BC}}{|\overrightarrow{BA}|\,|\overrightarrow{BC}|}\right) = \cos^{-1}\left(\frac{\langle -3, 3, -3 \rangle \cdot \langle -4, 1, -1 \rangle}{\sqrt{27}\sqrt{18}}\right)$$

$$= \cos^{-1}\left(\frac{18}{\sqrt{27}\sqrt{18}}\right) \approx 35.26°.$$

Then $\angle C = 180° - \angle A - \angle B = 90°$. As a check, note that

$$\overrightarrow{CA} \cdot \overrightarrow{CB} = \langle 1, 2, -2 \rangle \cdot \langle 4, -1, 1 \rangle = 0.$$

So angle C is, indeed, a right angle.

The **direction angles** of the nonzero vector $\mathbf{a} = \langle a_1, a_2, a_3 \rangle$ are the angles α, β, and γ that it makes with the vectors $\mathbf{i}$, $\mathbf{j}$, and $\mathbf{k}$, respectively, as shown in Figure 13.12. The cosines of these angles, $\cos \alpha$, $\cos \beta$, and $\cos \gamma$, are called the **direction cosines** of the vector $\mathbf{a}$. When we replace $\mathbf{b}$ in Equation (11) by $\mathbf{i}$, $\mathbf{j}$, and $\mathbf{k}$ in turn, we find that

$$\cos \alpha = \frac{\mathbf{a} \cdot \mathbf{i}}{|\mathbf{a}|\,|\mathbf{i}|} = \frac{a_1}{|\mathbf{a}|},$$

$$\cos \beta = \frac{\mathbf{a} \cdot \mathbf{j}}{|\mathbf{a}|\,|\mathbf{j}|} = \frac{a_2}{|\mathbf{a}|}, \qquad \text{and} \qquad (12)$$

$$\cos \gamma = \frac{\mathbf{a} \cdot \mathbf{k}}{|\mathbf{a}|\,|\mathbf{k}|} = \frac{a_3}{|\mathbf{a}|}.$$

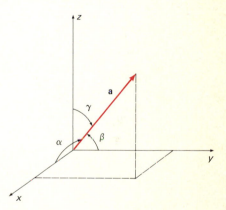

13.12 The direction angles of the vector $\mathbf{a}$.

That is, the direction cosines of $\mathbf{a}$ are the components of the **unit vector** $\mathbf{a}/|\mathbf{a}|$ with the same direction as $\mathbf{a}$. Consequently

$$\cos^2\alpha + \cos^2\beta + \cos^2\gamma = 1. \qquad (13)$$

EXAMPLE 3 Find the direction angles of the vector $\mathbf{a} = 2\mathbf{i} + 3\mathbf{j} - \mathbf{k}$.

Solution Since $|\mathbf{a}| = \sqrt{14}$, Equations (12) give $\alpha = \cos^{-1}(2/\sqrt{14}) \approx 57.69°$, $\beta = \cos^{-1}(3/\sqrt{14}) \approx 36.70°$, and $\gamma = \cos^{-1}(-1/\sqrt{14}) \approx 105.50°$.

Sometimes we need to find the component of one vector $\mathbf{a}$ in the direction of another (nonzero) vector $\mathbf{b}$. Think of the two vectors as located with the

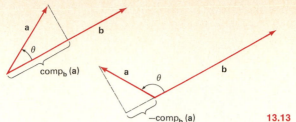

13.13 The component of **a** along **b**.

same initial point, as in Figure 13.13. Then the **component of a along b**, denoted by $comp_\mathbf{b}\ \mathbf{a}$, is numerically the length of the perpendicular projection of **a** onto the straight line determined by **b**. It is positive if the angle θ between **a** and **b** is acute (so that **a** and **b** point in the same general direction) and negative if $\theta > \pi/2$. Thus $comp_\mathbf{b}\ \mathbf{a} = |\mathbf{a}|\cos\theta$ in either case. Equation (10) then gives

$$comp_\mathbf{b}\ \mathbf{a} = \frac{|\mathbf{a}|\,|\mathbf{b}|\cos\theta}{|\mathbf{b}|} = \frac{\mathbf{a}\cdot\mathbf{b}}{|\mathbf{b}|}. \tag{14}$$

There is no real need to memorize this formula, for—in practice—we can always read $comp_\mathbf{b}\ \mathbf{a} = |\mathbf{a}|\cos\theta$ from the figure and then apply (10) to eliminate $\cos\theta$.

EXAMPLE 4 Given $\mathbf{a} = \langle 4, -5, 3\rangle$ and $\mathbf{b} = \langle 2, 1, -2\rangle$, write **a** as the sum of a vector $\mathbf{a}_\parallel$ parallel to **b** and a vector $\mathbf{a}_\perp$ perpendicular to **b**.

13.14 Construction of $\mathbf{a}_\parallel$ and $\mathbf{a}_\perp$ (see Example 4).

Solution Our method of solution is motivated by the diagram of Figure 13.14. We take

$$\mathbf{a}_\parallel = (comp_\mathbf{b}\ \mathbf{a})\frac{\mathbf{b}}{|\mathbf{b}|} = \frac{\mathbf{a}\cdot\mathbf{b}}{|\mathbf{b}|^2}\mathbf{b} = \frac{8-5-6}{9}\mathbf{b}$$

$$= -\tfrac{1}{3}\langle 2, 1, -2\rangle = \langle -\tfrac{2}{3}, -\tfrac{1}{3}, \tfrac{2}{3}\rangle,$$

and

$$\mathbf{a}_\perp = \mathbf{a} - \mathbf{a}_\parallel = \langle 4, -5, 3\rangle - \langle -\tfrac{2}{3}, -\tfrac{1}{3}, \tfrac{2}{3}\rangle = \langle \tfrac{14}{3}, -\tfrac{14}{3}, \tfrac{7}{3}\rangle.$$

The diagram makes our choice of $\mathbf{a}_\parallel$ plausible, and we have deliberately chosen $\mathbf{a}_\perp$ so that $\mathbf{a} = \mathbf{a}_\perp + \mathbf{a}_\parallel$. To verify that the vector $\mathbf{a}_\parallel$ is, indeed, parallel to **b**, we simply note that it is a scalar multiple of **b**. To verify that $\mathbf{a}_\perp$ is perpendicular to **b**, we compute the dot product:

$$\mathbf{a}_\perp \cdot \mathbf{b} = \tfrac{28}{3} - \tfrac{14}{3} - \tfrac{14}{3} = 0.$$

Thus $\mathbf{a}_\parallel$ and $\mathbf{a}_\perp$ have the required properties.

13.15 The vector force **F** is constant, but acts at an angle to the line of motion.

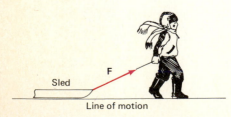

One important application of vector components is to the definition and computation of work. Recall that the work W done by a constant force F exerted along the line of motion in moving a particle a distance d is given by $W = Fd$. But what if the force is a constant vector **F** pointing in some direction other than the line of motion, as when a child pulls a sled against the resistance of friction (see Figure 13.15)? Suppose that **F** moves a particle along the line segment from P to Q, and let $\mathbf{u} = \overrightarrow{PQ}$. Then the **work** W done

CHAP. 13: Vectors, Curves, and Surfaces in Space

by **F** in moving the particle along the line from P to Q is *by definition* the product of the component of **F** along $\mathbf{u} = \overrightarrow{PQ}$ and the distance moved:

$$W = (\text{comp}_\mathbf{u}\,\mathbf{F})|\mathbf{u}|. \qquad (15)$$

EXAMPLE 5   Show that the work done by a constant force **F** in moving a particle along the line segment from P to Q is

$$W = \mathbf{F} \cdot \mathbf{D} \qquad (16)$$

where $\mathbf{D} = \overrightarrow{PQ}$ is the displacement vector. This formula is the natural vector generalization of the scalar formula $W = Fd$.

Solution   We combine Formulas (14) and (15). This gives

$$W = (\text{comp}_\mathbf{D}\,\mathbf{F})|\mathbf{D}| = \frac{\mathbf{F} \cdot \mathbf{D}}{|\mathbf{D}|}\,|\mathbf{D}| = \mathbf{F} \cdot \mathbf{D}.$$

13-1 PROBLEMS

In Problems 1–5 find: (i) $2\mathbf{a} + \mathbf{b}$; (ii) $3\mathbf{a} - 4\mathbf{b}$; (iii) $\mathbf{a} \cdot \mathbf{b}$; (iv) $|\mathbf{a} - \mathbf{b}|$; (v) $\mathbf{a}/|\mathbf{a}|$.

1 $\mathbf{a} = \langle 2, 5, -4 \rangle$ and $\mathbf{b} = \langle 1, -2, -3 \rangle$.
2 $\mathbf{a} = \langle -1, 0, 2 \rangle$ and $\mathbf{b} = \langle 3, 4, -5 \rangle$.
3 $\mathbf{a} = \mathbf{i} + \mathbf{j} + \mathbf{k}$ and $\mathbf{b} = \mathbf{j} - \mathbf{k}$.
4 $\mathbf{a} = 2\mathbf{i} - 3\mathbf{j} + 5\mathbf{k}$ and $\mathbf{b} = 5\mathbf{i} + 3\mathbf{j} - 7\mathbf{k}$.
5 $\mathbf{a} = 2\mathbf{i} - \mathbf{j}$ and $\mathbf{b} = \mathbf{j} - 3\mathbf{k}$.

6–10 Find the angle between the vectors **a** and **b** in Problems 1–5.

11–15 Find $\text{comp}_\mathbf{b}\,\mathbf{a}$ and $\text{comp}_\mathbf{a}\,\mathbf{b}$ for the vectors **a** and **b** given in Problems 1–5.

In Problems 16–20, write the equation of the indicated sphere.

16 Center $(3, 1, 2)$, radius 5.
17 Center $(-2, 1, -5)$, radius $\sqrt{7}$.
18 One diameter the segment joining $(3, 5, -3)$ and $(7, 3, 1)$.
19 Center $(4, 5, -2)$, passing through the point $(1, 0, 0)$.
20 Center $(3, -4, 3)$, tangent to the xz-plane.

In Problems 21–23, find the center and radius of the sphere having the given equation.

21 $x^2 + y^2 + z^2 + 4x - 6y = 0$.
22 $x^2 + y^2 + z^2 - 8x - 9y + 10z + 40 = 0$.
23 $3x^2 + 3y^2 + 3z^2 - 18z - 48 = 0$.

In Problems 24–30, describe the graph of the given equation in geometric terms.

24 $x = 0$.   **25** $z = 10$.
26 $xy = 0$.   **27** $xyz = 0$.
28 $x^2 + y^2 + z^2 + 7 = 0$.

29 $x^2 + y^2 + z^2 - 2x + 1 = 0$.
30 $x^2 + y^2 + z^2 - 6x + 8y + 25 = 0$.

In Problems 31–33, find the direction angles of the vector $\overrightarrow{PQ}$.

31 $P(1, -1, 0)$ and $Q(3, 4, 5)$.
32 $P(2, -3, 5)$ and $Q(1, 0, -1)$.
33 $P(-1, -2, -3)$ and $Q(5, 6, 7)$.

In Problems 34 and 35 find the work W done by the force **F** in moving a particle from P to Q.

34 $\mathbf{F} = \mathbf{i} - \mathbf{k}$; $P(0, 0, 0)$, $Q(3, 1, 0)$.
35 $\mathbf{F} = 2\mathbf{i} - 3\mathbf{j} + 5\mathbf{k}$; $P(5, 3, -4)$, $Q(-1, -2, 5)$.
36 Prove the **Cauchy-Schwarz inequality**: $|\mathbf{a} \cdot \mathbf{b}| \leq |\mathbf{a}|\,|\mathbf{b}|$ for all pairs of vectors **a** and **b**.
37 Given two arbitrary vectors **a** and **b**, prove that they must satisfy the **triangle inequality**

$$|\mathbf{a} + \mathbf{b}| \leq |\mathbf{a}| + |\mathbf{b}|.$$

38 Show that, if **a** and **b** are two arbitrary vectors, then

$$|\mathbf{a} - \mathbf{b}| \geq |\mathbf{a}| - |\mathbf{b}|.$$

(*Suggestion:* Write $\mathbf{a} = (\mathbf{a} - \mathbf{b}) + \mathbf{b}$; then apply the triangle inequality (Problem 37).)
39 Find the area of the triangle with vertices $A(1, 1, 1)$, $B(3, -2, 3)$, and $C(3, 4, 6)$.
40 Find the three angles of the triangle of Problem 39.
41 Find the angle between any longest diagonal of a cube and any edge it meets.
42 Show that the three points $P(0, -2, 4)$, $Q(1, -3, 5)$, and $R(4, -6, 8)$ lie on a single straight line. (*Suggestion:* Compare the vectors $\overrightarrow{PQ}$ and $\overrightarrow{PR}$.)

43 Show that the point M given in Formula (2) is indeed the midpoint of the line segment P_1P_2. (*Note:* You must show *both* that M is equally distant from P_1 and P_2 *and* that M lies on the line segment P_1P_2.)

44 Given vectors **a** and **b**, let $a = |\mathbf{a}|$ and $b = |\mathbf{b}|$. Show that the vector $\mathbf{c} = (b\mathbf{a} + a\mathbf{b})/(a + b)$ bisects the angle between **a** and **b**.

45 Let **a**, **b**, and **c** be three vectors in the xy-plane with **a** and **b** nonzero and not parallel. Show that there exist scalars α and β such that $\mathbf{c} = \alpha\mathbf{a} + \beta\mathbf{b}$. Begin by expressing **a**, **b**, and **c** in terms of **i** and **j**.

46 Let $ax + by + c = 0$ be the equation of the line L in the xy-plane with normal vector $\mathbf{n} = \langle a, b \rangle$. Let $P_0(x_0, y_0)$ be a point on this line and $P_1(x_1, y_1)$ a point not on L. Show that the (perpendicular) distance d from P_1 to L is

$$d = \frac{|\mathbf{n} \cdot \overrightarrow{P_0P_1}|}{|\mathbf{n}|} = \frac{|ax_1 + by_1 + c|}{\sqrt{a^2 + b^2}}.$$

47 Given the two points $A(3, -2, 4)$ and $B(5, 7, -1)$, write an equation in x, y, and z that says this: The point $P(x, y, z)$ is equally distant from the points A and B. Then simplify this equation and give a geometric description of the set of all such points $P(x, y, z)$.

48 Given the fixed point $A(1, 3, 5)$, the point $P(x, y, z)$, and the vector $\mathbf{n} = \mathbf{i} - \mathbf{j} + 2\mathbf{k}$, use the dot product to help you write an equation in x, y, and z that says this: **n** and $\overrightarrow{AP}$ are perpendicular. Then simplify this equation and give a geometric description of the set of all such points $P(x, y, z)$.

13-2

The Vector Product of Two Vectors

We often need to find a vector that is perpendicular to each of two given vectors **a** and **b**. A routine way of doing this is provided by the **vector product**, or **cross product**, $\mathbf{a} \times \mathbf{b}$ of the vectors **a** and **b**. This vector product is quite unlike the dot product $\mathbf{a} \cdot \mathbf{b}$: $\mathbf{a} \cdot \mathbf{b}$ is a scalar, while $\mathbf{a} \times \mathbf{b}$ is a vector.

The vector product of the vectors $\mathbf{a} = \langle a_1, a_2, a_3 \rangle$ and $\mathbf{b} = \langle b_1, b_2, b_3 \rangle$ can be defined by the formula

$$\mathbf{a} \times \mathbf{b} = \langle a_2b_3 - a_3b_2, a_3b_1 - a_1b_3, a_1b_2 - a_2b_1 \rangle. \tag{1}$$

Though this formula seems unmotivated, it has a redeeming feature: The product $\mathbf{a} \times \mathbf{b}$ is perpendicular both to **a** and to **b**.

> **Theorem 1** *Perpendicularity of the Vector Product*
> The vector $\mathbf{a} \times \mathbf{b}$ is perpendicular both to **a** and to **b**.

Proof We show that $\mathbf{a} \times \mathbf{b}$ is perpendicular to **a** by showing that the dot product of **a** and $\mathbf{a} \times \mathbf{b}$ is zero. With the components as in Formula (1), we find that

$$\begin{aligned}
\mathbf{a} \cdot (\mathbf{a} \times \mathbf{b}) &= a_1(a_2b_3 - a_3b_2) + a_2(a_3b_1 - a_1b_3) + a_3(a_1b_2 - a_2b_1) \\
&= a_1a_2b_3 - a_1a_3b_2 + a_2a_3b_1 - a_2a_1b_3 + a_3a_1b_2 - a_3a_2b_1 \\
&= 0.
\end{aligned}$$

A similar computation shows that $\mathbf{b} \cdot (\mathbf{a} \times \mathbf{b}) = 0$, so that $\mathbf{a} \times \mathbf{b}$ is perpendicular to the vector **b** as well.

Formula (1) need not be memorized, because there is an alternative version involving determinants that is easy to remember. Recall that a determinant of order two is defined by

$$\begin{vmatrix} a_1 & a_2 \\ b_1 & b_2 \end{vmatrix} = a_1b_2 - a_2b_1. \tag{2}$$

For example,

$$\begin{vmatrix} 2 & -1 \\ 3 & 4 \end{vmatrix} = (2)(4) - (-1)(3) = 11.$$

A determinant of order three can be defined in terms of determinants of order two:

$$\begin{vmatrix} a_1 & a_2 & a_3 \\ b_1 & b_2 & b_3 \\ c_1 & c_2 & c_3 \end{vmatrix} = a_1 \begin{vmatrix} b_2 & b_3 \\ c_2 & c_3 \end{vmatrix} - a_2 \begin{vmatrix} b_1 & b_3 \\ c_1 & c_3 \end{vmatrix} + a_3 \begin{vmatrix} b_1 & b_2 \\ c_1 & c_2 \end{vmatrix}. \tag{3}$$

Note the minus sign on the right-hand side in Formula (3).

Although a determinant of order three can be expanded along any row or column, we shall use only expansions along the first row, as in (3). For example,

$$\begin{vmatrix} 1 & 3 & -2 \\ 2 & -1 & 4 \\ -3 & 7 & 5 \end{vmatrix} = (1) \begin{vmatrix} -1 & 4 \\ 7 & 5 \end{vmatrix} - (3) \begin{vmatrix} 2 & 4 \\ -3 & 5 \end{vmatrix} + (-2) \begin{vmatrix} 2 & -1 \\ -3 & 7 \end{vmatrix}$$

$$= (1)[-5 - 28] + (-3)[10 + 12] + (-2)[14 - 3]$$

$$= -33 - 66 - 22 = -121.$$

Formula (1) for the vector product of the vectors $\mathbf{a} = a_1\mathbf{i} + a_2\mathbf{j} + a_3\mathbf{k}$ and $\mathbf{b} = b_1\mathbf{i} + b_2\mathbf{j} + b_3\mathbf{k}$ is equivalent to

$$\mathbf{a} \times \mathbf{b} = \begin{vmatrix} a_2 & a_3 \\ b_2 & b_3 \end{vmatrix} \mathbf{i} - \begin{vmatrix} a_1 & a_3 \\ b_1 & b_3 \end{vmatrix} \mathbf{j} + \begin{vmatrix} a_1 & a_2 \\ b_1 & b_2 \end{vmatrix} \mathbf{k}. \tag{4}$$

This is easy to verify by expanding the 2 by 2 determinants of the right-hand side, and noting that the three components of the right-hand side of Formula (1) result. Motivated by Formula (4), we write

$$\mathbf{a} \times \mathbf{b} = \begin{vmatrix} \mathbf{i} & \mathbf{j} & \mathbf{k} \\ a_1 & a_2 & a_3 \\ b_1 & b_2 & b_3 \end{vmatrix}. \tag{5}$$

The "symbolic determinant" in this equation is to be evaluated by expanding along its first row, just as in Equation (3) and just as though it were an ordinary determinant with real-number entries. The result of this expansion is the right-hand side of Formula (4). Note that the components of the *first* vector $\mathbf{a}$ in $\mathbf{a} \times \mathbf{b}$ constitute the *second* row of the 3 by 3 determinant, while the components of the *second* vector $\mathbf{b}$ constitute the *third* row of the determinant. The order in which the vectors $\mathbf{a}$ and $\mathbf{b}$ are written is important, because we shall see presently that $\mathbf{a} \times \mathbf{b}$ is generally *not* equal to $\mathbf{b} \times \mathbf{a}$: The vector product is *not commutative*.

Formula (5) for the vector product is the form that is most convenient for computational purposes. For example, if

$$\mathbf{a} = 3\mathbf{i} - \mathbf{j} + 2\mathbf{k} \quad \text{and} \quad \mathbf{b} = 2\mathbf{i} + 2\mathbf{j} - \mathbf{k}$$

then

$$\mathbf{a} \times \mathbf{b} = \begin{vmatrix} \mathbf{i} & \mathbf{j} & \mathbf{k} \\ 3 & -1 & 2 \\ 2 & 2 & -1 \end{vmatrix}$$

$$= \begin{vmatrix} -1 & 2 \\ 2 & -1 \end{vmatrix} \mathbf{i} - \begin{vmatrix} 3 & 2 \\ 2 & -1 \end{vmatrix} \mathbf{j} + \begin{vmatrix} 3 & -1 \\ 2 & 2 \end{vmatrix} \mathbf{k}$$

$$= (1 - 4)\mathbf{i} - (-3 - 4)\mathbf{j} + (6 - [-2])\mathbf{k};$$

thus

$$\mathbf{a} \times \mathbf{b} = -3\mathbf{i} + 7\mathbf{j} + 8\mathbf{k}.$$

You might pause to verify (using the dot product) that the vector $-3\mathbf{i} + 7\mathbf{j} + 8\mathbf{k}$ is perpendicular both to $\mathbf{a}$ and to $\mathbf{b}$.

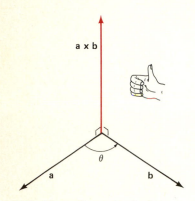

13.16 The vectors **a**, **b**, **a** × **b** form—in that order—a right-handed triple.

If the vectors **a** and **b** are located with the same initial point, then Theorem 1 tells us that $\mathbf{a} \times \mathbf{b}$ is normal to the plane determined by **a** and **b**, as indicated in Figure 13.16. There are still two possible directions for $\mathbf{a} \times \mathbf{b}$, but it turns out that if $\mathbf{a} \times \mathbf{b} \neq \mathbf{0}$, then the triple **a**, **b**, $\mathbf{a} \times \mathbf{b}$ is a **right-handed** triple in exactly the same sense as the triple **i**, **j**, **k**. Thus if the thumb of your right hand points in the direction of $\mathbf{a} \times \mathbf{b}$, your fingers curl in the direction of a rotation (less than 180°) from **a** to **b**.

Once the direction of $\mathbf{a} \times \mathbf{b}$ has been established, the vector product can be described in completely geometric terms by telling what the length $|\mathbf{a} \times \mathbf{b}|$ of the vector $\mathbf{a} \times \mathbf{b}$ is. This is given by the formula

$$|\mathbf{a} \times \mathbf{b}|^2 = |\mathbf{a}|^2 |\mathbf{b}|^2 - (\mathbf{a} \cdot \mathbf{b})^2. \tag{6}$$

This vector identity can be verified routinely (though tediously) by writing $\mathbf{a} = \langle a_1, a_2, a_3 \rangle$ and $\mathbf{b} = \langle b_1, b_2, b_3 \rangle$, computing both sides of Equation (6), and then verifying that the results are equal. That is,

$$(a_2 b_3 - a_3 b_2)^2 + (a_1 b_3 - a_3 b_1)^2 + (a_1 b_2 - a_2 b_1)^2$$
$$= (a_1^2 + a_2^2 + a_3^2)(b_1^2 + b_2^2 + b_3^2) - (a_1 b_1 + a_2 b_2 + a_3 b_3)^2.$$

Formula (6) tells us what $|\mathbf{a} \times \mathbf{b}|$ is, but the following theorem reveals the *geometric significance* of the length of the vector product.

Theorem 2 *Length of the Vector Product*

Let θ be the angle between the nonzero vectors **a** and **b** (measured so that $0 \leqq \theta \leqq \pi$). Then

$$|\mathbf{a} \times \mathbf{b}| = |\mathbf{a}| \, |\mathbf{b}| \sin \theta. \tag{7}$$

Proof We begin with Equation (6), and use the fact that $\mathbf{a} \cdot \mathbf{b} = |\mathbf{a}| \, |\mathbf{b}| \cos \theta$. Thus

$$|\mathbf{a} \times \mathbf{b}|^2 = |\mathbf{a}|^2 |\mathbf{b}|^2 - (\mathbf{a} \cdot \mathbf{b})^2$$
$$= |\mathbf{a}|^2 |\mathbf{b}|^2 - (|\mathbf{a}| \, |\mathbf{b}| \cos \theta)^2$$
$$= |\mathbf{a}|^2 |\mathbf{b}|^2 [1 - \cos^2 \theta]$$
$$= |\mathbf{a}|^2 |\mathbf{b}|^2 \sin^2 \theta.$$

Equation (7) now follows after we take the positive square root of both sides. (This is the correct root on the right-hand side because $\sin \theta \geqq 0$ for $0 \leqq \theta \leqq \pi$.)

Corollary *Parallel Vectors*
Two nonzero vectors **a** and **b** are parallel ($\theta = 0$ or $\theta = \pi$) if and only if **a** × **b** = **0**.

In particular, the cross product of any vector with itself is the zero vector. Also, Formula (1) shows immediately that the cross product of any vector with the zero vector is the zero vector again. Thus

$$\mathbf{a} \times \mathbf{a} = \mathbf{a} \times \mathbf{0} = \mathbf{0} \times \mathbf{a} = \mathbf{0} \tag{8}$$

for any vector **a**.

Equation (7) has an important geometric interpretation. Suppose that **a** and **b** are represented by adjacent sides of a parallelogram $PQRS$, with $\mathbf{a} = \overrightarrow{PQ}$ and $\mathbf{b} = \overrightarrow{PS}$, as indicated in Figure 13.17. The parallelogram then has base of length $|\mathbf{a}|$ and height $|\mathbf{b}| \sin \theta$, so its area is

$$A = |\mathbf{a}|\,|\mathbf{b}| \sin \theta = |\mathbf{a} \times \mathbf{b}|. \tag{9}$$

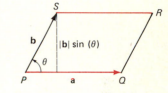

13.17 The area of the parallelogram *PQRS* is |**a** × **b**|.

Thus *the length of the vector product* **a** × **b** *is the same, numerically, as the area of the parallelogram determined by* **a** *and* **b**. It follows that the area of the triangle PQS—since it has half the area of the parallelogram—is

$$\tfrac{1}{2}A = \tfrac{1}{2}|\mathbf{a} \times \mathbf{b}| = \tfrac{1}{2}|\overrightarrow{PQ} \times \overrightarrow{PS}|. \tag{10}$$

Formula (10) gives a quick way to compute the area of a triangle—even one in space—without the need of finding any of its angles.

EXAMPLE 1 Find the area of the triangle with vertices $A(3, 0, -1)$, $B(4, 2, 5)$, and $C(7, -2, 4)$.

Solution $\overrightarrow{AB} = \langle 1, 2, 6 \rangle$ and $\overrightarrow{AC} = \langle 4, -2, 5 \rangle$, so

$$\overrightarrow{AB} \times \overrightarrow{AC} = \begin{vmatrix} \mathbf{i} & \mathbf{j} & \mathbf{k} \\ 1 & 2 & 6 \\ 4 & -2 & 5 \end{vmatrix} = 22\mathbf{i} + 19\mathbf{j} - 10\mathbf{k}.$$

So by Equation (10) the area of triangle ABC is

$$\tfrac{1}{2}\sqrt{(22)^2 + (19)^2 + (-10)^2} = \tfrac{1}{2}\sqrt{945} \approx 15.37.$$

Now let **u**, **v**, **w** be a right-handed triple of mutually perpendicular unit vectors. The angle between any two of these is $\theta = \pi/2$, so $|\mathbf{u}| = |\mathbf{v}| = |\mathbf{w}| = \sin \theta = 1$. Thus it follows from (7) that $\mathbf{u} \times \mathbf{v} = \mathbf{w}$. When we apply this observation to the basic unit vectors **i**, **j**, and **k**, we see that

$$\mathbf{i} \times \mathbf{j} = \mathbf{k}, \quad \mathbf{j} \times \mathbf{k} = \mathbf{i}, \quad \text{and} \quad \mathbf{k} \times \mathbf{i} = \mathbf{j}. \tag{11a}$$

But

$$\mathbf{j} \times \mathbf{i} = -\mathbf{k}, \quad \mathbf{k} \times \mathbf{j} = -\mathbf{i}, \quad \text{and} \quad \mathbf{i} \times \mathbf{k} = -\mathbf{j}. \tag{11b}$$

These relations, together with the fact that

$$\mathbf{i} \times \mathbf{i} = \mathbf{j} \times \mathbf{j} = \mathbf{k} \times \mathbf{k} = \mathbf{0}, \qquad (11c)$$

also follow directly from the original definition of the cross product (in the form of Equation (5)).

Note that $\mathbf{i} \times \mathbf{j} \neq \mathbf{j} \times \mathbf{i}$. *The vector product is not commutative.* Instead, it is *anticommutative*: $\mathbf{a} \times \mathbf{b} = -\mathbf{b} \times \mathbf{a}$. This is the first part of the following theorem.

Theorem 3 *Algebraic Properties of Vector Products*

If $\mathbf{a}$, $\mathbf{b}$, and $\mathbf{c}$ are vectors and k is a real number, then

(i)	$\mathbf{a} \times \mathbf{b} = -(\mathbf{b} \times \mathbf{a});$	(12)
(ii)	$(k\mathbf{a}) \times \mathbf{b} = \mathbf{a} \times (k\mathbf{b}) = k(\mathbf{a} \times \mathbf{b});$	(13)
(iii)	$\mathbf{a} \times (\mathbf{b} + \mathbf{c}) = (\mathbf{a} \times \mathbf{b}) + (\mathbf{a} \times \mathbf{c});$	(14)
(iv)	$\mathbf{a} \cdot (\mathbf{b} \times \mathbf{c}) = (\mathbf{a} \times \mathbf{b}) \cdot \mathbf{c};$ and	(15)
(v)	$\mathbf{a} \times (\mathbf{b} \times \mathbf{c}) = (\mathbf{a} \cdot \mathbf{c})\mathbf{b} - (\mathbf{a} \cdot \mathbf{b})\mathbf{c}.$	(16)

The proofs of Properties (12) through (15) are straightforward applications of the definition of the vector product in terms of components. See Problem 20 for an outline of the proof of (16).

Cross products of vectors expressed in terms of the basic unit vectors $\mathbf{i}$, $\mathbf{j}$, and $\mathbf{k}$ can be found by means of computations that closely resemble those of ordinary algebra. We simply apply the algebraic properties summarized in Theorem 3 together with the relations in (11) giving the various products of the basic unit vectors. Care must be taken to preserve the order of the factors, since vector multiplication is not commutative—although, of course, one should not hesitate to use Property (12). For example,

$$
\begin{aligned}
(\mathbf{i} - 2\mathbf{j} + 3\mathbf{k}) \times (3\mathbf{i} + 2\mathbf{j} - 4\mathbf{k}) = {} & 3(\mathbf{i} \times \mathbf{i}) + 2(\mathbf{i} \times \mathbf{j}) - 4(\mathbf{i} \times \mathbf{k}) \\
& - 6(\mathbf{j} \times \mathbf{i}) - 4(\mathbf{j} \times \mathbf{j}) + 8(\mathbf{j} \times \mathbf{k}) \\
& + 9(\mathbf{k} \times \mathbf{i}) + 6(\mathbf{k} \times \mathbf{j}) - 12(\mathbf{k} \times \mathbf{k}) \\
= {} & 3(\mathbf{0}) + 2\mathbf{k} - 4(-\mathbf{j}) - 6(-\mathbf{k}) - 4(\mathbf{0}) \\
& + 8\mathbf{i} + 9\mathbf{j} + 6(-\mathbf{i}) - 12(\mathbf{0}) \\
= {} & 2\mathbf{i} + 13\mathbf{j} + 8\mathbf{k}.
\end{aligned}
$$

SCALAR TRIPLE PRODUCTS

Let us examine the product $\mathbf{a} \cdot (\mathbf{b} \times \mathbf{c})$ appearing in Equation (15). Note first that the expression would not make sense were the parentheses around $\mathbf{a} \cdot \mathbf{b}$, since $\mathbf{a} \cdot \mathbf{b}$ is a scalar, and thus we could not form the cross product of $\mathbf{a} \cdot \mathbf{b}$ with the vector $\mathbf{c}$. This means that we may omit the parentheses; the expression $\mathbf{a} \cdot \mathbf{b} \times \mathbf{c}$ is not ambiguous. The dot product of the vectors $\mathbf{a}$ and $\mathbf{b} \times \mathbf{c}$ is a real number, called the **scalar triple product** of

the vectors **a**, **b**, and **c**. Equation (15) says that the operations · (dot) and
× (cross) can be interchanged without affecting the value of the expression:

$$\mathbf{a} \cdot \mathbf{b} \times \mathbf{c} = \mathbf{a} \times \mathbf{b} \cdot \mathbf{c}$$

for all vectors **a**, **b**, and **c**.

To compute the scalar triple product in terms of components, write
$\mathbf{a} = \langle a_1, a_2, a_3 \rangle$, $\mathbf{b} = \langle b_1, b_2, b_3 \rangle$, and $\mathbf{c} = \langle c_1, c_2, c_3 \rangle$. Then

$$\mathbf{b} \times \mathbf{c} = (b_2 c_3 - b_3 c_2)\mathbf{i} - (b_1 c_3 - b_3 c_1)\mathbf{j} + (b_1 c_2 - b_2 c_1)\mathbf{k},$$

so

$$\mathbf{a} \cdot (\mathbf{b} \times \mathbf{c}) = a_1(b_2 c_3 - b_3 c_2) - a_2(b_1 c_3 - b_3 c_1) + a_3(b_1 c_2 - b_2 c_1).$$

But the expression on the right is the value of the following 3 by 3 determinant:

$$\mathbf{a} \cdot \mathbf{b} \times \mathbf{c} = \begin{vmatrix} a_1 & a_2 & a_3 \\ b_1 & b_2 & b_3 \\ c_1 & c_2 & c_3 \end{vmatrix}. \tag{17}$$

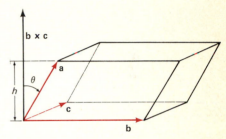

13.18 The volume of the parallelepiped is $|\mathbf{a} \cdot \mathbf{b} \times \mathbf{c}|$.

This is the quickest way to compute the scalar triple product. Its importance
for applications depends upon the following geometric interpretation.

Let **a**, **b**, and **c** be three vectors located with the same initial point. Figure
13.18 shows the parallelepiped determined by these vectors; that is, with
the vectors as adjacent edges. If the vectors **a**, **b**, and **c** are coplanar (lie in a
single plane) then the parallelepiped is *degenerate*; its volume is zero. Theorem 4 holds whether or not the three vectors are coplanar, but is of most
interest when they are not.

Theorem 4   *Scalar Triple Products and Volume*

The volume V of the parallelepiped determined by the vectors **a**, **b**, and
c is the absolute value of the scalar triple product $\mathbf{a} \cdot \mathbf{b} \times \mathbf{c}$; that is,

$$V = |\mathbf{a} \cdot \mathbf{b} \times \mathbf{c}|. \tag{18}$$

Proof   If the three vectors are coplanar, then **a** and $\mathbf{b} \times \mathbf{c}$ are perpendicular,
so $V = |\mathbf{a} \cdot \mathbf{b} \times \mathbf{c}| = 0$. Assume they are not coplanar. By Equation (9) the
area of the base (determined by **b** and **c**) of the parallelepiped is $A = |\mathbf{b} \times \mathbf{c}|$.

Now let α be the *acute* angle between **a** and the line through $\mathbf{b} \times \mathbf{c}$
that is perpendicular to the base. Then the height of the parallelepiped is
$h = |\mathbf{a}| \cos \alpha$. If θ is the angle between the vectors **a** and $\mathbf{b} \times \mathbf{c}$, then either
$\theta = \alpha$ or $\theta = \pi - \alpha$. Hence $\cos \alpha = |\cos \theta|$, so

$$V = Ah = |\mathbf{b} \times \mathbf{c}|\,|\mathbf{a}| \cos \alpha = |\mathbf{a}|\,|\mathbf{b} \times \mathbf{c}|\,|\cos \theta| = |\mathbf{a} \cdot \mathbf{b} \times \mathbf{c}|.$$

Thus we have verified Equation (18).

EXAMPLE 2   Use the scalar triple product to show that the points
$A(1, -1, 2)$, $B(2, 0, 1)$, $C(3, 2, 0)$, and $D(5, 4, -2)$ are coplanar.

Solution It is enough to show that the vectors $\overrightarrow{AB} = \langle 1, 1, -1 \rangle$, $\overrightarrow{AC} = \langle 2, 3, -2 \rangle$, and $\overrightarrow{AD} = \langle 4, 5, -4 \rangle$ are coplanar. But their scalar triple product is

$$\begin{vmatrix} 1 & 1 & -1 \\ 2 & 3 & -2 \\ 4 & 5 & -4 \end{vmatrix} = (1)(-2) - (1)(0) + (-1)(-2) = 0,$$

so Theorem 4 guarantees us that the parallelepiped determined by these three vectors has volume zero. Hence the four given points are coplanar.

The vector product occurs quite naturally in many scientific applications. For example, suppose that a body in space is free to rotate about the fixed point O. If a force $\mathbf{F}$ acts at the point P of the body, its effect is to cause rotation of the body. This effect is measured by the **torque vector** τ defined by the relation

$$\tau = \mathbf{r} \times \mathbf{F}$$

where $\mathbf{r} = \overrightarrow{OP}$. The straight line through O determined by τ is the axis of rotation, and the length $|\tau| = |\mathbf{r}|\,|\mathbf{F}|\sin\theta$ (see Figure 13.19) is the **moment** of the force $\mathbf{F}$ about this axis.

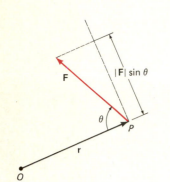

13.19 The torque vector τ is normal to both $\mathbf{r}$ and $\mathbf{F}$.

Another example is the force exerted on a moving charged particle by a magnetic field. This force is important in the cyclotron and in the television picture tube; controlling the paths of the ions is accomplished through the interplay of electric and magnetic fields. In such circumstances, the force $\mathbf{F}$ on the particle due to a magnetic field depends upon three things: The particle's charge q, its velocity vector $\mathbf{v}$, and the magnetic field vector $\mathbf{B}$ at the instantaneous location of the particle. And it turns out that

$$\mathbf{F} = (q\mathbf{v}) \times \mathbf{B}.$$

13-2 PROBLEMS

Find $\mathbf{a} \times \mathbf{b}$ in Problems 1–4.

1 $\mathbf{a} = \langle 5, -1, -2 \rangle$, $\mathbf{b} = \langle -3, 2, 4 \rangle$.
2 $\mathbf{a} = \langle 3, -2, 0 \rangle$, $\mathbf{b} = \langle 0, 3, -2 \rangle$.
3 $\mathbf{a} = \mathbf{i} - \mathbf{j} + 3\mathbf{k}$, $\mathbf{b} = -2\mathbf{i} + 3\mathbf{j} + \mathbf{k}$.
4 $\mathbf{a} = 4\mathbf{i} + 2\mathbf{j} - 2\mathbf{k}$, $\mathbf{b} = 2\mathbf{i} - 5\mathbf{j} + 5\mathbf{k}$.
5 Apply Equation (5) to verify Equations (11a).
6 Apply Equation (5) to verify Equations (11b).
7 Show that the vector product is not associative by calculating and comparing both $\mathbf{a} \times (\mathbf{b} \times \mathbf{c})$ and $(\mathbf{a} \times \mathbf{b}) \times \mathbf{c}$ with

$$\mathbf{a} = \mathbf{i}, \qquad \mathbf{b} = \mathbf{i} + \mathbf{j}, \quad \text{and} \quad \mathbf{c} = \mathbf{i} + \mathbf{j} + \mathbf{k}.$$

8 Find nonzero vectors $\mathbf{a}$, $\mathbf{b}$, and $\mathbf{c}$ such that $\mathbf{a} \times \mathbf{b} = \mathbf{a} \times \mathbf{c}$ but $\mathbf{b} \neq \mathbf{c}$.
9 Suppose that the three vectors $\mathbf{a}$, $\mathbf{b}$, and $\mathbf{c}$ are mutually perpendicular. Show that $\mathbf{a} \times (\mathbf{b} \times \mathbf{c}) = \mathbf{0}$.
10 Find the area of the triangle with vertices $P(1, 1, 0)$, $Q(1, 0, 1)$, and $R(0, 1, 1)$.

11 Find the area of the triangle with vertices $P(1, 3, -2)$, $Q(2, 4, 5)$, and $R(-3, -2, 2)$.
12 Find the volume of the parallelepiped with adjacent edges OP, OQ, and OR, where P, Q, and R are the points given in Problem 10.
13 (a) Find the volume of the parallelepiped with adjacent edges OP, OQ, and OR, where P, Q, and R are the points of Problem 11.
(b) Find the volume of the tetrahedron with vertices O, P, Q, and R.
14 Find a unit vector $\mathbf{n}$ perpendicular to the plane through the three points P, Q, and R of Problem 11. Then find the distance from the origin to this plane by computing $\mathbf{n} \cdot \overrightarrow{OP}$.
15 Apply Formula (5) to verify Equation (12), the anticommutativity of the vector product.
16 Apply Formula (17) to verify Identity (15) for scalar triple products.
17 Suppose that P and Q are points on a line L in space. Let A be a point not on L. Calculate in two ways the area

of the triangle APQ to show that the perpendicular distance from A to the line L is $d = |\overrightarrow{AP} \times \overrightarrow{AQ}|/|\overrightarrow{PQ}|$. Then use this formula to compute the distance from the point $A(1, 0, 1)$ to the line through the two points $P(2, 3, 1)$ and $Q(-3, 1, 4)$.

18 Suppose that A is a point not on the plane determined by the three points P, Q, and R. Calculate in two ways the volume of the tetrahedron $APQR$ to show that the perpendicular distance from A to this plane is

$$d = \frac{|\overrightarrow{AP} \cdot \overrightarrow{AQ} \times \overrightarrow{AR}|}{|\overrightarrow{PQ} \times \overrightarrow{PR}|}.$$

Use this formula to compute the distance from the point $A(1, 0, 1)$ to the plane through the points $P(2, 3, 1)$, $Q(3, -1, 4)$, and $R(0, 0, 2)$.

19 Suppose that P_1 and Q_1 are two points on the line L_1, and that P_2 and Q_2 are two points on the line L_2. If the lines L_1 and L_2 are not parallel, then the perpendicular distance d between them is the projection of $\overrightarrow{P_1P_2}$ onto a vector $\mathbf{n}$ that is perpendicular both to $\overrightarrow{P_1Q_1}$ and to $\overrightarrow{P_2Q_2}$. Show that

$$d = \frac{|\overrightarrow{P_1P_2} \cdot \overrightarrow{P_1Q_1} \times \overrightarrow{P_2Q_2}|}{|\overrightarrow{P_1Q_1} \times \overrightarrow{P_2Q_2}|}.$$

20 Use the following method to establish that the **vector triple product** $(\mathbf{a} \times \mathbf{b}) \times \mathbf{c}$ is equal to $(\mathbf{a} \cdot \mathbf{c})\mathbf{b} - (\mathbf{b} \cdot \mathbf{c})\mathbf{a}$.
(i) Let $\mathbf{I}$ be a unit vector in the direction of $\mathbf{a}$, and let $\mathbf{J}$ be a unit vector perpendicular to $\mathbf{I}$ and parallel to the plane of $\mathbf{a}$ and $\mathbf{b}$. Let $\mathbf{K} = \mathbf{I} \times \mathbf{J}$. Explain why there are scalars a_1, b_1, b_2, c_1, c_2, and c_3 such that

$$\mathbf{a} = a_1\mathbf{I},$$

$$\mathbf{b} = b_1\mathbf{I} + b_2\mathbf{J},$$

$$\mathbf{c} = c_1\mathbf{I} + c_2\mathbf{J} + c_3\mathbf{K}.$$

(ii) Now show that

$$(\mathbf{a} \times \mathbf{b}) \times \mathbf{c} = -a_1b_2c_2\mathbf{I} + a_1b_2c_1\mathbf{J}.$$

(iii) Finally, substitute for $\mathbf{I}$ and $\mathbf{J}$ in terms of $\mathbf{a}$ and $\mathbf{b}$.

21 By permutation of the vectors $\mathbf{a}$, $\mathbf{b}$, and $\mathbf{c}$, deduce from Problem 20 that

$$\mathbf{a} \times (\mathbf{b} \times \mathbf{c}) = (\mathbf{a} \cdot \mathbf{c})\mathbf{b} - (\mathbf{a} \cdot \mathbf{b})\mathbf{c}$$

[Formula (16)].

22 Deduce from the orthogonality properties of the vector product that the vector $(\mathbf{a} \times \mathbf{b}) \times (\mathbf{c} \times \mathbf{d})$ can be written both in the form $r_1\mathbf{a} + r_2\mathbf{b}$ and in the form $s_1\mathbf{c} + s_2\mathbf{d}$.

23 Consider the triangle in the xy-plane that has vertices $(x_1, y_1, 0)$, $(x_2, y_2, 0)$, and $(x_3, y_3, 0)$. Use the cross product to show that the area of this triangle is *half* the *absolute value* of the determinant

$$\begin{vmatrix} 1 & 1 & 1 \\ x_1 & x_2 & x_3 \\ y_1 & y_2 & y_3 \end{vmatrix}.$$

Lines and Planes in Space

A straight line in space is determined by any two points P_0 and P_1 on it. Alternatively, a line in space can be specified by giving a point P_0 on it *and* a vector, such as $\overrightarrow{P_0P_1}$, that determines the direction of the line.

To investigate equations describing lines in space, let us begin with a straight line L that passes through the point $P_0(x_0, y_0, z_0)$ and is parallel to the vector $\mathbf{v} = a\mathbf{i} + b\mathbf{j} + c\mathbf{k}$ (Figure 13.20). Then another point $P(x, y, z)$ lies on the line L if and only if the vectors $\mathbf{v}$ and $\overrightarrow{P_0P}$ are parallel, in which case

$$\overrightarrow{P_0P} = t\mathbf{v} \tag{1}$$

for some real number t. If $\mathbf{r}_0 = \overrightarrow{OP_0}$ and $\mathbf{r} = \overrightarrow{OP}$ are the position vectors of the points P_0 and P, respectively, then $\overrightarrow{P_0P} = \mathbf{r} - \mathbf{r}_0$. Hence (1) gives the *vector equation*

$$\mathbf{r} = \mathbf{r}_0 + t\mathbf{v} \tag{2}$$

describing the line L.

The left-hand and right-hand sides are equal in Equation (2), and each side is a vector. So corresponding components are also equal. When we write the resulting equations, we get a scalar description of the line L. Since

13.20 Finding the equation of the line L through the point P_0 parallel to the vector $\mathbf{v}$.

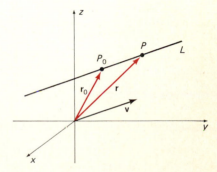

$\mathbf{r}_0 = \langle x_0, y_0, z_0 \rangle$ and $\mathbf{r} = \langle x, y, z \rangle$, Equation (2) thereby yields the three scalar equations

$$\left. \begin{aligned} x &= x_0 + at, \\ y &= y_0 + bt, \\ z &= z_0 + ct. \end{aligned} \right\} \qquad (3)$$

These are **parametric equations** of the line L through the point (x_0, y_0, z_0) parallel to the vector $\mathbf{v} = \langle a, b, c \rangle$.

Note that the parametric equations of a straight line are not unique. For example, if $P_1(x_1, y_1, z_1)$ is another point of L, then the equations

$$x = x_1 + a \tan \psi,$$

$$y = y_1 + b \tan \psi,$$

$$z = z_1 + c \tan \psi, \qquad -\frac{\pi}{2} < \psi < \frac{\pi}{2}$$

provide another parametric description of the same line, because $t = \tan \psi$ traverses the whole real line as ψ traverses the interval $(-\pi/2, \pi/2)$.

Given two straight lines L_1 and L_2 with parametric equations

$$x = x_1 + a_1 t, \qquad y = y_1 + b_1 t, \qquad z = z_1 + c_1 t \qquad (4)$$

and

$$x = x_2 + a_2 s, \qquad y = y_2 + b_2 s, \qquad z = z_2 + c_2 s, \qquad (5)$$

respectively, we can see at a glance whether or not L_1 and L_2 are parallel. Since L_1 is parallel to $\mathbf{v}_1 = \langle a_1, b_1, c_1 \rangle$ and L_2 is parallel to $\mathbf{v}_2 = \langle a_2, b_2, c_2 \rangle$, it follows that the lines L_1 and L_2 are parallel if and only if the vectors $\mathbf{v}_1$ and $\mathbf{v}_2$ are scalar multiples of each other. If the two lines are not parallel, we can attempt to find a point of intersection by solving the equations

$$x_1 + a_1 t = x_2 + a_2 s \quad \text{and} \quad y_1 + b_1 t = y_2 + b_2 s$$

simultaneously for s and t. If these values satisfy the equation $z_1 + c_1 t = z_2 + c_2 s$, then we have found a point of intersection (whose coordinates are obtained by substitution of the resulting value of t in Equations (4) or that of s in Equations (5)). Otherwise, the two lines L_1 and L_2 do not intersect; two nonparallel and nonintersecting lines in space are called **skew** lines.

If the coefficients a, b, and c in Equations (3) are all nonzero, then we can eliminate the parameter by equating the three expressions obtained by solving for t. This gives

$$\frac{x - x_0}{a} = \frac{y - y_0}{b} = \frac{z - z_0}{c}. \qquad (6)$$

These are called the **symmetric equations** of the line L. If one of a or b or c *is* zero, this means that L lies in a plane parallel to one of the coordinate planes, in which case the line does not have symmetric equations. For example, if $c = 0$, then L lies in the horizontal plane $z = z_0$. Of course, it is still possible to write equations for L not involving the parameter; if $c = 0$ while a and b are nonzero, we could describe the line L as the simultaneous solution of the equations

$$\frac{x - x_0}{a} = \frac{y - y_0}{b}, \qquad z = z_0.$$

EXAMPLE 1 Find both parametric equations and symmetric equations of the line L through the points $P_0(3, 1, -2)$ and $P_1(4, -1, 1)$. Also find the points in which L intersects the three coordinate planes.

Solution The line L is parallel to the vector $\mathbf{v} = \overrightarrow{P_0 P_1} = \langle 1, -2, 3 \rangle$, so we take $a = 1$, $b = -2$, and $c = 3$. Equations (3) then give the parametric equations

$$x = 3 + t, \qquad y = 1 - 2t, \qquad z = -2 + 3t$$

of L, while Equations (6) give the symmetric equations

$$\frac{x - 3}{1} = \frac{y - 1}{-2} = \frac{z + 2}{3}.$$

To find the point at which L intersects the xy-plane, we set $z = 0$ in the symmetric equations. This gives

$$\frac{x - 3}{1} = \frac{y - 1}{-2} = \frac{2}{3},$$

and so $x = \frac{11}{3}$, $y = -\frac{1}{3}$. Thus L meets the xy-plane at the point $(\frac{11}{3}, -\frac{1}{3}, 0)$. Similarly, $x = 0$ gives $(0, 7, -11)$ for the point where L meets the yz-plane, and $y = 0$ gives $(\frac{7}{2}, 0, -\frac{1}{2})$ for its intersection with the xz-plane.

A **plane** $\mathscr{P}$ in space is determined by a point $P_0(x_0, y_0, z_0)$ through which $\mathscr{P}$ passes and a line through P_0 that is normal to $\mathscr{P}$. Alternatively, we may be given P_0 on $\mathscr{P}$ and a normal vector $\mathbf{n} = \langle a, b, c \rangle$ to the plane $\mathscr{P}$. Then the point $P(x, y, z)$ lies on the plane $\mathscr{P}$ if and only if the vectors $\mathbf{n}$ and $\overrightarrow{P_0 P}$ are perpendicular (see Figure 13.21), in which case $\mathbf{n} \cdot \overrightarrow{P_0 P} = 0$. We write $\overrightarrow{P_0 P} = \mathbf{r} - \mathbf{r}_0$ where $\mathbf{r}$ and $\mathbf{r}_0$ are the position vectors $\mathbf{r} = \overrightarrow{OP}$ and $\mathbf{r}_0 = \overrightarrow{OP_0}$ of the points P and P_0, respectively. Thus we obtain a **vector equation**

$$\mathbf{n} \cdot (\mathbf{r} - \mathbf{r}_0) = 0 \qquad (7)$$

of the plane $\mathscr{P}$.

If we substitute $\mathbf{n} = \langle a, b, c \rangle$, $\mathbf{r} = \langle x, y, z \rangle$, and $\mathbf{r}_0 = \langle x_0, y_0, z_0 \rangle$ in Equation (7), we thereby obtain a **scalar equation**

$$a(x - x_0) + b(y - y_0) + c(z - z_0) = 0 \qquad (8)$$

of the plane through $P_0(x_0, y_0, z_0)$ **with normal vector** $\mathbf{n} = \langle a, b, c \rangle$. For example, an equation of the plane through $P_0(-1, 5, 2)$ with normal vector $\mathbf{n} = \langle 1, -3, 2 \rangle$ is

$$(1)(x + 1) + (-3)(y - 5) + (2)(z - 2) = 0;$$

that is,

$$x - 3y + 2z = -12.$$

Note that the coefficients of x, y, and z in the last equation are the components of the normal vector. This is always the case, for Equation (8)

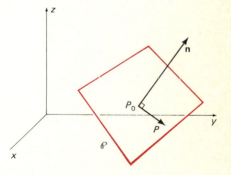

13.21 Because $\mathbf{n}$ is normal to $\mathscr{P}$, it follows that $\mathbf{n}$ is normal to $\overrightarrow{P_0 P}$ for all points P in $\mathscr{P}$.

can be written in the form

$$ax + by + cz = d \tag{9}$$

where $d = ax_0 + by_0 + cz_0$. Conversely, every *linear equation* in x, y, and z of the form in (9) represents a plane in space provided that the coefficients a, b, and c are not all zero. For if $c \neq 0$ (for instance), we can pick x_0 and y_0 arbitrarily and solve the equation $ax_0 + by_0 + cz_0 = d$ for z_0. With these values, Equation (9) takes the form

$$ax + by + cz = ax_0 + by_0 + cz_0$$

or

$$a(x - x_0) + b(y - y_0) + c(z - z_0) = 0,$$

so this equation represents the plane through (x_0, y_0, z_0) with normal vector $\langle a, b, c \rangle$.

EXAMPLE 2 Find an equation for the plane through the three points $P(2, 4, -3)$, $Q(3, 7, -1)$, and $R(4, 3, 0)$.

Solution We want to use Equation (8), so we first need a vector $\mathbf{n}$ that is normal to the plane in question. One easy way to obtain such a normal vector is through the cross product: Let

$$\mathbf{n} = \vec{PQ} \times \vec{PR} = \begin{vmatrix} \mathbf{i} & \mathbf{j} & \mathbf{k} \\ 1 & 3 & 2 \\ 2 & -1 & 3 \end{vmatrix} = 11\mathbf{i} + \mathbf{j} - 7\mathbf{k}.$$

Because $\vec{PQ}$ and $\vec{PR}$ are in the plane, their cross product $\mathbf{n}$ is normal to the plane. Hence our plane has equation

$$11(x - 2) + (y - 4) - 7(z + 3) = 0.$$

After simplifications, we write it as

$$11x + y - 7z = 47.$$

Two planes with normal vectors $\mathbf{n}$ and $\mathbf{m}$ are called **parallel** provided that $\mathbf{n}$ and $\mathbf{m}$ are parallel. Otherwise the two planes meet in a straight line. We define the angle between the two planes to be the angle between their normal vectors $\mathbf{n}$ and $\mathbf{m}$.

EXAMPLE 3 Find the angle θ between the planes with equations

$$2x + 3y - z = -3 \quad \text{and} \quad 4x + 5y + z = 1.$$

Then write symmetric equations of their line of intersection L.

Solution The vectors $\mathbf{n} = \langle 2, 3, -1 \rangle$ and $\mathbf{m} = \langle 4, 5, 1 \rangle$ are normal to the two planes, so

$$\cos \theta = \frac{\mathbf{n} \cdot \mathbf{m}}{|\mathbf{n}| \, |\mathbf{m}|} = \frac{22}{\sqrt{14}\sqrt{42}}.$$

Hence $\theta = \cos^{-1}(11\sqrt{3/21}) \approx 24.87°$.

To determine the line L of intersection of the two planes, we need first to find a point P_0 lying on L. We can do this by substituting an arbitrarily chosen value of x into the equations of the given planes and then solving the resulting two equations for y and z. With $x = 1$ we get the equations

$$2 + 3y - z = -3,$$

$$4 + 5y + z = 1.$$

The common solution is $y = -1$, $z = 2$; thus the point $P_0(1, -1, 2)$ lies on the line L.

Next we need a vector $\mathbf{v}$ parallel to L. The normal vectors $\mathbf{n}$ and $\mathbf{m}$ to the two planes are both perpendicular to L, so their cross product will be parallel to L. Hence we may choose

$$\mathbf{v} = \mathbf{n} \times \mathbf{m} = \begin{vmatrix} \mathbf{i} & \mathbf{j} & \mathbf{k} \\ 2 & 3 & -1 \\ 4 & 5 & 1 \end{vmatrix} = 8\mathbf{i} - 6\mathbf{j} - 2\mathbf{k}.$$

From (6) we now find the symmetric equations

$$\frac{x - 1}{8} = \frac{y + 1}{-6} = \frac{z - 2}{-2}$$

of the line of intersection of the two given planes.

In conclusion, we note that the symmetric equations of a line L exhibit the line as an intersection of planes. For we can rewrite Equations (6) in the form

$$b(x - x_0) - a(y - y_0) = 0,$$

$$c(x - x_0) - a(z - z_0) = 0, \quad \text{and} \tag{10}$$

$$c(y - y_0) - b(z - z_0) = 0.$$

These are the equations of three planes that intersect in the line L. The first has normal vector $\langle b, -a, 0 \rangle$, a vector parallel to the xy-plane. So the first plane is perpendicular to the xy-plane. The second plane is perpendicular to the xz-plane, and the third is perpendicular to the yz-plane.

Equations (10) are symmetric equations of the line through $P_0(x_0, y_0, z_0)$ parallel to $\mathbf{v} = \langle a, b, c \rangle$. They enjoy the advantage over Equations (6) of being meaningful whether or not the components a, b, and c of $\mathbf{v}$ are all nonzero. They have a special form, though, if one of these components is zero. If (for instance) $a = 0$, then the first two equations in (10) take the form $x = x_0$. The line is then the intersection of the two planes $x = x_0$ and $c(y - y_0) = b(z - z_0)$.

13-3 PROBLEMS

In Problems 1–6, write both parametric and symmetric equations for the indicated straight line.

1 Through $P(2, 3, -4)$ and parallel to $\mathbf{v} = \langle 1, -1, -2 \rangle$.
2 Through $P(2, 5, -7)$ and $Q(4, 3, 8)$.

3 Through $P(1, 1, 1)$ and perpendicular to the xy-plane.
4 Through the origin and perpendicular to the plane with equation $x + y + z = 1$.
5 Through $P(2, -3, 4)$ and perpendicular to the plane with equation $2x - y + 3z = 4$.

6 Through $P(2, -1, 5)$ and parallel to the line with parametric equations $x = 3t$, $y = 2 + t$, $z = 2 - t$.

In Problems 7–14, write an equation of the indicated plane.

7 Through $P(5, 7, -6)$ parallel to the xz-plane.
8 Through $P(1, 0, -1)$ with normal vector $\mathbf{n} = \langle 2, 2, -1 \rangle$.
9 Through $P(10, 4, -3)$ with normal vector $\mathbf{n} = \langle 7, 11, 0 \rangle$.
10 Through $P(1, -3, 2)$ with normal vector $\mathbf{n} = \vec{OP}$.
11 Through the origin and parallel to the plane with equation $3x + 4y = z + 10$.
12 Through $P(5, 1, 4)$ and parallel to the plane with equation $x + y - 2z = 0$.
13 Through the origin and the points $P(1, 1, 1)$ and $Q(1, -1, 3)$.
14 Through the points $A(1, 0, -1)$, $B(3, 3, 2)$, and $C(4, 5, -1)$.

In Problems 15–17, find the angle between the given planes.

15 $x = 10$ and $x + y + z = 0$.
16 $2x - y + z = 5$ and $x + y - z = 1$.
17 $x - y - 2z = 1$ and $x - y - 2z = 5$.
18 Find parametric equations of the line of intersection of the planes $2x + y + z = 4$ and $3x - y + z = 3$.
19 Write symmetric equations for the line through $P(3, 3, 1)$ that is parallel to the line of Problem 18.
20 Find an equation of the plane through $P(3, 3, 1)$ that is perpendicular to the planes $x + y = 2z$ and $2x + z = 10$.
21 Find an equation of the plane through $(1, 1, 1)$ that intersects the xy-plane in the same line as does the plane $3x + 2y - z = 6$.
22 Find an equation for the plane through the point $P(1, 3, -2)$ and the line of intersection of the planes
$$x - y + z = 1 \text{ and } x + y - z = 1.$$
23 Find an equation of the plane through the points $P(1, 0, -1)$ and $Q(2, 1, 0)$ that is also parallel to the line of intersection of the planes $x + y + z = 5$ and $3x - y = 4$.
24 Show that the lines $x - 1 = \frac{1}{2}(y + 1) = z - 2$ and $x - 2 = \frac{1}{3}(y - 2) = \frac{1}{2}(z - 4)$ intersect. Find an equation of the (only) plane containing them both.

25 Show that the line of intersection of the planes
$$x + 2y - z = 2 \text{ and } 3x + 2y + 2z = 7$$
is parallel to the line $x = 1 + 6t$, $y = 3 - 5t$, $z = 2 - 4t$. Find an equation of the plane determined by these two lines.
26 Show that the perpendicular distance D from the point $P_0(x_0, y_0, z_0)$ to the plane $ax + by + cz = d$ is
$$D = \frac{|ax_0 + by_0 + cz_0 - d|}{\sqrt{a^2 + b^2 + c^2}}.$$

(*Suggestion:* The line through P_0 perpendicular to the given plane has parametric equations $x = x_0 + at$, $y = y_0 + bt$, and $z = z_0 + ct$. Let $P_1(x_1, y_1, z_1)$ be the point of this line, corresponding to $t = t_1$, at which it intersects the given plane. Solve for t_1, and then compute $D = |\vec{P_0P_1}|$.)

In Problems 27 and 28, use the formula of Problem 26 to find the distance between the given point and the given plane.

27 The origin and the plane $x + y + z = 10$.
28 The point $P(5, 12, -13)$ and the plane with equation $3x + 4y + 5z = 12$.
29 Show that any two skew lines L_1 and L_2 lie in parallel planes.
30 Use the formula of Problem 26 to show that the perpendicular distance D between the two parallel planes $ax + by + cz + d_1 = 0$ and $ax + by + cz + d_2 = 0$ is
$$D = \frac{|d_1 - d_2|}{\sqrt{a^2 + b^2 + c^2}}.$$

31 The line L_1 is described by the equations
$$x - 1 = 2y + 2; \quad z = 4.$$

The line L_2 passes through the points $P(2, 1, -3)$ and $Q(0, 8, 4)$.
(a) Show that L_1 and L_2 are skew lines.
(b) Use the results of Problems 29 and 30 to find the perpendicular distance between L_1 and L_2.

13-4

Curves and Motion in Space

In Sections 10-4 and 10-6, we used vector-valued functions to discuss curves and motion in the plane. Much of that discussion applies, with only minor changes, to curves and motion in three-dimensional space. The principal difference is that our vectors now have three components rather than two.

Think of a point that moves along a curve in space. Its position at time t can be described by *parametric equations*

$$x = f(t), \quad y = g(t), \quad z = h(t)$$

that give its coordinates at time t. Alternatively, the location of the point can be given by its **position vector**

$$\mathbf{r}(t) = f(t)\mathbf{i} + g(t)\mathbf{j} + h(t)\mathbf{k} = x\mathbf{i} + y\mathbf{j} + z\mathbf{k} \qquad (1)$$

shown in Figure 13.22.

A three-dimensional vector-valued function like $\mathbf{r}(t)$ can be differentiated and integrated in a componentwise manner, just like a two-dimensional vector-valued function (see Theorem 1 in Section 10-4). Thus the **velocity** vector $\mathbf{v} = \mathbf{v}(t)$ of our moving point at time t is given by

$$\mathbf{v}(t) = \mathbf{r}'(t) = f'(t)\mathbf{i} + g'(t)\mathbf{j} + h'(t)\mathbf{k},$$

$$\mathbf{v} = \frac{d\mathbf{r}}{dt} = \frac{dx}{dt}\mathbf{i} + \frac{dy}{dt}\mathbf{j} + \frac{dz}{dt}\mathbf{k}. \qquad (2)$$

Its **acceleration** vector $\mathbf{a} = \mathbf{a}(t)$ is given by

$$\mathbf{a}(t) = \mathbf{v}'(t) = f''(t)\mathbf{i} + g''(t)\mathbf{j} + h''(t)\mathbf{k},$$

$$\mathbf{a} = \frac{d\mathbf{v}}{dt} = \frac{d^2x}{dt^2}\mathbf{i} + \frac{d^2y}{dt^2}\mathbf{j} + \frac{d^2z}{dt^2}\mathbf{k}. \qquad (3)$$

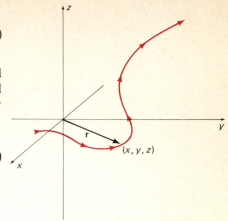

13.22 The position vector $\mathbf{r} = \langle x, y, z \rangle$ of a moving particle in space.

The **speed** $v(t)$ and **scalar acceleration** $a(t)$ of the moving point are the lengths of its velocity and acceleration vectors, respectively:

$$v(t) = |\mathbf{v}(t)| = \sqrt{\left(\frac{dx}{dt}\right)^2 + \left(\frac{dy}{dt}\right)^2 + \left(\frac{dz}{dt}\right)^2} \qquad (4)$$

and

$$a(t) = |\mathbf{a}(t)| = \sqrt{\left(\frac{d^2x}{dt^2}\right)^2 + \left(\frac{d^2y}{dt^2}\right)^2 + \left(\frac{d^2z}{dt^2}\right)^2}. \qquad (5)$$

EXAMPLE 1 The parametric equations of a moving point are

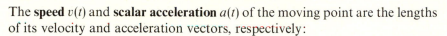

$$x = a \cos \omega t, \qquad y = a \sin \omega t, \qquad z = bt,$$

where a, b, and ω are positive. Describe the path of the point in geometric terms. Compute its velocity, speed, and acceleration at time t.

13.23 The particle of Example 1 moves in a helical path.

Solution Since

$$x^2 + y^2 = a^2 \cos^2\omega t + a^2 \sin^2\omega t = a^2,$$

the path of the moving point lies on the **cylinder** that stands above and below the circle $x^2 + y^2 = a^2$ in the xy-plane, extending infinitely far in both directions as in Figure 13.23. By our discussion of uniform circular motion in Section 10-5, the given parametric equations for x and y tell us that the point's projection in the xy-plane moves counterclockwise around the circle $x^2 + y^2 = a^2$ with angular speed ω. Meanwhile, since $z = bt$, the point itself is rising with vertical speed b. Its path on the cylinder is a spiral called a **helix** (also shown in Figure 13.23).

The derivative of the position vector

$$\mathbf{r}(t) = \mathbf{i}a \cos \omega t + \mathbf{j}a \sin \omega t + \mathbf{k}bt$$

of the moving point is its velocity vector

$$\mathbf{v}(t) = -\mathbf{i}a\omega \sin \omega t + \mathbf{j}a\omega \cos \omega t + b\mathbf{k}. \qquad (6)$$

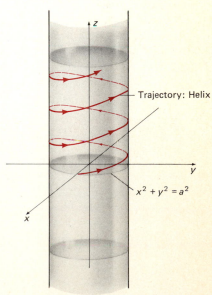

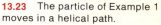

Trajectory: Helix

$x^2 + y^2 = a^2$

Another differentiation gives its acceleration vector

$$\mathbf{a}(t) = -\mathbf{i}a\omega^2 \cos \omega t - \mathbf{j}a\omega^2 \sin \omega t$$
$$= -a\omega^2(\mathbf{i} \cos \omega t + \mathbf{j} \sin \omega t). \qquad (7)$$

The speed of the moving point is a constant, for

$$v(t) = |\mathbf{v}(t)| = \sqrt{a^2\omega^2 + b^2}.$$

Note that the acceleration vector is a horizontal vector of length $a\omega^2$. Moreover, if we think of $\mathbf{a}(t)$ as attached to the moving point at the time t of evaluation—so that the initial point of $\mathbf{a}(t)$ is the terminal point of $\mathbf{r}(t)$—then $\mathbf{a}(t)$ points directly toward the point $(0, 0, bt)$ on the z-axis.

The helix of Example 1 is a typical trajectory of a charged particle in a constant magnetic field. Such a particle must satisfy both Newton's law $\mathbf{F} = m\mathbf{a}$ and the magnetic force law $\mathbf{F} = q\mathbf{v} \times \mathbf{B}$ mentioned in Section 13-2. Hence its velocity and acceleration vectors must satisfy the equation

$$q\mathbf{v} \times \mathbf{B} = m\mathbf{a}. \qquad (8)$$

If the constant magnetic field is vertical, $\mathbf{B} = B\mathbf{k}$, then with the velocity vector of Equation (6), we find that

$$q\mathbf{v} \times \mathbf{B} = q \begin{vmatrix} \mathbf{i} & \mathbf{j} & \mathbf{k} \\ -a\omega \sin \omega t & a\omega \cos \omega t & b \\ 0 & 0 & B \end{vmatrix}$$

$$= qa\omega B(\mathbf{i} \cos \omega t + \mathbf{j} \sin \omega t).$$

The acceleration vector of Equation (7) gives

$$m\mathbf{a} = -ma\omega^2(\mathbf{i} \cos \omega t + \mathbf{j} \sin \omega t).$$

When we compare these results, we see that the helix of Example 1 satisfies Equation (8) provided that

$$qa\omega B = -ma\omega^2, \qquad \text{or} \qquad \omega = -\frac{qB}{m}.$$

For example, this equation would determine the angular speed ω for the helical trajectory of electrons ($q < 0$) in a cathode-ray tube placed in a constant magnetic field parallel to the axis of the tube (see Figure 13.24).

Differentiation of three-dimensional vector-valued functions satisfies the same formal properties that we listed for two-dimensional vector-valued functions in Theorem 2 of Section 10-4:

13.24 A spiraling electron in a cathode-ray tube.

$$\begin{array}{lll} \text{(i)} & D_t[\mathbf{u}(t) + \mathbf{v}(t)] = \mathbf{u}'(t) + \mathbf{v}'(t), & \\ \text{(ii)} & D_t[h(t)\mathbf{u}(t)] \quad = h'(t)\mathbf{u}(t) + h(t)\mathbf{u}'(t), & \\ \text{(iii)} & D_t[\mathbf{u}(t) \cdot \mathbf{v}(t)] \ = \mathbf{u}'(t) \cdot \mathbf{v}(t) + \mathbf{u}(t) \cdot \mathbf{v}'(t). & \end{array} \right\} \quad (9)$$

Here, $\mathbf{u}(t)$ and $\mathbf{v}(t)$ are vector-valued functions with differentiable component functions, and $h(t)$ is a differentiable real-valued function. In addition, the expected product rule for the derivative of a cross product also holds:

$$\text{(iv)} \qquad D_t[\mathbf{u}(t) \times \mathbf{v}(t)] = \mathbf{u}'(t) \times \mathbf{v}(t) \ + \ \mathbf{u}(t) \times \mathbf{v}'(t). \qquad (10)$$

CHAP. 13: **Vectors, Curves, and Surfaces in Space**

Note that the order of the factors in Equation (10) *must* be preserved because the cross product is not commutative. Each of the formal properties in (9) and (10) can be verified routinely by componentwise differentiation, as in Section 10-4.

The arc length formula for parametric space curves is analogous to the arc length formula for parametric plane curves (Equation (6) in Section 10-2). The arc length s along the smooth curve with position vector

$$\mathbf{r}(t) = f(t)\mathbf{i} + g(t)\mathbf{j} + h(t)\mathbf{k} = x\mathbf{i} + y\mathbf{j} + z\mathbf{k}$$

from the point $\mathbf{r}(a)$ to the point $\mathbf{r}(b)$ is

$$
\begin{aligned}
s &= \int_a^b \sqrt{[x'(t)]^2 + [y'(t)]^2 + [z'(t)]^2}\, dt \\
&= \int_a^b \sqrt{\left(\frac{dx}{dt}\right)^2 + \left(\frac{dy}{dt}\right)^2 + \left(\frac{dz}{dt}\right)^2}\, dt.
\end{aligned}
\tag{11}
$$

From Equation (4) we see that the integrand is the speed of the moving point, so

$$s = \int_a^b v(t)\, dt. \tag{12}$$

EXAMPLE 2   Find the arc length of one turn of the helix of Example 1, from $t = 0$ to $t = 2\pi/\omega$.

Solution   In Example 1 we found that $v(t) = \sqrt{a^2\omega^2 + b^2}$. Hence Formula (12) gives

$$s = \int_0^{2\pi/\omega} \sqrt{a^2\omega^2 + b^2}\, dt = \frac{2\pi}{\omega}\sqrt{a^2\omega^2 + b^2}.$$

For instance, if $a = b = \omega = 1$, then $s = 2\pi\sqrt{2}$, which is $\sqrt{2}$ times the circumference of the circle in the xy-plane over which the helix lies.

Let $s(t)$ denote the arc length along a smooth curve from its initial point $\mathbf{r}(a)$ to the variable point $\mathbf{r}(t)$ (see Figure 13.25). Then from Formula (12) we obtain the **arc length function** $s(t)$ of the curve:

$$s(t) = \int_a^t v(u)\, du. \tag{13}$$

The Fundamental Theorem of Calculus then gives

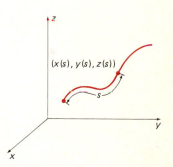

13.25 A curve parametrized by arc length s.

$$\frac{ds}{dt} = v. \tag{14}$$

Thus *the speed of the moving point is the time rate of change of its arc length function*. If $v(t) > 0$ for all t, then it follows that $s(t)$ is an increasing function of t and therefore has an inverse function $t(s)$. When we replace t by $t(s)$ in the curve's original parametric equations, we obtain the **arc length parametrization**

$$x = x(s), \qquad y = y(s), \qquad z = z(s).$$

This gives the position of the moving point as a function of arc length measured along the curve from its initial point (see Figure 13.25).

CURVATURE AND ACCELERATION
FOR SPACE CURVES

Consider now a moving particle in space with twice-differentiable position vector $\mathbf{r}(t)$. Suppose also that the velocity vector $\mathbf{v}(t)$ is always nonzero. The **unit tangent vector** at time t is defined (as in Section 10-6) to be

$$\mathbf{T}(t) = \frac{\mathbf{v}(t)}{|\mathbf{v}(t)|} = \frac{\mathbf{v}}{v}, \tag{15}$$

so

$$\mathbf{v} = v\mathbf{T}. \tag{16}$$

In Section 10-6 we defined the curvature of a plane curve to be $\kappa = d\phi/ds$, where ϕ is the angle of inclination of $\mathbf{T}$ from the positive x-axis. For a space curve, there is no single angle that determines the direction of $\mathbf{T}$, so we adopt the following approach. Differentiation of the identity $\mathbf{T} \cdot \mathbf{T} = 1$ with respect to arc length s gives

$$\mathbf{T} \cdot \frac{d\mathbf{T}}{ds} = 0.$$

It follows that the vectors $\mathbf{T}$ and $d\mathbf{T}/ds$ are always perpendicular.

Then we define the **curvature** κ of the curve at the point $r(t)$ to be

$$\kappa = \left| \frac{d\mathbf{T}}{ds} \right| = \left| \frac{d\mathbf{T}}{dt} \frac{dt}{ds} \right| = \frac{1}{v} \left| \frac{d\mathbf{T}}{dt} \right|. \tag{17}$$

At points where $\kappa \neq 0$, we define the **principal unit normal vector** $\mathbf{N}$ to be

$$\mathbf{N} = \frac{d\mathbf{T}/ds}{|d\mathbf{T}/ds|} = \frac{1}{\kappa} \frac{d\mathbf{T}}{ds}, \tag{18}$$

so

$$\frac{d\mathbf{T}}{ds} = \kappa \mathbf{N}. \tag{19}$$

Equation (19) shows that $\mathbf{N}$ has the same direction as $d\mathbf{T}/ds$, and Equation (18) shows that $\mathbf{N}$ is a unit vector. Intuitively, $\mathbf{N}$ is the normal vector that points in the direction in which the curve is bending (see Figure 13.26). Since Equation (19) is the same as Equation (11) of Section 10-6, we see that the present definitions of κ and $\mathbf{N}$ agree with those given earlier for the two-dimensional case.

EXAMPLE 3 Compute the curvature κ of the helix of Example 1.

Solution In Example 1 we computed the velocity vector

$$\mathbf{v} = \mathbf{i}(-a\omega \sin \omega t) + \mathbf{j}(a\omega \cos \omega t) + b\mathbf{k}$$

and speed $v = |\mathbf{v}| = \sqrt{a^2\omega^2 + b^2}$. Hence Equation (15) gives the unit tangent vector

$$\mathbf{T} = \frac{\mathbf{v}}{v} = \frac{\mathbf{i}(-a\omega \sin \omega t) + \mathbf{j}(a\omega \cos \omega t) + b\mathbf{k}}{\sqrt{a^2\omega^2 + b^2}}.$$

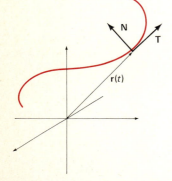

13.26 The principal unit normal vector **N** points in the direction in which the curve is turning.

CHAP. 13: Vectors, Curves, and Surfaces in Space

Then

$$\frac{d\mathbf{T}}{dt} = \frac{\mathbf{i}(-a\omega^2\cos\omega t) + \mathbf{j}(-a\omega^2\sin\omega t)}{\sqrt{a^2\omega^2 + b^2}},$$

so Equation (17) gives

$$\kappa = \frac{1}{v}\left|\frac{d\mathbf{T}}{dt}\right| = \frac{a\omega^2}{a^2\omega^2 + b^2}$$

for the curvature of the helix of Example 1. Note that the helix has constant curvature. Note also that, if $b = 0$ (so that the helix reduces to a circle of radius a in the xy-plane), our result reduces to $\kappa = 1/a$, in agreement with our computation of the curvature of a circle in Example 2 of Section 10-6.

Just as in Section 10-6, differentiation of both sides of Equation (16) yields

$$\mathbf{a} = \frac{d\mathbf{v}}{dt} = \frac{dv}{dt}\mathbf{T} + v\frac{d\mathbf{T}}{dt} = \frac{dv}{dt}\mathbf{T} + v\frac{d\mathbf{T}}{ds}\frac{ds}{dt}.$$

Since $ds/dt = v$ and $d\mathbf{T}/ds = \kappa\mathbf{N}$ by Equation (19), this gives the resolution

$$\mathbf{a} = \frac{dv}{dt}\mathbf{T} + \kappa v^2\mathbf{N} \qquad (20)$$

of the acceleration vector of a space curve into its **tangential and normal components.** As in Section 10-6, we write

$$a_T = \frac{dv}{dt} \quad\text{and}\quad a_N = \kappa v^2, \qquad (21)$$

so that

$$\mathbf{a} = a_T\mathbf{T} + a_N\mathbf{N}. \qquad (22)$$

Such a resolution is illustrated in Figure 13.27.

The resolution of the acceleration vector into its tangential and normal components is formally the same for a space curve as for a plane curve (Equation (12) in Section 10-6). But it remains for us to see how to compute a_T, a_N, and $\mathbf{N}$ effectively in the case of a space curve. We would like to have formulas that explicitly involve only the vectors $\mathbf{r}$, $\mathbf{v}$, and $\mathbf{a}$; that is, that involve only the position vector and its derivatives with respect to t.

If we take the scalar product of $\mathbf{v} = v\mathbf{T}$ with the acceleration $\mathbf{a}$ as given in Equation (20) and use the facts that $\mathbf{T}\cdot\mathbf{T} = 1$ and $\mathbf{T}\cdot\mathbf{N} = 0$, we get

$$\mathbf{v}\cdot\mathbf{a} = v\mathbf{T}\cdot\frac{dv}{dt}\mathbf{T} + v\mathbf{T}\cdot\kappa v^2\mathbf{N} = v\frac{dv}{dt}.$$

It follows that

$$a_T = \frac{dv}{dt} = \frac{\mathbf{v}\cdot\mathbf{a}}{v} = \frac{\mathbf{r}'(t)\cdot\mathbf{r}''(t)}{|\mathbf{r}'(t)|}. \qquad (23)$$

Similarly, when we compute the vector product of $\mathbf{v} = v\mathbf{T}$ with each side of Equation (20), we find that

$$\mathbf{v}\times\mathbf{a} = v\mathbf{T}\times\frac{dv}{dt}\mathbf{T} + v\mathbf{T}\times\kappa v^2\mathbf{N} = \kappa v^3\mathbf{T}\times\mathbf{N}.$$

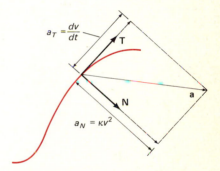

13.27 Resolution of the acceleration vector **a** into its tangential and normal components.

Since κ and v are nonnegative and since $\mathbf{T} \times \mathbf{N}$ is a unit vector, we may conclude that

$$\kappa = \frac{|\mathbf{v} \times \mathbf{a}|}{v^3} = \frac{|\mathbf{r}'(t) \times \mathbf{r}''(t)|}{|\mathbf{r}'(t)|^3}. \tag{24}$$

The curvature of a space curve is not often as easy to compute directly from the definition as we found in the case of the helix of Example 3. It is generally more convenient to use Formula (24). Once $\mathbf{a}$, $\mathbf{T}$, a_T, and a_N have been computed, we can rewrite Equation (22) as

$$\mathbf{N} = \frac{\mathbf{a} - a_T\mathbf{T}}{a_N} \tag{25}$$

to find the principal unit normal vector.

EXAMPLE 4 Compute $\mathbf{T}$, $\mathbf{N}$, κ, a_T, and a_N at the point $(1, \frac{1}{2}, \frac{1}{3})$ of the twisted cubic with parametric equations $x = t$, $y = \frac{1}{2}t^2$, $z = \frac{1}{3}t^3$.

Solution Differentiation of the position vector

$$\mathbf{r}(t) = t\mathbf{i} + \tfrac{1}{2}t^2\mathbf{j} + \tfrac{1}{3}t^3\mathbf{k}$$

gives

$$\mathbf{r}'(t) = \mathbf{i} + t\mathbf{j} + t^2\mathbf{k}$$

and

$$\mathbf{r}''(t) = \mathbf{j} + 2t\mathbf{k}.$$

When we substitute $t = 1$, we get the

velocity: $\qquad \mathbf{v}(1) = \mathbf{i} + \mathbf{j} + \mathbf{k},$

speed: $\qquad v(1) = |\mathbf{v}(1)| = \sqrt{3},$ $\qquad$ and

acceleration: $\qquad \mathbf{a}(1) = \mathbf{j} + 2\mathbf{k}$

at the point $(1, \frac{1}{2}, \frac{1}{3})$. Then Equation (23) gives the tangential component of acceleration:

$$a_T = \frac{\mathbf{v} \cdot \mathbf{a}}{v} = \frac{3}{\sqrt{3}} = \sqrt{3}.$$

Since

$$\mathbf{v} \times \mathbf{a} = \begin{vmatrix} \mathbf{i} & \mathbf{j} & \mathbf{k} \\ 1 & 1 & 1 \\ 0 & 1 & 2 \end{vmatrix} = \mathbf{i} - 2\mathbf{j} + \mathbf{k},$$

Equation (24) gives the curvature:

$$\kappa = \frac{|\mathbf{v} \times \mathbf{a}|}{v^3} = \frac{\sqrt{6}}{(\sqrt{3})^3} = \frac{\sqrt{2}}{3}.$$

The normal component of acceleration is $a_N = \kappa v^2 = \sqrt{2}$. The unit tangent vector is

$$\mathbf{T} = \frac{\mathbf{v}}{v} = \frac{\mathbf{i} + \mathbf{j} + \mathbf{k}}{\sqrt{3}}.$$

Finally, Equation (25) gives

$$\mathbf{N} = \frac{\mathbf{a} - a_T\mathbf{T}}{a_N} = \frac{(\mathbf{j} + 2\mathbf{k}) - (\mathbf{i} + \mathbf{j} + \mathbf{k})}{\sqrt{2}} = \frac{-\mathbf{i} + \mathbf{k}}{\sqrt{2}}.$$

Find the arc length of each of the curves in Problems 1–5.

1 $x = 3 \sin 2t$, $y = 3 \cos 2t$, $z = 8t$; from $t = 0$ to $t = \pi$.

2 $x = t$, $y = \dfrac{t^2}{\sqrt{2}}$, $z = \dfrac{1}{3}t^3$; from $t = 0$ to $t = 1$.

3 $x = 6e^t \cos t$, $y = 6e^t \sin t$, $z = 17e^t$; from $t = 0$ to $t = 1$.

4 $x = \frac{1}{2}t^2$, $y = \ln t$, $z = t\sqrt{2}$; from $t = 1$ to $t = 2$.

5 $x = 3t \sin t$, $y = 3t \cos t$, $z = 2t^2$; from $t = 0$ to $t = \frac{4}{5}$.

Find the curvature κ of each of the curves whose position vectors are given in Problems 6–10.

6 $\mathbf{r}(t) = t\mathbf{i} + (2t - 1)\mathbf{j} + (3t + 5)\mathbf{k}$.
7 $\mathbf{r}(t) = t\mathbf{i} + \mathbf{j} \sin t + \mathbf{k} \cos t$.
8 $\mathbf{r}(t) = t\mathbf{i} + t^2\mathbf{j} + t^3\mathbf{k}$.
9 $\mathbf{r}(t) = \mathbf{i}e^t \cos t + \mathbf{j}e^t \sin t + \mathbf{k}e^t$.
10 $\mathbf{r}(t) = \mathbf{i}t \sin t + \mathbf{j}t \cos t + \mathbf{k}t$.

11–15 Find the tangential and normal components of acceleration (a_T and a_N) for the curves of Problems 6–10, respectively.

In each of Problems 16–19, find the unit vectors $\mathbf{T}$ and $\mathbf{N}$ for the given curve at the indicated point.

16 The curve of Problem 8 at the point $(1, 1, 1)$.
17 The curve of Problem 7 at the point $(0, 0, 1)$.
18 The curve of Problem 3 at the point $(6, 0, 17)$.
19 The curve of Problem 9 at the point $(1, 0, 1)$.
20 Find $\mathbf{T}$, $\mathbf{N}$, a_T, and a_N as functions of t for the helix of Example 1.
21 A point moves on a sphere centered at the origin. Show that its velocity vector is always tangential to the sphere.
22 A particle moves along a space curve with constant speed. Show that its velocity and acceleration vectors are always perpendicular.
23 Verify that

$$D_t[\mathbf{u}(t) \times \mathbf{v}(t)] = \mathbf{u}'(t) \times \mathbf{v}(t) + \mathbf{u}(t) \times \mathbf{v}'(t).$$

24 Find the arc length parametrization of the circle $x = 2 \cos t$, $y = 2 \sin t$ in terms of the arc length s measured counterclockwise from the point $(2, 0)$.
25 Find the arc length parametrization of the helix $x = 3 \cos t$, $y = 3 \sin t$, $z = 4t$ in terms of arc length measured from the point $(3, 0, 0)$.
26 Substitute $x = t$, $y = f(t)$, $z = 0$ into Formula (24) to show that the curvature of the plane curve $y = f(x)$ is

$$\kappa = \frac{|f''(x)|}{[1 + (f'(x))^2]^{3/2}}.$$

27 Deduce from Equation (24) in this section that the curvature of the parametric plane curve $x = f(t)$, $y = g(t)$ is given by Equation (7) in Section 10-6.
28 (a) Show that $a_N = \sqrt{|\mathbf{a}|^2 - (a_T)^2}$. (b) Then use this formula to compute a_N for the helix of Example 1. (c) Finally, find the curvature $\kappa = a_N/v^2$ of that helix.
29 The **angular momentum** $\mathbf{L}(t)$ and **torque** $\boldsymbol{\tau}(t)$ of a moving particle of mass m with position vector $\mathbf{r}(t)$ are defined to be

$$\mathbf{L}(t) = m\mathbf{r}(t) \times \mathbf{v}(t), \quad \boldsymbol{\tau}(t) = m\mathbf{r}(t) \times \mathbf{a}(t).$$

Show that $\mathbf{L}'(t) = \boldsymbol{\tau}(t)$. It follows that $\mathbf{L}(t)$ must be constant if $\boldsymbol{\tau} \equiv \mathbf{0}$; this is the law of the conservation of angular momentum.
30 Suppose that a particle is moving under the influence of a *central* force field $\mathbf{F} = k\mathbf{r}$, where k is a scalar function of x, y, and z. Conclude that the particle's trajectory lies in a *fixed* plane through the origin.

13-5 Cylinders and Quadric Surfaces

Just as the graph of an equation $f(x, y) = 0$ is generally a curve in the xy-plane, the graph of an equation in three variables is generally a surface in space. A function F of three variables associates with each ordered triple (x, y, z) of real numbers a real number $F(x, y, z)$. The graph of the equation

$$F(x, y, z) = 0 \tag{1}$$

is the set of all points whose coordinates satisfy this equation. We will refer to the graph of such an equation as a **surface.** It should be noted, however, that the graph of Equation (1) does not always agree with the intuitive notion of a surface; for instance, the graph of the equation $(x^2 + y^2)(y^2 + z^2) = 0$ consists of the x- and z-axes. We leave for advanced calculus the precise

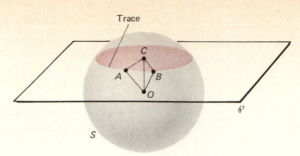

13.28 The intersection of the sphere S and the plane $\mathscr{P}$ is a circle.

definition of a surface, and the study of conditions sufficient to imply that the graph of (1) actually is a surface.

The simplest example of a surface is a plane with linear equation $Ax + By + Cz + D = 0$. This section is devoted to examples of other simple surfaces that frequently appear in multivariable calculus.

In order to sketch a surface S, it is often helpful to examine its intersections with various planes. The **trace** of the surface S in the plane $\mathscr{P}$ is the intersection of $\mathscr{P}$ and S. For example, if S is a sphere, then the methods of elementary geometry may be used to verify that the trace of S in the plane $\mathscr{P}$ is a circle (Figure 13.28), provided that $\mathscr{P}$ intersects the sphere but is not tangent to it (Problem 29).

Let C be a curve in a plane, and let L be a line not parallel to the plane. Then the set of all points on all lines parallel to L that intersect C is called a **cylinder.** This is a generalization of the familiar right circular cylinder, for which the curve C is a circle and the line L is perpendicular to the plane of the circle. Figure 13.29 shows this cylinder when C is the circle $x^2 + y^2 = a^2$ in the xy-plane. The trace of this cylinder in any horizontal plane $z = c$ is a circle of radius a and with center the point $(0, 0, c)$ on the z-axis. Thus the point (x, y, c) lies on the cylinder if and only if $x^2 + y^2 = a^2$. Hence this cylinder is the graph of the equation $x^2 + y^2 = a^2$, considered as an equation in *three* variables—even though the variable z does not appear explicitly in the equation.

Given a general cylinder generated by a plane curve C and a line L as in the definition above, the lines on the cylinder parallel to the line L are called **rulings** of the cylinder. Thus the rulings of the cylinder $x^2 + y^2 = a^2$ are vertical lines (parallel to the z-axis).

If the curve C in the xy-plane has equation

$$f(x, y) = 0, \qquad (2)$$

then the cylinder through C with vertical rulings has the same equation. This is so because the point $P(x, y, z)$ lies on the cylinder if and only if the point $Q(x, y, 0)$ lies on the curve C. Similarly, the graph of an equation $g(x, z) = 0$ is a cylinder with rulings parallel to the y-axis, and the graph of an equation $h(y, z) = 0$ is a cylinder with rulings parallel to the x-axis. Thus the graph in space of an equation involving only two of the three coordinate variables is always a cylinder; its rulings are parallel to the axis corresponding to the *missing* variable.

EXAMPLE 1 The graph of the equation $4y^2 + 9z^2 = 36$ is the **elliptic cylinder** shown in Figure 13.30. Its rulings are parallel to the x-axis, and

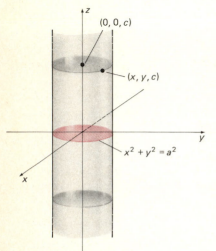

(0, 0, c)

(x, y, c)

$x^2 + y^2 = a^2$

13.29 A right circular cylinder.

13.30 An elliptic cylinder.

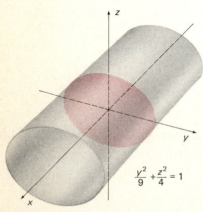

$\dfrac{y^2}{9} + \dfrac{z^2}{4} = 1$

CHAP. 13: **Vectors, Curves, and Surfaces in Space**

its trace in every plane perpendicular to the x-axis is an ellipse with semi-axes of lengths 3 and 2 (just like the pictured ellipse $y^2/9 + z^2/4 = 1$ in the yz-plane).

EXAMPLE 2 The graph of the equation $z = 4 - x^2$ is the **parabolic cylinder** shown in Figure 13.31. Its rulings are parallel to the y-axis, and its trace in every plane perpendicular to the y-axis is a parabola that is a parallel translate of the parabola $z = 4 - x^2$ in the xz-plane.

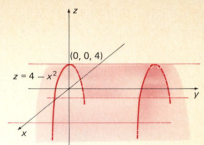

13.31 A parabolic cylinder.

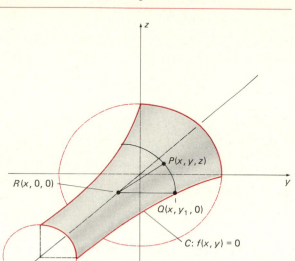

13.32 The surface generated by rotating C around the x-axis. (For clarity, only a quarter of the surface is shown.)

Another way to use a plane curve C to generate a surface is to revolve the curve (in space) around a line L in its plane. This gives a **surface of revolution** with **axis** L. For example, Figure 13.32 shows the surface generated by revolving the curve $f(x, y) = 0$ in the first quadrant of the xy-plane around the x-axis. The point $P(x, y, z)$ lies on the surface of revolution if and only if the point $Q(x, y_1, 0)$ lies on the curve, where

$$y_1 = |RQ| = |RP| = \sqrt{y^2 + z^2}.$$

Thus it is necessary that $f(x, y_1) = 0$, so the equation of the indicated surface of revolution around the x-axis is

$$f(x, \sqrt{y^2 + z^2}) = 0. \tag{3}$$

The equations of surfaces of revolution around other coordinate axes are obtained similarly. If the first quadrant curve $f(x, y) = 0$ above is revolved around the y-axis, we replace x by $(x^2 + z^2)^{1/2}$ to get the equation $f((x^2 + z^2)^{1/2}, y) = 0$ of the resulting surface of revolution. If the curve $g(y, z) = 0$ in the first quadrant of the yz-plane is revolved around the z-axis, we replace y by $(x^2 + y^2)^{1/2}$. Thus the equation of the resulting surface of revolution around the z-axis is $g((x^2 + y^2)^{1/2}, z) = 0$. These assertions are easily verified with the aid of diagrams similar to Figure 13.32.

EXAMPLE 3 Find an equation of the **ellipsoid of revolution** generated by revolving the ellipse $4y^2 + z^2 = 4$ around the z-axis (see Figure 13.33).

13.33 The ellipsoid of revolution of Example 3.

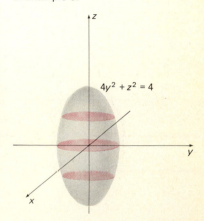

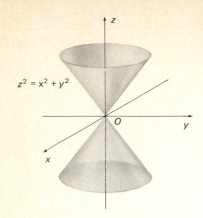

$z^2 = x^2 + y^2$

13.34 The cone of Example 4.

Solution We replace y by $(x^2 + y^2)^{1/2}$ in the given equation. This gives us $4x^2 + 4y^2 + z^2 = 4$ as an equation of the ellipsoid.

EXAMPLE 4 Determine the graph of the equation $z^2 = x^2 + y^2$.

Solution First we rewrite the given equation in the form $z = \pm(x^2 + y^2)^{1/2}$. Thus the surface is symmetric about the xy-plane, and the upper half has equation $z = +(x^2 + y^2)^{1/2}$. This last equation is obtained from the simple equation $z = y$ by replacing y by $(x^2 + y^2)^{1/2}$. Thus the upper half of the surface is obtained by revolving the line $z = y$ (for $y \geqq 0$) around the z-axis. Thus we have the **cone** shown in Figure 13.34. Its upper half has equation $z = +(x^2 + y^2)^{1/2}$, and its lower half has equation $z = -(x^2 + y^2)^{1/2}$. The whole cone $z^2 = x^2 + y^2$ is obtained by revolving the whole line $z = y$ around the z-axis.

QUADRIC SURFACES

Cones, spheres, circular and parabolic cylinders, and ellipsoids of revolution are all examples of surfaces with graphs that are second degree equations in x, y, and z. The graph of a second degree equation in three variables is called a **quadric surface**. We discuss here some important special cases of the equation

$$Ax^2 + By^2 + Cz^2 + Dx + Ey + Fz + G = 0. \tag{4}$$

This is a somewhat special second-degree equation in that it contains no terms involving the products xy, xz, or yz.

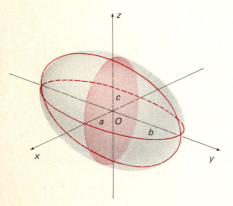

13.35 The ellipsoid of Example 5.

EXAMPLE 5 The **ellipsoid**

$$\frac{x^2}{a^2} + \frac{y^2}{b^2} + \frac{z^2}{c^2} = 1 \tag{5}$$

is symmetric about each of the three coordinate planes and has intercepts $(\pm a, 0, 0)$, $(0, \pm b, 0)$, and $(0, 0, \pm c)$ on the three coordinate axes. If $P(x, y, z)$ is a point of this ellipsoid, then $|x| \leqq a$, $|y| \leqq b$, and $|z| \leqq c$. Each trace in a plane parallel to one of the three coordinate axes is either a single point or an ellipse. For example, if $-c < z_0 < c$, then Equation (5) can be reduced to

$$\frac{x^2}{a^2} + \frac{y^2}{b^2} = 1 - \frac{z_0^2}{c^2} > 0$$

which is the equation of an ellipse with semiaxes $(a/c)(c^2 - z_0^2)^{1/2}$ and $(b/c)(c^2 - z_0^2)^{1/2}$. The ellipsoid is shown in Figure 13.35.

13.36 An elliptic paraboloid.

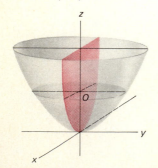

EXAMPLE 6 The **elliptic paraboloid**

$$\frac{x^2}{a^2} + \frac{y^2}{b^2} = \frac{z}{c} \tag{6}$$

is shown in Figure 13.36. Its trace in the horizontal plane $z = z_0 > 0$ is the ellipse $x^2/a^2 + y^2/b^2 = z_0/c$ with semiaxes $a\sqrt{z_0/c}$ and $b\sqrt{z_0/c}$. Its trace in any vertical plane is a parabola. For instance, its trace in the plane $y = y_0$ has equation $x^2/a^2 + y_0^2/b^2 = z/c$, which can be rewritten in the

CHAP. 13: Vectors, Curves, and Surfaces in Space

form $z - z_1 = k(x - x_1)^2$ by taking $z_1 = cy_0^2/b^2$ and $x_1 = 0$. The paraboloid opens upward if $c > 0$ and downward if $c < 0$. If $a = b$ the paraboloid is circular.

EXAMPLE 7   The **elliptic cone**

$$\frac{x^2}{a^2} + \frac{y^2}{b^2} = \frac{z^2}{c^2} \tag{7}$$

is shown in Figure 13.37. Its trace in the horizontal plane $z = z_0 \neq 0$ is an ellipse with semiaxes $a|z_0|/c$ and $b|z_0|/c$.

EXAMPLE 8   The **hyperboloid of one sheet** with equation

$$\frac{x^2}{a^2} + \frac{y^2}{b^2} - \frac{z^2}{c^2} = 1 \tag{8}$$

is shown in Figure 13.38. Its trace in the horizontal plane $z = z_0$ is the ellipse $x^2/a^2 + y^2/b^2 = 1 + z_0^2/c^2 > 0$. Its trace in a vertical plane is a hyperbola except when the vertical plane intersects the xy-plane in a line tangent to the ellipse $x^2/a^2 + y^2/b^2 = 1$. In this special case, the trace is a degenerate hyperbola consisting of two intersecting lines.

The graphs of the equations

$$\frac{y^2}{b^2} + \frac{z^2}{c^2} - \frac{x^2}{a^2} = 1 \quad \text{and} \quad \frac{x^2}{a^2} + \frac{z^2}{c^2} - \frac{y^2}{b^2} = 1$$

are also hyperboloids of one sheet, opening along the x- and y-axes, respectively.

EXAMPLE 9   The **hyperboloid of two sheets** with equation

$$\frac{z^2}{c^2} - \frac{x^2}{a^2} - \frac{y^2}{b^2} = 1 \tag{9}$$

consists of two connected pieces or sheets (Figure 13.39). The two sheets open along the positive and negative z-axis and intersect it at the points $(0, 0, \pm c)$. The trace of this hyperboloid in a horizontal plane $z = z_0$ with $|z_0| > c$ is the ellipse

$$\frac{x^2}{a^2} + \frac{y^2}{b^2} = \frac{z_0^2}{c^2} - 1 > 0.$$

Its trace in any vertical plane is a nondegenerate hyperbola.

The graphs of the equations

$$\frac{x^2}{a^2} - \frac{y^2}{b^2} - \frac{z^2}{c^2} = 1 \quad \text{and} \quad \frac{y^2}{b^2} - \frac{x^2}{a^2} - \frac{z^2}{c^2} = 1$$

are also hyperboloids of two sheets, opening along the x- and y-axes, respectively. Note that when the equation of a hyperboloid is written in standard form with $+1$ on the right-hand side (as in Equations (8) and (9)), the number of sheets is equal to the number of negative terms on the left-hand side.

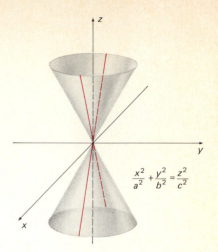

13.37   An elliptic cone (Example 7).

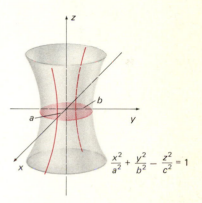

13.38   A hyperboloid of one sheet (Example 8).

13.39   A hyperboloid of two sheets (Example 9).

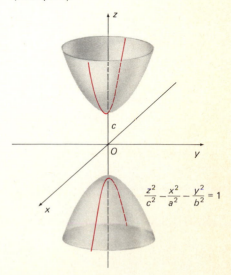

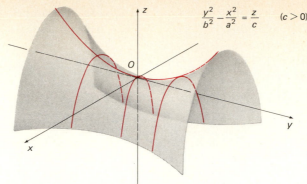

$$\frac{y^2}{b^2} - \frac{x^2}{a^2} = \frac{z}{c} \qquad (c > 0)$$

13.40 The hyperbolic paraboloid is a saddle-shaped surface.

EXAMPLE 10   The **hyperbolic paraboloid**

$$\frac{y^2}{b^2} - \frac{x^2}{a^2} = \frac{z}{c} \qquad (c > 0) \tag{10}$$

is saddle-shaped as indicated in Figure 13.40. Its trace in the horizontal plane $z = z_0$ is a hyperbola (or two intersecting lines if $z_0 = 0$). Its trace in a vertical plane parallel to the xz-plane is a parabola that opens downward, while its trace in a vertical plane parallel to the yz-plane is a parabola that opens upward. In particular, the trace of the hyperbolic paraboloid in the xz-plane is a parabola opening downward from the origin, while its trace in the yz-plane is a parabola opening upward from the origin. Thus the origin looks like a local maximum from one direction but like a local minimum from another direction. Such a point on a surface is called a **saddle point**.

*CONIC SECTIONS AS SECTIONS OF A CONE

The parabola, ellipse, and hyperbola that we studied in Chapter 9 were originally introduced by the ancient Greek mathematicians as plane sections (or traces) of a right circular cone. Here we show that the intersection of a plane and a cone is, indeed, one of the three conic sections as defined in Chapter 9.

Figure 13.41 shows the cone $z = (x^2 + y^2)^{1/2}$ and its intersection with a plane $\mathscr{P}$ that passes through the point $(0, 0, 1)$ and the line $x = c > 0$ in the xy-plane. The equation of $\mathscr{P}$ is

$$z = 1 - \frac{x}{c}. \tag{11}$$

The angle between $\mathscr{P}$ and the xy-plane is $\phi = \tan^{-1}(1/c)$. We want to show that the conic section obtained by intersecting the cone and the plane is:

a parabola if $\phi = 45°$ $(c = 1)$,

an ellipse if $\phi < 45°$ $(c > 1)$,   and

a hyperbola if $\phi > 45°$ $(c < 1)$.

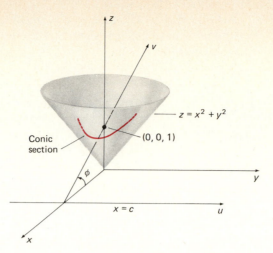

13.41 Finding an equation for a conic section.

We begin by introducing uv-coordinates in the plane $\mathscr{P}$ as follows. The u-coordinate of the point (x, y, z) of $\mathscr{P}$ is $u = y$. The v-coordinate of the same point is its perpendicular distance from the line $x = c$. This explains the u- and v-axes indicated in Figure 13.41. Figure 13.42 shows the cross section in the plane $y = 0$ exhibiting the relation between v, x, and z. We see that

$$z = v \sin \phi = \frac{v}{\sqrt{1 + c^2}}. \tag{12}$$

Equations (11) and (12) give

$$x = c(1 - z) = c\left(1 - \frac{v}{\sqrt{1 + c^2}}\right). \tag{13}$$

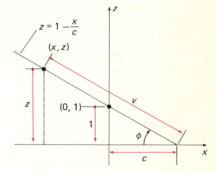

13.42 Computing coordinates in the uv-plane.

We had $z^2 = x^2 + y^2$ for the equation of the cone. We make the following substitutions in this equation: Replace y by u, and replace z and x by the expressions on the right-hand sides of Equations (12) and (13), respectively. This yields

$$\frac{v^2}{1 + c^2} = c^2\left(1 - \frac{v}{\sqrt{1 + c^2}}\right)^2 + u^2.$$

After simplification, this last equation takes the form

$$u^2 + \frac{c^2 - 1}{c^2 + 1}v^2 - \frac{2c^2}{\sqrt{1 + c^2}}v + c^2 = 0. \tag{14}$$

This is the equation of the curve of intersection in the uv-plane. We proceed to examine the three cases for the angle ϕ.

First suppose that $\phi = 45°$. Then $c = 1$, so that Equation (14) contains a u^2-term, a v-term, and a constant term. So the curve is a parabola; see Equation (6) of Section 9-2.

Next suppose that $\phi < 45°$. Then $c > 1$, and the coefficients of u^2 and v^2 in Equation (14) are both positive. So the curve is an ellipse; see Equation (6) of Section 9-3.

Finally, if $\phi > 45°$, then $c < 1$, and the coefficients of u^2 and v^2 in Equation (14) have different signs. So the curve is a hyperbola; see Equation (8) in Section 9-4.

13-5 PROBLEMS

Describe and sketch the graphs of the equations given in Problems 1–20.

1 $2z = x^2 + y^2$.	**2** $x = 1 + y^2 + z^2$.
3 $z^2 = 4(x^2 + y^2)$.	**4** $y^2 = 4x$.
5 $x^2 = 4z + 8$.	**6** $x = 9 - z^2$.
7 $4x^2 + y^2 = 4$.	**8** $x^2 + z^2 = 4$.
9 $x^2 = 4y^2 + 9z^2$.	**10** $x^2 - 4y^2 = z$.
11 $x^2 + y^2 + 4z = 0$.	**12** $x = \sin y$.
13 $x = 2y^2 - z^2$.	**14** $x^2 + 4y^2 + 2z^2 = 4$.
15 $x^2 + y^2 - 9z^2 = 9$.	**16** $x^2 - y^2 - 9z^2 = 9$.
17 $y = 4x^2 + 9z^2$.	**18** $y^2 + 4x^2 - 9z^2 = 36$.
19 $y^2 - 9x^2 - 4z^2 = 36$.	**20** $x^2 + 9y^2 + 4z^2 = 36$.

Each of Problems 21–28 gives the equation of a curve in one of the coordinate planes. Write an equation for the surface generated by revolving this curve around the indicated axis. Then sketch the surface.

21 $x = 2z^2$; the x-axis.
22 $4x^2 + 9y^2 = 36$; the y-axis.
23 $y^2 - z^2 = 1$; the z-axis.
24 $z = 4 - x^2$; the z-axis.
25 $y^2 = 4x$; the x-axis.

26 $yz = 1$; the z-axis.
27 $z = e^{-x^2}$; the z-axis.
28 $(y - z)^2 + z^2 = 1$; the z-axis.
29 Show that the triangles OAC and OBC in Figure 13.28 are congruent, and thereby conclude that the trace of a sphere in an intersecting plane is a circle.
30 Show that the projection into the yz-plane of the curve of intersection of the surfaces $x = 1 - y^2$ and $x = y^2 + z^2$ is an ellipse.
31 Show that the projection into the xy-plane of the intersection of the plane $z = 2y$ and the paraboloid $z = x^2 + y^2$ is a circle.
32 Show that the projection into the xz-plane of the intersection of the paraboloids $y = 2x^2 + 3z^2$ and $y = 5 - 3x^2 - 2z^2$ is a circle.
33 Show that the projection into the xy-plane of the intersection of the plane $x + y + z = 1$ and the ellipsoid $x^2 + 4y^2 + 4z^2 = 4$ is an ellipse.
34 Show that the curve of intersection of the plane $z = ky$ and the cylinder $x^2 + y^2 = 1$ is an ellipse. (*Suggestion:* Introduce uv-coordinates in the plane $z = ky$ as follows: Let the u-axis be the original x-axis, and let the v-axis be the line $z = ky, x = 0$.)

13-6

Cylindrical and Spherical Coordinates

Rectangular coordinates provide only one of several useful ways of describing points, curves, and surfaces in space. In this section we discuss two additional coordinate systems in three-dimensional space. Each is a generalization of polar coordinates in the plane.

Recall that the relationship between the rectangular coordinates (x, y) and the polar coordinates (r, θ) of a point in the plane is given by the equations

$$x = r \cos \theta, \qquad y = r \sin \theta \qquad (1)$$

and

$$r^2 = x^2 + y^2, \qquad \tan \theta = \frac{y}{x} \text{ if } x \neq 0. \qquad (2)$$

These relationships may be read directly from the triangle of Figure 13.43.

The cylindrical coordinates (r, θ, z) of a point P in space are a perfectly natural hybrid of its polar and rectangular coordinates. We use the polar co-

13.43 The relation between rectangular and polar coordinates in the xy-plane.

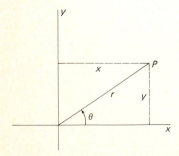

CHAP. 13: **Vectors, Curves, and Surfaces in Space**

ordinates (r, θ) of the point in the plane with rectangular coordinates (x, y), and use the same z-coordinate as in rectangular coordinates. (The cylindrical coordinates of a point P in space are illustrated in Figure 13.44.) This means that the relation between the rectangular coordinates (x, y, z) of the point P and its cylindrical coordinates (r, θ, z) is obtained by simply adding the identity $z = z$ to Equations (1) and (2):

$$x = r \cos \theta, \qquad y = r \sin \theta, \qquad z = z \qquad (3)$$

and

$$r^2 = x^2 + y^2, \qquad \theta = \tan^{-1}\left(\frac{y}{x}\right), \qquad z = z. \qquad (4)$$

These equations are used to convert from rectangular to cylindrical coordinates, and vice versa.

The term **cylindrical coordinates** arises from the fact that the graph in space of the equation $r = c$ (a constant) is a cylinder of radius c symmetric about the z-axis. This suggests that cylindrical coordinates be used when working problems involving circular symmetry about the z-axis; cylindrical coordinates exhibit a special simplicity in such cases. For instance, the sphere $x^2 + y^2 + z^2 = a^2$ and the cone $z^2 = x^2 + y^2$ have cylindrical coordinate equations $r^2 + z^2 = a^2$ and $z^2 = r^2$, respectively. It follows from our discussion of surfaces of revolution in Section 13-5 that, if the curve $f(y, z) = 0$ in the yz-plane is revolved around the z-axis, then the cylindrical coordinate equation of the surface generated is $f(r, z) = 0$.

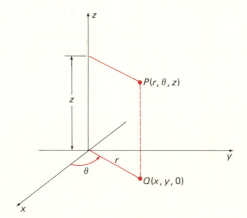

13.44 Finding the cylindrical coordinates of the point P.

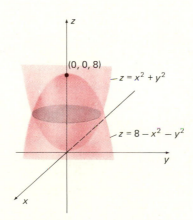

13.45 The two paraboloids of Example 1.

EXAMPLE 1 Sketch the region that is bounded by the graphs of the cylindrical coordinate equations $z = r^2$ and $z = 8 - r^2$.

Solution We first substitute $r^2 = x^2 + y^2$ from Equation (4) into the given equations. Thus the two surfaces of the example have rectangular coordinate equations $z = x^2 + y^2$ and $z = 8 - x^2 - y^2$. Their graphs are the two paraboloids shown in Figure 13.45. The region in question is bounded above by the paraboloid $z = 8 - x^2 - y^2$ and below by the paraboloid $z = x^2 + y^2$.

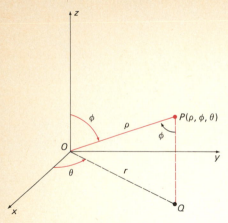

13.46 Finding the spherical coordinates of the point P.

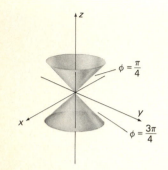

13.47 The two nappes of a 45° cone; $\phi = \pi/2$ is the xy-plane.

Figure 13.46 shows the **spherical coordinates** of the point P in space. The first spherical coordinate ρ is simply the distance $\rho = |OP|$ from the origin O. The second spherical coordinate ϕ is the angle between $\overrightarrow{OP}$ and the positive z-axis; thus $0 \leq \phi \leq \pi$. And θ is the familiar angle of cylindrical coordinates, the angular polar coordinate of the projection Q of the point P into the xy-plane. Both the angles ϕ and θ are always measured in *radians*.

The term *spherical coordinates* is used because the graph of the equation $\rho = c$ (constant) is a sphere of radius c centered at the origin. Note also that $\phi = c$ (constant) describes (one nappe of) a cone if $0 < c < \pi/2$ or $\pi/2 < c < \pi$. The spherical coordinate equation of the xy-plane is $\phi = \pi/2$ (see Figure 13.47).

From the right triangle OPQ of Figure 13.46, we see that

$$r = \rho \sin \phi, \qquad z = \rho \cos \phi. \tag{5}$$

Indeed, these equations are most easily remembered by visualizing this triangle. Substitution of Equations (5) into Equations (3) gives the equations

$$x = \rho \sin \phi \cos \theta,$$
$$y = \rho \sin \phi \sin \theta, \tag{6}$$
$$z = \rho \cos \phi.$$

These three equations give the relationship between rectangular and spherical coordinates. Also useful is the formula

$$\rho^2 = x^2 + y^2 + z^2, \tag{7}$$

a consequence of the distance formula.

EXAMPLE 2 Find a spherical coordinate equation for the paraboloid $z = x^2 + y^2$.

Solution We substitute $z = \rho \cos \phi$ from (5) and $x^2 + y^2 = r^2 = \rho^2 \sin^2\phi$ from (6). This gives $\rho \cos \phi = \rho^2 \sin^2\phi$. Cancellation of ρ gives $\cos \phi = \rho \sin^2\phi$, or $\rho = \csc \phi \cot \phi$ as the spherical coordinate equation of the paraboloid. We get the whole paraboloid by using ϕ in the range $0 < \phi \leq \pi/2$. Note that $\phi = \pi/2$ gives the point $\rho = 0$ that might otherwise have been lost by cancelling ρ.

EXAMPLE 3 Determine the graph of the spherical coordinate equation $\rho = \sin \phi \sin \theta$.

Solution We first multiply through by ρ and get $\rho^2 = \rho \sin \phi \sin \theta$. We then use Equations (6) and (7) and find that $x^2 + y^2 + z^2 = y$. This is the rectangular coordinate equation of a sphere with center $(0, \frac{1}{2}, 0)$ and with radius $\frac{1}{2}$.

*LATITUDE AND LONGITUDE

The spherical coordinates ϕ and θ are closely related to the latitude and longitude of points on the earth's surface. Assume that the earth is a sphere with radius $\rho = 3960$ miles. We begin with the **prime meridian** (a **meridian**

CHAP. 13: Vectors, Curves, and Surfaces in Space

is a north-south great semicircle) through Greenwich, just outside London. This is the point marked G in Figure 13.48.

We take the z-axis through the North Pole and the x-axis through the point where the prime meridian intersects the equator. The **latitude** α and (west) **longitude** β of a point P in the Northern Hemisphere are given by the equations

$$\alpha = 90° - \phi° \quad \text{and} \quad \beta = 360° - \theta°, \tag{8}$$

where $\phi°$ and $\theta°$ are the angular spherical coordinates, measured in *degrees*, of P. (That is, $\phi°$ and $\theta°$ denote the degree equivalents of the angles θ and ϕ, respectively, which are always measured in radians unless otherwise specified.) Thus the latitude α is measured northward from the equator, and the longitude β is measured westward from the prime meridian.

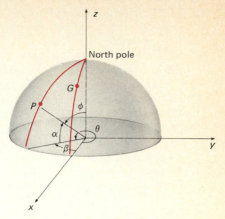

13.48 The relation between latitude, longitude, and spherical coordinates.

EXAMPLE 4 Find the great-circle distance between New York (latitude 40.75° north, longitude 74°) and London (latitude 51.5° north, longitude 0°).

Solution From Equations (8) we find that $\phi° = 49.25°$, $\theta° = 286°$ for New York, while $\phi° = 38.5°$, $\theta° = 360°$ (or 0°) for London. Hence the spherical coordinates of New York are $\phi = (49.25/180)\pi$, $\theta = (286/180)\pi$, while those of London are $\phi = (38.5/180)\pi$, $\theta = 0$. With these values of ϕ and θ and with $\rho = 3960$ (miles), Equations (6) give the rectangular coordinates

New York: $\quad P_1(826.90, -2883.74, 2584.93)$

and

London: $\quad P_2(2465.16, 0.0, 3099.13)$.

The angle γ between the radius vectors $\overrightarrow{OP_1}$ and $\overrightarrow{OP_2}$ in Figure 13.49 is given by

$$\cos \gamma = \frac{\overrightarrow{OP_1} \cdot \overrightarrow{OP_2}}{|\overrightarrow{OP_1}||\overrightarrow{OP_2}|}$$

$$= \frac{(826.90)(2465.16) + (-2883.74)(0.0) + (2584.93)(3099.13)}{(3960)^2}$$

$$\approx 0.641.$$

13.49 Finding the great circle distance d from New York to London.

Thus γ is approximately 0.875 radians. Hence the great circle distance between New York and London is close to $d = (3960)(0.875) \approx 3465$ miles.

13-6 PROBLEMS

In each of Problems 1–14, describe the graph of the given equation.

In each of Problems 15–20, convert the given equation both to cylindrical and to spherical coordinates.

1 $r = 5$.
2 $\theta = 3\pi/4$.
3 $\theta = \pi/4$.
4 $\rho = 5$.
5 $\phi = \pi/6$.
6 $\phi = 5\pi/6$.
7 $\psi = \pi/2$.
8 $\phi = \pi$.
9 $r = 2 \sin \theta$.
10 $\rho = 2 \sin \phi$.
11 $\cos \theta + \sin \theta = 0$.
12 $z = 10 - 3r^2$.
13 $\rho \cos \phi = 1$.
14 $\rho = \cot \phi$.

15 $x^2 + y^2 + z^2 = 25$.
16 $x^2 + y^2 = 2x$.
17 $x + y + z = 1$.
18 $x + y = 4$.
19 $x^2 + y^2 + z^2 = x + y + z$.
20 $z = x^2 - y^2$.
21 The parabola $z = x^2$, $y = 0$ is rotated about the z-axis. Write a cylindrical coordinate equation for the surface generated.

22 The hyperbola $y^2 - z^2 = 1$, $x = 0$ is rotated about the z-axis. Write a cylindrical coordinate equation for the surface generated.

23 A sphere of radius 2 is centered at the origin. A hole of radius 1 is drilled through the sphere, with the axis of the hole coinciding with the z-axis. Describe the solid region that remains in: (a) cylindrical coordinates; (b) spherical coordinates.

24 Find the great circle distance from Atlanta (latitude 33.75°, longitude 84.40°) to San Francisco (latitude 37.78°, longitude 122.42°).

25 Find the great circle distance from Fairbanks (latitude 64.80°, longitude 147.85°) to Leningrad (latitude 59.91°,

longitude 30.43° *east* of Greenwich; in our earlier notation, we would say that its longitude is 329.57°).

26 Since Fairbanks and Leningrad are at almost the same latitude, a plane could fly from one to the other roughly along the 62nd parallel of latitude. Estimate the length of such a trip.

27 In flying the great circle route from Fairbanks to Leningrad, how close to the North Pole would a plane come?

28 A right circular cone of radius R and height H is located with its vertex at the origin and its axis coincident with part of the nonnegative z-axis. Describe the solid cone in cylindrical coordinates.

29 Describe the cone of Problem 28 in spherical coordinates.

CHAPTER 13 REVIEW: Definitions, Concepts, Results

Use the list below as a guide to concepts that you may need to review.

1 Properties of addition of vectors in space, and of multiplication of vectors by scalars

2 The dot (scalar) product of vectors—definition and geometric interpretation

3 Use of the dot product to test perpendicularity of vectors

4 The cross (vector) product of vectors—definition and geometric interpretation

5 The scalar triple product of vectors—definition and geometric interpretation

6 The parametric and symmetric equations of the straight line through a given point and parallel to a given vector

7 Equation of the plane through a given point normal to a given vector

8 The velocity and acceleration vectors of a parametric space curve

9 Arc length of a parametric space curve

10 The curvature, unit tangent vector, and principal unit normal vector of a space curve

11 Tangential and normal components of the acceleration vector of a parametric space curve

12 Equations of cylinders and of surfaces of revolution

13 The standard examples of quadric surfaces

14 Definition of the cylindrical coordinate and spherical coordinate systems, and the equations relating cylindrical and spherical coordinates with rectangular coordinates

MISCELLANEOUS PROBLEMS

1 Suppose that M is the midpoint of the line segment PQ in space and that A is another point. Show that $\overrightarrow{AM} = \frac{1}{2}(\overrightarrow{AP} + \overrightarrow{AQ})$.

2 Let **a** and **b** be nonzero vectors, and let

$$\mathbf{a}_{||} = (\text{comp}_{\mathbf{b}} \mathbf{a}) \frac{\mathbf{b}}{|\mathbf{b}|} \quad \text{and} \quad \mathbf{a}_{\perp} = \mathbf{a} - \mathbf{a}_{||}.$$

Show that $\mathbf{a}_{\perp}$ is perpendicular to **b**.

3 Let P and Q be different points in space. Show that the point R lies on the line through P and Q *if and only if* there exist numbers a and b such that $a + b = 1$ and $\overrightarrow{OR} = a\overrightarrow{OP} + b\overrightarrow{OQ}$. Conclude that $\mathbf{r}(t) = t\overrightarrow{OP} + (1 - t)\overrightarrow{OQ}$ is a parametric equation of this line.

4 Conclude from the result of Problem 3 that the points P, Q, and R are collinear if and only if there exist numbers a, b, and c, not all zero, such that $a + b + c = 0$ and $a\overrightarrow{OP} + b\overrightarrow{OQ} + c\overrightarrow{OR} = \mathbf{0}$.

5 Let $P(x_0, y_0)$, $Q(x_1, y_1)$, and $R(x_2, y_2)$ be points in the xy-plane. Use the cross product to show that the area of

the triangle PQR is

$$A = \tfrac{1}{2}|(x_1 - x_0)(y_2 - y_0) - (x_2 - x_0)(y_1 - y_0)|.$$

6 Write both symmetric and parametric equations for the line through $P(1, -1, 0)$ parallel to $\mathbf{v} = \langle 2, -1, 3 \rangle$.

7 Write both symmetric and parametric equations of the line through $P_1(1, -1, 2)$ and $P_2(3, 2, -1)$.

8 Write an equation of the plane through $P(3, -5, 1)$ with normal vector $\mathbf{n} = \mathbf{i} + \mathbf{j}$.

9 Show that the lines with symmetric equations

$$x - 1 = 2(y + 1) = 3(z - 2)$$

and

$$x - 3 = 2(y - 1) = 3(z + 1)$$

are parallel. Then write an equation of the plane through these two lines.

10 Let the lines L_1 and L_2 have symmetric equations

$$\frac{x - x_i}{a_i} = \frac{y - y_i}{b_i} = \frac{z - z_i}{c_i}$$

for $i = 1, 2$. Show that L_1 and L_2 are skew lines if and only if

$$\begin{vmatrix} x_1 - x_2 & y_1 - y_2 & z_1 - z_2 \\ a_1 & b_1 & c_1 \\ a_2 & b_2 & c_2 \end{vmatrix} \neq 0.$$

11 Given the four points $A(2, 3, 2)$, $B(4, 1, 0)$, $C(-1, 2, 0)$, and $D(5, 4, -2)$, find an equation of the plane through A and B parallel to the line through C and D.

12 Given the points A, B, C, and D of Problem 11, find points P on the line AB and Q on the line CD such that the line PQ is perpendicular to both AB and CD. What is the perpendicular distance d between the lines AB and CD?

13 Let $P_0(x_0, y_0, z_0)$ be a point of the plane with equation

$$ax + by + cz + d = 0.$$

By projecting $\overrightarrow{OP_0}$ onto the normal vector $\mathbf{n} = \langle a, b, c \rangle$, show that the distance D from the origin to this plane is

$$D = \frac{|d|}{\sqrt{a^2 + b^2 + c^2}}.$$

14 Show that the distance D from the point $P_1(x_1, y_1, z_1)$ to the plane $ax + by + cz + d = 0$ is equal to the distance from the origin to the plane with equation

$$a(x + x_1) + b(y + y_1) + c(z + z_1) + d = 0.$$

Hence conclude from the result of Problem 13 that

$$D = \frac{|ax_1 + by_1 + cz_1 + d|}{\sqrt{a^2 + b^2 + c^2}}.$$

15 Find the perpendicular distance between the parallel planes $2x - y + 2z = 4$ and $2x - y + 2z = 13$.

16 Write an equation of the plane through the point $(1, 1, 1)$ that is normal to the twisted cubic $x = t$, $y = t^2$, $z = t^3$ at this point.

17 A particle moves in space with parametric equations $x = t$, $y = t^2$, $z = \frac{4}{3}t^{3/2}$. Find the curvature of its trajectory and the tangential and normal components of its acceleration when $t = 1$.

18 The **osculating plane** to a space curve at a point P of that curve is the plane through P that is parallel to the curve's unit tangent and principal unit normal vectors at P. Write an equation of the osculating plane to the curve of Problem 17 at the point $(1, 1, \frac{4}{3})$.

19 Show that the equation of the plane through the point $P_0(x_0, y_0, z_0)$ and parallel to the vectors $\mathbf{v}_1 = \langle a_1, b_1, c_1 \rangle$ and $\mathbf{v}_2 = \langle a_2, b_2, c_2 \rangle$ can be written in the form

$$\begin{vmatrix} x - x_0 & y - y_0 & z - z_0 \\ a_1 & b_1 & c_1 \\ a_2 & b_2 & c_2 \end{vmatrix} = 0.$$

20 Deduce from Problem 19 that the equation of the osculating plane (Problem 18) to the parametric curve

$\mathbf{r}(t)$ at the point $\mathbf{r}(t_0)$ can be written in the form

$$(\mathbf{R} - \mathbf{r}(t_0)) \cdot \mathbf{r}'(t_0) \times \mathbf{r}''(t_0) = 0$$

where $\mathbf{R} = \langle x, y, z \rangle$. Note first that the vectors $\mathbf{T}$ and $\mathbf{N}$ are coplanar with $\mathbf{r}'(t)$ and $\mathbf{r}''(t)$.

21 Use the result of Problem 20 to write an equation of the osculating plane to the twisted cubic $x = t$, $y = t^2$, $z = t^3$ at the point $(1, 1, 1)$.

22 Let a parametric space curve be described by equations $r = r(t)$, $\theta = \theta(t)$, $z = z(t)$ giving the cylindrical coordinates of a moving point on the curve for $a \leq t \leq b$. Use the equations relating rectangular and cylindrical coordinates to show that the arc length of the curve is

$$s = \int_a^b \left[\left(\frac{dr}{dt} \right)^2 + \left(r \frac{d\theta}{dt} \right)^2 + \left(\frac{dz}{dt} \right)^2 \right]^{1/2} dt.$$

23 A point moves on the *unit* sphere $\rho = 1$ with its spherical angular coordinates at time t given by $\phi = \phi(t)$, $\theta = \theta(t)$, $a \leq t \leq b$. Use the equations relating rectangular and spherical coordinates to show that the arc length of its path is

$$s = \int_a^b \left[\left(\frac{d\phi}{dt} \right)^2 + (\sin^2 \phi) \left(\frac{d\theta}{dt} \right)^2 \right]^{1/2} dt.$$

24 The cross product $\mathbf{B} = \mathbf{T} \times \mathbf{N}$ of the unit tangent vector and the principal unit normal vector is the *unit binormal vector* $\mathbf{B}$ of a curve.
(a) Differentiate $\mathbf{B} \cdot \mathbf{T} = 0$ to show that $\mathbf{T}$ is perpendicular to $d\mathbf{B}/ds$.
(b) Differentiate $\mathbf{B} \cdot \mathbf{B} = 1$ to show that $\mathbf{B}$ is perpendicular to $d\mathbf{B}/ds$.
(c) Conclude from (a) and (b) that $d\mathbf{B}/ds = -\tau \mathbf{N}$ for some number τ, called the **torsion** of the curve. The torsion of a curve measures the amount it twists at each point in space.

25 Show that the torsion of the helix of Example 1 in Section 13-4 is constant by showing that its value is

$$\tau = \frac{b\omega}{a^2 \omega^2 + b^2}.$$

26 Deduce from the definition of torsion (Problem 24) that $\tau \equiv 0$ for any curve such that $\mathbf{r}(t)$ lies in a fixed plane.

27 Write an equation in spherical coordinates of the sphere with radius 1 and center $x = 0 = y$, $z = 1$.

28 Let C be the circle in the yz-plane with radius 1 and center $y = 1$, $z = 0$. Write equations in both rectangular and cylindrical coordinates of the surface obtained by revolving C around the z-axis.

29 Let C be the curve in the yz-plane with equation $(y^2 + z^2)^2 = 2(z^2 - y^2)$. Write an equation in spherical coordinates of the surface obtained by revolving this curve around the z-axis. Then sketch this surface. (*Suggestion:* Remember that $r^2 = 2 \cos 2\theta$ is the polar equation of a figure-eight curve.)

30 Let A be the area of a parallelogram $PQRS$ in space determined by the vectors $\mathbf{a} = \overrightarrow{PQ}$ and $\mathbf{b} = \overrightarrow{PS}$. Let A' be the area of the perpendicular projection of *PQRS* into a plane that makes an acute angle γ with the plane of *PQRS*. Assuming that $A' = A \cos \gamma$ in this situation, show that the areas of the perpendicular projections of the parallelogram *PQRS* into the three coordinate planes are

$$|\mathbf{i} \cdot \mathbf{a} \times \mathbf{b}|, \qquad |\mathbf{j} \cdot \mathbf{a} \times \mathbf{b}|, \quad \text{and} \quad |\mathbf{k} \cdot \mathbf{a} \times \mathbf{b}|.$$

Conclude that the square of the area of a parallelogram in space is equal to the sum of the squares of its perpendicular projections into the three coordinate planes.

31 Take $\mathbf{a} = \langle a_1, a_2, a_3 \rangle$ and $\mathbf{b} = \langle b_1, b_2, b_3 \rangle$ in Problem 30. Show that

$$A^2 = \begin{vmatrix} a_2 & a_3 \\ b_2 & b_3 \end{vmatrix}^2 + \begin{vmatrix} a_3 & a_1 \\ b_3 & b_1 \end{vmatrix}^2 + \begin{vmatrix} a_1 & a_2 \\ b_1 & b_2 \end{vmatrix}^2.$$

32 Let C be a curve in a plane $\mathscr{P}$ that is not parallel to the z-axis. Suppose that the projection of C into the xy-plane is an ellipse. Introduce uv-coordinates in the plane $\mathscr{P}$ to prove that the curve C is itself an ellipse.

33 Conclude from Problem 32 that the intersection of a nonvertical plane and an elliptic cylinder with vertical axis is an ellipse.

34 Use the result of Problem 32 to show that the intersection of the plane $z = Ax + By$ and the paraboloid $z = a^2 x^2 + b^2 y^2$ is either empty, a single point, or an ellipse.

35 Use the result of Problem 32 to show that the intersection of the plane $z = Ax + By$ and the ellipsoid $x^2/a^2 + y^2/b^2 + z^2/c^2 = 1$ is either empty, a single point, or an ellipse.

Partial Differentiation

14

14-1

Introduction

In this and the following two chapters, we turn our attention to the calculus of functions of more than one variable. Many real-world functions depend upon two or more variables. For example:

- In physical chemistry the ideal gas law $PV = nRT$ (where n and R are constants) is used to express any one of the variables P, V, and T as a function of the other two.

- The altitude above sea level at a particular location on the earth's surface depends upon its latitude and longitude.

- A manufacturer's profit depends upon sales, overhead costs, the cost of each raw product that is used, and perhaps upon additional variables.

- The amount of useable energy a solar panel can gather depends upon its efficiency, its angle of inclination to the sun's rays, the angle of elevation of the sun above the horizon, and perhaps upon other factors.

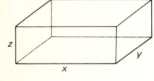

14.1

A typical application may call for an extreme value of a function of several variables. For example, suppose that we want to minimize the cost of making a rectangular box with a volume of 48 cubic feet, given that its front and back cost \$1/ft^2, its top and bottom cost \$2/ft^2, and its two ends cost \$3/ft^2. Figure 14.1 shows such a box with length x, width y, and height z. Under the conditions given, its total cost will be

$$C = 2xz + 4xy + 6yz \quad \text{(dollars)}.$$

But x, y, and z are not independent variables, because the box has fixed volume $V = xyz = 48$. We eliminate z from the first formula by using the second formula; since $z = 48/xy$, we find that the cost we want to minimize is given by

$$C = 4xy + \frac{288}{x} + \frac{96}{y}.$$

Because neither of the variables x and y can be expressed in terms of the other, the single-variable maximum-minimum techniques of Chapter 2 cannot be applied here. We need new optimization techniques applicable to functions of two or more independent variables. In Section 14-4 we shall return to this problem.

The problem of optimization is just a typical example. In this chapter, we shall see that the main ingredients of single-variable differential calculus—limits, derivatives and rates of change, chain-rule computations, and maximum-minimum techniques—can all be generalized to functions of two or more variables.

Functions of Several Variables

Recall that a real-valued *function* is a rule or correspondence f that associates a unique real number with each element of a set D. The set D is called the *domain* of definition of f. The domain D has always been a subset of the real line for the functions of a single variable that we have studied up to this point. If D is a subset of the plane, then f is a function of *two* variables—for, given a point P in D, we naturally associate with P its rectangular coordinates (x, y).

Definition *Functions of Two or Three Variables*

A **function of two variables,** defined on the **domain** D in the plane, is a rule f that associates with each point (x, y) in D a real number $f(x, y)$. A **function of three variables,** defined on the **domain** D in space, is a rule that associates with each point (x, y, z) in D a real number $f(x, y, z)$.

A function f of two (or three) variables is often defined by giving a formula that specifies $f(x, y)$ in terms of x and y (or $f(x, y, z)$ in terms of x, y, and z). In case the domain D of f is not explicitly specified, it is to be understood that D consists of all points for which the given formula is meaningful. For example, the domain of the function f with formula $f(x, y) = \sqrt{25 - x^2 - y^2}$ is the set of all (x, y) such that $25 - x^2 - y^2 \geqq 0$; that is, the circular disk $x^2 + y^2 \leq 25$ of radius 5 centered at the origin. Similarly, the function g defined as

$$g(x, y, z) = \frac{x + y + z}{\sqrt{x^2 + y^2 + z^2}}$$

is defined at all points of space where $x^2 + y^2 + z^2 \neq 0$; that is, everywhere except at the origin $(0, 0, 0)$.

In a geometric, physical, or economic situation, a function typically results from expressing one descriptive variable in terms of others. Thus, in Section 14-1, we saw that the cost C of the box discussed there was given by the formula

$$C = C(x, y) = 4xy + \frac{288}{x} + \frac{96}{y}$$

in terms of its length x and width y. The value C of this function is a variable that depends upon the values of x and y. Hence we call C a **dependent variable,** while x and y are **independent variables.** If the temperature T at the point (x, y, z) is given by some formula $T = h(x, y, z)$, then the dependent variable T is a function of the independent variables x, y, and z.

A function of four or more variables can be defined by giving a formula involving the appropriate number of independent variables. For example, if

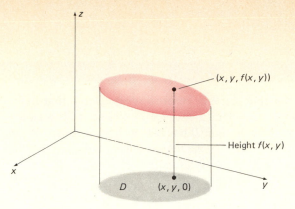

14.2 The graph of a function of two variables is frequently a surface "over" the domain of the function.

an amount A of heat is released at the origin at time $t = 0$ in a medium with conductivity k, then—under appropriate conditions—it turns out that the temperature at the point (x, y, z) at the time $t > 0$ is given by

$$T(x, y, z, t) = \frac{A}{(4\pi kt)^{3/2}} \, e^{-(x^2 + y^2 + z^2)/4kt}.$$

This formula gives the temperature T as a function of the four independent variables x, y, z, and t.

We shall see that the main differences between single-variable and multivariable calculus show up with examples involving only two independent variables. Hence most of our results will be stated in terms of functions of two variables. Many of these results will, however, readily generalize by analogy to the case of three or more independent variables.

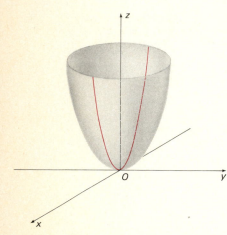

14.3 The paraboloid is [part of] the graph of the function $f(x, y) = x^2 + y^2$.

14.4 The upper half of an ellipsoid is the graph of a function of two variables.

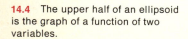

GRAPHS AND LEVEL CURVES

A function f of two variables x and y has the property that we can visualize how it "works" in terms of its graph. The **graph** of f is the graph of the equation $z = f(x, y)$; that is, the set of all points in space with coordinates (x, y, z) that satisfy the equation $z = f(x, y)$. See Figure 14.2.

You saw several examples of such graphs in Chapter 13. For example, the graph of the function $f(x, y) = x^2 + y^2$ is the paraboloid $z = x^2 + y^2$ shown in Figure 14.3. The graph of the function

$$g(x, y) = c \sqrt{1 - \frac{x^2}{a^2} - \frac{y^2}{b^2}}$$

is the *upper half* of the ellipsoid $x^2/a^2 + y^2/b^2 + z^2/c^2 = 1$ and is shown in Figure 14.4. Generally speaking, the graph of a function of two variables is a surface that lies above (or below, or both) its domain D in the xy-plane.

The intersection of the horizontal plane $z = k$ with the surface $z = f(x, y)$ is called the **contour curve** of height k on the surface, as in Figure 14.5. The vertical projection of this contour curve into the xy-plane is the **level curve** $f(x, y) = k$ of the function f. The level curves of f are simply the sets on which the value of f is constant. On a topographic map (like the one in Figure 14.6) the level curves are the curves of constant height above sea level.

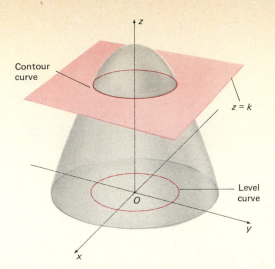

14.5 A contour curve and the corresponding level curve.

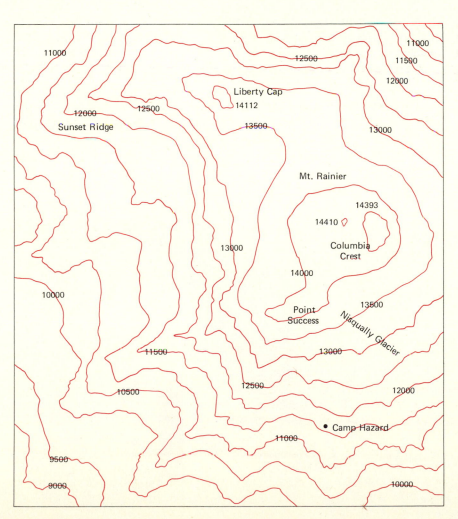

14.6 Area around Mt. Rainier, Washington, showing level curves at 500-foot intervals. [*Adapted from* U.S. Geological Survey Map N4645-W12145/7.5 (1971)]

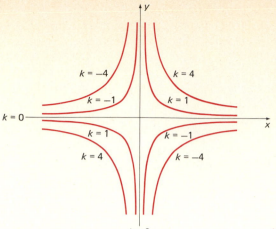

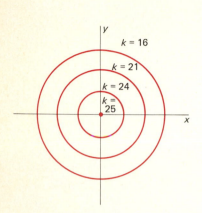

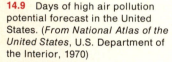

14.8 Some level curves of $f(x, y) = 25 - x^2 - y^2$.

14.9 Days of high air pollution potential forecast in the United States. (*From National Atlas of the United States*, U.S. Department of the Interior, 1970)

14.7 Level curves of $f(x, y) = xy$.

EXAMPLE 1 Some typical level curves of the function $f(x, y) = xy$ are the hyperbolas of the form $xy = k$ shown in Figure 14.7.

EXAMPLE 2 Some typical level curves of the function $f(x, y) = 25 - x^2 - y^2$ are the circles of the form $25 - x^2 - y^2 = k$ shown in Figure 14.8. The circle corresponding to the value $k < 25$ has radius $\sqrt{25 - k}$. The level curve of height $k = 25$ is the origin $(0, 0)$.

The graph of a function $f(x, y, z)$ of three variables cannot be drawn in three dimensions, but we can readily visualize its **level surfaces** of the form

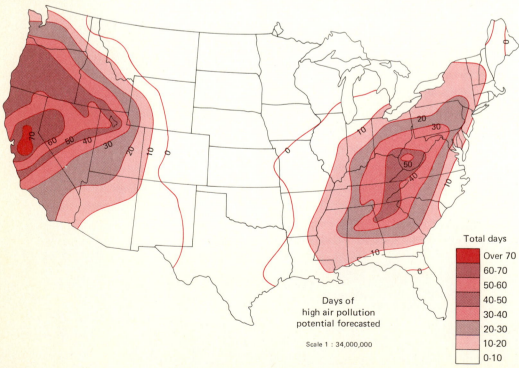

Days of
high air pollution
potential forecasted

Scale 1 : 34,000,000

Total days	
	Over 70
	60-70
	50-60
	40-50
	30-40
	20-30
	10-20
	0-10

$f(x, y, z) = k$. For example, the level surfaces of the function $f(x, y, z) = x^2 + y^2 + z^2$ are spheres centered at the origin. Thus the level surfaces of f are the sets in space on which the value $f(x, y, z)$ is constant.

If f is a temperature function, then its level curves or surfaces are called **isotherms**. A weather map often includes level curves of the atmospheric pressure; these are called **isobars**. Even though you may be able to construct the graph of a function of two variables, that graph might be so complicated that information about the function (or the situation it describes) is obscure. Frequently the level curves themselves give more information—as in weather maps. For example, Figure 14.9 shows level curves for the annual number of days of *high* air pollution forecast at different localities in the United States. The scale of this figure does not show local variations due to individual cities. But a glance indicates that western Colorado, southern Georgia, and central Illinois all expect about the same number (10, in this case) of high pollution days each year.

LIMITS AND CONTINUITY

We need limits of functions of several variables for the same reasons that we needed limits of functions of a single variable—so that we can discuss slopes and rates of change. Both the definition and the basic properties of limits of functions of several variables are essentially the same as those we stated in Section 1-8 for functions of a single variable. For simplicity, we shall state them here only for functions of two variables x and y; for a function of three variables, the pair (x, y) should be replaced by the triple (x, y, z).

We say that the number L is the *limit* of the function $f(x, y)$ as (x, y) approaches the point (a, b), and write

$$\lim_{(x,y) \to (a,b)} f(x, y) = L,$$

provided that the number $f(x, y)$ can be made as close as one pleases to L if the point (x, y) is chosen sufficiently close to—though not equal to—the point (a, b). More precisely:

Definition   *Limit of* $f(x, y)$
We say that **the limit of** $f(x, y)$ **as** (x, y) **approaches** (a, b) **is** L provided that, given $\varepsilon > 0$, there exists $\delta > 0$ such that there are points $(x, y) \neq (a, b)$ in the domain of f such that both $|x - a| < \delta$ and $|y - b| < \delta$, and for all such points the inequality $|f(x, y) - L| < \varepsilon$ holds.

We ordinarily shall rely upon continuity rather than the formal definition of the limit to evaluate limits of functions of several variables. We say that f is **continuous at the point** (a, b) provided that $f(a, b)$ exists and $f(x, y)$ approaches $f(a, b)$ as (x, y) approaches (a, b); that is,

$$\lim_{(x,y) \to (a,b)} f(x, y) = f(a, b).$$

Thus f is continuous at (a, b) if it is defined there and its limit there is equal to its value there. The function f is **continuous on the set** D if it is continuous at each point of D.

The limit laws of Section 1-8 have natural analogues for functions of several variables. If

$$\lim_{(x,y) \to (a,b)} f(x, y) = L \quad \text{and} \quad \lim_{(x,y) \to (a,b)} g(x, y) = M,$$

then the sum, product, and quotient laws say that

$$\lim_{(x,y) \to (a,b)} [f(x, y) + g(x, y)] = L + M, \tag{1}$$

$$\lim_{(x,y) \to (a,b)} [f(x, y)g(x, y)] = LM, \quad \text{and} \tag{2}$$

$$\lim_{(x,y) \to (a,b)} \frac{f(x, y)}{g(x, y)} = \frac{L}{M} \quad \text{if} \quad M \neq 0. \tag{3}$$

EXAMPLE 3 Show that $\lim\limits_{(x,y) \to (a,b)} xy = ab$.

Solution We take $f(x, y) = x$ and $g(x, y) = y$. Then it follows from the definition of limits that

$$\lim_{(x,y) \to (a,b)} f(x, y) = a \quad \text{and} \quad \lim_{(x,y) \to (a,b)} g(x, y) = b.$$

Hence the product law gives

$$\lim_{(x,y) \to (a,b)} xy = \lim_{(x,y) \to (a,b)} f(x, y)g(x, y) = ab.$$

More generally, suppose that $P(x, y)$ is a polynomial in the two variables x and y, so that it can be written in the form

$$P(x, y) = \Sigma c_{ij} x^i y^j.$$

Then the sum and product laws imply that

$$\lim_{(x,y) \to (a,b)} P(x, y) = P(a, b).$$

An immediate but important consequence is that every polynomial in two (or more) variables is a continuous function.

Just as in the single-variable case, any composition of continuous multivariable functions is also a continuous function. For instance, suppose that the functions f and g are each continuous at (a, b), and that h is continuous at the point $(f(a, b), g(a, b))$. Then the composite function $H(x, y) = h(f(x, y), g(x, y))$ is also continuous at (a, b). As a consequence, any finite combination involving sums, products, quotients, and compositions of the familiar elementary functions is continuous, except possibly at points where a denominator is zero or where the formula for the function is otherwise meaningless. This general rule suffices for the evaluation of most limits that we shall encounter. For example,

$$\lim_{(x,y) \to (1,2)} \left[e^{xy} \sin(\pi y/4) + xy \ln(\sqrt{y - x}) \right] = e^2 \sin\left(\frac{\pi}{2} \right) + 2 \ln 1 = e^2.$$

The following two examples illustrate techniques that are sometimes successful in handling cases with denominators approaching zero.

EXAMPLE 4 Show that $\displaystyle\lim_{(x,y)\to(0,0)} \frac{xy}{\sqrt{x^2 + y^2}} = 0.$

Solution Let (r, θ) be the polar coordinates of the point (x, y). Then

$$\frac{xy}{\sqrt{x^2 + y^2}} = \frac{(r\cos\theta)(r\sin\theta)}{\sqrt{r^2\cos^2\theta + r^2\sin^2\theta}}$$

$$= r\cos\theta\sin\theta \qquad \text{for} \quad r \neq 0.$$

Since $r = \sqrt{x^2 + y^2}$, it is clear that $r \to 0$ as x and y both approach zero. It therefore follows that

$$\lim_{(x,y)\to(0,0)} \frac{xy}{\sqrt{x^2 + y^2}} = \lim_{r\to 0} r\cos\theta\sin\theta = 0,$$

since $|\cos\theta\sin\theta| \leq 1$ for all values of θ.

EXAMPLE 5 Show that $\displaystyle\lim_{(x,y)\to(0,0)} \frac{xy}{x^2 + y^2}$ does not exist.

Solution Our plan is to show that $f(x, y) = xy/(x^2 + y^2)$ approaches different values as (x, y) approaches $(0, 0)$ from different directions. Suppose that (x, y) approaches $(0, 0)$ along the line of slope m through the origin. On this line we have $y = mx$. So on this line

$$f(x, y) = \frac{x(mx)}{x^2 + m^2 x^2} = \frac{m}{1 + m^2}$$

if $x \neq 0$. If we take $m = 1$, we see that $f(x, y) = \frac{1}{2}$ at every point of the line $y = x$ other than $(0, 0)$. If we take $m = -1$, we see that $f(x, y) = -\frac{1}{2}$ at every point of the line $y = -x$ other than $(0, 0)$. Thus $f(x, y)$ approaches two different values as (x, y) approaches $(0, 0)$ along these two lines, shown in Figure 14.10. Hence $f(x, y)$ cannot approach any *single* value as (x, y) approaches $(0, 0)$, and it follows that the limit in question does not exist.

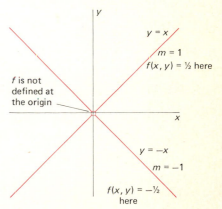

14.10 The function f of Example 5 takes on both values $+\frac{1}{2}$ and $-\frac{1}{2}$ at points arbitrarily close to the origin.

In order for $\displaystyle\lim_{(x,y)\to(a,b)} f(x, y)$ to exist, $f(x, y)$ must approach L for *any* mode of approach of (x, y) to (a, b). In Problem 29 below, we give an example of a function f such that $f(x, y) \to 0$ as $(x, y) \to (0, 0)$ along any straight line through the origin, but $f(x, y) \to 1$ as (x, y) approaches the origin along the parabola $y = x^2$. Fortunately, many important applications, including those we discuss in the remainder of this chapter, involve only functions that exhibit no such exotic behavior.

14-2 PROBLEMS

In each of Problems 1–5, state the largest possible domain of definition of the given function $f(x, y)$.

1 $f(x, y) = e^{-x^2-y^2}.$
2 $f(x, y) = \ln(x^2 + y^2 - 1).$

3 $f(x, y) = \dfrac{x + y}{x - y}.$

4 $f(x, y) = \sqrt{4 - x^2 - y^2}.$

5 $f(x, y) = \dfrac{1 + \sin xy}{xy}$.

In each of Problems 6–10, sketch the graph of the function f.

6 $f(x, y) = -10$.   **7** $f(x, y) = x$.
8 $f(x, y) = x + y$.   **9** $f(x, y) = \sqrt{4 - x^2 - 4y^2}$.
10 $f(x, y) = x^2 + y^2 - 10$.

In each of Problems 11–15, sketch some typical level curves of f.

11 $f(x, y) = x - y$.   **12** $f(x, y) = x^2 - y^2$.
13 $f(x, y) = x^2 + 4y^2$.   **14** $f(x, y) = y - x^2$.
15 $f(x, y) = y - x^3$.

In each of Problems 16–20, evaluate the limits

$$\lim_{h \to 0} \frac{f(x + h, y) - f(x, y)}{h} \quad \text{and} \quad \lim_{k \to 0} \frac{f(x, y + k) - f(x, y)}{k}.$$

16 $f(x, y) = x + y$.   **17** $f(x, y) = xy$.
18 $f(x, y) = x^2 + y^2$.   **19** $f(x, y) = xy^2 - 2$.
20 $f(x, y) = x^2 y^3 - 10$.

Use the limit laws and consequences of continuity to evaluate the limits in Problems 21–25.

21 $\displaystyle\lim_{(x,y) \to (1,-1)} e^{-xy}$.

22 $\displaystyle\lim_{(x,y) \to (0,0)} \frac{x + y}{1 + xy}$.

23 $\displaystyle\lim_{(x,y) \to (0,0)} \frac{e^{xy} \sin xy}{xy}$.

24 $\displaystyle\lim_{(x,y,z) \to (1,1,1)} (x + y + z) \ln xyz$.

25 $\displaystyle\lim_{(x,y,z) \to (1,1,0)} \frac{xy - z}{\cos xyz}$.

26 Use the method of Example 4 to show that

$$\lim_{(x,y) \to (0,0)} \frac{x^2 - y^2}{\sqrt{x^2 + y^2}} = 0.$$

27 Use the method of Example 5 to show that

$$\lim_{(x,y) \to (0,0)} \frac{x^2 - y^2}{x^2 + y^2}$$

does not exist.

28 Determine whether or not

$$\lim_{(x,y,z) \to (0,0,0)} \frac{xy + xz + yz}{x^2 + y^2 + z^2}$$

exists.

29 Let $f(x, y) = 2x^2 y/(x^4 + y^2)$. Show that
(a) $f(x, y) \to 0$ as $(x, y) \to (0, 0)$ along *any* (every) straight line through the origin; but that (b) $f(x, y) \to 1$ as $(x, y) \to (0, 0)$ along the parabola $y = x^2$. Hence conclude that the limit of $f(x, y)$ as $(x, y) \to (0, 0)$ does not exist.

30 Suppose that $f(x, y) = (x - y)/(x^3 - y)$ except at points of the curve $y = x^3$; we *define* $f(x, y)$ to be 1 at those points. Show that f is not continuous at the point $(1, 1)$. Evaluate the limits of $f(x, y)$ as $(x, y) \to (1, 1)$ along the vertical line $x = 1$ and along the horizontal line $y = 1$. (*Suggestion:* Recall that $a^3 - b^3 = (a - b)(a^2 + ab + b^2)$.)

31 Locate and identify the extrema (local or global, maximum or minimum) of the function $f(x, y) = x^2 - x + y^2 + 2y + 1$. (*Note:* The point of this problem is that you do *not* need calculus to work it.)

32 Sketch enough level curves of the function $h(x, y) = y - x^2$ to show that the function h has *no* extreme values— no local or global maxima or minima.

14-3

Partial Derivatives

Suppose that $y = f(x)$ is a function of *one* real variable. Its first derivative

$$\frac{dy}{dx} = D_x f(x) = \lim_{h \to 0} \frac{f(x + h) - f(x)}{h} \tag{1}$$

can be interpreted as the instantaneous rate of change of y with respect to x. For a function $z = f(x, y)$ of two variables, we need a similar understanding of the rate at which z changes as x and y vary (either singly or simultaneously). To reach this more complicated concept, we adopt a "divide and conquer" strategy.

First we hold y fixed and let x vary. The rate of change of z with respect to x is then denoted by $\partial z / \partial x$ and has the value

$$\frac{\partial z}{\partial x} = \lim_{h \to 0} \frac{f(x + h, y) - f(x, y)}{h}. \tag{2}$$

The value of this limit (assuming it exists) is called the **partial derivative of** f **with respect to** x. In like manner, we may hold x fixed and let y vary. The rate of change of z with respect to y is then the **partial derivative of** f **with respect to** y, defined to be

$$\frac{\partial z}{\partial y} = \lim_{k \to 0} \frac{f(x, y + k) - f(y)}{k} \tag{3}$$

for all (x, y) for which this limit exists. Note the symbol ∂ that is used instead of d to denote the partial derivatives of a function of two variables. A function of three or more independent variables has a partial derivative (defined similarly) with respect to each of its independent variables. Some other commonly used notations for partial derivatives are

$$\frac{\partial z}{\partial x} = \frac{\partial f}{\partial x} = f_x(x, y) = D_x f(x, y) = D_1 f(x, y),$$

$$\frac{\partial z}{\partial y} = \frac{\partial f}{\partial y} = f_y(x, y) = D_y f(x, y) = D_2 f(x, y).$$

Note that if the symbol y in Equation (2) is deleted, the result is the limit of Equation (1). This means that $\partial z / \partial x$ can be calculated as an "ordinary" derivative with respect to x, simply regarding y as a constant during the process of differentiation. Similarly, we can compute $\partial z / \partial y$ as an ordinary derivative, thinking of y as the *only* variable and treating x as a constant during the computation.

EXAMPLE 1 Compute the partial derivatives $\partial f / \partial x$ and $\partial f / \partial y$ of the function $f(x, y) = x^2 + 2xy^2 - y^3$.

Solution To compute the partial of f with respect to x, we regard y as a constant. Then we differentiate normally and find that

$$\frac{\partial f}{\partial x} = 2x + 2y^2.$$

When we regard x as a constant and differentiate with respect to y, we find that

$$\frac{\partial f}{\partial y} = 4xy - 3y^2.$$

To get an intuitive feel for the meaning of partial derivatives, we can think of $f(x, y)$ as the temperature at the point (x, y). Then $f_x(x, y)$ is the instantaneous rate of change of temperature at (x, y) per unit increase in x (with y held constant), while $f_y(x, y)$ is the rate of change of temperature per unit increase in y (with x held constant). For example, with the temperature function $f(x, y) = x^2 + 2xy^2 - y^3$ of Example 1, the rate of change of temperature at the point $(1, -1)$ is $+4°$ per unit distance in the positive x-direction and $-7°$ per unit distance in the positive y-direction.

EXAMPLE 2 The volume V (in cubic centimeters) of one mole of an ideal gas is given by

$$V = \frac{(82.06)T}{P},$$

where P is the pressure (in atmospheres) and T is the absolute temperature (in $°K = °C + 273$). Find the rates of change of the volume of one mole of an ideal gas with respect to pressure and with respect to temperature with $T = 300°K$ and $P = 5$ atm.

Solution The partial derivatives of V with respect to its two variables are

$$\frac{\partial V}{\partial P} = -\frac{(82.06)T}{P^2} \quad \text{and} \quad \frac{\partial V}{\partial T} = \frac{82.06}{P}.$$

With $T = 300°K$ and $P = 5$ atm, we have the values $\partial V/\partial P = -984.72$ cm³/atm and $\partial V/\partial T = 16.41$ cm³/°K. These partial derivatives allow us to estimate the effect of a change in temperature or in pressure on the volume V of gas in question, as follows. We are given $T = 300°K$ and $P = 5$ atm, so the volume of gas we are dealing with is $V = (82.06)(300)/5 = 4923.60$ cm³. We would expect an increase in pressure of one atmosphere (with temperature held constant) to decrease the volume of gas by roughly one liter (1000 cm³). An increase in temperature of 1°K (or 1°C) would, with pressure held constant, increase the volume by about 16 cm³.

GEOMETRIC INTERPRETATION OF PARTIAL DERIVATIVES

The partial derivatives f_x and f_y are the slopes of tangent lines to certain curves on the surface $z = f(x, y)$. Consider the point $P(a, b, f(a, b))$ on this surface. It lies directly above the point $Q(a, b, 0)$ in the xy-plane, as shown in Figure 14.11. The vertical plane $y = b$ parallel to the xz-plane intersects the surface in the curve $z = f(x, b)$ through the point P. Along this curve, x varies but y is constant; y is always equal to b, because the curve lies in the vertical plane with equation $y = b$.

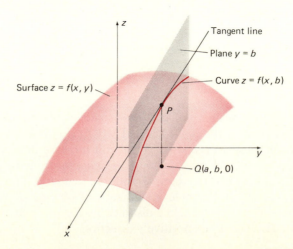

Surface $z = f(x, y)$
Tangent line
Plane $y = b$
Curve $z = f(x, b)$
P
$Q(a, b, 0)$

14.11 Geometric interpretation of $f_x(a, b)$ as the slope of a tangent line to a *curve*.

Let us project this curve and its tangent line at P into the xz-plane. We get exactly the same curve and tangent line, since this is a normal projection. But since we are now working in the xz-plane, we can actually "forget" the presence of y. In effect, as indicated in Figure 14.12, we are dealing with z as a function of the *single* variable x, and the projected curve is the graph of this function. Hence the slope of the tangent line to the original curve at the point $P(a, b, f(a, b))$ is equal to the slope of the tangent line of Figure 14.12. But by familiar single-variable calculus, this slope is

$$\lim_{h \to 0} \frac{f(a + h, b) - f(a, b)}{h} = f_x(a, b).$$

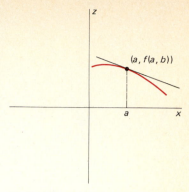

14.12 The projection of the curve and tangent line of Figure 14.11 into the xz-plane.

Thus $f_x(a, b)$ is the slope of the tangent line at P to the curve formed by intersecting the plane $y = b$ parallel to the xz-plane with the surface $z = f(x, y)$.

We proceed in much the same way to get a geometric interpretation of the partial derivative of f with respect to y. As Figure 14.13 suggests, $f_y(a, b)$ *is the slope of a line tangent to the curve of intersection of the surface $z = f(x, y)$ with the plane $x = a$ parallel to the yz-plane.*

We now define the **tangent plane** to the surface $z = f(x, y)$ at the point $P(a, b, f(a, b))$ to be the plane through P that contains the two tangent lines we have just found. In Section 14-7 we shall see that, if the partial derivatives f_x and f_y are continuous functions of x and y, then this plane contains the tangent line at P to *every* smooth curve on the surface passing through the point P.

In order to write the equation of this tangent plane, all we need is a vector $\mathbf{n}$ normal to the plane. One way to get such a normal vector is to form the cross product of tangent (or velocity) vectors to the curves $z = f(a, y)$ and $z = f(x, b)$ discussed above. The first curve has the parametrization $x = a$, $y = t$, $z = f(a, t)$ and passes through P when $t = b$ with velocity vector

$$\mathbf{v}_1 = \mathbf{j} + \mathbf{k}f_y(a, b).$$

The second curve has the parametrization $x = t$, $y = b$, $z = f(t, b)$ and passes through P when $t = a$ with velocity vector

$$\mathbf{v}_2 = \mathbf{i} + \mathbf{k}f_x(a, b).$$

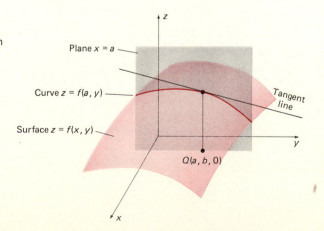

14.13 The partial derivative f_y is also the slope of a line tangent to a curve in the surface $z = f(x, y)$.

Plane $x = a$

Curve $z = f(a, y)$

Tangent line

Surface $z = f(x, y)$

$Q(a, b, 0)$

Using these two tangent vectors we obtain the normal vector

$$\mathbf{n} = \mathbf{v}_1 \times \mathbf{v}_2 = \begin{vmatrix} \mathbf{i} & \mathbf{j} & \mathbf{k} \\ 0 & 1 & f_y(a, b) \\ 1 & 0 & f_x(a, b) \end{vmatrix},$$

so that

$$\mathbf{n} = \mathbf{i}f_x(a, b) + \mathbf{j}f_y(a, b) - \mathbf{k}. \tag{4}$$

Note that

$$\mathbf{n} = \left\langle \frac{\partial z}{\partial x}, \frac{\partial z}{\partial y}, -1 \right\rangle \tag{5}$$

is a downward-pointing vector (because of its negative z-component); its negative $-\mathbf{n}$ is the normal vector shown in Figure 14.14.

Finally we use the normal vector $\mathbf{n}$ of Equation (4) to find the equation of the tangent plane to the surface $z = f(x, y)$ at the point $(a, b, f(a, b))$. This equation is

$$f_x(a, b)(x - a) + f_y(a, b)(y - b) - [z - f(a, b)] = 0. \tag{6}$$

EXAMPLE 3 Write an equation of the tangent plane to the paraboloid $z = x^2 + y^2$ at the point $(2, -1, 5)$.

Solution $\partial z/\partial x = 2x$ and $\partial z/\partial y = 2y$. The values of these partial derivatives when $x = 2$, $y = -1$ are 4 and -2, respectively, so Equation (6) gives

$$4(x - 2) - 2(y + 1) - (z - 5) = 0$$

or $4x - 2y - z = 5$ as an equation of the indicated tangent plane.

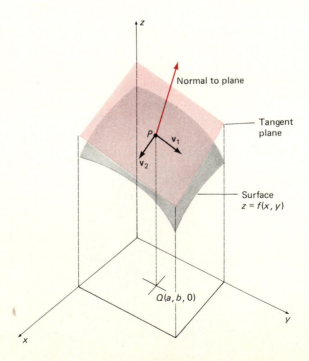

Normal to plane

Tangent plane

Surface $z = f(x, y)$

$Q(a, b, 0)$

14.14 A normal vector to the tangent plane can be found by forming the cross product of two tangent vectors.

HIGHER PARTIAL DERIVATIVES

The **first order** partial derivatives f_x and f_y are themselves functions of x and y, so they may be differentiable with respect to x and y. The partial derivatives of $f_x(x, y)$ and $f_y(x, y)$ are called the **second order partial derivatives** of f. There are four of them because there are four possibilities in the order of differentiation:

$$(f_x)_x = f_{xx} = \frac{\partial f_x}{\partial x} = \frac{\partial}{\partial x}\left(\frac{\partial f}{\partial x}\right) = \frac{\partial^2 f}{\partial x^2},$$

$$(f_x)_y = f_{xy} = \frac{\partial f_x}{\partial y} = \frac{\partial}{\partial y}\left(\frac{\partial f}{\partial x}\right) = \frac{\partial^2 f}{\partial y\, \partial x},$$

$$(f_y)_x = f_{yx} = \frac{\partial f_y}{\partial x} = \frac{\partial}{\partial x}\left(\frac{\partial f}{\partial y}\right) = \frac{\partial^2 f}{\partial x\, \partial y},$$

$$(f_y)_y = f_{yy} = \frac{\partial f_y}{\partial y} = \frac{\partial}{\partial y}\left(\frac{\partial f}{\partial y}\right) = \frac{\partial^2 f}{\partial y^2}.$$

If we write $z = f(x, y)$, we can replace each occurrence of the symbol f above by z.

Note that f_{xy} is the second order partial derivative of f with respect to x first, then y; f_{yx} is the result of differentiating with respect to x and y in the reverse order. It is proved in advanced calculus that these two "mixed" second partial derivatives are equal if they are continuous. More precisely, if f_{xy} and f_{yx} are continuous on a circular disk centered at the point (a, b), then

$$f_{xy}(a, b) = f_{yx}(a, b). \tag{7}$$

Since most functions of interest to us will have second order partial derivatives that are continuous everywhere they are defined, we shall ordinarily need deal with only three distinct second partial derivatives rather than with four. Similarly, if $f(x, y, z)$ is a function of three variables with continuous second order partial derivatives, then

$$\frac{\partial^2 f}{\partial x\, \partial y} = \frac{\partial^2 f}{\partial y\, \partial x}, \qquad \frac{\partial^2 f}{\partial x\, \partial z} = \frac{\partial^2 f}{\partial z\, \partial x}, \quad \text{and} \quad \frac{\partial^2 f}{\partial y\, \partial z} = \frac{\partial^2 f}{\partial z\, \partial y}.$$

Third order and higher order partial derivatives are defined similarly, and the order in which the differentiations are performed is unimportant so long as all derivatives involved are continuous. For instance, the distinct third order partial derivatives of a function $z = f(x, y)$ are

$$f_{xxx} = \frac{\partial}{\partial x}\left(\frac{\partial^2 f}{\partial x^2}\right) = \frac{\partial^3 f}{\partial x^3}, \qquad f_{xxy} = \frac{\partial}{\partial y}\left(\frac{\partial^2 f}{\partial x^2}\right) = \frac{\partial^3 f}{\partial y\, \partial x^2},$$

$$f_{xyy} = \frac{\partial}{\partial y}\left(\frac{\partial^2 f}{\partial y\, \partial x}\right) = \frac{\partial^3 f}{\partial y^2\, \partial x}, \qquad f_{yyy} = \frac{\partial}{\partial y}\left(\frac{\partial^2 f}{\partial y^2}\right) = \frac{\partial^3 f}{\partial y^3}.$$

EXAMPLE 4 Show that the partial derivatives of third order and higher of the function $f(x, y) = x^2 + 2xy^2 - y^3$ are all constant.

Solution We find that

$$f_x(x, y) = 2x + 2y^2 \quad \text{and} \quad f_y(x, y) = 4xy - 3y^2.$$

So

$$f_{xx}(x, y) = 2, \quad f_{xy}(x, y) = f_{yx}(x, y) = 4y, \quad \text{and} \quad f_{yy}(x, y) = 4x - 6y.$$

And finally,

$$f_{xxx}(x, y) = 0, \qquad f_{xxy}(x, y) = 0,$$

$$f_{xyy}(x, y) = 4, \quad \text{and} \quad f_{yyy}(x, y) = -6.$$

The function f is a polynomial, so all its partial derivatives are polynomials and are therefore continuous. Hence there is no need to check any other third order partial derivatives; each equals one of the last four above. Since the third order partial derivatives are all constant, all higher partial derivatives of f are zero.

14-3 PROBLEMS

Compute the first and second order partial derivatives of the functions in Problems 1–15.

1 $f(x, y) = x^4 - x^3 y + x^2 y^2 - xy^3 + y^4.$
2 $f(x, y) = x \sin y.$
3 $f(x, y) = e^x(\cos y - \sin y).$
4 $f(x, y) = x^2 e^{xy}.$
5 $f(x, y) = \dfrac{x + y}{x - y}.$
6 $f(x, y) = \dfrac{xy}{x^2 + y^2}.$ **7** $f(x, y) = \ln(x^2 + y^2).$
8 $f(x, y) = (x - y)^{14}.$ **9** $f(x, y) = x^y.$
10 $f(x, y) = \tan^{-1} xy.$ **11** $f(x, y, z) = x^2 y^3 z^4.$
12 $f(x, y, z) = x^2 + y^3 + z^4.$ **13** $f(x, y, z) = e^{xyz}.$
14 $f(x, y, z) = x^4 - 16yz.$ **15** $f(x, y, z) = x^2 e^y \ln z.$

In each of Problems 16–20, find an equation of the plane tangent to the given surface $z = f(x, y)$ at the given point P.

16 $z = 3x + 4y; \quad P = (1, 1, 7).$
17 $z = xy; \quad P = (1, -1, -1).$
18 $z = e^{-x^2 - y^2}; \quad P = (0, 0, 1).$
19 $z = x^2 - 4y^2; \quad P = (5, 2, 9).$
20 $z = \sqrt{x^2 + y^2}; \quad P = (3, -4, 5).$

21 Verify that the mixed second partial derivatives f_{xy} and f_{yx} are indeed equal if $f(x, y) = x^m y^n$ (m and n are positive integers).
22 Suppose that $z = e^{x+y}$. Show that the result of differentiating z m times with respect to x and n times with respect to y is

$$\frac{\partial^{m+n} z}{\partial x^m \partial y^n} = e^{x+y}.$$

23 Let $f(x, y, z) = e^{xyz}$. Calculate the six distinct second partial derivatives of f and also the third partial derivative f_{xyz}.

24 Suppose that $g(x, y) = \sin xy$. Verify that $g_{xy} = g_{yx}$ and that $g_{xxy} = g_{xyx} = g_{yxx}$.
25 In physics it is shown that the temperature $u(x, y)$ at the point x and the time t of a long thin insulated rod lying along the x-axis satisfies the *one-dimensional heat equation*

$$\frac{\partial u}{\partial t} = k \frac{\partial^2 u}{\partial x^2} \qquad (k \text{ is a constant}).$$

Show that the function

$$u = u(x, t) = e^{-n^2 kt} \sin nx$$

satisfies the heat equation for any choice of the constant n.
26 The *two-dimensional heat equation* for an insulated plane is

$$\frac{\partial u}{\partial t} = k \left(\frac{\partial^2 u}{\partial x^2} + \frac{\partial^2 u}{\partial y^2} \right).$$

Show that the function

$$u = u(x, y, t) = e^{-(m^2 + n^2)kt} \sin mx \cos ny$$

satisfies this equation for any choice of constants m and n.
27 A string is stretched along the x-axis, fixed at each end, then set in vibration. In physics it is shown that the displacement $y = y(x, t)$ of the point of the string at location x at time t satisfies the *one-dimensional wave equation*

$$\frac{\partial^2 y}{\partial t^2} = a^2 \frac{\partial^2 y}{\partial x^2}$$

where the constant a depends upon the density and tension of the string. Show that the following functions each satisfy the wave equation above:

(a) $y = \sin(x + at).$
(b) $y = \cosh(3[x - at]).$
(c) $y = \sin kx \cos kat$ (k is a constant).

28 A steady-state temperature function $u = u(x, y)$ for a thin flat plate satisfies *Laplace's equation*

$$\frac{\partial^2 u}{\partial x^2} + \frac{\partial^2 u}{\partial y^2} = 0.$$

Show that each of the following functions satisfies Laplace's equation.

(a) $u = \ln(\sqrt{x^2 + y^2})$.
(b) $u = (x^2 + y^2)^{1/2}$.
(c) $u = \arctan(y/x)$.
(d) $u = e^{-x} \sin y$.

29 The **ideal gas law** $PV = nRT$ (n is the number of moles of gas, R is a constant) determines each of the three variables P, V, and T (pressure, volume, and absolute temperature, respectively) as functions of the other two. Show that

$$\frac{\partial P}{\partial V} \frac{\partial V}{\partial T} \frac{\partial T}{\partial P} = -1.$$

30 It is geometrically clear that every tangent plane to the cone $z^2 = x^2 + y^2$ passes through the origin. Show this by methods of calculus.

31 There is only one point at which the tangent plane to the surface $z = x^2 + 2xy + 2y^2 - 6x + 8y$ is horizontal. Find it.

32 Show that the plane tangent to the paraboloid $z = x^2 + y^2$ at the point (a, b, c) intersects the xy-plane in the line with equation $2ax + 2by = a^2 + b^2$. Then apply the formula for the distance from a point to a line to conclude that this line is tangent to the circle $x^2 + y^2 = \frac{1}{4}(a^2 + b^2)$.

33 According to the van der Waals equation, one mole of a gas satisfies the equation

$$\left(P + \frac{a}{V^2}\right)(V - b) = (82.06)T$$

where P, V, and T are as in Example 2. For carbon dioxide, $a = 3.59 \times 10^6$ and $b = 42.7$, and V is 25,600 cm^3 when P is 1 atm and $T = 313°$K.

(a) Compute $\partial V/\partial P$ by differentiating the above equation with T held constant. Then estimate the change in volume that would result from an increase of 0.1 atm in pressure with T at 313°K.

(b) Compute $\partial V/\partial T$ by differentiating the above equation with P held constant. Then estimate the change in volume that would result from an increase of 1°K in temperature with P held at 1 atm.

14-4

Maxima and Minima of Functions of Several Variables

The single-variable maximum-minimum techniques of Chapter 2 generalize in a natural manner to functions of several variables. We consider first a function f of two variables. Suppose that we are interested in the extreme values attained by f on a plane region R consisting of the points on and within a simple closed curve C, as in Figure 14.15. We say that the function f attains its **absolute** (or **global**) **maximum value** M on R at the point (a, b) of R provided that

$$f(x, y) \leq M = f(a, b)$$

for all points (x, y) of R. Similarly, f attains its **absolute** (or **global**) **minimum value** m on R at the point (c, d) provided that $f(x, y) \geq m = f(c, d)$ for all points (x, y) of R. The following theorem, proved in advanced calculus courses, guarantees the existence of absolute maximum and minimum values in many situations of practical interest.

> **Theorem 1** *Existence of Extreme Values*
> Suppose that the function f is continuous on the region R which consists of the points on and within a simple closed curve C in the plane. Then f attains an absolute maximum value at some point (a, b) of R and attains an absolute minimum value at some point (c, d) of R.

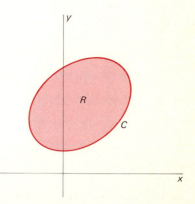

14.15 A bounded plane region R whose boundary is the simple closed curve C.

But how do we *find* these extrema? We are interested mainly in the case in which the function f attains its absolute maximum (or minimum) value at an interior point of R. The point (a, b) of R is called an **interior point** of

R provided that some circular disk centered at (a, b) lies wholly within R. The interior points of a region R of the sort described in Theorem 1 are precisely those that do *not* lie on the boundary curve C. Although this fact seems intuitively plausible, it is by no means obvious; we omit its proof.

An absolute extreme value attained by the function f at an *interior* point of R is necessarily a local extreme value. We say that $f(a, b)$ is a **local maximum value** of f if there is a circular disk D centered at (a, b) such that f is defined on D and $f(x, y) \leqq f(a, b)$ for all points (x, y) of D. If the inequality is reversed, then $f(a, b)$ is a **local minimum value** of the function f.

In Theorem 1 of Section 2-3, we saw that if the function $f(x)$ has a local maximum or local minimum value at a point $x = c$ where it is differentiable, then $f'(c) = 0$. We now shall show that an analogous event occurs in the two-dimensional situation: If $f(a, b)$ is either a local maximum value or a local minimum value of the function $f(x, y)$, then $f_x(a, b)$ and $f_y(a, b)$ are both zero, provided that these two partial derivatives exist at the point (a, b).

Suppose, for example, that $f(a, b)$ is a local maximum value of $f(x, y)$, and that the partial derivatives $f_x(a, b)$ and $f_y(a, b)$ exist. We look at cross-sections of the graph of $z = f(x, y)$ in the same way as when we defined these partial derivatives in Section 14-3. Let

$$G(x) = f(x, b) \quad \text{and} \quad H(y) = f(a, y).$$

Since f is defined on a circular disk centered at (a, b), it follows that $G(x)$ is defined on some open interval containing the point $x = a$, and that $H(y)$ is defined on some open interval containing the point $y = b$. Since f has a local maximum at (a, b), it follows that $G(x)$ has a local maximum at $x = a$ and that $H(y)$ has a local maximum at $y = b$. The single-variable maximum-minimum result cited above therefore implies that $G'(a) = 0$ and that $H'(b) = 0$. But

$$G'(a) = \lim_{h \to 0} \frac{G(a + h) - G(a)}{h} = \lim_{h \to 0} \frac{f(a + h, b) - f(a, b)}{h} = f_x(a, b)$$

and

$$H'(b) = \lim_{k \to 0} \frac{H(b + k) - H(b)}{k} = \lim_{k \to 0} \frac{f(a, b + k) - f(a, b)}{k} = f_y(a, b).$$

Hence we conclude that $f_x(a, b) = 0 = f_y(a, b)$. A similar argument gives the same conclusion if $f(a, b)$ is a local minimum value of f. This discussion establishes the following theorem.

Theorem 2   *Necessary Conditions for Local Extrema*

Suppose that $f(x, y)$ attains a local maximum value or a local minimum value at the point (a, b) and that the partial derivatives $f_x(a, b)$ and $f_y(a, b)$ both exist. Then

$$f_x(a, b) = 0 = f_y(a, b). \tag{1}$$

Equations (1) imply that the tangent plane to the surface $z = f(x, y)$ must be horizontal at any local maximum or local minimum point

$(a, b, f(a, b))$, just as in the single-variable case (in which the tangent line is horizontal at any local maximum or minimum point).

It is important to realize that the necessary condition in (1) for a local maximum or minimum is *not* a sufficient condition; that is, the fact that $f_x(a, b)$ and $f_y(a, b)$ are both zero is not by itself enough to imply that $f(a, b)$ is a local extreme value. To see this, look at the surfaces in Figure 14.16. The first of these is the graph of $f(x, y) = x^2 + y^2$. The second is the graph of $g(x, y) = -x^2 - y^2$, and the third is the graph of $h(x, y) = y^2 - x^2$. In the case of each of these three functions, both first-order partial derivatives vanish at the origin. But it is clear from the figure that $f(x, y)$ has a local minimum at $(0, 0)$, $g(x, y)$ has a local maximum at $(0, 0)$, and $h(x, y)$ has neither a local maximum nor a local minimum there. (Because of the shape of the graph of h, the point $(0, 0)$ is called a *saddle point*.) This shows that a point (a, b) that satisfies Equations (1) may be either a local maximum, a local minimum, or neither.

Nevertheless, Theorem 2 is a very useful tool for finding the absolute maximum and minimum values attained by a continuous function f on a region R of the type described in Theorem 1. If $f(a, b)$ is the absolute maximum value, for instance, then (a, b) is either an interior point of R or a point of the boundary curve C. If (a, b) is an interior point and the partial derivatives $f_x(a, b)$ and $f_y(a, b)$ both exist, then Theorem 2 implies that both partial derivatives must be zero. Thus we have the following result.

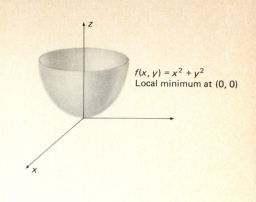

$f(x, y) = x^2 + y^2$
Local minimum at $(0, 0)$

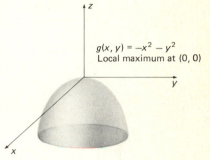

$g(x, y) = -x^2 - y^2$
Local maximum at $(0, 0)$

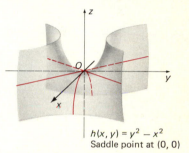

$h(x, y) = y^2 - x^2$
Saddle point at $(0, 0)$

14.16 When both partial derivatives are zero, there may be a maximum, a minimum, or neither.

Theorem 3 *Types of Absolute Extrema*

Suppose that f is continuous on the plane region R consisting of the points on and within a simple closed curve C. If $f(a, b)$ is either the absolute maximum or the absolute minimum value of $f(x, y)$ on R, then (a, b) is either:

(i) An interior point of R where

$$f_x(a, b) = 0 = f_y(a, b);$$

(ii) An interior point of R where not both first partial derivatives of f exist; or

(iii) A point of the boundary curve C.

Note the analogy with Theorem 2 in Section 2-3 which says that the maximum (or minimum) value of a single-variable function $f(x)$ on a closed interval I occurs either at an interior point c where $f'(c) = 0$, or at an interior point where f is not differentiable, or at an end point of I.

EXAMPLE 1 Let $f(x, y) = \sqrt{x^2 + y^2}$ on the region R consisting of the points on and within the circle $x^2 + y^2 = 1$ in the xy-plane. The graph of f is shown in Figure 14.17. We see that the minimum value 0 of f occurs at the origin $(0, 0)$, where the partial derivatives f_x and f_y fail to exist (why?), while the maximum value 1 of f on R occurs at *each and every* point of the boundary circle.

14.17 The graph of the function of Example 1.

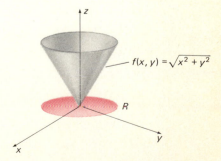

$f(x, y) = \sqrt{x^2 + y^2}$

R

In applied problems we frequently know in advance that the absolute maximum (or minimum) value of $f(x, y)$ occurs at an *interior* point of R where the partial derivatives of f exist. In this important case, Theorem 3 tells us that we can locate every possible point at which the maximum might occur by solving (simultaneously) the two equations

$$f_x(x, y) = 0 \quad \text{and} \quad f_y(x, y) = 0. \tag{2}$$

If we are so fortunate as to find that these equations have only one simultaneous solution (x, y) interior to R, then *it* must be the location of the desired maximum. If we find that Equations (2) have several solutions interior to R, then we simply evaluate the function f at each of them to determine which yields the largest value of $f(x, y)$ and is therefore the desired maximum point.

EXAMPLE 2 Find the highest point on the surface

$$z = 1 + 4x + 4y^3 - x^4 - y^4.$$

Solution Note that $z = 1$ at $(0, 0)$, so the maximum value of z—if any—is certainly positive. If we can find a region R such that z is negative at every point outside R and on the boundary of R, it will then follow that the maximum value of $z = z(x, y)$ must occur at some interior point of R. The fact that the highest-degree terms in z are of even degree and have negative coefficients suggests that such a region R exists; that is, that $z(x, y) < 0$ if $|x|$ or $|y|$ is sufficiently large. In particular, let R be the square shown in Figure 14.18, consisting of all points (x, y) such that $|x| \leq 10$ and $|y| \leq 10$. We claim that, if the point (x, y) is *not* interior to the set R, then $z(x, y) < 0$.

To see this, write

$$z = 1 + 4x + 4y^3 - x^4 - y^4 = 1 + f(x) + g(y)$$

where $f(x) = 4x - x^4$ and $g(y) = 4y^3 - y^4$. Then $f'(x) = 4(1 - x^3)$ and $f''(x) = -12x^2$. Since $f''(x) < 0$ for all $x \neq 0$ and since $x = 1$ is the only critical point of f, we may conclude that $f(1) = 3$ is the maximum value of f. Moreover,

$$f(x) \leq 4(10) - (10)^4 = -9960 \quad \text{if} \quad |x| \geq 10.$$

We can also show that 27 is the maximum value of $g(y)$, attained at $y = 3$, and that $g(y) \leq -6000$ if $|y| \geq 10$. It follows that

$$z \leq 1 - 9960 + 27 < 0 \quad \text{if} \quad |x| \geq 10$$

and that

$$z \leq 1 + 3 - 6000 < 0 \quad \text{if} \quad |y| \geq 10.$$

If (x, y) is not an interior point of R, at least one of the inequalities $|x| \geq 10$, $|y| \geq 10$ must hold, and so $z < 0$ if (x, y) is not an interior point of R. We saw at the outset that $z > 0$ at some interior points of R, so we now have enough evidence to conclude that the maximum value of z exists, and occurs at an interior point of R.

Since the partial derivatives of z with respect to x and y exist everywhere, Theorem 3 implies that we need only solve Equations (2); that is,

$$\frac{\partial z}{\partial x} = 4 - 4x^3 = 4(1 - x^3) = 0$$

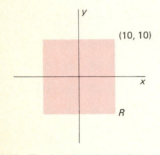

14.18 The function $z(x, y)$ of Example 2 is negative outside the square.

and

$$\frac{\partial z}{\partial y} = 12y^2 - 4y^3 = 4y^2(3 - y) = 0.$$

The first equation gives $x = 1$, the second gives $y = 0$ and $y = 3$. Thus there are two solutions for x and y: $(1, 0)$ and $(1, 3)$. Since $z(1, 0) = 4$ and $z(1, 3) = 31$, we conclude finally that $(1, 3, 31)$ is the highest point on the given surface.

EXAMPLE 3 Find the minimum cost of the rectangular box with volume 48 cubic feet discussed in Section 14-1. This is the box which has front and back that cost $1/ft^2, top and bottom that cost $2/ft^2, and two ends that cost $3/ft^2.

Solution In Section 14-1 we found that the cost C (in dollars) of this box is given by

$$C = 4xy + \frac{288}{x} + \frac{96}{y}$$

in terms of its length x and width y. Let R be a square like the one shown in Figure 14.19. Two sides of R are so close to the coordinate axes that $288/x > 1000$ on one and $96/y > 1000$ on the other. Also, the square is so large that $4xy > 1000$ at each point of the other two sides. This means that $C(x, y) > 1000$ at every point (x, y) in the first quadrant that lies on the boundary of or wholly outside the square R. Since $C(x, y)$ attains reasonably small values within R—at $C(1, 1) = 388$, for instance—it is clear that the absolute minimum of C must occur at an interior point of R.

We therefore proceed to solve the equations

$$\frac{\partial C}{\partial x} = 4y - \frac{288}{x^2} = 0, \qquad \frac{\partial C}{\partial y} = 4x - \frac{96}{y^2} = 0.$$

We multiply the first equation by x and the second one by y. This gives us

$$\frac{288}{x} = 4xy = \frac{96}{y},$$

so $x = (288/96)y = 3y$. We substitute $x = 3y$ into the second equation above and get

$$12y - \frac{96}{y^2} = 0, \quad \text{so that} \quad 12y^3 = 96.$$

Hence $y = \sqrt[3]{8} = 2$, and so $x = 6$. Therefore the minimum cost of this box is $C(6, 2) = 144$ (dollars). Since the volume of the box is $V = xyz = 48$, its height is $z = 48/(6)(2) = 4$ when $x = 6$ and $y = 2$. Thus our optimal box is 6 ft by 2 ft by 4 ft.

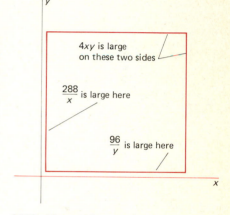

14.19 The cost function C of Example 3 takes on large positive values on the boundary of the square.

We have seen that, if $f_x(a, b) = 0 = f_y(a, b)$, then $f(a, b)$ may be either a maximum value, a local minimum value, or neither. In Section 14-10 we shall discuss sufficient conditions that $f(a, b)$ be either a local maximum or a local minimum; these will involve the second order partial derivatives of f at (a, b). Here we give a simple method that will sometimes tell us how the function behaves near the point (a, b). We write $x = a + h$, $y = b + k$, then express the difference

$$D = f(x, y) - f(a, b) \qquad (3)$$

in terms of h and k. If, for instance, we can then see by inspection that $D \geq 0$ for all h and k sufficiently small, it follows that $f(a, b)$ is a local minimum value.

EXAMPLE 4   Find the local maxima and minima of the function

$$f(x, y) = 3x^2 + 4xy + 4y^2 + 2x + 12y + 5.$$

Solution   We first apply the necessary condition (Theorem 2) for a local extremum—that the partial derivatives of f vanish. This gives

$$\frac{\partial f}{\partial x} = 6x + 4y + 2 = 0, \qquad \frac{\partial f}{\partial y} = 4x + 8y + 12 = 0.$$

The simultaneous solution of these equations is $x = 1$, $y = -2$. Hence the only possible local extreme value of f is $f(1, -2) = -6$.

Now we write $x = 1 + h$ and $y = -2 + k$ and calculate

$$\begin{aligned} D &= f(1 + h, -2 + k) - f(1, -2) \\ &= 3h^2 + 4hk + 4k^2 = (h + 2k)^2 + 2h^2. \end{aligned}$$

Our final expression for D, obtained by completing the square, makes it clear that $D > 0$ unless $h = k = 0$. Hence $f(x, y) \geq f(1, -2)$ for all x and y, so $f(1, -2) = -6$ is a local minimum value of f.

The methods of this section generalize readily to functions of three or more variables. For example, if the function $f(x, y, z)$ has a local maximum or a local minimum at the point (a, b, c) where its three first partial derivatives exist, then all three must vanish there; that is,

$$f_x(a, b, c) = f_y(a, b, c) = f_z(a, b, c) = 0. \qquad (4)$$

The following example illustrates a "line-through-the-point" method that can sometimes be used to show that a point (a, b, c) where the necessary conditions in (4) hold is neither a local maximum nor a local minimum point.

EXAMPLE 5   Determine whether the function $f(x, y, z) = xy + yz - xz$ has any local extrema.

Solution   The necessary conditions in (4) give the equations

$$f_x = y - z = 0, \qquad f_y = x + z = 0, \qquad f_z = y - x = 0.$$

We find easily that the only simultaneous solution of these equations is $x = y = z = 0$. On the line $x = y = z$ through $(0, 0, 0)$, the function $f(x, y, z)$ reduces to x^2, which is minimal at $x = 0$. But on the line $x = -y = z$ it reduces to $-3x^2$, which is maximal when $x = 0$. Hence f can have neither a local maximum nor a local minimum at $(0, 0, 0)$ and therefore has no local extrema at all.

In each of Problems 1–9, find the highest and lowest points (if any) on the given surface.

1 $z = x - 3y + 5$.
2 $z = x^2 + 6x + y^2 + 2y + 10$.
3 $z = 10 + 8x - 6y - x^2 - y^2$.
4 $z = xy + 25$.
5 $z = x^2 + x + y^3$.
6 $z = 3x^2 + 12x + 4y^3 + 6y^2 + 5$.
7 $z = 3x^4 + 4x^3 + 6y^4 - 16y^3 + 12y^2 + 7$.
8 $z = e^{-x^2 - 2x + y^2}$.

9 $z = \dfrac{1}{1 - x + y + x^2 + y^2}$.

For each of the functions in Problems 10–15, there is a single point (a, b) at which $f_x = 0 = f_y$. Use the method of Example 4 to determine whether $f(a, b)$ is a local maximum value, a local minimum value, or neither.

10 $f(x, y) = x^2 - 2x + y^2 + 2y + 3$.
11 $f(x, y) = 6x - 8y - x^2 - y^2$.
12 $f(x, y) = x^2 - y^2 + 2x - 4y + 17$.
13 $f(x, y) = xy - 2x - 3y - 4$.
14 $f(x, y) = x^2 - xy + y^2$.
15 $f(x, y) = x^2 + xy + y^2$.

16 Use the line-through-the-point method to determine the behavior of the function of Example 2 near the point $(1, 0)$. (*Suggestion:* Examine the values of f on straight lines through $(1, 0)$, including the vertical line $x = 1$.)

17 Find the dimensions x, y, and z of a rectangular box with fixed volume $V = 1000$ and minimum total surface area A.

18 Find the points of the surface $xyz = 1$ that are closest to the origin.

19 Find the dimensions of the rectangular box with maximum volume that has total surface area 600 square centimeters.

20 A rectangular box without a lid is to have fixed volume 4000 cubic centimeters. What should be its dimensions to minimize its total surface area?

21 A rectangular box is placed in the first octant, with one of its corners at the origin and three of its sides lying in the three coordinate planes. The vertex opposite the origin lies on the plane with equation $x + 2y + 3z = 6$. What is the maximum possible volume of such a box? What are the dimensions of that box?

22 The sum of three positive numbers is 120. What is the maximum possible value of their product?

23 A building in the shape of a rectangular box is to have a volume of 8000 cubic feet. Annual heating and cooling costs will amount to \$2 for each square foot of top, front, and back and \$4 for each square foot of the two end walls. What dimensions of the building will minimize these annual costs?

24 Use the maximum-minimum methods of this section to find the point of the plane $2x - 3y + z = 1$ that is closest to the point $(3, -2, 1)$. Note that the distance between two points is minimized when its square is minimized.

25 Find the maximum volume of a rectangular box that can be sent from a post office if the sum of its *length* and *girth* cannot exceed 84 inches.

26 Repeat Problem 25 for the case of a cylindrical box—one shaped like a hatbox or a fat mailing tube.

27 A rectangular box with its base in the xy-plane is inscribed under the graph of the paraboloid $z = 1 - x^2 - y^2$, $z \geq 0$. Find the maximum possible volume of the box.

28 What is the maximum possible volume of a rectangular box inscribed in a *hemisphere* of radius R? You may assume that one face of the box lies in the planar base of the hemisphere.

29 A wire 120 inches long is cut into three *or fewer* pieces, and each piece is bent into the shape of a square. How should this be done in order to minimize the total area of these squares? In order to maximize the total area?

30 A lump of putty of fixed volume V is to be divided into three or fewer pieces, and the pieces made into cubes. How should this be done so as to maximize the total surface area of the cubes? Or to minimize it?

31 Consider the function $f(x, y) = (y - x^2)(y - 3x^2)$. (a) Show that $f_x(0, 0) = 0 = f_y(0, 0)$. (b) Show that, for any straight line $y = mx$, the function $f(x, mx)$ has a local minimum at $x = 0$. (c) Examine the values of f at points of the parabola $y = 2x^2$ to show that f does *not* have a local minimum at $(0, 0)$. This shows that the line-through-the-point method cannot be used to show that a point *is* a local extremum.

---

*14-5

Least Squares and Applications to Economics

An important technique in science is the "fitting" of empirical data to equations, particularly linear equations. Suppose that the outcome of an experiment is a collection of pairs of data—points (x_1, y_1), (x_2, y_2), . . . , (x_n, y_n). For example, y_i might be the yield from an acre of cornfield to which x_i

pounds of fertilizer had been applied. Suppose that we have some reason to expect the relationship between the x-values and the y-values to be linear:

$$y = mx + b, \tag{1}$$

at least approximately, for some particular choice of slope m and y-intercept b.

In actuality, the empirically observed points are not likely to lie precisely on a straight line. Even in the best of theories there is inevitable discrepancy between prediction and reality. Hence a graph of the data points, together with a straight line that seems to fit them well, might look something like Figure 14.20.

We want to choose the constants m and b in Equation (1) so that the line "fits" the points as closely as possible. It is customary to measure how well a given line fits the empirical data points in terms of the *vertical* distances from the points to the line. Specifically, let $Q_i(x_i, mx_i + b)$ be the point on the line $y = mx + b$ that lies directly above or below the ith empirical point $P_i(x_i, y_i)$. We define the **deviation** between the ith point and the line to be

$$d_i = y_i - (mx_i + b) \tag{2}$$

for $1 \leqq i \leqq n$.

Our first inclination might be to choose the line so that the sum Σd_i of the deviations is as small as possible. But in this case positive deviations could cancel negative deviations, so that Σd_i is small while none of the points lies very close to the line. We avoid this by minimizing the sum of the *squares* of the deviations given in (2). Doing so is called the **least squares method** of fitting a straight line to empirical data; we choose m and b in Equation (1) so as to minimize the function

$$f(m, b) = d_1^2 + d_2^2 + d_3^2 + \cdots + d_n^2 = \sum_{i=1}^{n} [y_i - (mx_i + b)]^2. \tag{3}$$

The function f of Equation (3) is a function of the variables m and b; the numbers x_i and y_i are all constants. The partial derivatives of f are

$$\frac{\partial f}{\partial m} = \sum_{i=1}^{n} 2[y_i - (mx_i + b)](-x_i)$$

$$= 2m \sum_{i=1}^{n} x_i^2 + 2b \sum_{i=1}^{n} x_i - 2 \sum_{i=1}^{n} x_i y_i$$

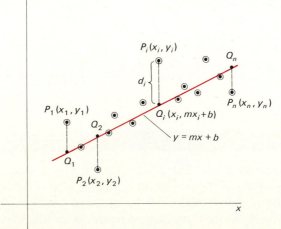

14.20 Fitting the best straight line to the data points (x_i, y_i), $1 \leqq i \leqq n$.

CHAP. 14: **Partial Differentiation**

and

$$\frac{\partial f}{\partial b} = \sum_{i=1}^{n} 2[y_i - (mx_i + b)](-1)$$

$$= 2m \sum_{i=1}^{n} x_i + 2b \sum_{i=1}^{n} 1 - 2 \sum_{i=1}^{n} y_i.$$

When we set both partial derivatives equal to zero, we get the pair of *linear* equations

$$m \sum_{i=1}^{n} x_i + nb \qquad = \sum_{i=1}^{n} y_i, \qquad (4)$$

$$m \sum_{i=1}^{n} x_i^2 + b \sum_{i=1}^{n} x_i = \sum_{i=1}^{n} x_i y_i. \qquad (5)$$

We can now solve these two linear equations simultaneously for m and b. It can be shown (using the method of Section 14-10) that the resulting line $y = mx + b$ always minimizes the sum of the squares of the deviations. Hence we need not worry in each individual problem about whether we have found a maximum, a minimum, or neither.

EXAMPLE 1   As a step toward determining the selling price that would maximize its profits, a perfume company wished to express its weekly sales y (in thousands of bottles) as a linear function of the price x (in dollars per bottle). In order to do this, it test-sold the perfume in four similar cities, with the following results:

	City 1	City 2	City 3	City 4
x	6.25	6.75	8.00	8.75
y	6.03	5.62	4.78	4.34

If the company has fixed weekly costs of $25,000, plus $1.25 per bottle, what selling price will maximize its weekly profit?

Solution   To find the sales-price relation $y = mx + b$, we first compute the coefficients in Equations (4) and (5). We find that

$\Sigma x_i = 6.25 + 6.75 + 8.00 + 8.75 = 29.75,$

$\Sigma x_i^2 = (6.25)^2 + (6.75)^2 + (8.00)^2 + (8.75)^2 = 225.19,$

$\Sigma y_i = 6.03 + 5.62 + 4.78 + 4.34 = 20.77,$ and

$\Sigma x_i y_i = (6.25)(6.03) + (6.75)(5.62) + (8.00)(4.78) + (8.75)(4.34) = 151.84$

(we carry two-place accuracy since the data are no better than this). Since $n = 4$ in this example, we then solve the equations [from (4) and (5)]

$$29.75m + \quad 4b = \quad 20.77,$$

$$225.19m + 29.75b = 151.84$$

and find that $m = -0.67$ and $b = 10.19$. Thus each increase of $1 in price will reduce weekly sales by 670 bottles, because the sales-price relation is

$$y = -0.67x + 10.19 \quad \text{(thousands)}.$$

The company's weekly production costs for y thousand bottles are $C = 25 + 1.25y$ (in thousands of dollars). So its weekly profits (still in thousands of dollars) are

$$P = xy - C = x(-0.67x + 10.19) - 25 - (1.25)(-0.67x + 10.19),$$

and so

$$P = P(x) = -0.67x^2 + 11.03x - 37.74.$$

Now $dP/dx = -1.34x + 11.03$, and $dP/dx = 0$ when $x = 8.23$. Thus $8.23 is the optimal selling price for each bottle of perfume. To complete the analysis, it follows that

$y = 4.68$ (thousand) bottles will be the weekly sales,   and

$P = 7.66$ (thousand) dollars will be the weekly profit,

with a profit of $1.64 on each bottle.

You can use the method of least squares to fit empirical data to non-linear equations or to equations with more than one independent variable. To fit the data $(x_1, y_1), (x_2, y_2), \ldots, (x_n, y_n)$ to the second degree equation

$$y = Ax^2 + Bx + C,$$

you need only to minimize the function of three variables

$$g(A, B, C) = \sum_{i=1}^{n} [y_i - (Ax_i^2 + Bx_i + C)]^2.$$

To fit the data $(x_1, y_1, z_1), (x_2, y_2, z_2), \ldots, (x_n, y_n, z_n)$ to the linear equation

$$z = Ax + By + C,$$

you need only to minimize the function

$$g(A, B, C) = \sum_{i=1}^{n} [z_i - (Ax_i + By_i + C)]^2.$$

Each case would call for solving the *three* simultaneous equations

$$\frac{\partial g}{\partial A} = \frac{\partial g}{\partial B} = \frac{\partial g}{\partial C} = 0.$$

The great German mathematician Carl Friedrich Gauss (1777–1855) invented the method of least squares when he was 18 years old. A short time later, he used it to determine the orbit of the first-discovered asteroid Ceres, initially observed on the first day of the nineteenth century, but lost from sight a few weeks later. Ceres was later rediscovered in the position that Gauss had predicted from the limited number of original observations.

COMPETITION AND COLLUSION

Here we give an economic application of maximum-minimum techniques. Suppose that companies 1 and 2 make competitive products, with the weekly sales of each product being determined jointly by the price of the product and the price of the competing product. For simplicity, let us suppose that the cost of production of each product is the same, $1 per unit.

The management of each company does some market research like that described in Example 1. They find that the weekly sales m and n of products 1 and 2 (in thousands of units) are given in terms of their respective prices x and y by the equations

$$m = 10 - 2x + y, \tag{6}$$

$$n = 15 + 2x - 3y. \tag{7}$$

Because

$$\frac{\partial m}{\partial x} = -2, \qquad \frac{\partial n}{\partial x} = +2,$$

$$\frac{\partial m}{\partial y} = +1, \quad \text{and} \quad \frac{\partial n}{\partial y} = -3,$$

we see that our model is at least qualitatively plausible. For an increase of $1 in the price of product 1 decreases its weekly sales by 2 (thousand) units, but increases the sales of product 2 by 2 (thousand) units. An increase of $1 in the price of product 2 decreases its weekly sales by 3 (thousand) units but increases the sales of product 1 by 1 (thousand) units. This is part of the basic nature of competition: Each price change in either product has opposite effects on their respective sales, as one's gain is the other's loss.

Suppose first that the two companies act independently—that is, in *competition*—to maximize their individual weekly profits. From the point of view of company 1, its own selling price x is a *variable* under its control, but the selling price y of product 2 is *fixed* by its competitor. The weekly profit of company 1 is then a function of x, $P = P(x)$, given by

$$P = mx - m = (10 - 2x + y)(x - 1);$$

$$P(x) = -2x^2 + 12x + xy - y - 10. \tag{8}$$

If the management of company 1 knew in advance the selling price y of product 2, then it would determine its own selling price x by solving the equation

$$P'(x) = -4x + y + 12 = 0 \tag{9}$$

for the purpose of maximizing its own profit.

From the point of view of company 2, the roles of the selling prices x and y are reversed. Its own selling price y is a *variable*, while x is *fixed*. The weekly profit of company 2 is then a function of y, $Q = Q(y)$, given by

$$Q = ny - n = (15 + 2x - 3y)(y - 1);$$

$$Q(y) = -3y^2 + 18y + 2xy - 2x - 15. \tag{10}$$

If the management of company 2 knew in advance its competitor's selling price x, then it would determine its own selling price y by solving the equation

$$Q'(y) = 2x - 6y + 18 = 0. \tag{11}$$

If the management of each company credits its competitor with a knowledge of elementary calculus, then each will solve the simultaneous linear equations in (9) and (11) and get

$$x = \tfrac{45}{11} \approx \$4.09, \qquad y = \tfrac{48}{11} \approx \$4.36.$$

Equations (6) and (7) will then give their weekly sales

$$m = 6.182, \qquad n = 10.091$$

(in thousands of units). Equations (8) and (10) finally give their respective weekly profits

$$P = 19.108, \qquad Q = 33.942$$

in thousands of dollars.

Now suppose instead that the managements of the two companies act in *collusion*. They enter into a price-fixing agreement (legal or otherwise) by which they plan to maximize their total weekly profit. (We suppose that they will divide the resulting profit in an equitable way, but the details of this intriguing problem are, for the moment, not the issue.)

The total weekly profit $T = T(x, y)$ of the two companies will be given by

$$T = (\text{revenue}) - (\text{cost}) = (mx + ny) - (m + n);$$
$$T(x, y) = 10x + 17y - 2x^2 + 3xy - 3y^2 - 25. \tag{12}$$

To maximize T they solve the two equations

$$\frac{\partial T}{\partial x} = 10 - 4x + 3y = 0,$$

$$\frac{\partial T}{\partial y} = 17 + 3x - 6y = 0$$

for $x = \$7.40$, $y = \$6.53$. These values give a total weekly profit of \$11,072 by the first company and \$56,461 by the second. The total weekly profit of the two companies is \$67,533, an increase of \$14,483 over their combined weekly profits when they acted independently. Figure 14.21 shows a comparison of the two situations—competition versus collusion.

Competition model		
Price 1: \$4.09	Company 1 profit/week:	\$19,108
Price 2: \$4.36	Company 2 profit/week:	\$33,942
	Total weekly profit:	\$53,050

Collusion model		
Price 1: \$7.40	Company 1 profit/week:	\$11,072
Price 2: \$6.53	Company 2 profit/week:	\$56,461
	Total weekly profit:	\$67,533

14.21 Comparison of the competition model with the collusion model.

In Problems 1–3, find the straight line $y = mx + b$ that best fits the given data points (x_i, y_i).

1 $(-1, 1.9), (1, 8.1), (4, 16.8)$.
2 $(-1.73, 4.61), (2.25, -3.23), (5.67, -2.13)$.
3 $(1.4, -11), (2.7, 11), (3.6, 26), (5.5, 58)$.
4 In order to find a linear relation $y = mx + b$ between the price x per box (in cents) and the weekly sales y (in thousands of cases), a soap manufacturer test-sold soap in three cities at different prices, with the following results.

x	79	89	99
y	110	95	85

What weekly sales should be expected if the price is set at 69¢ per box?

5 The per capita consumption of cigarettes in 1930 and the lung cancer death rate (deaths per million males) for 1950 in the Scandinavian countries were as follows:

Country	Cigarette Consumption	Deaths from Lung Cancer
Denmark	350	165
Finland	1100	350
Norway	250	95
Sweden	300	120

Fit these data to a straight line to estimate the 1950 male lung cancer death rate for Australia, in which the 1930 per capita cigarette consumption was 470. (The actual death rate was 170.) What lung cancer death rate might you expect in a country in which no cigarettes at all are smoked?

6 The total cost y (dollars) for printing x copies of a certain booklet is known for three lot sizes; here are the details:

x	700	2700	3700
y	2700	4200	5000

Express the total cost as the sum of a fixed cost and an additional cost per book printed. What, then, should 2000 copies cost?

7 The systolic blood pressure p (in millimeters of mercury) of a healthy child is essentially a linear function $p = m(\ln w) + b$ of the *logarithm* of his or her weight w (in pounds). The numbers m and b are constants. Determine them from the following experimental data.

w	41	67	78	93	125
p	89	100	103	107	110

What should the systolic blood pressure of a healthy 100-pound child be?

8 Consider the sum of squares

$$g(A, B, C) = \sum_{i=1}^{n} [z_i - (Ax_i + By_i + C)]^2$$

that you would use to fit the data $(x_1, y_1, z_1), (x_2, y_2, z_2), \ldots, (x_n, y_n, z_n)$ to the linear equation $z = Ax + By + C$. Show that the conditions $g_A = g_B = g_C = 0$ reduce to the simultaneous equations

$$A\Sigma x_i + B\Sigma y_i + nC = \Sigma z_i,$$
$$A\Sigma x_i^2 + B\Sigma x_i y_i + C\Sigma x_i = \Sigma x_i z_i,$$
$$A\Sigma x_i y_i + B\Sigma y_i^2 + C\Sigma y_i = \Sigma y_i z_i.$$

9 A company wishes to determine the weekly sales z (in thousands of units) as a linear function $z = Ax + By + C$ of the price x (in dollars) of its own product and the price y of a competitor's product. Given the data below, use the equations of Problem 8 to find A, B, and C. What sales should be expected if x is \$7.50 and y is \$6.50?

x	y	z
5	6	7.1
7	5	13.8
6	7	6.2
6	6	8.9

10 A company manufactures a product and sells it in two different countries. In country A, it can sell $1000 - 10x$ units at x dollars each, and in country B it can sell $2000 - 15y$ units at y dollars each. The production costs for the units are \$30,000 (fixed cost) plus \$20 for each unit. What prices x and y will maximize the total profit of the company?

11 A company manufactures and sells two competing products (I and II) that cost \$15 and \$25 per unit, respectively, to produce. The total revenue from making x units of product I and y units of product II is

$$55x + 45y - (0.02)xy - (0.15)x^2 - (0.05)y^2.$$

Find the values of x and y that will maximize the total profit (revenue minus cost) of the company.

12 The owner of an appliance store charges x dollars per videotape recorder and y dollars per service policy for each such machine. As a result he can sell

$$27,500 - 30x - 5y \quad \text{videotape recorders}$$

and

$$6500 - 5x - 20y \quad \text{service policies}.$$

What should be the values of x and y to maximize his total revenue?

13 A farmer can raise sheep, hogs, and cattle. She has space for 80 sheep or 120 hogs or 60 cattle or any combination using the same amount of space; that is, 8 sheep use as much space as 12 hogs or 6 cattle. The anticipated profits per animal are $10 per sheep, $8 per hog, and $20 for each head of cattle. State law requires that a farmer raise as many hogs as sheep and cattle combined. How does the farmer maximize her profit?

14 The profit of two firms depends in each case upon the production of both. Say the first firm has profit P, the second Q, the first has production x, and the second y. These quantities are thus related:

$$P = 48x - 2x^2 - 3y^2 + 1000;$$
$$Q = 60y - 3y^2 - 2x^2 + 2000.$$

(a) If each firm acts to maximize its profit independently (but while crediting the competition with intelligence), find P, Q, x, y, and $T = P + Q$. (b) If the firms form an agreement to maximize $T = P + Q$ and act accordingly, find x, y, and T.

15 Three firms—Ajax Products (AP), Behemoth Quicksilver (BQ), and Conglomerate Resources (CR)—produce products in quantities A, B, and C, respectively. The profits that accrue to each obey the following laws:

AP: $P = 1000A - A^2 - 2AB.$

BQ: $Q = 2000B - 2B^2 - 4BC.$

CR: $R = 1500C - 3C^2 - 6AC.$

(a) If each firm acts independently to maximize its profit, what will the profits be? (b) If firms AP and CR join to maximize their total profit while BQ continues to act alone, what effects will this have? Give a *complete* answer to this part. Assume that the fact of the merger of AP and CR *is* known to the management of BQ.

14-6

Increments and Differentials

Given a single-variable function $f(x)$, we defined (in Section 3-2) the *increment*

$$\Delta f = \Delta y = f(a + \Delta x) - f(a)$$

and the *differential*

$$df = dy = f'(a)\,\Delta x.$$

Each of these quantities is associated with the change in x from a to $a + \Delta x$. For a *fixed* value a, the differential df is a *linear* function of Δx. We saw that df is a good approximation to the actual change Δf in the sense that

$$\Delta f = df + \varepsilon\,\Delta x = f'(a)\,\Delta x + \varepsilon\,\Delta x,$$

where ε is a function of Δx that approaches zero as $\Delta x \to 0$. Hence

$$\Delta f - df = \varepsilon\,\Delta x$$

is "very *very* small" when Δx is "very small." The relation between Δf and df is shown in Figure 14.22; df is the change in the height of the *tangent line* to $y = f(x)$ at $x = a$.

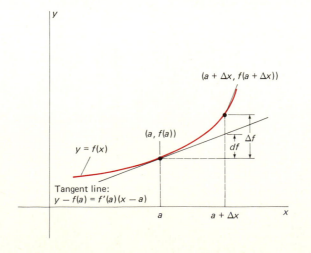

14.22 Increment and differential for a function of one variable.

We define the differential of a function $f(x, y)$ of two variables in much the same way, by using the tangent *plane* to the surface $z = f(x, y)$ in a way analogous to the way we use the tangent *line* for a single-variable function. Our goal is a simple approximation to the change in the value of f between $P(a, b)$ and $Q(a + \Delta x, b + \Delta y)$; that is, the **increment**

$$\Delta f = f(a + \Delta x, b + \Delta y) - f(a, b). \tag{1}$$

We consider the tangent plane at $(a, b, f(a, b))$, shown in Figure 14.23, with equation—by Formula (6) of Section 14-3—that may be written as

$$z - f(a, b) = f_x(a, b)(x - a) + f_y(a, b)(y - b). \tag{2}$$

The **differential** df of f at $P(a, b)$ we define to be the change in the height of the tangent plane from $P(a, b)$ to $Q(a + \Delta x, b + \Delta y)$. This is simply the value of $z - f(a, b)$ that we calculate from Equation (2), taking $x = a + \Delta x$ and $y = b + \Delta y$. So

$$df = f_x(a, b)\,\Delta x + f_y(a, b)\,\Delta y. \tag{3}$$

Note that, for a and b fixed, df is a *linear* function of Δx and Δy and that the coefficients $f_x(a, b)$ and $f_y(a, b)$ in this linear function depend upon a and b.

The differential df is a **linear approximation** to the actual increment Δf. At the end of this section, we shall show that df is a very good approximation to Δf when Δx and Δy are both small, in the sense that

$$\Delta f - df = \varepsilon_1 \,\Delta x + \varepsilon_2 \,\Delta y \tag{4}$$

where ε_1 and ε_2 are functions of Δx and Δy that approach zero as $\Delta x \to 0$ and $\Delta y \to 0$. Hence we write

$$\Delta f - df \approx 0, \quad \text{or} \quad \Delta f \approx df$$

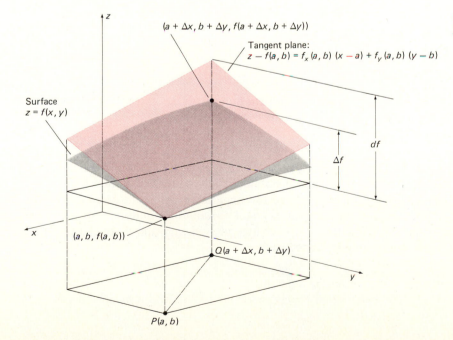

14.23 Increment and differential for $z = f(x, y)$.

when Δx and Δy are small. The approximation

$$f(a + \Delta x, b + \Delta y) = f(a, b) + \Delta f$$
$$\approx f(a, b) + df;$$
$$f(a + \Delta x, b + \Delta y) \approx f(a, b) + f_x(a, b)\,\Delta x + f_y(a, b)\,\Delta y \qquad (5)$$

may then be used to estimate the value of $f(a + \Delta x, b + \Delta y)$ when Δx and Δy are small and the values $f(a, b)$, $f_x(a, b)$, and $f_y(a, b)$ are known.

EXAMPLE 1 Use Equation (5) to estimate $\sqrt{2(2.02)^3 + (2.97)^2}$. Note that $\sqrt{2(2)^3 + (3)^2} = \sqrt{25} = 5$.

Solution We take $f(x, y) = (2x^3 + y^2)^{1/2}$, $a = 2$, $b = 3$, $\Delta x = 0.02$, and $\Delta y = -0.03$. Then $f(2, 3) = 5$, and

$$f_x(2, 3) = \frac{3(2)^2}{5} = \frac{12}{5}, \qquad f_y(2, 3) = \frac{3}{5}.$$

Hence Equation (5) gives

$$\sqrt{2(2.02)^3 + (2.97)^2} = f(2.02, 2.97)$$
$$\approx f(2, 3) + f_x(2, 3)(0.02) + f_y(2, 3)(-0.03)$$
$$= 5 + \tfrac{12}{5}(0.02) + \tfrac{3}{5}(-0.03) \approx 5.03.$$

The actual value, accurate to four decimal places, is 5.0305.

If $z = f(x, y)$, we often write dz in place of df. So the differential of the dependent variable z at the point (a, b) is $dz = f_x(a, b)\,\Delta x + f_y(a, b)\,\Delta y$. At the arbitrary point (x, y), the differential of z takes the form

$$dz = f_x(x, y)\,\Delta x + f_y(x, y)\,\Delta y.$$

More simply, we can write

$$dz = \frac{\partial f}{\partial x}\,\Delta x + \frac{\partial f}{\partial y}\,\Delta y. \qquad (6)$$

It is customary to write dx for Δx and dy for Δy in this formula; when this is done, Equation (6) takes the compact form

$$dz = \frac{\partial f}{\partial x}\,dx + \frac{\partial f}{\partial y}\,dy. \qquad (7)$$

When one uses this notation, it is important to realize that there is *no* connotation of dx and dy being "infinitesimal" or even small; dz is still simply a linear function of the ordinary real variables dx and dy, one that gives a linear approximation to the change in z when x and y are changed by the amounts dx and dy, respectively.

EXAMPLE 2 Find the differential dz if $z = x^2 + 3xy - 2y^2$, and estimate the change in z when x increases from 3 to 3.2 and y, meanwhile, decreases from 5 to 4.9.

Solution   Equation (7) gives

$$dz = (2x + 3y)\, dx + (3x - 4y)\, dy.$$

When we substitute $x = 3$, $y = 5$, $dx = 0.2$, and $dy = -0.1$, we find that

$$dz = (6 + 15)(0.2) + (9 - 20)(-0.1) = 5.3.$$

Remember that this is an approximation to the actual change

$$\Delta z = [(3.2)^2 + 3(3.2)(4.9) - 2(4.9)^2] - [9 + 45 - 50] = 5.26.$$

EXAMPLE 3   In Example 2 of Section 14-3, we considered a mole of an ideal gas—its volume V in cubic centimeters given in terms of its pressure P in atmospheres and temperature T in degrees Kelvin by the formula $V = 82.06\,T/P$. Approximate the change in its volume when P is increased from 5 atm to 5.2 atm and T is increased from $300°$K to $310°$K.

Solution   The differential of $V = V(P, T)$ is

$$dV = V_P\, dP + V_T\, dT = -\frac{82.06\,T}{P^2}\, dP + \frac{82.06}{P}\, dT.$$

With $P = 5$, $T = 300$, $dP = 0.2$, and $dT = 10$, we compute

$$dV = -\frac{(82.06)(300)}{(5)^2}(0.2) + \frac{82.06}{5}(10) = -32.8 \text{ cm}^3.$$

This indicates that the gas will decrease in volume by about 33 cm^3. The actual change is

$$\Delta V = \frac{(82.06)(310)}{5.2} - \frac{(82.06)(300)}{5} = 4892.0 - 4923.6 = -31.6$$

cubic centimeters.

Increments and differentials of functions of more than two variables are defined similarly. A function $w = f(x, y, z)$ has *increment*

$$\Delta w = \Delta f = f(x + \Delta x, y + \Delta y, z + \Delta z) - f(x, y, z)$$

and *differential*

$$dw = df = \frac{\partial f}{\partial x}\, \Delta x + \frac{\partial f}{\partial y}\, \Delta y + \frac{\partial f}{\partial z}\, \Delta z,$$

or

$$dw = w_x\, dx + w_y\, dy + w_z\, dz$$

if, as in Equation (7), we write dx for Δx, dy for Δy, and dz for Δz.

EXAMPLE 4   Suppose that three positive numbers, each less than 100, are rounded to the first decimal place. Estimate the largest error that this rounding might cause in computing their product $w = xyz$.

Solution The differential of $w = xyz$ is

$$dw = yz\,dx + xz\,dy + xy\,dz.$$

We substitute the largest possible values for x, y, and z: We take each to be 100. Because of the way the numbers are rounded, we take ± 0.05 for dx, dy, and dz. We then get

$$dw = (100)^2(\pm 0.05) + (100)^2(\pm 0.05) + (100)^2(\pm 0.05),$$

which shows that dw may be anywhere between -1500 and $+1500$. It is rather a surprise to find that an error of only 0.05 in each factor may cause an error of as much as 1500 in the product.

The differential $df = f_x\,dx + f_y\,dy$ is defined provided that the partial derivatives f_x and f_y both exist. The following theorem gives sufficient conditions that df be a good approximation to the increment Δf when Δx and Δy are small.

> **Theorem** *Linear Approximation*
>
> Suppose that $f(x, y)$ has continuous first order partial derivatives on a rectangular region with horizontal and vertical sides and containing the points $P(a, b)$ and $Q(a + \Delta x, b + \Delta y)$. Let Δf be the increment of Equation (1). Then
>
> $$\Delta f = f_x(a, b)\,\Delta x + f_y(a, b)\,\Delta y + \varepsilon_1\,\Delta x + \varepsilon_2\,\Delta y \qquad (8)$$
>
> where ε_1 and ε_2 are functions of Δx and Δy that approach zero as $\Delta x \to 0$ and $\Delta y \to 0$.

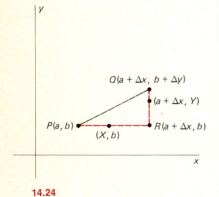

14.24

Proof If R is the point $(a + \Delta x, b)$ indicated in Figure 14.24, then

$$\Delta f = f(Q) - f(P) = [f(R) - f(P)] + [f(Q) - f(R)]$$
$$= [f(a + \Delta x, b) - f(a, b)] + [f(a + \Delta x, b + \Delta y) - f(a + \Delta x, b)]. \qquad (9)$$

We consider separately the two terms on the right in Equation (9). For the first term, we define the single-variable function

$$g(x) = f(x, b) \quad \text{for } x \text{ in } [a, a + \Delta x].$$

Then the Mean Value Theorem gives

$$f(a + \Delta x, b) - f(a, b) = g(a + \Delta x) - g(a)$$
$$= g'(X)\,\Delta x = f_x(X, b)\,\Delta x$$

for some number X in the open interval $(a, a + \Delta x)$.

For the second term on the right in Equation (9), we define the single-variable function

$$h(y) = f(a + \Delta x, y) \quad \text{for } y \text{ in } [b, b + \Delta y].$$

The Mean Value Theorem now yields

$$f(a + \Delta x, b + \Delta y) - f(a + \Delta x, b) = h(b + \Delta y) - h(b)$$
$$= h'(Y)\,\Delta y = f_y(a + \Delta x, Y)\,\Delta y$$

for some number Y in $(b, b + \Delta y)$.

When we substitute these two results into Equation (9), we find that

$$\Delta f = f_x(X, b)\,\Delta x + f_y(a + \Delta x, Y)\,\Delta y$$
$$= [f_x(a, b) + f_x(X, b) - f_x(a, b)]\,\Delta x$$
$$+ [f_y(a, b) + f_y(a + \Delta x, Y) - f_y(a, b)]\,\Delta y.$$

So

$$\Delta f = f_x(a, b)\,\Delta x + f_y(a, b)\,\Delta y + \varepsilon_1\,\Delta x + \varepsilon_2\,\Delta y,$$

where

$$\varepsilon_1 = f_x(X, b) - f_x(a, b), \qquad \varepsilon_2 = f_y(a + \Delta x, Y) - f_y(a, b).$$

Finally, since the points (X, b) and $(a + \Delta x, Y)$ both approach (a, b) as $\Delta x \to 0$ and $\Delta y \to 0$, it follows from the continuity of f_x and f_y that ε_1 and ε_2 both approach zero as Δx and Δy approach zero.

The function of Problem 29 illustrates the fact that a function $f(x, y)$ of two variables can have partial derivatives at a point without even being continuous there. Thus the mere existence of partial derivatives means less for a function of two (or more) variables than does differentiability for a single-variable function. But it follows from the linear approximation theorem above that a function with partial derivatives that are *continuous* at every point within a circle is itself continuous within that circle (see Problem 30).

A function f of two variables is called **differentiable** at the point (a, b) if $f_x(a, b)$ and $f_y(a, b)$ both exist *and also* there exist functions ε_1 and ε_2 of Δx and Δy that approach zero as Δx and Δy do and such that Equation (8) holds. We shall have no need of this concept of differentiability, since it will always suffice for us to assume that our functions have continuous partial derivatives.

The basic theorem above can be generalized in a natural way to functions of three or more variables. For example, if $w = f(x, y, z)$, the analogue of Equation (8) is

$$df = f_x(a, b, c)\,\Delta x + f_y(a, b, c)\,\Delta y + f_z(a, b, c)\,\Delta z$$
$$+ \varepsilon_1\,\Delta x + \varepsilon_2\,\Delta y + \varepsilon_3\,\Delta z,$$

where ε_1, ε_2, and ε_3 all approach zero as Δx, Δy, and Δz approach zero. The proof for the three-variable case is much like the one above for two variables.

14-6 PROBLEMS

Find the differential dw in Problems 1–10.

1 $w = 3x^2 + 4xy - 2y^3$. **2** $w = e^{-x^2 - y^2}$.

3 $w = \sqrt{1 + x^2 + y^2}$. **4** $w = xye^{x+y}$.

5 $w = \arctan\left(\dfrac{y}{x}\right)$. **6** $w = xz^2 - yx^2 + zy^2$.

7 $w = \ln(x^2 + y^2 + z^2)$. **8** $w = \sin xyz$.

9 $w = x \tan yz$. **10** $w = xye^{uv}$.

In each of Problems 11–15, use differentials to approximate $\Delta f = f(Q) - f(P)$.

11 $f(x, y) = (x^2 + y^2)^{1/2}$; $P(3, 4), Q(2.97, 4.04)$.

12 $f(x, y) = (x^2 - y^2)^{1/2}$; $P(13, 5), Q(13.2, 4.9)$.

13 $f(x, y) = \dfrac{1}{1 + x + y}$; $P(3, 6), Q(3.02, 6.05)$.

14 $f(x, y, z) = (xyz)^{1/2}$; $P(1, 3, 3), Q(0.9, 2.9, 3.1)$.

15 $f(x, y, z) = (x^2 + y^2 + z^2)^{1/2}$; $P(3, 4, 12)$,

                       $Q(3.03, 3.96, 12.05)$.

In each of Problems 16–19, use differentials to approximate the indicated number.

16 $\sqrt{26}\sqrt[3]{28}\sqrt[4]{17}$.

17 $(\sqrt{15} + \sqrt{99})^2$.

18 $\dfrac{\sqrt[3]{25}}{\sqrt[5]{30}}$.

19 $e^{0.4} = e^{(1.1)^2 - (0.9)^2}$.

20 The base and height of a rectangle are measured as 10 cm and 15 cm, respectively, with a possible error of as much as 0.1 cm in each. Use differentials to estimate the maximum resulting error in computing the area of the rectangle.

21 The base radius r and height h of a right circular cone are measured as 5 inches and 10 inches, respectively. There is a possible error of as much as $\frac{1}{16}$ inch in each measurement. Use differentials to estimate the maximum resulting error that might occur in computing the volume of the cone. (*Note:* The volume V is given by $V = \frac{1}{3}\pi r^2 h$.)

22 The dimensions of a closed rectangular box are found by measurement to be 10 cm, 15 cm, and 20 cm, but there is a possible error of 0.1 cm in each. Use differentials to estimate the maximum resulting error in computing the total surface area of the box.

23 A surveyor wants to find the area in acres (one acre is 43,560 square feet) of a certain field. She measures two adjacent sides, finding them to be $a = 500$ feet and $b = 700$ feet, with a possible error of as much as 1 foot in each. She finds the angle between these two sides to be $\theta = 30°$, with a possible error of 0.25°. The field is triangular, so its area is given by $A = \frac{1}{2}ab \sin \theta$. Use differentials to estimate the maximum resulting error, in acres, in computing the area of the field by this formula.

24 Use differentials to estimate the change in the volume of the gas of Example 3 if its pressure is decreased from 5 atm to 4.9 atm and its temperature is decreased from 300°K to 280°K.

25 The period of a pendulum of length L is given (approximately) by the formula $T = 2\pi(L/g)^{1/2}$. Estimate the change in the period of a pendulum if its length is increased from 2 ft to 2 ft 1 in. and if it is simultaneously moved from a location where g is exactly 32 ft/sec² to one where $g = 32.2$ ft/sec².

26 Show that, for the pendulum of Problem 25, the relative error in the determination of T is half the difference of the relative errors in measuring L and g; that is, that

$$\frac{dT}{T} = \frac{1}{2}\left(\frac{dL}{L} - \frac{dg}{g}\right).$$

27 The range of a projectile fired (in vacuum) with initial velocity v_0 and inclination angle α from the horizontal is $R = \frac{1}{32}v_0^2 \sin 2\alpha$. Use differentials to approximate the change in range if v_0 is increased from 400 ft/sec to 410 ft/sec and α is increased from 30° to 31°.

28 A horizontal beam is supported at both ends and supports a uniform load. The deflection or sag at its middle is given by $S = k/wh^3$ where w and h are the width and height of the beam, respectively, and k is a constant that depends on the length and composition of the beam and the amount of the load. Show that

$$dS = -S\left(\frac{dw}{w} + \frac{3\,dh}{h}\right).$$

If S is 1 inch when w is 2 inches and h is 4 inches, approximate the sag when $w = 2.1$ inches and $h = 4.1$ inches. Compare your approximation with the actual value you compute from the formula $S = k/wh^3$.

29 Let the function f be defined on the whole xy-plane by $f(x, y) = 1$ if $x = y \neq 0$, while $f(x, y) = 0$ otherwise. Show that (a) f is not continuous at $(0, 0)$; but (b) both first partial derivatives f_x and f_y exist at $(0, 0)$.

30 Deduce from Equation (8) that, under the hypotheses of the linear approximation theorem, $\Delta f \to 0$ as $\Delta x \to 0$ and $\Delta y \to 0$. What does this say about continuity of f at the point (a, b)?

14-7

The Chain Rule

The single-variable chain rule expresses the derivative of a composite function $f(g(t))$ in terms of the derivatives of f and g:

$$D_t f(g(t)) = f'(g(t))g'(t). \tag{1}$$

With $w = f(x)$ and $x = g(t)$, the chain rule says that

$$\frac{dw}{dt} = \frac{dw}{dx}\frac{dx}{dt}. \tag{2}$$

The simplest multivariable chain rule situation involves a function $w = f(x, y)$ where x and y are each functions of the same single variable $t: x = g(t)$

and $y = h(t)$. The composite function $f(g(t), h(t))$ is then a single-variable function of t, and the following theorem expresses its derivative in terms of the partial derivatives of f and the ordinary derivatives of g and h. We assume that the stated hypotheses hold on suitable domains such that the composite function is defined.

Theorem 1   *The Chain Rule*

Suppose that $w = f(x, y)$ has continuous first order partial derivatives, and that $x = g(t)$ and $y = h(t)$ are differentiable functions. Then w is a differentiable function of t and

$$\frac{dw}{dt} = \frac{\partial w}{\partial x}\frac{dx}{dt} + \frac{\partial w}{\partial y}\frac{dy}{dt}. \tag{3}$$

The variable notation of Formula (3) ordinarily will be more useful than functional notation. It must be remembered, however, that the partial derivatives in (3) are to be evaluated at the point $(g(t), h(t))$, so what (3) actually says is that

$$D_t[f(g(t), h(t))] = f_x(g(t), h(t))g'(t) + f_y(g(t), h(t))h'(t). \tag{4}$$

Proof of the Chain Rule   We choose a point t_0 at which we want to compute dw/dt, and write

$$a = g(t_0), \qquad b = h(t_0).$$

Let

$$\Delta x = g(t_0 + \Delta t) - g(t_0), \qquad \Delta y = h(t_0 + \Delta t) - h(t_0).$$

Then

$$g(t_0 + \Delta t) = a + \Delta x \quad \text{and} \quad h(t_0 + \Delta t) = b + \Delta y.$$

If

$$\begin{aligned} \Delta w &= f(g(t_0 + \Delta t), h(t_0 + \Delta t)) - f(g(t_0), h(t_0)) \\ &= f(a + \Delta x, b + \Delta y) - f(a, b), \end{aligned}$$

then what we want to compute is

$$\frac{dw}{dt} = \lim_{\Delta t \to 0} \frac{\Delta w}{\Delta t}.$$

The linear approximation theorem of Section 14-6 gives

$$\Delta w = f_x(a, b)\,\Delta x + f_y(a, b)\,\Delta y + \varepsilon_1\,\Delta x + \varepsilon_2\,\Delta y$$

where ε_1 and ε_2 approach zero as $\Delta x \to 0$ and $\Delta y \to 0$. We note that Δx and Δy approach zero as $\Delta t \to 0$ because the derivatives

$$\frac{dx}{dt} = \lim_{\Delta t \to 0} \frac{\Delta x}{\Delta t} \quad \text{and} \quad \frac{dy}{dt} = \lim_{\Delta t \to 0} \frac{\Delta y}{\Delta t}$$

both exist. Therefore

$$\frac{dw}{dt} = \lim_{\Delta t \to 0} \frac{\Delta w}{\Delta t}$$

$$= \lim_{\Delta t \to 0} \left[f_x(a, b) \frac{\Delta x}{\Delta t} + f_y(a, b) \frac{\Delta y}{\Delta t} + \varepsilon_1 \frac{\Delta x}{\Delta t} + \varepsilon_2 \frac{\Delta y}{\Delta t} \right]$$

$$= f_x(a, b) \frac{dx}{dt} + f_y(a, b) \frac{dy}{dt} + 0 \frac{dx}{dt} + 0 \frac{dy}{dt};$$

hence

$$\frac{dw}{dt} = \frac{\partial w}{\partial x} \frac{dx}{dt} + \frac{\partial w}{\partial y} \frac{dy}{dt}.$$

Thus we have established Formula (3), writing $\partial w/\partial x$ and $\partial w/\partial y$ for the partial derivatives $f_x(a, b)$ and $f_y(a, b)$, respectively, in the final step.

In the situation of Theorem 1, we may refer to w as the **dependent** variable, x and y as **intermediate** variables, and t as the **independent** variable. Then note that the right-hand side of Equation (3) has two terms, one for each intermediate variable, each like the right-hand side of the single-variable chain rule of Equation (2). If there are more than two intermediate variables, then there is still one term on the right-hand side for each intermediate variable. For example, if $w = f(x, y, z)$, with x, y, and z each a function of t, then the chain rule takes the form

$$\frac{dw}{dt} = \frac{\partial w}{\partial x} \frac{dx}{dt} + \frac{\partial w}{\partial y} \frac{dy}{dt} + \frac{\partial w}{\partial z} \frac{dz}{dt}. \tag{5}$$

The proof of (5) is essentially the same as the proof of (3); it uses the linear approximation theorem for three variables rather than for two variables.

EXAMPLE 1   Find dw/dt if $w = x^2 + ze^y + \sin xz$ and $x = t$, $y = t^2$, $z = t^3$.

Solution   Equation (5) gives

$$\frac{dw}{dt} = (2x + z \cos xz)(1) + (ze^y)(2t) + (e^y + x \cos xz)(3t^2)$$

$$= 2t + (3t^2 + 2t^4)e^{t^2} + 4t^3 \cos t^4.$$

In this example we could check the result given by the chain rule by first writing w as a function of t, then computing the ordinary single-variable derivative with respect to t.

There may be several independent variables as well as several intermediate variables. For example, if $w = f(x, y, z)$, where $x = g(u, v)$, $y = h(u, v)$, and $z = k(u, v)$, so that

$$w = f(x, y, z) = f(g(u, v), h(u, v), k(u, v)),$$

then we have the three intermediate variables x, y, and z and the two independent variables u and v. In this case we would want to compute the *partial* derivatives $\partial w/\partial u$ and $\partial w/\partial v$ of the composite function. The general chain rule below says that each partial derivative of the dependent variable w is given by a chain rule formula like (3) or (5) above. The only difference is that the derivatives with respect to the independent variables will be partial derivatives. For instance,

$$\frac{\partial w}{\partial u} = \frac{\partial w}{\partial x}\frac{\partial x}{\partial u} + \frac{\partial w}{\partial y}\frac{\partial y}{\partial u} + \frac{\partial w}{\partial z}\frac{\partial z}{\partial u}.$$

Theorem 2 *The General Chain Rule*

Suppose that w is a function of the variables $x_1, x_2, x_3, \ldots, x_m$, and that each of these is a function of the variables $t_1, t_2, \ldots, t_n$. If all these functions have continuous first order partial derivatives, then

$$\frac{\partial w}{\partial t_i} = \frac{\partial w}{\partial x_1}\frac{\partial x_1}{\partial t_i} + \frac{\partial w}{\partial x_2}\frac{\partial x_2}{\partial t_i} + \cdots + \frac{\partial w}{\partial x_m}\frac{\partial x_m}{\partial t_i} \qquad (6)$$

for each i, $1 \leq i \leq n$.

Thus there is a formula in (6) for *each* of the independent variables $t_1, t_2, \ldots, t_n$, and the right-hand side of each such formula contains one typical chain rule term for each of the intermediate variables $x_1, x_2, \ldots, x_m$.

EXAMPLE 2 Let $w = f(x, y)$ where x and y are given in polar coordinates by the equations $x = r \cos \theta$, $y = r \sin \theta$. Calculate

$$\frac{\partial w}{\partial r}, \qquad \frac{\partial w}{\partial \theta}, \quad \text{and} \quad \frac{\partial^2 w}{\partial \theta \, \partial r}$$

in terms of r and θ and the partial derivatives of w with respect to x and y.

Solution Here x and y are intermediate variables, while the independent variables are r and θ. First note that

$$\frac{\partial x}{\partial r} = \cos \theta, \qquad\qquad \frac{\partial y}{\partial r} = \sin \theta,$$

$$\frac{\partial x}{\partial \theta} = -r \sin \theta, \quad \text{and} \quad \frac{\partial y}{\partial \theta} = r \cos \theta.$$

Then

$$\frac{\partial w}{\partial r} = \frac{\partial w}{\partial x}\frac{\partial x}{\partial r} + \frac{\partial w}{\partial y}\frac{\partial y}{\partial r} = \frac{\partial w}{\partial x} \cos \theta + \frac{\partial w}{\partial y} \sin \theta \qquad (7)$$

and

$$\frac{\partial w}{\partial \theta} = \frac{\partial w}{\partial x}\frac{\partial x}{\partial \theta} + \frac{\partial w}{\partial y}\frac{\partial y}{\partial \theta} = -r \frac{\partial w}{\partial x} \sin \theta + r \frac{\partial w}{\partial y} \cos \theta. \qquad (8)$$

Next,

$$\frac{\partial^2 w}{\partial \theta \, \partial r} = \frac{\partial}{\partial \theta}\left(\frac{\partial w}{\partial x}\cos\theta + \frac{\partial w}{\partial y}\sin\theta\right)$$

$$= \left(\frac{\partial}{\partial \theta}\frac{\partial w}{\partial x}\right)\cos\theta - \frac{\partial w}{\partial x}\sin\theta + \left(\frac{\partial}{\partial \theta}\frac{\partial w}{\partial y}\right)\sin\theta + \frac{\partial w}{\partial y}\cos\theta$$

$$= \left[\left(\frac{\partial}{\partial x}\frac{\partial w}{\partial x}\right)\frac{\partial x}{\partial \theta} + \left(\frac{\partial}{\partial y}\frac{\partial w}{\partial x}\right)\frac{\partial y}{\partial \theta}\right]\cos\theta - \frac{\partial w}{\partial x}\sin\theta$$

$$+ \left[\left(\frac{\partial}{\partial x}\frac{\partial w}{\partial y}\right)\frac{\partial x}{\partial \theta} + \left(\frac{\partial}{\partial y}\frac{\partial w}{\partial y}\right)\frac{\partial y}{\partial \theta}\right]\sin\theta + \frac{\partial w}{\partial y}\cos\theta$$

$$= \left[\frac{\partial^2 w}{\partial x^2}(-r\sin\theta) + \frac{\partial^2 w}{\partial y \partial x}(r\cos\theta)\right]\cos\theta - \frac{\partial w}{\partial x}\sin\theta$$

$$+ \left[\frac{\partial^2 w}{\partial x \partial y}(-r\sin\theta) + \frac{\partial^2 w}{\partial y^2}(r\cos\theta)\right]\sin\theta + \frac{\partial w}{\partial y}\cos\theta.$$

We combine terms where we can and see finally that

$$\frac{\partial^2 w}{\partial \theta \, \partial r} = -r\frac{\partial^2 w}{\partial x^2}\sin\theta\cos\theta + r\frac{\partial^2 w}{\partial y \partial x}(\cos^2\theta - \sin^2\theta)$$

$$+ r\frac{\partial^2 w}{\partial y^2}\sin\theta\cos\theta - \frac{\partial w}{\partial x}\sin\theta + \frac{\partial w}{\partial y}\cos\theta.$$

Here we have used the chain rule to calculate the derivatives with respect to θ of $\partial w/\partial x$ and $\partial w/\partial y$, which are themselves composite functions.

EXAMPLE 3 Consider a parametric curve $x = x(t)$, $y = y(t)$, $z = z(t)$ lying on the surface $z = f(x, y)$ in space. Recall that if

$$\mathbf{T} = \left\langle \frac{dx}{dt}, \frac{dy}{dt}, \frac{dz}{dt}\right\rangle \quad \text{and} \quad \mathbf{N} = \left\langle \frac{\partial z}{\partial x}, \frac{\partial z}{\partial y}, -1\right\rangle,$$

then $\mathbf{T}$ is tangent to the curve and $\mathbf{N}$ is normal to the surface. Show that $\mathbf{T}$ and $\mathbf{N}$ are everywhere perpendicular.

Solution The chain rule of Equation (3) tells us that

$$\frac{dz}{dt} = \frac{\partial z}{\partial x}\frac{dx}{dt} + \frac{\partial z}{\partial y}\frac{dy}{dt}.$$

But this equation is equivalent to the vector equation

$$\left\langle \frac{\partial z}{\partial x}, \frac{\partial z}{\partial y}, -1\right\rangle \cdot \left\langle \frac{dx}{dt}, \frac{dy}{dt}, \frac{dz}{dt}\right\rangle = 0.$$

Thus $\mathbf{N} \cdot \mathbf{T} = 0$, and so $\mathbf{N}$ and $\mathbf{T}$ are indeed perpendicular.

EXAMPLE 4 Suppose that the function $z = f(x, y)$ is *implicitly* defined by means of the equation $F(x, y, z) = 0$, which means that $F(x, y, f(x, y)) \equiv 0$. Suppose that F has continuous first order partial derivatives and that

$F_z(x, y, z)$ is never zero. Show that

$$\frac{\partial z}{\partial x} = -\frac{F_x}{F_z} \quad \text{and} \quad \frac{\partial z}{\partial y} = -\frac{F_y}{F_z}. \tag{9}$$

Solution Since $w = F(x, y, f(x, y))$ is identically zero, differentiation with respect to x yields

$$0 = \frac{\partial w}{\partial x} = \frac{\partial F}{\partial x}\frac{\partial x}{\partial x} + \frac{\partial F}{\partial y}\frac{\partial y}{\partial x} + \frac{\partial F}{\partial z}\frac{\partial z}{\partial x}$$

$$= (F_x)(1) + (F_y)(0) + (F_z)\left(\frac{\partial z}{\partial x}\right);$$

and so

$$F_x + F_z\frac{\partial z}{\partial x} = 0.$$

This gives the first formula in (9); the second is obtained similarly by differentiating w with respect to y.

For instance, let's apply (9) with

$$F(x, y, z) = x^2 + y^2 + z^2 - 1 = 0$$

and

$$z = f(x, y) = \sqrt{1 - x^2 - y^2} \quad \text{if} \quad x^2 + y^2 < 1.$$

Then the first formula in (9) gives

$$\frac{\partial z}{\partial x} = -\frac{F_x}{F_z} = -\frac{2x}{2z} = -\frac{x}{\sqrt{1 - x^2 - y^2}},$$

the same result that would be obtained by direct partial differentiation of z with respect to x.

We close this section with an example that shows how we can use chain rule computations to *actually do something*. Consider a quantity of liquid which has pressure P, volume V, and temperature T that satisfy a "state equation" of the form

$$F(P, V, T) = 0, \tag{10}$$

analogous to the state equation $PV - nRT = 0$ of an ideal gas. We shall not need to know precisely what the function F is, but we shall assume that Equation (10) can be solved for each variable in terms of the other two. Let's write

$$P = g(V, T) \quad \text{and} \quad V = h(P, T). \tag{11}$$

The partial derivatives $\partial P/\partial V$, $\partial P/\partial T$, $\partial V/\partial P$, and $\partial V/\partial T$ of g and h are important in physical chemistry.

Suppose that V and T are changed to $V + \Delta V$ and $T + \Delta T$, while P is held constant. Then the fractional change in volume per unit change in temperature is $(\Delta V/V)/\Delta T$. The limit of this expression as $\Delta T \to 0$ gives the *thermal expansivity*

$$\alpha = \frac{1}{V} \frac{\partial V}{\partial T} \tag{12}$$

of our liquid. Similarly, if V and P change while T remains constant, the limit as $\Delta P \to 0$ of the negative of the fractional change in volume per unit change in pressure is the *isothermal compressivity* of our liquid,

$$\beta = -\frac{1}{V} \frac{\partial V}{\partial P}. \tag{13}$$

We may differentiate the equation $F(P, V, T) = 0$ with respect to P and T as in Example 4. This yields the results

$$\frac{\partial V}{\partial P} = -\frac{F_P}{F_V} \quad \text{and} \quad \frac{\partial V}{\partial T} = -\frac{F_T}{F_V}. \tag{14}$$

But when we differentiate Equation (10) with respect to V and T, we find that

$$\frac{\partial P}{\partial V} = -\frac{F_V}{F_P} \quad \text{and} \quad \frac{\partial P}{\partial T} = -\frac{F_T}{F_P}. \tag{15}$$

We finally substitute Equations (14) into the second of Equations (15) and obtain

$$\frac{\partial P}{\partial T} = -\frac{F_T}{F_P} = -\frac{-F_T/F_V}{-F_P/F_V} = -\frac{\partial V/\partial T}{\partial V/\partial P} = \frac{(1/V)(\partial V/\partial T)}{(-1/V)(\partial V/\partial P)};$$

consequently

$$\frac{\partial P}{\partial T} = \frac{\alpha}{\beta}. \tag{16}$$

The following example shows that this last equation has significant practical consequences.

EXAMPLE 5 The thermal expansivity and isothermal compressivity of liquid mercury are $\alpha = 1.8 \times 10^{-4}$ and $\beta = 3.9 \times 10^{-6}$, respectively, in liter-atm-°C units. Suppose that a thermometer bulb is exactly filled with mercury at 50°C. If the bulb can withstand an internal pressure of only 200 atm, can it be heated to 55°C without breaking?

Solution By Equation (16), we find that

$$\frac{\partial P}{\partial T} = \frac{\alpha}{\beta} = \frac{1.8 \times 10^{-4}}{3.9 \times 10^{-6}} \approx 46.$$

Thus each increase of 1 degree in temperature will generate an additional internal pressure of about 46 atm. Hence, if you heat the bulb to 55°C, you will generate a pressure of 230 atm, and therefore break the bulb.

In Problems 1–4, find dw/dt both by using the chain rule *and* by expressing w explicitly as a function of t before differentiation.

1 $w = e^{-x^2-y^2}, x = t, y = t^{1/2}$.

2 $w = \dfrac{1}{u^2 + v^2}, u = \cos 2t, v = \sin 2t$.

3 $w = \sin xyz, x = t, y = t^2, z = t^3$.
4 $w = \ln(u + v + z), u = \cos^2 t, v = \sin^2 t, z = t^2$.

In Problems 5–8, find $\partial w/\partial s$ and $\partial w/\partial t$.

5 $w = \ln(x^2 + y^2 + z^2), x = s - t, y = s + t, z = 2\sqrt{st}$.
6 $w = pq \sin r, p = 2s + t, q = s - t, r = st$.
7 $w = (u^2 + v^2 + z^2)^{1/2}, u = 3e^t \sin s, v = 3e^t \cos s, z = 4e^t$.
8 $w = yz + zx + xy, x = s^2 - t^2, y = s^2 + t^2, z = s^2 t^2$.

In Problems 9–10, find $\partial r/\partial x, \partial r/\partial y,$ and $\partial r/\partial z$.

9 $r = e^{u+v+w}, u = yz, v = xz, w = xy$.
10 $r = uvw - u^2 - v^2 - w^2, u = y + z, v = x + z, w = x + y$.

In Problems 11–15, find $\partial z/\partial x$ and $\partial z/\partial y$ as functions of $x, y,$ and z, assuming that $z = f(x, y)$ satisfies the given equation.

11 $x^{2/3} + y^{2/3} + z^{2/3} = 1$.   **12** $x^3 + y^3 + z^3 = xyz$.
13 $xe^{xy} + ye^{zx} + ze^{xy} = 3$.   **14** $z^5 + xy^2 + yz = 5$.

15 $\dfrac{x^2}{a^2} + \dfrac{y^2}{b^2} + \dfrac{z^2}{c^2} = 1$.

16 Differentiate the equation $F(V, P, T) = 0$ with respect to P and T to establish Equations (14) of this section.
17 Differentiate the equation $F(V, P, T) = 0$ with respect to V and T to establish Equations (15) of this section.
18 Suppose that $w = f(x, y), x = r \cos \theta,$ and $y = r \sin \theta$. Show that

$$\left(\frac{\partial w}{\partial x}\right)^2 + \left(\frac{\partial w}{\partial y}\right)^2 = \left(\frac{\partial w}{\partial r}\right)^2 + \frac{1}{r^2}\left(\frac{\partial w}{\partial \theta}\right)^2.$$

19 Suppose that $w = f(u)$ and that $u = x + y$. Show that $\partial w/\partial x = \partial w/\partial y$.
20 Suppose that $w = f(u)$ and that $u = x - y$. Show that $\partial w/\partial x = -\partial w/\partial y$ and that

$$\frac{\partial^2 w}{\partial x^2} = \frac{\partial^2 w}{\partial y^2} = -\frac{\partial^2 w}{\partial x\, \partial y}.$$

21 Suppose that $w = f(x, y),$ with $x = u + v$ and $y =$

$u - v$. Show that

$$\frac{\partial^2 w}{\partial x^2} - \frac{\partial^2 w}{\partial y^2} = \frac{\partial^2 w}{\partial u\, \partial v}.$$

22 Assume that $w = f(x, y)$ where $x = 2u + v$ and $y = u - v$. Show that

$$5\frac{\partial^2 w}{\partial x^2} + 2\frac{\partial^2 w}{\partial x\, \partial y} + 2\frac{\partial^2 w}{\partial y^2} = \frac{\partial^2 w}{\partial u^2} + \frac{\partial^2 w}{\partial v^2}.$$

23 In this problem, $w = f(x, y), x = r \cos \theta,$ and $y = r \sin \theta$. Show that

$$\frac{\partial^2 w}{\partial x^2} + \frac{\partial^2 w}{\partial y^2} = \frac{\partial^2 w}{\partial r^2} + \frac{1}{r}\frac{\partial w}{\partial r} + \frac{1}{r^2}\frac{\partial^2 w}{\partial \theta^2}.$$

24 If $w = (1/r)f(t - r/a)$ and $r = (x^2 + y^2 + z^2)^{1/2}$, show that

$$\frac{\partial^2 w}{\partial x^2} + \frac{\partial^2 w}{\partial y^2} + \frac{\partial^2 w}{\partial z^2} = \frac{1}{a^2}\frac{\partial^2 w}{\partial t^2}.$$

25 Suppose that $w = f(r)$ and that $r = (x^2 + y^2 + z^2)^{1/2}$. Show that

$$\frac{\partial^2 w}{\partial x^2} + \frac{\partial^2 w}{\partial y^2} + \frac{\partial^2 w}{\partial z^2} = \frac{d^2 w}{dr^2} + \frac{2}{r}\frac{dw}{dr}.$$

26 Suppose that $w = f(u) + g(v),$ that $u = x - at,$ and that $v = x + at$. Show that

$$\frac{\partial^2 w}{\partial t^2} = a^2\frac{\partial^2 w}{\partial x^2}.$$

27 Assume that $w = f(u, v)$ where $u = x + y$ and $v = x - y$. Show that

$$\frac{\partial w}{\partial x}\frac{\partial w}{\partial y} = \left(\frac{\partial w}{\partial u}\right)^2 - \left(\frac{\partial w}{\partial v}\right)^2.$$

28 If $w = f(x, y), x = e^u \cos v,$ and $y = e^u \sin v$, show that

$$\left(\frac{\partial w}{\partial x}\right)^2 + \left(\frac{\partial w}{\partial y}\right)^2 = e^{-2u}\left[\left(\frac{\partial w}{\partial u}\right)^2 + \left(\frac{\partial w}{\partial v}\right)^2\right].$$

29 Assume that $w = f(x, y)$ with α a constant such that $x = u \cos \alpha - v \sin \alpha$ and $y = u \sin \alpha + v \cos \alpha$. Show that

$$\left(\frac{\partial w}{\partial u}\right)^2 + \left(\frac{\partial w}{\partial v}\right)^2 = \left(\frac{\partial w}{\partial x}\right)^2 + \left(\frac{\partial w}{\partial y}\right)^2.$$

30 Suppose that $w = f(u)$ where $u = (x^2 - y^2)/(x^2 + y^2)$. Show that $xw_x + yw_y = 0$.

Directional Derivatives and the Gradient Vector

We know that the partial derivatives $f_x(x, y, z)$, $f_y(x, y, z)$, and $f_z(x, y, z)$ give the rates of change of $w = f(x, y, z)$ at the point $P(x, y, z)$ in the x-, y-, and z-directions, respectively. Now we want to say what is meant by the rate of change of w in the direction of an *arbitrary unit* vector $\mathbf{u} = \langle a, b, c \rangle$.

Let Q be a point on the ray in the direction of $\mathbf{u}$ from the point P (see Figure 14.25). The **average rate of change of w with respect to distance** between P and Q is

$$\frac{f(Q) - f(P)}{|\overrightarrow{PQ}|} = \frac{\Delta w}{\text{distance } PQ}.$$

We want to find the *instantaneous* rate of change of w in the direction $\mathbf{u}$ at the point P. So we take the limit of this average rate of change as Q approaches P along the ray:

$$\lim_{Q \to P} \frac{f(Q) - f(P)}{|\overrightarrow{PQ}|}. \tag{1}$$

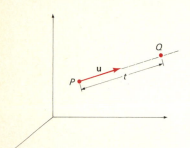

14.25 First step in computing the rate of change of $f(x, y, z)$ in the direction of the unit vector $\mathbf{u}$.

The value of this limit is called the **directional derivative of f at $P(x, y, z)$ in the direction $\mathbf{u}$**, and we denote it by $D_{\mathbf{u}} f(x, y, z)$.

To make the limit in (1) more precise, let $t = |\overrightarrow{PQ}|$. Then $\overrightarrow{PQ} = t\mathbf{u} = \langle at, bt, ct \rangle$, so $f(Q) = f(x + at, y + bt, z + ct)$. Now we can rewrite the limit in (1) as

$$D_{\mathbf{u}} f(x, y, z) = \lim_{t \to 0} \frac{f(x + at, y + bt, z + ct) - f(x, y, z)}{t}. \tag{2}$$

This is the formal definition of the directional derivative $D_{\mathbf{u}} f(x, y, z)$.

To ease the labor of computing directional derivatives, we use a form of the chain rule that involves the **gradient vector** ∇f of the function f, which is defined to be

$$\nabla f(x, y, z) = \mathbf{i} f_x(x, y, z) + \mathbf{j} f_y(x, y, z) + \mathbf{k} f_z(x, y, z). \tag{3}$$

We also write

$$\nabla f = \left\langle \frac{\partial f}{\partial x}, \frac{\partial f}{\partial y}, \frac{\partial f}{\partial z} \right\rangle = \frac{\partial f}{\partial x} \mathbf{i} + \frac{\partial f}{\partial y} \mathbf{j} + \frac{\partial f}{\partial z} \mathbf{k}.$$

For example, if $f(x, y, z) = x^2 + yz - 2xy - z^2$, then

$$\nabla f(x, y, z) = (2x - 2y)\mathbf{i} + (z - 2x)\mathbf{j} + (y - 2z)\mathbf{k},$$

and the value of ∇f at the point $(2, 1, 3)$ would, for instance, be $\nabla f(2, 1, 3) = 2\mathbf{i} - \mathbf{j} - 5\mathbf{k}$.

Suppose now that the first order partial derivatives of f are continuous and that

$$\mathbf{r}(t) = x(t)\mathbf{i} + y(t)\mathbf{j} + z(t)\mathbf{k}$$

is a differentiable vector-valued function. Then $f(\mathbf{r}(t)) = f(x(t), y(t), z(t))$ is a differentiable function of t, and its derivative is

$$D_t f(\mathbf{r}(t)) = \nabla f(\mathbf{r}(t)) \cdot \mathbf{r}'(t). \tag{4}$$

The operation on the right-hand side above is the *dot* product, because the gradient of f and the derivative of $\mathbf{r}$ are both *vector*-valued functions.

The rule of Equation (4) is the **vector chain rule**; it is a new way of writing the chain rule of Section 14-7. To see why this is so, recall that

$$\mathbf{r}'(t) = \frac{d\mathbf{r}}{dt} = \left\langle \frac{dx}{dt}, \frac{dy}{dt}, \frac{dz}{dt} \right\rangle.$$

Then

$$D_t f(\mathbf{r}(t)) = D_t f(x(t), y(t), z(t))$$

$$= \frac{\partial f}{\partial x}\frac{dx}{dt} + \frac{\partial f}{\partial y}\frac{dy}{dt} + \frac{\partial f}{\partial z}\frac{dz}{dt} = \nabla f \cdot \frac{d\mathbf{r}}{dt}.$$

Now we are ready to compute directional derivatives. Suppose that $\mathbf{r}(t) = \langle x + at, y + bt, z + ct \rangle$, so that

$$\mathbf{r}(0) = \langle x, y, z \rangle \quad \text{and} \quad \mathbf{r}'(0) = \langle a, b, c \rangle = \mathbf{u}.$$

Then, by (2), we find that

$$D_{\mathbf{u}} f(x, y, z) = \lim_{t \to 0} \frac{f(\mathbf{r}(t)) - f(\mathbf{r}(0))}{t}$$

$$= D_t f(\mathbf{r}(t)) \quad \text{at} \quad t = 0$$

$$= \nabla f(r(0)) \cdot \mathbf{r}'(0) \qquad \text{[by Equation (4)].}$$

Therefore

$$D_{\mathbf{u}} f(x, y, z) = \nabla f(x, y, z) \cdot \mathbf{u}. \tag{5}$$

This formula provides the quickest and easiest way of computing the directional derivative $D_{\mathbf{u}} f$; all we need is knowledge of the partial derivatives of f. In terms of the components a, b, and c of the unit vector $\mathbf{u}$, Formula (5) says simply that

$$D_{\mathbf{u}} f = a \frac{\partial f}{\partial x} + b \frac{\partial f}{\partial y} + c \frac{\partial f}{\partial z}. \tag{6}$$

If $a = \cos \alpha$, $b = \cos \beta$, and $c = \cos \gamma$ are the direction cosines of $\mathbf{u}$, then

$$D_{\mathbf{u}} f = \frac{\partial f}{\partial x} \cos \alpha + \frac{\partial f}{\partial y} \cos \beta + \frac{\partial f}{\partial z} \cos \gamma. \tag{7}$$

EXAMPLE 1   Compute the directional derivative of $f(x, y, z) = x^2 - 2yz$ in the direction of the vector $\mathbf{v} = \langle 2, 2, 1 \rangle$. At what rate is f increasing in the direction of $\mathbf{v}$ at the point $(1, 2, 3)$?

Solution   Since $\mathbf{v}$ is not a unit vector, we must replace it by a unit vector having the same direction before using the formulas above. So we take

$$\mathbf{u} = \frac{\mathbf{v}}{|\mathbf{v}|} = \left\langle \frac{2}{3}, \frac{2}{3}, \frac{1}{3} \right\rangle.$$

Then Formula (5) gives

$$D_{\mathbf{u}} f(x, y, z) = \langle 2x, -2z, -2y \rangle \cdot \langle \tfrac{2}{3}, \tfrac{2}{3}, \tfrac{1}{3} \rangle = \tfrac{1}{3}(4x - 2y - 4z).$$

And at the point $(1, 2, 3)$, we find that

$$D_{\mathbf{u}} f(1, 2, 3) = \tfrac{1}{3}(4 - 4 - 12) = -4.$$

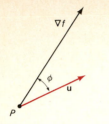

14.26 The angle ϕ between ∇f and the unit vector **u**.

Thus f is *decreasing* at the rate of 4 units per unit moved in the direction of **u** (or **v**) at $(1, 2, 3)$.

The gradient vector ∇f has an important interpretation that involves the *maximal* directional derivative of f. If ϕ is the angle between ∇f at the point P and the unit vector **u** (Figure 14.26), then Formula (5) gives

$$D_{\mathbf{u}} f(P) = \nabla f(P) \cdot \mathbf{u} = |\nabla f(P)| \cos \phi$$

because $|\mathbf{u}| = 1$. The maximum value of $\cos \phi$ is 1, and this occurs when $\phi = 0$. This is so when **u** is the particular unit vector $\nabla f(P)/|\nabla f(P)|$ that points in the direction of the gradient vector. In this case the formula above yields

$$D_{\mathbf{u}} f(P) = |\nabla f(P)|,$$

so that the value of the directional derivative is the length of the gradient vector. We have therefore proved the following theorem.

Theorem 1 *Significance of the Gradient Vector*
The maximum value of the directional derivative $D_{\mathbf{u}} f(P)$ is obtained when **u** is the unit vector in the direction of the gradient vector $\nabla f(P)$; that is, when $\mathbf{u} = \nabla f(P)/|\nabla f(P)|$. The value of this maximum directional derivative is $|\nabla f(P)|$, the length of the gradient vector.

Thus *the gradient vector ∇f points in the direction in which the function f increases most rapidly, and its length is the rate of increase of f (with respect to distance) in that direction.*

EXAMPLE 2 Suppose that the temperature W (in °C) at the point (x, y, z) is given by $W = 50 + xyz$. (a) Find the rate of change of temperature with respect to distance (in feet) at the point $P(3, 4, 1)$ in the direction of the vector $\mathbf{v} = \mathbf{i} + 2\mathbf{j} + 2\mathbf{k}$. (b) Find the maximal directional derivative $D_{\mathbf{u}} W$ at the point $P(3, 4, 1)$ and the direction **u** in which that maximum occurs.

Solution (a) $\nabla W = yz\mathbf{i} + xz\mathbf{j} + xy\mathbf{k}$, so $\nabla W(P) = 4\mathbf{i} + 3\mathbf{j} + 12\mathbf{k}$. The *unit* vector in the direction of **v** is

$$\mathbf{u} = \frac{\mathbf{v}}{|\mathbf{v}|} = \frac{1}{3}\mathbf{i} + \frac{2}{3}\mathbf{j} + \frac{2}{3}\mathbf{k}.$$

Hence the desired directional derivative is

$$D_{\mathbf{u}} W(3, 4, 1) = \nabla W(3, 4, 1) \cdot \mathbf{u}$$
$$= (4)(\tfrac{1}{3}) + (3)(\tfrac{2}{3}) + (12)(\tfrac{2}{3}) = \tfrac{34}{3}.$$

Thus the rate of change of temperature in the direction of the vector **v** is $11\frac{1}{3}$°C/ft.

(b) According to Theorem 1, the maximal directional derivative of W at P is

$$|\nabla W(P)| = |4\mathbf{i} + 3\mathbf{j} + 12\mathbf{k}| = 13°\text{C/ft}.$$

This is the directional derivative of W in the direction of the unit vector

$$\mathbf{u} = \frac{\nabla W(P)}{|\nabla W(P)|} = \frac{4\mathbf{i} + 3\mathbf{j} + 12\mathbf{k}}{13}.$$

Thus far, we only have discussed directional derivatives of functions of three variables. For a function of two variables the formulas are analogous:

$$\nabla f(x, y) = \left\langle \frac{\partial f}{\partial x}, \frac{\partial f}{\partial y} \right\rangle = \frac{\partial f}{\partial x}\mathbf{i} + \frac{\partial f}{\partial y}\mathbf{j} \tag{8}$$

and

$$D_{\mathbf{u}}f(x, y) = \nabla f(x, y) \cdot \mathbf{u} = a\frac{\partial f}{\partial x} + b\frac{\partial f}{\partial y} \tag{9}$$

where $\mathbf{u} = \langle a, b \rangle$ is a unit vector. If α is the inclination angle of $\mathbf{u}$, then $a = \cos\alpha$ and $b = \sin\alpha$, so Equation (9) takes the form

$$D_{\mathbf{u}}f(x, y) = \frac{\partial f}{\partial x}\cos\alpha + \frac{\partial f}{\partial y}\sin\alpha. \tag{10}$$

As in Example 2, change in temperature provides a tangible interpretation of the three-dimensional directional derivative. In the two-dimensional case, it is helpful to think of the directional derivative $D_{\mathbf{u}}f(x, y)$ as the slope of a curve on the surface $z = f(x, y)$. If $\mathbf{u} = \langle a, b \rangle$ is a (two-dimensional) unit vector, then the quotient

$$\frac{f(x + at, y + bt) - f(x, y)}{t} = \frac{\text{vertical rise}}{\text{horizontal run}}$$

is an average rate of change of height along the curve on $z = f(x, y)$ shown in Figure 14.27. The limit of this quotient as $t \to 0$ is (by definition) the value of the directional derivative $D_{\mathbf{u}}f(x, y)$. Hence $D_{\mathbf{u}}f(x, y)$ is equal to the slope (ratio of vertical rise to horizontal run) of the tangent line at $(x, y, f(x, y))$ to the indicated curve. So, if you think of the surface $z = f(x, y)$ as a hill, then $D_{\mathbf{u}}f(x, y)$ is your rate of climb (per unit of horizontal distance) as you proceed in the (*horizontal*) direction $\mathbf{u}$. And the angle at which you climb while

14.27 Geometry of $D_{\mathbf{u}}f(x, y)$.

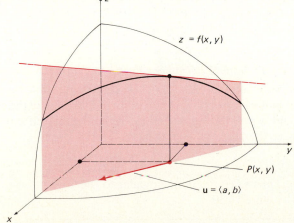

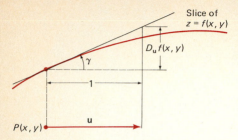

14.28 The cross section of the part of the graph above **u**.

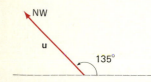

14.29 The unit vector in the northwest direction.

walking in this direction is $\gamma = \arctan(D_{\mathbf{u}} f(x, y))$, which appears in the vertical cross section of the graph shown in Figure 14.28.

EXAMPLE 3 You are standing at the point $(-100, -100, 430)$ on a hill that has the shape

$$z = 500 - (0.003)x^2 - (0.004)y^2.$$

What will be your rate of climb if you head northwest? At what angle from the horizontal will you be climbing?

Solution $\nabla z = -(0.006)x\mathbf{i} - (0.008)y\mathbf{j}$, so the gradient vector of $z(x, y)$ at $P(-100, -100)$ is $\nabla z(P) = 0.6\mathbf{i} + 0.8\mathbf{j}$. Since "north" means the direction of $\mathbf{j}$ and "east" that of $\mathbf{i}$, the unit vector in the northwest direction—shown in Figure 14.29—is

$$\mathbf{u} = \mathbf{i} \cos 135° + \mathbf{j} \sin 135° = -\frac{1}{\sqrt{2}}\mathbf{i} + \frac{1}{\sqrt{2}}\mathbf{j}.$$

Hence your rate of climb if you head northwest will be

$$D_{\mathbf{u}} z(P) = \nabla z(P) \cdot \mathbf{u} = (0.6)\left(\frac{-1}{\sqrt{2}}\right) + (0.8)\left(\frac{1}{\sqrt{2}}\right) \approx 0.14$$

feet vertically for each foot horizontally. You will then be climbing at the approximate angle $\tan^{-1}(0.14)$ or $8°$ from the horizontal.

THE GRADIENT VECTOR AS A NORMAL VECTOR

Consider the graph of the equation

$$F(x, y, z) = 0, \tag{11}$$

where F is a function with continuous first order partial derivatives. According to the **implicit function theorem** of advanced calculus, near every point where $\nabla F \neq \mathbf{0}$—that is, at least one of the partial derivatives of F is nonzero—the graph of Equation (11) agrees with the graph of an equation of one of the forms

$$z = f(x, y), \qquad y = g(x, z), \qquad x = h(y, z).$$

Because of this, we are justified in general in referring to the graph of Equation (11) as a "surface." The gradient vector ∇F is normal to this surface, in the sense of the following theorem.

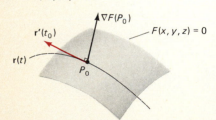

14.30 The gradient vector ∇F is normal to every curve in the surface $F(x, y, z) = 0$.

Theorem 2 *Gradient Vector as Normal Vector*

Suppose that $F(x, y, z)$ has continuous first order partial derivatives, and let $P_0(x_0, y_0, z_0)$ be a point of the graph of the equation $F(x, y, z) = 0$ at which $\nabla F(P_0) \neq \mathbf{0}$. If $\mathbf{r}(t)$ is a differentiable curve on this surface with $\mathbf{r}(t_0) = \langle x_0, y_0, z_0 \rangle$, then

$$\nabla F(P_0) \cdot \mathbf{r}'(t_0) = 0. \tag{12}$$

Thus $\nabla F(P_0)$ is perpendicular to the tangent vector $\mathbf{r}'(t_0)$, as indicated in Figure 14.30.

Proof The statement that $\mathbf{r}(t)$ lies on the surface $F(x, y, z) = 0$ means that $F(\mathbf{r}(t)) = 0$ for all t. Hence

$$0 = D_t F(\mathbf{r}(t_0)) = \nabla F(\mathbf{r}(t_0)) \cdot \mathbf{r}'(t_0) = \nabla F(P_0) \cdot \mathbf{r}'(t_0)$$

by the chain rule in the form of Equation (4). Therefore the vectors $\nabla F(P_0)$ and $\mathbf{r}'(t_0)$ are perpendicular.

Since $\nabla F(P_0) \neq \mathbf{0}$ is perpendicular to every curve on the surface $F(x, y, z) = 0$ through the point P_0, it is a normal vector to the surface at P_0,

$$\mathbf{n} = \frac{\partial F}{\partial x}\mathbf{i} + \frac{\partial F}{\partial y}\mathbf{j} + \frac{\partial F}{\partial z}\mathbf{k}. \tag{13}$$

Note that if the equation $z = f(x, y)$ is rewritten in the form $F(x, y, z) = f(x, y) - z = 0$, then

$$\left\langle \frac{\partial F}{\partial x}, \frac{\partial F}{\partial y}, \frac{\partial F}{\partial z} \right\rangle = \left\langle \frac{\partial f}{\partial x}, \frac{\partial f}{\partial y}, -1 \right\rangle.$$

Thus Equation (13) agrees with the definition of normal vector that we gave in Section 14-3 (Equation (4) there).

The **tangent plane** to the surface $F(x, y, z) = 0$ at the point $P_0(x_0, y_0, z_0)$ is the plane through P_0 that is perpendicular to the normal vector $\mathbf{n}$ of Equation (13). Its equation is

$$F_x(x_0, y_0, z_0)(x - x_0) + F_y(x_0, y_0, z_0)(y - y_0) + F_z(x_0, y_0, z_0)(z - z_0) = 0. \tag{14}$$

EXAMPLE 4 Write an equation of the tangent plane to the ellipsoid $2x^2 + 4y^2 + z^2 - 45 = 0$ at the point $(2, -3, -1)$.

Solution Here we have $\nabla F(x, y, z) = \langle 4x, 8y, 2z \rangle$, so

$$\nabla F(2, -3, -1) = 8\mathbf{i} - 24\mathbf{j} - 2\mathbf{k}.$$

Equation (14) then takes the form

$$8(x - 2) - 24(y + 3) - 2(z + 1) = 0;$$

$$4x - 12y - z = 45.$$

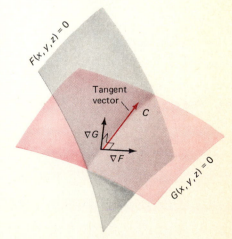

14.31 $\nabla F \times \nabla G$ is tangent to the curve C of intersection.

The intersection of the two surfaces $F(x, y, z) = 0$ and $G(x, y, z) = 0$ will generally be some sort of curve. By the implicit function theorem, this curve can be represented in parametric fashion near every point where the gradient vectors ∇F and ∇G are *not* parallel. This curve C is perpendicular to each of the normal vectors ∇F and ∇G. That is, if P is a point of C, then the tangent vector to C at P is perpendicular to each of the vectors $\nabla F(P)$ and $\nabla G(P)$, as indicated in Figure 14.31. It follows that the vector

$$\mathbf{T} = \nabla F \times \nabla G \tag{15}$$

is tangent to the curve of intersection of the surfaces $F(x, y, z) = 0$ and $G(x, y, z) = 0$.

EXAMPLE 5 The point $P(1, -1, 2)$ lies on both the paraboloid

$$F(x, y, z) = x^2 + y^2 - z = 0$$

and the ellipsoid

$$G(x, y, z) = 2x^2 + 3y^2 + z^2 - 9 = 0.$$

Write an equation of the plane through P that is normal to their curve of intersection.

Solution First we compute

$$\nabla F = 2x\mathbf{i} + 2y\mathbf{j} - \mathbf{k} \quad \text{and} \quad \nabla G = 4x\mathbf{i} + 6y\mathbf{j} + 2z\mathbf{k}.$$

At $P(1, -1, 2)$ these two vectors are

$$\nabla F(1, -1, 2) = 2\mathbf{i} - 2\mathbf{j} - \mathbf{k} \quad \text{and} \quad \nabla G(1, -1, 2) = 4\mathbf{i} - 6\mathbf{j} + 4\mathbf{k}.$$

Hence the tangent vector to the curve of intersection of the paraboloid and the ellipsoid is

$$\mathbf{T} = \nabla\mathbf{F} \times \nabla\mathbf{G} = \begin{vmatrix} \mathbf{i} & \mathbf{j} & \mathbf{k} \\ 2 & -2 & -1 \\ 4 & -6 & 4 \end{vmatrix} = -14\mathbf{i} - 12\mathbf{j} - 4\mathbf{k}.$$

This vector is normal to the desired plane through $(1, -1, 2)$, so the plane has equation

$$-14(x - 1) - 12(y + 1) - 4(z - 2) = 0;$$

$$7x + 6y + 2z = 5.$$

A result analogous to Theorem 2 holds in two dimensions. The graph of the equation $F(x, y) = 0$ looks like a curve near each point at which $\nabla F \neq \mathbf{0}$, and ∇F is normal to the curve in such cases.

EXAMPLE 6 Write an equation of the tangent line to the folium of Descartes

$$F(x, y) = 2x^3 + 2y^3 - 9xy = 0$$

at the point $(1, 2)$. See Figure 14.32.

Solution

$$\nabla F(x, y) = (6x^2 - 9y)\mathbf{i} + (6y^2 - 9x)\mathbf{j},$$

so the normal vector at $(1, 2)$ is $\nabla F(1, 2) = -12\mathbf{i} + 15\mathbf{j}$. Hence the tangent line has equation

$$-12(x - 1) + 15(y - 2) = 0,$$

or $4x - 5y + 6 = 0$.

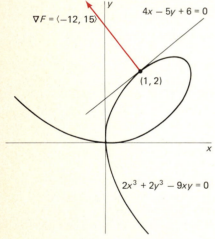

14.32 The foliium and its tangent (see Example 6).

14-8 PROBLEMS

In Problems 1–10, find the directional derivative of f at P in the direction of $\mathbf{v}$; that is, find

$$D_{\mathbf{u}}f(P) \quad \text{where} \quad \mathbf{u} = \frac{\mathbf{v}}{|\mathbf{v}|}.$$

1 $f(x, y) = x^2 + 2xy + 3y^2; \quad P(2, 1); \mathbf{v} = \mathbf{i} + \mathbf{j}.$

2 $f(x, y) = e^x \sin y; \quad P\left(0, \dfrac{\pi}{4}\right); \mathbf{v} = \mathbf{i} - \mathbf{j}.$

3 $f(x, y) = x^3 - x^2y + xy^2 + y^3; \quad P(1, -1); \mathbf{v} = 2\mathbf{i} + 3\mathbf{j}.$

4 $f(x, y) = \tan^{-1}\left(\dfrac{y}{x}\right); \quad P(-3, 3); \mathbf{v} = 3\mathbf{i} + 4\mathbf{j}.$

5 $f(x, y) = \sin x \cos y; \quad P\left(\dfrac{\pi}{3}, \dfrac{-2\pi}{3}\right); \mathbf{v} = 4\mathbf{i} - 3\mathbf{j}.$

6 $f(x, y, z) = xy + yz + zx; \quad P(1, -1, 2); \mathbf{v} = \mathbf{i} + \mathbf{j} + \mathbf{k}.$

7 $f(x, y, z) = (xyz)^{1/2}$; $P(2, -1, -2)$; $\mathbf{v} = \mathbf{i} + 2\mathbf{j} - 2\mathbf{k}$.
8 $f(x, y, z) = \ln(1 + x^2 + y^2 - z^2)$; $P(1, -1, 1)$; $\mathbf{v} = 2\mathbf{i} - 2\mathbf{j} + 3\mathbf{k}$.
9 $f(x, y, z) = e^{xyz}$; $P(4, 0, -3)$; $\mathbf{v} = \mathbf{j} - \mathbf{k}$.
10 $f(x, y, z) = (10 - x^2 - y^2 - z^2)^{1/2}$; $P(1, 1, -2)$; $\mathbf{v} = 3\mathbf{i} + 4\mathbf{j} - 12\mathbf{k}$.

In Problems 11–15, find the maximum directional derivative of f at P and the direction in which it occurs.

11 $f(x, y) = 2x^2 + 3xy + 4y^2$; $P(1, 1)$.

12 $f(x, y) = \arctan\left(\dfrac{y}{x}\right)$; $P(1, -2)$.

13 $f(x, y, z) = 3x^2 + y^2 + 4z^2$; $P(1, 5, -2)$.
14 $f(x, y, z) = e^{x-y-z}$; $P(5, 2, 3)$.
15 $f(x, y, z) = (xy^2z^3)^{1/2}$; $P(2, 2, 2)$.

In Problems 16–20, write an equation of the tangent line (or plane) to the given curve (or surface) at the given point P.

16 $2x^2 + 3y^2 = 35$; $P(2, 3)$.
17 $x^4 + xy + y^2 = 19$; $P(2, -3)$.
18 $3x^2 + 4y^2 + 5z^2 = 73$; $P(2, 2, 3)$.
19 $x^{1/3} + y^{1/3} + z^{1/3} = 1$; $P(1, -1, 1)$.
20 $xyz + x^2 - 2y^2 + z^3 = 14$; $P(5, -2, 3)$.

21 Show that the gradient operator has the following formal properties that show its close analogy with the single-variable derivative operator D.

(a) $\nabla(au + bv) = a\,\nabla u + b\,\nabla v$ (a, b constants).

(b) $\nabla(uv) = u\,\nabla v + v\,\nabla u$.

(c) $\nabla\left(\dfrac{u}{v}\right) = \dfrac{v\,\nabla u - u\,\nabla v}{v^2}$.

(d) $\nabla u^n = nu^{n-1}\,\nabla u$.

22 Suppose that f is a function of the three independent variables x, y, and z. Show that $D_{\mathbf{i}}f = f_x$, $D_{\mathbf{j}}f = f_y$, and $D_{\mathbf{k}}f = f_z$.

23 Show that the equation of the line tangent to the conic section $Ax^2 + Bxy + Cy^2 = D$ at the point (x_0, y_0) is

$$(Ax_0)x + \tfrac{1}{2}B(y_0x + x_0y) + (Cy_0)y = D.$$

24 Show that the equation of the tangent plane to the quadric surface $Ax^2 + By^2 + Cz^2 = D$ at the point (x_0, y_0, z_0) is

$$(Ax_0)x + (By_0)y + (Cz_0)z = D.$$

25 Assume that the temperature at the point (x, y, z) in space is given by the function $W = 50 + xyz$, as in Example 2. You are standing at the point $P(3, 4, 1)$ on a mountain ridge shaped like the surface $z = x^2 - 2y$. Suppose that you start up the hill with unit speed and a northeast compass heading. What initial rate of change of temperature do you observe?

26 Suppose that the temperature at the point (x, y, z) in space is given by the formula $W = 100 - x^2 - y^2 - z^2$.
(a) Find the rate of change of temperature at $P(3, -4, 5)$ in the direction of the vector $\mathbf{v} = 3\mathbf{i} - 4\mathbf{j} + 12\mathbf{k}$.
(b) In what direction does W increase most rapidly at P? What is the value of the maximal directional derivative?

27 You're standing at the point $(30, 20, 50)$ on a hill with the shape of the surface

$$z = 1000 \exp(-[x^2 + 3y^2]/701).$$

(a) In what direction (that is, with what compass heading) should you go in order to climb most steeply? At what angle from the horizontal will you initially be climbing?
(b) If instead of climbing as in Part (a), you head directly toward the summit of the hill, then at what angle will you be climbing initially?

28 Find an equation for the plane tangent to the paraboloid $z = 2x^2 + 3y^2$ and, simultaneously, parallel to the plane $4x - 3y - z = 10$.

29 The cone $z^2 = x^2 + y^2$ and the plane $2x + 3y + 4z + 2 = 0$ intersect in an ellipse. Write an equation of the plane normal to this ellipse at the point $(3, 4, -5)$.

30 It is geometrically apparent that the highest and lowest points of the ellipse of Problem 29 are those points where its tangent line is horizontal. Find those points.

31 Show that the sphere $x^2 + y^2 + z^2 = r^2$ and the elliptical cone $z^2 = a^2x^2 + b^2y^2$ are orthogonal (that is, have perpendicular tangent planes) at every point of their intersection.

14-9

In Section 14-4 we discussed the problem of finding the maximum and minimum values attained by a function $f(x, y)$ at points of the plane region R, in the simple case in which R consists of the points on and within the simple closed curve C. We saw that any local maximum or minimum in the *interior* of R occurs at a point where $f_x = 0 = f_y$. In this section we discuss the very different matter of finding the maximum and minimum values attained by f at points of the *boundary* curve C.

Lagrange Multipliers and Constrained Maximum-Minimum Problems

If the curve C is the graph of the equation $g(x, y) = 0$, then our task is to maximize or minimize the function $f(x, y)$ subject to the **constraint** or **side condition**

$$g(x, y) = 0. \tag{1}$$

We could in principle try to solve this constraint equation for $y = \phi(x)$, and then maximize or minimize the single-variable function $f(x, \phi(x))$ by the standard method of finding where its derivative is zero. But what if it is impractical or impossible to solve Equation (1) explicitly for y in terms of x? An alternative approach that does not require that we first solve this equation is the **method of Lagrange multipliers.** It is named for its discoverer, the French mathematician Joseph Louis Lagrange (1736–1813). The method is based on the following theorem.

Theorem 1   *Lagrange Multipliers (one constraint)*

Let $f(x, y)$ and $g(x, y)$ be functions with continuous first order partial derivatives. If the maximum (or minimum) value of f subject to the condition

$$g(x, y) = 0 \tag{1}$$

occurs at a point P where $\nabla g(P) \neq \mathbf{0}$, then

$$\nabla f(P) = \lambda \, \nabla g(P) \tag{2}$$

for some constant λ.

Proof   By the implicit function theorem mentioned in Section 14-8, the fact that $\nabla g(P) \neq \mathbf{0}$ allows us to represent the curve $g(x, y) = 0$ near P by a parametric curve $\mathbf{r}(t)$ and in such fashion that $\mathbf{r}$ has a nonzero tangent vector near P: $\mathbf{r}'(t) \neq \mathbf{0}$. See Figure 14.33. Let t_0 be the value of t such that $\mathbf{r}(t_0) = \overrightarrow{OP}$. If $f(x, y)$ attains its maximum value at P, then the composite function $f(\mathbf{r}(t))$ attains its maximum value at $t = t_0$, and so

$$D_t f(\mathbf{r}(t)) = \nabla f(\mathbf{r}(t_0)) \cdot \mathbf{r}'(t_0)$$
$$= \nabla f(P) \cdot \mathbf{r}'(t_0) = 0. \tag{α}$$

Here we have used the vector chain rule—Equation (4) of Section 14-8.

Because $\mathbf{r}(t)$ lies on the curve $g(x, y) = 0$, the composite function $g(\mathbf{r}(t))$ is a constant function. Therefore

$$D_t g(\mathbf{r}(t)) = \nabla g(\mathbf{r}(t_0)) \cdot \mathbf{r}'(t_0)$$
$$= \nabla g(P) \cdot \mathbf{r}'(t_0) = 0. \tag{β}$$

Equations (α) and (β) tell us that the vectors $\nabla f(P)$ and $\nabla g(P)$ are both perpendicular to the tangent vector $\mathbf{r}'(t_0)$. Hence $\nabla f(P)$ must be a scalar multiple of $\nabla g(P)$, and this is what Equation (2) says. Thus we have proved Theorem 1.

The Lagrange multiplier method based on Theorem 1 goes like this. Suppose that we want to maximize (or minimize) $z = f(x, y)$ subject to the constraint or side condition $g(x, y) = 0$. Equation (1) and the two scalar

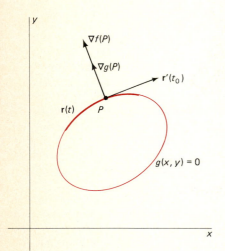

14.33   The conclusion of Theorem 1 illustrated.

CHAP. 14:   **Partial Differentiation**

components of Equation (2) yield the three equations

$$g(x, y) = 0, \qquad (1)$$

$$f_x(x, y) = \lambda g_x(x, y), \qquad \text{and} \qquad (2a)$$

$$f_y(x, y) = \lambda g_y(x, y). \qquad (2b)$$

Thus we have three equations which we can attempt to solve for the three unknowns x, y, and λ. The points (x, y) that we find (assuming that our efforts are successful) are the only possible locations for the extrema of f subject to the constraint $g(x, y) = 0$. The associated values of λ, called **Lagrange multipliers**, may come out in the wash but are usually not of much interest to us. Finally we calculate the value $f(x, y)$ at each of the solution points (x, y) so as to spot its maximum and minimum values.

We must bear in mind the additional possibility that the maximum or minimum (or both) values might occur at a point where $g_x = 0 = g_y$. The Lagrange multiplier method may fail to locate these exceptional points, but they can usually be recognized as points at which the graph $g(x, y) = 0$ fails to be a smooth curve.

EXAMPLE 1 Let us return to the sawmill problem of Example 5, Section 2-4: To maximize the cross-sectional area of a rectangular beam cut from a circular log with radius 1 foot. We want to show that the optimal beam has a square cross section.

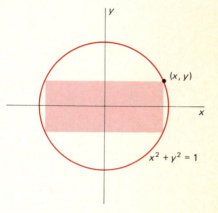

14.34 Cutting a rectangular beam from a circular log.

Solution With the coordinate system indicated in Figure 14.34, we want to maximize the area $A = f(x, y) = 4xy$ subject to the constraint

$$g(x, y) = x^2 + y^2 - 1 = 0.$$

Equations (2a) and (2b) take the forms

$$4y = 2\lambda x \quad \text{and} \quad 4x = 2\lambda y.$$

It is clear that neither $x = 0$ nor $y = 0$ gives the maximum area. Hence we may divide the first equation by $4x$ and the second by $4y$ to obtain

$$\frac{\lambda}{2} = \frac{y}{x} = \frac{x}{y}.$$

We forget λ, and see that $x^2 = y^2$. We're looking for a solution point in the first quadrant, so we conclude that $x = 1/\sqrt{2} = y$ gives the maximum. This corresponds to a square beam of edge $\sqrt{2}$ feet, which has cross-sectional area of 2 square feet—about 64% of the total cross-sectional area π of the log.

Note that $f(x, y) = 4xy$ attains its maximum value of 2 at both $(1/\sqrt{2}, 1/\sqrt{2})$ and $(-1/\sqrt{2}, -1/\sqrt{2})$ and its minimum value of -2 at both $(-1/\sqrt{2}, 1/\sqrt{2})$ and $(1/\sqrt{2}, -1/\sqrt{2})$. The Lagrange multiplier method actually locates all four of these points for us.

EXAMPLE 2 After the square beam of Example 1 has been cut from our circular log of radius 1 ft, let us cut 4 planks from the remaining pieces, each of dimensions u by $2v$, as shown in Figure 14.35. How should this be done in order to maximize the combined cross-sectional area of all 4 planks,

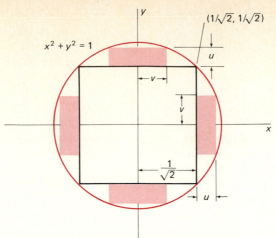

$x^2 + y^2 = 1$

$(1/\sqrt{2}, 1/\sqrt{2})$

$\dfrac{1}{\sqrt{2}}$

14.35 Cutting four more planks from the log.

thereby using what might otherwise be scrap lumber as efficiently as possible?

Solution Since the point $(u + 1/\sqrt{2}, v)$ lies on the unit circle, we want to maximize the function $f(u, v) = 8uv$ subject to the condition

$$g(u, v) = \left(u + \frac{1}{\sqrt{2}}\right)^2 + v^2 - 1 = 0.$$

The conditions $f_u = \lambda g_u$ and $f_v = \lambda g_v$ give us the equations

$$8v = 2\lambda\left(u + \frac{1}{\sqrt{2}}\right) \quad \text{and} \quad 8u = 2\lambda v.$$

We solve each of these equations for $\lambda/4$ and find that

$$\frac{\lambda}{4} = \frac{v}{u + 1/\sqrt{2}} = \frac{u}{v}.$$

Thus $v^2 = u(u + 1/\sqrt{2})$.

Finally we apply the one equation that we've not used yet—the constraint equation $g(u, v) = 0$. We substitute our last equation into that equation and get the quadratic equation

$$\left(u + \frac{1}{\sqrt{2}}\right)^2 + u\left(u + \frac{1}{\sqrt{2}}\right) - 1 = 2u^2 + \frac{3}{\sqrt{2}}u - \frac{1}{2} = 0.$$

The only positive root of this equation is $u = 0.199$ (to three-place accuracy), and this in turn tells us that $v = 0.424$. Thus our planks should be 0.199 ft thick and 0.848 ft wide. Their combined cross-sectional area will be $f(0.199, 0.424) \approx 0.673$ square feet, or about 21% of the original log.

LAGRANGE MULTIPLIERS IN THREE DIMENSIONS

Now suppose that $f(x, y, z)$ and $g(x, y, z)$ have continuous first order partial derivatives, and that we want to find the points of the *surface*

$$g(x, y, z) = 0 \tag{3}$$

at which the function $f(x, y, z)$ attains its maximum and minimum values. With functions of three rather than two variables, Theorem 1 holds precisely as we stated it above. We leave the details to Problem 33, but an argument similar to the proof of Theorem 1 shows that, at a maximum-minimum point P of $f(x, y, z)$ on the surface in (3), the gradient vectors $\nabla f(P)$ and $\nabla g(P)$ are both normal vectors to the surface, as indicated in Figure 14.36. It follows that

$$\nabla f(P) = \lambda \, \nabla g(P) \tag{4}$$

for some scalar λ. This vector equation corresponds to three scalar equations, and so we can attempt to solve simultaneously the four equations

$$g(x, y, z) = 0, \tag{3}$$

$$f_x(x, y, z) = \lambda g_x(x, y, z), \tag{4a}$$

$$f_y(x, y, z) = \lambda g_y(x, y, z), \quad \text{and} \tag{4b}$$

$$f_z(x, y, z) = \lambda g_z(x, y, z) \tag{4c}$$

for the four unknowns x, y, z, and λ. If successful, we then evaluate $f(x, y, z)$ at each of the solution points (x, y, z) to see at which it attains its maximum and its minimum values. Thus the Lagrange multiplier method with one side condition is essentially the same in dimension three as in dimension two.

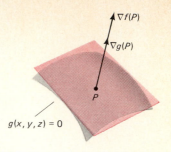

14.36 The natural generalization of Theorem 1 also holds for functions of three variables.

EXAMPLE 3 Find the maximum volume of a rectangular box inscribed in the ellipsoid $x^2/a^2 + y^2/b^2 + z^2/c^2 = 1$ with its faces parallel to the coordinate planes.

Solution Let (x, y, z) be the vertex of our box that lies in the first octant (where x, y, and z are all positive). We want to maximize the volume $V = f(x, y, z) = 8xyz$ subject to the constraint

$$g(x, y, z) = \frac{x^2}{a^2} + \frac{y^2}{b^2} + \frac{z^2}{c^2} - 1 = 0.$$

Equations (4a), (4b), and (4c) give

$$8yz = \frac{2\lambda x}{a^2}, \qquad 8xz = \frac{2\lambda y}{b^2}, \qquad 8xy = \frac{2\lambda z}{c^2}.$$

Part of the art of mathematics lies in pausing for a moment to find an elegant way to solve a problem, rather than by rushing in headlong with brute force methods. Here, if we multiply the first equation by x, the second by y, and the third by z, we find that

$$2\lambda \frac{x^2}{a^2} = 2\lambda \frac{y^2}{b^2} = 2\lambda \frac{z^2}{c^2} = 8xyz.$$

Now $\lambda \neq 0$ because (at maximum volume) x, y, and z are nonzero. We conclude that

$$\frac{x^2}{a^2} = \frac{y^2}{b^2} = \frac{z^2}{c^2}.$$

The sum of the last three expressions is 1—that's the constraint condition for this problem—and so each must be equal to $\frac{1}{3}$. All three of x, y, and z

are positive, and so

$$x = \frac{a}{\sqrt{3}}, \qquad y = \frac{b}{\sqrt{3}}, \quad \text{and} \quad z = \frac{c}{\sqrt{3}}.$$

Therefore the box of maximum volume has volume $V = 8abc/3\sqrt{3}$.

Check the ratio of box volume to ellipsoid volume ($4\pi abc/3$) to see if the above answer is plausible.

PROBLEMS WITH TWO CONSTRAINTS

Suppose that we want to find the maximum and minimum values of the function $f(x, y, z)$ at points of the curve of intersection of the two surfaces

$$g(x, y, z) = 0 \quad \text{and} \quad h(x, y, z) = 0. \tag{5}$$

This is a maximum-minimum problem with *two* constraints or side conditions. The Lagrange multiplier method for such situations is based upon the following theorem.

Theorem 2   *Lagrange Multipliers (two constraints)*

Let $f(x, y, z)$, $g(x, y, z)$, and $h(x, y, z)$ be functions with continuous first order partial derivatives. If the maximum (or minimum) value of f subject to the two conditions

$$g(x, y, z) = 0 \quad \text{and} \quad h(x, y, z) = 0 \tag{5}$$

occurs at a point P where the vectors $\nabla g(P)$ and $\nabla h(P)$ are nonzero and nonparallel, then

$$\nabla f(P) = \lambda_1 \nabla g(P) + \lambda_2 \nabla h(P) \tag{6}$$

for some two constants λ_1 and λ_2.

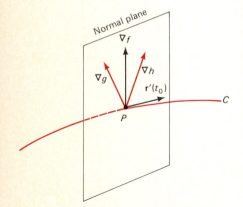

14.37 Relation between the gradients in the proof of Theorem 2.

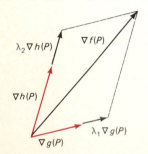

14.38 Geometry of the equation $\nabla f(P) = \lambda_1 \nabla g(P) + \lambda_2 \nabla h(P)$.

Outline of Proof   By an appropriate version of the implicit function theorem, the curve C of intersection of the two surfaces (Figure 14.37) may be represented near P by a parametric curve $\mathbf{r}(t)$ with nonzero tangent vector $\mathbf{r}'(t)$. Let t_0 be the value of t such that $\mathbf{r}(t_0) = P$. We compute the derivatives at t_0 of the composite functions $f(\mathbf{r}(t))$, $g(\mathbf{r}(t))$, and $h(\mathbf{r}(t))$. We find—exactly as in the proof of Theorem 1—that

$$\nabla f(P) \cdot \mathbf{r}'(t_0) = 0, \qquad \nabla g(P) \cdot \mathbf{r}'(t_0) = 0, \quad \text{and} \quad \nabla h(P) \cdot \mathbf{r}'(t_0) = 0.$$

These three equations say that all three gradient vectors are perpendicular to the curve C at P and thus all lie in a single plane, the normal plane to the curve C at the point P.

Now $\nabla g(P)$ and $\nabla h(P)$ are nonzero and nonparallel, so $\nabla f(P)$ is the sum of its projections onto $\nabla g(P)$ and $\nabla h(P)$ (see Problem 45 of Section 13-1). As illustrated in Figure 14.38, this fact implies Equation (6).

In examples we prefer to avoid subscripts by writing λ and μ for the Lagrange multipliers λ_1 and λ_2 of Theorem 2. Equations (5) and the three scalar components of the vector Equation (6) then give us the five simultaneous equations

$$g(x, y, z) = 0, \tag{5a}$$

$$h(x, y, z) = 0, \tag{5b}$$

$$f_x(x, y, z) = \lambda g_x(x, y, z) + \mu h_x(x, y, z), \tag{6a}$$

$$f_y(x, y, z) = \lambda g_y(x, y, z) + \mu h_y(x, y, z), \quad \text{and} \tag{6b}$$

$$f_z(x, y, z) = \lambda g_z(x, y, z) + \mu h_z(x, y, z). \tag{6c}$$

EXAMPLE 4 The plane $2x + 3y - 5z + 6 = 0$ intersects the cone $z^2 = x^2 + y^2$ in an ellipse. Find the highest and lowest points on this ellipse, and thereby determine its semiaxes.

Solution See Figure 14.39. The reason why the curve of intersection is an ellipse (rather than a parabola or hyperbola) is that the angle between the plane's normal vector $\langle -2, -3, 5 \rangle$ and the z-axis is

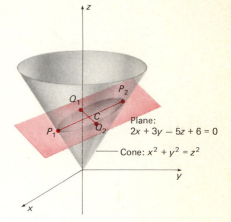

14.39 The ellipse of Example 4.

$$\theta = \arccos\left(\frac{\langle -2, -3, 5 \rangle \cdot \langle 0, 0, 1 \rangle}{\sqrt{(-2)^2 + (-3)^2 + (5)^2}}\right) \approx 36°,$$

while the elements of this familiar right cone make an angle of 45° with the z-axis. We want to find the maximum and minimum values of $f(x, y, z) = z$ subject to the conditions

$$g(x, y, z) = x^2 + y^2 - z^2 = 0 \tag{7}$$

and

$$h(x, y, z) = 2x + 3y - 5z + 6 = 0. \tag{8}$$

Equations (6a), (6b), and (6c) give

$$0 = 2\lambda x + 2\mu, \quad 0 = 2\lambda y + 3\mu, \quad 1 = -2\lambda z - 5\mu.$$

From these equations we find that

$$x = -\frac{2\mu}{2\lambda}, \quad y = -\frac{3\mu}{2\lambda}, \quad \text{and} \quad z = -\frac{5\mu + 1}{2\lambda}. \tag{9}$$

We substitute these values into Equation (7) and then multiply by $4\lambda^2$. This results in the quadratic equation

$$12\mu^2 + 10\mu + 1 = 0,$$

with roots $\mu = -0.116$ and $\mu = -0.717$ (approximately).

Then we substitute Equations (9) into the constraint equation in (8) and find that

$$-\frac{4\mu}{2\lambda} - \frac{9\mu}{2\lambda} + \frac{5(5\mu + 1)}{2\lambda} + 6 = 0.$$

Finally we solve for λ, and find that $\lambda = -\mu - \frac{5}{12}$. Equations (9) then give the (approximate) values of x, y, and z shown in the following table.

μ	λ	x	y	z
-0.116	-0.300	-0.39	-0.58	0.70
-0.717	0.300	2.39	3.58	4.30

Thus the lowest and highest points on the ellipse are evident: The lowest is $P_1(-0.39, -0.58, 0.70)$ and the highest is $P_2(2.39, 3.58, 4.30)$. When we average these coordinates, we get the center of the ellipse; it is $C(1.00, 1.50, 2.50)$. The major semiaxis is $|CP_1| = |CP_2| = 3.08$.

The endpoints Q_1 and Q_2 of the (horizontal) minor axis can be found by routinely solving Equations (7) and (8) with $z = 2.50$; it turns out that they are $Q_1(-0.44, 2.46, 2.50)$ and $Q_2(2.44, 0.54, 2.50)$. The minor semiaxis of our ellipse is then $|CQ_1| = |CQ_2| = 1.73$.

ECONOMIC APPLICATIONS

The method of Lagrange multipliers is applicable to economic problems in which total production (or profit, or what-have-you) is to be maximized subject to the constraint of fixed available resources. For example, let P be the number of units of a certain product that is being manufactured. Then P may be given in terms of the number of units x of labor and y of capital utilized. In particular, if

$$P = f(x, y) = kx^\alpha y^\beta \qquad (\alpha + \beta = 1),$$

then this function—known as the Cobb-Douglas production function—has the convenient property that $f(cx, cy) = cf(x, y)$. Thus doubling (or tripling) each of the resources utilized will double (or triple) production.

If the cost of each unit of labor is A dollars, and that of capital is B dollars, and a total of C dollars is available, then we want to maximize the production $f(x, y)$ subject to the constraint

$$g(x, y) = Ax + By - C = 0.$$

In Problem 27 we ask you to show that production is maximized when $x = \alpha C/A$ and $y = \beta C/B$. In Problem 28 we ask you to minimize the cost function $Ax + By$ subject to the constraint of fixed production.

14-9 PROBLEMS

In each of Problems 1–10, find the maximum and minimum values—if any—of the given function f subject to the given constraint or constraints.

1 $f(x, y) = x^2 - y^2$; $x^2 + y^2 = 4$.
2 $f(x, y) = x^2 + y^2$; $2x + 3y = 6$.
3 $f(x, y) = xy$; $4x^2 + 9y^2 = 36$.
4 $f(x, y) = 4x^2 + 9y^2$; $x^2 + y^2 = 1$.
5 $f(x, y, z) = x^2 + y^2 + z^2$; $3x + 2y + z = 6$.
6 $f(x, y, z) = 3x + 2y + z$; $x^2 + y^2 + z^2 = 1$.
7 $f(x, y, z) = x + y + z$; $x^2 + 4y^2 + 9z^2 = 36$.
8 $f(x, y, z) = xyz$; $x^2 + y^2 + z^2 = 1$.
9 $f(x, y, z) = x^2 + y^2 + z^2$; $x + y + z = 1$ and $x + 2y + 3z = 6$.
10 $f(x, y, z) = z$; $x^2 + y^2 = 1$ and $2x + 2y + z = 5$.

In each of Problems 11–20, use Lagrange multipliers to solve the indicated problem.

11 Section 14-4, Problem 17.
12 Section 14-4, Problem 18.

13 Section 14-4, Problem 19.
14 Section 14-4, Problem 20.
15 Section 14-4, Problem 21.
16 Section 14-4, Problem 23.
17 Section 14-4, Problem 24.
18 Section 14-4, Problem 27.
19 Section 14-4, Problem 28.
20 Section 14-4, Problem 29.
21 Find the point(s) of the surface $z = xy + 5$ closest to the origin. (*Suggestion:* Minimize the *square* of the distance.)
22 A triangle with sides x, y, and z has fixed perimeter $2s = x + y + z$. Its area A is given by Heron's formula:

$$A^2 = s(s - x)(s - y)(s - z).$$

Use the method of Lagrange multipliers to show that, among all triangles with the given perimeter, the one of largest area is the equilateral one.
23 Use the method of Lagrange multipliers to show that, of all triangles inscribed in the unit circle, the one of greatest

area is equilateral. (*Suggestion:* Use Figure 14.40 and the fact that the area of a triangle with sides a and b and included angle θ is given by the formula $A = \frac{1}{2} ab \sin \theta$.)

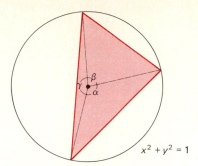

$x^2 + y^2 = 1$

14.40 A triangle inscribed in a circle (see Problem 23).

24 Find the points on the rotated ellipse $x^2 + xy + y^2 = 3$ that are nearest to and farthest from the origin. (*Suggestion:* Write the Lagrange multiplier equations in the form

$$ax + by = 0 \quad \text{and} \quad cx + dy = 0.$$

These equations have a nontrivial solution *only if* $ad - bc = 0$. Use this fact to solve first for λ.)

25 Use the method of Problem 24 to find the points of the rotated hyperbola $x^2 + 12xy + 6y^2 = 130$ that are closest to the origin.

26 Find the points of the ellipse $4x^2 + 9y^2 = 36$ that are closest to, and those farthest from, the point $(1, 1)$.

27 Show that the production function $P = P(x, y) = kx^\alpha y^\beta$ (where $\alpha + \beta = 1$) is maximized subject to fixed costs $Ax + By = C$ by $x = \alpha C/A$, $y = \beta C/B$.

28 Show that the cost function $C = C(x, y) = Ax + By$ is minimized subject to fixed production $P = kx^\alpha y^\beta$ (where $\alpha + \beta = 1$) by

$$x = \frac{P}{k}\left(\frac{\alpha B}{\beta A}\right)^\beta, \qquad y = \frac{P}{k}\left(\frac{\beta A}{\alpha B}\right)^\alpha.$$

29 Find the highest and lowest points on the ellipse of intersection of the cylinder $x^2 + y^2 = 1$ and the plane $2x + y - z = 4$.

30 Apply the method of Example 4 to find the highest and lowest points on the ellipse of intersection of the cone $z^2 = x^2 + y^2$ and the plane $x + 2y + 3z = 3$.

31 Find the points on the ellipse of Problem 30 that are nearest to, and those farthest from, the origin.

32 The ice tray shown in Figure 14.41 is to be made of material that costs 1¢ per square inch. Minimize the cost function $f(x, y, z) = xy + 3xz + 7yz$ subject to the constraints that each of the 12 compartments is to be square and the total volume (ignoring partitions) is to be 12 cubic inches.

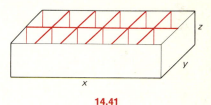

14.41

33 Prove Theorem 1 for functions of three variables by showing that each of the vectors $\nabla f(P)$ and $\nabla g(P)$ is perpendicular at P to every curve on the surface $g(x, y, z) = 0$.

14-10

The Second Derivative Test for Functions of Two Variables

We saw in Section 14-4 that, in order for the differentiable function $f(x, y)$ to have either a local minimum or a local maximum at the point $P(a, b)$, it is a *necessary* condition that P be a *critical point* of f; that is, that

$$f_x(a, b) = 0 = f_y(a, b).$$

In this section we give *sufficient* conditions that f have a local extremum at a critical point. The criterion stated below involves the second order partial derivatives of f at (a, b) and plays the role of the single-variable second derivative test (Section 3-5) for functions of two variables. To simplify the statement of this result, we shall use the following abbreviations:

$$A = f_{xx}(a, b), \qquad B = f_{xy}(a, b), \qquad C = f_{yy}(a, b), \qquad (1)$$

and

$$\Delta = AC - B^2 = f_{xx}(a, b)f_{yy}(a, b) - [f_{xy}(a, b)]^2. \qquad (2)$$

We shall prove the following theorem at the end of this section.

Theorem *Sufficient Conditions for Local Extrema*

Let (a, b) be a critical point of the function $f(x, y)$, and assume that f has continuous first order and second order partial derivatives in some circular disk centered at (a, b). Then:

(i) If $\Delta > 0$ and $A > 0$, then f has a local minimum at (a, b).
(ii) If $\Delta > 0$ and $A < 0$, then f has a local maximum at (a, b).
(iii) If $\Delta < 0$ then f has neither a local minimum nor a local maximum at (a, b).

Thus f has *either* a local maximum *or* a local minimum at the critical point (a, b) provided that the **discriminant** $\Delta = AC - B^2$ is *positive*. In this case $A = f_{xx}(a, b)$ plays the role of the second derivative of a single-variable function: There is a local minimum at (a, b) if $A > 0$, a local maximum if $A < 0$.

If $\Delta < 0$ then f has *neither* a local maximum *nor* a local minimum at (a, b). In this case we call (a, b) a **saddle point** for f, thinking of the appearance of the hyperbolic paraboloid $f(x, y) = x^2 - y^2$ (Figure 14.42), a typical example of this case.

Note that the theorem is silent on the question of what happens when $\Delta = 0$. In this case the two-variable second derivative test fails (to apply) and *anything* can happen, ranging from the local minimum of $f(x, y) = x^4 + y^4$ at $(0, 0)$ to the "monkey saddle" of Example 2 below.

In the case of a function $f(x, y)$ with several critical points, the quantities A, B, C, and Δ must be computed separately at each critical point in order to apply the above test.

EXAMPLE 1 Locate and classify the critical points of $f(x, y) = 3x - x^3 - 3xy^2$.

Solution This function is a polynomial, so all its partial derivatives exist and are continuous everywhere. When we equate its first partial derivatives

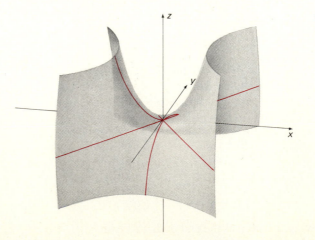

14.42 The origin is a saddle point of $z = x^2 - y^2$.

Critical point	A	B	C	Δ	Type of extremum
(1, 0)	−6	0	−6	36	Local maximum
(−1, 0)	6	0	6	36	Local minimum
(0, 1)	0	−6	0	−36	Not an extremum
(0, −1)	0	6	0	−36	Not an extremum

14.43 Critical point analysis for the function of Example 1.

to zero (in order to locate the critical points), we get

$$f_x(x, y) = 3 - 3x^2 - 3y^2 = 0, \qquad f_y(x, y) = -6xy = 0.$$

The second of these equations implies that one of the variables x and y must be zero, and then the first equation implies that the other must be ± 1. Thus there are four critical points: $(0, \pm 1)$ and $(\pm 1, 0)$.

The second order partial derivatives of f are

$$A = f_{xx} = -6x, \qquad B = f_{xy} = -6y, \qquad C = f_{yy} = -6x,$$

so $\Delta = 36(x^2 - y^2)$ at each of the critical points. The table in Figure 14.43 summarizes the situation at each of the four critical points. Figure 14.44 shows the graph of the function f.

EXAMPLE 2 Find and classify the critical points of the function $f(x, y) = 6xy^2 - 2x^3 - 3y^4$.

Solution Equating the first order partial derivatives to zero, we get the equations

$$f_x(x, y) = 6y^2 - 6x^2 = 0 \quad \text{and} \quad f_y(x, y) = 12xy - 12y^3 = 0;$$

that is,

$$x^2 = y^2 \quad \text{and} \quad xy - y^3 = 0.$$

The first of these equations gives $x = \pm y$. If $x = y$, the second equation gives $y = 0$ or $y = 1$. If $x = -y$, the second equation gives $y = 0$ or $y = -1$. Hence there are three critical points: $(0, 0)$, $(1, 1)$, and $(1, -1)$.

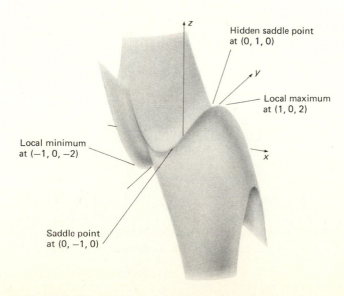

Hidden saddle point at (0, 1, 0)

Local maximum at (1, 0, 2)

Local minimum at (−1, 0, −2)

Saddle point at (0, −1, 0)

14.44 Graph of the function of Example 1.

14.45 Critical point analysis for the function of Example 2.

The second order partial derivatives of f are

$$A = f_{xx} = -12x, \qquad B = f_{xy} = 12y, \qquad C = f_{yy} = 12x - 36y^2.$$

These give the data shown in the table of Figure 14.45. Note that our test fails at the point $(0, 0)$, so we must find another way to test this point. We observe that $f(x, 0) = -2x^3$ and that $f(0, y) = -3y^4$. Hence, as we move away from the origin in the

positive x-direction:	f decreases,
negative x-direction:	f increases,
positive y-direction:	f decreases,
negative y-direction:	f decreases.

Consequently f has neither a local maximum nor a local minimum at the origin. Note (Figure 14.46) that if a monkey sits with its rump at the origin, facing the negative x-direction, then the three directions in which $f(x, y)$ decreases provide places for both its tail and its two legs to hang. That's why this particular surface is called a *monkey saddle.*

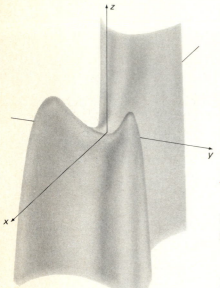

14.46 The graph of the function of Example 2 is a "monkey saddle."

Now we turn to the proof of the theorem above. It happens that the behavior of $f(x, y)$ near the critical point (a, b) is determined by the behavior near $(0, 0)$ of the *quadratic form*

$$q(h, k) = Ah^2 + 2Bhk + Ck^2. \tag{3}$$

Here, $A = f_{xx}(a, b)$, $B = f_{xy}(a, b)$, and $C = f_{yy}(a, b)$, as before. Also writing $\Delta = AC - B^2$ as before, we find upon completing the square that

$$q(h, k) = \frac{1}{A}\left[(Ah + Bk)^2 + \Delta k^2\right]. \tag{4}$$

This form makes clear the following properties of $q(h, k)$:

(i) If $\Delta > 0$ and $A > 0$, then $q(h, k) > 0$ for all h, k not both zero.

(ii) If $\Delta > 0$ and $A < 0$, then $q(h, k) < 0$ for all h, k not both zero.

(iii) If $\Delta < 0$, then there are some values of h and k (not both zero) arbitrarily close to zero such that $q(h, k) > 0$, and other such values of h and k with $q(h, k) < 0$.

Assertions (i) and (ii) are obvious upon consideration of signs. The reason for (iii) in the case $A > 0$, $B \neq 0$ is this: If $\Delta < 0$, then $q(h, 0) = Ah^2 > 0$ while $q(h, -Ah/B) = \Delta k^2/A < 0$.

We want to analyze the consequences of the assumption that

$$\Delta = AC - B^2 = f_{xx}(a, b)f_{yy}(a, b) - [f_{xy}(a, b)]^2 \neq 0$$

at the critical point (a, b) of the function f. Draw a circular disk centered at (a, b), as in Figure 14.47. Since the second order partial derivatives of f are continuous, we can make the radius of the disk small enough so that $f_{xx}f_{yy} - (f_{xy})^2$ has the same sign as Δ at every point of the disk.

Now consider the single-variable function

$$G(t) = f(a + th, b + tk), \quad t \text{ in } [0, 1].$$

We apply Taylor's formula to $G(t)$ on the interval $[0, 1]$ and find that

$$G(1) = G(0) + G'(0) + \tfrac{1}{2}G''(\hat{t}) \tag{5}$$

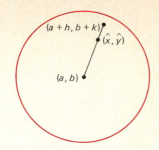

14.47 The circular disk used in proving the theorem.

for some $\hat{t}$ in $(0, 1)$. But the chain rule gives

$$G'(t) = \frac{\partial f}{\partial x}\frac{dx}{dt} + \frac{\partial f}{\partial y}\frac{dy}{dt} = hf_x + kf_y$$

and

$$G''(t) = \frac{\partial}{\partial x}(hf_x + kf_y)\frac{dx}{dt} + \frac{\partial}{\partial y}(hf_x + kf_y)\frac{dy}{dt}$$

$$= h^2 f_{xx} + 2hk f_{xy} + k^2 f_{yy},$$

where the partial derivatives of f are to be evaluated at the point $(a + th, b + tk)$. Consequently $G'(0) = 0$ because $f_x(a, b) = f_y(a, b) = 0$, and

$$G''(\hat{t}) = \hat{A}h^2 + 2\hat{B}hk + \hat{C}k^2$$

where $\hat{A}$, $\hat{B}$, and $\hat{C}$ denote the values of f_{xx}, f_{xy}, and f_{yy}, respectively, at the point $(a + \hat{t}h, b + \hat{t}k)$. Because $G(0) = f(a, b)$ and $G(1) = f(a + h, b + k)$, Equation (5) says that

$$f(a + h, b + k) = f(a, b) + \tfrac{1}{2}(\hat{A}h^2 + 2\hat{B}hk + \hat{C}k^2). \tag{6}$$

Now $\hat{\Delta} = \hat{A}\hat{C} - \hat{B}^2$ has the same sign as $\Delta = AC - B^2$. And if $A \neq 0$, we assume that our circular disk is so small that A has the same sign as $\hat{A}$. So the quadratic form

$$\hat{q}(h, k) = \hat{A}h^2 + 2\hat{B}hk + \hat{C}k^2$$

that appears in Equation (6) exhibits the same behavior as the quadratic form $q(h, k)$ of Equation (3). The theorem now follows from our analysis of $q(h, k)$. For example, if Δ and A are both positive, then so are $\hat{\Delta}$ and $\hat{A}$. Hence $\hat{q}(h, k) > 0$ for all h, k not both zero. So (6) shows us that $f(a + h, b + k) > f(a, b)$ for each point $(a + h, b + k)$—other than (a, b) itself—in the circular disk. Thus $f(a, b)$ is a local minimum in this case. The other two cases follow from similar arguments.

14-10 PROBLEMS

Find and classify the critical points of each of the functions in Problems 1–15.

1 $f(x, y) = 2x^2 + y^2 + 4x - 4y + 5$.
2 $f(x, y) = 10 + 12x - 12y - 3x^2 - 2y^2$.
3 $f(x, y) = 2x^2 - 3y^2 + 2x - 3y + 7$.
4 $f(x, y) = xy + 3x - 2y + 4$.

5 $f(x, y) = 2x^2 + 2xy + y^2 + 4x - 2y + 1$.
6 $f(x, y) = x^2 + 4xy + 2y^2 + 4x - 8y + 3$.
7 $f(x, y) = x^3 + y^3 + 3xy + 3$.
8 $f(x, y) = x^2 - 2xy + y^3 - y$.
9 $f(x, y) = 6x - x^3 - y^3$.
10 $f(x, y) = 3xy - x^3 - y^3$.
11 $f(x, y) = x^4 + y^4 - 4xy$.

12 $f(x, y) = (x^2 + y^2) \exp(x^2 - y^2)$.

13 $f(x, y) = 1 - x^4 - y^4$.

14 $f(x, y) = x^3 + y^3$.

15 $f(x, y) = y^3 - 3x^2 y$.

16 Let $f(s, t)$ denote the *square* of the distance between a typical point of the line $x = t$, $y = t + 1$, $z = 2t$ and a typical point of the line $x = 2s$, $y = s - 1$, $z = s + 1$. Show that the single critical point of f is a local minimum. Hence find the closest points on these two skew lines.

17 Let $f(x, y)$ denote the square of the distance from $(0, 0, 2)$ to a typical point of the surface $z = xy$. Find and classify the critical points of f.

18 Show that the surface $z = (x^2 + 2y^2) \exp(1 - x^2 - y^2)$ looks like two mountain peaks joined by two ridges with a pit between them.

19 A wire 120 inches long is cut into three pieces of lengths x, y, and $120 - x - y$, and each piece is bent into the shape of a square. Let $f(x, y)$ denote the sum of the areas of these squares. Show that the single critical point of f is a local minimum. But surely it is possible to *maximize* the sum of the areas. Explain.

20 Apply the test of this section to prove that the method of least squares (Section 14-5) always gives a minimum. That is, show that the function

$$f(m, b) = \sum_{i=1}^{n} [y_i - (mx_i + b)]^2$$

has a single critical point, and it is a local minimum. You may use the fact that $n\Sigma x_i^2 > (\Sigma x_i)^2$ unless the numbers $x_1, x_2, \ldots, x_n$ are all equal (see Miscellaneous Problem 37).

CHAPTER 14 REVIEW: Definitions, Concepts, Results

Use the list below as a guide to concepts that you may need to review.

1 Graphs and level curves of functions of two variables

2 Limits and continuity for functions of two and three variables

3 Partial derivatives—definition and computation

4 Geometric interpretation of partial derivatives and the tangent plane to the surface $z = f(x, y)$

5 Absolute and local maxima and minima

6 Necessary conditions for a local extremum

7 The method of least squares

8 Increments and differentials for functions of two and three variables

9 The linear approximation theorem

10 The chain rule for functions of several variables

11 Directional derivatives—definition and computation

12 The gradient vector and the vector chain rule

13 Significance of the length and direction of the gradient vector

14 The gradient vector as a normal vector; tangent plane to a surface $F(x, y, z) = 0$

15 Constrained maximum-minimum problems and the Lagrange multiplier method

16 Sufficient conditions for a local extremum of a function of two variables

MISCELLANEOUS PROBLEMS

1 Use the method of Example 4 in Section 14-2 to show that

$$\lim_{(x,y)\to(0,0)} \frac{x^2 y^2}{x^2 + y^2} = 0.$$

2 Use spherical coordinates to show that

$$\lim_{(x,y,z)\to(0,0,0)} \frac{x^3 + y^3 - z^3}{x^2 + y^2 + z^2} = 0.$$

3 Suppose that $g(x, y) = xy/(x^2 + y^2)$ unless $x = 0 = y$; let $g(0, 0)$ be defined to be zero. Show that g is not continuous at $(0, 0)$.

4 Compute $g_x(0, 0)$ and $g_y(0, 0)$ for the function of Problem 3.

5 Find a function $f(x, y)$ such that

$$f_x(x, y) = 2xy^3 + e^x \sin y$$

and

$$f_y(x, y) = 3x^2 y^2 + e^x \cos y + 1.$$

6 Prove that there is *no* function with continuous second order partial derivatives such that $f_x(x, y) = 6xy^2$ and $f_y(x, y) = 8x^2 y$.

7 Find the points on the paraboloid $z = x^2 + y^2$ at which the normal line passes through the point $(0, 0, 1)$.

8 Write an equation of the tangent plane to the surface $\sin xy + \sin yz + \sin xz = 1$ at the point $(1, \pi/2, 0)$.

9 Prove that every normal line to the cone $z = (x^2 + y^2)^{1/2}$ intersects the z-axis.

10 Show that the function

$$u(x, t) = (4\pi kt)^{-1/2} \exp\left(\frac{-x^2}{4kt}\right)$$

satisfies the one-dimensional heat equation of Problem 25 in Section 14-3.

11 Show that the function

$$u(x, y, t) = (4\pi kt)^{-1} \exp\left(\frac{-[x^2 + y^2]}{4kt}\right)$$

satisfies the two-dimensional heat equation of Problem 26 in Section 14-3.

12 Let $f(x, y) = xy(x^2 - y^2)/(x^2 + y^2)$ unless $x = 0 = y$, in which case we have $f(0, 0) = 0$. Show that the second-order partial derivatives f_{xx}, f_{xy}, f_{yx}, and f_{yy} all exist at $(0, 0)$ but that $f_{xy}(0, 0) \neq f_{yx}(0, 0)$.

13 Define the partial derivatives $\mathbf{r}_x$ and $\mathbf{r}_y$ of the vector-valued function $\mathbf{r}(x, y) = x\mathbf{i} + y\mathbf{j} + f(x, y)\mathbf{k}$ by component-wise partial differentiation. Then show that the vector $\mathbf{r}_x \times \mathbf{r}_y$ is normal to the surface $z = f(x, y)$.

14 An open-topped rectangular box is to have a total surface area of 300 in². Find the dimensions that maximize its volume.

15 A rectangular shipping crate is to have a volume of 60 ft³. Its sides cost $1/ft², its top costs $2/ft², and its bottom costs $3/ft². What dimensions will minimize the cost of the box?

16 A pyramid is bounded by the three coordinate planes and the tangent plane to the surface $xyz = 1$ at a point in the first octant. Find the volume of this pyramid (it is independent of the point of tangency).

17 The total resistance R of two resistances R_1 and R_2 connected in parallel is given by the formula

$$\frac{1}{R} = \frac{1}{R_1} + \frac{1}{R_2}.$$

Suppose that R_1 and R_2 are measured to be 300 and 600 ohms, respectively, with a maximum error of 1% in each measurement. Use differentials to estimate the maximum error (in ohms) in the calculated value of R.

18 Consider the gas of Problem 33 in Section 14-3, a gas satisfying the van der Waals equation. Use differentials to approximate the change in its volume if P is increased from 1 atm to 1.1 atm and T is decreased from 313°K to 303°K.

19 The semiaxes a, b, and c of an ellipsoid with volume $V = \frac{4}{3}\pi abc$ are each measured with a maximum percentage error of 1%. Use differentials to estimate the maximum percentage error in the calculated value of V.

20 Two spheres have radii a and b and the distance between their centers is $c < a + b$, so that the spheres meet in a common circle. Let P be a point on this circle and let $\mathscr{P}_1$ and $\mathscr{P}_2$ be the tangent planes at P to the two spheres. Find the angle between $\mathscr{P}_1$ and $\mathscr{P}_2$ in terms of a, b, and c. (*Note:* The angle between two planes in space is by definition the angle between their normal vectors.)

21 Find every point on the surface of the ellipsoid $x^2 + 4y^2 + 9z^2 = 16$ at which the normal line at the point passes through the center $(0, 0, 0)$ of the ellipsoid.

22 Suppose that

$$F(x) = \int_{g(x)}^{h(x)} f(t)\, dt.$$

Show that $F'(x) = f(h(x))h'(x) - f(g(x))g'(x)$. (*Suggestion:* Write $w = \int_u^v f(t)\, dt$ with $u = g(x)$ and $v = h(x)$.)

23 Suppose that $\mathbf{a}$, $\mathbf{b}$, and $\mathbf{c}$ are mutually perpendicular unit vectors, and that f is a function of the three independent variables x, y, and z. Show that

$$\nabla f = (D_{\mathbf{a}}f)\mathbf{a} + (D_{\mathbf{b}}f)\mathbf{b} + (D_{\mathbf{c}}f)\mathbf{c}.$$

24 Let $\mathbf{R} = \langle \cos\theta, \sin\theta, 0 \rangle$ and $\mathbf{\Theta} = \langle -\sin\theta, \cos\theta, 0 \rangle$ be the polar coordinate unit vectors. Given $f(x, y, z) = w(r, \theta, z)$, show that

$$D_{\mathbf{R}}f = \frac{\partial w}{\partial r} \quad \text{and} \quad D_{\mathbf{\Theta}}f = \frac{1}{r}\frac{\partial w}{\partial \theta}.$$

Then conclude from Problem 23 that the gradient vector is given in cylindrical coordinates by

$$\nabla f = \frac{\partial w}{\partial r}\mathbf{R} + \frac{1}{r}\frac{\partial w}{\partial \theta}\mathbf{\Theta} + \frac{\partial w}{\partial z}\mathbf{k}.$$

25 Suppose that you are standing at the point $(-100, -100, 430)$ on the hill of Example 3 in Section 14-8. (a) In what (horizontal) direction should you move in order to climb the most steeply? (b) What will be your resulting rate of climb (rise/run)?

26 Suppose that the blood concentration in the water at the point (x, y) is given by $f(x, y) = A\,\exp(-k[x^2 + 2y^2])$ where A and k are positive constants. A shark always swims in the direction of ∇f. (Why?) Show that its path is a parabola $y = cx^2$. (*Suggestion:* Show that the condition that $\langle dx/dt, dy/dt \rangle$ be a multiple of ∇f implies that $x'/x = y'/2y$. Then antidifferentiate this equation.)

27 Consider a tangent plane to the surface $x^{2/3} + y^{2/3} + z^{2/3} = 1$. Show that the sum of the squares of the x-, y-, and z-intercepts of this plane is 1.

28 Find the points on the ellipse $x^2/a^2 + y^2/b^2 = 1$ (with $a \neq b$) where the normal line passes through the origin.

29 (a) Show that the origin $(0, 0)$ is a critical point of the function f of Problem 12. (b) Show that f does not have a local extremum at $(0, 0)$.

30 Find the point of the surface $z = xy + 1$ that is closest to the origin.

31 Use the method of Problem 24 in Section 14-9 to find the semiaxes of the rotated ellipse $73x^2 + 72xy + 52y^2 = 100$.

32 Use the Lagrange multiplier method to show that the longest chord of the sphere $x^2 + y^2 + z^2 = 1$ has length 2. (*Suggestion:* There is no loss of generality in assuming that $(1, 0, 0)$ is one end point of the chord.)

33 Use the method of Lagrange multipliers, the law of cosines, and Figure 14.40 to find the triangle of minimum perimeter inscribed in the unit circle.

34 When a current I enters two resistances R_1 and R_2 connected in parallel, it splits into two currents I_1 and I_2 to minimize $R_1 I_1^2 + R_2 I_2^2$ (the sum of the energies). Express I_1 and I_2 in terms of R_1, R_2, and I.

35 Use the method of Lagrange multipliers to find the points of the ellipse $x^2 + 2y^2 = 1$ that are closest to and farthest from the line $x + y = 2$. (*Suggestion:* Let $f(x, y, u, v)$ denote the square of the distance between the point (x, y) of the ellipse and the point (u, v) of the line.)

36 (a) Show that the maximum value of $f(x, y, z) = x + y + z$ at points of the sphere $x^2 + y^2 + z^2 = a^2$ is $a\sqrt{3}$.

(b) Conclude from the result of (a) that

$$(x + y + z)^2 \leq 3(x^2 + y^2 + z^2)$$

for any three numbers x, y, and z.

37 Generalize the method of Problem 36 to show that

$$\left(\sum_{i=1}^{n} x_i \right)^2 \leq n \sum_{i=1}^{n} x_i^2$$

for any n real numbers $x_1, x_2, \ldots, x_n$.

38 Find the minimum and maximum values of $f(x, y) = xy - x - y$ at points on and within the triangle with vertices $(0, 0)$, $(0, 1)$, and $(3, 0)$.

39 Find the maximum and minimum values of $f(x, y, z) = x^2 - yz$ at points of the sphere $x^2 + y^2 + z^2 = 1$.

40 Find the maximum and minimum values of $f(x, y) = x^2 y^2$ at points of the ellipse $x^2 + 4y^2 = 24$.

Locate and classify the critical points (local maxima, local minima, saddle points, and other points at which the tangent plane is horizontal) of the functions in Problems 41–50.

41 $f(x, y) = x^3 y - 3xy + y^2$.

42 $f(x, y) = x^2 + xy + y^2 - 6x + 2$.

43 $f(x, y) = x^3 - 6xy + y^3$.

44 $f(x, y) = x^2 y + xy^2 + x + y$.

45 $f(x, y) = x^3 y^2 (1 - x - y)$.

46 $f(x, y) = x^4 - 2x^2 + y^2 + 4y + 3$.

47 $f(x, y) = e^{xy} - 2xy$.

48 $f(x, y) = x^3 - y^3 + x^2 + y^2$.

49 $f(x, y) = (x - y)(xy - 1)$.

50 $f(x, y) = (2x^2 + y^2)e^{-x^2 - y^2}$.

Multiple Integrals

15

Double Integrals

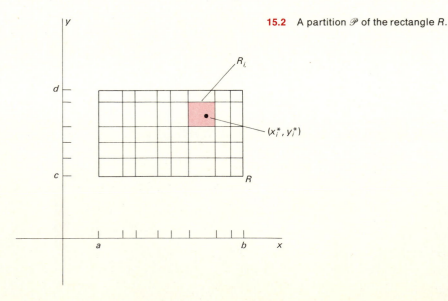

15.1 We will use a double integral to compute the volume V.

This chapter is devoted to integrals of functions of two and three variables. Such integrals are called **multiple integrals.** The applications of multiple integrals include computations of area, volume, mass, and surface area in a wider variety of situations than can be handled with the single integral of Chapters 4 and 5.

The simplest sort of multiple integral is the *double integral*

$$\iint\limits_{R} f(x, y) \, dA$$

of a continuous function $f(x, y)$ over the *rectangle*

$$R = [a, b] \times [c, d] = \{(x, y) \mid a \leqq x \leqq b, \quad c \leqq y \leqq d\}$$

in the xy-plane. Just as the definition of the single integral is motivated by the problem of computing areas, the definition of the double integral is motivated by the problem of computing the volume V of the solid of Figure 15.1: A solid that is bounded above by the graph $z = f(x, y)$ of the non-negative function f and that lies above the rectangle R in the xy-plane.

To define the double integral $\iint_{R} f(x, y) \, dA$ which has value V, we begin with an *approximation* to V. To obtain this approximation, our first step is to construct a **partition** $\mathscr{P}$ of R into subrectangles $R_1, R_2, \ldots, R_k$ determined by the partitions

$$a = x_0 < x_1 < x_2 < \cdots < x_m = b$$

of $[a, b]$ and

$$c = y_0 < y_1 < y_2 < \cdots < y_n = d$$

of $[c, d]$. Such a partition of R into $k = mn$ subrectangles is shown in Figure 15.2. The order in which these rectangles are labeled makes no difference.

15.2 A partition $\mathscr{P}$ of the rectangle R.

CHAP. 15:   Multiple Integrals

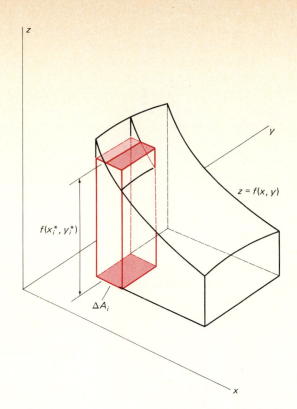

15.3 Approximating the volume under the surface by summing volumes of towers with rectangular bases.

Next, we choose an arbitrary point (x_i^*, y_i^*) of the ith subrectangle R_i for each i $(1 \leq i \leq k)$. The collection of points $S = \{(x_i^*, y_i^*)\}_{i=1}^k$ is called a **selection** for the partition $\mathscr{P} = \{R_i\}_{i=1}^k$. As a measure of the size of the subrectangles of the partition $\mathscr{P}$, we define its **mesh** $|\mathscr{P}|$ to be the maximum of the lengths of the diagonals of the rectangles $\{R_i\}$.

Figure 15.3 shows the rectangular column with base the subrectangle R_i and height the value $f(x_i^*, y_i^*)$ of f at the selected point (x_i^*, y_i^*) of R_i. If ΔA_i denotes the area of R_i, then the volume of this column is $f(x_i^*, y_i^*) \, \Delta A_i$. The sum of the volumes of all these columns is the **Riemann sum**

$$\sum_{i=1}^k f(x_i^*, y_i^*) \, \Delta A_i, \tag{1}$$

an approximation to the volume V of the solid region that lies over the rectangle R and under the graph $z = f(x, y)$.

We would expect to get the exact volume V by taking the limit of the Riemann sum in (1) as the mesh $|\mathscr{P}|$ of the partition $\mathscr{P}$ approaches zero. We therefore define the **(double) integral** of the function f over the rectangle R by

$$\iint_R f(x, y) \, dA = \lim_{|\mathscr{P}| \to 0} \sum_{i=1}^k f(x_i^*, y_i^*) \, \Delta A_i, \tag{2}$$

provided that this limit (we discuss its meaning in more detail below) exists. In advanced calculus it is proved that the limit in (2) *does* exist if f is continuous on R. In order to motivate the introduction of the Riemann sum in (1), we assumed that f was nonnegative on R, but Equation (2) serves to define the double integral whether or not f is nonnegative.

The direct evaluation of the limit in Equation (2) is generally even less practical than the direct evaluation of the limit we used in Section 4-3 to define the single-variable integral. In practice, we shall calculate double integrals over rectangles by means of the **iterated integrals** that appear in the following theorem.

Theorem 1 *Iterated Double Integrals*

Suppose that $f(x, y)$ is continuous on the rectangle $R = [a, b] \times [c, d]$. Then

$$\iint\limits_{R} f(x, y)\, dA = \int_a^b \left(\int_c^d f(x, y)\, dy \right) dx = \int_c^d \left(\int_a^b f(x, y)\, dx \right) dy. \quad (3)$$

This theorem tells us how to compute a double integral by means of two successive (or *iterated*) single-variable integrations, each of which can be carried out using the Fundamental Theorem of Calculus (provided that the function f is nice enough).

The meaning of the parentheses in the iterated integral

$$\int_a^b \int_c^d f(x, y)\, dy\, dx = \int_a^b \left(\int_c^d f(x, y)\, dy \right) dx \quad (4)$$

is this: First we hold x constant and integrate with respect to y, from $y = c$ to $y = d$. The result of this first integration is the **partial integral of f with respect to y**, denoted by

$$\int_c^d f(x, y)\, dy,$$

and it is a function of x alone. The final step is to integrate this latter function with respect to x, from $x = a$ to $x = b$.

Similarly, the iterated integral

$$\int_c^d \int_a^b f(x, y)\, dx\, dy = \int_c^d \left(\int_a^b f(x, y)\, dx \right) dy \quad (5)$$

is calculated by first integrating from a to b with respect to x (while holding y fixed) and then integrating the result from c to d with respect to y. Note that the order of integration (either first with respect to x and then with respect to y or the reverse) is determined by the order in which the differentials dx and dy appear in the iterated integrals in (4) and (5). We always work "from the inside out." Theorem 1 guarantees that the value obtained is independent of the order of integration provided that f is continuous.

EXAMPLE 1 Compute the iterated integrals in (4) and (5) for the function $f(x, y) = 4x^3 + 6xy^2$ on the rectangle $R = [1, 3] \times [-2, 1]$.

Solution

$$\int_1^3 \left(\int_{-2}^1 (4x^3 + 6xy^2)\, dy \right) dx = \int_1^3 \left[4x^3 y + 2xy^3 \right]_{-2}^1 dx$$

$$= \int_1^3 [(4x^3 + 2x) - (-8x^3 - 16x)]\, dx$$

$$= \int_1^3 (12x^3 + 18x)\, dx$$

$$= \left[3x^4 + 9x^2 \right]_1^3 = 312.$$

On the other hand,

$$\int_{-2}^1 \left(\int_1^3 (4x^3 + 6xy^2)\, dx \right) dy = \int_{-2}^1 \left[x^4 + 3x^2 y^2 \right]_1^3 dy$$

$$= \int_{-2}^1 [(81 + 27y^2) - (1 + 3y^2)]\, dy$$

$$= \int_{-2}^1 (80 + 24y^2)\, dy$$

$$= \left[80y + 8y^3 \right]_{-2}^1 = 312.$$

An outline of the proof of Theorem 1 exhibits an instructive relationship between iterated integrals and the method of cross sections (for computing volumes) discussed in Section 5-2. First, we subdivide $[a, b]$ into n equal subintervals each with length $\Delta x = (b - a)/n$, and we also subdivide $[c, d]$ into n equal subintervals each with length $\Delta y = (d - c)/n$. This gives n^2 subrectangles, each of which has area $\Delta A = \Delta x\, \Delta y$. Pick a point x_i^* in $[x_{i-1}, x_i]$ for each i, $1 \le i \le n$. Then the average value theorem for single integrals (Section 4-4) gives a point y_{ij}^* in $[y_{j-1}, y_j]$ such that

$$\int_{y_{j-1}}^{y_j} f(x_i^*, y)\, dy = f(x_i^*, y_{ij}^*)\, \Delta y.$$

This gives our selected point (x_i^*, y_{ij}^*) in the subrectangle $[x_{i-1}, x_i] \times [y_{j-1}, y_j]$. Then

$$\iint_R f(x, y)\, dA \approx \sum_{i,j=1}^n f(x_i^*, y_{ij}^*)\, \Delta A$$

$$= \sum_{i=1}^n \sum_{j=1}^n f(x_i^*, y_{ij}^*)\, \Delta y\, \Delta x$$

$$= \sum_{i=1}^n \left(\sum_{j=1}^n \int_{y_{j-1}}^{y_j} f(x_i^*, y)\, dy \right) \Delta x$$

$$= \sum_{i=1}^n \left(\int_c^d f(x_i^*, y)\, dy \right) \Delta x = \sum_{i=1}^n A(x_i^*)\, \Delta x$$

where $A(x) = \int_c^d f(x, y)\, dy$. This last sum is a Riemann sum for the integral $\int_a^b A(x)\, dx$, so the result of our computation is that

$$\iint_R f(x, y)\, dA \approx \sum_{i=1}^n A(x_i^*)\, \Delta x$$

$$\approx \int_a^b A(x)\, dx = \int_a^b \left(\int_c^d f(x, y)\, dy \right) dx.$$

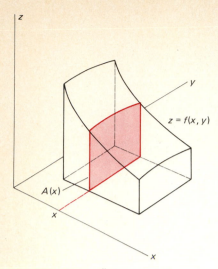

z = f(x, y)

A(x)

15.4 $A(x) = \int_a^b f(x, y)\, dy$ is the area of a cross section.

This outline can be converted into a complete proof of Theorem 1 by showing that the approximations above become equalities when we take limits as $n \to \infty$.

In case the function f is nonnegative on R, the function $A(x)$ introduced above gives the area of the vertical cross section perpendicular to the x-axis shown in Figure 15.4. Thus the iterated integral in (4) expresses the volume V as the integral from $x = a$ to $x = b$ of the cross-sectional area function $A(x)$. Similarly, the iterated integral in (5) expresses V as the integral from $y = c$ to $y = d$ of the function

$$A(y) = \int_a^b f(x, y)\, dx,$$

which gives the area of a vertical cross section in a plane perpendicular to the y-axis. (Although it is suggestive to use $A(y)$ here, note that $A(y)$ and $A(x)$ are not the same function.)

DOUBLE INTEGRALS
OVER MORE GENERAL REGIONS

Now we want to define and compute double integrals over regions more general than rectangles. Let the function f be defined on the plane region R, and suppose that R is **bounded**—that is, suppose that R lies within some rectangle S. To define the (double) integral of f over the region R, we begin with a partition $\mathcal{Q}$ of the rectangle S into subrectangles. Some of the rectangles of $\mathcal{Q}$ will lie inside R, some will lie outside R, and some will lie partly inside and partly outside. We consider the collection $\mathcal{P} = \{R_1, R_2, \ldots, R_k\}$ of all those subrectangles of $\mathcal{Q}$ that lie *completely within* the region R. This collection $\mathcal{P}$ is called the **inner partition** of the region R determined by the partition $\mathcal{Q}$ of the rectangle S (see Figure 15.5). By the **mesh** $|\mathcal{P}|$ of the inner partition $\mathcal{P}$ is meant the mesh of the partition $\mathcal{Q}$ that determines $\mathcal{P}$. (Note that $|\mathcal{P}|$ depends not only upon $\mathcal{P}$ but upon $\mathcal{Q}$ as well.)

Using the inner partition $\mathcal{P}$ of the region R, we can proceed in much the same way as before. By choosing an arbitrary point (x_i^*, y_i^*) in the ith subrectangle R_i of $\mathcal{P}$ for $i = 1, 2, 3, \ldots, k$, we obtain a **selection** for the inner partition $\mathcal{P}$. Let us denote by ΔA_i the area of R_i. Then this selection gives the **Riemann sum**

$$\sum_{i=1}^{k} f(x_i^*, y_i^*)\, \Delta A_i$$

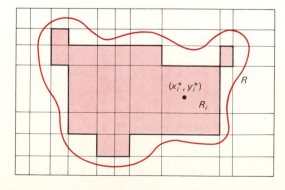

15.5 The rectangular partition of S produces an associated inner partition (shown shaded) of the region R.

associated with the inner partition $\mathscr{P}$. In case f is nonnegative on R, this Riemann sum approximates the volume of the three-dimensional region that lies under the surface $z = f(x, y)$ and above the region R in the xy-plane. We therefore define the double integral of f over the region R by taking the limit of this Riemann sum as the mesh $|\mathscr{P}|$ approaches zero. Thus

$$\iint\limits_{R} f(x, y) \, dA = \lim_{|\mathscr{P}| \to 0} \sum_{i=1}^{k} f(x_i^*, y_i^*) \, \Delta A_i \tag{6}$$

provided that this limit exists in the sense of the following definition.

Definition   *The Double Integral*

The **(double) integral** of the bounded function f over the plane region R is the number

$$I = \iint\limits_{R} f(x, y) \, dA$$

provided that, for every $\varepsilon > 0$, there exists a number $\delta > 0$ such that

$$\left| \sum_{i=1}^{k} f(x_i^*, y_i^*) \, \Delta A_i - I \right| < \varepsilon$$

for every inner partition $\mathscr{P} = \{R_1, R_2, R_3, \ldots, R_k\}$ of R having mesh $|\mathscr{P}| < \delta$ and every selection of points (x_i^*, y_i^*) in R_i $(i = 1, 2, 3, \ldots, k)$.

Thus the meaning of the limit in (6) is that the Riemann sum can be made arbitrarily close to $I = \iint_R f(x, y) \, dA$ by choosing the mesh of the inner partition $\mathscr{P}$ sufficiently small.

Note that, if R is a rectangle and we choose $S = R$ (so that an inner partition of R is simply a partition of R), then the definition above reduces to our earlier definition of a double integral over a rectangle. In advanced calculus it is proved that the double integral of the function f over the bounded plane region R exists provided that f is continuous on R and the *boundary* of R is reasonably nice. In particular, it suffices for the boundary of R to consist of finitely many piecewise smooth simple closed curves (that is, each boundary curve consists of finitely many smooth arcs).

For certain common types of regions, we can evaluate double integrals by using iterated integrals, in much the same way as when the region is a rectangle. The region R is called **vertically simple** if it is described by means of the inequalities

$$a \leqq x \leqq b, \qquad g_1(x) \leqq y \leqq g_2(x), \tag{7}$$

where g_1 and g_2 are continuous on $[a, b]$. Such a region appears in Figure 15.6. The region R is called **horizontally simple** if it is described by the inequalities

$$c \leqq y \leqq d, \qquad h_1(y) \leqq x \leqq h_2(y), \tag{8}$$

where h_1 and h_2 are continuous on $[c, d]$. The region of Figure 15.7 is horizontally simple.

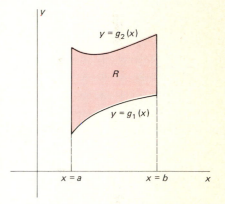

15.6   A vertically simple region R.

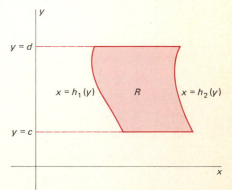

15.7   A horizontally simple region R.

The following theorem tells us how to compute by iterated integration a double integral over a region R that is either vertically simple or horizontally simple.

Theorem 2 *General Iterated Double Integrals*

Suppose that $f(x, y)$ is continuous on the region R. If R is the vertically simple region given in (7), then

$$\iint_R f(x, y)\, dA = \int_a^b \int_{g_1(x)}^{g_2(x)} f(x, y)\, dy\, dx. \tag{9}$$

If R is the horizontally simple region given in (8), then

$$\iint_R f(x, y)\, dA = \int_c^d \int_{h_1(y)}^{h_2(y)} f(x, y)\, dx\, dy. \tag{10}$$

This theorem includes Theorem 1 as a special case (when R is a rectangle), and it can be proved by a generalization of the argument we outlined for the proof of Theorem 1.

EXAMPLE 2   Compute in two different ways the integral $\iint_R xy^2\, dA$, where R is the first-quadrant region bounded by the two curves $y = x^2$ and $y = x^3$.

Solution   *Always sketch the region R of integration before attempting to evaluate a double integral.* The region R bounded by $y = x^2$ and $y = x^3$ is shown in Figure 15.8. This region is vertically simple with $a = 0$, $b = 1$, $g_1(x) = x^3$, and $g_2(x) = x^2$. Therefore (9) yields

$$\iint_R xy^2\, dA = \int_0^1 \int_{x^3}^{x^2} xy^2\, dy\, dx = \int_0^1 \left[\tfrac{1}{3}xy^3 \right]_{x^3}^{x^2} dx$$

$$= \int_0^1 \left(\tfrac{1}{3}x^7 - \tfrac{1}{3}x^{10} \right) dx = \tfrac{1}{24} - \tfrac{1}{33} = \tfrac{1}{88}.$$

The region R is also horizontally simple with $c = 0$, $d = 1$, $h_1(y) = y^{1/2}$, and $h_2(y) = y^{1/3}$, so we can reverse the order of integration, and integrate first with respect to x. Then (10) gives

$$\iint_R xy^2\, dA = \int_0^1 \int_{y^{1/2}}^{y^{1/3}} xy^2\, dx\, dy = \int_0^1 \left[\tfrac{1}{2}x^2 y^2 \right]_{y^{1/2}}^{y^{1/3}} dy$$

$$= \int_0^1 \left(\tfrac{1}{2}y^{8/3} - \tfrac{1}{2}y^3 \right) dy = \tfrac{1}{88}.$$

EXAMPLE 3   Evaluate $\iint_R (6x + 2y^2)\, dA$, where R is the region bounded by the parabola $x = y^2$ and the straight line $x + y = 2$.

Solution   The region R appears in Figure 15.9. It is both horizontally and vertically simple. If we wished to integrate first with respect to y and then with respect to x, we would need to evaluate two integrals:

$$\iint_R f(x, y)\, dA = \int_0^1 \int_{-\sqrt{x}}^{\sqrt{x}} (6x + 2y^2)\, dy\, dx + \int_1^4 \int_{-\sqrt{x}}^{2-x} (6x + 2y^2)\, dy\, dx.$$

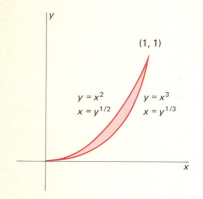

$y = x^2$
$x = y^{1/2}$
$y = x^3$
$x = y^{1/3}$
$(1, 1)$

15.8 The region R of Example 2.

15.9 The region of Example 3.

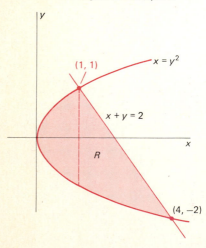

$(1, 1)$
$x = y^2$
$x + y = 2$
R
$(4, -2)$

The reason is that the "top" function $g_2(x)$ changes its formula at the point $(1, 1)$ (see the sketch of the region R).

To avoid this extra work, we prefer to integrate in the opposite order:

$$\iint\limits_{R} (6x + 2y^2)\, dA = \int_{-2}^{1} \int_{y^2}^{2-y} (6x + 2y^2)\, dx\, dy$$

$$= \int_{-2}^{1} \left[3x^2 + 2xy^2 \right]_{y^2}^{2-y} dy$$

$$= \int_{-2}^{1} \left[3(2 - y)^2 + 2(2 - y)y^2 - 3(y^2)^2 - 2(y^2)y^2 \right] dy$$

$$= \int_{-2}^{1} (12 - 12y + 7y^2 - 2y^3 - 5y^4)\, dy$$

$$= \left[12y - 6y^2 + \tfrac{7}{3}y^3 - \tfrac{1}{2}y^4 - y^5 \right]_{-2}^{1} = \tfrac{99}{2}.$$

Example 3 indicates that, even when the region R is both vertically and horizontally simple, it may be simpler to integrate in one order rather than the other, because of the shape of R. We naturally prefer the easier route. The choice of the preferable order of integration may also be influenced by the nature of the function $f(x, y)$. For it may be difficult—or even impossible—to compute a given iterated integral, but easy to do *after reversing the order of integration*. The following example shows that the key to reversing the order of integration is this: Find (and sketch) the region R over which the integration is performed.

EXAMPLE 4 Evaluate $\int_{0}^{2} \int_{y/2}^{1} y e^{x^3}\, dx\, dy$.

Solution We cannot integrate first with respect to x, as indicated, because it happens that e^{x^3} has no elementary antiderivative. So we decide to try to find the value of the above integral by reversing the order. To do this, we first sketch the region of integration.

We first note that $0 \le y \le 2$ for all points (x, y) of R. In addition, all the points of R lie between the two graphs $x = y/2$ and $x = 1$. We draw the four straight lines $y = 0$, $y = 2$, $x = y/2$, and $x = 1$ and find that the region of integration is the triangle that appears in Figure 15.10.

Integrating first with respect to y, from $g_1(x) = 0$ to $g_2(x) = 2x$, we obtain

$$\int_{0}^{2} \int_{y/2}^{1} y e^{x^3}\, dx\, dy = \int_{0}^{1} \int_{0}^{2x} y e^{x^3}\, dy\, dx$$

$$= \int_{0}^{1} \left[\tfrac{1}{2}y^2 \right]_{0}^{2x} e^{x^3}\, dx = \int_{0}^{1} 2x^2 e^{x^3}\, dx$$

$$= \left[\tfrac{2}{3} e^{x^3} \right]_{0}^{1} = \tfrac{2}{3}(e - 1).$$

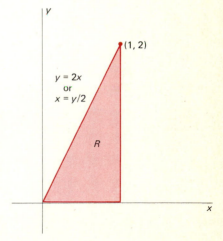

15.10 The region R of Example 4.

We conclude this section by listing some useful formal properties of double integrals. Let c be a constant and f and g continuous functions on a region R on which $f(x, y)$ attains a minimum value m and a maximum value M. Let $a(R)$ denote the area of the region R. If the indicated integrals all exist, then

$$\iint_R cf(x, y)\, dA = c \iint_R f(x, y)\, dA. \tag{11}$$

$$\iint_R [f(x, y) + g(x, y)]\, dA = \iint_R f(x, y)\, dA + \iint_R g(x, y)\, dA. \tag{12}$$

$$m \cdot a(R) \le \iint_R f(x, y)\, dA \le M \cdot a(R). \tag{13}$$

$$\iint_R f(x, y)\, dA = \iint_{R_1} f(x, y)\, dA + \iint_{R_2} f(x, y)\, dA. \tag{14}$$

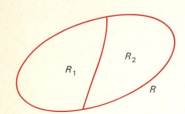

15.11 The regions of Formula (14).

In (14), R_1 and R_2 are two nonoverlapping regions (regions with disjoint interiors) with union R, as in Figure 15.11. In Problems 25–28, we indicate proofs of Properties (11) through (14) for the special case in which R is a rectangle.

Property (14) enables us to evaluate $\iint_R f(x, y)\, dA$ when the region R is neither horizontally simple nor vertically simple but can be decomposed into finitely many simple regions $R_1, R_2, R_3, \ldots, R_n$. Then we may write

$$\iint_R f(x, y)\, dA = \sum_{i=1}^{n} \iint_{R_i} f(x, y)\, dA$$

and use Theorem 2 to evaluate each of the right-hand integrals by iterated integration.

15-1 PROBLEMS

In Problems 1–4, verify that the values of $\iint_R f(x, y)\, dA$ given by the iterated integrals in (4) and (5) are indeed equal.

1 $f(x, y) = 2xy - 3y^2$; $R = [-1, 1] \times [-2, 2]$.

2 $f(x, y) = \sin x \cos y$; $R = [0, \pi] \times \left[-\dfrac{\pi}{2}, \dfrac{\pi}{2}\right]$.

3 $f(x, y) = \sqrt{x + y}$; $R = [0, 1] \times [0, 2]$.

4 $f(x, y) = e^{x+y}$; $R = [0, \ln 2] \times [0, \ln 3]$.

Evaluate the iterated integrals in Problems 5–14.

5 $\displaystyle\int_0^1 \int_0^{x^2} xy \, dy \, dx.$

6 $\displaystyle\int_0^1 \int_y^{\sqrt{y}} (x + y) \, dx \, dy.$

7 $\displaystyle\int_0^4 \int_x^{\sqrt{x}} (2x - y) \, dy \, dx.$

8 $\displaystyle\int_0^2 \int_{-\sqrt{2y}}^{\sqrt{2y}} (3x + 2y) \, dx \, dy.$

9 $\displaystyle\int_0^1 \int_{x^4}^x (y - x) \, dy \, dx.$

10 $\displaystyle\int_{-1}^2 \int_{-y}^{y+2} (x + 2y^2) \, dx \, dy.$

11 $\displaystyle\int_0^1 \int_0^{x^3} e^{y/x} \, dy \, dx.$

12 $\displaystyle\int_0^\pi \int_0^{\sin x} y \, dy \, dx.$

13 $\displaystyle\int_0^3 \int_0^y \sqrt{y^2 + 16} \, dx \, dy.$

14 $\displaystyle\int_1^{e^2} \int_0^{1/y} e^{xy} \, dx \, dy.$

In Problems 15–24, reverse the order of integration and then evaluate the resulting integral. Also sketch each region of integration.

15 $\displaystyle\int_{-2}^2 \int_{x^2}^4 x^2 y \, dy \, dx.$

16 $\displaystyle\int_0^1 \int_{x^4}^x (x - 1) \, dy \, dx.$

17 $\displaystyle\int_{-1}^3 \int_{x^2}^{2x+3} x \, dy \, dx.$

18 $\displaystyle\int_{-2}^2 \int_{y^2-4}^{4-y^2} y \, dx \, dy.$

19 $\displaystyle\int_0^2 \int_{2x}^{4x-x^2} 1 \, dy \, dx.$

20 $\displaystyle\int_0^1 \int_y^1 e^{-x^2} \, dx \, dy.$

21 $\displaystyle\int_0^\pi \int_x^\pi \frac{\sin y}{y} \, dy \, dx.$

22 $\displaystyle\int_0^{\sqrt{\pi}} \int_y^{\sqrt{\pi}} \sin x^2 \, dx \, dy.$

23 $\displaystyle\int_0^1 \int_y^1 \frac{dx \, dy}{1 + x^4}.$

24 $\displaystyle\int_0^1 \int_{\tan^{-1}y}^{\pi/4} \sec x \, dx \, dy.$

25 Use Riemann sums to prove (11) for the case in which R is a rectangle.

26 Use iterated integrals and familiar properties of single integrals to prove (12) for the case in which R is a rectangle.

27 Use Riemann sums to prove (13) when R is a rectangle.

28 Use iterated integrals and familiar properties of single integrals to prove (14) when the right edge of the rectangle R_1 is the left edge of the rectangle R_2.

29 Use Riemann sums to show that

$$\iint\limits_R f(x, y)\, dA \le \iint\limits_R g(x, y)\, dA$$

if $f(x, y) \le g(x, y)$ at each point of the rectangle R.

30 Suppose that the continuous function f is integrable on the plane region R and that f attains a minimum value m and a maximum value M at points of R. Assume also that R is *connected* in the following sense: For any two points (x_0, y_0) and (x_1, y_1) of R, there is a continuous parametric curve $\mathbf{r}(t)$ in R with $\mathbf{r}(0) = \langle x_0, y_0 \rangle$ and $\mathbf{r}(1) = \langle x_1, y_1 \rangle$. Then deduce from (13) the *average value property* of double integrals:

$$\iint\limits_R f(x, y)\, dA = f(\hat{x}, \hat{y}) \cdot \text{area}(R)$$

for some point $(\hat{x}, \hat{y})$ of R. [*Suggestion:* If $m = f(x_0, y_0)$ and $M = f(x_1, y_1)$, then you may apply the Intermediate Value Property of the function $f(\mathbf{r}(t))$.]

15-2

Area and Volume by Double Integration

In Section 15-1 our definition of $\iint_R f(x, y)\, dA$ was *motivated* by the problem of computing the volume of the solid

$$T = \{(x, y, z) \mid (x, y) \in R \quad \text{and} \quad 0 \le z \le f(x, y)\}$$

that lies under the surface $z = f(x, y)$ and above the region R in the xy-plane. Such a solid T appears in Figure 15.12. Despite this geometric motivation, the actual definition of the double integral as a limit of Riemann sums does not depend upon the concept of volume. We may therefore turn matters around and use the double integral to *define* volume.

> **Definition** *Volume under* $z = f(x, y)$
>
> Suppose that the function f is continuous and nonnegative on the bounded plane region R. Then the **volume** V of the solid that lies under the surface $z = f(x, y)$ and above the region R is given by
>
> $$V = \iint\limits_R f(x, y)\, dA \qquad (1)$$
>
> provided this integral exists.

15.12 A solid region with vertical sides and base R in the xy-plane.

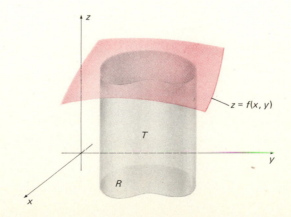

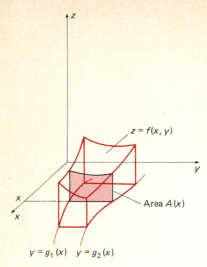

$z = f(x, y)$

Area $A(x)$

$y = g_1(x)$ $y = g_2(x)$

15.13 The cross-sectional area is $A(x) = \int_{g_1(x)}^{g_2(x)} f(x, y)\, dy$.

It is of interest to note the connection between this definition and the cross-sections approach to volume that we discussed in Section 5-2. If, for instance, the region R is vertically simple, then the volume integral in (1) takes the form

$$V = \int_a^b \int_{g_1(x)}^{g_2(x)} f(x, y)\, dy\, dx$$

in terms of iterated integrals. The inner integral

$$A(x) = \int_{g_1(x)}^{g_2(x)} f(x, y)\, dy$$

is simply the area of the cross section of the solid region T in a plane perpendicular to the x-axis (see Figure 15.13). Thus

$$V = \int_a^b A(x)\, dx,$$

and so in this case Formula (1) reduces to "volume is the integral of cross-sectional area."

Suppose now that the solid region T lies above the plane region R, as before, but *between* the surfaces $z = f_1(x, y)$ and $z = f_2(x, y)$, where $f_1(x, y) \leqq f_2(x, y)$ for all (x, y) in R (see Figure 15.14). Then we get the volume V of T by subtracting the volume under $z = f_1(x, y)$ from the volume under $z = f_2(x, y)$, so

$$V = \iint\limits_R [f_2(x, y) - f_1(x, y)]\, dA. \tag{2}$$

This is a natural generalization of the formula for the area of the plane region between the curves $y = f_1(x)$ and $y = f_2(x)$ over the interval $[a, b]$.

The special case $f(x, y) \equiv 1$ in Formula (1) above gives the area

$$A = a(R) = \iint\limits_R 1\, dA = \iint\limits_R dA \tag{3}$$

of the plane region R. In this case the solid region T is a flat-topped "mesa"; that is, a solid cylinder with base area R and height 1. And the volume of any cylinder—not necessarily circular—is the product of its height and the area of its base.

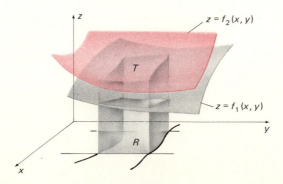

15.14 The solid T has vertical sides and is bounded above and below by surfaces.

$z = f_2(x, y)$

$z = f_1(x, y)$

T

R

714

CHAP. 15: **Multiple Integrals**

A three-dimensional region T is typically described in terms of the surfaces that bound it. The first step in applying either Formula (1) or Formula (2) to compute its volume V is to determine the region R in the xy-plane over which T lies. The second step is to determine the appropriate order of iterated integration. This may be done in the following way.

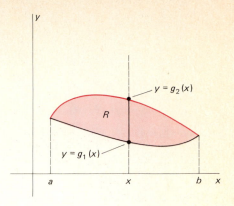

> If each vertical line in the xy-plane meets R in a *single* line segment, then R is vertically simple, and you may integrate first with respect to y. The limits on y will be the y-coordinates $g_1(x)$ and $g_2(x)$ of the end points of this line segment (as indicated in Figure 15.15). The limits on x will be the end points a and b of the interval on the x-axis onto which R projects. Theorem 2 of Section 15-1 then gives
>
> $$V = \iint\limits_R f(x, y)\, dA = \int_a^b \int_{g_1(x)}^{g_2(x)} f(x, y)\, dy\, dx. \qquad (4)$$

15.15   A vertically simple region.

Alternatively:

> If each horizontal line in the xy-plane meets R in a *single* line segment, then R is horizontally simple, and you may integrate with respect to x first. In this case,
>
> $$V = \iint\limits_R f(x, y)\, dA = \int_c^d \int_{h_1(y)}^{h_2(y)} f(x, y)\, dx\, dy. \qquad (5)$$
>
> As indicated in Figure 15.16, $h_1(y)$ and $h_2(y)$ are the x-coordinates of the end points of this horizontal line segment, and c and d are the end points of the corresponding interval on the y-axis.

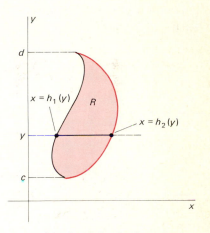

15.16   A horizontally simple region.

If the region R is both vertically simple and horizontally simple, you have the pleasant option of choosing the order of integration that will lead to the simplest subsequent computations. If R is neither vertically simple nor horizontally simple, then you must first subdivide into simple regions before proceeding with iterated integration.

15.17   The region R of Example 1.

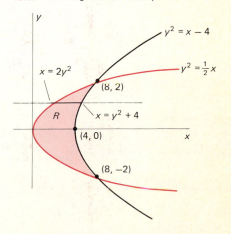

EXAMPLE 1   Compute the area of the region R in the xy-plane that is bounded by the two parabolas $y^2 = x/2$ and $y^2 = x - 4$.

Solution   The region R is shown in Figure 15.17. It's horizontally simple but not vertically simple, so we integrate first with respect to x. We read from Figure 15.17 that $h_1(y) = 2y^2$ and $h_2(y) = y^2 + 4$. So

$$a(R) = \int_{-2}^{2} \int_{2y^2}^{y^2+4} 1\, dx\, dy = \int_{-2}^{2} \left[x \right]_{2y^2}^{y^2+4} dy$$

$$= \int_{-2}^{2} (4 - y^2)\, dy = \left[4y - \tfrac{1}{3}y^3 \right]_{-2}^{2} = \tfrac{32}{3}.$$

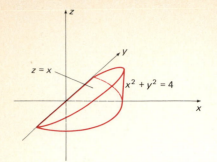

15.18 The wedge of Example 2.

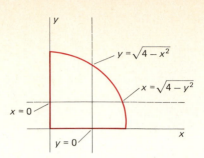

15.19 *Half* of the base R of the wedge.

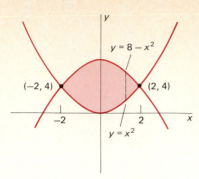

15.20 The region R of Example 3.

EXAMPLE 2   Find the volume of the wedge-shaped solid T that lies above the xy-plane, under the plane $z = x$, and within the cylinder $x^2 + y^2 = 4$ (as shown in Figure 15.18).

Solution   The base region R is a semicircle of radius 2, but by symmetry we may integrate over the first-quadrant quarter-circle alone, then double the result. A sketch of the quarter-circle—see Figure 15.19—helps establish the limits of integration. We could integrate in either order, but integration with respect to x first gives a slightly simpler computation:

$$V = 2 \int_0^2 \int_0^{\sqrt{4-y^2}} x \, dx \, dy = 2 \int_0^2 \left[\tfrac{1}{2}x^2 \right]_0^{\sqrt{4-y^2}} dy$$

$$= \int_0^2 (4 - y^2) \, dy = \left[4y - \tfrac{1}{3}y^3 \right]_0^2 = \tfrac{16}{3}.$$

You should integrate in the other order and then compare the results.

EXAMPLE 3   Find the volume V of the solid T that is bounded by the parabolic cylinders $z = x^2$, $z = 2x^2$, $y = x^2$, and $y = 8 - x^2$.

Solution   We think of T as lying between the surfaces $z = x^2$ and $z = 2x^2$ and above the region R in the xy-plane that is bounded by the parabolas $y = x^2$ and $y = 8 - x^2$. These parabolas intersect at the points $(-2, 4)$ and $(2, 4)$, and they and R are shown in Figure 15.20.

The figure indicates that we should integrate first with respect to y, for otherwise we would need two integrals. Integrating $2x^2 - x^2 = x^2$ first from $y = x^2$ to $y = 8 - x^2$ and then from $x = -2$ to $x = 2$, we get

$$V = \int_{-2}^2 \int_{x^2}^{8-x^2} x^2 \, dy \, dx = \int_{-2}^2 \left[x^2 y \right]_{x^2}^{8-x^2} dx$$

$$= \int_{-2}^2 (8x^2 - 2x^4) \, dx = 2 \left[\tfrac{8}{3}x^3 - \tfrac{2}{5}x^5 \right]_0^2 = \tfrac{256}{15}.$$

15-2   PROBLEMS

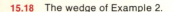

In each of Problems 1–10, use double integration to find the area of the region in the xy-plane that is bounded by the given curves.

1 $y = x$, $y^2 = x$.
2 $y = x$, $y = x^4$.

3 $y = x^2$, $y = 2x + 3$.
4 $y = 2x + 3$, $y = 6x - x^2$.
5 $y = x^2$, $x + y = 2$, $y = 0$.
6 $y = (x - 1)^2$, $y = (x + 1)^2$, $y = 0$.

7 $y = x^2 + 1$, $y = 2x^2 - 3$.

8 $y = x^2 + 1$, $y = 9 - x^2$.

9 $y = x$, $y = 2x$, $xy = 2$.

10 $y = x^2$, $y = \dfrac{2}{1 + x^2}$.

In each of Problems 11–16, find the volume of the solid that lies under the surface $z = f(x, y)$ and above the region in the xy-plane that is bounded by the given curves.

11 $z = x^2 + y^2$; $x = 0$, $y = 0$, $x = 1$, $y = 2$.

12 $z = 1 + x^2 + y^2$; $y = x$, $y = 2 - x^2$.

13 $z = 9 - x - y$; $y = 0$, $x = 3$, $y = \dfrac{2x}{3}$.

14 $z = 10 + y - x^2$; $y = x^2$, $x = y^2$.

15 $z = 4x^2 + y^2$; $x = 0$, $y = 0$, $2x + y = 2$.

16 $z = 2x + 3y$; $y = x^2$, $y = x^3$.

17 Use double integration to find the volume of the tetrahedron in the first octant that is bounded by the coordinate planes and the plane $x/a + y/b + z/c = 1$. The numbers a, b, and c are positive constants.

18 Show that the volume of the solid bounded by the cylinder $x^2 + y^2 = a^2$, the plane $z = 0$, and the plane $z = x + h$ (where $h > a > 0$) is $\pi a^2 h$.

19 Find the volume of the first octant part of the solid bounded by the cylinders $x^2 + y^2 = 1$ and $y^2 + z^2 = 1$. (*Suggestion:* One order of integration is considerably easier than the other.)

20 Find the areas of the two regions bounded by the parabola $y = x^2$ and the curve $y(2x - 7) = -9$, a translated rectangular hyperbola. (*Suggestion:* $x = -1$ is a root of the cubic equation that you need to solve.)

In the following problems you may consult Chapter 8 or the integral table inside the covers of this book to find such antiderivatives as $\int \sqrt{a^2 - u^2}\, du$ and $\int (a^2 - u^2)^{3/2}\, du$.

21 Find the volume of a sphere of radius a by double integration.

22 Use double integration to find the formula $V = V(a, b, c)$ for the volume of an ellipsoid with semiaxes a, b, and c.

23 Find the volume of the solid that is bounded by the xy-plane and the paraboloid $z = 25 - x^2 - y^2$.

24 Find the volume of the solid bounded by the paraboloids $z = x^2 + 2y^2$ and $z = 12 - 2x^2 - y^2$.

25 Find the volume removed when a vertical square hole of edge length R is cut directly through the center of a long horizontal cylinder of radius R.

26 Find the volume of the solid bounded by the two surfaces $z = x^2 + 3y^2$ and $z = 4 - y^2$.

15-3

Double Integrals in Polar Coordinates

A double integral may be easier to evaluate after it has been "transformed" from rectangular xy-coordinates to polar $r\theta$-coordinates. In particular, this is likely to be the case when the region R of integration is a *polar rectangle*. A **polar rectangle** is a region described in polar coordinates by the inequalities

$$a \leqq r \leqq b, \qquad \alpha \leqq \theta \leqq \beta. \tag{1}$$

This polar rectangle is shown in Figure 15.21. If $a = 0$, it is a sector of a circular disk of radius b. If $0 < a < b$ and $\alpha = 0$, $\beta = 2\pi$, it is an annular ring of inner radius a and outer radius b. Since the area of a circular sector with radius r and central angle θ is $\frac{1}{2}r^2\theta$, the area of the polar rectangle (1) is

$$A = \tfrac{1}{2}b^2(\beta - \alpha) - \tfrac{1}{2}a^2(\beta - \alpha)$$
$$= \tfrac{1}{2}(a + b)(b - a)(\beta - \alpha) = \bar{r}\,\Delta r\,\Delta\theta, \tag{2}$$

where $\Delta r = b - a$, $\Delta\theta = \beta - \alpha$, and $\bar{r} = \frac{1}{2}(a + b)$ is the *average radius* of the polar rectangle.

Now suppose that we want to compute the double integral $\iint_R f(x, y)\, dA$, where R is the polar rectangle in (1). In Section 15-1 we defined the double integral as a limit of Riemann sums associated with partitions consisting of ordinary rectangles. It can also be defined in terms of *polar partitions*, consisting of polar rectangles. We begin with a partition

$$a = r_0 < r_1 < r_2 < \cdots < r_m = b$$

15.21 A "polar rectangle."

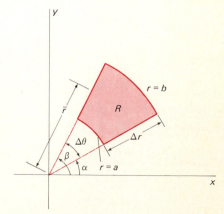

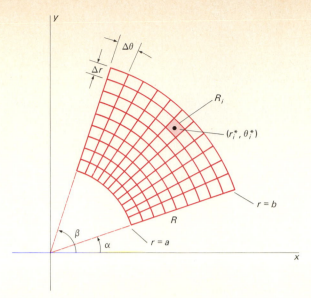

15.22 A polar partition of the polar rectangle R.

of $[a, b]$ into m equal subintervals of length $\Delta r = (b - a)/m$, and a partition

$$\alpha = \theta_0 < \theta_1 < \theta_2 < \cdots < \theta_n = \beta$$

of $[\alpha, \beta]$ into n equal subintervals of length $\Delta\theta = (\beta - \alpha)/n$. This gives the **polar partition** $\mathscr{P}$ of R into the $k = mn$ polar rectangles $R_1, R_2, R_3, \ldots, R_k$ indicated in Figure 15.22. The **mesh** $|\mathscr{P}|$ of this polar partition is the maximum of the lengths of the diagonals of its polar subrectangles.

Let the center point of R_i have polar coordinates (r_i^*, θ_i^*), where r_i^* is the average radius of R_i; then the rectangular coordinates of this point are $x_i^* = r_i^* \cos \theta_i^*$ and $y_i^* = r_i^* \sin \theta_i^*$. Therefore the Riemann sum for the function $f(x, y)$ associated with the polar partition $\mathscr{P}$ is $\sum\limits_{i=1}^{k} f(x_i^*, y_i^*)\, \Delta A_i$, where $\Delta A_i = r_i^* \Delta r\, \Delta\theta$ is the area of the polar rectangle R_i (in part a consequence of Formula (2)). When we express this Riemann sum in polar coordinates, we obtain

$$\sum_{i=1}^{k} f(x_i^*, y_i^*)\, \Delta A_i = \sum_{i=1}^{k} f(r_i^* \cos \theta_i^*, r_i^* \sin \theta_i^*) r_i^* \, \Delta r\, \Delta\theta$$

$$= \sum_{i=1}^{k} g(r_i^*, \theta_i^*)\, \Delta r\, \Delta\theta,$$

where $g(r, \theta) = f(r \cos \theta, r \sin \theta)r$. This last sum is simply a Riemann sum for the double integral

$$\int_{\alpha}^{\beta} \int_{a}^{b} g(r, \theta)\, dr\, d\theta = \int_{\alpha}^{\beta} \int_{a}^{b} f(r \cos \theta, r \sin \theta)\, r\, dr\, d\theta,$$

so it follows finally that

$$\iint\limits_{R} f(x, y)\, dA = \lim_{|\mathscr{P}| \to 0} \sum_{i=1}^{k} f(x_i^*, y_i^*)\, \Delta A_i$$

$$= \lim_{\Delta r, \Delta\theta \to 0} \sum_{i=1}^{k} g(r_i^*, \theta_i^*)\, \Delta r\, \Delta\theta = \int_{\alpha}^{\beta} \int_{a}^{b} g(r, \theta)\, dr\, d\theta.$$

That is,

$$\iint\limits_{R} f(x, y)\, dA = \int_{\alpha}^{\beta} \int_{a}^{b} f(r \cos \theta, r \sin \theta)\, r\, dr\, d\theta. \tag{3}$$

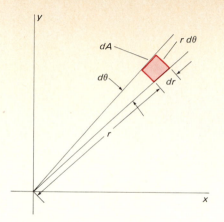

15.23 The dimensions of the small polar rectangle suggest that $dA = r\,dr\,d\theta$.

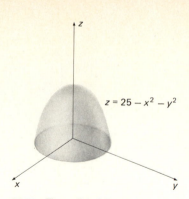

15.24 The solid of Example 1.

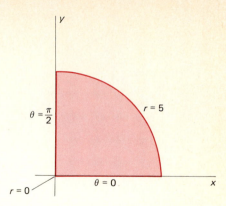

15.25 One-fourth of the domain of the integral of Example 1.

Thus we formally transform a double integral over a polar rectangle (1) into polar coordinates by replacing x by $r \cos \theta$, y by $r \sin \theta$, $dA = dx\,dy$ by $r\,dr\,d\theta$, and inserting the appropriate limits on r and θ. In particular, *note the "extra" r on the right-hand side in Formula (3).* It may be remembered by visualizing the "infinitesimal polar rectangle" of Figure 15.23, with "area" $dA = r\,dr\,d\theta$ (formally).

EXAMPLE 1 Find the volume of the solid shown in Figure 15.24. This is the figure bounded below by the xy-plane and above by the paraboloid $z = 25 - x^2 - y^2$.

Solution The paraboloid intersects the xy-plane in the circle $x^2 + y^2 = 25$. We can compute the volume of the solid by integrating over the quarter of that circle in the first quadrant (see Figure 15.25) and then multiplying the result by 4. Thus

$$V = 4 \int_0^5 \int_0^{\sqrt{25-x^2}} (25 - x^2 - y^2)\,dy\,dx.$$

There is no difficulty in performing the integration with respect to y, but then we are confronted with the integrals

$$\int \sqrt{25 - x^2}\,dx, \qquad \int x^2 \sqrt{25 - x^2}\,dx, \quad \text{and} \quad \int (25 - x^2)^{3/2}\,dx.$$

Let us instead transform the integral into polar coordinates. Since $25 - x^2 - y^2 = 25 - r^2$ and since the first quadrant of our circular disk is described by $0 \le r \le 5$, $0 \le \theta \le \pi/2$, Formula (3) yields

$$V = 4 \int_0^{\pi/2} \int_0^5 (25 - r^2) r\,dr\,d\theta$$

$$= 4 \int_0^{\pi/2} \left[\frac{25}{2} r^2 - \frac{1}{4} r^4 \right]_0^5 d\theta$$

$$= 4 \left(\frac{625}{4} \right)\left(\frac{\pi}{2} \right) = \frac{625\pi}{2}.$$

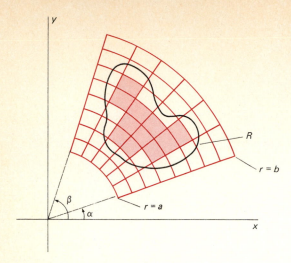

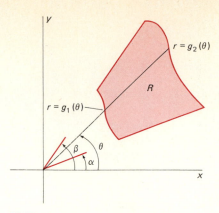

15.26 A "polar inner partition" of the region R.

15.27 A radially simple region R.

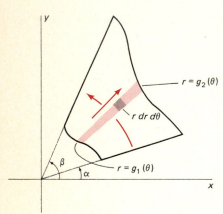

15.28 Integrating first with respect to r, then with respect to θ.

15.29 An angularly simple region R.

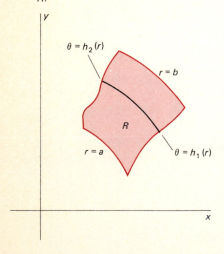

If R is a more general region, then the double integral $\iint_R f(x, y)\, dA$ can be transformed into polar coordinates by expressing it as a limit of Riemann sums associated with "polar inner partitions" of the sort indicated in Figure 15.26. Instead of giving the detailed derivation—a generalization of the above derivation of Formula (3)—we shall simply give the results in two special cases of practical importance. These correspond to the two types of plane regions that play the same role in polar coordinates that horizontally simple and vertically simple regions play in rectangular coordinates.

Figure 15.27 shows a *radially simple* region R consisting of those points which have polar coordinates that satisfy the inequalities

$$\alpha \leqq \theta \leqq \beta, \qquad g_1(\theta) \leqq r \leqq g_2(\theta).$$

In this case the formula

$$\iint_R f(x, y)\, dA = \int_\alpha^\beta \int_{g_1(\theta)}^{g_2(\theta)} f(r \cos \theta, r \sin \theta)\, r\, dr\, d\theta \qquad (4)$$

gives the evaluation in polar coordinates of a double integral over R (under the usual assumption that the indicated integrals exist). Note that we integrate first with respect to r, with the limits $g_1(\theta)$ and $g_2(\theta)$ being the r-coordinates of the end points of a typical radial segment in R, as indicated in Figure 15.27.

Figure 15.28 indicates how the iterated integral on the right-hand side of (4) can be set up in a formal way. A typical area element $dA = r\, dr\, d\theta$ is first swept radially from $r = g_1(\theta)$ to $r = g_2(\theta)$. The resulting strip is then rotated from $\theta = \alpha$ to $\theta = \beta$ to sweep out the region R.

Figure 15.29 shows an *angularly simple* region R described in polar coordinates by the inequalities

$$a \leqq r \leqq b, \qquad h_1(r) \leqq \theta \leqq h_2(r).$$

In this case

$$\iint_R f(x, y)\, dA = \int_a^b \int_{h_1(r)}^{h_2(r)} f(r \cos \theta, r \sin \theta)\, r\, d\theta\, dr. \qquad (5)$$

CHAP. 15: **Multiple Integrals**

Here we integrate first with respect to θ, with the limits $h_1(r)$ and $h_2(r)$ being the θ-coordinates of the end points of a typical circular arc in R, as indicated in Figure 15.29. Figure 15.30 indicates how the iterated integral in (5) can be set up in a formal manner.

Observe that Formulas (3), (4), and (5) for the evaluation of a double integral in polar coordinates all take the form

$$\iint_R f(x, y)\, dA = \iint_S f(r\cos\theta, r\sin\theta)\, r\, dr\, d\theta. \qquad (6)$$

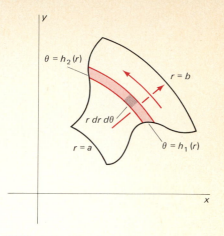

The letter S on the right-hand side corresponds to choosing appropriate limits on r and θ (once the order of integration has been determined) so that the region R is swept out in the manner of Figures 15.28 and 15.30. With $f(x, y) \equiv 1$, Formula (6) reduces to the formula

$$A = a(R) = \iint_S r\, dr\, d\theta \qquad (7)$$

15.30 Integrating first with respect to θ, last with respect to r.

for computing the area of R by double integration in polar coordinates.

EXAMPLE 2 Find the volume of the solid region that is interior to both the sphere $x^2 + y^2 + z^2 = 4$ of radius 2 and the cylinder $(x - 1)^2 + y^2 = 1$ of radius 1. This is the volume of material removed when an off-center hole of radius 1 is bored all the way through a sphere of radius 2. The upper hemisphere is shown in Figure 15.31.

Solution We need to integrate the function $f(x, y) = (4 - x^2 - y^2)^{1/2}$ over the disk R bounded by the circle with center $(1, 0)$ and radius 1, shown in Figure 15.32. The desired volume is twice that of the part above the xy-plane:

$$V = 2 \iint_R \sqrt{4 - x^2 - y^2}\, dA.$$

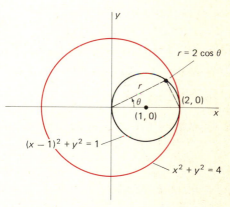

15.32 The small circle is the domain of the integral for Example 2.

15.31 The *upper half* of the sphere-with-hole of Example 2.

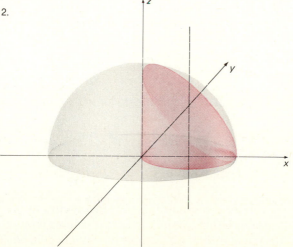

But this integral would be particularly troublesome to evaluate in rectangular coordinates, so we change to polar coordinates.

The unit circle in Figure 15.32 is familiar from Section 9-6 (Equation (3)); its polar coordinate equation is $r = 2 \cos \theta$. Therefore the region R is described by the inequalities

$$0 \leq r \leq 2 \cos \theta, \qquad -\pi/2 \leq \theta \leq \pi/2.$$

We shall integrate over only the upper half of R, taking advantage of the symmetry of the sphere-with-hole; this involves doubling for a *second* time the integral we write. So—using Formula (4)—we find that

$$V = 4 \int_0^{\pi/2} \int_0^{2 \cos \theta} \sqrt{4 - r^2}\, r\, dr\, d\theta$$

$$= 4 \int_0^{\pi/2} \left[-\tfrac{1}{3}(4 - r^2)^{3/2} \right]_0^{2 \cos \theta} d\theta$$

$$= \tfrac{32}{3} \int_0^{\pi/2} (1 - \sin^3 \theta)\, d\theta.$$

Now we see from Formula (113) inside the back cover that $\int_0^{\pi/2} \sin^3 \theta\, d\theta = \tfrac{2}{3}$, and so $V = 16\pi/3 - \tfrac{64}{9} \approx 9.64405$.

EXAMPLE 3 Here we apply a standard polar coordinates technique to show that

$$I = \int_0^\infty e^{-x^2}\, dx = \frac{\sqrt{\pi}}{2}. \tag{8}$$

This important improper integral converges because

$$\int_1^b e^{-x^2}\, dx \leq \int_1^b e^{-x}\, dx \leq \int_1^\infty e^{-x}\, dx = 1.$$

(The first inequality is valid because $e^{-x^2} \leq e^{-x}$ for $x \geq 1$.) It follows that $\int_1^b e^{-x^2}\, dx$ is a bounded, increasing function of b.

Solution Let V_b denote the volume of the region that lies under the surface $z = e^{-x^2 - y^2}$ and above the square with vertices $(\pm b, \pm b)$ in the xy-plane. Then

$$V_b = \int_{-b}^b \int_{-b}^b e^{-x^2 - y^2}\, dx\, dy = \int_{-b}^b e^{-y^2} \left(\int_{-b}^b e^{-x^2}\, dx \right) dy$$

$$= \left(\int_{-b}^b e^{-x^2}\, dx \right) \left(\int_{-b}^b e^{-y^2}\, dy \right) = \left(\int_{-b}^b e^{-x^2}\, dx \right)^2 = 4 \left(\int_0^b e^{-x^2}\, dx \right)^2.$$

It follows that the volume under $e^{-x^2 - y^2}$ above the entire xy-plane is

$$V = \lim_{b \to \infty} V_b = \lim_{b \to \infty} 4 \left(\int_0^b e^{-x^2}\, dx \right)^2$$

$$= 4 \left(\int_0^\infty e^{-x^2}\, dx \right)^2 = 4I^2.$$

Now we compute V by another method—the use of polar coordinates. We take the limit, as $b \to \infty$, of the volume under $z = e^{-x^2 - y^2} = e^{-r^2}$ above the circular disk with center $(0, 0)$ and radius b. This disk is described by

$0 \leq r \leq b, 0 \leq \theta \leq 2\pi$, so we obtain

$$V = \lim_{b \to \infty} \int_0^{2\pi} \int_0^b e^{-r^2} r \, dr \, d\theta = \lim_{b \to \infty} \int_0^{2\pi} \left[-\tfrac{1}{2} e^{-r^2} \right]_0^b d\theta$$

$$= \lim_{b \to \infty} \int_0^{2\pi} \tfrac{1}{2}(1 - e^{-b^2}) \, d\theta = \lim_{b \to \infty} \pi(1 - e^{-b^2}) = \pi.$$

We equate our two values of V, and it follows that $4I^2 = \pi$. Therefore $I = \tfrac{1}{2}\sqrt{\pi}$, as desired.

In Problems 1–7, find the indicated area by double integration in polar coordinates.

1 The area bounded by the circle $r = 1$.
2 The area bounded by the circle $r = 3 \sin \theta$.
3 The area bounded by the cardioid $r = 1 + \cos \theta$.
4 The area bounded by one loop of $r = 2 \cos 2\theta$.
5 The area within both the circles $r = 1$ and $r = 2 \sin \theta$.
6 The area within $r = 2 + \cos \theta$ and outside the circle $r = 2$.
7 The area within the smaller loop of $r = 1 - 2 \sin \theta$.

In each of Problems 8–12, find by double integration in polar coordinates the volume of the solid that lies under the given surface and over the plane region R bounded by the given curve.

8 $z = x^2 + y^2$; $r = 3$.
9 $z = \sqrt{x^2 + y^2}$; $r = 2$.
10 $z = x^2 + y^2$; $r = 2 \cos \theta$.
11 $z = 10 + 2x + 3y$; $r = \sin \theta$.
12 $z = a^2 - x^2 - y^2$; $r = a$.

In Problems 13–18, evaluate the given integral by first changing to polar coordinates.

13 $\int_0^1 \int_0^{\sqrt{1-y^2}} \dfrac{dx \, dy}{1 + x^2 + y^2}$.

14 $\int_0^1 \int_0^{\sqrt{1-x^2}} \dfrac{dy \, dx}{\sqrt{4 - x^2 - y^2}}$.

15 $\int_0^2 \int_0^{\sqrt{4-x^2}} (x^2 + y^2)^{3/2} \, dy \, dx$.

16 $\int_0^1 \int_x^1 x^2 \, dy \, dx$.

17 $\int_0^1 \int_0^{\sqrt{1-y^2}} \sin(x^2 + y^2) \, dx \, dy$.

18 $\int_1^2 \int_0^{\sqrt{2x-x^2}} \dfrac{dy \, dx}{\sqrt{x^2 + y^2}}$.

Solve Problems 19–28 by double integration in polar coordinates.

19 Problem 21, Section 15-2.
20 Problem 24, Section 15-2.
21 Problem 18, Section 15-2.
22 Find the volume of the wedge-shaped solid described in Example 2 of Section 15-2.
23 Find the volume bounded by the paraboloids $z = x^2 + y^2$ and $z = 4 - 3x^2 - 3y^2$.
24 Find the volume that is bounded by the paraboloids $z = x^2 + y^2$ and $z = 2x^2 + 2y^2 - 1$.
25 Find the volume of the "ice cream cone" bounded by the sphere $x^2 + y^2 + z^2 = a^2$ and the cone $z = (x^2 + y^2)^{1/2}$.
26 Find the volume that is bounded by the paraboloid $z = r^2$, the cylinder $r = 2a \sin \theta$, and the plane $z = 0$.
27 Find the volume that lies under the paraboloid $z = r^2$ and above one loop of the lemniscate $r^2 = 2 \sin 2\theta$.
28 Find the volume that lies inside both the cylinder $x^2 + y^2 = 4$ and the ellipsoid $2x^2 + 2y^2 + z^2 = 18$.
29 If $0 < h < a$, then the plane $z = a - h$ cuts off a spherical segment of height h and radius b from the sphere $x^2 + y^2 + z^2 = a^2$. (a) Show that $b^2 = 2ah - h^2$. (b) Show that the volume of this spherical segment is $V = (\pi h/6)(3b^2 + h^2)$.
30 Show by the method of Example 3 that

$$\int_0^{\infty} \int_0^{\infty} \frac{dx \, dy}{(1 + x^2 + y^2)^2} = \frac{\pi}{4}.$$

15-4

The double integral can be used to find the mass M and the centroid $(\bar{x}, \bar{y})$ of a thin plate or plane *lamina* that occupies a bounded region R in the xy-plane. We suppose that the density of the lamina (in units of mass per unit *area*) at the point (x, y) is given by the continuous function $\rho(x, y)$.

Applications of Double Integrals

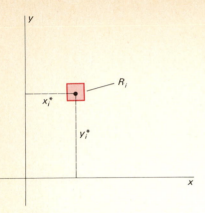

15.33 The moment of an infinitesimal rectangle about the coordinate axes.

Let $\mathcal{P} = \{R_1, R_2, R_3, \ldots, R_n\}$ be an inner partition of R, and choose a point (x_i^*, y_i^*) in each subrectangle R_i (see Figure 15.33). Then the mass of the piece of the lamina occupying R_i will be approximately $\rho(x_i^*, y_i^*)\,\Delta A_i$, where ΔA_i denotes the area $a(R_i)$ of R_i. Hence the mass of the entire lamina is given approximately by

$$M \approx \sum_{i=1}^{n} \rho(x_i^*, y_i^*)\,\Delta A_i.$$

Recall from Section 5-6 that the moment of a particle about the x-axis is the product of its mass and its directed distance from the x-axis; its moment about the y-axis is the product of its mass and its directed distance from the y-axis. So the moments M_x and M_y of our lamina about the x- and y-axes, respectively, are given approximately by

$$M_x \approx \sum_{i=1}^{n} y_i^* \rho(x_i^*, y_i^*)\,\Delta A_i \quad \text{and} \quad M_y \approx \sum_{i=1}^{n} x_i^* \rho(x_i^*, y_i^*)\,\Delta A_i.$$

As the mesh $|\mathcal{P}|$ of the inner partition $\mathcal{P}$ approaches zero, these Riemann sums approach the corresponding double integrals over R, so we may *define* the **mass** M and the **moments** M_x and M_y be means of the formulas

$$M = \iint\limits_{R} \rho(x, y)\,dA, \tag{1}$$

$$M_x = \iint\limits_{R} y\rho(x, y)\,dA, \qquad \text{and} \tag{2}$$

$$M_y = \iint\limits_{R} x\rho(x, y)\,dA. \tag{3}$$

As in Section 5-6, the coordinates $(\bar{x}, \bar{y})$ of the **centroid** or center of mass of the lamina are given by the formulas

$$\bar{x} = \frac{M_y}{M} = \frac{1}{M} \iint\limits_{R} x\rho(x, y)\,dA, \tag{4}$$

$$\bar{y} = \frac{M_x}{M} = \frac{1}{M} \iint\limits_{R} y\rho(x, y)\,dA. \tag{5}$$

15.34 The lamina of Example 1.

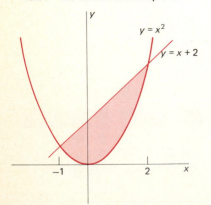

EXAMPLE 1 A lamina occupies the region bounded by the line $y = x + 2$ and the parabola $y = x^2$, as in Figure 15.34. The density of the lamina at the point $P(x, y)$ is proportional to the square of the distance of P from the y-axis: $\rho(x, y) = kx^2$. Find the mass and the centroid of this lamina.

Solution The line and the parabola intersect in the two points $(-1, 1)$ and $(2, 4)$, so Formula (1) gives

$$M = \int_{-1}^{2} \int_{x^2}^{x+2} kx^2\,dy\,dx = k \int_{-1}^{2} \left[x^2 y \right]_{x^2}^{x+2} dx$$

$$= k \int_{-1}^{2} (x^3 + 2x^2 - x^4)\,dx = \frac{63k}{20}.$$

Then Formulas (4) and (5) give

$$\bar{x} = \frac{20}{63k} \int_{-1}^{2} \int_{x^2}^{x+2} kx^3 \, dy \, dx = \frac{20}{63} \int_{-1}^{2} \left[x^3 y \right]_{x^2}^{x+2} dx$$

$$= \frac{20}{63} \int_{-1}^{2} (x^4 + 2x^3 - x^5) \, dx = \frac{20}{63} \cdot \frac{18}{5} = \frac{8}{7};$$

$$\bar{y} = \frac{20}{63k} \int_{-1}^{2} \int_{x^2}^{x+2} kx^2 y \, dy \, dx = \frac{20}{63} \int_{-1}^{2} \left[\frac{1}{2} x^2 y^2 \right]_{x^2}^{x+2} dx$$

$$= \frac{10}{63} \int_{-1}^{2} (x^4 + 4x^3 + 4x^2 - x^6) \, dx = \frac{10}{63} \cdot \frac{531}{35} = \frac{118}{49}.$$

Thus the lamina of our example has mass $63k/20$ and its centroid is located at the point $(\frac{8}{7}, \frac{118}{49})$.

EXAMPLE 2 A lamina is shaped like the first-quadrant quarter-circle of radius a, shown in Figure 15.35. Its density is proportional to distance from the origin; that is, $\rho(x, y) = k\sqrt{x^2 + y^2} = kr$. Find its mass and its centroid.

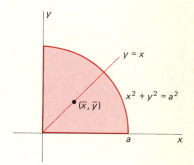

15.35 Finding mass and centroid—see Example 2.

Solution First we change to polar coordinates—the shape of the boundary of the lamina suggests that this will make the problem easier. Then Formula (1) yields the result

$$M = \iint_R \rho \, dA = \int_0^{\pi/2} \int_0^a kr^2 \, dr \, d\theta$$

$$= k \int_0^{\pi/2} \left[\tfrac{1}{3} r^3 \right]_0^a d\theta = k \int_0^{\pi/2} \tfrac{1}{3} a^3 \, d\theta = \frac{k\pi a^3}{6}.$$

By symmetry of the lamina and its density function, the centroid lies on the line $y = x$. So Formula (5) gives

$$\bar{x} = \bar{y} = \frac{1}{M} \iint_R y\rho \, dA = \frac{6}{k\pi a^3} \int_0^{\pi/2} \int_0^a kr^3 \sin\theta \, dr \, d\theta$$

$$= \frac{6}{\pi a^3} \int_0^{\pi/2} \left[\frac{1}{4} r^4 \sin\theta \right]_0^a d\theta = \frac{6}{\pi a^3} \frac{a^4}{4} \int_0^{\pi/2} \sin\theta \, d\theta = \frac{3a}{2\pi}.$$

Thus the given lamina has mass $\frac{1}{6} k\pi a^3$ and its centroid is located at $(3a/2\pi, 3a/2\pi)$.

An important physical concept, the *moment of inertia*, is defined this way: Consider a finite system of particles with masses $m_1, m_2, m_3, \ldots, m_n$, all revolving in uniform circular motion with angular speed ω (in radians per second) about a given fixed axis. Let r_i denote the (perpendicular) distance of the ith particle from this axis. Then the linear speed of this particle is $v_i = r_i \omega$. Hence the total kinetic energy of this system of particles will be

$$\text{K.E.} = \sum_{i=1}^{n} \tfrac{1}{2} m_i v_i^2 = \sum_{i=1}^{n} \tfrac{1}{2} m_i r_i^2 \omega^2,$$

so that

$$\text{K.E.} = \tfrac{1}{2} I \omega^2 \tag{6}$$

where I denotes the **moment of inertia** of the system of particles about the given axis and is defined by

$$I = \sum_{i=1}^{n} m_i r_i^2. \tag{7}$$

Since linear kinetic energy has the formula $\text{K.E.} = \tfrac{1}{2} m v^2$, the result of Formula (6) suggests that moment of inertia is the rotational analogue of mass.

The **radius of gyration** of the system is defined by means of the equation

$$\hat{r} = \sqrt{\frac{I}{M}}, \tag{8}$$

where $M = m_1 + m_2 + \cdots + m_n$. Then

$$I = M \hat{r}^2, \quad \text{so that} \quad \text{K.E.} = \tfrac{1}{2} M (\hat{r} \omega)^2.$$

Thus the kinetic energy of the rotating system is the same as that of a single particle of mass M revolving at the distance $\hat{r}$ from the given axis.

Formula (7) gives the definition of the moment of inertia about a given axis of a *discrete* system of mass particles. We want to generalize this formula, to define the moment of inertia of a *continuous* distribution of mass in the form of a plane lamina occupying the region R in the xy-plane. The moment of inertia of this lamina *about the z-axis* is called its **polar moment of inertia** and is denoted by I_0.

To define I_0, we consider our original inner partition $\mathscr{P} = \{R_1, R_2, R_3, \ldots, R_n\}$ of the region R. Denote by r_i^* the distance from the origin (or from the z-axis) of the selected point (x_i^*, y_i^*) in R_i. Then, by analogy with Formula (7), I_0 is given approximately by

$$I_0 \approx \sum_{i=1}^{n} (r_i^*)^2 \rho(x_i^*, y_i^*)\, \Delta A_i.$$

We take the limit as $|\mathscr{P}| \to 0$ and obtain the integral formula

$$I_0 = \iint_R r^2 \rho(x, y)\, dA = \iint_R (x^2 + y^2) \rho(x, y)\, dA, \tag{9}$$

which serves to define I_0. Note that

$$I_0 = I_x + I_y$$

where

$$I_x = \iint_R y^2 \rho(x, y)\, dA \tag{10}$$

and

$$I_y = \iint\limits_R x^2 \rho(x, y)\, dA. \tag{11}$$

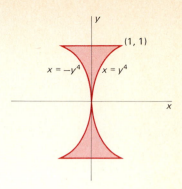

15.36 The lamina of Example 3.

Here I_x is the lamina's **moment of inertia about the x-axis**, while I_y is its **moment of inertia about the y-axis**.

EXAMPLE 3 Compute I_x for a lamina of constant density $\rho = 1$ occupying the region bounded by the curves $x = \pm y^4$, $-1 \leq y \leq 1$. See Figure 15.36.

Solution Formula (10) gives

$$I_x = \int_{-1}^{1} \int_{-y^4}^{y^4} y^2\, dx\, dy$$

$$= \int_{-1}^{1} \left[xy^2 \right]_{-y^4}^{y^4} dy = \int_{-1}^{1} 2y^6\, dy = \tfrac{4}{7}.$$

The region of Example 3 resembles the cross section of an I beam. It is known that the stiffness, or resistance to bending, of a horizontal beam is proportional to the moment of inertia of its cross section with respect to a horizontal axis through the centroid of the cross section. Let's compare our I beam with a rectangular beam of equal height 2 and equal area

$$A = \int_{-1}^{1} \int_{-y^4}^{y^4} 1\, dx\, dy = \tfrac{4}{5}.$$

The cross section of such a rectangular beam is shown in Figure 15.37. Its width will be $\tfrac{2}{5}$, and the moment of inertia of its cross section will be

$$I_x = \int_{-1}^{1} \int_{-1/5}^{1/5} y^2\, dx\, dy = \tfrac{4}{15}.$$

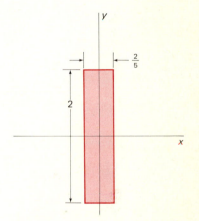

15.37 A rectangular beam for comparison with the I beam of Example 3.

Since the ratio of $\tfrac{4}{7}$ to $\tfrac{4}{15}$ is $\tfrac{15}{7}$, we see that the I beam is more than twice as strong as a rectangular beam of the same cross-sectional area. This is why I beams frequently are used in construction.

EXAMPLE 4 Find the polar moment of inertia of a circular lamina of radius a and constant density ρ centered at the origin.

Solution Formula (9) gives

$$I_0 = \iint\limits_{x^2+y^2 \leq a^2} r^2 \rho\, dA$$

$$= \int_0^{2\pi} \int_0^a \rho r^3\, dr\, d\theta = \frac{\rho \pi a^4}{2} = \frac{1}{2} Ma^2,$$

where $M = \rho \pi a^2$ is the mass of the circular lamina.

To show the moment of inertia in action, we apply the result of Example 4 to show that a Hula-Hoop will roll down an incline more slowly than a quarter will. You can easily perform an experiment to verify this assertion.

Consider a circular object of mass M and polar moment of inertia I. Suppose that it starts from rest and rolls down an incline under the influence

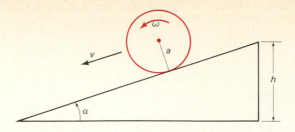

of gravity. Suppose that the incline has vertical height h and makes an angle α with the horizontal, as shown in Figure 15.38.

At any instant t after the start at $t = 0$, the linear speed v and angular speed ω of the rolling mass M are related by $v = a\omega$, where a is the radius of M. After a vertical descent of h down the incline, the rolling mass has lost potential energy Mgh (g is the acceleration of gravity). But it arrives at the bottom with translational kinetic energy $\frac{1}{2}Mv^2$ and rotational kinetic energy

$$\frac{1}{2} I\omega^2 = \frac{Iv^2}{2a^2}$$

(by the natural generalization of (6)). Hence

$$Mgh = \frac{Mv^2}{2} + \frac{Iv^2}{2a^2} \tag{12}$$

because mechanical energy is conserved.

Now for the Hula-Hoop and the quarter. First, a quarter has polar moment $I = \frac{1}{2}Ma^2$ by Example 4. So

$$Mgh = \frac{Mv^2}{2} + \frac{Ma^2v^2}{4a^2} = \frac{3}{4} Mv^2.$$

We cancel M and solve for v:

$$v = 2\sqrt{\frac{gh}{3}}.$$

For instance, $v \approx 46.2$ ft/sec if $h = 50$ feet and $g \approx 32$ ft/sec^2.

The Hula-Hoop can be thought of as a circular object with mass M all concentrated in its rim at distance a from its center, so that $I = Ma^2$. We substitute this formula into (12), and we find that

$$Mgh = \frac{Mv^2}{2} + \frac{Ma^2v^2}{2a^2} = Mv^2.$$

So $v = \sqrt{gh}$. If, as before, we take $h = 50$ feet, then we find that v will be approximately 40 feet per second. This establishes our assertion about which of the two rolls faster, because h is actually arbitrary.

Finally, the **radius of gyration** of a lamina about an axis is defined by Equation (8), with I being its moment of inertia about that axis. For example, the radii of gyration $\hat{x}$ and $\hat{y}$ about the y-axis and the x-axis, respectively, are given by

$$\hat{x} = \left(\frac{I_y}{M}\right)^{1/2} \quad \text{and} \quad \hat{y} = \left(\frac{I_x}{M}\right)^{1/2} \qquad (13)$$

Now suppose that this lamina lies in the right half-plane $x > 0$ and is symmetric about the x-axis. If it represents the "face" of a racquet whose handle (considered weightless) extends along the x-axis from the origin to the face, then the point $(\hat{x}, 0)$ is a plausible candidate for the tennis player's "sweet spot."

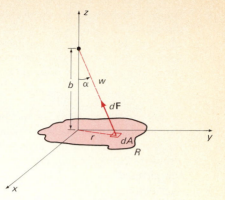

15.39 Computing gravitational attraction between a point mass and a lamina.

EXAMPLE 5 Suppose that a lamina of uniform density ρ occupies the region R in the xy-plane, and that a mass m is located at the point $(0, 0, b)$ on the z-axis. From Figure 15.39 we see that the z-component dF_z of the force of gravitational attraction between m and an infinitesimal piece dA of the lamina is

$$dF_z = \frac{Gm\rho \, dA}{w^2} \cos \alpha = \frac{Gm\rho b \, dA}{w^3}$$

where $w = \sqrt{r^2 + b^2}$. It follows that the vertical component F_z of the force of gravitational attraction between the point mass and the lamina is

$$F_z = \iint_R \frac{Gm\rho b \, dA}{(r^2 + b^2)^{3/2}}.$$

In Problem 28 we ask you to compute this integral in the case in which R is a circular disk.

15-4 PROBLEMS

In each of Problems 1–10, find the mass and centroid of a plane lamina with the indicated shape and density.

1 The square with vertices $(0, 0)$, $(0, a)$, (a, a), $(a, 0)$; $\rho = x + y$.

2 The triangular region bounded by the coordinate axes and the line $x + y = a$; $\rho = x^2 + y^2$.

3 The region bounded by $y = x^2$ and $y = 4$; $\rho = y$.

4 The region bounded by $y = x^2$ and $y = 2x + 3$; $\rho = x^2$.

5 The region between the x-axis and $y = \sin x$ over the interval $0 \leqq x \leqq \pi$; $\rho = x$.

6 The semicircular region $x^2 + y^2 \leqq a^2$, $y \geqq 0$; $\rho = y$.

7 The semicircular region $x^2 + y^2 \leqq a^2$, $y \geqq 0$; $\rho = r$.

8 The region bounded by the cardioid $r = 1 + \cos \theta$; $\rho = r$.

9 The region within the circle $r = 2 \sin \theta$ and outside the circle $r = 1$; $\rho = y$.

10 The region within the limaçon $r = 1 + 2 \cos \theta$ and outside the circle $r = 2$; $\rho = r$.

In each of Problems 11–15, find the polar moment of inertia I_0 of the indicated lamina.

11 The region bounded by the circle $r = a$; $\rho = r^n$ where n is a fixed positive integer.

12 The lamina of Problem 6.

13 The disk bounded by $r = 2 \cos \theta$; $\rho = k$ (a constant).

14 The lamina of Problem 9.

15 The region bounded by the right-hand loop of the lemniscate $r^2 = \cos 2\theta$; $\rho = r^2$.

In each of Problems 16–20, find the radii of gyration $\hat{x}$ and $\hat{y}$ of the indicated lamina about the coordinate axes.

16 The lamina of Problem 1.

17 The lamina of Problem 3.

18 The lamina of Problem 4.

19 The lamina of Problem 7.

20 The lamina of Problem 13.

21 Show that the moment of inertia of a uniform rectangular lamina with dimensions a and b about a perpendicular axis through its centroid is given by $I = \frac{1}{12}\rho(a^3b + ab^3)$.

22 Show that the moment of inertia of a uniform plane lamina in the xy-plane with centroid at $(0, 0)$, about an axis perpendicular to the xy-plane at (x_0, y_0), is

$$I = I_0 + M(x_0^2 + y_0^2).$$

23 Suppose that a plane lamina consists of two nonoverlapping laminae. Show that its polar moment of inertia is the sum of theirs. Use this fact, together with the results of Problems 21 and 22, to find the polar moment of inertia of the pictured T-shaped lamina of constant density $\rho = k$ shown in Figure 15.40.

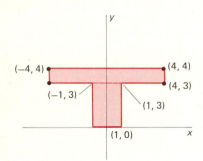

15.40 One lamina made of two simple ones—see Problem 23.

24 Find the polar moment of inertia of a uniform elliptical lamina with semiaxes a and b centered at the origin.

25 A homogeneous washer-shaped lamina is bounded by the circles $r = a$ and $r = b$, where $b > a$. (a) Compute its polar moment of inertia. (b) Suppose that $a = 1$ ft and $b = 2$ ft. Compute the speed with which this washer will arrive at the bottom of the incline of Figure 15.38.

26 A racquet consists of a uniform lamina that occupies the region within the right-hand loop of $r^2 = \cos 2\theta$, on the end of a handle (which we'll assume weightless) corresponding to the interval $-1 \leq x \leq 0$ (as in Figure 15.41). Find the radius of gyration of the racquet about the line $x = -1$. Where is its sweet spot?

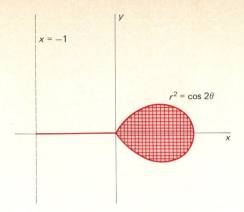

15.41 The racquet of Problem 26.

27 Find the centroid of the unbounded region under the graph of $y = e^{-x}$, $x \geq 0$.

28 (a) Suppose that the lamina of Example 5 is a uniform circular disk of radius a centered at the origin. Show by integration in polar coordinates that

$$F_z = 2\pi Gm\rho \left[1 - \left(1 + \frac{a^2}{b^2} \right)^{-1/2} \right].$$

(b) Suppose that b is very large in comparison with a. Use the binomial series (Section 12-7, Example 6) to show that $F_z \approx GmM/b^2$ where $M = \pi\rho a^2$ is the mass of the disk.
(c) Note that $F_z \to 2\pi Gm\rho$ as $a \to \infty$. Thus the gravitational field of an infinite plane with uniform mass density is *constant*—independent of distance from the plane.

29 Consider a solid sphere of radius R centered at the origin, with constant density ρ, and with mass $M = \frac{4}{3}\pi\rho R^3$. A point mass m is located at the point $(0, 0, D)$, where $D > R$. Think of the slice of the sphere between z and $z + dz$ as a circular lamina with density $\rho \, dz$ and radius $a = (R^2 - z^2)^{1/2}$. The vertical force it exerts on m is then given by the formula of Problem 28 (a) with $b = D - z$. Integrate from $z = -R$ to $z = R$ to show that the total force exerted by the solid sphere on the point mass is $F = GMm/D^2$. Thus the solid sphere acts as if all its mass were concentrated at its center.

15-5

Triple Integrals

The definition of the triple integral is a three-dimensional version of the definition of the double integral in Section 15-1. Let $f(x, y, z)$ be continuous on the bounded space region T, and suppose that T lies inside the rectangular block R determined by the inequalities $a \leq x \leq b$, $c \leq y \leq d$, and $p \leq z \leq q$. We subdivide $[a, b]$ into equal subintervals of length Δx, $[c, d]$ into equal subintervals of length Δy, and $[p, q]$ into equal subintervals of length Δz.

This generates a partition of R into smaller rectangular blocks (as in Figure 15.42), each with volume $\Delta V = \Delta x\, \Delta y\, \Delta z$. Let $\mathscr{P} = \{T_1, T_2, T_3, \ldots, T_n\}$ be the collection of these smaller blocks that lie wholly within T. Then $\mathscr{P}$ is called an **inner partition** of the region T. The **mesh** $|\mathscr{P}|$ of $\mathscr{P}$ is the length of a longest diagonal of any of the blocks T_i. If (x_i^*, y_i^*, z_i^*) is an arbitrarily selected point of T_i (for each $i = 1, 2, 3, \ldots, n$), then the **Riemann sum**

$$\sum_{i=1}^{n} f(x_i^*, y_i^*, z_i^*)\, \Delta V$$

is an approximation to the triple integral of f over the region T.

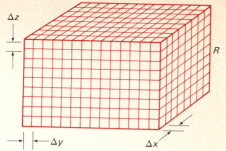

15.42 A partition of a large block into many small ones.

For instance, if T is a solid body with density function f, then the above sum approximates its total mass. We define the **triple integral of f over T** by means of the equation

$$\iiint_T f(x, y, z)\, dV = \lim_{|\mathscr{P}| \to 0} \sum_{i=1}^{n} f(x_i^*, y_i^*, z_i^*)\, \Delta V. \tag{1}$$

It is proved in advanced calculus that the limit of the Riemann sums exists as the mesh $|\mathscr{P}|$ approaches zero, provided that f is continuous on T and the boundary of the region T is reasonably nice. For instance, it suffices for the boundary of T to consist of finitely many pieces of smooth surfaces.

The applications of double integrals that we saw in earlier sections generalize immediately to triple integrals. If T is a solid body with density function $\rho(x, y, z)$, then its **mass M** is given by

$$M = \iiint_T \rho\, dV. \tag{2}$$

The case $\rho \equiv 1$ gives the **volume**

$$V = \iiint_T dV$$

of T. The **moments** of T about the three coordinate planes are defined to be

$$M_{yz} = \iiint_T x\rho\, dV, \qquad M_{xz} = \iiint_T y\rho\, dV, \qquad M_{xy} = \iiint_T z\rho\, dV. \tag{3}$$

The coordinates of the **centroid** of M are given by

$$\bar{x} = \frac{M_{yz}}{M} = \frac{1}{M} \iiint_T x\rho\, dV, \tag{4a}$$

$$\bar{y} = \frac{M_{xz}}{M} = \frac{1}{M} \iiint_T y\rho\, dV, \qquad \text{and} \tag{4b}$$

$$\bar{z} = \frac{M_{xy}}{M} = \frac{1}{M} \iiint_T z\rho\, dV. \tag{4c}$$

The **moments of inertia** of T about the three coordinate axes are defined to be

$$I_x = \iiint_T (y^2 + z^2)\rho\, dV, \tag{5a}$$

$$I_y = \iiint_T (x^2 + z^2)\rho\, dV, \qquad \text{and} \tag{5b}$$

$$I_z = \iiint_T (x^2 + y^2)\rho\, dV. \tag{5c}$$

Like double integrals, triple integrals are almost always evaluated by iterated single integration. Suppose that the nice region T is **z-simple,** in that each line parallel to the z-axis intersects T (if at all) in a single line segment. In effect, this means that T can be described by the inequalities

$$g_1(x, y) \leqq z \leqq g_2(x, y), \quad (x, y) \text{ in } R,$$

where R is the vertical projection of T into the xy-plane. Then

$$\iiint\limits_{T} f(x, y, z) \, dV = \iint\limits_{R} \left(\int_{g_1(x,y)}^{g_2(x,y)} f(x, y, z) \, dz \right) dA. \tag{6}$$

In Formula (6), we take $dA = dx\,dy$ or $dy\,dx$, depending upon the preferable order of integration over the set R. The limits $g_1(x, y)$ and $g_2(x, y)$ for z will be the z-coordinates of the end points of the line segment in which the vertical line at (x, y) meets T.

If the region R has the description

$$\alpha(y) \leqq x \leqq \beta(y), \qquad c \leqq y \leqq d,$$

and we integrate *last* with respect to y, then

$$\iiint\limits_{T} f(x, y, z) \, dV = \int_c^d \int_{\alpha(y)}^{\beta(y)} \int_{g_1(x,y)}^{g_2(x,y)} f(x, y, z) \, dz \, dx \, dy.$$

Thus the triple integral on the left above reduces in this case to three iterated single integrals; these can (in principle) be evaluated using the Fundamental Theorem of Calculus.

If the solid T is bounded by the *two* surfaces $z = g_1(x, y)$ and $z = g_2(x, y)$, as in Figure 15.43, we can find the region R of Equation (6) as follows. The equation $g_1(x, y) = g_2(x, y)$ determines a vertical cylinder that passes through the curve of intersection of these two surfaces. So the cylinder intersects the xy-plane in the boundary curve of R. For instance, the solid of Example 2 below is bounded by the paraboloid $z = x^2 + y^2$ and the plane $z = y + 2$. And the graph of the equation $x^2 + y^2 = y + 2$ in the xy-plane is a circle that bounds the region R above which the solid T lies.

We proceed similarly if T is either **x-simple** or **y-simple.** Such situations, as well as a z-simple solid, appear in Figure 15.44. For example, suppose

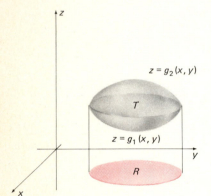

15.43 To find the boundary of R, solve $g_1(x, y) = g_2(x, y)$.

15.44 Solids that are z-simple, y-simple, and x-simple.

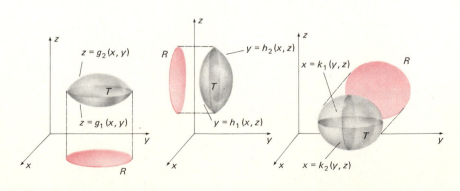

T is z-simple T is y- simple T is x-simple

CHAP. 15: **Multiple Integrals**

that T is y-simple, so that it has a description of the form

$$h_1(x, z) \leq y \leq h_2(x, z), \quad (x, z) \text{ in } R,$$

where R is the projection of T into the xz-plane. Then

$$\iiint_T f(x, y, z)\, dV = \iint_R \left(\int_{h_1(x,z)}^{h_2(x,z)} f(x, y, z)\, dy \right) dA, \tag{7}$$

where $dA = dx\, dz$ or $dz\, dx$, and the limits $h_1(x, z)$ and $h_2(x, z)$ are the y-coordinates of the end points of the line segment in which a typical line parallel to the y-axis intersects T. If T is x-simple (with the obvious meaning), we have

$$\iiint_T f(x, y, z)\, dV = \iint_R \left(\int_{k_1(y,z)}^{k_2(y,z)} f(x, y, z)\, dx \right) dA \tag{8}$$

where $dA = dy\, dz$ or $dz\, dy$ and R is the projection of T into the yz-plane.

EXAMPLE 1   Compute by triple integration (in three ways) the volume of the region T that is bounded by the parabolic cylinder $x = y^2$ and the planes $z = 0$ and $x + z = 1$. Also find the centroid of T (under the assumption that its density ρ is the constant 1).

Solution   The three segments in Figure 15.45 parallel to the coordinate axes indicate that the region T is simultaneously x-simple, y-simple, and z-simple. We may therefore integrate in any order we choose. The projection of T into the xy-plane is the region shown in Figure 15.46, bounded by $x = y^2$ and $x = 1$, so Formula (6) gives

$$V = \int_{-1}^{1} \int_{y^2}^{1} \int_{0}^{1-x} dz\, dx\, dy = 2 \int_{0}^{1} \int_{y^2}^{1} (1 - x)\, dx\, dy$$

$$= 2 \int_{0}^{1} \left[x - \tfrac{1}{2}x^2 \right]_{y^2}^{1} dy = 2 \int_{0}^{1} (\tfrac{1}{2} - y^2 + \tfrac{1}{2}y^4)\, dy = \tfrac{8}{15}.$$

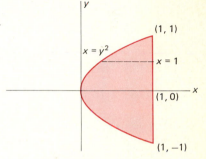

15.45   The region T of Example 1 is x- , y- , and z-simple.

15.46   The vertical projection of the solid region T into the xy-plane.

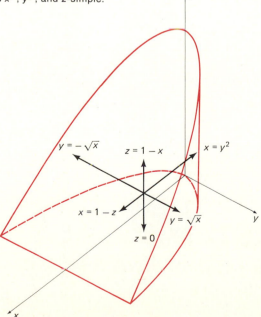

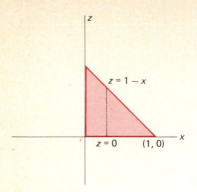

15.47 The vertical projection of the solid region T into the xz-plane.

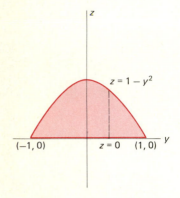

15.48 The vertical projection of the solid region T into the yz-plane.

The projection of T into the xz-plane is the triangle bounded by the coordinate axes and the line $x + z = 1$ (Figure 15.47), so Formula (7) gives

$$V = \int_0^1 \int_0^{1-x} \int_{-\sqrt{x}}^{\sqrt{x}} dy \, dz \, dx = 2 \int_0^1 \int_0^{1-x} \sqrt{x} \, dz \, dx$$

$$= 2 \int_0^1 (x^{1/2} - x^{3/2}) \, dx = \tfrac{8}{15}.$$

The projection of T into the yz-plane is bounded by the y-axis and the parabola $z = 1 - y^2$ (as in Figure 15.48), so Formula (8) yields

$$V = \int_{-1}^1 \int_0^{1-y^2} \int_{y^2}^{1-z} dx \, dz \, dy,$$

and evaluation of this multiple integral again gives $V = \tfrac{8}{15}$.

Now for the centroid of T. Since the region T is symmetric about the xz-plane, its centroid lies in this plane, and so $\bar{y} = 0$. We compute $\bar{x}$ and $\bar{z}$ by integrating first with respect to y:

$$\bar{x} = \frac{1}{V} \iiint_T x \, dV = \frac{15}{8} \int_0^1 \int_0^{1-x} \int_{-\sqrt{x}}^{\sqrt{x}} x \, dy \, dz \, dx$$

$$= \frac{15}{4} \int_0^1 \int_0^{1-x} x^{3/2} \, dz \, dx = \frac{15}{4} \int_0^1 (x^{3/2} - x^{5/2}) \, dx = \frac{3}{7},$$

and similarly

$$\bar{z} = \frac{1}{V} \iiint_T z \, dV = \frac{15}{8} \int_0^1 \int_0^{1-x} \int_{-\sqrt{x}}^{\sqrt{x}} z \, dy \, dz \, dx = \frac{2}{7}.$$

Thus the centroid of T is located at the point $(\tfrac{3}{7}, 0, \tfrac{2}{7})$.

EXAMPLE 2 Find the volume of the *oblique segment of a paraboloid* bounded by the paraboloid $z = x^2 + y^2$ and the plane $z = y + 2$. See Figure 15.49.

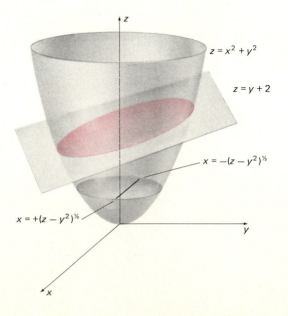

15.49 An oblique segment of a paraboloid (Example 2).

Solution The given region T is z-simple, but its projection into the xy-plane is bounded by the graph of the equation $x^2 + y^2 = y + 2$, which is a translated circle. Consequently it would be rather inconvenient to integrate first with respect to z.

The region T is also x-simple, so we may integrate first with respect to x. The projection of T into the yz-plane is bounded by the line $z = y + 2$ and the parabola $z = y^2$, which intersect at the points $(-1, 1)$ and $(2, 4)$, as shown in Figure 15.50. The end points of a segment in T parallel to the x-axis have x-coordinates $x = \pm\sqrt{z - y^2}$. Since T is symmetric about the yz-plane, we can integrate from $x = 0$ to $x = \sqrt{z - y^2}$ and double the result. Hence

$$V = 2 \int_{-1}^{2} \int_{y^2}^{y+2} \int_{0}^{(z-y^2)^{1/2}} dx\, dz\, dy = 2 \int_{-1}^{2} \int_{y^2}^{y+2} (z - y^2)^{1/2}\, dz\, dy$$

$$= 2 \int_{-1}^{2} \left[\frac{2}{3}(z - y^2)^{3/2}\right]_{z=y^2}^{y+2} dy = \frac{4}{3} \int_{-1}^{2} (2 + y - y^2)^{3/2}\, dy$$

$$= \frac{4}{3} \int_{-3/2}^{3/2} \left(\frac{9}{4} - u^2\right)^{3/2} du \qquad [\text{Complete the square; } u = y - \tfrac{1}{2}.]$$

$$= \frac{27}{4} \int_{-\pi/2}^{\pi/2} \cos^4\theta\, d\theta \qquad [u = \tfrac{3}{2}\sin\theta]$$

$$= \frac{27}{4} \cdot 2 \cdot \frac{1 \cdot 3}{2 \cdot 4} \frac{\pi}{2} = \frac{81\pi}{32}.$$

In the final evaluation, we have used Formula (113) (inside the back cover) for $\int_0^{\pi/2} \cos^n\theta\, d\theta$.

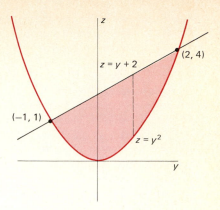

15.50 Projection of the segment of the paraboloid into the yz-plane.

In Problems 1–10, sketch the solid bounded by the graphs of the given equations, and then find its volume by triple integration.

1 $2x + 3y + z = 6$, $x = 0$, $y = 0$, $z = 0$.
2 $z = y$, $y = x^2$, $y = 4$, $z = 0$.
3 $y + z = 4$, $y = 4 - x^2$, $y = 0$, $z = 0$.
4 $z = x^2 + y^2$, $z = 0$, $x = 0$, $y = 0$, $x + y = 1$.
5 $z = 10 - x^2 - y^2$, $y = x^2$, $x = y^2$, $z = 0$.
6 $x = z^2$, $x = 8 - z^2$, $y = -1$, $y = -3$.
7 $z = x^2$, $y + z = 4$, $y = 0$.
8 $z = 1 - y^2$, $z = y^2 - 1$, $x + z = 1$, $x = 0$.
9 $y = z^2$, $z = y^2$, $x + y + z = 2$, $x = 0$.
10 $y = 4 - x^2 - z^2$, $x = 0$, $y = 0$, $z = 0$, $x + z = 2$.

In each of Problems 11–20, assume that the indicated solid has constant density $\rho = 1$.

11 Find the centroid of the solid of Problem 2.
12 Find the centroid of the hemisphere $x^2 + y^2 + z^2 \leq R^2$, $z \geq 0$.
13 Find the centroid of the solid of Problem 7.

14 Find the centroid of the solid bounded by $z = 1 - x^2$, $z = 0$, $y = -1$, and $y = 1$.
15 Find the centroid of the solid bounded by $z = \cos x$, $x = -\pi/2$, $x = \pi/2$, $y = 0$, $z = 0$, $y + z = 1$.
16 Find the moment of inertia about the z-axis of the solid of Problem 2.
17 Find the moment of inertia about the y-axis of the solid of Problem 14.
18 Find the moment of inertia about the z-axis of the solid cylinder $x^2 + y^2 \leq R^2$, $0 \leq z \leq H$.
19 Find the moment of inertia about the z-axis of the solid bounded by $x = 0$, $y = 0$, $z = 0$, and $x + y + z = 1$.
20 Find the moment of inertia about the z-axis of the cube with vertices $(\pm\frac{1}{2}, \frac{7}{2} \pm \frac{1}{2}, \pm\frac{1}{2})$.

In the following problems the indicated solid has uniform density $\rho \equiv 1$ unless otherwise indicated.

21 Find the moment of inertia about one of its edges of a cube with edge length a.
22 The first octant cube with edge a, faces parallel to the coordinate planes, and opposite vertices $(0, 0, 0)$ and (a, a, a),

has density proportional to the square of the distance from the origin. Find the coordinates of its centroid.

23 Find the moment of inertia about the z-axis of the cube of Problem 22.

24 The cube bounded by the coordinate planes and the planes $x = 1$, $y = 1$, and $z = 1$ has density $\rho = kz$ (where k is a positive constant). Find its centroid.

25 Find the moment of inertia about the z-axis of the cube of Problem 24.

26 Find the moment of inertia about a diameter of a solid sphere of radius a.

27 Find the centroid of the first octant region that is interior to the two cylinders $x^2 + z^2 = 1$ and $y^2 + z^2 = 1$.

28 Find the moment of inertia about the z-axis of the solid of Problem 27.

29 Find the volume bounded by the elliptic paraboloids $z = 2x^2 + y^2$ and $z = 12 - x^2 - 2y^2$. Note that this solid projects onto a circular disk in the xy-plane.

30 Find the volume bounded by the elliptic paraboloid $y = x^2 + 4z^2$ and the plane $y = 2x + 3$.

31 Find the volume of the elliptical cone bounded by $z = \sqrt{x^2 + 4y^2}$ and $z = 1$. (*Suggestion:* Integrate first with respect to x.)

32 Find the volume of the region that is bounded by the paraboloid $x = y^2 + 2z^2$ and the parabolic cylinder $x = 2 - y^2$.

Integration in Cylindrical and Spherical Coordinates

Suppose that $f(x, y, z)$ is a continuous function defined on the z-simple region T, which—because it is z-simple—can be described by

$$g_1(x, y) \leq z \leq g_2(x, y) \quad \text{for } (x, y) \text{ in } R.$$

We saw in Section 15-5 that

$$\iiint\limits_{T} f(x, y, z)\, dV = \iint\limits_{R} \left(\int_{g_1(x,y)}^{g_2(x,y)} f(x, y, z)\, dz \right) dA. \tag{1}$$

If the region R is described more naturally in polar coordinates than in rectangular coordinates, then it is likely that the integration over the plane region R will be simpler if it is carried out in polar coordinates.

We first express the inner partial integral in (1) in terms of r and θ by writing

$$\int_{g_1(x,y)}^{g_2(x,y)} f(x, y, z)\, dz = \int_{G_1(r,\theta)}^{G_2(r,\theta)} F(r, \theta, z)\, dz \tag{2}$$

where

$$F(r, \theta, z) = f(r \cos \theta, r \sin \theta, z) \tag{3a}$$

and

$$G_i(r, \theta) = g_i(r \cos \theta, r \sin \theta) \quad \text{for} \quad i = 1, 2. \tag{3b}$$

Substitution of (2) into (1) with $dA = r\, dr\, d\theta$ gives

$$\iiint\limits_{T} f(x, y, z)\, dV = \iint\limits_{S} \left(\int_{G_1(r,\theta)}^{G_2(r,\theta)} F(r, \theta, z)\, dz \right) r\, dr\, d\theta \tag{4}$$

where F, G_1, and G_2 are the functions given in (3) and S represents the appropriate limits on r and θ needed to describe the plane region R in polar coordinates (as discussed in Section 15-3). The limits on z are simply the z-coordinates (in terms of r and θ) of a typical line segment joining the lower and upper boundary surfaces of T, as indicated in Figure 15.51.

15.51

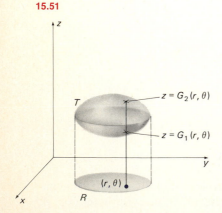

CHAP. 15: **Multiple Integrals**

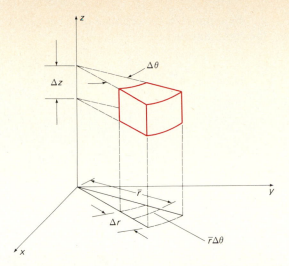

15.52 The volume of the "cylindrical block" is $\Delta V = \bar{r}\, \Delta z\, \Delta r\, \Delta\theta$.

Thus the general formula for **triple integration in cylindrical coordinates** is

$$\iiint\limits_{T} f(x,\, y,\, z)\, dV = \iiint f(r\cos\theta,\, r\sin\theta,\, z) r\, dz\, dr\, d\theta, \qquad (5)$$

with limits on z, r, and θ appropriate to describe the space region T in cylindrical coordinates. Before integrating, the variables x and y are replaced by $r\cos\theta$ and $r\sin\theta$, respectively, while z is left unchanged. The cylindrical coordinates volume element

$$dV = r\, dz\, dr\, d\theta$$

may be regarded formally as the product of dz and the polar coordinates area element $dA = r\, dr\, d\theta$. It is a consequence of the formula $\Delta V = \bar{r}\, \Delta z\, \Delta r\, \Delta\theta$ for the volume of the *cylindrical block* shown in Figure 15.52.

Integration in cylindrical coordinates is particularly useful for computations associated with solids of revolution. The solid should be placed so that the axis of revolution is the z-axis.

EXAMPLE 1 Find the volume and centroid of the solid T that is bounded by the paraboloid $z = b(x^2 + y^2)$ $(b > 0)$ and the plane $z = h$ $(h > 0)$.

15.53 The paraboloid of Example 1.

Solution Figure 15.53 makes it clear that we get the radius of the circular top by equating $z = b(x^2 + y^2) = br^2$ and $z = h$. This gives $a = (h/b)^{1/2}$ for the radius of the circle over which the solid lies. Hence Formula (4), with $f(x,\, y,\, z) \equiv 1$, gives the volume:

$$V = \iiint\limits_{T} dV = \int_0^{2\pi} \int_0^a \int_{br^2}^h r\, dz\, dr\, d\theta$$

$$= \int_0^{2\pi} \int_0^a (hr - br^3)\, dr\, d\theta$$

$$= 2\pi\left(\frac{1}{2} ha^2 - \frac{1}{4} ba^4\right) = \frac{\pi h^2}{2b} = \frac{1}{2}\pi a^2 h$$

(because $a^2 = h/b$).

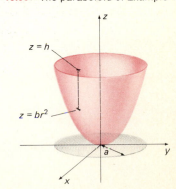

By symmetry, the centroid of T lies on the z-axis, so all that remains is the computation of $\bar{z}$:

$$\bar{z} = \frac{1}{V} \iiint_T z \, dV = \frac{2}{\pi a^2 h} \int_0^{2\pi} \int_0^a \int_{br^2}^h rz \, dz \, dr \, d\theta$$

$$= \frac{2}{\pi a^2 h} \int_0^{2\pi} \int_0^a \left(\frac{1}{2} h^2 r - \frac{1}{2} b^2 r^5 \right) dr \, d\theta$$

$$= \frac{4}{a^2 h} \left(\frac{1}{4} h^2 a^2 - \frac{1}{12} b^2 a^6 \right) = \frac{2}{3} h,$$

again using $a^2 = h/b$. Note that this answer is dimensionally correct; $\bar{z}$ *must* be in units of length.

The results of Example 1 can be summarized as follows: The volume of a right circular paraboloid is *half* that of the circumscribed cylinder (see Figure 15.54), and its centroid lies on its axis of symmetry *two-thirds* of the way from the "vertex" $(0, 0, 0)$ to the base (top).

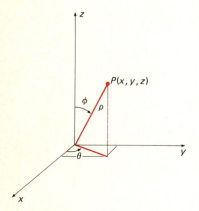

15.54 Volume and centroid of a right circular paraboloid in terms of the circumscribed cylinder.

SPHERICAL COORDINATES

When the boundary surfaces of the region T of integration are spheres, cones, or other surfaces with simple descriptions in spherical coordinates, it is generally advantageous to transform the integral into spherical coordinates. Recall from Section 13-6 that the relationship between spherical coordinates (ρ, ϕ, θ) (shown in Figure 15.55) and rectangular coordinates (x, y, z) is given by the equations

$$x = \rho \sin \phi \cos \theta, \qquad y = \rho \sin \phi \sin \theta, \qquad z = \rho \cos \phi. \qquad (6)$$

Suppose, for example, that T is the **spherical block** determined by the inequalities

$$\rho_1 \leq \rho \leq \rho_2 = \rho_1 + \Delta\rho,$$

$$\phi_1 \leq \phi \leq \phi_2 = \phi_1 + \Delta\phi, \qquad (7)$$

$$\theta_1 \leq \theta \leq \theta_2 = \theta_1 + \Delta\theta.$$

15.55 The spherical coordinates (ρ, ϕ, θ) of the point P.

As indicated by the dimensions labeled in Figure 15.56, this spherical block is (if $\Delta\rho$, $\Delta\phi$, and $\Delta\theta$ are small) *approximately* a rectangular block with dimensions $\Delta\rho$, $\rho_1 \Delta\phi$, and $\rho_1 \sin \phi_2 \Delta\theta$. Thus its volume is approximately $\rho_1^2 \sin \phi_2 \Delta\rho \Delta\phi \Delta\theta$. It can be shown (see Problem 15 in Section 15-7) that the *exact* volume of the spherical block (7) is

$$\Delta V = \hat{\rho}^2 \sin \hat{\phi} \, \Delta\rho \, \Delta\phi \, \Delta\theta \qquad (8)$$

for appropriate numbers $\hat{\rho}$ and $\hat{\phi}$ such that $\rho_1 < \hat{\rho} < \rho_2$ and $\phi_1 < \hat{\phi} < \phi_2$.

Now suppose that we partition each of the intervals $[\rho_1, \rho_2]$, $[\phi_1, \phi_2]$, and $[\theta_1, \theta_2]$ into n equal subintervals with lengths

$$\Delta\rho_i = \frac{\rho_1 - \rho_2}{n}, \qquad \Delta\phi_i = \frac{\phi_2 - \phi_1}{n}, \qquad \text{and} \qquad \Delta\theta_i = \frac{\theta_2 - \theta_1}{n},$$

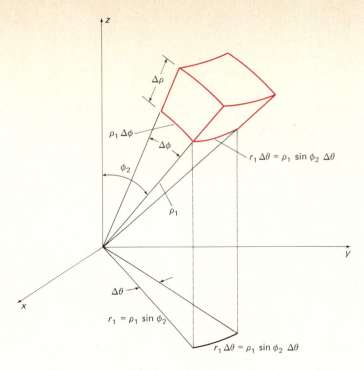

respectively. This produces a **spherical partition** $\mathcal{P}$ of the spherical block T into $k = n^3$ smaller spherical blocks $T_1, T_2, T_3, \ldots, T_k$. By Formula (8) there exists a point $(\hat\rho_i, \hat\phi_i, \hat\theta_i)$ of the spherical block T_i such that its volume is $\Delta V_i = \hat\rho_i^2 \sin \hat\theta_i \, \Delta\rho_i \, \Delta\phi_i \, \Delta\theta_i$. The mesh $|\mathcal{P}|$ of $\mathcal{P}$ is the length of the longest diagonal of any of the small spherical blocks $T_1, T_2, T_3, \ldots, T_k$.

If (x_i^*, y_i^*, z_i^*) are the rectangular coordinates of the point with spherical coordinates $(\hat\rho_i, \hat\phi_i, \hat\theta_i)$, then the definition of the triple integral as a limit of Riemann sums as the mesh $|\mathcal{P}|$ approaches zero gives

$$\iiint\limits_T f(x, y, z) \, dV = \lim_{|\mathcal{P}| \to 0} \sum_{i=1}^{n} f(x_i^*, y_i^*, z_i^*) \, \Delta V_i$$

$$= \lim_{|\mathcal{P}| \to 0} \sum_{i=1}^{k} F(\hat\rho_i, \hat\phi_i, \hat\theta_i) \hat\rho_i^2 \sin \hat\phi_i \, \Delta\rho_i \, \Delta\phi_i \, \Delta\theta_i, \qquad (9)$$

where

$$F(\rho, \phi, \theta) = f(\rho \sin \phi \cos \theta, \rho \sin \phi \sin \theta, \rho \cos \phi). \qquad (10)$$

But the sum in limit (9) is simply a Riemann sum for the triple integral

$$\int_{\theta_1}^{\theta_2} \int_{\phi_1}^{\phi_2} \int_{\rho_1}^{\rho_2} F(\rho, \phi, \theta) \, \rho^2 \sin \phi \, d\rho \, d\phi \, d\theta.$$

It therefore follows that

$$\iiint\limits_T f(x, y, z) \, dV = \int_{\theta_1}^{\theta_2} \int_{\phi_1}^{\phi_2} \int_{\rho_1}^{\rho_2} F(\rho, \phi, \theta) \, \rho^2 \sin \phi \, d\rho \, d\phi \, d\theta. \qquad (11)$$

Thus we transform the integral $\iiint_T f(x, y, z) \, dV$ into spherical coordinates by replacing the rectangular coordinate variables x, y, z by their expressions in (6) in terms of the spherical coordinate variables ρ, ϕ, θ and formally writing

$$dV = \rho^2 \sin\phi \, d\rho \, d\phi \, d\theta$$

for the volume element in spherical coordinates.

More generally, we can transform the triple integral $\iiint_T f(x, y, z) \, dV$ into spherical coordinates whenever the region T is **centrally simple**; that is, whenever it has a spherical coordinates description of the form

$$G_1(\phi, \theta) \leq \rho \leq G_2(\phi, \theta), \qquad \phi_1 \leq \phi \leq \phi_2, \qquad \theta_1 \leq \theta \leq \theta_2. \quad (12)$$

If so, then

$$\iiint_T f(x, y, z) \, dV = \int_{\theta_1}^{\theta_2} \int_{\phi_1}^{\phi_2} \int_{G_1(\phi,\theta)}^{G_2(\phi,\theta)} F(\rho, \phi, \theta) \rho^2 \sin\phi \, d\rho \, d\phi \, d\theta. \quad (13)$$

The limits on ρ are simply the ρ-coordinates (in terms of ϕ and θ) of the end points of a typical radial segment joining the "inner" and "outer" parts of the boundary of T (see Figure 15.57). Thus the general formula for **triple integration in spherical coordinates** is

$$\iiint_T f(x, y, z) \, dV$$

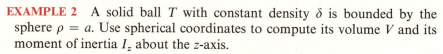

$$= \iiint f(\rho \sin\phi \cos\theta, \rho \sin\phi \sin\theta, \rho \cos\phi)\rho^2 \sin\phi \, d\rho \, d\phi \, d\theta \quad (14)$$

with limits on ρ, ϕ, θ appropriate to describe the region T in spherical coordinates.

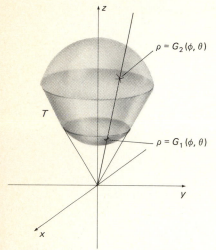

15.57 A centrally simple region.

EXAMPLE 2 A solid ball T with constant density δ is bounded by the sphere $\rho = a$. Use spherical coordinates to compute its volume V and its moment of inertia I_z about the z-axis.

Solution The points of the ball are described by the inequalities

$$0 \leq \rho \leq a, \qquad 0 \leq \phi \leq \pi, \qquad 0 \leq \theta \leq 2\pi.$$

We take $f = F \equiv 1$ in Formula (11) and thereby obtain

$$V = \iiint_T dV = \int_0^{2\pi} \int_0^{\pi} \int_0^a \rho^2 \sin\phi \, d\rho \, d\phi \, d\theta$$

$$= \frac{a^3}{3} \int_0^{2\pi} \int_0^{\pi} \sin\phi \, d\phi \, d\theta = \frac{a^3}{3} \int_0^{2\pi} \left[-\cos\phi \right]_0^{\pi} d\theta$$

$$= \frac{2a^3}{3} \int_0^{2\pi} d\theta = \frac{4}{3}\pi a^3.$$

The distance from the typical point (ρ, ϕ, θ) to the z-axis is $r = \rho \sin\phi$, so the moment of inertia of the sphere about that axis is

$$I_z = \iiint_T r^2 \delta \, dV = \int_0^{2\pi} \int_0^{\pi} \int_0^a \delta \rho^4 \sin^3\phi \, d\rho \, d\phi \, d\theta$$

$$= \frac{\delta a^5}{5} \int_0^{2\pi} \int_0^{\pi} \sin^3\phi \, d\phi \, d\theta = \frac{2\pi\delta a^5}{5} \int_0^{\pi} \sin^3\phi \, d\phi$$

$$= \frac{2\pi\delta a^5}{5} \cdot 2 \cdot \frac{2}{3} = \frac{2}{5} Ma^2.$$

where $M = \frac{4}{3}\pi a^3 \delta$ is the mass of the ball. In Problem 22 we ask you to apply this result to continue the discussion associated with Equation (12) in Section 15-4 and show that a marble rolls down an inclined plane faster than a quarter does.

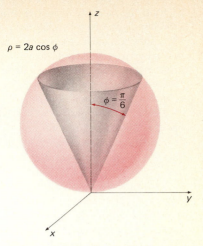

$\rho = 2a \cos \phi$

$\phi = \dfrac{\pi}{6}$

EXAMPLE 3 Find the volume and centroid of the ice cream cone bounded by the cone $\phi = \pi/6$ and the sphere $\rho = 2a \cos \phi$ of radius a and tangent to the xy-plane at the origin. The sphere and the part of the cone inside it are shown in Figure 15.58.

Solution The ice cream cone is described by the inequalities

$$0 \leq \theta \leq 2\pi, \qquad 0 \leq \phi \leq \frac{\pi}{6}, \qquad 0 \leq \rho \leq 2a \cos \phi.$$

Using Formula (13) to compute its volume, we get

$$V = \int_0^{2\pi} \int_0^{\pi/6} \int_0^{2a \cos \phi} \rho^2 \sin \phi \, d\rho \, d\phi \, d\theta$$

$$= \tfrac{8}{3} a^3 \int_0^{2\pi} \int_0^{\pi/6} \cos^3 \phi \sin \phi \, d\phi \, d\theta$$

$$= \frac{16}{3} \pi a^3 \left[-\frac{1}{4} \cos^4 \phi \right]_0^{\pi/6} = \frac{7\pi a^3}{12}.$$

Now for the centroid. It is clear by symmetry that $\bar{x} = 0 = \bar{y}$. Since $z = \rho \cos \phi$, the moment of the ice cream cone about the xy-plane is

$$M_{xy} = \int_0^{2\pi} \int_0^{\pi/6} \int_0^{2a \cos \phi} \rho^3 \cos \phi \sin \phi \, d\rho \, d\phi \, d\theta$$

$$= 4a^4 \int_0^{2\pi} \int_0^{\pi/6} \cos^5 \phi \sin \phi \, d\phi \, d\theta = 8\pi a^4 \left[-\frac{1}{6} \cos^6 \phi \right]_0^{\pi/6} = \frac{37\pi a^4}{48}.$$

Hence the z-coordinate of the centroid is $\bar{z} = M_{xy}/V = 37a/28$.

15.58 The "ice cream cone" of Example 3 is the part of the cone that lies within the sphere.

15-6 PROBLEMS

Solve Problems 1–10 by integration in cylindrical coordinates.

1 Find the volume and centroid of the region bounded by the plane $z = 0$ and the paraboloid $z = 9 - x^2 - y^2$.

2 Find the volume and centroid of the region bounded by the paraboloids $z = x^2 + y^2$ and $z = 12 - 2x^2 - 2y^2$.

3 Find the volume of the region bounded by the paraboloids $z = 2x^2 + y^2$ and $z = 12 - x^2 - 2y^2$.

4 Find the volume of the region that is bounded below by the paraboloid $z = x^2 + y^2$ and above by the plane $z = 2x$.

5 Find the volume of the region that is bounded above by the sphere $x^2 + y^2 + z^2 = 2$ and below by the paraboloid $z = x^2 + y^2$.

6 A homogeneous solid cylinder has mass M and radius a. Show that its moment of inertia about its axis of symmetry is $\frac{1}{2}Ma^2$.

7 Find the moment of inertia I of a homogeneous solid cylinder about a diameter of its base. Express I in terms of the radius a, the height h, and the density δ of the cylinder.

8 Find the centroid of a homogeneous solid hemisphere of radius a.

9 Find the volume of the region bounded by the plane $z = 1$ and the cone $z = r$.

10 Show that the centroid of a homogeneous solid circular cone lies on its axis $\frac{3}{4}$ of the way from its vertex to its base.

Solve Problems 11–20 by integration in spherical coordinates.

11 Find the centroid of a homogeneous solid hemisphere of radius a.

12 Find the mass and centroid of a solid hemisphere of radius a if its density δ is proportional to distance z from its base: $\delta = kz$.

13 Problem 9 by integration in spherical coordinates.

14 Problem 10 by integration in spherical coordinates.

15 Find the volume and centroid of the solid that lies within the sphere $\rho = a$ and above the cone $r = z$.

16 Find the moment of inertia I_z of the solid of Problem 15 under the assumption that it has constant density δ.

17 Find the moment of inertia about a tangent line of a solid homogeneous sphere of radius a and total mass M.

18 A spherical shell of mass M is bounded by the spheres $\rho = a$ and $\rho = 2a$, and its density function is $\delta = \rho^2$. Find its moment of inertia about a diameter.

19 Describe the surface $\rho = 2a \sin \phi$, and compute the volume of the region it bounds.

20 Describe the surface $\rho = 1 + \cos \phi$, and compute the volume of the region it bounds.

21 Find the moment of inertia about the x-axis of the region that lies inside both the cylinder $r = a$ and the sphere $\rho = 2a$.

22 Substitute the result of Example 2 of this section into Equation (12) of Section 15-4, and thereby find the speed with which a homogeneous solid spherical ball will reach the bottom of the inclined plane of Figure 15.38.

23 A homogeneous spherical shell is bounded by the concentric spheres $\rho = a$ and $\rho = b > a$. (a) Compute its moment of inertia about the z-axis. (b) Assume that $a = 1$ ft, $b = 2$ ft and find the speed with which it will arrive at the bottom of the inclined plane of Figure 15.38.

24 Consider a homogeneous spherical ball of radius a centered at the origin, with density δ and mass $M = \frac{4}{3}\pi a^3 \delta$. Show that the gravitational force $\mathbf{F}$ exerted by this ball upon a point mass m located at the point $(0, 0, c)$, with $c > a$ (Figure 15.59) is the same as though all the mass of the ball were concentrated at its center $(0, 0, 0)$. That is, show that $|\mathbf{F}| = GMm/c^2$. (*Suggestion:* By symmetry you may assume that the force is vertical, so that $\mathbf{F} = F_z\mathbf{k}$. Set up the integral

$$F_z = -\int_0^{2\pi} \int_0^a \int_0^\pi \frac{Gm\delta \cos \alpha}{w^2} \rho^2 \sin \phi \, d\phi \, d\rho \, d\theta.$$

Change the first variable of integration from ϕ to w by using

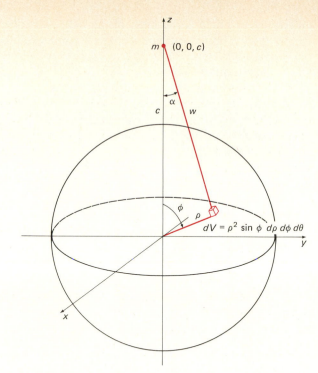

15.59 The point-and-mass system of Problem 24.

the law of cosines: $w^2 = \rho^2 + c^2 - 2\rho c \cos \phi$. Then $2w \, dw = 2\rho c \sin \phi \, d\phi$ and $w \cos \alpha + \rho \cos \phi = c$. (Why?))

25 Consider now the spherical shell $a \leqq \rho \leqq b$ with uniform density δ. Show that this shell exerts *no* force on a point mass m located at the point $(0, 0, c)$ *inside* it; that is, with $c < a$. The computation will be the same as in Problem 24, except for the limits of integration on ρ and w.

15-7

Surface Area

Until now our concept of a surface has been the graph $z = f(x, y)$ of a function of two variables. Occasionally we have seen such a surface defined implicitly by an equation of the form $F(x, y, z) = 0$. Now we want to introduce the more precise concept of a parametric surface—a two-dimensional analogue of a parametric curve.

A **parametric surface** S is the image of a function or transformation $\mathbf{r}$ defined on a region R in the uv-plane and with values in xyz-space. The **image** under $\mathbf{r}$ of each point (u, v) in R is the point in xyz-space with position vector

$$\mathbf{r}(u, v) = \langle x(u, v), y(u, v), z(u, v) \rangle. \tag{1}$$

We shall assume throughout this section that the component functions of $\mathbf{r}$ have continuous partial derivatives with respect to u and v, and also that the vectors

$$\mathbf{r}_u = \frac{\partial \mathbf{r}}{\partial u} = \langle x_u, y_u, z_u \rangle = \frac{\partial x}{\partial u}\mathbf{i} + \frac{\partial y}{\partial u}\mathbf{j} + \frac{\partial z}{\partial u}\mathbf{k} \tag{2}$$

and

$$\mathbf{r}_v = \frac{\partial \mathbf{r}}{\partial v} = \langle x_v, y_v, z_v \rangle = \frac{\partial x}{\partial v} \mathbf{i} + \frac{\partial y}{\partial v} \mathbf{j} + \frac{\partial z}{\partial v} \mathbf{k} \qquad (3)$$

are nonzero and nonparallel at each interior point of R.

We call the variables u and v the *parameters* for the surface S (analogous to the single parameter t for a parametric curve). The graph $z = f(x, y)$ of a function may be regarded as a parametric surface with parameters x and y. In this case the transformation $\mathbf{r}$ from the xy-plane to xyz-space has the component functions

$$x = x, \qquad y = y, \qquad z = f(x, y). \qquad (4)$$

Similarly, a surface given in cylindrical coordinates as the graph of $z = g(r, \theta)$ may be regarded as a parametric surface with parameters r and θ; the transformation $\mathbf{r}$ from the $r\theta$-plane to xyz-space is then given by

$$x = r \cos \theta, \qquad y = r \sin \theta, \qquad z = g(r, \theta). \qquad (5)$$

A surface given in spherical coordinates by $\rho = h(\phi, \theta)$ may be regarded as a parametric surface with parameters ϕ and θ, and the corresponding transformation from the $\phi\theta$-plane to xyz-space is then given by

$$x = h(\phi, \theta) \sin \phi \cos \theta, \qquad y = h(\phi, \theta) \sin \phi \sin \theta, \qquad z = h(\phi, \theta) \cos \phi. \qquad (6)$$

The concept of a parametric surface lets us treat all these cases, and others, with the same techniques.

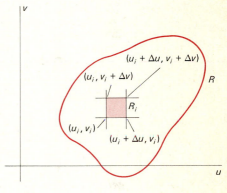

15.60

Now we want to define the *surface area* of the general parametric surface given by Equation (1). We begin with an inner partition of the region R—the domain of $\mathbf{r}$ in the uv-plane—into rectangles $R_1, R_2, R_3, \ldots, R_n$, each with dimensions Δu and Δv. Let (u_i, v_i) be the lower left-hand corner of R_i (as in Figure 15.60). The image S_i of R_i under $\mathbf{r}$ generally won't be a rectangle in xyz-space; it will look more like a *curvilinear figure* on the image surface S, with $\mathbf{r}(u_i, v_i)$ as one "vertex." See Figure 15.61. Let ΔS_i denote the area of this curvilinear figure S_i.

15.61 The image of R_i is a curvilinear figure.

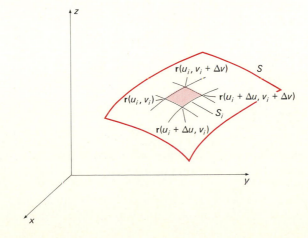

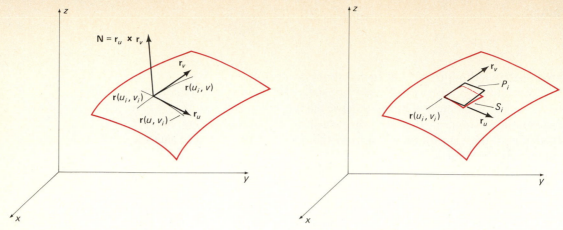

15.62 Normal **N** to the surface at $\mathbf{r}(u_i, v_i)$.

15.63 The area of the real parallelogram P_i is an approximation to the area of the curvilinear figure S_i.

The parametric curves $\mathbf{r}(u, v_i)$ and $\mathbf{r}(u_i, v)$—with parameters u and v, respectively—lie on the surface S and meet at the point $\mathbf{r}(u_i, v_i)$. At this point of intersection, these two curves have the tangent vectors $\mathbf{r}_u(u_i, v_i)$ and $\mathbf{r}_v(u_i, v_i)$ shown in Figure 15.62. Hence their cross product

$$\mathbf{N}(u_i, v_i) = \mathbf{r}_u(u_i, v_i) \times \mathbf{r}_v(u_i, v_i) \tag{7}$$

is a normal vector to S at the point $\mathbf{r}(u_i, v_i)$.

Now suppose that Δu and Δv are both small. Then the area ΔS_i of the curvilinear figure S_i will be approximately equal to the area ΔP_i of the parallelogram with adjacent sides $\mathbf{r}_u(u_i, v_i) \, \Delta u$ and $\mathbf{r}_v(u_i, v_i) \, \Delta v$ (see Figure 15.63). But the area of this parallelogram is

$$\Delta P_i = \left| \mathbf{r}_u(u_i, v_i) \, \Delta u \times \mathbf{r}_v(u_i, v_i) \, \Delta v \right| = \left| \mathbf{N}(u_i, v_i) \right| \Delta u \, \Delta v.$$

This means that the area of the surface S is given approximately by

$$\text{area}(S) = \sum_{i=1}^{n} \Delta S_i \approx \sum_{i=1}^{n} \Delta P_i,$$

so that

$$\text{area}(S) \approx \sum_{i=1}^{n} \left| \mathbf{N}(u_i, v_i) \right| \Delta u \, \Delta v.$$

But this last sum is a Riemann sum for the double integral $\iint_R |\mathbf{N}(u, v)| \, du \, dv$. We are therefore motivated to *define* the **surface area** of the parametric surface S by

$$A = \text{area}(S) = \iint_R \left| \mathbf{N}(u, v) \right| du \, dv = \iint_R \left| \frac{\partial \mathbf{r}}{\partial u} \times \frac{\partial \mathbf{r}}{\partial v} \right| du \, dv. \tag{8}$$

In the case of the surface $z = f(x, y)$, for (x, y) in the region R in the xy-plane, the component functions of $\mathbf{r}$ are given by Equations (4) with

744   CHAP. 15:   **Multiple Integrals**

parameters x and y (in place of u and v). Then

$$\mathbf{N} = \frac{\partial \mathbf{r}}{\partial x} \times \frac{\partial \mathbf{r}}{\partial y} = \begin{vmatrix} \mathbf{i} & \mathbf{j} & \mathbf{k} \\ 1 & 0 & \dfrac{\partial f}{\partial x} \\ 0 & 1 & \dfrac{\partial f}{\partial y} \end{vmatrix} = -\frac{\partial f}{\partial x}\mathbf{i} - \frac{\partial f}{\partial y}\mathbf{j} + \mathbf{k},$$

so Formula (8) becomes

$$A = \text{area } (S) = \iint\limits_{R} \sqrt{1 + \left(\frac{\partial f}{\partial x}\right)^2 + \left(\frac{\partial f}{\partial y}\right)^2}\, dx\, dy$$

$$= \iint\limits_{R} \sqrt{1 + z_x^2 + z_y^2}\, dx\, dy. \tag{9}$$

EXAMPLE 1   Find the area of the ellipse cut from the plane $z = 2x + 2y + 1$ by the cylinder $x^2 + y^2 = 1$.

Solution   Here, R is the unit circle in the xy-plane with area $\iint_R 1\, dx\, dy = \pi$, so Formula (9) gives

$$A = \iint\limits_{R} \sqrt{1 + \left(\frac{\partial z}{\partial x}\right)^2 + \left(\frac{\partial z}{\partial y}\right)^2}\, dx\, dy = \iint\limits_{R} 3\, dx\, dy = 3\pi.$$

Now consider a cylindrical coordinates surface $z = g(r, \theta)$ parametrized by Equations (5), for (r, θ) in a region R of the $r\theta$-plane. Then the normal vector is

$$\mathbf{N} = \frac{\partial \mathbf{r}}{\partial r} \times \frac{\partial \mathbf{r}}{\partial \theta} = \begin{vmatrix} \mathbf{i} & \mathbf{j} & \mathbf{k} \\ \cos\theta & \sin\theta & \dfrac{\partial z}{\partial r} \\ -r\sin\theta & r\cos\theta & \dfrac{\partial z}{\partial \theta} \end{vmatrix}$$

$$= \mathbf{i}\left(\frac{\partial z}{\partial \theta}\sin\theta - r\frac{\partial z}{\partial r}\cos\theta\right) - \mathbf{j}\left(\frac{\partial z}{\partial \theta}\cos\theta + r\frac{\partial z}{\partial r}\sin\theta\right) + r\mathbf{k}.$$

After some simplification, we find that

$$|\mathbf{N}| = \sqrt{r^2 + r^2\left(\frac{\partial z}{\partial r}\right)^2 + \left(\frac{\partial z}{\partial \theta}\right)^2}.$$

Thus Formula (8) yields the formula

$$A = \iint\limits_{R} \sqrt{r^2 + r^2\left(\frac{\partial z}{\partial r}\right)^2 + \left(\frac{\partial z}{\partial \theta}\right)^2}\, dr\, d\theta \tag{10}$$

for surface area in cylindrical coordinates.

EXAMPLE 2   Find the surface area cut from the paraboloid $z = r^2$ by the cylinder $r = 1$.

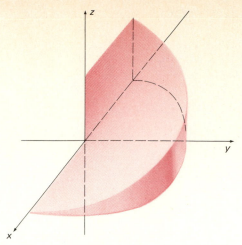

15.64 The "spiral ramp" of Example 3.

Solution Formula (10) gives

$$A = \int_0^{2\pi} \int_0^1 \sqrt{r^2 + r^2(2r)^2} \, dr \, d\theta = 2\pi \int_0^1 r\sqrt{1 + 4r^2} \, dr$$

$$= 2\pi \left[\frac{2}{3} \cdot \frac{1}{8} (1 + 4r^2)^{3/2} \right]_0^1 = \frac{\pi}{6}(5\sqrt{5} - 1) \approx 5.3304.$$

In Example 2, you would get the same result if you first wrote $z = x^2 + y^2$, used Equation (9), which gives

$$A = \iint_R \sqrt{1 + 4x^2 + 4y^2} \, dx \, dy,$$

and then changed to polar coordinates. In the following example it would be less convenient to begin with rectangular coordinates.

EXAMPLE 3 Find the area of the *spiral ramp* $z = \theta, 0 \leq r \leq 1, 0 \leq \theta \leq \pi$: the upper surface of the solid shown in Figure 15.64.

Solution Formula (10) gives

$$A = \int_0^\pi \int_0^1 \sqrt{r^2 + 1} \, dr \, d\theta = \frac{\pi}{2} \left[\sqrt{2} + \ln(1 + \sqrt{2}) \right] \approx 3.6059;$$

we avoided a trigonometric substitution by using the table of integrals inside the front cover.

EXAMPLE 4 Find the surface area of the torus generated by revolving the circle $(x - b)^2 + z^2 = a^2$ $(a < b)$ in the xz-plane around the z-axis.

Solution With the ordinary polar coordinate θ and the angle ψ of Figure 15.65, the torus is described for $0 \leq \theta \leq 2\pi$ and $0 \leq \psi \leq 2\pi$ by the parametric equations

$$x = r \cos \theta = (b + a \cos \psi) \cos \theta,$$

$$y = r \sin \theta = (b + a \cos \psi) \sin \theta,$$

$$z = a \sin \psi.$$

15.65 The circle that generates the torus of Example 4.

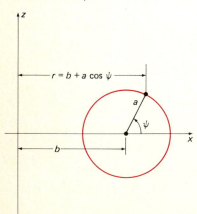

CHAP. 15: **Multiple Integrals**

When we compute $\mathbf{N} = \mathbf{r}_\theta \times \mathbf{r}_\psi$ and simplify, we find that

$$|\mathbf{N}| = a(b + a \cos \psi).$$

Hence the general surface area formula (Formula (8)) gives

$$A = \int_0^{2\pi} \int_0^{2\pi} a(b + a \cos \psi)\, d\theta\, d\psi$$

$$= 2\pi a \left[b\psi + a \sin \psi \right]_0^{2\pi} = 4\pi^2 ab.$$

We obtained the same result in Section 5-6 with the aid of Pappus' theorem.

15-7 PROBLEMS

1 Find the area of the surface that is the graph of $z = x + y^2$ for $0 \leq x \leq 1, 0 \leq y \leq 2$.

2 Find the area of that part of the surface of Problem 1 that lies above the triangle in the xy-plane with vertices $(0, 0)$, $(0, 1)$, and $(1, 1)$.

3 Find by integration the area of the part of the plane $2x + 3y + z = 6$ that lies in the first octant.

4 Find the area of the ellipse that is cut from the plane of Problem 3 by the cylinder $x^2 + y^2 = 2$.

5 Find the area that is cut from the saddle-shaped surface $z = xy$ by the cylinder $x^2 + y^2 = 1$.

6 Find the area that is cut from the surface $z = x^2 - y^2$ by the cylinder $x^2 + y^2 = 4$.

7 Find the surface area of the part of the paraboloid $z = 16 - x^2 - y^2$ that lies above the xy-plane.

8 Show by integration that the surface area of the conical surface $z = br$ between the planes $z = 0$ and $z = h = ab$ is given by $A = \pi aL$, where L is the slant height $(a^2 + h^2)^{1/2}$ and a is the radius of the base of the cone.

9 Let the part of the cylinder $x^2 + y^2 = a^2$ between the planes $z = 0$ and $z = h$ be parametrized by $x = a \cos \theta$, $y = a \sin \theta$, $z = z$. Apply Formula (8) to show that its surface area is $A = 2\pi ah$.

10 Consider the meridianal zone of height $h = c - b$ lying on the sphere $r^2 + z^2 = a^2$ between the planes $z = b$ and $z = c$, where $0 \leq b < c \leq a$. Apply Formula (10) to show that the area of this zone is $A = 2\pi ah$.

11 Find the area of the part of the cylinder $x^2 + z^2 = a^2$ that lies within the cylinder $r^2 = x^2 + y^2 = a^2$.

12 Find the area of the part of the sphere $r^2 + z^2 = a^2$ that lies within the cylinder $r = a \sin \theta$.

13 (a) Apply Formula (8) to show that the surface area of the surface $y = f(x, z)$, for (x, z) in the region R of the xz-plane, is given by

$$A = \iint_R \sqrt{1 + \left(\frac{\partial f}{\partial x}\right)^2 + \left(\frac{\partial f}{\partial z}\right)^2}\; dx\, dz.$$

(b) State and derive a similar formula for the area of the surface $x = f(y, z)$, (y, z) in R.

14 Suppose that R is a region in the $\phi\theta$-plane. Consider the part of the sphere $\rho = a$ that corresponds to (ϕ, θ) in R, parametrized by Equations (6) with $h(\phi, \theta) = a$. Apply (8) to show that the surface area of this part of the sphere is $A = \iint_R a^2 \sin \phi\, d\phi\, d\theta$.

15 (a) Consider the "spherical rectangle" defined by $\rho = a$, $\phi_1 \leq \phi \leq \phi_2 = \phi_1 + \Delta\phi$, $\theta_1 \leq \theta \leq \theta_2 = \theta_1 + \Delta\theta$. Apply the formula of Problem 14 and the average value property (Problem 30 in Section 15-1) to show that the area of this spherical rectangle is $A = a^2 \sin \hat{\phi}\, \Delta\phi\, \Delta\theta$ for some $\hat{\phi}$ in (ϕ_1, ϕ_2).

(b) Conclude from the result of Part (a) that the volume of the spherical block defined by $\rho_1 \leq \rho \leq \rho_2 = \rho_1 + \Delta\rho$ and $\phi_1 \leq \phi \leq \phi_2, \theta_1 \leq \theta \leq \theta_2$ is

$$\Delta V = \tfrac{1}{3}(\rho_2^3 - \rho_1^3) \sin \hat{\phi}\, \Delta\phi\, \Delta\theta.$$

Finally derive Formula (8) in Section 15-6 by applying the Mean Value Theorem to the function $f(\rho) = \rho^3$ on $[\rho_1, \rho_2]$.

16 Describe the surface $z = 2a \sin \phi$. Why is it called a pinched torus? It is parametrized by Equations (6) with $h(\phi, \theta) = 2a \sin \phi$. Show that its surface area is $A = 4\pi^2 a^2$.

17 The surface of revolution obtained when one revolves the curve $x = f(z)$, $a \leq z \leq b$ around the z-axis, is parametrized in terms of θ in $[0, 2\pi]$ and z in $[a, b]$ by $x = f(z) \cos \theta$, $y = f(z) \sin \theta$, $z = z$. Derive from (8) the surface area formula

$$A = \int_0^{2\pi} \int_a^b f(z) \sqrt{1 + [f'(z)]^2}\; dz\, d\theta.$$

Note that this agrees with surface area of revolution as defined in Section 5-4.

Change of Variables in Multiple Integrals

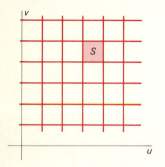

We have seen in preceding sections that it is easier to evaluate certain multiple integrals by transforming them into polar or spherical coordinates. The technique of changing coordinate systems in order to evaluate a multiple integral is the multivariable analogue of substitution in a single integral. Recall from Section 4-5 that if $x = g(u)$, then

$$\int_a^b f(x)\, dx = \int_c^d f(g(u))g'(u)\, du \tag{1}$$

where $a = g(c)$ and $b = g(d)$. The method of substitution involves a "change of variables" that is tailored to the evaluation of a given integral.

Suppose that we want to evaluate the double integral $\iint_R F(x, y)\, dx\, dy$. A change of variables for this integral is determined by a **transformation** T from the uv-plane to the xy-plane; that is, a function that associates with the point (u, v) a point $T(u, v) = (x, y)$ given by equations of the form

$$x = f(u, v), \qquad y = g(u, v). \tag{2}$$

The point (x, y) is called the **image** of the point (u, v) under the transformation T. If no two different points in the uv-plane have the same image point in the xy-plane, then the transformation T is said to be **one-to-one.** In this case it may be possible to solve Equations (2) for u and v in terms of x and y and thus obtain the equations

$$u = h(x, y), \qquad v = k(x, y) \tag{3}$$

of the **inverse transformation** T^{-1} from the xy-plane to the uv-plane.

It is often convenient to visualize the transformation T geometrically in terms of its u-curves and v-curves. The **u-curves** of T are the images in the xy-plane of horizontal lines in the uv-plane; the **v-curves** are the images of vertical lines in the uv-plane. Note that the image under T of a rectangle that is bounded by horizontal and vertical lines in the uv-plane is a *curvilinear figure* bounded by u-curves and v-curves in the xy-plane. This phenomenon is shown in Figure 15.66. If we know the Equations (3) of the inverse transformation, then we can find the u-curves and v-curves quite simply by writing the equations

$$k(x, y) = C_1 \quad \text{and} \quad h(x, y) = C_2,$$

respectively, where C_1 and C_2 are constants.

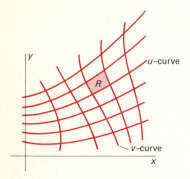

15.66 The transformation T turns the rectangle S into the curvilinear figure R.

15.67 The u-curves and v-curves of Example 1.

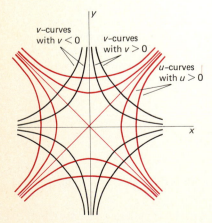

EXAMPLE 1 Determine the u-curves and the v-curves of the transformation T whose inverse T^{-1} is specified by the equations $u = xy$, $v = x^2 - y^2$.

Solution The v-curves are the rectangular hyperbolas

$$xy = u = C_1 \qquad \text{(constant)},$$

while the u-curves are the hyperbolas

$$x^2 - y^2 = v = C_2 \qquad \text{(constant)}.$$

These two families of hyperbolas are shown in Figure 15.67.

CHAP. 15: Multiple Integrals

Now we describe the change of variables in a double integral that corresponds to the transformation T specified by Equations (2). Let the region R in the xy-plane be the image under T of the region S in the uv-plane. Let $F(x, y)$ be continuous on R. Let $\{S_1, S_2, S_3, \ldots, S_n\}$ be an inner partition of S into rectangles each having dimensions Δu by Δv. Each rectangle S_i is transformed by T into a curvilinear figure R_i in the xy-plane, as shown in Figure 15.68. The images $\{R_1, R_2, R_3, \ldots, R_n\}$ under T of the S_i then constitute an inner partition of the region R (into curvilinear figures rather than rectangles).

Let (u_i^*, v_i^*) be the lower left-hand corner point of S_i and write $(x_i^*, y_i^*) = (f(u_i^*, v_i^*), g(u_i^*, v_i^*))$ for its image under T. The u-curve through (x_i^*, y_i^*) has velocity vector

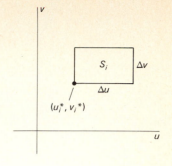

$$\mathbf{t}_u = \mathbf{i}f_u(u_i^*, v_i^*) + \mathbf{j}g_u(u_i^*, v_i^*) = \frac{\partial x}{\partial u}\mathbf{i} + \frac{\partial y}{\partial u}\mathbf{j},$$

while the v-curve through (x_i^*, y_i^*) has velocity vector

$$\mathbf{t}_v = \mathbf{i}f_v(u_i^*, v_i^*) + \mathbf{j}g_v(u_i^*, v_i^*) = \frac{\partial x}{\partial v}\mathbf{i} + \frac{\partial y}{\partial v}\mathbf{j}.$$

Thus we can approximate the curvilinear figure R_i by a parallelogram P_i with edges that are "copies" of the vectors $\mathbf{t}_u \Delta u$ and $\mathbf{t}_v \Delta v$. These edges and the approximating parallelogram P_i also appear in Figure 15.68.

Now the area ΔA_i of R_i is also approximated by the area of the parallelogram P_i, and we can compute the latter. Indeed,

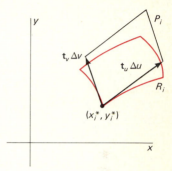

15.68 The effect of the transformation T; we estimate the area of $R_i = T(S_i)$ by computing the area of P_i.

$$\Delta A_i \approx \text{area}\,(P_i) = |\mathbf{t}_u \Delta u \times \mathbf{t}_v \Delta v| = |\mathbf{t}_u \times \mathbf{t}_v|\, \Delta u\, \Delta v.$$

But

$$\mathbf{t}_u \times \mathbf{t}_v = \begin{vmatrix} \mathbf{i} & \mathbf{j} & \mathbf{k} \\ \dfrac{\partial x}{\partial u} & \dfrac{\partial y}{\partial u} & 0 \\ \dfrac{\partial x}{\partial v} & \dfrac{\partial y}{\partial v} & 0 \end{vmatrix} = \begin{vmatrix} \dfrac{\partial x}{\partial u} & \dfrac{\partial x}{\partial v} \\ \dfrac{\partial y}{\partial u} & \dfrac{\partial y}{\partial v} \end{vmatrix}\mathbf{k}.$$

The two-by-two determinant above is called the **Jacobian** of the transformation T, after the German mathematician Carl Jacobi (1804–1851), who first investigated general changes of variables in multiple integrals. The Jacobian of the transformation T is a function of u and v, and we denote it by $J_T = J_T(u, v)$. Thus

$$J_T(u, v) = \begin{vmatrix} f_u(u, v) & f_v(u, v) \\ g_u(u, v) & g_v(u, v) \end{vmatrix}. \tag{4}$$

A common and particularly suggestive notation for the Jacobian is

$$J_T = \frac{\partial(x, y)}{\partial(u, v)}.$$

The computation above shows that the area ΔA_i of R_i is given approximately by

$$\Delta A_i \approx |J_T(u_i^*, v_i^*)|\, \Delta u\, \Delta v.$$

Therefore, when we set up Riemann sums for approximating double integrals, we find that

$$\iint\limits_{R} F(x, y)\, dx\, dy \approx \sum_{i=1}^{n} F(x_i^*, y_i^*)\, \Delta A_i$$

$$\approx \sum_{i=1}^{n} F(f(u_i^*, v_i^*), g(u_i^*, v_i^*)) \left| J_T(u_i^*, v_i^*) \right| \Delta u\, \Delta v$$

$$\approx \iint\limits_{S} F(f(u, v), g(u, v)) \left| J_T(u, v) \right| du\, dv.$$

This discussion is, in fact, an outline of a proof of the following general **change of variables** theorem. We assume that T transforms the bounded region S in the uv-plane into the bounded region R in the xy-plane, and that T is one-to-one from the interior of S to the interior of R. Suppose that the function $F(x, y)$ and the first order partial derivatives of the component functions of T are continuous functions. Finally, in order to assure the existence of the indicated double integrals, we assume that the boundary of each of the regions S and R consists of finitely many piecewise-smooth simple closed curves.

Theorem *Change of Variables*

If the transformation T with component functions $x = f(u, v)$, $y = g(u, v)$ satisfies the above conditions, then

$$\iint\limits_{R} F(x, y)\, dx\, dy = \iint\limits_{S} F(f(u, v), g(u, v)) \left| J_T(u, v) \right| du\, dv. \qquad (5)$$

If we write $G(u, v) = F(f(u, v), g(u, v))$, then the change of variables formula in (5) becomes

$$\iint\limits_{R} F(x, y)\, dx\, dy = \iint\limits_{S} G(u, v) \left| \frac{\partial(x, y)}{\partial(u, v)} \right| du\, dv. \qquad (5a)$$

Thus we formally transform $\iint_R F(x, y)\, dA$ by replacing the variables x and y by $f(u, v)$ and $g(u, v)$, respectively, and writing

$$dA = \left| \frac{\partial(x, y)}{\partial(u, v)} \right| du\, dv$$

for the area element in terms of u and v. Note the analogy between (5a) and the single-variable formula (1). In fact, if $g'(x) \neq 0$ on $[c, d]$, and we denote by α the smaller and by β the larger of the two limits c and d in (1), then it becomes

$$\int_a^b f(x)\, dx = \int_\alpha^\beta f(g(u)) \left| g'(u) \right| du. \qquad (1a)$$

Thus the Jacobian in (5a) plays the role of the derivative $g'(u)$ in Formula (1a).

For example, suppose that the transformation T from the $r\theta$-plane to the xy-plane is determined by the polar coordinates equations

$$x = f(r, \theta) = r\cos\theta, \qquad y = g(r, \theta) = r\sin\theta.$$

Then the Jacobian of T is

$$\frac{\partial(x, y)}{\partial(r, \theta)} = \begin{vmatrix} \cos\theta & -r\sin\theta \\ \sin\theta & r\cos\theta \end{vmatrix} = r > 0,$$

so Equation (5) or (5a) reduces to the familiar formula

$$\iint\limits_{R} F(x, y)\, dx\, dy = \iint\limits_{S} F(r\cos\theta, r\sin\theta)\, r\, dr\, d\theta.$$

Given a particular double integral $\iint_R f(x, y)\, dx\, dy$, how do we find a *productive* change of variables? One standard approach is to choose a transformation T such that the boundary of R consists of u-curves and v-curves. In case it is more convenient to express u and v in terms of x and y, we can first compute $\partial(u, v)/\partial(x, y)$ explicitly and then find the needed Jacobian $\partial(x, y)/\partial(u, v)$ from the formula

$$\frac{\partial(x, y)}{\partial(u, v)} \cdot \frac{\partial(u, v)}{\partial(x, y)} = 1. \tag{6}$$

Formula (6) is a consequence of the chain rule (see Problem 18).

EXAMPLE 2 Suppose that R is the plane region that is bounded by the hyperbolas

$$xy = 1, \quad xy = 3 \quad \text{and} \quad x^2 - y^2 = 1, \quad x^2 - y^2 = 4.$$

Find the polar moment of inertia $I_0 = \iint_R (x^2 + y^2)\, dx\, dy$ of this region.

Solution The hyperbolas bounding R are v-curves and u-curves if $u = xy$ and $v = x^2 - y^2$, as in Example 1. We can most easily write $x^2 + y^2$ in terms of u and v by first noting that

$$4u^2 + v^2 = 4x^2y^2 + (x^2 - y^2)^2 = (x^2 + y^2)^2,$$

so that $x^2 + y^2 = \sqrt{4u^2 + v^2}$. Now

$$\frac{\partial(u, v)}{\partial(x, y)} = \begin{vmatrix} y & x \\ 2x & -2y \end{vmatrix} = -2(x^2 + y^2).$$

Hence Equation (6) gives

$$\frac{\partial(x, y)}{\partial(u, v)} = -\frac{1}{2(x^2 + y^2)} = -\frac{1}{2\sqrt{4u^2 + v^2}}.$$

We are now ready to apply the theorem, with the regions S and R as shown in Figure 15.69. With $F(x, y) = x^2 + y^2$, Formula (5a) gives

$$I_0 = \iint\limits_{R} (x^2 + y^2)\, dx\, dy = \int_1^4 \int_1^3 \sqrt{4u^2 + v^2} \cdot \frac{1}{2\sqrt{4u^2 + v^2}}\, du\, dv$$

$$= \int_1^4 \int_1^3 \frac{1}{2}\, du\, dv = 3.$$

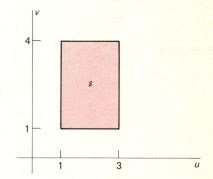

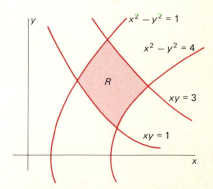

15.69 The transformation T and the new region S constructed in Example 2.

Our next example has an important application as its motivation. Consider an engine with an operating cycle that consists of alternate expansion and compression of the gas in a piston. During a cycle the point (P, V)

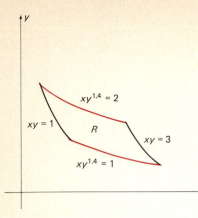

15.70 Find the area of the region R (Example 3).

giving the pressure and volume of this gas traces a closed curve in the *PV*-plane. The work done by the engine—ignoring friction and related losses—is then equal (in appropriate units) to the *area enclosed by this curve*, called the *indicator diagram* of the engine. In an ideal Carnot engine, the indicator diagram consists of two *isothermals* $xy = a$, $xy = b$ and two *adiabatics* $xy^\gamma = c$, $xy^\gamma = d$, where γ is the heat capacity ratio of the working gas in the piston. A typical value is $\gamma = 1.4$.

EXAMPLE 3   Find the area of the region R bounded by the curves $xy = 1$, $xy = 3$, and $xy^{1.4} = 1$, $xy^{1.4} = 2$ (see Figure 15.70).

Solution   To force the given curves to be *u*-curves and *v*-curves, we define our change of variables transformation by means of the equations $u = xy$ and $v = xy^{1.4}$. Then

$$\frac{\partial(u, v)}{\partial(x, y)} = \begin{vmatrix} y & x \\ y^{1.4} & (1.4)xy^{0.4} \end{vmatrix} = (0.4)xy^{1.4} = (0.4)v.$$

So

$$\frac{\partial(x, y)}{\partial(u, v)} = \frac{1}{\partial(u, v)/\partial(x, y)} = \frac{2.5}{v}.$$

Consequently, the change of variables theorem gives the formula

$$A = \iint_R 1 \, dx \, dy = \int_1^2 \int_1^3 \frac{2.5}{v} \, du \, dv = 5 \ln 2.$$

The change of variables formula for triple integrals is similar to Formula (5). Let S and R be regions that correspond under the one-to-one transformation T from *uvw*-space to *xyz*-space, where the coordinate functions that comprise T are

$$x = f(u, v, w), \qquad y = g(u, v, w), \qquad z = h(u, v, w). \tag{7}$$

The Jacobian of T is

$$J_T(u, v, w) = \frac{\partial(x, y, z)}{\partial(u, v, w)} = \begin{vmatrix} \dfrac{\partial x}{\partial u} & \dfrac{\partial x}{\partial v} & \dfrac{\partial x}{\partial w} \\[2mm] \dfrac{\partial y}{\partial u} & \dfrac{\partial y}{\partial v} & \dfrac{\partial y}{\partial w} \\[2mm] \dfrac{\partial z}{\partial u} & \dfrac{\partial z}{\partial v} & \dfrac{\partial z}{\partial w} \end{vmatrix}. \tag{8}$$

Then the change of variables formula for triple integrals is

$$\iiint_R F(x, y, z) \, dx \, dy \, dz = \iiint_S G(u, v, w) \left| \frac{\partial(x, y, z)}{\partial(u, v, w)} \right| du \, dv \, dw, \tag{9}$$

where $G(u, v, w) = F(f(u, v, w), g(u, v, w), h(u, v, w))$ is the function obtained from $F(x, y, z)$ by expressing the variables x, y, and z in terms of u, v, and w.

For example, if T is the spherical coordinates transformation given by

$$x = \rho \sin \phi \cos \theta, \qquad y = \rho \sin \phi \sin \theta, \qquad z = \rho \cos \phi,$$

CHAP. 15:   Multiple Integrals

then the Jacobian of T is

$$\frac{\partial(x, y, z)}{\partial(\rho, \phi, \theta)} = \begin{vmatrix} \sin\phi\cos\theta & \rho\cos\phi\cos\theta & -\rho\sin\phi\sin\theta \\ \sin\phi\sin\theta & \rho\cos\phi\sin\theta & \rho\sin\phi\cos\theta \\ \cos\phi & -\rho\sin\phi & 0 \end{vmatrix} = \rho^2\sin\phi.$$

Thus (9) reduces to the familiar formula

$$\iiint_R F(x, y, z)\, dx\, dy\, dz = \iiint_S G(\rho, \phi, \theta)\, \rho^2\sin\phi\, d\rho\, d\phi\, d\theta$$

because $\rho^2\sin\phi \geqq 0$ for ϕ in $[0, \pi]$.

EXAMPLE 4 Find the volume of the solid torus R that is obtained by revolving around the z-axis the circular disk $(x - b)^2 + z^2 \leqq a^2$, $a < b$, in the xz-plane.

Solution This is the torus of Example 4 in Section 15-7. Let us write u for the ordinary polar coordinate angle θ, v for the angle ψ of Figure 15.65, and w for distance from the center of the circular disk described by the above inequality. We then define the transformation T by means of the equations

$$x = (b + w\cos v)\cos u, \qquad y = (b + w\cos v)\sin u, \qquad z = w\sin v.$$

Then the solid torus R is the image under T of the region S in uvw-space described by the inequalities $0 \leqq u \leqq 2\pi$, $0 \leqq v \leqq 2\pi$, $0 \leqq w \leqq a$. By a routine computation, we find that the Jacobian of T is

$$\frac{\partial(x, y, z)}{\partial(u, v, w)} = w(b + w\cos v).$$

Hence Formula (9) with $F(x, y, z) \equiv 1$ gives

$$V = \iiint_R dx\, dy\, dz = \int_0^{2\pi}\int_0^{2\pi}\int_0^a (bw + w^2\cos v)\, dw\, du\, dv$$

$$= 2\pi\int_0^{2\pi} (\tfrac{1}{2}a^2 b + \tfrac{1}{3}a^3\cos v)\, dv = 2\pi^2 a^2 b,$$

which agrees with the value $V = (2\pi b)(\pi a^2)$ given by Pappus' theorem (Section 5-6).

15-8 PROBLEMS

In each of Problems 1–6, solve for x and y in terms of u and v and then compute the Jacobian $\partial(x, y)/\partial(u, v)$.

1 $u = x + y$, $v = x - y$.

2 $u = x - 2y$, $v = 3x + y$.

3 $u = xy$, $v = \dfrac{y}{x}$.

4 $u = 2(x^2 + y^2)$, $v = 2(x^2 - y^2)$.

5 $u = x + 2y^2$, $v = x - 2y^2$.

6 $u = \dfrac{2x}{x^2 + y^2}$, $v = \dfrac{-2y}{x^2 + y^2}$.

7 Let R be the parallelogram bounded by the lines $x + y = 1$, $x + y = 2$ and $2x - 3y = 2$, $2x - 3y = 5$. Substitute $u = x + y$, $v = 2x - 3y$ to find the area $A = \iint_R dx\, dy$ of R.

8 Substitute $u = xy$, $v = y/x$ to find the area of the first quadrant region bounded by the lines $y = x$, $y = 2x$ and the hyperbolas $xy = 1$, $xy = 2$.

9 Substitute $u = xy$, $v = xy^3$ to find the area of the region in the first quadrant bounded by the curves $xy = 2$, $xy = 4$ and $xy^3 = 3$, $xy^3 = 6$.

10 Find the area of the region in the first quadrant bounded by the curves $y = x^2$, $y = 2x^2$ and $x = y^2$, $x = 4y^2$. (*Suggestion:* Let $y = ux^2$ and $x = vy^2$.)

11 Use the method of Problem 10 to find the area of the region in the first quadrant bounded by the curves $y = x^3$, $y = 2x^3$ and $x = y^3$, $x = 4y^3$.

12 Let R be the region in the first quadrant bounded by the circles $x^2 + y^2 = 2x$, $x^2 + y^2 = 6x$ and the circles $x^2 + y^2 = 2y$, $x^2 + y^2 = 8y$. Use the transformation $u = 2x/(x^2 + y^2)$, $v = 2y/(x^2 + y^2)$ to evaluate the integral $\iint_R (x^2 + y^2)^{-2} \, dx \, dy$.

13 Use elliptical coordinates $x = 3r \cos \theta$, $y = 2r \sin \theta$ to find the volume of the region that is bounded by the xy-plane, the paraboloid $z = x^2 + y^2$, and the elliptical cylinder $x^2/9 + y^2/4 = 1$.

14 Let R be the solid ellipsoid with outer boundary surface $x^2/a^2 + y^2/b^2 + z^2/c^2 = 1$. Use the transformation $x = au$, $y = bv$, $z = cw$ to show that the volume of this ellipsoid is $V = \iiint_R 1 \, dx \, dy \, dz = \frac{4}{3}\pi abc$.

15 Find the volume of the region in the first octant that is bounded by the hyperbolic cylinders $xy = 1$, $xy = 4$; $xz = 1$, $xz = 9$; $yz = 4$, $yz = 9$. (*Suggestion:* Let $u = xy$, $v = xz$, $w = yz$, and note that $uvw = x^2y^2z^2$.)

16 Use the transformation $x = (r/t) \cos \theta$, $y = (r/t) \sin \theta$, $z = r^2$ to find the volume of the region R that lies between the paraboloids $z = x^2 + y^2$, $z = 4(x^2 + y^2)$ and also between the planes $z = 1$, $z = 4$.

17 Let R be the rotated elliptical region bounded by the graph of $x^2 + xy + y^2 = 3$. Let $x = u + v$ and $y = u - v$. Show that

$$\iint_R e^{-(x^2 + xy + y^2)} \, dx \, dy = 2 \iint_S e^{-(3u^2 + v^2)} \, du \, dv.$$

Then substitute $u = r \cos \theta$, $v = \sqrt{3}(r \sin \theta)$ to evaluate the latter integral.

18 Derive Relation (6) between the Jacobians of a transformation and its inverse from the chain rule and the following property of determinants:

$$\begin{vmatrix} a_1 & b_1 \\ c_1 & d_1 \end{vmatrix} \cdot \begin{vmatrix} a_2 & b_2 \\ c_2 & d_2 \end{vmatrix} = \begin{vmatrix} a_1 a_2 + b_1 c_2 & a_1 b_2 + b_1 d_2 \\ a_2 c_1 + c_2 d_1 & b_2 c_1 + d_1 d_2 \end{vmatrix}.$$

19 Change to spherical coordinates to show that, for $k > 0$,

$$\int_{-\infty}^{+\infty} \int_{-\infty}^{+\infty} \int_{-\infty}^{+\infty} (x^2 + y^2 + z^2)^{1/2} e^{-k(x^2 + y^2 + z^2)} \, dx \, dy \, dz$$

$$= \frac{2\pi}{k^2}.$$

CHAPTER 15 REVIEW: Definitions, Concepts, Results

Use the list below as a guide to concepts that you may need to review.

1 Definition of the double integral as a limit of Riemann sums

2 Evaluation of double integrals by iterated single integrals

3 Use of the double integral to find the volume between two surfaces above a given plane region

4 Transformation of the double integral $\iint_R f(x, y) \, dA$ into polar coordinates

5 Application of double integrals to find mass, centroids, and moments of inertia of plane laminae

6 Definition of the triple integral as a limit of Riemann sums

7 Evaluation of triple integrals by iterated single integrals

8 Application of triple integrals to find volume, mass, centroids, and moments of inertia

9 Transformation of the triple integral $\iiint_T f(x, y, z) \, dV$ into cylindrical and spherical coordinates

10 The surface area of a parametric surface

11 The area of a surface $z = f(x, y)$ for (x, y) in the plane region R

12 The Jacobian of a transformation of coordinates

13 The transformation of a double or triple integral corresponding to a given change of variables

MISCELLANEOUS PROBLEMS

In each of Problems 1–5, evaluate the given integral by first reversing the order of integration.

1 $\displaystyle\int_0^1 \int_{y^{1/3}}^1 \frac{dx \, dy}{\sqrt{1 + x^2}}$.

2 $\displaystyle\int_0^1 \int_y^1 \frac{\sin x}{x} \, dx \, dy$.

3 $\displaystyle\int_0^1 \int_x^1 e^{-y^2} \, dy \, dx$.

4 $\displaystyle\int_0^8 \int_{x^{2/3}}^4 x \cos y^4 \, dy \, dx$.

5 $\displaystyle\int_0^4 \int_{\sqrt{y}}^2 \frac{y e^{x^2}}{x^3} \, dx \, dy$.

6 The double integral $\int_0^\infty \int_x^\infty y^{-1} e^{-y} \, dy \, dx$ is an improper integral over the unbounded region in the first quadrant bounded by the lines $y = x$ and $x = 0$. Assuming the validity of reversing the order of integration, evaluate this integral by integrating first with respect to x.

7 Find the volume of the solid T that lies under the paraboloid $z = x^2 + y^2$ and over the triangle R in the xy-plane having vertices $(0, 0, 0)$, $(1, 1, 0)$, and $(2, 0, 0)$.

8 Find by integration in cylindrical coordinates the volume bounded by the paraboloids $z = 2x^2 + 2y^2$ and $z = 48 - x^2 - y^2$.

9 Use integration in spherical coordinates to find the volume and centroid of the solid region that is inside the sphere $\rho = 3$, under the cone $\phi = \pi/3$, and above the xy-plane $\phi = \pi/2$.

10 Find the volume of the solid bounded by the elliptic paraboloids $z = x^2 + 3y^2$ and $z = 8 - x^2 - 5y^2$.

11 Find the volume bounded by the paraboloid $y = x^2 + 3z^2$ and the parabolic cylinder $y = 4 - z^2$.

12 Find the volume of the region that is bounded by the parabolic cylinders $z = x^2$, $z = 2 - x^2$, and the planes $y = 0$, $y + z = 4$.

13 Find the volume of the region bounded by the elliptical cylinder $y^2 + 4z^2 = 4$ and the planes $x = 0$, $x = y + 2$.

14 Show that the volume of the solid bounded by the elliptic cylinder $x^2/a^2 + y^2/b^2 = 1$ and the planes $z = 0$, $z = h + x$ (where $h > a > 0$) is $V = \pi abh$.

15 Let R be the first-quadrant region bounded by the line $y = x$ and the curve $x^4 + x^2y^2 = y^2$. Use polar coordinates to evaluate the integral $\iint_R (1 + x^2 + y^2)^{-2}\, dA$.

In each of Problems 16–20, find the mass and centroid of a plane lamina having the indicated shape and density ρ.

16 The region bounded by $y = x^2$ and $x = y^2$; $\rho = x^2 + y^2$.

17 The region bounded by $y^2 = \frac{1}{2}x$ and $y^2 = x - 4$; $\rho = y^2$.

18 The region between $y = \ln x$ and the x-axis over the interval $1 \leq x \leq 2$; $\rho = 1/x$.

19 The circle bounded by $r = 2\cos\theta$; $\rho = k$ (a constant).

20 The region of Problem 19; $\rho = r$.

21 Find the centroid of the region in the xy-plane bounded by the x-axis and the parabola $y = 4 - x^2$.

22 Find the volume of the solid that lies under the parabolic cylinder $z = x^2$ and over the triangle in the xy-plane bounded by the x-axis, the y-axis, and the line $x + y = 1$.

23 Use cylindrical coordinates to find the volume of the ice cream cone bounded above by the sphere $x^2 + y^2 + z^2 = 5$ and below by the cone $z = 2(x^2 + y^2)^{1/2}$.

24 Find the volume and centroid of the ice cream cone that is bounded above by the sphere $\rho = a$ and below by the cone $\phi = \pi/3$.

25 Find the moment of inertia about its natural axis of a homogeneous solid circular cone with mass M and base radius a.

26 Find the mass of the first octant of the ball $\rho \leq a$ if its density function is $\delta = xyz$.

27 Find the moment of inertia about the x-axis of the homogeneous solid ellipsoid bounded by $x^2/a^2 + y^2/b^2 + z^2/c^2 = 1$.

28 Find the volume of the region in the first octant that is bounded by the sphere $\rho = a$, the cylinder $r = a$, the plane $z = a$, the xz-plane, and the yz-plane.

29 Find the moment of inertia about the z-axis of the homogeneous region that lies inside both the sphere $\rho = 2$ and the cylinder $r = 2\cos\theta$.

In each of Problems 30–32, a volume is generated by revolving a plane region R around an axis. To find this volume, set up a *double* integral over R by revolving an area element dA around the indicated axis to generate a volume element dV.

30 Find the volume of the solid that is obtained by revolving around the y-axis the area within the circle $r = 2a\cos\theta$.

31 Find the volume of the solid obtained by revolving the area enclosed by the cardioid $r = 1 + \cos\theta$ around the x-axis.

32 Find the volume of the solid torus that is obtained by revolving the circle $r \leq a$ about the line $x = -b$, $|b| \geq a > 0$.

33 This problem deals with the oblique segment of a paraboloid discussed in Example 2 of Section 15-5 (shown in Figure 15.49).

(a) Show that its centroid is the point $C(0, \frac{1}{2}, \frac{7}{4})$.

(b) Show that the center of the elliptical upper "base" of the solid paraboloid is the point $Q(0, \frac{1}{2}, \frac{5}{2})$.

(c) Verify that the point $V(0, \frac{1}{2}, \frac{1}{4})$ is the point where the tangent plane to the paraboloid is parallel to the upper base. The point V is called the *vertex* of the oblique segment, and the line segment VQ is its *principal axis*.

(d) Show that C lies on the principal axis *two-thirds* of the way from the vertex to the upper base. Archimedes showed that this is true for any segment cut off from a paraboloid by a plane. This was a key step in his determination of the equilibrium position of a floating right circular paraboloid, in terms of its density and dimensions. The possible positions are shown in Figure 15.71. The principles he introduced

15.71 How a uniform solid paraboloid might float.

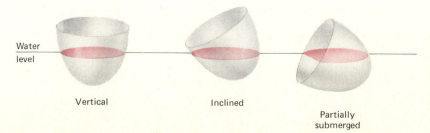

Water level

Vertical   Inclined   Partially submerged

for the solution of this problem are still important in naval architecture.

Problems 34–40 deal with average distance. The **average distance** $\bar{d}$ of the point (x_0, y_0) from the points of the plane region R with area A is defined to be

$$\bar{d} = \frac{1}{A} \iint_R [(x - x_0)^2 + (y - y_0)^2]^{1/2} \, dA.$$

The average distance of a point (x_0, y_0, z_0) from the points of a space region is defined similarly.

34 Show that the average distance of the points of a disk of radius a from its center is $2a/3$.

35 Show that the average distance of the points of a disk of radius a from a fixed point on its boundary is $\bar{d} = 32a/9\pi$.

36 A circle of radius 1 is interior to and tangent to a circle of radius 2. Find the average distance of the point of tangency from the points that lie between the two circles.

37 Show that the average distance of the points of a spherical ball of radius a from its center is $3a/4$.

38 Show that the average distance of the points of a spherical ball of radius a from a fixed point on its surface is $6a/5$.

39 A sphere of radius 1 is interior to and tangent to a sphere of radius 2. Find the average distance of the point of tangency from the set of all points between the two spheres.

40 A right circular cone has radius R and height H. Find the average distance of the points of the cone from its vertex.

41 Find the surface area of the part of the paraboloid $z = 10 - r^2$ that lies between the two planes $z = 1$ and $z = 6$.

42 Find the surface area of the part of the surface $z = y^2 - x^2$ that is inside the cylinder $x^2 + y^2 = 4$.

43 Deduce from the formula of Problem 14 in Section 15-7 that the surface area of the zone on the sphere $\rho = a$ between the planes $z = z_1$ and $z = z_2$ (where $-a \leq z_1 < z_2 \leq a$) is $A = 2\pi a h$ where $h = z_2 - z_1$.

44 Find the surface area of the part of the sphere $\rho = 2$ that is inside the cylinder $x^2 + y^2 = 2x$.

45 A square hole with side length 2 is cut through a cone of height and base radius 2; the center line of the hole is the axis of symmetry of the cone. Find the area of the surface removed from the cone.

46 Numerically approximate the surface area of the part of the parabolic cylinder $z = \frac{1}{2}x^2$ that lies inside the cylinder $x^2 + y^2 = 1$.

47 A "fence" of variable height $h(t)$ stands above the plane curve $(x(t), y(t))$. Thus the fence has the parametrization $x = x(t), y = y(t), z = z$ for $a \leq t \leq b, 0 \leq z \leq h(t)$. Apply Formula (8) of Section 15-7 to show that the area of the

fence is

$$A = \int_a^b \int_0^{h(t)} \left[\left(\frac{dx}{dt} \right)^2 + \left(\frac{dy}{dt} \right)^2 \right]^{1/2} dz \, dt.$$

48 Apply the formula of Problem 47 to compute the area of the part of the cylinder $r = a \sin \theta$ that lies inside the sphere $r^2 + z^2 = a^2$.

49 Find the polar moment of inertia of the first-quadrant region of constant density δ that is bounded by the hyperbolas $xy = 1$, $xy = 3$ and $x^2 - y^2 = 1$, $x^2 - y^2 = 4$.

50 Substitute $u = x - y, v = x + y$ to evaluate

$$\iint_R \exp\left(\frac{x - y}{x + y} \right) dx \, dy,$$

where R is bounded by the coordinate axes and the line $x + y = 1$.

51 Use ellipsoidal coordinates $x = a\rho \sin \phi \cos \theta$, $y = b\rho \sin \phi \sin \theta$, $z = c\rho \cos \phi$ to find the mass of the solid ellipsoid $x^2/a^2 + y^2/b^2 + z^2/c^2 = 1$ if its density at the point (x, y, z) is $\delta = 1 - x^2/a^2 - y^2/b^2 - z^2/c^2$.

52 Let R be the first-quadrant region bounded by the lemniscates $r^2 = 3 \cos 2\theta$, $r^2 = 4 \cos 2\theta$, and $r^2 = 3 \sin 2\theta$, $r^2 = 4 \sin 2\theta$ (see Figure 15.72). Show that its area is $A = (10 - 7\sqrt{2})/4$. (*Suggestion:* Define the transformation T from the uv-plane to the $r\theta$-plane by $r^2 = u^{1/2} \cos 2\theta$, $r^2 = v^{1/2} \sin 2\theta$. Show first that

$$r^4 = \frac{uv}{u + v}, \qquad \theta = \frac{1}{2} \tan^{-1} \frac{u^{1/2}}{v^{1/2}}.$$

Then show that $\partial(r, \theta)/\partial(u, v) = -1/[16r(u + v)^{3/2}]$.)

15.72 The region R of Problem 52.

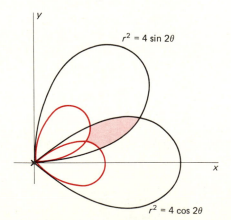

$r^2 = 4 \sin 2\theta$

$r^2 = 4 \cos 2\theta$

Vector Analysis

16

Introduction

This chapter is devoted to topics in the calculus of vector fields that are important in science and engineering. A **vector field** defined on a region T in space is a vector-valued function $\mathbf{F}$ that associates with each point (x, y, z) of T a vector

$$\mathbf{F}(x, y, z) = \mathbf{i}P(x, y, z) + \mathbf{j}Q(x, y, z) + \mathbf{k}R(x, y, z).$$

Here are some typical examples of such vector fields:

1 A force field, such as the gravitational force

$$\mathbf{F}(x, y, z) = -\frac{GM}{r^3}\,\mathbf{r} = \frac{-GM(x\mathbf{i} + y\mathbf{j} + z\mathbf{k})}{(x^2 + y^2 + z^2)^{3/2}}$$

that a particle of mass M at the origin exerts upon a unit mass particle located at the point $\mathbf{r} = \langle x, y, z \rangle$. (See Figure 16.1.)

2 A velocity field $\mathbf{v}(x, y, z)$ of a steady fluid flow; $\mathbf{v}(x, y, z)$ is the velocity *vector* of the fluid at the point (x, y, z).

3 The gradient vector field of a temperature function $T(x, y, z)$, in which case the gradient vector $\nabla T(x, y, z)$ points *opposite* to the direction in which heat is flowing at (x, y, z). Its length $|\nabla T|$ is proportional to the rate of heat flow. In general, any scalar-valued function f has associated with it a vector field $\nabla f(x, y, z)$.

In Section 16-2 we define line integrals, which are used (for example) to compute the work done by a force field in moving a particle along a curved path. In Section 16-4 we discuss surface integrals, which are used (for example) to compute the rate at which a fluid with a known velocity vector field is moving across a given surface. The three basic integral theorems of vector analysis—Green's theorem (Section 16-3), the divergence theorem (Section 16-5), and Stokes' theorem (Section 16-6)—play much the same role for line and surface integrals that the Fundamental Theorem of Calculus plays for ordinary single-variable integrals.

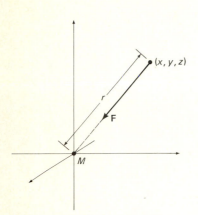

16.1 A mass M at the origin produces a gravitational force field $\mathbf{F}$ elsewhere in space.

Line Integrals

To motivate the definition of the line integral, we imagine a thin wire, shaped like the smooth curve C with end points A and B, as in Figure 16.2. Suppose that the wire has variable density given at the point (x, y, z) by the known continuous function $f(x, y, z)$, in units such as grams per (linear) centimeter. Let

$$x = x(t), \quad y = y(t), \quad z = z(t), \qquad t \text{ in } [a, b], \tag{1}$$

be a smooth parametrization of the curve C, with $t = a$ corresponding to the initial point A of the curve and $t = b$ to its terminal point B.

In order to *approximate* the total mass M of our curved wire, we begin with a partition

$$a = t_0 < t_1 < t_2 < \cdots < t_{n-1} < t_n = b$$

of $[a, b]$ into n equal subintervals, each with length $\Delta t = (b - a)/n$. These subdivision points of $[a, b]$ produce, via our parametrization, a physical subdivision of the wire into short curved segments, as shown in Figure 16.3.

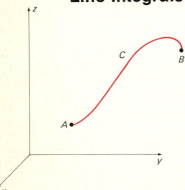

16.2 A wire of variable density in the shape of the smooth curve C.

CHAP. 16: Vector Analysis

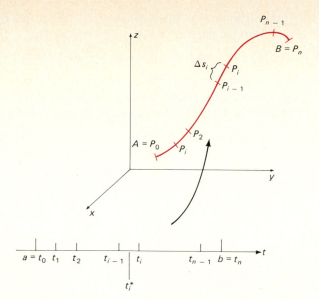

16.3 The partition of the interval $[a, b]$ determines a related partition of the curve C into short arcs.

We let $P_i = (x(t_i), y(t_i), z(t_i))$ (for $i = 0, 1, 2, \ldots, n$) be the resulting subdivision points of C. Note that $A = P_0$ and that $B = P_n$.

From our study of arc length in Sections 10-2 and 13-4, we know that the arc length Δs_i of the segment of C from P_{i-1} to P_i is

$$\Delta s_i = \int_{t_{i-1}}^{t_i} \sqrt{[x'(t)]^2 + [y'(t)]^2 + [z'(t)]^2} \, dt$$
$$= \sqrt{[x'(t_i^*)]^2 + [y'(t_i^*)]^2 + [z'(t_i^*)]^2} \, \Delta t \qquad (2)$$

for some number t_i^* in the interval $[t_{i-1}, t_i]$—a consequence of the average value theorem for integrals (Section 4-4).

If we multiply the density at the point (x_i^*, y_i^*, z_i^*) by the length Δs_i of the segment of C containing it, we obtain an estimate of the mass of that segment of C. So, after we sum over all the segments, we have an estimate of the total mass of the wire:

$$M \approx \sum_{i=1}^{n} f(x(t_i^*), y(t_i^*), z(t_i^*)) \, \Delta s_i.$$

The limit of this sum as $\Delta t \to 0$ should be the actual mass M. This is our motivation for the definition of the line integral of the function f along the curve C, denoted by $\int_C f(x, y, z) \, ds$.

Definition *Line Integral with respect to Arc Length*

Suppose that the function $f(x, y, z)$ is continuous at each point of the smooth parametric curve C from A to B, as given in (1). Then the **line integral of f along C from A to B with respect to arc length** is defined to be

$$\int_C f(x, y, z) \, ds = \lim_{\Delta t \to 0} \sum_{i=1}^{n} f(x(t_i^*), y(t_i^*), z(t_i^*)) \, \Delta s_i. \qquad (3)$$

When we substitute (2) into (3), we recognize the result as the limit of a Riemann sum; therefore

$$\int_C f(x, y, z) \, ds = \int_a^b f(x(t), y(t), z(t))([x'(t)]^2 + [y'(t)]^2 + [z'(t)]^2)^{1/2} \, dt. \qquad (4)$$

Thus we may evaluate the line integral $\int_C f(x, y, z)\, ds$ by expressing everything in terms of the parameter t, including the symbolic arc length element

$$ds = \sqrt{[x'(t)]^2 + [y'(t)]^2 + [z'(t)]^2}\, dt. \tag{5}$$

The result—the right-hand side of Equation (4)—is an **ordinary integral with respect to the single real variable** t. Line integrals along plane curves are obtained by deleting z wherever it appears in the formulas above.

Let us now return to our physical wire, and denote its density function by the more usual $\rho(x, y, z)$. The mass of a small piece of length Δs is $\Delta m = \rho\, \Delta s$, so we write

$$dm = \rho(x, y, z)\, ds$$

for its (symbolic) element of mass. Then the **mass** M of the wire and its **moments** M_{yz}, M_{xz}, and M_{xy} about the respective coordinate planes are given by the line integrals

$$M = \int_C dm = \int_C \rho\, ds, \qquad\qquad M_{yz} = \int_C x\, dm,$$

$$M_{xz} = \int_C y\, dm, \qquad\qquad M_{xy} = \int_C z\, dm. \tag{6}$$

The **centroid** of the wire is the point $(\bar{x}, \bar{y}, \bar{z})$ where

$$\bar{x} = \frac{M_{yz}}{M}, \qquad \bar{y} = \frac{M_{xz}}{M}, \quad \text{and} \quad \bar{z} = \frac{M_{xy}}{M}.$$

Note the analogy with Equations (3) and (4) in Section 15-5. The **moment of inertia** of the wire about a given axis is

$$I = \int_C w^2\, dm \tag{7}$$

where $w = w(x, y, z)$ denotes the perpendicular distance from the point (x, y, z) to the axis in question.

EXAMPLE 1 Find the centroid of a wire with density $\rho = kz$ if it has the shape of the helix C with parametrization

$$x = 3\cos t, \quad y = 3\sin t, \quad z = 4t, \quad 0 \leq t \leq \pi.$$

Solution The mass element of the wire is

$$dm = \rho\, ds = kz\, ds$$

$$= 4kt\sqrt{(-3\sin t)^2 + (3\cos t)^2 + (4)^2}\, dt = 20kt\, dt.$$

Hence Formulas (6) yield

$$M = \int_C \rho\, ds = \int_0^\pi 20kt\, dt = 10k\pi^2;$$

$$M_{yz} = \int_C \rho x\, ds = \int_0^\pi 60kt\cos t\, dt$$

$$= 60k\Big[\cos t + t\sin t\Big]_0^\pi = -120k;$$

$$M_{xz} = \int_C \rho y\, ds = \int_0^\pi 60kt\sin t\, dt$$

$$= 60k\Big[\sin t - t\cos t\Big]_0^\pi = 60k\pi;$$

and

$$M_{xy} = \int_C \rho z \, ds = \int_0^\pi 80kt^2 \, dt = \frac{80k\pi^3}{3}.$$

Hence

$$\bar{x} = \frac{-120k}{10k\pi^2} \approx -1.22, \qquad \bar{y} = \frac{60k\pi}{10k\pi^2} \approx 1.91, \qquad \bar{z} = \frac{80k\pi^3}{3(10k\pi^2)} \approx 8.38.$$

So the centroid of the wire is approximately the point $(-1.22, 1.91, 8.38)$.

A different type of line integral is obtained by replacing Δs_i in (3) by

$$\Delta x_i = x(t_i) - x(t_{i-1}) = x'(t_i^*) \, \Delta t.$$

The **line of integral f along C with respect to** x is defined to be

$$\int_C f(x, y, z) \, dx = \lim_{\Delta t \to 0} \sum_{i=1}^n f(x(t_i^*), y(t_i^*), z(t_i^*)) \, \Delta x_i;$$

$$\int_C f(x, y, z) \, dx = \int_a^b f(x(t), y(t), z(t)) \, x'(t) \, dt. \tag{8a}$$

Similarly, the line integrals of f along C with respect to y and with respect to z are given by

$$\int_C f(x, y, z) \, dy = \int_a^b f(x(t), y(t), z(t)) \, y'(t) \, dt \tag{8b}$$

and

$$\int_C f(x, y, z) \, dz = \int_a^b f(x(t), y(t), z(t)) \, z'(t) \, dt. \tag{8c}$$

The three integrals in (8) often occur together. If P, Q, and R are continuous functions of the variables x, y, and z, then we write (indeed, *define*)

$$\int_C P \, dx + Q \, dy + R \, dz = \int_C P \, dx + \int_C Q \, dy + \int_C R \, dz. \tag{9}$$

The line integrals in (8) and (9) are evaluated by expressing x, y, z, dx, dy, and dz in terms of t as determined by a parametrization of the curve C; the result is an ordinary single-variable integral.

There is an important difference between the line integral in (4) with respect to arc length s and the line integral in (9) with respect to x, y, and z. Suppose that the *orientation* of the curve C (the direction in which it is traced as t increases) is reversed. Then, because of the terms $x'(t)$, $y'(t)$, and $z'(t)$ in (8), the *sign* of the line integral (9) is changed. But this reversal of orientation does *not* change the value of the line integral in (4). We express this by writing

$$\int_{-C} f \, ds = \int_C f \, ds, \tag{10}$$

in contrast with the formula

$$\int_{-C} P \, dx + Q \, dy + R \, dz = -\int_C P \, dx + Q \, dy + R \, dz. \tag{11}$$

Here the symbol $-C$ denotes the curve C with its orientiation reversed (from B to A, rather than from A to B). It is proved in advanced calculus that for

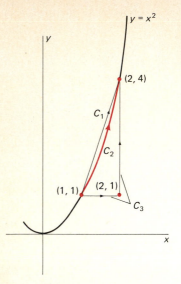

$y = x^2$

$(2, 4)$

C_1

C_2

$(1, 1)$ $(2, 1)$

C_3

16.4 The three arcs of Example 2.

either type of line integral, two one-to-one parametrizations of the curve C that *agree in orientation* will give the same value.

If the curve C consists of finitely many smooth curves joined at corner points, we say that C is **piecewise smooth.** The value of a line integral along C is then defined to be the sum of its values along the smooth segments of C.

EXAMPLE 2 Evaluate the line integral $\int_C y\,dx + 2x\,dy$ for C each of the three curves C_1, C_2, and C_3 from $A(1, 1)$ to $B(2, 4)$ shown in Figure 16.4.

Solution The straight line segment C_1 from A to B can be parametrized by $x = 1 + t$, $y = 1 + 3t$, $0 \leq t \leq 1$. So

$$\int_{C_1} y\,dx + 2x\,dy = \int_0^1 (1 + 3t)(dt) + 2(1 + t)(3\,dt)$$

$$= \int_0^1 (7 + 9t)\,dt = \tfrac{23}{2}.$$

Next, the arc C_2 of the parabola $y = x^2$ from A to B has the parametrization $x = x$, $y = x^2$, $1 \leq x \leq 2$, so

$$\int_{C_2} y\,dx + 2x\,dy = \int_1^2 (x^2)(dx) + 2(x)(2x\,dx) = \int_1^2 5x^2\,dx = \tfrac{35}{3}.$$

Finally, along the straight line segment from $(1, 1)$ to $(2, 1)$ we have $y = 1$ and $dy = 0$. Along the vertical segment from $(2, 1)$ to $(2, 4)$ we have $x = 2$ and $dx = 0$. Therefore

$$\int_{C_3} y\,dx + 2x\,dy = \int_1^2 \left[(1)(dx) + (2x)(0)\right] + \int_1^4 \left[(y)(0) + (4)(dy)\right]$$

$$= \int_1^2 1\,dx + \int_1^4 4\,dy = 13.$$

Example 2 also shows that we may obtain different values for the line integral from A to B if we evaluate it along different curves from A to B. In Theorem 2 below, we give a sufficient condition for the line integral $\int_C P\,dx + Q\,dy + R\,dz$ to have the same value for *all* smooth curves C from A to B and thus that the line integral be *independent of path.*

LINE INTEGRALS AND WORK

Suppose now that $\mathbf{F} = P\mathbf{i} + Q\mathbf{j} + R\mathbf{k}$ is a force field that is defined on a region containing the curve C from A to B. Suppose also that our parametrization

$$x = x(t), \qquad y = y(t), \qquad z = z(t), \qquad t \text{ in } [a, b]$$

of C has a *nonzero* velocity vector

$$\mathbf{v} = \mathbf{i}\frac{dx}{dt} + \mathbf{j}\frac{dy}{dt} + \mathbf{k}\frac{dz}{dt}.$$

The speed associated with this velocity vector is

$$v = |\mathbf{v}| = \sqrt{\left(\frac{dx}{dt}\right)^2 + \left(\frac{dy}{dt}\right)^2 + \left(\frac{dz}{dt}\right)^2}.$$

Recall that the *unit tangent vector* to the curve C is

$$\mathbf{T} = \frac{\mathbf{v}}{v} = \frac{1}{v}\left(\frac{dx}{dt}\mathbf{i} + \frac{dy}{dt}\mathbf{j} + \frac{dz}{dt}\mathbf{k}\right).$$

We want to approximate the work W done by the force field $\mathbf{F}$ in moving a particle along the curve C from A to B. Subdivide C as indicated in Figure 16.5. Think of $\mathbf{F}$ moving the particle from P_{i-1} to P_i, two consecutive division points of C. The work ΔW_i done is approximately the product of the distance Δs_i from P_{i-1} to P_i (measured along C) and the tangential component $\mathbf{F} \cdot \mathbf{T}$ of the force $\mathbf{F}$ at a typical point $(x(t_i^*), y(t_i^*), z(_i^*))$ between P_{i-1} and P_i. Thus

$$\Delta W_i \approx \mathbf{F}(x(t_i^*), y(t_i^*), z(t_i^*)) \cdot \mathbf{T}(t_i^*)\,\Delta s_i,$$

so the total work W is given approximately by

$$W \approx \sum_{i=1}^{n} \mathbf{F}(x(t_i^*), y(t_i^*), z(t_i^*)) \cdot \mathbf{T}(t_i^*)\,\Delta s_i.$$

This approximation suggests that we define the **work** W as

$$W = \int_C \mathbf{F} \cdot \mathbf{T}\, ds. \tag{12}$$

Thus *work is the integral with respect to arc length of the tangential component of the force.*

It is customary to write formally

$$\mathbf{r} = x\mathbf{i} + y\mathbf{j} + z\mathbf{k}, \qquad d\mathbf{r} = \mathbf{i}\,dx + \mathbf{j}\,dy + \mathbf{k}\,dz,$$

and

$$\mathbf{T}\,ds = \left(\frac{dx}{ds}\mathbf{i} + \frac{dy}{ds}\mathbf{j} + \frac{dz}{ds}\mathbf{k}\right)ds = d\mathbf{r}.$$

With this notation, (12) takes the form

$$W = \int_C \mathbf{F} \cdot d\mathbf{r} \tag{13}$$

that is common in physics and engineering texts.

16.5 The component of **F** along C from P_{i-1} to P_i is **F · T**.

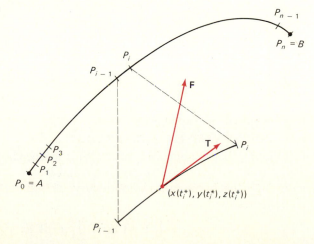

To evaluate the line integral $\int_C \mathbf{F} \cdot \mathbf{T}\, ds$, we express its integrand in terms of the parameter t, as usual. Thus

$$W = \int_C \mathbf{F} \cdot \mathbf{T}\, ds = \int_a^b (P\mathbf{i} + Q\mathbf{j} + R\mathbf{k}) \cdot \frac{1}{v}\left(\frac{dx}{dt}\mathbf{i} + \frac{dy}{dt}\mathbf{j} + \frac{dz}{dt}\mathbf{k}\right) v\, dt$$

$$= \int_a^b \left(P\frac{dx}{dt} + Q\frac{dy}{dt} + R\frac{dz}{dt}\right) dt$$

and so

$$W = \int_C P\, dx + Q\, dy + R\, dz. \tag{14}$$

This computation yields an important relation between the two types of line integrals we have defined in this section.

Theorem 1

Suppose that the vector field $\mathbf{F} = P\mathbf{i} + Q\mathbf{j} + R\mathbf{k}$ has continuous component functions and that $\mathbf{T}$ is the unit tangent to the smooth curve C. Then

$$\int_C \mathbf{F} \cdot \mathbf{T}\, ds = \int_C P\, dx + Q\, dy + R\, dz. \tag{15}$$

Note that, if the orientation of the curve C is reversed, then the sign of the right-hand integral is changed according to Equation (11), while the sign of the left-hand integral is changed because $\mathbf{T}$ is replaced by $-\mathbf{T}$.

EXAMPLE 3 Find the work done by the inverse-square law force field

$$\mathbf{F}(x, y, z) = \frac{k\mathbf{r}}{r^3} = \frac{k(x\mathbf{i} + y\mathbf{j} + z\mathbf{k})}{(x^2 + y^2 + z^2)^{3/2}}$$

in moving a particle along the straight line segment C from $(0, 4, 0)$ to $(0, 4, 3)$.

Solution Along C we have $x = 0$, $y = 4$, and $z = z$. We choose z as the parameter. Since $dx = 0 = dy$, Formula (14) gives

$$W = \int_C \frac{k(x\, dx + y\, dy + z\, dz)}{(x^2 + y^2 + z^2)^{3/2}}$$

$$= \int_0^3 \frac{kz\, dz}{(16 + z^2)^{3/2}} = \left[\frac{-k}{(16 + z^2)^{1/2}}\right]_0^3 = \frac{k}{20}.$$

**INDEPENDENCE OF PATH
AND CONSERVATIVE FORCE FIELDS**

We say that the line integral in (15) is **independent of path in the region** D provided that, given any two points A and B of D, the integral has the same value along every piecewise smooth curve or **path** in D from A to B. In this case, we may write

$$\int_C \mathbf{F} \cdot \mathbf{T}\, ds = \int_A^B \mathbf{F} \cdot \mathbf{T}\, ds$$

because the value of the integral depends only upon the points A and B and not upon the particular choice of path C joining them.

> **Theorem 2** *Independence of Path*
> The line integral $\int_C \mathbf{F} \cdot \mathbf{T}\, ds$ is independent of path in D if and only if $\mathbf{F} = \nabla f$ for some function f defined on D.

Proof Suppose first that $\mathbf{F} = \nabla f = \langle \partial f/\partial x, \partial f/\partial y, \partial f/\partial z \rangle$ and that C is a path from A to B parametrized as usual with parameter t in $[a, b]$. Then, by Equation (15),

$$\int_C \mathbf{F} \cdot \mathbf{T}\, ds = \int_C \frac{\partial f}{\partial x}\, dx + \frac{\partial f}{\partial y}\, dy + \frac{\partial f}{\partial z}\, dz$$

$$= \int_a^b \left(\frac{\partial f}{\partial x}\frac{dx}{dt} + \frac{\partial f}{\partial y}\frac{dy}{dt} + \frac{\partial f}{\partial z}\frac{dz}{dt} \right) dt$$

$$= \int_a^b D_t(f(x(t), y(t), z(t)))\, dt$$

$$= f(x(b), y(b), z(b)) - f(x(a), y(a), z(a)).$$

And so

$$\int_C \mathbf{F} \cdot \mathbf{T}\, ds = f(B) - f(A). \tag{16}$$

The last step follows from the Fundamental Theorem of Calculus. Equation (16) shows that the value of the line integral depends only upon the points A and B and is therefore independent of choice of the particular path C. This proves the "if" part of Theorem 2.

To outline the "only if" part, we suppose that the line integral is independent of path in D. Choose a *fixed* point $A_0 = A_0(x_0, y_0, z_0)$ in D, and let $B = B(x, y, z)$ be an arbitrary point in D. Given any path C from A_0 to B in D, we *define* the function f by means of the equation

$$f(x, y, z) = \int_C \mathbf{F} \cdot \mathbf{T}\, ds = \int_{(x_0, y_0, z_0)}^{(x, y, z)} \mathbf{F} \cdot \mathbf{T}\, ds.$$

Because of our hypothesis of independence of path, the resulting value of $f(x, y, z)$ depends only upon (x, y, z) and not upon the particular path C used. We shall omit the verification that $\nabla f = \mathbf{F}$. (See Problem 19 in Section 16-6.)

It follows from Theorem 2 that the vector field $\mathbf{F} = y\mathbf{i} + 2x\mathbf{j}$ of Example 2 is not the gradient of any scalar function f, because we found different values for its line integral along different paths from $(1, 1)$ to $(2, 4)$. Alternatively, we may conclude that $\mathbf{F}$ is not the gradient of a function f by noting that the equations $\partial f/\partial x = y$ and $\partial f/\partial y = 2x$ would imply that $\partial^2 f/\partial y \partial x = 1$ while $\partial^2 f/\partial x \partial y = 2$.

The force field $\mathbf{F}$ is called **conservative** if it is the gradient of a scalar function. It is customary to introduce a minus sign and write $\mathbf{F} = -\nabla V$. Then $V(x, y, z)$ is called the **potential energy** at the point (x, y, z). With $f = -V$ in Equation (16), we have

$$W = \int_A^B \mathbf{F} \cdot \mathbf{T}\, ds = V(A) - V(B), \tag{17}$$

and this means that the work done by $\mathbf{F}$ in moving a particle from A to B is equal to the *decrease* in potential energy.

For example, a brief computation shows that

$$\nabla\left(-\frac{k}{\sqrt{x^2 + y^2 + z^2}}\right) = \frac{k(x\mathbf{i} + y\mathbf{j} + z\mathbf{k})}{(x^2 + y^2 + z^2)^{3/2}} = \mathbf{F}$$

for the inverse-square force field of Example 3. With $V = k/(x^2 + y^2 + z^2)^{1/2}$, Equation (16) then gives

$$\int_{(0,4,0)}^{(0,4,3)} \mathbf{F} \cdot \mathbf{T}\, ds = \frac{k}{(0^2 + 4^2 + 0^2)^{1/2}} - \frac{k}{(0^2 + 4^2 + 3^2)^{1/2}} = \frac{k}{20},$$

as we also found by direct integration.

Here's the reason why the expression *conservative field* is used. Suppose that a particle of mass m moves from A to B under the influence of the conservative force $\mathbf{F}$, with position vector $\mathbf{r}(t)$, $a \le t \le b$. Then Newton's law $\mathbf{F}(\mathbf{r}(t)) = m\mathbf{r}''(t) = m\mathbf{v}'(t)$ gives

$$\int_A^B \mathbf{F} \cdot \mathbf{T}\, ds = \int_a^b m\mathbf{v}'(t) \cdot \frac{\mathbf{v}(t)}{v}\, v\, dt$$

$$= \int_a^b mD_t[\tfrac{1}{2}\mathbf{v}(t) \cdot \mathbf{v}(t)]\, dt = \left[\frac{1}{2}m[v(t)]^2\right]_a^b.$$

Thus, with the abbreviations v_A for $v(a)$ and v_B for $v(b)$, we see that

$$\int_A^B \mathbf{F} \cdot \mathbf{T}\, ds = \tfrac{1}{2}m(v_B)^2 - \tfrac{1}{2}m(v_A)^2. \qquad (18)$$

By equating the right-hand sides of Equations (17) and (18), we get the formula

$$\tfrac{1}{2}m(v_A)^2 + V(A) = \tfrac{1}{2}m(v_B)^2 + V(B). \qquad (19)$$

This is the law of **conservation of energy** for a particle moving under the influence of a *conservative* force field—the sum of its kinetic energy and potential energy remains constant.

16-2   PROBLEMS

In Problems 1–5, evaluate the line integrals $\int_C f(x, y)\, ds$, $\int_C f(x, y)\, dx$, and $\int_C f(x, y)\, dy$ along the indicated parametric curve.

1 $f(x, y) = x^2 + y^2$; $x = 4t - 1, y = 3t + 1, -1 \le t \le 1$.

2 $f(x, y) = x$; $x = t, y = t^2, 0 \le t \le 1$.

3 $f(x, y) = x + y$; $x = e^t + 1, y = e^t - 1, 0 \le t \le \ln 2$.

4 $f(x, y) = 2x - y$; $x = \sin t, y = \cos t, 0 \le t \le \pi/2$.

5 $f(x, y) = xy$; $x = 3t, y = t^4, 0 \le t \le 1$.

6 Evaluate $\int_C xy\, dx + (x + y)\, dy$ where C is the part of the graph of $y = x^2$ from $(-1, 1)$ to $(2, 4)$.

7 Evaluate $\int_C y^2\, dx + x\, dy$ where C is the part of the graph of $x = y^3$ from $(-1, -1)$ to $(1, 1)$.

8 Evaluate $\int_C y\sqrt{x}\, dx + x\sqrt{x}\, dy$ where C is the part of the graph of $y^2 = x^3$ from $(1, 1)$ to $(4, 8)$.

9 Evaluate the line integral $\int_C x^2y\, dx + xy^3\, dy$ where C consists of the line segments from $(-1, 1)$ to $(2, 1)$ and from $(2, 1)$ to $(2, 5)$.

10 Evaluate $\int_C (x + 2y)\, dx + (2x - y)\, dy$ where C consists of the line segments from $(3, 2)$ to $(3, -1)$ and from $(3, -1)$ to $(-2, -1)$.

In Problems 11–15, evaluate the line integral $\int_C \mathbf{F} \cdot \mathbf{T}\, ds$ along the indicated curve C.

11 $\mathbf{F} = z\mathbf{i} + x\mathbf{j} - y\mathbf{k}$; $x = t, y = t^2, z = t^3, t$ in $[0, 1]$.

12 $\mathbf{F} = yz\mathbf{i} + xz\mathbf{j} + xy\mathbf{k}$; C is the straight line segment from $(2, -1, 3)$ to $(4, 2, -1)$.

13 $\mathbf{F} = y\mathbf{i} - x\mathbf{j} + z\mathbf{k}$; $x = \sin t, y = \cos t, z = 2t, 0 \le t \le \pi$.

14 $\mathbf{F} = (2x + 3y)\mathbf{i} + (3x + 2y)\mathbf{j} + 3z^2\mathbf{k}$; C is the path from $(0, 0, 0)$ to $(4, 2, 3)$ that consists of three line segments parallel to the x-axis, the y-axis, and the z-axis, in that order.

15 $\mathbf{F} = yz^2\mathbf{i} + xz^2\mathbf{j} + 2xyz\mathbf{k}$; C is the path from $(-1, 2, -2)$ to $(1, 5, 2)$ consisting of three line segments parallel to the z-axis, the x-axis, and the y-axis, in that order.

16 Find $\int_C xyz \, ds$ if C is the line segment from $(1, -1, 2)$ to $(3, 2, 5)$.

17 Find $\int_C (2x + 9xy) \, ds$ given that C is the curve $x = t$, $y = t^2$, $z = t^3$, $0 \le t \le 1$.

18 Evaluate $\int_C xy \, ds$ where C is the elliptical helix $x = 4 \cos t$, $y = 9 \sin t$, $z = 7t$, $0 \le t \le 5\pi/2$.

19 Find the centroid of a uniform thin wire shaped like the semicircle $x^2 + y^2 = a^2$, $y \ge 0$.

20 Find the moments of inertia about the x- and y-axes of the wire of Problem 19.

21 Find the mass and centroid of a wire with constant density $\rho = k$ and shaped like the helix $x = 3 \cos t$, $y = 3 \sin t$, $z = 4t$, t in $[0, 2\pi]$.

22 Find the moment of inertia about the z-axis of the wire of Example 1 of this section.

23 A wire shaped like the first-quadrant portion of the circle $x^2 + y^2 = a^2$ has density $\rho = kxy$ at the point (x, y). Find its mass, its centroid, and its moment of inertia about each coordinate axis.

24 Find the work done by the inverse-square force field of Example 3 in moving a particle from $(1, 0, 0)$ to $(0, 3, 4)$. Integrate first along the line segment from $(1, 0, 0)$ to $(5, 0, 0)$ and then along a path on the sphere $x^2 + y^2 + z^2 = 25$, and note that the second integral is automatically zero (why?).

25 Imagine an infinitely long and uniformly charged wire that coincides with the z-axis. The electric force that it exerts on a unit charge at the point $(x, y) \ne (0, 0)$ in the xy-plane is $\mathbf{F} = k(x\mathbf{i} + y\mathbf{j})/(x^2 + y^2)$. Find the work done by $\mathbf{F}$ in moving a unit charge along the straight line segment from: (a) $(1, 0)$ to $(1, 1)$; (b) $(1, 1)$ to $(0, 1)$.

26 Let $\mathbf{F} = (-y\mathbf{i} + x\mathbf{j})/(x^2 + y^2)$ for x and y not both zero. Calculate the values of $\int_C \mathbf{F} \cdot \mathbf{T} \, ds$ along both the upper and lower halves of the circle $x^2 + y^2 = 1$ from $(1, 0)$ to $(-1, 0)$. Is there a function $f = f(x, y)$ defined for x, y not both zero such that $\nabla f = \mathbf{F}$? Why?

27 Show that if the force field $\mathbf{F} = P\mathbf{i} + Q\mathbf{j}$ is conservative, then $\partial P/\partial y = \partial Q/\partial x$. Show that the force field of Problem 26 satisfies the condition $\partial P/\partial y = \partial Q/\partial x$, but nevertheless is *not* conservative.

28 Suppose that the force field $\mathbf{F} = P\mathbf{i} + Q\mathbf{j} + R\mathbf{k}$ is conservative. Show that $\partial P/\partial y = \partial Q/\partial x$, $\partial P/\partial z = \partial R/\partial x$, and $\partial Q/\partial z = \partial R/\partial y$.

29 Apply the result of Problem 28 together with Theorem 2 to show that $\int_C 2xy \, dx + x^2 \, dy + y^2 \, dz$ is not independent of path.

30 Let $\mathbf{F} = \mathbf{F}(x, y, z) = yz\mathbf{i} + (xz + y)\mathbf{j} + (xy + 1)\mathbf{k}$. Define the function f by $f(x, y, z) = \int_C \mathbf{F} \cdot \mathbf{T} \, ds$ where C is the line segment from $(0, 0, 0)$ to (x, y, z). Determine f by evaluating this line integral, and then show that $\nabla f = \mathbf{F}$.

16-3 Green's Theorem

Green's theorem relates a line integral around a simple closed plane curve C to an ordinary double integral over the plane region R bounded by C. Suppose that the curve C is piecewise smooth—it consists of finitely many parametric arcs with continuous nonzero velocity vectors. Then C has a unit tangent vector $\mathbf{T}$ except possibly at finitely many *corner points*. The **positive** or **counterclockwise** direction along C is the one determined by a parametrization $\mathbf{r}(t)$ of C such that the region R remains on the *left* as the point $\mathbf{r}(t)$ traces the boundary curve C. That is, the vector obtained from the unit tangent vector $\mathbf{T}$ by a counterclockwise rotation through $90°$ always points *into* the region R, as shown in Figure 16.6. The symbol

$$\oint_C P \, dx + Q \, dy$$

denotes a line integral along or around C in this positive direction. A reversed arrow on the circle through the integral sign indicates a line integral around C in the opposite direction, which we naturally call either the **negative** or the **clockwise** direction.

The following result first appeared (in an equivalent form) in a booklet on the applications of mathematics to electricity and magnetism, published

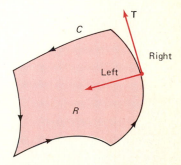

16.6 Positive orientation of the curve C: The region R within C is to the *left* of the unit tangent vector $\mathbf{T}$.

privately in 1828 by the self-taught English mathematical physicist George Green (1793–1841).

Green's Theorem

Let C be a piecewise-smooth simple closed curve that bounds the region R in the plane. Suppose that the functions $P(x, y)$ and $Q(x, y)$ are continuous and have continuous first order partial derivatives on R. Then

$$\oint_C P\,dx + Q\,dy = \iint_R \left(\frac{\partial Q}{\partial x} - \frac{\partial P}{\partial y} \right) dA. \tag{1}$$

Proof First we give a proof for the case in which the region R is both horizontally simple and vertically simple. Then we shall indicate how to extend the result to more general regions.

Recall that if R is vertically simple, then it has a description of the form $g_1(x) \leqq y \leqq g_2(x)$, $a \leqq x \leqq b$. The boundary curve C is then the union of the four arcs C_1, C_2, C_3, and C_4 of Figure 16.7, oriented as indicated there. Hence

$$\oint_C P\,dx = \int_{C_1} P\,dx + \int_{C_2} P\,dx + \int_{C_3} P\,dx + \int_{C_4} P\,dx.$$

The integrals along C_2 and C_4 are both zero, because on these two curves $x(t)$ is constant, so that $dx = x'(t)\,dt = 0$. It remains to compute the integrals along C_1 and C_3.

The point $(x, g_1(x))$ traces C_1 as x increases from a to b, while the point $(x, g_2(x))$ traces C_3 as x *decreases* from b to a. Hence

$$\oint_C P\,dx = \int_a^b P(x, g_1(x))\,dx + \int_b^a P(x, g_2(x))\,dx$$

$$= -\int_a^b (P(x, g_2(x)) - P(x, g_1(x)))\,dx$$

$$= -\int_a^b \int_{g_1(x)}^{g_2(x)} \frac{\partial P}{\partial y}\,dy\,dx$$

by the Fundamental Theorem of Calculus. Thus

$$\oint_C P\,dx = -\iint_R \frac{\partial P}{\partial y}\,dA. \tag{2}$$

In Problem 24 we ask you to show in a precisely similar way that

$$\oint_C Q\,dy = +\iint_R \frac{\partial Q}{\partial x}\,dA \tag{3}$$

if the region R is horizontally simple. We then obtain Equation (1), the conclusion of Green's theorem, simply by adding Equations (2) and (3).

The complete proof of Green's theorem for more general regions is beyond the scope of an introductory text. But the typical region R that appears in practice can be subdivided into smaller regions $R_1, R_2, R_3, \ldots, R_k$

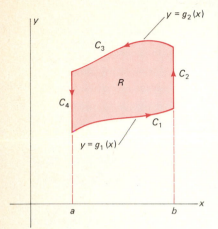

16.7 The boundary curve C is the union of the four arcs C_1, C_2, C_3, and C_4.

CHAP. 16: Vector Analysis

that are both vertically and horizontally simple. Green's theorem for the region R then follows from the fact that it applies to each of the regions $R_1, R_2, R_3, \ldots, R_k$ (see Problem 25).

For example, the horseshoe-shaped region R of Figure 16.8 can be subdivided into the regions R_1 and R_2, each of which is both horizontally simple and vertically simple. We also subdivide the boundary C of R and write $C_1 \cup D_1$ for the boundary of R_1 and $C_2 \cup D_2$ for the boundary of R_2 (see Figure 16.8). Applying Green's theorem separately to the regions R_1 and R_2, we get

$$\oint_{C_1 \cup D_1} P\,dx + Q\,dy = \iint_{R_1} \left(\frac{\partial Q}{\partial x} - \frac{\partial P}{\partial y} \right) dA$$

and

$$\oint_{C_2 \cup D_2} P\,dx + Q\,dy = \iint_{R_2} \left(\frac{\partial Q}{\partial x} - \frac{\partial P}{\partial y} \right) dA.$$

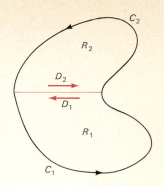

16.8 Decomposing the region R into two horizontally and vertically simple regions by means of a crosscut.

When we add these two equations, the result is Equation (1), Green's theorem for the region R, because the two line integrals along D_1 and D_2 cancel. This occurs because D_1 and D_2 represent the same curve with opposite orientations, so that

$$\int_{D_2} P\,dx + Q\,dy = -\int_{D_1} P\,dx + Q\,dy$$

by Equation (11) in Section 16-2. It therefore follows that

$$\int_{C_1 \cup D_1 \cup C_2 \cup D_2} P\,dx + Q\,dy = \int_{C_1 \cup C_2} P\,dx + Q\,dy = \oint_C P\,dx + Q\,dy.$$

Similarly, Green's theorem for the region shown in Figure 16.9 could be established by subdividing it into four simple regions as indicated.

EXAMPLE 1   Use Green's theorem to evaluate the line integral

$$\oint_C (2y + \sqrt{9 + x^3})\,dx + (5x + e^{\arctan y})\,dy,$$

where C is the circle $x^2 + y^2 = 4$.

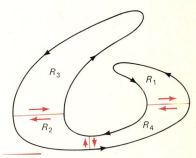

16.9 Many important regions can be decomposed into simple regions by using two or more crosscuts.

Solution   With $P(x, y) = 2y + (9 + x^3)^{1/2}$ and $Q(x, y) = 5x + \exp(\arctan y)$, we see that $\partial Q/\partial x - \partial P/\partial y = 5 - 2 = 3$. Since R is a circular disk with area 4π, Green's theorem therefore tells us that the given line integral is equal to

$$\iint_R 3\,dA = 3(4\pi) = 12\pi.$$

In Example 1 we found the double integral much easier to evaluate directly than the line integral. Sometimes the situation is the reverse. The following consequence of Green's theorem illustrates the technique of evaluating a double integral $\iint_R f(x, y)\,dA$ by converting it into a line integral $\oint_C P\,dx + Q\,dy$. To do this, we must be able to find functions $P(x, y)$ and $Q(x, y)$ such that $\partial Q/\partial x - \partial P/\partial y = f(x, y)$. As in the proof of the following result, this is sometimes easy.

The area A of the region R bounded by the piecewise-smooth simple closed curve C is given by

$$A = \tfrac{1}{2}\oint_C -y\,dx + x\,dy = -\oint_C y\,dx = \oint_C x\,dy. \qquad (4)$$

Proof With $P(x, y) = -y$ and $Q(x, y) \equiv 0$, Green's theorem gives

$$-\oint_C y\,dx = \iint_R 1\,dA = A.$$

Similarly, with $P(x, y) \equiv 0$ and $Q(x, y) = x$, we obtain $A = \oint_C x\,dy$. With $P(x, y) = -y/2$ and $Q(x, y) = x/2$, Green's theorem gives

$$\tfrac{1}{2}\oint_C -y\,dx + x\,dy = \iint_R (\tfrac{1}{2} + \tfrac{1}{2})\,dA = A.$$

EXAMPLE 2 Apply the corollary above to find the area bounded by the ellipse $x^2/a^2 + y^2/b^2 = 1$.

Solution With the parametrization $x = a \cos t$, $y = b \sin t$, $0 \leqq t \leqq 2\pi$, Equation (4) gives

$$A = \oint_{\text{ellipse}} x\,dy = \int_0^{2\pi} (a \cos t)(b \cos t)\,dt$$

$$= \tfrac{1}{2}ab \int_0^{2\pi} (1 + \cos 2t)\,dt = \pi ab.$$

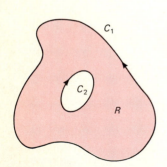

16.10 An annular region—the boundary consists of two simple closed curves, one within the other.

16.11 Two crosscuts convert the annular region into the union of two ordinary regions.

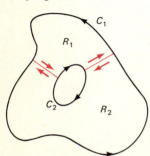

By using the technique of subdividing a region into simpler ones, Green's theorem may be extended to regions which have boundaries that consist of two or more simple closed curves. For example, consider the annular region R of Figure 16.10, with boundary C consisting of the two simple closed curves C_1 and C_2. The positive direction along C—the direction for which the region R always lies on the left—is counterclockwise on the outer curve C_1 but clockwise on the inner curve C_2.

We subdivide R into two regions R_1 and R_2 by using two crosscuts, as shown in Figure 16.11. Applying Green's theorem to each of these subregions, we get

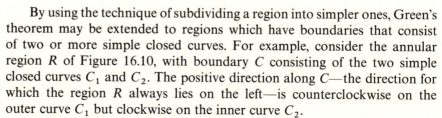

$$\iint_R \left(\frac{\partial Q}{\partial x} - \frac{\partial P}{\partial y}\right) dA = \iint_{R_1} \left(\frac{\partial Q}{\partial x} - \frac{\partial P}{\partial y}\right) dA + \iint_{R_2} \left(\frac{\partial Q}{\partial x} - \frac{\partial P}{\partial y}\right) dA$$

$$= \int_{C_1} P\,dx + Q\,dy + \int_{C_2} P\,dx + Q\,dy$$

$$= \oint_C P\,dx + Q\,dy.$$

Thus we obtain Green's theorem for the given region R. What makes this proof work is that the opposite line integrals along the two crosscuts cancel each other.

EXAMPLE 3 Suppose that C is a smooth simple closed curve that encloses the origin $(0, 0)$. Then show that

$$\oint_C \frac{-y\,dx + x\,dy}{x^2 + y^2} = 2\pi,$$

but that this integral is zero if C does *not* enclose the origin.

Solution With $P(x, y) = -y/(x^2 + y^2)$ and $Q(x, y) = x/(x^2 + y^2)$, a brief computation gives $\partial Q/\partial x - \partial P/\partial y = 0$ when x and y are not both zero. If the region R bounded by C does not contain the origin, then P and Q and their derivatives are continuous on R. Hence Green's theorem gives

$$\oint_C \frac{-y\,dx + x\,dy}{x^2 + y^2} = \iint_R 0\,dA = 0.$$

If C does enclose the origin, then we enclose the origin in a small circle C_a interior to C (as in Figure 16.12), and parametrize this small circle by $x = a\cos t$, $y = a\sin t$. Then Green's theorem, applied to the region R between C and C_a, gives

$$\oint_C \frac{-y\,dx + x\,dy}{x^2 + y^2} + \oint_{C_a} \frac{-y\,dx + x\,dy}{x^2 + y^2} = \iint_R 0\,dA = 0.$$

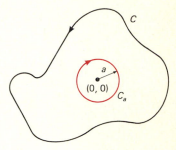

16.12 Use the small circle C_a if C encloses the origin.

Therefore

$$\oint_C \frac{-y\,dx + x\,dy}{x^2 + y^2} = \oint_{C_a} \frac{-y\,dx + x\,dy}{x^2 + y^2}$$

$$= \int_0^{2\pi} \frac{(-a\sin t)(-a\sin t\,dt) + (a\cos t)(a\cos t\,dt)}{(a\cos t)^2 + (a\sin t)^2}$$

$$= \int_0^{2\pi} 1\,dt = 2\pi.$$

The result of Example 3 can be interpreted in terms of the polar coordinate angle $\theta = \tan^{-1}(y/x)$. Since

$$d\theta = \frac{-y\,dx + x\,dy}{x^2 + y^2},$$

the line integral of Example 3 measures the net change in θ as we go around the curve C in a counterclockwise direction. This net change is 2π if C encloses the origin and is zero otherwise.

16.13 The area of the parallelogram approximates the fluid flow across Δs_i in unit time.

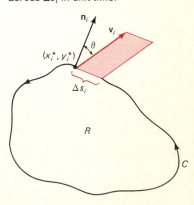

THE DIVERGENCE AND FLUX OF A VECTOR FIELD

Now let us consider the steady flow of a thin layer of fluid in the plane (perhaps like a sheet of water spreading across the floor). Let $\mathbf{v}(x, y)$ be its velocity vector field and $\rho(x, y)$ the density of the fluid at the point (x, y). The term *steady flow* means that $\mathbf{v}$ and ρ depend only upon x and y and *not* upon time t. We want to compute the rate at which the fluid flows out of the region R bounded by a given simple closed curve C (Figure 16.13). We seek the net rate of outflow—the actual outflow minus the inflow.

Let Δs_i be a short segment of the curve C, and let (x_i^*, y_i^*) be an end point of Δs_i. Then the area of the portion of fluid that flows out of R across Δs_i per unit time is approximately the area of the parallelogram of Figure 16.13; this parallelogram is spanned by the segment Δs_i and the vector $\mathbf{v}_i = \mathbf{v}(x_i^*, y_i^*)$. Suppose that $\mathbf{n}_i$ is the unit normal vector to C at the point (x_i^*, y_i^*), the normal pointing *out* of R. Then the area of this parallelogram is

$$(|\mathbf{v}_i| \cos \theta) \, \Delta s_i = \mathbf{v}_i \cdot \mathbf{n}_i \, \Delta s_i$$

where θ is the angle between $\mathbf{n}_i$ and $\mathbf{v}_i$.

We multiply by the density $\rho_i = \rho(x_i^*, y_i^*)$ then add these terms over those values of i that correspond to a subdivision of the entire curve C. This gives the (net) total mass of fluid leaving R per unit of time; it is approximately

$$\sum_{i=1}^{n} \rho_i \mathbf{v}_i \cdot \mathbf{n}_i \, \Delta s_i = \sum_{i=1}^{n} \mathbf{F}_i \cdot \mathbf{n}_i \, \Delta s_i$$

where $\mathbf{F} = \rho \mathbf{v}$. The line integral around C that this sum approximates is called the *flux of the vector field* $\mathbf{F}$ *across the curve* C,

$$(\text{flux of } F \text{ across } C) = \oint_C \mathbf{F} \cdot \mathbf{n} \, ds, \tag{5}$$

where $\mathbf{n}$ is the *outer* unit normal to C.

In the present case of fluid flow with velocity vector $\mathbf{v}$, the flux of $\mathbf{F} = \rho \mathbf{v}$ is the rate at which the fluid is flowing out of R across the boundary curve C, in units of mass per unit of time. But the same terminology is used for an arbitrary vector field $\mathbf{F} = P\mathbf{i} + Q\mathbf{j}$. For example, we may speak of the flux of an electric or gravitational field across a curve C.

From Figure 16.14 we see that the outer unit normal $\mathbf{n}$ is equal to $\mathbf{T} \times \mathbf{k}$. The unit tangent $\mathbf{T}$ to the curve C is given by

$$\mathbf{T} = \frac{1}{v}\left(\mathbf{i} \frac{dx}{dt} + \mathbf{j} \frac{dy}{dt}\right) = \mathbf{i} \frac{dx}{ds} + \mathbf{j} \frac{dy}{ds}$$

since $v = ds/dt$. Hence

$$\mathbf{n} = \mathbf{T} \times \mathbf{k} = \left(\mathbf{i} \frac{dx}{ds} + \mathbf{j} \frac{dy}{ds}\right) \times \mathbf{k}.$$

Since $\mathbf{i} \times \mathbf{k} = -\mathbf{j}$ and $\mathbf{j} \times \mathbf{k} = \mathbf{i}$, we find that

$$\mathbf{n} = \mathbf{i} \frac{dy}{ds} - \mathbf{j} \frac{dx}{ds}. \tag{6}$$

Substitution of Expression (6) into the flux integral of Equation (5) gives

$$\oint_C \mathbf{F} \cdot \mathbf{n} \, ds = \oint_C (P\mathbf{i} + Q\mathbf{j}) \cdot \left(\mathbf{i} \frac{dy}{ds} - \mathbf{j} \frac{dx}{ds}\right) ds = \oint_C -Q \, dx + P \, dy.$$

Applying Green's theorem to this last line integral, we get

$$\oint_C \mathbf{F} \cdot \mathbf{n} \, ds = \iint_R \left(\frac{\partial P}{\partial x} + \frac{\partial Q}{\partial y}\right) dA \tag{7}$$

for the flux of $\mathbf{F} = P\mathbf{i} + Q\mathbf{j}$ across C.

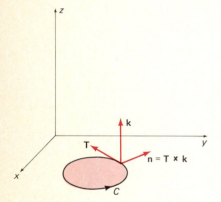

16.14 Computing the outer unit normal $\mathbf{n}$ from the unit tangent vector $\mathbf{T}$.

The scalar function $\partial P/\partial x + \partial Q/\partial y$ that arises in Equation (7) is called the **divergence** of the two-dimensional vector field $\mathbf{F} = P\mathbf{i} + Q\mathbf{j}$ and is denoted by

$$\text{div } \mathbf{F} = \nabla \cdot \mathbf{F} = \frac{\partial P}{\partial x} + \frac{\partial Q}{\partial y}. \tag{8}$$

The notation $\nabla \cdot \mathbf{F}$ for the divergence is suggested by the formal expression

$$\nabla = \mathbf{i}\frac{\partial}{\partial x} + \mathbf{j}\frac{\partial}{\partial y},$$

which is consistent both with Equation (8) and with the use of the symbol ∇ in the notation

$$\nabla f = \mathbf{i}\frac{\partial f}{\partial x} + \mathbf{j}\frac{\partial f}{\partial y}$$

for the gradient vector.

When we substitute (8) into (7), we obtain a **vector form of Green's theorem**:

$$\oint_C \mathbf{F} \cdot \mathbf{n}\, ds = \iint_R \nabla \cdot \mathbf{F}\, dA, \tag{9}$$

with the understanding that $\mathbf{n}$ is the *outer* unit normal to C. Thus the flux of a vector field across a simple closed curve C is equal to the double integral of its divergence over the region R bounded by C.

If the region R is bounded by a circle C_r of radius r centered at the point (x_0, y_0), then

$$\oint_{C_r} \mathbf{F} \cdot \mathbf{n}\, ds = \iint_R \nabla \cdot \mathbf{F}\, dA = (\pi r^2)\nabla \cdot \mathbf{F}(\bar{x}, \bar{y})$$

for some point $(\bar{x}, \bar{y})$ in R; this is a consequence of the average value property of double integrals (see Problem 30 in Section 15-1). We divide by πr^2 and then let r approach zero. Thus we find that

$$\nabla \cdot \mathbf{F}(x_0, y_0) = \lim_{r \to 0} \frac{1}{\pi r^2} \oint_{C_r} \mathbf{F} \cdot \mathbf{n}\, ds \tag{10}$$

because $(\bar{x}, \bar{y}) \to (x_0, y_0)$ as $r \to 0$.

In the case of our original fluid flow, with $\mathbf{F} = \rho\mathbf{v}$, Equation (10) tells us that the value of $\nabla \cdot \mathbf{F}$ at (x_0, y_0) is a measure of the rate at which the fluid is "diverging away" from the point (x_0, y_0).

EXAMPLE 4 The vector field $\mathbf{F} = -y\mathbf{i} + x\mathbf{j}$ is the velocity field of a steady-state counterclockwise rotation about the origin. Show that the flux of $\mathbf{F}$ across any simple closed curve C is zero.

Solution This follows immediately from Equation (9), because $\nabla \cdot \mathbf{F} = (\partial/\partial x)(-y) + (\partial/\partial y)(x) = 0$.

In each of Problems 1–8, apply Green's theorem to evaluate the indicated line integral.

1 $\oint_C (x + y^2)\,dx + (y + x^2)\,dy$, where C is the square with vertices $(\pm 1, \pm 1)$.

2 $\oint_C (x^2 + y^2)\,dx - 2xy\,dy$, where C is the boundary of the triangle bounded by the lines $x = 0$, $y = 0$, and $x + y = 1$.

3 $\oint_C (y + e^x)\,dx + (2x^2 + \cos y)\,dy$, where C is the boundary of the triangle with vertices $(0, 0)$, $(1, 1)$, and $(2, 0)$.

4 $\oint_C (x^2 - y^2)\,dx + xy\,dy$, where C is the boundary of the region that is bounded by the line $y = x$ and the parabola $y = x^2$.

5 $\oint_C (-y^2 + e^{e^x})\,dx + (\tan^{-1} y)\,dy$, where C is the boundary of the region between the parabolas $y = x^2$ and $x = y^2$.

6 $\oint_C y^2\,dx + (2x - 3y)\,dy$, where C is the circle $x^2 + y^2 = 9$.

7 $\oint_C (x - y)\,dx + y\,dy$, where C is the boundary of the region between the x-axis and the graph $y = \sin x$ for $0 \leqq x \leqq \pi$.

8 $\oint_C e^x \sin y\,dx + e^x \cos y\,dy$, where C is the right-hand loop of the lemniscate $r^2 = 4 \cos \theta$.

In each of Problems 9–12, use the corollary to Green's theorem to find the area of the indicated region.

9 The circle bounded by $x = a \cos t$, $y = a \sin t$, $0 \leqq t \leqq 2\pi$.

10 The region under one arch of the cycloid with parametric equations $x = a(t - \sin t)$ and $y = a(1 - \cos t)$.

11 The region bounded by the astroid $x = \cos^3 t$, $y = \sin^3 t$, $0 \leqq t \leqq 2\pi$.

12 The region bounded by $y = x^2$ and $x = y^3$.

13 Suppose that f is a twice-differentiable scalar function of x and y. Show that

$$\mathbf{V}^2 f = \operatorname{div}(\mathbf{V}f) = \frac{\partial^2 f}{\partial x^2} + \frac{\partial^2 f}{\partial y^2}.$$

14 Show that $f(x, y) = \ln(x^2 + y^2)$ satisfies **Laplace's equation** $\mathbf{V}^2 f = 0$ (except at the point $(0, 0)$).

15 Suppose that f and g are twice-differentiable functions. Show that $\mathbf{V}^2(fg) = f\mathbf{V}^2 g + g\mathbf{V}^2 f + 2\mathbf{V}f \cdot \mathbf{V}g$.

16 Suppose that $\mathbf{F}(x, y)$ is a vector-valued function and that $g(x, y)$ is scalar-valued. Show that

$$\mathbf{V} \cdot (g\mathbf{F}) = g(\mathbf{V} \cdot \mathbf{F}) + (\mathbf{V}g) \cdot \mathbf{F}.$$

17 Let R be the plane region with area A enclosed by the piecewise-smooth simple closed curve C. Use Green's theo-

rem to show that the coordinates of the centroid of R are

$$\bar{x} = \frac{1}{2A}\oint_C x^2\,dy \quad \text{and} \quad \bar{y} = -\frac{1}{2A}\oint_C y^2\,dx.$$

18 Use the result of Problem 17 to find the centroid of: (a) a semicircular region of radius a; (b) a quarter-circular region of radius a.

19 Suppose that a lamina shaped like the region of Problem 17 has constant density ρ. Show that its moments of inertia about the coordinate axes are

$$I_x = -\frac{\rho}{3}\oint_C y^3\,dx \quad \text{and} \quad I_y = \frac{\rho}{3}\oint_C x^3\,dy.$$

20 Use the result of Problem 19 to show that the polar moment of inertia $I_0 = I_x + I_y$ of a circular lamina of radius a, centered at the origin, and of constant density ρ, is $\frac{1}{2}Ma^2$ where M is the mass of the lamina.

21 The loop of the folium of Descartes (with equation $x^3 + y^3 = 3xy$) appears in Figure 16.15. Apply the corollary to Green's theorem to find the area of this loop. (*Suggestion:* Set $y = tx$ to discover the parametrization $x = 3t/(1 + t^3)$, $y = 3t^2/(1 + t^3)$. To obtain the area of the loop, use values of t in the interval $[0, 1]$. This gives the half of the loop lying below the line $y = x$.)

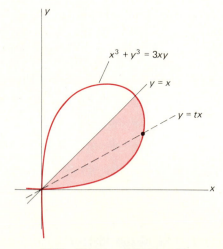

16.15 The loop of Problem 21.

22 Find the area bounded by one loop of the curve $x = \sin 2t$, $y = \sin t$.

23 Let f and g be functions with continuous second order partial derivatives in the region R bounded by the curve C.

Apply Green's theorem in vector form to show that

$$\oint_C f\nabla g \cdot \mathbf{n} \, ds = \iint_R [(f)(\nabla \cdot \nabla g) + \nabla f \cdot \nabla g] \, dA.$$

It was this formula rather than "Green's theorem" itself that appeared in Green's book.

24 Complete the proof of the simple case of Green's theorem by showing directly that $\oint_C Q \, dy = \iint_R (\partial Q/\partial x) \, dA$ if the region R is horizontally simple.

25 Suppose that the bounded plane region R is subdivided into the nonoverlapping subregions $R_1, R_2, R_3, \ldots, R_k$. If

Green's theorem, Equation (1), holds for each of these subregions, explain why it follows that Green's theorem holds for R. State carefully any assumptions that you need to make.

In Problems 26–28, find the flux of the given vector field $\mathbf{F}$ outward across the ellipse $x = 4 \cos t$, $y = 3 \sin t$, $0 \leq t \leq 2\pi$.

26 $\mathbf{F} = x\mathbf{i} + 2y\mathbf{j}$.

27 $\mathbf{F} = \dfrac{x\mathbf{i} + y\mathbf{j}}{x^2 + y^2}$.

28 $\mathbf{F} = -y\mathbf{i} + x\mathbf{j}$.

16-4

Surface Integrals

A surface integral is to surfaces in space what a line (or "curve") integral is to curves in the plane. Consider a curved thin metal sheet shaped like the surface S. Suppose that this sheet has a variable density, given at the point (x, y, z) by the known continuous function $f(x, y, z)$, in units such as grams per square centimeter of surface. We want to define the surface integral

$$\iint_S f(x, y, z) \, dS$$

in such a way that—upon evaluation—it gives the total mass of the thin metal sheet. In case $f(x, y, z) \equiv 1$, the numerical value of the integral should also equal the surface area of S.

As in Section 15-7, we assume that S is a parametric surface described by the function or transformation

$$\mathbf{r}(u, v) = \langle x(u, v), y(u, v), z(u, v) \rangle$$

for (u, v) in a region D in the uv-plane. We suppose throughout that the component functions of $\mathbf{r}$ have continuous partial derivatives, and also that the vectors $\mathbf{r}_u = \partial \mathbf{r}/\partial u$ and $\mathbf{r}_v = \partial \mathbf{r}/\partial v$ are nonzero and nonparallel at each interior point of D.

Recall now how we computed the surface area A of S in Section 15-7. We began with an inner partition of D consisting of n rectangles $R_1, R_2, R_3, \ldots, R_n$, each Δu by Δv in size. The images under $\mathbf{r}$ of these rectangles are curvilinear figures filling up most of the surface S, and these pieces of S are themselves approximated by parallelograms P_i of the sort shown in Figure 16.16. This gave us the approximation

$$A \approx \sum_{i=1}^{n} \Delta P_i = \sum_{i=1}^{n} |\mathbf{N}(u_i, v_i)| \, \Delta u \, \Delta v. \tag{1}$$

Here, $\Delta P_i = |\mathbf{N}(u_i, v_i)| \, \Delta u \, \Delta v$ is the area of the parallelogram P_i that is tangent to the surface S at the point $\mathbf{r}(u_i, v_i)$. The vector

$$\mathbf{N} = \frac{\partial \mathbf{r}}{\partial u} \times \frac{\partial \mathbf{r}}{\partial v} \tag{2}$$

is normal to S at $\mathbf{r}(u, v)$. If the surface S now has a density function $f(x, y, z)$, we can approximate the total mass M of the surface by first multiplying each

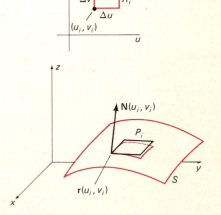

16.16 Approximating surface area with parallelograms.

parallelogram area ΔP_i in Equation (1) by the density $f(\mathbf{r}(u_i, v_i))$ at $\mathbf{r}(u_i, v_i)$, and then summing these estimates over all such parallelograms. Thus we obtain the approximation

$$M \approx \sum_{i=1}^{n} f(\mathbf{r}(u_i, v_i)) \, \Delta P_i = \sum_{i=1}^{n} f(\mathbf{r}(u_i, v_i)) |\mathbf{N}(u_i, v_i)| \, \Delta u \, \Delta v.$$

This approximation is a Riemann sum for the **surface integral of the function** f **over the surface** S, defined by

$$\iint_S f(x, y, z) \, dS = \iint_D f(\mathbf{r}(u, v)) |\mathbf{N}(u, v)| \, du \, dv \tag{3}$$

$$= \iint_D f(\mathbf{r}(u, v)) \left| \frac{\partial \mathbf{r}}{\partial u} \times \frac{\partial \mathbf{r}}{\partial v} \right| \, du \, dv.$$

To evaluate the surface integral $\iint_S f(x, y, z) \, dS$, we simply use the parametrization $\mathbf{r}$ to express the variables x, y, and z in terms of u and v and formally replace the **surface area element** dS by

$$dS = |\mathbf{N}(u, v)| \, du \, dv = \left| \frac{\partial \mathbf{r}}{\partial u} \times \frac{\partial \mathbf{r}}{\partial v} \right| \, du \, dv. \tag{4}$$

This converts the surface integral into an ordinary double integral over the region D in the uv-plane.

In the important special case of a surface S described by $z = h(x, y)$, (x, y) in D, we may use x and y as the parameters (rather than u and v). The surface area element takes the form

$$dS = \sqrt{1 + (\partial h/\partial x)^2 + (\partial h/\partial y)^2} \, dx \, dy \tag{5}$$

corresponding to Equation (9) in Section 15-7. The surface integral of f over S is then given by

$$\iint_S f(x, y, z) \, dS = \iint_D f(x, y, h(x, y)) \sqrt{1 + \left(\frac{\partial h}{\partial x}\right)^2 + \left(\frac{\partial h}{\partial y}\right)^2} \, dx \, dy. \tag{6}$$

Moments, centroids, and moments of inertia for surfaces are computed in the same way as for curves (Section 16-2), using surface integrals in place of line integrals. For example, suppose that the surface S has density $\rho(x, y, z)$ at the point (x, y, z). Then its moment M_{xy} about the xy-plane and its moment of inertia I_z about the z-axis are given by

$$M_{xy} = \iint_S z\rho(x, y, z) \, dS \quad \text{and} \quad I_z = \iint_S (x^2 + y^2)\rho(x, y, z) \, dS.$$

EXAMPLE 1 Find the centroid of the unit-density hemispherical surface $z = \sqrt{a^2 - x^2 - y^2}$, $x^2 + y^2 \leqq a^2$.

Solution By symmetry, $\bar{x} = 0 = \bar{y}$. A simple computation gives $\partial z/\partial x = -x/z$ and $\partial z/\partial y = -y/z$, so Equation (5) yields

$$dS = \sqrt{1 + \left(\frac{\partial z}{\partial x}\right)^2 + \left(\frac{\partial z}{\partial y}\right)^2} \, dx \, dy = \frac{a}{z} \, dx \, dy.$$

Hence

$$\bar{z} = \frac{M_{xy}}{A} = \frac{1}{2\pi a^2} \iint_D (z)\left(\frac{a}{z}\right) dx\,dy = \frac{1}{2\pi a} \iint_D 1\,dx\,dy = \frac{a}{2}.$$

Note in the final step that D is a circle of radius a in the xy-plane; this simplifies the computation.

EXAMPLE 2 Find the moment of inertia about the z-axis of the spherical surface $x^2 + y^2 + z^2 = a^2$, assuming that it has constant density $\rho = k$.

Solution The sphere of radius a is parametrized in spherical coordinates by

$$x = a \sin \phi \cos \theta, \qquad y = a \sin \phi \sin \theta, \qquad z = a \cos \phi$$

for $0 \le \phi \le \pi$, $0 \le \theta \le 2\pi$. By Problem 14 in Section 15-7, the surface element is then $dS = a^2 \sin \phi \, d\phi \, d\theta$. We first note that

$$x^2 + y^2 = a^2 \sin^2\phi \cos^2\theta + a^2 \sin^2\phi \sin^2\theta = a^2 \sin^2\phi,$$

and it then follows that

$$I_z = \iint_S (x^2 + y^2)\rho \, dS = \int_0^{2\pi} \int_0^{\pi} k(a^2 \sin^2\phi)a^2 \sin \phi \, d\phi \, d\theta$$

$$= 2\pi ka^4 \int_0^{\pi} \sin^3\phi \, d\phi = \tfrac{8}{3}\pi ka^4 = \tfrac{2}{3}Ma^2.$$

In the last step we use the fact that the mass M of the spherical surface with density k is $4\pi ka^2$.

The surface integral $\iint_S f(x, y, z) \, dS$ is analogous to the line integral $\int_C f(x, y) \, ds$. There is a second type of surface integral which is analogous to the line integral of the form $\int_C P \, dx + Q \, dy$. To define the surface integral

$$\iint_S f(x, y, z) \, dx \, dy,$$

with $dx \, dy$ in place of dS, we replace the parallelogram area ΔP_i in Equation (1) by its projection into the xy-plane (see Figure 16.17). To see how this works out, consider the *unit* normal to S,

$$\mathbf{n} = \frac{\mathbf{N}}{|\mathbf{N}|} = \mathbf{i} \cos \alpha + \mathbf{j} \cos \beta + \mathbf{k} \cos \gamma. \tag{7}$$

Since

$$\mathbf{N} = \begin{vmatrix} \mathbf{i} & \mathbf{j} & \mathbf{k} \\ \dfrac{\partial x}{\partial u} & \dfrac{\partial y}{\partial u} & \dfrac{\partial z}{\partial u} \\ \dfrac{\partial x}{\partial v} & \dfrac{\partial y}{\partial v} & \dfrac{\partial z}{\partial v} \end{vmatrix}$$

or—in the Jacobian notation of Section 15-8—

$$\mathbf{N} = \mathbf{i} \frac{\partial(y, z)}{\partial(u, v)} + \mathbf{j} \frac{\partial(z, x)}{\partial(u, v)} + \mathbf{k} \frac{\partial(x, y)}{\partial(u, v)},$$

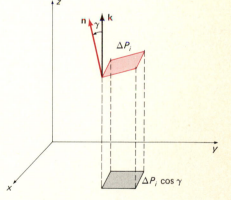

16.17 Finding the area of the projected parallelogram.

we see that the components of the unit normal $\mathbf{n}$ are

$$\cos \alpha = \frac{1}{|\mathbf{N}|} \frac{\partial(y, z)}{\partial(u, v)}, \qquad \cos \beta = \frac{1}{|\mathbf{N}|} \frac{\partial(z, x)}{\partial(u, v)}, \qquad \cos \gamma = \frac{1}{|\mathbf{N}|} \frac{\partial(x, y)}{\partial(u, v)}. \qquad (8)$$

From Figure 16.17, we see that the (signed) projection of the area ΔP_i into the xy-plane is $\Delta P_i \cos \gamma$. The corresponding Riemann sum motivates the *definition*

$$\iint_S f(x, y, z) \, dx \, dy = \iint_S f(x, y, z) \cos \gamma \, dS = \iint_D f(\mathbf{r}(u, v)) \frac{\partial(x, y)}{\partial(u, v)} \, du \, dv, \qquad (9)$$

where the last integral is obtained by substituting for $\cos \gamma$ from (8) and for dS from (4). Similarly, we *define*

$$\iint_S f(x, y, z) \, dy \, dz = \iint_S f(x, y, z) \cos \alpha \, dS = \iint_D f(\mathbf{r}(u, v)) \frac{\partial(y, z)}{\partial(u, v)} \, du \, dv \qquad (10)$$

and

$$\iint_S f(x, y, z) \, dz \, dx = \iint_S f(x, y, z) \cos \beta \, dS = \iint_D f(\mathbf{r}(u, v)) \frac{\partial(z, x)}{\partial(u, v)} \, du \, dv. \qquad (11)$$

Note that the symbols z and x in Formula (11) appear in the reverse of alphabetical order. It is important to write them in the correct order because the fact that $\partial(x, z)/\partial(u, v) = -\partial(z, x)/\partial(u, v)$ implies that

$$\iint_S f(x, y, z) \, dx \, dz = -\iint_S f(x, y, z) \, dz \, dx.$$

In an ordinary *double integral*, the order in which the differentials are written simply indicates the order of integration. But in a *surface integral*, it instead indicates the order of appearance of the corresponding variables in the Jacobians of Formulas (9) through (11).

The general surface integral of the second type is the sum

$$\iint_S P \, dy \, dz + Q \, dz \, dx + R \, dx \, dy = \iint_S (P \cos \alpha + Q \cos \beta + R \cos \gamma) \, dS \qquad (12)$$

$$= \iint_D \left(P \frac{\partial(y, z)}{\partial(u, v)} + Q \frac{\partial(z, x)}{\partial(u, v)} + R \frac{\partial(x, y)}{\partial(u, v)} \right) du \, dv. \qquad (13)$$

Here, P, Q, and R are continuous functions of x, y, and z.

Suppose that $\mathbf{F} = P\mathbf{i} + Q\mathbf{j} + R\mathbf{k}$. Then the integrand in Equation (12) is simply $\mathbf{F} \cdot \mathbf{n}$, so we obtain the basic relation

$$\iint_S \mathbf{F} \cdot \mathbf{n} \, dS = \iint_S P \, dy \, dz + Q \, dz \, dx + R \, dx \, dy \qquad (14)$$

between our two types of surface integrals. This formula is analogous to the earlier formula

$$\int_C \mathbf{F} \cdot \mathbf{T} \, ds = \int_C P \, dx + Q \, dy + R \, dz$$

for line integrals. On the other hand, Formula (13) gives the evaluation procedure for the surface integral of (12)—substitute for x, y, z and their derivatives in terms of u and v and then integrate over the appropriate region D in the uv-plane.

It turns out that only the *sign* of the right-hand surface integral in Equation (14) depends upon the parametrization of S. The unit normal on the left-hand side is the one that is provided by the parametrization of S via Equations (8). In the case of a surface given by $z = h(x, y)$, with x and y used as the parameters u and v, this will be the *upper* normal, as you will see in the computation of the following example.

EXAMPLE 3 Suppose that S is the surface $z = h(x, y)$, (x, y) in D. Then show that

$$\iint_S P \, dy \, dz + Q \, dz \, dx + R \, dx \, dy = \iint_D \left(-P \frac{\partial z}{\partial x} - Q \frac{\partial z}{\partial y} + R \right) dx \, dy, \quad (15)$$

where P, Q, and R in the second integral are evaluated at $(x, y, h(x, y))$.

Solution This is simply a matter of computing the three Jacobians in (13) with the parameters x and y. We note first that $\partial x/\partial x = 1 = \partial y/\partial y$ and that $\partial x/\partial y = 0 = \partial y/\partial x$. Hence

$$\frac{\partial(y, z)}{\partial(x, y)} = \begin{vmatrix} \dfrac{\partial y}{\partial x} & \dfrac{\partial y}{\partial y} \\[2mm] \dfrac{\partial z}{\partial x} & \dfrac{\partial z}{\partial y} \end{vmatrix} = -\frac{\partial z}{\partial x},$$

$$\frac{\partial(z, x)}{\partial(x, y)} = \begin{vmatrix} \dfrac{\partial z}{\partial x} & \dfrac{\partial z}{\partial y} \\[2mm] \dfrac{\partial x}{\partial x} & \dfrac{\partial x}{\partial y} \end{vmatrix} = -\frac{\partial z}{\partial y},$$

and

$$\frac{\partial(x, y)}{\partial(x, y)} = \begin{vmatrix} \dfrac{\partial x}{\partial x} & \dfrac{\partial x}{\partial y} \\[2mm] \dfrac{\partial y}{\partial x} & \dfrac{\partial y}{\partial y} \end{vmatrix} = 1.$$

Equation (15) is an immediate consequence.

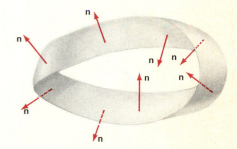

16.18 A Möbius strip—a one-sided surface.

16.19 Inner and outer normals to a two-sided closed surface.

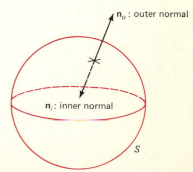

One of the most important applications of surface integrals is to the computation of the flux of a vector field. In order to define the flux of the vector field **F** across the surface S, it is necessary that S have a unit normal vector **n** that varies *continuously* from point to point of S. This condition excludes from our consideration one-sided, or *nonorientable* surfaces, such as the Möbius strip of Figure 16.18. If S is a two-sided, or *orientable*, surface, then there are two possible choices for **n**. For example, if S is a closed surface (such as a sphere or torus) that separates space, then we may choose for **n** either the outer normal or the inner normal. These two options are shown in Figure 16.19. The unit normal determined by Formula (7) may be either the outer normal or the inner normal; which of the two it is depends upon how S has been parametrized.

Now we turn our attention to the flux of a vector field. Suppose that we are given the vector field $\mathbf{F}$, the orientable surface S, and a continuous unit normal vector $\mathbf{n}$ on S. We define the **flux of F across S in the direction of n** in analogy with Equation (5) of Section 16-3:

$$\text{(flux of } \mathbf{F}) = \iint_S \mathbf{F} \cdot \mathbf{n}\, dS. \tag{16}$$

For example, if $\mathbf{F} = \rho\mathbf{v}$ where $\mathbf{v}$ is the velocity vector field corresponding to the steady flow in space of a fluid of density ρ and $\mathbf{n}$ is the *outer* normal for a closed surface S bounding a space region T, then the flux determined by the above equation is the rate of flow of the fluid *out of* T across its boundary surface S, in such units as grams per second.

A similar application is to the flow of heat, which is mathematically quite similar to the flow of a fluid. Suppose that a body has temperature $u = u(x, y, z)$ at the point (x, y, z). Experiments indicate that the flow of heat in the body is described by the heat flow vector

$$\mathbf{q} = -K\,\nabla u. \tag{17}$$

The number K—normally, but not always, a constant—is the *heat conductivity* of the body. The vector $\mathbf{q}$ points in the direction of heat flow, and its length is the rate of flow of heat across a unit area normal to $\mathbf{q}$. This flow rate is measured in such units as calories per second per square centimeter. If S is a closed surface within the body bounding a region T and $\mathbf{n}$ denotes the outer unit normal to S, then

$$\iint_S \mathbf{q} \cdot \mathbf{n}\, dS = -\iint_S K\,\nabla u \cdot \mathbf{n}\, dS \tag{18}$$

is the rate of flow of heat (in calories per second, for example) out of the region T across its boundary surface S.

EXAMPLE 4   Find the flux of the vector field $\mathbf{F} = x\mathbf{i} + y\mathbf{j} + 3\mathbf{k}$ out of the region T bounded by the paraboloid $z = x^2 + y^2$ and the plane $z = 4$ (see Figure 16.20).

16.20 The surface of Example 4.

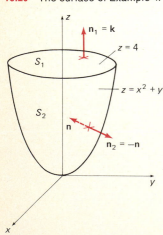

Solution   Let S_1 denote the circular top, which has outer unit normal $\mathbf{n}_1 = \mathbf{k}$; let S_2 be the parabolic part of the surface, with outer unit normal $\mathbf{n}_2$. The flux across S_1 is

$$\iint_{S_1} \mathbf{F} \cdot \mathbf{n}_1\, dS = \iint_{S_1} 3\, dS = 12\pi$$

because S_1 is a circle of radius 2.

The computation of Example 3 gives

$$\mathbf{N} = \left\langle -\frac{\partial z}{\partial x}, -\frac{\partial z}{\partial y}, 1 \right\rangle = \langle -2x, -2y, 1 \rangle$$

for a normal vector to the paraboloid $z = x^2 + y^2$. Then $\mathbf{n} = \mathbf{N}/|\mathbf{N}|$ is an upper, and thus an *inner*, unit normal to the surface S_2. The unit *outer* normal is therefore $\mathbf{n}_2 = -\mathbf{n}$, opposite to the direction of $\mathbf{N} = \langle -2x, -2y, 1 \rangle$. With parameters (x, y) in the circular disk $x^2 + y^2 \leq 4$ in the xy-plane, the surface area element is $dS = |\mathbf{N}| \, dx \, dy$. Therefore the outward flux across S_2 is

$$\iint_{S_2} \mathbf{F} \cdot \mathbf{n}_2 \, dS = -\iint_{S_2} \mathbf{F} \cdot \mathbf{n} \, dS = -\iint_{D} \mathbf{F} \cdot \frac{\mathbf{N}}{|\mathbf{N}|} \, |\mathbf{N}| \, dx \, dy$$

$$= -\iint_{D} [(x)(-2x) + (y)(-2y) + (3)(1)] \, dx \, dy.$$

We change to polar coordinates and find that

$$\iint_{S_2} \mathbf{F} \cdot \mathbf{n}_2 \, dS = \int_0^{2\pi} \int_0^2 (2r^2 - 3) \, r \, dr \, d\theta = 2\pi \left[\tfrac{1}{2} r^4 - \tfrac{3}{2} r^2 \right]_0^2 = 4\pi.$$

Hence the total flux of $\mathbf{F}$ out of T is 16π, or approximately 50.27.

16-4 PROBLEMS

In each of Problems 1–5, evaluate the surface integral $\iint_S f(x, y, z) \, dS$.

1 $f(x, y, z) = xyz$; S is the first-octant part of the plane $x + y + z = 1$.

2 $f(x, y, z) = x + y$; S is the part of the plane $z = 2x + 3y$ that lies within the cylinder $x^2 + y^2 = 9$.

3 $f(x, y, z) = z$; S is the part of the paraboloid $z = r^2$ that lies within the cylinder $r = 2$.

4 $f(x, y, z) = (x^2 + y^2)z$; S is the hemispherical surface $\rho = 1$, $z \geq 0$. (*Suggestion:* Use spherical coordinates.)

5 $f(x, y, z) = z^2$; S is the cylindrical surface parametrized by $x = \cos \theta$, $y = \sin \theta$, $z = z$, $0 \leq \theta \leq 2\pi$, $0 \leq z \leq 2$.

In Problems 6–10, use Formulas (13) and (14) to evaluate the surface integral $\iint_S \mathbf{F} \cdot \mathbf{n} \, dS$, where $\mathbf{n}$ is the upward-pointing unit normal to the given surface S.

6 $\mathbf{F} = x\mathbf{i} + y\mathbf{j} + z\mathbf{k}$; S is the first-octant part of the plane $2x + 2y + z = 3$.

7 $\mathbf{F} = 2y\mathbf{i} + 3z\mathbf{k}$; S is the part of the plane $z = 3x + 2$ that lies within the cylinder $x^2 + y^2 = 4$.

8 $\mathbf{F} = z\mathbf{k}$; S is the upper half of the spherical surface $\rho = 2$. (*Suggestion:* Use spherical coordinates.)

9 $\mathbf{F} = y\mathbf{i} - x\mathbf{j}$; S is the part of the cone $z = r$ that lies within the cylinder $r = 3$.

10 $\mathbf{F} = 2x\mathbf{i} + 2y\mathbf{j} + 3\mathbf{k}$; S is the paraboloid $z = 4 - x^2 - y^2$ above the xy-plane.

11 The first-octant part of the spherical surface $\rho = a$ has unit density. Find its centroid.

12 The conical surface $z = r$, $r \leq a$, has constant density $\delta = k$. Find its centroid and its moment of inertia about the z-axis.

13 The paraboloid $z = r^2$, $r \leq a$, has constant density δ. Find its centroid and its moment of inertia about the z-axis.

14 Find the centroid of the part of the spherical surface $\rho = a$ that lies within the cone $r = z$.

15 Find the centroid of the part of the spherical surface $x^2 + y^2 + z^2 = 4$ that lies inside the cylinder $x^2 + y^2 = 2x$ and above the xy-plane.

16 Suppose that the toroidal surface of Example 4 in Section 15-7 has uniform density and total mass M. Show that its moment of inertia about the z-axis is $M(3a^2 + 2b^2)/2$.

17 Let S be the boundary of the region bounded by the coordinate planes and the plane $2x + 3y + z = 6$. Find the flux of $\mathbf{F} = x\mathbf{i} - y\mathbf{j}$ across S in the direction of its outer normal.

18 Let S be the boundary of the solid that is bounded by the paraboloid $z = 4 - x^2 - y^2$ and the xy-plane. Find the flux across S in the direction of the outer normal of the vector field $\mathbf{F} = 2x\mathbf{i} + 2y\mathbf{j} + 3\mathbf{k}$.

19 Let S be the boundary of the solid bounded by the paraboloids $z = x^2 + y^2$ and $z = 18 - x^2 - y^2$. Find the flux of $\mathbf{F} = z^2\mathbf{k}$ across S in the direction of the outer normal.

20 Let S be the surface $z = h(x, y)$ for (x, y) in the region D in the xy-plane, and let γ be the angle between $\mathbf{k}$ and the upper normal vector $\mathbf{N}$ to S. Show that

$$\iint_S f(x, y, z) \, dS = \iint_D f(x, y, h(x, y)) \sec \gamma \, dx \, dy.$$

21 Consider a homogeneous thin spherical shell S of radius a centered at the origin, with density δ and total mass $M = 4\pi a^2 \delta$. A particle of mass m is located at the point $(0, 0, c)$ with $c > a$. Use the method and notation of Problem 24 in Section 15-6 to show that the gravitational force of attraction between the particle and the spherical shell is

$$F = \iint_S \frac{Gm\delta \, dS}{w^2} = \frac{GMm}{c^2}.$$

The Divergence Theorem

The *divergence theorem* is for surface integrals what Green's theorem is for line integrals. It lets us convert a surface integral over a closed surface into a triple integral over the enclosed region, or vice versa. The divergence theorem is also known as *Gauss's theorem*, and in some eastern European countries it is called *Ostrogradski's theorem*. The German "prince of mathematics" Carl Friedrich Gauss (1777–1855) used it to study inverse-square force fields, while the Russian Michel Ostrogradski (1801–1861) used it to study heat flow. The work of each was done in the 1830s.

The surface S is called **piecewise smooth** if it consists of finitely many smooth parametric surfaces. It is called **closed** if it is the boundary of a bounded space region. For example, the boundary of a cube is a closed piecewise-smooth surface, as are the boundary of a pyramid and the boundary of a solid cylinder.

The Divergence Theorem

Suppose that S is a closed piecewise smooth surface bounding the space region T. Let $\mathbf{F} = P\mathbf{i} + Q\mathbf{j} + R\mathbf{k}$ be a vector field with component functions that have continuous first-order partial derivatives on T. Let $\mathbf{n}$ be the *outer* unit normal to S. Then

$$\iint_S \mathbf{F} \cdot \mathbf{n}\, dS = \iiint_T \nabla \cdot \mathbf{F}\, dV. \tag{1}$$

Equation (1) is a three-dimensional analogue of the vector form of Green's theorem (Equation (9) in Section 16-3). Note that the left-hand side of Equation (1) is the flux of $\mathbf{F}$ across S in the direction of the outer normal $\mathbf{n}$. The **divergence** $\nabla \cdot \mathbf{F}$ of the vector field $\mathbf{F}$ is given in three dimensions by

$$\operatorname{div} \mathbf{F} = \nabla \cdot \mathbf{F} = \frac{\partial P}{\partial x} + \frac{\partial Q}{\partial y} + \frac{\partial R}{\partial z}. \tag{2}$$

If $\mathbf{n}$ is given in terms of its direction cosines, as $\mathbf{n} = \langle \cos\alpha, \cos\beta, \cos\gamma \rangle$, then we can write the divergence theorem in scalar form:

$$\iint_S (P\cos\alpha + Q\cos\beta + R\cos\gamma)\, dS = \iiint_T \left(\frac{\partial P}{\partial x} + \frac{\partial Q}{\partial y} + \frac{\partial R}{\partial z} \right) dV. \tag{3}$$

It is best to parametrize S so that the normal vector given by the parametrization is the outer normal. For then Equation (3) can be written entirely in Cartesian form:

$$\iint_S P\, dy\, dz + Q\, dz\, dx + R\, dx\, dy = \iiint_T \left(\frac{\partial P}{\partial x} + \frac{\partial Q}{\partial y} + \frac{\partial R}{\partial z} \right) dV. \tag{4}$$

Proof of the Divergence Theorem We shall prove the theorem only in the case in which the region T is simultaneously x-simple, y-simple, and z-simple.

This guarantees that every straight line parallel to a coordinate axis intersects T, if at all, in a single point or a line segment. It will suffice to derive separately the equations

$$\iint_S P \, dy \, dz = \iiint_T \frac{\partial P}{\partial x} \, dV,$$

$$\iint_S Q \, dz \, dx = \iiint_T \frac{\partial Q}{\partial y} \, dV, \quad \text{and}$$

$$\iint_S R \, dx \, dy = \iiint_T \frac{\partial R}{\partial z} \, dV. \tag{5}$$

For the sum of the Equations (5) is Equation (4).

Since T is z-simple, it has the description $h_1(x, y) \leq z \leq h_2(x, y), (x, y)$ in D. As in Figure 16.21, we denote the lower surface $z = h_1(x, y)$ of T by S_1, the upper surface $z = h_2(x, y)$ by S_2, and the lateral surface between S_1 and S_2 by S_3. In the case of some simple surfaces—such as a sphere—there may be *no* S_3 to consider. But if there is, we see that

$$\iint_{S_3} R \, dx \, dy = \iint_{S_3} R \cos \gamma \, dS = 0 \tag{6}$$

because $\gamma = 90°$ at each point of the vertical cylinder S_3.

On the upper surface S_2, the upper unit normal corresponding to the parametrization $z = h_2(x, y)$ is the given outer normal $\mathbf{n}$, so Formula (15) of Section 16-4 tells us that

$$\iint_{S_2} R \, dx \, dy = \iint_D R(x, y, h_2(x, y)) \, dx \, dy. \tag{7}$$

But on the lower surface S_1, the unit upper normal corresponding to the parametrization $z = h_1(x, y)$ is the inner normal $-\mathbf{n}$, so we must reverse the sign. Thus

$$\iint_{S_1} R \, dx \, dy = -\iint_D R(x, y, h_1(x, y)) \, dx \, dy. \tag{8}$$

16.21 A z-simple space region bounded by the surfaces S_1, S_2, and S_3.

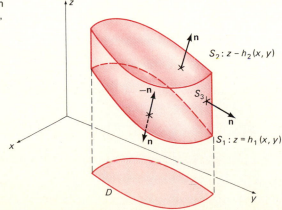

We add Equations (6), (7), and (8), and get

$$\iint\limits_{S} R\, dx\, dy = \iint\limits_{D} \Big(R(x, y, h_2(x, y)) - R(x, y, h_1(x, y)) \Big)\, dx\, dy$$

$$= \iint\limits_{D} \left(\int_{h_1(x,y)}^{h_2(x,y)} \frac{\partial R}{\partial z}\, dz \right) dx\, dy.$$

Therefore

$$\iint\limits_{S} R\, dx\, dy = \iiint\limits_{T} \frac{\partial R}{\partial z}\, dV.$$

This is the third of the three equations in (5), and the other two may be derived in much the same way.

The divergence theorem is established for more general regions by the device of subdivision of T into simpler regions, ones for which the above proof holds. For example, suppose that T is the shell between the concentric spheres S_a and S_b of radii a and b, with $0 < a < b$. The coordinate planes separate T into eight regions $T_1, T_2, T_3, \ldots, T_8$, each shaped like the one of Figure 16.22. Let Σ_i be the boundary of T_i, and let $\mathbf{n}_i$ be the outer unit normal to Σ_i. We apply the divergence theorem to each of these eight regions, and obtain

$$\iiint\limits_{T} \nabla \cdot \mathbf{F}\, dV = \sum_{i=1}^{8} \iiint\limits_{T_i} \nabla \cdot \mathbf{F}\, dV$$

$$= \sum_{i=1}^{8} \iint\limits_{i} \mathbf{F} \cdot \mathbf{n}_i\, dS \qquad \text{(divergence theorem)}$$

$$= \iint\limits_{S_a} \mathbf{F} \cdot \mathbf{n}_a\, dS + \iint\limits_{S_b} \mathbf{F} \cdot \mathbf{n}_b\, dS.$$

16.22 One octant of the shell between S_a and S_b.

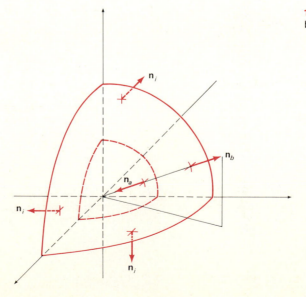

Here we write $\mathbf{n}_a$ for the inner normal on S_a and $\mathbf{n}_b$ for the outer normal for S_b. The last equality holds because the surface integrals over the internal boundary surfaces (the ones in the coordinate planes) cancel in pairs—the normals are oppositely oriented there. Since the boundary S of T is the union of the spherical surfaces S_a and S_b, it follows that

$$\iiint_T \mathbf{V} \cdot \mathbf{F}\, dV = \iint_S \mathbf{F} \cdot \mathbf{n}\, dS.$$

This is the divergence theorem for the spherical shell T.

EXAMPLE 1   Apply the divergence theorem to rework Example 4 in Section 16-4; that is, to find the flux of the vector field $\mathbf{F} = x\mathbf{i} + y\mathbf{j} + 3\mathbf{k}$ out of the region bounded by the paraboloid $z = x^2 + y^2$ and the plane $z = 4$.

Solution   We must compute $\iint_S \mathbf{F} \cdot \mathbf{n}\, dS$ where S is the surface of the region in question and $\mathbf{n}$ is the outer unit normal to S. The divergence theorem lets us replace this rather troublesome integral by an easy triple integral over the region T enclosed by S; the integral is easy because div $\mathbf{F} = 2$. Thus the flux of $\mathbf{F}$ out of T is

$$\iint_S \mathbf{F} \cdot \mathbf{n}\, dS = \iiint_T \operatorname{div} \mathbf{F}\, dV = \int_0^{2\pi} \int_0^2 \int_{r^2}^4 (\operatorname{div} \mathbf{F})\, r\, dz\, dr\, d\theta$$

$$= \int_0^{2\pi} \int_0^2 \int_{r^2}^4 2r\, dz\, dr\, d\theta = 16\pi.$$

EXAMPLE 2   Suppose that the region T is bounded by the closed surface S with a parametrization that gives the outer unit normal. Show that the volume V of T is given by

$$V = \tfrac{1}{3} \iint_S x\, dy\, dz + y\, dz\, dx + z\, dx\, dy. \tag{9}$$

Solution   Equation (9) follows immediately from Equation (4) if we take $P(x, y, z) = x$, $Q(x, y, z) = y$, and $R(x, y, z) = z$. For example, if S is the spherical surface $x^2 + y^2 + z^2 = a^2$ with outer unit normal

$$\mathbf{n} = \langle \cos \alpha, \cos \beta, \cos \gamma \rangle = \left\langle \frac{x}{a}, \frac{y}{a}, \frac{z}{a} \right\rangle,$$

then Equation (9) gives

$$V = \tfrac{1}{3} \iint_S x\, dy\, dz + y\, dz\, dx + z\, dx\, dy$$

$$= \tfrac{1}{3} \iint_S (x \cos \alpha + y \cos \beta + z \cos \gamma)\, dS$$

$$= \tfrac{1}{3} \iint_S \frac{x^2 + y^2 + z^2}{a}\, dS = \frac{a}{3} \iint_S 1\, dS = \frac{1}{3} aA.$$

This result is consistent with the familiar formulas $V = \tfrac{4}{3}\pi a^3$ and $A = 4\pi a^2$.

EXAMPLE 3 Show that the divergence of the vector field $\mathbf{F}$ at the point P is given by

$$[\text{div } \mathbf{F}](P) = \lim_{r \to 0} \frac{1}{V_r} \iint\limits_{S_r} \mathbf{F} \cdot \mathbf{n} \, dS \qquad (10)$$

where S_r is the sphere of radius r centered at P and $V_r = \frac{4}{3}\pi r^3$ is the volume of the ball B_r that it bounds.

Solution The divergence theorem gives

$$\iint\limits_{S_r} \mathbf{F} \cdot \mathbf{n} \, dS = \iiint\limits_{B_r} \text{div } \mathbf{F} \, dV.$$

Then we apply the average value property of triple integrals, a result analogous to the double integral result of Problem 30 in Section 15-1. This yields

$$\iiint\limits_{B_r} \text{div } \mathbf{F} \, dV = V_r \cdot [\text{div } \mathbf{F}](P^*)$$

for some point P^* of B_r (here we write $[\text{div } \mathbf{F}](P^*)$ for the value of div $\mathbf{F}$ at P^*). We assume that the component functions of $\mathbf{F}$ have continuous first order partial derivatives at P, so it follows that

$$[\text{div } \mathbf{F}](P^*) \to [\text{div } \mathbf{F}](P) \quad \text{as } P^* \to P.$$

Equation (10) follows after we divide both sides by V_r and then take the limit as $r \to 0$.

For instance, suppose that $\mathbf{F} = \rho \mathbf{v}$ is a fluid flow vector field. We can interpret Equation (10) as saying that $[\text{div } \mathbf{F}](P)$ is the net rate per unit volume that fluid mass is flowing away (or "diverging") from the point P. For this reason the point P is called a **source** if $[\text{div } \mathbf{F}](P) > 0$ but a **sink** if $[\text{div } \mathbf{F}](P) < 0$.

*GAUSS'S LAW FOR INVERSE-SQUARE FORCE FIELDS

Suppose that point masses $m_1, m_2, m_3, \ldots, m_k$ are located at the points $\mathbf{r}_1, \mathbf{r}_2, \mathbf{r}_3, \ldots, \mathbf{r}_k$, respectively. We write the endpoint of $\mathbf{r}_i$ as $\langle x_i, y_i, z_i \rangle$. By Newton's law of gravitation, the force exerted by m_i on a unit mass at the point $\mathbf{r} = \langle x, y, z \rangle$ is

$$\mathbf{F}_i = -Gm_i \frac{\mathbf{r} - \mathbf{r}_i}{|\mathbf{r} - \mathbf{r}_i|^3}.$$

Hence the total gravitational force the particles exert on the unit mass is

$$\mathbf{F} = \sum_{i=1}^{k} \mathbf{F}_i = -G \sum_{i=1}^{k} m_i \frac{\mathbf{r} - \mathbf{r}_i}{|\mathbf{r} - \mathbf{r}_i|^3}.$$

Gauss's law gives the (outer) flux of $\mathbf{F}$ across a closed surface S enclosing these masses as

$$\iint\limits_{S} \mathbf{F} \cdot \mathbf{n} \, dS = -4\pi G(m_1 + m_2 + m_3 + \ldots + m_k). \qquad (11)$$

To prove this result, we let $B_1, B_2, B_3, \ldots, B_k$ be small spherical balls centered at the points $\mathbf{r}_1, \mathbf{r}_2, \mathbf{r}_3, \ldots, \mathbf{r}_k$, with no two intersecting and so small that all lie completely within the surface S. This is represented schematically in Figure 16.23. Let S_i denote the boundary of B_i, and let $\mathbf{n}_i$ be the outer unit normal to the spherical surface S_i. Let a_i be the radius of S_i. In Problem 11 we ask you to verify that div $\mathbf{F}_i = 0$ except at the point $\mathbf{r}_i$. Granted this, it follows that

$$\iint\limits_{S_i} \mathbf{F} \cdot \mathbf{n}_i \, dS = \sum_{j=1}^{k} \iint\limits_{S_i} \mathbf{F}_j \cdot \mathbf{n}_i \, dS = \iint\limits_{S_i} \mathbf{F}_i \cdot \mathbf{n}_i \, dS$$

$$= -Gm_i \iint\limits_{S_i} \frac{\mathbf{r} - \mathbf{r}_i}{|\mathbf{r} - \mathbf{r}_i|^3} \cdot \frac{\mathbf{r} - \mathbf{r}_i}{|\mathbf{r} - \mathbf{r}_i|} \, dS = -Gm_i \iint\limits_{S_i} \frac{1}{a_i^2} \, dS.$$

Therefore

$$\iint\limits_{S_i} \mathbf{F} \cdot \mathbf{n}_i \, dS = -4\pi G m_i \qquad (12)$$

because the area of S_i is $4\pi a_i^2$.

Now div $\mathbf{F} = 0$ on the region T bounded by S and the spheres $S_1, S_2, S_3, \ldots, S_k$ (Problem 11). So the divergence theorem yields

$$\iint\limits_{S} \mathbf{F} \cdot \mathbf{n} \, dS - \sum_{i=1}^{k} \iint\limits_{S_i} \mathbf{F} \cdot \mathbf{n}_i \, dS = \iiint\limits_{T} \text{div } \mathbf{F} \, dV = 0.$$

Hence

$$\iint\limits_{S} \mathbf{F} \cdot \mathbf{n} \, dS = \sum_{i=1}^{k} \iint\limits_{S_i} \mathbf{F} \cdot \mathbf{n}_i \, dS = -4\pi G(m_1 + m_2 + \cdots + m_k)$$

by (12), and Equation (11) follows.

Gauss's law can be generalized to the case of a total mass M distributed continuously throughout a body B, with gravitational force field $\mathbf{F}$. If B is subdivided into small pieces $\Delta m_1, \Delta m_2, \Delta m_3, \ldots, \Delta m_k$, having the force fields $\Delta\mathbf{F}_1, \Delta\mathbf{F}_2, \Delta\mathbf{F}_3, \ldots, \Delta\mathbf{F}_k$, and these pieces are approximated by point masses, then

$$\iint\limits_{S} \mathbf{F} \cdot \mathbf{n} \, dS = \sum_{i=1}^{k} \iint\limits_{S} (\Delta\mathbf{F}_i) \cdot \mathbf{n} \, dS \approx \sum_{i=1}^{k} (-4\pi G \, \Delta m_i) = -4\pi G M$$

for any closed surface S containing the body in its interior. A limit argument then gives

$$\iint\limits_{S} \mathbf{F} \cdot \mathbf{n} \, dS = -4\pi G M \qquad (13)$$

for the flux across the surface S in terms of the total mass M enclosed by S.

As an application of Gauss's law, we deduce Newton's important result that the gravitational force $\mathbf{F}$ exerted on an exterior particle of unit mass by a uniform solid sphere is the same as if all its mass M were concentrated at its center. Arrange coordinates so that the spherical mass is centered at the origin, and let S be a large sphere of radius r centered at the origin, as shown

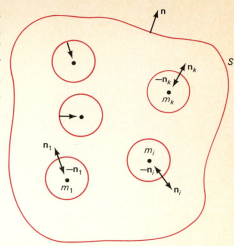

16.23 A two-dimensional diagram of the three-dimensional situation in Gauss's law.

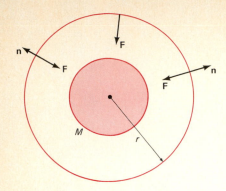

16.24 A two-dimensional diagram: Using Gauss's law to derive Newton's result on the gravitational attraction of a uniform three-dimensional solid ball.

in Figure 16.24. By symmetry, $\mathbf{F}$ points towards the origin and has the same magnitude $|\mathbf{F}|$ at each point of S. So we need only verify that $|\mathbf{F}| = GM/r^2$, because this would be the force were M contracted to a point mass at the origin. But Gauss's law in (13) gives

$$-4\pi GM = \iint_S \mathbf{F} \cdot \mathbf{n} \, dS = \iint_S (-|\mathbf{F}|) \, dS = -4\pi r^2 |\mathbf{F}|,$$

so we obtain the desired value $|\mathbf{F}| = GM/r^2$.

Gauss's law also applies to electric force fields. The force field (exerted on a unit positive charge) of a charge q at the origin is

$$\mathbf{E} = \frac{q}{4\pi\varepsilon_0} \frac{\mathbf{r}}{|\mathbf{r}|^3}$$

where $\varepsilon_0 \approx 8.85 \times 10^{-12}$ in mks units (charge in coulombs). We replace G above by $-1/(4\pi\varepsilon_0)$ and thus obtain

$$\iint_S \mathbf{E} \cdot \mathbf{n} \, dS = \frac{Q}{\varepsilon_0} \tag{14}$$

for the electric field flux across the surface S in terms of the total charge Q enclosed by S.

This is only a taste of the extensive physical applications of the divergence theorem. In Miscellaneous Problems 25–27 (at the end of this chapter), we ask you to apply it to show that—if $u = u(x, y, z, t)$ is the temperature at the point (x, y, z) at the time t in a body through which heat is flowing—then the function u must satisfy the equation

$$\frac{\partial^2 u}{\partial x^2} + \frac{\partial^2 u}{\partial y^2} + \frac{\partial^2 u}{\partial z^2} = \frac{1}{k} \frac{\partial u}{\partial t} \tag{15}$$

where k is a constant (the *thermal diffusivity* of the body). This is a *partial differential equation* called the **heat equation.** If both the initial temperature $u(x, y, z, 0)$ and the temperatures on the boundary of the body are given, then its interior temperatures at future times are determined by the heat equation. A large part of advanced applied mathematics consists of techniques for solving such partial differential equations.

Yet another impressive consequence of the divergence theorem is Archimedes' law of buoyancy; see Problem 19 below and also Problem 20 in Section 16-6.

16-5 PROBLEMS

In each of Problems 1–5, verify the divergence theorem by direct computation of both the surface integral and the triple integral in Equation (1).

1 $\mathbf{F} = x\mathbf{i} + y\mathbf{j} + z\mathbf{k}$; S is the spherical surface with equation $x^2 + y^2 + z^2 = 1$.
2 $\mathbf{F} = |\mathbf{r}|\mathbf{r}$ where $\mathbf{r} = x\mathbf{i} + y\mathbf{j} + z\mathbf{k}$; S is the spherical surface with equation $x^2 + y^2 + z^2 = 9$.
3 $\mathbf{F} = x\mathbf{i} + y\mathbf{j} + z\mathbf{k}$; S is the surface of the cube that is bounded by the three coordinate planes and the three planes $x = 2, y = 2$, and $z = 2$.

4 $\mathbf{F} = xy\mathbf{i} + yz\mathbf{j} + xz\mathbf{k}$; S is the surface of Problem 3.
5 $\mathbf{F} = (x + y)\mathbf{i} + (y + z)\mathbf{j} + (z + x)\mathbf{k}$; S is the surface of the tetrahedron bounded by the three coordinate planes and the plane $x + y + z = 1$.

In Problems 6–10, use the divergence theorem to find $\iint_S \mathbf{F} \cdot \mathbf{n} \, dS$ where $\mathbf{n}$ is the outer unit normal to the surface S.

6 $\mathbf{F} = x^2\mathbf{i} + y^2\mathbf{j} + z^2\mathbf{k}$; S is the surface of Problem 3.
7 $\mathbf{F} = x^3\mathbf{i} + y^3\mathbf{j} + z^3\mathbf{k}$; S is the surface of the cylinder that is bounded by $x^2 + y^2 = 9$, $z = -1$, and $z = 4$.

8 $\mathbf{F} = |\mathbf{r}|\mathbf{r}$ where $\mathbf{r} = x\mathbf{i} + y\mathbf{j} + z\mathbf{k}$; S is the surface of the region bounded by the plane $z = 0$ and the paraboloid $z = 25 - x^2 - y^2$.

9 $\mathbf{F} = (x^2 + e^{-yz})\mathbf{i} + (y + \ln xz)\mathbf{j} + (\ln xy)\mathbf{k}$; S is the surface of Problem 5.

10 $\mathbf{F} = (xy^2 + e^{-y}\sin z)\mathbf{i} + (x^2y + e^{-x}\cos z)\mathbf{j} + (\tan^{-1}xy)\mathbf{k}$; S is the surface of the region bounded by the paraboloid $z = x^2 + y^2$ and the plane $z = 9$.

11 Let $\mathbf{r} = \langle x, y, z \rangle$, $\mathbf{r}_0 = \langle x_0, y_0, z_0 \rangle$, and suppose that $\mathbf{F} = (\mathbf{r} - \mathbf{r}_0)/|\mathbf{r} - \mathbf{r}_0|^3$. Show that div $\mathbf{F} = 0$ except at the point $\mathbf{r}_0$.

12 Suppose that $\mathbf{F}$ is a vector-valued function, g is a scalar-valued function, and both are differentiable. Show that

$$\text{div}(g\mathbf{F}) = (g)(\text{div } \mathbf{F}) + (\text{grad } g) \cdot \mathbf{F}.$$

13 The **Laplacian** of the twice-differentiable scalar function f is defined by means of the equation $\nabla^2 f = \text{div}(\text{grad } f) = \nabla \cdot \nabla f$. Show that

$$\nabla^2 f = \frac{\partial^2 f}{\partial x^2} + \frac{\partial^2 f}{\partial y^2} + \frac{\partial^2 f}{\partial z^2}.$$

14 Let $\partial f/\partial \mathbf{n} = \nabla f \cdot \mathbf{n}$ denote the directional derivative of the scalar function f in the direction of the outer unit normal $\mathbf{n}$ to the surface S that bounds the region T. Show that

$$\iint_S \frac{\partial f}{\partial \mathbf{n}}\, dS = \iiint_T \nabla^2 f\, dV.$$

15 Suppose that $\nabla^2 f \equiv 0$ in the region T with boundary S. Show that

$$\iint_S f \frac{\partial f}{\partial \mathbf{n}}\, dS = \iiint_T |\nabla f|^2\, dV.$$

(See Problems 13 and 14 for the notation.)

16 Apply the divergence theorem to $\mathbf{F} = f \nabla g$ to establish **Green's first identity**

$$\iint_S f \frac{\partial g}{\partial \mathbf{n}}\, dS = \iiint_T (f\nabla^2 g + \nabla f \cdot \nabla g)\, dV.$$

17 Interchange f and g in Green's first identity (Problem 16) to establish **Green's second identity**

$$\iint_S \left(f \frac{\partial g}{\partial \mathbf{n}} - g \frac{\partial f}{\partial \mathbf{n}} \right) dS = \iiint_T (f\nabla^2 g - g\nabla^2 f)\, dV.$$

18 Suppose that f is a differentiable scalar function defined on the region T of space and that S is the boundary of T. Show that

$$\iint_S f\mathbf{n}\, dS = \iiint_T \nabla f\, dV.$$

(*Suggestion:* Apply the divergence theorem to $\mathbf{F} = f\mathbf{a}$ where $\mathbf{a}$ is an arbitrary constant vector. Note: Integrals of vector-valued functions are defined by componentwise integration.)

19 (Archimedes' law of buoyancy) Let S be the surface of a body T submerged in a fluid of constant density ρ. Set up coordinates so that $z = 0$ corresponds to the surface of the fluid and so that positive values of z are measured *downward* from the surface. Then the pressure at depth z is $p = \rho g z$. The buoyant force exerted on the body by the fluid is $\mathbf{B} = -\iint_S p\mathbf{n}\, dS$ (why?). Apply the result of Problem 18 to show that $\mathbf{B} = -W\mathbf{k}$, where W is the weight of fluid displaced by the body. Since z is measured downward, the vector $\mathbf{B}$ is directed upward.

20 Let $\mathbf{F}$ be the gravitational force field due to a uniform distribution of mass in the shell bounded by the concentric spherical surfaces $\rho = a$ and $\rho = b > a$. Apply Gauss's law in (13), with S the sphere $\rho = r < a$, to show that $\mathbf{F}$ is zero at points interior to this shell.

21 Consider a solid sphere of radius a and constant density ρ. Apply Gauss's law to show that the gravitational force on a unit mass particle at a distance $r < a$ from the center of the sphere is given by $F = GM_r/r^2$, where $M_r = \frac{4}{3}\pi\rho r^3$ is the mass enclosed by the sphere of radius r.

22 Consider an infinitely long vertical straight wire with a uniform positive charge of q coulombs per meter. By symmetry, the electric field vector $\mathbf{E}$ is a radial vector directed away from the wire. Apply Gauss's law (14), with S a cylinder with the wire as its axis, to show that $|\mathbf{E}| = q/2\pi\varepsilon_0 r$. That is, the electric field intensity is inversely proportional to distance r from the wire.

16-6

Stokes' Theorem

In Equation (9) of Section 16-3, we gave a vector form of Green's theorem

$$\oint_C P\, dx + Q\, dy = \iint_R \left(\frac{\partial Q}{\partial x} - \frac{\partial P}{\partial y} \right) dA \qquad (1)$$

that amounted to a two-dimensional version of the divergence theorem. There is another vector form of Green's theorem, one that involves the **curl** of a vector field. If $\mathbf{F} = P\mathbf{i} + Q\mathbf{j} + R\mathbf{k}$ is a vector field, then curl $\mathbf{F}$

is the vector field with *definition*

$$\text{curl } \mathbf{F} = \nabla \times \mathbf{F} = \begin{vmatrix} \mathbf{i} & \mathbf{j} & \mathbf{k} \\ \dfrac{\partial}{\partial x} & \dfrac{\partial}{\partial y} & \dfrac{\partial}{\partial z} \\ P & Q & R \end{vmatrix}$$

$$= \left(\frac{\partial R}{\partial y} - \frac{\partial Q}{\partial z} \right) \mathbf{i} + \left(\frac{\partial P}{\partial z} - \frac{\partial R}{\partial x} \right) \mathbf{j} + \left(\frac{\partial Q}{\partial x} - \frac{\partial P}{\partial y} \right) \mathbf{k}. \qquad (2)$$

Note first that the **k**-component of $\nabla \times \mathbf{F}$ is the integrand of the double integral of Equation (1). We know from Section 16-1 that the line integral in (1) can also be written as $\oint_C \mathbf{F} \cdot \mathbf{T} \, ds$, where $\mathbf{T}$ is the positive-directed unit tangent. Consequently Green's theorem can be rewritten in the form

$$\oint_C \mathbf{F} \cdot \mathbf{T} \, ds = \iint_R (\text{curl } \mathbf{F}) \cdot \mathbf{k} \, dA. \qquad (3)$$

Stokes' theorem is the generalization of Formula (3) that we get by replacing the plane region R with a floppy two-dimensional version: an oriented bounded surface S in three-dimensional space with boundary C that consists of one or more simple closed space curves.

Recall that an *oriented* surface is one with a continuous unit normal vector **n**. The positive orientation of the boundary C of an oriented surface S corresponds to the unit tangent vector **T** such that $\mathbf{n} \times \mathbf{T}$ always points *into* S (see Figure 16.25). You should check that, for a plane region with unit normal **k**, the positive orientation of its outer boundary curve is counterclockwise.

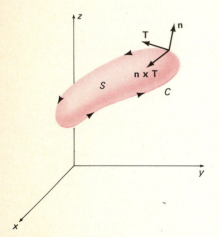

16.25 Vectors, surface, and boundary curve mentioned in the statement of Stokes' theorem.

Stokes' Theorem

Let S be an oriented, bounded piecewise smooth surface in space with positively oriented boundary C. Suppose that the components of the vector field **F** have continuous first order partial derivatives in a space region containing S. Then

$$\oint_C \mathbf{F} \cdot \mathbf{T} \, ds = \iint_S (\text{curl } \mathbf{F}) \cdot \mathbf{n} \, dS. \qquad (4)$$

Thus Stokes' theorem says that *the line integral around the boundary curve of the tangential component of* **F** *is equal to the surface integral of the normal component of curl* **F**. Compare Equations (3) and (4).

This result first appeared publicly as a problem posed by George Stokes (1819–1903) on a prize examination for Cambridge University students in 1854. It had been stated in an 1850 letter to Stokes from the physicist William Thomson (Lord Kelvin, 1824–1907).

In terms of the components of $\mathbf{F} = P\mathbf{i} + Q\mathbf{j} + R\mathbf{k}$ and those of curl **F**, Stokes' theorem can—with the aid of Equation (14) of Section 16-4—be recast into its scalar form:

$$\oint_C P\,dx + Q\,dy + R\,dz = \iint_S \left(\frac{\partial R}{\partial y} - \frac{\partial Q}{\partial z}\right)dy\,dz + \left(\frac{\partial P}{\partial z} - \frac{\partial R}{\partial x}\right)dz\,dx$$

$$+ \left(\frac{\partial Q}{\partial x} - \frac{\partial P}{\partial y}\right)dx\,dy. \qquad (5)$$

Here we need the usual stipulation that the parametrization of S corresponds to the given unit normal $\mathbf{n}$.

To prove Stokes' theorem, it is therefore enough to establish the equation

$$\oint_C P\,dx = \iint_S \left(\frac{\partial P}{\partial z}\,dz\,dx - \frac{\partial P}{\partial y}\,dx\,dy\right) \qquad (6)$$

and the corresponding two equations that are the Q and R "components" of Equation (5). Equation (5) itself then follows by adding the three results.

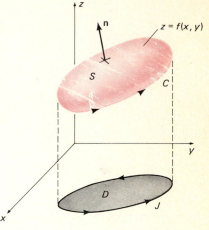

16.26

Partial Proof Suppose first that S is the graph of a function $z = f(x, y)$, (x, y) in D, with an upper unit normal and with D a region in the xy-plane that is bounded by the simple closed curve J. See Figure 16.26. Then

$$\oint_C P\,dx = \oint_J P(x, y, f(x, y))\,dx$$

$$= \oint_J p(x, y)\,dx \qquad \text{(where } p(x, y) \equiv P(x, y, f(x, y)))$$

$$= -\iint_D \frac{\partial p}{\partial y}\,dx\,dy \qquad \text{(by Green's theorem).}$$

We now use the chain rule to compute $\partial p/\partial y$ and find that

$$\oint_C P\,dx = -\iint_D \left(\frac{\partial P}{\partial y} + \frac{\partial P}{\partial z}\frac{\partial z}{\partial y}\right)dx\,dy. \qquad (7)$$

Next we use Equation (15) of Section 16-4:

$$\iint_S P\,dy\,dz + Q\,dz\,dx + R\,dx\,dy = \iint_D \left(-P\frac{\partial z}{\partial x} - Q\frac{\partial z}{\partial y} + R\right)dx\,dy.$$

In this equation we replace P by 0, Q by $\partial P/\partial z$, and R by $-\partial P/\partial y$. This gives

$$\iint_S \left(\frac{\partial P}{\partial z}\,dz\,dx - \frac{\partial P}{\partial y}\,dx\,dy\right) = \iint_D \left(-\frac{\partial P}{\partial z}\frac{\partial z}{\partial y} - \frac{\partial P}{\partial y}\right)dx\,dy. \qquad (8)$$

Finally we compare Equations (7) and (8) and see that we have established Equation (6). If the surface S can also be written in the forms $y = g(x, z)$ and $x = h(y, z)$, then the Q and R "components" of Equation (5) can be derived in essentially the same way. This proves Stokes' theorem for the special case of a surface S that can be represented as a graph in all three coordinate directions. Stokes' theorem may then be extended to a more general oriented surface in the now-familiar way—by subdividing it into simpler surfaces, to each of which the above proof is applicable.

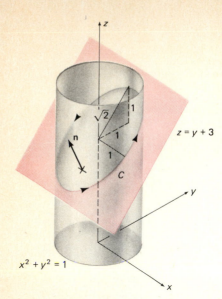

16.27 The ellipse of Example 1.

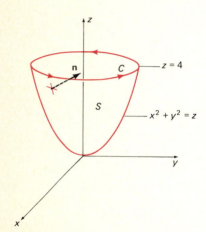

16.28 The parabolic surface of Example 2.

16.29 A physical interpretation of the curl of a vector field.

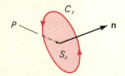

EXAMPLE 1 Apply Stokes' theorem to evaluate $\oint_C \mathbf{F} \cdot \mathbf{T}\, ds$, where C is the ellipse in which the plane $z = y + 3$ intersects the cylinder $x^2 + y^2 = 1$. Orient the ellipse counterclockwise as viewed from above. Take $\mathbf{F}(x, y, z) = 3z\mathbf{i} + 5x\mathbf{j} - 2y\mathbf{k}$.

Solution The plane, cylinder, and ellipse appear in Figure 16.27. The given orientation of C corresponds to the upward unit normal $\mathbf{n} = (-\mathbf{j} + \mathbf{k})/\sqrt{2}$ to the elliptical region S in the plane $z = y + 3$ bounded by C. Now

$$\text{curl } \mathbf{F} = \begin{vmatrix} \mathbf{i} & \mathbf{j} & \mathbf{k} \\ \dfrac{\partial}{\partial x} & \dfrac{\partial}{\partial y} & \dfrac{\partial}{\partial z} \\ 3z & 5x & -2y \end{vmatrix} = -2\mathbf{i} + 3\mathbf{j} + 5\mathbf{k}.$$

So, by Stokes' theorem,

$$\oint_C \mathbf{F} \cdot \mathbf{T}\, ds = \iint_S (\text{curl } \mathbf{F}) \cdot \mathbf{n}\, dS$$

$$= \iint_S \frac{2}{\sqrt{2}}\, dS = \frac{2}{\sqrt{2}} \text{ area } (S) = 2\pi,$$

since we can see from Figure 16.27 that S is an ellipse with semiaxes 1 and $\sqrt{2}$; thus its area is $\pi\sqrt{2}$.

EXAMPLE 2 Apply Stokes' theorem to evaluate $\iint_S (\nabla \times \mathbf{F}) \cdot \mathbf{n}\, dS$ where $\mathbf{F} = 3z\mathbf{i} + 5x\mathbf{j} - 2y\mathbf{k}$ and S is the part of the parabolic surface $z = x^2 + y^2$ that lies below the plane $z = 4$ and has orientation given by the upper unit normal vector (see Figure 16.28).

Solution We parametrize the boundary circle C of S by $x = 2\cos t$, $y = 2\sin t$, $z = 4$ for $0 \leq t \leq 2\pi$. Then $dx = -2\sin t\, dt$, $dy = 2\cos t\, dt$, and $dz = 0$. So Stokes' theorem yields

$$\iint_S (\nabla \times \mathbf{F}) \cdot \mathbf{n}\, dS = \oint_C \mathbf{F} \cdot \mathbf{T}\, ds$$

$$= \oint_C 3z\, dx + 5x\, dy - 2y\, dz$$

$$= \int_0^{2\pi} (-24\sin t + 20\cos^2 t)\, dt = 20\pi.$$

Just as the divergence theorem yields a physical interpretation of div $\mathbf{F}$ (Formula (10) in Section 16-5), Stokes' theorem yields a physical interpretation of curl $\mathbf{F}$. Let S_r be a circular disk of radius r, centered at the point P in space, and perpendicular to the unit vector $\mathbf{n}$. Let C_r be the boundary circle of S_r; see Figure 16.29. Then Stokes' theorem and the average value property of double integrals together give

$$\oint_{C_r} \mathbf{F} \cdot \mathbf{T}\, ds = \iint_{S_r} (\text{curl } \mathbf{F}) \cdot \mathbf{n}\, dS = \pi r^2 [\{\text{curl } \mathbf{F}\}(P^*)] \cdot \mathbf{n}$$

for some point P^* of S_r. We divide this equality by πr^2 and then take the limit as $r \to 0$. This gives

$$[\{\text{curl } \mathbf{F}\}(P)] \cdot \mathbf{n} = \lim_{r \to 0} \frac{1}{\pi r^2} \oint_{C_r} \mathbf{F} \cdot \mathbf{T} \, ds. \qquad (9)$$

This last formula has a natural physical meaning. Suppose that $\mathbf{F} = \rho \mathbf{v}$, where $\mathbf{v}$ is the velocity vector field of a steady-state fluid flow with constant density ρ. Then the value of the line integral

$$\Gamma(C) = \oint_C \mathbf{F} \cdot \mathbf{T} \, ds \qquad (10)$$

measures the rate of flow of fluid mass *around* the curve C and is therefore called the **circulation** of $\mathbf{F}$ around C. From (9) we see that

$$[\{\text{curl } \mathbf{F}\}(P)] \cdot \mathbf{n} \approx \frac{\Gamma(C_r)}{\pi r^2}$$

if C_r is a circle of very small radius r centered at P and perpendicular to $\mathbf{n}$. If $\{\text{curl } \mathbf{F}\}(P) \neq \mathbf{0}$, it follows that $\Gamma(C_r)$ is greatest (for r fixed and small) when the unit vector $\mathbf{n}$ points in the direction of $\{\text{curl } \mathbf{F}\}(P)$. Hence the line through P determined by $\{\text{curl } \mathbf{F}\}(P)$ is the axis about which the fluid near P is revolving the most rapidly. A tiny paddle wheel placed in the fluid at P would rotate fastest if its axis lay along this line. In Miscellaneous Problem 32 (at the end of the chapter), we ask you to show that, in the case of a fluid revolving steadily about a fixed axis with constant angular speed ω (in radians per second), $|\text{curl } \mathbf{F}| = 2\rho\omega$. Thus $\{\text{curl } \mathbf{F}\}(P)$ indicates both the direction *and* rate of rotation of the fluid near P. Because of this interpretation, the notation rot $\mathbf{F}$ for the curl is sometimes seen in older books, an abbreviation that we are happy has disappeared from general use.

If curl $\mathbf{F} = \mathbf{0}$ everywhere, the fluid flow and the vector field $\mathbf{F}$ itself are said to be **irrotational.** An infinitesimal straw placed in an irrotational fluid flow would be translated parallel to itself without rotating. It turns out that a vector field $\mathbf{F}$ defined on a simply connected region D is irrotational if and only if it is conservative, which in turn is true if and only if the line integral $\int_C \mathbf{F} \cdot \mathbf{T} \, ds$ is independent of the path in D. (The region D is said to be **simply connected** if every simple closed curve in D can be continuously shrunk to a point while staying within D. The interior of a torus is an example of a space region that is *not* simply connected. It is true, though not obvious, that any piecewise-smooth simple closed curve in a simply connected region D is the boundary of a piecewise smooth oriented surface in D.)

> **Theorem** *Conservative and Irrotational Fields*
>
> Let $\mathbf{F}$ be a vector field with continuous first order partial derivatives in a simply connected space region D. Then the vector field $\mathbf{F}$ is irrotational if and only if it is conservative; that is, if and only if $\mathbf{F} = \nabla\phi$ for some scalar function ϕ defined on D.

Partial Proof A complete proof of the "if" part of the theorem is easy; in Problem 11 we ask you to show that $\nabla \times (\nabla\phi) = \mathbf{0}$ for any twice-differentiable scalar function ϕ.

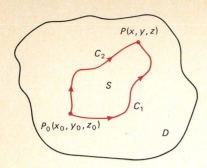

16.30 Two paths from P_0 to P in the simply connected space region D.

Here is a sketch of how one might show the "only if" part of the proof of the theorem. Assume that $\mathbf{F}$ is irrotational. Let $P_0(x_0, y_0, z_0)$ be a fixed point of D. Given an arbitrary point $P(x, y, z)$ of D, we would like to define

$$\phi(x, y, z) = \int_{C_1} \mathbf{F} \cdot \mathbf{T}\, ds \tag{11}$$

where C_1 is a path in D from P_0 to P. But it is essential that we show that any *other* path C_2 from P_0 to P would give the *same* value for $\phi(x, y, z)$.

We may assume, as suggested by Figure 16.30, that C_1 and C_2 intersect only at their endpoints. Then the simple closed curve $C = C_1 \cup (-C_2)$ bounds an oriented surface S, and Stokes' theorem gives

$$\int_{C_1} \mathbf{F} \cdot \mathbf{T}\, ds - \int_{C_2} \mathbf{F} \cdot \mathbf{T}\, ds = \oint_C \mathbf{F} \cdot \mathbf{T}\, ds$$

$$= \iint_S (\nabla \times \mathbf{F}) \cdot \mathbf{n}\, dS = 0$$

because of our hypothesis that $\nabla \times \mathbf{F} \equiv \mathbf{0}$. This shows that the line integral $\int_C \mathbf{F} \cdot \mathbf{T}\, ds$ is *independent of path*, just as desired. In Problem 19 we ask you to complete this proof by showing that the function ϕ of Equation (11) is the one whose existence is claimed by the theorem. That is, if ϕ is defined as in Equation (11), then $\mathbf{F} = \nabla\phi$.

APPLICATION   Suppose that $\mathbf{v}$ is the velocity vector field of a steady fluid flow that is both irrotational and incompressible—the density ρ of the fluid is constant. Suppose that S is any closed surface bounding a region T. Then, because of conservation of mass, the flux of $\mathbf{v}$ across S must be zero; the mass of fluid within S remains constant. Hence the divergence theorem gives

$$\iiint_T \text{div } \mathbf{v}\, dV = \iint_S \mathbf{v} \cdot \mathbf{n}\, dS = 0.$$

Since this holds for *any* region T, it follows from the usual average value property argument that div $\mathbf{v} = 0$ everywhere. The scalar function ϕ provided by the theorem above, for which $\mathbf{v} = \nabla\phi$, is called the **velocity potential** of the fluid flow. We substitute $\mathbf{v} = \nabla\phi$ into the equation div $\mathbf{v} = 0$ and thereby obtain

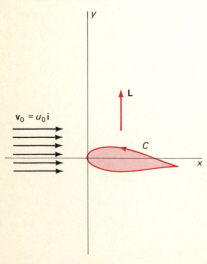

16.31 An airfoil.

$$\text{div}(\nabla\phi) = \frac{\partial^2\phi}{\partial x^2} + \frac{\partial^2\phi}{\partial y^2} + \frac{\partial^2\phi}{\partial z^2} = 0. \tag{12}$$

Thus the velocity potential ϕ of an irrotational and incompressible fluid flow satisfies *Laplace's equation*.

Now consider an *airfoil*, shaped like a (noncircular) cylinder perpendicular to the xy-plane and placed in a steady flow that is initially parallel to the x-axis, $\mathbf{v}_0 = u_0\mathbf{i}$, as shown in cross section in Figure 16.31. The assumption of incompressibility is valid in the range of subsonic aerodynamics (Mach number less than 0.3). To determine the resulting flow around the airfoil, it suffices to solve Laplace's equation in (12) for the velocity potential ϕ. The solution ϕ must satisfy the *boundary condition* that $\mathbf{v} = \nabla\phi \approx u_0\mathbf{i}$ at points far from the airfoil, and that $\mathbf{v}$ is tangential to the boundary curve C of the airfoil. (Since the z-coordinate is not involved, we may project everything into the xy-plane.)

The solution of Laplace's equation ordinarily requires advanced techniques, but suppose that it has been carried out and that we now know the velocity vector field $\mathbf{v} = \nabla\phi$ of the fluid flow around the airfoil. Then a standard theorem of fluid dynamics says that the net force or *lift* $\mathbf{L}$ exerted on the airfoil by the fluid is perpendicular to the *free stream* velocity $\mathbf{v}_0 = u_0\mathbf{i}$, and its magnitude is $|\mathbf{L}| = \rho u_0 \Gamma$, where $\Gamma = \oint_C \mathbf{v} \cdot \mathbf{T} \, ds$ is the circulation of $\mathbf{v}$ around C (the boundary curve of the airfoil). Thus the question of "whether it'll fly" boils down to the evaluation of a line integral! For the "standard theorem" to which we refer above, see page 122 in W. F. Hughes and J. A. Brighton, *Theory and Problems of Fluid Dynamics*, Schaum's Outline (New York: McGraw-Hill, 1967).

This brief outline of a single application should indicate how the mathematics of this chapter forms the starting point for investigations in a number of areas, including acoustics, aerodynamics, electromagnetism, meteorology, and oceanography, among others. Indeed, the entire subject of vector analysis stems historically from its scientific applications rather than from abstract mathematical considerations. The modern form of the subject is due primarily to J. Willard Gibbs (1839–1903), the first great American physicist, and the English electrical engineer Oliver Heaviside (1850–1925).

16-6 PROBLEMS

In Problems 1–5, use Stokes' theorem for the evaluation of $\iint_S [\text{curl } \mathbf{F}] \cdot \mathbf{n} \, dS$.

1 $\mathbf{F} = 3y\mathbf{i} - 2x\mathbf{j} + xyz\mathbf{k}$, and S is the hemispherical surface $z = +(4 - x^2 - y^2)^{1/2}$ with upper unit normal.

2 $\mathbf{F} = 2y\mathbf{i} + 3x\mathbf{j} + e^z\mathbf{k}$, and S is the part of the paraboloid $z = x^2 + y^2$ below the plane $z = 4$ with upper unit normal.

3 $\mathbf{F} = \langle xy, -2, \tan^{-1}x^2 \rangle$, and S is the part of the paraboloid $z = 9 - x^2 - y^2$ above the xy-plane with upper unit normal.

4 $\mathbf{F} = yz\mathbf{i} + xz\mathbf{j} + xy\mathbf{k}$, and S is the part of the cylinder $x^2 + y^2 = 1$ between the two planes $z = 1$ and $z = 3$ with outer unit normal.

5 $\mathbf{F} = \langle yz, -xz, z^3 \rangle$, and S is the part of the cone $z = (x^2 + y^2)^{1/2}$ between the two planes $z = 1$ and $z = 3$ with upper unit normal.

In Problems 6–10, use Stokes' theorem to evaluate $\oint_C \mathbf{F} \cdot \mathbf{T} \, ds$.

6 $\mathbf{F} = 3y\mathbf{i} - 2z\mathbf{j} + 4x\mathbf{k}$, and C is the circle $x^2 + y^2 = 9$, $z = 4$ oriented counterclockwise as viewed from above.

7 $\mathbf{F} = 2z\mathbf{i} + x\mathbf{j} + 3y\mathbf{k}$, and C is the ellipse in which the plane $z = x$ meets the cylinder $x^2 + y^2 = 4$, oriented counterclockwise as viewed from above.

8 $\mathbf{F} = y\mathbf{i} + z\mathbf{j} + x\mathbf{k}$, and C is the boundary of the triangle with vertices $(0, 0, 0)$, $(2, 0, 0)$, and $(0, 2, 2)$, oriented counterclockwise as viewed from above.

9 $\mathbf{F} = \langle y - x, x - z, x - y \rangle$, and C is the boundary of the part of the plane $x + 2y + z = 2$ in the first octant, oriented counterclockwise as viewed from above.

10 $\mathbf{F} = y^2\mathbf{i} + z^2\mathbf{j} + x^2\mathbf{k}$, and C is the intersection of the plane $z = y$ and the cylinder $x^2 + y^2 = 2y$, oriented counterclockwise as viewed from above.

11 Suppose that $\mathbf{F}$ is a vector field and that ϕ is a scalar function. Show that

(a) $\text{curl}(\nabla\phi) \equiv \mathbf{0}$; (b) $\nabla \times (\phi\mathbf{F}) = (\phi)(\nabla \times \mathbf{F}) + (\nabla\phi) \times \mathbf{F}$.

12 Suppose that $\mathbf{F}$ and $\mathbf{G}$ are vector fields. Show that

(a) $\text{div}(\text{curl } \mathbf{F}) \equiv 0$;
(b) $\nabla \cdot (\mathbf{F} \times \mathbf{G}) = (\nabla \times \mathbf{F}) \cdot \mathbf{G} - \mathbf{F} \cdot (\nabla \times \mathbf{G})$.

13 Suppose that $\mathbf{r} = x\mathbf{i} + y\mathbf{j} + z\mathbf{k}$ and that $\mathbf{a}$ is a constant vector. Show that

(a) $\nabla \cdot (\mathbf{a} \times \mathbf{r}) = 0$;   (b) $\nabla \times (\mathbf{a} \times \mathbf{r}) = 2\mathbf{a}$;
(c) $\nabla \cdot ([\mathbf{r} \cdot \mathbf{r}]\mathbf{a}) = 2\mathbf{r} \cdot \mathbf{a}$;   (d) $\nabla \times ([\mathbf{r} \cdot \mathbf{r}]\mathbf{a}) = 2(\mathbf{r} \times \mathbf{a})$.

14 Show that $\iint_S (\text{curl } \mathbf{F}) \cdot \mathbf{n} \, dS$ has the same value for all oriented surfaces S that have the same oriented boundary curve C.

15 Suppose that S is a closed surface. Show that

$$\iint_S (\text{curl } \mathbf{F}) \cdot \mathbf{n} \, dS = 0$$

in two different ways: (a) by using the divergence theorem, with T the region bounded outside by S; (b) by using Stokes' theorem, with the aid of a nice simple closed curve C on S.

Line integrals, surface integrals, and triple integrals of vector-valued functions are defined by componentwise integration. Such integrals appear in the following three problems.

16 Suppose that C and S are as in the statement of Stokes' theorem and that ϕ is a scalar function. Show that

$$\oint_C \phi\mathbf{T}\, ds = \iint_S \mathbf{n} \times \nabla\phi\, dS.$$

(*Suggestion:* Apply Stokes' theorem with $\mathbf{F} = \phi\mathbf{a}$, where $\mathbf{a}$ is an arbitrary constant vector.)

17 Suppose that $\mathbf{a}$ and $\mathbf{r}$ are as in Problem 13. Show that

$$\oint_C \mathbf{a} \times \mathbf{r} \cdot \mathbf{T}\, ds = 2\mathbf{a} \cdot \iint_S \mathbf{n}\, dS.$$

18 Suppose that S is a closed surface that bounds the region T. Show that

$$\iint_S \mathbf{n} \times \mathbf{F}\, dS = \iiint_T \nabla \times \mathbf{F}\, dV.$$

(*Suggestion:* Apply the divergence theorem to $\mathbf{F} \times \mathbf{a}$ where $\mathbf{a}$ is an arbitrary constant vector. Note that the formulas of this problem, the divergence theorem, and Problem 18 in Section 16-5 all fit the pattern

$$\iint_S \mathbf{n} * (\ )\, dS = \iiint_T \nabla * (\ )\, dV,$$

where $*$ may denote either ordinary multiplication, the dot product, or the cross product, and either a scalar function or a vector-valued function is placed within the parentheses, as appropriate.)

19 Suppose that the line integral $\int_C \mathbf{F} \cdot \mathbf{T}\, ds$ is independent of path. If $\phi(x, y, z) = \int_{P_0}^P \mathbf{F} \cdot \mathbf{T}\, ds$ as in Equation (11), show that $\nabla\phi = \mathbf{F}$. (*Suggestion:* Note that if L is the line segment from (x, y, z) to $(x + \Delta x, y, z)$, then

$$\phi(x + \Delta x, y, z) - \phi(x, y, z) = \int_L \mathbf{F} \cdot \mathbf{T}\, ds = \int_x^{x + \Delta x} P\, dx.)$$

20 Let T be the submerged body of Problem 19 in Section 16-5, with centroid

$$\mathbf{r}_0 = \frac{1}{V} \iiint_T \mathbf{r}\, dV.$$

The torque about $\mathbf{r}_0$ of Archimedes' buoyant force $\mathbf{B} = -W\mathbf{k}$ is given by

$$\mathbf{L} = \iint_S (\mathbf{r} - \mathbf{r}_0) \times (-\rho g z \mathbf{n})\, dS$$

(why?). Apply the result of Problem 18 above to show that $\mathbf{L} = 0$. It follows that $\mathbf{B}$ acts along the vertical line through the centroid $\mathbf{r}_0$ of the submerged body (why?).

CHAPTER 16 REVIEW: Definitions, Concepts, Results

Use the list below as a guide to concepts that you may need to review.

1 Definition and evaluation of the line integral $\int_C f(x, y, z)\, ds$

2 Definition and evaluation of the line integral

$$\int_C P\, dx + Q\, dy + R\, dz$$

3 Relationship between the two types of line integrals, and the line integral of the tangential component of a vector field

4 Line integrals and independence of path

5 Green's theorem

6 Flux and the vector form of Green's theorem

7 The divergence of a vector field

8 Definition and evaluation of the surface integral $\iint_S f(x, y, z)\, dS$

9 Definition and evaluation of the surface integral $\iint_S P\, dy\, dz + Q\, dz\, dx + R\, dx\, dy$

10 Relationship between the two types of surface integrals, and the flux of a vector field across a surface

11 The divergence theorem, in vector and in scalar notation

12 Gauss's law and inverse-square force fields

13 The curl of a vector field

14 Stokes' theorem, in vector and in scalar notation

15 The circulation of a vector field around a simple closed curve

16 Physical interpretations of the divergence and the curl of a vector field

MISCELLANEOUS PROBLEMS

In each of Problems 1–5, evaluate the given line integral.

1 $\int_C (x^2 + y^2)\, ds$, where C is the straight line segment from $(0, 0)$ to $(3, 4)$.

2 $\int_C y^2\, dx + x^2\, dy$, where C is the graph of $y = x^3$ from $(-1, -1)$ to $(1, 1)$.

3 $\int_C \mathbf{F} \cdot \mathbf{T}\, ds$ where $\mathbf{F} = x\mathbf{i} + y\mathbf{j} + z\mathbf{k}$ and C is the curve $x = e^{2t}, y = e^t, z = e^{-t}, 0 \leq t \leq \ln 2$.

4 $\int_C xyz\, ds$, where C is the path from $(1, 1, 2)$ to $(2, 3, 6)$ consisting of three straight line segments, the first parallel to the x-axis, the second parallel to the y-axis, and the third parallel to the z-axis.

5 $\int_C \sqrt{z}\, dx + \sqrt{x}\, dy + y^2\, dz$, where C is the curve $x = t$, $y = t^{3/2}, z = t^2, 0 \leq t \leq 4$.

6 Apply Theorem 2 in Section 16-2 to show that the line integral $\int_C y^2\, dx + 2xy\, dy + z\, dz$ is independent of the path C from A to B.

7 Apply Theorem 2 in Section 16-2 to show that the line integral $\int_C x^2 y\, dx + xy^2\, dy$ is not independent of the path C from $(0, 0)$ to $(1, 1)$.

8 A wire shaped like the circle $x^2 + y^2 = a^2, z = 0$, has constant density and total mass M. Find its moment of inertia about: (a) the z-axis; (b) the x-axis.

9 A wire shaped like the parabola $y = \frac{1}{2}x^2, 0 \leq x \leq 2$, has density function $\rho = x$. Find its mass and moment of inertia about the y-axis.

10 Find the work done by the force field $\mathbf{F} = z\mathbf{i} - x\mathbf{j} + y\mathbf{k}$ in moving a particle from $(1, 1, 1)$ to $(2, 4, 8)$ along the curve $y = x^2, z = x^3$.

11 Apply Green's theorem to evaluate the line integral $\oint_C x^2 y dx + xy^2\, dy$, where C is the boundary of the region between the curves $y = x^2$ and $y = 8 - x^2$.

12 Evaluate the line integral $\oint_C x^2\, dy$, where C is the cardioid $r = 1 + \cos\theta$, by first applying Green's theorem and then changing to polar coordinates.

13 Let C_1 be the circle $x^2 + y^2 = 1$ and C_2 the circle $(x - 1)^2 + y^2 = 9$. Show that, if $\mathbf{F} = x^2 y\mathbf{i} - xy^2\mathbf{j}$, then $\oint_{C_1} \mathbf{F} \cdot \mathbf{n}\, ds = \oint_{C_2} \mathbf{F} \cdot \mathbf{n}\, ds$.

14 (a) Let C be the straight line segment from (x_1, y_1) to (x_2, y_2). Show that

$$\frac{1}{2}\int_C -y\, dx + x\, dy = \frac{1}{2}(x_1 y_2 - x_2 y_1).$$

(b) Suppose that the vertices of a polygon are (x_1, y_1), $(x_2, y_2), \ldots, (x_n, y_n)$, named in counterclockwise order around the polygon. Apply Part (a) to show that the area of the polygon is

$$A = \frac{1}{2}\sum_{i=1}^{n}(x_i y_{i+1} - x_{i+1} y_i)$$

where x_{n+1} means x_1 and y_{n+1} means y_1.

15 Suppose that the line integral $\int P\, dx + Q\, dy$ is independent of path in the plane region D. Show that $\oint_C P\, dx + Q\, dy = 0$ for every piecewise smooth simple closed curve C in D.

16 Use Green's theorem to show that $\oint_C P\, dx + Q\, dy = 0$ for every piecewise smooth simple closed curve C in the plane region D if and only if $\partial P/\partial y = \partial Q/\partial x$ in D.

17 Evaluate the surface integral $\iint_S (x^2 + y^2 + 2z)\, dS$ if S is the part of the paraboloid $z = 2 - x^2 - y^2$ above the xy-plane.

18 Suppose that $\mathbf{F} = (x^2 + y^2 + z^2)(x\mathbf{i} + y\mathbf{j} + z\mathbf{k})$ and that S is the spherical surface $x^2 + y^2 + z^2 = a^2$. Evaluate $\iint_S \mathbf{F} \cdot \mathbf{n}\, dS$ without actually performing an antidifferentiation.

19 Let T be the solid bounded by the paraboloids $z = x^2 + 2y^2$ and $z = 12 - 2x^2 - y^2$ and suppose that $\mathbf{F} =$

$x\mathbf{i} + y\mathbf{j} + z\mathbf{k}$. Find (by evaluation of surface integrals) the outward flux of $\mathbf{F}$ across the boundary of T.

20 Give a reasonable definition—as a surface integral—of the average distance of the point P from points of the surface S. Then show that the average distance of a fixed point of a spherical surface of radius a from all points of the surface is $4a/3$.

21 Suppose that the surface S is the graph of the equation $x = g(y, z)$ for (y, z) in the region D of the yz-plane. Show that

$$\iint_S P\, dy\, dz + Q\, dz\, dx + R\, dx\, dy$$

$$= \iint_D \left(P - Q\frac{\partial x}{\partial y} - R\frac{\partial x}{\partial z}\right) dy\, dz.$$

22 Suppose that the surface S is the graph of the equation $y = g(x, z)$ for (x, z) in the region D of the xz-plane. Show that

$$\iint_S f(x, y, z)\, dS = \iint_D f(x, g(x, z), z) \sec\beta\, dx\, dz$$

where $\sec\beta = \sqrt{1 + (\partial y/\partial x)^2 + (\partial y/\partial z)^2}$.

23 Let T be a space region with volume V, boundary surface S, and centroid $(\bar{x}, \bar{y}, \bar{z})$. Use the divergence theorem to show that $\bar{z} = (1/2V) \iint_S z^2\, dx\, dy$.

24 Apply the result of Problem 23 to find the centroid of the solid hemisphere $x^2 + y^2 + z^2 \leq a^2, z \geq 0$.

Problems 25–27 outline the derivation of the heat equation for a body having temperature $u = u(x, y, z, t)$ at the point (x, y, z) at time t. Denote by K its heat conductivity and by c its heat capacity, both assumed constant, and let $k = K/c$. Let B be a small solid ball within the body, and let S denote the boundary sphere of B.

25 Deduce from the divergence theorem and Equation (18) of Section 16-4 that the rate of flow of heat across S *into* B is $R = \iiint_B K\nabla^2 u\, dV$.

26 The meaning of heat capacity is that, if Δu is small, then $(c\,\Delta u)\,\Delta V$ calories of heat are required to raise the temperature of the volume ΔV by Δu degrees. It follows that the rate at which the volume ΔV is absorbing heat is $c(\partial u/\partial t)\,\Delta V$ (why?). Conclude that the rate of flow of heat into B is $R = \iiint_B c(\partial u/\partial t)\, dV$.

27 Equate the results of Problems 25 and 26, apply the average value property of triple integrals, and then take the limit as the radius of the ball B approaches zero. You should thereby obtain the heat equation $\partial u/\partial t = k\nabla^2 u$.

28 For a *steady state* temperature function—one that is independent of time t—the heat equation reduces to Laplace's equation

$$\nabla^2 u = \frac{\partial^2 u}{\partial x^2} + \frac{\partial^2 u}{\partial y^2} + \frac{\partial^2 u}{\partial z^2} = 0.$$

(a) Suppose that u_1 and u_2 are two solutions of Laplace's equation in the region T, and that u_1 and u_2 agree on its

boundary surface S. Apply Problem 15 in Section 16-5 to $f = u_1 - u_2$ to conclude that $\nabla f = \mathbf{0}$ at each point of T.
(b) From the fact that $\nabla f \equiv \mathbf{0}$ in T and $f \equiv 0$ on S, conclude that $f \equiv 0$, so that $u_1 \equiv u_2$. Thus the steady-state temperatures within a region are *determined* by the boundary value temperatures.

29 Suppose that $\mathbf{r} = x\mathbf{i} + y\mathbf{j} + z\mathbf{k}$ and that $\phi(r)$ is a scalar function of $r = |\mathbf{r}|$. Compute: (a) $\nabla\phi(r)$; (b) $\text{div}[\phi(r)\mathbf{r}]$; (c) $\text{curl}[\phi(r)\mathbf{r}]$.

30 Let S be the upper half of the torus obtained by revolving the circle $(y - a)^2 + z^2 = b^2$ in the yz-plane around the z-axis, with upper unit normal. Describe how to subdivide S to establish Stokes' theorem for it. How are the two boundary circles oriented?

31 Explain why the method of subdivision does not suffice to establish Stokes' theorem for the Möbius strip of Figure 16.18.

32 (a) Suppose that a fluid or a rigid body is rotating with angular speed ω radians per second about the line through the origin determined by the unit vector $\mathbf{u}$. Show that the velocity of the point with position vector $\mathbf{r}$ is $\mathbf{v} = \boldsymbol{\omega} \times \mathbf{r}$, where $\boldsymbol{\omega} = \omega\mathbf{u}$ is the angular velocity vector. Note first that $|\mathbf{v}| = \omega|\mathbf{r}|\sin\theta$ where θ is the angle between $\mathbf{r}$ and $\boldsymbol{\omega}$.
(b) Use the fact that $\mathbf{v} = \boldsymbol{\omega} \times \mathbf{r}$ (established in Part (a)) to show that $\text{curl } \mathbf{v} = 2\boldsymbol{\omega}$.

33 Consider the steady flow past a circular cylinder as shown in Figure 16.32. Let

$$\phi(x, y) = u_0 x\left(1 + \frac{a^2}{x^2 + y^2}\right) = u_0 x\left(1 + \frac{a^2}{r^2}\right).$$

(a) Show that $\phi(x, y)$ satisfies Laplace's equation.

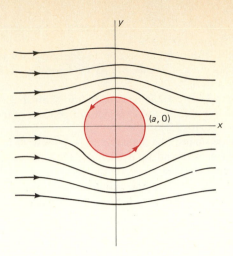

16.32 Fluid flow around a cylindrical airfoil.

(b) Show that, in polar coordinates,

$$\mathbf{v} = \nabla\phi = \frac{u_0 a^2}{r^2}\left[\left(\frac{r^2}{a^2} - \cos 2\theta\right)\mathbf{i} - (\sin 2\theta)\mathbf{j}\right].$$

(c) Show that $\lim\limits_{r \to \infty} (\mathbf{v} - u_0\mathbf{i}) = \mathbf{0}$ and that $\mathbf{v}$ is tangential to the circle $r = a$.
(d) Parts (a), (b), and (c) above establish that $\phi(x, y)$ as given is the velocity potential for the indicated flow. Now show that the lift

$$|\mathbf{L}| = \rho u_0 \oint_C \mathbf{v} \cdot \mathbf{T} \, ds$$

is zero, as symmetry indicates it should be.

Differential Equations

17

Introduction

A **differential equation** is one that involves one or more derivatives of an "unknown function" or dependent variable. If the dependent variable is a function of a *single* independent variable, so that the derivatives involved are *ordinary* single-variable derivatives, then the equation is called an **ordinary differential equation.** If the dependent variable is a function of two or more independent variables, so that partial derivatives are involved, then the equation is a **partial differential equation.** In this chapter we concentrate on ordinary differential equations; more advanced techniques are required for the solution of most partial differential equations.

The **order** of a differential equation is the order of the highest derivative that appears. Thus the general form of an **nth order** ordinary differential equation is

$$F(x, y, y', \ldots, y^{(n)}) = 0, \tag{1}$$

where F is a given function of $n + 2$ variables. For instance, the equations

$$\left(\frac{d^4 y}{dx^4}\right)^2 + \left(\frac{d^3 y}{dx^3}\right)\left(\frac{d^2 y}{dx^2}\right) + x^2 y^2 = 0$$

and

$$y^{(4)}(x) + x^2 y''(x) + x^4 y(x) = 0$$

are both fourth order differential equations. The second of these two equations is **linear** in the dependent variable y and its derivatives (only their first powers are involved), but the first equation above is not linear. The function $y = \phi(x)$ is called a **solution** of the differential equation in (1) on the interval (a, b) provided that

$$F(x, \phi(x), \phi'(x), \ldots, \phi^{(n)}(x)) = 0$$

for all x in (a, b).

In order to solve an nth order differential equation we must, in principle, integrate n times (to go from $y^{(n)}$ to y). Consequently the result of solving Equation (1) may be expected to involve n *arbitrary constants* $c_1, c_2, c_3, \ldots, c_n$ (of integration). Hence the solution we obtain will be of the form

$$y = \phi(x, c_1, c_2, c_3, \ldots, c_n). \tag{2}$$

Such a function, one which (as a function of x) is a solution of (1) for each choice of the constants $c_1, c_2, c_3, \ldots, c_n$, is called a **general solution** of (1). The function of x alone that results from assigning specific values to these constants is called a **particular solution** of (1).

For example, we saw in Section 7-4 that the general solution of the second order equation $y'' + y = 0$ is

$$y = c_1 \cos x + c_2 \sin x.$$

The specific choice $c_1 = 3$ and $c_2 = -2$ yields the particular solution $y = 3 \cos x - 2 \sin x$.

It is convenient to refer to *the* general solution of a differential equation, despite the fact that a differential equation can have two or more different general solutions. Moreover, the concept of a function containing n arbitrary constants requires a more careful examination than is feasible here. Finally, it is possible for a differential equation to have a *singular solution* that is not a special case of any general solution and so is not a particular solution as defined above. Nevertheless, this is an exceptional circumstance and the terminology of general and particular solutions is both suggestive and helpful.

An applied problem frequently calls for a particular solution of an nth order equation that satisfies n given **initial conditions** of the form

$$y(a) = b_0, \quad y'(a) = b_1, \quad \ldots, \quad y^{(n-1)}(a) = b_{n-1}$$

imposed at the point $x = a$. That is, we want the solution and its first $n - 1$ derivatives to have preassigned values at $x = a$. If we can find the general solution $y = \phi(x, c_1, c_2, \ldots, c_n)$, then we can attempt to solve the n equations

$$\phi(a, c_1, c_2, \ldots, c_n) = b_0,$$
$$\phi'(a, c_1, c_2, \ldots, c_n) = b_1,$$
$$\vdots$$
$$\phi^{(n-1)}(a, c_1, c_2, \ldots, c_n) = b_{n-1}$$

for the n constants $c_1, c_2, c_3, \ldots, c_n$.

For example, suppose that we want to find the particular solution of the second order equation $y'' = y$ that satisfies the initial conditions $y(0) = 1$, $y'(0) = 3$. Now $y = e^x$ and $y = e^{-x}$ are obviously solutions of $y'' = y$; the general solution is

$$y = c_1 e^x + c_2 e^{-x}$$

with derivative $y' = c_1 e^x - c_2 e^{-x}$. The given initial conditions yield the equations

$$c_1 + c_2 = 1, \quad c_1 - c_2 = 3$$

which has solution $c_1 = 2$, $c_2 = -1$. Thus the desired particular solution is $y = 2e^x - e^{-x}$.

In a brief introduction to the subject of differential equations, we can only scratch the surface of a vast and venerable area of advanced mathematics with applications that permeate modern science and technology. Even so, we go far enough in this chapter to show in several concrete applications— rocket propulsion (Section 17-2), mechanical and electrical vibrations (Section 17-6), underground temperature variations (Section 17-7)—what a wonderful amount of useful information can sometimes be extracted from the solution of a differential equation.

Linear First Order Equations and Applications

The first order differential equation $F(x, y, y') = 0$ is called **linear** if it is linear in the dependent variable y and its derivative y'; that is, only the first powers of y and y' appear. It need not be linear in x. Thus the general **linear first order differential equation** can be written in the form

$$\frac{dy}{dx} + P(x)y = Q(x), \tag{1}$$

where the coefficients P and Q are functions of x alone.

There is a standard technique for solving (1) on an interval on which P and Q are continuous functions. We multiply (1) by the **integrating factor**

$$\rho = \rho(x) = e^{\int P(x)\,dx}. \tag{2}$$

The result is

$$e^{\int P\,dx}\frac{dy}{dx} + P(x)e^{\int P\,dx}y = Q(x)e^{\int P\,dx}.$$

Since $D_x[\int P(x)\,dx] = P(x)$, the left-hand side is the derivative of the product $ye^{\int P\,dx}$, so we have

$$D_x\left(ye^{\int P\,dx}\right) = Q(x)e^{\int P\,dx}.$$

Integration of both sides of this equation gives

$$ye^{\int P\,dx} = \int\left(Q(x)e^{\int P(x)\,dx}\right)dx + C.$$

Finally solving for y, we obtain the general solution of the linear first order equation in (1):

$$y = y(x) = e^{-\int P(x)\,dx}\left[\int\left(Q(x)e^{\int P(x)\,dx}\right)dx + C\right]. \tag{3}$$

Formula (3) need *not* be memorized. In a specific problem it generally is simpler to use the *method* by which we derived this formula. Begin by calculating the integrating factor given by (2), $\rho = e^{\int P(x)\,dx}$. Then multiply the differential equation by ρ, recognize the left-hand side of the resulting equation as the derivative of a product, integrate this equation, and finally solve for y.

EXAMPLE 1 Find the particular solution of

$$x\frac{dy}{dx} + 3y - 2x^5 = 0$$

for $x > 0$ such that $y(2) = 1$.

Solution The given equation may be rewritten in the form

$$\frac{dy}{dx} + \frac{3y}{x} = 2x^4,$$

and in this form we recognize it as a linear first order equation with $P = 3/x$ and $Q = 2x^4$. The integrating factor is

$$\rho = e^{\int P\, dx} = e^{\int (3/x)\, dx} = e^{3\ln x} = x^3.$$

Multiplication of the above differential equation by this integrating factor gives

$$x^3 \frac{dy}{dx} + 3x^2 y = 2x^7,$$

which we recognize as

$$\frac{d}{dx}(x^3 y) = 2x^7.$$

Hence integration with respect to x gives

$$x^3 y = \tfrac{1}{4}x^8 + C,$$

so the general solution is

$$y = \frac{x^5}{4} + \frac{C}{x^3}.$$

The condition that $y = 1$ when $x = 2$ gives $1 = 32/4 + C/8$, and we solve for $C = -56$. Thus the desired particular solution is

$$y = \frac{x^5}{4} - \frac{56}{x^3}.$$

Linear first order differential equations have numerous scientific applications. In Section 6-6 we discussed applications of the linear first order equation $dx/dt = ax + b$ with *constant* coefficients, which we solved there by separating the variables. The remainder of this section is devoted to some typical applications that require the more powerful solution method described above.

*A MIXTURE PROBLEM

Consider two brine tanks arranged as shown in Figure 17.1. The volumes in gallons of brine (water containing dissolved salt) in tanks 1 and 2 are V_1 and V_2, respectively. Suppose that pure water flows into tank 1, the completely mixed solution in tank 1 flows into tank 2, and the completely mixed solution in tank 2 flows out, with each of these three flow rates being k gal/min (so that V_1 and V_2 remain constant).

Let $x(t)$ and $y(t)$ denote the amounts of salt (in pounds) present in tanks 1 and 2, respectively, at time t (min). Given the initial amounts $x(0) = x_0$ and $y(0) = y_0$, we want to solve for $x(t)$ and $y(t)$. We must consider the two tanks separately.

TANK 1 We need to compute the change Δx in the amount of salt in tank 1 during the short time interval from t to $t + \Delta t$. Writing x instead of $x(t)$ for simplicity, the concentration of salt in the tank is approximately x/V_1

17.1 Two brine tanks.

Volume V_1 Tank 1
Amount x

Tank 2 Volume V_2
Amount y

lbs/gal, and $k\,\Delta t$ gal of the mixed solution flows out during this time interval. Since the inflow to tank 1 is pure water, none of the salt in the outflow is replaced. Hence

$$\Delta x \approx -\frac{x}{V_1}\,k\,\Delta t = -\alpha x\,\Delta t$$

where $\alpha = k/V_1$. We divide by Δt, then take the limit as $\Delta t \to 0$. This gives the differential equation

$$\frac{dx}{dt} = -\alpha x \qquad \left(\alpha = \frac{k}{V_1}\right), \tag{4}$$

with the familiar solution

$$x(t) = x_0 e^{-\alpha t}. \tag{5}$$

TANK 2 Similarly, we compute the change Δy in the salt content of tank 2 during the same short time interval Δt. The salt concentration in this tank is approximately y/V_2 lbs/gal, and $k\,\Delta t$ gal of the mixed solution leaves during the time interval Δt. But at the same time the amount $(x/V_1)(k\,\Delta t)$ of salt from tank 1 (computed above) enters tank 2. Hence the net change in the amount of salt in tank 2 from time t to time $t + \Delta t$ is

$$\Delta y \approx \frac{x}{V_1}\,k\,\Delta t - \frac{y}{V_2}\,k\,\Delta t = \alpha x\,\Delta t - \beta y\,\Delta t$$

where $\beta = k/V_2$. We divide by Δt and take the limit as $\Delta t \to 0$, which gives the differential equation

$$\frac{dy}{dt} = \alpha x - \beta y \qquad \left(\beta = \frac{k}{V_2}\right). \tag{6}$$

Substituting $x = x_0 e^{-\alpha t}$ from (5), we get the linear first order differential equation

$$\frac{dy}{dt} + \beta y = \alpha x_0 e^{-\alpha t} \tag{7}$$

for y as a function of t. In Problem 11 we ask you to derive its solution, which (in the case $\alpha \neq \beta$) is

$$y(t) = y_0 e^{-\beta t} + \frac{\alpha x_0}{\beta - \alpha}\,(e^{-\alpha t} - e^{-\beta t}). \tag{8}$$

You could use this solution, for instance, to find when the amount of salt in tank 2 is maximal (Problem 12).

The arrangement of tanks shown in Figure 17.1 is called a **two-step cascade.** An industrial chemical process may involve an n-step cascade.

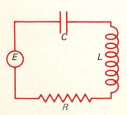

17.2 The simple RLC series circuit.

*ELECTRICAL CIRCUITS

Consider the simple electrical circuit shown in Figure 17.2. It contains a *resistor* with a resistance of R ohms, an *inductor* with an inductance of L henries, and a *capacitor* with a capacitance of C farads, in series with a source of electromotive force (a battery or a generator) that supplies a voltage of $E(t)$ volts at time t. This results in a current of $I(t)$ amperes in the circuit and a charge of $Q(t)$ coulombs on the capacitor at time t. The relation between Q and I is

$$I = \frac{dQ}{dt}. \tag{9}$$

According to elementary principles of electricity, the *voltage drops* across the three circuit elements are those shown in the table of Figure 17.3. **Kirchoff's law** says that the (algebraic) sum of the voltage drops across the elements in a simple loop of an electric circuit is equal to the voltage supplied. Applied to the circuit of Figure 17.2, this gives the differential equation

$$L\frac{dI}{dt} + RI + \frac{1}{C}Q = E(t). \tag{10}$$

Circuit element	Voltage drop
INDUCTOR	$L\frac{dI}{dt}$
RESISTOR	RI
CAPACITOR	$\frac{1}{C}Q$

17.3 Voltage drops across circuit elements.

In Examples 2 and 3 below, we consider the still simpler circuit of Figure 17.4, containing only two circuit elements—a resistor and an inductor—in addition to the battery or generator. Then the Q/C term is missing from Equation (10), so we have the linear first order differential equation

$$L\frac{dI}{dt} + RI = E(t) \tag{11}$$

for the current I as a function of time t.

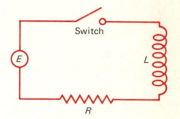

17.4 An RL series circuit.

EXAMPLE 2 In the circuit of Figure 17.4, suppose that a battery supplies a constant voltage of $E = 40$ volts, and that $L = 2$ henries and $R = 8$ ohms. The switch S is closed at time $t = 0$ and current begins to flow, with $I(0) = 0$. Find $I(t)$.

Solution With the given values $L = 2$, $R = 8$, and $E = 40$, Equation (11) is

$$2\frac{dI}{dt} + 8I = 40, \quad \text{or} \quad \frac{dI}{dt} + 4I = 20.$$

Multiplication by the integrating factor $e^{\int 4\,dt} = e^{4t}$ gives

$$e^{4t}\frac{dI}{dt} + 4e^{4t}I = 20e^{4t};$$

$$\frac{d}{dt}(e^{4t}I) = 20e^{4t};$$

so

$$e^{4t}I = 5e^{4t} + C.$$

The fact that $I = 0$ when $t = 0$ gives $C = -5$, so

$$I(t) = 5(1 - e^{-4t}).$$

Thus the current in the circuit builds toward a maximum current of 5 amperes (without ever quite reaching that value). When $t = 1$ (sec) the current is $I(1) = 5(1 - e^{-4}) \approx 4.91$ amp.

EXAMPLE 3   The battery in Example 2 is replaced by an alternating current generator that supplies a voltage of $E(t) = 40 \sin 8t$ volts. With everything else the same, now find $I(t)$.

Solution   After replacement of $E = 40$ with $E = 40 \sin 8t$, our differential equation takes the form

$$2 \frac{dI}{dt} + 8I = 40 \sin 8t, \quad \text{or} \quad \frac{dI}{dt} + 4I = 20 \sin 8t.$$

Multiplying by the same integrating factor e^{4t}, we get

$$\frac{d}{dt}(e^{4t}I) = 20e^{4t} \sin 8t.$$

We integrate with the aid of Formula (67) (inside the back cover) for $\int e^{at} \sin bt \, dt$; we get

$$e^{4t}I = 20 \int e^{4t} \sin 8t \, dt = e^{4t}(\sin 8t - 2 \cos 8t) + C.$$

The condition that $I = 0$ when $t = 0$ gives $C = 2$, so the current at time t is

$$I(t) = \sin 8t - 2 \cos 8t + 2e^{-4t}.$$

Note that the term $2e^{-4t}$ is a *transient* current, which dies out as $t \to \infty$; this leaves "in the limit" an alternating current $\sin 8t - 2 \cos 8t$ with the same frequency as the imposed voltage.

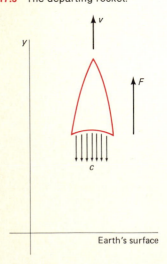

17.5 The departing rocket.

*ROCKET PROPULSION

Suppose that the rocket of Figure 17.5 blasts off straight upward from the surface of the earth at time $t = 0$. We want to compute its height y and velocity $v = dy/dt$ at time t. The rocket is propelled by exhaust gases that exit with constant speed c (relative to the rocket). Because of the combustion of its fuel, the mass $m = m(t)$ of the rocket is variable.

To derive the equation of motion of the rocket, we need Newton's second law in the form

$$\frac{dP}{dt} = F \tag{12}$$

where $P = mv$ is momentum and F is the net external force (gravity, air resistance, and so on). If m is constant, then (12) reduces to the standard form $F = m(dv/dt)$.

Suppose that m changes to $m + \Delta m$ and v changes to $v + \Delta v$ during the short time interval Δt. The change in the momentum of the *rocket it-*

CHAP. 17: **Differential Equations**

self is

$$(m + \Delta m)(v + \Delta v) - mv = m\,\Delta v + v\,\Delta m + \Delta m\,\Delta v.$$

But the system also includes the exhaust gases expelled during this time interval, with mass $-\Delta m$ and velocity $v - c$. Hence the total change in momentum during the time interval Δt is

$$\Delta P \approx (m\,\Delta v + v\,\Delta m + \Delta m\,\Delta v) + (-\Delta m)(v - c)$$
$$= m\,\Delta v + c\,\Delta m + \Delta m\,\Delta v.$$

Dividing by Δt and taking the limit as $\Delta t \to 0$, we get

$$\frac{dP}{dt} = m\frac{dv}{dt} + c\frac{dm}{dt}$$

because Δm and Δv approach zero as $\Delta t \to 0$. We substitute this expression for dP/dt into Newton's law (12) and thereby obtain the **rocket propulsion equation**

$$m\frac{dv}{dt} + c\frac{dm}{dt} = F. \tag{13}$$

We shall assume that $F = F_G + F_R$, where $F_G = -mg$ is a constant force of gravity and $F_R = -kv$ is a force of resistance proportional to velocity. This gives

$$m\frac{dv}{dt} + c\frac{dm}{dt} = -mg - kv. \tag{14}$$

Now suppose that the rocket's fuel is consumed at a constant rate β (burn rate) during the time interval $[0, \tau]$, during which time the rocket's mass decreases from m_0 to m_1. Thus

$$m(0) = m_0, \qquad m(\tau) = m_1,$$

$$m(t) = m_0 - \beta t, \qquad \frac{dm}{dt} = -\beta \qquad \text{for} \quad t \le \tau, \tag{15}$$

with burnout occurring at time $t = \tau$.

We proceed to compute the velocity at time $t \le \tau$. If we substitute the expressions in (15) into Equation (14), the result is

$$(m_0 - \beta t)\frac{dv}{dt} - \beta c = -(m_0 - \beta t)g - kv,$$

which we can rewrite in the form

$$\frac{dv}{dt} + \frac{k}{m_0 - \beta t}v = -g + \frac{\beta c}{m_0 - \beta t}. \tag{16}$$

This is a linear first order equation for v as a function of t, with integrating factor

$$\rho = \exp\left(\int \frac{k\,dt}{m_0 - \beta t}\right) = (m_0 - \beta t)^{-k/\beta}.$$

The details are somewhat tedious (Problem 22), but the result of integrating Equation (16) is

$$v(t) = v_0 M^{k/\beta} + \frac{\beta c}{k}(1 - M^{k/\beta}) + \frac{gm_0}{\beta - k}(M - M^{k/\beta}), \tag{17}$$

where $v_0 = v(0)$ and

$$M = \frac{m_0 - \beta t}{m_0} = \frac{m(t)}{m_0}$$

is the *fractional mass* of the rocket at time t.

NO RESISTANCE Let us simplify the situation by ignoring the force of resistance to the rocket's motion. Then $k = 0$. In Problem 20 we ask you to take the limit of the right-hand side of (17) as $k \to 0$, using l'Hôpital's rule, to obtain

$$v(t) = v_0 - gt + c \ln\left(\frac{m_0}{m_0 - \beta t}\right) \tag{18}$$

for the case of no resistance. You can get the same result by setting $k = 0$ in the differential equation in (16) and then integrating (Problem 21). Since $m_0 - \beta\tau = m_1$, the velocity of the rocket at burnout ($t = \tau$) is

$$v(\tau) = v_0 - g\tau + c \ln\left(\frac{m_0}{m_1}\right). \tag{19}$$

We can compute the height of the rocket at time $t \leqq \tau$ by integrating Equation (18):

$$y(t) = (v_0 + c \ln m_0)t - \tfrac{1}{2}gt^2 - c\int_0^t \ln(m_0 - \beta t)\, dt,$$

assuming that $y_0 = y(0) = 0$. Using the integration formula $\int \ln u\, du = u \ln u - u + C$, we find after some simplification that

$$y(t) = (v_0 + c)t - \tfrac{1}{2}gt^2 - \frac{c}{\beta}(m_0 - \beta t)\ln\left(\frac{m_0}{m_0 - \beta t}\right). \tag{20}$$

When we substitute $t = \tau$, we find the height at burnout to be

$$y(\tau) = (v_0 + c)\tau - \tfrac{1}{2}g\tau^2 - \frac{cm_1}{\beta}\ln\left(\frac{m_0}{m_1}\right). \tag{21}$$

FREE SPACE Suppose finally that the rocket is accelerating in free space, where there is neither gravity nor resistance, so that $g = k = 0$. With $g = 0$ in Equation (19), we see that the rocket's increase in velocity as its mass decreases from m_0 to m_1 is

$$\Delta v = v(\tau) - v_0 = c \ln\left(\frac{m_0}{m_1}\right). \tag{22}$$

Note that Δv depends only upon the exhaust gas speed c and the initial to final mass ratio m_0/m_1 but does *not* depend upon the burn rate β. For example, if $c = 2$ mi/sec and $m_0/m_1 = 10$, then $\Delta v = 2 \ln 10 \approx 4.61$ mi/sec. Thus, if a rocket initially consists predominately of fuel, then it can attain speeds significantly greater than the (relative) speed of its exhaust gases.

Solve the differential equations in Problems 1–10.

1 $xy' + y = 3xy$; $y(1) = 0$.
2 $y' + y = e^x$; $y(0) = 1$.
3 $xy' + 2y = 3x$; $y(1) = 4$.
4 $xy' - 3y = x^3$; $y(1) = 10$.
5 $y' + 2xy = x$; $y(0) = -2$.
6 $y' = (1 - y) \cos x$; $y(\pi) = 2$.
7 $(x + 1)y' + 5y = 10$.
8 $xy' = 2y + x^3 \cos x$.
9 $y' + y \cot x = \cos x$.
10 $y' + y = \sin x$.
11 Derive the solution (8) of Equation (7), assuming that $\alpha \neq \beta$.
12 In a two-step cascade of the sort shown in Figure 17.1, suppose that the volume of tank 1 is 50 gal and that of tank 2 is 100 gal. At time $t = 0$, tank 1 contains 10 lb of salt while tank 2 contains pure water. The flow rates are all 10 gal/min. What is the maximum amount of salt in tank 2, and when does it occur?
13 A tank initially contains 60 gallons of pure water. Brine containing 1 lb of salt per gallon enters the tank at 2 gal/min, and the (perfectly mixed) solution leaves the tank at 3 gal/min; the tank is empty after exactly 1 hour. (a) Find the amount of salt in the tank after t minutes. (b) What is the maximum amount of salt in the tank?
14 In the two-step cascade of Figure 17.1, each tank contains 100 gal of salt water and each initially contains 20 lb of salt. The flow rates are all 5 gal/min. How much salt remains in each tank after 5 minutes?
15 An electric circuit contains a voltage source E, a resistor R (constant), and a capacitor C (also constant). By Equations (9) and (10) with $L = 0$, the charge $Q(t)$ on the capacitor satisfies the equation

$$R \frac{dQ}{dt} + \frac{1}{C} Q = E(t).$$

If $E(t) = E_0 \cos \omega t$ and $Q(0) = 0$, find $Q(t)$.
16 In the RC circuit of Problem 15, assume that $R = 10$ ohms, $C = 0.02$ farads, and $E = 100e^{-5t}$ volts. (a) Find the charge Q and the current $I = dQ/dt$ at time t. (b) What is the maximum charge on the capacitor, and when does it occur?
17 A rocket has an initial weight of 25 tons, of which 20 tons consists of fuel mixture which burns at the rate of 1 ton per second. Its exhaust gas speed is 1 mi/sec. It blasts off at time $t = 0$ with $y_0 = v_0 = 0$. Find its height and velocity (mph) at burnout. Ignore air resistance and use $g = 32$ ft/sec^2.
18 For the rocket of Problem 17, how large must its exhaust speed be in order for it to get off the ground?
19 For a rocket in free space, show that Equation (13) can be written in the form $dv/dm = -c/m$. Integrate this equation to obtain the velocity of the rocket as given in Equation (22).
20 Derive Equation (18) by taking the limit as $k \to 0$ in Equation (17).
21 Derive Equation (18) by solving Equation (16) with $k = 0$.
22 Derive Equation (17) by solving Equation (16) in the case $k > 0$.
23 The V-2 rocket of World War II had an initial weight of 28,300 lb (so $m_0 = 878.88$ slugs, using $g = 32.2$ ft/sec^2), of which 68.5% was fuel. This fuel burned uniformly for 70 seconds with an exhaust velocity of 1.25 mi/sec. Assuming an air resistance force of $(0.1)v$ lb (with v in ft/sec) find the velocity and height of the V-2 at burnout. Begin by solving Equation (16) with the given numerical parameters.

17-3

Aside from simple second order equations that can be reduced by elementary substitutions to familiar first order equations, higher order differential equations ordinarily require specialized techniques for their solution. The remainder of this chapter is devoted to *linear* ordinary differential equations of order two and higher. The general form of a **linear nth order differential equation** is

Higher Order Linear Equations

$$a_n(x) \frac{d^n y}{dx^n} + a_{n-1}(x) \frac{d^{n-1} y}{dx^{n-1}} + \cdots + a_1(x) \frac{dy}{dx} + a_0(x)y = F(x) \quad (1)$$

where the *coefficients* $a_0(x)$, $a_1(x)$, ..., $a_n(x)$ and the function $F(x)$ are given. Equation (1) is called *linear* because only the *first* powers of the dependent variable $y(x)$ and of its derivatives are involved. It is linear in y but not necessarily in the independent variable x—the coefficients and the function $F(x)$ need not be linear in x.

The linear differential equation (1) is called **homogeneous** provided that $F(x) \equiv 0$; otherwise it is said to be **inhomogeneous.** Thus the general form of a homogeneous linear equation is

$$a_n(x)\frac{d^n y}{dx^n} + a_{n-1}(x)\frac{d^{n-1} y}{dx^{n-1}} + \cdots + a_1(x)\frac{dy}{dx} + a_0(x)y = 0. \qquad (2)$$

If the function $F(x)$ in Equation (1) is not identically zero, then Equation (2) is the homogeneous equation *associated with* the inhomogeneous Equation (1). Thus

$$x^2 y'' + 4xy' + x^3 y = 0$$

is the homogeneous equation associated with the inhomogeneous equation

$$x^2 y'' + 4xy' + x^3 y = 5 \sin 2x.$$

The study of linear differential equations is greatly simplified by the following fact: If $y_1(x)$ and $y_2(x)$ are solutions of the homogeneous equation in (2), then so is the **linear combination**

$$y(x) = c_1 y_1(x) + c_2 y_2(x)$$

for any choice of the *constants* c_1 and c_2. This is so because the linearity of the operation of differentiation implies that

$$a_k y^{(k)}(x) = c_1 a_k y_1^{(k)}(x) + c_2 a_k y_2^{(k)}(x) \qquad (3)$$

for each $k = 0, 1, 2, \ldots, n$. So

$$\sum_{k=0}^{n} a_k y^{(k)}(x) = c_1 \sum_{k=0}^{n} a_k y_1^{(k)}(x) + c_2 \sum_{k=0}^{n} a_k y_2^{(k)}(x) = 0$$

provided that y_1 and y_2 are both solutions of (2).

It is known that (on an interval where the coefficient functions are continuous) every *homogeneous* linear equation (2) has n particular solutions $y_1(x), y_2(x), \ldots, y_n(x)$ such that *every* solution $y(x)$ of (2) is a linear combination of these particular solutions [see Kaplan, *Ordinary Differential Equations* (Reading, Mass.: Addison-Wesley, 1958), p. 115]. That is,

$$y(x) = c_1 y_1(x) + c_2 y_2(x) + \cdots + c_n y_n(x) \qquad (4)$$

for an appropriate choice of the constants $c_1, c_2, \ldots, c_n$. In this case the linear combination in (4) is the *general solution* of Equation (2).

For example, we saw in Section 7-4 that the general solution of the homogeneous second order linear differential equation $y'' + \omega^2 y = 0$ (where ω is a constant) is

$$y(x) = c_1 \cos \omega x + c_2 \sin \omega x, \qquad (5)$$

so here $y_1(x) = \cos \omega x$ and $y_2(x) = \sin \omega x$. In Section 17-5 we will see how to find the general solution of any *homogeneous* linear equation with *constant* coefficients.

In order to solve the *inhomogeneous* linear equation (1), it suffices to find *both* a single particular solution $y_p(x)$ of (1) *and* the general solution

$$y_h(x) = c_1 y_1(x) + \cdots + c_n y_n(x)$$

of the associated homogeneous equation in (2). Then a computation similar to (3)—see Problem 21—shows that *every* solution $y(x)$ of (1) can be written in the form

$$y(x) = y_h(x) + y_p(x)$$

$$= c_1 y_1(x) + c_2 y_2(x) + \cdots + c_n y_n(x) + y_p(x) \tag{6}$$

for an appropriate choice of the constants $c_1, c_2, \ldots, c_n$. In this case the sum $y(x) = y_h(x) + y_p(x)$ is the *general solution* of the inhomogeneous equation (1).

METHOD OF UNDETERMINED COEFFICIENTS

The **method of undetermined coefficients** is a simple method of attempting to find a particular solution $y_p(x)$ of the inhomogeneous linear equation

$$a_n y^{(n)} + a_{n-1} y^{(n-1)} + \cdots + a_1 y' + a_0 y = F(x). \tag{7}$$

This method often succeeds when the coefficients a_i are *constants* and the function $F(x)$ is sufficiently simple that we can make a reasonable guess as to the general form of $y_p(x)$. We shall illustrate the method of undetermined coefficients with some examples involving simple second order equations, but the method can be applied in similar fashion to the nth order equation (7).

Suppose first that $F(x)$ is a polynomial of degree k. Then, since the derivatives of a polynomial are polynomials of lower degree, it is reasonable to suspect a particular solution

$$y_p(x) = b_k x^k + b_{k-1} x^{k-1} + \cdots + b_1 x + b_0$$

that is a polynomial of degree k as well. We therefore substitute this form of $y_p(x)$ into our differential equation, and attempt to determine the coefficients $b_0, b_1, \ldots, b_{k-1}, b_k$ so that it will be a solution.

EXAMPLE 1   Find a particular solution of $y'' + y' + 2y = 4x^2$.

Solution   Here $F(x) = 4x^2$, a second degree polynomial, so we try

$$y_p(x) = Ax^2 + Bx + C.$$

Then $y_p' = 2Ax + B$ and $y_p'' = 2A$, so substitution of y_p into the given differential equation yields

$$(2A) + (2Ax + B) + 2(Ax^2 + Bx + C) = 4x^2.$$

As in the method of partial fractions, we may equate coefficients of like powers of x. Thus we obtain the system of equations

$$2A = 4, \qquad 2A + 2B = 0, \qquad 2A + B + 2C = 0,$$

whose solution is $A = 2, B = -2, C = -1$. Thus our particular solution is

$$y_p(x) = 2x^2 - 2x - 1.$$

If $F(x)$ is of the form Ae^{rx}, then we suspect that $y_p(x)$ is of the same form, because derivatives of e^{rx} are constant multiples of e^{rx}. If $F(x)$ is a sum of terms of two different types, then we naturally try as $y_p(x)$ a sum of terms corresponding to these different types.

EXAMPLE 2   Find a particular solution of

$$y'' + y' + 3y = 6e^{2x} - 2.$$

Solution   We try $y_p(x) = Ae^{2x} + B$, so that $y_p' = 2Ae^{2x}$ and $y_p'' = 4Ae^{2x}$. Substitution of y_p into the given differential equation gives

$$(4Ae^{2x}) + (2Ae^{2x}) + 3(Ae^{2x} + B) = 6e^{2x} - 2.$$

We equate coefficients of e^{2x} and of the constant terms, and get the equations $9A = 6$ and $3B = -2$, so that $A = \frac{2}{3}$ and $B = -\frac{2}{3}$. Thus our particular solution is

$$y_p = \tfrac{2}{3}e^{2x} - \tfrac{2}{3} = \tfrac{2}{3}(e^{2x} - 1).$$

If $F(x) = A \cos kx + B \sin kx$, then we expect a particular solution of this same form, because any derivative of a linear combination of $\cos kx$ and $\sin kx$ is again a linear combination of $\cos kx$ and $\sin kx$.

EXAMPLE 3   Find the general solution of $y'' + 4y = 10 \sin 3x$.

Solution   The associated homogeneous equation is $y'' + 4y = 0$, and by (5) *its* general solution is

$$y_h(x) = c_1 \cos 2x + c_2 \sin 2x.$$

For the particular solution we try

$$y_p(x) = A \cos 3x + B \sin 3x.$$

Since $y_p'' = -9A \cos 3x - 9B \sin 3x$, substitution of y_p into the given inhomogeneous equation gives

$$-5A \cos 3x - 5B \sin 3x = 10 \sin 3x,$$

so that $A = 0$ and $B = -2$. Thus the particular solution is

$$y_p(x) = -2 \sin 3x.$$

Indeed, we might well have anticipated that A would be zero, because the second derivative of $\sin 3x$ is a multiple of $\sin 3x$. Had we done so, we could have tried the simpler particular solution $y_p(x) = B \sin 3x$ to begin with. At any rate, by Equation (6) the general solution of our inhomogeneous equation is the sum of the particular solution and the general solution of the associated homogeneous equation. This yields the general solution

$$y = c_1 \cos 2x + c_2 \sin 2x - 2 \sin 3x. \tag{8}$$

EXAMPLE 4   Find the particular solution of the differential equation of Example 3 for which $y(0) = 1$ and $y'(0) = 0$.

Solution   From the general solution in (8), we see immediately that the condition $y(0) = 1$ implies that $c_1 = 1$, so

$$y'(x) = -2 \sin 2x + 2c_2 \cos 2x - 6 \cos 3x.$$

The condition $y'(0) = 0$ therefore implies that $2c_2 - 6 = 0$, so $c_2 = 3$. Hence the desired particular solution is

$$y = \cos 2x + 3 \sin 2x - 2 \sin 3x.$$

Occasionally the natural trial solution turns out to be a solution of the associated *homogeneous* equation and therefore cannot possibly produce a solution of our inhomogeneous equation. Various rules as to what to try next can be stated, but often a bit of experimentation suffices.

EXAMPLE 5 Find a particular solution of $y'' + 4y = 10 \sin 2x$.

Solution Here it will do no good to try $A \sin 2x$ as a trial solution, because this function satisfies the associated homogeneous equation $y'' + 4y = 0$. We need a function $f(x)$ with second derivative that involves *both* $f(x)$ itself *and* $\sin 2x$, so that cancellation may occur and leave the $\sin 2x$ term we require. A good candidate is

$$y_p(x) = Ax \cos 2x,$$

because then

$$y_p'' = -4Ax \cos 2x - 4A \sin 2x.$$

Substitution into the original differential equation gives the equation

$$(-4Ax \cos 2x - 4A \sin 2x) + 4(Ax \cos 2x) = 10 \sin 2x,$$

which is satisfied if $A = -\frac{5}{2}$. Thus the desired particular solution is

$$y_p(x) = -\frac{5}{2}x \cos 2x.$$

17-3 PROBLEMS

In each of Problems 1–6, a differential equation and its general solution are given. Find the particular solution satisfying the indicated initial conditions.

1 $y'' + y' = 0$; $y = c_1 + c_2 e^{-x}$; $y(0) = 1$, $y'(0) = -1$.
2 $y'' + y = 0$; $y = c_1 \cos x + c_2 \sin x$; $y(0) = 0$, $y'(0) = 2$.
3 $y'' + y' - 6y = 0$; $y = c_1 e^{2x} + c_2 e^{-3x}$; $y(0) = 0$, $y'(0) = 1$.
4 $y'' - 2y' + y = 0$; $y = c_1 e^x + c_2 x e^x$; $y(0) = y'(0) = 3$.
5 $y'' + 2y' + 2y = 0$; $y = c_1 e^{-x} \cos x + c_2 e^{-x} \sin x$; $y(0) = 2$, $y'(0) = 3$.
6 $y''' - 4y'' + 4y' = 0$; $y = c_1 + c_2 e^{2x} + c_3 x e^{2x}$; $y(0) = 0$, $y'(0) = y''(0) = 1$.

In each of Problems 7–10, find all values of r such that $y = e^{rx}$ is a solution of the given differential equation.

7 $y'' = 4y$.
8 $y'' + 4y' + 4y = 0$.
9 $y'' - 5y' - 6y = 0$.
10 $y''' - 9y' = 0$.

In Problems 11–20, use the method of undetermined coefficients to find a particular solution of the given equation.

11 $y' + y = 3x$.
12 $y'' + y' + 2y = x^2 - 1$.
13 $y'' - 4y = e^x$.
14 $y'' + 3y' = \sin 2x$.
15 $y'' + 2y' + 2y = 3 \cos 2x$.
16 $y''' + y'' - y = x^3$.
17 $y'' + 9y = 5 \cos 3x$.
18 $y'' - 9y = 5e^{-3x}$.
19 $y''' - 2y'' + y' = x$.
20 $y'' - 2y = x e^x$.
21 Let $y_p(x)$ be a particular solution of the inhomogeneous equation in (1), and let

$$y_h(x) = \sum_{i=1}^{n} c_i y_i(x)$$

be the general solution of the associated homogeneous equation in (2). If $y(x)$ is a second particular solution of (1), show that $y(x) = y_h(x) + y_p(x)$ for an appropriate choice of the constants $c_1, c_2, \ldots, c_n$. (*Suggestion:* Show first that $y(x) - y_p(x)$ is a solution of the homogeneous equation.)

Complex Numbers and Functions

We include in this section the elementary algebra and calculus of complex numbers and functions—as much as is needed to solve homogeneous linear differential equations with constant coefficients (Section 17-5). Recall that a **complex number** is one of the form

$$z = a + bi \tag{1}$$

where a and b are real numbers, and i denotes the (imaginary) square root of -1; that is,

$$i = \sqrt{-1}; \qquad i^2 = -1. \tag{2}$$

Complex numbers may appear, for example, when we solve a quadratic equation by use of the quadratic formula.

The **real part** $Re(z)$ of the complex number $z = a + bi$ is a, and its **imaginary part** $Im(z)$ is b (not bi):

$$Re(a + bi) = a, \qquad Im(a + bi) = b. \tag{3}$$

The **conjugate** of $z = a + bi$ is the complex number $\bar{z} = a - bi$, with the same real part as z but with imaginary part opposite in sign from that of z. Complex numbers are added (and subtracted) by adding their real and imaginary parts:

$$(a + bi) \pm (c + di) = (a \pm c) + (b \pm d)i. \tag{4}$$

For instance, $(2 + 7i) + (3 - 5i) = 5 + 2i$. Complex numbers are multiplied just as binomials are multiplied, using the fact that $i^2 = -1$:

$$(a + bi)(c + di) = ac + (ad + bc)i + bdi^2$$
$$= (ac - bd) + (ad + bc)i. \tag{5}$$

To find the quotient of two complex numbers, we can use the familiar technique of rationalization:

$$\frac{a + bi}{c + di} = \frac{a + bi}{c + di} \cdot \frac{c - di}{c - di} = \frac{(ac + bd) + (bc - ad)i}{c^2 + d^2}. \tag{6}$$

EXAMPLE 1 Express the complex number $(2 - 3i)/(3 + 4i)$ in the standard form of (1).

Solution

$$\frac{2 - 3i}{3 + 4i} = \frac{2 - 3i}{3 + 4i} \cdot \frac{3 - 4i}{3 - 4i} = -\frac{6}{25} - \frac{17}{25}i.$$

We can represent complex numbers geometrically by regarding the x-axis as the *real axis* and the y-axis as the *imaginary axis*. We then identify the complex number $z = x + iy$ with the point $P(x, y)$ in the plane. In terms of the polar coordinates r and θ of P we have

$$z = x + iy = r(\cos \theta + i \sin \theta). \qquad (7)$$

We call the right-hand side in Equation (7) the **polar form** of the complex number z. Note that the polar form is expressed in terms of the **magnitude** $r = |z|$ of the complex number z and its **argument** $\theta = \arg z$. From Figure 17.6 we see that

$$r = |z| = \sqrt{x^2 + y^2} \quad \text{and} \quad \tan \theta = \frac{y}{x}. \qquad (8)$$

In terms of the common and useful abbreviation

$$\operatorname{cis} \theta = \cos \theta + i \sin \theta, \qquad (9)$$

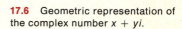

17.6 Geometric representation of the complex number $x + yi$.

the polar form of z is

$$z = r \operatorname{cis} \theta. \qquad (10)$$

For example,

$$1 + i = \sqrt{2}\left(\cos \frac{\pi}{4} + i \sin \frac{\pi}{4}\right) = \sqrt{2} \operatorname{cis} \frac{\pi}{4}.$$

The polar form is especially convenient for multiplying and dividing complex numbers. For if

$$z_1 = r_1 \operatorname{cis} \theta_1 \quad \text{and} \quad z_2 = r_2 \operatorname{cis} \theta_2,$$

then a simple computation (Problem 17) using the addition formulas for sine and cosine shows that

$$z_1 z_2 = (r_1 \operatorname{cis} \theta_1)(r_2 \operatorname{cis} \theta_2) = r_1 r_2 \operatorname{cis}(\theta_1 + \theta_2). \qquad (11)$$

Thus the magnitudes are multiplied while the arguments are added. Similarly,

$$\frac{z_1}{z_2} = \frac{r_1 \operatorname{cis} \theta_1}{r_2 \operatorname{cis} \theta_2} = \frac{r_1}{r_2} \operatorname{cis}(\theta_1 - \theta_2). \qquad (12)$$

For example,

$$\frac{1 + i}{1 - i} = \frac{\sqrt{2} \operatorname{cis} \pi/4}{\sqrt{2} \operatorname{cis}(-\pi/4)} = \operatorname{cis} \pi/2 = i.$$

From Equation (11) we obtain $(r \operatorname{cis} \theta)^2 = r^2 \operatorname{cis} 2\theta$ and

$$(r \operatorname{cis} \theta)^3 = (r \operatorname{cis} \theta)(r^2 \operatorname{cis} 2\theta) = r^3 \operatorname{cis} 3\theta.$$

We proceed by induction, and find that

$$(r \operatorname{cis} \theta)^n = r^n \operatorname{cis} n\theta \qquad (13)$$

for all positive integers n. The special case $r = 1$ is **DeMoivre's formula**

$$(\cos \theta + i \sin \theta)^n = \cos n\theta + i \sin n\theta. \qquad (14)$$

EXAMPLE 2 Write $(1 + i\sqrt{3})^{10}$ in standard form.

Solution If $z = 1 + i\sqrt{3}$, then $r = |z| = 2$ and $\theta = \tan^{-1}\sqrt{3} = \pi/3$. Hence (14) gives

$$(1 + i\sqrt{3})^{10} = \left(2 \operatorname{cis} \frac{\pi}{3}\right)^{10} = 2^{10} \operatorname{cis}\left(\frac{10\pi}{3}\right)$$

$$= 1024 \operatorname{cis}\left(2\pi + \frac{4\pi}{3}\right) = 1024 \operatorname{cis}\left(\frac{4\pi}{3}\right)$$

$$= 1024\left(-\frac{1}{2} - \frac{\sqrt{3}}{2}i\right) \approx -512 - (886.81)i.$$

DeMoivre's formula can also be used to compute roots of complex numbers. Let $r^{1/n}$ denote the ordinary positive real nth root $\sqrt[n]{r}$ of the positive real number r. Then Equation (13) gives

$$\left[r^{1/n} \operatorname{cis} \frac{\theta}{n}\right]^n = r \operatorname{cis} \theta.$$

So $r^{1/n} \operatorname{cis}(\theta/n)$ is an nth root of $z = r \operatorname{cis} \theta$. But the number z can also be written in the forms

$$z = r \operatorname{cis}(\theta + 2\pi k), \qquad k = 0, \pm 1, \pm 2, \pm 3, \ldots.$$

The values $k = 0, 1, 2, \ldots, n-1$ give, as above, the n distinct nth roots

$$\sqrt[n]{z} = r^{1/n} \operatorname{cis}\left(\frac{\theta}{n} + \frac{2\pi k}{n}\right), \qquad k = 0, 1, 2, \ldots, n - 1. \tag{15}$$

These n complex numbers are, indeed, nth roots of z. But no other nth roots of z can be obtained by using other values of k. For if m is a multiple of n, then

$$r^{1/n} \operatorname{cis}\left(\frac{\theta}{n} + \frac{2\pi(k + m)}{n}\right) = r^{1/n} \operatorname{cis}\left(\frac{\theta}{n} + \frac{2\pi k}{n}\right).$$

The n distinct nth roots of $z = r \operatorname{cis} \theta$ given in (15) are evenly spaced around the circle of radius $r^{1/n}$, the first (corresponding to $k = 0$) having argument θ/n, and with the arguments of the others differing from θ/n by integral multiples of $2\pi/n$. For instance, the eight eighth roots of 256 are shown in Figure 17.7.

EXAMPLE 3 Find the three cube roots $z_0, z_1,$ and z_2 of $z = 27i = 27 \operatorname{cis} \pi/2$.

Solution We take $n = 3$, $r = 27$, $\theta = \pi/2$, and $k = 0, 1,$ and 2 in Formula (15). We find that

$$z_0 = 3 \operatorname{cis} \frac{\pi}{6} = \frac{3}{2}(\sqrt{3} + i),$$

$$z_1 = 3 \operatorname{cis}\left(\frac{\pi}{6} + \frac{2\pi}{3}\right) = 3 \operatorname{cis} \frac{5\pi}{6} = \frac{3}{2}(-\sqrt{3} + i),$$

$$z_2 = 3 \operatorname{cis}\left(\frac{\pi}{6} + \frac{4\pi}{3}\right) = 3 \operatorname{cis} \frac{3\pi}{2} = -3i.$$

These three roots are shown in Figure 17.8.

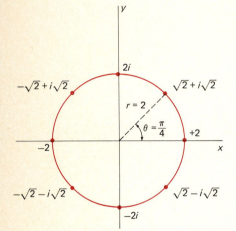

17.7 The eight eighth roots of 256.

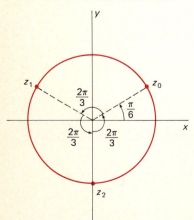

17.8 The three cube roots of $27i$.

COMPLEX SEQUENCES AND SERIES

Suppose that $\{a_n + ib_n\}_{n=1}^{\infty}$ is an infinite sequence of complex numbers. We say that

$$\lim_{n \to \infty} (a_n + ib_n) = A + Bi$$

provided that both

$$\lim_{n \to \infty} a_n = A \quad \text{and} \quad \lim_{n \to \infty} b_n = B.$$

When we apply this definition to the sequence of partial sums of an infinite series of complex numbers, we find that we have the definition of the sum of such a series. We thus say that

$$\sum_{n=1}^{\infty} (a_n + ib_n) = A + Bi$$

provided that both

$$\sum_{n=1}^{\infty} a_n = A \quad \text{and} \quad \sum_{n=1}^{\infty} b_n = B.$$

Thus we can sum an infinite series of complex numbers by separately summing the real and imaginary parts of its terms.

For instance, if we put $t = i\theta$ in the Taylor series for e^t as a way to *define* $e^{i\theta}$, we obtain

$$e^{i\theta} = \sum_{n=0}^{\infty} \frac{(i\theta)^n}{n!} = 1 + i\theta - \frac{\theta^2}{2!} - \frac{i\theta^3}{3!} + \frac{\theta^4}{4!} + \frac{i\theta^5}{5!} - \cdots$$

$$= \left(1 - \frac{\theta^2}{2!} + \frac{\theta^4}{4!} - \cdots\right) + i\left(\theta - \frac{\theta^3}{3!} + \frac{\theta^5}{5!} - \cdots\right);$$

$$e^{i\theta} = \cos\theta + i\sin\theta = \operatorname{cis}\theta. \tag{16}$$

This result is known as **Euler's formula.** Because of it we *define* the exponential function e^z, for $z = x + iy$, by

$$e^z = e^{x+iy} = e^x e^{iy} = e^x(\cos y + i\sin y). \tag{17}$$

Note that if

$$z_1 = x_1 + iy_1 \quad \text{and} \quad z_2 = x_2 + iy_2,$$

then Formula (17) implies that

$$e^{z_1}e^{z_2} = e^{x_1}\operatorname{cis} y_1\, e^{x_2}\operatorname{cis} y_2 = e^{x_1}e^{x_2}\operatorname{cis}(y_1 + y_2)$$

$$= e^{x_1 + x_2}e^{i(y_1 + y_2)} = e^{z_1 + z_2}.$$

This shows that the complex exponential function satisfies one important law of exponents; by like methods, it can be shown to satisfy the others as well.

Euler's formula leads to important expressions for the sine and cosine functions in terms of the exponential function. The equations

$$e^{ix} = \cos x + i\sin x \quad \text{and} \quad e^{-ix} = \cos x - i\sin x,$$

which follow from (16) (put $\theta = x$, then $\theta = -x$), imply (add them, subtract them) the formulas

$$\cos x = \frac{e^{ix} + e^{-ix}}{2} \quad \text{and} \quad \sin x = \frac{e^{ix} - e^{-ix}}{2i}. \tag{18}$$

COMPLEX-VALUED FUNCTIONS

A **complex-valued function** F of the *real* variable x associates with each real number x (in its domain of definition) the complex number

$$z = F(x) = f(x) + ig(x). \tag{19}$$

The real-valued functions $f(x)$ and $g(x)$ are called the real and imaginary parts, respectively, of $F(x)$. If they are differentiable, we can define the **derivative** of F to be

$$F'(x) = f'(x) + ig'(x). \tag{20}$$

Thus we merely differentiate the real and imaginary parts of F separately. Similarly, we integrate a complex-valued function by separately integrating its real and imaginary parts—assuming the integrals exist. That is,

$$\int_a^b F(x)\,dx = \int_a^b f(x)\,dx + i\int_a^b g(x)\,dx. \tag{21}$$

A particular *complex-valued* function that plays an important role in the solution of *real* differential equations is the exponential function $F(x) = e^{rx}$ where $r = a + bi$, a complex number. By (17), we have

$$e^{rx} = e^{ax}(\cos bx + i \sin bx). \tag{22}$$

So the real and imaginary parts of e^{rx} are $e^{ax} \cos bx$ and $e^{ax} \sin bx$, respectively.

EXAMPLE 4   Suppose that $r = a + bi$. Show that

$$De^{rx} = re^{rx}, \tag{23}$$

just as if r were a real number.

Solution   From (22) we obtain

$$\begin{aligned}
De^{rx} &= D(e^{ax} \cos bx) + iD(e^{ax} \sin bx) \\
&= [ae^{ax} \cos bx - be^{ax} \sin bx] + i[ae^{ax} \sin bx + be^{ax} \cos bx] \\
&= (a + bi)(e^{ax} \cos bx + ie^{ax} \sin bx) = re^{rx}.
\end{aligned}$$

THE FUNDAMENTAL THEOREM OF ALGEBRA

The fundamental theorem of algebra, easy to state but difficult to prove, is the underlying reason for the importance—indeed, the very unavoidability—of complex numbers in mathematics. The theorem tells us that every polynomial equation of degree n,

$$p(x) = a_n x^n + a_{n-1} x^{n-1} + \cdots + a_1 x + a_0 = 0, \tag{24}$$

has at least one (complex) root r_1. By the **factor theorem**, we can then write $p(x) = (x - r_1)q(x)$, where $q(x)$ is a polynomial of degree $n - 1$. By induction on n, it then follows that the nth degree polynomial $p(x)$ has the factorization

$$p(x) = a_n(x - r_1)(x - r_2) \cdots (x - r_n),$$

and the n roots $r_1, r_2, r_3, \ldots, r_n$ (not necessarily distinct). Even if the coefficients $a_0, a_1, a_2, \ldots, a_n$ of the polynomial $p(x)$ are all real, the roots will be, in general, complex; this is one reason for the necessity of complex numbers. But if the coefficients of the polynomial $p(x)$ are all real, then it can be shown that the complex roots (if any) occur in **conjugate pairs** of the form $a \pm bi$.

Express the complex numbers given in Problems 1–6 both in the form $x + iy$ and in the form r cis θ.

1 $(3 - 4i)(3 + 4i)$.

2 $(1 + 2i)(2 - 3i)$.

3 $\dfrac{5 + 7i}{3 - 2i}$.

4 $\dfrac{(1 + i)^2}{(1 - i\sqrt{3})^2}$.

5 $(1 + i)^6$.

6 $(-3 + 4i)^3$.

7 Find the four fourth roots of -1.

8 Find the three cube roots of 27.

9 Find the six sixth roots of -64.

10 Find the five fifth roots of $-4 - 4i$.

11 Find the four roots of the equation $x^4 - 16 = 0$.

12 Find the four roots of the equation $x^4 - 4x^2 + 6 = 0$.

13 Find the six roots of the equation $x^6 + 4x^3 + 8 = 0$.

14 Expand $(\cos \theta + i \sin \theta)^3$. Then compare the result with that given by DeMoivre's formula, and thereby discover formulas for $\cos 3\theta$ and $\sin 3\theta$ in terms of $\sin \theta$ and $\cos \theta$.

15 Use DeMoivre's formula as in Problem 14 to discover formulas for $\cos 4\theta$ and $\sin 4\theta$ in terms of $\sin \theta$ and $\cos \theta$.

16 Suppose that a is a positive real number. Find the 4 roots of the equation $x^4 + a^4 = 0$.

17 Prove that $\operatorname{cis}(\alpha + \beta) = \operatorname{cis} \alpha \operatorname{cis} \beta$.

18 If $e^w = z$ then $z^i = e^{wi}$ by definition. Choose w so that $e^w = i$, then conclude that

$$i^i = e^{-\pi/2} \approx 0.20788.$$

19 Obtain formulas for $\int e^{ax} \cos bx \, dx$ and $\int e^{ax} \sin bx \, dx$ by equating real and imaginary parts in the formula

$$\int e^{(a + bi)x} \, dx = \frac{e^{(a + bi)x}}{a + bi} + C.$$

17-5

Our goal here is to solve the nth order homogeneous linear differential equation

$$a_n \frac{d^n y}{dx^n} + a_{n-1} \frac{d^{n-1} y}{dx^{n-1}} + \cdots + a_1 \frac{dy}{dx} + a_0 y = 0 \qquad (1)$$

Homogeneous Linear Equations with Constant Coefficients

with real *constant* coefficients $a_0, a_1, a_2, \ldots, a_n$. As we saw in Section 17-3, the general solution of (1) will be a linear combination

$$y(x) = c_1 y_1(x) + c_2 y_2(x) + \cdots + c_n y_n(x) \qquad (2)$$

of n particular solutions $y_1, y_2, \ldots, y_n$. In the general theory of linear differential equations it is shown that the linear combination in (2) is the general solution of (1) if and only if the particular solutions $y_1, y_2, \ldots, y_n$ are **linearly independent,** meaning that no one of them can be expressed as a linear combination of the others.

So we need a way to find n independent solutions of Equation (1). We begin with the observation that

$$\frac{d^k(e^{rx})}{dx^k} = r^k e^{rx}. \tag{3}$$

Thus the kth derivative of e^{rx} is a constant multiple of e^{rx}. This suggests that we attempt to determine r so that $y = e^{rx}$ is a solution of Equation (1).

To find such a value of r, we merely substitute $y = e^{rx}$ into Equation (1), then choose r so that this particular function *is* a solution. By Formula (3), the result of this substitution is

$$a_n r^n e^{rx} + a_{n-1} r^{n-1} e^{rx} + \cdots + a_1 r e^{rx} + a_0 e^{rx} = 0,$$

or

$$e^{rx}(a_n r^n + a_{n-1} r^{n-1} + \cdots + a_1 r + a_0) = 0. \tag{4}$$

Consequently, $y = e^{rx}$ will be a solution of Equation (1) precisely when r is a root of the equation

$$a_n r^n + a_{n-1} r^{n-1} + \cdots + a_1 r + a_0 = 0. \tag{5}$$

This equation is called the **characteristic equation** of the differential equation in (1). If it has n *distinct* (unequal) roots $r_1, r_2, \ldots, r_n$, then the functions

$$e^{r_1 x}, e^{r_2 x}, \ldots, e^{r_n x}$$

turn out to be linearly independent. This gives the following result.

Theorem 1 *Distinct Roots*

If the roots $r_1, r_2, r_3, \ldots, r_n$ of the characteristic equation in (5) are distinct, then the general solution of (1) is

$$y = c_1 e^{r_1 x} + c_2 e^{r_2 x} + \cdots + c_n e^{r_n x}. \tag{6}$$

EXAMPLE 1 Find the general solution of $y''' - y'' - 6y' = 0$.

Solution The characteristic equation of this differential equation is

$$r^3 - r^2 - 6r = 0,$$

which we solve by factoring:

$$r(r - 3)(r + 2) = 0.$$

The three roots are 0, 3, and -2. They are distinct, and therefore—because $e^{0x} = 1$—the general solution of our differential equation is

$$y = c_1 + c_2 e^{3x} + c_3 e^{-2x}.$$

Note the two surprises: First, that the exponential function is so important in solving linear differential equations with constant coefficients; second, that *no calculus* is used in any explicit way. Theorem 1 makes a problem in differential equations into a problem involving the fundamental theorem of algebra.

If the roots of the characteristic equation in (5) are *not* distinct—there are *repeated* roots—then we don't yet have n independent solutions of (1). The problem, then, is to produce the "missing" independent solutions.

Suppose, for example, that $n = 2$ and that the characteristic equation has the double root $r = r_1$. This will happen precisely when the characteristic equation is a constant multiple of the equation

$$(r - r_1)^2 = r^2 - 2r_1 r + r_1^2 = 0.$$

The differential equation with this characteristic equation is

$$y'' - 2r_1 y' + r_1^2 y = 0. \tag{7}$$

But it is easy to verify by direct substitution that $y = xe^{r_1 x}$ is a second solution (in addition to $y = e^{r_1 x}$) of Equation (7). It is clear that $e^{r_1 x}$ and $xe^{r_1 x}$ are independent functions, so the general solution of (7) is

$$y = c_1 e^{r_1 x} + c_2 x e^{r_1 x} = (c_1 + c_2 x)e^{r_1 x}.$$

Similarly, it is easy to verify by direct substitution that $e^{r_1 x}$, $xe^{r_1 x}$, and $x^2 e^{r_1 x}$ are all solutions of the third order equation

$$y''' - 3r_1 y'' + 3r_1^2 y' - r_1^3 y = 0, \tag{8}$$

whose characteristic equation has the triple root $r = r_1$. Thus the general solution of (8) derived from this triple root is

$$y = (c_1 + c_2 x + c_3 x^2)e^{r_1 x}.$$

By a more general argument, the following theorem can be established.

Theorem 2 *Repeated Roots*

If the characteristic equation in (5) has a repeated root r_1 of multiplicity k, then the part of the general solution of (1) that corresponds to r_1 is

$$(c_1 + c_2 x + \cdots + c_k x^{k-1})e^{r_1 x}. \tag{9}$$

EXAMPLE 2 Find the general solution of the differential equation
$$y^{(4)} + 3y^{(3)} + 3y'' + y' = 0.$$

Solution The characteristic equation of this differential equation is

$$r^4 + 3r^3 + 3r^2 + r = r(r + 1)^3 = 0.$$

It has the simple root 0 and the triple root -1. Hence the general solution of the differential equation is

$$y = c_1 + (c_2 + c_3 x + c_4 x^2)e^{-x}.$$

The roots of the characteristic equation may be either real or complex. Since we are assuming that the coefficients are real numbers, the nonreal complex roots will occur in *conjugate pairs*, of the form $\alpha \pm \beta i$. The terms in the general solution of (1) corresponding to the two roots $r_1 = \alpha + \beta i$ and $r_2 = \alpha - \beta i$ are

$$c_1 e^{(\alpha + \beta i)x} + c_2 e^{(\alpha - \beta i)x} = c_1 e^{\alpha x}(\cos \beta x + i \sin \beta x) + c_2 e^{\alpha x}(\cos \beta x - i \sin \beta x)$$

$$= e^{\alpha x}(C_1 \cos \beta x + C_2 \sin \beta x)$$

where $C_1 = c_1 + c_2$ and $C_2 = (c_1 - c_2)i$. This computation gives the following result.

> **Theorem 3** *Complex Roots*
>
> If the characteristic equation in (5) has an unrepeated pair of complex conjugate roots $\alpha \pm \beta i$, then the corresponding part of the general solution of (1) is
>
> $$e^{\alpha x}(C_1 \cos \beta x + C_2 \sin \beta x). \qquad (10)$$

In the case of a complex conjugate pair of roots $\alpha \pm \beta i$ of multiplicity k, the corresponding part of the general solution is a linear combination of the functions

$$x^p e^{\alpha x} \cos \beta x \quad \text{and} \quad x^p e^{\alpha x} \sin \beta x$$

for $p = 0, 1, 2, \ldots, k - 1$.

EXAMPLE 3 Find the general solution of $y'' + \beta^2 y = 0 \ (\beta > 0)$.

Solution The characteristic equation is $r^2 + \beta^2 = 0$, and its roots are $\pm \beta i$. Hence we have the two solutions in (10) with $\alpha = 0$. This gives the familiar solution

$$y = C_1 \cos \beta x + C_2 \sin \beta x$$

to the given differential equation.

EXAMPLE 4 Find the particular solution of $y'' - 4y' + 5y = 0$ for which $y(0) = 1$ and $y'(0) = 5$.

Solution The characteristic equation is $r^2 - 4r + 5 = 0$, with roots $r = 2 \pm i$. Hence the general solution is

$$y = e^{2x}(c_1 \cos x + c_2 \sin x).$$

To find the particular solution required in this example, we need to know dy/dx too; it is

$$y' = 2e^{2x}(c_1 \cos x + c_2 \sin x) + e^{2x}(-c_1 \sin x + c_2 \cos x).$$

Then our initial conditions give

$$y(0) = c_1 = 1, \qquad y'(0) = 2c_1 + c_2 = 5.$$

So $c_2 = 3$, and the desired particular solution is

$$y = e^{2x}(\cos x + 3 \sin x).$$

EXAMPLE 5 Find the general solution of

$$y^{(4)} + \gamma^4 y = 0 \qquad (\gamma > 0).$$

Solution The characteristic equation of this fourth order linear differential equation is

$$r^4 + \gamma^4 = (r^2 + \gamma^2 i)(r^2 - \gamma^2 i) = 0$$

and its four roots are $r = \pm\gamma\sqrt{\pm i}$. Now $i = \operatorname{cis} \pi/2$ and $-i = \operatorname{cis}(3\pi/2)$, so

$$\sqrt{i} = \operatorname{cis} \frac{\pi}{4} = \frac{1+i}{\sqrt{2}}, \qquad \sqrt{-i} = \operatorname{cis} \frac{3\pi}{4} = \frac{-1+i}{\sqrt{2}}.$$

It follows that the four distinct roots of our characteristic equation are $r = \pm(1 \pm i)\alpha$, where $\alpha = \gamma/\sqrt{2}$. These two pairs of complex conjugate roots, $\alpha \pm \alpha i$ and $-\alpha \pm \alpha i$, give the general solution

$$y = e^{\alpha x}(c_1 \cos \alpha x + c_2 \sin \alpha x) + e^{-\alpha x}(c_3 \cos \alpha x + c_4 \sin \alpha x)$$

of the equation $y^{(4)} + \gamma^4 y = 0 \quad (\gamma = \alpha\sqrt{2})$.

HOMOGENEOUS LINEAR SECOND ORDER EQUATIONS

For reference in the following section, we here record the general solution of the homogeneous second order linear differential equation

$$y'' + 2py' + qy = 0 \tag{11}$$

with constant coefficients p and q. The characteristic equation is

$$r^2 + 2pr + q = 0,$$

with roots

$$r = \frac{-2p \pm \sqrt{4p^2 - 4q}}{2} = -p \pm \sqrt{p^2 - q}.$$

If $p^2 > q$: There are distinct roots r_1 and r_2. The general solution is

$$y = c_1 e^{r_1 x} + c_2 e^{r_2 x}. \tag{12}$$

Note that if p and q are both positive, then r_1 and r_2 are both negative.

If $p^2 = q$: Then $r = -p$ is a double root, and the general solution is

$$y = e^{-px}(c_1 + c_2 x). \tag{13}$$

If $p^2 < q$: Then $r = -p \pm \omega i$ where

$$\omega = \sqrt{q - p^2}.$$

In this case, the general solution of (11) is

$$y = e^{-px}(c_1 \cos \omega x + c_2 \sin \omega x). \tag{14}$$

Find the general solution of each of the differential equations given in Problems 1-18.

1 $y'' - 4y = 0.$

2 $y'' - 3y' = 0.$

3 $y'' + 3y' - 10y = 0.$

4 $y'' - 5y' + 4y = 0.$

5 $y'' + 5y' + 5y = 0.$

6 $y'' + 4y' + 4y = 0.$

7 $y'' - 6y' + 9y = 0.$

8 $y'' - 6y' + 13y = 0.$

9 $y'' + 8y' + 25y = 0.$

10 $y^{(4)} + 2y^{(3)} = 0.$

11 $y^{(4)} - 6y^{(3)} + 9y'' = 0.$

12 $y^{(4)} - 3y^{(3)} + 3y'' - y' = 0.$

13 $y^{(3)} + 2y'' + 2y' = 0.$

14 $y^{(4)} + 3y'' - 4y = 0.$

15 $y^{(4)} - 8y'' + 16y = 0.$

16 $y^{(4)} + 8y'' + 16y = 0.$

17 $y^{(4)} = \gamma^4 y.$

18 $y^{(4)} + 2y^{(3)} + 3y'' + 2y' + y = 0.$ (*Suggestion:* Compute $(1 + r + r^2)^2.$)

19 Show that the substitution $x = e^z$ reduces the *Euler equation* $ax^2 y'' + bxy' + cy = 0$ to a homogeneous linear equation with *constant* coefficients (for y as a function of z).

Use the substitution of Problem 19 to solve the Euler equations in Problems 20-22.

20 $x^2 y'' - 6y = 0.$

21 $x^2 y'' + 5xy' + 4y = 0.$

22 $x^2 y'' + xy' + 4y = 0.$

---

*17-6

Mechanical Vibrations

In Section 7-4 we discussed the motion of a mass m on the end of a spring that exerts a restoring force $F_S = -kx$ on the mass when its displacement from the equilibrium position is x (Figure 17.9). Newton's law then gives $m(d^2x/dt^2) = F_S = -kx$, or

$$\frac{d^2x}{dt^2} + \omega^2 x = 0 \qquad \left(\omega^2 = \frac{k}{m}\right). \tag{1}$$

The general solution of (1) is

$$x(t) = A \cos \omega t + B \sin \omega t. \tag{2}$$

If the mass is released from rest at $x = x_0$, then the initial conditions $x(0) = x_0$ and $x'(0) = 0$ give

$$x(t) = x_0 \cos \omega t \tag{3}$$

as the motion that results. This equation describes a periodic oscillation or vibration with **amplitude** x_0, **period**

$$T = \frac{2\pi}{\omega} = 2\pi \sqrt{\frac{m}{k}}, \tag{4}$$

and **frequency**

$$f = \frac{1}{T} = \frac{\omega}{2\pi} = \frac{1}{2\pi}\sqrt{\frac{k}{m}}. \tag{5}$$

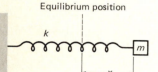

Equilibrium position

17.9 The undamped mass-and-spring system.

DAMPED VIBRATIONS

Now suppose that the mass above is attached, as indicated in Figure 17.10, to a dashpot or shock absorber—as in the suspension system of an automobile—that exerts on it a force F_R of resistance to motion that is proportional to its velocity v. Then $F_R = -cv$, where the proportionality constant c will depend upon the viscosity of the fluid in the dashpot; the more

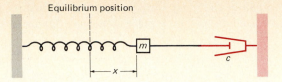

17.10 A spring-and-mass system with damping force provided by a dashpot.

Equilibrium position

viscous this fluid is, the larger $c > 0$ is. To take this resistive force into account, we replace Equation (1) by

$$m \frac{d^2x}{dt^2} = F_S + F_R = -kx - c \frac{dx}{dt}$$

or

$$\frac{d^2x}{dt^2} + 2p \frac{dx}{dt} + qx = 0 \tag{6}$$

where

$$p = \frac{c}{2m} \quad \text{and} \quad q = \frac{k}{m}. \tag{7}$$

Now Equation (6) is a homogeneous second order linear differential equation, so its solution depends upon the relative magnitudes of p and q as we described at the end of Section 17-5. We want to determine the motion that results if the mass is released at time $t = 0$ with initial conditions $x(0) = x_0 > 0$ and $x'(0) = 0$.

OVERDAMPED CASE $p^2 > q$ or $c^2 > 4km$.

Since c is relatively large in this case, we are dealing with a *strong* resistive or frictional force compared with a relatively weak spring and small mass. This case corresponds to Equation (12) in Section 17-5. In our present notation, the general solution of the differential equation in (6) is $x = c_1 e^{r_1 t} + c_2 e^{r_2 t}$ where the distinct roots

$$r_1 = -p + \sqrt{p^2 - q}, \qquad r_2 = -p - \sqrt{p^2 - q}$$

are both negative. The initial conditions $x(0) = x_0$ and $x'(0) = 0$ give us the equations

$$c_1 + c_2 = x_0, \qquad r_1 c_1 + r_2 c_2 = 0.$$

It follows that

$$c_1 = \frac{r_2 x_0}{r_2 - r_1} \quad \text{and} \quad c_2 = -\frac{r_1 x_0}{r_2 - r_1}.$$

Hence the resulting motion of the mass is described by

$$x(t) = \frac{x_0}{r_2 - r_1} \left(r_2 e^{r_1 t} - r_1 e^{r_2 t} \right). \tag{8}$$

Since r_1 and r_2 are both negative, it is not difficult to verify that $x(t)$ is strictly decreasing for $t > 0$ and that $\lim_{t \to 0} x(t) = 0$. As indicated by the graph of $x(t)$ in Figure 17.11, the large viscosity of the dashpot fluid in this case

17.11 Variation of displacement x with time t: the overdamped case.

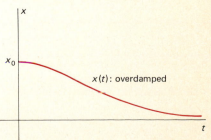

causes the mass simply to recede toward its equilibrium position, with no oscillations taking place—they are "damped out."

CRITICALLY DAMPED CASE $p^2 = q$ or $c^2 = 4km$.

This case corresponds to Equation (13) in Section 17-5; the general solution is

$$x = e^{-pt}(c_1 + c_2 t).$$

Since

$$x'(t) = -pe^{-pt}(c_1 + c_2 t) + c_2 e^{-pt},$$

the initial conditions $x(0) = x_0$, $x'(0) = 0$ give the equations

$$c_1 = x_0, \qquad -pc_1 + c_2 = 0.$$

So $c_1 = x_0$ and $c_2 = px_0$. The resulting motion of the mass m is then described by

$$x(t) = x_0 e^{-pt}(1 + pt). \tag{9}$$

The graph of this function is shown in Figure 17.12. In this case the viscosity is just large enough to damp out all vibrations, but even a slight decrease in viscosity will yield the following case, the one that shows the most dramatic behavior.

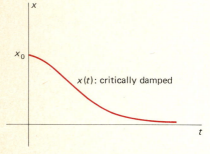

17.12 Variation of displacement x with time t: the critically damped case.

UNDERDAMPED CASE $p^2 < q$ or $c^2 < 4km$.

By Equation (14) of Section 17-5, the general solution now is

$$x = e^{-pt}(c_1 \cos \omega t + c_2 \sin \omega t)$$

where

$$\omega = \sqrt{q - p^2} = \sqrt{\frac{k}{m} - \frac{c^2}{4m^2}}. \tag{10}$$

Because

$$x'(t) = -pe^{-pt}(c_1 \cos \omega t + c_2 \sin \omega t) + \omega e^{-pt}(-c_1 \sin \omega t + c_2 \cos \omega t),$$

the initial conditions $x(0) = x_0$, $x'(0) = 0$ give the equations

$$c_1 = x_0, \qquad -pc_1 + \omega c_2 = 0.$$

So $c_1 = x_0$ and $c_2 = px_0/\omega$. Hence the resulting motion is described by

$$x(t) = \frac{x_0}{\omega} e^{-pt}(\omega \cos \omega t + p \sin \omega t).$$

With the use of the cosine addition formula, this solution can be rewritten in the form

$$x(t) = Ae^{-pt} \cos(\omega t - \alpha) \tag{11}$$

where

$$A = \frac{x_0}{\omega} \sqrt{\omega^2 + p^2} \quad \text{and} \quad \alpha = \tan^{-1}\left(\frac{p}{\omega}\right).$$

CHAP. 17: **Differential Equations**

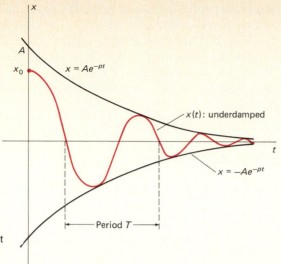

17.13 Variation of displacement x with time t: the underdamped case.

The graph of the function $x(t)$ shown in Figure 17.13 indicates an *exponentially damped vibration* with "amplitude" Ae^{-pt} at time t. The frequency of oscillation (with T as in the figure) is

$$f = \frac{1}{T} = \frac{\omega}{2\pi} = \frac{1}{2\pi}\left(\frac{k}{m} - \frac{c^2}{4m^2}\right)^{1/2}. \tag{12}$$

In comparing this last equation with Equation (5), we see that the frequency in this underdamped case is *less* than in the case of free vibrations. In addition, there is a *phase delay* that corresponds to the *phase angle* α of Equation (11).

FORCED VIBRATIONS

Let us return to the undamped case—no resistance—but suppose that the motion of our mass on a spring is modified by an external force. Of special interest is the case in which that external force is itself periodic, say of the form $F_E = F_0 \sin \omega_0 t$. Such periodic forces are typical in machines involving rotating components. From Newton's law and the diagram of Figure 17.14, we get the differential equation of motion:

$$m\frac{d^2x}{dt^2} = F_S + F_E = -kx + F_0 \sin \omega_0 t$$

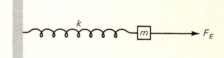

17.14 Mass-and-spring system with externally impressed force F_E.

or

$$\frac{d^2x}{dt^2} + \omega^2 x = \frac{F_0}{m} \sin \omega_0 t \tag{13}$$

where $\omega = \sqrt{k/m}$. We shall now make the important assumption that $\omega_0 \neq \omega$.

The general solution of the homogeneous equation $x'' + \omega^2 x = 0$ associated with (13) is

$$x_h(t) = c_1 \cos \omega t + c_2 \sin \omega t. \tag{14}$$

From Section 17-3 we know that, if we can find a particular solution x_p of the inhomogeneous equation in (13), then its general solution will be $x = x_h + x_p$. But by the method of undetermined coefficients (as in Example 3 of Section 17-3), we can find a particular solution of the form $x_p = A \sin \omega_0 t$. When we substitute this trial solution into Equation (13), we find that $A = F_0/m(\omega^2 - \omega_0^2)$. Thus our particular solution of Equation (13) is

$$x_p(t) = \frac{F_0}{m(\omega^2 - \omega_0^2)} \sin \omega_0 t, \tag{15}$$

and its general solution is

$$x(t) = c_1 \cos \omega t + c_2 \sin \omega t + \frac{F_0}{m(\omega^2 - \omega_0^2)} \sin \omega_0 t. \tag{16}$$

This equation describes the forced motion of the mass as a combination or *superposition* of vibrations with two different frequencies

$$f = \frac{\omega}{2\pi} \quad \text{and} \quad f_0 = \frac{\omega_0}{2\pi}.$$

The frequency $f = \omega/2\pi$ corresponding to the first two terms in the solution (16) is the **natural frequency** of the oscillations that would occur if there were no impressed external force. The last term in (16) represents a vibration with the **impressed frequency** $f_0 = \omega_0/2\pi$ of the external force.

An interesting phenomenon occurs if the difference between ω and ω_0 is very small. Then the coefficient of $\sin \omega_0 t$ in (16) is very large. In this case the impressed vibrations tend to be reinforced by the natural vibrations of the system, and the amplitude of the resulting vibrations is much larger than would otherwise occur. This is the phenomenon of **resonance** (see also Problem 11).

A spectacular example of resonance can occur when a column of soldiers marches in step over a bridge. A bridge is a complicated structure with many natural frequencies of vibration. If the frequency of the soldiers' cadence is approximately equal to one of the bridge's natural frequencies, then—just as in our simple example of a mass on a spring—resonance may occur. Indeed, the resulting resonance vibrations can be of such large amplitude that the bridge will collapse. This has actually happened in the past (for example, the collapse of the Broughton Bridge near Manchester, England in 1831), and it's the reason for the now standard practice of "breaking cadence" when crossing a bridge. Similarly, the avoidance of destructive resonance vibrations is a standard factor in the design of machinery of all types.

On the other hand, many common electrical devices could not function properly without taking advantage of the phenomenon of resonance. For example, consider a radio. A highly simplified model of its tuning circuit is the *RLC* circuit discussed in Section 17-2 (see Figure 17.2), with L and R constant but C variable. Suppose we want to "pick up" a particular radio station that is broadcasting at frequency ω_0, and thereby (in effect) provides a voltage $E(t) = E_0 \sin \omega_0 t$ to the tuning circuit of our radio. If we differentiate Equation (10) in Section 17-2, remembering that $dQ/dt = I$, we obtain the differential equation

$$L\frac{d^2I}{dt^2} + R\frac{dI}{dt} + \frac{1}{C}I = \omega_0 E_0 \cos \omega_0 t \tag{17}$$

for the current I in the circuit. The associated homogeneous equation corresponds to Equation (6) with $p = R/2L$ and $q = 1/LC$. Assuming that $R^2 < 4L/C$ (as is usual in practice), the homogeneous solution is therefore of the form

$$I = Ae^{-pt}\cos(\omega t - \alpha)$$

given by Equation (11). This is a transient solution so it follows that, when t is large, the current in the circuit is effectively described by the particular solution of (17) of the form

$$I(t) = B_1 \cos \omega_0 t + B_2 \sin \omega_0 t \tag{18}$$

(which we can find by using the method of undetermined coefficients). In Problem 10 we ask you to show that the *amplitude* $|I|$ of this steady-state current is

$$|I| = \sqrt{B_1^2 + B_2^2} = \frac{E_0}{(R^2 + [\omega_0 L - (1/\omega_0 C)]^2)^{1/2}}. \tag{19}$$

Thus I is maximal (recalling that R is constant) when $\omega_0 L = 1/\omega_0 C$; this is **electrical resonance.** And so we "tune in" the station that is broadcasting at frequency ω_0 by adjusting our radio's variable capacitor so that $C = 1/L\omega_0^2$, thus maximizing $|I|$ and hence the "volume" of sound that we hear.

17-6   PROBLEMS

Problems 1–3 deal with a highly simplified model of a car with weight 3200 lb and hence mass $m = \frac{3200}{32} = 100$ slugs (a **slug** is the unit of mass in the fps system). Assume that the suspension system of the car acts like a single spring and its shock absorbers like a single dashpot, so that its vertical vibrations satisfy Equation (6) with appropriate values of the constant coefficients.

1 Find the stiffness coefficient k of the spring if the car undergoes free vibrations at 80 cycles per minute when its shock absorbers are disconnected.

2 The car with shock absorbers disconnected is subjected to a periodic force by driving it over a "washboard" road surface that has one sinusoidal wave every 20 feet. At what speed (in mph) will resonance vibrations occur?

3 With the shock absorbers connected the car is set into vibration by driving it over a bump, and the resulting damped vibrations have a frequency of 78 cycles per minute. After how long will their amplitude be 1% of its original value?

4 A 12-lb weight (for which $m = \frac{12}{32}$ slugs) is attached to the end of a spring; this spring is stretched 1 ft by a force of 24 lb. (a) If the weight is displaced 1 ft from equilibrium and then released from rest at time $t = 0$, find its position

function $x(t)$. (b) Find the amplitude, period, and frequency of the motion.

5 Suppose that the mass-and-spring system of Problem 4 is attached to a dashpot that provides a resistance of $3v$ lb. (a) Find $x(t)$ if the mass is released as in Problem 4. (b) What are the new frequency and period of motion?

6 If in Problem 5 the resistive force of the dashpot is $15v/2$, show that no oscillations occur.

7 Show that resonance occurs if the mass-and-spring system of Problem 4 (no dashpot) is subjected to an external force of $F(t) = 20 \sin 8t$ lb. Assume that $x(0) = 0 = x'(0)$.

8 Finally suppose that the mass-and-spring system of Problem 4 is subject both to a dashpot resistance of $3v$ lb and an external force of $F(t) = 20 \sin 8t$ lb. (a) Find $x(t)$, assuming that $x(0) = 1$ and $x'(0) = 0$. (b) Find the amplitude and period of the resulting steady periodic oscillations (when t is so large that the transient terms are negligible).

9 An inductor ($L = 1$ henry), a resistor ($R = 12$ ohms), and a capacitor ($C = 0.01$ farad) are connected in series with a generator providing a voltage of $E(t) = 30 \sin 10t$ volts. (a) Find the charge $Q(t)$ on the capacitor, given that $Q(0) = I(0) = 0$. (b) What is the steady periodic value of

$Q(t)$ when t is large? (*Suggestion:* Equation (10) in Section 17-2 can be written in the form $LQ'' + RQ' + (1/C)Q = E$.)

10 Verify that, if the coefficients B_1 and B_2 in (18) are chosen by the method of undetermined coefficients to give

a particular solution of (17), then $|I| = (B_1^2 + B_2^2)^{1/2}$ is given by (19).

11 Determine A so that $x_p(t) = At \cos \omega t$ is a particular solution of Equation (13) in the resonance case, $\omega_0 = \omega$. Note that the amplitude is unbounded as $t \to \infty$.

---

*17-7

Periodic Temperature Oscillations

The underground cellar is an ancient solution to the problem of keeping wine or buttermilk (according to taste) at a cool and steady temperature year-round. Nowadays we hear of buildings and homes built underground for energy-saving purposes. In this final section we investigate the mathematics of these strategies, and we will find that most of the content of this book is involved, either explicitly or implicitly. Our discussion will explain the common experience of entering a cavern on a hot summer day and noting that the temperature there is appreciably cooler than at the surface, as well as the subtler fact that in some cellars it is coolest in the summer and warmest in the winter.

Our physical experience indicates that the underground temperature at a given location is a function both of time t and of the depth x beneath the surface. Let $u = u(x, t)$ denote this function. According to Equation (15) in Section 16-5, any temperature function satisfies the heat equation

$$\frac{\partial u}{\partial t} = k\left(\frac{\partial^2 u}{\partial x^2} + \frac{\partial^2 u}{\partial y^2} + \frac{\partial^2 u}{\partial z^2}\right).$$

In our cellar-temperature problem, the variables y and z are not involved, since the function u depends only upon time t and the depth x (measured with the positive direction downward). So the heat equation takes here the simpler form

$$\frac{\partial u}{\partial t} = k\frac{\partial^2 u}{\partial x^2}. \tag{1}$$

We will make the simple but reasonably accurate assumption that k, the thermal diffusivity of soil, is constant.

We should like to solve the partial differential equation in (1) above, but any useful solution must be a *particular* solution rather than its general solution. This means that we need to find the appropriate boundary conditions, and then write them in mathematical form.

To begin, we may regard the temperature $u(0, t)$ at time t at the surface $x = 0$ of the ground as known. In fact, the periodic seasonal variation of monthly average surface temperatures, with a maximum in midsummer (July) and a minimum in midwinter (January), is very close to a sine or cosine oscillation. We shall therefore assume that

$$u(0, t) = T_0 + A_0 \cos \omega t, \tag{2}$$

where we take $t = 0$ at midsummer, with T_0 the annual average temperature, A_0 the amplitude of seasonal temperature variation, and ω chosen to make the period of the above function 1 year. (In cgs units, for instance, ω would be 2π divided by the number of seconds in a year.)

n	0	1	2	3	4	5	6	7	8	9	10	11
Month	July	Aug.	Sept.	Oct.	Nov.	Dec.	Jan.	Feb.	Mar.	Apr.	May	June
Actual	79.1	78.4	72.6	62.6	51.6	44.1	43.4	45.6	51.8	61.8	70.0	76.8
Predicted	79.2	76.8	70.3	61.3	52.4	45.8	43.4	45.8	52.4	61.3	70.3	76.8

17.15

To illustrate the accuracy of our assumption about the cosine-like behavior of surface temperature, we have shown in Figure 17.15 the actual monthly average temperature data for Athens, Georgia, compared with the values predicted by Formula (2). These data are plotted as the points in Figure 17.16, and they are close to the values of $u(0, t)$, whose graph appears in the same figure. (For Athens, Georgia, we find that $T_0 = 61.3$ and $A_0 = 17.9$.)

Let us next consider the *qualitative* character of underground temperature variations. It is reasonable to assume that the temperature at depth x varies in the same general fashion as on the surface—that it oscillates cyclically with a period of 1 year, but with an annual average temperature $T(x)$ and amplitude $A(x)$ that may depend upon x. If geothermal heat is ignored, then the average annual temperature at depth x will be the same as at the surface: $T(x) \equiv T_0$. But there is no reason to assume that the seasons underground coincide with those on the surface, so we introduce a "phase shift" $\rho(x)$ to indicate the amount that the seasons at depth x are delayed.

When we put all this together, we are in effect assuming that the general form of our temperature function $u(x, t)$ is

$$u(x, t) = T_0 + A(x) \cos(\omega t - \rho(x)). \qquad (3)$$

We have yet to determine the **amplitude function** $A(x)$ and the **phase shift function** $\rho(x)$. And, in order that Equations (2) and (3) are consistent when $x = 0$, we must also impose the initial conditions

$$A(0) = A_0 \qquad (4)$$

17.16 Predicted and actual average monthly temperatures in Athens, Georgia.

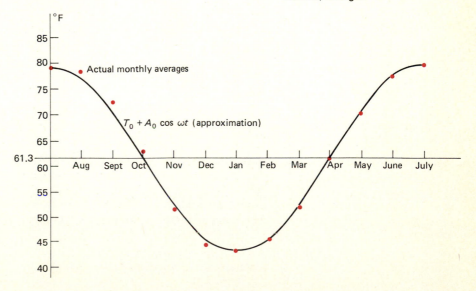

and

$$\rho(0) = 0. \tag{5}$$

Finally, we assume that the function $A(x)$ is *bounded* as $x \to \infty$; otherwise, our mathematical model would seem physically unrealistic.

Now we are ready to solve the mathematical model consisting of Equations (1) through (5). The arithmetic will be a bit simpler if we temporarily shift the temperature scale to make $T_0 = 0$; at the conclusion we shall add T_0 to the solution to restore our original scale.

So, with this change, Equation (3) becomes

$$u(x, t) = A(x) \cos(\omega t - \rho(x)). \tag{3'}$$

By the cosine addition formula, we can rewrite this as

$$u(x, t) = V(x) \cos \omega t + W(x) \sin \omega t \tag{6}$$

where

$$V(x) = A(x) \cos \rho(x) \quad \text{and} \quad W(x) = A(x) \sin \rho(x). \tag{7}$$

Next, we substitute (6) into the heat equation (1) and then equate the coefficients of $\cos \omega t$ and of $\sin \omega t$ on the two sides of the resulting equation. We find that

$$V''(x) = \frac{\omega}{k} W(x) \quad \text{and} \quad W''(x) = -\frac{\omega}{k} V(x). \tag{8}$$

We translate the initial conditions of (4) and (5) into the initial conditions

$$V(0) = A_0 \quad \text{and} \quad W(0) = 0. \tag{9}$$

Note what has happened: We now have the pair of *ordinary* differential equations in (8) to solve, rather than the *partial* differential equation in (1).

We differentiate the first equation in (8) twice, and this gives

$$V^{(4)}(x) = \frac{\omega}{k} W''(x) = \frac{\omega}{k} \left(-\frac{\omega}{k} V(x) \right).$$

Thus $V(x)$ satisfies the fourth order differential equation

$$V^{(4)}(x) + \gamma^4 V(x) = 0 \tag{10}$$

where $\gamma = \sqrt{\omega/k}$. By Example 5 in Section 17-5, the general solution of (10) is

$$V = e^{-\alpha x}(c_1 \cos \alpha x + c_2 \sin \alpha x) + e^{\alpha x}(c_3 \cos \alpha x + c_4 \sin \alpha x)$$

where

$$\alpha = \frac{\gamma}{\sqrt{2}} = \sqrt{\frac{\omega}{2k}}. \tag{11}$$

Now we have assumed that $A(x)$, and hence $V(x)$, is bounded as $x \to \infty$. This implies that $c_3 = 0 = c_4$. So, finally,

$$V(x) = ae^{-\alpha x} \cos \alpha x + be^{-\alpha x} \sin \alpha x \tag{12}$$

(with a and b in place of c_1 and c_2 to avoid subscripts).

We must yet find $W(x)$. But routine differentiation of (12) gives

$$V''(x) = -2b\alpha^2 e^{-\alpha x} \cos \alpha x + 2a\alpha^2 e^{-\alpha x} \sin \alpha x$$

$$= -\frac{b\omega}{k} e^{-\alpha x} \cos \alpha x + \frac{a\omega}{k} e^{-\alpha x} \sin \alpha x$$

because $\alpha^2 = \omega/2k$ by (11). The first equation in (8) now gives

$$W(x) = -be^{-\alpha x} \cos \alpha x + ae^{-\alpha x} \sin \alpha x. \tag{13}$$

But the fact that $W(0) = 0$ implies that $b = 0$, and then $V(0) = A_0$ implies that $a = A_0$. So Equations (12) and (13) can be simplified to the forms

$$V(x) = A_0 e^{-\alpha x} \cos \alpha x \quad \text{and} \quad W(x) = A_0 e^{-\alpha x} \sin \alpha x. \tag{14}$$

Finally, we substitute Equations (14) into Equation (6) and add the constant T_0 to recover our original temperature scale. Thus we have the solution

$$u(x, t) = T_0 + A_0 e^{-\alpha x} \cos \alpha x \cos \omega t + A_0 e^{-\alpha x} \sin \alpha x \sin \omega t,$$

or

$$u(x, t) = T_0 + A_0 e^{-\alpha x} \cos(\omega t - \alpha x). \tag{15}$$

This, then, with $\alpha = \sqrt{\omega/2k}$, is the temperature at depth x at time t. Thus the amplitude of the annual temperature variation is subject to an exponential damping factor $e^{-\alpha x}$ at depth x:

$$A(x) = A_0 e^{-\alpha x}. \tag{16}$$

The phase shift or seasonal delay factor at depth x is

$$\rho(x) = \alpha x. \tag{17}$$

We leave for you the pleasure of deriving in the following problems some interesting numerical consequences from the above underground temperature solution. As a typical value of the thermal diffusivity of soil, you may use $k = 0.002$ (cgs units). The frequency constant ω is

$$\omega = \frac{2\pi}{365 \times 24 \times 3600} \approx 1.992 \times 10^{-7} \quad (\text{sec}^{-1}),$$

and so

$$\alpha = \sqrt{\frac{\omega}{2k}} \approx \left(\frac{1.992 \times 10^{-7}}{(2)(0.002)}\right)^{1/2} \approx 0.00706 \quad (\text{cm}^{-1}).$$

17-7 PROBLEMS

1 At what depth (in centimeters) is the amplitude of annual temperature variation half the surface amplitude?

2 Suppose that the surface amplitude is $A_0 = 16°C$. Determine x so that $A(x) = 1$ (in degrees Celsius). In a cellar at this depth x, the year-round temperature would be

essentially constant. Is this a practical depth for a home-owner to seek?

3 (a) Determine x so that $\rho(x) = \pi$. Why are the *seasons reversed* at this depth? (b) Suppose that the surface temperature extremes are 27°C in July (time $t = 0$) and 5°C in January. Find the July and January temperatures at the depth that you found in Part (a).

4 Given surface temperature extremes as in Problem 3(b), how deep should a cellar be in order that the temperature never exceed 20°C at any time during the year?

5 In this problem you are to calculate the depth to which winter freezing temperatures penetrate the soil, in a certain locale where the annual surface temperature range is from $-8°C$ to $22°C$ and the thermal diffusivity of the soil is $k = 0.004$ (cgs units) instead of the value we mentioned above. Assume that the minimum surface temperature of $-8°C$ occurs annually on January 1. What is the maximum depth at which the temperature is $0°C$ sometime during the

year? On what date does this occur? (*Answer:* Valentine's Day.)

Daily Fluctuations. So far, we have assumed that $u(x, t)$ is the daily *average* temperature and have ignored daily fluctuations. For the daily cycle of temperature variations we have $\omega = 2\pi/(24 \times 3600) \approx 7.27 \times 10^{-5}$ sec^{-1}, so that $\alpha = \sqrt{\omega/2k} \approx 0.135$ cm^{-1} if $k = 0.002$ (cgs units). Use these values in Problems 6 and 7. They show that only a *thin surface layer* is affected by *daily* temperature fluctuations.

6 Show that for daily temperature variations the "half-depth" of Problem 1 is about 5 centimeters.

7 Show that the "reversal of day and night," analogous to the reversal of seasons in Problem 3, occurs at a depth of about 23 centimeters.

CHAPTER 17 REVIEW: Concepts and Methods

Use the list below as a guide to concepts that you may need to review.

1 Solution of linear first order equations
2 General and particular solutions of linear differential equations: homogeneous and inhomogeneous
3 The method of undetermined coefficients

4 Complex-valued functions and Euler's formula
5 Solution of homogeneous linear equations with constant coefficients; the characteristic equation and the solutions corresponding to distinct roots, repeated roots, and complex roots

MISCELLANEOUS PROBLEMS

Find general solutions of the differential equations in Problems 1–18.

1 $xy' = 2x^2y^2 + 3y^2.$
2 $yy' = xe^{x^2 - y^2}.$
3 $xy' + y = x^5.$
4 $y' + y \tan x = \cos x.$
5 $(x^2 + 1)y' + y = 1.$
6 $(x + 1)y' = x + y + 2.$
7 $2y'' + 3y' + y = 0.$
8 $4y'' + 5y' - 6y = 0.$
9 $y''' - 4y' = 0.$
10 $4y''' + 4y'' + y' = 0.$
11 $y''' + 3y'' + 3y' + y = 0.$
12 $y''' + 9y' = 0.$
13 $y'' + 5y' + 4y = 0.$
14 $y'' + 13y' + 12y = 0.$
15 $y^{(4)} = 16y.$
16 $y^{(4)} + 16y = 0.$
17 $y^{(4)} + 2y'' + y = 0.$
18 $y^{(4)} - 3y'' - 4y = 0.$

Find a particular solution of each of the differential equations in Problem 19–24.

19 $y'' + 2y = x^2 - 4.$
20 $y'' + y' - 3y = e^{2x}.$

21 $y'' + 2y = \sin x.$
22 $y'' + 2y' + 2y = \sin x.$
23 $y'' + 4y = \cos 4x.$
24 $y'' - y' - 2y = e^{2x}.$

Use the substitution $x = e^z$ to solve the Euler equations in Problems 25–27.

25 $2x^2y'' + 5xy' + y = 0.$
26 $4x^2y'' + 8xy' + y = 0.$
27 $x^2y'' + 6xy' + 4y = 0.$

The differential equation $y' + P(x)y = Q(x)y^n$ is called a *Bernoulli equation* if $n \neq 0$ or 1. Use the substitution $v = y^{1-n}$ to solve the Bernoulli equations in Problems 28–30.

28 $xy' + y = x^3y^3.$
29 $y' + y = y^2e^x.$
30 $y' = ay(b - y)$ (with a and b positive constants; this is the logistic equation—compare your solution with Equation (6) in Section 6-8).

*References
for
Further Study*

References 2, 3, 6, and 11 below may be consulted for historical topics pertinent to calculus. Reference 14 provides a more theoretical treatment of single-variable calculus than ours. References 4, 5, 7, 8, 9, and 15 include advanced topics in multivariable calculus. Reference 12 is a standard work on infinite series. References 1, 10, and 13 are differential equations textbooks.

1   BOYCE, W. E., and R. C. DiPRIMA, *Elementary Differential Equations* (3rd ed.) New York: John Wiley, 1976.

2   BOYER, C. B., *A History of Mathematics*. New York: John Wiley, 1968.

3   BOYER, C. B., *The History of the Calculus*. New York: Dover Publications, 1959.

4   BUCK, R. CREIGHTON, *Advanced Calculus* (3rd ed.) New York: McGraw-Hill, 1978.

5   COURANT, RICHARD, and FRITZ JOHN, *Introduction to Calculus and Analysis*. Vol. I, New York: Interscience, 1965. Vol. II, New York: Wiley-Interscience, 1974, with the assistance of Albert A. Blank and Alan Solomon.

6   EDWARDS, C. H., JR., *The Historical Development of the Calculus*. New York: Springer-Verlag, 1979.

7   EDWARDS, C. H., JR., *Advanced Calculus of Several Variables*. New York: Academic Press, 1973.

8   HILDEBRAND, FRANCIS B., *Advanced Calculus for Applications* (2nd ed.) Englewood Cliffs, N.J.: Prentice-Hall, 1976.

9   HURLEY, JAMES F., *Intermediate Calculus*. Philadelphia: Saunders, 1980.

10 KAPLAN, WILFRED, *Ordinary Differential Equations*. Reading, Mass.: Addison-Wesley, 1958.

11 KLINE, MORRIS, *Mathematical Thought from Ancient to Modern Times*. New York: Oxford University Press, 1972.

12 KNOPP, KONRAD, *Theory and Application of Infinite Series*. New York: Hafner, 1971.

13 SIMMONS, GEORGE F., *Differential Equations with Applications and Historical Notes*. New York: McGraw-Hill, 1972.

14 SPIVAK, MICHAEL, *Calculus*. Menlo Park, Calif.: Benjamin, 1967.

15 TAYLOR, ANGUS E., *Advanced Calculus*. Boston: Ginn and Company, 1955.

Appendices

Units of measurement and conversion factors

MKS SCIENTIFIC UNITS

Length in meters (m); *mass* in kilograms (kg); *time* in seconds (sec).

Force in newtons (N); a force of 1 newton provides an acceleration of 1 m/sec^2 to a mass of 1 kg.

Work in joules (J); 1 joule is the work done by a force of 1 newton acting through a distance of 1 meter.

Power in watts (W); a watt is one joule per second.

BRITISH ENGINEERING UNITS

Length in feet (ft); *force* in pounds (lb); *time* in seconds (sec).

Mass in slugs; 1 lb of force provides an acceleration of 1 ft/sec^2 to a mass of 1 slug. A mass of m slugs at the surface of the earth has a *weight* of $w = mg$ pounds, where $g \approx 32.17$ ft/sec^2.

Work in ft-lb; *power* in ft-lb/sec.

CONVERSION FACTORS

1 in. = 2.54 cm = 0.0254 m, 1 m $\approx$ 3.2808 ft

1 mile = 5280 ft; 60 mph = 88 ft/sec

1 lb $\approx$ 4.4482 newtons; 1 slug $\approx$ 14.594 kg

1 horsepower = 550 ft-lb/sec $\approx$ 745.7 watts

Gravitational acceleration: $g \approx 32.17$ ft/sec^2 ≈ 9.807 m/sec^2

Atmospheric pressure: 1 atm is the pressure exerted by a column of 76 cm of mercury; 1 atm $\approx$ 14.70 lb/in^2 $\approx 1.013 \times 10^5$ N/m^2.

Heat energy: 1 BTU $\approx$ 778 ft-lb $\approx$ 252 cal, 1 cal $\approx$ 4.184 joules

1 *Laws of Exponents*

$a^m a^n = a^{m+n}$, $(a^m)^n = a^{mn}$, $(ab)^n = a^n b^n$, $a^{m/n} = \sqrt[n]{a^m}$; in particular, $a^{1/2} = \sqrt{a}$.

If $a \neq 0$, then $a^{m-n} = \dfrac{a^m}{a^n}$, $a^{-n} = \dfrac{1}{a^n}$, and $a^0 = 1$.

Formulas from algebra and geometry

2 *Quadratic Formula*

The quadratic equation

$$ax^2 + bx + c = 0 \qquad (a \neq 0)$$

has solutions

$$x = \frac{-b \pm \sqrt{b^2 - 4ac}}{2a}.$$

3 *Factoring*

$a^2 - b^2 = (a - b)(a + b)$.

$a^3 - b^3 = (a - b)(a^2 + ab + b^2)$.

$a^4 - b^4 = (a - b)(a^3 + a^2 b + ab^2 + b^3) = (a - b)(a + b)(a^2 + b^2)$.

$a^5 - b^5 = (a - b)(a^4 + a^3 b + a^2 b^2 + ab^3 + b^4)$.

(The pattern continues.)

$a^3 + b^3 = (a + b)(a^2 - ab + b^2)$.

$a^5 + b^5 = (a + b)(a^4 - a^3 b + a^2 b^2 - ab^3 + b^4)$.

(The pattern continues for odd exponents.)

4 *Binomial Formula*

$$(a + b)^n = a^n + na^{n-1}b + \frac{n(n - 1)}{1 \cdot 2} a^{n-2}b^2$$

$$+ \frac{n(n - 1)(n - 2)}{1 \cdot 2 \cdot 3} a^{n-3}b^3 + \cdots + nab^{n-1} + b^n$$

if n is a positive integer.

5 *Areas and Volumes* (See the figure and formulas below: A denotes area; b, base; B, area of base; C, circumference; h, height; ℓ, length; r, radius; V, volume; w, width.)

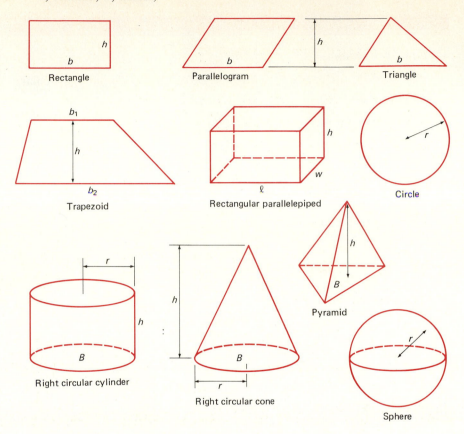

Rectangle: $A = bh$.

Parallelogram: $A = bh$.

Triangle: $A = \frac{1}{2}bh$.

Trapezoid: $A = \frac{1}{2}(b_1 + b_2)h$.

Circle: $C = 2\pi r$ and $A = \pi r^2$.

Rectangular parallelepiped: $V = \ell wh$.

Right circular cylinder: $V = \pi r^2 h = Bh$.

Right circular cone: $V = \frac{1}{3}\pi r^2 h = \frac{1}{3}Bh$.

Pyramid: $V = \frac{1}{3}Bh$.

Sphere: $V = \frac{4}{3}\pi r^3$, $A = 4\pi r^2$.

6 *Pythagorean Theorem*

In a right triangle with legs a and b and hypotenuse c, $a^2 + b^2 = c^2$.

$\sin(-\theta) = -\sin\theta, \cos(-\theta) = \cos\theta.$

$\sin^2\theta + \cos^2\theta = 1.$

$\sin 2\theta = 2\sin\theta\cos\theta$ and $\cos 2\theta = \cos^2\theta - \sin^2\theta.$

$\sin(\alpha + \beta) = \sin\alpha\cos\beta + \cos\alpha\sin\beta.$

$\cos(\alpha + \beta) = \cos\alpha\cos\beta - \sin\alpha\sin\beta.$

$\tan(\alpha + \beta) = \dfrac{\tan\alpha + \tan\beta}{1 - \tan\alpha\tan\beta}.$

$\sin^2\dfrac{\theta}{2} = \dfrac{1}{2}(1 - \cos\theta).$

$\cos^2\dfrac{\theta}{2} = \dfrac{1}{2}(1 + \cos\theta).$

For an arbitrary triangle:

Law of Sines $\qquad \dfrac{\sin A}{a} = \dfrac{\sin B}{b} = \dfrac{\sin C}{c}.$

Law of Cosines $\qquad c^2 = a^2 + b^2 - 2ab\cos C.$

Formulas from trigonometry

Greek alphabet

A	α	alpha	I	ι	iota	P	ρ	rho
B	β	beta	K	κ	kappa	Σ	σ	sigma
Γ	γ	gamma	Λ	λ	lambda	T	τ	tau
Δ	δ	delta	M	μ	mu	Υ	υ	upsilon
E	ε	epsilon	N	ν	nu	Φ	ϕ	phi
Z	ζ	zeta	Ξ	ξ	xi	X	χ	chi
H	η	eta	O	o	omicron	Ψ	ψ	psi
Θ	θ	theta	Π	π	pi	Ω	ω	omega

TABLE 1   *Trigonometric Functions in Radians*

t	sin t	cos t	tan t	t	sin t	cos t	tan t
.0	.00000	1.00000	.00000	3.1	.04158	−.99914	−.04162
.1	.09983	.99500	.10033	π	.00000	−1.00000	.00000
.2	.19867	.98007	.20271	3.2	−.05837	−.99829	.05847
.3	.29552	.95534	.30934	3.3	−.15775	−.98748	.15975
.4	.38942	.92106	.42279	3.4	−.25554	−.96680	.26432
.5	.47943	.87758	.54630	3.5	−.35078	−.93646	.37459
.6	.56464	.82534	.68414	3.6	−.44252	−.89676	.49347
.7	.64422	.76484	.84229	3.7	−.52984	−.84810	.62473
.8	.71736	.69671	1.02964	3.8	−.61186	−.79097	.77356
.9	.78333	.62161	1.26016	3.9	−.68777	−.72593	.94742
1.0	.84147	.54030	1.55741	4.0	−.75680	−.65364	1.15782
1.1	.89121	.45360	1.96476	4.1	−.81828	−.57482	1.42353
1.2	.93204	.36236	2.57215	4.2	−.87158	−.49026	1.77778
1.3	.96356	.26750	3.60210	4.3	−.91617	−.40080	2.28585
1.4	.98545	.16997	5.79788	4.4	−.95160	−.30733	3.09632
1.5	.99749	0.7074	14.10142	4.5	−.97753	−.21080	4.63733
π/2	1.00000	.00000	********	4.6	−.99369	−.11215	8.86017
1.6	.99957	−.02920	−34.23253	4.7	−.99992	−.01239	80.71269
1.7	.99166	−.12884	−7.69660	3π/2	−1.00000	.00000	********
1.8	.97385	−.22720	−4.28626	4.8	−.99616	.08750	−11.38487
1.9	.94630	−.32329	−2.92710	4.9	−.98245	.18651	−5.26749
2.0	.90930	−.41615	−2.18504	5.0	−.95892	.28366	−3.38052
2.1	.86321	−.50485	−1.70985	5.1	−.92581	.37798	−2.44939
2.2	.80850	−.58850	−1.37382	5.2	−.88345	.46852	−1.88564
2.3	.74571	−.66628	−1.11921	5.3	−.83227	.55437	−1.50127
2.4	.67546	−.73739	−.91601	5.4	−.77276	.63469	−1.21754
2.5	.59847	−.80114	−.74702	5.5	−.70554	.70867	−.99558
2.6	.51550	−.85689	−.60160	5.6	−.63127	.77557	−.81394
2.7	.42738	−.90407	−.47273	5.7	−.55069	.83471	−.65973
2.8	.33499	−.94222	−.35553	5.8	−.46460	.88552	−.52467
2.9	.23925	−.97096	−.24641	5.9	−.37388	.92748	−.40311
3.0	.14112	−.98999	−.14255	6.0	−.27942	.96017	−.29101
—	—	—	—	6.1	−.18216	.98327	−.18526
—	—	—	—	6.2	−.08309	.99654	−.08338
—	—	—	—	2π	.00000	1.00000	.00000

TABLE 2 *The Natural Exponential Function*

x	e^x	e^{-x}	x	e^x	e^{-x}	x	e^x	e^{-x}
.00	1.00000	1.00000	.40	1.49182	.67032	.80	2.22554	.44933
.01	1.01005	.99005	.41	1.50682	.66365	.81	2.24791	.44486
.02	1.02020	.98020	.42	1.52196	.65705	.82	2.27050	.44043
.03	1.03045	.97045	.43	1.53726	.65051	.83	2.29332	.43605
.04	1.04081	.96079	.44	1.55271	.64404	.84	2.31637	.43171
.05	1.05127	.95123	.45	1.56831	.63763	.85	2.33965	.42741
.06	1.06184	.94176	.46	1.58407	.63128	.86	2.36316	.42316
.07	1.07251	.93239	.47	1.59999	.62500	.87	2.38691	.41895
.08	1.08329	.92312	.48	1.61607	.61878	.88	2.41090	.41478
.09	1.09417	.91393	.49	1.63232	.61263	.89	2.43513	.41066
.10	1.10517	.90484	.50	1.64872	.60653	.90	2.45960	.40657
.11	1.11628	.89583	.51	1.66529	.60050	.91	2.48432	.40252
.12	1.12750	.88692	.52	1.68203	.59452	.92	2.50929	.39852
.13	1.13883	.87810	.53	1.69893	.58860	.93	2.53451	.39455
.14	1.15027	.86936	.54	1.71601	.58275	.94	2.55998	.39063
.15	1.16183	.86071	.55	1.73325	.57695	.95	2.58571	.38674
.16	1.17351	.85214	.56	1.75067	.57121	.96	2.61170	.38289
.17	1.18530	.84366	.57	1.76827	.56553	.97	2.63794	.37908
.18	1.19722	.83527	.58	1.78604	.55990	.98	2.66446	.37531
.19	1.20925	.82696	.59	1.80399	.55433	.99	2.69123	.37158
.20	1.22140	.81873	.60	1.82212	.54881	1.00	2.71828	.36788
.21	1.23368	.81058	.61	1.84043	.54335	1.01	2.74560	.36422
.22	1.24608	.80252	.62	1.85893	.53794	1.02	2.77319	.36059
.23	1.25860	.79453	.63	1.87761	.53259	1.03	2.80107	.35701
.24	1.27125	.78663	.64	1.89648	.52729	1.04	2.82922	.35345
.25	1.28403	.77880	.65	1.91554	.52205	1.05	2.85765	.34994
.26	1.29693	.77105	.66	1.93479	.51685	1.06	2.88637	.34646
.27	1.30996	.76338	.67	1.95424	.51171	1.07	2.91538	.34301
.28	1.32313	.75578	.68	1.97388	.50662	1.08	2.94468	.33960
.29	1.33643	.74826	.69	1.99372	.50158	1.09	2.97427	.33622
.30	1.34986	.74082	.70	2.01375	.49659	1.10	3.00417	.33287
.31	1.36343	.73345	.71	2.03399	.49164	1.11	3.03436	.32956
.32	1.37713	.72615	.72	2.05443	.48675	1.12	3.06485	.32628
.33	1.39097	.71892	.73	2.07508	.48191	1.13	3.09566	.32303
.34	1.40495	.71177	.74	2.09594	.47711	1.14	3.12677	.31982
.35	1.41907	.70469	.75	2.11700	.47237	1.15	3.15819	.31664
.36	1.43333	.69768	.76	2.13828	.46767	1.16	3.18993	.31349
.37	1.44773	.69073	.77	2.15977	.46301	1.17	3.22199	.31037
.38	1.46228	.68386	.78	2.18147	.45841	1.18	3.25437	.30728
.39	1.47698	.67706	.79	2.20340	.45384	1.19	3.28708	.30422

TABLE 2 *The Natural Exponential Function (continued)*

x	e^x	e^{-x}	x	e^x	e^{-x}	x	e^x	e^{-x}
1.20	3.32012	.30119	**1.60**	4.95303	.20190	**4.0**	54.598	.01832
1.21	3.35348	.29820	1.61	5.00281	.19989	4.1	60.340	.01657
1.22	3.38719	.29523	1.62	5.05309	.19790	4.2	66.686	.01500
1.23	3.42123	.29229	1.63	5.10387	.19593	4.3	73.700	.01357
1.24	3.45561	.28938	1.64	5.15517	.19398	4.4	81.451	.01228
1.25	3.49034	.28650	1.65	5.20698	.19205	4.5	90.017	.01111
1.26	3.52542	.28365	1.66	5.25931	.19014	4.6	99.484	.01005
1.27	3.56085	.28083	1.67	5.31217	.18825	4.7	109.947	.00910
1.28	3.59664	.27804	1.68	5.36556	.18637	4.8	121.510	.00823
1.29	3.63279	.27527	1.69	5.41948	.18452	4.9	134.290	.00745
1.30	3.66930	.27253	**1.70**	5.47395	.18268	**5.0**	148.41	.00674
1.31	3.70617	.26982	1.71	5.52896	.18087	5.1	164.02	.00610
1.32	3.74342	.26714	1.72	5.58453	.17907	5.2	181.27	.00552
1.33	3.78104	.26448	1.73	5.64065	.17728	5.3	200.34	.00499
1.34	3.81904	.26185	1.74	5.69734	.17552	5.4	221.41	.00452
1.35	3.85743	.25924	1.75	5.75460	.17377	5.5	244.69	.00409
1.36	3.89619	.25666	1.80	6.04965	.16530	5.6	270.43	.00370
1.37	3.93535	.25411	1.85	6.35982	.15724	5.7	298.87	.00335
1.38	3.97490	.25158	1.90	6.68589	.14957	5.8	330.30	.00303
1.39	4.01485	.24908	1.95	7.02869	.14227	5.9	365.04	.00274
1.40	4.05520	.24660	**2.0**	7.3891	.13534	**6.0**	403.43	.00248
1.41	4.09596	.24414	2.1	8.1662	.12246	6.1	445.86	.00224
1.42	4.13712	.24171	2.2	9.0250	.11080	6.2	492.75	.00203
1.43	4.17870	.23931	2.3	9.9742	.10026	6.3	544.57	.00184
1.44	4.22070	.23693	2.4	11.0232	.09072	6.4	601.85	.00166
1.45	4.26311	.23457	2.5	12.1825	.08208	6.5	665.14	.00150
1.46	4.30596	.23224	2.6	13.4637	.07427	6.6	735.10	.00136
1.47	4.34924	.22993	2.7	14.8797	.06721	6.7	812.41	.00123
1.48	4.39295	.22764	2.8	16.4446	.06081	6.8	897.85	.00111
1.49	4.43710	.22537	2.9	18.1741	.05502	6.9	992.27	.00101
1.50	4.48169	.22313	**3.0**	20.086	.04979	**7.0**	1096.6	.00091
1.51	4.52673	.22091	3.1	22.198	.04505	7.5	1808.0	.00055
1.52	4.57223	.21871	3.2	24.533	.04076	8.0	2981.0	.00034
1.53	4.61818	.21654	3.3	27.113	.03688	8.5	4914.8	.00020
1.54	4.66459	.21438	3.4	29.964	.03337	9.0	8103.1	.00012
1.55	4.71147	.21225	3.5	33.115	.03020	9.5	13360	.00007
1.56	4.75882	.21014	3.6	36.598	.02732	10.0	22026	.00005
1.57	4.80665	.20805	3.7	40.447	.02472	10.5	36316	.00003
1.58	4.85496	.20598	3.8	44.701	.02237	11.0	59874	.00002
1.59	4.90375	.20393	3.9	49.402	.02024	11.5	98716	.00001

TABLE 3 *Natural Logarithms*

x	ln x	x	ln x	x	ln x	x	ln x	x	ln x
		0.50	−0.69315	1.00	0.00000	1.50	0.40547	2.00	0.69315
0.01	−4.60517	0.51	−0.67334	1.01	0.00995	1.51	0.41211	2.10	0.74194
0.02	−3.91202	0.52	−0.65393	1.02	0.01980	1.52	0.41871	2.20	0.78846
0.03	−3.50656	0.53	−0.63488	1.03	0.02956	1.53	0.42527	2.30	0.83291
0.04	−3.21888	0.54	−0.61619	1.04	0.03922	1.54	0.43178	2.40	0.87547
0.05	−2.99573	0.55	−0.59784	1.05	0.04879	1.55	0.43825	2.50	0.91629
0.06	−2.81341	0.56	−0.57982	1.06	0.05827	1.56	0.44469	2.60	0.95551
0.07	−2.65926	0.57	−0.56212	1.07	0.06766	1.57	0.45108	2.70	0.99325
0.08	−2.52573	0.58	−0.54473	1.08	0.07696	1.58	0.45742	2.80	1.02962
0.09	−2.40795	0.59	−0.52763	1.09	0.08618	1.59	0.46373	2.90	1.06471
0.10	−2.30259	0.60	−0.51083	1.10	0.09531	1.60	0.47000	3.00	1.09861
0.11	−2.20727	0.61	−0.49430	1.11	0.10436	1.61	0.47623	3.10	1.13140
0.12	−2.12026	0.62	−0.47804	1.12	0.11333	1.62	0.48243	3.20	1.16315
0.13	−2.04022	0.63	−0.46204	1.13	0.12222	1.63	0.48858	3.30	1.19392
0.14	−1.96611	0.64	−0.44629	1.14	0.13103	1.64	0.49470	3.40	1.22378
0.15	−1.89712	0.65	−0.43078	1.15	0.13976	1.65	0.50078	3.50	1.25276
0.16	−1.83258	0.66	−0.41552	1.16	0.14842	1.66	0.50682	3.60	1.28093
0.17	−1.77196	0.67	−0.40048	1.17	0.15700	1.67	0.51282	3.70	1.30833
0.18	−1.71480	0.68	−0.38566	1.18	0.16551	1.68	0.51879	3.80	1.33500
0.19	−1.66073	0.69	−0.37106	1.19	0.17395	1.69	0.52473	3.90	1.36098
0.20	−1.60944	0.70	−0.35667	1.20	0.18232	1.70	0.53063	4.00	1.38629
0.21	−1.56065	0.71	−0.34249	1.21	0.19062	1.71	0.53649	4.10	1.41099
0.22	−1.51413	0.72	−0.32850	1.22	0.19885	1.72	0.54232	4.20	1.43508
0.23	−1.46968	0.73	−0.31471	1.23	0.20701	1.73	0.54812	4.30	1.45862
0.24	−1.42712	0.74	−0.30111	1.24	0.21511	1.74	0.55389	4.40	1.48160
0.25	−1.38629	0.75	−0.28768	1.25	0.22314	1.75	0.55962	4.50	1.50408
0.26	−1.34707	0.76	−0.27444	1.26	0.23111	1.76	0.56531	4.60	1.52606
0.27	−1.30933	0.77	−0.26136	1.27	0.23902	1.77	0.57098	4.70	1.54756
0.28	−1.27297	0.78	−0.24846	1.28	0.24686	1.78	0.57661	4.80	1.56862
0.29	−1.23787	0.79	−0.23572	1.29	0.25464	1.79	0.58222	4.90	1.58924
0.30	−1.20397	0.80	−0.22314	1.30	0.26236	1.80	0.58779	5.00	1.60944
0.31	−1.17118	0.81	−0.21072	1.31	0.27003	1.81	0.59333	5.10	1.62924
0.32	−1.13943	0.82	−0.19845	1.32	0.27763	1.82	0.59884	5.20	1.64866
0.33	−1.10866	0.83	−0.18633	1.33	0.28518	1.83	0.60432	5.30	1.66771
0.34	−1.07881	0.84	−0.17435	1.34	0.29267	1.84	0.60977	5.40	1.68640
0.35	−1.04982	0.85	−0.16252	1.35	0.30010	1.85	0.61519	5.50	1.70475
0.36	−1.02165	0.86	−0.15082	1.36	0.30748	1.86	0.62058	5.60	1.72277
0.37	−0.99425	0.87	−0.13926	1.37	0.31481	1.87	0.62594	5.70	1.74047
0.38	−0.96758	0.88	−0.12783	1.38	0.32208	1.88	0.63127	5.80	1.75786
0.39	−0.94161	0.89	−0.11653	1.39	0.32930	1.89	0.63658	5.90	1.77495
0.40	−0.91629	0.90	−0.10536	1.40	0.33647	1.90	0.64185	6.00	1.79176
0.41	−0.89160	0.91	−0.09431	1.41	0.34359	1.91	0.64710	6.10	1.80829
0.42	−0.86750	0.92	−0.08338	1.42	0.35066	1.92	0.65233	6.20	1.82455
0.43	−0.84397	0.93	−0.07257	1.43	0.35767	1.93	0.65752	6.30	1.84055
0.44	−0.82098	0.94	−0.06188	1.44	0.36464	1.94	0.66269	6.40	1.85630
0.45	−0.79851	0.95	−0.05129	1.45	0.37156	1.95	0.66783	6.50	1.87180
0.46	−0.77653	0.96	−0.04082	1.46	0.37844	1.96	0.67294	6.60	1.88707
0.47	−0.75502	0.97	−0.03046	1.47	0.38526	1.97	0.67803	6.70	1.90211
0.48	−0.73397	0.98	−0.02020	1.48	0.39204	1.98	0.68310	6.80	1.91692
0.49	−0.71335	0.99	−0.01005	1.49	0.39878	1.99	0.68813	6.90	1.93152

TABLE 3 *Natural Logarithms* (*continued*)

x	$\ln x$	x	$\ln x$	x	$\ln x$	x	$\ln x$	x	$\ln x$
7.00	1.94591	30.0	3.40120	80.0	4.38203	400.	5.99146	900.	6.80239
7.10	1.96009	31.0	3.43399	81.0	4.39445	410.	6.01616	910.	6.81344
7.20	1.97408	32.0	3.46574	82.0	4.40672	420.	6.04025	920.	6.82437
7.30	1.98787	33.0	3.49651	83.0	4.41884	430.	6.06379	930.	6.83518
7.40	2.00148	34.0	3.52636	84.0	4.43082	440.	6.08677	940.	6.84588
7.50	2.01490	35.0	3.55535	85.0	4.44265	450.	6.10925	950.	6.85646
7.60	2.02815	36.0	3.58352	86.0	4.45435	460.	6.13123	960.	6.86693
7.70	2.04122	37.0	3.61092	87.0	4.46591	470.	6.15273	970.	6.87730
7.80	2.05412	38.0	3.63759	88.0	4.47734	480.	6.17379	980.	6.88755
7.90	2.06686	39.0	3.66356	89.0	4.48864	490.	6.19441	990.	6.89770
8.00	2.07944	40.0	3.68888	90.0	4.49981	500.	6.21461	1000.	6.90776
8.10	2.09186	41.0	3.71357	91.0	4.51086	510.	6.23441	—	—
8.20	2.10413	42.0	3.73767	92.0	4.52179	520.	6.25383	—	—
8.30	2.11626	43.0	3.76120	93.0	4.53260	530.	6.27288	—	—
8.40	2.12823	44.0	3.78419	94.0	4.54329	540.	6.29157	—	—
8.50	2.14007	45.0	3.80666	95.0	4.55388	550.	6.30992	—	—
8.60	2.15176	46.0	3.82864	96.0	4.56435	560.	6.32794	—	—
8.70	2.16332	47.0	3.85015	97.0	4.57471	570.	6.34564	—	—
8.80	2.17475	48.0	3.87120	98.0	4.58497	580.	6.36303	—	—
8.90	2.18605	49.0	3.89182	99.0	4.59512	590.	6.38012	—	—
9.00	2.19722	50.0	3.91202	100.	4.60517	600.	6.39693	—	—
9.10	2.20827	51.0	3.93183	110.	4.70048	610.	6.41346	—	—
9.20	2.21920	52.0	3.95124	120.	4.78749	620.	6.42972	—	—
9.30	2.23001	53.0	3.97029	130.	4.86753	630.	6.44572	—	—
9.40	2.24071	54.0	3.98898	140.	4.94164	640.	6.46147	—	—
9.50	2.25129	55.0	4.00733	150.	5.01064	650.	6.47697	—	—
9.60	2.26176	56.0	4.02535	160.	5.07517	660.	6.49224	—	—
9.70	2.27213	57.0	4.04305	170.	5.13580	670.	6.50728	—	—
9.80	2.28238	58.0	4.06044	180.	5.19296	680.	6.52209	—	—
9.90	2.29253	59.0	4.07754	190.	5.24702	690.	6.53669	—	—
10.0	2.30259	60.0	4.09434	200.	5.29832	700.	6.55108	—	—
11.0	2.39790	61.0	4.11087	210.	5.34711	710.	6.56526	—	—
12.0	2.48491	62.0	4.12713	220.	5.39363	720.	6.57925	—	—
13.0	2.56495	63.0	4.14313	230.	5.43808	730.	6.59304	—	—
14.0	2.63906	64.0	4.15888	240.	5.48064	740.	6.60665	—	—
15.0	2.70805	65.0	4.17439	250.	5.52146	750.	6.62007	—	—
16.0	2.77259	66.0	4.18965	260.	5.56068	760.	6.63332	—	—
17.0	2.83321	67.0	4.20469	270.	5.59842	770.	6.64639	—	—
18.0	2.89037	68.0	4.21951	280.	5.63479	780.	6.65929	—	—
19.0	2.94444	69.0	4.23411	290.	5.66988	790.	6.67203	—	—
20.0	2.99573	70.0	4.24850	300.	5.70378	800.	6.68461	—	—
21.0	3.04452	71.0	4.26268	310.	5.73657	810.	6.69703	—	—
22.0	3.09104	72.0	4.27667	320.	5.76832	820.	6.70930	—	—
23.0	3.13549	73.0	4.29046	330.	5.79909	830.	6.72143	—	—
24.0	3.17805	74.0	4.30407	340.	5.82895	840.	6.73340	—	—
25.0	3.21888	75.0	4.31749	350.	5.85793	850.	6.74524	—	—
26.0	3.25810	76.0	4.33073	360.	5.88610	860.	6.75693	—	—
27.0	3.29584	77.0	4.34381	370.	5.91350	870.	6.76849	—	—
28.0	3.33220	78.0	4.35671	380.	5.94017	880.	6.77992	—	—
29.0	3.36730	79.0	4.36945	390.	5.96615	890.	6.79122	—	—

Answers
to
Odd-Numbered
Problems

CHAPTER 1

1 $(-\infty, 2)$ **3** $(-\frac{7}{3}, \frac{2}{3}]$ **5** $[-\frac{7}{5}, \frac{1}{5}]$ **7** $(-\frac{2}{7}, 0) \cup (\frac{2}{7}, \frac{4}{7})$ **9** $(-\frac{1}{2}, -\frac{1}{3})$

	Domain	Range
11	$\mathscr{R}$	$(-\infty, 10]$
13	$\mathscr{R}$	$[0, +\infty)$
15	$[\frac{5}{3}, +\infty)$	$[0, +\infty)$
17	$(-\infty, \frac{1}{2}]$	$[0, +\infty)$
19	$(-\infty, 3) \cup (3, +\infty)$	$(-\infty, 0) \cup (0, +\infty)$
21	$\mathscr{R}$	$[3, \infty)$
23	$[0, 16]$	$[0, 2]$
25	$(-\infty, 0) \cup (0, +\infty)$	$\{-1, 1\}$

27 $C(A) = 2\sqrt{\pi A},\ A \geqq 0$ **29** $C(F) = \frac{5}{9}(F - 32),\ F \geqq F_{\min}$ **31** $A(x) = x\sqrt{16 - x^2},\ 0 \leqq x \leqq 4$

SECTION 1-3 (pages 13–14)

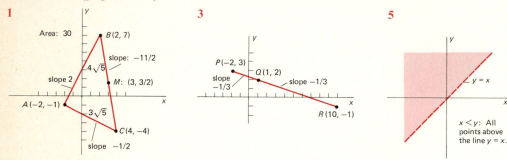

15 $y = -5$ **17** $y = 2x - 7$ **19** $y = -x + 6$ **21** $y = -2x + 7$ **23** $y = -\frac{1}{2}x + \frac{13}{2}$
25 $\frac{4}{13}\sqrt{26}$ **27** $K = \frac{1}{9}(5F + 2297);\ -459.4$ **29** 1136

SECTION 1-4 (pages 18–19)

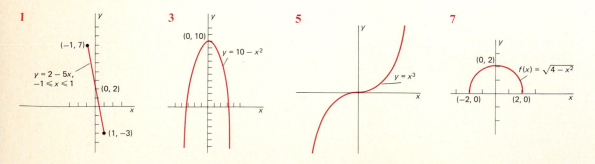

Answers to Odd-Numbered Problems

9

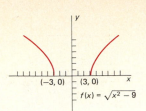

$(-3, 0)$ $(3, 0)$
$f(x) = \sqrt{x^2 - 9}$

11

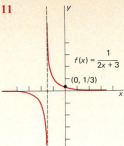

$f(x) = \dfrac{1}{2x + 3}$
$(0, 1/3)$

13

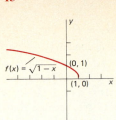

$f(x) = \sqrt{1 - x}$
$(0, 1)$
$(1, 0)$

15

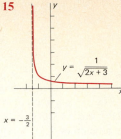

$y = \dfrac{1}{\sqrt{2x + 3}}$
$x = -\dfrac{3}{2}$

17

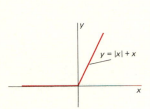

$y = |x| + x$

19

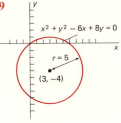

$x^2 + y^2 - 6x + 8y = 0$
$r = 5$
$(3, -4)$

21 There are no points on the graph

23

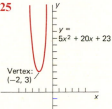

$y = x^2 + 2x + 4$
$(-1, 3)$

25

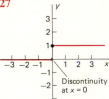

$y = -5x^2 + 20x + 23$
Vertex: $(-2, 3)$

27

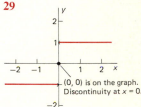

Discontinuity at $x = 0$

29

$(0, 0)$ is on the graph.
Discontinuity at $x = 0$.

31

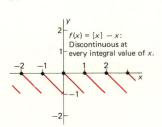

$f(x) = [x] - x$:
Discontinuous at every integral value of x.

33

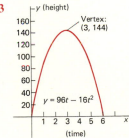

y (height)
Vertex: $(3, 144)$
$y = 96t - 16t^2$
(time)

SECTION 1-5 (page 27)

1 $f'(x) \equiv 3$; $y = 3x - 1$ **3** $f'(x) = 6x^2 - 3$; $y = 21x - 27$ **5** $f'(x) = 2x - 2$; $y = 2x - 3$ **7** $f'(x) = -1/x^2$; $y = -\frac{1}{4}x + 1$
9 $f'(x) = -2/x^3$; $y = -\frac{1}{4}x + \frac{3}{4}$ **11** $f'(x) = -2/(x - 1)^2$; $y = -2x + 6$ **13** Tangent line: $4x + y + 4 = 0$. Normal line:
$4y - x = 18$ **15** Tangent line: $y = 11x - 13$; Normal line: $x + 11y = 101$ **17** $(5, 25)$ **19** $y = 12x - 36$ **21** $y = 2x + 1$;
$y + 2x = 9$ **23** 4500 barrels per day, achieved when $x = 10$ wells

SECTION 1-6 (pages 31–32)

1 $s = 100$ **3** $s = 99$ **5** $s = 120$ **7** $dA/dC = C/2\pi$ **9** It skids for 10 seconds, a total distance of 500 feet. **11** (a) 2.5 months
(b) 50 rodents per month **13** $v(20) \approx 73$ ft/sec, or about 50 mph, $v(40) \approx 91$ ft/sec, or about 62 mph **17** When $t = 30$,
$dV/dt = -(\pi/3)(6.25)$, so the air is leaking out at approximately 6.545 cubic inches per second.

SECTION 1-7 (page 38)

1 $(f + g)(x) = x^2 + 3x - 2$, domain $\mathscr{R}$; $(fg)(x) = x^3 + 3x^2 - x - 3$, domain $\mathscr{R}$; $(f/g)(x) = (x + 1)/(x^2 + 2x - 3)$, domain $x \neq -3, x \neq 1$ **3** $(f + g)(x) = \sqrt{x} + \sqrt{x - 2}$, domain $[2, +\infty)$; $(fg)(x) = (x^2 - 2x)^{1/2}$, domain $[2, +\infty)$; $(f/g)(x) = \sqrt{x/(x - 2)}$, domain $(2, +\infty)$ **5** $(f + g)(x) = (x^2 - 1)^{1/2} + 1/(4 - x^2)^{1/2}$; $(fg)(x) = (x^2 - 1)^{1/2}/(4 - x^2)^{1/2}$; $(f/g)(x) = (x^2 - 1)^{1/2}(4 - x^2)^{1/2}$; Domain in each case: $(-2, -1] \cup [1, 2)$ **7** $f(g(x)) \equiv -17$, domain $\mathscr{R}$; $g(f(x)) \equiv 17$, domain $\mathscr{R}$ **9** $f(g(x)) = 1 + 1/(x^2 + 1)^2$, domain $\mathscr{R}$; $g(f(x)) = 1/[1 + (x^2 + 1)^2]$, domain $\mathscr{R}$ **11** $g(x) = x + 1$ **13** $g(x) = (x^4 + 1)^{1/2}$ or $g(x) = -(x^4 + 1)^{1/2}$, etc. **15** $g(x) = x + 4$ **17** No, because $f(g(0)) \neq 0$. **19** Yes, if the domains are suitably restricted: Use $(-1, +\infty)$ for the domain of f, $(0, +\infty)$ for the domain of g. Not if the "natural" domains are used, because $f(g(-0.5)) = +0.5$ **21** $g(x) = (x + 5)/13$, domain $\mathscr{R}$ **23** $g(x) = ((x - 1)/3)^{1/3}$, domain $\mathscr{R}$ **25** $g(x) = (1 + x^{1/17})^{1/3}$, $-1 \leq x \leq 0$ **27** Let $g(x) = (x - b)/a$. Show that $f(g(x)) = x$ for all x in $\mathscr{R}$ and that $g(f(x)) = x$ for all x in $\mathscr{R}$. **29** The domain of g_1 is the range of f, and so is that of g_2, so g_1 and g_2 have the same domain D. Choose x in D. It now suffices to show that $g_1(x) = g_2(x)$. Use the fact that $x = f(u)$ for some number u in the domain of f. **31** If $x_1 < x_2$ then $x_1^3 < x_2^3$ and $2x_1 < 2x_2$. So $f(x_1) < f(x_2)$. Therefore f is an increasing function.

SECTION 1-8 (pages 47–48)

1 $\lim_{x \to 0} (3x^2 + 7x - 12) = \lim_{x \to 0} (3x^2) + \lim_{x \to 0} (7x) - \lim_{x \to 0} 12 = 3(\lim_{x \to 0} x)^2 + 7(\lim_{x \to 0} x) - 12 = (3)(0)^2 + (7)(0) - 12 = -12$ **3** -1 **5** 128 **7** $(x + 1)/(x^2 - x - 2) = (x + 1)/(x + 1)(x - 2) = 1/(x - 2) \to -\frac{1}{3}$ as $x \to -1$ **9** 6 **11** $16\sqrt{2}$ **13** 1 **15** $(\sqrt{x + 4} - 2)/x = (x + 4 - 4)/x(\sqrt{x + 4} + 2) = 1/(\sqrt{x + 4} + 2) \to \frac{1}{4}$ as $x \to 0$ **17** $-\frac{1}{54}$ **19** $-|x| \leq x \cos x \leq |x|$. So $x \cos x \to 0$ as $x \to 0$ by the squeeze law.

21 x	$(\sqrt{x + 9} - 3)/x$	**23** x	$(1 - \cos x)/x^2$	**25** x	$(3^x - 1)/x$
0.1	0.16621	0.1	0.49958	0.01	1.10467
0.01	0.16662	0.01	0.50000	0.001	1.09922
0.001	0.16666	0.001	0.50000	0.0001	1.09867
-0.001	0.16667	-0.001	0.50000	-0.0001	1.09855
0.0001	0.16667			0.00001	1.09860
-0.0001	0.16667			-0.00001	1.09861

The limit seems to be $\frac{1}{6}$ The limit seems to be 0.5   The limit seems to be approximately 1.0986

27 $f'(x) = -1/2x\sqrt{x}$ **29** $f'(x) = 1/(2x + 1)^2$ **31** $\theta^2/\sin\theta = \theta (\theta/\sin\theta) \to (0)(1) = 0$ as $\theta \to 0$ **33** Use the identity $(1 - \cos 2x)/2 = \sin^2 x$ **35** $2x/((\sin x) - x) = 2/((\sin x)/x - 1) \to -\infty$ as $x \to 0$ (because $(\sin x)/x < 1$ if $x \neq 0$) **39** All real numbers other than integers

SECTION 1-9 (pages 54–55)

1 As $x \to 5^-$, $5 - x \to 0$. So their product approaches zero by the product law. Therefore $\sqrt{x(5 - x)} \to 0$ by the root law **3** As $x \to 4^+$, $x - 4 \to 0$ through positive values and also $4x \to 16$. So their quotient approaches zero through positive values. So the given limit is zero by the root law **5** As $x \to -4$ from the right, $x^2 \to 16$ from the left. So $1/(x^2 - 16) \to -\infty$; i.e., the limit does not exist **7** The fraction can be simplified to $|x - 3|/(x - 3)$, which is equal to -1 for all $x < 3$. So -1 is the limit **9** Continuous where defined; cannot be made continuous at $x = -3$ **11** Continuous where defined; cannot be made continuous at $x = -2$ **13** Continuous where defined; cannot be made continuous at $x = \pm 1$ **15** Continuous if $x > 3$ and if $x < -3$ (no isolated points of discontinuity) **17** Continuous on the interval $(-1, 1]$ (no isolated points of discontinuity) **19** Continuous where defined; cannot be made continuous at $x = 17$ **21** Continuous at each real number x not an integer; cannot be made continuous at any integer **23** Continuous at no point **25** Minimum value 0 at $x = 0$ **27** Minimum value $\frac{1}{2}$ at $x = -1$ **29** Minimum value -2 at $x = -1$; maximum value 2 at $x = 1$ **31** Since $x^2 - 5 = -1$ when $x = 2$ and $x^2 - 5 = 4$ when $x = 3$, it follows that $x^2 - 5 = 0$ for some x between 2 and 3 **35** Suggestion: Let $f(x) = x^2 - a$; compare with Problem 31 **37** Suggestion: Let $f(x) = x - \cos x$. Show that $f(x) = 0$ for some x between 0 and $\pi/2$

SECTION 1-10 (page 60)

1 Given $\varepsilon > 0$, choose $\delta = \varepsilon$ **3** Given $\varepsilon > 0$, let δ be the minimum of 1 and $\varepsilon/6$ **5** It suffices to consider the case $a \geq 0$. The case $a = 0$ isn't much trouble. If $a > 0$, then—given $\varepsilon > 0$—choose δ to be the minimum of $\varepsilon/(2a + 1)$ and 1. **9** Suppose that x is very close to a. What is an upper bound on the size of $|x^{n-1} + x^{n-2}a + \cdots + a^{n-1}|$?

1 $(-\frac{1}{2}, 2]$ **3** $(-\infty, -\frac{1}{8}] \cup [\frac{7}{16}, +\infty)$ **5** $(-\infty, \frac{2}{3}]$ **7** $[2, 4]$ **9** $y = 2x + 11$ **11** $y = \frac{1}{2}x - 5$ **13** $x + 2y = 11$

15

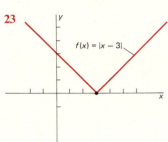

17

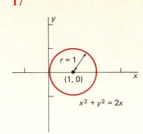

19

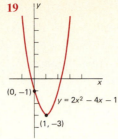

21

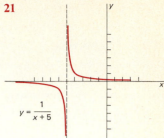

23

25 $f'(x) = 4x + 3$ **27** $f'(x) = 1/(3 - x)^2$

29 $f'(x) = -1/(2x - 1)^{3/2}$ **31** $y - 14 - 6\sqrt{5} = (6 + 2\sqrt{5})(x - 3 - \sqrt{5})$ and $y - 14 + 6\sqrt{5} = (6 - 2\sqrt{5})(x - 3 + \sqrt{5})$
33 Touchdown at time $t = 2$; the speed then is zero. **35** (a) When $t = 6$, $V'(t) = -144\pi$, so the volume is then *decreasing* at approximately 452.39 cc/hr. (b) The average is -156π, so the volume is decreasing at the *average* rate of about 490.09 cc/hr over the given time interval. **37** $g(x) = (5 - x^2)/2$, $x \geqq 0$ **39** $g(x) = (1 + x^{-1/5})^{1/3}$, $x \neq 0$ **41** $g(x) = x^2 - x$ **43** $\frac{1}{6}$ **45** $-\frac{1}{54}$
47 -1 **49** The function is continuous where defined, but we cannot assign a value to $f(2)$ so that f will be continuous at $x = 2$.
51 The function is continuous where it is defined. It cannot be made continuous at $x = \pm 1$ by assignment of values at those numbers.
53 Suggestion: Find values of x for which $f(x) = x^5 + x - 1$ is positive and negative. **55** Choose $\varepsilon = f(a)/2$.

CHAPTER 2

SECTION 2-2 (pages 73–74)

1 $f'(x) = 6x - 1$ **3** $f'(x) = 2(3x - 2) + 3(2x + 3)$ **5** $h'(x) = 3x^2 + 6x + 3$ **7** $f'(y) = (2y - 1)(2y + 1) + 2y(2y + 1) + 2y(2y - 1)$ **9** $g'(x) = -1/(x + 1)^2 + 1/(x - 1)^2$ **11** $h'(x) = -3(2x + 1)/(x^2 + x + 1)^2$ **13** $g'(t) = 2t(t^3 + t^2 + 1) + (3t^2 + 2t)(t^2 + 1)$ **15** $g'(z) = -1/2z^2 + 2/3z^3$ **17** $g'(y) = 2(3y^2 - 1)(y^2 + 2y + 3) + 6y(2y)(y^2 + 2y + 3) + (2y + 2)(2y)(3y^2 - 1)$ **19** $g'(t) = ((t^2 + 2t + 1) - (t - 1)(2t + 2))/(t^2 + 2t + 1)^2$ **21** $v'(t) = -3/(t - 1)^4$ **23** $g'(x) = (3(x^3 + 7x - 5) - (3x)(3x^2 + 7))/(x^3 + 7x - 5)^2$ **25** $g'(x) = ((2x - 3)(3x^2 - 4x) - 2(x^3 - 2x^2))/(2x - 3)^2$ **27** $(0, 7)$
29 $(-1, 27), (2, 0)$ **31** $(0, 1)$ **33** $v > 0: (-\infty, -3), (2, +\infty)$; $v < 0: (-3, 2)$ **35** (a) It contracts (b) 0.06427 cc/°C
37 $14,400\pi \approx 45,239$ cc/cm **39** $y = 3x + 2$ **41** Suppose that such a line L is tangent at the points (a, a^2) and (b, b^2). Show that it follows that $a = b$. **43** $((n - 1)/n)x_0$ for $x_0 \neq 0$ **49** $g'(x) = 17(x^3 - 17x + 35)^{16}(3x^2 - 17)$

SECTION 2-3 (pages 82–83)

1 Minimum: 1; Maximum: 5 **3** Minimum: -16; Maximum: 9 **5** Minimum: -22; Maximum: 10 **7** Minimum: -56; Maximum: 56 **9** Minimum: 5; Maximum: 13 **11** Minimum: 0; Maximum: 17 **13** Minimum: 0; Maximum: $\frac{3}{4}$ **15** Minimum: $-\frac{1}{6}$ at $x = 3$; Maximum: $\frac{1}{2}$ at $x = -1$ **17** 1152 **19** Approximately 3.9665 °C **21** 1000 cc **23** 0.25 cubic meters (all cubes, no open-top boxes) **25** 30,000 sq. meters **27** $\frac{250000}{27} \approx 9259.26$ cubic inches **29** Five presses **31** The minimizing value of x is $-2 + 10\sqrt{6}/3$ inches. To the nearest integer, use $x = 6$ inches of insulation for an annual savings of \$285. **33** Charge either \$1.10 or \$1.15 for the largest revenue. **35** Radius $2R/3$, height $H/3$

1 $f'(x) = 3x^2/2\sqrt{x^3 + 1}$ **3** $f'(x) = 2x/\sqrt{2x^2 + 1}$ **5** $f'(t) = \frac{3}{2}\sqrt{2t}$ **7** $f'(x) = \frac{3}{2}(2x^2 - x + 7)^{1/2}(4x - 1)$ **9** $g'(x) = -4(1 - 6x^2)/3(x - 2x^3)^{7/3}$ **11** $f'(x) = (1 - 2x^2)/\sqrt{1 - x^2}$ **13** $f'(t) = -2t/(t^2 + 1)^{1/2}(t^2 - 1)^{3/2}$ **15** $f'(x) = 3(x^2 + 1)(x^2 - 1)^2/x^4$ **17** $f'(v) = -(v + 2)/2v^2\sqrt{v + 1}$ **19** $f'(x) = -2x/3(1 - x^2)^{2/3}$ **21** If R is the radius of the circle and $2x$ is one side of the inscribed rectangle, then its perimeter is $P(x) = 4x + 4(R^2 - x^2)^{1/2}, 0 \leq x \leq R$ **23** Each plank is $(\sqrt{34} - 3\sqrt{2})/8 \approx 0.19854$ by $(7 - \sqrt{17})^{1/2}/2 \approx 0.84807$ **25** Strike the shore $2\sqrt{3}/3 \approx 1.1547$ km from the point nearest the island—that is, $(18 - 2\sqrt{3})/3 \approx 4.8453$ km from the town **27** $\sqrt{3}/3$ **29** Walk directly to the junction of the river and the pipeline and back **31** The actual value of x is approximately 3.45246232

1 $f(0) = 0 = f(2), f'(x) = 2x - 2, c = 1$ **3** $f(-1) = 0 = f(1), f'(x) = -4x/(1 + x^2)^2, c = 0$ **5** $f'(0)$ does not exist **7** $f(0) \neq f(1)$ **9** $c = -\frac{1}{2}$ **11** $c = \frac{35}{27}$ **13** The average slope is $\frac{1}{3}$, but $f'(x) = \pm 1$ wherever f' exists **15** The average slope is 1, but $f'(x) = 0$ wherever it exists **17** Increasing: $(-\infty, -1)$ and $(1, +\infty)$, decreasing: $(-1, 0)$ and $(0, 1)$ **19** Increasing: $(-\infty, -\frac{1}{2})$ and $(\frac{1}{2}, +\infty)$, decreasing: $(-\frac{1}{2}, 0)$ and $(0, \frac{1}{2})$ **21** If $g(x) = x^5 + 2x - 3$, then $g'(x) > 0$ for all x in $[0, 1]$ and $g(1) = 0$. So $x = 1$ is the only root of the equation in the given interval **23** If $g(x) = x^4 - 3x - 20$, then $g(2) = -10$ and $g(3) = 52$. Since $g'(x) = 4x^3 - 3 \geq 4(2^3) - 3 = 29 > 0$, g is an increasing function, and thus has exactly one zero in the interval $[2, 3]$ **25** Note that $f'(x) = \frac{3}{2}(-1 + \sqrt{x + 1})$ **27** Assume that $f'(x)$ has the form $a_0 + a_1 x + \cdots + a_{n-1}x^{n-1}$. Construct a polynomial $p(x)$ such that $p'(x) = f'(x)$. Conclude that $f(x) = p(x) + C$ on $[a, b]$

1 Minimum: $(0.75, -10.125)$, increasing: $(0.75, +\infty)$, decreasing: $(-\infty, 0.75)$

3 Minimum: $(1, -7)$, maximum: $(-2, 20)$, decreasing: $(-2, 1)$, increasing: $(-\infty, -2)$ and $(1, +\infty)$

5 Maximum: $(0.6, 16.2)$, minimum: $(0.8, 16.0)$, increasing: $(-\infty, \frac{3}{5})$, and $(\frac{4}{5}, +\infty)$ decreasing: $(\frac{3}{5}, \frac{4}{5})$

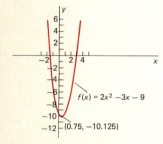

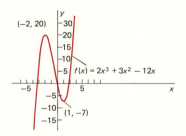

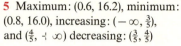

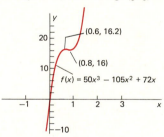

7 Increasing: $(-1, 0)$, and $(2, +\infty)$, decreasing: $(-\infty, -1)$ and $(0, 2)$, Maximum: $(0, 8)$, Minima: $(-1, 3)$ and $(2, -24)$

9 Maximum: $(-2, 64)$, minimum: $(2, -64)$, increasing: $(-\infty, -2)$ and $(2, +\infty)$, decreasing: $(-2, 2)$

11 Increasing: $(-\infty, +\infty)$, decreasing: Nowhere, Extrema: None

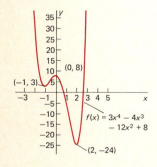

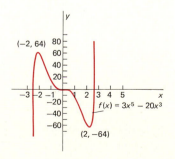

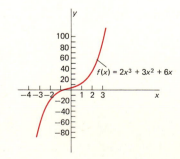

13 Maxima: $(\pm\sqrt{2}, 16)$, minimum: (0, 0), increasing: $(-\infty, -\sqrt{2})$ and $(0, \sqrt{2})$, decreasing: $(-\sqrt{2}, 0)$ and $(\sqrt{2}, +\infty)$

15 Increasing: $(-\infty, 1)$, decreasing: $(1, +\infty)$, maximum: (1, 3)

17 Decreasing: (0.6, 1), increasing: $(-\infty, 0.6)$ and $(1, +\infty)$, minimum: (1, 0), maximum: (0.6, 0.3257) (ordinate approximate)

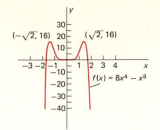

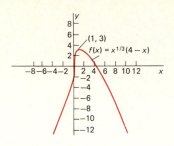

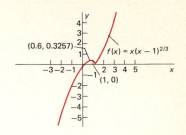

21 Base: 5 by 5 inches; height: 2.5 inches   **23** Radius: $(25/\pi)^{1/3} \approx 1.9965$ inches, height: 4 times the radius   **25** The nearest points are $(\pm\sqrt{6}/2, \frac{3}{2})$; (0, 0) is *not* the nearest point   **27** 8 inches   **29** $L = [20 + 12\sqrt[3]{4} + 24\sqrt[3]{2}]^{1/2} \approx 8.324$ meters

SECTION 2-7 (pages 111–112)

$f(x)$	$g(x)$	$h'(x)$
1 $x^2 + 1$	$1/x$	$-(1/(x^2 + 1)^2)(2x)$
3 $x^2 + 25$	$x^{-1/2}$	$-\frac{1}{2}(x^2 + 25)^{-3/2}(2x)$
5 $x - 1$	$x^{5/3}$	$\frac{5}{3}(x - 1)^{2/3}(1)$
7 $x^2 + 1$	$\cos x$	$[-\sin(x^2 + 1)](2x)$

9 $dy/dx = -4x(x^2 + 1)/([x^2 + 1]^2 + 1)^2$ **11** $-1/(x + 1)^{1/2}(x - 1)^{3/2}$ **13** $f'(x) = x \cos x + \sin x$
15 $g'(t) = 2t^3 \cos(t^4 + 1)^{1/2}/(t^4 + 1)^{1/2}$ **17** $f'(x) = 2[\cos(\cos x)][-\sin(\cos x)][-\sin x]$ **19** 10   **21** -18
25 $400\pi \approx 1256.64$ cc/sec   **27** 5 cm   **29** Total melting time: $2/(2 - \sqrt[3]{4}) \approx 4.8473$ hours. Completely melted at about 2:50:50 P.M. of the same day   **31** $17/x$ **33** $2x/(x^2 + 1)$ **35** $-1/x[L(x)]^2$ **37** $\cot x$

SECTION 2-8 (pages 116–117)

1 x/y **3** $-16x/25y$ **5** $-(y/x)^{1/2}$ **7** $-(y/x)^{1/3}$ **9** $dy/dx = (3x^2 - 2xy - y^2)/(3y^2 + 2xy + x^2)$ **11** Tangent: $y = 2 - x$.
Normal: $y = x$ **13** Tangent: $y = x - 2$. Normal: $y = -x$ **15** $d^2y/dx^2 = (y^2 - x^2)/y^3 = -1/y^3$ **17** $d^2y/dx^2 = 2y^{1/3}(x^{1/3} + y^{1/3})/3x^{5/3}$ **19** Tangent line equations: $y = \frac{1}{2}x \pm 3$ **21** Closest points: $(\sqrt{3}, -\sqrt{3})$ and $(-\sqrt{3}, \sqrt{3})$. Farthest points: (3, 3) and $(-3, -3)$ **23** Farthest points: (1, 1) and $(-1, -1)$ **25** Maximum volume: 500 cubic inches   **29** Maximum volume: All sphere, no cube, volume $(5000/3)\sqrt{10/\pi} \approx 2973.54$ cubic inches. Minimum volume: Edge of cube twice the radius of the sphere, total volume $(1000 + (500\pi/3))(10/(\pi + 6))^{3/2}$ or about 1743.16 cubic inches   **31** Base 5 by 5 inches; height: 2.5 inches   **33** Radius: $r = (25/\pi)^{1/3}$, height $4r$

SECTION 2-9 (pages 124–126)

1 (a) $\frac{2}{9}\pi$ (b) $-\frac{3}{2}\pi$ (c) $\frac{7}{4}\pi$ (d) $\frac{7}{6}\pi$ (e) $-\frac{5}{6}\pi$ **3**

x	(a), (d) $-\pi/3, 5\pi/3$	(b) $3\pi/4$	(c) $7\pi/6$
$\sin x$	$-\sqrt{3}/2$	$\sqrt{2}/2$	$-1/2$
$\cos x$	$\frac{1}{2}$	$-\sqrt{2}/2$	$-\sqrt{3}/2$
$\tan x$	$-\sqrt{3}$	-1	$\sqrt{3}/3$
$\sec x$	2	$-\sqrt{2}$	$-\frac{2}{3}\sqrt{3}$
$\csc x$	$-\frac{2}{3}\sqrt{3}$	$\sqrt{2}$	-2
$\cot x$	$-\frac{1}{3}\sqrt{3}$	-1	$\sqrt{3}$

5 (a) The odd multiples of $\pi/2$: $x = (2k + 1)\pi/2$ where k is an integer   (b) The even multiples of π: $x = 2k\pi$ (c) The odd multiples of π: $x = (2k + 1)\pi$ **7** $\sin x = -\frac{3}{5}$, $\cos x = -\frac{4}{5}$, etc.   **13** $\frac{1}{2}$ **15** 1   **17** $\frac{1}{3}$ **19** $\frac{1}{4}$ **21** $2 \cos 2x$ **23** $3x^2 \cos x^3$

25 $\sin\sqrt{x}\cos\sqrt{x}/\sqrt{x}$ **27** $2x\cos(3x^2-1)-6x^3\sin(3x^2-1)$ **29** $(2\cos 2x)(\cos 3x)-(3\sin 3x)(\sin 2x)$
31 $(-3\sin 5x\sin 3x-5\cos 5x\cos 3x)/\sin^2 5x$ **33** $x(\sin\sqrt{1-x^2})\cos(\cos\sqrt{1-x^2})/\sqrt{1-x^2}$ **37** $\alpha=\pi/4$
39 $(2000\pi)/27$ ft/sec, or about 158.67 mph **41** $\theta=\pi/3$ (60°) **43** $V=(8\pi/3)R^3$ **45** $A=3\sqrt{3}/4\approx 1.299$
47 Suggestion: $A(\theta)=s^2(\theta-\sin\theta)/2\theta^2$

CHAPTER 2 MISCELLANEOUS PROBLEMS (pages 127–130)

1 $2x-6/x^3$ **3** $1/2\sqrt{x}-1/3x^{4/3}$ **5** $7(x-1)^6(3x+2)^9+27(x-1)^7(3x+2)^8$ **7** $4(3x-1/2x^2)^3(3+1/x^3)$
9 $-y/x=-9/x^2$ **11** $-\frac{3}{2}(x^3-x)^{-5/2}(3x^2-1)$ **13** $4x(1+x^2)/(x^4+2x^2+2)^2$ **15** $\frac{7}{3}(\sqrt{x}+\sqrt[3]{2x})^{4/3}(1/2\sqrt{x}+\sqrt[3]{2}/3x^{2/3})$
17 $-1/\sqrt{x+1}\,(\sqrt{x+1}-1)^2$ **19** $dy/dx=(-2xy^2+1)/(2x^2y-1)$
21 $\frac{1}{2}[x+(2x+\sqrt{3x})^{1/2}]^{-1/2}[1+\frac{1}{2}(2x+\sqrt{3x})^{-1/2}(2+\sqrt{3}/2\sqrt{x})]$ **23** $-(y/x)^{2/3}$ **25** $-18(1+2/[1+x]^3)^2/[1+x]^4$
27 $\frac{1}{2}((1+\cos x)/\sin^2 x)^{1/2}(2(1+\cos x)\sin x\cos x+\sin^3 x)/(1+\cos x)^2$ **29** $-(4\sin 2x\sin 3x+3\cos 2x\cos 3x)/2(\sin 3x)^{3/2}$
31 $(6\sin^2 2x\cos 3x)(\cos 2x\cos 3x-\sin 2x\sin 3x)$ **33** $(5\sin^4(x+1/x)\cos(x+1/x))(1-1/x^2)$
35 $(3\cos^2\sqrt[3]{x^4+1})(-\sin\sqrt[3]{x^4+1})\frac{1}{3}(x^4+1)^{-2/3}(4x^3)$ **37** $x=1$ **39** $x=0$ **41** $\frac{1}{2}$ ft/min
43 Minimum at $(3,-5)$, decreasing on $(-\infty,3)$, increasing on $(3,+\infty)$ **45** Increasing everywhere, no extrema **47** Increasing on $(-\infty,\frac{1}{4})$, decreasing on $(\frac{1}{4},+\infty)$, maximum: $(\frac{1}{4},3/4\sqrt[3]{4})$

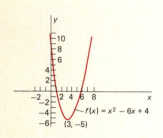
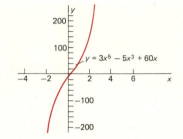
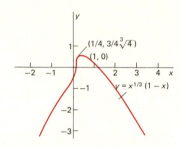

49 $dV/dS=\frac{1}{4}\sqrt{S/\pi}$ **51** $(\pi/18)\sec^2 50°\approx 0.4224$ mi/sec, or about 1521 mph, or about 2230 ft/sec **53** The maximum area is R^2
55 Minimum area: $(36\pi V^2)^{1/3}$, obtained by making *one* sphere of radius $(3V/4\pi)^{1/3}$. Maximum area: $(8\pi)^{1/3}(3V)^{2/3}$, obtained by making two equal spheres, each of radius $\frac{1}{2}(3V/\pi)^{1/3}$ **57** $(32\pi/81)R^3$ **59** Radius: $(1/5\pi)^{1/3}$ ft, about 4.79 inches; Height: $(25/\pi)^{1/3}$ ft, or about 23.96 inches **61** The rectangle should be twice as long as it is wide (dimensions $P/3$ by $P/6$) and should be rotated about its *short* side.
63 The minimizing angle is $\tan^{-1}\sqrt[3]{2}\approx 0.9$, or about $51.56°$. The minimum length is $4(1+\sqrt[3]{4})^{3/2}\approx 16.648$ feet.
67 Two miles from the point on the shore nearest the first town
69 (a) The maximum height is $m^2 v_0^2/64(m^2+1)$ (b) The maximum range occurs when $m=1$; thus when $\alpha=\pi/4$ **71** $x=\frac{2}{3}H$

CHAPTER 3

SECTION 3-2 (page 137)

1 $\Delta y=3$, $dy=3$ **3** $\Delta y=\sqrt{11}-\sqrt{10}\approx 0.15435$, $dy=1/2\sqrt{10}\approx 0.15811$ **5** $\Delta y=-1+1/(1+0.04)=-\frac{1}{26}\approx -0.03846$, $dy=0$
7 $dy=3(4x-x^2)^{1/2}(2-x)\,dx$ **9** $dy=-2\,dx/(x-1)^2$ **11** $dy=(2x\cos\sqrt{x}-\frac{1}{2}x^{3/2}\sin\sqrt{x})dx$ **13** $dy/dx=-x/y$
15 $(y-x^2)/(y^2-x)$ **17** $-(1/x)\tan y$ **19** $10+\frac{1}{60}\approx 10.0167$ **21** 132.5 **23** $2+\frac{1}{32}\approx 2.0313$ **25** 0.1 **27** $\Delta y=(2x-1)\Delta x+(\Delta x)^2$, $dy=(2x-1)\Delta x$, $\varepsilon=\Delta x$ **29** $\Delta y=-\Delta x/x(x+\Delta x)$, $dy=(-1/x^2)\Delta x$, $\varepsilon=\Delta x/x^2(x+\Delta x)$ **31** 7.5 **33** $10\pi\approx 31.416$ cubic inches **35** $\pi/96\approx 0.0327$ sec **37** $25\pi\approx 78.54$ cubic inches **39** $4\pi\approx 12.57$ square meters

SECTION 3-3 (pages 140–141)

1 2 **3** 5 **5** $(\pi-3\sqrt{3})/(\pi+6\sqrt{3})$ **7** $4/5\pi\approx 0.25465$ ft/sec **9** $32\pi/125\approx 0.80425$ meters/hr **11** $\frac{11}{15}\sqrt{21}\approx 3.36$ ft/sec downward
13 They are closest at $t=12$ minutes, and the distance between them then is $32\sqrt{13}\approx 115.38$ miles. **15** $1/162\pi\approx 0.001965$ ft/sec
17 $10/81\pi\approx 0.0393$ inches/minute **19** $300\sqrt{2}\approx 424.26$ mph

SECTION 3-4 (pages 149–150)

1 $x_0=0.25$; $x_{n+1}=(1-x_n^2)/3$. The solution (after seven iterations) is 0.302776 **3** 1.236130 **5** -0.740222 **7** 2.154435

9 1.802191   **11** (b) 1.25992   **13** 1.145048   **15** $x_{n+1} = \sqrt{\sqrt{\sqrt{ax}}}$; 1.389495   **17** 3.4525   **19** 0.2261   **21** $\alpha_1 \approx 2.0287578 \approx$ (1.29/2)π, $\alpha_2 \approx 4.9131804 \approx$ (3.13/2)π, $\alpha_3 \approx 7.9786657 \approx$ (5.08/2)π, $\alpha_4 \approx 11.085538 \approx$ (7.06/2)π

SECTION 3-5 (pages 157–158)

1 $3x^2 - 2, 6x, 6$ **3** $-(t^{-2}) + 2(2t + 1)^{-2}, 2t^{-3} - 8(2t + 1)^{-3}, -6t^{-4} + 48(2t + 1)^{-4}$ **5** $3t^{1/2} - 4t^{1/3}, \frac{3}{2}t^{-1/2} - \frac{4}{3}t^{-2/3},$
$-\frac{3}{4}t^{-3/2} + \frac{8}{9}t^{-5/3}$ **7** $-4(t - 2)^{-2}, 8(t - 2)^{-3}, -24(t - 2)^{-4}$ **9** $-\frac{4}{3}(5 - 4x)^{-2/3}, -\frac{32}{9}(5 - 4x)^{-5/3}, -\frac{640}{27}(5 - 4x)^{-8/3}$
11 $y' = x/y, y'' = -1/y^3$ **13** $y' = -(x + 2y)/(2x + 3y), y'' = 5/(2x + 3y)^3$ **15** $y' = (y - x^2)/(y^2 - x), y'' = 2xy/(x - y^2)^3$

17 Local maximum at $(-1, 10)$, local minimum at $(2, -17)$, inflection point at $(0.5, -3.5)$

19 Local maximum at $(-2, 22)$ and at $(2, 22)$, local minimum at $(0, 6)$. Inflection points at $(\pm 2\sqrt{3}/3, \frac{134}{9})$

21 Minima at $(-1, -6)$ and $(2, -32)$, maximum at $(0, -1)$. Inflection points at $(x, f(x))$ where $x = (1 \pm \sqrt{7})/3$ [approximately $(1.22, -19.36)$ and $(-0.55, -3.68)$]

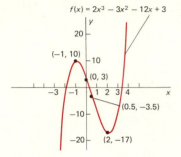

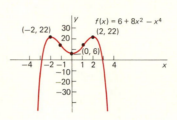

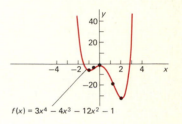

23 Local maximum at $(\frac{3}{7}, 0.008)$ (ordinate approximate), local minimum at $(1, 0)$. Inflection points at $(0, 0)$ and at $(x, f(x))$ where $x = (3 \pm \sqrt{2})/7$. The general shape of the graph is shown in the upper left; the lower right shows the behavior of f on and near $[0, 1]$ with the vertical scale greatly magnified

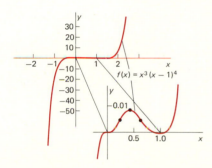

25 Inflection point at $(0, 1)$; no extrema

27 Minimum at $(0, 0)$, inflection point at $(1, 4)$. Concave up for $x > 1$

29 Maximum at $(1, 3)$, inflection points at $(0, 0)$ and $(-2, -7.56)$ (ordinate approximate)

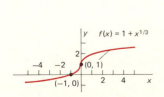

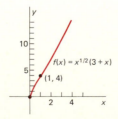

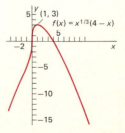

31 $5(x + 1)^4, 20(x + 1)^3, 60(x + 1)^2, 120(x + 1), 120$ **39** $a = 3PV^2 \approx 3583858.8, b = \frac{1}{3}V \approx 42.7, R = 8PV/3T \approx 81.80421$

1 1 **3** 3 **5** 2 **7** 1 **9** 4 **11** 0 **13** 2 **15** $+\infty$ or "does not exist."

17 No critical points, no inflection points. Vertical asymptote: $x = 3$. Horizontal asymptote: $y = 0$ (the x-axis). No x-intercept; the y-intercept is $(0, -\frac{2}{3})$

19 No critical points, no inflection points. Vertical asymptote: $x = -2$. Horizontal asymptote: $y = 0$

21 No critical points, no inflection points. Vertical asymptote: $x = \frac{3}{2}$. Horizontal asymptote: $y = 0$

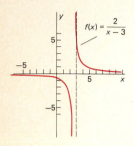

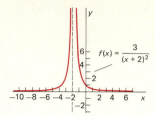

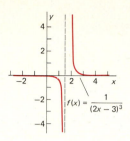

23 Minimum at $(0, 0)$, inflection points at $(\pm\frac{1}{3}\sqrt{3}, \frac{1}{4})$. Horizontal asymptote: $y = 1$

25 Local maximum at $(0, -\frac{1}{9})$, no inflection points. Vertical asymptotes: $x = \pm 3$. Horizontal asymptote: $y = 0$

27 Local maximum: $(-\frac{1}{2}, -\frac{4}{25})$. Horizontal asymptote: $y = 0$. Vertical asymptotes: $x = -3$ and $x = 2$. No inflection points.

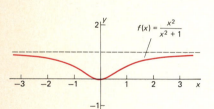

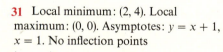

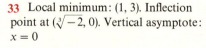

29 Local minimum at $(1, 2)$, local maximum at $(-1, -2)$. No inflection points. Vertical asymptote: $x = 0$. The line $y = x$ is also an asymptote

31 Local minimum: $(2, 4)$. Local maximum: $(0, 0)$. Asymptotes: $y = x + 1$, $x = 1$. No inflection points

33 Local minimum: $(1, 3)$. Inflection point at $(\sqrt[3]{-2}, 0)$. Vertical asymptote: $x = 0$

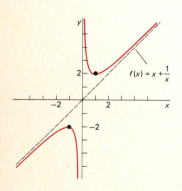

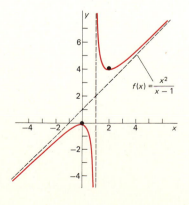

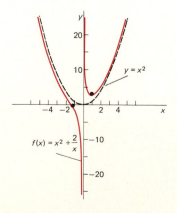

SECTION 3-7 (pages 172–173)

1 $x^3 + x^2 + x + C$ **3** $x - \frac{2}{3}x^3 + \frac{3}{4}x^4 + C$ **5** $-\frac{3}{2}x^{-2} + \frac{4}{5}x^{5/2} - x + C$ **7** $t^{3/2} + 7t + C$ **9** $\frac{3}{5}x^{5/3} - 16x^{-1/4} + C$
11 $\frac{1}{12}(2t - 1)^6 + C$ **13** $\frac{3}{8}(x^3 + 7)^{8/3} + C$ **15** $\sqrt{x^2 + 1} + C$ **17** $\frac{3}{5}x^{5/3} + 2x + 3x^{1/3} + C$ **19** $\frac{1}{2}\sin 2x + C$
21 $\frac{1}{2}\sin x^2 + C$ **23** $\frac{1}{2}\sin(2t + 1) + C$ **25** $\frac{1}{4}\sin^2 2x + C$ or $-\frac{1}{4}\cos^2 2x + C$ **27** (a) 250 tons   (b) $50\sqrt{29} \approx 269.258$ tons
(c) $\cos^{-1}(5/\sqrt{29}) \approx 0.38$, or about $21°48'5''$ **29** (b) 225 animals   (c) $20(\sqrt{2} - 1) \approx 8.284$

SECTION 3-8 (pages 182–183)

1 $y = f(x) = x^2 + x + 3$ **3** $\frac{2}{3}(x^{3/2} - 8)$ **5** $2\sqrt{x + 2} - 5$ **7** $x(t) = 25t^2 + 10t + 20$ **9** $x(t) = \frac{1}{2}t^3 + 5t$ **11** $x(t) = $
$\frac{1}{3}(t + 3)^4 - 37t - 26$ **13** $x(t) = 1/2(t + 1) + t/2 - \frac{1}{2}$ **15** $y(x) = 1/(1 - x)$ **17** $y(x) = (x + 1)^{1/4}$ **19** $y(x) = (2 + \frac{1}{3}x^{3/2})^2$
21 5 sec; -160 ft/sec **23** 400 ft; 10 sec **25** $(-5 + 2\sqrt{145})/4 \approx 4.77$ sec; $16\sqrt{145} \approx 192.6655$ ft/sec **27** 544/3 ft/sec
29

	m/sec	km/hr	mph
Moon	2376	8554	5315
Mars	4983	17940	11147
Jupiter	58996	212387	131971
Sun	614267	2,211,361	1,374,076
Ganymede	2562	9222	5730

31 Approximately 2.78 meters   **33** Approximately 0.19 seconds

SECTION 3-9 (page 186)

	(a) $C(400)$	(b) $C(400)/400$	(c) $C'(400)$	(d) $C(401) - C(400)$
1	$2200.00	$5.50	$3.00	$3.00
3	$24640.00	$61.60	$43.20	$43.20
5	$2705.00	$6.76	$4.99	$4.99

7 Minimum value $51.31 for $x = 1414$ **11** 417 per week, to be sold at $35.83 each, for maximum profit of $3208.33   **15** $C(x) = $
$50x + (0.001)x^2 + 990$; $\frac{1}{200}C(200) = \$55.15$ **17** (b) 5 years

CHAPTER 3 MISCELLANEOUS PROBLEMS (pages 187–188)

1 $\Delta y = 123^{2/3} - 25 \approx -0.2674$, $dy = -\frac{4}{15} \approx -0.2667$ **3** $373/75 \approx 4.9733$ **5** $323/108 \approx 2.9907$ **7** 4 sq. in. per sec.
9 $-50/9\pi \approx -1.7684$ ft/min **11** 1 inch per minute **13** Approximately 1.54785 feet **15** To five places, the three real solutions
are -2.72249, 0.80126, and 2.30998. The other two are the two complex numbers $-0.19437 \pm (1.93204)i$ **17** $y' = -(y/x)^{2/3}$,
$y'' = \frac{2}{3}(y/x^5)^{1/3}$ **19** $y' = 1/(2(5y^4 - 4)\sqrt{x})$, $y'' = -(20\sqrt{x}y^3 + (5y^4 - 4)^2)/(4x^{3/2}(5y^4 - 4)^3)$

21 Minimum at $(2, -48)$, x-intercepts 0 and $2\sqrt[3]{4} \approx 3.1748$. Concave up everywhere, no inflection points, no asymptotes

23 Local maximum at $(0, 0)$, minima at $(\pm\frac{2}{3}\sqrt{3}, -\frac{32}{27})$ (about $(\pm 1.15, -1.19)$). Inflection points at $(\pm\frac{2}{5}\sqrt{5}, -\frac{96}{125})$ (approximately $(\pm 0.89, -0.77)$). No asymptotes

25 Maximum at $(3, 3)$, inflection points at $(4, 0)$ and $(6, -6\sqrt[3]{2})$ (the ordinate is approximately -7.56). No asymptotes; vertical tangent at $(4, 0)$

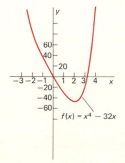

$f(x) = x^4 - 32x$

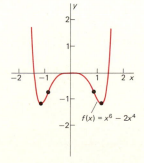

$f(x) = x^6 - 2x^4$

27 Local maximum at $(0, -\frac{1}{4})$, horizontal asymptote $y = 1$, vertical asymptotes $x = \pm 2$. No inflection points.

29 The inflection point has abscissa the only real solution of $x^3 + 6x^2 + 4 = 0$, which is approximately -6.10724.

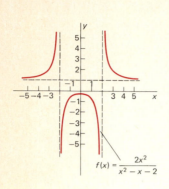

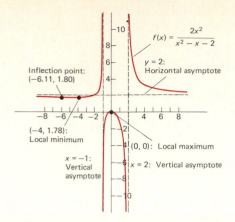

31 $\frac{1}{3}x^3 + 2/x - 5/2x^2 + C$ **33** $\frac{3}{4}(x^2 + 2x)^{2/3} + C$ **35** $\frac{2}{3}(1 + \sin x)^{3/2} + C$ **37** $-1/[3(1 + 2x^{3/2})] + C$ **39** $y(x) = \frac{1}{12}(2x + 1)^6 + \frac{23}{12}$ **41** $y(x) = \frac{3}{2}x^{2/3} - \frac{1}{2}$ **43** $y(x) = \frac{1}{16}(x + 1)^4$ **45** Impact time: $t = 2\sqrt{10} \approx 6.3246$ sec. Speed at impact: $20\sqrt{10} \approx 63.2456$ ft/sec **47** 176 ft **49** $100(\frac{20}{9})^{0.4} \approx 54.79$ mph **51** Horizontal tangents when $x = 1 - \frac{1}{3}\sqrt{3} \approx 0.42$, $y = \pm\frac{1}{3}(12)^{1/4} \approx \pm 0.62$. Intercepts: $(0, 0)$, $(1, 0)$, and $(2, 0)$. Vertical tangent at each intercept. Inflection points corresponding to the [only] positive solution of $3x^2(x - 2)^2 = 4$: $x \approx 2.46789$, at which $y \approx \pm 1.30191$. No asymptotes

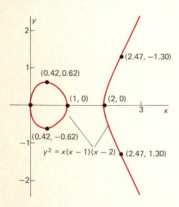

CHAPTER 4

SECTION 4-2 (page 198)

1 190 **3** 1165 **5** 224 **7** 1834 **9** 333,833,500 **11** $\frac{1}{2}$ **13** $\frac{1}{5}$ **15** $\frac{1}{3}n(2n - 1)(2n + 1)$ **17** 24 **19** $\frac{1}{3}$ **21** 132 **23** 90 **25** $\frac{1}{3}b^3$

SECTION 4-3 (pages 206–207)

1 0.44, 0.24, 0.33 (exact) **3** $\frac{29}{20} = 1.45$, $\frac{137}{60} \approx 2.283333$, $\frac{6086}{3465} \approx 1.756421$ **5** 19.5, 16.5, 18 (exact) **7** 58.8, 26.4, 40.575 (exact) **9** 2 **11** 12 **13** 30 **15** $\frac{17}{6}$ **17** $\frac{28}{3}$ **19** $\frac{52}{5}$ **21** $(13\sqrt{13} - 8)/3 \approx 12.9574$ **23** $\pi/2$ **25** 0 **27** $\frac{1}{2}$ **29** $\frac{2}{3}$ **31** $25\pi/4 \approx 19.635$ **37** $1000 + \int_0^{30} v'(t)\,dt = 160$ gal

SECTION 4-4 (pages 213–214)

1 $\frac{125}{4}$ **3** $\frac{14}{9}$ **5** $\frac{1}{2}$ **7** 4 **9** $\frac{1}{3}$ **11** $-\frac{22}{81} \approx -0.271605$ **13** 0 **15** $\frac{35}{24} \approx 1.4583$ **17** 0 **19** 4 **21** $f'(x) = (x^2 + 1)^{17}$ **23** $h'(z) = (z - 1)^{1/3}$ **25** $f'(x) = -x - 1/x$ **27** $f'(x) = 3\sin 9x^2$ **29** The integral does not exist (look at the graph of $y = 1/x^2$) **33** Average height: $\frac{800}{3} \approx 266.67$ ft. Average velocity: -80 ft/sec **35** $\frac{5000}{3} \approx 1666.67$ gal

1 $\frac{1}{24}(4x - 3)^6 + C$ **3** $-\frac{1}{9}(2 - 3x^2)^{3/2} + C$ **5** $-1/9(x^3 + 5)^3 + C$ **7** $-\frac{1}{16}(2 - 4x^3)^{4/3} + C$ **9** $-3 \cos t/3 + C$ **11** $-\frac{3}{8}$
13 2 **15** $\frac{1192}{15} \approx 79.46667$ **17** $\frac{15}{128} \approx 0.11719$ **19** $\frac{2}{5}(4\sqrt{2} - 1) \approx 1.86274$ **21** $\frac{3}{4}(x^4 - 4x)^{1/3} + C$ **23** $\frac{4}{15}(6t - t^3)^{5/4} + C$

SECTION 4-6 (pages 225–226)

1 $\frac{500}{3}$ **3** $\frac{64}{3}$ **5** $\frac{500}{3}$ **7** $\frac{4}{3}$ **9** $\frac{1}{12}$ **11** $\frac{16}{3}$ **13** $\frac{16}{3}$ **15** $\frac{27}{2}$ **17** $\frac{320}{3}$ **19** $\frac{8}{15}$ **21** $\frac{37}{12}$ **25** 1

SECTION 4-7 (pages 235–236)

1 $L_5 = 2.04$, $R_5 = 2.64$; $n \geq 300$ **3** $L_4 \approx 1.07$, $R_4 \approx 1.17$; $n \geq 42$ **5** $L_{10} \approx 0.95$, $R_{10} \approx 0.94$; $n \geq 16$ **7** $M_{10} \approx 0.9462$
9 (a) $T_8 \approx 1.0892$ (b) $S_8 \approx 1.0901$ **11** (a) 1.2846 (b) 1.2854 **13** (a) 3.0200 (b) 3.0717 **15** Both answers should be near
2435 **17** $L_{10} \approx 0.71877$, $R_{10} \approx 0.66877$ **19** $M_{10} \approx 0.692835$, $T_{10} \approx 0.693771$ **21** $n \geq 14$

CHAPTER 4 MISCELLANEOUS PROBLEMS (pages 237–238)

1 1700 **3** 2845 **5** $2(\sqrt{2} - 1) \approx 0.82843$ **7** $(2\pi/3)(2\sqrt{2} - 1) \approx 3.82945$ **13** $2x\sqrt{2x}/3 + 2/\sqrt{3x} + C$ **15** $-2/x - x^2/4 + C$
17 $\frac{2}{3} \sin x^{3/2} + C$ **19** $\cos(1/t) + C$ **21** $-\frac{3}{8}(1 + u^{4/3})^{-2} + C$ **23** $\frac{38}{3}$ **25** $(4x^2 - 1)^{3/2}/3x^3 + C$ (use $u = 1/x$) **27** $\frac{1}{30}$ **29** $\frac{44}{15}$
31 $\frac{125}{6}$ **33** Semicircle, center (1, 0), radius 1: area $\pi/2$ **35** $f(x) = \sqrt{4x^2 - 1}$ **37** $n \geq 9$. $L_{10} \approx 1.12767$, $R_{10} \approx 1.16909$. An
estimate: their average is A = 1.1483 ± 0.05 **39** $M_5 \approx 0.28667$, $T_5 \approx 0.28971$. They bound the true value of the integral because
the second derivative of the integrand is positive for all $x > 0$.

CHAPTER 5

SECTION 5-1 (pages 244–245)

1 -320; 320 **3** -50; 106.25 **5** 65; 97 **7** 1; 1 **9** 0; $4/\pi$ **11** 1 **13** $2/\pi$ **15** $\frac{98}{3}$ **17** 550 gal **19** 385,000 **21** 109.5 in.

SECTION 5-2 (pages 250–251)

1 $\pi/5$ **3** 8π **5** $\pi^2/2$ **7** $3\pi/10$ **9** $512\pi/3$ **11** 9π **13** $\frac{4}{3}\pi a^2 b$ **15** $\frac{16}{3}a^3$ **17** $(4\sqrt{3}/3)a^3$ **23** $\frac{16}{3}a^3$

SECTION 5-3 (page 255)

1 8π **3** $625\pi/2$ **5** 16π **7** π **9** $6\pi/5$ **11** $256\pi/15$ **15** $\frac{4}{3}\pi a^2 b$ **17** $V = 2\pi^2 a^2 b$ **19** $V = 2\pi^2 a^3$ **21** (a) $V = (\pi/6)h^3$

SECTION 5-4 (pages 262–263)

1 $\frac{22}{3}$ **3** $\frac{14}{3}$ **5** $\frac{123}{32} = 3.84375$ **7** $\frac{1}{27}[104\sqrt{13} - 125] \approx 9.25842$ **9** $(\pi/6)(5\sqrt{5} - 1) \approx 5.3304$ **11** $339\pi/16 \approx 66.5625$
13 $(\pi/9)(82\sqrt{82} - 1) \approx 258.8468$ **15** 4π **17** 3.8194 (the true value is approximately 3.8202) **21** Avoid the problem when
$x = 0$: $L = 8 \int_{1/2\sqrt{2}}^{1} x^{-1/3} \, dx = 6$

SECTION 5-5 (pages 268–269)

1 30 **3** 9 **5** 0 **7** 15 ft-lb **9** 2.816×10^9 ft-lb (with $R = 4000$ mi, $g = 32$ ft/sec^2) **11** $13000\pi \approx 40840.7$ ft-lb
13 $125,000\pi/3 \approx 130,899.69$ ft-lb **15** $156,000\pi \approx 490,088.54$ ft-lb **17** $4,160,000\pi \approx 13,069,025.44$ ft-lb **19** 8750 ft-lb
21 11250 ft-lb **23** $25000(1 - [0.1]^{0.4})$ in-lb, or approximately 1253.9434 ft-lb

SECTION 5-6 (pages 276–277)

1 $(\bar{x}, \bar{y}) = (0, \frac{8}{5})$ **3** $(\bar{x}, \bar{y}) = (\frac{3}{4}, \frac{9}{10})$ **5** $(\bar{x}, \bar{y}) = (-\frac{1}{2}, 2)$ **7** $(\bar{x}, \bar{y}) = (\frac{3}{5}, \frac{12}{35})$ **9** $(\bar{x}, \bar{y}) = (4r/3\pi, 4r/3\pi)$ **11** $(\bar{x}, \bar{y}) = (2r/\pi, 2r/\pi)$
19 $\bar{y} = (4a^2 + 3\pi ab + 6b^2)/(12b + 3\pi a)$, $\bar{x} = 0$, $V = (\pi a/3)(4a^2 + 3\pi ab + 6b^2)$

SECTION 5-7 (page 282)

1 $(\frac{4}{3}, 0, 0)$ **3** $(1, 0, 0)$ **5** $(\frac{15}{8}, 0, 0)$ **7** $(\frac{48}{11}, 0, 0)$ **11** $(0, 3b/8, 0)$ **15** 249.6 lb **17** 748.8 lb **19** 19,500 lb **21** $700\rho/3 \approx 14560$ lb
23 About 32,574 tons **25** (a) $(62.4)(12.5)(25) = 19,500$ lb (b) $(62.4)(13)(9\pi)$ or about 22936.14 lb (c) $(62.4)(10 + 16/3\pi)(8\pi) \approx$
18345.231 lb

1 $\frac{1}{3}$ 3 $\frac{8}{3}$ 5 $\frac{1}{2}$ 7 $\frac{52}{9}$

CHAPTER 5 MISCELLANEOUS PROBLEMS (pages 286–288)

1 $-\frac{3}{2}; \frac{31}{6}$ 3 $1; 3$ 5 $\frac{124}{3}$ 7 1 9 12 in. 11 $\frac{41}{105}\pi$ 13 $10.625\pi \approx 33.379$ gm 19 $f(x) = \sqrt{1 + 3x}$ 21 $(24 - 2\pi^2)/3\pi \approx 0.4521$
23 $\frac{10}{3}$ 25 $\frac{63}{8}$ 27 $52\pi/5$ 31 $\bar{x} = \frac{393}{352}, \bar{y} = \frac{268}{165}$ 33 $\bar{x} = \frac{111}{112}, \bar{y} = \frac{1136}{245}$ 35 $\bar{x} = \frac{16}{21}, \bar{y} = 0$ 37 $\bar{y} = 4b/3\pi$ 43 1 ft
45 $W = 4\pi R^4 \rho$ 47 $10,454,400$ ft-lb 49 $36,400$ tons 51 Approximately 42031 tons

CHAPTER 6

SECTION 6-2 (pages 300–301)

1 $3/(3x - 1)$ 3 $1/(1 + 2x)$ 5 $(3x^2 - 1)/3(x^3 - x)$ 7 $-(1/x)\sin(\ln x)$ 9 $-1/x(\ln x)^2$ 11 $6/(2x + 1) + 8x/(x^2 - 4)$
13 $-x/(4 - x^2) - x/(9 + x^2)$ 15 $-15/4(1 - 5x) - 20x^3/3(1 + x^4)$ 17 $\frac{1}{6}\ln(1 + 3x^2) + C$ 19 $\frac{1}{4}\ln|2x^2 + 4x + 1| + C$
21 $\frac{1}{3}(\ln x)^3 + C$ 25 $m \approx -0.2479, k \approx 291.7616$

31 $y \to 0$ as $x \to 0^+$; also $y' \to 0$ as $x \to 0^+$. The point $(0, 0)$ is not on the graph. Intercept at $(1, 0)$, minimum at $(1/\sqrt{e}, -1/2e) \approx$ $(0.61, -0.18)$, inflection point at $(1/e\sqrt{e}, -3/2e^3) \approx (0.22, -0.07)$. The figure is *not drawn to scale*.

33 $y \to -\infty$ as $x \to 0^+$; $y \to 0$ as $x \to +\infty$. Maximum at $(e^2, 2/e)$, inflection point at $(e^{8/3}, 8/3e^{4/3})$. The x-axis is a horizontal asymptote and the y-axis is a vertical asymptote. The only intercept is $(1, 0)$. The figure is *not drawn to scale*.

37 Midpoint estimate: Approximately 872.47. Trapezoidal estimate: Approximately 872.60. The true value of the integral is approximately 872.5174.

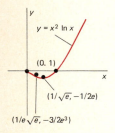

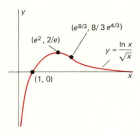

SECTION 6-3 (pages 306–307)

1 $-3e^{-3x}$ 3 $-xe^{-x^2/2}$ 5 $11x^{10}$ 7 $(x - 2)/e^x$ 9 $e^x e^{e^x}$ 11 $2e^x \cos 2e^x$ 13 $-((x + 1)\ln(x + 1) + 1)/(x + 1)e^x[\ln(x + 1)]^2$
15 $(\sqrt{x} - 1)e^{\sqrt{x}}/2x^{3/2}$ 17 $-e^{-t^2/2} + C$ 19 $2e^{\sqrt{x}} + C$ 21 $\ln(1 + e^x) + C$

23 Minimum and intercept at $(0, 0)$, local maximum at $(2, 4/e^2)$. Inflection points where $x = 2 \pm \sqrt{2}$. The x-axis is an asymptote. The figure is *not drawn to scale*.

25 Maximum at $(0, 1)$, the only intercept. The x-axis is the only asymptote. Inflection points where $x = \pm\sqrt{2}/2$

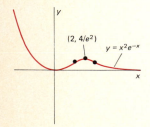

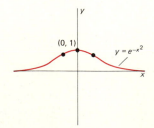

27 $(\pi/2)(e^2 - 1) \approx 10.0359$ **29** $(e^2 - 1)/2e \approx 1.1752$ **31** The solution is approximately 1.278464543. Note that if $f(x) = e^{-x} - x + 1$, then $f'(x) < 0$ for all x.   **33** $f'(x) = 0$ only for $x = 0$ and $x = n$; f is increasing on $(0, n)$ and decreasing on $(n, +\infty)$. Thus $x = n$ yields the absolute maximum value of f for $x \geq 0$. The x-axis is a horizontal asymptote, and there are inflection points above $x = n \pm \sqrt{n}$. The graph is *not drawn to scale*.   **37** $y = e^{-2x} + 4e^x$

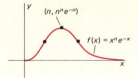

SECTION 6-4 (pages 311–312)

1 $17^x \ln 17$ **3** $-(10^{1/x}/x^2) \ln 10$ **5** $(2^{2x} \ln 2)(2^x \ln 2)$ **7** $(1/\ln 3)(x/(x^2 + 4))$ **9** $\ln 2/\ln 3$ or $\log_3 2$ **11** $1/x(\ln x)(\ln 2)$
13 $\exp(\log_{10} x)/(x \ln 10)$ **15** $3^{2x}/(2 \ln 3) + C$ **17** $2(2^{\sqrt{x}}/\ln 2) + C$ **19** $7^{x^3+1}/(3 \ln 7) + C$ **21** $(\ln x)^2/(2 \ln 2) + C$
23 $dy/dx = (x/(x^2 - 4) + 1/2(2x + 1))\sqrt{(x^2 - 4)}\sqrt{2x + 1}$ **25** $dy/dx = 2^x \ln 2$ **27** $dy/dx = (x^{\ln x})(2 \ln x)/x$ **29** $\frac{1}{3}(1/(x + 1) + 1/(x + 2) - 2x/(x^2 + 1) - 2x/(x^2 + 2))\sqrt[3]{(x + 1)(x + 2)/(x^2 + 1)(x^2 + 2)}$ **31** $dy/dx = (\ln x)^{\sqrt{x}}(\ln(\ln x)/2\sqrt{x} + 1/(\sqrt{x} \ln x))$
37 Note that $\ln(x^x/e^x) = x \ln(x/e)$

SECTION 6-5 (pages 319–320)

1 \$119.35; \$396.24   **3** Approximately 3.8685 hours   **5** The sample is about 686 years old.   **7** (a) 9.308%   (b) 9.381%
(c) 9.409%   (d) 9.416%   (e) 9.417%   **9** \$44.52   **11** After an additional 32.26 days   **13** Approximately 35 years   **15** About
4.2521×10^9 years old   **17** 2.40942 minutes   **19** (a) 20.486 inches; 9.604 inches   (b) 3.4524 miles, or about 18,230 feet

SECTION 6-6 (pages 327–328)

1 $y(x) = -1 + 2e^x$ **3** $y(x) = \frac{1}{2}(e^{2x} + 3)$ **5** 4,870,238   **9** About 46 days after the rumor starts   **13** $400/\ln 2$ feet, or about
577 feet   **17** \$1,587,337   **19** The maximum height is $1600(1 - 2 \ln 1.5) \approx 302.5$ feet, attained after $10 \ln 1.5 \approx 4.055$ seconds
aloft. The total time aloft is approximately 8.7422 seconds, and the impact speed is about 119.75 ft/sec   **21** After five minutes the
boat has coasted $400 \ln 31 \approx 1373.6$ feet, and is then moving at $40/31 \approx 1.29$ ft/sec

SECTION 6-7 (page 332)

1 $20 \ln 3 \approx 22$ years   **3** $50 \ln 1.8 \approx 29.4$ years   **5** (a) \$43196.96   (b) \$7634.94   **7** At \$21672.83 per year   **9** (a) \$1,334,109
(b) \$1,265,014

SECTION 6-8 (pages 337–338)

1 $x(t) = t^2 - t + 3$ **3** $x(t) = [5e^t - 1]^{1/2}$ **5** $x(t) = (18t - 3)/(6t + 1)$ **7** $x(t) = \sec t$ **9** (b) 256   **11** $x(t) = (3 + e^{2t})/(3 - e^{2t})$
13 $x(t) = 2(e^t - 1)/(2 - e^t)$ **15** After another 9.24 days

CHAPTER 6 MISCELLANEOUS PROBLEMS (pages 339–340)

1 $1/2x$ **3** $(1 - e^x)/(x - e^x)$ **5** $\ln 2$ **7** $(2 + 3x^2)e^{-1/x^2}$ **9** $(1 + \ln(\ln x))/x$ **11** $(1/x) 2^{\ln x} \ln 2$
13 $-(2/(x - 1)^2) \exp([x + 1]/[x - 1])$ **15** $\frac{3}{2}(1/(x - 1) + 8x/(3 - 4x^2))$ **17** $(\sin x \cos x) \exp(1 + \sin^2 x)^{1/2}/(1 + \sin^2 x)^{1/2}$
19 $\cot x + \ln 3$ **21** $dy/dx = x^{1/x}(1 - \ln x)/x^2$ **23** $[(1 + \ln(\ln x))/x](\ln x)^{\ln x}$ **25** $-\frac{1}{2} \ln|1 - 2x| + C$ **27** $\frac{1}{2} \ln|1 + 6x - x^2| + C$
29 $-\ln|2 + \cos x| + C$ **31** $(2/\ln 10)10^{\sqrt{x}} + C$ **33** $\frac{2}{3}(1 + e^x)^{3/2} + C$ **35** $6^x/\ln 6 + C$ **37** $x(t) = t^2 + 17$ **39** $x(t) = 1 + e^t$
41 $x(t) = \frac{1}{3}(2 + 7e^{3t})$ **43** $x(t) = \sqrt{2} \, e^{\sin t}$

45 Horizontal asymptote: The x-axis. Maximum at $(\frac{1}{2}, 1/\sqrt{2e})$, inflection point above $x = (1 + \sqrt{2})/2$—approximately $(1.21, 0.33)$. Minimum and intercept at $(0, 0)$, with a vertical tangent there as well. The graph is *not drawn to scale*.

47 Minimum at $(4, 2 - \ln 4)$, inflection point at $(16, 1.23)$ (ordinate approximate). The y-axis is a vertical asymptote. The graph continues to rise for large increasing x; there is no horizontal asymptote. The graph is *not drawn to scale*.

49 Inflection point at $(\frac{1}{2}, 1/e^2)$. The horizontal line $y = 1$ and the y-axis are asymptotes. The point $(0, 0)$ is *not* on the graph. As $x \to 0^+$, $y \to 0$; as $x \to 0^-$, $y \to +\infty$. As $|x| \to \infty$, $y \to 1$. The graph is *not drawn to scale*.

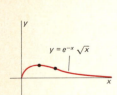

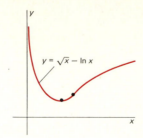

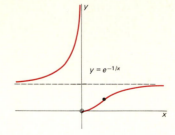

51 Sell immediately! **53** (b) The minimizing value is about 10.516. But since the batch size must be an integer, it turns out that 11 (rather than 10) minimizes $f(x)$. Thus the answer to part (c) is $977.85 **57** 20 weeks. **59** (a) $231.30 (b) $315.72 **61** About 22.567 hours after the power failure; that is, at about 9:34 P.M. on the following evening. **63** (a) 2000 gm (b) $\frac{5}{4}\ln(1.56/0.76) \approx$ 0.8989 sec. **65** (b) $k = \frac{1}{600}$; 240 sec (c) 120 ln 3 $\approx$ 131.8 sec

CHAPTER 7

SECTION 7-2 (pages 346–347)

1 $2\cos(2x + 3)$ **3** $\frac{2}{3}\sec^2(2x/3)$ **5** $-(1/x^2)\sec(1/x)\tan(1/x)$ **7** $12\sec^2 4x\tan^2 4x$ **9** $\cot x$ **11** $(2/x)\tan(\ln x)\sec^2(\ln x)$ **13** $-\csc x$ **15** $-3[\sin^2(\csc x)][\cos(\csc x)][\csc x \cot x]$ **17** $\sec x \tan x$ **19** $e^{\sin x}(1 + \sec x \tan x)$ **21** $2\tan(x/2) + C$ **23** $x/2 + \frac{1}{12}\sin 6x + C$ **25** $-\cos x + C$ **27** $\ln|1 + \sec x| + C$ **29** $\frac{1}{2}\sec x^2 + C$ **31** $\frac{1}{2}e^{\sin 2x} + C$ **33** $\sec x + C$ **35** $\ln|\tan e^x| + C$ **37** $\frac{1}{6}\sin^6 x + C$ **39** $\frac{1}{8}\sec^4 2x + C$ **41** $f'(x) = \sec^2 x$, which is always positive. $f''(x) = 2\sec^2 x \tan x$, which is zero at $(0, 0)$, $(\pi, 0)$, $(-\pi, 0)$, etc.; these are all x-intercepts and inflection points. There are vertical asymptotes at the odd multiples of $\pi/2$.
43 Minima at $(-3\pi/4, -\sqrt{2})$, $(5\pi/4, -\sqrt{2})$, etc.; maxima at $(\pi/4, \sqrt{2})$. **45** $\pi/4$ **47** $\ln 4$ **51** About 0.86033359; about 3.42561846 $(11\pi/4, \sqrt{2})$, etc.; x-intercepts at $3\pi/4$, $7\pi/4$, $-\pi/4$, etc.; y-intercept is $(0, 1)$.

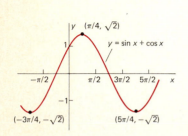

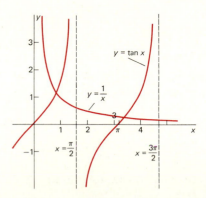

SECTION 7-3 (pages 353–354)

1 $e^x/\sqrt{1 - e^{2x}}$ **3** $-2/\sqrt{1 - x^2}$ **5** $-2/x\sqrt{x^4 - 1}$ **7** $-1/(1 + x^2)(\arctan x)^2$ **9** $1/x(1 + [\ln x]^2)$ **11** $2e^x/(1 + e^{2x})$ **13** $\cos(\arctan x)/(1 + x^2)$ **15** $[((1 + 9x^2)(\arctan 3x) - 3)/(1 + 9x^2)(\arctan 3x)^2]e^x$ **17** $(1 - 4x \arctan x)/(1 + x^2)^3$

19 $1/(a^2 + x^2)$ **21** $\pi/4$ **23** $\pi/3$ **25** $\pi/12$ **27** $\frac{1}{2}\arcsin 2x + C$

29 $\frac{1}{5}\operatorname{arcsec}|x/5| + C$ **31** $\arctan(e^x) + C$ **33** $\frac{1}{15}\operatorname{arcsec}|x^3/5| + C$

35 $\sin^{-1}(2x - 1) + C$ *or* $2\sin^{-1}(\sqrt{x}) + C$ **39** 8 feet   **43** π

45 $A = 1 - \pi/3,\, B = 1 + 2\pi/3$

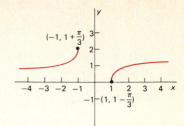

SECTION 7-4 (pages 361–362)

1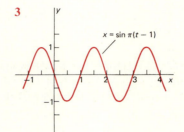

3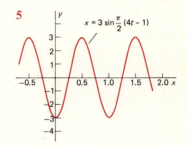

5

7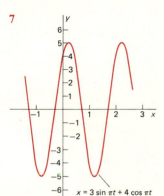

9 1   **11** $\frac{1}{2}\sqrt{5}$ **13** 5π **15** Period 2π, amplitude 5

19 Approximately 1.9795 miles, or about 10,452 feet

SECTION 7-5 (pages 366–367)

1 $3\sinh(3x - 2)$ **3** $2x\tanh(1/x) - \operatorname{sech}^2(1/x)$ **5** $-12\coth^2 4x\,\operatorname{csch}^2 4x$ **7** $-e^{\operatorname{csch} x}\operatorname{csch} x\coth x$ **9** $(\cosh x)\cos(\sinh x)$

11 $4x^3\cosh x^4$ **13** $-(1 + \operatorname{sech}^2 x)/(x + \tanh x)^2$ **15** $\frac{1}{2}\cosh x^2 + C$ **17** $x - \frac{1}{3}\tanh 3x + C$ **19** $\frac{1}{6}\sinh^3 2x + C$

21 $-\frac{1}{2}\operatorname{sech}^2 x + C$ **23** $-\frac{1}{2}\operatorname{csch}^2 x + C$ **25** $\ln(1 + \cosh x) + C$ **27** $\frac{1}{4}\tanh x + C$ **33** $\sinh a$

SECTION 7-6 (page 374)

1 $2/\sqrt{4x^2 + 1}$ **3** $1/2\sqrt{x}(1 - x)$ **5** $1/\sqrt{x^2 - 1}$ **7** $3\sqrt{\sinh^{-1} x}/2\sqrt{x^2 + 1}$ **9** $1/((1 - x^2)\tanh^{-1} x)$ **11** $\sinh^{-1}(x/3) + C$

13 $\frac{1}{4}\ln\frac{9}{5} \approx 0.14695$ **15** $-\frac{1}{2}\operatorname{sech}^{-1}|3x/2| + C$ **17** $\sinh^{-1}(e^x) + C$ **19** $-\operatorname{sech}^{-1}(e^x) + C$ **29** Time: Approximately 485 sec.

Impact speed: About 20.656 ft/sec   **35** (a) $T_0 \approx 259.16$ lb   (b) $T_{\max} \approx 278.70$ lb   (c) $H \approx 19.5$ ft

CHAPTER 7 MISCELLANEOUS PROBLEMS (pages 375–376)

1 $\cos\sqrt{x}/2\sqrt{x}$ **3** $\sec^2(\ln x)/x$ **5** $-10(\csc 2x)(\csc 2x + \cot 2x)^5$ **7** $6\tan 3x$ **9** $e^{\arctan x}/(1 + x^2)$ **11** $1/(2\sqrt{x}\sqrt{1 - x})$

13 $2x/(x^4 + 2x^2 + 2)$ **15** $e^x\sinh e^x + e^{2x}\cosh e^x$ **17** 0   **19** $x/|x|\sqrt{x^2 + 1}$ **21** $\frac{1}{2}\tan x^2 + C$ **23** $-\frac{1}{4}\csc^2 2x + C$

25 $1/x - \tan(1/x) + C$ **27** $\frac{1}{2}\ln|\sec(x^2 + 1)| + C$ **29** $\sin^{-1}(e^x) + C$ **31** $\frac{1}{2}\arcsin(2x/3) + C$ **33** $\frac{1}{3}\arctan(x^3) + C$

35 $\sec^{-1}|2x| + C$ **37** $\sec^{-1}(e^x) + C$ **39** $2\cosh\sqrt{x} + C$ **41** $\frac{1}{2}(\tan^{-1} x)^2 + C$ **43** $\frac{1}{2}\sinh^{-1}(2x/3) + C$ **45** $\pi^2/6$

47 The zeros shown are at (approximately) -0.197, 1.373, and 2.944. The maximum shown is at $(0.588, 13)$ and the minima at $(-0.983, -13)$ and $(2.159, -13)$ (abscissas approximate)

51 $x \approx 4.730041$ **55** Approximately 8 feet
57 (d) $\sqrt{Rg}$, or—in miles per hour—about 17,691

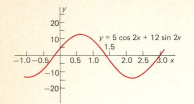

CHAPTER 8

SECTION 8-2 (pages 380–381)

1 $-\frac{1}{15}(2 - 3x)^5 + C$ **3** $\frac{1}{9}(2x^3 - 4)^{3/2} + C$ **5** $\frac{9}{8}(2x^2 + 3)^{2/3} + C$ **7** $-2 \csc\sqrt{y} + C$ **9** $\frac{1}{6}(1 + \sin\theta)^6 + C$ **11** $e^{-\cot x} + C$
13 $\frac{1}{11}(\ln t)^{11} + C$ **15** $\frac{1}{3}\sin^{-1}3t + C$ **17** $\frac{1}{2}\arctan(e^{2x}) + C$ **19** $\frac{3}{2}\arcsin(x^2) + C$ **21** $\frac{1}{15}\tan^5 3x + C$ **23** $\tan^{-1}(\sin\theta) + C$
25 $\frac{2}{5}(1 + \sqrt{x})^5 + C$ **27** $\ln|\arctan t| + C$ **29** $\sec^{-1}e^x + C$ **31** $\frac{2}{7}(x - 2)^{7/2} + \frac{8}{5}(x - 2)^{5/2} + \frac{8}{3}(x - 2)^{3/2} + C$ *or* $\frac{2}{105}(x - 2)^{3/2}(15x^2 + 24x + 32) + C$ **33** $\frac{1}{3}(2x + 3)^{1/2}(x - 3) + C$ **35** $\frac{3}{10}(x + 1)^{2/3}(2x - 3) + C$ **37** The substitution $u = e^x$
leads to an integral of the form of #44 in the Table of Integrals (see end papers); the answer is $\frac{1}{2}e^x\sqrt{9 + e^{2x}} + \frac{9}{2}\ln(e^x + \sqrt{9 + e^{2x}}) + C$
39 With $u = x^2$ and #47 in the Table of Integrals: $\frac{1}{2}(x^4 - 1)^{1/2} - \sec^{-1}x^2 + C$ **41** With $u = \ln x$ and #48 in the Table of
Integrals: $\frac{1}{8}([\ln x](2[\ln x]^2 + 1)([\ln x]^2 + 1)^{1/2} - \ln|[\ln x] + ([\ln x]^2 + 1)^{1/2}|) + C$ **45** $\sin^{-1}(x - 1) + C$

SECTION 8-3 (pages 384–385)

1 $\frac{1}{3}\cos^3 x - \cos x + C$ **3** $\frac{1}{3}\sin^3\theta - \frac{1}{5}\sin^5\theta + C$ **5** $\frac{1}{5}\sin^5 x - \frac{2}{3}\sin^3 x + \sin x + C$ **7** $\frac{2}{5}(\cos x)^{5/2} - 2(\cos x)^{1/2} + C$
9 $-\frac{1}{14}\cos^7 2z + \frac{1}{5}\cos^5 2z - \frac{1}{6}\cos^3 2z + C$ **11** $\frac{1}{4}(\sec 4x + \cos 4x) + C$ **13** $\frac{1}{3}\tan^3 t + \tan t + C$ **15** $-\frac{1}{4}\csc^2 2x - \frac{1}{2}\ln|\sin 2x| + C$
17 $\frac{1}{12}\tan^6 2x + C$ **19** $-\frac{1}{10}\cot^5 2t - \frac{1}{3}\cot^3 2t - \frac{1}{2}\cot 2t + C$ **21** $\frac{1}{4}\cos^4\theta - \frac{1}{2}\cos^2\theta + C$ **23** $\frac{2}{3}(\sec t)^{3/2} + 2(\sec t)^{-1/2} + C$
25 $\frac{1}{3}\sin^3\theta + C$ **27** $\frac{1}{4}\sec^4 x + C$ and $\frac{1}{4}\tan^4 x + \frac{1}{2}\tan^2 x + C$. The latter is equal to $\frac{1}{4}(\tan^4 x + 2\tan^2 x + 1) + \bar{C}$
29 $\frac{1}{4}\cos 2x - \frac{1}{16}\cos 8x + C$ **31** $\frac{1}{6}\sin 3x + \frac{1}{10}\sin 5x + C$

SECTION 8-4 (page 389)

1 $\frac{1}{2}xe^{2x} - \frac{1}{4}e^{2x} + C$ **3** $-t\cos t + \sin t + C$ **5** $(x/3)\sin 3x + \frac{1}{9}\cos 3x + C$ **7** $\frac{1}{4}x^4\ln x - \frac{1}{16}x^4 + C$ **9** $x\tan^{-1}x - \frac{1}{2}\ln(1 + x^2) + C$ **11** $\frac{2}{3}y^{3/2}\ln y - \frac{4}{9}y^{3/2} + C$ **13** $t(\ln t)^2 - 2t\ln t + 2t + C$ **15** $\frac{2}{3}x(x + 3)^{3/2} - \frac{4}{15}(x + 3)^{5/2} + C = \frac{2}{5}(x + 3)^{3/2}(x - 2) + C$ **17** $\frac{2}{9}x^3(x^3 + 1)^{3/2} - \frac{4}{45}(x^3 + 1)^{5/2} + C = \frac{2}{45}(x^3 + 1)^{3/2}(3x^3 - 2) + C$ **19** $-\frac{1}{2}(\csc\theta\cot\theta + \ln|\csc\theta + \cot\theta|) + C$ **21** $\frac{1}{3}x^3\tan^{-1}x + \frac{1}{6}x^2 + \frac{1}{6}\ln(1 + x^2) + C$ **23** $x\sec^{-1}\sqrt{x} - \sqrt{x - 1} + C$ **25** $(x + 1)\tan^{-1}\sqrt{x} - \sqrt{x} + C$
27 $\pi(e - 2) \approx 2.25655$ **29** $\bar{x} = (e^2 - 1)/4(e - 2) \approx 2.22373$ **37** $6 - 2e \approx 0.563436$ **39** $6 - 2e \approx 0.563436$

SECTION 8-5 (page 394)

1 $-(1/x)\sqrt{1 - x^2} - \sin^{-1}x + C$ **3** $\ln|x + \sqrt{x^2 - 1}| - (1/x)\sqrt{x^2 - 1} + C$ **5** $\frac{1}{80}[(9 + 4x^2)^{5/2} - 15(9 + 4x^2)^{3/2}] + C =$
$((2x^2 - 3)/40)(9 + 4x^2)^{3/2} + C$ **7** $\sqrt{1 - 4x^2} - \ln|(1 + \sqrt{1 - 4x^2})/2x| + C$ **9** $\frac{1}{2}\ln|2x + \sqrt{9 + 4x^2}| + C$ **11** $\frac{25}{2}\sin^{-1}(x/5) -$
$(x/2)\sqrt{25 - x^2} + C$ **13** $(x/2)\sqrt{x^2 + 1} - \frac{1}{2}\ln|x + \sqrt{1 + x^2}| + C$ **15** $\frac{1}{18}x\sqrt{4 + 9x^2} - \frac{2}{27}\ln|3x + \sqrt{4 + 9x^2}| + C$
17 $x/\sqrt{1 + x^2} + C$ **19** $\frac{1}{256}(16x/(4 - x^2)^2 + 6x/(4 - x^2) + 3\ln|(2 + x)/\sqrt{4 - x^2}|) + C = (20x - 3x^3)/128(4 - x^2)^2 +$
$\frac{3}{256}\ln|(2 + x)/(2 - x)| + C$ **21** $\sinh^{-1}(x/5) + C$ **23** $\cosh^{-1}(x/2) - (1/x)\sqrt{x^2 - 4} + C$ **25** $\frac{1}{8}(x(1 + 2x^2)\sqrt{1 + x^2} - \sinh^{-1}x) + C$
27 $(\pi/32)(18\sqrt{5} - \ln(2 + \sqrt{5})) \approx 3.8097$ **29** $\sqrt{5} - \sqrt{2} + \ln(2(1 + \sqrt{2})/(1 + \sqrt{5})) \approx 1.222016$
33 $2\pi(\sqrt{2} + \ln(1 + \sqrt{2})) \approx 14.4236$

SECTION 8-6 (page 398)

1 $\frac{1}{3}\tan^{-1}((x + 2)/3) + C$ **3** $\frac{1}{4}\ln|(1 + x)/(3 - x)| + C$ **5** $\ln(x^2 + 2x + 2) - 7\tan^{-1}(x + 1) + C$ **7** $\frac{2}{9}\sin^{-1}(x - \frac{2}{3}) -$
$\frac{1}{9}\sqrt{5 + 12x - 9x^2} + C$ **9** $\frac{75}{4}\sin^{-1}(\frac{2}{5}[x - 2]) + \frac{3}{2}(x - 2)\sqrt{9 + 16x - 4x^2} + \frac{1}{6}(9 + 16x - 4x^2)^{3/2} + C$
11 $(7x - 12)/9\sqrt{6x - x^2} + C$ **13** $-1/4(4x^2 + 12x + 13) + C$ **15** $\frac{3}{2}\ln(x^2 + x + 1) - (5\sqrt{3}/3)\tan^{-1}((\sqrt{3}/3)[2x + 1]) + C$
17 $\frac{1}{32}\ln|(x + 2)/(x - 2)| - \frac{1}{8}(x/(x^2 - 4)) + C$

n: $2x + y = 13$

17 $(x - 6)^2 + (y - 11)^2 = 18$

19 $(x/5)^2 + (y/3)^2 = 1$

$+ 13$

$(7, 4)$

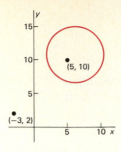

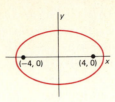

$4\sqrt{3} = (4 - 2\sqrt{3})(x - 2 + \sqrt{3}), y - 7 - 4\sqrt{3} = (4 + 2\sqrt{3})(x - 2 - \sqrt{3})$ **23** $y - 1 = 4(x - 4)$ and $y + 1 =$
5 $h = p(e^2 + 1)/(1 - e^2), a = \sqrt{|h^2 - p^2|}, b = a\sqrt{e^2 - 1}$

2 (pages 425–426)

3 $(x - 2)^2 = -8(y - 3)$

5 $(y - 3)^2 = -8(x - 2)$

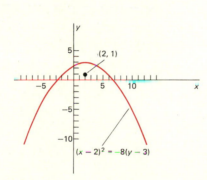

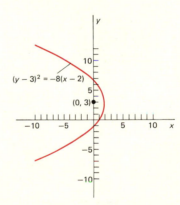

$+ \frac{3}{2})$

9 $x^2 = 4(y + 1)$

11 $y^2 = 12x$, vertex: $(0, 0)$, axis: The x-axis

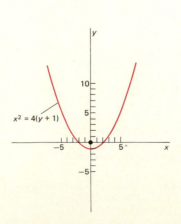

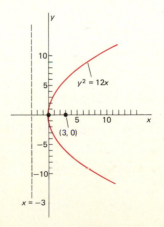

Answers to Odd-Numbered Problems

SECTION 8-7 (pages 405–406)

1 $x - 2\ln|x + 1| + C$ **3** $x + 1/(x + 1) + C$ **5** $\frac{1}{4}\ln|(x - 2)/(x + 2)| + C$ **7** $\frac{3}{2}\ln|2x - 1|$
9 $\ln|x| + 2/(x + 1) + C$ **11** $\frac{3}{2}\ln|x^2 - 4| + \frac{1}{2}\ln|x^2 - 1| + C$ **13** $\ln|x + 2| + 4/(x + 2) - 2$
15 $\frac{1}{2}\ln(x^2/(x^2 + 1)) + C$ **17** $\frac{1}{2}\ln(x^2/(x^2 + 4)) + \frac{1}{2}\arctan(x/2) + C$ **19** $-\frac{1}{2}\ln|x + 1| + \frac{1}{4}\ln$
21 $\tan^{-1}(x/2) - (3\sqrt{2}/2)\tan^{-1}(x\sqrt{2}) + C$ **23** $-\frac{1}{8}[\tan^{-1}((x + 1)/2) + (2x + 14)/(x^2 + 2x +$
$\frac{1}{8}((3x^2 + 4x + 1)/(x^2 + 2x + 5) - \tan^{-1}((x + 1)/2)) + C$, depending on the technique used t
25 $x + \frac{1}{2}\ln|x - 1| - \frac{5}{2}/(x - 1) + \frac{3}{4}\ln(x^2 + 1) + 2\tan^{-1}x + C$ **27** $(1 - 2e^{2t})/4(e^{2t} - 1)^2 +$
29 $\frac{1}{4}(\ln|3 + 2\ln t| + 1/(3 + 2\ln t)) + C$ **31** $\frac{1}{2}x^2 + \ln|x - 1| + \frac{1}{2}\ln|x^2 + x + 1| + (\sqrt{3}/3)$
33 $\frac{1}{4}\ln(x^4 + x^2 + 1) - (\sqrt{3}/2)\arctan(\sqrt{3}/(2x^2 + 1)) + C$

SECTION 8-8 (pages 410–411)

1 $\frac{1}{81}(\frac{2}{9}(3x - 2)^{9/2} + \frac{12}{7}(3x - 2)^{7/2} + \frac{24}{5}(3x - 2)^{5/2} + \frac{16}{3}(3x - 2)^{3/2}) + C$ **3** $2\sqrt{x} - 2\ln(1$
$3(x^2 - 3)/4(x^2 - 1)^{1/3} + C$ **7** $[x = u^4]\ x - \frac{4}{5}x^{5/4} + C$ **9** $[u = 1 + x^3]\frac{2}{9}(1 + x^3)^{3/2} - \frac{2}{}$
$3x^{1/3} - 3\tan^{-1}x^{1/3} + C$ **13** $[u = \sqrt{x + 4}]\ 2\sqrt{x + 4} - 2\ln(1 + \sqrt{x + 4}) + C$ **15** $(\sin\theta$
17 $[u = \tan(\theta/2)]\ \sqrt{2}\ln|(\sqrt{2} - 1 + \tan(\theta/2))/\sqrt{1 + 2\tan(\theta/2) - \tan^2(\theta/2)}| + C$ **19** $-\ln|2$
25 $\sqrt{2} + \ln(1 + \sqrt{2}) \approx 2.295587$

CHAPTER 8 MISCELLANEOUS PROBLEMS (pages 412–414)

Note: Different techniques of integration may produce answers that appear to differ from th
only be a constant.

1 $2\arctan\sqrt{x} + C$ **3** $\ln|\sec x| + C$ **5** $\frac{1}{2}\sec^2\theta + C$ **7** $x\tan x - \frac{1}{2}x^2 + \ln|\cos x| + C$ **9**
11 $(x/2)\sqrt{25 + x^2} - \frac{25}{2}\ln|x + \sqrt{25 + x^2}| + C$ **13** $(2\sqrt{3}/3)\tan^{-1}((\sqrt{3}/3)[2x - 1]) + C$ **15** (103)
$\frac{5}{6}\ln(9x^2 - 12x + 33) + C$ **17** $\frac{2}{3}\arctan\frac{1}{3}\tan(\theta/2) + C$ **19** $\sin^{-1}(\frac{1}{2}\sin x) + C$ **21** $-\ln|$
23 $(1 + x)\ln(1 + x) - x + C$ **25** $(x/2)\sqrt{x^2 + 9} + \frac{9}{2}\ln|x + \sqrt{x^2 + 9}| + C$ **27** $\frac{1}{2}(x - 1$
29 $\frac{1}{3}x^3 + 2x - \sqrt{2}\ln|(x + \sqrt{2})/(x - \sqrt{2})| + C$ **31** $\frac{1}{2}((x^2 + x)/(x^2 + 2x + 2) - \tan^{-1}($
$\sin 2\theta/2(1 + \cos 2\theta) + C$ **35** $\frac{1}{5}\sec^5x - \frac{1}{3}\sec^3x + C$ **37** $(x^2/8)[4(\ln x)^3 - 6(\ln x)^2 + 6($
39 $\frac{1}{2}(e^x\sqrt{1 + e^{2x}} + \ln[e^x + \sqrt{1 + e^{2x}}]) + C$ **41** $\frac{1}{54}\sec^{-1}(x/3) + \sqrt{x^2 - 9}/18x^2 + C$ **4**
45 $\frac{1}{2}(\sec x\tan x - \ln|\sec x + \tan x|) + C$ **47** $\ln|x + 1| - (2/3x^3) + C$ **49** $\frac{1}{9}(\ln|(x^2 + x$
$2\sqrt{3}\tan^{-1}((\sqrt{3}/3)[2x + 1]) + (x - 1)/(x^2 + x + 1)) + C$ **51** $\frac{1}{3}\ln|(3\sin\theta + 1 - \cos\theta)$
53 $\frac{1}{3}(\sin^{-1}x)^3 + C$ **55** $\frac{1}{2}\sec^2z + \ln|\cos z| + C$ **57** $\frac{1}{2}\arctan(e^{x^2}) + C$ **59** $-((x^2 + 1)/2$
$\ln|(1 + \sqrt{1 - x^2})/x| + C$ **63** $\frac{1}{8}(\sin^{-1}x + x\sqrt{1 - x^2}(2x^2 - 1)) + C$ **65** $\frac{1}{4}(\ln|2x + 1| +$
67 $\frac{1}{2}\ln|e^{2x} - 1| + C$ **69** $2\ln|x + 1| + 3/(x + 1) - 5/(3(x + 1)^3) + C$ **71** $\frac{1}{2}\ln(x^2 + 1)$
73 $\frac{1}{45}(x^3 - 1)^{3/2}(6x^3 + 4) + C$ **75** $\frac{2}{3}(1 + \sin x)^{3/2} + C$ **77** $\frac{1}{2}\ln|\sec x + \tan x| + C$ **7**
81 $-2x + \sqrt{3}\tan^{-1}((\sqrt{3}/3)[2x + 1]) + ((2x + 1)/2)\ln|x^2 + x + 1| + C$ **83** $-(1/x)\tan$
85 $\frac{1}{2}\ln(x^2 + 1) + 1/2(x^2 + 1) + C$ **87** $(x - 6)/2\sqrt{x^2 + 4} + C$ **89** $\frac{1}{3}(1 + \sin^2x)^{3/2} +$
91 $(e^x/2)(x\sin x - x\cos x + \cos x) + C$ **93** $-\frac{1}{2}(x - 1)^{-2}\tan^{-1}x + \frac{1}{8}[(x^2 + 1)(x - 1)^{-2}$
95 $\frac{1}{9}(11\sin^{-1}((3x - 1)/2) - 2\sqrt{3 + 6x - 9x^2}) + C$ **97** $\frac{1}{2}\cos^2\theta + \cos\theta + C$ **99** $x\sec^{-1}$
101 $(\pi/4)(e^2 - e^{-2} + 4)$ **103** (a) $A_b = \pi(\sqrt{2} - e^{-b}\sqrt{1 + e^{-2b}} + \ln|(1 + \sqrt{2})/(e^{-b} +$
(b) $\pi[\sqrt{2} + \ln(1 + \sqrt{2})] \approx 7.2118$ **105** $(\pi\sqrt{2}/2)[2\sqrt{14} - \sqrt{2} + \ln((1 + \sqrt{2})/(2\sqrt{2} +$
111 The value of the integral is $\frac{1}{630}$. **113** $((5\sqrt{3} - 3)/2)\sqrt{2} + \frac{1}{2}\ln((1 + \sqrt{2})/(\sqrt{3} + \sqrt{2}$
115 The substitution is $u = e^x$. (a) $(2\sqrt{3}/3)\tan^{-1}((\sqrt{3}/3)[1 + 2e^x]) + C$

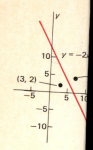

21 $y - 7 +$
$4(x + 4)$ **2**

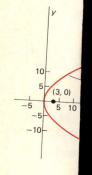

SECTION 9-

1 $y^2 = 12x$

7 $x^2 = -6(y$

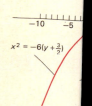

$x^2 = -6(y + \frac{3}{2})$

SECTION 9-1 (pages 420–421)

1 $x + 2y + 3 = 0$ **3** $4y + 25 = 3x$ **5** $x + y = 1$ **7** Center: $(-1, 0)$, radius: $\sqrt{5}$ **9**
11 $(x + 1)^2 + (y + 2)^2 = 34$ **13** $(x - 6)^2 + (y - 6)^2 = \frac{4}{5}$

13 $y^2 = -6x$, vertex: $(0, 0)$, axis: The x-axis

15 $x^2 - 4x - 4y = 0$, vertex: $(2, -1)$, axis: $x = 2$

17 $4x^2 + 4x + 4y + 13 = 0$, vertex: $(-\frac{1}{2}, -3)$, axis: $x = -\frac{1}{2}$

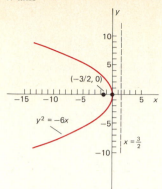

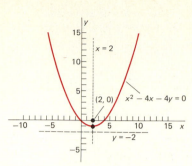

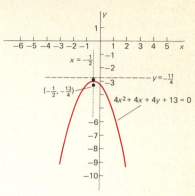

23 About 0.693 days, or about 16 hr 38 min   **27** $\alpha = \pi/12$ or $5\pi/12$

SECTION 9-3 (pages 430–431)

1 $(x/4)^2 + (y/5)^2 = 1$ **3** $(x/15)^2 + (y/17)^2 = 1$ **5** $x^2/16 + y^2/7 = 1$ **7** $x^2/100 + y^2/75 = 1$ **9** $x^2/16 + y^2/12 = 1$
11 $(x - 2)^2/16 + (y - 3)^2/4 = 1$ **13** $(x - 1)^2/25 + (y - 1)^2/16 = 1$ **15** $(x - 1)^2/81 + (y - 2)^2/72 = 1$
17 Center: $(0, 0)$, foci: $(\pm 2\sqrt{5}, 0)$, major axis length: 12, minor axis length: 8

19 Center: $(0, 4)$, foci: $(0, 4 \pm \sqrt{5})$, major axis length: 6, minor axis length: 4

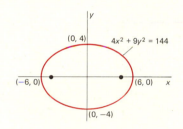

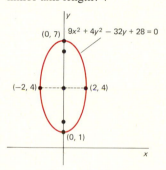

21 About 3466.54 A.U., or 3.22×10^{11} miles, or 20 light-days   **27** $(x - 1)^2/4 + y^2/(\frac{16}{3}) = 1$

SECTION 9-4 (pages 437–438)

1 $x^2/1 - y^2/15 = 1$ **3** $x^2/16 - y^2/9 = 1$ **5** $y^2/25 - x^2/25 = 1$ **7** $y^2/9 - x^2/27 = 1$ **9** $x^2/4 - y^2/12 = 1$
11 $(x - 2)^2/9 - (y - 2)^2/27 = 1$ **13** $(y + 2)^2/9 - (x - 1)^2/4 = 1$
15 Center: $(1, 2)$, foci: $(1 \pm \sqrt{2}, 2)$, asymptotes: $y - 2 = \pm(x - 1)$

17 Center: $(0, 3)$, foci: $(0, 3 \pm 2\sqrt{3})$, asymptotes: $y = 3 \pm x\sqrt{3}$

19 Center: $(-1, 1)$, foci: $(-1 \pm \sqrt{13}, 1)$, asymptotes: $y = \frac{3}{2}x + \frac{5}{2}$ and $y = -\frac{3}{2}x - \frac{1}{2}$

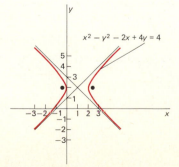

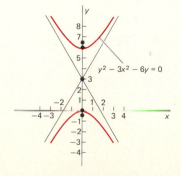

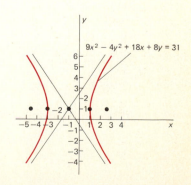

21 There are no points on the graph if $c > 15$. **25** $16x^2 + 50xy + 16y^2 = 369$ **27** The plane is about 16.42 miles north of B and 8.66 miles west of B; alternatively, it is about 18.56 miles from B at a bearing of $27°\,48'$ west of north.

SECTION 9-5 (page 442)

1 $B^2 - 4AC = 0$: A parabola (or a degenerate case). Rotation angle: $\alpha = \pi/4$. Transformed equation: $(x')^2 = 1$

3 $B^2 - 4AC = 5 > 0$: Hyperbola. Rotation angle: $\alpha = \pi/4$. Transformed equation: $(x')^2/2 - (y')^2/10 = 1$

5 $B^2 - 4AC = 24 > 0$: Hyperbola. Rotation angle: $\alpha = \tan^{-1}(\tfrac{1}{2}) \approx 26°\,33'\,54''$. Transformed equation: $(x')^2/9 - (y')^2/6 = 1$

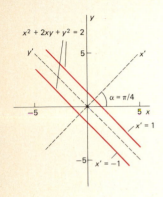

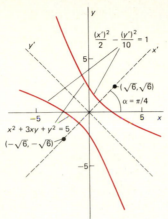

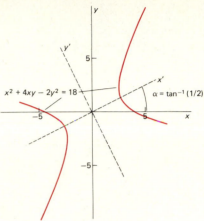

7 $B^2 - 4AC = -3 < 0$: Ellipse. Rotation angle: $\alpha = \pi/4$. Transformed equation: $(x'-1)^2/4 + (y'+1)^2/(\tfrac{4}{3}) = 1$

9 $B^2 - 4AC = 5000 > 0$: Hyperbola. Rotation angle: $\alpha = \tan^{-1}(\tfrac{4}{3}) \approx 0.9273$. Transformed equation: $(y'-1)^2/(\tfrac{1}{2}) - (x'-2)^2/1 = 1$

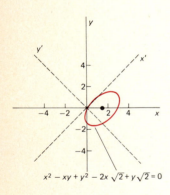

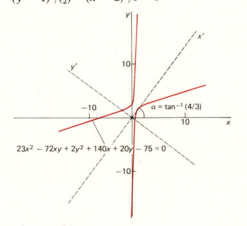

11 $B^2 - 4AC = 334084 > 0$: Hyperbola. Rotation angle: $\alpha = \tan^{-1}(\tfrac{8}{15}) \approx 0.49$ (a little over $28°$). Transformed equation: $(x'-1)^2/1 - (y')^2/1 = 1$

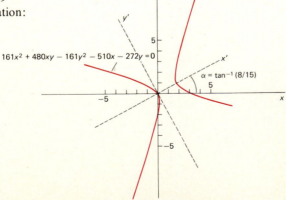

17 $\alpha = \frac{1}{2}\cot^{-1}(-B/2)$; rectangular hyperbola   **19** $\alpha \approx \tan^{-1}(0.7656)$; closest points $(\pm 0.1238, \pm 0.0947)$, farthest points $(\pm 0.3610, \mp 0.4715)$

SECTION 9-6 (pages 447–448)

1 (a) $(\sqrt{2}/2, \sqrt{2}/2)$ (b) $(1, -\sqrt{3})$ (c) $(1/2, -\sqrt{3}/2)$
(d) $(0, -3)$ (e) $(\sqrt{2}, \sqrt{2})$ (f) $(\sqrt{3}, -1)$ (g) $(-\sqrt{3}, 1)$

3 $r\cos\theta = 4$ **5** $\theta = \tan^{-1}(\frac{1}{3})$ **7** $r^2\cos\theta\sin\theta = 1$ **9** $r = \tan\theta\sec\theta$ **11** $x^2 + y^2 = 9$ **13** $x^2 + 5x + y^2 = 0$
15 $(x^2 + y^2)^3 = 4y^4$ **17** $x = 3$

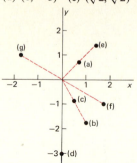

19 Symmetric about the x-axis

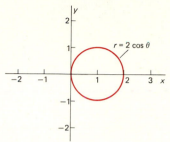

21 Symmetric about the x-axis

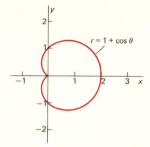

23 Symmetric about the x-axis

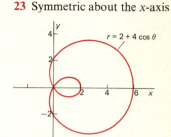

25 Symmetric about the origin

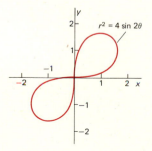

27 Symmetric about both axes and about the origin

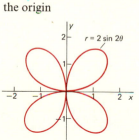

29 Symmetric about the x-axis

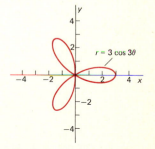

31 Symmetric about the y-axis

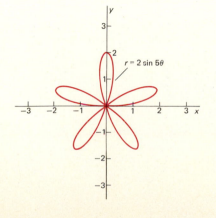

33 No points of intersection   **35** $(0, 0)$, $(\frac{1}{2}, \pi/6)$, $(\frac{1}{2}, 5\pi/6)$, $(1, \pi/2)$ **37** Four points: The pole, the point $r = 2$, $\theta = \pi$, and the two points $r = 2(\sqrt{2} - 1)$, $\theta = \pm\cos^{-1}(3 - 2\sqrt{2})$ **39** (a) $r\cos(\theta - \alpha) = p$

SECTION 9-7 (pages 451–452)

1 Parabola, directrix $x = 6$, vertex $(3, 0)$

3 Parabola, directrix $y = -3$, vertex $(0, -\frac{3}{2})$

5 Ellipse, eccentricity $\frac{1}{2}$, directrix $y = -6$

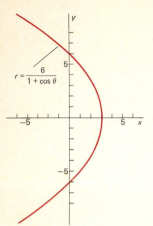

$r = \dfrac{6}{1 + \cos \theta}$

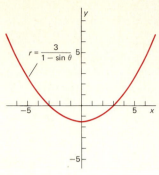

$r = \dfrac{3}{1 - \sin \theta}$

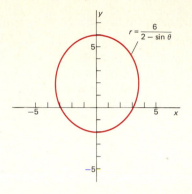

$r = \dfrac{6}{2 - \sin \theta}$

7 Hyperbola, eccentricity $\frac{3}{2}$

9 Hyperbola, eccentricity $\sqrt{2}$, rotated $\pi/4$ from the "standard position"

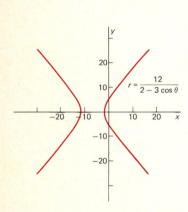

$r = \dfrac{12}{2 - 3 \cos \theta}$

$r = \dfrac{2}{1 - \sin \theta - \cos \theta}$

11 $r = 3/(1 + \cos \theta)$ **13** $r = 2/(3 + \sin \theta)$ **15** $r = (3\sqrt{2}/2)/(1 + \cos(\theta - \pi/4))$ **17** $3x^2 + 32x + 64 = y^2$

19 $9x^2 + 5y^2 = 24y + 36$

21 $A = \sqrt{2}$, $\alpha = \pi/4$. Equation: $r = 2/(1 - \cos(\theta - \pi/4))$.   **25** 2000 miles

Parabola, focus at the pole, rotated $45°$ from such a parabola with directrix $x' = -2$

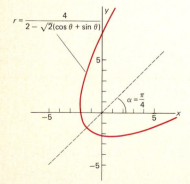

$r = \dfrac{4}{2 - \sqrt{2}(\cos \theta + \sin \theta)}$

$\alpha = \dfrac{\pi}{4}$

1 π **3** $\frac{3}{2}\pi$ **5** $\frac{9}{2}\pi$ **7** $\pi/2$ (one of *four* loops)   **9** $\pi/4$ (one of *eight* loops)   **11** 2 (one of *two* loops)   **13** $(2\pi + 3\sqrt{3})/6$
15 $(5\pi - 6\sqrt{3})/24$ **17** $(39\sqrt{3} - 10\pi)/6 \approx 6.02234$ **19** $(2 - \sqrt{2})/2$ **21** $\pi/2$

CHAPTER 9 MISCELLANEOUS PROBLEMS (pages 455–456)

1 Circle, center $(1, -1)$, radius 2   **3** Circle, center $(3, -1)$, radius 1   **5** Parabola, vertex $(4, -2)$, opening downward
7 Ellipse, center $(2, 0)$, major axis 6, minor axis 4   **9** Hyperbola, center $(-1, 1)$, vertical axis, foci at $(-1, 1 \pm \sqrt{3})$
11 There are no points on the graph   **13** Hyperbola, axis inclined at $22.5°$ $(\pi/8)$ from horizontal   **15** Ellipse, major axis $2\sqrt{2}$, minor
axis 1, rotated through angle $\alpha = \pi/4$, center at the origin   **17** Parabola, vertex $(0, 0)$, opening to the "northeast," axis at angle $\alpha =$
$\tan^{-1}(\frac{3}{4})$ from the horizontal   **19** Circle, center $(1, 0)$, radius 1   **21** Straight line $y = x + 1$ **23** Horizontal line $y = 3$
25 Two ovals tangent to each other and to the y-axis at $(0, 0)$ **27** Apple-shaped curve, symmetric about y-axis

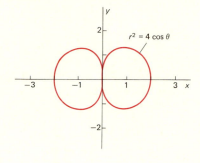

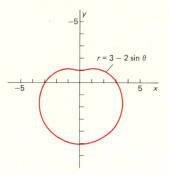

29 Ellipse, one focus at $(0, 0)$, directrix $x = 4$, eccentricity $e = \frac{1}{2}$ **31** $(\pi - 2)/2$ **33** $(39\sqrt{3} - 10\pi)/6 \approx 6.02234$ **35** 2   **37** $5\pi/4$
39 $r = 2p\cos(\theta - \alpha)$ **41** If $a > b$, the maximum is $2a$ and the minimum is $2b$.   **43** $b^2 y = 4hx(b - x)$ or $r = b\sec\theta -$
$(b^2/4h)\sec\theta\tan\theta$ **45** Suggestion: Let θ be the angle that QR makes with the x-axis.   **49** Suggestion: $4x^2 + 12x + 9 = x^2 + y^2$,
so $(2x + 3)^2 = r^2$. The curve is a hyperbola with one focus at the origin, directrix $x = -\frac{3}{2}$, and eccentricity $e = 2$.   **51** $\frac{3}{2}$
53 (a) $r = 31.11/(1 + 0.83\cos\theta)$ (b) About 18.7 days

CHAPTER 10

1 $y = 2x - 3$ **3** $y^2 = x^3$ **5** $y = 2x^2 - 5x + 2$ **7** $y = 4x^2, x > 0$ **9** $(x/5)^2 + (y/3)^2 = 1$ **11** $x^2 - y^2 = 1$ **13** $y - 5 =$
$\frac{9}{4}(x - 3)$ **15** $y = -(\pi/2)(x - \pi/2)$ **17** $x + y = 3$ **19** $\psi = \pi/6$ [constant]   **21** $\psi = \pi/2$ **25** $x = p/m^2, y = 2p/m$

1 $\frac{22}{5}$ **3** $\frac{4}{3}$ **5** $(1 + e^\pi)/2 \approx 12.0703$ **7** $358\pi/35 \approx 32.13400$ **9** $16\pi/15 \approx 3.35103$ **11** $\frac{74}{3}$ **13** $\pi\sqrt{2}/4 \approx 1.11072$
15 $(e^{2\pi} - 1)\sqrt{5} \approx 1195.1597$ **17** $(8\pi/3)(5\sqrt{5} - 2\sqrt{2}) \approx 69.96882$ **19** $(2\pi/27)(13\sqrt{13} - 8) \approx 9.04596$ **21** $16\pi^2$ **23** $5\pi^2 a^3$
25 (a) $A = \pi ab$ (b) $V = \frac{4}{3}\pi ab^2$ **27** $\pi\sqrt{1 + 4\pi^2} + \frac{1}{2}\ln(2\pi + \sqrt{1 + 4\pi^2}) \approx 21.25629$ **29** $\frac{3}{8}\pi a^2$ **31** $\frac{12}{5}\pi a^2$

1 $\sqrt{5}; 2\sqrt{13}; 4\sqrt{2}; \langle -2, 0 \rangle; \langle 9, -10 \rangle$; no   **3** $2\sqrt{2}; 10; \sqrt{5}; \langle -5, -6 \rangle; \langle 0, 2 \rangle$; no   **5** $\sqrt{10}; 2\sqrt{29}; \sqrt{65}; 3\mathbf{i} - 2\mathbf{j}; -\mathbf{i} + 19\mathbf{j}$; no
7 $4; 14; \sqrt{65}; 4\mathbf{i} - 7\mathbf{j}; 12\mathbf{i} + 14\mathbf{j}$; yes   **9** $\mathbf{u} = \langle -\frac{3}{5}, -\frac{4}{5} \rangle, \mathbf{v} = \langle \frac{3}{5}, \frac{4}{5} \rangle$ **11** $\mathbf{u} = \frac{8}{17}\mathbf{i} + \frac{15}{17}\mathbf{j}, \mathbf{v} = -\frac{8}{17}\mathbf{i} - \frac{15}{17}\mathbf{j}$ **13** $-4\mathbf{j}$
15 $8\mathbf{i} - 14\mathbf{j}$ **17** Yes   **19** No   **21** (a) $15\mathbf{i} - 21\mathbf{j}$ (b) $\frac{5}{3}\mathbf{i} - \frac{7}{3}\mathbf{j}$ **23** (a) $(35\sqrt{58}/58)\mathbf{i} - (15\sqrt{58}/58)\mathbf{j}$ (b) $-(40\sqrt{89}/89)\mathbf{i} -$
$(25\sqrt{89}/89)\mathbf{j}$ **25** $c = 0$ **33** $\mathbf{v}_a = (500 + 25\sqrt{2})\mathbf{i} + 25\sqrt{2}\mathbf{j}$ **35** $\mathbf{v}_a = -225\sqrt{2}\mathbf{i} + 275\sqrt{2}\mathbf{j}$

1 $0; 0$ **3** $2\mathbf{i} - \mathbf{j}; 4\mathbf{i} + \mathbf{j}$ **5** $6\pi\mathbf{i}; 12\pi^2\mathbf{j}$ **7** $\mathbf{j}; \mathbf{i}$ **9** $((2 - \sqrt{2})/2)\mathbf{i} + \sqrt{2}\mathbf{j}$ **11** $\frac{484}{15}\mathbf{i}$ **13** 11   **15** 0   **17** $\mathbf{i} + 2t\mathbf{j}; t\mathbf{i} + t^2\mathbf{j}$
19 $\frac{1}{2}t^2\mathbf{i} + \frac{1}{3}t^3\mathbf{j}; (1 + \frac{1}{6}t^3)\mathbf{i} + \frac{1}{12}t^4\mathbf{j}$ **25** Begin with $(d/dt)(\mathbf{v} \cdot \mathbf{v}) = 0$.   **27** A repulsive force acting directly away from the origin,
with magnitude proportional to distance from the origin

1 v_0 411.047 ft/sec   **5** (a) $y_m = 100$ ft, $R = 400\sqrt{3}$ ft   (b) $y_m = 200$, $R = 800$ (c) $y_m = 300$, $R = 400\sqrt{3}$ **7** $v_0 = 1056$ ft/sec
9 $v_0 = 4\sqrt{4181} \approx 258.64$ ft/sec; $\alpha = \tan^{-1}(0.82)$, approximately $39°21'6''$ **11** $M_S \approx 1.98885 \times 10^{30}$ kg; the ratio is about
333252 to 1   **13** Approximately 238,020 miles   **15** $\alpha = \tan^{-1}(55/108) \approx 26°59'16''$ **17** $20\sqrt{5}$ ft/sec (about 30.5 mph)

1 0   **3** 1   **5** $40\sqrt{2}/41\sqrt{41} \approx 0.2155$ **7** At $(-\frac{1}{2}\ln 2, \frac{1}{2}\sqrt{2})$ **9** $\kappa_{\max} = 5/9$ at $(\pm 5, 0)$; $\kappa_{\min} = 3/25$ at $(0, \pm 3)$.
11 $\mathbf{T} = (\sqrt{10}/10)\mathbf{i} + (3\sqrt{10}/10)\mathbf{j}$; $\mathbf{N} = (3\sqrt{10}/10)\mathbf{i} - (\sqrt{10}/10)\mathbf{j}$ **13** $\mathbf{T} = (\sqrt{57}/57)(3\mathbf{i} - 4\sqrt{3}\mathbf{j})$; $\mathbf{N} = -(\sqrt{57}/57)(4\sqrt{3}\mathbf{i} + 3\mathbf{j})$
15 $\mathbf{T} = -(\sqrt{2}/2)\mathbf{i} - (\sqrt{2}/2)\mathbf{j}$; $\mathbf{N} = (\sqrt{2}/2)\mathbf{i} - (\sqrt{2}/2)\mathbf{j}$ **17** $a_T = 18t/\sqrt{9t^2 + 1}$, $a_N = 6/\sqrt{9t^2 + 1}$ **19** $a_T = t/(1 + t^2)^{1/2}$,
$a_N = (2 + t^2)/(1 + t^2)^{1/2}$ **21** $\kappa = 1/a$ **23** $x^2 + (y - \frac{1}{2})^2 = \frac{1}{4}$ **25** $(x - \frac{3}{2})^2 + (y - \frac{3}{2})^2 = 2$ **27** Begin with $(d/dt)(\mathbf{v} \cdot \mathbf{v})$
29 $\kappa = 1/|t|$ **31** $y = 3x^5 - 8x^4 + 6x^3$

1 $\mathbf{v} = a\mathbf{u}_\theta$, $\mathbf{a} = -a\mathbf{u}_r$ **3** $\mathbf{v} = \mathbf{u}_r + t\mathbf{u}_\theta$, $\mathbf{a} = -t\mathbf{u}_r + 2\mathbf{u}_\theta$ **5** $\mathbf{v} = (12 \cos 4t)\mathbf{u}_r + (6 \sin 4t)\mathbf{u}_\theta$, $\mathbf{a} = (-60 \sin 4t)\mathbf{u}_r + (48 \cos 4t)\mathbf{u}_\theta$
9 36.65 mi/sec; 24.13 mi/sec   **11** 0.672 mi/sec; 0.602 mi/sec   **13** (a) 5 h 23 m 48 s   (b) 8.755 sec

CHAPTER 10 MISCELLANEOUS PROBLEMS (pages 499–500)

1 The straight line $y = x + 2$ **3** The circle $(x - 2)^2 + (y - 1)^2 = 1$
5 Equation: $y^2 = (x - 1)^3$

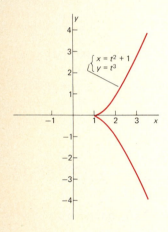

7 $y - 2\sqrt{2} = -\frac{4}{3}(x - (3\sqrt{2}/2))$ **9** $y + (2/\pi)x = \pi/2$ **11** 24   **13** 3π **15** $(13\sqrt{13} - 8)/27 \approx 1.4397$ **17** $\frac{43}{6}$
19 $(4\pi - 3\sqrt{3})/8 \approx 0.92128$ **21** $471295\pi/1024 \approx 1445.915$ **23** $(\pi\sqrt{5}/2)(e^\pi + 1) \approx 84.7919$
25 $x = a\theta - b \sin \theta$, $y = a - b \cos \theta$. The graph shows the case $a = 1$, $b = 0.7$

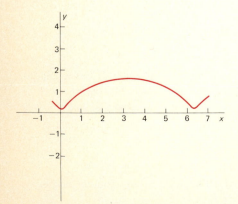

27 Suggestion: Compute $r^2 = x^2 + y^2$ 29 $6\pi^3 a^3$ 35 There are two solutions: $\alpha \approx 0.033364$ radians (about $1°54'53''$) and $\alpha \approx 1.29116$ radians (about $73°58'40''$) 37 The curvature is zero when x is a multiple of π, and reaches the maximum value 1 when x is an odd multiple of $\pi/2$ 39 $\mathbf{T} = -(\pi/(\pi^2 + 4)^{1/2})\mathbf{i} + (2/(\pi^2 + 4)^{1/2})\mathbf{j}$, $\mathbf{N} = -(2/(\pi^2 + 4)^{1/2})\mathbf{i} - (\pi/(\pi^2 + 4)^{1/2})\mathbf{j}$
43 $y = \frac{15}{8}x - \frac{5}{4}x^3 + \frac{3}{8}x^5$ 45 $f(x) = (\omega^2/2g)x^2$

CHAPTER 11

SECTION 11-1 (pages 508–509)

1 $\frac{1}{2}$ 3 $\frac{2}{5}$ 5 0 7 0 9 $\frac{1}{2}$ 11 2 13 0 15 1 17 1 19 0 21 The limit does not exist. 23 $-\infty$ (also correct to say that the limit does not exist) 25 Does not exist 27 $-\frac{1}{2}$ 29 0 31 1 33 1 35 $e^{-1/6}$

SECTION 11-2 (page 518)

1 $e^{-x} = 1 - x/1! + x^2/2! - x^3/3! + x^4/4! - x^5/5! + (e^{-z}/6!)x^6$ for some z between 0 and x 3 $\cos x = 1 - x^2/2! + x^4/4! - ((\sin z)/5!)x^5$ for some z between 0 and x 5 $\sqrt{1 + x} = 1 + (x/1!2) - (x^2/2!4) + (3x^3/3!8) - (5x^4/128)(1 + z)^{7/2}$ for some z between 0 and x 7 $\tan x = x/1! + 2x^3/3! + ((16 \sec^4 z \tan z + 8 \sec^2 z \tan^3 z)/4!)x^4$ for some z between 0 and x 9 $\sin^{-1} x = x/1! + (x^3/3!)(1 + 2z^2)/(1 - z^2)^{5/2}$ for some z between 0 and x 11 $e^x = e + (e/1!)(x - 1) + (e/2!)(x - 1)^2 + (e/3!)(x - 1)^3 + (e/4!)(x - 1)^4 + (e^z/5!)(x - 1)^5$ for some z between 1 and x 13 $\sin x = \frac{1}{2} + (\sqrt{3}/1!2)(x - \pi/6) - (1/2!2)(x - \pi/6)^2 - (\sqrt{3}/3!2)(x - \pi/6)^3 + ((\sin z)/4!)(x - \pi/6)^4$ for some z between $\pi/6$ and x 15 $1/(x - 4)^2 = 1 - 2(x - 5) + 3(x - 5)^2 - 4(x - 5)^3 + 5(x - 5)^4 - 6(x - 5)^5 + (7/(z - 4)^8)(x - 5)^6$ for some z between 5 and x 17 Six-place accuracy
19 Five-place accuracy 21 Problem 17 predicts $e^{1/3} \approx 1.3955826$; the true value is about 1.3956124 (the error is about 0.00003)
25 The third-degree polynomial gives the estimate 0.81915 27 The third-degree polynomial gives the estimate 0.8829476

SECTION 11-3 (page 523)

1 50 [exact] 3 Bounds: 0.631813, 0.632341. True value: About 0.63212056 5 Bounds: 1.974231, 2.014605. True value: Exactly 2
7 Bounds: 1.8627403, 1.8627423. True value: About 1.8627417 9 Bounds: 0.3862936, 0.3862968. True value: About 0.3862944
11 $n \geq 104$ 13 $n \geq 5$ 15 $n \geq 26$ 17 $n \geq 4$ 19 $V = (h/6)(0 + 4\pi(r/2)^2 + \pi r^2) = \frac{1}{3}\pi r^2 h$ 21 $V = (a/6)(0 + 4\pi a/2 + \pi a) = \pi a^2/2$ 23 $n = 10$; $T_{10} \approx 0.6937714$, so $\ln 2 \approx 0.69$ 25 Use $n = 10$; $T_{10} \approx 1.112332$ (the true value of the integral is approximately 1.11144797) 27 $S_2 - ES_2 = \frac{1}{6}(0 + \frac{4}{16} + 1) - 24/(180)(2^4) = \frac{1}{5}$

SECTION 11-4 (pages 530–531)

1 1 3 $+\infty$ 5 $+\infty$ 7 1 9 $+\infty$ 11 $-\frac{1}{2}$ 13 $\frac{9}{2}$ 15 $+\infty$ 17 Does not exist 19 $2(e - 1)$ 21 $+\infty$ 23 $\frac{1}{4}$

SECTION 11-5 (page 536)

1 $x - \frac{1}{2}x^2 + \frac{1}{3}x^3 - \frac{1}{4}x^4 + \cdots$ 3 $1 - x + x^2/2! - x^3/3! + x^4/4! - \cdots$ 5 $(x - 1) - \frac{1}{2}(x - 1)^2 + \frac{1}{3}(x - 1)^3 - \frac{1}{4}(x - 1)^4 + \cdots$
7 $\sqrt{2}/2 - (\sqrt{2}/1!2)(x - \pi/4) - (\sqrt{2}/2!2)(x - \pi/4)^2 + (\sqrt{2}/3!2)(x - \pi/4)^3 + (\sqrt{2}/4!2)(x - \pi/4)^4 - (\sqrt{2}/5!2)(x - \pi/2)^5 - (\sqrt{2}/6!2)(x - \pi/2)^6 + (\sqrt{2}/7!2)(x - \pi/4)^7 + \cdots$ 11 Five terms give the estimate 0.7966
13 Three terms give the estimate 0.2315266 15 Four terms give the estimate 0.487222 19 Seven terms give the estimate 0.2877

CHAPTER 11 MISCELLANEOUS PROBLEMS (pages 537–538)

1 $\frac{1}{4}$ 3 $\frac{1}{2}$ 5 $\frac{7}{45}$ 7 1 9 $-\infty$ 11 $+\infty$ 13 e^2 15 $-e/2$ 17 $+\infty$ 19 1 21 2 23 $e^{x+1} = e + ex/1 + ex^2/2! + ex^3/3! + ex^4/4! + ex^5/5! + (e^{z+1}/6!)x^6$ for some number z between 0 and x 25 $e^{-x^2} = 1 - (2/2!)x^2 + ((16z^4 - 48z^2 + 12)e^{-z^2}/4!)x^4$ for some number z between 0 and x 27 $x^{1/3} = 5 + (x - 125)/75 - (x - 125)^2/28125 + (x - 125)^3/6328125 - 10(x - 125)^4/243z^{11/3}$ for some z between 125 and x 29 $x^{-1/2} = 1 - (x - 1)/1!2 + 3(x - 1)^2/2!4 - 15(x - 1)^3/3!8 + 105(x - 1)^4/4!16 - 945(x - 1)^5/5!32z^{11/2}$ for some z between 1 and x 31 $x^4 = 1 + 4(x - 1) + 6(x - 1)^2 + 4(x - 1)^3 + (x - 1)^4 + 0$—the remainder ("0") is in fact zero in this case 33 $\sinh x = x + x^3/3! + x^5/5! + x^7/7! + \cdots$ 35 $1/x = 1 - (x - 1) + (x - 1)^2 - (x - 1)^3 + \cdots$ 37 $\sin x = \sqrt{2}/2 + \sqrt{2}(x - \pi/4)/1!2 - \sqrt{2}(x - \pi/4)^2/2!2 - \sqrt{2}(x - \pi/4)^3/3!2 + \sqrt{2}(x - \pi/4)^4/4!2 + \cdots$
39 Five terms of the series for $f(x) = x^{1/10}$, center 1024, give the estimate 1.99526. The true value is approximately 1.995262315
41 Not even one-place accuracy, since the error may be as large as 0.2

CHAPTER 12

SECTION 12-2 (pages 546–547)

1 $\frac{2}{5}$ 3 $\frac{1}{2}$ 5 1 7 Does not converge 9 0 11 0 13 0 15 1 17 0 19 0 21 0 23 0 25 e 27 $1/e^2$ 29 2 31 1
33 Does not converge 35 0 41 (b) 4

SECTION 12-3 (pages 554–555)

1 $\frac{3}{2}$ **3** Diverges **5** Diverges **7** 6 **9** Diverges **11** Diverges **13** Diverges **15** $\sqrt{2}/(\sqrt{2}-1)$ **17** Diverges **19** $\frac{1}{12}$ **21** $\frac{47}{99}$ **23** $\frac{41}{333}$ **25** $\frac{314156}{99999}$ **27** $S_n = \ln(n+1)$; diverges **29** $S_n = \frac{3}{2} - 1/n - 1/(n+1)$; the sum is $\frac{3}{2}$ **35** Computations with S are meaningless since S is not a number. **37** 4.5 seconds **39** (a) $M_n = (0.95)^n M_0$ (b) 0 **41** $\frac{1}{12}$

SECTION 12-4 (page 561)

1 Diverges **3** Diverges **5** Converges **7** Diverges **9** Converges **11** Converges **13** Converges **15** Converges **17** Converges **19** Converges **23** With $n = 6$, $1.03689 < S < 1.03699$ **25** About a million centuries **27** Results with $n = 10$: $S_{10} \approx 1.08203658$, and $3.141566 < \pi < 3.141627$

SECTION 12-5 (page 566)

1 Converges **3** Diverges **5** Converges **7** Diverges **9** Converges **11** Converges **13** Diverges **15, 17, 19, 21, 23,** and **25** Converges

SECTION 12-6 (pages 573–574)

1 Converges **3, 5,** and **7** Converges **9** Diverges **11** Converges absolutely **13** Converges conditionally **15** Converges absolutely **17** Converges absolutely **19** Diverges **21** Converges absolutely **23** Converges conditionally **25** Diverges **27** Diverges **29** Converges absolutely **31** Converges absolutely **33** $n = 1999$ **35** $n = 6$ **37** The first six terms give the estimate 0.6065 **39** The first three terms give the estimate 0.0953 **41** Results using the first 10 terms: $S_{10} \approx 0.8179622$ and $3.1329 < \pi < 3.1488$

SECTION 12-7 (pages 583–584)

1 $[-1, 1)$ **3** $(-1, 1)$ **5** $[-1, 1]$ **7** $(\frac{2}{5}, \frac{4}{5})$ **9** $[\frac{5}{2}, \frac{7}{2}]$ **11** Converges only for $x = 0$ **13** $(-4, 2)$ **15** $[2, 4]$ **17** Converges only for $x = 5$ **19** $(-1, 1)$ **21** $f(x) = x^2 - 3x^3/1! + 3^2 x^4/2! - 3^3 x^5/3! + 3^4 x^6/4! - \cdots$; the radius of convergence is $+\infty$ **23** $f(x) = x^2 - x^6/3! + x^{10}/5! - x^{14}/7! + x^{18}/9! - \cdots$; the radius of convergence is $+\infty$ **25** $(1-x)^{1/3} = 1 - \frac{1}{3}x - (2/2!3^2)x^2 - (2 \cdot 5/3!3^3)x^3 - (2 \cdot 5 \cdot 8/4!3^4 - (2 \cdot 5 \cdot 8 \cdot 11/5!3^5)x^5 - \cdots$; $r = 1$ **27** $f(x) = 1 - 3x + (4 \cdot 3/2)x^2 - (5 \cdot 4/2)x^3 + (6 \cdot 5/2)x^4 + \cdots$; $r = 1$ **29** $f(x) = 1 - x/2 + x^2/3 - x^3/4 + x^4/5 - \cdots$; $r = 1$ **31** $f(x) = x^4/4 - x^{10}/3!10 + x^{16}/5!16 - x^{22}/7!22 + \cdots = \Sigma_{n=0}^{\infty} (-1)^n x^{6n+4}/(2n+1)!(6n+4)$ **33** $f(x) = x - x^4/1!4 + x^7/2!7 - x^{10}/3!10 + x^{13}/4!13 - \cdots = \Sigma_{n=0}^{\infty} (-1)^n x^{3n+1}/n!(3n+1)$ **35** $f(x) = x/1! - x^3/2!3 + x^5/3!5 - x^7/4!7 + x^9/5!9 - \cdots = \Sigma_{n=1}^{\infty} (-1)^{n+1} x^{2n-1}/n!(2n-1)$ **37** By using six terms we find that $3.14130878 < \pi < 3.1416744$

SECTION 12-8 (pages 588–589)

1 $65^{1/3} = 4(1 + \frac{1}{64})^{1/3}$. The first four terms of the binomial series give $\sqrt[3]{65} \approx 4.020726$; answer: 4.021 **3** Three terms of the usual sine series give 0.479427 with error less than 0.000002; answer: 0.479 **5** Five terms of the usual arctangent series give 0.463684 with error less than 0.000045; answer: 0.464 **7** 0.916 **9** 0.0976 **11** $-1/2$ **13** 1/2 **15** 0 **19** $1, 0, \frac{1}{2}, 0, 5/24$

SECTION 12-9 (pages 591–592)

1 19 min 9 sec **3** 26556.97 ft

CHAPTER 12 MISCELLANEOUS PROBLEMS (pages 592–594)

1 1 **3** 10 **5** 0 **7** 0 **9** No limit **11** 0 **13** $+\infty$ or "No limit" **15** 1 **17** Converges **19** Converges **21** Converges **23** Diverges **25** Converges **27** Converges **29** Diverges **31** $(-\infty, +\infty)$ **33** $[-2, 4)$ **35** $[-1, 1]$ **37** It converges only for $x = 0$ **39** $(-\infty, +\infty)$ **41** It converges for *no* x **43** It converges for all x **45** Arithmetic computations involving x are meaningless because x is not a number. **51** Seven terms of the binomial series give 1.084 **53** Six terms of the "obvious" series give 0.747 **55** Four terms give 0.444

CHAPTER 13

SECTION 13-1 (pages 603–604)

1 $\langle 5, 8, -11 \rangle$; $\langle 2, 23, 0 \rangle$; 4; $\sqrt{51}$; $(\sqrt{5}/15)\langle 2, 5, -4 \rangle$ **3** $2\mathbf{i} + 3\mathbf{j} + \mathbf{k}$; $3\mathbf{i} - \mathbf{j} + 7\mathbf{k}$; 0; $\sqrt{5}$; $(\sqrt{3}/3)(\mathbf{i} + \mathbf{j} + \mathbf{k})$ **5** $4\mathbf{i} - \mathbf{j} - 3\mathbf{k}$; $6\mathbf{i} - 7\mathbf{j} + 12\mathbf{k}$; -1; $\sqrt{17}$; $(\sqrt{5}/5)(2\mathbf{i} - \mathbf{j})$ **7** $\theta = \cos^{-1}(-13\sqrt{10}/50) \approx 2.536$ **9** $\theta = \cos^{-1}(-34/\sqrt{3154}) \approx 2.221$ **11** $\text{comp}_\mathbf{b}\,\mathbf{a} = \frac{2}{7}\sqrt{14}$; $\text{comp}_\mathbf{a}\,\mathbf{b} = \frac{4}{15}\sqrt{5}$ **13** $\text{comp}_\mathbf{b}\,\mathbf{a} = 0 = \text{comp}_\mathbf{a}\,\mathbf{b}$ **15** $\text{comp}_\mathbf{b}\,\mathbf{a} = -\frac{1}{10}\sqrt{10}$; $\text{comp}_\mathbf{a}\,\mathbf{b} = -\frac{1}{5}\sqrt{5}$

17 $(x + 2)^2 + (y - 1)^2 + (z + 5)^2 = 7$ **19** $(x - 4)^2 + (y - 5)^2 + (z + 2)^2 = 38$ **21** Center: $(-2, 3, 0)$; radius: $\sqrt{13}$
23 Center: $(0, 0, 3)$; radius: 5   **25** A plane perpendicular to the z-axis at $z = 10$ **27** All points in the coordinate planes
29 The point $(1, 0, 0)$ **31** $\alpha = \cos^{-1}(\sqrt{6}/9) \approx 74.21°$, $\beta = \cos^{-1}(5\sqrt{6}/18) \approx 47.12°$, $\gamma = \beta$ **33** $\alpha = \cos^{-1}(3\sqrt{2}/10) \approx 64.90°$,
$\beta = \cos^{-1}(2\sqrt{2}/5) \approx 55.55°$, $\gamma = \cos^{-1}(\sqrt{2}/2) = 45°$ **35** 48   **39** $A = \frac{3}{2}\sqrt{69} \approx 12.46$ **41** $\cos^{-1}(\sqrt{3}/3) \approx 0.9553$, or about
$54.7356°$ **47** $2x + 9y - 5z = 23$; the plane through the midpoint of the segment AB perpendicular to AB

SECTION 13-2 (pages 610–611)

1 $\langle 0, -14, 7 \rangle$ **3** $\langle -10, -7, 1 \rangle$ **7** $(\mathbf{a} \times \mathbf{b}) \times \mathbf{c} = \langle -1, 1, 0 \rangle$, $\mathbf{a} \times (\mathbf{b} \times \mathbf{c}) = \langle 0, 0, -1 \rangle$ **11** $A = \frac{1}{2}\sqrt{2546} \approx 25.229$
13 (a) $V = 55$ (b) $V = 55/6$ **17** $\frac{1}{38}\sqrt{9842} \approx 2.6107$

SECTION 13-3 (pages 615–616)

1 $x = 2 + t$, $y = 3 - t$, $z = -4 - 2t$. $x - 2 = -y + 3 = (-z - 4)/2$ **3** $x = 1$, $y = 1$, $z = 1 + t$. $x - 1 = 0 = y - 1$,
z arbitrary   **5** $x = 2 + 2t$, $y = -3 - t$, $z = 4 + 3t$. $(x - 2)/2 = -y - 3 = (z - 4)/3$ **7** $y = 7$ **9** $7x + 11y = 114$
11 $3x + 4y - z = 0$ **13** $2x - y - z = 0$ **15** $\theta = \cos^{-1}(1/\sqrt{3}) \approx 54.736°$ **17** The planes are parallel—$\theta = 0$ **19** $(x - 3)/2 =$
$y - 3 = (-z + 1)/5$ **21** $3x + 2y + z = 6$ **23** $7x - 5y - 2z = 9$ **25** $x - 2y + 4z = 3$ **27** $\frac{10}{3}\sqrt{3}$ **31** $D = 133/\sqrt{501} \approx 5.942$

SECTION 13-4 (page 623)

1 10π **3** $19(e - 1) \approx 32.647$ **5** $2 + \frac{9}{10}\ln 3 \approx 3.5588$ **7** $\kappa = \frac{1}{2}$ **9** $\kappa = (\sqrt{2}/3)e^{-t}$ **11** $a_T = 0$, $a_N = 0$ **13** $a_T = (4t + 18t^3)/$
$(1 + 4t^2 + 9t^4)^{1/2}$, $a_N = 2(1 + 9t^2 + 9t^4)^{1/2}/(1 + 4t^2 + 9t^4)^{1/2}$ **15** $a_T = t/(t^2 + 2)^{1/2}$, $a_N = (t^4 + 5t^2 + 8)^{1/2}/(t^2 + 2)^{1/2}$
17 $\mathbf{T} = (\sqrt{2}/2)(\mathbf{i} + \mathbf{j}\cos t - \mathbf{k}\sin t)$, $\mathbf{N} = -\mathbf{j}\sin t - \mathbf{k}\cos t$; at $(0, 0, 1)$, $\mathbf{T} = (\sqrt{2}/2)(\mathbf{i} + \mathbf{j})$, $\mathbf{N} = -\mathbf{k}$.   **19** $\mathbf{T} = (\sqrt{3}/3)(\mathbf{i} + \mathbf{j} + \mathbf{k})$,
$\mathbf{N} = (\sqrt{2}/2)(-\mathbf{i} + \mathbf{j})$ **21** Suggestion: Compute $(d/dt)(\mathbf{r} \cdot \mathbf{r})$.   **25** $x(s) = 3\cos(s/5)$, $y(s) = 3\sin(s/5)$, $z(s) = 4s/5$

SECTION 13-5 (page 630)

1 Paraboloid, vertex at $(0, 0, 0)$, axis the z-axis, opening upward   **3** Cone, vertex $(0, 0, 0)$, axis the z-axis (both nappes)
5 Parabolic cylinder, perpendicular to xz-plane, its trace there the parabola opening upward with axis the z-axis and vertex $(0, -2)$
7 Elliptical cylinder, perpendicular to the xy-plane, its trace there the ellipse with center $(0, 0)$ and intercepts $(\pm 1, 0)$ and $(0, \pm 2)$
9 Elliptical cone, vertex $(0, 0, 0)$, axis the x-axis   **11** Paraboloid, vertex $(0, 0, 0)$, its axis the z-axis, opening downward
13 Hyperbolic paraboloid, saddle point at the origin, meets the xz-plane in a parabola with vertex $(0, 0, 0)$ and opening downward,
meets the xy-plane in a parabola with vertex $(0, 0, 0)$ and opening upward, meets planes parallel to the yz-plane in hyperbolas with
directrices parallel to the y-axis   **15** Hyperboloid of one sheet, axis the z-axis, trace in the xy-plane the circle with center $(0, 0)$ and
radius 3, its traces in parallel planes are larger circles, and its traces in planes parallel to the z-axis are hyperbolas   **17** Elliptic
paraboloid, axis the y-axis, vertex $(0, 0, 0)$ **19** Hyperboloid of two sheets, axis the y-axis   **21** Paraboloid, axis the x-axis, vertex
$(0, 0, 0)$, equation $x = 2(y^2 + z^2)$ **23** Hyperboloid of one sheet; see Figure 13.38; equation $x^2 + y^2 - z^2 = 1$ **25** Paraboloid,
vertex $(0, 0, 0)$, axis the x-axis, equation $y^2 + z^2 = 4x$ **27** The surface resembles a rug with a basketball underneath. Its highest
point is $(0, 0, 1)$; $z \to 0$ from above as x or y (or both) increase without bound. Its equation is $z = e^{-(x^2 + y^2)}$.   **31** The projection of
the intersection has equation $x^2 + y^2 = 2y$; it is the circle with center $(0, 1)$ and radius 1.   **33** Equation: $5x^2 + 8xy + 8y^2 - 8x -$
$8y = 0$. Since $B^2 - 4AC < 0$, it is an ellipse. (In a uv-plane rotated approximately $55°\ 16'\ 41''$ from the xy-plane, the ellipse has
center $(0.517, -0.453)$, minor axis 0.352 in the u-direction, major axis 0.774 in the v-direction.)

SECTION 13-6 (pages 633–634)

1 Cylinder, radius 5, axis the z-axis   **3** The *plane* $y = x$ **5** The upper nappe of the cone $x^2 + y^2 = 3z^2$ **7** The xy-plane
9 Cylinder, axis the vertical line $x = 0$, $y = 1$; its trace in the xy-plane the circle $x^2 + (y - 1)^2 = 1$ **11** The vertical plane whose
trace in the xy-plane is the line $y = -x$ **13** The horizontal plane $z = 1$ **15** $r^2 + z^2 = 25$; $\rho = 5$ **17** $r(\cos \theta + \sin \theta) + z = 1$;
$\rho(\sin \phi \cos \theta + \sin \phi \sin \theta + \cos \phi) = 1$ **19** $r^2 + z^2 = r(\cos \theta + \sin \theta) + z$; $\rho = \sin \phi \cos \theta + \sin \phi \sin \theta + \cos \phi$ **21** $z = r^2$
23 (a) $1 \leqq r^2 \leqq 4 - z^2$ (b) $\csc \phi \leqq \rho \leqq 2$ (and, as a consequence, $\pi/6 \leqq \phi \leqq 5\pi/6$)   **25** About 3821 miles   **27** Just under
31 miles   **29** $0 \leqq \rho \leqq H\sec \phi$, $0 \leqq \phi \leqq \tan^{-1}(R/H)$, θ arbitrary

CHAPTER 13 MISCELLANEOUS PROBLEMS (pages 634–636)

7 $x = 1 + 2t$, $y = -1 + 3t$, $z = 2 - 3t$; $(x - 1)/2 = (y + 1)/3 = (-z + 2)/3$ **9** $-13x + 22y + 6z = -23$ **11** $x - y +$
$2z = 3$ **15** 3   **17** $\kappa = 1/9$, $a_T = 2$, $a_N = 1$ **21** $3x - 3y + z = 1$ **25** $\rho = 2\cos \phi$ **29** $\rho^2 = 2\cos 2\phi$; shape: like an hour-
glass with rounded ends

CHAPTER 14

SECTION 14-2 (pages 645–646)

1 $\mathscr{R}^2(\mathscr{R} \times \mathscr{R})$ 3 $y \neq x$ 5 $x \neq 0, y \neq 0$ 7 The plane through the y-axis making a 45° angle with the x-axis   9 The upper half of an ellipsoidal surface; the graph meets the coordinate axes at $(\pm 2, 0, 0)$, $(0, \pm 1, 0)$, and $(0, 0, 2)$ 11 Straight lines of slope 1   13 Ellipses, centered at $(0, 0)$, major axis twice the minor axis and lying on the x-axis   15 Vertical translates of the graph of $y = x^3$ 17 $y; x$ 19 $y^2; 2xy$ 21 e 23 1   25 1   31 Note that $f(x, y) = (x - \frac{1}{2})^2 + (y + 1)^2 - \frac{1}{4} \geqq -\frac{1}{4}$ for all (x, y)

SECTION 14-3 (pages 652–653)

1 $\partial f/\partial x = 4x^3 - 3x^2 y + 2xy^2 - y^3$, $\partial f/\partial y = -x^3 + 2x^2 y - 3xy^2 + 4y^3$, $\partial^2 f/\partial x^2 = 12x^2 - 6xy + 2y^2$, $\partial^2 f/\partial x\, \partial y = \partial^2 f/\partial y\, \partial x = -3x^2 + 4xy - 3y^2$, $\partial^2 f/\partial y^2 = 2x^2 - 6xy + 12y^2$ 3 $\partial f/\partial x = e^x(\cos y - \sin y)$, $\partial f/\partial y = -e^x(\sin y + \cos y)$, $\partial^2 f/\partial x^2 = e^x(\cos y - \sin y)$, $\partial^2 f/\partial y^2 = e^x(\sin y - \cos y)$, $\partial^2 f/\partial x\, \partial y = \partial^2 f/\partial y\, \partial x = -e^x(\sin y + \cos y)$ 5 $\partial f/\partial x = -2y/(x - y)^2$, $\partial f/\partial y = 2x/(x - y)^2$, $\partial^2 f/\partial x^2 = 4y/(x - y)^3$, $\partial^2 f/\partial x\, \partial y = \partial^2 f/\partial y\, \partial x = -2(x + y)/(x - y)^3$, $\partial^2 f/\partial y^2 = 4x/(x - y)^3$ 7 $f_x(x, y) = 2x/(x^2 + y^2)$, $f_y(x, y) = 2y/(x^2 + y^2)$, $f_{xx}(x, y) = 2(y^2 - x^2)/(x^2 + y^2)^2$, $f_{yy}(x, y) = 2(x^2 - y^2)/(x^2 + y^2)^2$, $f_{xy}(x, y) = -4xy/(x^2 + y^2)^2$ 9 $z_x = yx^{y-1}$, $z_y = x^y \ln x$, $z_{xx} = y(y - 1)x^{y-2}$, $z_{xy} = z_{yx} = x^{y-1}(1 + y \ln x)$, $z_{yy} = x^y(\ln x)^2$ 11 $\partial f/\partial x = 2xy^3 z^4$, $\partial f/\partial y = 3x^2 y^2 z^4$, $\partial f/\partial z = 4x^2 y^3 z^3$, $\partial^2 f/\partial x^2 = 2y^3 z^4$, $\partial^2 f/\partial x\, \partial y = \partial^2 f/\partial y\, \partial x = 6xy^2 z^4$, $\partial^2 f/\partial y^2 = 6x^2 yz^4$, $\partial^2 f/\partial y\, \partial z = \partial^2 f/\partial z\, \partial y = 12x^2 y^2 z^3$, $\partial^2 f/\partial x\, \partial z = \partial^2 f/\partial z\, \partial x = 8xy^3 z^3$, $\partial^2 f/\partial z^2 = 12x^2 y^3 z^2$ 13 $f_x(x, y, z) = yze^{xyz}$, $f_y(x, y, z) = xze^{xyz}$, $f_z(x, y, z) = xye^{xyz}$, $f_{xx}(x, y, z) = y^2 z^2 e^{xyz}$, $f_{yy}(x, y, z) = x^2 z^2 e^{xyz}$, $f_{zz}(x, y, z) = x^2 y^2 e^{xyz}$. $f_{xy}(x, y, z) = f_{yx}(x, y, z) = (xyz + 1)ze^{xyz}$, $f_{xz}(x, y, z) = f_{zx}(x, y, z) = (xyz + 1)ye^{xyz}$, $f_{yz}(x, y, z) = f_{zy}(x, y, z) = (xyz + 1)xe^{xyz}$ 15 $\partial f/\partial x = 2xe^y \ln z$, $\partial f/\partial y = x^2 e^y \ln z$, $\partial f/\partial z = x^2 e^y/z$. $\partial^2 f/\partial x^2 = 2e^y \ln z$, $\partial^2 f/\partial x\, \partial y = \partial^2 f/\partial y\, \partial x = 2xe^y \ln z$, $\partial^2 f/\partial y^2 = x^2 e^y \ln z$, $\partial^2 f/\partial y\, \partial z = \partial^2 f/\partial z\, \partial y = x^2 e^y/z$, $\partial^2 f/\partial z^2 = -x^2 e^y/z^2$, $\partial^2 f/\partial x\, \partial z = \partial^2 f/\partial z\, \partial x = 2xe^y/z$ 17 $x - y + z = 1$ 19 $10x - 16y - z = 9$ 23 See answer to 13; $f_{xyz}(x, y, z) = (x^2 y^2 z^2 + 3xyz + 1)e^{xyz}$.   31 $(10, -7, -58)$ 33 (a) A decrease of about 2570 cc, (b) An increase of approximately 82.5 cc

SECTION 14-4 (page 659)

1 None   3 Highest point: $(4, -3, 35)$, lowest point: none   5 None   7 Lowest point: $(-1, 0, 6)$, highest point: none   9 Lowest point: none. Highest point: $(\frac{1}{2}, -\frac{1}{2}, 2)$ 11 If $x = 3 + h$, $y = -4 + k$, then $z = 25 - h^2 - k^2$. The maximum value of z is 25, at $(x, y) = (3, -4)$.   13 No extremum at $(3, 2)$ 15 Minimum value 0 at $(0, 0)$ 17 Dimensions: 10 by 10 by 10   19 10 by 10 by 10 cm   21 Maximum value: $\frac{4}{3}$; $x = 2$, $y = 1$, $z = \frac{2}{3}$ 23 Height: 10 ft; front, back 40 ft wide, sides 20 ft deep   25 $V_{max} = 5488$ in.^{3} 27 $V_{max} = \frac{1}{2}$ 29 Maximum area: 900 [one square]; minimum: 300 [three equal squares]

SECTION 14-5 (pages 665–666)

1 $m = \frac{113}{38} \approx 2.9737$, $b = \frac{472}{95} \approx 4.9684$ 3 $m = \frac{1497}{89} \approx 16.8202$, $b \approx -\frac{30711}{890} \approx -34.5067$ 5 174; 41   7 107 ($p \approx (19.38) \times \ln w + 17.95$)   9 $A = \frac{29}{15}$, $B = -\frac{17}{6}$, $C = \frac{72}{5}$, Answer: 10.5   11 $x = 122$, $y = 176$, Profit: \$4189.16   13 Raise 40 head of cattle, 40 hogs. Profit: \$1120.   15 (a) $A = 125$, $B = 375$, $C = 125$, $P = \$15625$, $Q = \$281{,}250$, $R = \$46875$. Note that $P + R = \$62500$ (b) $A = 0$, $B = 250$, $C = 250$, $P = 0$, $Q = \$125{,}000$, $R = \$187{,}500$, $P + R = \$187{,}500$.

SECTION 14-6 (pages 671–672)

1 $dw = (6x + 4y)\, dx + (4x - 6y^2)\, dy$ 3 $dw = (x\, dx + y\, dy)/\sqrt{1 + x^2 + y^2}$ 5 $dw = (-y\, dx + x\, dy)/(x^2 + y^2)$ 7 $dw = (2x\, dx + 2y\, dy + 2z\, dz)/(x^2 + y^2 + z^2)$ 9 $dw = (\tan yz)\, dx + xz(\sec^2 yz)\, dy + xy(\sec^2 yz)\, dz$ 11 $\Delta f \approx 0.014$ (true value: about 0.01422975)   13 $\Delta f \approx -0.0007$ 15 $\Delta f \approx \frac{53}{1300} \approx 0.04077$ 17 191.1   19 1.4   21 8.18   23 0.022 acres   25 The period increases by about 0.0278 sec   27 Approximately 303.8 ft

SECTION 14-7 (page 679)

1 $-(2t + 1)e^{-t^2 - t}$ 3 $6t^5 \cos t^6$ 5 $\partial w/\partial s = \partial w/\partial t = 2/(s + t)$ 7 $\partial w/\partial s = 0$, $\partial w/\partial t = 5e^t$ 9 $\partial r/\partial x = (y + z)\exp(yz + xz + xy)$, $\partial r/\partial y = (x + z)\exp(yz + xz + xy)$, $\partial r/\partial z = (x + y)\exp(yz + xz + xy)$ 11 $z_x = -(z/x)^{1/3}$, $z_y = -(z/y)^{1/3}$ 13 $z_x = -(yz(e^{xy} + e^{xz}) + (xy + 1)e^{xy})/(e^{xy} + xye^{yz})$, $z_y = -(x(x + z)e^{xy} + e^{xz})/(xye^{xz} + e^{xy})$ 15 $z_x = -c^2 x/a^2 z$, $z_y = -c^2 y/b^2 z$ 19 $\partial w/\partial x = f'(u)[\partial u/\partial x] = f'(u)$, etc.   21 Show that $w_u = w_x + w_y$. Then note that $w_{uv} = (\partial/\partial v)(w_u) = (\partial w_u/\partial x)(\partial x/\partial v) + (\partial w_u/\partial y)(\partial y/\partial v)$.

SECTION 14-8 (pages 686–687)

1 $8\sqrt{2}$ 3 $\frac{12}{13}\sqrt{13}$ 5 $-\frac{13}{20}$ 7 $-\frac{1}{6}$ 9 $-6\sqrt{2}$ 11 Maximum: $\sqrt{170}$; direction: $7\mathbf{i} + 11\mathbf{j}$ 13 Maximum: $14\sqrt{2}$; direction: $3\mathbf{i} + 5\mathbf{j} - 8\mathbf{k}$ 15 Maximum: $2\sqrt{14}$; direction: $\mathbf{i} + 2\mathbf{j} + 3\mathbf{k}$ 17 $29(x - 2) - 4(y + 3) = 0$ 19 $x + y + z = 1$ 25 $55\sqrt{2}/6$ 27 (a) Direction: $-\mathbf{i} - 2\mathbf{j}$. Angle: about $84°2'4''$ (b) Angle: about $83°8'14''$ 29 $x - 2y + z + 10 = 0$

1 Maximum: 4, at $(\pm 2, 0)$, minimum: -4, at $(0, \pm 2)$ **3** Maximum: 3, at $(\pm\frac{3}{2}\sqrt{2}, \pm\sqrt{2})$ [same signs]; minimum: -3, at $(\pm\frac{3}{2}\sqrt{2},$ $\mp\sqrt{2})$ [opposite signs]   **5** $\frac{126}{49}$ at $(\frac{9}{7}, \frac{6}{7}, \frac{3}{7})$ is the minimum; no maximum   **7** Maximum: 7, at $(\frac{36}{7}, \frac{9}{7}, \frac{4}{7})$, minimum: -7, at $(-\frac{36}{7},$ $-\frac{9}{7}, -\frac{4}{7})$ **9** Minimum: $\frac{25}{3}$, at $(-\frac{5}{3}, \frac{1}{3}, \frac{7}{3})$.   **21** Closest points: $(2, -2, 1)$ and $(-2, 2, 1)$.   **25** $(2, 3)$ and $(-2, -3)$.
29 Highest point: $(\frac{2}{5}\sqrt{5}, \frac{1}{5}\sqrt{5}, \sqrt{5} - 4)$. Lowest point: $(-\frac{2}{5}\sqrt{5}, -\frac{1}{5}\sqrt{5}, -\sqrt{5} - 4)$.
31 Farthest: $((-15 - 9\sqrt{5})/20, (-15 - 9\sqrt{5})/10, (9 + 3\sqrt{5})/4)$, nearest: $((-15 + 9\sqrt{5})/20, (-15 + 9\sqrt{5})/10, (9 - 3\sqrt{5})/4)$

1 Minimum: $(-1, 2, -1)$, no other extrema   **3** Saddle point: $(-\frac{1}{2}, -\frac{1}{2}, \frac{29}{4})$, no extrema   **5** Minimum: $(-3, 4, -9)$, no other extrema   **7** Saddle point at $(0, 0, 3)$; local maximum at $(-1, -1, 4)$ **9** No extrema   **11** Saddle point at $(0, 0, 0)$, local minima at $(-1, -1, -2)$ and $(1, 1, -2)$ **13** Maximum at $(0, 0, 1)$ **15** No extrema   **17** $(1, 1)$ and $(-1, -1)$ minimize f; its minimum value is 3
19 See Problem 29, Section 14-4, and its answer.

3 On the line $y = x$, $g(x, y) \equiv \frac{1}{2}$, except that $g(0, 0) = 0$ **5** $f(x, y) = x^2y^3 + e^x \sin y + y + C$ **7** All points of the form $(a, b, \frac{1}{2})$ (so that $a^2 + b^2 = \frac{1}{2}$) together with $(0, 0, 0)$.   **9** The normal to the cone at $(a, b, [a^2 + b^2]^{1/2})$ meets the z-axis at $z = 2[a^2 + b^2]^{1/2}$.
15 Base: $2\sqrt[3]{3}$ by $2\sqrt[3]{3}$, height $5\sqrt[3]{3}$ **17** 200 ± 2 ohms   **19** 3%   **21** $(\pm 4, 0, 0)$, $(0, \pm 2, 0)$, $(0, 0, \pm\frac{4}{3})$ **25** $3\mathbf{i} + 4\mathbf{j}$; 1
27 The plane tangent to the graph at (a, b, c) has x-intercept $a^{1/3}$ **31** Semiaxes: 1 and 2   **33** There is no such triangle of minimum perimeter, unless one is willing to consider as a triangle the figure with all sides of length zero—a single point on the circumference of the circle. The triangle of *maximum* perimeter is equilateral, with perimeter $3\sqrt{3}$.   **35** Closest point: $(\frac{1}{3}\sqrt{6}, \frac{1}{6}\sqrt{6})$, farthest point: $(-\frac{1}{3}\sqrt{6}, -\frac{1}{6}\sqrt{6})$ **39** Maximum: 1, minimum: $-\frac{1}{2}$ **41** Local minima: -1, at $(1, 1)$ and at $(-1, -1)$; horizontal tangent plane (not extrema): $(0, 0, 0)$, $(\pm\sqrt{3}, 0, 0)$ **43** Local minimum: -8 at $(2, 2)$, horizontal tangent plane at $(0, 0, 0)$
45 Local maximum $\frac{1}{432}$ at $(\frac{1}{2}, \frac{1}{3})$. The points on the intervals $(-\infty, 0)$ and $(1, +\infty)$ on the x-axis are all local minima (value 0), and those on the interval $(0, 1)$ on the x-axis are local maxima (value 0). Saddle point at $(0, 1, 0)$.   **47** Saddle point at $(0, 0, 1)$; each point on the hyperbola $xy = \ln 2$ yields a global minimum.   **49** No extrema; saddle points at $(1, 1, 0)$ and $(-1, -1, 0)$.

CHAPTER 15

1 -32 **3** $\frac{4}{15}(9\sqrt{3} - 4\sqrt{2} - 1)$ **5** $\frac{1}{12}$ **7** $-\frac{52}{5}$ **9** $-\frac{1}{18}$ **11** $(e - 2)/2$ **13** $\frac{61}{3}$ **15** $\int_0^4 \int_{-y^{1/2}}^{y^{1/2}} x^2 y \, dx \, dy = \frac{512}{21}$
17 $\int_0^1 \int_{-y^{1/2}}^{y^{1/2}} x \, dx \, dy + \int_1^9 \int_{(1/2)(y-3)}^{y^{1/2}} x \, dx \, dy = \frac{32}{3}$ **19** $\int_0^4 \int_{2-(4-y)^{1/2}}^{y/2} 1 \, dx \, dy = \frac{4}{3}$ **21** $\int_0^\pi \int_0^y (\sin y/y) \, dx \, dy = 2$
23 $\int_0^1 \int_0^x (1/(1 + x^4)) \, dy \, dx = \pi/8$

1 $\frac{1}{6}$ **3** $\frac{32}{3}$ **5** $\frac{5}{6}$ **7** $\frac{32}{3}$ **9** $2\ln 2$ **11** $\frac{10}{3}$ **13** 19   **15** $\frac{4}{3}$ **17** $abc/6$ **19** $\frac{2}{3}$ **23** $625\pi/2 \approx 981.748$ **25** $(R^3/6)(2\pi + 3\sqrt{3}) \approx (1.913)R^3$

3 $3\pi/2$ **5** $\frac{1}{6}(4\pi - 3\sqrt{3}) \approx 1.22837$ **7** $\frac{1}{2}(2\pi - 3\sqrt{3}) \approx 0.5435$ **9** $16\pi/3$ **11** $23\pi/8$ **13** $(\pi/4)\ln 2$ **15** $16\pi/5$
17 $(\pi/4)(1 - \cos 1) \approx 0.36105$ **23** 2π **25** $(\pi a^3/3)(2 - \sqrt{2}) \approx (0.6134)a^3$ **27** $\pi/4$

1 $M = a^3$, $\bar{x} = \bar{y} = 7a/12$ **3** $M = \frac{128}{5}$, $\bar{x} = 0$, $\bar{y} = \frac{20}{7}$ **5** $M = \pi$, $\bar{x} = (\pi^2 - 4)/\pi$, $\bar{y} = \pi/8$ **7** $M = \pi a^3/3$, $\bar{x} = 0$, $\bar{y} = 3a/2\pi$
9 $M = \frac{2}{3}\pi + \frac{1}{4}\sqrt{3}$, $\bar{x} = 0$, $\bar{y} = (36\pi + 33\sqrt{3})/(32\pi + 12\sqrt{3}) \approx 1.4034$ **11** $I_0 = 2\pi a^{n+4}/(n + 4)$ **13** $I_0 = \frac{3}{2}\pi k$ **15** $I_0 = \frac{1}{9}$
17 $\hat{x} = \frac{2}{21}\sqrt{105}$, $\hat{y} = \frac{4}{3}\sqrt{5}$ **19** $\hat{x} = \hat{y} = (a/10)\sqrt{30}$ **23** $484k/3$ **25** (i) $(\pi\rho/2)(b^4 - a^4)$, (ii) $v = 4\sqrt{gh/13}$ **27** $(1, \frac{1}{4})$

1 $V = \int_0^3 \int_0^{2-(2/3)x} \int_0^{6-2x-3y} 1 \, dz \, dy \, dx = 6$ **3** $\frac{128}{5}$ **5** $\frac{332}{105}$ **7** $\frac{256}{15}$ **9** $\frac{11}{30}$ **11** $(0, \frac{20}{7}, \frac{10}{7})$ **13** $(0, \frac{8}{7}, \frac{12}{7})$ **15** $\bar{x} = 0$, $\bar{y} = (44 - 9\pi)/(72 - 9\pi)$, $\bar{z} = (9\pi - 16)/(72 - 9\pi)$ **17** $\frac{8}{7}$ **19** $\frac{1}{30}$ **21** $\frac{2}{3}a^5$ **23** $\frac{38}{45}ka^7$ **25** $k/3$ **27** $(9\pi/64, 9\pi/64, \frac{3}{8})$ **29** 24π **31** $\pi/6$

1 $V = 81\pi/2$, centroid $(0, 0, 3)$ **3** $V = 24\pi$ **5** $V = (\pi/6)(8\sqrt{2} - 7)$ **7** $I_x = (\pi\delta a^2 h/12)(3a^2 + 4h^2)$ **9** $\pi/3$ **13** $\pi/3$
15 $V = (\pi a^3/3)(2 - \sqrt{2})$; centroid: $\bar{x} = 0 = \bar{y}$, $\bar{z} = (3a/16)(2 + \sqrt{2}) \approx (0.6402)a$ **17** $I_x = \frac{7}{5}Ma^2$ **19** Description: The surface obtained by rotating the circle in the xz-plane with center $(a, 0)$ and radius a about the z-axis—a doughnut with an infinitesimal hole. $V = 2\pi^2 a^3$ **21** $I_x = (2\pi a^5/15)(128 - 51\sqrt{3})$ **23** (i) $I_z = (8\pi\delta/15)(b^5 - a^5)$,   (ii) $v = (140gh/101)^{1/2}$. With $h = 50$ and $g = 32$, v 47.094 ft/sec: slower than a marble, faster than a quarter

1 $A = \frac{1}{2}(6\sqrt{2} + \ln(3 + 2\sqrt{2})) \approx 5.124$ **3** $A = 3\sqrt{14}$ **5** $A = (2\pi/3)(2\sqrt{2} - 1) \approx 3.829$ **7** $(\pi/6)(65\sqrt{65} - 1)$
11 $A = 8a^2$

1 $x = (u + v)/2$, $y = (u - v)/2$, $J = -\frac{1}{2}$ **3** $x = (u/v)^{1/2}$, $y = (uv)^{1/2}$, $J = 1/2v$ **5** $x = (u + v)/2$, $y = \frac{1}{2}(u - v)^{1/2}$, $J = -1/4(u - v)^{1/2}$ **7** $A = \frac{3}{5}$ **9** $A = \ln 2$ **11** $A = (2 - \sqrt{2})/8$ **13** $V = 39\pi/2$ **15** $V = 8$
17 S is the region $3u^2 + v^2 \le 3$; the value of the integral is $2\pi\sqrt{3}(e^3 - 1)/3e^3$

1 $(2 - \sqrt{2})/3$ **3** $(e - 1)/2e$ **5** $(e^4 - 1)/4$ **7** $\frac{4}{3}$ **9** $V = 9\pi$; centroid: $(0, 0, \frac{9}{16})$ **11** 4π **13** 4π **15** $(\pi - 2)/16$ **17** Mass: $\frac{128}{15}$, centroid: $(\frac{32}{7}, 0)$ **19** Mass: $k\pi$, centroid: $(1, 0)$ **21** $(0, \frac{8}{5})$ **23** $(10\pi/3)(\sqrt{5} - 2) \approx 2.4721$ **25** $\frac{3}{10}Ma^2$ **27** $\frac{1}{5}M(b^2 + c^2)$
29 $128\delta(15\pi - 26)/225 \approx (12.017)\delta$ (δ is [constant] density)   **31** $8\pi/3$ **39** $\frac{18}{7}$ **41** $(\pi/6)(37\sqrt{37} - 17\sqrt{17}) \approx 81.1418$ **45** $4\sqrt{2}$
46 Approximately 3.49608   **49** $I_0 = 3\delta$ (δ is [constant] density)   **51** $M = \frac{8}{15}\pi abc$

CHAPTER 16

1 $\frac{310}{3}, \frac{248}{3}, 62$ **3** $3\sqrt{2}, 3, 3$ **5** $\frac{49}{24}, \frac{3}{2}, \frac{4}{3}$ **7** $\frac{6}{5}$ **9** 315 **11** $\frac{19}{60}$ **13** $\pi + 2\pi^2$ **15** 28 **17** $\frac{1}{6}(14\sqrt{14} - 1) \approx 8.563867$ **19** $(0, 2a/\pi)$
21 $10k\pi$; centroid: $(0, 0, 4\pi)$ **23** $M = \frac{1}{2}ka^3$; centroid $(2a/3, 2a/3)$; $I_x = I_y = \frac{1}{4}ka^5$ **25** (i) $\frac{1}{2}k \ln 2$, (ii) $-\frac{1}{2}k \ln 2$ **27** $\mathbf{F}$ is not conservative on any region containing $(0, 0)$ **29** $Q_z = 0 \neq 2y = R_y$

1 0 **3** 3 **5** $\frac{3}{10}$ **7** 2 **9** $A = \oint_C x \, dy = \int_0^{2\pi} a^2 \cos^2 t \, dt = \pi a^2$ **11** $3\pi/8$ **21** $\frac{3}{2}$ **27** 0

1 $\sqrt{3}/120$ **3** $(\pi/60)(1 + 391\sqrt{17}) \approx 84.4635$ **5** $16\pi/3$ **7** 24π **9** 0 **11** $(a/2, a/2, a/2)$ **13** $\bar{x} = 0 = \bar{y}$, $\bar{z} = (1 + (24a^4 + 2a^2 - 1)\sqrt{1 + 4a^2})/10[(1 + 4a^2)^{3/2} - 1]$, $I_z = (\pi\delta/60)[1 + (24a^4 + 2a^2 - 1)\sqrt{1 + 4a^2}]$ **15** $\bar{x} = 4/3(\pi - 2) \approx 1.16796$, $\bar{y} = 0$, $\bar{z} = \pi/2(\pi - 2)$ 1.37597 **17** Net flux: 0 **19** Net flux: 1458π

1 Both values: 4π **3** Both values: 24 **5** Both values: $\frac{1}{2}$ **7** $2385\pi/2$ **9** $\frac{1}{4}$

1 -20π **3** 0 **5** -52π **7** -8π **9** -2

1 $\frac{125}{3}$ **3** (Use the fact that $\int \mathbf{F} \cdot \mathbf{T} \, ds$ is independent of the path.) $\frac{69}{8}$ **5** $\frac{2148}{5}$ **9** $M = (5\sqrt{5} - 1)/3 \approx 3.3934$; $I_y = (50\sqrt{5} + 2)/15 \approx 7.5869$ **11** $\frac{2816}{7}$ **17** $371\pi/30$ **19** 72π **29** (a) $\phi'(r)(\mathbf{r}/r)$, (b) $3\phi(r) + r\phi'(r)$, (c) $\mathbf{0}$

CHAPTER 17

1 $y \equiv 0$ **3** $y = x + 3/x^2$ **5** $y = \frac{1}{2}(1 - 5e^{-x^2})$ **7** $y = 2 + C/(x + 1)^5$ **9** $y = \frac{1}{2}\sin x + C\csc x$ **13** (a) $60 - t - \frac{1}{3600} \times (60 - t)^3$ (b) The maximum is $\frac{40}{3}\sqrt{3}$ lb, about 23.094 lb   **15** $Q(t) = [E_0 C/(1 + (RC\omega)^2)](\cos \omega t + RC\omega \sin \omega t - e^{-t/RC})$

17 $v(t) = -32t + 5280 \ln(25/(25 - t))$ [ft/sec]; $y(t) = -16t^2 + 5280[t - (25 - t)\ln(25/(25 - t))]$ [ft]. At burnout (when $t = 20$), $y = 99200 - 26400 \ln 5$, or about 56711 ft, or about 10.74 mi. At that time $v = -640 + 5280 \ln 5 \approx 7857.83$ ft/sec, or about 5358 mph.   **23** $v(70) = 5337$ ft/sec; $y(70) = 136{,}925$ ft

SECTION 17-3 (page 813)

1 $y = e^{-x}$ **3** $y = \frac{1}{5}e^{2x} - \frac{1}{5}e^{-3x}$ **5** $y = 2e^{-x}\cos x + 5e^{-x}\sin x$ **7** $r = \pm 2$ **9** $r = -1, r = 6$ **11** $y = 3x - 3$ **13** $y = -\frac{1}{3}e^x$ **15** $y = -\frac{3}{10}\cos 2x + \frac{3}{5}\sin 2x$ **17** $y = \frac{5}{6}x \sin 3x$ **19** $y = \frac{1}{2}x^2 + 2x$

SECTION 17-4 (page 819)

1 $25 + (0)i$; $25 \operatorname{cis} 0$ **3** $\frac{1}{13} + \frac{31}{13}i$; $(\sqrt{962}/13) \operatorname{cis}(\tan^{-1}31)$ **5** $0 + (-8)i$; $8 \operatorname{cis}(3\pi/2)$ **7** $\pm\sqrt{2}/2 \pm (\sqrt{2}/2)i$ **9** $\pm 2i, \pm\sqrt{3} \pm i$ **11** $\pm 2, \pm 2i$ **13** Let $P = \frac{1}{2}\sqrt{2 + \sqrt{3}}$ and $Q = \frac{1}{2}\sqrt{2 - \sqrt{3}}$; let $R = (2\sqrt{2})^{1/3}$. The solutions are then given by $x = R(-Q + Pi)$, $R(Q - Pi), R(Q + Pi), R(-P - Qi)$, and $R(\sqrt{2}/2 \pm (\sqrt{2}/2)i)$ **15** $\cos 4\theta = \cos^4\theta - 6\cos^2\theta \sin^2\theta + \sin^4\theta$, $\sin 4\theta = 4\cos^3\theta \sin\theta - 4\cos\theta \sin^3\theta$

SECTION 17-5 (page 824)

1 $y = C_1 e^{2x} + C_2 e^{-2x}$ **3** $y = Ae^{2x} + Be^{-5x}$ **5** $y = A\exp[((-5 + \sqrt{5})/2)x] + B\exp[((-5 - \sqrt{5})/2)x]$ **7** $y = C_1 e^{3x} + C_2 x e^{3x}$ **9** $y = e^{-4x}(C_1 \cos 3x + C_2 \sin 3x)$ **11** $y = C_1 e^{3x} + C_2 x e^{3x} + C_3 + C_4 x$ **13** $y = C_1 + e^{-x}(C_2 \cos x + C_3 \sin x)$ **15** $y = (A + Bx)e^{2x} + (C + Dx)e^{-2x}$ **17** $y = C_1 e^{\gamma x} + C_2 e^{-\gamma x} + C_3 \cos \gamma x + C_4 \sin \gamma x$ **21** $y = A/x^2 + (B/x^2)\ln x$

SECTION 17-6 (pages 829–830)

1 $k = (80\pi/3)^2 \approx 7018.385$ lb/ft   **3** $(30\sqrt{79}/316\pi) \ln((16 \times 10^6)/1521)$ sec, or about 2.487 sec
5 (a) $x(t) = \frac{2}{3}\sqrt{3}e^{-4t}\cos(4\sqrt{3}t - \frac{1}{6}\pi)$;   (b) Frequency $2\sqrt{3}/\pi$ hz, period about 0.91 sec
9 (a) $Q(t) = -\frac{1}{4}\cos 10t + \frac{1}{16}e^{-6t}(4\cos 8t + 3\sin 8t)$ (b) When t is large, $Q(t) \approx -\frac{1}{4}\cos 10t$ **11** $A = -F_0/2m\omega$

SECTION 17-7 (pages 833–834)

1 98 cm   **3** (a) 445 cm   (b) July: 15.5°C, January: 16.5°C   **5** 153 cm   **7** 23.27 cm

CHAPTER 17 MISCELLANEOUS PROBLEMS (page 834)

1 $y(x) = 1/(C - x^2 - 3\ln|x|)$ **3** $v(x) = \frac{1}{6}x^5 + C/x$ **5** $y(x) = 1 + C\exp(-\tan^{-1} x)$ **7** $y(x) = Ae^{-x} + Be^{-x/2}$
9 $y(x) = C_1 + C_2 e^{2x} + C_3 e^{-2x}$ **11** $y(x) = (A + Bx + Cx^2)e^{-x}$ **13** $y(x) = C_1 e^{-x} + C_2 e^{-4x}$ **15** $y(x) = C_1 e^{2x} + C_2 e^{-2x} + C_3 \cos 2x + C_4 \sin 2x$ **17** $y(x) = (C_1 + C_2 x)\cos x + (C_3 + C_4 x)\sin x$ **19** $y(x) = \frac{1}{2}x^2 - \frac{5}{2}$ **21** $y(x) = \sin x$
23 $y(x) = -\frac{1}{12}\cos 4x$ **25** $y(x) = A/x + B/\sqrt{x}$ **27** $y(x) = A/x + B/x^4$ **29** $y(x) = 1/(C - x)e^x$

Index

Change of variables in double integral, 749
Change of variables theorem, 750
Characteristic equation, 820
    complex roots case, 822
    distinct roots case, 820
    repeated roots case, 821
Charged wire, 767
Chemical reactions, 336, 338
Cheops (see Khufu)
Circle:
    of curvature, 489
    equation, 14, 15, 417
    osculating, 489
Circular motion, 483–485
Circulation of a field, 793
cis θ, 815
Clepsydra, 251
Clock problem, 591
Closed curve, 458
Closed interval, 4
Closed interval maximum-minimum
    method, 77
Closed surface, 782
Cobb–Douglas production function, 694
Collusion and competition, 662–664
Combinations of functions, 32
Comet(s):
    examples, 429, 450
    Halley's, 429
    Kahoutek, 430
    problems, 425, 452, 455
Communications satellite, 485
Comparison property, 204
Comparison test for series, 562
Competition and collusion, 662–664
Completeness property of real numbers,
    544, 546
Completing the square, 15, 395
Complex exponential function, 817
Complex number(s), 814–816
    conjugate, 814
    DeMoivre's formula, 815
    imaginary part, 814
    magnitude, 815
    real part, 814
    polar form, 815
    powers of, 815–816
    products of, 815
    quotients of, 815
    roots of, 816
    sequences, 817
    series, 817
Complex-valued function(s), 818
    derivative, 818
    integral, 818
Component of **a** along **b**, 602
Components:
    of acceleration vector, 621
    of vectors, 470
Componentwise differentiation, 477
Componentwise integration, 479
Composition of functions, 34
    derivative of, 105
Compound interest, 317, 319
Computation:
    of e, 292, 307, 515, 568
    of ln 2, 555, 561, 573

of π, 192, 521, 533, 534, 536, 559, 560,
    574, 583
Concave upward and downward, 152–153
    test, 153
Conditionally convergent series, 570
Cone:
    area, 259
    centroid problem, 282
    circumscribed about sphere, 125
    elliptic, 627
    inscribed in inverted cone, 129
    inscribed in sphere, 128
    lateral surface area, 277
    slant height, 259
Conic section(s), 416, 418–420, 438, 628
    classification, 442
    directrices, 421, 426, 431
    discriminant, 442
    eccentricity, 420, 426, 431
    foci, 420, 421, 426, 431
    general equation, 438–442
    in polar coordinates, 448–451
    summary of properties, 455
Conjugate axis of hyperbola, 432
Conjugate complex numbers, 814
Conjugate pairs of roots, 819
Conservation of energy, 359, 766
Conservative force field, 765, 766, 793
Conservative and irrotational fields, 793
Constant:
    integral of, 203
    of integration, 166
    law, 42
    law for limits, 57
    multiple property, 203
Constraint(s), 688
    in maximum-minimum problems, 114
    two, in lagrange multiplier method, 692,
    693
Consumption of a commodity, 328–329
Conte, S. D., 520
Continued fraction, 593
Continuity, 48
    of compositions, 49
    of functions of two variables, 285, 643
    on an interval, 51
    of inverse functions, 50
    of rational functions, 49
    of vector functions, 477
Continuous function, 48
Continuously compounded interest, 317
Contour curve, 640
Convergence (see also Convergence of
    series)
    of continued fraction, 593
    of improper integrals, 524
    of infinite product, 593
    interval of, 575, 577
    radius of, 575, 577
    of sequence, 142
Convergence of series, 547–554
    absolute, 569
    alternating, 567
    comparison test, 562
    conditional, 570
    integral test, 556
    introduction, 556

    limit comparison test, 564
    nth term test, 551
    power series, 575
    p-series, 558
    ratio test, 571
    root test, 572
Conversion factors, 838
Cooling, Newton's law of, 165
Coordinate(s):
    angular, 443
    Cartesian (rectangular), 9, 596–597
    cylindrical, 631
    ellipsoidal, 756
    polar, 443, 630
    radial, 443
    spherical, 631
    translation of, 439
Copernicus, Nicholas (1473–1543), 493
Cork ball example, 143
Corner point, 26
Corrugated sheet example, 258
Cosecant, 117, 118
Cosine, 117, 118
    direction, 601
    power series of, 532, 574
    properties, 118, 119
    Taylor polynomial of, 518
Coordinate axes, 9, 596
Coordinate functions of a curve, 458
Coordinate line, 4
Coordinate planes, 596
Coordinates in the plane:
    polar, 443, 630
    rectangular (Cartesian), 9
Coordinates in space:
    cylindrical, 631, 736
    rectangular, 596
    spherical, 631, 738
Cost:
    average, 183
    inflation of, 331
    marginal, 183
Cost function, 80, 183
Cotangent, 117, 118
Courant, Richard, 836
Critical point, 76
Critically damped vibrations, 826
Cross (vector) product of vectors, 604–608
Cross section(s), 246
    method of, 245, 707
Cross-sectional area function, 708
Curl of a vector field, 789–790
    physical interpretation, 792
Curvature, 486–489, 510, 620
    center of, 489
    of circle, 488
    problems, 500
    radius of, 489
    of smooth parametric curve, 488
    zero, 486
Curve (see also Curve in three dimensions)
    approach, 491
    banking of, 484
    closed, 458
    concavity of, 153
    length of, 257, 466
    orientation of, 761

Force (*cont.*)
  on submerged plate, 280, 282
  properties of, 263
Force field, 758
  conservative, 765
  electric, 788
  gravitational, 787–788
  inverse-square, 786
Force function, 263
Forced vibrations, 827
Formulas from elementary mathematics,
      839–841
Four-leaved rose, 446
Fractional mass of rocket, 808
Free stream velocity, 795
Frequency, 358, 824
  impressed, 828
  natural, 828
From the Earth to the Moon, 529
Frustum of cone:
  area of, 259
  volume of, 277, 287
ft-lb, 838
Function(s), 5
  algebraic, 342
  average value, 208
  a^x, 307
  complex-valued, 818
  composition of, 33–34
  continuous, 48, 643
  cross-sectional area, 246
  decreasing, 91, 96
  definition, 5
  derivative of, 23, 64, 65
  difference, 32
  differentiable, 26, 65
  discontinuous, 18, 48, 645
  domain, 6
  elementary, 226
  with equal derivatives, 95
  even, 218
  exponential (e^x), 290
  exponential with base a, 291
  force, 263
  gamma, 528
  graph of, 15, 18
  greatest integer, 18
  hyperbolic, 362–364
  implicitly defined, 112
  increasing, 30, 91, 96
  integrable, 200
  increment of, 132, 666–667
  inverse, 35–36
  inverse cosine (arccosine), 348
  inverse cosecant (arccosecant), 352
  inverse cotangent (arccotangent), 350
  inverse hyperbolic, 367–369
  inverse secant (arcsecant), 351
  inverse sine (arcsine), 347
  inverse tangent (arctangent), 349
  inverse trigonometric, 347–352
  limit of, 40, 643
  linear, 97
  logarithmic, 290
  logarithmic to base a, 291, 309
  monotone (*see* increasing; decreasing)

natural exponential, 293, 301
notation, 65
odd, 218
one-to-one, 37
periodic, 120, 343
polynomial, 33
power, 299
product, 32
production, 694
quadratic, 24
quotient, 32
radial probability density, 305
range, 6
rational, 33
scalar multiple, 32
smooth, 256
sum, 32
of three variables, 639
of two variables, 639
transcendental, 342
vector-valued, 477
trigonometric, 117, 342
with zero derivative, 95
Fundamental identity of trigonometry, 119
Fundamental theorem of algebra, 818–819
Fundamental Theorem of Calculus, 208,
      706, 765, 768
  proof, 210–211
  statement, 210
  vector form, 480

G

Gabriel's horn, 530
Gabriel's trumpet, 301
Galilei, Galileo (1564–1642), 424
Gamma function, 528
  properties, 530–531, 537, 538
Ganymede, 486
Gas:
  adiabatic expansion, 301
  ideal, 301
Gas piston engine, 751–752
Gauss, Carl Friedrich (1777–1855), 301,
      662, 782
Gauss's law, 786–788
Gauss's theorem, 782
General chain rule, 675
General exponential function, 291, 307
General logarithm function, 291
General iterated double integrals, 710
General solution, 333, 800, 810
Generalized Mean Value Theorem, 504
  proof, 507–508
Generalized power rule, 84, 105
  proof, 108
Generalized root rule, 85, 108
Geocentric system, 493
Geological fault problem, 90-91
Geometric interpretation of partial
      derivatives, 648–649
Geometric series, 549
  sum, 549
Geometry formulas, 840

Gibbs, J. Willard (1839–1903), 795
Gizeh, pyramid at, 268
Global extrema:
  of $f(x)$, 42, 76
  of $f(x,y)$, 653
Gradient and curl, 793
Gradient field, 758
Gradient vector, 680
  in chain rule, 681
  and directional derivative, 681
  as normal vector, 684
  significance, 682
Grain warehouse problem, 339
Graph(s):
  of equation, 14
  of equation in three variables, 598
  of function, 15
  of function of two variables, 640
  of parametric curve, 458
  in polar coordinates, 444
  of trigonometric functions, 343
Graph sketching, 99–100, 155–157,
      161–165
Gravitation law, 175, 494, 500
Gravitational acceleration, 174–175
Gravitational attraction:
  of disk, 730
  of lamina, 729
  of plane, 730
  of spherical shell, 742, 781, 789
  of solid sphere, 730, 787–788, 789
  of uniform rod, 531
Greatest integer function, 18
Greek alphabet, 841
Green, George (1793–1841), 768
Green's first identity, 789
Green's second identity, 789
Green's theorem, 768
  corollary, 770
  vector form, 773
Greenwich, 633, 634
Grouping terms of series, 565
Growth, natural, 312, 314
Gyration, radius of, 726

H

Hailstone problem, 112
Half-angle formulas, sine and cosine, 119
Half life, 315, 319–320
Halley, Edmund (1656–1742), 429
Halley's comet, 429
Hanging cable, 371–373, 374, 376
Harmonic motion, 335–336
  amplitude, 356
  frequency, 358
  period, 356
  phase delay, 356
Harmonic series, 551, 557
  alternating, 533, 566, 582
  partial sums, 553, 561
Hastings, Battle of, 429
Hay bale example, 482

Newton's first law, 481
Newton's law of cooling, 165, 320,
   322–323, 327, 340
Newton's law of gravitation, 175, 265, 485,
   494, 500
Newton's second law of motion, 176
Newton's wine barrel problem, 251, 538
Newton's method, 146–148
   modified, 148
   problems, 187
Nonorientable surface, 779
Nonrepeating decimal, 550
Normal line, 25
Normal component of acceleration, 490,
   620
Normal vector:
   to curve, 619
   to a plane, 613
   principal unit, 487, 620
   to surface, 684, 779
   to $z = f(x,y)$, 650
nth term test for divergence, 551
Nuclear carrier problem, 332
Nuclear reactor example, 115
Number:
   complex, 814
   real, 3–4
Numerical integration methods (*see also*
   Approximation)
   by Simpson's approximation, 231, 232,
      234, 519
   with Taylor polynomials, 534–535
   by trapezoidal approximation, 229, 234,
      519

O

Oblique segment of paraboloid, 734–735,
   755
Octant, first, 597
Oil can problem, 105
Oil field, 8, 27, 128
Oil slick problem, 140
One-to-one function, 37
One-to-one transformation, 748
One-sided limits, 51
Open interval, 4
Operator notation, 65
Optics:
   Fermat's principle, 88
   Snell's law, 90
Optical property:
   of ellipse, 430, 431
   of hyperbola, 436
   of parabola, 36, 423
Orbit, planetary, 496, 498
   of earth, 429
   of Icarus, 456
   of Mercury, 431
Orbital, 305
   1s, 306
Ordinate, 9
Order of differential equation, 800
Order of integration (reversing), 711

Order of magnitude, 304
   of $x^n/n!$, 536
Ordinary differential equation, 800
Ordinary Differential Equations, 836
Orientable surface, 779
Orientation of a curve, 761
Origin, 9
Osculating circle, 489, 492
Osculating plane, 635
Ostrogradski, Michel (1801–1861), 782
Ostrogradski's theorem, 782
Overdamping, 825
Ozone problem, 340

P

Pappus (ca. A.D. 300):
   first theorem, 274
   second theorem, 275
Parabola, 15, 418–419, 421
   axis, 422
   as conic section, 418
   construction, 425
   directrix, 421
   equation, 422
   focus, 421
   and hanging cable, 172
   polar form, 448
   reflection property, 423
   semicubical, 258
   tangent line property, 425
   vertex, 16
Parabolic approximation, 233
Parabolic bridge arch problem, 456
Parabolic cylinder, 625
Parabolic mirror, 424
Parabolic orbit, 425
Parabolic trajectories, 424
Paraboloid:
   elliptic, 626
   floating, 756
   of revolution, 251, 262, 279
   surface area, 262
   volume of oblique segment, 734–735
   volume problem, 251
Parachute, 325, 326, 328
Parallel lines' slopes, 12
Parallel planes, 614
Parallel resistance problem, 701, 702
Parallel vectors, 471, 600, 607
Parallelepiped volume, 609
Parallelogram area, 607
Parallelogram law, 471, 599
Parameter(s), 458, 743
   elimination, 459
Parametric curve, 458
   arc length, 466
   graph, 458
   smooth, 461
   tangent line to, 462
Parametric equations, 458, 616
   of cycloid, 460
   and derivatives, 461
   of line, 612

in polar coordinates, 468
Parametric surface, 742
Parametrization of a curve, 459
Partial derivative(s):
   chain rule, 672–678
   equality of mixed, 651
   of $f(x,y)$, 646–649
   of $f(x,y,z)$, 651
   geometric interpretation, 648–649
   higher order, 651
   of implicitly defined function, 676–677
Partial differential equation, 800
Partial-fraction decomposition, 400
Partial fractions method, 399
Partial sums of series, 547
*Partial Sums of Infinite Series and How
   They Grow,* 553
Partial integral, 706
Particular solution, 333, 800
Partition, 194, 199
   mesh, 199, 705
   inner, 708
   of interval, 194, 199
   polar, 717–718
   of rectangle, 704
   regular, 202
   selection for, 199, 705
   spherical, 739
Parts, integration by, 385–389
Path, 764 (*see also* Curve in three
   dimensions)
Path-independent integral, 762, 764
Pendulum, 359, 589–592
   length, 359–361, 589–591, 672
   period, 128, 137, 672
Pendulum clocks, 359–361
Period, 356, 824
   of pendulum, 128, 137, 589–591
   of sine and cosine, 120
   of trigonomic functions, 343
Perpendicular lines' slopes, 12
Perpendicular vectors, 473–475, 600
Perpendicularity of vector product, 604
Petroleum example, 329
Perpetual annuity, 529
   problems, 531
Perpetuity (*see* Perpetual annuity)
*Philosophiae Naturalis Principia
   Mathematica,* 176, 493
Phase angle, 827
Phase delay, 356, 827
Phase shift function, 831
Phenylethylamine, 327
π :
   computation of, 192, 533–534, 536, 560,
      561, 574, 583
   estimate by Simpson's approximation,
      521
   series for, 559, 583
   and 22/7, 413
Piecewise smooth curve, 762
Piecewise smooth surface, 782
Piston:
   example, 357
   problem, 269
Piston engine, 751–752

T

Tangent line(s) (*cont.*)
  vertical, 88
Tangent plane:
  to surface $z = f(x,y)$, 649
  to $f(x,y,z) = 0$, 685
Tangent reduction formula, 384
Tangent vector, unit, 620
Tangential component of acceleration,
    490, 621
Taylor, Angus E., 836
Taylor, Brook (1685–1731), 514
Taylor polynomial, 512
  remainder, 514
Taylor series, 532, 574, 578
  for binomial, 579, 584
  for cos x, 532
  for e^x, 532, 563
  for inverse hyperbolic tangent, 538
  for inverse sine, 582
  for inverse tangent, 533
  for log$(1 + x)$, 582
  remainder, 514
  for sin x, 533
Taylor's formula, 514
  proof, 517–518
  with remainder, 514
  and second derivative test, 535
Telescope design, 437
Telescoping sum, 549
Television picture tube, 610
Temperature oscillations, 830–833
Terminal speed, 324, 325, 367
Terms of a sequence, 540
Termwise addition and multiplication of
    infinite series, 550
Termwise differentiation and integration
    of power series, 581
Test for concavity, 153
Test for local extrema, 98
Tests for convergence of series (*see*
    Convergence of series)
Tetrahedron volume problem, 611
Theory and Application of Infinite Series,
    587, 836
Thermal diffusivity, 788
Thermal expansivity, 678
Thermometer bulb example, 678
Thomson, William (Lord Kelvin,
    1824–1907), 790
Three-leaved rose, 448
Timber problem, 339
Time rate of change, 313
Torque vector, 610
Torricelli, Evangelista (1608–1647), 28
Torricelli's law, 28, 166, 413
  problems, 141, 172, 173
Torsion, 635
Torus, 251
  surface area, 469
  volume example, 753
Trace of a surface, 624
Trajectory of projectile, 424
Transcendental functions, 342
Transfer orbits, 497, 498
Transformation, 748
Transient current, 806
Transient solution, 829

Translation of coordinates, 439
Translation principle, 14, 417
Transmission line problem, 374
Transverse axis of hyperbola, 432
Transverse unit vector, 493
Trapezoidal approximation, 229
  error estimate, 234, 519
Triangle area problem, 611
Triangle inequality, 4, 603
Triangle law, 471
Trigonometric functions, 117–118
  antiderivatives, 344–345
  basic identities, 119
  definition, 117–118, 342
  derivatives, 121–122, 342–343
  graphs, 343
  periodicity, 343
  series expansions (*see specific function*)
  table of values, 842
Trigonometric formulas, 841
Trigonometric integrals, 381–384
Trigonometric substitution, 390–393
Trigonometry review, 117–121
Triple integrals, 730–736
Triple integration:
  in cylindrical coordinates, 736–738
  in rectangular coordinates, 730–735
  in spherical coordinates, 741–748
Triple scalar product, 608
Triple vector product, 608, 611
Trochoid, 499
T-shaped lamina, 730
Touchdown of spaceship, 62
Tuning circuit, 828
Two-point equation of line, 11
Two-step cascade, 804

U

u-curves and v-curves, 748
Underdamping, 827
Underground cellar, 830
Undetermined coefficients method,
    811–813
Uniform circular motion, 483–485
Unit binormal vector, 635
Unit normal vector, 620
Unit tangent vector, 486, 620
Unit vector, 472, 599
  basic, 599
  polar form, 493
Units of measurement, 838
United States population, 335
Universe, age of, 316
Upper and lower sums, 202
Upper bound for sequence, 545
Uranium isotopes, 316
Uranium-238 dating, 316

V

V-2 rocket, 809
Vacuum pump, 555

Value of a function, 6
van der Waals' equation, 158, 653, 701
Variable(s):
  change in double integral, 749
  dependent, 6, 639, 674
  independent, 6, 639, 674
  intermediate, 674
v-curves, 748
Vector(s), 470
  acceleration, 478–479
  addition, 471
  angle between two, 475, 600
  apparent velocity, 476
  arrow representation, 470
  basic unit, 599
  binormal, 635
  components, 470
  cross (vector) product, 604
  difference of two, 471
  differentiation formulas, 478
  direction, 471
  direction angles, 601
  direction cosines, 601
  dot product, 474, 599
  field, 758
  force, 473
  function, 477
  gradient, 680
  i and **j**, 472
  i, **j**, and **k**, 599
  length, 470, 598, 606
  magnitude, 470
  multiplication by scalar, 471
  normal, 487, 619, 620, 684, 779
  notation, 470–473, 598–599
  negative of, 471
  parallel, 600, 607
  parallelogram law, 599
  perpendicular, 473, 475, 600
  polar form, 493
  position, 470, 598
  principal unit normal, 487, 620
  product, 604, 606, 608
  product of three, 608, 611
  product with scalar, 471, 599
  projection onto another vector, 602
  radial unit, 493
  scalar multiple, 471, 599
  scalar product, 474, 599
  scalar triple product, 608
  subtraction, 471
  sum, 471, 599
  in three dimensions, 598–603
  torque, 610
  transverse unit, 493
  triple scalar product, 608
  triple vector product, 608, 611
  true velocity, 476
  unit, 472, 599
  unit normal, 620
  unit tangent, 486, 620
  vector (cross) product, 604
  velocity, 470, 478, 617
  zero, 470
Vector chain rule, 681
Vector equation of line in space, 611
Vector fields, 758

INVERSE TRIGONOMETRIC FORMS

69 $\displaystyle\int \sin^{-1} u \, du = u \sin^{-1} u + \sqrt{1 - u^2} + C$

70 $\displaystyle\int \tan^{-1} u \, du = u \tan^{-1} u - \frac{1}{2} \ln (1 + u^2) + C$

71 $\displaystyle\int \sec^{-1} u \, du = u \sec^{-1} u - \ln|u + \sqrt{u^2 - 1}| + C$

72 $\displaystyle\int u \sin^{-1} u \, du = \frac{1}{4} (2u^2 - 1) \sin^{-1} u + \frac{u}{4} \sqrt{1 - u^2} + C$

73 $\displaystyle\int u \tan^{-1} u \, du = \frac{1}{2} (u^2 + 1) \tan^{-1} u - \frac{u}{2} + C$

74 $\displaystyle\int u \sec^{-1} u \, du = \frac{u^2}{2} \sec^{-1} u - \frac{1}{2} \sqrt{u^2 - 1} + C$

75 $\displaystyle\int u^n \sin^{-1} u \, du = \frac{u^{n+1}}{n + 1} \sin^{-1} u - \frac{1}{n + 1} \int \frac{u^{n+1}}{\sqrt{1 - u^2}} \, du + C \quad \text{if } n \neq -1$

76 $\displaystyle\int u^n \tan^{-1} u \, du = \frac{u^{n+1}}{n + 1} \tan^{-1} u - \frac{1}{n + 1} \int \frac{u^{n+1}}{1 + u^2} \, du + C \quad \text{if } n \neq -1$

77 $\displaystyle\int u^n \sec^{-1} u \, du = \frac{u^{n+1}}{n + 1} \sec^{-1} u - \frac{1}{n + 1} \int \frac{u^n}{\sqrt{u^2 - 1}} \, du + C \quad \text{if } n \neq -1$

HYPERBOLIC FORMS

78 $\displaystyle\int \sinh u \, du = \cosh u + C$

79 $\displaystyle\int \cosh u \, du = \sinh u + C$

80 $\displaystyle\int \tanh u \, du = \ln (\cosh u) + C$

81 $\displaystyle\int \coth u \, du = \ln|\sinh u| + C$

82 $\displaystyle\int \text{sech } u \, du = \tan^{-1} |\sinh u| + C$

83 $\displaystyle\int \text{csch } u \, du = \ln \left| \tanh \frac{u}{2} \right| + C$

84 $\displaystyle\int \sinh^2 u \, du = \frac{1}{4} \sinh 2u - \frac{u}{2} + C$

85 $\displaystyle\int \cosh^2 u \, du = \frac{1}{4} \sinh 2u + \frac{u}{2} + C$

86 $\displaystyle\int \tanh^2 u \, du = u - \tanh u + C$

87 $\displaystyle\int \coth^2 u \, du = u - \coth u + C$

88 $\displaystyle\int \text{sech}^2 u \, du = \tanh u + C$

89 $\displaystyle\int \text{csch}^2 u \, du = -\coth u + C$

90 $\displaystyle\int \text{sech } u \tanh u \, du = -\text{sech } u + C$

91 $\displaystyle\int \text{csch } u \coth u \, du = -\text{csch } u + C$

MISCELLANEOUS ALGEBRAIC FORMS

92 $\displaystyle\int u(au + b)^{-1} \, du = \frac{u}{a} - \frac{b}{a^2} \ln|au + b| + C$

93 $\displaystyle\int u(au + b)^{-2} \, du = \frac{1}{a^2} \left[\ln|au + b| + \frac{b}{au + b} \right] + C$

94 $\displaystyle\int u(au + b)^n \, du = \frac{(au + b)^{n+1}}{a^2} \left(\frac{au + b}{n + 2} - \frac{b}{n + 1} \right) + C \quad \text{if } n \neq -1, -2$

95 $\displaystyle\int \frac{du}{(a^2 \pm u^2)^n} = \frac{1}{2a^2(n - 1)} \left(\frac{u}{(a^2 \pm u^2)^{n-1}} + (2n - 3) \int \frac{du}{(a^2 \pm u^2)^{n-1}} \right) \quad \text{if } n \neq 1$